ENCYCLOPEDIA OF

BIOPROCESS TECHNOLOGY: FERMENTATION, BIOCATALYSIS, AND BIOSEPARATION

VOLUME 4

WILEY BIOTECHNOLOGY ENCYCLOPEDIAS

Encyclopedia of Bioprocess Technology: Fermentation, Biocatalysis, and Bioseparation
Edited by Michael C. Flickinger and Stephen W. Drew

Encyclopedia of Molecular Biology
Edited by Thomas E. Creighton

Encyclopedia of Cell Technology
Edited by Raymond E. Spier

Encyclopedia of Ethical, Legal, and Policy Issues in Biotechnology
Edited by Thomas J. Murray and Maxwell J. Mehlman

ENCYCLOPEDIA OF BIOPROCESS TECHNOLOGY: FERMENTATION, BIOCATALYSIS, AND BIOSEPARATION
EDITORIAL BOARD

ENCYCLOPEDIA OF

BIOPROCESS TECHNOLOGY: FERMENTATION, BIOCATALYSIS, AND BIOSEPARATION

VOLUME 4

Michael C. Flickinger
University of Minnesota
St. Paul, Minnesota

Stephen W. Drew
Merck and Co., Inc.
Rahway, New Jersey

A Wiley-Interscience Publication
John Wiley & Sons, Inc.
New York / Chichester / Weinheim / Brisbane / Singapore / Toronto

This book is printed on acid-free paper. ♾

For ordering and customer service, call 1-800-CALL-WILEY.

Library of Congress Cataloging-in-Publication Data:

Flickinger, Michael C.
The encyclopedia of bioprocess technology : fermentation, biocatalysis, and bioseparation / Michael C. Flickinger, Stephen W. Drew.
p. cm.
Includes index.
ISBN 0-471-13822-3 (alk. paper)
1. Biochemical engineering--Encyclopedias. I. Drew, Stephen W., 1945- . II. Title.
TP248.3.F57 1999
660.6′03--dc21 99-11576
CIP

Printed in the United States of America.

10 9 8 7 6 5 4 3 2 1

ENCYCLOPEDIA OF

BIOPROCESS TECHNOLOGY: FERMENTATION, BIOCATALYSIS, AND BIOSEPARATION

VOLUME 4

MICROCARRIER CULTURE

MADHUSUDAN V. PESHWA
Dendreon Corp.
Mountain View, California
KATHY WONG
WEI-SHOU HU
University of Minnesota
Minneapolis, Minnesota

KEY WORDS

Anchorage-dependent cells
Bioreactor
Cell aggregate
Cell attachment
Cell immobilization
Mammalian cells
Microcarrier
rDNA protein
Shear damage
Vaccine

OUTLINE

INTRODUCTION

Most recombinant DNA–derived mammalian cell culture products are manufactured using cells grown in suspension. Commonly used cells include Chinese hamster ovary (CHO), baby hamster kidney (BHK), hybridoma, and myeloma cells. Those cells can be cultivated in fermentors similarly to microorganisms. However, the vast majority of cells grown in research laboratories, and many of those used in the manufacturing of viral vaccines for human and veterinary use, are anchorage-dependent or anchorage-preferred cells. This means that they need to attach to a compatible surface to survive and multiply, or, if given a compatible surface, they will readily attach to it and multiply rather than stay in suspension. Frequently used anchorage-dependent cells include MDCK, vero, mouse L, fibroblasts such as MRC-5 and WI-38, and almost all differentiated primary cells from nonblood tissues, such as hepatocytes, neuronal cells, and chondrocytes. These cells are grown on petri dishes, flasks (T-flasks), and roller bottles in research laboratories. Traditionally roller bottles are used in the production of biologics, such as cytokines and viruses, derived from those cells. In a typical production plant of those biologics, it is common to see tens of thousands of roller bottles being used for cell cultivation. The amount of labor requirement for roller bottle handling is daunting. Many of those processes manufacture products that were approved by regulatory agencies long ago. If the products were developed more recently, many, if not most, would have been manufactured with an alternative and improved method.

The need to establish a better method of culturing anchorage-dependent cells in large scale prompted the development of microcarriers. The first microcarrier culture was demonstrated by van Wezel in 1967 with diethylaminoethyl (DEAE) Sephadex A50 particles originally designed as column packing for ion exchange chromatography (1). The first industrial-scale product (inactivated polio vaccine) produced in microcarrier cultures was subsequently introduced by van Wezel in 1972 (2). This new way of culturing anchorage-dependent cells spurred other investigators to discover novel techniques to improve cell attachment to the beads, to test new types of microcarriers, and to increase the cell density. Serum coating of DEAE Sephadex beads (3), electric corona discharge for treatment of plastic (polystyrene) beads, and use of dextran microcarriers with a lower charge density (4) led to improvements in microcarrier technology. Since then, a variety of microcarriers have been developed and much research has provided insights into optimization of culture conditions. Table 1 presents a list of the most commonly used commercial microcarriers. The advantages of microcarrier culture over conventional monolayer culture methods (e.g., roller bottles, stacking flat plates) are numerous. Using microcarriers to provide a surface for cell attachment, a large surface area can be contained in a given reactor volume with a relatively uniform environment. Using a reactor similar to conventional fermentors, environmental monitoring and control become possible, thus opening up the possibility of further process optimization.

Most conventional microcarriers have a diameter of approximately 200–250 μm when suspended in medium and

Table 1. Commercially Available Conventional and Macroporous Microcarriers

Microcarrier type	Supplier
Conventional microcarriers	
Biosilon (polystyrene based)	Nunclon
Cytodex 1 (DEAE dextran)	Pharmacia
Cytodex 2 (DEAE dextran)	Pharmacia
Cytodex 3 (gelatin)	Pharmacia
Glass-, collagen-, and hydroxyapatite-coated plastics	Solohil Engineering
Macroporous microcarriers	
Cultisphere (collagen or gelatin)	Hyclone
Cytoline (polyethylene)	Pharmacia
Cytopore (cellulose)	Pharmacia

a specific density of 1.02–1.03 g/cm^3, slightly higher than that of the medium. This higher density allows a minimum agitation to be used to keep the microcarriers in suspension. It also allows for ready settlement after agitation is turned off, which is ideal for culture harvest or exchange of medium for supporting higher cell growth. A typical culture uses a microcarrier concentration equivalent to a settled bead volume 5–15% of total culture volume. This gives a surface area of about 0.5–1.5 m^2/L culture volume. Each liter of culture is equivalent to 5–150 roller bottles. The potential of increasing cell growth surface area was further advanced by the development of macroporous microcarriers (5). These are highly porous convoluted spheres with internal cavities for cell attachment and growth. Macroporous microcarriers can be used for culturing both anchorage-dependent and suspension cells.

Over the past 30 years, microcarrier culture technology has developed and matured to its present-day status. It allows for an easy way of producing a large quantity of cells both in laboratories and for industrial manufacturing. The ease of cell retention in microcarrier cultures also facilitates the employment of a perfusion system, which increases the maximum achievable cell density significantly. For large-scale cultivation of anchorage-dependent cells, microcarrier culture is still the best method.

CONVENTIONAL MICROCARRIERS

The microcarriers developed in the 1970s were initially based on cross-linked dextran. Soon after their introduction they were used successfully for the production of various vaccines and biologicals (6). Further development led to the use of other materials as microcarrier supports. These materials include polystyrene (7), glass (8), cellulose (9), and gelatin (10). The original dextran-based microcarriers require tedious procedures for detachment of cells (11) for subsequent inoculation, which is not the case for some of the other materials developed (8). Several studies have been directed at the effect of surface charge on cell attachment (12,13) and the kinetics of attachment (12).

Cell Surface Interactions and Attachment Kinetics

A microcarrier culture is typically initiated by trypsinizing cells that are adherent on a surface, either on tissue culture flasks or on microcarriers. Sometimes cells for inoculation are cultivated in suspension, and the production scale is conducted with microcarriers. Inoculation from a suspension culture eliminates the need for trypsinization. Once inoculated into a microcarrier suspension, the cells can either attach to the microcarriers or agglomerate to each other to form aggregates. It is essential that the rate of cell attachment to the microcarriers follows rapid kinetics to prevent them from agglomerating. The need for rapid attachment is even more pronounced in cases of primary cells such as hepatocytes or in case of other adherent cell lines that lose their viability if maintained in suspension over extended periods of time. For single cell populations, several studies have demonstrated that the initial rate of attachment of cells to microcarriers is essentially a first-order process, meaning the rate of increase in attached cells is proportional to the concentration of unattached cells, provided that the surface area available for attachment is not limiting (12). The process of cell attachment to any foreign surface can be viewed as a combination of two separate processes: initial adsorption of cells to the microcarrier surface followed by adhesion molecule–mediated attachment and cell spreading. The affinity of the binding of the cells to the microcarriers depends on the cell line, the microcarrier characteristics, the growth phase of the culture, the medium composition, and the cell loading per microcarrier. Cells initially adsorbed can possibly desorb from the microcarrier surface without really developing cell adhesion; the surface should therefore have sufficient affinity so as to allow the cells to remain attached to develop a firm grip by cell adhesion molecule–mediated interactions.

The initial adsorption of cells onto the microcarrier surface is a physical process and is believed to be facilitated by various attractive or affinitive interactions between the cell surface and the microcarriers. Cells in general fail to attach to unmodified crossed-dextran beads such as Sephadex but attach readily to DEAE Sephadex, although the optimal charge density for cell spreading and growth is less than that in DEAE Sephadex (4). Plastics such as polypropylene and polystyrene, in their native state, have a lower surface energy and are therefore incompatible for cell attachment. When treated with an electric corona discharge, their surface energy rises to that of tissue culture dishes, which enhances their ability for cell attachment. It has been demonstrated that corona discharge leads to the unmasking of carboxylic groups on the polystyrene surface, leading to this increase in surface charge density (14). Also, the attachment of cells has been shown to be affected by the length of the carbon chain on the unmasked surface groups (15). The cell surface is negatively charged due to the presence of negatively charged residues on the side chains of membrane-associated cell surface proteins. The interaction of the cell surface with positively charged surfaces can be attributed to the electrostatic interactions. The attachment of cells to negatively charged surfaces is likely due to Lewis acid type reactions in which short-range electron donor–acceptor interactions lead to coupling. In some cases microcarriers coated with cell adhesion factors, such as fibronectin, collagen, and laminin (16), have also been used to facilitate cell adhesion. It is interesting to note that for many cell types the initial attachment rate to charged microcarriers such as DEAE-modified cross-linked dextran beads (Cytodex-1) is faster than to those coated with collagen, which is an excellent surface for cell adhesion on petri dishes.

Furthermore, although serum provides many factors that promote cell adhesion, the initial attachment to charged microcarriers is often faster in the absence of serum (12). These observations suggest that electrostatic interactions between the surfaces of cells and microcarriers is a dominant factor for initial cell attachment. The provision of adhesion molecules, either covalently bound to microcarriers (16) or in soluble form in medium (17), promotes subsequent development of firm grips of cells on microcarriers and facilitates cell spreading. In many cases, no exogenous adhesion molecules are necessary because cells synthesize their own. For some cell types the omission of these factors impedes cell spreading despite successful cell attachment to microcarriers (16). It is possible then to coimmobilize charged groups and cell adhesion molecules on the microcarriers to promote both initial physical attachment and subsequent biological events of development of adhesion plaques and cell spreading. Irrespective of the surface charge and the hydrophobicity or hydrophilicity of the microcarriers, it is essential that the surface have good wettability characteristics (18).

It has also been shown that a correlation exists between the exchange capacity of the microcarriers and the rate of cell attachment. A higher exchange capacity of the microcarriers leads to a higher first-order rate constant for cell attachment. However, at physiological pH, a higher exchange capacity is not correlated to a higher charge density. Moreover, a difference in cell spreading is observed at different exchange capacities (12). It appears that higher exchange capacities lead to increased cell attachment, but one must be cautioned that a higher exchange capacity may not lead to optimal growth of cells. The optimum exchange capacities for cell attachment and for cell growth are thus very different, and it is advisable to perform preliminary experiments to determine what exchange capacity is best suited for a particular cell type and a particular microcarrier type.

Inoculum Cell Density

After inoculation, cells attach to microcarriers randomly and the number of cells per microcarrier is distributed over a range. Once the adhesion plaques are developed and spreading occurs, most cells do not detach readily and reattach to other microcarriers. They grow until they cover the surface of the bead. For cells subject to contact inhibition, growth ceases then. Many cell types can form multiple layers, but growth rate often slows down after reaching confluence. Therefore, cells on beads with more cells initially reach confluence and stop growing or slow their growth rate faster while others are still growing. Once this happens the overall growth rate in the culture slows. The initial cell distribution on microcarriers thus affects the overall growth kinetics of microcarrier culture. The narrower the initial cell distribution is, the better the growth kinetics will be. Furthermore, microcarriers that do not receive any cells initially will remain barren. Thus, the initial cell distribution after inoculation also affects the maximum cell concentration attainable in the culture.

The initial random distribution of cells on the microcarriers can be simulated by a Poisson distribution, which predicts the probability of a microcarrier having a certain number of cells given the average number of cells per microcarrier (19). Ideally the average number of cells that attach per bead should be such that the fraction of beads without any attached cells is kept at a minimum. It has been shown that a minimal average of about five to six cells per microcarrier is a necessary inoculum size (19,20). With this inoculum ratio, the percentage of microcarriers without any cells attached to them is less than 1%, as estimated from the Poisson distribution.

Cell Shear

Adequate agitation in microcarrier cultures is critical for providing a homogeneous environment and thus facilitates oxygen transfer. The hydrodynamic forces that result from agitation can be detrimental to cells grown on microcarriers. Cells grown microcarrier culture are more shear sensitive than those grown in suspension. The dominant cell damage mechanism in microcarrier bioreactors is believed to be due to forces generated by the collision of microcarrier beads with each other and by small turbulent eddies, which are motions generated by the random variations of fluid direction and velocity in a turbulent flow field (21). In general, except in the spinner flasks used in laboratories, a pitch-blade or marine-type impeller is used for agitation. The ratio of impeller diameter to tank diameter is typically larger than that used in microbial fermentation. The larger impeller provides a larger capacity in liquid pumping for suspending the microcarriers. Because of the low solubility and mass transfer rate of oxygen in culture medium, external aeration is needed at larger scale. In many cases oxygen is supplied indirectly through membrane or silicone rubber tubing, which has a high permeability for oxygen. Use of such tubing is practical even up to a reactor of hundreds of liters. Direct sparging with air or oxygen is used in large stirred tank bioreactors. It is well known in suspension cultures that shear force from rupturing bubbles in a sparged system is the most dominant cell damaging mechanism. This is also likely the case in microcarrier cultures (22). With direct sparging surface active agents, such as Pluronic F-68, are used to reduce the attachment of cell-laden microcarriers to rising bubbles. Attaching to these bubbles and rising with them at a high velocity is often detrimental to cells attached to microcarriers. Furthermore, it is necessary to employ antifoaming agents. Otherwise, cell-laden microcarriers may rise to the top of the foam layer and accumulate there, eventually losing viability.

MACROPOROUS MICROCARRIERS

The cell concentration achievable in a conventional microcarrier system is limited by the available surface for growth. Increasing the microcarrier concentration beyond a certain limit does not lead to an increase in cell concentration because the increased agitation needed to suspend the microcarrier slurry becomes damaging to cells. Macroporous microcarriers were developed with an aim of circumventing some of these limitations of conventional microcarriers.

Macroporous beads are essentially convoluted spheres with relatively large pores for cell attachment and growth. The commonly used materials are gelatin (5), collagen (23), cellulose, glass, polyethylene (24), and polystyrene. The desirable properties of the materials of construction for macroporous microcarriers were discussed earlier. After inoculation of cultures, the cells attach to the external surface of the microcarriers and subsequently migrate inward to populate the interior of the beads. If large open channels leading to internal pores exist, cells may also populate the interior soon after inoculation. Thus, these macroporous microcarriers provide extended surface area in the interior and potentially support a higher cell concentration with an equivalent number of beads. Also, the cells in the interior of the macroporous microcarriers are shielded from fluid flow and are hence not susceptible to detrimental shear effects (22). However, there are two areas for potential concern. First, in contrast to conventional microcarriers the surface of the macroporous microcarriers is full of cavities and is highly irregular. This irregularity interrupts the flow pattern in the bulk fluid in the immediate vicinity of the beads and gives rise to local microeddies that may damage the cells attached on the surface. Therefore, to avoid shear damage, once attached cells need to migrate inward at a relatively fast rate or the internal pore must be relatively open to allow for cells be carried into the interior by fluid convection during the cell attachment period after inoculation. The second cause for concern is the potential gradients of nutrients, oxygen, and metabolites along the radius of the microcarrier which may cause growth limitation. As a result, necrotic core could be induced within the beads (24,25). Use of intraparticle convection has been postulated as a method to overcome the formation of heterogeneous environment. Numerous cell types have been successfully cultivated on macroporous microcarriers for purposes of product production and virus production (26). Apart from stirred tank reactors, fluidized-bed reactors (27) are also used for cultivating cells in macroporous microcarriers.

Spatial Distribution of Cells and Cell Viability

To ensure efficient use of macroporous microcarriers the internal pores should be open to the exterior through open channels that allow for cell movement into the interior. Some collagen-based microcarriers have leaflike interiors with open and flat surfaces, and others have open pores interconnected by tortuous channels. The materials that the microcarriers are made of also affect the rate of cell movement. Thus, both the structure and the material may affect the performance of the culture. Ideally, macroporous microcarriers should have very open structures to allow cell penetration into the interior immediately after inoculation. Architectural staining of paraffin-embedded thin sections of cell-laden macroporous microcarriers with hemotoxylin and eosin (H&E) indicate that the distribution of cells in the interior is usually not uniform. Some of the internal cavities are packed to confluency, whereas others are devoid of cells.

Traditional methods of examining the pore structure and cell behavior in the interior require fixation and sectioning, which cause aberation. Furthermore, reconstruction of the three-dimensional image is difficult. By digitizing sectioned macroporous culture in three colors, it is possible to classify pixels as conforming to either background, cytoplasmic, microcarrier matrix, or nuclear parameters, based on a set of classification rules determined by statistical analysis (28). When relating the number of pixels occupied by cellular material to the total number of pixels in the sectioned microcarrier, quantification of cell growth can be generated. To examine the accessibility and geometrical features of the pores, confocal laser scanning microscopy (CLSM) can be used to optically section the beads (29). CLSM requires fluorescent samples. Most microcarriers can be rendered fluorescent by appropriate staining techniques. It is also possible to stain cells with a vital dye that has different excitation wavelengths than the dye used for microcarrier staining. This dual stain allows simultaneous observation of cells and microcarriers. Individual beads can be optically sectioned serially through their depths. A three-dimensional image can be reconstructed by combining the sections for examination of the openness and connectivity of the pores. To further evaluate the geometry of the pores, the solid structure of the beads can be removed from the reconstructed three-dimensional images to allow for direct visualization of the pores. Cell distribution, divisional, and other morphological characteristics can be observed directly. This confocal microscopic technique for direct evaluation of pore structure of macroporous beads is noninvasive and requires minimal sample preparation. It can easily be adapted for in situ observation of cell behaviors and to provide information to optimize the structure of those microcarriers.

MICROSPHERE-INDUCED AGGREGATES

A possible way to increase cell concentration in conventional microcarrier systems is by increasing the microcarrier concentration. However, a potential drawback of this scheme is the large ratio of settled bead volume to the culture volume. This large volume of settled beads calls for sophisticated methods of agitation to keep the microcarriers in suspension. Compared with these systems, the settled volume of cell aggregate with an equivalent cell concentration is relatively small, and the agitation requirements parallel those of simple suspension cultures. Also, some adherent cell lines have been engineered to grow in suspension or as aggregates by processes of adaptation or media adaptation. It has been demonstrated that agglomeration can be induced by lectin, concanavalin A, and wheat germ agglutanin (30). However, these processes are relatively slow, and a significant number of the cells initially inoculated lose their viability because of the duration required for aggregate induction.

Microspheres made of DEAE-derivatized Sephadex beads with diameters of 20–50 μm have been demonstrated to increase the rate of aggregate formation (31). Hence, this process can be thought of as a process similar to nucleation and growth in crystallization. After the initial attachment phase, these cell-laden microspheres come into contact with each other to form larger clumps, which serve as nucleation sites for aggregate formation via growth of these attached cells. Over time, aggregates emerge, each consisting of a few microspheres surrounded by multiple layers of cells.

A variety of cells, including CHO, vero, human embryonic kidney cells (293 cells), and swine testicular cells have been successfully grown on these microspheres (31) for virus and protein C productions (31). The drawback of employing aggregate cultures is that little is known about controlling the size of aggregates. As the aggregate size increases, nutrient limitation and a decreased viability can be expected.

CONCLUSION

Microcarrier processes, because of their ease of operation and control, are ideal for cultivating anchorage-dependent cells in both the laboratory and in manufacturing. The bioreactor operation requirements are virtually identical to those for a typical suspension culture in fermentors except that the upper limit for agitation is lower. Operating modes such as batch, fed batch (32), and perfusion are routinely being used in microcarrier cultures. The prevailing use of microcarriers in industrial scale is for the manufacture of viral vaccines and reagents. Depending on the cell type, serum-free cultivation is also possible. Because most newly developed cell culture processes for rDNA products employ suspension cells, microcarrier culture processes will see only moderate growth in the future. However, with increased efforts in research and product development employing differentiated cells, we may see microcarriers play an increasingly important role in small to moderate scales. This technology—developed three decades ago—is still intriguing to cell culture technologists. Its versatility in supporting the growth of a variety of anchorage-dependent cells ensures it a unique place in bioprocessing.

BIBLIOGRAPHY

1. A.L. van Wezel, *Nature* **216**, 64–65 (1967).
2. A.L. van Wezel, *Progr. Immunobiol. Stand.* **5**, 187–192 (1972).
3. R.E. Spier and J.P.W. Whiteside, *Biotechnol. Bioeng.* **18**, 659–667 (1976).
4. D.W. Levine, D.I.C. Wang, and W.G. Thilly, *Biotechnol. Bioeng.* **21**, 821–845 (1979).
5. K. Nilsson, F. Buzsaky, and K. Mosbach, *Bio / Technology* **4**, 989–990 (1986).
6. B. Meignier, *Dev. Biol. Stand.* **42**, 141–145 (1978).
7. A. Johansson and V. Nielsen, *Dev. Biol. Stand.* **46**, 125–129 (1980).
8. J. Varani, M. Dame, T.F. Beals, and J.A. Wass, *Biotechnol. Bioeng.* **25**, 1359–1372 (1983).
9. S. Reuveny, L. Silberstein, A. Shahar, E. Freeman, and A. Mizrahi, *Dev. Biol. Stand.* **50**, 115–123 (1982).
10. J. Varani, M. Dame, J. Fediske, T.F. Beals, and W. Hillegas, *J. Biol. Stand.* **13**, 67–76 (1985).
11. T.-Y. Tao, G.-Y. Ji, and W.-S. Hu, *Biotechnol. Bioeng.* **32**, 1037–1052 (1988).
12. V.B. Himes and W.-S. Hu, *Biotechnol. Bioeng.* **29**, 1155–1163 (1987).
13. S. Reuveny, A. Mizrahi, M. Kotler, and A. Freeman, *Biotechnol. Bioeng.* **25**, 469–480 (1983).
14. W.S. Ramsey, W. Hertl, E.D. Nowlan, and N.J. Binkowski, *In Vitro* **20**, 802–808 (1984).
15. M.J. Kuo, C.J. Lewis, R.A. Martin, R.E. Miller, R.A. Schoenfeld, J.M. Scheck, and B.S. Wildi, *In Vitro* **17**, 901–906 (1981).
16. J. Varani, S.E.G. Fligiel, D.R. Inman, T.F. Beals, and W.J. Hillegas, *Journal of Biomedical Materials Research* **29**, 993–997 (1995).
17. M. Kan, Y. Miyamoto, S. Sunami, I. Yamane, and M. Umeda, *Cell Structure and Function* **7**, 245–252 (1982).
18. M. Kiremitci and E. Piskin, *Biomat., Art. Cells, Art. Org.* **18**, 599–603 (1990).
19. W.-S. Hu, J. Meier, and D.I.C. Wang, *Biotechnol. Bioeng.* **27**, 585–595 (1985).
20. W.-S. Hu and D.I.C. Wang, *Biotechnol. Bioeng.* **27**, 1466–1476 (1985).
21. E.T. Papoutsakis, *TIBTECH* **9**, 427–437 (1991).
22. D.E. Martens, E.A.A. Nollen, M. Hardeveld, C.A.M. Vander Velden-de Groot, C.D. de Gooijer, E.C. Beuvery, and J. Tramper, *Cytotechnology* **21**, 45–59 (1996).
23. R.C. Dean, Jr., S.B. Karkare, P.G. Phillips, N.G. Ray, and P.W. Runstadler, Jr., in B.J. Lydersen ed., *Large Scale Mammalian Cell Culture Technology*, Hanser, New York, 1987, pp. 145–167.
24. A. Preissmann, R. Wiesmann, R. Buchholz, R.G. Werner, and W. Noe, *Cytotechnology* **24**, 121–134 (1997).
25. M.L. Kennard and J.M., *Biotechnol. Bioeng.* **47**, 550–556 (1995).
26. T.J. Nikolai and W.S. Hu, *Enzymol. Microb. Technol.* **14**, 203–208 (1992).
27. P.W. Runstadler, Jr., A.S. Tung, E.G. Hayman, N.G. Ray, S.G. Sample, and D.E. De Lucia, *Bioprocess Technol.* **10**, 363–391 (1990).
28. D.J. Foran, F. Cahn, and E.F. Eikenberry, *Analytical & Quantitative Cytology & Histology* **13**, 215–222 (1991).
29. S. Bancel and W.S. Hu, *Biotechnol. Prog.* **12**, 398–402 (1996).
30. P. Whur, H. Koppel, C. Urguhart, and D.C. Williams, *J. Cell Sci.* **23**, 193–209 (1977).
31. S. Goetghebeur and W.S. Hu, *Appl. Microbiol. Biotechnol.* **34**, 735–741 (1991).
32. R.Z. Mendonca and C.A. Pereira, *Bioproc. Eng.* **12**, 279–282 (1995).

See also MAMMALIAN CELL BIOREACTORS; MAMMALIAN CELL CULTURE REACTORS, SCALE-UP.

MICROENCAPSULATION

HOLGER HUEBNER
R. BUCCHOLZ
Technical University of Berlin
Berlin, Germany

KEY WORDS

Alginate
Beads
Cellulose sulfate
Coated beads
Encapsulation techniques
Hollow microspheres
Immobilization
Microencapsulation
Polyvinyl alcohol

OUTLINE

INTRODUCTION

The immobilization of active biological substances or living cells has become a universal tool in biotechnology over the past decades. Immobilization can be defined as any procedure that confines substances or cells inside a given system and limits its free diffusion or migration. In biotechnology, the concept of microencapsulation is related to a special immobilization system, where the biological material is confined inside particles, beads, or hollow spheres.

Spherical particles can be considered to be the best-suited immobilization system for use in bioreactors, due to their optimal hydrodynamic properties and abrasion resistance. Because microencapsulation is a very powerful and universal tool, the applications for microencapsulated systems cover a very wide field. To date, it has been used to improve the relatively outdated biotechnological processes in the food industry (1,2), wastewater treatment (3), leaching (4), and environmental detoxification (5). In modern biotechnology it has been applied to cell culture processes for the production of high-value substances such as antibodies (6,7), erythropoetin (8), and the anticancer drug taxol (9), as well as for the production of artificial seeds (10,11) and cryopreservation (12). In medicine, microencapsulated cells have been used as artificial organs to treat diabetes (13,14), as a delivery system for gene therapy (15,16), as an intermediate step in cancer therapy (17), and for the treatment of Parkinson's disease (18,19), to name a few. Furthermore, microencapsulation can serve as a controlled release system for drugs in medicine (20–22) and pesticides in agriculture (23,24).

MICROCAPSULES

Although there are a variety of encapsulation techniques and suitable materials, the resulting microcapsules can be divided into only three main groups: beads, coated beads, and hollow spheres.

Beads

Beads can be produced by cooling liquid drops of gelling agents (gelatine, agarose) under their melting point, thus transforming the polymer solution into a stable cryogel by internal hydrogen bonding, or by chemically or ionically cross-linking polymers to produce a hydrogel. Because bead formation is a simple one-step process, it is relatively easy to develop a large-scale production process (25). The most common bead formation system is alginate cross-linked with calcium ions or any other divalent metal ions.

The fully developed three-dimensional internal network structure enables the beads to withstand extreme mechanical stress. Permeability of hydrogel beads is excellent, and they are often used for culturing cells and microorganisms. The disadvantage is their lack of a real barrier on the surface, so cells can expand during cultivation.

Coated Beads

Providing beads with one or several additional walls leads to the formation of coated beads, which have been developed to overcome the problems associated with the open porous structure of bead surfaces. To produce a membrane, the beads from charged polymers (e.g., alginate, agarose) are treated with a diluted solution of an oppositely charged polyelectrolyte (e.g., poly-L-lysine, poly(ethylene imine), poly(*N*-vinylamine), chitosan) thus forming a simplex membrane on the bead surface. The use of several alternatively charged electrolytes leads to multilayer membranes (26–28). Alternatively, beads can be soaked with a photosensitive substance that cross-links when irradiated (29–31).

A proper coating process produces an additional membrane without incorporation of the material to be immobilized and has the advantage that the membrane properties can be engineered independently with respect to diffusion, molecular weight cut-off, cell retention, or immunoprotection from the internal bead structure and material. Possibly the most well-investigated clinical application of microencapsulation is based on poly-L-lysine-coated calcium alginate beads in the treatment of diabetes (32,33), which has recently been applied to humans (34,35).

Hollow Spheres

Hollow spheres can be produced in a one-step process using two membrane-forming materials that cannot penetrate each other due to diffusional limitations. The cross-linking reaction is then limited to the interface area of the hollow sphere, forming drops and producing a stable membrane around a liquid core. Typical hollow-sphere-forming materials are all polyelectrolyte combinations, for example, cellulose sulfate (CS) and poly(diallyl dimethyl ammonium chloride) (PDADMAC), and CS and poly(ethylene imine) (PEI), able to produce a simplex membrane (36). Another method for the production of hollow spheres is the use of two materials such that the cross-linking material of the drop can diffuse into the surrounding solution. The membrane starts forming at the surface of the droplets and then proceeds outward, creating cell-free membranes. The membrane thickness can be controlled by the electrolyte concentration of the droplet. In principle it is a reversal of the bead-forming process (37). Of course, hollow spheres can also be obtained from a multistep process in which the core of the coated beads is solubilized. This method has been used for controlled release systems in agriculture and the pharmaceutical industry. It is applied on shear-sensitive compounds such as animal cells in particular (38,39).

Recently simplex membrane systems have gained importance for medical applications such as artificial liver support (40) and gene therapy (41). Considerable progress has been made through a U.S. National Aeronautics and Space Administration study in which over a thousand combinations of polyanions and polycations were tested to identify new polymer candidates that could be suitable for the encapsulation of living cells, especially pancreatic islet cells for treatment of diabetes in humans (42,43).

FUNDAMENTAL METHODS IN MICROENCAPSULATION

The basis of all microencapsulation processes is the formation of a liquid drop followed by gelation (44,45), a cross-linking reaction, or membrane formation. The liquid drops can be obtained from extrusion of a liquid through a small needle or orifice, or from emmulsification of the drop-forming solution in a second immiscible solution by use of dynamic or static mixers. Scale-up ability of the process and a uniform microcapsule size and shape are the main parameters in encapsulation technology (46).

Dropping Methods

A liquid ejected with low velocity from a needle will break into individual drops. If the velocity is increased, drop formation increases until the maximum velocity is reached and the liquid begins to form a jet. The maximum velocity can be calculated from the following formula using the liquid velocity, v, interfacial tension, γ, liquid density, ρ, and the inner needle diameter, d_i:

$$v < 2\ (\gamma/\rho d_i)^{0.5}$$

Simple Dropping. The two main factors affecting drop size are the force of gravity trying to tear the drop from the needle tip, and the resisting product of the interfacial tension and the tip perimeter. Other forces interacting with drop formation are resistance power and inertial force, but these can be neglected in computations. The drop mass can be calculated from the equilibrium of the two main forces with the mass, m, gravity acceleration constant, g, external needle diameter, d_e, interfacial tension, γ, and the liquid density, ρ:

$$mg = \pi d_e \gamma \text{ and } m = \frac{\rho \pi d^3}{6}$$

The drop mass must however be corrected by a factor of 0.85 to get the true mass, because drops stretch out and leave a small portion behind when dropping from the needle. Even if very thin needles are used, it is very difficult to obtain droplets with a diameter of 1 mm or less (47,48).

Dropping with a Superposed Air Jet. The droplet size obtained from simply dropping can be reduced by superposing the drop forming process with an additional air jet. For calculating the droplet mass, the dragging forces of the airflow have to be taken into account. The viscous drag force is the laminar effect of the fluid, and the kinetic energy dissipation term represents the impact of the turbulence on the drag. Assuming that the drop diameter is larger than the tip, that there are no lateral fluid effects, and that the nascent drop has a spherical shape, the mass, m, of the generated droplet can be calculated using the gravity acceleration constant, g, liquid viscosity, μ, fluid velocity, v, droplet diameter, d_D, external needle diameter, d_e, the surface tension, γ, and the density of the fluid, ρ (49):

$$mg = \pi d_e \gamma - (3\pi \mu v d_D + 0.055\ \rho v^2 d_D^2)$$

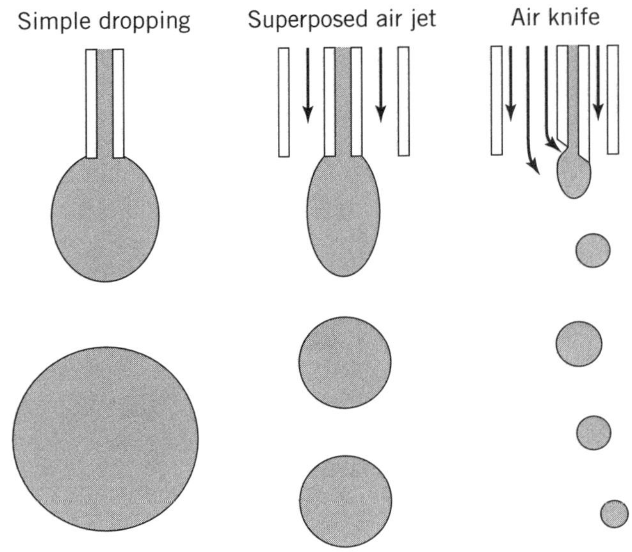

Figure 1. Monoaxial extrusion technologies based on simple dropping.

Depending on the immobilization material used, uniform droplets from 2 to 6 mm can be produced. Droplet production can be scaled up by use of several needles at once and has been successfully performed for the encapsulation of IgG-producing hybridoma cell cultures and yeasts (50). However, the increasing airflow influences the line of flight of the droplets, which becomes important if several adjacent needles are used simultaneously. As long as the superposed airflow is laminar ($Re < 0.3$, Stokes' area), droplets will fall in a perfectly vertical line. For turbulent superposed airflow producing periodical eddies ($1{,}000 < Re < 10^5$, Kármán vortex street), the diversion can be neglected up to $Re < 1{,}080$. Stronger turbulence may cause collisions of the falling drops but can be avoided if the distance between the needles is increased. At any rate, small satellite drops are formed and site distribution increases significantly.

This method uses an airstream that is concentric with respect to the droplet-forming needle. Excentric air sleeve positioning combined with a beveled droplet-forming needle facilitates the generation of very small droplets. The distal end of the needle is beveled at an angle of about 15 to 45°; the beveled end is disposed facing the central axis of the sleeve for the airstream. This arrangement leads to the formation of an air knife that generates droplets of 25–300 μm in diameter. The distribution of size is quite high due to the fact that the airstream not only drags the nascent drops but also breaks up the drop into smaller droplets. The rate of production is limited to 0.1–3.0 mL/min; the preferable limit is 0.3 mL/min (51,52).

Coaxial Dropping with a Superposed Air Jet. All of the systems mentioned so far include the mixing of the microcapsule-forming polymer with the material to be immobilized. This material is distributed evenly in the immobilization matrix and affects the membrane-forming process. In particular, if animal cells are immobilized in hollow spheres from polyelectrolyte complexes, the cells will be incorporated into the membrane. Because these membranes are very thin (1–5 μm) compared with those of animal cells (10–18 μm), several gaps result when these cells disintegrate. Complete cell and product retention or immunprotection cannot therefore be achieved. Coaxial dropping can be used to overcome this problem. This system uses one needle for the membrane-forming polymer solution and a second one, which is fixed in the center of the first, for the cell suspension. Both needles are of the same length. The nascent drop formed at their ends has a core containing the cell suspension and is enveloped with the polymer solution. While falling, the polymer solution completely surrounds the cells that comprise the core. The membrane-forming process in the precipitation bath is not affected. The core may have a volume of up to 25% of the whole drop. The droplet diameter can be varied from 1 to 5 mm. All the equations of the previous section are still valid.

This system has been scaled up to a productivity of 4 L/H by the simultaneous use of 24 concentric needles. To date this system has been successfully used for the immobilization of hybridomas (53), insect cell cultures (54), and the treatment of rheumatoid disease in mice, where microcapsules have been used intraperitoneously in mice containing a hybridoma cell line secreting antirheumatoid monoclonal antibodies (55).

Coaxial Dropping with a Superposed Liquid Jet. Superpositioning is not restricted to the use described in the last section; it can be used with any liquid immiscible with a drop-forming solution along with a precipitation bath. There are two main advantages to this method. Fluids still flow in a laminar fashion at higher velocities as compared with gases, and the viscous drag force is much higher. Both effects enable the production of hollow spheres from polyelectrolytes with a diameter ranging from 500 to 1,000 μm. This is considerably less compared with the previous system. Unfortunately this system is restricted to a productivity of 150 mL/h, and scale-up by use of multiple devices is problematic.

Electrostatic Extrusion. The application of a high static potential between the capillary and collecting solution has been used to improve droplet formation with respect to reducing the droplet size. The electric field force effectively pulls the forming droplet off the tip of the needle at a much lower mass (and hence size) compared with the simple dropping method. A series of small droplets (as small as

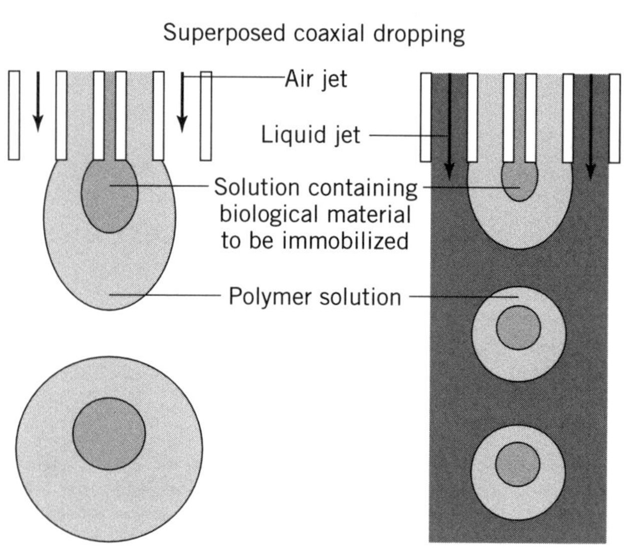

Figure 2. Coaxial extrusion technologies based on simple dropping.

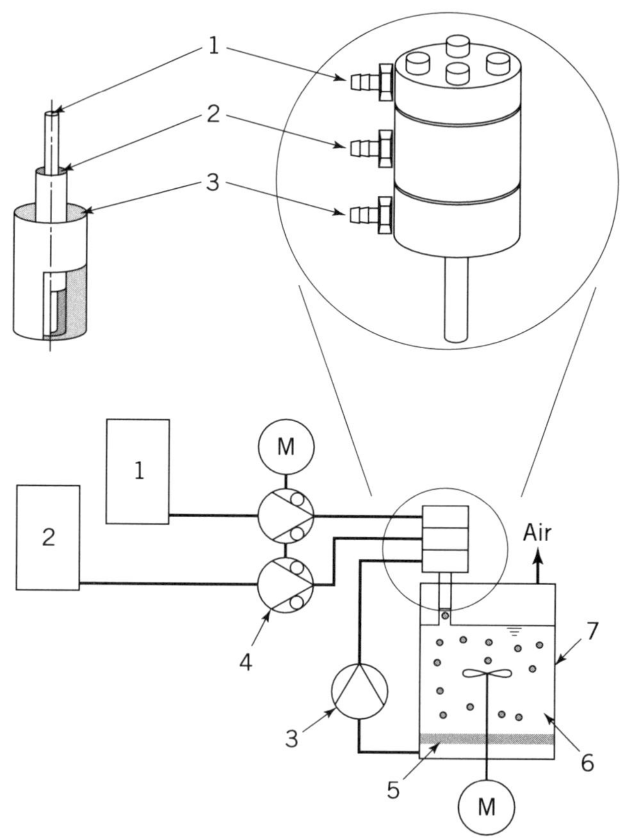

Figure 3. Schematic representation of the coaxial encapsulation apparatus (Torsten Steinau Verfahrenstechnik Berlin): (1) cell suspension, (2) cellulose sulfate solution, (3) inert carrier fluid, (4) multichannel peristaltic pump, (5) sieve for capsule retention, (6) polyDADMAC precipitation bath, and (7) stirred tank reactor.

26 μm in diameter) is generated with a comparably low standard deviation. The average microbead size can be easily adjusted from 2.5 to 0.2 mm by decreasing the concentration of the polymer solution, by using a higher-gauge needle, and by increasing the applied voltage. However there is a limit to how much the size can be reduced. In very high electric fields, droplets are no longer formed, and the solution pours from the needle in a steady stream. Charged molecules accumulate at the nascent drop's surface and counteract the surface tension. According to Lippmann's theory, the size of the drops can be determined with the electrostatic force, Fe, electric permitivity, ε, droplet diameter, d, the applied electrostatic potential, U, surface tension, γ, mass, m, and gravity acceleration constant, g:

$$mg = \pi d\gamma - Fe$$

$$mg = \pi d(\gamma - (k\varepsilon U^2/d))$$

$$d = 1/(6\ m/\pi\rho)^{1/3}$$

The correction factor, k, is difficult to evaluate. It depends on system design and may be a function of the form of the pending drop and of the electrostatic potential. This system has so far been applied to the encapsulation of cells in alginate beads (56,57), pancreatic islets in coated beads to study the influence of diameter and membrane compactness integrity and immunoprotection (58), and to the production of protein-loaded ethylene vinyl acetate beads for controlled-release purposes (59).

Rotating Disk. A liquid can be sprayed into a precipitation bath by using a rotating disk or a dish rotating at a high speed. The liquid for dispersion emerges from an orifice at the center of the axisymmetric rotating disk. The liquid leaves the disk in the form of small jets or ligaments, which form drops prior to falling into the precipitation bath. Droplet diameters, d, can be determined from the surface tension, γ, rotation speed (per second) of the disk, w, disk radius, r, and the liquid density, σ (60,61):

$$d = 0.425\ (\gamma/(w^2 r\sigma))^{0.5}$$

This system can be improved by applying a well-designed wave on the liquid flowing on the spinning disk. The ligaments are broken into uniform droplets with a standard deviation lower than 5%. Typically, a disk of 1 cm in diameter rotating at 2,000 rpm forms about 60 ligaments, which then form droplets of 500 μm at a flow rate of 6 L/h. Equations describing this process are much more complex than those for classical jet rupture. Physical properties of the liquid, design of the rotating disk, and wave frequency and amplitude must be adjusted to achieve correct particle size. This method is currently being used for experimental research and agriculture (62).

Rotating Nozzles. Instead of a rotating disk, a perforated rotating cylinder can be used. This cylinder can be placed in a reactor, with the reactor serving as the precipitation bath and the culturing vessel. Monodisperse and spherical drops are produced by rotating the cylinder at a velocity that hurls the drops directly down into the reactor liquid. The size of the beads is a function of the cylinder's rotation velocity, with a size of 1–3 mm and a deviation of up to 10%. The diameter and number of holes drilled in the cylinder can be varied according to the desired capacity, which can be up to hundreds of liters of alginate beads per hour. The volume, V, of the droplets can be calculated from the hole diameter, d, surface tension, γ, density of the liquid, ρ, gravity acceleration, g, and a correction factor, k, which depends on the shape of the holes:

$$V = (2\pi d)/(k\rho g)$$

The variation in bead diameter, d, for gelled alginate beads can be expressed as a function of the rotation velocity, ω:

$$d = (1/\omega^2)^{1/3}$$

As has been observed for simple dropping, the diameter, d, has to be corrected by a factor of 0.84 to 0.95, depending on the rotation velocity used, because the drop leaves behind a portion of the pendant drop upon breaking. The landing radius of the drops on the liquid surface must be taken into account to avoid splattering of the drops on the reactor walls. The landing radius, r, is dependent on the velocity, v, of the drops when leaving the hurler, the height, h, of the holes above the liquid surface, and the acceleration force, g:

$$r = v(2h/g)^{0.5}$$

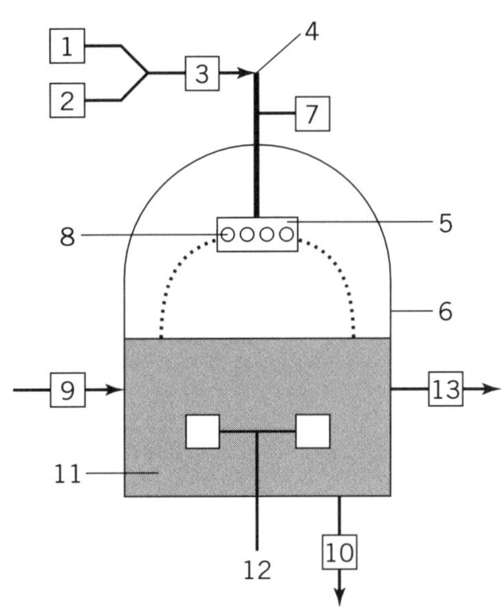

Figure 4. Schematic representation of the Biosphere (Landteknikk A/L): (1) biocatalyst solution, (2) polymer solution, (3) mixing chamber, (4) rigid tube, (5) perforated cylinder, (6) bioreactor, (7) velocity regulation unit, (8) nozzles, (9) feed pipe, (10) product outlet, (11) precipitation bath, (12) stirring device, and (13) reactor outlet.

Liquid-Jet-Based Methods

Dropping methods can be scaled up only by the parallel use of several needles, because drop formation depends on a relatively slow liquid velocity. Hence, dropping methods are limited to about 200 mL/h per needle. To overcome this problem, several methods for producing droplets from liquid jets have been developed, and several liters of microcapsules can be produced. The maximum productivity of these systems is limited due to the fact that the pressure needed to ensure flow increases with the second power of the liquid velocity, and with increasing liquid viscosity and decreasing orifice diameter. Pressure can be calculated using the following equations with pressure, P, resistance coefficient, χ, liquid density, ρ, liquid velocity, v, needle diameter, d, needle length, I, Reynold's number, Re, and the liquid viscosity, η:

$$\Delta P = (\chi \rho v^2/2)(l/d)$$

$$Re = v\rho d/\eta$$

and

$$\chi = 64/Re \text{ for } Re < 2{,}320 \text{ (laminar flow)}$$

or

$$\chi = 0.3164/Re^{0.25} \text{ for } 3{,}000 < Re < 10{,}000 \text{ (turbulent flow)}$$

or

$$\chi = 0.0054 + 0.3964/Re^{0.3} \text{ for } 20{,}000 < Re < 10^6 \text{ (turbulent flow)}$$

Dipping Jet. The dipping jet method is the simplest process for producing a large amount of microcapsules. It depends on the fact that a liquid jet at high velocity disintegrates into many drops when penetrating a fluid. Drop formation depends on irregular dynamic processes and leads to drops with a large size distribution (63,64). This method has been used to immobilize the fungi *Claviceps purpurea* in alginate particles for further production of alcaloids in a gas–solid fluidized-bed reactor. Researchers produced 110-L capsules with a mean diameter of 2.5 mm (65).

Vibrating Jet Breakage. Lord Raleigh influenced liquid jets with tuning forks in the late nineteenth century. He observed that vibration can force a liquid jet to disintegrate into equal drops if the wavelength of the vibration is greater than the perimeter of the liquid jet. He showed that the frequency, f, for maximum instability is related to the jet velocity, v, and the wavelength, λ (66):

$$f = v/\lambda$$

When a laminar-flowing jet is mechanically disturbed at this frequency, uniform drops are formed. The optimum wavelength, λ_{opt}, for breakup can be calculated with the needle diameter, d, dynamic viscosity, η, liquid density, ρ, and the surface tension, σ (67):

$$\lambda_{opt} = 2^{0.5}\, \pi d(1 + (3\eta/(\rho\sigma d)^{0.5}))^{0.5}$$

By considering that the jet will break into liquid cylinders with the wavelength λ, each of which will form uniform droplets, the diameter of the droplets, d, can be calculated:

$$d = (1.5\, d^2\lambda)^{1/3}$$

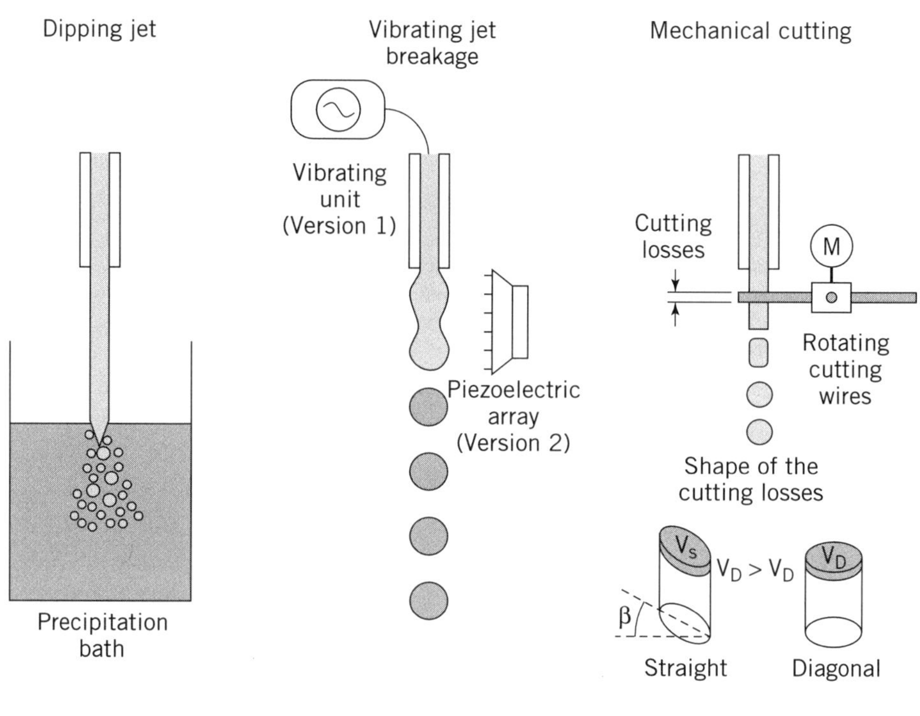

Figure 5. Liquid-jet-based extrusion technologies.

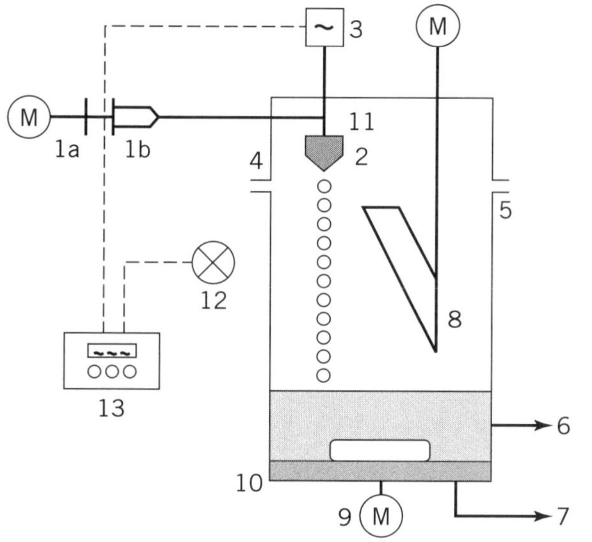

Figure 6. Schematic representation of the Encapsulator AP (INOTECH AG): (1a) syringe pump, (1b) syringe containing polymer solution and cell suspension, (2) vibrating nozzle, (3) vibration unit, (4) pulsation unit, (5) reaction vessel, (6) reactor outlet for capsule harvesting, (7) waste outlet, (8) bypass unit (used to eliminate undesired beads at the beginning and the end of bead formation), (9) magnetic stirrer, (10) sieve, (11) capillary, (12) stroboscope lamp, and (13) frequency generator.

Mechanical Cutting. The principle of the mechanical cutting encapsulation method is to cut a liquid jet with a rotating wire into a series of uniform liquid cylinders behind the cutting wire. Due to surface tension, the liquid cylinders form drops immediately as they fall down into the precipitation bath. The bead build volume, V, of the cylinder can be calculated from the cutting angle, β, which is a function of the liquid velocity, v, and the local velocity of the wire, w, at the cutting point. Other relevant parameters are the thickness of the wire, d_w, the number of revolutions, n, of the cutting wire, the number of wires, z, and the nozzle diameter, d.

$$V = 0.25\ \pi d^2((v/(nz) - (d_w + d\ \sin\beta)/\cos\beta))$$
$$\text{with } \beta = \arctan(\text{v/w})$$

Because the wire used for cutting cannot be infinitesimally small, and its size is also determined by the material, it slings away about 10–20% of the liquid jet; this can be considered lost material. This cutting loss, V_L, can be determined as follows:

$$V_L = 0.25\ \pi d^2(d_w/\cos\beta + d\ \tan\beta)$$

When the cutting device is arranged at an optimal angle, α, depending on the liquid jet's velocity, the the drop volume is affected, and cutting loss can be reduced further:

$$\alpha = \arcsin(v/w)$$
$$V = 0.25\ \pi d^2((v/nz) - d_w)$$
$$V_L = 0.25\ \pi d^2 d_w$$

The cutting loss can also be reduced by use of a thinner wire for cutting. However, if a stainless steel wire becomes thinner than 0.2 mm, distortion effects of the wire can be observed when using fluids of high viscosity. Therefore, special stabilized wires 0.06 mm in diameter have been designed; in combination with diagonal cutting, they reduce cutting loss to only 3.3% (68,69).

This method is especially suited to produce beads from high-viscosity polymer solutions, with high productivity rates of up to 20 kg/h for 1-mm beads. This method has been used to immobilize a variety of microorganisms for further use in wastewater treatment experiments.

Liquid–Liquid Emulsification

Microcapsules can be obtained from dispersion of one liquid into another followed by a membrane-forming process. Generally, dispersions can be obtained from static and dynamic mixers. However dynamic mixers are preferred for microencapsulation because of the need to stabilize the dispersion until microcapsule formation has taken place, otherwise coalescence leads to increasing drop size.

Dynamic Mixer. It is advantageous to use an impeller that disperses the lighter liquid in both the center and the periphery at the same impeller speed. The optimum impeller diameter, d, depends on the vessel diameter, D, and presence of baffles:

$$d = D/3 \text{ (impeller)}$$

or

$$d = 0.4\ D \text{ (impeller in baffled vessels)}$$

The minimum agitation speed, n, required to disperse a dispersed (index D) liquid into a continuous (index C) liquid by using a four-paddled stirrer centered in a flat-bottomed cylindrical vessel with a liquid height equal to the vessel diameter, D, can be calculated from the liquid densities, ρ, and the viscosity, η_H, of the continuous phase:

$$n = 750\ D^{-2/3}\ (\eta_c/\rho_c)^{1/9}\ ((\rho_c - \rho_D)/\rho_c)^{0.26}$$

The viscosity of the dispersed phase and the interfacial tensions can be neglected (70). Calculating the mean particle size is fairly difficult and depends on type and diameter of the stirrer, agitation rate, and the physical properties of the liquids and their relative amounts (71). Generally, droplet size and size distribution decreases with increasing agitator speed. Droplet sizes ranging from a few micrometers to several hundreds of micrometers have been reported. A variety of correlations have been published relating the Sauter mean diameter, d_{32}, of the droplets to the vessel geometry and the physical properties of the dispersion system. The most well known correlation is derived from the impeller diameter, d_I and speed, n, the Weber number, N_{We}, liquid density, ρ, interfacial tension, η, the dispersed phase volume fraction, φ, and two correlation factors, b and c (72):

$$d_{32}/d_l = b(1 + c\varphi)\, N_{\mathrm{We}}^{-0.6}$$
$$N_{\mathrm{We}} = \rho n^2 d_l^3/\eta$$

Due to the considerable internal sheer forces applied during dispersion, this method is especially suited for microencapsulation of inanimated matter (73–75) or viruses and microorganisms (76,77), but it can also be adjusted to the needs of cell cultures (78,79).

SELECTED MICROCAPSULE-FORMING MATERIALS

There are many microcapsule-forming materials available, indicating that there is no such thing as an all-purpose material. Hence, various materials have been proposed as carriers. They are based on either natural polymers, such as the polysaccharides alginate, carrageenan, agarose (80), and chitosan (81,82), and proteins such as gelatin (83,84) and collagen, or synthetic polymers such as polyacrylates, polyurethane and its less-toxic derivates (85), and polyethers. Each material has its pros and cons with respect to physicochemical stability, biodegradability or degradation into units small enough for renal excretion (86,87), and influence on the entrapped cells. Sulfated polyanions, for example, have been successfully used to obtain and maintain stable single-cell suspensions of insect cells (88), whereas these polymers exhibit the opposite effect: the formation of multicellular spheroids on mammalian cells (89). Technical considerations such as simplicity of preparation, availability, and cost also have to be taken into account.

Calcium Alginate

Alginic acid is a polyuronic acid extracted from brown seaweeds and some bacteria, mostly derived from the genus *Pseudomonas*. Chemically, alginates are block copolymers of 1,4-linked β-D-mannuronic acid (M) and α-L-guluronic acid (G). The mannuronate residuals of the bacterial alginate are acetylate to some extent at positions O-2 and/or O-3. The residues, occuring in varying proportions, depend on the source, and the residues are arranged in block patterns comprising homopolymeric regions (MM blocks and GG blocks) linked with alternating regions of heteropolymeric regions (MG blocks). Divalent cations, such as Ca^{2+}, bind preferentially between GG blocks, which appear to have preferred binding sites. Therefore, the composition (20–75% G content) and molecular size of the polymer impacts on the porosity, stability, gel strength, and swelling behavior (90–93).

Interfacial polymerization of alginate and Ca^{2+} is instantaneous, with precipitation of calcium alginate followed by a more gradual gelation of the interior as calcium ions penetrate through the alginate. During formation of gel beads, the biopolymer localized at the bead center moves gradually to the bead surface by attracting the ionotropic cations, which induce gelation. Consequently, the immobilized material flows with the macromolecules to the periphery of the bead, producing a superficial crust. At the same time the bead volume is reduced due to cross-linking of the macromolecules and expulsion of water. The interior of the bead consists of a fluffy core surrounded by a medium gelled zone containing supramolecular structures such as shafts and cavities. These structural inhomogeneities originate from gradients of local concentrations of components during the gelling process or the presence of chemically inert substances such as PEG. They remarkably affect the strength of beads and their permeability for low and high molecular weight components (94,95).

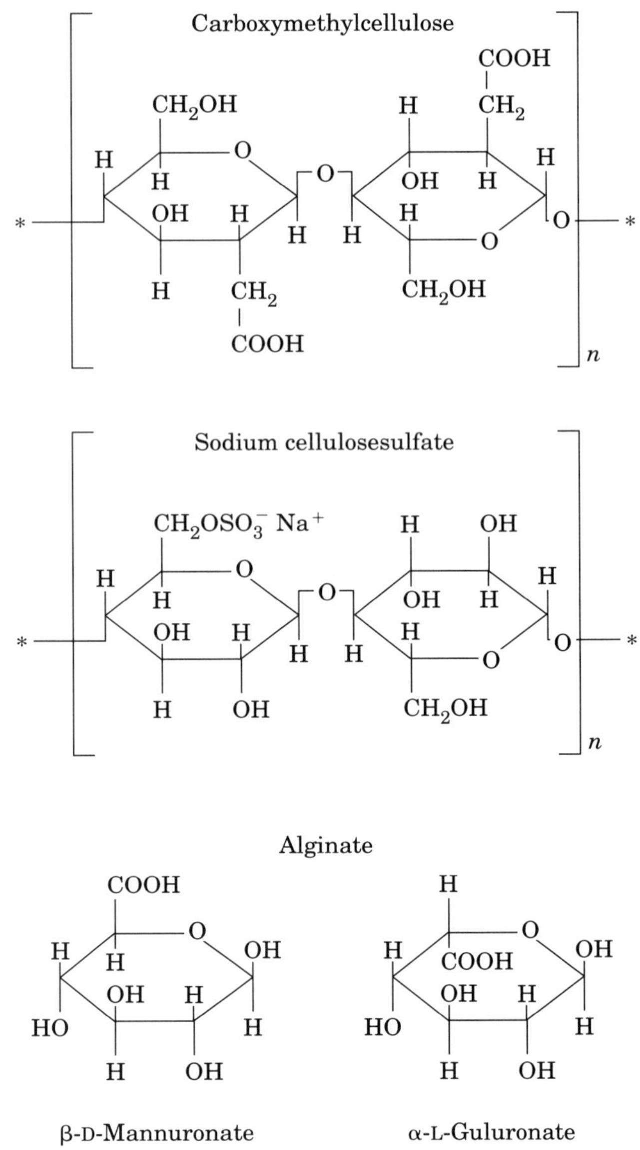

Figure 7. Polysaccharides used for encapsulation (I).

Calcium alginate beads are very stable for long periods and can tolerate a pH range from 3 to 10 at 25 °C. Thermal stability is also good, and beads can withstand incubation temperatures of up to 85 °C at pH 7.0. Destabilization of calcium alginates is promoted by chelating agents, which remove divalent cations. Phosphate, citrate, or EDTA should be avoided. The breaking of beads during fermentation of microorganisms is mainly caused by low initial alginate and/or high initial cell concentration, whereas the length of incubation has only a minor impact (96,97).

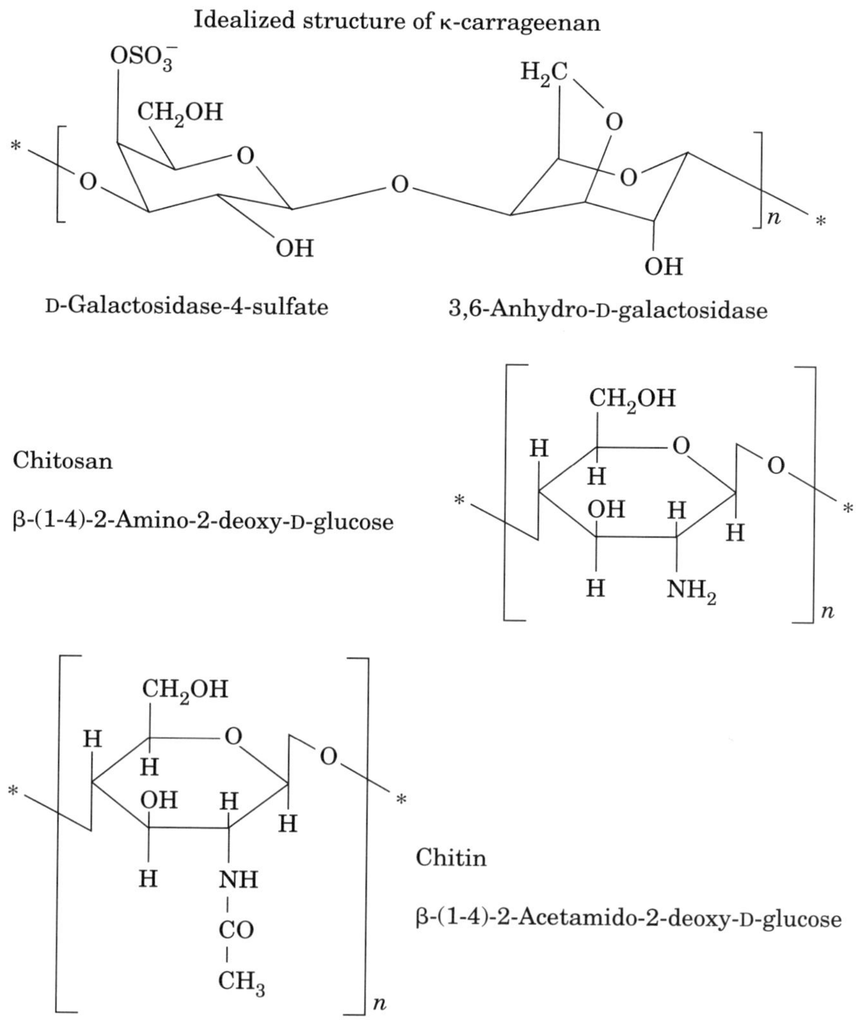

Figure 8. Polysaccharides used for encapsulation (II).

Microencapsulation in alginate beads is an inexpensive, versatile, and nontoxic method. It is used in food process industries to enhance fermentation productivity and cell stability. Costs for downstream processing are reduced by facilitating cell recovery and recycling (98). In the beverage industry, several processes including several different microorganism species have been developed for producing beer (99), wine (100), cider, (101) and ethanol (102).

Coating with Poly-L-Lysine

To prevent cell leakage from alginate beads in general and to provide immobilized cells with a semipermeable immunisolation barrier in particular, the coating of alginate beads with poly-L-lysine (PLL) has been introduced. This system is currently the best investigated and by far the most widespread microencapsulation system used in medicine.

First, viable cells are immobilized within an alginate bead. The bead is washed and transferred to another precipitation bath containing a PLL solution. Because the alginate is a negatively charged polymer, simplex membrane formation with the positively charged PLL is instant, decreasing the molecular weight cut-off of the system to about 100 kDa. Finally, the now-coated capsules are washed again and transferred to another precipitation bath containing a dilute solution of more oppositely charged polymers, with respect to the outer PLL surface. This final polyanion may be PEI (103) or alginate (104). The purpose of the additional coating is to neutralize nonreacted PLL and to mask the PLL that otherwise might provoke unwanted immune or fibrotic response. Because the outmost layer consist of alginate, purification of the alginate prior to use improves biocompatibility. Bioincompatibility is influenced by the chemical composition and, to a greater extent, by individual imperfections of the capsules; it has been clearly shown that fibrotic response against implanted capsules is associated with the presence of PLL around the capsules (105,106).

The stability of the PLL layer is governed mainly by the composition of the alginate, where alginate with a high M content ensures both more rapid binding of the PLL during the coating step and much thicker PLL layers. Because the reaction between the polymers is very fast, the rate-limiting step is the diffusion of PLL into the alginate gel. The free diffusion of the PLL is determined by gel porosity and the relative molecular weight of the polycation. Therefore the PLL layer can be improved through an increased concentration of PLL and by using a lower molecular weight sample of PLL. However, a certain minimum polymer length is necessary to obtain stable capsules, so optimum PLL molecular weight is believed to be around 18 kDa (107,108).

Improper performance of the microcapsules can be caused by a distorted coating process due to the presence of living cells on the surface area of the alginate bead. As previously pointed out, immobilized material has the tendency to be concentrated in the outer areas of alginate beads. The consequence is that small alginate beads are even more liable to improper coating caused by protruding cells than large beads, while cells in large beads are more likely to suffer from mass transport limitations. The additional entrapment of microspheres within a larger sphere can cover undesirable cell protuberances and debris integrated or exposed on the surface of the microsphere with a complete spherical shell of calcium alginate prior to the following coating process (109). Because only a small fraction of the alginate may diffuse throughout the coacervate membrane, the capsules are liable to osmotic swelling under physiological conditions because the cross-linking divalent ions of the bead are exchanged with monovalent ions. This swelling can also lead to capsule breakage. Strategies to reduce swelling include strengthening of the PLL layer by simply letting more polycations bind to the capsule surface, exchanging of a fraction of the calcium ions with ions with higher affinity toward the G blocks such as strontium or barium, using more alternate PLL–alginate layers, and decreasing capsule diameter. PLL-coated alginate capsules have been successfully used to treat diabetes in humans for periods of over one year (110–114).

Polyelectrolytes

Polyelectrolytes are polymers that contain a charged group in their monomer repeating unit. They have been used as a flocculent in water purification, protein precipitation, and membrane technology. When brought into contact, aqueous solutions of oppositely charged polyelectrolytes immediately form water-insoluble but highly swollen precipitates or membranes. Membrane formation is completed within minutes.

Although there are is a vast number of possible polyanion polycation combinations, only a minority can be used for microencapsulation. For the encapsulation of living cells, it is strongly recommended that a polyanion be used for the inner polymer. This is in contact with the cells during capsule formation. The reason for this is that polycations are cytotoxic, especially those that contain quarternary ammonium groups such as PDADMAC. Polycations interact with the negatively charged phosphate groups of the cytoplasmic membrane, which finally leads to membrane rupture and cell death. This has been extensively studied with liposomes (115). Additionally, polycations form stable complexes with DNA and proteins (116,117). Membrane-bound polycations do not show these effects at all.

The best-suited polyanions to form a cell suspension containing drop are highly charged polysaccharide derivates with a molar mass in the range of 10^6 Da, with a cyclic backbone and pH-dependent charge. Typical examples are cellulose sulfate, carboxymethyl cellulose, and alginate. The corresponding polycation should contain either permanently charged quarternary amonium groups (PDADMAC) or a tertiary amine. The charge density should be between 50 and 100 mol%. The molecular mass may range from 10^3 to 10^6 Da, with high molecular weight species requiring a higher charge density to compensate for the lower diffusion coefficient (118–121).

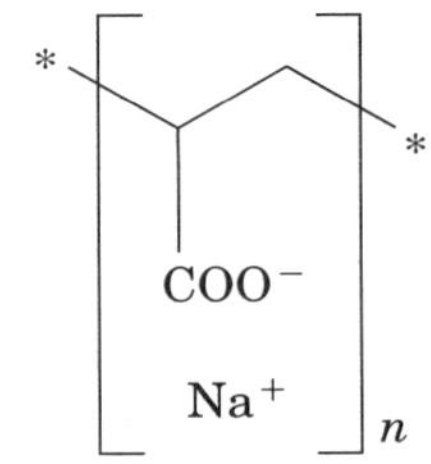

Poly(acrylic acid)

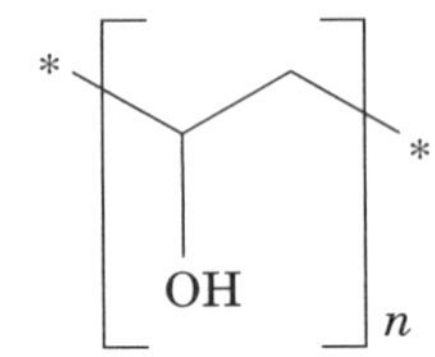

Poly(vinyl alcohol)

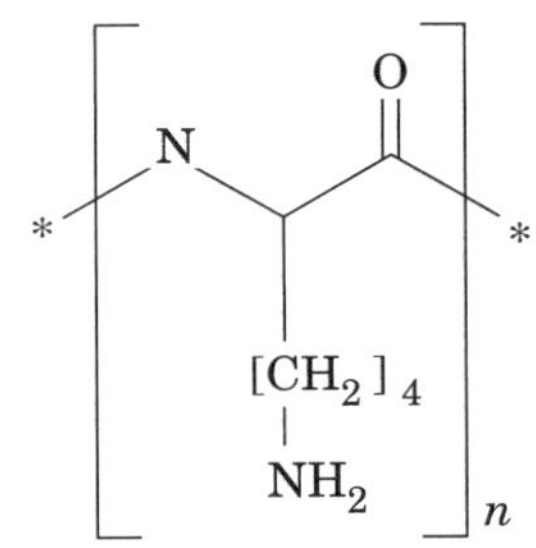

Poly(L-lysine)

Figure 9. Polymers used for encapsulation.

Microcapsules from CS and PDADMAC exhibit immense mechanical stability up to 7 N per 3-mm capsule (122). About 0.3 N is required for this type of capsule to withstand the stress occurring in a heavily stirred tank reactor. The capsules typically possess a thin but very compact primary layer (a few microns) followed by a less densely packed region joining the inner layers of high porosity. The membrane properties mainly depend on the primary layer; therefore distortion of the membrane-forming reaction leads to microcapsules with very different properties. The molecular weight cut-off can be finely tuned according to the needs of the application. It can be as low as 5,500 Da for complete product recovery (123) or immunoprotection (40,124), in the range of 4.4×10^5 to 2.0×10^6 Da for continuous production of monoclonal antibodies

(53,125), or within the range of several microns to allow free virus passage for some gene therapy (126) and insect cell culture (54) applications.

Poly(Vinyl Alcohol)

An aqueos poly(vinyl) alcohol (PVA) can be cross-linked through hydrogen bonding by a freeze–thaw process with hydroxyl groups on PVA molecules. The typical freezing temperature is around −20 °C. The mechanical strength of the cryogel greatly depends on the thawing rate, and shear modulus increases with decreasing thawing rate (preferably slower than 0.2 °C/min). During slow thawing, the system resides in the temperature range from −4 °C to −1 °C for a long time, and numerous hydrogen bonds are linked. It is important to consider that such interactions occur relatively slowly due to the high viscosity of the liquid microphase and the low temperature of the frozen system. Stability can be further increased by using a more highly concentrated PVA solution and storage in a frozen state. The ability of the gel to melt depends strongly on the preparation process. Thermoresistance is excellent, and immobilized thermophiles in PVA beads have operated at 65 °C for a period of 1 year and longer (127).

As a result of the formation of ice crystals during freezing, the cryogel has a high micro- and macroporous structure, which is necessary for nonhindered mass transport of any solutes for cultivation purposes. The polymer itself is nontoxic and biocompatible and is not degraded by most enzymes and microbes (128). The gel beads are stable in any biotechnology liquid media; swelling or softening can occur only under harsh conditions such as extreme pH or high concentration of chaotropic substances such as urea. Physiological concentrations of most compounds do not affect the gel or cause it to shrink, which increases stability (129,130). Apart from the freeze–thaw method, PVA can be cross-linked chemically by additions of compounds such as aldehydes or borates to form a hydrogel. Chemical cross-linking creates a gel of a comparable low strength; it can be described as having an appearance and strength similar to common gelatine (131). It is a very suitable method and, with the exception of the cross-linking reaction, nontoxic biomaterial has been studied extensively as membranes in artificial kidneys and as materials for skin replacement and contact with blood (132). It is therefore especially suited for coating beads. Microcapsules from PVL cryogels seem to be especially useful for use in wastewater treatment for removal of nitrogen (133,134), heavy metals (135), and environmental decontamination (136,137).

Poly(Methylmethacrylate-Hydoxyethylmethacrylate)

Poly(methylmethacrylate-hydroxyethylmethacrylate) (HEMA-MMA), which is 75–80% HEMA and 25–20% MMA, is a nonionic random copolymer. The polymer is water insoluble, which makes encapsulation more difficult. However water insolubility is considered an advantage because this is presumed to be a key factor leading to greater biocompatibility.

The cell suspension and the polymer are mixed together in a biocompatible solvent such as polyethylene glycol (PEG), and droplets are formed by jet extrusion or emulsification in another liquid. The membrane formation is initiated upon removal of the polymer solvent. This can be achieved easily by transferring the droplets in a phosphate-buffered saline (PBS) precipitation bath. The PEG is also soluble in the PBS solution, thus rapidly depleting the polymer solvent at the interface of the droplets, leading to a dense outer layer that provides all the resistance to the mass transfer. Below the dense layer are macroporous "trabecular" layers, similar to asymmetric phase-intervention membranes. Typically the membrane thickness ranges from 20 to 150 μm, with an average of 60 μm. The molecular weight cut-off of the membrane is above 52 kDa, possibly above 340 kDa, but can be influenced by the polymer composition, whereas the permeability of the membrane increases with higher HEMA content of the polymer and when the solubility parameters of the solvent and the precipitate are both high. Furthermore, the precipitation media influences the permeability of the microcapsule membrane with Tris-buffered glycerol, leading to denser

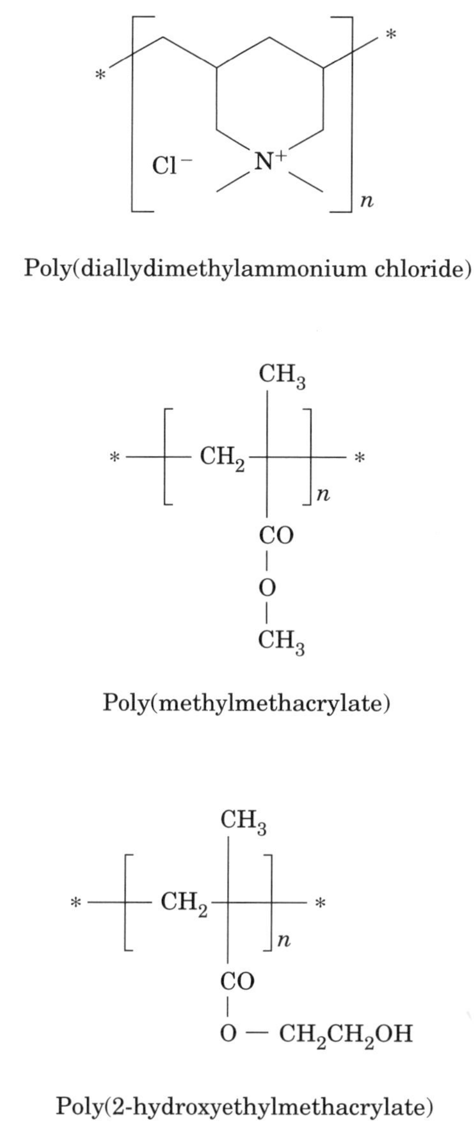

Figure 10. Synthetic polymers used for encapsulation.

membranes that reflect the slower rate of precipitation in PBS or unbuffered glycerol. The membrane-forming process is not absolutely reliable with respect to complete membrane formation, and procedures to screen and remove capsules containing pinholes or flaws might be necessary for certain applications (138–140).

BIBLIOGRAPHY

1. A. Groboillot, D.K. Boadi, D. Poncelet, and R.J. Neufeld, *Crit. Rev. Biotechnol.* **14**(2), 75–107 (1994).
2. Z.M. Sun, D. Poncelet, J. Conway, and R.J. Neufeld, *J. Microencapsulation* **12**(5), 495–504 (1995).
3. V.A. Santos, J. Tramper, and R.H. Wijffels, *Biomater. Art. Cells Immob. Biotech.* **21**(3), 317–322 (1993).
4. Y. Andrès, H.J. MacCordick, and J.C. Hubert, *Appl. Microbiol. Biotechnol.* **44**, 271–276 (1995).
5. N.G. Wilson and G. Bradley, *J. Appl. Bacteriol.* **80**, 99–104 (1996).
6. E. Poscillico, *Biotechnology* **4**, 114–117 (1986).
7. U.S. Pat. 5,116,747 (May 26, 1992), M. Moo-Young, N.C. Bols, S.E. Overgaard, and J.M. Scharer (to University of Waterloo).
8. Y. Shirai, R. Sasaki, K. Hashimoto, H. Hitomi, and H. Chiba, *Appl. Microbiol. Biotechnol.* **29**, 544–549 (1988).
9. M. Seki and S. Furusaki, *Chemtech* **26**(3), 41–45 (1996).
10. K. Redenbaugh, *Synseeds: Applications of Synthetic Seeds to Crop Improvement*, CRC Press, Boca Raton, Fla., 1993, pp. 3–7.
11. J.I. Shigeta, *Biotechnol. Techn.* **9**(10), 771–776 (1995).
12. Y. Bachiri, C. Gazeau, J. Hansz, C. Morisset, and J. Dereuddre, *Plant Cell Tissue and Organ Culture* **43**, 241–248 (1995).
13. F. Lim and A.M. Sun, *Science* **210**, 908–910 (1980).
14. U.S. Pat. 5,578,314 (November 26, 1996), K.C. Cochrum, R.E. Dorian, and S.A. Jemtrud (to the Regents of the University of California).
15. I.T. Tai and A.M. Sun, *FASEB J.* **7**, 1061–1069 (1993).
16. P. Aebischer, N.A.-M. Pochon, B. Heyd, N. Déglon, J.-M. Joseph, A.D. Zurn, E.E. Baetge, J.P. Hammang, M. Goddard, M. Lysaght, F. Kaplan, A.C. Kato, M. Schluep, L. Hirt, F. Porchet, and N. De Tribolet, *Hum. Gene Ther.* **7**, 851–860 (1996).
17. E.B. Reilly, G. Antognetti, J.S. Wesolowski, Jr., and P. Sakorafas, *J. Immunol. Methods* **126**, 273–279 (1990).
18. S.R. Win, P.A. Tresco, B. Zielinski, L.A. Greene, C.B. Jaeger, and P. Aebischer, *Exp. Neurol.* **113**, 322–329 (1991).
19. L. Christenson, D.F. Emerich, and P.R. Sanberg, *Mov. Disord.* **7**, 185–186 (1995).
20. J.K. Embleton and B.J. Tighe, *J. Microencapsulation* **10**(3), 341–351 (1993).
21. K.A. Brown, C.A. Moser, T.J. Speaker, C.A. Khoury, J.E. Kim, and P.A. Offit, *J. Infect. Dis.* **171**, 1334–1338 (1995).
22. D. Chickering, J. Jacob, and E. Mathiowitz, *Biotechnol. Bioeng.* **52**(1), 96–101 (1996).
23. B. Katzbauer, M. Narodoslawsky, and A. Moser, *Bioprocess Eng.* **12**, 173–179 (1995).
24. C. Cokmus and Y.M. Elçin, *Pestic. Sci.* **45**, 351–355 (1995).
25. PCT WO 93/15190 (August 5, 1993), M. Lysberg, I. Storrø, H. Svendsen, O. Eide, and B. Oterholm (to Norske Meierier).
26. U.S. Pat. RE28779 (April 20, 1976), S. Katayama, H. Matsukawa, J. Matsuyama, and M. Yamamoto (to Fuji Foto Film Co., Ltd.).
27. R. Pommersheim, J. Schrezenmeir, and W. Vogt, *Macromol. Chem. Phys.* **195**, 1557–1567 (1994).
28. M.L. Huguet and E. Dellacherie, *Process Biochem.* **31**(8), 745–751 (1996).
29. U.S. Pat. 5,410,016 (April 25, 1995), J.A. Hubbell, C.P. Pathak, A.S. Sawhney, and N.P. Hill (to Board of Regents, the University of Texas System).
30. U.S. Pat. 5,464,932 (November 7, 1995), H.R. Allcock, C.G. Cameron, and D.E. Smith (to the Penn State Research Foundation).
31. U.S. Pat. 5,529,914 (June 25, 1996), J.A. Hubbell, C.P. Pathak, A.S. Sawhney, N.P. Desai, and S.F.A. Hossainy (to the Board of Regents, the University of Texas Systems).
32. U.S. Patent 4,352,883 (October 5, 1978), F. Lim (to Damon Corp.).
33. F. Lim and A. Sun, *Science* **210**, 908–910 (1980).
34. P. Soon-Shiong, P. Sandford, R. Heintz, T. Zheng, Q. Yao, Z. Yao, N. Meredith, M. Schmehl, M. Moloney, and M. Murphey, *Trans. Soc. Biomater.* **356** (1994).
35. P. Soon-Shiong and P. Sandford, *Surg. Technol. Int. IV,* 93–99 (1995).
36. H. Dautzenberg, F. Loth, K. Fechner, B. Mehlis, and K. Pommering, *Macromol. Chem. Suppl.* **9**, 203–210 (1985).
37. P. Spiekermann, K.-D. Vorlop, and J. Klein, *Proc. 4th Eur. Cong. Biotechnol.* **3**, 590–593 (1987).
38. J.E. Tyler, *Bioprocess Technol.* **10**, 343–361 (1990).
39. L.C. Colombi, *Feed Mix* **4**, 12–16 (1995).
40. J. Stange, S. Mitzner, H. Dautzenberg, W. Ramlow, M. Knippel, M. Steiner, B. Ernst, R. Klinkmann, and H. Klinkmann, *Biomater. Art. Cells Immob. Biotechnol.* **21**, 343–352 (1993).
41. R.M. Saller, J. Stange, S. Mitzner, K. von Rombs, P. Hutzler, H. Dautzenberg, W.H. Güntzburg, and B. Salmons, in H. Dautzenberg and D. Poncelet eds., *Bioencapsulation Research Group, International Workshop, Bioencapsulation V: From Fundamentals to Industrial Applications,* Potsdam, Germany, September 23–25, 1996.
42. D. Hunkeler, A. Prokop, S. DiMari, M.A. Haralson, and T.G. Wang, in H. Dautzenberg and D. Poncelet eds., *Bioencapsulation Research Group, International Workshop, Bioencapsulation V: From Fundamentals to Industrial Applications,* Potsdam, Germany, September 23–25, 1996.
43. T. Wang, I. Lacík, M. Brissová, A.V. Anilkumar, A. Prokop, D. Hunkeler, R. Green, K. Shahrokhi, and A. Powers, *Nat. Biotechnol.* **15**, 358–362 (1997).
44. U.S. Pat. 5,053,332 (October 1, 1991), R.B. Cook, R.B. Provonchee, and S. Nochumson.
45. U.S. Pat. 5,160,745 (November 3, 1992), P.P. DeLuca and F. Rypacek (to the University of Kentucky Research Foundation).
46. D. Poncelet, C. Dulieu, and R.J. Neufed, in D. Poncelet and V. Santos eds., *Bioencapsulation Research Group, International Workshop, Bioencapsulation VI: From Fundamentals to Industrial Applications,* Barcelona, Spain, August 30–September 1, 1997.
47. W.D. Harkins and F.E. Brown, *J. Am. Chem. Soc.* **41**, 499 (1919).
48. M.C. Levy and D. Poncelet, *Biofuture* **132**, 16–21 (1994).
49. W. Lane, *Rev. Sci. Instrum.* **24**, 98–101 (1947).
50. U.S. Pat. 4,814,274 (March 21, 1989), T. Shioya and Y. Shiinoki (to Snow Brand Milk Products Co., Ltd.).
51. M. Van Dyke, *An Album of Fluid Motion,* Parabolic Press, Stanford, 1982, p. 86.
52. U.S. Pat. 5,521,079 (May 28, 1996), R.E. Dorian and K.C. Cochrum (to the Regents of the University of California).

53. M.G. Cho, *Verfahrenstechnische Auslegung einer Apparatur zur Herstellung mikroverkapselter Biokatalysatoren mit getrennter Zuführung von Katalysatorlösung und Kapselgrundsubstanz,* VDI Fortschrittsberichte, VDI Verlag, Düsseldorf, 1994, pp. 93–102.
54. H. Hübner and R. Buchholz, in K. Maramorosch and J. Mitsuhashi eds., *Invertebrate Cell Culture: Novel Directions and Biotechnology Applications,* Science Publishers, Enfield, N.H., 1997, pp. 123–130.
55. C. Kloth, C. Riese, M. Cho, R. Buchholz, and U. Marx, in E.C. Beuvery, J.B. Griffiths, and W.P. Zeijlemaker eds., *Animal Cell Technology: Developments towards the 21st Century,* Kluwer Academic, Dordrecht, The Netherlands, 1995, pp. 765–769.
56. B. Bugarski, B. Amsden, R.J. Neufeld, D. Poncelet, and M.F.A. Goosen, *Can. J. Chem. Eng.* **72**, 517–521 (1994).
57. B. Bugarski, Q. Li, M.F.A. Goosen, D. Poncelet, R.J. Neufeld, and G. Vunjak, *MchE J.* **40**, 1026–1031 (1994).
58. B.R.S. Hsu, Y.S. Ho, S.H. Fu, Y.Y. Huang, S.C. Chiou, and H.S. Huang, *Transplant. Proc.* **27**(6), 3227–3231 (1995).
59. B.G. Amsden and M.F.A. Goosen, *J. Controlled Release* **43**, 183–196 (1997).
60. R.P. Fraser and P. Eisenklamm, *Trans. Inst. Chem. Eng.* **34**, 294–319 (1956).
61. H. Brauer, *Grundlagen der Einphasen- und Mehrphasenströmung,* Verlag Sauerländer, Frankfurt, 1971, pp. 272–352.
62. Eur. Pat. EP0446134 (September 11, 1991), J.-P. Renaudeaux and J.-M. Chicheportiche (to Government of France, represented by Delegue General pour l'Armement).
63. W. Ohnesorge, *Z. angew. Math. Mech.* **16**(6), 355–358 (1936).
64. R.M. Christiansen and A.N. Hixon, *Ind. Eng. Chem.* **49**(6), 1017–1024 (1957).
65. K. Merkel, *Vergleichende Untersuchungen verschiedener Kultivierungssysteme zur Produktion von Ergotalkaloiden mit* Claviceps purpurea *Dl-nic,* VDI Fortschrittsberichte, VDI Verlag, Düsseldorf, 1996, pp. 70–73.
66. Lord Raleigh, *Proc. Lond. Math. Soc.* **10**(4), 4–13 (1878).
67. C. Weber, *Z. angew. Math.* **11**, 136 (1936).
68. Ger. Pat. DE 4,424,998 (July 20, 1996), K.-D. Vorlop and J. Breford.
69. U. Prüße, B. Fox, M. Kirchhoff, F. Bruske, J. Breford, and K.-D. Vorlop, *Chem. Eng. Technol.* **21**(1), 29–33 (1998).
70. S. Nagata, *Mixing: Principles and applications,* John Wiley & Sons, New York, 1975, pp. 297–315.
71. A. Mersmann and H. Grossmann, *Chem. Ing. Technol.* **52**(8), 621–628 (1980).
72. E.G. Chatzi, A.D. Gavrielidis, and C. Kiparissides, *Ind. Eng. Chem. Res.* **28**, 1704 (1989).
73. U.S. Pat. 3,869,406 (March 4, 1975), H. Matsukawa and K. Saeki (to Fuji Foto Film Co., Ltd.).
74. U.S. Pat. 4,822,535 (April 18, 1989), B. Ekman and A. Lindahl (to Norsk Hydro A.S.).
75. U.S. Pat. 5,271,961 (December 21, 1993), E. Mathiowitz, H. Bernstein, E. Morrel, and K. Schwaller (to Alkermes Controlled Therapeutics, Inc.).
76. U.S. Pat. 5,500,161 (March 19, 1996), A.K. Andrianov and R.S. Langer (to Massachusetts Institute of Technology and Virus Research Institute).
77. U.S. Pat. 4,774,178 (September 27, 1988), P. Egerer, W. Haese, H. Perrey, and G. Schmidt-Kastner (to Bayer Aktiengesellschaft).
78. U.S. Pat. 4,353,888 (October 12, 1982), M.V. Sefton.
79. U.S. Pat. 4,647,536 (March 3, 1987), K. Mosbach and K. Nilsson.
80. A.C. Jen, M.C. Wake, and A.G. Mikos, *Biotechnol. Bioeng.* **50**, 357–364 (1996).
81. S.-K. Kim and C.K. Rha, in G. Skjåk-Bræk, T. Anthonson, and P. Sandford eds., *Chitin and Chitosan*, Elsvier Applied Science, London, 1990, pp. 617–626.
82. C.-H. Chu, H. Kumagai, and K. Nakamura, *J. Appl. Polym. Sci.* **60**, 1041–1047 (1996).
83. E. Esposito, R. Cortesi, and C. Nastruzzi, *Biomaterials* **17**, 2009–2020 (1996).
84. H. Akin and N. Hasirci, *J. Appl. Polym. Sci.* **58**, 95–100 (1995).
85. M.Z.-C. Hu and M. Reeves, *Biotechnol. Prog.* **13**, 60–70 (1997).
86. S. Pulapura and J. Kohn, *J. Biomater. Appl.* **6**, 216–250 (1992).
87. B. Jeong, Y.H. Bae, D.S. Lee, and S.W. Kim, *Nature* **388**, 860–862 (1997).
88. K.U. Dee, M.L. Shuler, and H.A. Wood, *Biotechnol. Bioeng.* **54**, 191–205 (1997).
89. W. Handschack, J. Kopp, and C. Pitra, *Acta Biol. Med. Germ.* **36**, 1429–1434 (1977).
90. G. Skjåk-Bræk, H. Grasdalen, and B. Larsen, *Carbohydr. Res.* **154**, 239–250 (1986).
91. O. Smidsrød, *Faraday Discuss. Chem. Soc.* **57**, 263–274 (1974).
92. A. Martinsen, G. Skjåk-Bræk, and O. Smidsrød, *Biotechnol. Bioeng.* **33**, 79–89 (1989).
93. B.H.A. Rehm and S. Valla, *Appl. Microbiol. Biotechnol.* **48**, 281–288 (1997).
94. U.S. Pat. 5,175,093 (December 29, 1992), D.B. Seifert (to Leight University of Bethlehem).
95. J.E. Nava Saucedo, C. Roisin, and J.-N. Barbotin, in R.H. Wijffels, R.M. Buitelaar, C. Bucke, and J. Tramper eds., *Immobilized Cells: Basics and Applications*, Elsevier Science, B.V., 1996, pp. 39–46.
96. J.C. Ogbonna, Y. Amano, and K. Nakamura, *J. Ferment. Bioeng.* **67**(2), 92–96 (1988).
97. O. Smidsrød and G. Skjåk-Bræk, *Trends Biotechnol.* **8**, 71–78 (1990).
98. A. Grobillot, D.K. Boadi, D. Poncelet, and R.J. Neufeld, *Crit. Rev. Biotechnol.* **14**(2), 75–107 (1994).
99. C.A. Masschelein, D.S. Ryder, and J.-P. Simon, *Crit. Rev. Biotechnol.* **14**(2), 155–177 (1994).
100. C.R. Davis, D.J. Wibowo, T.H. Lee, and G.H. Fleet, *Appl. Environ. Microbiol.* **51**(3), 539–545 (1986).
101. A. Durieux, X. Nicolay, J.-M. Jourdain, C. Depaepe, A. Piercelie, A.-M. Plaisant, and J.-P. Simon, in D. Poncelet and V. Santos eds., *Bioencapsulation Research Group, International Workshop, Bioencapsulation VI: From Fundamentals to Industrial Applications,* Barcelona, Spain, August 30–September 1, 1997.
102. B.J.M. Hannoun and G. Stephanopoulos, *Biotechnol. Prog.* **6**, 341–345 (1987).
103. U.S. Pat. 4,407,957 (October 4, 1983), F. Lim (to Damon Corporation).
104. U.S. Pat. 4,673,566 (June 16, 1987), M.F.A. Goosen, G.M. O'Shea, and A.M. Sun (to Connaught Laboratories Limited).
105. P. De Vos, B. De Haan, and R. Van Schilfgaarde, *Biomaterials* **18**, 273–278 (1997).

106. P. De Vos, B.J. De Haan, G.H.J. Wolters, J.H. Strubbe, and R. Van Schilfgaarde, *Diabetologica* **40**, 262–270 (1997).
107. M.F.A. Goosen, G.M. O'Shea, H.M. Charapetian, S. Chu, and A.M. Sun, *Biotechnol. Bioeng.* **27**, 146–150 (1984).
108. B. Thu, P. Brunheim, T. Espevik, O. Smidsrød, P. Soon-Shiong, and G. Skjåk-Bræk, *Biomaterials* **17**, 1031–1040 (1996).
109. H. Wong and T.M.S. Chang, *Biomat. Art. Cells Immob. Biotechnol.* **19**(4), 687–697 (1991).
110. H.A. Clayton, R.F.L. James, and N.J.M. London, *Acta Diabetol.* **30**, 181–189 (1993).
111. P. Soon-Shiong and P.A. Sandford, *Surg. Technol. Int. IV* 93–99 (1995).
112. D.E. Awrey, M. Tse, G. Hortelano, and P.L. Chang, *Biotechnol. Bioeng.* **52**, 472–484 (1996).
113. B. Thu, P. Brunheim, T. Espevik, O. Smidrød, P. Soon-Shiong, and G. Skjåk-Bræk, *Biomaterials* **17**, 1069–1079 (1996).
114. P. De Vos, B.J. De Haan, J. Pater, and R. Van Schilfgaarde, *Transplantation* **62**, 893–899 (1996).
115. V.A. Kabanov, A.A. Yaroslavov, and S.A. Sukhishvili, *J. Controlled Release* **39**, 173–189 (1996).
116. V.A. Kabanov, in P. Dubin, J. Bock, R. Davis, D.N. Schulz, and C. Thies eds., *Macromolecular Complexes in Chemistry and Biology*, Springer-Verlag, Berlin, 1994, pp. 151–174.
117. J. Xia and P.L. Dubin, in P. Dubin, J. Bock, R. Davis, D.N. Schulz, and C. Thies eds., *Macromolecular complexes in chemistry and biology*, Springer-Verlag, Berlin, 1994, pp. 247–271.
118. D.J. Burgess, in P. Dubin, J. Bock, R. Davis, D.N. Schulz, and C. Thies eds., *Macromolecular Complexes in Chemistry and Biology*, Springer-Verlag, Berlin, 1994, pp. 285–300.
119. H. Dautzenberg, G. Arnoldt, B. Tiersch, B. Lukanoff, and U. Eckert, *Prog. Colloid. Polym. Sci.* **101**, 149–156 (1996).
120. D. Hunkeler, A. Prokop, S. DiMari, M.A. Haralson, and T.G. Wang, in H. Dautzenberg and D. Poncelet eds., *Bioencapsulation Research Group, International Workshop, Bioencapsulation V: From Fundamentals to Industrial Applications,* Potsdam, Germany, September 23–25, 1996.
121. R. Darvari and V. Hasirci, *J. Microencapsulation* **13**(1), 9–24 (1996).
122. H. Dautzenberg, B. Lukanoff, U. Schuldt, and B. Tiersch, in D. Poncelet and V. Santos eds., *Bioencapsulation Research Group, International Workshop, Bioencapsulation VI: From Fundamentals to Industrial Applications,* Barcelona, Spain, August 30–September 1, 1997.
123. H. Dautzenberg, F. Loth, K. Fechner, B. Mehlis, and K. Pommering, *Makromol. Chem. Suppl.* **9**, 211–217 (1985).
124. K.L. Lafferty, S.E. Prowse, C.J. Simeonovic, and H. Warren, *Annu. Rev. Immunol.* **1**, 143 (1983).
125. J. Mansfels and H. Dautzenberg, in G.F. Bickerstaff ed., *Methods in Biotechnology*, vol. 1, Humana Press, Totowa, N.J., 1997, pp. 309–317.
126. I.T. Tai and A.M. Sun, *FASEB J.* **7**, 1061–1069 (1993).
127. E.I. Rainina, M.A. Pusheva, A.M. Ryabokon, N.P. Bolotina, V.I. Lozinsky, and S.D. Varfolomeyev, *Biotechnol. Appl. Biochem.* **32**, 321–329 (1994).
128. U.S. Patent (March 29, 1988), T. Tanabe and M. Nambu (to Nippon Oil Company, Ltd.).
129. N.A. Pepas and S.R. Stauffer, *J. Controlled Release* **16**, 302–310 (1991).
130. V.I. Lozinsky and A.L. Zubov, in D. Poncelet and V. Santos eds., *Bioencapsulation Research Group, International Workshop, Bioencapsulation VI: From Fundamentals to Industrial Applications,* Barcelona, Spain, August 30–September 1, 1997.
131. P.K. Watler, C.H. Cholakis, and M.V. Sefton, *Biomaterials* **9**, 150–154 (1988).
132. N.A. Peppas, in N.A. Peppas ed., *Hydrogels in Medicine and Pharmacy*, CRC Press, Boca Raton, Fla., 1987.
133. K. Furukawa, *J. Ferment. Bioeng.* **76**(6), 515–520 (1993).
134. M. Jekel, A. Buhr, T. Wilke, and K.-D. Vorlop, in D. Poncelet and V. Santos eds., *Bioencapsulation Research Group, International Workshop, Bioencapsulation VI: From Fundamentals to Industrial Applications,* Barcelona, Spain, August 30–September 1, 1997.
135. P.K. Robinson, A.L. Mak, and M.D. Trevan, *Bioprocess Chem.* 122–127 (August 1986).
136. S.F. Nishino and J.C. Spain, *Environ. Sci. Technol.* **27**, 489–494 (1993).
137. E.I. Rainina, M.A. Pusheva, A.M. Ryabokon, N.P. Bolotina, V.I. Lozinsky, and S.D. Varfolomeyev, *Proceedings of the 1993 ERDEC Scientific Conference on Chemical Defense Research*, 895–900, (1994).
138. C.A. Crooks, J.A. Douglas, R.L. Broughton, and M.V. Sefton, *J. Biomed. Mater. Res.* **24**, 1241–1262 (1990).
139. G.D.M. Wells, M.M. Fisher, and M.V. Sefton, *Biomaterials* **14**(8), 615–620 (1993).
140. H. Uludag, V. Horvath, J.P. Black, and M.V. Sefton, *Biotechnol. Bioeng.* **44**, 1199–1204 (1994).

See also FREEZE DRYING, PHARMACEUTICALS.

MIXED CULTURE

SUTEAKI SHIOYA
Osaka University
Osaka, Japan

KEY WORDS

Activated sludge process
Bacteriocin
Bulking
Competition
Cooperation
Lactobacilli
Mixed culture
Mutualism
Nisin
Yeast

OUTLINE

INTRODUCTION

Most fermented foods are produced by mixed cultures acting on various substrates. Cheese, yogurt, pickles, whiskey, and sake (Japanese rice wine) are some examples of fermented foods (1). Numerous interactions occur between microbial communities, plants, and animals. The field of mixed culture is too broad to be reviewed completely in a short article. Of course, it is our belief that models, simulations, and stability analysis used for mixed culture systems are important aids in planning various experiments and interpreting the results (2). Pioneering works in this direction had been done by Henry Tsuchiya and Arnold Fredrickson. For example, we can see their careful results by referring to their review articles in this field (3,4). However, in spite of their extensive efforts, when using these aids general close agreement with data on the dynamic behavior of a complicated mixed culture has yet to be obtained. Thus we stress here applications of mixed culture and discuss case studies that demonstrate how these techniques can contribute to the design, operation, and control of mixed culture systems.

INTERACTION BETWEEN TWO SPECIES

Interactions between two species can be classified into three types: neutral (0), positive, and negative (5). Moreover, positive interactions are characterized into two interactions in detail: one is necessary for the survival of microorganisms (+*) whereas the other has an enhancing effect but is not obligatory for the life of these microorganisms (+). Negative interactions are also divided into two interactions: one directly affects other microorganisms (−*) and the other indirectly affects them (−).

Combination of these interaction categories results in more than nine groups classified under the following headings: neutralism (0, 0), mutual inhibition competition (−*, −*), competition—resource use type (−, −), amensalism (−, 0), parasitism (+, −), predation (+, −), commensalism (+, 0), proto-cooperation (+, +), mutualism (+*, +*). The first symbol in the parentheses shows the relationship of species 1 to species 2 and the second symbol shows the reverse relationship. The difference between parasitism and predation is only with respect to the size of species 1; that is, in parasitism, the size of species 1 is smaller than that of species 2 whereas in predation species 1 is larger than species 2.

Sometimes, of course, the actual relationship between two microorganisms cannot be classified into these nine groups because the relationship cannot be defined clearly. Nevertheless, the nine-group classification is useful for building a mathematical model and understanding the relationship between two species through simulation because a rough overview of the interaction between two microorganisms is possible based on this classification.

There are several examples of interactions belonging to these nine groups, but here only three examples that have been studied by the author are given. One is a mixed culture system of lactobacilli and propionibacteria in a bioreactor, the interaction of which is classified under proto-cooperation (+, +); the second example is a mixed culture system of lactobacilli and yeast classified as mutualism under a limited initial medium condition; and the last example is that of an activated sludge process consisting not of two microorganisms but of two competing groups of microorganism populations because of the convenience of the stability analysis and application of the result to the operation of the system.

One of the aims of this article is to show how models, simulations, and stability analysis are important aids in interpreting results of laboratory experiments or interacting phenomena and for the operation and control of a mixed culture system. Another aim is to demonstrate that the mixed culture system can be utilized for effective bioproduction. Mixed culture systems are encountered frequently in nature, but artificially constructed mixed culture systems can also contribute to the microbial production of various bioproducts such as bacteriocins.

MIXED CULTURE OF LACTOBACILLI AND PROPIONIBACTERIA

Bacteriocins are peptides produced from bacteria such as lactate-producing microorganisms, that is, lactobacilli, and effective antibiotics. Nisin is a typical bacteriocin effective against gram-positive microorganisms (6). Lactobacilli grow at pH higher than 4.0 and their growth is affected by the lactate produced, that is, product inhibition. Thus, the removal of lactate during fermentation is an effective way of optimizing nisin production, but the method of removal of lactate is a difficult problem to solve to assure optimal nisin production. In this study, we try to remove lactate by coculture of lactobacilli with propionibacteria, which can assimilate lactate efficiently.

Growth and Lactate Fermentation Characteristics of *Lactococcus lactis*

Figure 1 shows a typical result of the anaerobic batch culture of *Lactococcus lactis* subsp. *lactis* ATCC11454, a homolactate-producing bacterium. Without pH control, the pH decreased from 7.0 to 5.0 after 2 h of culture and to 4.5 after 3 h. Cell growth as well as nisin production stopped

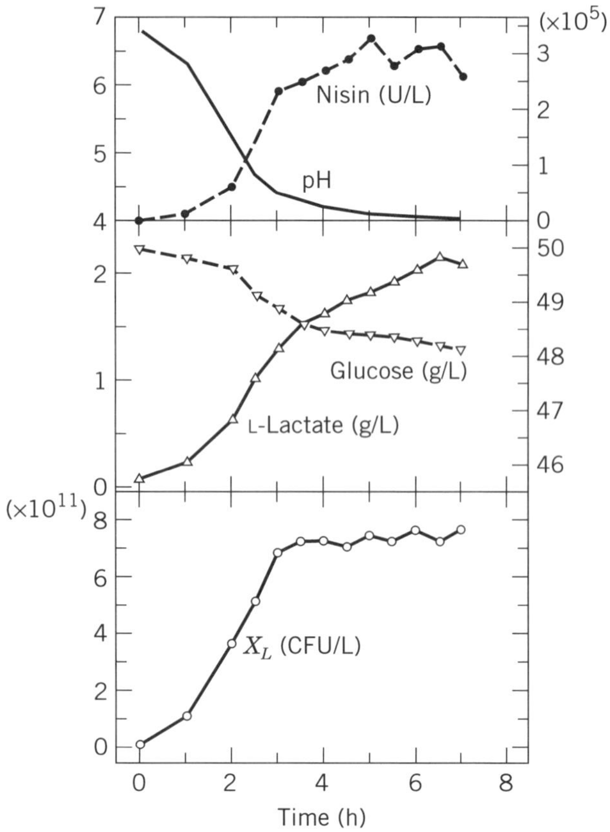

Figure 1. Time course of a batch culture of *Lactococcus lactis* without pH control. *Circles*, cell concentration; *inverted triangles*, glucose concentration; *triangles*, lactate concentration; *solid circles*, nisin concentration.

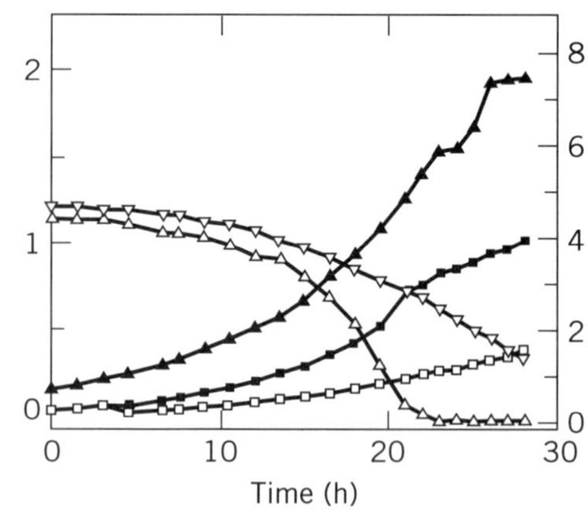

Figure 2. Time course of a batch culture of *Propionibacterium freudenreichii*. *Solid triangles*, cell concentration; *inverted triangles*, glucose concentration; *triangles*, lactate concentration; *open squares*, acetate concentration; *solid squares*, propionate concentration.

at 3 h after the start of culture. Lactate was produced at a level of approximately 1.5 g/L and was still produced at a decreased rate until 7 h later. Thus, control of pH to a value greater than 4.5 is essential for achieving a high cell growth rate. Also, nisin production is strictly associated with cell growth. Thus, for effective nisin production, the pH of the medium should be controlled to a value greater than 4.5. Finally, 300 U/mL nisin was produced in this batch culture without pH control.

To improve this low productivity of nisin, pH was controlled and lactate concentration was kept at a fairly low level by diluting the cultivation medium by adding fresh medium. This cultivation method is called repeated dilution culture. Using the method of repeated dilution culture, about 2,000 U/mL of nisin was produced after 10 h of operation. Cell growth, however, stopped at about 15 g/L of lactate. Thus, the control of lactate concentration kept to a level less than 15 g/L is also required for effective continuous production of nisin.

Growth Characteristics of Propionate Bacteria

Propionibacterium freudenreichii IFO12424 was used as a propionate-producing microorganism that can assimilate lactate. Figure 2 shows a typical result of anaerobic batch culture of *P. freudenreichii*. Lactate and glucose were used as carbon sources. From the figure, it can be seen that this microorganism assimilates lactate. However, glucose was also assimilated, which is disadvantageous for the control of lactate concentration because glucose, the main carbon source for *L. lactis*, can be assimilated by *P. freudenreichii*. The glucose is competitively assimilated by both *L. lactis* and *P. freudenreichii*. If the population balance of these microorganisms is destroyed by the deficiency of glucose for *L. lactis* or by accumulation of lactate, stable operation of the mixed culture system will be difficult. Acetate was also produced by *P. freudenreichii*.

Interaction between Two Microorganisms

The interaction between the two microorganisms, *L. lactis* and *P. freudenreichii*, is illustrated in Figure 3. Lactate produced by *L. lactis* can be assimilated by *P. freudenreichii*. Without removing lactate, *L. lactis* cannot grow to a high density because of inhibition by low pH and by lactate itself. In this mixed culture system, the interaction between the two microorganisms is termed proto-cooperation.

Propionate and acetate produced by *P. freudenreichii* contribute somewhat to microbial contaminants in their

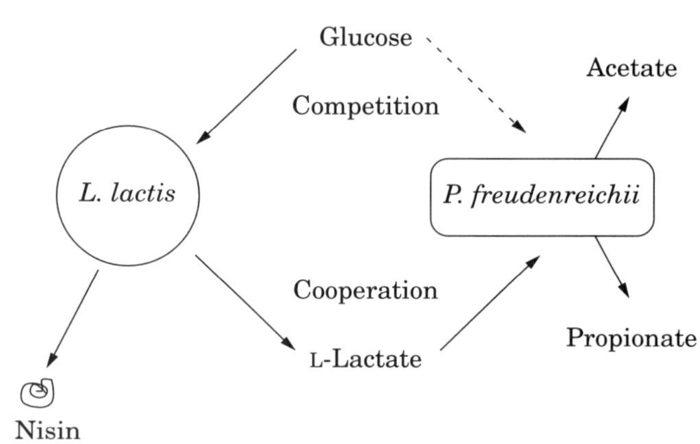

Figure 3. Interaction between *L. lactis* and *P. freudenreichii*.

capacities as antibiotics when they are used for food additives. However, these organic acids also contribute to the decrease in medium pH. Glucose, a carbon source for *L. lactis*, can be assimilated competitively by *P. freudenreichii*. In this sense the interaction between the two microorganisms is classified under competition. Thus, the interaction between the two microorganisms is a mix of cooperation and competition, that is, (±, ±), in glucose-rich medium. However, if we can control the glucose concentration to a level at which it is utilized by *L. lactis* but not by *P. freudenreichii*, the system can operate as a cooperation system. The most important issue is to manipulate or manage the system so as to represent the interaction (+, +), in practice, and to maintain a low supply of glucose, as well as preventing pH decrease of the medium. For this, mathematical modeling and simulation of the system are important tools.

Simulation Model of the Mixed Culture

Based on the kinetic data for cultivation, the following mathematical model was proposed under the assumption that the mixed culture can be described by use of the kinetic data of individual pure cultures.

Dynamics of the mixed culture system can be written by the following equations. To control the growth rate of *L. lactis* at a level comparable to that of *P. freudenreichii* by limiting glucose concentration, a fed-batch culture was employed. The cell growth rate for *L. lactis* in fed-batch culture is written as

$$\frac{dX_L}{dt} = \left(\mu_L - \frac{F}{V}\right)X_L \qquad (1)$$

and the rate for *P. freudenreichii* is written as

$$\frac{dX_P}{dt} = \left(\mu_P - \frac{F}{V}\right)X_p \qquad (2)$$

where X_i ($i = L,P$) and μ_i are viable cell concentration and specific growth rate, respectively. In these equations, the subscripts L and P represent *L. lactis* and *P. freudenreichii*, respectively, and F and V indicate the flow rate of the feeding medium and the liquid volume in the tank, respectively. Nisin (N, U/mL) production by *L. lactis* is written by

$$\frac{dN}{dt} = \rho_N X_L - \frac{F}{V}N \qquad (3)$$

Propionate (P, g/L) production by *P. freudenreichii* is written as

$$\frac{dP}{dt} = \rho_P X_P - \frac{F}{V}P \qquad (4)$$

Also, acetate (A, g/L) production by *P. freudenreichii* is written as

$$\frac{dA}{dt} = \rho_A X_P - \frac{F}{V}A \qquad (5)$$

where ρ_k is the specific production rate of product k ($k = A,L,N,P$). Lactate production by *L. lactis* and its assimilation by *P. freudenreichii* can be summarized as

$$\frac{dL}{dt} = \rho_L X_L - \nu_{LP} X_P - \frac{F}{V}L \qquad (6)$$

As a result of glucose assimilation by both organisms, the dynamics of glucose concentration (S, g/L) can be written as

$$\frac{dS}{dt} = -\nu_{SL}X_L - \nu_{SP}X_P + \frac{F}{V}(S_F - S) \qquad (7)$$

where ν_{ji} is a specific uptake rate of substrate j by microorganism i ($j = L,S$; $i = L,P$).

The dynamic pH change should be described by a mathematical model in which the following three types of equations are necessary. For example, for lactate concentration, one equation describes the equilibrium between dissociation and association for ionization of lactate such as

$$K_L = \frac{[L^-][H^+]}{[LH]} \qquad (8)$$

where K_k is a dissociation constant for product k ($k = A,L,P$).

The second equation describes the total balance between ionized lactate and nonionized lactate $[LH]$ as

$$L_{MT} = [L^-] + [LH] \qquad (9)$$

where L_{MT} is the mole-based total lactate concentration. From equation 8, nonionized lactate $[LH]$ is described by other variables, $[H^+]$ and $[L^-]$, and is substituted into equation 9 to give

$$L_{MT} = [L^-] + \frac{[L^-][H^+]}{K_L} = [L^-]\left\{1 + \frac{[H^+]}{K_L}\right\} \qquad (10)$$

Then,

$$[L^-] = \frac{L_{MT}}{\left\{1 + \frac{[H^+]}{K_L}\right\}} = \frac{L_{MT}}{(1 + 10^{-pH+pK_L})} \qquad (11)$$

where pK_L is defined in the same way as pH and $pK_L = -\log_{10}(K_L)$. In a similar way, ionized propionate $[P^-]$ can be written as

$$[P^-] = \frac{P_{MT}}{(1 + 10^{-\mathrm{kp}H+pK_P})} \qquad (12)$$

Also, the ionized acetate $[A^-]$ can be written as

$$[A^-] = \frac{A_{MT}}{(1 + 10^{-\mathrm{kp}H+pK_A})} \qquad (13)$$

These mole-based concentrations can be related to the weight-based concentrations as

$$L_{MT} = \frac{L}{MW_L},\ P_{MT} = \frac{P}{MW_P} \text{ and } A_{MT} = \frac{A}{MW_A} \quad (14)$$

Here MW_k ($k = A,L,P$) is the molecular weight of chemicals k.

Finally, the following total ion balance should be considered.

$$[H^+] + X_B = [OH^-] + [L^-] + [P^-] + [A^-] + X_A \quad (15)$$

where X_A and X_B are the concentrations of strong acid and base in the medium, respectively. By substituting equations 11, 12, and 13 into equation 15 and using the definition of pH:

$$X_A - X_B = 10^{-pH} - 10^{pH-14} - \frac{L_{MT}}{1 + 10^{-pH+pK_L}} - \frac{P_{MT}}{1 + 10^{-pH+pK_P}} - \frac{A_{MT}}{1 + 10^{-pH+pK_A}} \quad (16)$$

By differentiating this equation with respect to time t:

$$\frac{d(X_A - X_B)}{dt} = \frac{\partial f}{\partial pH}\frac{dpH}{dt} + \frac{\partial f}{\partial L_{MT}}\frac{dL_{MT}}{dt} + \frac{\partial f}{\partial P_{MT}}\frac{dP_{MT}}{dt} + \frac{\partial f}{\partial A_{MT}}\frac{dA_{MT}}{dt} \quad (17)$$

where f represents the right-hand side of equation 16. If we can assume that

$$\frac{d(X_A - X_B)}{dt} = 0 \quad (18)$$

then

$$\frac{dpH}{dt} = \frac{1}{\left(\frac{\partial f}{\partial pH}\right)}\left\{-\left(\frac{\partial f}{\partial L_{MT}}\right)\frac{dL_{MT}}{dt} - \left(\frac{\partial f}{\partial P_{MT}}\right)\frac{dP_{MT}}{dt} - \left(\frac{\partial f}{\partial A_{Mt}}\right)\frac{dA_{MT}}{dt}\right\} \quad (19)$$

Now

$$\frac{\partial f}{\partial L_{MT}} = \frac{-1}{(1 + 10^{-pH+pK_L})},\ \frac{\partial f}{\partial P_{MT}} = \frac{-1}{(1 + 10^{-pH+pK_P})},\ \frac{\partial f}{\partial A_{MT}} = \frac{-1}{(1 + 10^{-pH+pK_A})} \quad (20)$$

and

$$\frac{\partial f}{\partial pH} = \text{Ln}10\left[-10^{-pH} - 10^{pH-14} - \frac{L_{MT}\cdot 10^{-pH+pK_L}}{(1 + 10^{-pH+pK_L})^2} - \frac{P_{MT}\cdot 10^{-pH+pK_P}}{(1 + 10^{-pH+pK_P})^2} - \frac{A_{MT}\cdot 10^{-pH+pK_A}}{(1 + 10^{-pH+pK_A})^2}\right] \quad (21)$$

can be easily derived. Taking into account that equation 8 is not always valid, a correction term was introduced into the denominator of equation 19 as 1/230, which was determined using experimental data of the mixed culture system and resulted in the following equation (22). Finally, the pH change can be described by

$$\frac{dpH}{dt} = \left\{-\frac{1}{(1 + 10^{-pH+pK'_L})}\frac{dL_{MT}}{dt} - \frac{1}{(1 + 10^{-pH+pK_P})}\cdot\frac{dP_{MT}}{dt} - \frac{1}{(1 + 10^{-pH+pK_{AL}})}\frac{dA_{MT}}{dt}\right\} \div \left\{\frac{1}{230} + \text{Ln}10\left[10^{-pH} + 10^{pH-14} + \frac{L_{MT}\cdot 10^{-pH+pK_L}}{(1 + 10^{-pH+pK_L})^2} + \frac{P_{MT}\cdot 10^{-pH+pK_P}}{(1 + 10^{-pH+pK_P})^2} + \frac{A_{MT}\cdot 10^{-\&pH+pK_A}}{(1 + 10^{-pH+pKA})^2}\right]\right\} \quad (22)$$

where the dynamic change of L_{MT}, P_{MT}, and A_{MT} can be calculated using equations 4, 5, 6, and 14. Finally, the kinetic parameters such as μ_i, ν_{ji}, and ρ_k ($i = L,P, j = L,S, k = A,L,N,P$) are given as a function of pH as well as substrate concentration, and the dynamic behavior of the fed-batch culture can be simulated using the mathematical model described by equations 1 through 7, 14, and 22.

Control of pH by Adjusting an Appropriate Initial Population

To maintain the culture pH at a value higher than 5.0 for a long period, *P. freudenreichii* should grow at a fast rate so that lactate is assimilated well. By comparing the specific growth rates of both microorganisms, it can be seen that *L. lactis* grows at a much faster rate than *P. freudenreichii*. This result means that lactate produced in correlation with *L. lactis* growth will accumulate in the fermentor if there is no control of the growth of *L. lactis*. For control of the growth rate of *L. lactis*, a fed-batch culture was employed. By manipulating the exponential feeding rate of glucose (in this case, μ_L was set at 0.2 h^{-1}; initial tank volume at 1.5 L, and initial glucose concentration at 2 g/L), the growth of *L. lactis* can be controlled. The mathematical model proposed here can be used for simulation of this exponential fed-batch mixed culture system. The kinetic parameters of *L. lactis* that depend on the pH were determined from the data for the pure batch culture. Other kinetic parameters were also determined from the experimental data.

By changing the initial cell populations of *L. lactis* and *P. freudenreichii*, the dynamic behavior of the mixed culture can also be altered. In this way the most appropriate initial cell populations were selected. After several simulation studies, an appropriate initial ratio of cell concentration of *L. lactis* to that of *P. freudenreichii* was determined as 1:14. The simulation result is shown in Figure 4. Here, to easily compare the simulation and experimental data, the cell concentration of *L. lactis* is represented by cell number whereas that for *P. freudenreichii* is by dry cell weight, because in the experiment the cell concentration of *P. freudenreichii* was calculated by subtraction of the

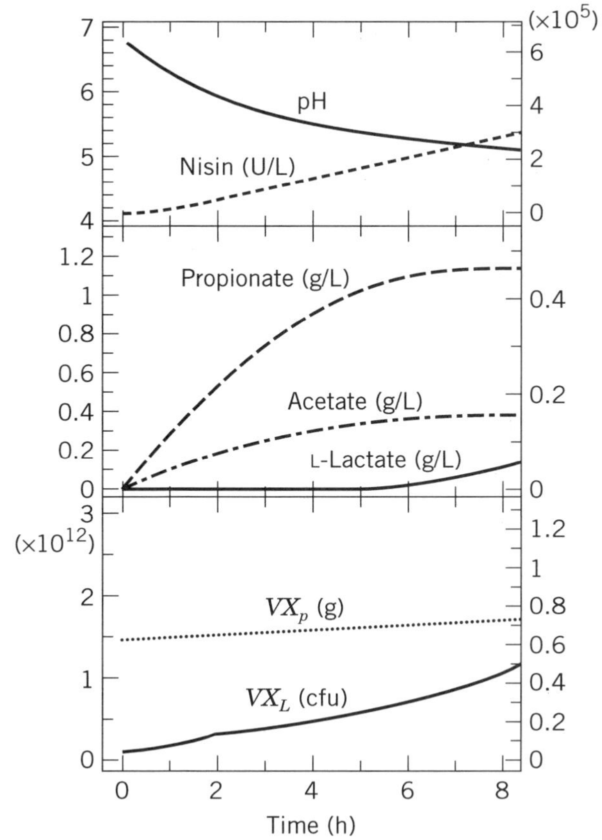

Figure 4. Simulation of a mixed culture for effective nisin production.

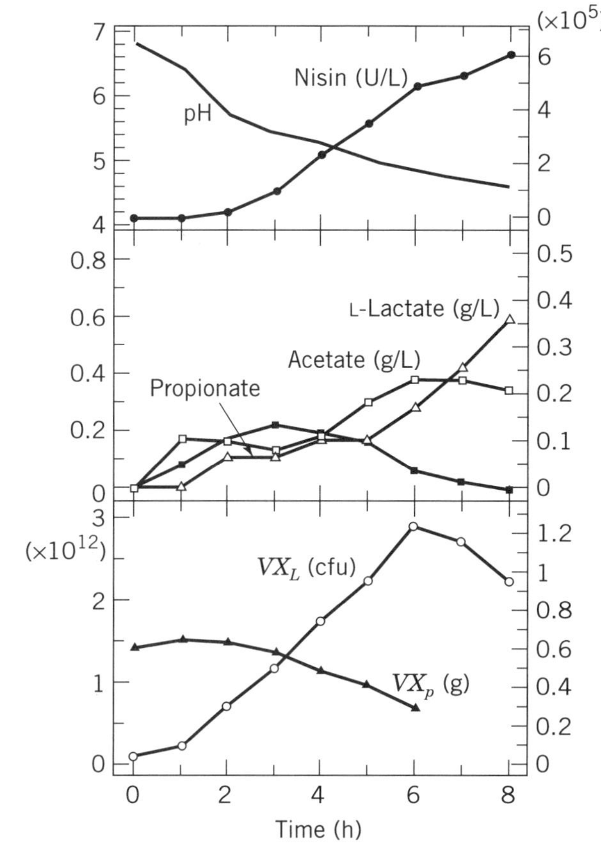

Figure 5. Experimental result of a mixed culture of *L. lactis* and *P. freudenreichii*. *Circles*, cell concentration of *L. lactis*; *solid triangles*, cell concentration of *P. freudenreichii*; *triangles*, lactate concentration; *open squares*, acetate concentration; *solid squares*, propionate concentration; *solid circles*, nisin concentration.

amount of dry cell weight of *L. lactis* converted from the cell number counted by plate colony from the total dry cell weight. It is shown that pH will be maintained at a value higher than 5 for a period of 5 h and that nisin will be produced much more effectively than for a given specific growth rate of *L. lactis*.

Experimental Result for the Mixed Culture

Figure 5 shows an experimental result of the mixed fed-batch culture with the same initial cell concentrations of both microorganisms, *L. lactis* and *P. freudenreichii*, as those in Figure 4. The pH could be controlled until 5 h from the start of cultivation. However, under these conditions *P. freudenreichii* could not grow well, whereas *L. lactis* grew faster than the estimated rate shown in Figure 4. The reason why there is a large difference between dynamic simulation using the mathematical model and experimental data is not clear at this point. However, the simulation model at least can be used for forecasting the dynamics of the mixed culture and for adjusting the initial cell concentrations of both species so as to achieve effective production of nisin by this mixed culture. Finally, the experimental data for the mixed culture show that the final nisin concentration was improved twofold.

In the near future, the process will be improved further after determination of the reason for the difference between the experimental data and the simulation data.

MIXED CULTURE OF LACTOBACILLI AND YEAST

To enhance bacteriocin productivity, a mixed culture system was employed as outlined in the previous section. However, the propionibacteria can assimilate glucose, which in the system described is the main carbon source of lactobacilli. Thus the glucose was competitively assimilated by both microorganisms, which may result in the need to use a fed-batch operation. One improvement that can be made to the system is use a main carbon source of lactobacilli which cannot be assimilated by another microorganism. Only the lactate produced by the lactobacilli should be a carbon source of the second microorganism. From this aspect, maltose was chosen as a carbon source of lactobacilli, and a microorganism that can assimilate lactate but not maltose was selected. From kefir, a type of yogurt, such a microorganism, *Kluyveromyces marxianus*, was isolated. The following mixed culture system was then constructed.

Interaction between *Lactococcus lactis* and *Kluyveromyces marxianus*

Kluyveromyces marxianus MS1, isolated from a type of yogurt (kefir), can assimilate lactate as a carbon source. *K.*

marxianus cannot assimilate starch and maltose directly whereas *L. lactis* can grow on maltose. Whether *L. lactis* could grow and produce nisin under aerobic conditions had to be determined because *K. marxianus* grows under aerobic conditions. The results indicated that there is no significant difference in the productivity of nisin and growth rate of *L. lactis* between aerobic and anaerobic conditions. Thus, the mixed culture system can be operated under aerobic conditions. The interaction of the two microorganisms is illustrated in Figure 6. *L. lactis* can assimilate the maltose and produce lactate, which can be assimilated by the yeast, *K. marxianus*. This cooperative interaction results in a very low or undetectable lactate concentration, which may enhance the growth of *L. lactis* as well as nisin production. Thus, the relationship between these microorganisms can be termed proto-cooperation or mutualism because their relation can be summarized by (+, +*).

Mixed Culture of *L. lactis* and *K. marxianus*

From the characteristics of the interaction, it is expected that a high growth rate of *L. lactis* as well as high productivity of nisin will be achieved. In Figure 7, the experimental data indicated that the system worked as desired because the lactate concentration can be kept at almost zero or at least at an undetectable level and *L. lactis* grew significantly. The monoculture of *L. lactis* using maltose as a carbon source stopped at 7 h because of low pH (Fig. 7a). Under mixed culture, pH could be maintained well around 6.0 (Fig. 7b) even by a manual dissolved oxygen (DO) concentration control. The role of DO control is explained later. Also, *L. lactis* as well as *K. marxianus* could grow at a level fourfold higher than the monoculture of *L. lactis*. The yeast also grew well and the nisin yield reached 1,400 U/mL at the end of fermentation after only 8 h operation (Fig. 7c), similar to the amount obtained in the pH-controlled repeated batch culture, which was fourfold higher than the level in the monoculture of *L. lactis*.

The pH level of the culture medium can be controlled by the lactate uptake rate linked with the growth of *K. marxianus*. On the other hand, the growth of *K. marxianus* can be controlled by the DO level. Then, the DO level can control the pH level in this mixed culture system. Finally, it is clear that a setpoint for the DO control by changing the agitation speed can control pH level. This is a so-called cascade controller. Based on this cascade controller concept, the pH level could be maintained almost completely automatically at a constant level by manipulating the agitation speed of the fermentor. Finally, the result of this mixed culture experiment with cascade control of pH is shown in Figure 8; the product nisin concentration was enhanced more and reached about 4,000 U/mL after 11 h of operation, which was almost 10 fold that by anaerobic monoculture of *L. lactis* without pH control and about twice as much as by anaerobic monoculture of *L. lactis* with pH control by NaOH addition.

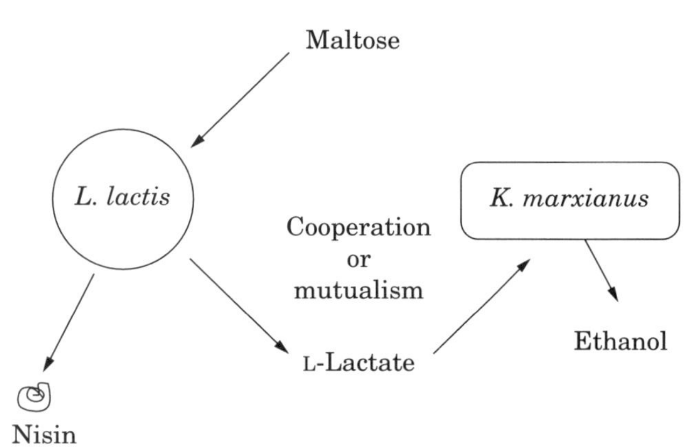

Figure 6. Interaction between *L. lactis* and *Kluyveromyces marxianus*.

The system developed here functions successfully. However, it should be stressed that a simple and reliable method for the measurement of individual cell concentrations should be developed. One way is to use a flow cytometer that can count individual cells if the cell size of each population is significantly different. One example of flow cytometry measurement of the concentration of lactobacilli and yeast in the mixed culture is shown in Figure 9. The abscissa of the figure represents the cell size, and one dot indicates one cell. Thus, densely clustered points means there are many cells. In this case, the lactobacilli cells are small compared with yeast cells. Using the scale of the abscissa, it may be possible to count each population. However, to identify the two populations of microorganisms more effectively, staining of their DNA using a dye such as propidium iodine (PI) also aids in discriminating the yeast and lactobacillus by cell size as well as by DNA content under two-dimensional plotting as shown in the figure. The ordinate indicates the DNA contents represented by strength of PI staining. In Figure 9, the clustering between lactobacilli and yeast cells becomes evident because the two populations separate like two islands.

DYNAMICS OF AN ACTIVATED SLUDGE PROCESS AS A MIXED CULTURE SYSTEM

An activated sludge process is a typical mixed culture system. The formation of activated sludge flocs as a result of the overgrowth of filamentous microorganisms results in a phenomenon known as bulking, which is a major problem in the operation of the activated sludge process. Wastewater composition, sludge loading, DO concentration, the feeding pattern of the aerator, and pH are some of the factors involved in the bulking phenomenon. The objective of this work is to develop a simple mathematical model describing the bulking phenomenon caused by high sludge loading, including the growth kinetics of microorganisms in an aeration tank in terms of a mixed culture system and the ability of the sludge to settle, and to obtain some means of preventing and controlling bulking. In our experiment, only the sludge loading rate was changed; the other operating conditions such as pH and temperature were kept constant. The DO concentration in the aeration tank was maintained at a high level so as not to induce the bulking phenomenon. The model system selected for examination is a mixed culture of two groups of microorganisms, one having the ability to induce the formation of flocs and the

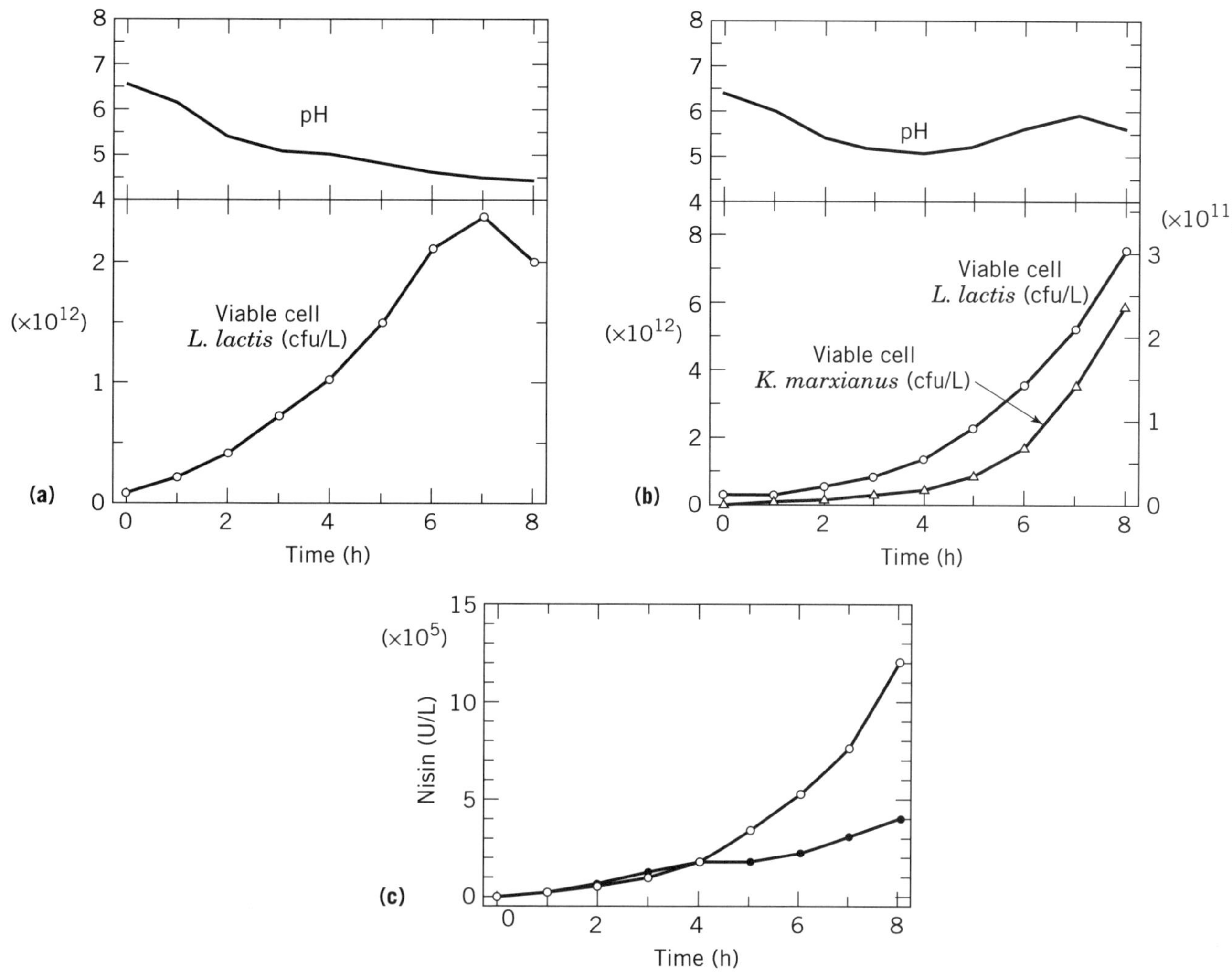

Figure 7. (**a**) Monoculture of *L. lactis* without pH control or pH changes. *Circles*, cell concentration of *L. lactis*. (**b**) Mixed culture of *L. lactis* and *K. marxianus* by manual dissolved oxygen (DO) control. *Circles*, cell concentration of *L. lactis*; *triangles*, cell concentration of *K. marxianus*. (**c**) Production of nisin under a mono- and mixed cultures by manual DO control. *Solid circles*, nisin concentration by anaerobic monoculture without pH control; *open circles*, nisin concentration by (aerobic) mixed culture.

other inducing bulking. The interaction between the two groups is classified under competition with respect to the utilization of the same substrate (7).

A continuous flow biochemical reactor in which a mixed culture of microorganisms is grown very often exhibits multiple steady states. The analysis of the stability of these steady states has been the subject of research. The techniques most frequently employed in determining the stability characteristics of a steady state are based on Liapunov's first method, which consists mainly of a linearization of the differential equations around the steady state and an examination of the eigenvalue of the Jacobian matrix. Here, the dynamic behavior of the activated sludge process, in which competition occurs with respect to assimilation of the substrate, is analyzed based on a mathematical model utilizing phase plane analysis and computer simulation. The results including those pertaining to the conditions that result in the system reaching the bulking steady state are very informative and useful for the practical operation of activated sludge processes.

Kinetics of Substrate Uptake and Cell Growth

The kinetics in two synthetic media, one in which the carbon source is glucose and polypeptone (medium A) and the other in which the carbon source is glucose (medium B), were studied experimentally using batch culture. In this system, the chemical oxygen demand (COD) concentration, S, is considered to be a limiting substrate for describing the kinetics of the process. Both groups of microorganism populations, that is, the floc-forming and the bulking-inducing groups, were cultured by feeding different types of synthetic medium into a batch reactor on a fill-and-draw basis. Two synthetic media were used to purify and main-

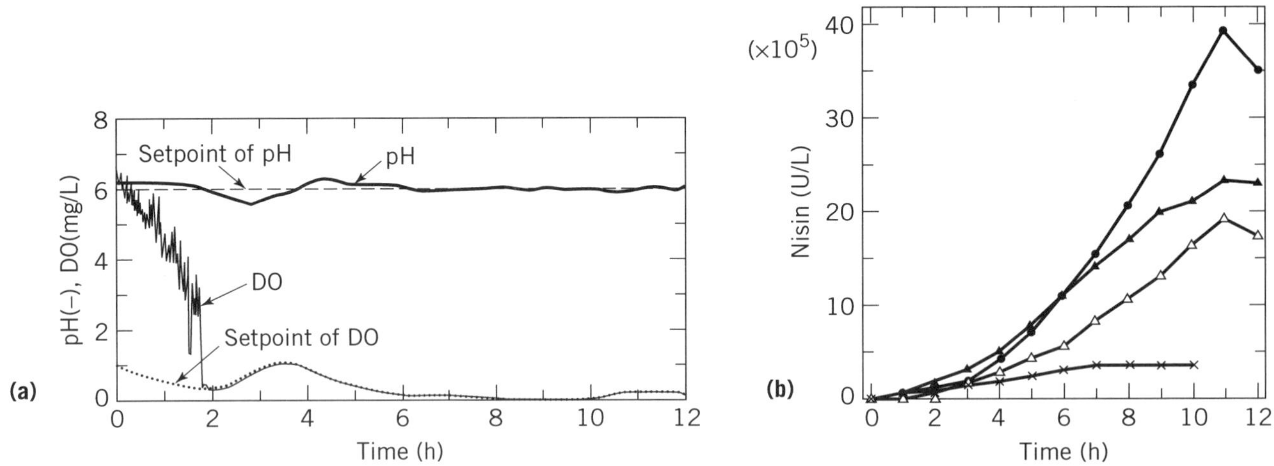

Figure 8. (**a**) pH control by an automatic cascade controller coupled with DO control in the mixed culture system. (**b**) Production of nisin by cascade control of pH compared with other pH control strategies. ×, anaerobic monoculture without pH control; *solid triangles*, anaerobic monoculture with pH control by NaOH addition; *open triangles*, aerobic monoculture with pH control by NaOH addition; *solid circles*, mixed culture with cascade control of pH.

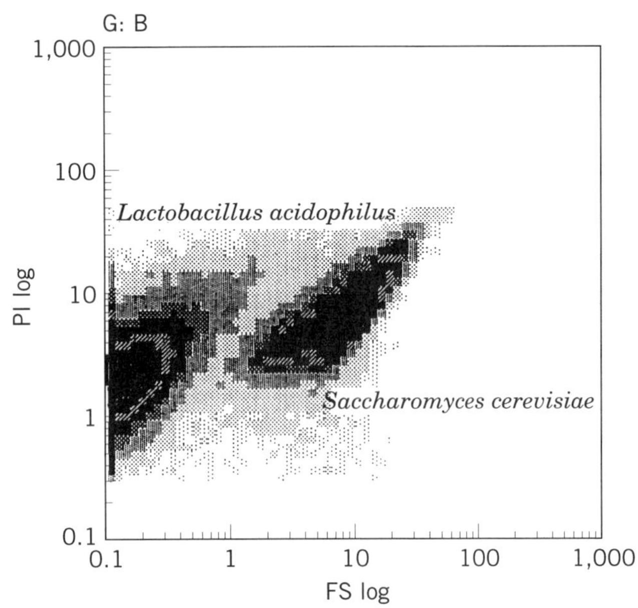

Figure 9. Two-dimensional plot of flow cytometry for a mixed culture system of *Lactobacillus acidophilus* and *Saccharomyces cerevisiae*.

tain each population of microorganisms for more than 2 months. Using medium A in which the initial COD concentration was about 110 mg/L, the sludge settled well and was called floc-forming sludge. On the other hand, using medium B in which the initial COD concentration was about 450 mg/L, the sludge did not settle well, that is, it had a high sludge volume index. The sludge could be kept stable for more than 1 month by the feeding of medium B and was called bulking sludge. From the results of microscopic observation, it was determined that the bulking sludge was composed of filamentous microorganisms such as *Sphaphaeroutilus* sp., while the floc-forming sludge did not contain such microorganisms.

A 1-L jar fermentor equipped with automatic temperature control, an air flow rate indicator, DO sensors, and a magnetic agitator was used. The kinetics of floc-forming and bulking populations were analyzed from the batch experiments using medium A or B with various initial COD concentrations. A typical result (Fig. 10) shows the time courses of floc-forming sludge concentration X_F and the COD concentration S with medium A. From this figure, it was shown that the specific growth rate μ_F and the substrate uptake rate ν_F could be represented by a linear relationship with respect to substrate (COD) concentration S and that the yield coefficient of the sludge from COD was constant. Similar results for growth rates and uptake rates as described here were obtained for both types of sludge and for both types of media. The mathematical forms of growth rates and uptake rates thus are as follows:

$$\mu_j = K_j(S - S_j) \quad (23)$$

and

$$\nu_j = \frac{-1}{Y_j}\mu_j \quad j = F, B \quad (24)$$

where F and B mean floc-forming and bulking sludge, respectively. The solid lines in Figure 10 show the calculated values using the estimated parameters. The kinetic model described by equations 23 and 24 is generally acceptable in the low COD concentration range.

The ability of the activated sludge to settle in the sedimentation vessel was also studied experimentally. An empirical description of the concentration of the sedimenting sludge related to the ratio of X_F/X_B was obtained. The ratio of the concentration of the settled sludge, X_g, to the con-

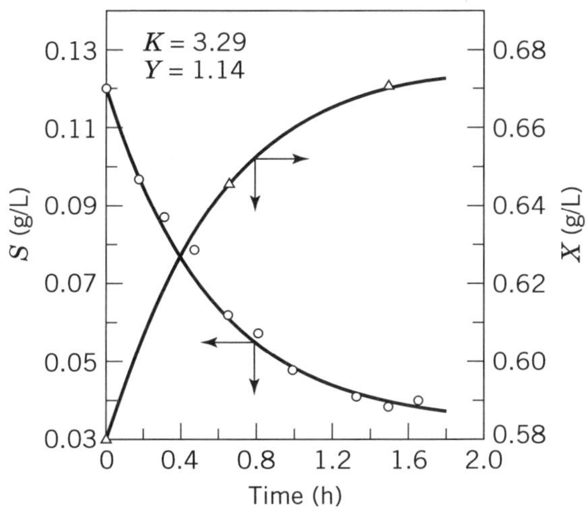

Figure 10. Batch culture of floc-forming sludge using medium A. *Circles*, chemical oxygen demand (COD) concentration; *triangles*, cell concentration.

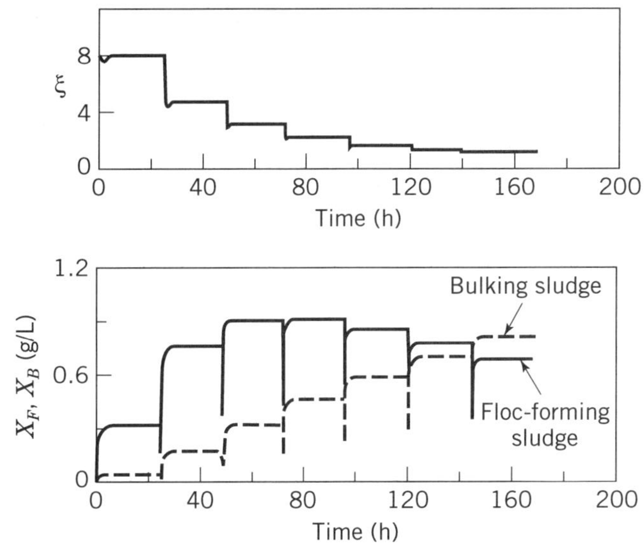

Figure 11. Simulation of fill-and-draw-type batch culture using medium B. After more than 4 days, the concentration ratio χ approached 1.0. After that, the bulking phenomenon was observed.

centration of the sludge to be settled, X, which is denoted by χ, can be represented by

$$\chi = \frac{X_R}{X} = \alpha\left(\frac{X_B}{X_F} + \beta\right)^{-\gamma} + 1 \tag{25}$$

Interaction between Bulking and Floc-Forming Sludges

It is assumed that there is competitive use of substrate between the bulking and floc-forming sludge. This assumption could be partially verified by the following experiment and computer simulation. When a fill-and-draw-type batch experiment is conducted from the point at which the ratio of the bulking to the floc-forming sludge is very small, the ratio of the bulking sludge to the total sludge increases gradually using medium B, and its ability to settle becomes poorer after more than 4 days from the start of the experiment. This experimental result can be simulated as shown in Figure 11 by the mathematical model given in equations 23 and 24. After 4 or 5 days from the start of the experiment, χ becomes almost 1.0; that is, the state of the sludge approaches bulking, which is reflected by the experimental data.

Mathematical Model of the Activated Sludge Process

For an activated sludge process in a completely mixed aeration tank and sedimentation vessel, a mathematical model of the system is derived. The mathematical model of the activated sludge process considered here is as follows:

$$\frac{dX_j}{dt} = \{(\chi - 1)q_r - q_i + \mu_j\}X_j, \quad j = F, B \tag{26}$$

$$\frac{dS}{dt} = q_i(S_i - S) - \frac{\mu_F}{Y_F}X_F - \frac{\mu_B}{Y_B}X_B \tag{27}$$

$$\mu_j = \begin{cases} K_j(S - S_j) & \text{for } S \geq S_j \\ 0 & \text{for } S < S_j, \end{cases} \quad j = F, B \tag{28}$$

and

$$\chi = \alpha\left(\frac{X_B}{X_F} + \beta\right)^{-\gamma} + 1 \tag{29}$$

where X_F, X_B, and S are the outlet concentration of the floc-forming sludge and the bulking sludge and the COD in the aeration tank, respectively, and S_i is the inlet COD concentration in the system; q_i and q_r are the inlet flow rate and recycled flow rate divided by the tank volume, respectively. Equations 26 through 29 are analyzed to obtain information about the bulking phenomenon caused by high organic loading. One of the important characteristics of process equations is that the growth rates of the bulking and floc-forming sludges intersect (Fig. 12). This result means that the interaction between the two sludges can be classified as competition for assimilation of the same substrate (COD). Using medium B, the relationship between the specific growth rates of both sludges becomes similar to those using medium A.

The intersection in Figure 12 defines the value of the COD concentration S_c at which both types of sludge, bulking and floc-forming, can coexist. As shown in Figure 12, in the region where COD is smaller than S_c, the specific growth rate of the floc-forming sludge is higher than that of the bulking sludge; that is, the floc-forming sludge may become dominant. On the other hand, in the region where COD is larger than S_c, the bulking sludge is dominant.

Another distinguishable feature of the mixed culture system employed here is the existence of the cell recycle stream. Reflecting on this feature, the coexistence of both sludges should be taken into account at a certain dilution rate, although the coexistence state does not occur without the cell recycle stream.

Dynamics of an Activated Sludge Process

Equilibrium Points. We next analyzed the dynamic behavior of equations 26 through 29 when the organic load-

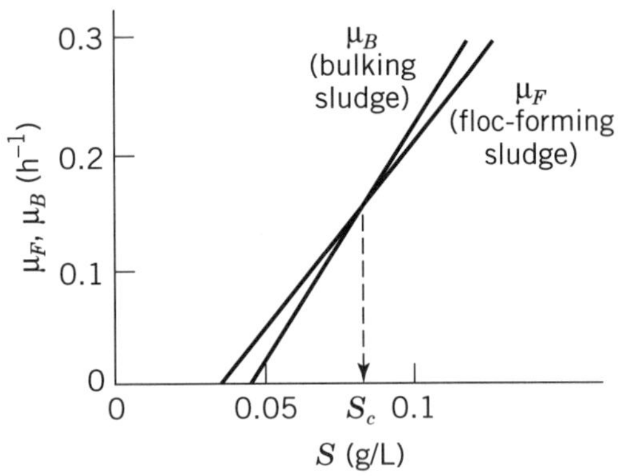

Figure 12. The specific growth rates of floc-forming sludge and bulking sludge using medium A.

ing rates S_i and q_i are changed. As the characteristics of a mixed culture system using medium B are similar to those using medium A, analysis is based on the parameters obtained using medium A. First, the possible steady states should be clarified.

It can be shown that there are, in general, four equilibrium points: normal state E-1 ($X_F \neq 0$, $X_B = 0$); bulking state E-2 ($X_F = 0$, $X_B \neq 0$); coexistence state E-3 ($X_F \neq 0$, $X_B \neq 0$); and washout state E-4 ($X_F = 0$, $X_B = 0$). The normal state, E-1, is the only desirable state for stable operation, and the bulking state, E-2, should be avoided. Figure 13 shows the relationship between these four equilibrium points in the state space. These points are given as intersections of three planes, called "isoclines," and the plane $X_B = 0$ or $X_F = 0$. An isocline is defined here as the curve or the plane along which the first derivative of the state variables with respect to time t is zero excluding $X_B = 0$ and $X_F = 0$. Then, the isoclines are three planes that satisfy $dX_F/dt = 0$, $dX_B/dt = 0$, and $dS/dt = 0$ in equations 26 through 29. From the definition, the intersections of these planes are equilibrium points. For example, point E-1 is given as the intersection of both isoclines of $dX_F/dt = 0$ and $dS/dt = 0$, and the plane $X_B = 0$. The number of possible equilibrium points under a given operating condition is counted using the shapes and interconnections of these isoclines.

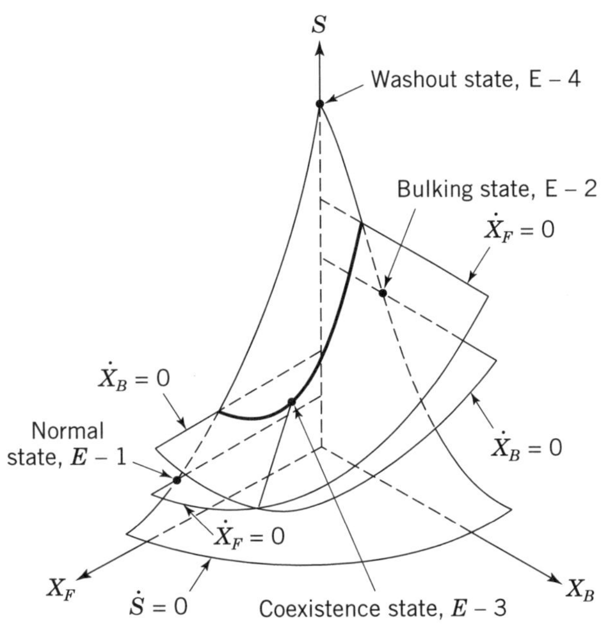

Figure 13. The possible independent equilibrium points, which are shown as intersection of isoclines and $X_F = 0$ or $X_B = 0$. Four possible points are shown as E-1 to E-4.

The trajectory of the state from a given initial condition to a final state is roughly estimated using these three isoclines because an isocline divides the gradient of a state variable into two regions having negative and positive values. From the gradient of the three state variables, the gradient of the trajectory in the state space is estimated. Finally, the maximum number of possible equilibrium points is four, as shown in Figure 13, and normal state E-1 is the only desirable state for the stable operation of the activated sludge process. The other states should be avoided. To make it possible for normal state E-1 to exist, parameters such as input flow rate q_i or recycled flow rate q_r and input COD concentration S_i must satisfy the following equations:

$$S_F \leq S_e \leq S_i \tag{30}$$

where

$$S_e = \frac{\{q_i - (\chi^* - 1)q_r + K_F S_F\}}{K_F} \tag{31}$$

Here χ^* is the value of χ at $X_B/X_F = 0$. This constraint was deduced from the assumption that normal state E-1 really exists. The same type of analysis can be performed with respect to the other equilibrium points.

Locally Stable Conditions of the Activated Sludge Process. The activated sludge process should be operated under conditions guaranteeing that the equilibrium point E-1 exists and is locally stable. The term locally stable means that the equilibrium point is asymptotically stable in the linearized system. The occurrence of the locally stable condition can be determined by checking the sign of the real part of the eigenvalues of the characteristic equation in the linearized system (8). Stability is guaranteed only in the vicinity of the equilibrium point. Of course, in practical operation, whether the system reaches the state E-1 from a given initial state depends on global stability, based not only on local stability but also on the global dynamic behavior. However, locally stable conditions provide useful information with respect to the stable operation of the activated sludge process. Using the locally stable condition, the domain of stable operation in the operating space consisting of S_i, q_i, and q_r can be obtained.

The locally stable conditions of four equilibrium points were deduced analytically as functions of the operating conditions S_i, q_i, and q_r. We showed that the coexistence state E-3 is unstable under all operating conditions. The locally stable conditions and the existence conditions of the four equilibrium points divide the domain of the operating conditions into many regions. Such an analysis can be used to identify the existence and stability of an equilibrium point.

Figure 14 shows the region described by q_i, q_r, and S_i in which E-1 exists and is locally stable. Under the locally stable condition at E-1, S_e defined in equation 31 is as follows:

$$S_F \leq S_e \leq S_c \tag{32}$$

where S_c is the intersection of μ_F and μ_B as shown in Figure 12; that is, $\mu_F \geqq \mu_B$ in this region, and the floc-forming sludge becomes dominant. The well-known fact that high organic loading will cause the bulking phenomenon is verified using the model employed here (Fig. 14). This figure can also be utilized to determine the strategy required to obtain locally stable conditions. When the state of both sludges and COD concentrations does not change substantially from E-1, the system will be operated in a stable way as long as the operating condition is maintained in the region indicated in Figure 14. If the flow rate of wastewater q_i increases, the recycled flow rate q_r should be increased, as in Figure 14, to maintain the conditions in this locally stable region.

Figure 15 shows the region in which the bulking equilibrium E-2 is unstable or does not exist. In the domain of the operating conditions indicated in Figure 15, only E-1 is locally stable. Even though the bulking sludge increases slightly, based on the local stability it can be said that the desired operating state will be restored as long as the operating condition is maintained in this region. These conditions coincide with the well-known rule that the sludge loading rate should be kept low. The kind of deviation in the operating and initial conditions that renders the normal state E-2 unstable was clarified.

The locally stable conditions are not sufficient to maintain a stable operation that prevents the occurrence of the bulking phenomenon. This is because in practice locally stable conditions are valid only in the vicinity of equilibrium points. Therefore, the manipulating variables should be changed dynamically to maintain the system in normal state E-1 when disturbances have been introduced. The control problem has been solved (7). Finally, the interaction between the two sludges has been classified as competitive, and a mathematical model has been proposed and simulations as well as a stability analysis have been performed to determine the manner in which to operate the activated sludge process in a stable way.

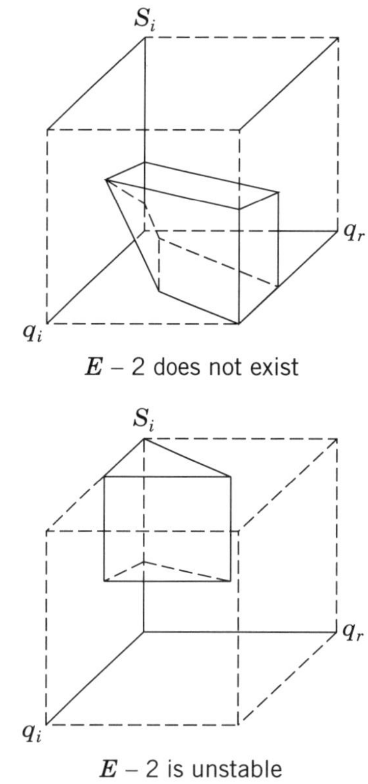

Figure 15. The desirable operating domain so that E-1 is stable and the other equilibrium points do not exist or are unstable.

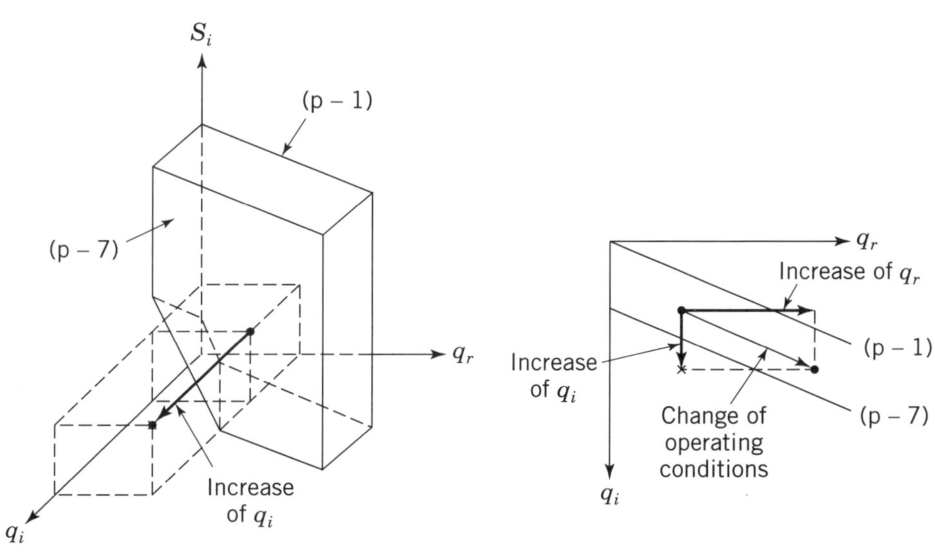

Figure 14. The domain of operating conditions in which E-1 exists and is stable.

CONCLUSION

In this article, practical examples that show typical interactions such as those classified under the headings competition, cooperation, and mutualism between two microorganisms have been described so as to demonstrate how the models, simulations, and stability analysis can contribute to better understanding of the dynamic behavior as well as the practical phenomena of the processes. The examples also demonstrate that the mixed culture system can be utilized for solving practical problems such as the avoidance of product inhibition.

Only a few examples, however, are described here, but there are others indicating interesting characteristics of mutual interactions. For example, it is speculated that many mixed culture systems in which the microorganisms exhibit synergistic relationships exist but that isolation of the strains is very difficult because of the strict mutualism. It is likely that during the coming decade many examples of such systems will be found using modern molecular biological techniques such as the PCR method. The application of mixed culture systems to practical problems so as to enhance the productivity should then increase.

BIBLIOGRAPHY

1. M.E. Bushell and J.H. Slater, eds., *Mixed Culture Fermentations*, Academic Press, London, 1981.
2. J.E. Bailey and D.F. Ollis, *Biochemical Engineering Fundamentals*, McGraw-Hill, New York, 1977.
3. A.G. Fredrickson and H.M. Tsuchiya, in N.R. Amundson and L. Lapidus, eds., *Chemical Reactor Theory: A Review*, Prentice-Hall, Englewood Cliffs, N.J., 1977.
4. A.G. Fredrickson, *Annu. Rev. Microbiol.* **31**, 63–87 (1977).
5. E.P. Odum, *Basic Ecology*, W.B. Saunders, Philadelphia, Pa., 1983.
6. M. Taniguchi, K. Hoshino, H. Urasaki, and M. Fujii, *J. Ferment. Bioeng.* **77**, 704–708 (1994).
7. T. Takamastu, S. Shioya, and H. Kurome, *AIChE J.* **30**, 368–376 (1984).
8. M. Athans and P.L. Falb, *Optimal Control: An Introduction to the Theory and Its Application*, McGraw-Hill, New York, 1966.

See also BIOREMEDIATION; CULTURE COLLECTIONS.

MONOACYLGLYCEROLS

TSUNEO YAMANE
Nagoya University
Nagoya, Japan

KEY WORDS

Antimicrobial
Esterification
Glycerol
Glycerolysis
Lipase
MAG
Monoacylglycerols
Monoglycerides
Protected glycerol

OUTLINE

INTRODUCTION

Monoacylglycerols (MAGs), popularly called monoglycerides, is a generic name for a group of esters whose sole glycerol hydroxyl group is esterified with a fatty acid carboxyl group. These are compounds of the most popular emulsifiers or surfactants used in food, cosmetics, and the pharmaceuticals industry. Because of their wide application and importance for humans, research and development concerning MAGs have been carried out in both the academic and industrial sectors.

STRUCTURE AND NOMENCLATURE

Glycerol has a symmetrical plane that centers at the C-2 position, and it is achiral (i.e. optically inactive). Because glycerol has three hydroxyl groups, three types of structurally different MAGs can exist, depending on which hydroxyl group, C-1, C-2, or C-3, is esterified. In the older literature these three types have been often differentiated by the use of α and β (Scheme 1a). Introduction of a different substituent group at C-1 or C-3 makes glycerol chiral. Due to the recommendation of the IUPAC-IUB Commission and according to Buchnea (1), the stereospecific numbering (*sn*) system has been widely adopted. Thus enantiomers of MAG are named as 1-*O*-acyl-*sn*-glycerol and 3-*O*-acyl-*sn*-glycerol, or more simply 1-acyl-*sn*-glycerol and 3-acyl-*sn*-glycerol. The absolute configulations of 1-*O*-acyl-*sn*-glycerol and 3-*O*-acyl-*sn*-glycerol are, (*S*) and (*R*), respectively. An equimolar mixture of these enantiomers gives a racemate and is described as *rac*-monoacylglycerol. If the secondary hydroxyl group is shown to the left of C-2 in a Fischer projection, then these enantiomers are depicted as in Scheme 1b. There exist quite a number of MAGs corresponding to various fatty acids with which a glycerol hydroxyl group reacts to yield a monoester. MAGs whose fatty acid moieties are popular have particular but

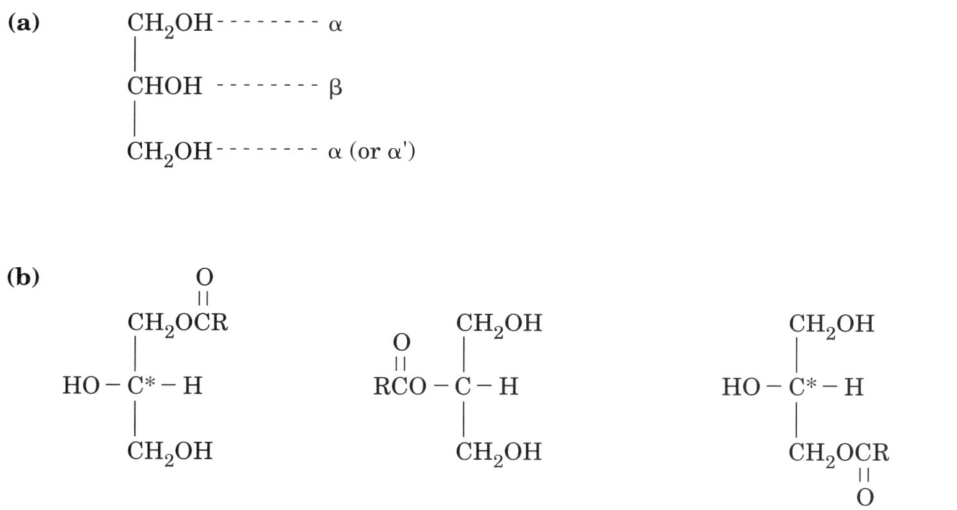

1-*O*-Monoacyl-*sn*-glycerol (1-MAG) 2-*O*-Monoacyl-*sn*-glycerol (2-MAG) 3-*O*-Monoacyl-*sn*-glycerol (3-MAG)

Scheme 1. Structure and nomenclature of MAGs.

nonscientific names such as monopalmitin, nomostearin, and monoolein.

ANALYSIS

MAGs are determined by various analytical methods including titration (volumetric analysis), thin-layer chromatography (TLC), high-performance liquid chromatography (HPLC), and gas chromatography (GC).

Classical determination of α-MAG is a volumetric titration using periodic acid (H_5IO_6). This method utilizes the oxidative cleavage of vicinal hydroxyl groups by H_5IO_6 (Scheme 2, reaction 1). In case of α-MAG, the oxidative reaction proceeds as shown in Scheme 2, reaction 2. In the actual volumetric analysis, α-MAG is completely oxidized by excess H_5IO_6, followed by the reduction of remaining unreacted H_5IO_6 with KI solution; the iodine formed is quantitatively determined with sodium thiosulfate ($Na_2S_2O_3$) (Scheme 2, reactions 3 and 4). This method is accepted as a standard analytical method for MAGs by both the American Oil Chemists' Society and Japan Oil Chemists' Society. It must be noted that the above method is in principle applicable only to the determination of α-MAG; β-MAG (2-MAG) cannot be determined due to inert reactivity with H_5IO_6. Commercial MAG contains usually 5–9% β-MAG. In the presence of a small amount of perchloric acid, α- and β-MAGs quickly reach an equilibrium (86.9% α-MAG and 13.1% β-MAG) through acyl migration started either from pure α-MAG or pure β-MAG (2). The application of this behavior enables the determination of total MAG as well as α-MAG by titration after pretreatment of the sample in both the absence and presence of perchloric acid.

TLC enables separation of a mixture of α-MAG, β-MAG, 1,2-diacylglycerols (1,2-DAGs), 1,3-diacylglycerols (1,3-DAGs), triacylglycerols (TAGs), and free fatty acids (FFAs) by applying the proper solvent system (e.g., chloroform: acetone, 96:4 v/v), but requires additional spectrophotometry or gas–liquid chromatography in order to determine MAG content precisely. However, a unique and very convenient analytical instrument called a TLC–flame ionization detector (TLC/FID) (Iatroscan®) is commercially available for the quantitative analysis of neutral lipids and partial glycerides. For details of TLC/FID analysis of partial glycerides, see Ref. 3; Figure 1a shows a chromatogram of a glycerides mixture in TLC/FID.

MAGs are also able to be quantitatively determined with gel permeation chromatography (GPC) using a refractive index (RI) detector. Silica gel or a related gel is used as column packing. Because a chromatographic column has very high number of theoretical plates (NTP), GPC-HPLC makes it possible to separate and determine MAGs having different molecular size. This means that one can use GPC-HPLC to determine MAGs whose fatty acids are different. MAGs having different fatty acid moieties can be also separated as their di-3,5-ditnitrophenylurethane derivatives with normal phase HPLC on nitrile-bonded phase (4).

As noted earlier, there are three isomers of MAG, including a pair of enantiomers. Resolutions of these isomers using chiral column chromatography after derivatizing 3,5-dinitrophenylurethane were reported by Takagi and Itabashi (5) and reviewed by Takagi (6). A typical HPLC chromatogram of a mixture of monopalmitoylglycerols is shown in Figure 1b (7). More recently, Meguro et al. reported a highly sensitive HPLC method to discriminate enantiomeric MAGs based on fluorescent chiral derivatization (7); its details are summarized in a monograph by Kim et al. (8).

MAGs used widely as surfactants or emulsifiers in food and cosmetics are mixtures of esters containing different fatty acids. Direct separation of MAGs having different fatty acids is feasible with HPLC if the number of the fatty acids is small, but is difficult if it is very large. Fatty acid composition of MAGs can be most successfully determined with a capillary GC after derivatization of fatty acids into

Oxidative cleavage of vicinal hydroxyl groups by periodic acid:

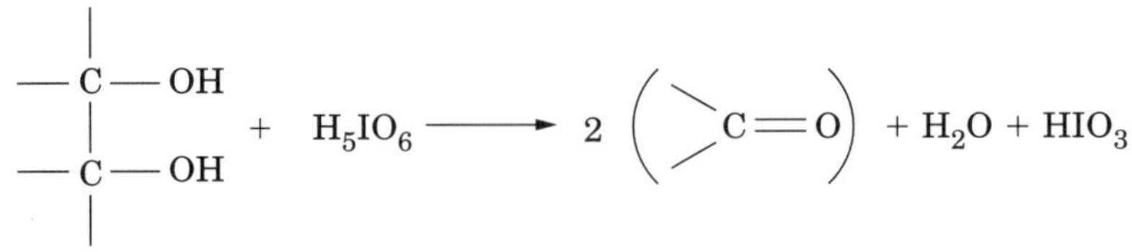

(1)

Oxidation of α-MAG by periodic acid:

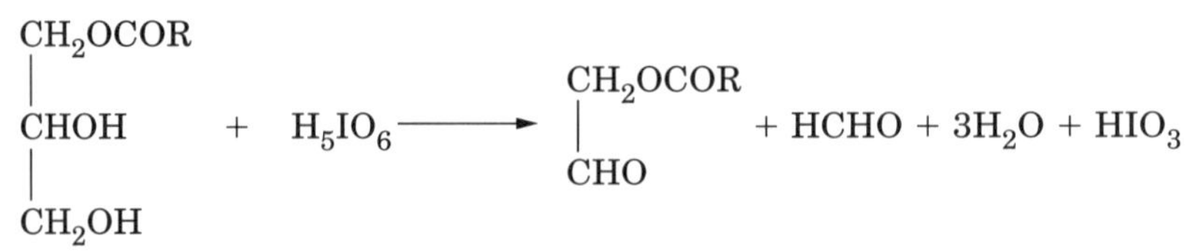

(2)

Reduction of periodic acid by potassium iodide:

$$H_5IO_6 + 2KI \longrightarrow HIO_3 + 2KOH + I_2 + H_2O \quad \textbf{(3)}$$

Reduction of iodide by sodium thiosulfate:

$$I_2 + 2Na_2S_2O_3 \longrightarrow 2NaI + Na_2S_4O_6 \quad \textbf{(4)}$$

Scheme 2. Reactions involved in MAG determination by titration.

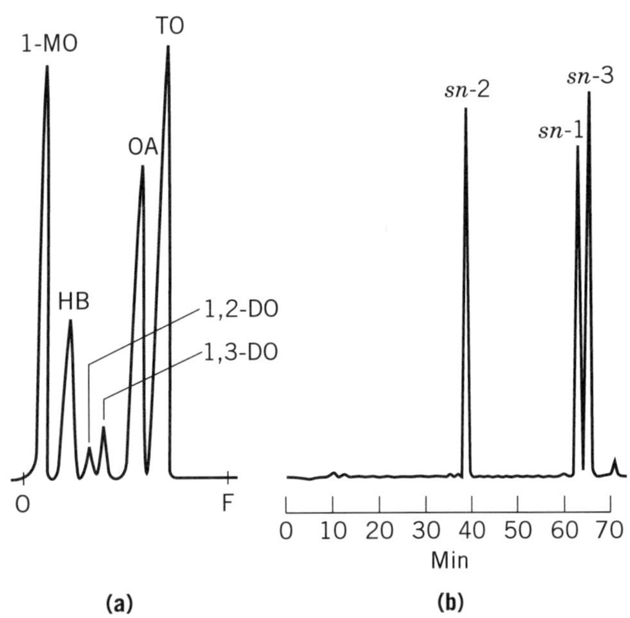

Figure 1. Chromatogram of a glycerides mixture. **(a)** TLC/FID chromatogram of oleic acid glycerides; 1-MO, 1-monooleylglycerol; HB, *p*-hydroxybenzoic acid; 1,2-DO, 1,2-dioleylglycerol; 1,3-DO, 1,3-dioleylglycerol; OA, oleic acid; TO, trioleylglycerol. *Source*: Ref. 3. **(b)** HPLC chromatogram of a mixture of monopalmitoylglycerols. *Source*: Ref. 7.

methyl esters by concentrated KOH/methanol (in a solvent if necessary).

SYNTHESIS

MAGs are synthesized by various methods. Various MAGs can be prepared biochemically (i.e., by the action of lipase) on a laboratory scale. Industrially they have been produced by chemical methods.

Chemical Synthesis

Chemical methods are used for the large-scale production of MAGs for food additives and cosmetics. Annual production in the world is estimated to be more than 150 thousand tons. A widely applied reaction is the glycerolysis of fats and oils, which is a reaction between glycerol and fats or oils (Scheme 3). To accelerate the reaction and to avoid undesirable side reactions such as oxidation of fatty acids and polymerization, the reaction is performed at high temperature (220–250 °C), employing an inorganic alkaline catalyst under a nitrogen gas atmosphere. Glycerol and fat that are immiscible at ambient temperature become miscible at such high temperature, which is also favorable for the reaction. The reaction is operated in either a batch or continuous method. Continuous operation gives better quality MAGs than the batch method because there is less damage to the fat with shorter heating and reaction time. The product immediately after the glycerolysis reaction contains DAGs, TAGs, and glycerol as well as MAGs, according to the random distribution rule. This rule states that relative composition of these four components de-

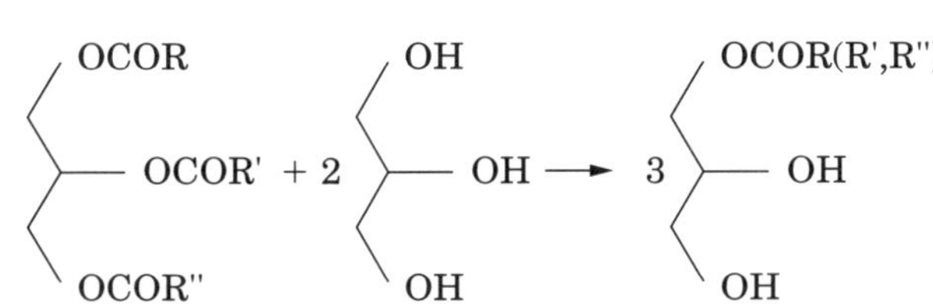

(5)

Scheme 3. Glycerolysis of fats.

pends only on the ratio of hydroxyl and carboxyl groups if three hydroxyl groups of glycerol react randomly with one carboxyl group of any fatty acid. The probability calculation is shown in Figure 2. The composition of the product is determined by the initial mixing ratio of glycerol and fat. Usually 40 wt. part of glycerol is reacted with 100 wt. part of fat, resulting in a product containing 40–50% MAG. The reaction mixture is then subjected to molecular distillation or short-path vacuum distillation to increase MAG content and also to remove undesirable byproducts. The molecular distillation is a kind of fractionation under very high vacuum (2–4 μmHg). In the equipment, clearance between vaporization and condensation surfaces is narrow so that molecules form a thin film of oil on the heated vaporization surface, which vaporizes and subsequently condenses at the cooled condensation surface without colliding with other molecules (O_2 etc.). Therefore, unstable compounds having high melting points can be purified. Two types of molecular distillation equipment are known: falling film and centrifugal. After the molecular distillation, MAG content is ca. 90%, and the rest is mostly DAGs; the residue is recycled as raw material to subsequent glycerolysis. Commercial MAGs produced by this method are called distilled monoglycerides. The isomeric composition of the MAGs is 91–95% α-MAG and 5–9% β-MAG. Raw fats or oils used for the glycerolysis include hydrogenated beef tallow, lard, soybean oil, and palm oil. Conventional chemical glycerolysis has been well reviewed by Sonntag (9). Glycerolysis of fat has been carried out without any solvent. Chemical glycerolysis of soybean oil was recently conducted in supercritical CO_2 media without alkali catalyst (10). A maximum MAG content of 50% was achieved at 250 °C, 20.7 MPa, a glycerol/oil ratio of 25, and 4% water, in 4 h.

Although industrial demand is much less, it is possible to chemically synthesize α- and β-MAGs using the protected hydroxyl group of glycerol. To synthesize α-MAG, α'- and β-hydroxyl groups of glycerol are first protected by acetone to form 1,2-O-*rac*-isopropyridene-*sn*-glycerol (1,2-*O*-IPG; 2,2-dimethyl-1,3-dioxolane-4-methanol; acetonide; acetal), followed by acylation of the α-hydroxyl group using acid chloride or free fatty acid, finally followed by deacetonization of the product using acid (Scheme 4). In order to protect the α- and α'-hydroxyl group for the purpose of 2-MAG synthesis, benzaldehyde or triphenylchloromethane (triphenylmethylchoride, $(C_6H_5)_3CCl$) is used. Reaction of glycerol with benzaldehyde gives rise to a mixture of α,β- and α,α'- benzylideneglycerols. α,α'-Isomer (1,3-*O*-benzylidene-*sn*-glycerol) is isolated by fractional crystallization utilizing difference in their solubilities. 1,3-*O*-Benzylideneglycerol is then esterified by acid chloride in the presence of pyridine or quinoline. Care must be taken here to deprotect the product to yield 2-MAG in order to avoid acyl migration from the 2-position to the 1- or 3-position. Only catalytic reduction (11) or boric acid (12) makes it possible to get 2-MAG. The complete scheme of 2-MAG synthesis is shown in Scheme 5. Overall yield of 2-MAG by this method is 15–20% because of difficulty in the preparation of 1,3-*O*-benzylidene-*sn*-glycerol.

Glycidol (2,3-epoxypropanol) is available in racemic and in (*R*) or (*S*) form, and these are also convenient starting compounds for the preparation of racemic and enantiomeric MAGs. Burgos et al. reported high yield preparation of MAGs from glycidol using titanium (IV) *tert*-butoxide as a catalyst at low temperature (13).

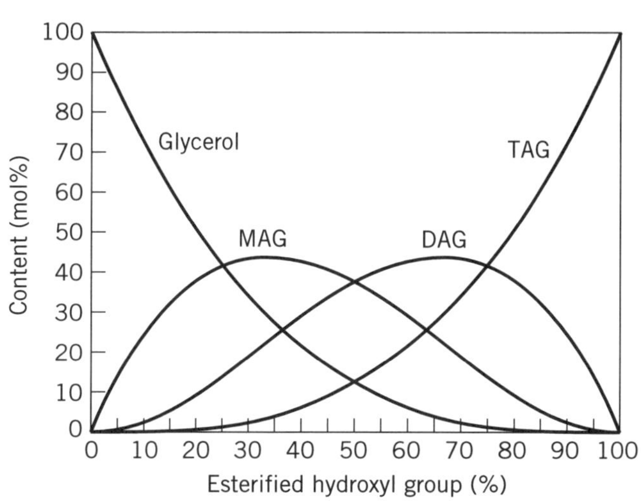

Figure 2. Composition of glycerides produced by glycerolysis, calculated according to the random distribution rule.

Biochemical Synthesis

A great deal of research and development has been done for the past 15 years in enzymatic synthesis of MAGs. The enzyme lipase (and esterase, in some research) has been exclusively utilized because lipase was employed successfully in the area of research and technology of fats. In fact lipase-catalyzed reaction for modifications of fats and lipids is a current trend in biotechnology. The incentive for the recent extensive research on lipase-catalyzed MAG synthesis is problems associated with the conventional chemical glycerolysis. The high temperatures involved in the reaction often result in a dark-colored product with a burnt odor. In addition, the presence of FFAs and their metallic soaps can lead to associated odor problems in food products into which they are incorporated. Operation at much lower temperature using the lipase-catalyzed MAG synthesis is feasible and can minimize these odor and color problems. Moreover, various specificities (stereo, positional, regional, and fatty acid) of lipase lead researchers to expect to synthesize unique MAGs that cannot be prepared by chemical methods. Also, it has been expected that immobilization of lipase and bioreactor technology, both of which have been well investigated over the last decade, can reduce the cost of enzymatic synthesis substantially, even though lipase is a rather expensive enzyme. The latest state of the art for the bioreactors with immobilized lipase was reviewed by Balcão et al. (14). Lipase and esterase catalyze three types of reaction: ester hydrolysis, esterification, and transesterification (15). Transesterification is categorized into four subdivisions according to the chemical species with which the ester reacts; acidolysis, alcoholysis, ester exchange (interestrification), and aminolysis.

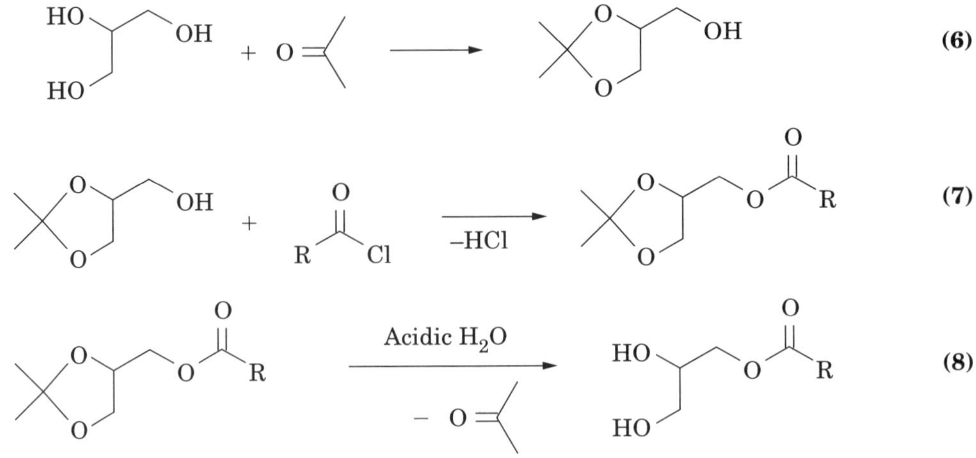

Scheme 4. Synthesis of α-MAG via protected glycerol by acetonization.

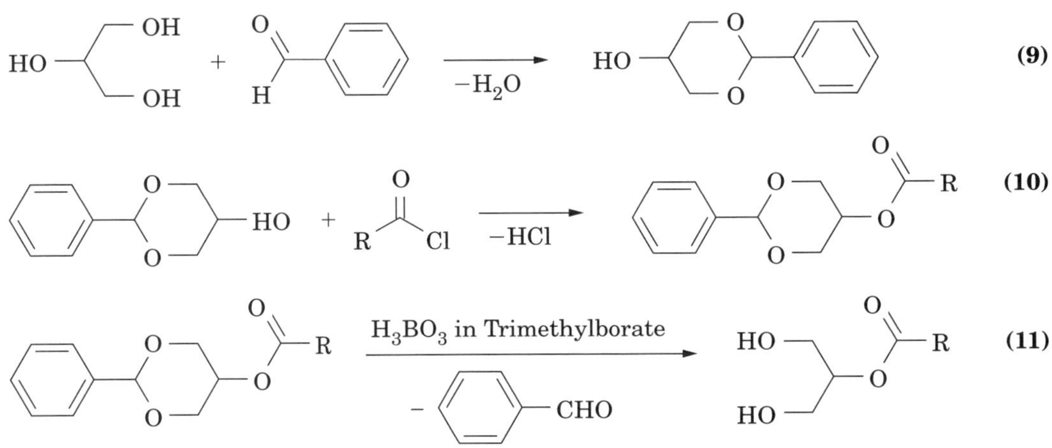

Scheme 5. Synthesis of β-MAG via protected glycerol with benzaldehyde.

All these types of reaction can be applied to MAG synthesis.

Bornscheuer comprehensively reviewed lipase-catalyzed syntheses of MAG through 1993 (16). Most references (123 articles) in the area of the enzymatic MAGs syntheses through that year were quoted in his article. He classified the lipase-catalyzed syntheses of MAG into four categories: selective hydrolysis, esterification of fatty acids, transesterification of fatty acid esters with glycerol, and glycerolysis of fats and oils. In this article, the latest achievements in this field are described according to three categories: hydrolysis, esterification, and alcoholysis.

Hydrolysis. MAG formation by enzymatic hydrolysis of TAG seems limited for two reasons: (*1*) low overall yield of MAG because 2 mol FFA and 1 mol MAG are liberated, and (*2*) the hydrolysis needs to be carefully controlled to avoid complete hydrolysis. Some studies have been done on the hydrolyses of several seed oils, such as mustard, crambe, honesty, and castor. Under ideal conditions, in which the lipase used is strictly 1,3-specific and no acyl migration occurs, one will be able to get 2-MAG with high yield (Scheme 6). Holmbery and Osterberg reported 80% yield of 2-MAG by the hydrolysis of palm oil with *Rhizopus delemar* lipase (1,3-specific lipase) in microemulsion (reversed micelle) prepared from Aerosol OT (Sulfobutanedioic acid 1,4-bis(2-ethylhexyl) ester sodium salt), isooctane, and aqueous buffer (pH7) (17). Attempts have been made to increase the content of polyunsaturated fatty acids (PUFAs) such as arachidonic acid (AA), eicosapentaenoic acid (EPA), and docosahexanoic acid (DHA) in the hydrolysate by enzymatic partial hydrolysis of fish oil. The earliest literature concerning this attempt is a Japanese patent (18). *Candida cylindracea* (18–21), *Geotrichum candidum* (22,23), *Pseudomonas* sp. (24), or porcine pancreas (25) have provided the enzyme. One might expect a PUFA-concentrated MAG fraction to result from selective hydrolysis because PUFAs are usually involved in the β-hydroxyl position, 1,3-specific lipase does not attack the β-acylglycerol bond, and lipase hardly hydrolyzes PUFA esters, but the actual MAG content of the hydrolysate was quite low (2–5%) (19,22). This may have been due to both intermolecular acyl migration between partial glycerides, and esterification between the free PUFAs and the partial glycerides, resulting in the formation of PUFA-enriched TAGs.

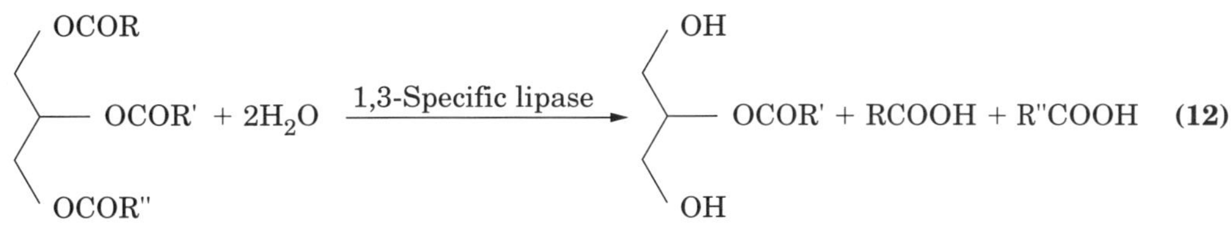

Scheme 6. 2-MAG formation by hydrolysis of TAG with a 1,3-specific lipase.

Esterification. Lipase-catalyzed esterification of fatty acids with a hydroxyl compound is possible only in microaqueous (low-water) systems with just trace amounts of water. Water liberated during the esterification must be removed to raise the ester yield. For this purpose, various methods have been proposed, including the addition of molecular sieves, reduced pressure, an organic solvent, and reversed micelles (microemulsion). The achievements through 1994 of MAG synthesis by biochemical esterification are summarized in Ref. 16. Using a selected fatty acid, MAG with a single fatty acid moiety can be synthesized. To achieve a high-yield formation of MAG, protected glycerol (1,2-isopropylidene-glycerol or 1,3-benzylidene glycerol, see Schemes 4 and 5) is desirable. Hess et al. synthesized monostearoylglycerol in organic solvents from 1,2-*O*-isopropylidene glycerol using *Pseudomonas cepacia* lipase (26). However, when mono- and diacylglycerol lipase from *Penillium camembertii* was used, a high yield of MAG was synthesized from FFAs and free glycerol (27). It is claimed that the enzyme specifically catalyzes hydrolysis of MAG and DAG but not TAG.

Alcoholysis. Here, alcoholysis is defined as the reaction between fatty acid esters, including TAG, and glycerol (free or protected). Alcoholysis of TAG with free glycerol is specifically called glycerolysis (Scheme 3) .

Since the work of Yamane et al. (28), lipase-catalyzed glycerolysis has received considerable attention, and much research related to this type of reaction has been carried out. Work before 1994 is reviewed in Ref. 15. Among this research it is notable that McNeill et al. found that glycerolysis of fats at low temperature increased MAG yield due to solidification of the reaction mixture (29). In one of serial study on the lipase-catalyzed solid-phase glycerolysis of fat, Yamane et al. demonstrated recently that immobilized lipase worked effectively as a catalyst on the solid-phase glycerolysis of olive oil (30). *Pseudomonas* lipase adsorbed on $CaCO_3$ fine powder can be reused eight times with 80–90% MAG yield in each run (Fig. 3). Myrnes et al. carried out glycerolysis of marine TAGs (cod liver oil and oils from the blubber of the harp seal and mink whale) at low temperature using *Pseudomonas* lipase without any solvent (31). MAG yield and *n*-3-PUFA content of the MAG were 42–53% and 12–20%, respectively.

In order to increase MAG yield, alcoholysis was conducted on a fatty acid ester with a protected glycerol (32). As ester vinyl ester gave the best results due to tautomeric conversion of liberated vinyl alcohol into acetaldehyde. Almost 100% conversion was achieved after a short time (Scheme 7). Afterward, acetonide was deprotected with boric acid, as shown in Scheme 4.

As just described, there are a number of possible strategies to synthesize MAG enzymatically, such as ethyl esters in fatty acid glycerolysis, TAG glycerolysis, and direct esterification; free and protected glycerol; solvent-free and solvent systems; and native, modified, and immobilized lipases. Some scientists carried out comparative studies to determine which is best (33–35). Their conclusion was that glycerolysis of TAG and direct esterification in the absence of solvents are less costly and more selective.

When free glycerol is used, glycerolysis, either chemical or biochemical, gives rise to *rac*-MAG formation. It is interesting to know whether alcoholysis of a fatty acid ester with *rac-O*-isopropylidene glycerol is enantioselective. Hess showed that in the case of ethyl stearate, the product, 1-stearoyl-2,3-*O*-isopropylidene glycerol, was found at only 11% ee (26).

Biochemical glycerolysis of fats yields α-MAG (mixture of 1-*sn*-MAG and 3-*sn*-MAG) in most cases. However, Millqvist et al. attempted to produce 2-MAG by lipase-catalyzed alcoholysis of TAG in organic media (36). When tripalmitin, ethanol, and methyl-*tert*-butyl ether were used as substrate, the fatty acid acceptor, and the solvent, respectively, at a water activity of 0.11 with *Rhizopus arrhizus* lipase immobilized on Celite, the highest yield (97%) was achieved. They emphasize that the reaction is not equilibrium controlled, but kinetically controlled, and that the enzymatic reaction must be much faster than spontaneous acyl migration to obtain high production of 2-MAG.

PHYSICOCHEMICAL PROPERTIES

The properties of distilled MAG depend on the fatty acid composition of the fat used as the raw material. The most influential parameter is the degree of unsaturation (i.e., iodine value). Saturated fatty acid MAG is a white, hard powder, but the higher iodine value, the softer the MAG. Unsaturated fatty acid MAG is soft and elastic. Table 1 shows some properties of distilled MAG. MAG is surface active because the molecule has both hydrophilic and lipophilic groups; it is widely used as an emulsifier in the food industry and is generally recognized as safe (GRAS). The hydrophile–lipophile balance (HLB) value of distilled MAG is 3–4. Table 2 shows the reduction in interfacial tension by MAG addition. A MAG content of several percent in oil reduces the oil's surface tension significantly. Solubility of MAG in water is very low, but in fat is quite high. At higher temperature, MAG and fat are miscible. MAG is

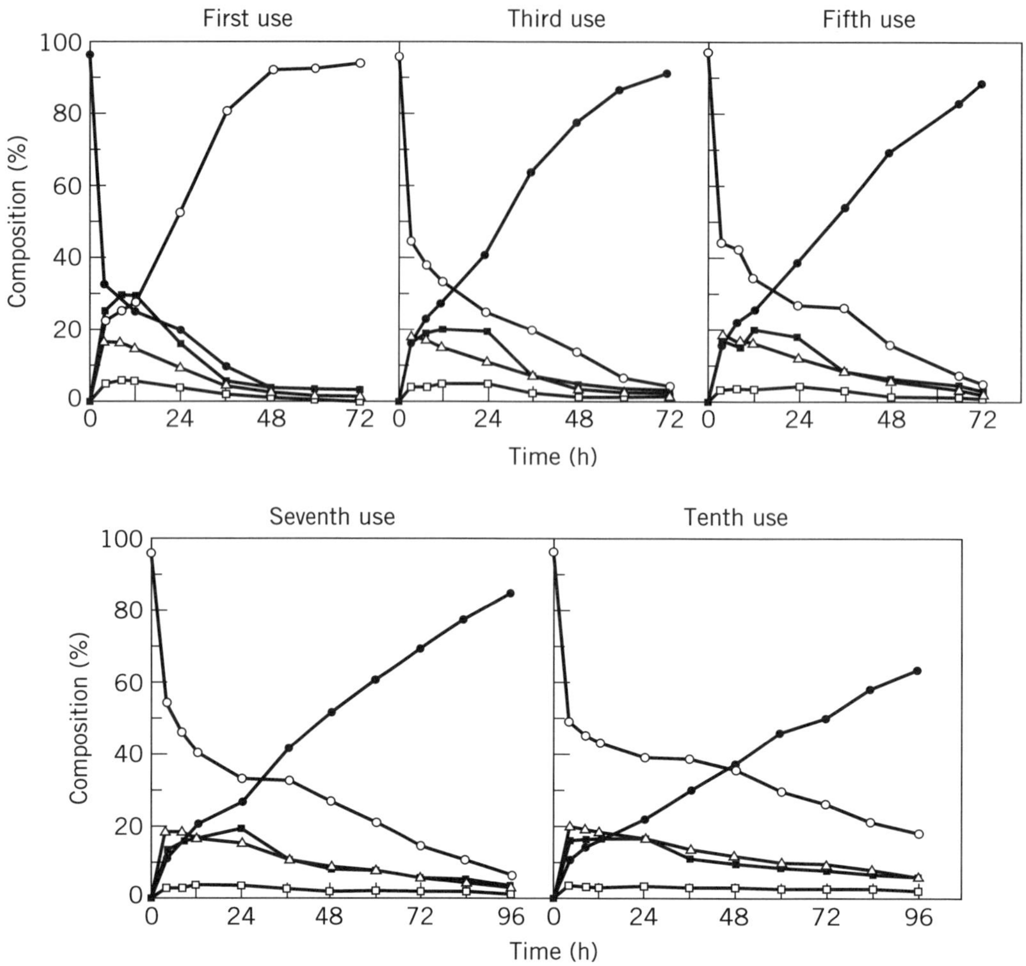

Figure 3. Time course of glycerolysis of olive oil by $CaCO_3$-immobilized lipase during reuses. MAG (○), 1,3-DAG (■), 1,2-DAG (△), TAG (●), and FFA (□). *Source*: Ref. 30.

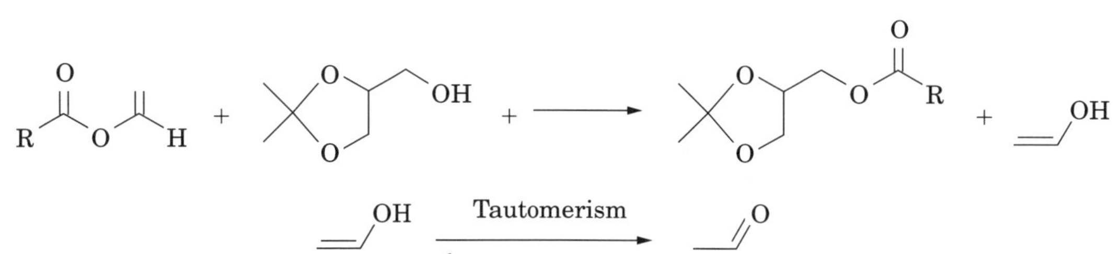

(13)

(14)

Scheme 7. Reaction of fatty acid vinyl ester and protected glycerol.

Table 1. Some Properties of Distilled MAG

	Hydrogenated lard	Hydrogenated soybean oil	Lard	Safflower oil
MAG (%)	>90	>90	>90	>90
Free fatty acids (%)	<1	<1	<1	<1
Free glycerol (%)	<1	<1	<1	<1
Saponification value	160–165	155–160	160–165	155–160
Iodine value	<1	<2	40	100
Melting point (°C)	70	72	56	40

Table 2. Reduction in the Interfacial Tension by MAG Addition (0.5% MAG in Oil)

Oil	Without MAG	1-Monostearin	2-Monostearin
Soybean oil	29.6	22.4	21.4
Benzene	35.6	22.1	20.4
Petroleum	50.0	5.1	2.9
n-Heptane	45.8	6.9	5.3
Liquid paraffin	48.0	2.6	2.2

soluble in organic solvents, the degree of solubility depending on the solvent (Table 3) (37). MAG has four different crystal forms: sub-α, α, β', and β. Each form has its characteristic melting point (Table 4) (37). The β-form is the most closely packed, so it has the highest melting point. Densities of MAG with a longer-chain fatty acid are smaller than unity, but those having shorter-chain fatty acids are heavier than unity.

NUTRITION AND BIOLOGICAL EFFECTS

There is evidence that distilled MAGs are nutritionally equal to the fats from which they were prepared in terms of absorption from intestinal tract, support of body growth, reproduction, and lactation (38).

Dietary TAG is acted upon by lipolytic enzymes in the stomach and the proximal small intestine, releasing FFAs and MAG as the first intermediate products. Enzymes involved in the TAG lipolysis are gastric and pancreatic colipase-dependent lipases, and the lipolytic products are FFAs from the *sn*-1,3 position and *sn*-2 MAGs. *sn*-2 MAGs are then absorbed through the intestinal mucosa and hydrolyzed by monoacylglycerol lipase (MAGL) in the blood, the hydrolysis proceeding to completion. A membrane-bound MAGL, previously demonstrated in intact human erythrocytes, has been recently purified and characterized (39). Distilled MAGs, composed mainly of α-MAG (as described in the section "Synthesis") and supplied as food additives, are hydrolyzed by MAGL because it has no positional specificity. Thus, MAGs having common fatty acid moieties are completely utilized in humans.

Human breast milk contains 20–30% palmitic acid (16:00), and ca. 70% of the 16:00 is esterified to the *sn*-2 position. This is in contrast to other edible oils, all of which have unsaturated fatty acids esterified to *sn*-2 position. Although the reason for the preferential esterification of 16:0 to the *sn*-2 position during TAG synthesis in the mammary gland and its physiological importance to the infant are not yet clear, it has been hypothesized that *sn*-2 monopalmitoylglycerol is more effectively absorbed than free 16:0, because the coefficient of absorption of free 16:0 is relatively low due in part to its melting point, which is above body temperature, and due to the tendency of the unesterified 16:0 to form hydrated fatty acid soaps at the pH of the intestine (40).

Table 3. Solubility of Monostearin in Several Solvents

Solvent	Temperature (°C)	Solubility (g/100 g solvent)
Acetone	8.83	0.482
	17.20	0.990
	24.08	1.957
	31.33	4.070
	41.19	9.883
Isopropyl alcohol	11.40	1.002
	18.23	1.973
	22.99	3.215
	27.25	4.953
	34.94	9.999
	43.30	24.779
	47.22	40.009
	49.01	50.000
Methanol	19.65	0.999
	25.46	2.002
	34.23	4.960
	39.49	9.809
	44.82	25.000
Ethanol	29.35	5.000
	36.72	9.960
	45.34	19.802
	45.66	29.965
	49.31	49.770
n-Hexane	44.28	0.227
	53.19	0.783

Table 4. Melting Points of Polymorphic α-MAG

	Melting point of polymorphic form (°C)			
	β-Form	β'-Form	α-Form	Sub-α-form
Monocaploin	19.4			
Monocaprylin	39.6			
Monocaprin	53	49	27	8
Monolaurin	63	50.5	44	15
Monomyristin	70.5	67.5	56	24
Monopalmitin	77	74	66.9	34
Monostearin	81.5	79	74	47.5
Monoarakidin	83.5	81.5	77	
Monoolein	35	32	25	12.5
Monoelaidin	58.5	56	42	29.5
Monolinolein	14.0			−22.8
Monolinolenin	16.0			−13.5

Medium-chain MAGs are known for bactericidal or antimicrobial effects (41). For example, Petschow et al. demonstrated recently that saturated MAGs, ranging in carbon chain length from C-10:00 to C-14:00, at 1 mM caused a 4-log-unit or greater reduction in the number of viable *Helicobacter pylori* after exposure for 1 h (42). Isaacs et al. found that MAGs with chain lengths varying from 8 to 12 carbons were more strongly antiviral and antibacterial when added to milk and formula than long-chain MAGs (43).

NATURAL OCCURRENCE

Some common MAGs have been found in the plant kingdom. Monoolein, monolinolein, and monopalmitin have been identified as active ingredients in immature seeds of rice (*Oryza sativa*), monoolein ond monopalmitin in immature seeds of *Perilla frutenscens*, and monopalmitin as a major active ingredient in cultured cells of *Nicotiana tabacum* (44).

Some kinds of MAG whose fatty acids are unique are found in plants and microbial cultures, although only at trace levels. Recent literature shows that 1-(28-hydroxyoctacosanoyl)glycerol and 1-(24-hydroxytetracosanoyl)-glycerol have been isolated from the bark of *Cinnamomum camphora* (45), and 1-(26-hydroxyhexacosanoyl)glycerol and 1-(26-ferulyloxyhexacosanoyl)glycerol have been isolated from the root bark of *Pentaclethra eetveldeana* (46). From the mycelial cake of *Streptomyces* sp., monoglycerides having branched-chain fatty acids have been isolated, and they exhibited both antibacterial and antifungal activities (47).

BIBLIOGRAPHY

1. D. Buchnea, in A. Kuksis ed., *Handbook of Lipid Research*, vol. 1, Plenum, New York, 1978, p. 233.
2. G.Y. Brokaw, E.S. Perry, and W.C. Lyman, *J. Am. Oil Chem. Soc.* **32**, 194–000 (1955).
3. T. Tatara, T. Fujii, T. Kawase, and M. Minagawa, *Lipids* **18**, 732–736 (1983).
4. T. Takagi and Y. Ando, *J. Am. Oil Chem. Soc.* **71**, 459–460 (1994).
5. T. Takagi and Y. Itabashi, *Yukagaku* **34**, 962–963 (1985).
6. T. Takagi, *Prog. Lipid Res.* **29**, 277–298 (1990).
7. J.-H. Kim, Y. Nishida, H. Ohrui, and H. Meguro, *J. Chromatogr. A* **709**, 375–380 (1995).
8. J.-H. Kim, H. Ohrui, and H. Meguro, *Kagaku-to-Seibutsu* **33**, 680–684 (1995).
9. N.O.V. Sonntag, *J. Am. Oil Chem. Soc.* **59**, 795A–802A (1982).
10. F. Temelli, J.W. King, and C.R. List, *J. Am. Oil Chem. Soc.* **73**, 699–706 (1996).
11. B.F. Daubert, H.H. Fricke, and H.E. Logenecker, *J. Am. Chem. Soc.* **65**, 1718–000 (1943).
12. J.B. Martin, *J. Am. Chem. Soc.* **75**, 5482–5483 (1953).
13. C.E. Burgos, D.E. Ayer, and R.A. Johnson, *J. Org. Chem.* **52**, 4973–4977 (1987).
14. V.M. Balcao, A.L. Paiva, and F.X. Malcata, *Enzyme Microb. Technol.* **18**, 392–419 (1996).
15. T. Yamane, *J. Am. Oil Chem. Soc.* **64**, 1657–1662 (1987).
16. U.T. Bornscheuer, *Enzyme Microb. Technol.* **17**, 578–586 (1995).
17. K. Holmbery and E. Osterberg, *J. Am. Oil Chem. Soc.* **65**, 1544–1548 (1988).
18. Y. Murase, K. Suzuki, and F. Saito, *Japan Kokai Tokkyo Koho* 58,165,796 (1983).
19. T. Hoshino, T. Yamane, and S. Shimizu, *Agric. Biol. Chem.* **54**, 1459–1467 (1990).
20. Y. Tanaka, J. Hirono, and T. Funada, *J. Am. Oil Chem. Soc.* **69**, 1210–1214 (1992).
21. Y. Shimada, A. Sugihara, K. Maruyama, T. Nagao, S. Nakayama, H. Nakano, and Y. Tominaga, *J. Am. Oil Chem. Soc.* **72**, 1323–1327 (1995).
22. Y. Shimada, K. Maruyama, S. Okazaki, M. Nakamura, A. Sugihara, and Y. Tominaga, *J. Am. Oil Chem. Soc.* **71**, 951–954 (1994).
23. Y. Shimada, K. Maruyama, M. Nakamura, S. Nakayama, A. Sugihara, and Y. Tominaga, *J. Am. Oil Chem. Soc.* **72**, 1577–1581 (1995).
24. H. Maehr, G. Zenchoff, and D.L. Coffen, *J. Am. Oil Chem. Soc.* **71**, 463–467 (1994).
25. F.J. Plou, M. Barandiarán, M.V. Calvo, A. Ballesteros, and E. Pastor, *Enzyme Microb. Technol.* **18**, 66–71 (1996).
26. R. Hess, U. Bornscheuer, A. Capewell, and T. Scheper, *Enzyme Microb. Technol.* **17**, 725–728 (1995).
27. S. Yamaguchi and T. Mase, *J. Ferment. Bioeng.* **72**, 162–167 (1991).
28. T. Yamane, M.M. Hoq, S. Itoh, and S. Shimizu, *J. Jpn. Oil Chem. Soc.* **35**, 625–631 (1986).
29. G.P. McNeill, S. Shimizu, and T. Yamane, *J. Am. Oil Chem. Soc.* **67**, 779–783 (1990).
30. R. Rosu, Y. Uozaki, Y. Iwasaki, and T. Yamane, *J. Am. Oil Chem. Soc.* **74**, 445–450 (1996).
31. B. Myrnes, H. Barstad, R.L. Oken, and E.O. Elvevoll, *J. Am. Oil Chem. Soc.* **72**, 1339–1344 (1995).
32. U.T. Bornscheuer and T. Yamane, *J. Am. Oil Chem. Soc.* **72**, 193–197 (1995).
33. U. Bornscheuer, H. Stamatis, A. Xenakis, T. Yamane, and F.N. Kolisis, *Biotechnol. Lett.* **16**, 697–702 (1994).
34. C. Otero and A. Ballesteros, *Appl. Biochem. Biotechnol.* **50**, 251–263 (1995).
35. J.A. Arcos and C. Otero, *J. Am. Oil Chem. Soc.* **73**, 673–682 (1996).
36. A. Millqvist, P. Adlercreutz, and B. Mattiason, *Enzyme Microb. Technol.* **16**, 1042–1047 (1994).
37. S. Tsuda, *Monoglycerides: Production and Application* (in Japanese), Maki Bookstore, Tokyo, 1957.
38. F.H. Mattson, F.J. Baur, and L.W. Beck, *J. Am. Oil Chem. Soc.* **28**, 386–390 (1951).
39. C. Sommadelpero, A. Valette, J. Lepetitthevenin, O. Nobili, J. Boyer, and A. Verine, *Biochem. J.* **312**, 519–525 (1995).
40. D.M. Small, *Annu. Rev. Nutr.* **11**, 413–434 (1991).
41. J.J. Kabara, in J.J. Kabara ed., *The Pharmacological Effects of Lipids*, American Oil Chemists' Society, Champaign, Ill., 1978, pp. 1–14.
42. B.W. Petschow, R.P. Batema, and L.L. Ford, *Antimicrob. Agents Chemother.* **40**, 302–306 (1996).
43. C.E. Isaacs, R.E. Litov, and H. Thormar, *J. Nutr. Biochem.* **6**, 362–366 (1995).
44. K.H. Park, J.D. Park, M. Hakayama, and T. Yokota, *Biosci. Biotechnol. Biochem.* **58**, 2241–2243 (1994).
45. R.K. Mukherjee, Y. Fujimoto, and K. Kakinuma, *Phytochemistry* **37**, 1641–1643 (1994).
46. H.W. Babadybyla, *Phytochemistry* **42**, 501–504 (1996).
47. Y. Akeda, K. Shibata, X. Ping, T. Tanaka, and M. Taniguchi, *J. Antibiotics* **48**, 363–368 (1995).

MUTAGENESIS

RICHARD H. BALTZ
CognoGen Enterprises
Indianapolis, Indiana

KEY WORDS

Antibiotic biosynthesis
Gene duplication
Mutagenesis
Secondary metabolite
Streptomyces
Transposon

OUTLINE

INTRODUCTION

Mutations are heritable changes in the genome that arise "spontaneously" or are induced by chemical or physical means. Mutational changes include deletions, duplications, translocations (transpositions), and base-pair substitutions. Base-pair substitutions are subdivided into transitions (a purine substituted for a purine, or a pyrimidine substituted for a pyrimidine) and transversions (a purine substituted for a pyrimidine and vise versa). There are two types of transitions (GC to AT and AT to GC), and four types of transversions (GC to CG, GC to TA, AT to CG, and AT to TA). Mutagenesis is an important aspect of fermentation process development for the production of primary metabolites (e.g., amino acids and vitamins) and secondary metabolites (e.g., antibiotics, antitumor agents, antihelmintic agents, and immunomodulators). Mutagenesis can be used to increase the levels of important precursors, cofactors, and enzymes involved in the biosynthesis of the desired product. It can also be used to eliminate competing nonessential biosynthetic pathways or to modify other parameters important in the fermentation process (e.g., culture viscosity, oxygen transfer, culture stability). There are two general approaches to mutagenesis, both of which are important at different stages of process development. The first approach is random chemical or physical mutagenesis. This is an indispensible approach to improve the production of complex secondary metabolites, because a priori it is essentially impossible to deduce what factors will be most important in the fermentation to improve the yield. The key to success with random mutagenesis is to identify procedures that give the optimum yields of mutants exhibiting the desired features. The second type of approach is more directed and can be focused specifically at rate-limiting steps in biosynthesis or at key regulatory elements. This generally requires more sophisticated molecular genetic tools and more knowledge about the genetics and biosynthesis of the product of interest. I discuss some key elements of the random and directed mutagenesis approaches, using secondary metabolite production in actinomycetes as models. Many of the methods and approaches can be adapted for other prokaryotes and for lower eukaryotes that have a haploid sporulation stage.

APPROACHES TO RANDOM BASE-PAIR SUBSTITUTION MUTAGENESIS

Comparative Mutagenesis

A number of chemical mutagens and UV light have been compared for their efficiencies of induction of mutations to spectinomycin resistance in *Streptomyces fradiae* (1,2). Of the mutagens tested, *N*-methyl-*N*′-nitro-*N*-nitrosoguanide (MNNG) was the most efficient, and UV light (UV) was the least efficient. The mutagens ranked in efficiency of mutation induction as follows: MNNG > 4-nitroquinoline-1-oxide (NQO) > methyl methanesulfonate (MMS) > ethyl methanesulfonate (EMS) > hydroxylamine (HA) > UV (Fig. 1). The efficiency of MNNG mutagenesis was enhanced by carrying out mutagenesis in the presence of chloramphenicol (4).

Optimizing Mutagenesis

The objective in optimizing mutagenesis conditions is to obtain the highest frequency of survivors containing single mutations altering secondary metabolite production. Higher levels of mutagenesis are not desirable because most mutations that alter product yields have negative effects (2). Therefore higher levels of mutagenesis will result in the coupling of gain mutations with loss-of-yield mutations. The optimum level of mutagenesis can be identified by statistical analysis of the progeny of mutagenesis. Because mutations, under ideal conditions, will be distributed among individual cells according to the Poisson distribution (2), the optimum condition giving a multiplicity of one mutation affecting secondary metabolite production per cell is obtained when 37% of the survivors of mutagenesis produce control yields (i.e., the null fraction of the Poisson distribution, $P_0 = e^{-m} = 0.37$, where the multiplicity $(m) = 1$). At this multiplicity, the fraction of cells containing only one mutation is also 37% (i.e., $P_1 = 1 \times e^{-1}/1!$). In *S. fradiae* the only mutagen capable of achieving

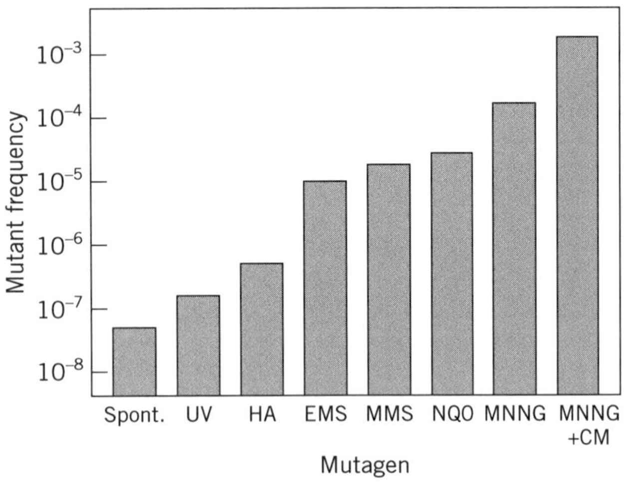

Figure 1. Frequency of spectinomycin-resistant mutants induced by mutagenic agents in *Streptomyces fradiae*. Spont., spontaneous; UV, ultraviolet light; HA, hydroxylamine; EMS, ethyl methanesulfonate; MMS, methyl methanesulfonate; NQO; 4-nitroquinoline-1 oxide; MNNG, *N*-methyl-*N'*-nitro-*N*-nitrosoguanidine; CM, chloramphenicol. *Source*: Data from Refs. 2 and 3.

a mutational multiplicity of as high as 1 is MNNG. However, MNNG plus CM (4) gives much too high levels of mutagenesis to be a useful procedure for random mutagenesis.

Mutagen Specificity and Limitations of Current Protocols

The specificity of base-pair substitutions induced by mutagenic agents has been studied extensively in *Escherichia coli* (5–7). MNNG induces mutations by the transition pathway 99% of the time, and 95% of these are GC to AT transitions (5,7). EMS induces GC to AT transitions almost exclusively, and NQO induces mutations by this pathway about 90% of the time (5). UV induces a wider variety of transition and transversion mutations, but it appears to be too weak to be very useful in actinomycetes. Because MNNG, NQO, and EMS are among the most potent mutagens to choose from for streptomycetes, the spectrum of base-pair substitutions obtainable by chemical mutagenesis is almost completely limited to GC to AT transitions.

Streptomycetes and other actinomycetes have DNA containing very high G + C content (~70%). The G + C content in the third position in codons is generally >90%. Recently, the spectrum of potential amino acid substitutions induced by different transition or transversion mutation pathways at dominant streptomycete codons (those containing G or C in the third position) has been analysed (3). The GC to AT transition pathway cannot induce mutations at phenylalanine-, isoleucine-, asparagine-, or lysine-prevalent codons, nor can it generate codons for alanine, glycine, or proline. Of the other base-pair substitution pathways, the AT to CG transversion pathway is strikingly complementary to the GC to AT transition pathway. Thus AT to CG mutations can occur at phenylalanine-, isoleucine-, asparagine-, or lysine-prevalent codons and can also generate alanine, glycine, and proline codons (3). Twenty-six different amino acid substitutions can be generated by GC to AT transitions at prevalent codons, and 25 can be generated by AT to CG transversions. Importantly, none of the specific amino acid substitutions induced by one pathway can be induced by the other. Therefore, the simplest way to broaden the spectrum of base-pair substitution mutations for yield improvement would be to develop a mutagenic protocol that is specific for AT to CG transversions.

Approach to Expand the Range of Base-Pair Substitution Mutagenesis

No chemical mutagen induces AT to CG transversions as the primary pathway. However, in *E. coli* the *mutT* mutation causes ~1,000-fold enhancement in spontaneous AT to CG transversions (5,8). Because the *mutT* gene is highly conserved in bacteria, it should be present in streptomycetes. A system for reversible gene disruption has been proposed that would allow for the screening of mutants containing spontaneous AT to CG mutations in a *mutT* background, followed by direct selection of $mutT^+$ recombinational revertants that would have normal spontaneous mutation rates (3). The implementation of such a system awaits the cloning of an actinomycete *mutT* gene.

OTHER APPROACHES TO MUTAGENESIS

Gene Duplications by Genetic Engineering

Although gene duplications can arise as spontaneous or induced mutations at low frequencies, the process has not been harnessed as a general technique to improve product yields. In some cases, gene duplications or amplifications have been recognized after the fact as being associated with yield enhancement (9–12). Gene cloning methodologies and chromosomal insertion methods have provided the means to carry out directed gene duplications. These approaches can be used to alleviate rate limitations in secondary metabolite biosynthetic pathways, thus improving product yields and product quality (3,13–15).

The general approach for directed gene duplication is to identify a neutral site in the chromosome where genes can be inserted without changing the product yield or the fermentation properties of the production strain. The neutral site can be cloned, and a gene of interest can be inserted within the neutral site on a plasmid vector. The vector is introduced into the streptomycete host by transformation or by conjugation from *E. coli*, and the gene is inserted into the chromosomal neutral site by a homologous double crossover (Fig. 2), resulting in a gene duplication and loss of the delivery plasmid (13–15). In practice, the individual single crossover can be selected sequentially. The first crossover, which integrates the plasmid into the chromosome, is selected using the antibiotic resistance marker (AR) on the plasmid. This can be accomplished with a temperature-sensitive plasmid replicon that can be readily cured at elevated temperature in the absence of homologous recombination into the chromosome. The second crossover can be screened or selected using a counterselectable marker (CSM), such as the *rpsL* gene (16). In this

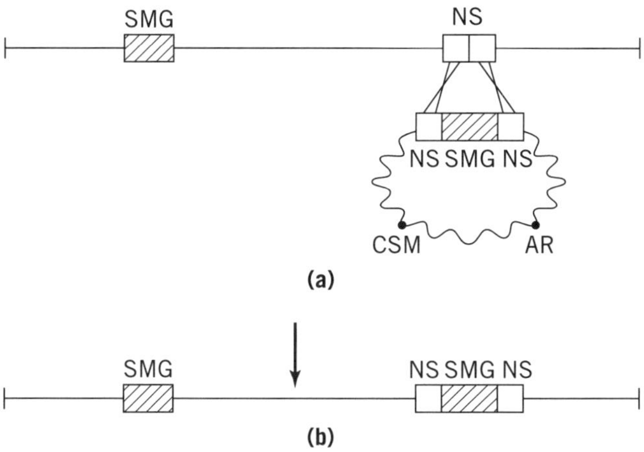

Figure 2. Gene duplication by homologous recombination of a secondary metabolite gene (SMG; crosshatched) into a predetermined neutral site (NS) from a plasmid (wavy line) containing markers for antibiotic resistance (AR) and counterselection (CSM).

case the wild-type *rpsL* gene encoding dominant streptomycin sensitivity is located on the plasmid, and the recessive streptomycin-resistant mutant allele of *rpsL* is located on the chromosome. Thus recombinants that have undergone a second crossover, deleting the plasmid, can be selected for streptomycin resistance at the temperature nonpermissive for plasmid replication. A good example of this neutral site cloning approach was the duplication of the *tylF* gene that encodes a rate-limiting step in the biosynthesis of the antibiotic tylosin in *S. fradiae*. In this case the neutral site was identified by transposon mutagenesis and cloned by selecting for the transposon marker that was expressed in streptomycetes and in *E. coli* (17,18).

Another variation of this method is to introduce the cloned gene into a plasmid that can insert site specifically into a plasmid or phage chromosomal insertion site, providing that the insertion site is neutral (15). A good example of this was the duplication of a gene involved in a rate-limiting step in pristinamycin biosynthesis by inserting a copy of the *snaAB* genes under the transcriptional control of the strong promoter *ermEp** into the plasmid pSAM2 *attB* site (19).

Duplication of much larger blocks of genes (~40 kb) can be accomplished by delivering the genes on a conjugal cosmid cloning vector lacking actinomycete replication functions (15,20). Recombinants are formed by homologous single crossovers of circular plasmids, or double crossovers of linear plasmid concatemers (Fig. 3), duplicating the complete ~40-kb region of interest. In this case, it is not predictable a priori if the insertion will be neutral, but in many cases it is, because ~50% of chromosomal sites in streptomycetes may be neutral (17). With this method of gene cluster duplication, the recombinants retain the plasmid sequences in the chromosome.

These methods also have the advantage that the recombinants are generally stable and lack self-replicating plasmids, which often reduce the productivity of the fermentation (13–15). Furthermore, the neutral site insertion

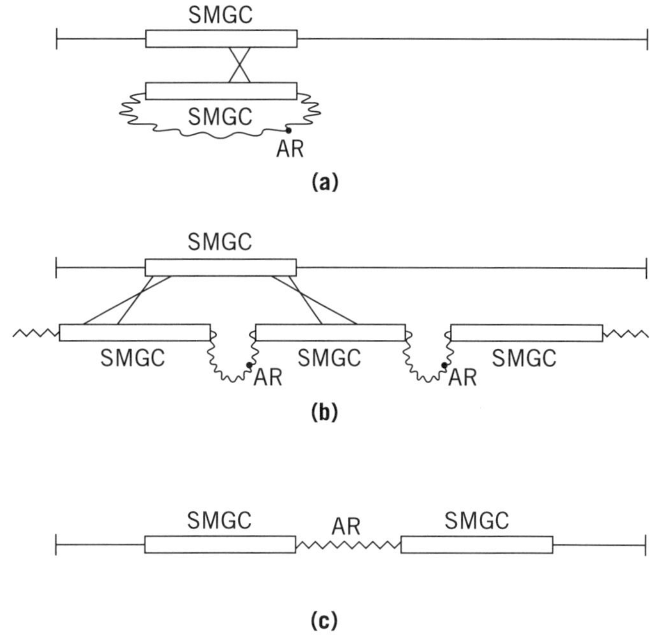

Figure 3. Gene duplication by homologous recombination of plasmid DNA into a secondary metabolite biosynthetic gene cluster (SMGC). A, Single crossover of a circular plasmid containing cloned SMGC into the chromosomal SMGC, selecting for an antibiotic resistance marker (AR); B, double crossover of a linear plasmid concatemer containing cloned SMGC into the chromosomal SMGC; C, recombinant containing SMGC duplication flanking plasmid DNA (wavy line).

method by double crossover results in a recombinant containing no heterologous DNA, so this method is in effect a directed gene duplication and directed translocation procedure that can give a very high frequency of the desired mutant that might otherwise occur randomly at an exceedingly low frequency, if at all.

Insertional Inactivation of Regulatory Elements by Transposons

Both positive and negative regulatory elements can influence the levels of production of secondary metabolites (21,22). Because transposons can be used to insertionally inactivate genes, they should be useful to disrupt regulatory genes. Disruption of negative regulatory genes should cause enhanced production of the desired products, whereas disruption of positive regulatory genes should cause reduced production. In both cases, the regulatory gene can be isolated by cloning the transposon and the flanking chromosomal DNA. This approach has been facilitated in IS*493*-derived transposons (17,23,24) by using markers that are selectable in streptomycetes and in *E. coli* (e.g., hygromycin resistance or apramycin resistance). In the case of negative regulatory genes, the cloned genes can then be used to make deletion mutants. Cloned positive regulatory genes can be used to make directed gene duplications to enhance production, as discussed in the previous section.

Random Insertion of Promoters by Transposition

The concept of using transposons as portable promoters was first proposed by Kleckner et al. (25). Several transposons are now available for streptomycetes, and those derived from IS*493* have outward reading promoter activity (18). Therefore this concept can now be tested using the natural promoter activity or by cloning stronger or regulatable promoters into a transposon. A preliminary study in the *Streptomyces roseosporus* indicated that Tn*5099* can induce mutations that cause enhanced daptomycin production (26). It has not been determined if the phenotype of these mutants is due to the portable promoter, to the inactivation of negative regulatory elements, or to both mechanisms. The data indicate, however, that transposon mutagenesis could become an important adjunct to robust industrial mutagenesis programs.

Biosynthetic Pathway Amplification

A somewhat more speculative approach to yield enhancement is by the duplication or amplification of the complete biosynthetic pathway. Because many secondary metabolite biosynthetic gene clusters are very large (e.g., 50 to 100 kb), it is not possible to duplicate them using standard cosmid cloning vectors. Recent genetic mapping experiments indicate that a number of secondary metabolite gene clusters are located close to one end of their respective linear chromosomes (3). In *Streptomyces rimosus*, which has a linear chromosome and a linear plasmid (11,27), the oxytetracycline biosynthetic genes are located about 600 kb from one end of the chromosome. One strain with improved oxytetracycline production contained 3 to 4 copies of a 1-Mb chimeric plasmid containing one plasmid end, one chromosomal end, and the complete oxytetracycline gene cluster. The strain also had one copy of the oxytetracycline gene cluster on the chromosome. In terms of classical mutagenesis, this strain harbors a duplication, translocation, and amplification because of the multicopy nature of the chimeric plasmid. This example of secondary metabolite pathway amplification has suggested a more directed homology-driven recombinational approach that might be applied to other secondary metabolite biosynthetic pathways clustered near the ends of linear chromosomes (3). Because linear plasmids are relatively common in streptomycetes (21,28), this could be a fruitful approach to assure that the secondary metabolite biosynthetic enzymes are not limiting, thus allowing other beneficial mutations influencing the flow of precursors into the pathway to be fully expressed phenotypically. Much more work is needed to fully develop this approach, but the potential benefits would appear to justify the additional studies needed to test this concept in other actinomycetes.

BIBLIOGRAPHY

1. J. Stonesifer and R.H. Baltz, *Proc. Natl. Acad. Sci. U.S.A.* **82**, 1180–1183 (1985).
2. R.H. Baltz, *Industrial Microbiology and Biotechnology*, American Society of Microbiology, Washington, D.C., 1986, pp. 184–190.
3. R.H. Baltz, *J. Ind. Microbiol. Biotechnol*, **20**, 360–363 (1998).
4. R.H. Baltz and J. Stonesifer, *J. Bacteriol.* **164**, 944–946 (1985).
5. C. Coulondre and J.H. Miller, *J. Mol. Biol.* **117**, 577–606 (1977).
6. J.H. Miller, *Annu. Rev. Genet.* **17**, 215–238 (1983).
7. C.G. Cupples and J.H. Miller, *Proc. Natl. Acad. Sci. U.S.A.* **86**, 5345–5349 (1989).
8. R.M. Schaaper, B.I. Bond, and R.G. Fowler, *Mol. Gen. Genet.* **219**, 256–262 (1989).
9. D.J. Smith, J.H. Bull, J. Edwards, and G. Turner, *Mol. Gen. Genet.* **216**, 492–497 (1989).
10. J.L. Barredo, B. Diez, E. Alvarez, and J.F. Martin, *Curr. Genet.* **16**, 453–459 (1989).
11. B. Gravius, D. Glocker, J. Pigac, K. Pandza, D. Hranueli, and J. Cullum, *Microbiology* **140**, 2271–2277 (1994).
12. U. Peschke, H. Schmidt, H.-Z. Zhang, and W. Piepersberg, *Mol. Microbiol.* **16**, 1137–1156 (1995).
13. R.H. Baltz and T.J. Hosted, *Trends Biotechnol.* **14**, 245–250 (1996).
14. R.H. Baltz, *Drugs Pharmaceut. Sci.* **82**, 49–62 (1997).
15. R.H. Baltz, *Trends Microbiol.* **6**, 76–83 (1998).
16. T.J. Hosted and R.H. Baltz, *J. Bacteriol.* **170**, 180–186 (1997).
17. P.J. Solenberg, C.A. Cantwell, A.J. Tietz, D. Mc Gilvray, S.W. Queener, and R.H. Baltz, *Gene* **168**, 67–72 (1996).
18. R.H. Baltz, M.A. McHenney, C.A. Cantwell, S.W. Queener, and P.J. Solenberg, *Antonie van Leeuenhoek*, **71**, 179–187 (1997).
19. G. Sezonov, V. Blanc, N. Bamas-Jacques, A. Friedman, J.-L. Pernodet, and M. Guerineau, *Nat. Biotechnol.* **15**, 349–353 (1997).
20. P. Matsushima, M.C. Broughton, J.R. Turner, and R.H. Baltz, *Gene* **146**, 39–45 (1994).
21. R.H. Baltz, *Bioprocess Technol.* **22**, 309–381 (1995).
22. M. Bibb, *Microbiology* **142**, 1335–1344 (1996).
23. R.H. Baltz, D.H. Hahn, M.A. McHenney, and P.J. Solenberg, *Gene* **115**, 61–65 (1992).
24. P.J. Solenberg and R.H. Baltz, *Gene* **147**, 47–54 (1994).
25. N. Kleckner, J. Roth, and D. Botstein, *J. Mol. Biol.* **116**, 125–159 (1977).
26. M.A. McHenney and R.H. Baltz, *Microbiology* **142**, 2363–2373 (1996).
27. K. Pandza, G. Pfalzer, J. Cullum, and D. Hranueli, *Microbiology* **143**, 1493–1501 (1997).
28. H. Kinashi, *Actinomycetologica* **8**, 87–96 (1994).

See also GENE TRANSFER, GRAM POSITIVE BACTERIA; GLUTAMIC ACID PRODUCING MICROORGANISMS; INSECT CELLS AND LARVAE, GENE EXPRESSION SYSTEMS; METABOLITES, PRIMARY AND SECONDARY; SECONDARY METABOLITE PRODUCTION, ACTINOMYCETES, OTHER THAN STREPTOMYCES.

MYXOBACTERIA

Hans Reichenbach
Gesellschaft für Biotechnologische Forschung
Braunschweig, Germany

KEY WORDS

Adsorber resin
Antibiotics fermentation
Cellulose degraders
Immobilization
Liquid cultures
Myxobacteria
Myxovirescin

OUTLINE

INTRODUCTION

Myxobacteria are Gram-negative, rod-shaped, strictly aerobic soil bacteria that move by gliding or creeping and are famous for their cell–cell interactions, culminating in fruiting body formation. Until now, no actual technical process has been performed with myxobacteria, but they have already been grown in industrial-scale bioreactors for the production of certain secondary metabolites. Myxobacteria certainly have a potential for technical applications. They could be used as producers of enzymes, for example, proteases to be applied for milk clotting (1,2); enzymes degrading xylan, chitin, cellulose, or yeast cell wall β-glucans (sometimes with unusual cleavage patterns, such as converting starch into trisaccharides) (3); or unusual enzymes for biochemical research such as restriction endonucleases (4,5), reversed transcriptases (6,7), or protein kinases (8). All these possibilities have been very little explored because relatively few scientists have studied myxobacteria in the past, and very few study them at present on a broader scale for such applications. As myxobacteria have efficient export systems for exoenzymes and proteins, *Myxococcus xanthus* was investigated for some time as a host of foreign genes coding for useful proteins (9,10). The advantage in this application was seen in the fact that myxobacteria are nonpathogenic and apparently free of endotoxin. Also, less obvious applications are conceivable. Thus, an anticoagulant, myxalin, has been described from *M. xanthus* that could be useful in medicine (11,12). Lectins have been isolated from myxobacteria and might be further explored (13,14). However, the most promising application of myxobacteria may be in the field of secondary metabolites. So far, about 80 different basic structures have been isolated and elucidated from these organisms, with roughly 350 structural variants (1,15–18). Two compounds, soraphen and epóthilon, proceeded to the stage of an industrial development project. Although soraphen had to be given up as a fungicide for plant protection because of its intolerable side effects (19), epóthilon still is under investigation as a cytotoxin for antitumor application (20).

WHAT KIND OF ORGANISMS ARE MYXOBACTERIA?

Several up-to-date reviews about myxobacteria in general (21,22) and, more specifically, about their isolation, cultivation, preservation, and taxonomy are available (23–25). These facts, therefore, need be only briefly summarized here. Also, citations of the older literature are given in the recent reviews. Only fermentation of myxobacteria is discussed in some detail in this article.

Distinguishing Characteristics of the Myxobacteria

The myxobacterial cells are relatively large, between 3.5 and 12 μm long and from 0.6 to 1 μm wide, depending on the species and, to some extent, on growth conditions. Two morphological types, representatives of two suborders, can readily be distinguished under the microscope: the cells are either slender with tapering ends (*Cystobacterineae*) or are cylindrical with rounded ends; the latter are usually in the shorter size range and often more stout (*Sorangineae*).

The cells move by gliding over or within the substrate. Therefore the colonies spread, sometimes over the whole culture plate within a couple of days, and are named swarm colonies or swarms. Gliding motility is much reduced or totally suppressed on media with a high nutrient content, such as those containing 1% or more peptone. When transferred into a liquid medium, myxobacteria grow at first as a film along the walls of the container or, when shaken, as nodules or flakes suspended in the medium. After several transfers, often homogeneously growing strains can be obtained. To achieve this, transfers should be made from the supernatant, avoiding clumps as much as possible, at short intervals and using small volumes (20 mL in 100-mL flasks). Also, the composition of the medium may play a role. In general, strains belonging to the suborder *Sorangineae* are more recalcitrant than those of the *Cystobacterineae*. For these, weeks and months may be required before a specific strain grows as single cells, and sometimes this stage is never obtained. For this reason, until the 1960s it was generally believed that myxobacteria cannot be grown in liquid media, an opinion that definitely delayed their research.

While gliding bacteria are found in most branches of the bacterial phylogenetic system, two other characteristics of the myxobacteria are exceptional and even unique. Myxobacterial cells seem to communicate with one another with the aim to stay together. This can be seen in the behavior of the swarms. While individual cells can move much faster than the swarm expands, they still keep together, returning whenever they push beyond the swarm edge or waiting for the other cells to catch up. Small swarms may even migrate as a whole, like a pseudoplasmodium. Within the swarm, patterns of veins, rings, and parallel rows of cells in a ripplelike arrangement may develop. The cells in the ripples appear to move back and forth, creating in this way the illusion of oscillating waves, particularly striking when seen in time-lapse films but also recognizable by the patient observer under the microscope. The culmination of intercellular communication is the production of fruiting bodies. Under unfavorable conditions, especially starvation, hundreds of thousands to millions of cells aggregate at certain sites within the swarm. They often mill around for some time on a ring-shaped path and pile up. From such cell masses, fruiting bodies may differentiate. The simplest types are just knobs of cells and soft slime (*Myxococcus*); the most sophisticated ones consist of an excreted slime stalk, which may even be branched, with clusters of sporangioles at the end (*Chondromyces*). These sporangioles consist of a tough wall enclosing hardened slime and cells. In other genera, sporangioles are also found single on a stalk or sitting directly on or within the substrate, often in large, dense clusters. The size of the fruiting bodies varies between 10 and 1,000 μm, and most of them can readily be recognized with the naked eye. They usually are brightly colored (yellow, orange, red, brown, black), and are often produced in large numbers. The biological significance of the fruiting bodies is apparently to guarantee that a new life cycle is started with a community rather than a single cell. Why this strategy would be favorable will be seen when the physiology of the myxobacteria is discussed.

Inside the maturing fruiting body, the vegetative cells convert into myxospores. In the suborder *Cystobacterineae*, the vegetative cells always change their shape dramatically. They shorten and fatten—in the family *Myxococcaceae* they even become spherical—and turn optically refractile. They surround themselves with a capsule, which in most cases, however, is seen only under the electron microscope. In the suborder *Sorangineae*, the shape change is less spectacular; no capsule seems ever to be developed, but the cells also become optically refractile. Myxospores are fully desiccation resistant, in contrast to vegetative cells that are rather sensitive to drying. Temperature resistance is only mildly increased: wet myxospores may survive at 58–60 °C for 30–60 min, just long enough to get rid of contaminants in many cases. Dry myxospores may be heated to 140 °C for 1 h and still germinate. Thus, myxospores clearly are the myxobacteria's survival strategy, and fruiting must have some other significance.

There Are Two Physiological Types of Myxobacteria

All myxobacteria are specialized in the decomposition of biomacromolecules. Most species are bacteriolytic, that is, they excrete a host of hydrolytic enzymes—proteases, nucleases, lipases, polysaccharidases—that decompose living and dead, sometimes only dead, bacteria, yeasts, and other microorganisms, and of course also organic matter coming from higher organisms. One group, the genus *Sorangium* and a second as yet unnamed related genus, decompose cellulose. These organisms seem to attack only dead microorganisms, such as autoclaved *E. coli* and baker's yeast. Because a population obviously is more efficient than a single cell in the decomposition of environmental macromolecules by creating high enzyme levels, minimizing diffusion losses, and maximizing recovery of solubilized monomers, it becomes understandable why myxobacteria developed a social behavior.

Myxobacteria of both physiological types can, as a rule, be cultivated, and most of them in pure culture. A few, seemingly rare myxobacteria have never been grown in culture, mainly the *Haploangium* species, *Polyangium parasiticum* (discovered on the freshwater green alga *Cladophora*), and a few other species that have never been seen again by anybody after their original description from collected samples or crude cultures. A case of obligatory symbiosis has recently been reported for *Chondromyces crocatus* (26,27). Most strains of that species in our collection grow only if a companion is present. As the companion, too, cannot grow without *Chondromyces*, both could be cultivated together in large bioreactors, where they keep themselves automatically in balance (28). The companion could be characterized by 16*S* rRNA sequencing. It belonged, in all our *Chondromyces* strains, to the same species, a new member of the *Flavobacterium* group, regardless from where the *Chondromyces* had been isolated. The intimacy of the relationship is emphasized by the fact that the companion can even be seen within the sporangioles of the *Chondromyces* fruiting body (27), whereas normally the interior of myxobacterial fruiting bodies is free of contaminants for some time after their production.

All myxobacteria, the cellulose degraders included, can be grown on yeast agar, such as VY/2 agar (baker's yeast, 0.5% by fresh weight of yeast cake; $CaCl_2 \cdot 2H_2O$, 0.1%; vitamin B_{12}, 0.5 mg/L; agar, 0.5%; pH adjusted to 7.2 with KOH; autoclaved). On this medium, spreading colonies are obtained; the cultures stay alive for at least 2–3 weeks (30 °C) and often produce fruiting bodies. Fruiting bodies are useful when the strain is to be preserved by lyophilization, because vegetative cells cannot reliably be dried. Further, fruiting bodies are required for classifying an isolate (see following). Fresh isolates usually form fruiting bodies, but during cultivation fruiting body formation usually is lost after three to eight transfers. Fruiting body formation can be maintained, if cultures are started from fruiting body preserves. Those can be prepared by streaking swarm material onto small pieces of filter paper placed on water agar. After incubation for 2–3 days at 30 °C, often fruiting bodies are produced. They are allowed to mature for another 4–6 days. The filter papers are then inserted into sterile medicine bottles and dried in a desiccator under vacuum for a week. From dry filter paper preserves stored at room temperature, fruiting cultures can still be started after 8–20 years. The fruiting cultures must then be used after one to two transfers to make fresh preserves; otherwise, fruiting may be lost.

The bacteriolytic myxobacteria usually also grow well on peptone-containing media, for example, CY agar (Casitone, Difco, 0.3%; yeast extract, Difco, 0.1%; $CaCl_2 \cdot 2H_2O$, 0.1%; agar 1.5%; pH adjusted with KOH to 7.2; autoclaved). On this medium growth often is heavier, but the swarms stay more compact, are often more slimy, rarely produce fruiting bodies, and die sooner (after 8–10 days) than on VY/2 agar. The reason for early death could be the production of ammonia and the concomitant shift of pH to alkaline conditions. The pH range of myxobacteria is 6.8 to 7.8. Of course, the media could be buffered. A useful buffer is HEPES, which is tolerated without adverse effects up to at least 100 mM. Several bacteriolytic myxobacteria seem to grow only on peptone, proteins, or amino acids and seem not to utilize carbohydrates at all (the genera *Myxococcus, Corallococcus*, and *Nannocystis*). However, there may be some unorthodox ways of carbohydrate utilization. Thus, *Myxococcus* degrades yeast cell wall β-glucan, and *Nannocystis* corrodes agar, so why should they leave the valuable decomposition products to other organisms? At least some *Corallococcus* strains decompose starch to tri- and oligosaccharides; those are utilized, while glucose and maltose remain untouched (3). (For more media recipes, see Ref. 24.)

Incidentally, in spite of their name of "slime bacteria," myxobacteria are poor producers of slime on plates and in liquid media. Only rarely colonies are somewhat raised and liquid media become recognizably viscose. Ropy liquids have never been reported. The slime of the swarm sheet, however, may become extremely tough, rubberlike, and almost impossible to divide with the inoculation loop.

For the cellulose degraders, too, alternative media are available (24), for example, filter paper on mineral salts agar with KNO_3 as the only nitrogen source (e.g., ST21 agar), on which they often form huge numbers of fruiting bodies, or mineral salts–glucose agar (e.g, CA2 agar). The organisms seem to prefer NO_3^- to NH_4^+ as the nitrogen source, grow even better in the presence of low peptone concentrations, but often are absolutely dependent on a carbohydrate (e.g., glucose, starch, cellulose) for carbon and energy. All myxobacteria can be grown in liquid media. This aspect is discussed in connection with fermentation in the next section.

Taxonomy and Genetics of the Myxobacteria

The taxonomy of the myxobacteria rests almost exclusively on morphological characteristics. Only very few useful physiological traits have been found; thus, cellulose degradation in *Sorangium* and synthesis of lanosterol in *Polyangium* (29) distinguish the two genera. A bright yellow fluorescence at 366 nm of the swarm colonies caused by the production of phenalamid (30,31) helps to distinguish *Myxococcus stipitatus* from *M. fulvus* when typical fruiting bodies are lacking. While suborders, families, and most genera can usually be identified without problems because the morphology of their vegetative cells, myxospores, and swarm colonies is sufficiently typical and can normally reliably be registered, differentiation of species is often not possible, because it requires typical fruiting bodies and those may be degenerate or completely lacking. In such cases, one can only guess what species one has. Today, molecular taxonomy could be applied to solve the problem (DNA–DNA hybridization).

Presently about 40 different species are recognized, but there certainly are more species in nature, at least another 20 to 30. They are classified in 12 genera, 4 families, and 2 suborders in the order *Myxococcales*. As already mentioned, the suborders can easily be distinguished by the shape of the vegetative cells. As demonstrated by 16*S* rRNA sequencing studies, the order is phylogenetically coherent and belongs to the δ-branch of the Proteobacteria (32). Complete 16*S* rRNA sequences are available for about 40 strains of many different genera and species, but the resulting picture does not much modify the established structure of the order. Clearly, the resolution of the technique is not particularly helpful at the species and genus level.

It should be mentioned that formerly, before it was appreciated that gliding motility is not a useful taxonomic characteristic, the genera *Cytophaga, Sporocytophaga*, the *Cytophaga*-like bacteria, and *Flexibacter* were classified as nonfruiting myxobacteria. In fact, they belong to a phylum of their own and are not related to the myxobacteria at all. This is of some practical importance, because several fish-pathogenic and allergenic species are known in that group, while no pathogenic myxobacteria have ever been reported. Another group, later recognized as a new genus, *Lysobacter*, was even classified as *Myxobacter* and *Sorangium* for some time. Those names can still be found occasionally in the current literature. *Lysobacter* comprises biotechnologically useful strains—suppliers of well-known antibiotics, such as myxin, and of interesting enzymes—but has nothing to do with myxobacteria. Phylogenetically the genus belongs into the γ-branch of the Proteobacteria.

The DNA of myxobacteria has a high G+C content of 66 to 72 mol%. The genomes are large for bacteria, between 9,500 and 10,000 kbp. Physical genome maps are available for *Myxococcus xanthus* (33) and *Stigmatella aurantiaca* (34). Two myxobacteria, *Myxococcus xanthus* and *Stigmatella aurantiaca*, have been analyzed with genetic methods as model systems for questions of morphogenesis, cell signaling, gliding motility, carotenoid synthesis, protein export, etc. (for details, see Refs. 21,22). Several gene transfer techniques have been developed: *E. coli* phage P1 suicide vectors, conjugation with *E. coli*, myxobacterial generalized transducing phages, and electroporation (21). Transposon mutagenesis has successfully been applied to gene tagging (35). In a few cases, those techniques have also been used for other myxobacteria, for example, *Sorangium cellulosum* (36,37), but there is always need of an adaptation.

Myxobacteria Are Very Common Soil Organisms

One might expect that organisms that are so unusual and fascinating as the myxobacteria, yet are so little investigated, are rare and difficult to obtain. The contrary is the case. Myxobacteria are found worldwide in all climate zones and vegetation belts and are very common everywhere. They live in soil, on rotting plant material, on decaying wood, on bark of living and dead trees, and on dung of various herbivores, especially rabbits, hares, deer, goats, and sheep; in general, wherever there is a rich microbial flora. The most diversified myxobacterial communities are found in warm, semiarid areas, like Egypt, the southwestern United States, or northern India, probably because, with their myxospores and fruiting bodies, they are beautifully adapted to such environments. Myxobacteria also seem to live in freshwater and to survive for some time in marine sediments; they can be isolated from marine sediments but are not halotolerant and thus probably cannot live there. They seem not to be able to grow at a low pH, and acid soils are the one habitat free of myxobacteria. Alkalophilic myxobacteria, however, have been obtained from alkaline lakes in Africa. Myxobacterial are typical mesophiles, although psychrophilic strains have been isolated from Antarctic samples.

Isolation and Maintenance of Myxobacteria

The relevant methods of maintenance of these bacteria have relatively recently been described in great detail (24), so a brief summary may suffice here. Because Myxobacteria have a rather generalized metabolism, a highly selective enrichment is not possible. The isolation techniques make use of gliding motility, the ability to attack living bacteria or cellulose, and resistance to certain antibiotics. There are essentially three ways to start enrichment cultures. In the first procedure, materials collected in nature are incubated in a moist chamber at 30 °C. Suitable sources are rotting wood, bark, or the dung of herbivores. To suppress growth of fungi, cycloheximide (50 mg/L) may be added to the water used to wet the source material. The cultures are inspected under a dissecting microscope with incident light at magnifications of 20–80×. Fruiting bodies may form within 2–8 days on the surface of the substrate and can be used to make transfers. From soft slimy *Myxococcus* fruiting bodies, myxospores can be transferred directly to a growth medium such as CY agar by cautiously touching the top of the fruiting body with the sharp tip of a needle on a 1-mL syringe. The young fruiting bodies are normally free of contaminants, so that often a pure culture is obtained in one step. In a similar way, sporangioles can be removed from the ends of the long stalks of *Chondromyces* fruiting bodies. The fruiting bodies of all other myxobacteria cannot be harvested without touching the heavily contaminated substrate and transferring pieces of it together with the fruiting bodies. Therefore, in these cases the fruiting bodies are transferred to streaks of living *E. coli* on water agar plus cycloheximide for further enrichment. The developing swarms usually expand quickly beyond the streak of the feed bacteria, and either uncontaminated sections from the swarm edge or fruiting bodies can be inoculated on growth medium. Often the procedure has to be repeated several times before a pure culture is obtained, which typically takes 2 weeks to 3 months. Sometimes special purification steps have to be included, such as heating fruiting bodies, suspended in water, at 58–60 °C for 10–60 min; shaking fruiting bodies overnight in a rich medium (in which they will not germinate) with high doses of suitable antibiotics; or gassing cultures with ammonia vapors to eliminate soil amoebae.

In the second approach, cultures are started by inoculating circular spots of living *E. coli* (about 2 cm in diameter) on water agar plus cycloheximide with small samples of soil or rotting plant material. After 2–3 days at 30 °C, swarm colonies and soon also fruiting bodies may develop, and transfers and purification steps are performed as described in the first approach. With this method, different types of samples can be processed, and myxobacteria with small, inconspicuous fruiting bodies are also obtained. Incidentally, on such plates often other gliding bacteria such as *Cytophaga*-like bacteria, *Flexibacter, Taxeobacter*, and *Herpetosiphon* also develop and can be recognized by their spreading colonies. In the third method, the cellulose degraders are isolated by inoculating filter paper pads on mineral salts plus cycloheximide agar with soil, rotting wood, decaying plant material, or dung of herbivores. After 10–14 days at 30 °C, bright yellow, orange, brown, or black spots around the inoculum indicate growth of *Sorangium* strains. Those usually produce huge numbers of densely packed fruiting bodies, which in fact are responsible for the colored areas. Transfers are made first to small pieces of filter paper on mineral salts agar, then on VY/2 agar. All *Sorangium* strains appear to be resistant to high doses of kanamycin, so that the antibiotic can be used for purification (50 mg/L kanamycin sulfate). Cellulose degraders are more tedious to purify, and it may take 1–2 years before the last contaminant is eliminated. Fortunately, pure strains are usually obtained within 1–3 months. In crude cultures for cellulose degraders, noncellulolytic myxobacteria always also develop. The inoculum, the solubilized cellulose, and other saprophytic bacteria supply sufficient nutrients for their growth. In particular, *Nannocystis, Polyangium*, and *Chondromyces* strains are often found in this way. They must, of course, be transferred to *E. coli* streaks on water agar.

Preservation of strains is done best by freezing 0.5- to 1-ml suspensions of cells in a 1% Casitone (Difco) solution at −70 °C in a deep freeze or in liquid nitrogen. The organisms appear to stay alive for many years (about 20 years has been experimentally proven). Freezing at −20 to −25 °C in ordinary household freezers seems not to work. In our experience, myxobacteria survive that at best for a couple of weeks. Drying of fruiting bodies has already been mentioned.

FERMENTATION OF MYXOBACTERIA

Fermentation of myxobacteria at a 5- to 50-m^3 scale has only been performed for the production of secondary metabolites. Data are available from many publications. Process descriptions are also found in a large number of published patents. Consequently, the following discussion centers upon this topic.

General Principles

With homogeneously growing strains, typical growth curves can be obtained. The generation times are long, typically between 4 and 14 h. The faster growing species are normally those of the suborder *Cystobacterineae*. On the other hand, those species usually show short stationary phases; that is, soon after the end of the log phase is reached, the culture breaks down. This collapse can be really dramatic. Cultures of *Stigmatella aurantiaca* strains in peptone medium without carbohydrate become pitch black within 2–3 h, probably because of the activity of phenol oxidases from lysed cells, and are then dead. Therefore, transfers are best made in the upper log phase, and extended production in the stationary phase is usually not possible. Species of the suborder *Sorangineae*, in contrast, often are under stationary conditions for days and may continue to produce secondary metabolites during this period. With these species it is also advisable to make transfers in the upper log phase, because later the inoculum may result in a very long lag phase or will no longer start growing. Obtainable cell densities typically are 10^8–10^{10} cells/mL depending on the strain and the medium. It should be remembered, however, that myxobacterial cells are larger than those of most other myxobacteria. The biomass thus is often between 4 and 12 g wet weight/L, corresponding to 1–3 g dry weight. As already mentioned, slime production is normally not seen.

Liquid media suitable for bacteriolytic myxobacteria are all based on peptones or proteins, while those for the cellulose degraders can be varied in wider limits. Enzymatically digested casein peptones are the substrates of choice, for example, Difco Casitone or Marcor peptone from casein, tryptically digested. Peptones from meat and acid-hydrolyzed peptones are much less useful or outright inhibitory, possibly because of elevated phosphate concentrations (38) or an imbalance in amino acid composition. Yet some myxobacteria can also be grown on casamino acids (e.g., Ref. 39). Examples of simple media suitable for many bacteriolytic myxobacteria are CAS lm (Casitone, Difco, 1%; $MgSO_4 \cdot 7H_2O$, 0.1%; pH 6.8; autoclaved); if the peptone concentration is too high, which is the case with quite a few myxobacteria, MD1 lm may be used (Casitone, Difco, 0.3%; $CaCl_2 \cdot 2H_2O$, 0.07%; $MgSO_4 \cdot 7H_2O$, 0.2%; cyanocobalamin, 0.5 mg/L; standard trace element solution; pH 7.0; autoclaved). For liquid cultures of *Sorangium* strains, AMB lm (40) could be used (Casitone, Difco, 0.25%; soluble starch, 0.5%; $MgSO_4 \cdot 7H_2O$, 0.05%; K_2HPO_4, 0.025%; pH 6.8; autoclaved). Those organisms also grow, however, on much simpler media containing only KNO_3 as a nitrogen source and glucose as a carbon source in a mineral salts solution (e.g., CK1 and CK6 lm).

Casein peptones contain sufficient phosphate, so that extra phosphate need not be added. Thus, 0.5 mM phosphate was determined in a 0.2% casein peptone solution (38). In general, myxobacteria are rather sensitive to phosphate. Also it may have been realized that the media mentioned contain relatively high concentrations of $MgSO_4$; the optimum is indeed often at 5 to 10 mM. Calcium is normally present in adequate amounts or is not specifically required. In contrast, the cellulose degraders would not grow without an addition of $CaCl_2$ (41). Vitamins are usually not required, but an occasional strain may depend on vitamin B_{12}.

A peptone medium may be complemented with 0.2–1% of a carbohydrate, for example, glucose, sucrose, or starch. The latter is usually readily hydrolyzed. A carbohydrate is a must for most strains of *Sorangium*. Normally cultures of bacteriolytic myxobacteria do not grow faster in the presence of carbohydrate but rather achieve higher cell density. Also, the pH does not rise as fast and as high as in a pure peptone medium. The details of carbohydrate metabolism have not yet been studied for a wide variety of myxobacteria. Some appear not to utilize carbohydrates because they lack essential glycolytic enzymes (hexokinase and pyruvate kinase in *Myxococcus xanthus*; 42). Others utilize only polysaccharides, because they can only ingest tri- and oligosaccharides (*Corallococcus coralloides*; 3). In this case it was shown that the glucose produced within the cell is mainly utilized via the pentose phosphate pathway, probably only to produce ribose and deoxyribose for nucleic acid syntheses. Many bacteriolytic myxobacteria, however, metabolize also mono- and disaccharides under production of acid and radioactive CO_2, if labeled sugars are supplied (e.g., *Stigmatella aurantiaca*; 43). Nothing is known about the biochemistry of sugar metabolism in those organisms. The cellulose degraders utilize sugars very efficiently for energy and all kinds of syntheses, as is suggested by their growth in the minimal media just described. Fully synthetic, defined media have also been reported for several bacteriolytic myxobacteria. They contain various amino acids and can in fact be very simple. *Cystobacter fuscus* and *Cystobacter ferrugineus* strains grow, for example, in HP16 lm with (per liter) L-Gln·Na, 1 g; L-Ile, 40 mg; L-Phe, 40 mg; glucose, 5 g; $MgSO_4 \cdot 7H_2O$, 1 g; standard trace elements; in phosphate buffer, 0.5 mM, pH 6.5. Some strains become vitamin dependent on thiamine or biotin, when grown in this minimal medium, while in peptone media they are not. For most strains, defined media are, however, more complex, containing more amino acids and still other ingredients. Apparently the relative concentrations of amino acids are as important as their kind. Also, every strain may require a medium of its own. Essential amino acids often are the branched-chain and aromatic amino acids. In any case, growth in those defined media is always substantially slowed down, and cell yields are much reduced.

When an organism relies solely on amino acids for nitrogen, carbon and energy, elimination of NH_3/NH_4^+ becomes a problem. This problem is indeed a serious one with bacteriolytic myxobacteria. The intracellular ammonia pools may become very high under such circumstances: in *Myxococcus virescens* 80–140 mM has been measured (44). Although the pH shift by ammonia can easily be controlled in the bioreactor by titration and in shake flasks by buffering with HEPES or MOPS, for example, the regulatory effects of ammonium still may interfere substantially with growth and production of antibiotics. This was demonstrated by pumping the contents of a small bioreactor through a module with hollow fibers made of a hydrophobic membrane. The ammonia in the culture supernatant could thus be reduced from 35–42 mM to 3–7.5 mM. In consequence, the generation time fell from 4 to 2 h, the cell density more than doubled, and myxovirescin production rose from 8 to 115 mg/L (K. Gerth, personal communication). A technical solution of this problem is presently not available.

For large-scale fermentations to be cheaper, technical substrates may be desirable. A wide variety of such substrates have been tested with myxobacteria and several have been found useful: soybean flour and meal (must be defatted), single cell protein (Probion, Hoechst, no longer available), skim milk, maize gluten, corn steep powder, or suspensions of baker's yeast (45). Which substrate should be used in a specific case must be tested. Also, as is typical for such poorly defined materials, considerable differences are observed from batch to batch and from supplier to supplier. Production of secondary metabolites may vary substantially with the particular substrate used (39,45,46). An unexpected discovery was that soybean meal may be free of iron, so that the addition of Na·Fe^{III} EDTA (8 mg/L) becomes essential (47).

The temperature range for growth of myxobacteria is rather wide: they still grow in the refrigerator at 6 °C, and quite a few strains still grow at 38 or even 40 °C. The temperature curves are skewed as usual, and the optimum is close to the maximum by 1–3 °C. Useful temperatures for fermentations are much narrower, normally 28–32 °C. At least with respect to secondary metabolites, production often goes down or stops completely at higher temperatures even if growth still is satisfactory. Also, at higher temperatures the generation time may rise (47), and the cultures become more sensitive to unfavorable pH and oxygen conditions and quickly break down as soon as the stationary phase is reached. At temperatures below 28 °C, growth, production, and process performance slow down and become unfavorable.

The pH range of myxobacteria is narrow, as a rule 6.8 to 7.8. Cellulose degraders tolerate a slightly lower pH, down to about 6.2, but with ever more reduced growth rates. If a utilizable carbohydrate is present, the pH drops in unbuffered media but can be controlled by titration with a 10% KOH solution. Without carbohydrate or after carbohydrate is used up, the pH always rises and usually becomes decidedly alkaline (pH 8.0–8.5). This change can be prevented by adding 5% sulfuric acid or 30% acetic acid. In the latter case care must be taken because acetate concentrations exceeding 0.1% may soon become inhibitory for production and growth. In a process, pH can often be stabilized by feeding glucose whenever pH reaches a set limit.

Although myxobacteria are strict aerobes, their oxygen demand is moderate because of slow growth and modest biomass. This requirement may become a problem if the aeration rate cannot be lowered unrestrictedly in a bioreactor. In fermentations for secondary metabolites, the oxygen level may play a crucial but unpredictable role for production. There are examples of a high as well as a low requirement of oxygen for optimal biosynthesis. In Table 1, the aeration schedules of a selection of such fermentations are listed. It should be understood that in no case have strictly standardized and optimized fermentations been performed. Maximum total oxygen consumption of 1.8 g O_2/L during 22 h of fermentation on a single cell protein (Probion) medium has been reported for *Myxococcus fulvus* strain Mx f16 (45).

Special Problems with Secondary Metabolite Production

In principle, the situation with secondary metabolite-producing myxobacteria is not much different from that with other antibiotics-producing bacteria. Yields initially are usually, but not always, low, typically between 0.5 and 20 mg/L. However, wild strains with initial yields of 70 mg/L soraphen or 120 mg/L myxalamid (39,58) have also been isolated. In several cases, yields could be much improved, for example, to nearly 1 g/L with sorangicin or close to 2 g/L with soraphen. In other instances, yields remained low in spite of much effort, as with saframycin Mx (59).

The methods of yield improvement are the usual ones. The influence of medium components can be substantial, as already mentioned. Thus, in fermentations of *Stigmatella aurantiaca* yields varied by a factor of 3, between 3 and 11 mg/L stigmatellin and between 47 and 120 ml/L myxalamid, depending on whether corn steep powder or maize gluten (Zein) was used as the main medium ingredient (39). At the same time growth was rather poor on the Zein medium. An improvement of yield and production stability is almost always achieved if the strain is cloned and if low and nonproducing clones are eliminated. Unfortunately this is less trivial as it seems, because initially all strains grow in clumps when transferred to a liquid medium, and if homogeneous cell suspensions are available a suitable plating medium giving high numbers of single cell colonies must first be found. This can take weeks and months, because normally no colonies are formed on the usual culture media that allow excellent growth when massively inoculated. Also, every strain needs its own plating medium. In developing a plating medium, the kind and concentration of the medium components are crucial. Inclusion of autoclaved old liquid cultures may help (36). Development of single-cell colonies takes 8–12 days for most bacteriolytic myxobacteria and 14–21 days for *Sorangium* strains (30 °C).

As soon as single-cell colonies can be produced, a mutation program becomes feasible. Mutations can be induced in myxobacteria with high yields using UV irradiation or the mutagen NTG (*N*-methyl-*N'*-nitro-*N*-nitrosoguanidine, about 40 μg/mL) (60). When inducing mutations, photoreactivation has to be taken in account (61). The size of the inoculum should be 5–10% (v/v). Lower ratios, down to 1–2%, have been found possible in some cases, but this may be paid for with a longer lag phase and a lower biomass attained at the end of the fermentation. Also, a certain minimal cell density may be necessary to start solubilization of polymeric substrates, such as casein (62). Under these circumstances a substantial decrease of the mean doubling time was observed when the size of the inoculum was increased, for example, with two *Myxococcus xanthus* strains on a casein medium from 15 to 8 h and from 26 to 8.5 h, respectively, when initial cell densities were raised from 10^4 to 10^7 cells/mL (62).

A very useful strategy to improve yields is the addition to the cultures of a neutral adsorber resin such as Amberlite XAD-16 (formerly XAD-1180; Rohm & Haas). All myxobacteria appear to tolerate the resin in the culture without adverse effects on growth. So far, virtually all excreted myxobacterial metabolites have been adsorbed by the resin (46,47,49,52,55,58). Only strictly hydrophilic compounds would remain unadsorbed, but such substances appear not to be produced by the organisms or only

Table 1. Examples of Oxygen Demand during Myxobacterial Fermentations for Secondary Metabolites

Metabolite	Producer[a]	Main components of medium[b]	Volume (L)	Stirrer speed (rpm)	Aeration rate[c] (L L broth^{-1} h^{-1})	Oxygen saturation[d] (%)	Reference
Rhizopodin	Mx s	pep	300	200	6		48
Angiolam	An d	Prob	650	300–450	1.8–2.2	90–20	49
Gephyronic acid	Ar g	Prob/yex/starch	300	200	6		50
Stigmatellin	Sg a	maize/gluten	55	200	1.5–5.4		39
			265	600	2.4–3.9	90–5	
Crocacin	Cm c	Prob/starch	55	200	3.3	90–65	28
Chondramid	Cm c	Prob/starch	90	150	3.3		46
			600	50	8	90–40	
Thiangazol	Pl fu	Prob/starch	65	200	3.3	100–85	51
Phenoxan	Pl spec.	Prob/starch	65	200	3.3	90–50	52
Ripostatin	So ce	Soy/starch/Fruc	300	300	1	100–15	53
Soraphen	So ce	Soy/yex/starch	230	400	6.5		47
			4400	230	4.5		
Epothilon	So ce	Soy/yex/starch	60	250	5		54
			230	350	4.3		54
Tartrolon	So ce	Soy/yex/starch	65	150	2.3		55
Jerangolid	So ce	Soy/yex/starch	60	250	12		56
Nannochelin	Na e	pep	65	200	3.3	90–70	57

[a]Mx s, *Myxococcus stipitatus*; An d, *Angiococcus disciformis*; Ar g, *Archangium gephyra*; Sg a, *Stigmatella aurantiaca*, Cm c, *Chondromyces crocatus*; Pl fu, *Polyangium fumosum*; So ce, *Sorangium cellulosum*; Na e, *Nannocystis exedens*.

[b]pep, casein peptone; yex, yeast extract; Prob, Probion (an experimental single-cell protein of the Hoechst Company); soy, soybean meal defatted; starch is often complemented with glucose.

[c]Stirrer speed is sometimes changed during fermentation to adjust dissolved oxygen.

[d]Dissolved oxygen is sometimes increased or held constant in the later stage of fermentation.

very rarely so. The advantages of the resin are obvious: by removing the metabolites from the supernatant, end product inhibitions are prevented, the adsorbed substances are stabilized and concentrated in a small volume, and harvest is much facilitated. If the strain grows as a homogeneous cell suspension, harvest can be done simply by collecting the resin on a sieve. The metabolites are recovered by eluting the washed resin with methanol or methanol–water mixtures. Sometimes a metabolite may crystallize from the eluate. The regulatory and protective effects of the resin are clearly demonstrated when fermentations are done without it. Almost always the yields decrease dramatically, and variants of the metabolite appear that have not been seen in cultures with resin. Normally an addition of 1–2% (v/v) is sufficient. If yields go up, and depending on the metabolite, higher amounts of resin may have to be added. Freshly bought resin is strongly alkaline and has to be washed until it becomes neutral. The resin is added to the culture medium and autoclaved together with it. Resin has also been used with excellent results in large-scale fermentations (47). There may be a loss of about 10% in the bioreactor from mechanical attrition, but the recovered resin can be regenerated by washing with NaOH solution after elution of the metabolites.

Most myxobacterial substances do not inhibit the producer. There are, however, exceptions. Myxovalargin, a peptide antibiotic that destroys the cell membranes (63), blocks growth of the producer, *Myxococcus fulvus* strain Mx f65, at 6 mg/L (64). Myxobacterial saframycin has a minimal inhibitory concentration of 15 mg/L for the producing strain, *Myxococcus xanthus* Mx x48 (58). In both cases mutants with somewhat increased resistance could be obtained, but removal of the inhibitor by adding adsorber resin to the cultures is an easy solution. As foaming is often a serious problem, an antifoam agent must always be added to myxobacterial fermentations. Silicon antifoams usually have been applied because they do not interfere with the chemical purification steps. For the organisms, however, other antifoam agents are also acceptable.

Feeding of precursors was successful only in a few fermentations. Thus, the addition of 0.3% sodium acetate stimulated production of myxalamid by *Myxococcus xanthus* strain Mx x12 (58). Normally, acetate is not a limiting factor, and its addition may even be inhibitory. This situation appears to be different with propionate. Added propionate is readily incorporated into polyketides containing propionate units and may indeed stimulate their production. Propionate is only tolerated in low concentrations, however, so that real yield improvements require repeated addition of small quantities. We have never observed a stimulation of secondary metabolite production by adding certain amino acids, probably because our peptone and protein media already contain an excess supply. It has been reported however that Ala, Ser, and Gly (0.9 mg/mL each) when given to *Myxococcus xanthus* strain TA result in higher yields of antibiotic TA (identical with myxovirescin). All three amino acids are indeed incorporated into the antibiotic, Ala and Ser serving as precursors for acetate, which seems to be not as readily taken up, while Gly is a building block of the molecule directly (65). The removal of inhibitory components may also be a prerequisite for production. The catechole iron chelator, myxochelin, is only produced when the iron content of the medium is about 10^{-7} M. Production is achieved by removing the iron in the medium by treatment with the chelating resin, Chelex 100 (Bio-Rad), and adding the optimal amount of $FeCl_3$ (66).

While feeding of specific precursors is problematical, fed-batch fermentations may be very useful (e.g., 47,67). Myxobacteria prefer moderate media concentrations: 1% peptone–protein and 0.5–1% carbohydrate are the upper limits allowing unrestricted growth, and optimal concentrations for production may be even lower. As mentioned, carbohydrate metabolism normally leads to acid production, so that feeding strategies can be developed using pH shifts or titration curves, respectively, as indicators. For, as soon as carbohydrate is used up, the metabolism switches to amino acids under ammonia production with a concomitant rise of pH. Sometimes production is directly connected to carbohydrate consumption, for example, that of tartrolon (55) and soraphen (47). Fed-batch fermentations often result in much higher yields; for example, in the case of disorazol a rise from 18 to 292 mg/L was achieved, with a production rate of 21 mg/L per day for 13 days (67). A fully standardized fed-batch process has still to be reported, however.

Often certain medium components have specific effects on production without much affecting growth. The type and concentration of peptone or protein is often critical (28,46,47,56,68). Individual sugars have differential effects on yields of myxalamid and stigmatellin even when produced biomass is comparable (46). Ripostatin synthesis is stimulated when 0.3% fructose and 0.01% pyruvic acid are added (53). *Sorangium cellulosum* strain So ce12 synthesizes concomitantly sorangicin, disorazol, chivosazol, and sorangiolid. In the presence of 0.1 mg/L sodium barbital the production rate of disorazol was reduced from 21 to about 2.5 mg/L per day irrespective of whether barbital was added in early or late log phase (67). Sorangicin production was not impeded.

As is well known for other producers, individual strains of myxobacteria may show a completely different production behavior for the same compound. A comparison of four wild strains producing soraphen revealed that all four synthesized the antifungal compound with reasonable yields at 30 °C in the presence of XAD resin (47). When the temperature was raised to 36 °C, however, one strain doubled soraphen yield while another one reduced it to 25%. When 0.2% casein peptone was added to the XAD culture at 30 °C, the yields of two strains dropped to near zero, while one strain increased its yield by more than 100%. In the absence of XAD the strains decreased their yields by 50% to 100%.

Efforts have been made to optimize myxovirescin production by *Myxococcus virescens* strain Mx v48 in a medium containing 1% casein peptone (69). The antibiotic is only synthesized under good oxygen supply. In normal batch cultures, antibiotic production started at the end of logarithmic growth at about 20 h after inoculation and continued until the stationary phase at about 50 h, that is, production took place in a typical idiophase. At the beginning of the production phase, the oxygen uptake and CO_2 production rates (both about 100 mg L^{-1} h^{-1}) slowed down, and the respiratory quotient, RQ, started to rise, up to 0.8 during the production phase. After 50 h, 3.1 g/L wet biomass was reached. The cell yield was 0.31 g dry biomass/g peptone, or 0.5 g dry biomass/g oxygen. Total oxygen consumption over 55 h was 6.25 g oxygen/L. Production of myxovirescin was very low, the final yield being 0.04 mg/L. By feeding peptone at different rates to a peptone-limited culture (0.16% peptone), it was demonstrated that the peptone consumption rate closely correlated with the myxovirescin production rate. In the batch culture, this rate reached a maximum of 0.4 g peptone g dry weight^{-1} h^{-1} after 8 h, but myxovirescin synthesis started only much later, after 26 h, when the peptone consumption rate was down at about 0.12 g peptone g dry weight^{-1} h^{-1}.

The feeding experiments revealed a very sharp optimum of the specific peptone uptake rate for maximal myxovirescin production. Under the chosen experimental conditions the optimal rate was about 1.8 g peptone g dry weight^{-1} h^{-1} and resulted in a specific myxovirescin production rate of about 0.05 mg myxovirescin g dry weight^{-1} h^{-1}. For optimal production, the specific peptone uptake rate had to be maintained between 0.15 and 0.2 g peptone g dry weight^{-1} h^{-1}. The final yield of myxovirescin still was very low, 2 mg/L. In another experiment under chemostat conditions, the myxovirescin production rate was raised from 0.12 to 0.55 mg L^{-1} h^{-1} under constant biomass by passing the culture supernatant through a vortex chamber containing XAD resin and thus removing myxovirescin continuously from the culture (70). Cell dry mass and antibiotic concentration reached a steady state after 170 h. The latter was reduced from 0.95 to 0.1 mg myxovirescin/L after the vortex chamber had been connected to the bioreactor. The effect on myxovirescin production of removal of NH_3 from the culture has already been mentioned. Incidently, in chemostat experiments myxobacteria turned out to respond very slowly to changed conditions, perhaps because they have a high homeostasis. For the same antibiotic, a scale-up procedure in a 0.5% Casitone (Difco) medium is also reported, although with a different producer, *Myxococcus xanthus* strain TA (71). The best titers were obtained at a relatively low oxygen absorption rate, which in fact was too low to be maintained in the largest bioreactor used. No direct correlation between growth and antibiotic production was seen. Final biomass yield was 3–4 g wet weight/L.

Several studies have reported immobilization of myxobacterial cells. The objectives were to retain a relatively high biomass, to perform chemostat experiments (not possible with a strain growing in flakes and nodules), and to investigate the influence of immobilization on the production rates of enzymes (72–74) and secondary metabolites (38,75). A more theoretical study considered the role of hydrophobic and electrostatic properties of bacterial cell surfaces on adsorbtion to porous SiO_2 (glass) and Al_2O_3 particles (76). This study included *Sorangium cellulosum*, which showed a relatively low zeta potential (-33 mV at pH 7 and in the presence of 1 mM NaCl) comparable to that of *E. coli*, but a very high hydrophobicity in the range typical for gram-positive rather than gram-negative bacteria. The contact angle, the measure of hydrophobicity, increased from 52° at the beginning of the growth curve to 75° at its end. Both values—low zeta potential and high contact angle—are contrary to good adsorption.

Immobilized cells of *Sorangium cellulosum* strain So ce12 were used to study the limitations of sorangicin synthesis (38). As immobilizing agent, calcium alginate (Protanal LF 20/60, Carroux) was found most suitable. Beads 1 mm in diameter containing 5% biomass (v/w) were prepared and used in a 1.6-L bioreactor with 1 L medium in continuous culture under chemostat conditions. Phosphate (0.23 mM) was used as the limiting factor preventing growing out of the enclosed cells. Higher phosphate concentrations also repressed sorangicin production. Consequently peptone from casein could only be added to such a concentration that its phosphate content did not exceed the optimal phosphate concentration; if that was the case, peptone stimulated sorangicin production. With the applied medium, the optimum dilution rate was 0.016/h, the optimum pH 7.8, and the optimum p_{O_2} 80% saturation (with free cells, 40% were optimal). For sorangicin production, the cells within the beads had to be maintained under growth conditions. Their generation time was 19 h (14 h with free cells). Productivity began to decline after 50 days and came to an end after 75 days when the cells within the beads degenerated. An increase of the biocatalyst wet weight beyond 60 g/L, or a liquid to catalyst ratio of 10:1 (v/v), resulted in a decrease of growth and production, partly because of oxygen limitation. Productivity could be stimulated by the addition of glycerol (9 mM) and pyruvate (5 mM), while the direct precursors, acetate and propionate, were inhibitory. A shift to enzyme activities of the pentose phosphate pathway in the immobilized cells compared with free cells was seen. The maximum volumetric productivity obtained with immobilized cells was 0.8 to 5 mg sorangicin $L^{-1} h^{-1}$ and the specific production rate was 0.25 mg sorangicin g biomass^{-1} h^{-1}, compared with 0.17 mg $L^{-1} h^{-1}$ and 0.12 mg g dry biomass^{-1} h^{-1}, respectively, with free cells.

In another project, cells of *Cystobacter* and *Corallococcus* strains were immobilized by either adsorbing them to nylon pads or by enclosing them in κ-carrageenan beads (75). The preparations were tested in shake flasks for production of chemically unknown antibiotics. Yields with adsorbed cells were 20–30% and those with enclosed cells 45–49% higher than with free cells which, however, also grew in small pellets. A substantial increase in the excretion of proteolytic enzymes has been reported for *Myxococcus xanthus* cells immobilized in calcium alginate beads (72,73,74). Yield improvements as high as eightfold have been found.

CONCLUSION

Myxobacteria may be at the threshold of biotechnological application as producers of secondary metabolites and of enzymes. No large-scale technical process has as yet been worked out for any one of them, but it already is obvious that they do not behave fundamentally different from other bacteria used in those fields. Solutions of problems met there may well be applicable or adaptable to myxobacteria.

ACKNOWLEDGMENT

I thank my secretary, Ms. Karin Rahn, for patiently and diligently typing the manuscript. I gratefully acknowledge financial support by the Fond der Chemischen Industrie.

BIBLIOGRAPHY

1. J.R. Carias, J. Raingeaud, C. Mazaud, G. Vachon, N. Lucas, Y. Cenatiempo, and R. Julien, *FEBS Lett.* **262**, 97–100 (1990).
2. E. Petit and J.F. Guespin-Michel, *J. Bacteriol.* **174**, 5136–5140 (1992).
3. H. Irschik and H. Reichenbach, *Arch. Microbiol.* **142**, 40–44 (1985).
4. D.W. Morris and J.H. Parish, *Arch. Microbiol.* **108**, 227–230 (1976).
5. H. Mayer and H. Reichenbach, *J. Bacteriol.* **136**, 708–713 (1978).
6. S. Inouye, M.Y. Hsu, S. Eagle, and M. Inouye, *Cell* **56**, 709–717 (1989).
7. S. Inouye, P.J. Herzer, and M. Inouye, *Proc. Natl. Acad. Sci. USA* **87**, 952–945 (1990).
8. W. Zhang, J. Munoz-Dorado, M. Inouye, and S. Inouye, *J. Bacteriol.* **174**, 5450–5453 (1992).
9. A.M. Breton and J.F. Guespin-Michel, *FEMS Microbiol. Lett.* **40**, 183–188 (1987).
10. A.M. Breton, G. Younes, F. van Gijsegem, and J. Guespin-Michel, *J. Biotechnol.* **4**, 303–311 (1986).
11. A. El Akoum, M. Vijayalakshmi, P. Cardon, B. Fournet, M. Sigot, and J.F. Guespin-Michel, *Enzyme Microb. Technol.* **9**, 426–429 (1987).
12. P.J. Masson and J.F. Guespin-Michel, *J. Gen. Microbiol.* **134**, 801–806 (1988).
13. D.R. Nelson, M.G. Cumsky, and D.R. Zusman, *J. Biol. Chem.* **256**, 12589–12595 (1981).
14. M.G. Cumsky and D.R. Zusman, *J. Biol. Chem.* **256**, 12596–12599 (1981).
15. H. Reichenbach and G. Höfle, *Biotechnol. Adv.* **11**, 219–277 (1993).
16. H. Reichenbach and G. Höfle, in M. Dworkin and D. Kaiser eds., *Myxobacteria II*, American Soc. for Microbiology, Washington, D.C., 1993, pp. 347–397.
17. G. Höfle and H. Reichenbach, in W. Kuhn and H.P. Fiedler, eds., *Sekundärmetabolismus bei Mikroorganismen, Beiträge zur Forschung*, Atempto Verlag, Tübingen, 1995, pp. 61–78.
18. H. Reichenbach and G. Höfle, in S. Grabley and R. Thiericke eds., *Drug Discovery from Nature*, Springer-Verlag, Berlin, 1999, pp. 149–179.
19. H. Reichenbach and G. Höfle, in *Scientific Annual Report of the GBF, Gesellschaft für Biotechnologische Forschung*, Braunschweig, 1994, pp. 5–22.
20. H. Reichenbach, in *Scientific Annual Report of the GBF, Gesellschaft für Biotechnologische Forschung*, Braunschweig, 1996, pp. 25–30.
21. M. Dworkin and D. Kaiser, eds., *Myxobacteria II*, American Soc. for Microbiology, Washington, D.C., 1993.
22. M. Dworkin, *Microbiol. Rev.* **60**, 70–102 (1996).
23. H.D. McCurdy, E.R. Brockman, H. Reichenbach, and D. White, in J.T. Staley, M.P. Briant, N. Pfennig, and J.G. Holt eds., *Bergey's Manual of Systematic Bacteriology, vol. 3*, Williams & Wilkins, Baltimore, 1989, pp. 2139–2170.
24. H. Reichenbach and M. Dworkin, in A. Balows, H.G. Trüper, M. Dworkin, W. Harder, and K.H. Schleifer eds., *The Prokaryotes*, 2nd ed., *vol. IV*, Springer-Verlag, New York, 1992, pp. 3416–3487.
25. H. Reichenbach, in J.G. Holt, N.R. Krieg, P.H.A. Sneath, J.T. Staley, and S.T. Williams eds., *Bergey's Manual of Determi-*

native Bacteriology, 9th ed., Williams & Wilkins, Baltimore, 1994, pp. 515–525.
26. C.A. Jacobi, H. Reichenbach, B.J. Tindall, and E. Stackebrandt, *Int. J. Syst. Bacteriol.* **46**, 119–122 (1996).
27. C.A. Jacobi, B. Assmuss, H. Reichenbach, and E. Stackebrandt, *Appl. Environ. Microbiol.* **63**, 719–723 (1997).
28. B. Kunze, R. Jansen, G. Höfle, and H. Reichenbach, *J. Antibiot. (Tokyo)* **47**, 881–886 (1994).
29. B. Zeggel, Doctoral Thesis, TU Braunschweig, 1993.
30. J.R. Lampky and E.R. Brockman, *Int. J. Syst. Bacteriol.* **27**, 161 (1977).
31. W. Trowitzsch-Kienast, E. Forche, V. Wray, H. Reichenbach, E. Jurkiewicz, G. Hunsmann, and G. Höfle, *Liebigs Ann. Chem.* **1992**, 659–664 (1992).
32. L. Shimkets and C. Woese, *Proc. Natl. Acad. Sci. USA* **89**, 9459–9463 (1992).
33. H.W. Chen, A. Kuspa, I.M. Keseler, and L.J. Shimkets, *J. Bacteriol.* **173**, 2109–2115 (1991).
34. B. Neumann, A. Pospiech, and H.U. Schairer, *Mol. Microbiol.* **10**, 1087–1099 (1993).
35. M. Kalos and J. Zissler, *Proc. Natl. Acad. Sci. USA* **87**, 8316–8320 (1990).
36. S. Jaoua, S. Neff, and T. Schupp, *Plasmid* **28**, 157–165 (1992).
37. T. Schupp, C. Toupet, B. Cluzel, S. Neff, S. Hill, J.J. Beck, and J.M. Liggon, *J. Bacteriol.* **177**, 3673–3679 (1995).
38. K. Becker, Doctoral Thesis, TU Braunschweig, 1990.
39. B. Kunze, T. Kemmer, G. Höfle, and H. Reichenbach, *J. Antibiot. (Tokyo)* **37**, 454–461 (1984).
40. S.M. Ringel, R.C. Greenough, S. Roemer, D. Connor, A.L. Gutt, B. Blair, G. Kanter, and M. von Strandtmann, *J. Antibiot. (Tokyo)* **30**, 371–375 (1977).
41. P. Coucke and J.P. Voets, *Z. Allg. Mikrobiol.* **7**, 175–182 (1967).
42. B.F. Watson and M. Dworkin, *J. Bacteriol.* **96**, 1465–1473 (1968).
43. K. Gerth and H. Reichenbach, *Arch. Microbiol.* **117**, 173–182 (1978).
44. K. Gerth and H. Reichenbach, *Anal. Biochem.* **152**, 78–82 (1986).
45. K. Gerth, W. Trowitzsch, G. Piehl, R. Schultze, and J. Lehmann, *Appl. Microbiol. Biotechnol.* **19**, 23–28 (1984).
46. B. Kunze, R. Jansen, F. Sasse, G. Höfle, and H. Reichenbach, *J. Antibiot. (Tokyo)* **48**, 1262–1266 (1995).
47. K. Gerth, N. Bedorf, H. Irschik, G. Höfle, and H. Reichenbach, *J. Antibiot. (Tokyo)* **47**, 23–31 (1994).
48. F. Sasse, H. Steinmetz, G. Höfle, and H. Reichenbach, *J. Antibiot. (Tokyo)* **46**, 741–748 (1993).
49. B. Kunze, W. Kohl, G. Höfle, and H. Reichenbach, *J. Antibiot. (Tokyo)* **38**, 1649–1654 (1985).
50. F. Sasse, H. Steinmetz, G. Höfle, and H. Reichenbach, *J. Antibiot. (Tokyo)* **48**, 21–25 (1995).
51. B. Kunze, R. Jansen, L. Pridzun, E. Jurkiewicz, G. Hunsmann, G. Höfle, and H. Reichenbach, *J. Antibiot. (Tokyo)* **46**, 1752–1755 (1993).
52. B. Kunze, R. Jansen, L. Pridzun, E. Jurkiewicz, G. Hunsmann, G. Höfle, and H. Reichenbach, *J. Antibiot. (Tokyo)* **45**, 1549–1552 (1992).
53. H. Irschik, H. Augustiniak, K. Gerth, G. Höfle, and H. Reichenbach, *J. Antibiot. (Tokyo)* **48**, 787–792 (1995).
54. K. Gerth, N. Bedorf, G. Höfle, H. Irschik, and H. Reichenbach, *J. Antibiot. (Tokyo)* **49**, 560–563 (1996).
55. H. Irschik, D. Schummer, K. Gerth, G. Höfle, and H. Reichenbach, *J. Antibiot. (Tokyo)* **48**, 26–30 (1995).
56. K. Gerth, P. Washausen, G. Höfle, H. Irschik, and H. Reichenbach, *J. Antibiot. (Tokyo)* **49**, 71–74 (1996).
57. B. Kunze, W. Trowitzsch-Kienast, G. Höfle, and H. Reichenbach, *J. Antibiot. (Tokyo)* **45**, 147–150 (1992).
58. K. Gerth, R. Jansen, G. Reifenstahl, G. Höfle, H. Irschik, B. Kunze, H. Reichenbach, and G. Thierbach, *J. Antibiot. (Tokyo)* **36**, 1150–1156 (1983).
59. H. Irschik, W. Trowitzsch-Kienast, K. Gerth, G. Höfle, and H. Reichenbach, *J. Antibiot. (Tokyo)* **41**, 993–998 (1988).
60. K. Grimm, *J. Bacteriol.* **135**, 748–753 (1978).
61. K. Grimm, *Z. Allg. Mikrobiol.* **18**, 399–407 (1978).
62. E. Rosenberg, K.H. Keller, and M. Dworkin, *J. Bacteriol.* **129**, 770–777 (1977).
63. H. Irschik and H. Reichenbach, *J. Antibiot. (Tokyo)* **38**, 1237–1245 (1985).
64. H. Irschik, K. Gerth, T. Kemmer, H. Steinmetz, and H. Reichenbach, *J. Antibiot. (Tokyo)*, **36**, 6–12 (1983).
65. S. Fytlovitch, P.D. Nathan, D. Zafriri, and E. Rosenberg, *J. Antibiot. (Tokyo)* **36**, 1525–1530 (1983).
66. B. Kunze, N. Bedorf, W. Kohl, G. Höfle, and H. Reichenbach, *J. Antibiot. (Tokyo)* **42**, 14–17 (1989).
67. H. Irschik, R. Jansen, K. Gerth, G. Höfle, and H. Reichenbach, *J. Antibiot. (Tokyo)* **48**, 31–35 (1995).
68. H. Irschik, R. Jansen, K. Gerth, G. Höfle, and H. Reichenbach, *J. Antibiot. (Tokyo)* **40**, 7–13 (1987).
69. J.N. Nigam, J. Lehmann, K. Gerth, H. Piehl, R. Schultze, and W. Trowitzsch, *Appl. Microbiol. Biotechnol.* **19**, 157–160 (1984).
70. V. Hecht, J. Vorlop, H. Kalbitz, K. Gerth, and J. Lehmann, *Biotechnol. Bioeng.* **29**, 222–227 (1987).
71. A. Mizrahi, J. Arnan, G. Miller, Z. Liron, M. Manai, Y. Batus, and E. Rosenberg, *J. Appl. Chem. Biotechnol.* **26**, 160–166 (1976).
72. G. Younes, A.M. Breton, and J. Guespin-Michel, *Appl. Microbiol. Biotechnol.* **25**, 507–512 (1987).
73. G. Younes, J.M. Nicaud, and J. Guespin-Michel, *Appl. Microbiol. Biotechnol.* **19**, 67–69 (1984).
74. J.C. Vuillemard, S. Terré, S. Benoit, and J. Amiot. *Appl. Microbiol. Biotechnol.* **27**, 423–431 (1988).
75. F.N. Yasouri and H.A. Foster, *Biotechnol. Tech.* **5**, 149–152 (1991).
76. C. Krekeler, H. Ziehr, and J. Klein, *Appl. Microbiol. Biotechnol.* **35**, 484–490 (1991).

N

NITRILE HYDRATASE

YOSHIRO ASHINA
MASARU SUTO
TAKAKAZU ENDO
Nitto Chemical Industry Co., Ltd.
Tokyo, Japan

KEY WORDS

Acrylamide
Bioreactor
Flocculant
Inducer
Metalloenzyme
Nicotinamide
Nitrilase
Nitrile hydratase
Rhodococcus chlororaphis B-23
Rhodococcus rhodochrous J-1
Rhodococcus sp. N-774

OUTLINE

MANUFACTURE OF ACRYLAMIDE

Manufacture of acrylamide is the most important commercial application of nitrile hydratase. Acrylamide is used as a starting material for a variety of polymer products such as flocculant, paper-strength agent, coating materials, and EOR (enhanced oil recovery) polymers. Current world production of acrylamide exceeds 200,000 metric tons per annum (MTA). Nitto Chemical Industry Co., Ltd. (Japan) started commercial production of acrylamide in 1986 using a catalyst based on nitrile hydratase. This was one of the first commercial bioconversion processes that made a commodity chemical. The very first acrylamide process based on hydration of acrylonitrile with sulfuric acid (Fig. 1) was commercialized in 1950, which was replaced by the copper-catalyst process 20 years later. This copper-catalyzed process currently prevails in worldwide acrylamide production. Although the copper process gives an excellent efficiency in comparison with the acid hydration process, it still has some drawbacks, such as formation of impurities that are difficult to remove and adversely influence polymerization reactions. These impurities include ethylene cyanohydrin, β-hydroxypropionamide, and nitrilotrispropionamide, which may interfere with the chain reactions lowering the molecular weight of polymers and deteriorate the stability of the monomer product. Overcoming these limitations was one of the important motivations for the development of the enzymatic process using nitrile hydratase, which would minimize the side reactions caused by high specificity of enzymes and mild conditions under which they work.

Strain N-774

Because microorganisms capable of converting nitrile compounds to corresponding amides had been reported in the 1960s (1) many studies were conducted to find a microorganism with an activity high enough for industrial applications. *Rhodococcus* sp. N-774, discovered in 1980 (2), was the first strain that was put into commercial use to produce acrylamide from acrylonitrile. The nitrile hydratase enzyme is produced in this strain without an inducer. Thiamine is essential for the cell growth. Fe^{2+} and Fe^{3+} remarkably promote production of nitrile hydratase in the cell (3). Table 1 and Figure 2 show the effect of metal ions on the activity and the typical time course of the cell growth and the enzyme activity, respectively. The electron micrograph of the strain N-774 is shown in Figure 3.

Reaction Conditions

According to the semi-batchwise reaction test using immobilized cells of N-774 with a serial addition of acrylonitrile, a reaction temperature of 0 °C is optimum to obtain the product in high concentration, as shown in Figure 4 (4). Enzyme activity of this cell is light-inductive (5). Figure 5 shows the effect of light irradiation on the activity, which should be properly reflected on the reactor design.

Immobilization of the Cell

For mass production of biologically converted products, immobilized cells give wider options in selection and designing of the reactor. The bacteria cells entrapped in a polyacrylamide-based hydrogel are used in the production of acrylamide. To maintain the enzyme activity and improve the physical characteristics of the immobilized cells, the following techniques are used: (*1*) Swelling of the gel is controlled by copolymerization with cationic monomers (6); (*2*) Shearing force applied while mixing the cells with monomers or polymers is carefully controlled; (*3*) polymerization conditions are adjusted to minimize diffusion resistance within the gel, which not only reduces the apparent

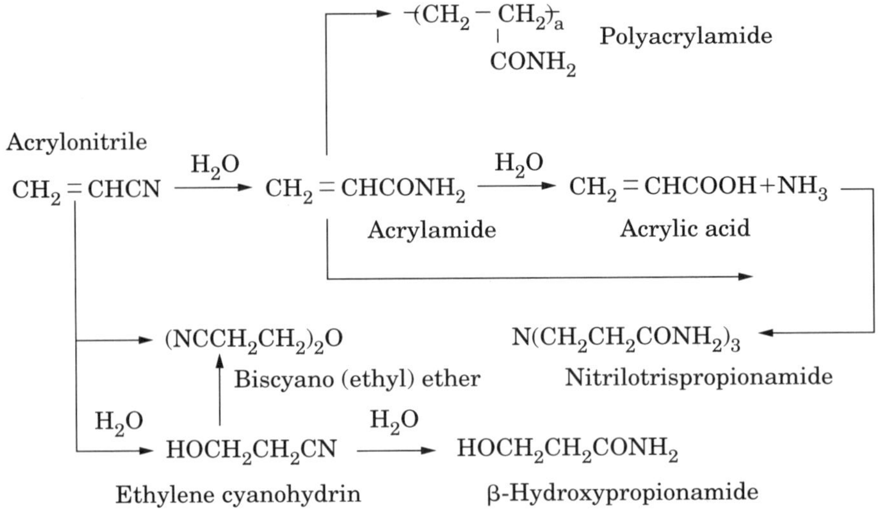

Figure 1. Products of acrylonitrile hydration.

Table 1. Effect of Metal Ions on Activity

Inorganic salts	(mg/L)	Call growth (g/L)	pH (−)	Specific activity (units/mg-dry cell)
$FeSO_4$ $7H_2O$	10	4.30	7.58	50.0
$MnSO_4$ $4H_2O$	10	4.75	6.94	0.70
$NiSO_4$ $6H_2O$	10	4.42	7.53	0.92
$CoSO_4$ $7H_2O$	10	4.34	7.37	1.27
$BaCl_2$ $2H_2O$	10	4.63	7.42	1.16
Na_2MoO_4 $2H_2O$	10	4.05	7.60	0.70
$CuSO_4$ $5H_2O$	2	4.54	7.30	1.24
$ZnSO_4$ $7H_2O$	2	4.38	7.45	0.96
None	—	4.42	7.47	1.05

Note: Basal medium: 10 g glucose, 3 g $(NH_4)_2SO_4$, 0.5 g KH_2PO_4, 0.5 g K_2HPO_4, 0.5 g $MgSO_4$ $7H_2O$, 0.1 g NaCl, 0.1 g $CaCl_2$ $2H_2O$, 5 g casamino acid and 0.002 g thiamine HCl in 1 L of distilled water, pH 7.5. Culture: at 30 °C for 55 hrs by shaking.

activity but also degenerates the activity by increasing the amide concentration near the active site (Figs. 6 and 7); and (*4*) Physical stability of the gel is improved by adjusting the polymer composition and polymerization conditions to prevent the cells from escaping the gel. An electron micrograph of the entrapped state of the cell is shown in Figure 8. The immobilization is carried out on a continuous basis at the commercial plant.

Bioreactor

Various types of the reactors—fixed bed, moving bed, fluidized bed, and stirred type—can be used (Membrane type does not seem suitable because of its flow-through resistance). The reactor is designed in consideration of (*1*) inhibition of polymerization; (*2*) replacement of degenerated catalyst; (*3*) control of reaction temperature; (*4*) control of substrate concentration; and (*5*) irradiation of light.

Comparison of the Processes

In comparison with the copper-catalyst process the biological acrylamide process features (*1*) 100% one-pass conversion; (*2*) no metal ions; (*3*) minimized organic impurities; (*4*) mild reaction conditions; and (*5*) a simplified process. The block diagram of the two processes is shown in Figure 9. Table 2 compares the typical product analysis.

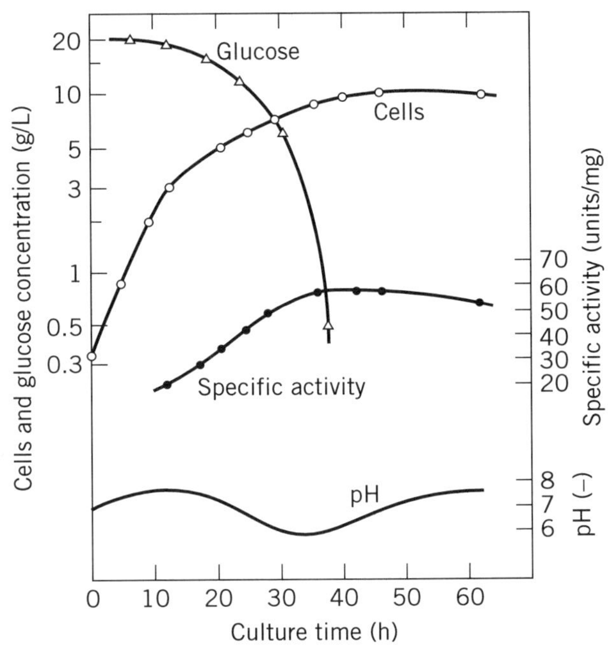

Figure 2. Time courses of cell growth and enzyme production. Culture conditions: medium consisting of 20 g glucose, 10 g peptone, 6 g yeast extract, and 6 g malt extract in 1 L of distilled water, pH, 7.2; aeration, 0.75 v/v/m; temperature, 30 °C; stirring, 350 rpm; culture volume, 13 L in a 20-L jar fermenter.

Strain B-23

Pseudomonas chlororaphis B-23, discovered by Yamada et al. of Kyoto University (7) (Table 3), potentially gives higher activity than N-774. The tolerance to the product is also higher than that of N-774. Sucrose, glutamic acid, cystine, and proline are used as medium components to grow the enzyme with minimized catabolite repression (8).

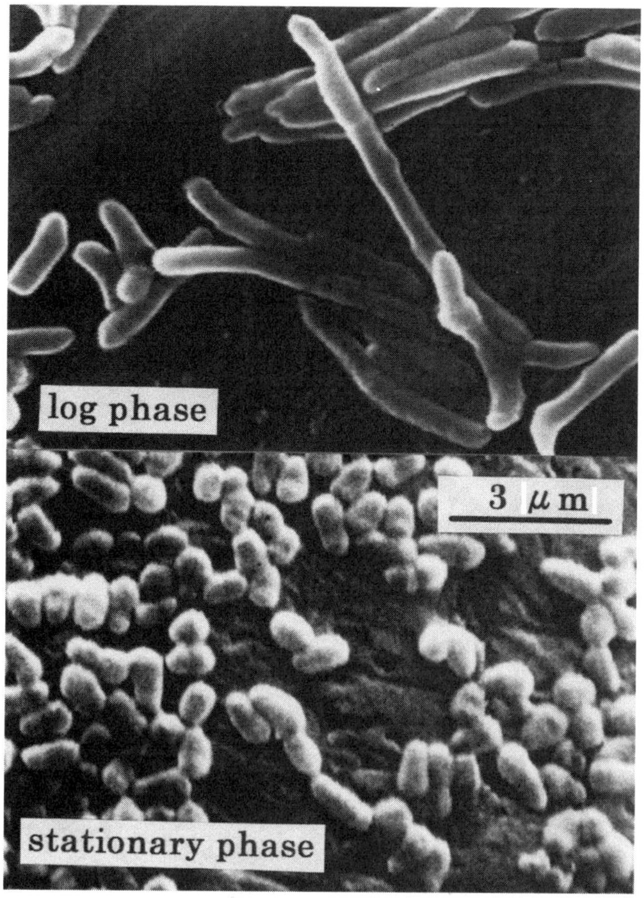

Figure 3. Electron micrograph of strain N-774.

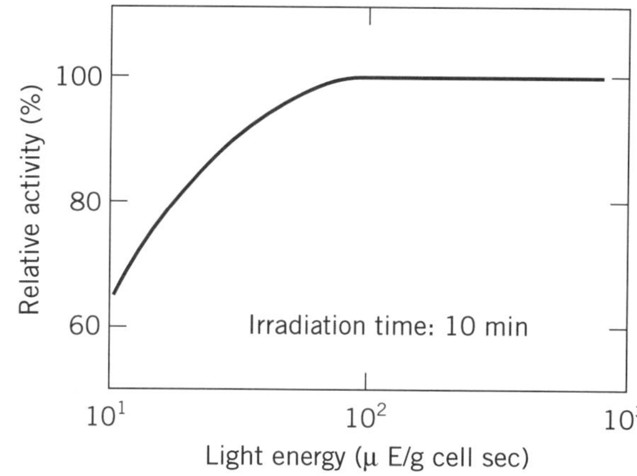

Figure 5. Activation of enzyme by light energy.

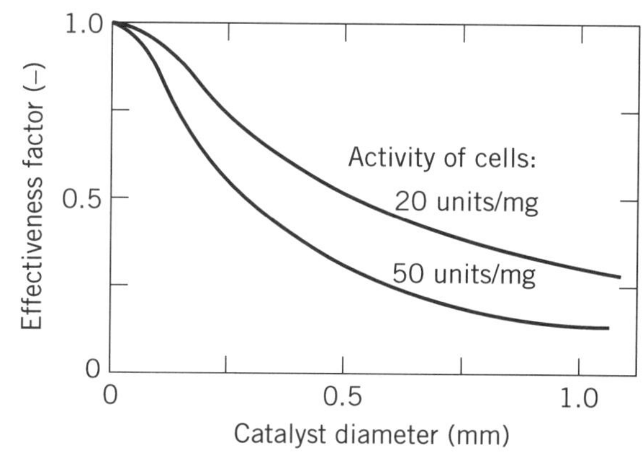

Figure 6. Catalyst particle size and effectiveness factor.

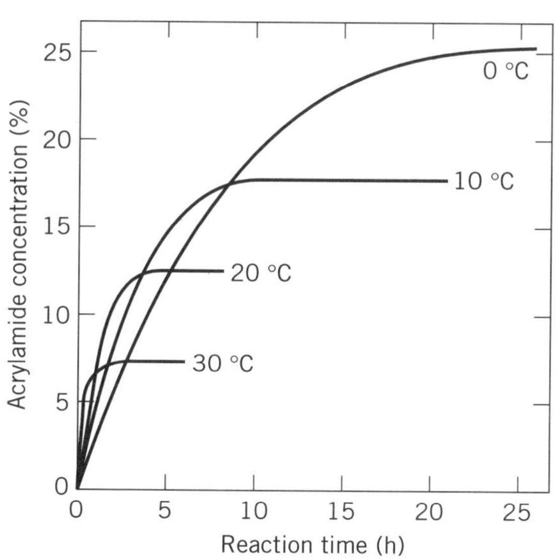

Figure 4. Effect of temperature on acrylamide production. The reaction was carried out in a semi-batchwise fashion in a stirred-tank reactor. Acrylonitrile was added consecutively. Reaction conditions: pH, 8.5; catalyst concentration, 0.5%; acrylonitrile concentration, 2.0%; temperature, as indicated in the figure.

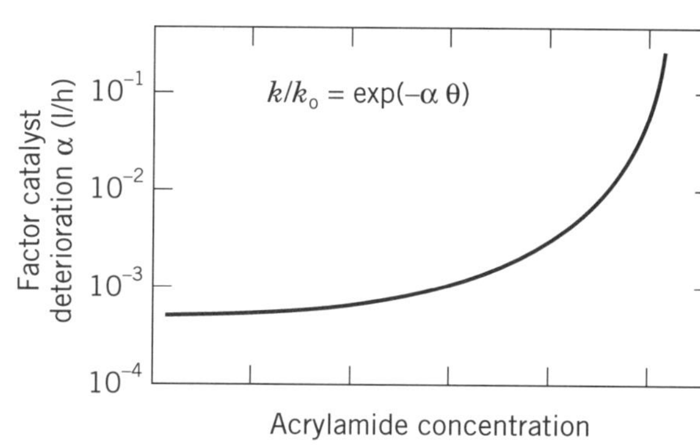

Figure 7. Catalyst deterioration by amide without reaction.

Methacrylamide works remarkably well as an enzyme inducer for this strain. The activity is improved by means of mutation using chemicals such as nitrosoguanidine. The mutant Am-324, shown in Figure 10, gives the activity 3,000 times as high as the original cell from nature. The catalyst based on this mutant, when used in the commercial acrylamide plant, increased the production capacity to 6,000 MTA from 4,000 MTA with the N-774 catalyst.

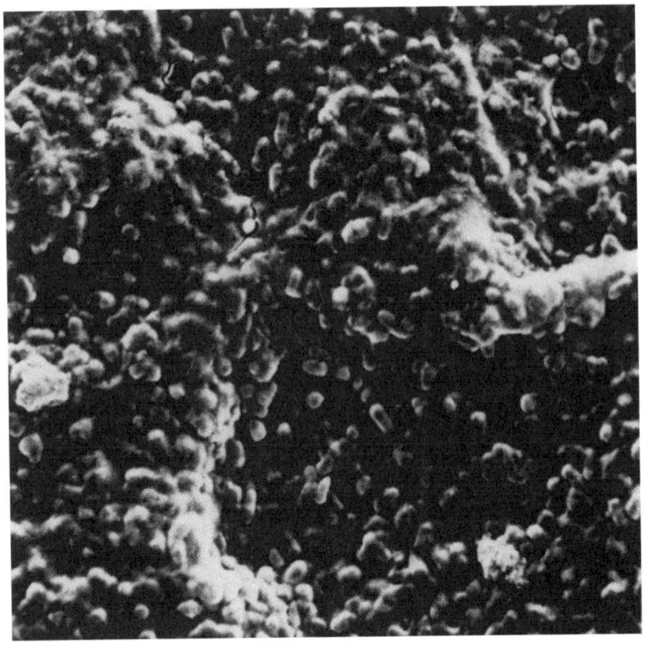

Figure 8. Electron micrograph of immobilized cells.

Table 2. Typical Product Analysis

Impurities	Acrylamide from enzymatic process	Acrylamide from copper-catalyzed process
Acrylonitrile	N.D. (<5)	100 ~ 200
Acrylic acid	<5.0	100 ~ 200
Ethylene cyanohydrin	N.D. (<10)	300 ~ 400
β-Hydroxypropionamide	N.D. (<10)	600 ~ 1000
Nitrilotrispropionamide	N.D. (<10)	~100
Propionamide	N.D. (<10)	20 ~ 50

Note: ppm in 50% aq. solution.

Strain J-1

The enzyme induction mechanism of *Rhodococcus rhodochrous* J-1 was also found by Yamada et al. (9). At the beginning of their discovery, J-1 produced two different kinds of nitrile-converting enzymes (nitrilase and nitrile hydratase), and the reaction activity of both enzymes was very low. They found addition of cobalt ions into the cell-cultured medium improved the nitrile hydratase activity, and further increase of the activity was affected by action of urea or its derivatives as an inducer in the presence of cobalt ions. This strain has very high tolerance to acrylamide. Table 4 shows properties of the nitrile hydratase from the strains N-774, B-23, and J-1. All three enzymes belong to the metalloenzyme group but are different in kind of the coordinated metals to their active sites (i.e., iron for N-774 and B-23 and cobalt for J-1). Nitrile hydratase enzymes having nickel as the coordinated metal have also been found. The protein chemistry of nitrile hydratase, including molecular weight and light inducibility, differs widely among different genuses and species. Reaction performance of the three strains, in laboratory tests and at the commercial plant, are compared in Figure 11 and

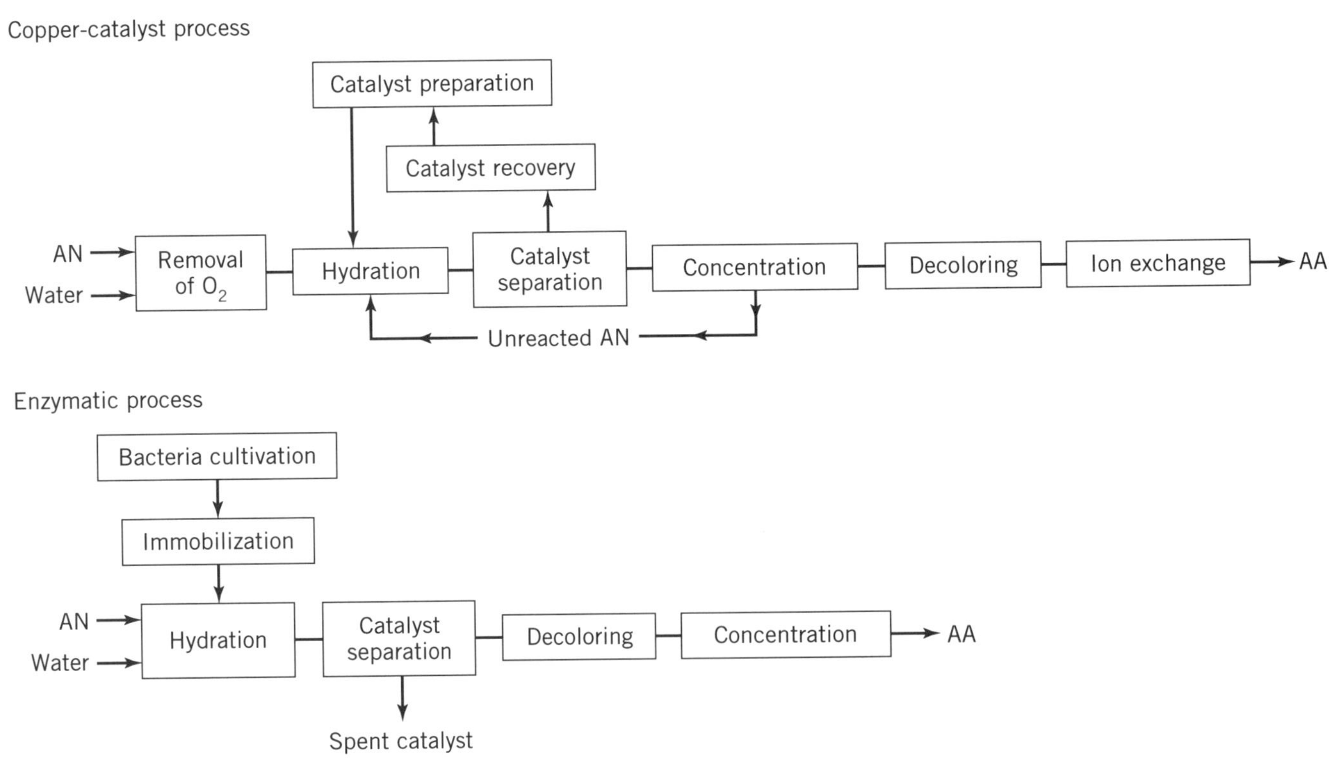

Figure 9. Comparison of enzymatic process with conventional process. AN = acrylonitrile; AA = Acrylamide (50% aqueous solution).

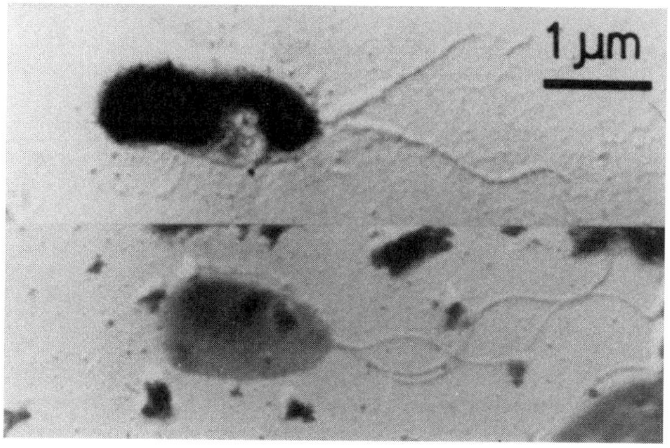

Figure 10. Electron micrograph of strain B-23.

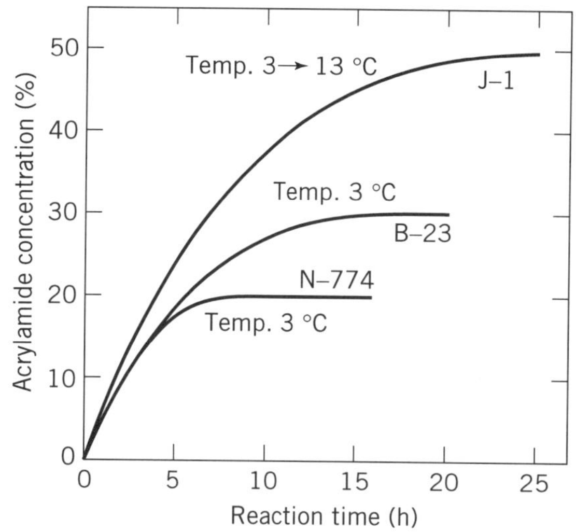

Figure 11. Comparison of reaction performance. The reaction was carried out in a semi-batchwise fashion in a stirred-tank reactor. Acrylonitrile was added consecutively. Reaction conditions: pH, 6.5 ~ 8.5; temperature, as indicated in the figure; acrylonitrile concentration, 2.0%.

Table 5. The optimized J-1 exhibits a magnitude of order higher productivity than the N-774. Production capacity of the acrylamide plant using the J-1 catalyst is by four to five times greater than that of the same plant with the N-774 catalyst.

APPLICATION TO OTHER AMIDES

Another important industrial application of nitrile hydratase is the production of nicotinamide, which is widely used as a vitamin supplement. At the end of 1997, Lonza Ltd. (China) was scheduled to manufacture 3,000 MTA of nicotinamide by the continuous biotransformation of 3-cyanopyridine using immobilized cells of *Rhodococcus rhodochrous*. Considerable quantity of nicotinic acid is inevitably formed as a by-product in alkaline hydrolysis of 3-cyanopyridine. Under the biological route, nicotinic acid and other contaminating by-products are not produced. This route is also competitive with the methylethylpyridine route.

NITRILASE

Nitrile hydratase is one of the two types of the enzymes that catalyze hydration or hydrolysis of nitrile compounds. The other one, nitrilase, converts nitriles to corresponding carboxylic acids. Some of the nitrilases are capable of hydrolyzing nitriles stereospecifically. Optical resolution of α-substituted nitriles using a nitrilase to obtain one of the corresponding acid enantiomers was reported (10). The yield of the proposed process, however, was limited to 50% with the other nitrile enantiomer remaining intact. The biotransformation process to produce *R*-mandelic acid using a *R*-specific nitrilase (equation 1) has been commercialized by Nitto Chemical (11). In this process, the unconverted *S*-nitrile enantiomer is racemized in one-pot equilibrium dissociation-association reactions (equation 2), which eventually gives nearly 100% conversion to *R*-acid enantiomer at an optical purity over 99%. Addition of sodium sulfate in the reaction mixture enhances the productivity sixfold by reversibly forming the benzaldehyde-sulfite complex to promote racemization of *S*-mandellonitrile (equation 3) (12).

$$\text{RS-}C_6H_5CH\,(OH)\,CN \rightarrow \text{S-}C_6H_5CH\,(OH)\,CN + \text{R-}C_6H_5CH\,(OH)\,COOH + NH_3 \quad (1)$$

$$C_6H_5CH\,(OH)\,CN \leftrightarrow C_6H_5CHO + HCN \quad (2)$$

$$C_6H_5CHO + Na_2SO_3 + H_2O \leftrightarrow C_6H_5CH\,(OH)\,SO_3Na + NaOH \quad (3)$$

Although R-specific nitrilase is found in a variety of bacteria, such as *Pseudomonas* sp., *Alcaligenes* sp., *Acineto-*

Table 3. The Development of B-23 Strain Am-324

Strain	Mutation	Medium	SA (U/mg)	BA (U/mL)	Property
Wild	—	M-A	0.72	0.4	Mucilaginous bacterium
↓	—	M-R	66	363	
Mutant I (Am-3)	NTG	M-R	65	465	Unmucilaginous
↓					
Mutant II (Am-324)	NTG	M-R	125	952	High activity
↓	—	Consecutive addition of inducer	141	1260	

Note: NTG = *N*-methyl-*N'*-nitro-*N*-nitrosoguanidine; SA = specific activity, BA = enzyme activity of culture broth.

Table 4. Properties of Nitrile Hydratase from N-774, B-23, and J-1

	Rhodococcus sp N-774	*Pseudomonas chlororaphis* B-23	*Rhodococcus rhodochrous* J-1
Molecular weight	70,000	100,000	520,000
Subunit molecular weight	α: 27,000 β: 27,500	25,000	α: 26,000 β: 29,000
Metal	Fe	Fe	Co
Activation by light	Positive	Negative	Negative
Tolerance for acrylamide	High	Very high	Very high
Production of organic acid	Very low	Negligible	Low
Optimum temperature [°C]	35	20	
Heat stability	Unstable	Unstable	Stable
Optimum pH	7.7	7.5	
Stable pH	7.0 ~ 8.5	6.0 ~ 7.5	6.0 ~ 8.5
Enzyme formation	Constitutive	Inducible	Inducible

Table 5. Reaction Performance in Commercial Production

	Reaction conditions		Reaction performance	
N-774	pH	7.5 ~ 8.5	Conversion of AN	99.9 %
	Temperature	0 ~ 5 °C	Selectivity of AA	99.9 %
	AN concentration	1.5 ~ 2.0 %	AA concentration at outlet of reactor	20%
B-23	pH	7.5 ~ 8.5	Conversion of AN	99.97 %
	Temperature	0 ~ 5 °C	Selectivity of AA	99.98 %
	AN concentration	1.5 ~ 2.0 %	AA concentration at outlet of reactor	30 %
J-1	pH	6.5 ~ 8.0	Conversion of AN	99.97 %
	Temperature	0 ~ 15 °C;	Selectivity of AA	99.98 %
	AN concentration	1.5 ~ 2.0 %	AA concentration at outlet of reactor	50 %

Note: AN = acrylonitrile, AA = acrylamide.

bacter sp., *Caseobacter* sp., *Nocardia* sp., *Brevibacterium* sp., *Bacillus* sp., *Aureobacterium* sp., and *Rhodococcus* sp. (13), *S*-specific nitrilase has hardly been isolated.

BIBLIOGRAPHY

1. A. Miura et al., *J. Ferment. Technol.* **47**, 631 (1969).
2. Jpn. Pat. 17918 (1981), I. Watanabe et al. (to Nitto Chemical Industry Co., Ltd.).
3. Jpn. Pat. 4987 (1984), I. Watanabe et al. (to Nitto Chemical Industry Co., Ltd.).
4. Jpn. Pat. 38118 (1981), I. Watanabe et al. (to Nitto Chemical Industry Co., Ltd.).
5. Jpn. Pat. 162195 (1986), K. Enomoto et al. (to Nitto Chemical Industry Co., Ltd., Mitsubishi Rayon).
6. Jpn. Pat. 35078 (1983), I. Watanabe et al. (to Nitto Chemical Industry Co., Ltd.).
7. Jpn. Pat. 37951 (1984), H. Yamada et al. (to H. Yamada).
8. Jpn. Pat. 43996 (1986), H. Yamada et al. (to H. Yamada, Nitto Chemical Industry Co., Ltd.); Jpn. Pat. 43997 (1986), H. Yamada et al. (to H. Yamada, Nitto Chemical Industry Co., Ltd.); Jpn. Pat. 43998 (1986), H. Yamada et al. (to H. Yamada, Nitto Chemical Industry Co., Ltd.); Jpn. Pat. 43999 (1986), H. Yamada et al. (to H. Yamada, Nitto Chemical Industry Co., Ltd.).
9. Jpn. Pat. 55148 (1986), H. Yamada et al. (to H. Yamada).
10. Eur. Pat. 0,348,901 B, K. Yamamoto et al. (to Asahi Kasei Kogyo).
11. Eur. Pat. 0,610,048 A, Y. Hashimoto et al. (to Nitto Chemical Industry Co., Ltd.).
12. U.S. Pat. 5,326,702, T. Endo et al. (to Nitto Chemical Industry Co., Ltd.).
13. U.S. Pat. 5,223,416, T. Endo et al. (to Nitto Chemical Industry Co., Ltd.).

O

OILS, MICROBIAL PRODUCTION

SAKAYU SHIMIZU
JUN OGAWA
Kyoto University
Kyoto, Japan

KEY WORDS

Arachidonic acid
Docosahexaenoic acid
Eicosadienoic acid
Eicosapentaenoic acid
Fatty acid
Linoleic acid
Mortierella
Polyunsaturated fatty acid
Single cell oil
Triacylglycerol

OUTLINE

INTRODUCTION

As are microbial carbohydrates, microbial lipids are a diverse group of compounds with important biochemical functions. The exploitation of microbial oil (triacylglycerol) has been attempted since the 1980s. The term *single cell oil* is used for unique oils produced by microorganisms (1,2). It is probable that single cell oil could compete against plant seed oils or fish oils. Some yeasts and molds have long been known to accumulate very high levels of triacylglycerols. A lipid content in excess of 40% (w/w) is not exceptional, and values of 70% and even 80% have been reported (3). The oil produced by these oleaginous species not only has the same triacylglycerol structure as plant oils but also shows a similar distribution of fatty acids at the three positions: saturated fatty acids are almost totally excluded from the central (*sn*-2) carbon atom.

The fatty acyl groups themselves are the same as those found in plant seed oils or, in the highly polyunsaturated fatty acids, are the same as those found in fish oils or other animal products. Many attempts have been made to produce useful fatty acids or oils containing them by microbial processes. One example is the production of cocoa-butter-equivalent oil by a yeast that abundantly produces stearic acid (4,5). The most successful example occurs in the production of oils containing polyunsaturated fatty acids (PUFAs). Certain microorganisms may also produce prostaglandin precursors, or even prostaglandins themselves from PUFAs, as well as cerebroside lipids and other unusual lipids that are not normally regarded as being of microbial origin (6). Here, microbial PUFA production is discussed mainly as a typical example of a single cell oil.

OUTLINE OF MICROBIAL PUFA PRODUCTION

PUFAs are fatty acids having more than one double bond. Among these, 5(*Z*),8(*Z*),11(*Z*)-eicosatrienoic acid (20:3ω9; Mead acid), dihomo-γ-linolenic acid (20:3ω6; DGLA), arachidonic acid (20:4ω6; AA), and 5(*Z*),8(*Z*),11(*Z*),14(*Z*),17(*Z*)-eicosapentaenoic acid (20:5ω3; EPA) (see Fig. 1 for structures and biosynthetic pathways) are of interest because they are precursors for prostaglandins, thromboxanes, leukotrienes, prostacyclins, and so on (7–9), all of which have hormonelike activities. Possibly as a result of the physiological activities of these compounds, PUFAs exhibit several unique biological activities, such as lowering of plasma cholesterol level and prevention of thrombosis. The inclusion of EPA and 4(*Z*),7(*Z*),10(*Z*),13(*Z*),16(*Z*),19(*Z*)-docosahexaenoic acid (22:6ω3; DHA) as dietary supplements has been recommended for the prevention of heart diseases (10), and γ-linolenic acid (18:3ω6; GLA) has been recommended for the relief of eczema (11). Accordingly, PUFAs are highly important substances in the pharmaceutical, medical, and nutritional fields. Because food sources rich in these PUFAs are limited to a few seed oils that contain GLA and fish oil, which contains EPA and DHA, recent investigations have focused on microorganisms as alternative sources of oils containing these PUFAs.

Microorganisms are potentially very promising lipid sources because of their extremely high growth rates in simple media and the simplicity of their manipulation.

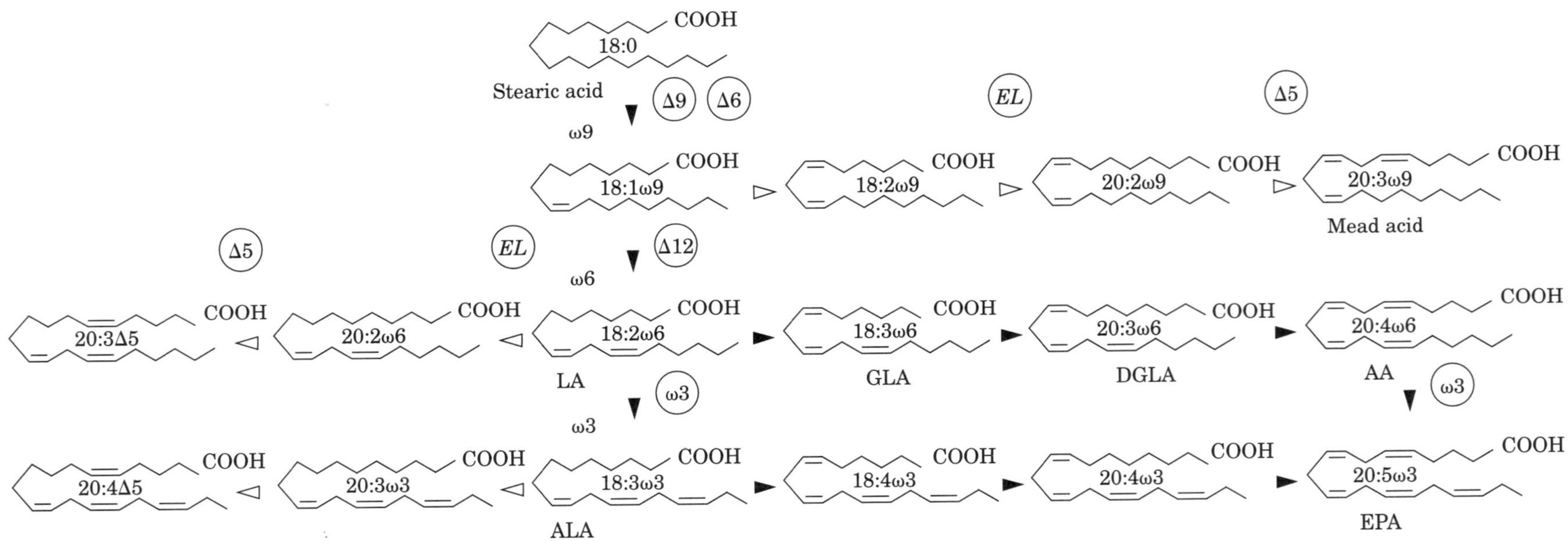

Figure 1. Proposed pathways for polyunsaturated fatty acid (PUFA) biosynthesis in *Mortierella alpina* 1S-4. *Open arrowheads* show bypaths through which *n*-9 fatty acids and nonmethylene-interrupted PUFAs are formed in Mut48 and Mut49, respectively. **Δ***n*, *Δ*n desaturation; *EL*, elongation.

Furthermore, microbial PUFA-containing oils are usually characterized by the absence of other PUFAs, making purification of individual fatty acids from microorganisms an easier task than it is from other sources. A variety of PUFAs have been detected in microorganisms including bacteria, fungi, algae, mosses, and protozoa (Table 1). Erwin and Bloch (12) suggested that lower classes of organisms including microorganisms can be classified into several groups based on their ability to produce PUFAs. Shaw (13) pointed out that some fungi belonging to the Mucorales can accumulate relatively large amounts of GLA in their mycelia. Based on these early observations, several groups have been screening for microorganisms capable of accumulating lipids containing PUFAs so as to obtain more suitable sources for large-scale preparation of important nutritional components.

Suzuki (14,15) found several potent producers of GLA through extensive screening of filamentous fungi. They reported that a strain of *Mortierella isabellina* accumulates about 3 g/L of GLA when grown in a medium containing glucose as the major carbon source. Based on this finding, commercial production of GLA with a *Mortierella* fungus recently commenced in Japan.

Shimizu et al. (16,17) found that several *Mortierella* fungi are potent producers for triacylglycerol containing AA, DGLA, or EPA. A soil isolate strain, *Mortierella alpina* 1S-4, produced AA on cultivation in a medium containing glucose and yeast extract. It also produced EPA when grown at low temperature (20 °C) or when grown in a medium supplemented with α-linolenic acid (18:3ω3; ALA). When grown in the presence of sesame seed oil, the same fungus accumulated DGLA in place of AA because of specific inhibition of Δ5 desaturase by sesamin and related lignan compounds present in the oil. Mutants that were considered to be defective (or to have low activity) in Δ5, Δ6, Δ12, Δ9, and ω3 desaturase were derived from *M. alpina* 1S-4 (Fig. 2) (18,19). Mutant Mut44 and S14 had low Δ5-desaturase activity and accumulated large amounts of DGLA. Mutant Mut48 was completely defective in Δ12 desaturase and produced three *n*-9 PUFAs: 6(*Z*),9(*Z*)-

Table 1. Microbial Polyunsaturated Fatty Acids

Polyenoic acid[a]	Major source	Typical source	Content (%) in total FA
14:2	Green algae	*Chrorella pyrenoidosa*	tr[b]
14:3	Green algae	*Chrorella* sp.	tr
15:2(7,10)	Algae	*Euglena gracilis*	1
15:4(4,7,10,13)	Algae	*Euglena gracilis*	1
16:2(5,9)	Slime molds	*Dictyostelium* sp.	10
16:2(7,10)	Green algae	*Chrorella yulgaria*	3
16:2(9,12)	Diatoms	*Bidulphia sinensis*	28
16:3(4,7,10)	Green algae	*Scenedesmus* sp.	tr[b]
16:3(6,9,12)	Algae (diatoms, etc.)	*Prorocentrum* sp.	12
16:3(7,10,13)	Green algae, *Euglena*, etc.	*Chrorella pyrenoidosa*	8
16:4(4,7,10,13)	Green algae, *Euglena*, etc.	*Chrorella pyrenoidosa*	3
16:4(6,9,12,15)	Diatoms	*Bidulphia sinensis*	9
17:2	Yeasts	*Candida tropicalis*	5
17:2(5,9)	Slime molds	*Dictyostelium* sp.	tr[b]
18:2(5,9)	Slime molds	*Dictyostelium* sp.	tr[b]
18:2(5,11)	Slime molds	*Dictyostelium* sp.	41
18:2(9,12)LA	Many microorganisms	*Agricus bisporus*	86
18:3(6,9,12)GLA	Filamentous fungi, green algae, blue-green algae, red algae, ciliates, etc.	*Mucor genevenis*	32
18:3(9,12,15)ALA	Yeasts, filamentous fungi, basidiomycetes, green algae, brown algae, etc.	*Chrorella* sp.	41
18:4(6,9,12,15)	Green algae, brown algae	*Scenedesmus* sp.	36
19:2(9,12)	Algae	*Euglena gracilis*	tr
19:2(11,14)	Algae	*Euglena gracilis*	tr
19:4(5,8,11,14)	Algae	*Euglena gracilis*	6
20:2(11,14)	Algae (Chrysophyceae, *Euglena*)	*Achanthamoeba* sp.	7
20:3(8,11,14)DGLA	Algae (Chrysophyceae, Xanthophyceae, red algae, *Euglena*, flagellates, etc.)	*Ocheromonas* sp.	5
20:3(11,14,17)	Algae (flagellates, etc.)	*Leishmania* sp.	2
20:4(5,8,11,14)AA	Filamentous fungi, many algae, protozoa	*Conidiobolus* sp.	41
20:4(8,11,14,17)	Algae (Crytomphyceae, brown algae, flagellates, *Euglena*)	*Cryptomonas ovata*	6
20:5(5,8,11,14,17)EPA	Filamentous fungi, marine bacteria, algae	*Monodus* sp.	31
22:4(7,10,13,16)	Algae (Crytomphyceae), ciliates	*Cryptomonas ovata*	2
22:5(4,7,10,13,16)	Algae (Haptophyceae), ciliates	*Critthidia* sp.	14
22:5(7,10,13,16,19)	Algae (Dinophyceae, Chrysophyceae)	*Amiphidinium* sp.	25
22:6(4,7,10,13,16,19)DHA	Algae (Dinophyceae, Haaptomphyceae)	*Exuviella* sp.	16

[a]All fatty acids are *cis*-form. Numbers in parentheses indicate double-bond position.
[b]tr, trace.

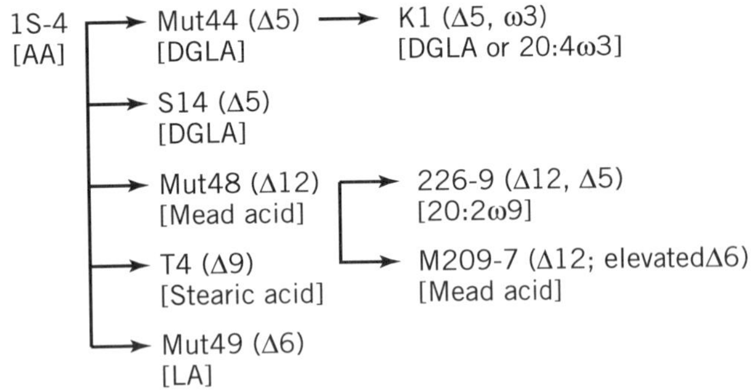

Figure 2. Mutants derived from *M. alpina* 1S-4. Desaturases missing are shown in *parentheses*; major fatty acids produced are shown in *square brackets*.

octadecadienoic acid (18:2ω9), 8(*Z*),11(*Z*)-eicosadienoic acid (20:2ω9), and Mead acid. Two nonmethylene-interrupted PUFAs, that is, 5(*Z*),11(*Z*),14(*Z*)-eicosatrienoic acid (20:3Δ5) and 5(*Z*),11(*Z*),14(*Z*),17(*Z*)-eicosatetraenoic acid (20:4Δ5), and accumulation of linoleic acid (18:2ω6; LA) were found in the Δ6 desaturase-defective mutant Mut 49. Two other mutants, T4 and K1, were considered to have defects in Δ9 and ω3 desaturase, respectively. T4 accumulated a high level of stearic acid (18:0) and K1 did not produce *n*-3 fatty acids, which were usually produced by the parental strain at low temperature. Above all, *M. alpina* 1S-4 and its mutants are new and promising sources of C_{20} PUFAs. In addition, because of intensive investigations done worldwide, all PUFAs listed in Table 2 can now be obtained from a variety of microorganisms.

PRODUCTION OF C_{18} FATTY ACIDS

γ-Linolenic Acid

Suzuki (14,15) reported that several *Mortierella* strains belonging to the subgenus *Micromucor* produce lipids, mainly triacylglycerol rich in GLA, in their mycelia when grown in a liquid medium containing glucose or sugar cane molasses as the major carbon source. These strains grow rapidly in a medium containing a high concentration of glucose (200 g/L) under the conditions of 2 ppm dissolved oxygen, pH 4.0, and 30 °C, yielding a mycelial mass of 40–156 g/L. The lipid content of GLA was 18–41 mg/g dry mycelia, corresponding to a GLA yield of 1.2–3.4 g/L. The purified oil obtained from the mycelia contained myristic acid (0.7%, by weight), palmitic acid (27.2), palmitoleic acid (0.9), stearic acid (5.7), oleic acid (43.9), linoleic acid (12.0), GLA (8.3), arachidic acid (0.6), eicocenoic acid (0.4), behenic acid (0.1), and erucic acid (0.2). The GLA content of the oil was comparable to that of *Oenothera biennis* seed oil. The low linoleic acid content of the oil may be advantageous for obtaining a highly pure GLA preparation with a high recovery. Recently, *Mortierella ramanniana*, *Mucor rouxii*, and *Mucor ambiguus* were evaluated as to their productivities of GLA. *M. ambiguus* immobilized in porous support particles was shown to excrete lipids containing GLA into the culture broth or onto the surface of the cell wall in the presence of a nonionic surfactant.

Linoleic Acid

The mutant Mut49 derived from *M. alpina* 1S-4 was considered to be defective in Δ6 desaturase. This mutant accumulated high amounts of LA with low levels of GLA, DGLA, and AA in its mycelia (20) (Table 3).

Stearic Acid

Mutant T4 derived from *M. alpina* 1S-4 is considered to be defective in Δ9 desaturation (19). Its mycelial fatty acids included 38% stearic acid (18:0), the level being only 5% for the wild type (Table 3). When grown at 24–28 °C, its mycelial lipids included a markedly high level (up to 50 mol% of total mycelial lipids) of free fatty acids, of which about 90 mol% was stearic acid. However, the levels of free fatty acids were markedly decreased with a concomitant increase in the triacylglycerol level when the mutant was grown at 20 °C or lower. When mutant T4 was grown in a culture medium supplemented with LA or linseed oil that included about 60% ALA, decreased levels of free fatty acid and increased levels of triacylglycerol were also found, suggesting that PUFA is needed for the synthesis of triacylglycerol in this mutant.

PRODUCTION OF C_{20} FATTY ACIDS

Arachidonic Acid

Shimizu et al. have assayed the C_{20} PUFA productivity of a wide variety of microorganisms (16,17). Most C_{20} PUFA producers were found to be filamentous fungi belonging to the orders Mucorales and Entomophthorales. Mainly they produce C_{20} PUFAs of the *n*-6 family (i.e., AA and DGLA) together with C_{18} PUFAs of the same family (i.e., GLA). Most of the PUFA-producing isolates from natural sources were found to belong to the genus *Mortierella*. The genus *Mortierella* is subdivided into two subgenera, *Micromucor* and *Mortierella*. Interestingly, all the strains found as C_{20} PUFA producers belong to the subgenus *Mortierella*. Neither the stock cultures nor isolates belonging to the subgenus *Micromucor* showed any detectable accumulation of C_{20} PUFA, although they were good producers of GLA (51). As a result of this screening effort, a soil isolate was obtained that was taxonomically identified as *Mortierella alpina* 1S-4, as a potent producer of AA (16,17). The fungus effectively utilized not only glucose but glycerol, maltose, palmitate, stearate, oleate, *n*-hexadecane, and *n*-octadecane as carbon sources for AA production (Fig. 3).

Using a 10-m^3 fermentor, a mycelial mass of 20 kg/m^3 of medium (dry wt) containing 60% lipid, and 40% of AA in the total fatty acids, was produced by *M. alpina* 1S-4. This fermentation involved intermittent feeding of glucose and 7 days cultivation at 28 °C. The concentration of AA in the harvested mycelia increased to nearly 70% when allowed to stand at room temperature for a further 6 days. The lipids containing AA could be recovered as a triacylglycerol (80–90% recovery) from the mycelia using a seven-step process: *(1)* separation of the mycelia by filtration, *(2)* drying, *(3)* crushing by ball mill, *(4)* extraction of lipids with *n*-hexane, *(5)* removal of insoluble materials by centrifugation, *(6)* decolorization and deodorization with ac-

Table 2. Sources of Polyunsaturated Fatty Acids (PUFAs) and Their Occurrence in Microorganisms

Fatty acid	Representative sources from: Microorganisms	Others
	n-6 series	
18:2ω6	*Mortierella alpina* 1S-4 Mut49 (20)	
GLA	*Mucor circinelloides* (3), *Mo. isabellina* (14), *Mo. ramanniana* var. *anglulispora* (21) *Cunninghamella echinucalta* (22), *Mu. mucedo* (23), *Cunninghamella elegans* (24), *Rhizopus arrhizus* (25), *Rhizopus* spp. (26) *Thamnidium elegans* (24)	Soybean, sunflower, evening primrose, borage, black currant
DGLA	*Mo. alpina* 1S-4 (27), *Mo. alpina* 1S-4 Mut44 (28), *Mo. alpina* 1S-4 S14 (29) *Saprolegnia ferax* (30)	Human milk, animal organs, mosses
AA	*Mo. alpina* 1S-4 and other *Mortierella* subgenus (16,17) *Entomophthora exitalis* (32), *Blastocladiella emersonii* (30) *Conidiobulos nanodes*	Algae (31), porcine liver, moss (33)
DPA(n-6)	*Schyzochytrium* sp. (34)	
	n-3 series	
ALA	Most microorganisms	Linseed, perilla
20:4ω3	*Mo. alpina* 1S-4 S14 (35)	
EPA	*Mortierella* sp. (36), *Pythium* sp. (37), *Shewanella putrefaciens* (38), *Mo. alpina* 1S-4 Mut48 (39) *Phythium irregulare (40)*	Fishes, algae (41)
DHA	*Thraustochytrium aureum* (42) *Schyzochytrium* sp. (34)	Fishes, algae (43)
DPA(n-3)	*Thraustochytrium aureum* (32)	
	n-9 series	
18:2ω9	*Mo. alpina* 1S-4 Mut48 (44)	Animal organs
20:2ω9	*Mo. alpina* 1S-4 Mut48 (44) *Mo. alpina* 1S-4 M226-9 (unpublished observation)	Animal organs
Mead acid	*Mo. alpina* 1S-4 Mut48 (44) *Mo. alpina* 1S-4 M209-7 (45)	Animal organs
	Others	
19:4ω5	*Mortierella* sp. (46)	Mullet
19:5ω2	*Saprolegnia* sp. (47)	Mullet
20:5ω1	*Mo. alpina* 1S-4 (48)	None
20:3Δ5	*Mo. alpina* 1S-4 Mut49 (18)	Plant seeds (49)
20:4Δ5	*Mo. alpina* 1S-4 Mut49 (18)	Plant seeds (50)

AA, arachidonic acid (20:4ω6); ALA, α-linolenic acid (18:3ω3); DGLA, dihomo-γ-linolenic acid (20:3ω6); DHA, 4(Z), 7(Z), 10(Z), 13(Z), 16(Z), 19(Z)-docosahexaenoic acid (22:6ω3); DPA(*n*-3), 7(Z), 10(Z), 13(Z), 16(Z, 19(Z)-docosapentaenoic acid (22:5ω3); DPA(*n*-6), 4(Z), 7(Z), 10(Z), 13(Z), 16(Z)-docosapentaenoic acid (22:5ω6); EPA, 5(Z), 8(Z), 11(Z), 14(Z), 17(Z)-eicosapentaenoic acid (20:5ω3); GLA, γ-linolenic acid (18:3ω6); LA, linoleic acid (18:2ω6); Mead acid, 5(Z), 8(Z), 11(Z)-eicosatrienoic acid (20:3ω9); *Mo., Mortierella; Mu., Mucor.*

tive charcoal, and *(7)* concentration (Fig. 4). If necessary, the AA in the purified oil could be isolated as the methyl or ethyl ester after successive transesterification, liquid–liquid partition chromatography, and high-performance liquid chromatography.

Dihomo-γ-Linolenic Acid

Most AA-producing fungi can accumulate small amounts of DGLA when they are cultivated under the conditions optimized for AA production (52). Because AA is derived from DGLA through the *n*-6 route, all AA-producing fungi potentially should produce large amounts of DGLA if one can block its conversion to AA (Δ5 desaturation).

The first approach to blocking the Δ5 desaturation began with screening for Δ5-desaturase inhibitors. Among various substances tested, sesame oil and peanut oil were found to cause a marked decrease in AA content (27,53) when added to the culture medium. On addition of 3% sesame seed oil to a glucose–yeast extract medium, the production of DGLA with *M. alpina* 1S-4 reached 1.7 g/L, whereas the production of AA was only 0.7 g/L. The effective factors responsible for this phenomenon were isolated from sesame seeds extracts or sesame oil and identified to be lignan compounds, that is, (+)-sesamin, (+)-episesamin, (+)-sesaminol, and (+)-episesaminol (Fig. 5) (54). The results obtained from experiments using either cell-free extracts of *M. alpina* 1S-4 or a rat liver microsomal fraction clearly demonstrated that these lignan compounds specifically inhibited Δ5 desaturase at low concentrations.

Table 3. Comparison of Fatty Acid Compositions of the Fatty Acid Desaturation-Defective Mutants of *Mortierella alpina* 1S-4

	Growth temperature (°C)								
	28						12		
Fatty acid (wt%)	1S-4	Mut48	Mut49	Mut44	S14	T4	Mut49	Mut44	K1
16:0	11.1	9.9	12.7	10.8	16.9	8.2	5.8	10.4	10.7
Stearic acid	5.3	6.1	10.7	7.8	7.5	38.0	10.2	9.8	6.5
18:1ω9	13.7	49.3	15.3	11.1	10.7	5.2	8.7	13.7	21.9
LA	7.4	—[a]	31.8	5.6	2.9	3.9	25.5	4.3	5.5
18:2ω9	—	12.1	—	—	—	—	—	—	—
GLA/20:0[b]	4.6	0.7	1.7	9.6	6.7	11.9	2.3	10.0	11.6
ALA/20:1ω9[c]	1.1	2.7	1.0	1.1	0.6	—	2.7	1.0	1.2
18:4ω3	—	—	—	—	—	—	—	1.3	—
20:2ω6	0.6	—	2.1	1.0	0.5	—	6.1	0.6	0.8
20:2ω9	—	2.1	—	—	—	—	—	—	—
DGLA/22:0[b]	6.5	2.2	2.9	32.8	48.8	1.1	1.7	30.8	32.7
Mead acid	—	9.3	—	—	—	—	—	—	—
20:3Δ5	—	—	7.8	—	—	—	6.4	—	—
AA/20:3ω3[c]	44.4	—	6.8	11.7	0.9	31.7	22.7	7.1	8.1
20:4ω3	—	—	—	—	—	—	—	6.4	—
20:4Δ5	—	—	—	—	—	—	3.2	—	—
24:0	5.2	5.6	7.8	8.5	4.2	—	3.3	1.5	1.0
EPA	—	—	—	—	—	—	1.5	3.1	—

Note: All strains were grown in medium of 2% (w/v) glucose and 1% (w/v) yeast extract (pH 6.0) at either 28 °C for 7 days (28 °C) or at 12 °C for 6 days after preincubation for 1 day at 28 °C (12 °C), except that K1 was grown at 12 °C for 10 days after 1-day cultivation at 28 °C (12 °C).

AA, arachidonic acid (20:4ω6); ALA, α-linolenic acid (18:3ω3); DGLA, dihomo-γ-linolenic acid (20:3ω6); EPA, 5(*Z*), 8(*Z*), 11(*Z*), 14(*Z*), 17(*Z*)-eicosapentaenoic acid (20:5ω3); GLA, γ-linolenic acid (18:3ω6); LA, linoleic acid (18:2ω6); Mead acid, 5(*Z*), 8(*Z*), 11(*Z*)-eicosatrienoic acid (20:3ω9); stearic acid, 18:0.

[a]—, undetectable.

[b]Both were found in all strains except for Mut48, in which GLA and DGLA were not detected. 20:0, arachidic acid; 22:0, behenic acid.

[c]ALA and 11(*Z*), 14(*Z*), 17(*Z*)-eicosatrienoic acid (20:3ω3) were found as an additional fatty acid on growth at 12 °C except for Mut48.

Figure 3. Photomicrographs of *M. alpina* 1S-4 colony grown on potato-dextrose agar (*left*) and oil droplets accumulated in mycelia (*right*).

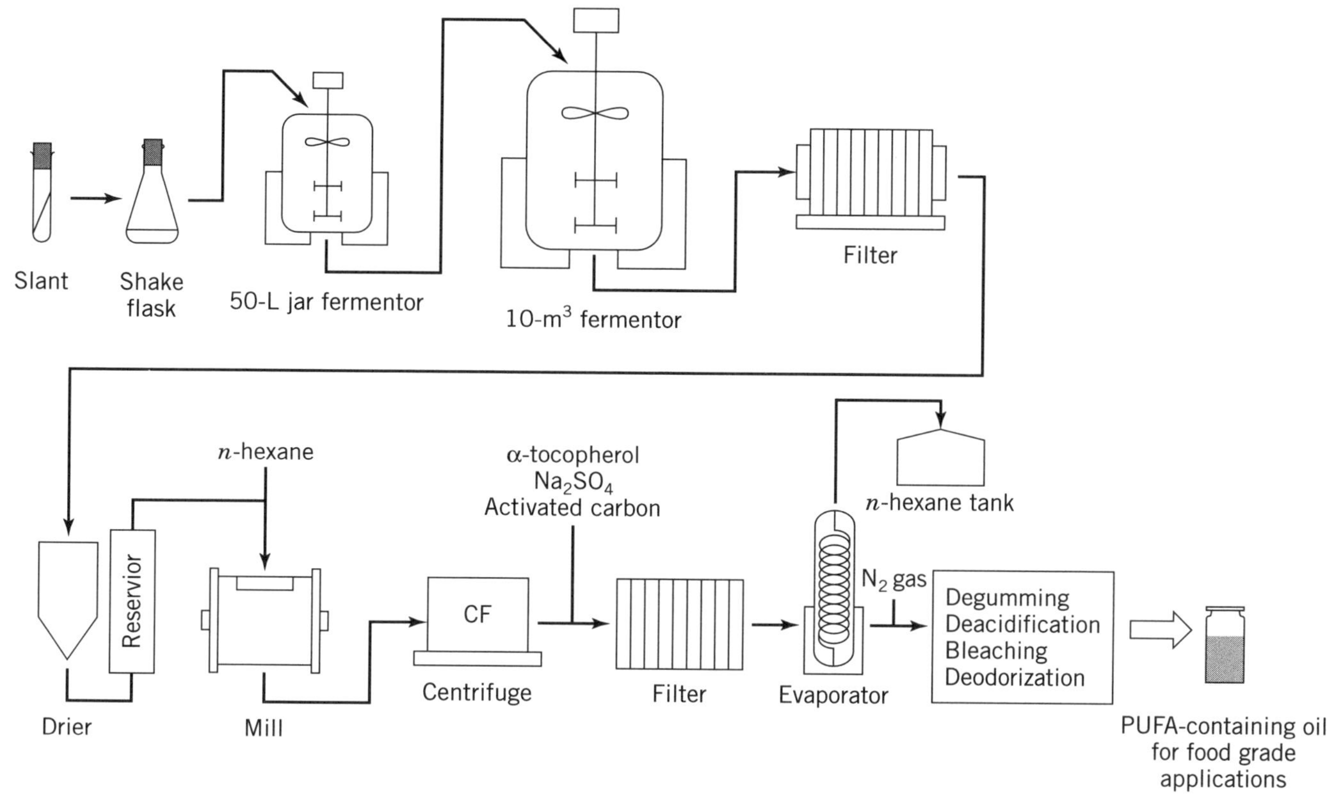

Figure 4. Flowchart for the production of a fungal single cell oil with a high PUFA content.

At these concentrations, none of the Δ6, Δ9, and Δ12 desaturases were inhibited by these lignan compounds. Kinetic analysis showed that sesamin is a noncompetitive inhibitor (K_i = 155 μM for rat liver Δ5 desaturase) (54). (−)-Asarinin and (−)-epiasarinin, the stereoisomers of (+)-episesamin and (+)-sesamin, respectively, which were isolated from a Chinese crude medicine "Saishin" (*Asiasari radix*), also showed specific noncompetitive inhibition of Δ5 desaturase (55). Another desaturase inhibitor, curcumin, was found in turmeric. Curcumin inhibited both Δ5 and Δ6 desaturase in vitro, but it was not as effective as sesamin in stimulation of DGLA production (56,57). Recently, an antioxidant, alkyl gallate (Fig. 5), and the Ca channel blockers nicardipine and nifedipine (Fig. 5) were also shown to inhibit both Δ5 and Δ6 desaturases (58). In a study on optimization of the culture conditions for the production of DGLA by *M. alpina* 1S-4, a medium containing glucose, yeast extract, and the non-oil fraction of sesame oil was found to be suitable. Under optimal conditions in a 50-L fermentor, the fungus produced DGLA at 2.2 g/L (107 mg/g dry mycelia) (Table 4). This value accounted for 23.1% of the total mycelial fatty acids.

The second approach to blocking the Δ5 desaturation began with screening for Δ5-desaturase-defective mutants. Several mutants of this type, which is characterized by a high DGLA level and a reduced AA level, were obtained, and two (Mut44 and S14) of them were studied in detail (28,29). The mycelial fatty acid composition of the two mutants is compared in Table 3. The mutant S14 has a higher DGLA level than Mut44; its AA level is about one-tenth that in Mut44. On cultivation for 6 days at 28 °C in a 10-L fermentor, Mut44 produced 3.2 g DGLA per liter of culture broth (123 mg/g dry mycelia), which accounted for 23.4% of the total mycelial fatty acids (28). The mycelial AA amounted to only 19 mg/g dry mycelia (0.5 g/L culture broth), which accounted for 3.7% of the total mycelial fatty acids (Table 4). The level of AA in Mut44 could be further decreased by growing the fungus in the presence of a mixture of Δ5-desaturase inhibitors, sesamin and episesamin. Using a culture medium supplemented with 0.075 g/L sesamin mixture, the production of DGLA was 2.1 g/L culture broth (16.1% of total fatty acids) and the AA production was 0.2 g/L (1.3%) (59). The DGLA production of S14 was 2.4 g/L culture broth (43.3% of total fatty acids) when grown at 28 °C for 7 days in a 5-L jar fermentor (29). The production of DGLA by using mutants was superior to the aforementioned method of using inhibitors because it did not require inhibitors and the concentration of DGLA in the mycelial oil was higher (Table 4).

By using mutant K1 derived from strain Mut44, which was considered to be defective in the conversion of *n*-6 to *n*-3 fatty acids (ω3 desaturation) (60), a DGLA-rich oil without *n*-3 fatty acids was obtained (Table 3). The mutant could not produce 8(*Z*),11(*Z*),14(*Z*), 17(*Z*)-eicosatetraenoic acid (20:4ω3) or any other *n*-3 fatty acids, of which about 10% was found in its parental strain on cultivation at 12 °C. The growth rate of the mutant was comparable to that of the parental strain when grown at 28 °C, but it became much slower when the mutant was grown at 12 °C, at

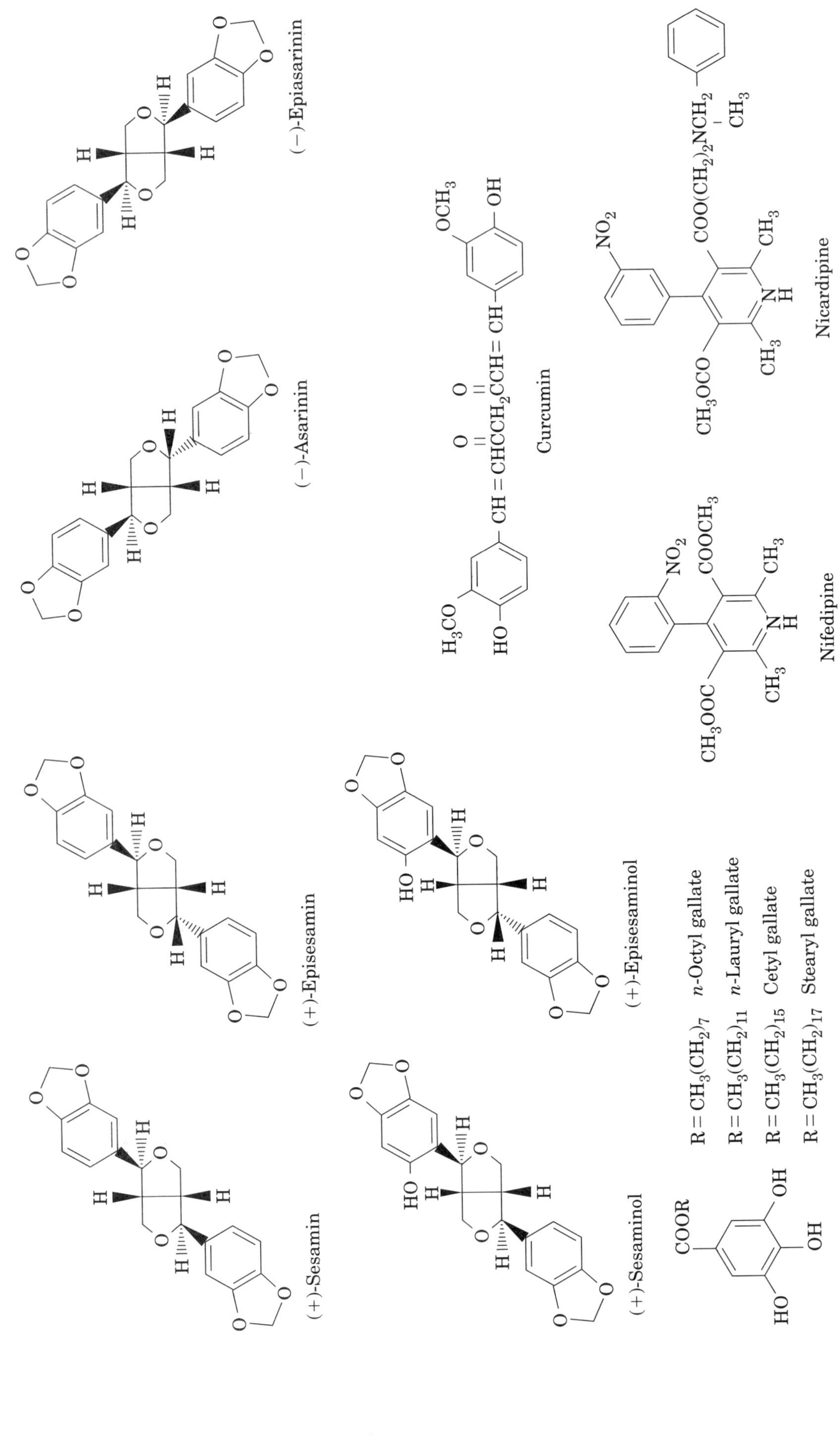

Figure 5. Sesamin and related compounds as desaturase inhibitor.

Table 4. Comparison of DGLA Productivities and Mycelial Fatty Acid Profiles of Some *Mortierella* Strains under Different Culture Conditions

Strain and conditions	Mycelial mass (g/L)	DGLA content (mg/g dry mycelia)	DGLA yield (g/L)	Mycelial fatty acid composition (wt%)			
				LA	GLA	DGLA	AA
		M. alpina 1S-4					
+Sesame oil	20.3	107	2.2	6.6	4.1	23.1	11.2
		Mutant Mut44					
−Inhibitors[a]	26.0	123	3.2	8.9	6.5	23.4	3.7
+inhibitors[a]	28.4	74	2.1	10.0	6.8	16.1	1.3
		Mutant S14					
−inhibitors[a]	17.1	140	2.4	4.4	5.8	43.3	1.5

AA, arachidonic acid (20:4ω6); DGLA, dihomo-γ-linolenic acid (20:3ω6); GLA, γ-linolenic acid (18:3ω6); LA, linoleic acid (18:2ω6).
[a]Inhibitors, a mixture of sesamin and episesamin.

which point the lag phase for Mut44 was about 2 days but 5 days for the mutant.

Eicosapentaenoic Acid

Shimizu et al. (36,52) found that lowering the cultivation temperature caused the additional accumulation of EPA by all the AA producers tested. Most of these AA producers grew well at low temperature (6–16 °C) and produced enough mycelia in simple growth media. Experiments with cell-free extracts of *M. alpina* 1S-4 demonstrated that the enzyme that catalyzed the formation of EPA was produced even when the fungus was grown at high temperature, but that the reaction(s) yielding EPA did not take place at high temperature (52). The Δ17 (or ω3) desaturation of AA to EPA was observed in *Saprolegnia parasitica* by Gellerman and Schlenk (61). If this is also the case for *Mortierella* fungi, an enzyme or enzyme system catalyzing the methyl end-directed desaturation of AA (Δ17 (or ω3) desaturation) may be activated by cold treatment and the resultant EPA may be necessary for maintaining the proper membrane fluidity in a low-temperature environment. As shown in Table 5, *M. alpina* 1S-4 produced EPA at levels of more than 0.3 g/L at 12 °C.

It was further demonstrated that several AA-producing *Mortierella* strains accumulated detectable amounts of EPA in their mycelia when grown in media containing ALA (62). This observation suggests that the conversion of ALA to EPA through the *n*-3 route occurs in *Mortierella* fungi as well as in animals. This conversion is independent of the growth temperature, because EPA production takes place even at 28 °C. The ability of the *Mortierella* fungi to convert added ALA to EPA is very promising from a biotechnological point of view because various kinds of easily available natural oils contain ALA. The potential of such natural oils as precursors of EPA was examined. Linseed oil, in which ALA amounts to about 60% of the total fatty acids, was found to be the most suitable for EPA production. Under optimal culture conditions, *M. alpina* 1S-4 converted 3.8% of the ALA in the added linseed oil into EPA, and EPA production reached 0.99 g/L (41 mg/g dry mycelia). This value is 3.3-fold higher than that obtained under low-temperature growth conditions (Table 5). Another advantage of this conversion is that it can be carried out under normal growth temperature conditions (20–30 °C). Under such conditions, the fungal growth is rapid and dense, and the energy costs should be lower than those for low-temperature cultivation.

EPA production was further stimulated when AA-producing fungi were grown in a medium containing linseed oil at low temperature. This phenomenon was suggested to be mainly caused by the low temperature-dependent production of EPA from AA formed through the *n*-6 route, and the conversion of the 18:3ω3 in the added linseed oil to EPA through the n-3 route took place at the same time (63). Stimulation of the *n*-3 route itself at low temperature was also suggested to contribute to the increased EPA production. The amount of EPA accumulated reached 1.88 g/L (66.6 mg/g dry mycelia) on cultivation of *M. alpina* 1S-4 with 3% linseed oil at 12 °C (Table 5).

By using the Δ12 desaturase-defective mutant Mut48 derived from *M. alpina* 1S-4, an EPA-rich oil with a low level of AA was obtained. Like the wild-type strain, Mut48 converted exogenous ALA to EPA (39). On cultivation at 20 °C for 10 days in a 5-L fermentor containing medium supplemented with linseed oil, the production of EPA was about 1 g/L culture broth (64 mg/g dry mycelia), accounting for approximately 20% of the total mycelial fatty acids. The mycelial AA content was 26 mg/g dry mycelia (0.4 g/L), accounting for 7.8% of the total mycelial fatty acids. As shown in Table 5, the EPA concentrations were lower than AA concentrations in the oils produced by all strains, except for Mut48 in which the concentration of EPA was 2.5-fold that of AA because of the Δ12 desaturase defect.

Recently, *Saprolegnia* sp. 28YTF-1, isolated from a freshwater sample, was shown to be a potent producer of EPA. Accompanying production of AA and other *n*-6 PUFAs was markedly low in this organism (64).

8(*Z*),11(*Z*),14(*Z*),17(*Z*)-Eicosatetraenoic Acid

Δ5 Desaturase-defective mutants of *M. alpina* 1S-4 accumulate 8(*Z*),11(*Z*),14(*Z*),17(*Z*)-eicosatetraenoic acid (20:4ω3) when grown with linseed oil. One of the mutants,

Table 5. Comparison of EPA Productivities and Mycelial Fatty Acid Profiles of Some *Mortierella* Strains under Different Culture Conditions

Strain and conditions	Mycelial mass (g/L)	EPA content (mg/g dry mycelia)	EPA yield (g/L)	Mycelial fatty acid composition (wt%)					
				GLA	ALA	DGLA	AA	EPA	Others
M. alpina 1S-4									
12 °C	11.1	27.0	0.30	4.5	—	2.9	63.8	10.9	17.9
28 °C + linseed oil	24.1	41.0	0.99	1.4	29.3	1.5	13.8	9.2	44.8
12 °C + linseed oil	28.2	66.6	1.88	1.6	28.2	3.0	15.7	12.0	39.5
M. alpina 2O-17									
12 °C	17.0	29.0	0.49	3.5	—	2.3	60.0	13.5	20.7
28 °C + linseed oil	32.4	41.5	1.35	0.9	38.5	1.4	12.3	7.1	39.8
M. hygrophila									
12 °C	15.2	20.4	0.31	3.5	—	1.6	7.2	6.7	81.0
Mutant Mut48									
20 °C + linseed oil	15.1	64.0	0.97	0.3	20.3	0.8	7.8	19.5	51.3

AA, arachidonic acid (20:4ω6); ALA, α-linolenic acid (18:3ω3); DGLA, dihomo-γ-linolenic acid (20:3ω6); EPA, 5(*Z*), 8(*Z*), 11(*Z*), 14(*Z*), 17(*Z*)-eicosapentaenoic acid (20:5ω3); GLA, γ-linolenic acid (18:3ω6).

M. alpina S-14, produced about 1.7 g/L (66 mg/g dry mycelia; 11.6 wt% of total mycelial fatty acids) of this PUFA per liter of culture medium (35).

Mead Acid

A Δ12 desaturase-defective mutant, Mut48, derived from *M. alpina* 1S-4 is characterized by a high oleic acid (18:1ω9) level and the absence of *n*-6 PUFA in its mycelia. In contrast to the wild type, this mutant produces three *n*-9 PUFAs, that is, 6(*Z*),9(*Z*)-octadecadienoic acid (18:2ω9), 8(*Z*),11(*Z*)-eicosadienoic acid (20:2ω9), and Mead acid (44). Significantly high levels of these fatty acids were produced on growth of the mutant at low temperature (12–20 °C). On cultivation at 20 °C for 10 days in a 5-L fermentor, the production of Mead acid reached approximately 0.8 g/L (56 mg/g dry mycelia), accounting for 15 wt% of the total mycelial fatty acids. The other major fatty acids were palmitic acid (6%), stearic acid (11%), oleic acid (45%), 18:2ω9 (12%), and 20:2ω9 (3%). About 70 mol% of Mead acid was present in the triacylglycerol and the remainder in the phospholipid fraction, especially in phosphatidylcholine. The formation of Mead acid could occur through the same sequential reactions suggested by Fulco and Mead (65), that is, desaturation 18:1ω9 at the Δ6 position into 18:2ω9 followed by elongation and Δ5 desaturation to Mead acid (see Fig. 1).

Recently, enhanced production of Mead acid was reported by a mutant fungus, *M. alpina* M209-7, derived from Δ12 desaturase-defective Mut48 strain. The production of Mead acid by M209-7 strain was 1.3 times greater than in Mut48, thought to result from its enhanced Δ6 desaturation activity, which was 1.4 times higher than that of Mut48 (45).

8(*Z*),11(*Z*)-Eicosadienoic Acid

A mutant strain, *M. alpina* M226-9, was derived from the Δ12 desaturase-defective mutant *M. alpina* Mut48. In the M226-9 strain, Δ5 and Δ12 desaturases were completely defective and the microorganism could produce large amounts of 8(*Z*),11(*Z*)-eicosadienoic acid (20:2ω9) (1.68 g/L; 101 mg/g dry mycelia). No detectable Mead acid occurred in the resultant fungal oil. The major fatty acids other than 20:2ω9 (17.6%) were palmitic acid (6.1), stearic acid (7.0), oleic acid (44.0), and 6(*Z*),9(*Z*)-octadecadienoic acid (18:2ω9, 17.6). Addition of olive oil, in which oleic acid is abundantly present, to the growth medium enhanced the production of this unique fatty acid (S. Shimizu, unpublished observation, 1998).

Nonmethylene-Interrupted C_{20} PUFAs

The occurrence in plants of two nonmethylene-interrupted PUFAs, that is, 5(*Z*),11(*Z*),14(*Z*)-eicosatrienoic acid (20:3Δ5) and 5(*Z*),11(*Z*),14(*Z*),17(*Z*)-eicosatetraenoic acid (20:4Δ5), has been reported (40,41). Microorganisms could be a good source of nonmethylene-interrupted PUFAs. These nonmethylene-interrupted PUFAs were found in the Δ6 desaturase-defective mutant Mut49 (18). The amount of 20:3Δ5 was greatest (27 mg/g dry mycelia) on growth at 20 °C, and it accounted for about 7% of the total mycelial fatty acids. 20:4Δ5 was detected only when the mutant was grown at a temperature lower than 24 °C or in a culture medium supplemented with either ALA or 20:3ω3. 20:3Δ5 is believed to be derived from LA by two subsequent reactions without Δ6 desaturation, that is, elongation and Δ5 desaturation (see Fig. 1).

PRODUCTION OF C_{22} FATTY ACIDS

Docosahexanoic Acid

Nakahara et al. (66) have reported that a thraustochytrid isolated from the marine environment grew well in conventional seawater medium and accumulated DHA intracellularly. The microorganism was thought to be a strain belonging to *Thraustochytrium* / *Shizochytrium*, because heterkont zoospores were observed in the early stage of growth in its life cycle. A dry cell weight of 40 g/L, lipid content of 2 g/L, and DHA content of about 35% were obtained in a stirred tank fermentor at 28 °C when grown in a medium that contained 90 g glucose, 10 g polypeptone, and 10 g corn steep liquor in 1 L diluted artificial seawater (one-half concentration of seawater) in 14 days. The lipids extracted were composed mostly of triacylglycerol (53.5% of total lipids). The main fatty acids in the total lipids were 16:0 (44.6%), DHA (34.0%), and 22:5 (8.5%); the EPA content was only 0.4%. The organisms may be an excellent source of DHA.

PRODUCTION OF NOVEL PUFAS

Odd-Numbered PUFAs

The occurrence of straight-chain odd-numbered PUFAs in the lipids of ruminants, fish, and several animals has been reported. Shimizu et al. found that *Mortierella* and several related filamentous fungi convert *n*-alkanes or fatty acids of 15 or 17 carbon atoms to unusual C_{19} PUFAs efficiently (Fig. 6) (46,47,67).

n-1 Fatty Acid

Mortierella alpina 1S-4 accumulates a novel 5(*Z*),8(*Z*),11(*Z*),-14(*Z*),19-eicosapentaenoic acid (20:5ω1) together with additional *n*-1 fatty acids, such as 15-hexadecenoic acid, 17-octadecenoic acid, and 8(*Z*),11(*Z*), 14(*Z*),19-eicosatetraenoic acid (20:4ω1) on growth with 1-hexadecene or 1-octadecene (Fig. 6) (48). Based on the hypothesis that a terminal double bond has no effect on PUFA biosynthesis, it is suggested that the 20:5ω1 is formed through the *n*-6 route. The ω methyl group, which is opposite the double bond end, may be first oxidized to a carboxyl group, and the resultant *n*-1 fatty acid may be introduced to the *n*-6 route.

PRODUCTION OF COCOA-BUTTER-EQUIVALENT YEAST OIL

Cocoa butter, the essential ingredient of chocolate, owes its unique physical properties to the fatty acyl groups on the three available positions of the glycerol moiety, these being palmitoyl, oleoyl, and stearoyl residues, respectively. Thus,

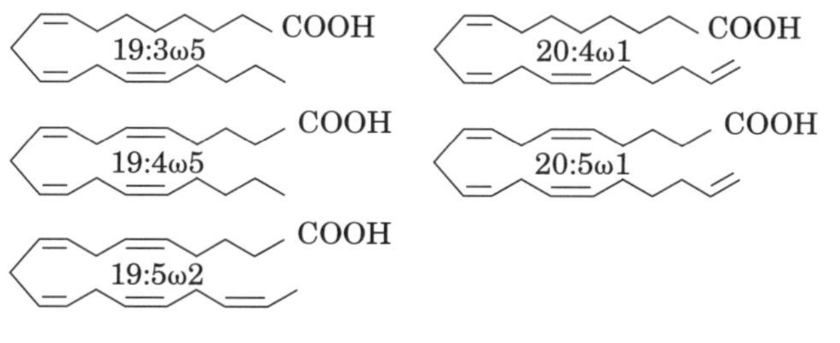

Figure 6. Novel PUFAs produced by *M. alpina* 1S-4 and related filamentous fungi.

an ideal oil substitute, or cocoa butter equivalent, should consist of two-thirds saturated fatty acid (palmitate and stearic acid) and one-third monounsaturated acid (oleic acid; 18:1ω9). When sterculic acid, a known inhibitor of the Δ9 desaturase, was added to growing yeast cultures, stearic acid accumulated to greater than 40% (w/w) of the total fatty acid in the yeast oil. The resulting oil was good for a cocoa butter equivalent (68). Based on this result, a number of oleic acid–requiring yeasts that had a functionally inactive Δ9 desaturase were screened and found to produce considerable amounts of stearic acid good for a cocoa butter equivalent (4). Decreasing the O_2 supply to culture of the yeasts could also increase the stearic acid content of the cell (69). As a result, the large-scale development of a yeast cocoa butter equivalent was established (5).

CLOSING REMARKS

Filamentous fungi of the genus *Mortierella* are good producers of several useful single cell oils, especially those containing PUFAs such as AA, DGLA, and EPA. Furthermore, mutants of these fungi have expanded abilities for producing new oils and PUFAs or increased productivities for existing ones. A soil isolate, *M. alpina* 1S-4, and its mutants are useful not only as producers of some useful fatty acids and PUFAs such as AA, DGLA, and Mead acid, but they also provide useful information on PUFA biosynthesis. The enzymes, that is, desaturases and elongase, that are involved in the PUFA biosynthesis in this fungus seem to have a wide substrate specificity. For example, Δ5 desaturase can act on odd-numbered fatty acids (i.e., 19:3ω5), PUFA with an ω-terminal double bond (i.e., 20:4ω1), *n*-3 PUFA (i.e., 20:4ω3), *n*-6 PUFA (i.e., DGLA), *n*-9 PUFA (i.e., 20:2ω9), and PUFAs with no C-8 double bond (i.e., 11(*Z*),14(*Z*)-eicosadienoic acid and 11(*Z*),14(*Z*), 17(*Z*)-eicosatrienoic acid). However, the first double bond needs to be inserted at the C-9 position. These mutants may also be used as tools for cloning the desaturase genes that may be introduced into other organisms to produce either transgenic plants or animals that have more desirable fatty acid compositions.

Of longer-term interest is the prospect of using microorganisms to produce the end products of PUFA metabolism found in animals, the prostaglandins, leukotrienes, and thromboxanes, which are extraordinarily expensive to produce by chemical synthetic routes. Some microorganisms appear to have the ability to carry out a cascade of reactions, similar to those found in animals, leading to the formation of hydroxy-PUFAs and perhaps even to the prostaglandins (70). This concept would be a logical extension of the work already accomplished in the production of the PUFAs. The ability of microorganisms to catalyze the formation of products not found in plants, or that are difficult to extract from animals, is also promising. Microorganisms may be a source of these materials, and perhaps also some unusual lipids, ranging from the cerebrosides found in yeasts (6) to the ether lipids of archaebacteria, which approximate the platelet-activating factor found in animals (6).

BIBLIOGRAPHY

1. R.S. Moreton, ed., *Single Cell Oil*, Longman, Harlow, U.K., 1988.
2. D.J. Kyle and C. Ratlege, eds., *Industrial Applications of Single Cell Oils*, American Oil Chemists Soc., Illinois, 1992.
3. C. Ratledge, *Microbial Lipids*, vol. 2, Academic Press, London, 1989, pp. 567–668.
4. R.J. Davis, *Lipid Technol.* **4**, 6–13 (1992).
5. R.J. Davis and J.E. Holdworth, *Adv. Appl. Lipid Res.* **1**, 119–159 (1992).
6. C. Ratledgs, *Trends Biotechnol.* **11**, 278–284 (1993).
7. S. Bergstrom, H. Danielsson, and B. Samuelsson, *Biochim. Biophys. Acta* **90**, 207–210 (1964).
8. D.A. Van Dorp, R.K. Beerthuis, D.H. Nugteren, and H. Vonkeman, *Biochim. Biophys. Acta* **90**, 204–207 (1964).
9. B.A. Jakschik, A.R. Sams, H. Sprecher, and P. Needleman, *Prostaglandins* **20**, 401–410 (1980).
10. P.M. Herold and J.K. Kinsella, *Am. J. Clin. Nutr.* **43**, 566–598 (1986).
11. L. Oxdale, *Chem. Br.* **21**, 813 (1990).
12. J. Erwin and K. Bloch, *Science* **143**, 1006–1012 (1964).
13. R. Shaw, *Biochim. Biophys. Acta* **98**, 230–237 (1965).
14. O. Suzuki, *Hakko to Kogyo* **43**, 1024–1031 (1985).
15. O. Suzuki, *Proc. of the World Conf. on Biotechnology for the Fats and Oils Industry, Hamburg*, American Oil Chemists' Soc., 1988, p. 110.
16. H. Yamada, S. Shimizu, and Y. Shinmen, *Agric. Biol. Chem.* **51**, 785–790 (1987).
17. Y. Shinmen, S. Shimizu, K. Akimoto, H. Kawashima, and H. Yamada, *Appl. Microbiol. Biotechnol.* **31**, 11–16 (1989).
18. S. Jareonkitmongkol, S. Shimizu, and H. Yamada, *J. Gen. Microbiol.* **138**, 997–1002 (1992).
19. S. Jareonkitmongkol, E. Sakuradani, H. Kawashima, and S. Shimizu, *Nippon Nogeikagakukaishi* **67**, 531 (1993).
20. S. Jareonkitmongkol, S. Shimizu, and H. Yamada, *Biochim. Biophys. Acta* **1167**, 137–141 (1993).
21. O. Hiruta, K. Yamamura, H. Takebe, T. Futamura, K. Iinuma, and H. Tanaka, *J. Ferment. Bioeng.* **83**, 79–86 (1997).
22. M. Certik, S. Sereke-Berhan, and J. Sajbidor, *Acta Biotechnol.* **13**, 193–196 (1993).
23. M. Certik, P. Andrasi, and J. Sajbidor, *J. Am. Oil Chem. Soc.* **73**, 357–365 (1996).
24. M. Certik, L. Balteszova, and J. Sajbidor, *Lett. Appl. Microbiol.* **25**, 101–105 (1997).
25. L. Kristofikova, M. Rosenberg, A. Vlnova, J. Sajbidor, and M. Certik, *Folia Microbiol.* **36**, 451–455 (1991).
26. M. Certik and J. Sajbidor, *Microbios* **85**, 151–160 (1996).
27. S. Shimizu, K. Akimoto, H. Kawashima, Y. Shinmen, and H. Yamada, *J. Am. Oil Chem. Soc.* **66**, 237–244 (1989).
28. S. Jareonkitmongkol, H. Kawashima, N. Shirasaka, S. Shimizu, and H. Yamada, *Appl. Environ. Microbiol.* **58**, 2196–2200 (1992).
29. S. Jareonkitmongkol, E. Sakuradani, and S. Shimizu, *Appl. Environ. Microbiol.* **59**, 4300–4304 (1993).
30. D.M. Lösel, in C. Ratledge and S.G. Wilkinson eds., *Microbial Lipids*, Academic Press, London, 1988, pp. 699–806.
31. T.J. Ahern, S. Katoh, and E. Sada, *Biotechnol. Bioeng.* **25**, 1057–1070 (1983).
32. A. Kendrick and C. Ratledge, *Appl. Microbiol. Biotechnol.* **37**, 18–22 (1992).
33. Y. Shinmen, K. Katoh, S. Shimizu, S. Jareonkitmongkol, and H. Yamada, *Phytochemistry* **30**, 3255–3260 (1991).
34. T. Yaguchi, S. Tanaka, T. Yokochi, T. Nakahara, and T. Higashihara, *J. Am. Chem. Soc.* **74**, 1431–1434 (1997).
35. H. Kawashima, N. Kamada, E. Sakuradani, S. Jareonkitmongkol, K. Akimoto, and S. Shimizu, *J. Am. Oil Chem. Soc.* **74**, 455–459 (1997).
36. S. Shimizu, H. Kawashima, Y. Shinmen, K. Akimoto, and H. Yamada, *J. Am. Oil Chem. Soc.* **65**, 1455–1459 (1988).
37. S.R. Gandai and J.D. Weete, *J. Gen. Microbiol.* **137**, 1825–1830 (1991).
38. K. Yazawa, K. Watanabe, C. Ishikawa, K. Kondo, and S. Kimura, in D.J. Kyle and C. Ratledge eds., *Industrial Applications of Single Oils*, American Oil Chemists' Soc., Illinois, 1992, pp. 29–51.
39. S. Jareonkitmongkol, S. Shimizu, and H. Yamada, *J. Am. Oil Chem. Soc.* **70**, 119–123 (1993).
40. D.J. O'Brien, M.J. Kurantz, and R. Kwoczak, *Appl. Microbiol. Biotechnol.* **40**, 211–214 (1993).
41. Z. Cohen and Y.M. Heimer, in D.J. Kyle and C. Ratledge eds., *Industrial Applications of Single Oils*, American Oil Chemists' Soc., Illinois, 1992, pp. 243–273.
42. P.K. Bajpai, P. Bajpai, and O.P. Ward, *J. Am. Oil Chem. Soc.* **68**, 509–514 (1991).
43. D.J. Kyle, V.J. Sicotte, and S.E. Reet, in D.J. Kyle and C. Ratledge eds., *Industrial Applications of Single Oils*, American Oil Chemists' Soc., Illinois, 1992, pp. 287–300.
44. S. Jareonkitmongkol, H. Kawashima, S. Shimizu, and H. Yamada, *J. Am. Oil Chem. Soc.* **69**, 939–944 (1992).
45. H. Kawashima, M. Nishihara, Y. Hirano, N. Kamada, and S. Shimizu, *Appl. Environ. Microbiol.* **63**, 1820–1825 (1997).
46. S. Shimizu, H. Kawashima, K. Akimoto, Y. Shinmen, and H. Yamada, *J. Am. Oil Chem. Soc.* **68**, 254–258 (1991).
47. N. Shirasaka, T. Yokochi, and S. Shimizu, *Biosci. Biotechnol. Biochem.* **59**, 1963–1965 (1995).
48. S. Shimizu, S. Jareonkitmongkol, H. Kawashima, K. Akimoto, and H. Yamada, *Arch. Microbiol.* **156**, 163–166 (1991).
49. C.R. Smith, Jr., R. Kleiman, and I.A. Wolff, *Lipids* **3**, 37–42 (1968).
50. J.L. Gellerman and H. Schlenk, *J. Am. Oil Chem. Soc.* **42**, 504–511 (1965).
51. N. Amano, Y. Shinmen, K. Akimoto, H. Kawashima, T. Amachi, S. Shimizu, and H. Yamada, *Mycotaxon* **44**, 257–265 (1992).
52. S. Shimizu, Y. Shinmen, H. Kawashima, K. Akimoto, and H. Yamada, *Biochem. Biophys. Res. Commun.* **150**, 335–341 (1988).
53. S. Shimizu, K. Akimoto, H. Kawashima, Y. Shinmen, S. Jareonkitmongkol, and H. Yamada, *Agric. Biol. Chem.* **53**, 1437–1438 (1989).
54. S. Shimizu, K. Akimoto, Y. Shinmen, H. Kawashima, M. Sugano, and H. Yamada, *Lipids* **26**, 512–516 (1991).
55. S. Shimizu, H. Kawashima, K. Akimoto, Y. Shinmen, H. Sugano, and H. Yamada, *Phytochemistry* **31**, 757–760 (1992).
56. S. Shimizu, S. Jareonkitmongkol, H. Kawashima, K. Akimoto, and H. Yamada, *Lipids* **27**, 509–512 (1992).
57. H. Kawashima, K. Akimoto, S. Jareonkitmongkol, N. Shirasaka, and S. Shimizu, *Biosci. Biotechnol. Biochem.* **60**, 108–110 (1996).
58. H. Kawashima, K. Akimoto, S. Jareonkitmongkol, N. Shirasaka, and S. Shimizu, *Biosci. Biotechnol. Biochem.* **60**, 1672–1676 (1996).

59. S. Jareonkitmongkol, H. Kawashima, and H. Yamada, *J. Ferment. Bioeng.* **76**, 406–407 (1993).
60. S. Jareonkitmongkol, E. Sakuradani, and S. Shimizu, *Arch. Microbiol.* **161**, 316–319 (1994).
61. J.L. Gellerman and H. Schlenk, *Biochim. Biophys. Acta* **573**, 23–30 (1979).
62. S. Shimizu, H. Kawashima, K. Akimoto, Y. Shinmen, and H. Yamada, *J. Am. Oil Chem. Soc.* **66**, 342–347 (1989).
63. S. Shimizu, H. Kawashima, K. Akimoto, Y. Shinmen, and H. Yamada, *Appl. Microbiol. Biotechnol.* **32**, 1–4 (1989).
64. N. Shirasaka and S. Shimizu, *J. Am. Oil. Chem. Soc.* **72**, 1545–1549 (1995).
65. A.J. Fulco and J.F. Mead, *J. Biol. Chem.* **234**, 1411–1416 (1959).
66. T. Nakahara, T. Yokochi, K. Takeda, K. Higashihara, and O. Suzuki, *Abstracts of 21th World Congress and Exhibition of the International Society for Fat Research*, 13 (1995).
67. S. Shimizu, H. Kawashima, K. Akimoto, Y. Shinmen, and H. Yamada, *Proceedings of Session Lectures and Scientific Presentation on ISF-JOCS World Congress 1988*, vol. 2, Japan Oil Chemists' Soc., 1989, pp. 1102–1106.
68. R.S. Moreton, *Appl. Microbiol. Biotechnol.* **33**, 41–45 (1985).
69. R.J. Davies, J.E. Holdworth, and S.L. Reader, *Appl. Microbiol. Biotechnol.* **33**, 569–573 (1990) [Erratum: **43**, 832–833].
70. J.L.F. Kock and C. Ratledge, *J. Gen. Microbiol.* **139**, 459–464 (1993).

OPINE DEHYDROGENASE, SECONDARY AMINE DICARBOXYLIC ACIDS

Yasuo Kato
Yasuhisa Asano
Toyama Prefectural University
Toyama, Japan

KEY WORDS

Amino acid dehydrogenase
Enzymatic synthesis
Enzyme purification and properties
Opine compound
Opine dehydrogenase
Secondary amine dicarboxylic acid

OUTLINE

NATURALLY OCCURRING OPINES

Opines are unusual compounds found in crown gall tumor tissues in plants induced by an infection of *Agrobacterium tumefaciens* (1–6), and in muscle tissue of marine invertebrates (2,3,7–8). They can be categorized as secondary amine dicarboxylic acids (octopine and nopaline families), mannityl opines (agropine), and phosphorylated sugars (agrocinopine). Synthesis of opines in infected plants occurs when the infecting bacterium (i.e., *A. tumefaciens*) transfers a tumor-inciting (Ti) plasmid to the plant, causing tumorigenesis (9). A gene present in this plasmid induces opine synthesis in the plants, while a chromosomal gene encodes for the utilization of the synthesized opine. In the muscle of marine invertebrates, secondary amine dicarboxylic acids are the biochemical equivalent of lactate and represent the terminal products of the glycolytic pathway (10) due to an absence of lactate dehydrogenase (LDH) in their tissues.

SECONDARY AMINE DICARBOXYLIC ACIDS

Of the naturally occurring opines, secondary amine dicarboxylic acids (Fig. 1) are the most extensively studied. These compounds may be regarded as two amino acids linked via a common imino nitrogen, or as secondary amines. One group can be categorized as an octopine family (2,3,11), which has an N-substituted D-alanine moiety, and the other, representing N-substituted D-glutamate derivatives, is the nopaline family (2,3,12). The members of the former family are formed by the reductive condensation reaction between pyruvate and the α-NH_2 group of L-arginine, L-lysine, L-ornithine, L-histidine, and L-methionine, and the corresponding opines are named as octopine (13,14), lysinopine (4,15), octopinic acid (14,16), histopine (6), and methiopine (17), respectively. Octopine is also found in marine invertebrates, together with structurally related alanopine (18), strombine (19), β-alanopine (20), and tauropine (21), which have L-alanine, glycine, β-alanine, and taurine moieties, respectively. Opines of the nopaline family are derived from the condensation between α-ketoglutarate and the α-NH_2 group of L-arginine and L-ornithine, to give the corresponding opines nopaline (22) and nopalinic acid (23), respectively. Saccharopine, N^6-(1,3-L-dicarboxypropyl)-L-lysine, was isolated from brewer's yeast (24) and has the L,L-stereochemistry. In the amino acid pool of *Lactococcus lactis* subsp. lactis (2,3), Thompson et al. reported the existence of N^5-(1-L-carboxyethyl)-L-ornithine and N^5-(1-L-carboxyethyl)-L-lysine, the condensation products between alanine and the ω-NH_2 group of L-ornithine and L-lysine, respectively. Recently, Fushiya et al. (25) isolated four new opines, valinopine,

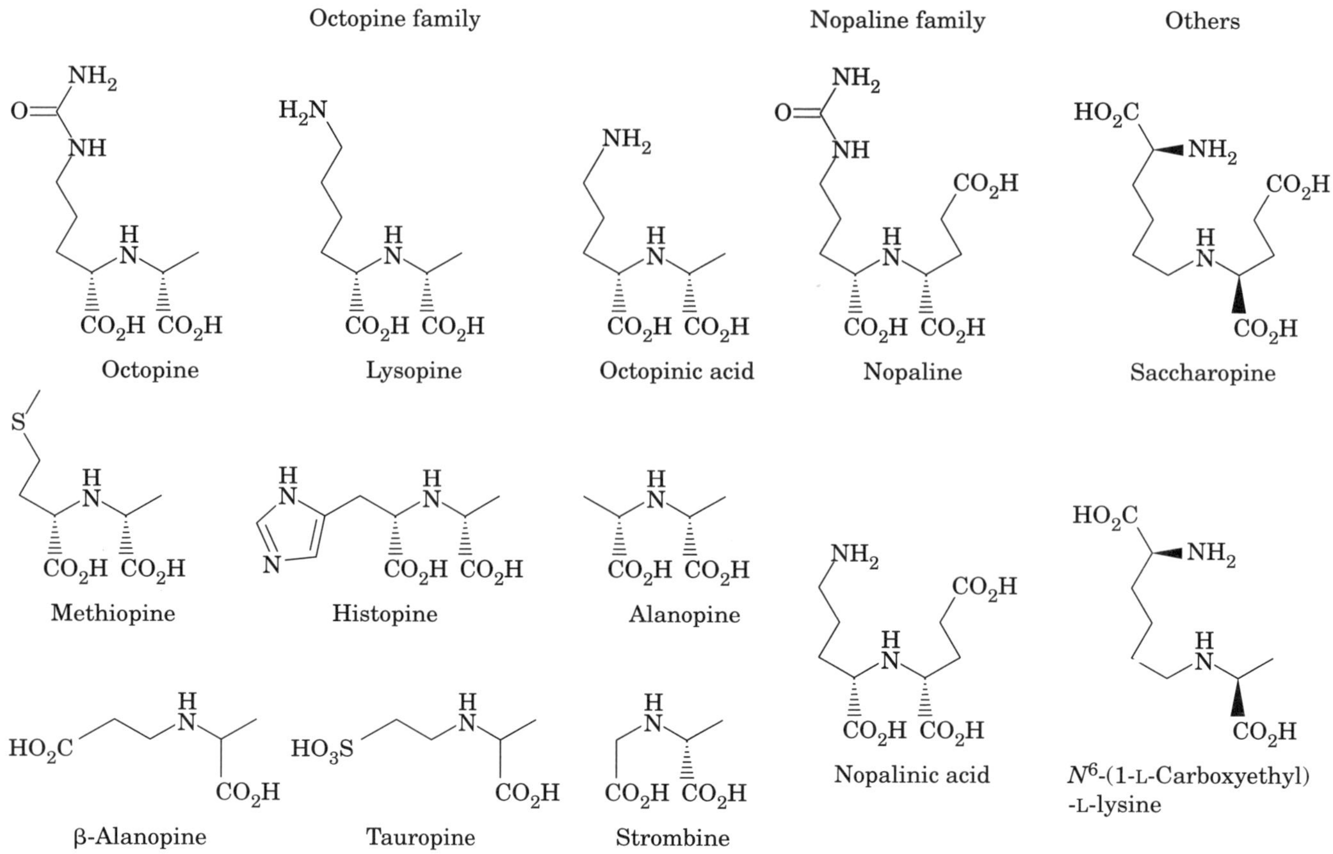

Figure 1. Naturally occurring secondary amine dicarboxylic acids.

epileucinopine, isoleucinopine, and phenylalanopine, which were derived from α-ketoglutarate and L-valine, L-leucine, L-isoleucine, and L-phenylalanine, respectively, from a poisonous mushroom, *Clitocybe acromelalga*.

ENZYMES ACTING ON SECONDARY AMINE DICARBOXYLIC ACIDS

Enzymes that can synthesize secondary amine dicarboxylic acids by the reductive condensation of an α-keto acid with the NH_2-group of an amino acid have been isolated from plants, marine invertebrates, and bacteria (2,3). These enzymes are all NAD(P)H-dependent oxidoreductases and appear to be responsible for the biosynthesis of opines.

Octopine dehydrogenase [N^2-(1-D-carboxyethyl)-L-arginine: $NAD(P)^+$ oxidoreductase, EC 1.5.1.1.1] has been purified from crown gall (26), *Pecten maximus*, and other marine invertebrates (27). The enzyme is a monomeric enzyme and has a molecular weight of 37,000–46,000. The enzyme from crown gall catalyzes the NAD(P)H-dependent condensation reaction between pyruvate and several amino acids such as basic and short-chain neutral amino acids (26), whereas that from *Pecten maximus* exhibits a marked preference for pyruvate, L-arginine, and NADH. The gene encoding the enzyme (*ocs*) has been cloned and sequenced from *A. tumefaciens* (28,29).

Lysopine dehydrogenase [N^2-(1-D-carboxyethyl)-L-lysine: $NADP^+$ oxidoreductase, EC 1.5.1.16] has been purified from crown gall tumor tissues (30). The enzyme catalyzes pyruvate-dependent synthesis of secondary amine dicarboxylic acids of L-lysine, L-arginine, L-ornithine, L-histidine, L-methionine, L-glutamine, and L-citrulline. The enzyme is a monomeric enzyme with a molecular weight of 35,000 and prefers NADP(H) to NAD(H) as the cofactor. This makes it different from the analogous enzyme, octopine dehydrogenase, which utilizes both NAD(H) and NADP(H).

Nopaline dehydrogenase [N^2-(1,3-D-dicarboxypropyl)-L-arginine: $NADP^+$ oxidoreductase, EC 1.5.1.19] catalyzes the α-ketoglutarate-dependent condensation reaction with L-arginine and L-ornithine, producing nopaline and nopalinic acid (ornaline), respectively. The enzyme has been purified from tumor tissue of *Nicotiania tabacum* (31) and sunflower crown gall (32). The enzyme is a tetrameric protein with a subunit molecular weight of 40,000 and requires NADP(H) as a specific cofactor. L-Lysine and L-histidine, good substrates for octopine and lysopine dehydrogenases, do not act as substrates for the enzyme. The nopaline dehydrogenase gene (*nos*) from *A. tumefaciens* has been cloned and sequenced, and the enzyme was expressed in *Escherichia coli* (33,34).

Alanopine dehydrogenase [*meso*-*N*-(1-carboxyethyl)-alanine: NAD^+ oxidoreductase, EC 1.5.1.17] has been purified to homogeneity from adductor muscle of oyster (35) and partially from the lungworm (36) and the foot muscle of the mussel (37). The monomeric enzyme with a molecular weight of 38,500–46,000 requires NADH as a cofactor and catalyzes the reversible condensation reaction between hydrophilic α-keto acids, such as pyruvate, glyoxylate, α-ketoglutarate, and hydroxypyruvate, and amino acids, such as L-alanine, glycine, L-cysteine, L-serine, and L-threonine.

Strombine dehydrogenase [*N*-(1-carboxymethyl)-D-alanine: NAD^+ oxidoreductase, EC 1.5.1.22] has been purified from *Mytilus edulis* (37). The enzyme (M_r = 38,000) catalyzes the NAD(H)-dependent reversible condensation reaction between glycine and pyruvate. Although the enzyme and alanopine dehydrogenase exhibit similar substrate specificities with respect to glycine and alanine, the enzyme prefers glycine as its amino donor.

Tauropine dehydrogenase [*N*-(2-sulfonylethyl)-D-alanine: NAD^+ oxidoreductase, EC 1.5.1.-] has been purified from *Arabella iricolor* (Polychaeta: Errantia) (38). The monomeric enzyme (M_r = 43,500) catalyzes the NAD(H)-dependent reversible condensation reaction between taurine and hypotaurine, and hydrophilic α-keto acids, such as pyruvate, oxalacetate, α-ketobutyrate, hydroxypyruvate, and glyoxylate.

These $NAD(P)^+$-dependent dehydrogenases appear to share a similar reaction mechanism with the $NAD(P)^+$-dependent amino acid dehydrogenases. It is peculiar that the $NAD(P)^+$-dependent secondary amino acid dehydrogenases will form a D-configuration from α-keto acid in the reductive amine-forming reaction because no $NAD(P)^+$-dependent amino acid dehydrogenase forms a D-amino acid from the ketoanalog in the reductive amination reaction, with $NADP^+$-dependent *meso*-α, ε-diaminopimelate D-dehydrogenase as an exception (39).

Saccharopine dehydrogenase, isolated from yeasts and fungi (2,3), catalyze the reversible oxidoreduction reaction of N^6-(1,3-L-dicarboxypropyl)-L-lysine, and the enzyme gene from *Yarrowia lipolytica* has been cloned and sequenced (40). Thompson purified an NADPH-dependent oxidoreductase responsible for the biosynthesis of both N^5-(L-1-carboxyethyl)-L-ornithine and N^6-(L-1-carboxyethyl)-L-lysine from *Lactococcus lactis* subsp. lactis (2,3). This enzyme, N^5-(L-1-carboxyethyl)-L-ornithine: $NADP^+$ oxidoreductase (EC 1.5.1.24), is a tetramer composed of identical subunits (M_r = 35,000) and mediates the reductive condensation reaction between pyruvate and the ω-NH_2 group of L-ornithine and L-lysine, but not an oxidation reaction of the secondary amine dicarboxylic acids. Recently, the gene encoding for the enzyme was cloned and sequenced, and the enzyme was expressed in *E. coli.* (41). Both enzymes are different from the already-mentioned dehydrogenases because opines synthesized by the enzymes arise via alkylation of the ω-NH_2 group of amino acids, and the stereochemistry of the newly formed asymmetric center is of L-configuration.

OPINE DEHYDROGENASE FROM *ARTHROBACTER* SP. STRAIN 1C

Asano et al. have purified phenylalanine dehydrogenases in crystalline forms or to homogeneity and characterized those from *Sporosarcina ureae* SCRC-R04 (42), *Bacillus sphaericus* SCRC-R79a (42), and *B. badius* IAM 11059 (43). The *pdh* genes have been cloned into plasmids and expressed in *E. coli*, and the *pdh* gene has been sequenced (44,45). They have used phenylalanine dehydrogenase and other amino acid dehydrogenases to synthesize optically pure, natural and unnatural amino acids from their corresponding α-keto acids (46,47). Much attention has been paid to a new amino acid dehydrogenase acting on secondary amine dicarboxylic acid. Comparison of the primary structure of the dehydrogenase with that of the known amino acid dehydrogenases would give a valuable information on the structure and function of this class of enzymes. Furthermore, the reversible nature of the enzyme-catalyzed reaction will make it possible to synthesize new chiral opine derivatives, the chemical synthesis of which is rather laborious.

Isolation of (1-D-Carboxyethyl)-L-Phenylalanine-Assimilating Bacteria

Several opine degraders, such as *A. tumefaciens, Pseudomonas* sp., and *Arthrobacter* sp., were isolated from soil and crown gall tumors (48,49). Degradation of opines (i.e., lysopine by *A. tumefaciens*) has been attributed by Jubier (50) to a membrane-bound enzyme complex that is responsible for the oxidation of lysopine, forming lysine and pyruvate.

In an attempt to utilize an NAD^+-dependent dehydrogenase for the stereospecific in vitro enzymatic synthesis of a wide variety of opines, Asano et al. (51,52) attempted to isolate several microorganisms that can grow on chemically synthesized opines. They isolated the bacterial strain YEO8, which utilizes *N*-(1-D-carboxy-2-phenylethyl)-D-phenylalanine as its carbon source from soil and is identified as a glucose-nonfermenting Gram-negative bacteria under CDC group IVc-2 (52). The cell-free extract of the strain contained enzymes that catalyzed the oxidation of the opine into phenylpyruvate and D-phenylalanine, with DCPIP but not NAD(P) as its hydrogen acceptor. They isolated a novel NAD^+-dependent secondary amine dicarboxylic acid dehydrogenase producer, strain 1C, from soil with *N*-(1-D,L-carboxyethyl)-L-phenylalanine as a substrate (51). The taxonomical studies of the strain 1C indicate that it belongs to the genus *Arthrobacter* because it is nonmotile and never fermentative, occurs in rod-coccus shape, its peptidoglycan contains lysine as the diamino acid, and no mycelium formation is observed. *Arthrobacter* strains in the type culture collections did not grow on a medium containing *N*-(1-D-carboxyethyl)-L-phenylalanine as the sole source of carbon.

Purification of Opine Dehydrogenase from Arthrobacter sp. Strain 1C

The enzyme activity found in *Arthrobacter* sp. strain 1C was inducible in the presence of *N*-(1-D,L-carboxyethyl)-L-phenylalanine; only one tenth of total enzyme activity per culture was obtained in the absence of the substrate. The enzyme was purified about 57-fold, with a 36% yield from the cell-free extract of the strain by a procedure involving ammonium sulfate fractionation, and DEAE-Toyopearl, butyl-Toyopearl, and Sephadex G-200 column chromatographies (53). The enzyme has a molecular weight of about 70,000 and consists of two identical subunits with an M_r of 36,000. Typical properties of the enzyme are shown in Table 1. The enzyme catalyzes a reversible oxidation–reduction reaction of opine-type secondary amine dicarboxylic acids. The enzyme is unique because the only similar microbial enzyme thus far reported (2,3) is N^5-(L-1-carboxyethyl)-L-ornithine: $NADP^+$ oxidoreductase from *L. lactis* subsp. lactis, which catalyzes only the reductive secondary amine forming reaction with NADPH, not the oxidative degradation reaction. The velocity of opine dehydrogenase for the secondary amine–forming reaction is about six times greater than that for the oxidative reaction. In the oxidative deamination reaction, the enzyme is active toward synthetic opines such as *N*-(1-D-carboxyethyl)-L-methionine and *N*-(1-D-carboxyethyl)-L-phenylalanine, and its K_m values for them are 9.0 and 14 mM, respectively. On the other hand, the enzyme does not act on naturally occurring opines, such as saccharopine, D-octopine, and L-allooctopine, or stereoisomers of *N*-(1-D-carboxyethyl)-L-phenylalanine, such as *N*-(1-L-carboxyethyl)-L-phenylalanine, *N*-(1-L-carboxyethyl)-D-phenylalanine, and *N*-(1-D-carboxyethyl)-D-phenylalanine. In the reaction that forms secondary amines using pyruvate as a fixed amino acceptor, the enzyme is active toward short-chain aliphatic L-amino acids and those substituted with acyloxy, phosphonooxy, and halogen groups, as shown in Table 2. Hydrophilic α-keto acids, such as pyruvate, oxaloacetate, glyoxylate, and α-ketobutyrate, are accepted as amino acceptors (Table 3). Based on its results of substrate specificity, the enzyme is tentatively named *N*-(1-D-carboxyethyl)-L-norvaline: NAD^+ oxidoreductase (L-norvaline-forming: EC 1.5.1.-, opine dehydrogenase [ODH]). Other substrates for the enzyme are 3-aminobutyric acid and L-phenylalaninol.

The $NADP^+$-dependent D-lysopine dehydrogenase (31,32) and D-nopaline dehydrogenase (30) catalyze a reversible oxidation–reduction reaction similar to the reaction catalyzed by ODH. However, it is a monomeric protein and the substrate and coenzyme specificities are totally different from those of ODH. In the reductive secondary amine–forming reactions with NADPH and NADH (20% the velocity with NADPH) as cofactors, the enzyme does not accept L-phenylalanine, L-alanine, and L-isoleucine, and so on, which are good substrates for ODH.

Table 1. Properties of Opine Dehydrogenase from *Arthrobacter* sp. Strain 1C

Molecular weight	
SDS	36,000
Gene sequencing	37,935
Molecular structure	Dimer
Maximum absorption wavelength	280 nm
Absorption coefficient	$\varepsilon_{280} = 75.87\ mM^{-1}\ cm^{-1}$
Specific activity[a]	140 U/mg
Optimum pH	8.0 (reduction) 10.0 (oxidation)
Optimum temperature	55 °C
pH stability[b]	pH 5.0 ~ 9.5
Temperature stability[c]	~50 °C
K_m for NAD^+	0.76 mM
Inhibitors	Sulfuhydryl reagents Heavy metals

[a]Measured by an oxidation of 10 mM of *N*-(1-D-carboxyethyl)-L-phenylalanine at pH 9.5.
[b]The enzyme activity was measured after incubation at various temperatures and pH 8.0 for 10 min.
[c]The enzyme activity was measured after incubation at 30 °C for 1 h with buffers at various pHs.

Table 2. Substrate Specificity of Opine Dehydrogenase in the Reductive Secondary Amine–Forming Reaction with Pyruvate as a Fixed Substrate

Amino donor[a]	Relative activity (%)[b]	K_m (mM^{-1})
L-Norvaline	100	2.17
L-α-Aminobutyric acid	83.7	20.0
L-Norleucine	72.5	3.72
β-Chloro-L-alanine	53.1	3.19
O-Acetyl-L-serine	35.8	5.94
L-Methionine	23.9	4.10
L-Isoleucine	22.3	6.20
L-Valine	21.9	3.00
L-Phenylalanine	21.7	8.70
L-Leucine	21.2	2.90
L-Alanine	16.0	5.10
O-Phospho-L-serine	15.8	8.69
D,L-α,γ-Diaminopropionic acid	9.67	11.9
L-Phenylglycine	4.40	28.3
D,L-β-Aminobutyric acid	0.683	7.73
O-Phospho-L-threonine	0.640	9.01
L-Phenylalaninol	0.631	46.0

Note: Assayed with fixed concentrations of pyruvate and NADH at 10 and 0.1 mM, respectively.
[a]The following compounds were inert as substrates: D-amino acids, γ-amino-β-hydroxybutyric acid, γ-aminobutyric acid, ε-aminohexanoic acid, L-phenylalanine methylester, L-phenylglycine methylester, L-norvaline methylester, L-alanine methylester, L-phenylalanine amide, L-phenylglycine amide, L-alanine amide, L-isoleucinol, L-valinol, L-alaninol, L-serinol, 1-amino-2,3-propanediol, 1-amino-2-propanol, (2*S*,3*S*)-2-amino-1-phenyl-1,3-propanediol, 3-amino-4-phenyl-4-butanol, ethanolamine, 2-amino-2-methyl-1,3-propanediol, methylamine, ethylamine, isopropylamine, butylamine, benzylamine, 3-aminopentane, 2-phenylethylamine, methoxyamine, benzyloxyamime dimethylamine, hydrazine, phenylhydrazine, and hydroxylamine.
[b]The enzyme activity for L-norvaline was taken as 100%.

Table 3. Substrate Specificity of Opine Dehydrogenase in the Reductive Secondary Amine–Forming Reaction with L-Methionine as a Fixed Substrate

Keto acid[a]	Relative activity (%)[b]	K_m (mM^{-1})
Pyruvate	100	3.0
Oxaloacetate	30	3.5
Glyoxylate	11	8.8
α-Ketobutyrate	11	28
Hydroxypyruvate	6.1	N.D.[c]
α-Ketocaproate	2.1	N.D.[c]

Note: Assayed with fixed concentrations of L-methionine and NADH at 10 and 0.1 mM, respectively.
[a]α-Keto-γ-phenylbutyrate, glyoxal, α-ketoglutarate, phenylpyruvate, and acetone were inert as a substrate.
[b]The enzyme activity for pyruvate was taken as 100%.
[c]Not determined.

Cloning, Nucleotide Sequencing, and Expression of Opine Dehydrogenase

The gene coding for ODH was cloned into plasmid p-Bluescript KS(−), and the nucleotide sequence of the 1,077-bp open reading frame consisting of 359 codons was identified as the *odh* gene coding for the enzyme (54). About 6.6-fold total activity per liter of culture was detected with a feeding of 1 mM IPTG, as compared with the wild-type strain *Arthrobacter* sp. 1C (51). The enzyme was purified and crystallized from the cell-free extracts of the recombinant strain (Fig. 2). The chemical and physical properties of the purified enzyme showed a complete agreement with those of the wild type.

The enzyme (359 amino acids) is similar to the sequence of the 40-kDa protein (27.4% identity among 359 amino acids) of the Ti plasmid of *A. tumefaciens* (55), D-lysopine dehydrogenase (24.6% identity among 358 amino acids) (28) and D-nopaline dehydrogenase (20.8% identity among 168 amino acids) (34) of *A. tumefaciens*, D-nopaline dehydrogenase (21.8% identity among 371 amino acids) of *A. vitis* (56), and phenylalanine dehydrogenases of *B. sphaericus* (21.8% identity among 78 amino acids) (44) and of *B. badius* (25.6% identity among 60 amino acids) (45). There is no apparent homology between ODH and saccharopine dehydrogenase (21) or N^5-(L-1-carboxyethyl)-L-ornithine: $NADP^+$ oxidoreductase (41). The common similarities between the 40-kDa protein, D-lysopine dehydrogenase, and the N-terminal region of ODH and the phenylalanine dehydrogenase (PheDH) are in the glycine-rich nucleotide-binding domain G-X-G-X-X-(G or A), connecting the β-strand with the α-helix in the region of adenine dinucleotide phosphate (ADP)–binding $\beta\alpha\beta$ folds, which is strongly conserved among $NAD(P)^+$-dependent dehydrogenases and FAD-containing oxidoreductases (57).

It can be suggested that ^{34}D of ODH corresponds to ^{214}D of the NAD^+-dependent *B. sphaericus* phenylalanine dehydrogenase, the position of which is considered to be important for the NADH binding discriminating NADPH among $NAD(P)^+$-dependent dehydrogenases (57). The equivalent positions with ^{34}D of the enzyme in the 40-kDa protein and D-lysopine dehydrogenase (which utilizes both $NADP^+$ and NAD^+) are neutral ^{30}S and ^{32}A, respectively, which are found in $NADP^+$-dependent dehydrogenases. There is also a similarity among these enzymes in the conserved ^{238}D-X-X-^{241}R residue, which is thought to be important for the proton-relay mechanism (3). The nucleotide-binding domain, as can be seen in typical lactate dehydrogenases, is located at the N-terminal region of these secondary amine dicarboxylic acid dehydrogenases, whereas other amino acid dehydrogenases have one more domain in the N-terminus important for accommodating the amino acid substrate (58). Octopine and nopaline dehydrogenases show 26% amino acid identity (3), but no similarity was seen with saccharopine dehydrogenase and N^5-(L-1-carboxyethyl)-L-ornithine: $NADP^+$ oxidoreductase. The former enzymes synthesize opines having D,L-configuration and are linked via the α-NH_2 group of amino acids, but the latter ones produce opines with L,L-configuration and are linked via the α- and ε-NH_2 group of amino acids. These differences in substrate specificities reflect the amino acid sequence similarity between these enzymes.

Although N^5-(L-1-carboxyethyl)-L-ornithine: $NADP^+$ oxidoreductase (41) has a similarity with lactate and alanine dehydrogenases (59) at pyruvate-binding sites, including active arginine (which is a catalytically relevant residue), ODH might have a different binding site with pyruvate because the enzyme had no similarity with these enzymes.

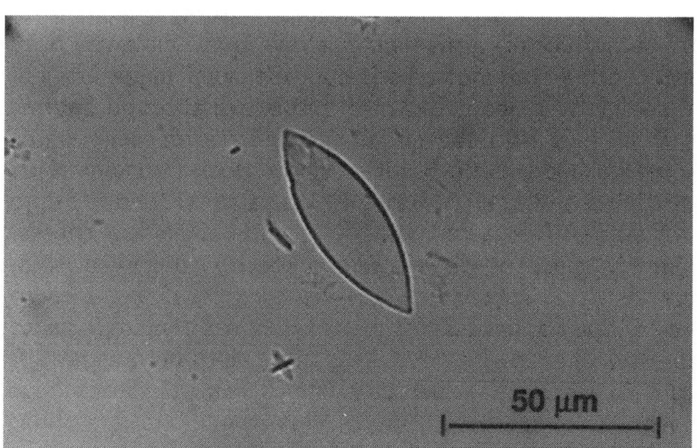

Figure 2. Photomicrograph of crystalline opine dehydrogenase.

Stereoselective Synthesis of Opine-Type Secondary Amine Carboxylic Acids by Opine Dehydrogenase

Opine-type secondary amine dicarboxylic acids are useful chiral intermediates of angiotensin-converting enzyme (ACE) inhibitors such as enalapril and lysinopril (60). They have been chemically synthesized by reductive condensation reactions of α-keto acids or their esters and amino acid derivatives (61,62) or SN_2 reactions of optically active 2-halo- and 2-trifluoromethanesulfonyloxy esters with amino acid derivatives (63,64). However, these methods generally require protection of the functional groups, and the stereoselectivity is not always high. Kato et al. (53) constructed a new system for the synthesis of secondary amine dicarboxylic acids by ODH coupled with NADH regeneration by formate dehydrogenase (FDH) from *Moraxella* sp. C-1 (65), without protection of the substrates (Fig. 3).

As shown in Table 4, several secondary amine dicarboxylic acids have been synthesized from amino acids and their analogues, such as L-methionine, L-isoleucine, L-leucine, L-valine, L-phenylalanine, L-alanine, L-threonine, and L-serine, and α-keto acids, such as glyoxylate, pyruvate, and 2-ketobutyrate, using ODH. A diastereo and enantiomerically pure secondary amine carboxylic acid from (*S*)-phenylalaninol and pyruvate was also quantitatively synthesized. The absolute configuration of the nascent asymmetric center of the opines was of the D-stereochemistry with $>$ 99.9% ee. This method seems to be the simplest for stereoselectively synthesizing opines without protection of the substrates, because chemical processes generally require five to eight steps.

On the other hand, one-pot synthesis of *N*-[1-D-(carboxyl)ethyl]-L-phenylalanine from phenylpyruvate

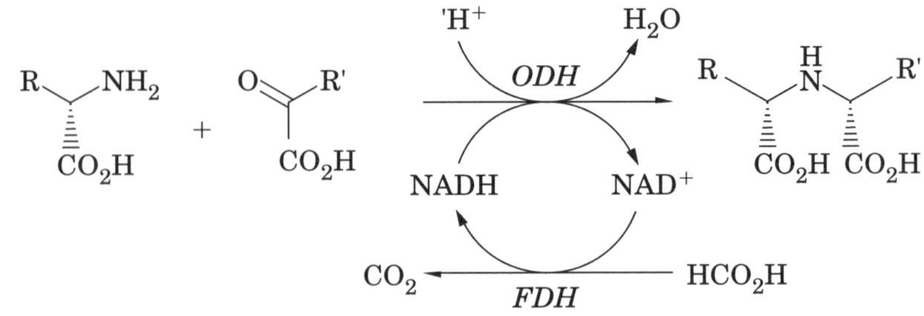

Figure 3. Synthesis of secondary amine dicarboxylic acids from L-amino acids and α-keto acids by opine dehydrogenase with regeneration of NADH by formate dehydrogenase. ODH, opine dehydrogenase from *Arthrobacter* sp. 1C; FDH, formate dehydrogenase from *Moraxella* sp. C-1.

Table 4. Synthesis of Opine-Type Secondary Amine Dicarboxylic Acids from l-Amino Acids and α-Keto Acids Using Opine Dehydrogenase and Formate Dehydrogenase

	α-Keto Acid						
	Glyoxylic acid	Pyruvic acid			2-Ketobutyric acid		
			Purity			Purity	
Amino acid	Yield[a] (%)	Yield[a] (%)	(% d.e.)[b]	(% e.e.)[a]	Yield[a] (%)	(% d.e.)[b]	(% e.e.)[a]
L-Methionine	–	96	>99.9	>99.9	–	–	–
L-Isoleucine	>99	>99	>99.9	>99.9	97	>99.9	>99.9
L-Leucine	>99	>99	>99.9	>99.9	–	–	–
L-Valine	>99	>99	>99.9	>99.9	>99	>99.9	>99.9
L-Phenylalanine	–	95	>99.9	>99.9	–	–	–
L-Alanine	–	98[c]	>99.9	>99.9	–	–	–
L-Threonine	–	>99	>99.9	>99.9	–	–	–
L-Serine	–	97	>99.9	>99.9	–	–	–

[a]Determined by HPLC.
[b]Determined by 400 MHz ^{1}H- and ^{13}C-NMR.
[c]The obtained compound was in the *meso*-form.

and pyruvate by using ODH, FDH, and PheDH from *B. sphaericus* (43,46) is also described, as shown in Figure 4. This reaction can be carried out well because of the substrate specificities of the enzymes thus used; ODH acts on pyruvate but not on phenylpyruvate (51), whereas PheDH acts on phenylpyruvate but not on pyruvate (43). The opine compound was obtained in a yield of 87.5% with high (>99.9%) enantio- and diastereomeric excess.

CONCLUSION

Both secondary amine dicarboxylic acid and amino acid dehydrogenases use NAD(P)H, α-keto acids, and amino donors as substrates and have a common consensus motif in their primary structures. The structural and catalytic similarities of these enzymes suggest that these genes have evolved from a common progenitor. On the other hand, differences have been found in the substrate specificities: *(1)* secondary amine dicarboxylic acid dehydrogenases utilize the α- or ε-amino group of amino acids, but amino acid dehydrogenases use ammonia as amino donors; and *(2)* the asymmetric center newly formed by secondary amine dicarboxylic acid dehydrogenases is of D-stereochemistry, except saccharopine dehydrogenase and N^5-(L-1-carboxyethyl)-L-ornithine: NADP$^+$ oxidoreductase, but that by formed amino acid dehydrogenases is exclusively of L-

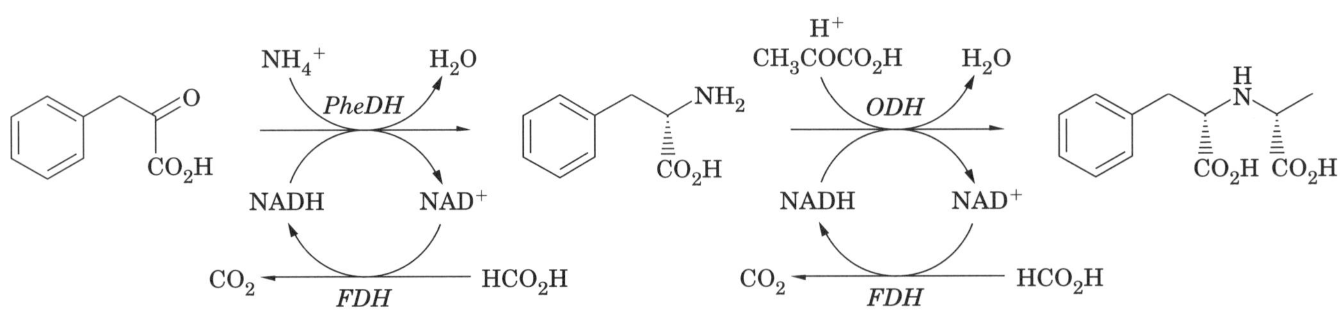

Figure 4. One-pot synthesis of *N*-(1-D-carboxyethyl)-L-phenylalanine from phenylpyruvate and pyruvate by phenylalanine and opine dehydrogenases with NADH regeneration by formate dehydrogenase. PheDH, phenylalanine dehydrogenase from *Bacillus sphaericus* R79a; ODH, opine dehydrogenase from *Arthrobacter* sp. 1C; FDH, formate dehydrogenase from *Moraxella* sp. C-1.

stereochemistry. Among secondary amine dicarboxylic acid dehydrogenases, some differences found in the substrate specificity include octopine, lysopine, and nopaline dehydrogenases characteristically promoting reductive alkylation at the α-NH_2 of basic amino acids; ODH, alanopine, and strombine dehydrogenases acting on that of neutral amino acids; and saccharopine dehydrogenase and N^5-(L-1-carboxyethyl)-L-ornithine: $NADP^+$ oxidoreductase accommodating the ε-NH_2 of basic amino acids. Octopine, lysopine, alanopine, and strombine dehydrogenases, N^5-(L-1-carboxyethyl)-L-ornithine: $NADP^+$ oxidoreductase, and ODH accept pyruvate as an amino acceptor, whereas nopaline and saccharopine dehydrogenases act on α-ketoglutarate. A comparison of the structure of ODH with other amino acid dehydrogenases would provide insight to the catalytic, structural, and evolutionary relationships among them. To determine the high-dimensional structural elucidation of ODH, X-ray crystallographic studies are in progress (66,67) and should provide details of the structure of the catalytic domains of ODH. The findings might help in understanding the mechanism and evolutionary relationship among the amino acid dehydrogenases.

BIBLIOGRAPHY

1. E.W. Nester, M.P. Gordon, R.M. Amasino, and M.F. Yanofsky. *Annu. Rev. Plant Physiol.* **35**, 387–413 (1984).
2. J. Thompson and S.P.F. Miller, *Adv. Enzymol.* **64**, 317–399 (1991).
3. J. Thompson and J.A. Donkarsloot, *Annu. Rev. Biochem.* **61**, 517–557 (1992).
4. W.S. Chilton, J. Tempe, M. Matzke, and M.D. Chilton, *J. Bacteriol.* **158**, 650–658 (1984).
5. E. Szegedi, M. Czako, L. Otten, and C.S. Koncz, *Physiol. Mol. Plant Pathol.* **32**, 237–248 (1988).
6. C.C. Chang, R.K. Japaswal, C.M. Chen, and S.B. Gelvin, *J. Bacteriol.* **171**, 5922–5927 (1989).
7. Y. Dessaux, A. Petit, and J. Tempe, in D.P.S. Verma ed., *Molecular Signals in Plant–Microbe Communications*, CRC Press, Boca Raton, Fla., 1992, pp. 109–136.
8. W.S. Chilton, A.M. Stomp, V. Beringue, H. Bouzar, and V. Vaudequin-Dransart, *Phytochemistry* **40**, 619–628 (1995).
9. A. Petit and J. Tempe, in L. van Floten-Doting, G.P.S. Groot, and T.C. Hall eds., *Molecular Form and Function of the Plant Genome*, Plenum Press, New York, 1984, pp. 625–636.
10. J. Baldwin and A.M. Opie, *Comp. Biochem. Physiol. B* **61**, 85–92 (1978).
11. E. Hack and J.D. Kemp, *Biochem. Biophys. Res. Commun.* **78**, 785–791 (1977).
12. S. Hatanaka, S. Atsumi, K. Furukawa, and Y. Ishida. *Phytochemistry* **21**, 225–227 (1982).
13. N. Izumiya, R. Wade, M. Winitz, M.C. Otey, S.M. Birnbaum, R.J. Koegel, and J.P. Greenstein, *J. Am. Chem. Soc.* **79**, 652–658 (1957).
14. A. Menage and G. Morel, *Coll. R. Acad. Sci.* **259**, 4795–4796 (1964).
15. C. Lioret, *Bull. Soc. Fr. Physiol. Veg.* **2**, 76 (1956).
16. K. Goto, M. Waki, N. Mitsuyasu, Y. Kitajima, and N. Izumiya, *Bull. Chem. Soc. Jpn.* **55**, 261–265 (1982).
17. J.L. Firmin, I.M. Stewart, and K.E. Wilson, *Biochem. J.* **232**, 431–434 (1985).
18. M. Sato, Y. Sato, and Y. Tsuchiya, *Nippon Suisan Gakkaishi*, **43**, 1077–1079 (1977).
19. A.W. Sangster, S.E. Thomas, and N.L. Tingling, *Tetrahedron*, **31**, 1135–1137 (1975).
20. M. Sato, M. Takahara, N. Kanno, Y. Sato, and W.R. Ellington, *Comp. Biochem. Physiol. B*, **88**, 803–806 (1987).
21. M. Kuriyama, *Nature* **192**, 969 (1961).
22. A. Goldmann, D.W. Thomas, and G. Morel, *Coll. R. Acad. Sci.* **268**, 852–854 (1969).
23. J.L. Firmin and R.G. Fenwick, *Phytochemistry* **16**, 761–762 (1977).
24. A. Kjaer and P.O. Larsen, *Acta Chem. Scand.* **15**, 750–759 (1961).
25. S. Fushiya, M. Matsuda, S. Yamada, and S. Nozoe, *Tetrahedron* **52**, 877–886 (1996).
26. E. Hack and J.D. Kemp, *Plant Physiol.* **65**, 949–955 (1980).
27. K.B. Storey and P.R. Dando, *Comp. Biochem. Physiol.* **73B**, 521–528 (1982).
28. H. De Greve, P. Dhaese, J. Seurinck, M. Lemmers, M. van Montagu, and J. Schell, *J. Mol. Appl. Genet.* **1**, 499–511 (1983), NBRF-PIR database accession number A00399.
29. R.F. Barker, K.B. Idler, D.V. Thompson, and J.D. Kemp, *Plant. Mol. Biol.* **2**, 335–350 (1983), NBRF-PIR database accession number S28683.
30. L.A. Otten, D. Vreugdenhil, and R.A. Schilperoort, *Biochim. Biophys. Acta* **485**, 268–277 (1977).
31. A. Goldmann, *Plant Sci. Lett.* **10**, 49–58 (1977).
32. D.W. Sutton, J.D. Kemp, and E. Hack, *Plant Physiol.* **62**, 363–367 (1978).
33. Y. Gafni and M.-D. Chilton, *Gene* **39**, 141–146 (1985).
34. C. Drevet, A.C.M. Brasileired, L. Jouanin, *Plant. Mol. Biol.* **25**, 83–89 (1994), NBRF-PIR database accession number S46512.
35. J.H.A. Fields and P.W. Hochachka, *Eur. J. Biochem.* **114**, 615–621 (1981).
36. B. Siegmund, M. Griesharker, M. Reitze, and E. Zebe, *Comp. Biochem. Physiol.* **82**, 337–345 (1985).
37. P.R. Dando, *Biochem. Soc. Trans.* **9**, 297–298 (1981).
38. N. Kanno, M. Sato, E. Nagashita, and Y. Sato, *Comp. Biochem. Physiol.* **114B**, 409–416 (1996).
39. H. Misono and K. Soda, *J. Biol. Chem.* **255**, 10599–10605 (1980).
40. J.-W. Xuan, P. Fournier, N. Declerk, M. Chasles, and C. Gaillardin, *Mol. Cell. Biol.* **10**, 4795–4806 (1990), NBRF-PIR database accession number A36467.
41. J.A. Donkersloot and J. Thompson, *J. Biol. Chem.* **21**, 12226–12234 (1995), NBRF-PIR database accession number A57499.
42. Y. Asano, A. Nakazawa, and K. Endo, *J. Biol. Chem.* **262**, 10346–10354 (1987).
43. Y. Asano, A. Nakazawa, K. Endo, Y. Hibino, M. Ohmori, N. Numao, and K. Kondo, *Eur. J. Biochem.* **168**, 153–159 (1987).
44. N. Okazaki, Y. Hibino, Y. Asano, M. Ohmori, N. Numao, and K. Kondo, *Gene* **63**, 337–341 (1988), Swiss-Plot database accession number P23307.
45. A. Yamada, T. Dairi, Y. Ohno, X.-L. Huang, and Y. Asano, *Biosci. Biotechnol. Biochem.* **59**, 1994–1995 (1995), NBRF-PIR database accession number JC4328.
46. Y. Asano, A. Yamada, Y. Kato, K. Yamaguchi, Y. Hibino, K. Hirai, and K. Kondo, *J. Org. Chem.* **55**, 5567–5571 (1990).

47. Y. Kato, K. Fukumoto, and Y. Asano, *Appl. Microbiol. Biotechnol.* **39**, 301–304 (1993).
48. G. Tremblay, R. Gagliardo, W.S. Chilton, and P. Dion, *Appl. Environ. Microbiol.* **53**, 1519–1524 (1987).
49. C.J. Beauchamp, J.W. Kloepper, R. Lifshitz, P. Dion, and H. Antoun, *Can. J. Microbiol.* **37**, 158–164 (1990).
50. M.F. Jubier, *FEBS Lett.* **28**, 129–132 (1972).
51. Y. Asano, K. Yamaguchi, and K. Kondo, *J. Bacteriol.* **171**, 4466–4471 (1989).
52. B.R.B.-D. Guzman, Y. Asano, and H. Yamada, *Annu. Rep. Int. Center Coop. Res. Biotechnol. Jpn.* **14**, 167–178 (1991)
53. Y. Kato, H. Yamada, and Y. Asano, *J. Mol. Catal. B: Enzymatic*, **1**, 151–160 (1996).
54. T. Dairi and Y. Asano, *Appl. Environ. Microbiol.* **61**, 3196–3171 (1995).
55. A. Schrell and J. Schröder. *Biochim. Biophys. Acta* **1174**, 303–304 (1993), NBRF-PIR database accession number S38355, PIR protein sequence database, National Biomedical Research Foundation, Washington, URL: *http://www.psc.edu/general/software/packages/nbrf-pir/nbrf.html*.
56. L. Otten and P. Ruffray, NBRF-PIR database accession number S41895, PIR protein sequence database, National Biomedical Research Foundation, Washington, URL: *http://www.psc.edu/general/software/packages/nbrf-pir/nbrf.html*.
57. K.L. Britton, P.J. Barker, P.C. Engel, D.W. Rice, and T.J. Stillman, *J. Mol. Biol.* **234**, 938–945 (1993).
58. D. Garmyn, T. Ferain, N. Bernard, P. Hols, and J. Delcour, *Appl. Environ. Microbiol.* **61**, 266–272 (1995).
59. A.R. Clarke, T. Atkinson, and J.J. Holbrook, *Trends Biochem. Sci.* **14**, 101–105 (1989).
60. M.T. Wu, A.W. Douglas, D.L. Ondeyka, L.G. Payne, T.L. Ikeler, H. Joshua, and A.A. Patchett, *J. Pharm. Sci.* **74**, 352–354 (1985).
61. T.J. Blacklock, R.F. Shuman, J.W. Butcher, W.E. Shearin, Jr., J. Budavari, and V.J. Grenda, *J. Org. Chem.* **53**, 836–844 (1988).
62. G. Iwasaki, R. Kimura, N. Numao, and K. Kondo, *Chem. Lett.* 1691–1694 (1988).
63. G. Iwasaki, R. Kimura, N. Numao, and K. Kondo, *Chem. Pharm. Bull.* **37**, 280–283 (1989).
64. F. Effenberger and U. Burkard, *Liebigs Ann. Chem.* 334–358 (1986).
65. Y. Asano, T. Sekigawa, M. Inukai, and A. Nakazawa, *J. Bacteriol.* **170**, 3189–3193 (1988).
66. K.L. Britton, H.F. Rogers, Y. Asano, T. Dairi, Y. Kato, T.J. Stillman, and D.W. Rice, *Acta Cryst.* **D54**, 124–126 (1998)
67. K.L. Britton, Y. Asano, and D.W. Rice, *Nat. Struc. Biol.* **5**, 593–601 (1998).

OPTICAL RESOLUTION, BIOCATALYSIS

TETSUO OMATA
SHUJI SENDA
YASUO NINOMIYA
Nitto Denko Company
Osaka, Japan

KEY WORDS

Asymmetric microbial degradation
Backer's yeast
Enzymatic hydrolysis
Enzymatic transesterification
E-value
Immobilized microorganisms
Lipase-catalyzing esterification
Microbial oxidation
Prelog's rule
ρ-Ketoesters

OUTLINE

INTRODUCTION

Enzyme proteins are produced by living organisms as biochemical catalysts. A typical feature of biocatalysts is the ability to catalyze stereo- and regiospecific reactions under mild conditions (e.g., neutral pH, normal temperature, and pressure).

Increasingly, enzymes are used in organic synthesis, especially for optical resolution of racemic organic compounds. This article provides an overview of some of the newer optical resolution reactions that use enzymes and microbial cells.

OPTICAL RESOLUTION WITH LIPASE-CATALYZED TRANSESTERIFICATION

Transesterifications catalyzed by lipases in organic solvent are most widely applied for optical resolution of racemic alcohols. Various kinds of racemic alcohols have been successfully resolved by lipase-catalyzed enantioselective transesterification reactions (1,2). Ward reported several examples of optical resolution accompanying the epimerization or racemization of starting materials at high conversion rates (3). Ward further introduced the enzyme-catalyzed acylating reaction of a racemic cyanohydrin under racemization of the residual enantiomer by a basic anion with extremely high conversion rates (Scheme 1).

A method of kinetic resolution of *cis*-1-diethylphosphomethyl-2-hydroxymethyl cyclohexane **1** by lipase has been developed (Scheme 2 and Table 1). The transesterification of (±)-**1** with vinyl acetate in the presence of lipase AK without solvent resulted in (+)-**1** with 99% ee and 41% yield and the corresponding acetate with 93% ee and 35% yield (the E value as reported by Sih was 152) (4).

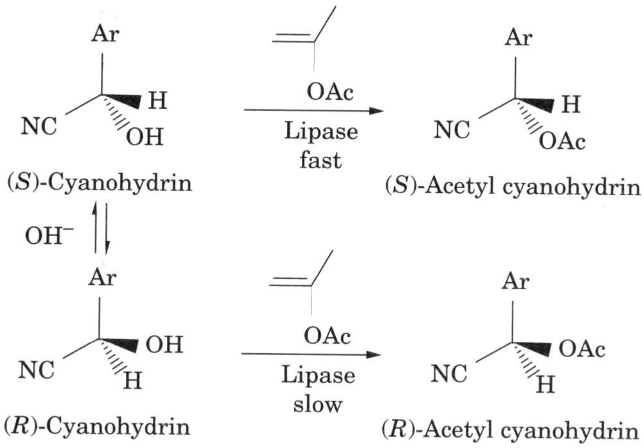

Scheme 1. Lipase-catalyzed asymmetric acylation of racemic cyanohydrin.

Enantioselective synthesis of both enantiomers of 1-arylethanols with a condensed aromatic ring has been achieved through acetylation of the racemic alcohols with vinyl acetate in the presence of lipase from *Pseudomonas aeruginosa* (Toyobo, LIP). The lipase catalyzed the reaction of (±)-(3-phenanthrenyl)ethanol with vinyl acetate at high enantioselectivity (99% ee and conversion; 49% of (±)-substrate). The remaining alcohol had a high optical purity (94% ee). The E value was 100 as determined by Sih's method (5).

Yamada et al. (6) reported an expedient route to optically pure four-carbon diols and triols using a *meso*-symmetric four-carbon precursor and lipase-mediated kinetic resolution as a key step in the transesterification reaction with vinyl acetate in an organic solvent (Scheme 3).

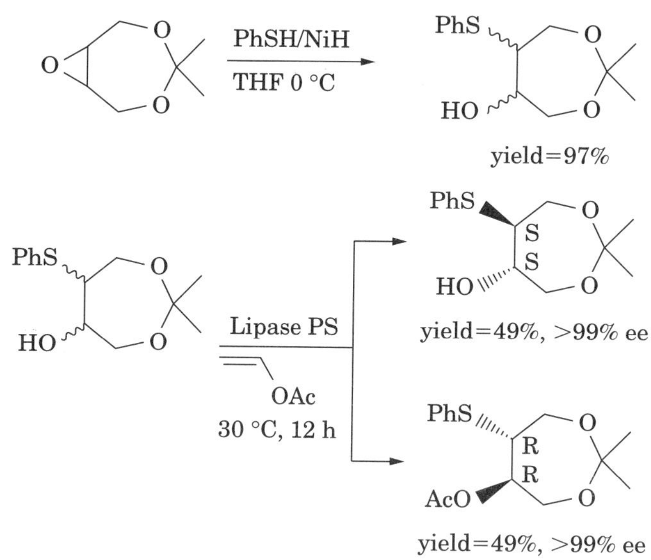

Scheme 3. *meso*-Mediated preparation of optically pure four-carbon triols with transesterification by lipase PS.

Tanaka et al. (7) reported a facilitated route to optically pure 3-*endo*-hydroxy dicyclopentadiene via lipase-mediated kinetic resolution in a transesterification reaction with vinyl acetate in an organic solvent (Scheme 4).

Chromobacterium viscosum lipase (CVL) catalyzes selectively the acylation of the C-5 hydroxyl of the three stereoisomeric vitamin D A-ring precursors **1a**, **2a**, and **2b** (Fig. 1). It stereoselectively acylated the hydroxyl group at the carbon of the 3-position of the fourth stereoisomer, **1b**,

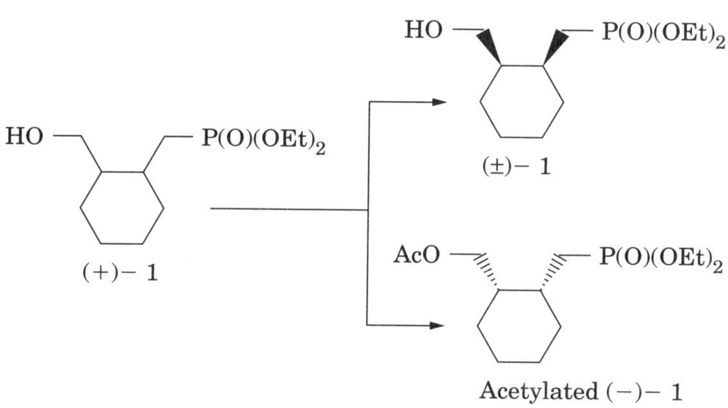

Scheme 2. Transesterification of phosphono alcohol 1 catalyzed lipases.

Table 1. Transesterification of (+)-1 Catalyzed Lipases

Lipase mg/100 mg of 1	Condition solv.	Condition time (h)	Alcohol (+)1 Yield (%)	Alcohol (+)1 ee (%)	Acetyl-(−)1 Yield (%)	Acetyl-(−)1 ee (%)	E value
PS (200)	THF	15	38	60	55	36	3.7
PS (100)	THF	15	60	62	35	97	62
PS (100)	*t*-BuOMe	12	38	14	25	14	1.3
PS (50)	None	52	42	98	54	64	20
AK (50)	None	6.5	41	>99	35	93	152

Note: All reactions were carried out at 37 °C. PS = *Pseudomonas cepacia*, Amano; AK = *Pseudomonas fluoresence*, Amano.

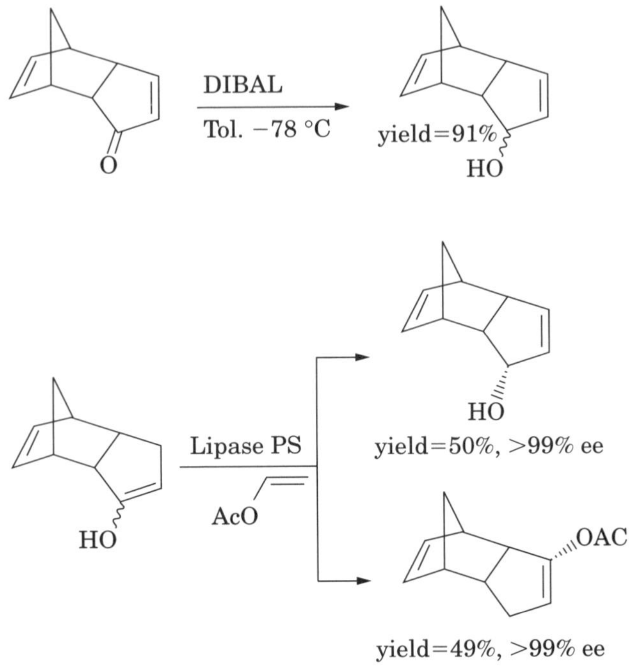

Scheme 4. Preparation route to optically pure 3-*endo*-hydroxy-dicyclopentadiene.

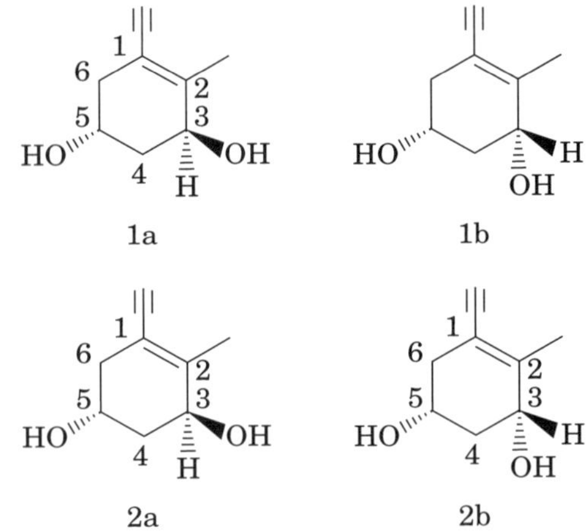

Figure 1. Structure of precursors of vitamin D A-ring.

with vinyl acetate under the same conditions in organic solvent (95% isolated yield, 90% ee) (8).

Izumi et al. (9) reported that *trans*-2,3-dihydro-3-hydroxy-2-phenyl-4H-1-benzopyran-4-one (*trans*-flavanon-3-ol) was resolved to acetate of (2*S*,3*S*)-(−)-*trans*-flavanon-3-ol with 96% ee and (2*R*,3*R*)-(−)-*trans*-flavanon-3-ol by an enzymatic transesterification with vinyl acetate in the presence of *P. capacia* lipase (Scheme 5).

Lundth et al. (10) proposed that enzyme-catalyzed transesterification of racemic alcohol with vinyl acetate by lipase from *P. fluorescens* (PFL) was a reversible process, although it is commonly referred to as irreversible. When racemic *β*-methyl-(2-thiophene)propanol **1** was resolved via a transesterification reaction catalyzed by lipase from PFL using an excess vinyl acetate in chloroform, the enantioselectivity was a high E = 200 (conversion, 39%), as calculated from enantioexcess (ee) of (*S*)-ester, (*S*)-**1Ac**. However, based on the ee of the remaining (*R*)-alcohol (*R*)-**1**, assuming irreversibility, and when measured at a conversion rate of 57%, the value E = 37 was obtained. Similarly, when the 5-(1-ethoxyethoxy)-3-pentyn-2-ol **2** was resolved using immobilized lipase B from *Candida antarctica* (Novozym SP435) under similar conditions, the (*R*)-ester product (*R*)-**2Ac** was obtained at 22% conversion with E = 140, compared with 65% conversion and E = 15 for the remaining alcohol (*S*)-**2**. These results were interpreted as the consequence of a reversible process occurring in transesterifications of this type, which were referred as irreversible (Schemes 6 and 7).

In the total synthesis of *endo*-brevicomin **1**, a pheromone of bark beetles such as *Dendrocutonus frontalis* and *Dryocoetes aoutograghus*, Kim et al. (11) examined *anti*-trityloxy-3,4-diol **2** as a substrate with lipase PS (LPS) from *P. capacia* in transesterification. They found that it was optically resolved by LPS with high enantioselectivity, thus allowing preparation of each enantiomer in optically pure form (>98% ee). The synthesis of optically pure (+)-

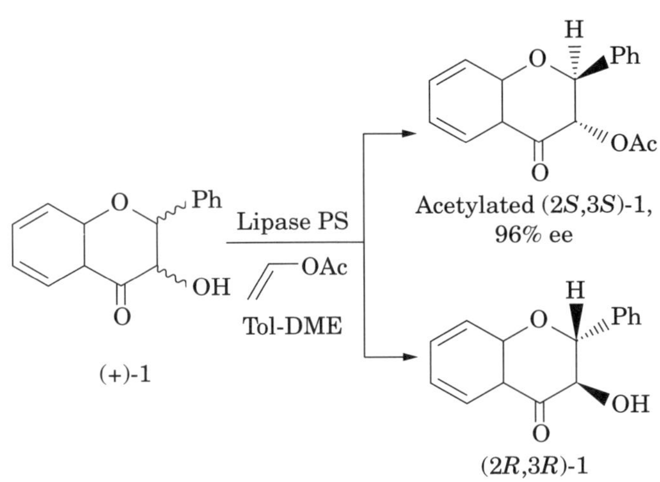

Scheme 5. Enzymatic resolution of *trans*-flavanon-3-ol by lipase.

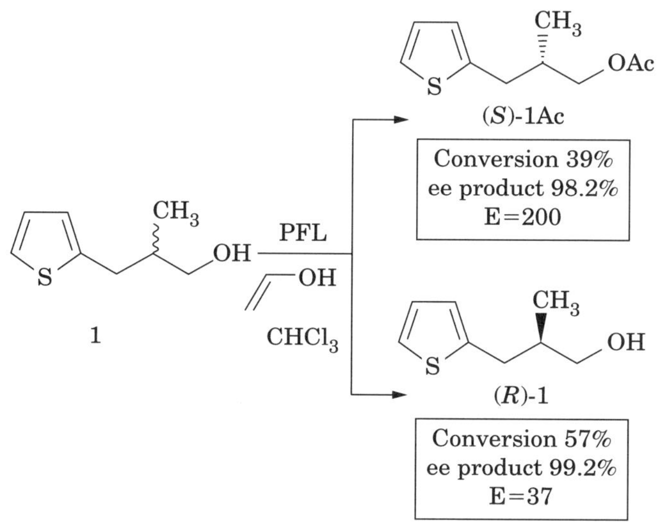

Scheme 6. Transesterification of 2-methyl-3-(2′-thienyl)-1-propanol.

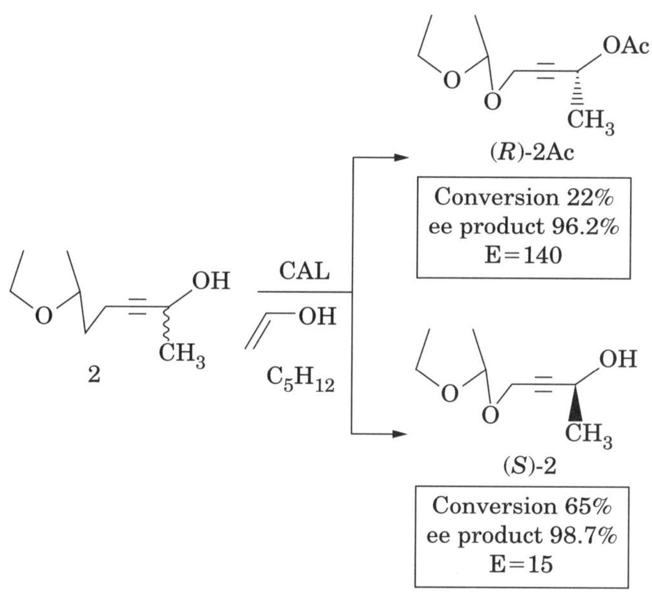

Scheme 7. Transesterification of 5-(1-ethoxyethoxy)-3-pentyn-2-ol by lipase.

endo-brevicomin **1** was achieved from the resolved (−)-2 by transesterification with vinyl acetate using LPS (Scheme 8).

Yamane (12) reported the total synthesis of optically pure (+)-Mitsugashiwa lactone by lipase-catalyzed transesterification under the presence of vinyl acetate in organic solvents (Scheme 9).

OPTICAL RESOLUTION WITH LIPASE-CATALYZING HYDROLYSIS

The stereoselective hydrolysis of 2-(3′-methylphenoxy)propionyl and 2-(2,3-dimethylphenoxy)propionyl esters by carboxy esterase NP was estimated under different experimental conditions to verify whether their optical resolution could be performed using this procedure. The reaction was performed on methyl, ethyl, isopropyl, isobutyl, and benzyl esters of 2-(3-methylphenoxy)propionic and 2-(2,3-dimethylphenoxy)propionic acids in aqueous media. The best substrate for the biocatalytic resolution was the *S* enantiomer for the monomethyl-substituted compounds and the *R* enantiomer for the dimethyl derivative. The extent of 2-(3-methylphenoxy) propionic ester hydrolysis was greater than the 2-(2,3-dimethyl phenoxy) propionic esters. The hydrolysis of isobutyl and benzyl derivatives suffered from poor enantioselectivity and long reaction time; this may be attributable to steric hindrance (caused by the acidic moiety) of the substrates' interaction with the active sites of the enzymes. Changes in pH, temperature, and substrate/enzyme ratio were investigated.

Optimum conditions for these biocatalytic resolutions were obtained when methyl 2-(3-methylphenoxy)-propionate and isopropyl 2-(2,3-dimethyl phenoxy)propionate were hydrolyzed at pH 7.2, 23 °C, and substrate/enzyme ratio 50. The scaling-up to preparative amounts of the enzymatic hydrolysis of these compounds succeeded in producing satisfactory quantities of the corresponding acids and residual esters with a good-enough optical yield (13). An enzymatic method for preparation of the enantiomers of chiral diols **1** and their monoacetate **2** with 100% enantiomeric purity was established using lipase-catalyzed esterification and/or hydrolysis.

The (*R*)-enantiomers were more susceptible in both acetylation of (±)-**1** and hydrolysis of (±)-**2** catalyzed by lipase OF from *Candida cylindracea* (14) (Scheme 10).

Tan et al. (15) envisioned that thioesters of certain carboxylic acids having the chiral center at the α-carbon might be susceptible to sufficiently acidic deprotonation, resulting in racemization under conditions of enzymatic resolution. Although thioesters have been used as substrates in enzymatic resolution, their α-proton acidity has apparently never been exploited for resolution coupled with substrate racemization. To demonstrate that thioester could be used for racemization coupled with enzymatic resolution, ethyl thioester of α-(phenylthio)propionate **1** was used (Scheme 11).

The corresponding (*R*)-acid **2** was subjected to diastereoselective oxidation for conversion to sulfoxide in the synthesis of a chiral vinyl sulfoxide. Because of the insolubility in water of **1** and most other substrates of interest, it was decided to use a lipase as catalyst in a biphasic system (water and water-immiscible organic solvent) while inducing racemization of the substrate in the organic phase.

A novel enzymatic process for the optical resolution of racemic pantolactone through the stereospecific hydrolysis of D-pantolactone by lactonohydrolase of *Fusarium oxysporum* was reported (16). *F. oxysporum* cells were found to catalyze the stereoselective hydrolysis of the D-enantiomer of racemic pantolactone. With 135g/L of DL-pantolactone as the substrate, 41% of the substrate was hydrolyzed to produce pantoic acid with optical purity of 90% enantiomer excess (for D-pantoic acid) (Scheme 12).

Bommarius et al. (17) reported the synthesis of enantiometrically pure *tert*-leucine, which was prepared using

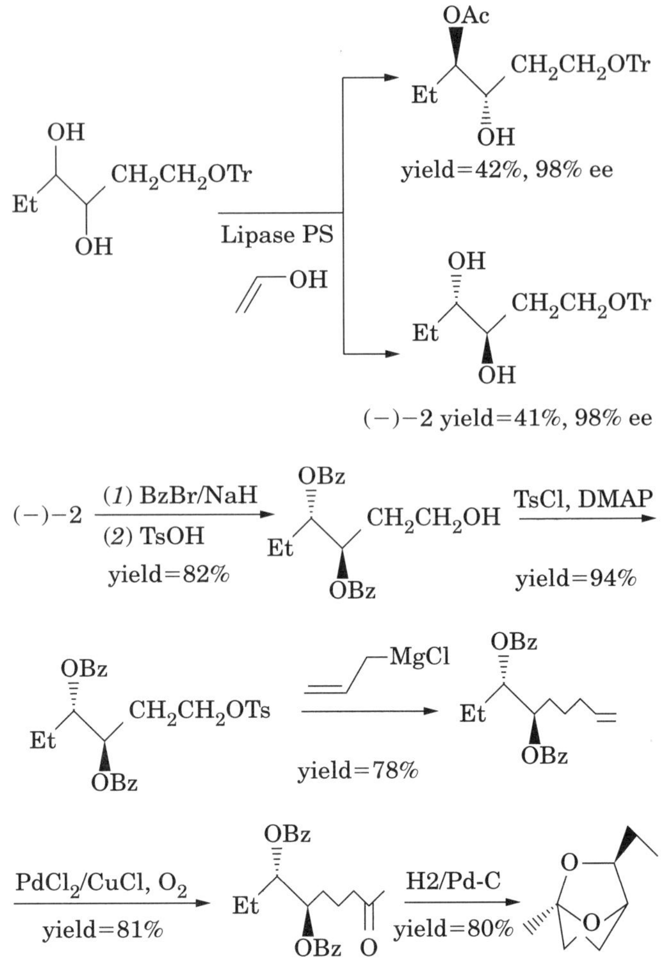

Scheme 8. Total synthesis of (+)-*endo*-brevicomin via transesterification reaction by lipase PS.

various methods using penicillin-G acylase, hydantoinase, and dehydrogenase.

Owing to its bulky, inflexible, and hydrophobic *tert*-butyl side chain, the use of *tert*-leucine has increased in templates and catalysts in asymmetric synthesis as well as in medicinal peptide compounds. (*S*)-*tert*-Leucine, available through a large-scale enzymatic reductive amination process, has been incorporated into a variety of anti-AIDS and anticancer compounds.

Optically pure inden-1-ol **1** has potential utility as a key starting material for the preparation of HIV-1 protease inhibitors such as L-735,524 and L-754,394. Synthesis of optically active inden-1-ol **1**, however, has been little studied, and only one enantiotopical preparation has been reported so far. Takahashi et al. (18) first examined the kinetic acetylation of racemic inden-1-ol, (±)-**1** with vinyl acetate in organic solvent in the presence of a lipase. However, no practically satisfactory conditions could be established, although partial optical resolution was observed using lipase PS-on-Celite (*Pseudomonas* sp., Amano) (22% ee). They next examined the kinetic deacetylation of racemic 1-acetoxyinden (±)-**2** in a phosphate buffer-acetone solution in the presence of lipases. Of the enzymes tested, lipase PS (*Pseudomonas* sp., Amano) gave the best result. From 10 mg/mmol of the racemic acetate (±)-**2**, 94% ee of (*R*)-inden-1-ol (*R*)-**1** was obtained in 46% yield with recovery of optically pure (*S*)-1-acetoxyinden (*S*)-**2** in 45% yield after 48 h (Scheme 13).

Ibuprofen is a nonsteroidal anti-inflammatory drug with annual sales of about $100 million (U.S.) in 1994. Sales are expected to reach $600 million by the year 2000. The physiological performance of the *S* form of ibuprofen is 28-fold higher than that of the *R* form. The price of the *S* form is three times that of its racemic mixture. Therefore, pharmaceutical companies have devoted much effort to developing production of the *S* form. Ibuprofen can be optically resolved by chemical and enzymatic methods and asymmetric synthesis. Many enzymatic resolutions are reported with lipase-catalyzed hydrolysis or transesterifications. Tai et al. (19) reported that enzymatic hydrolysis of ibuprofen esters produced 99% optically pure (*S*)-(+)-ibuprofen by immobilized lipase MY entrapped in chitosan-*N*-(3-dimethyl aminopropyl)-*N*′-ethylcarbodiimide-glutaraldehyde.

Sulfoxides are valuable chiral auxiliaries as precursors because they carry out carbon–carbon bond formation. As a route to enantiomerically pure sulfoxides, Serreqi et al. (20) examined hydrolase-catalyzed kinetic resolution by

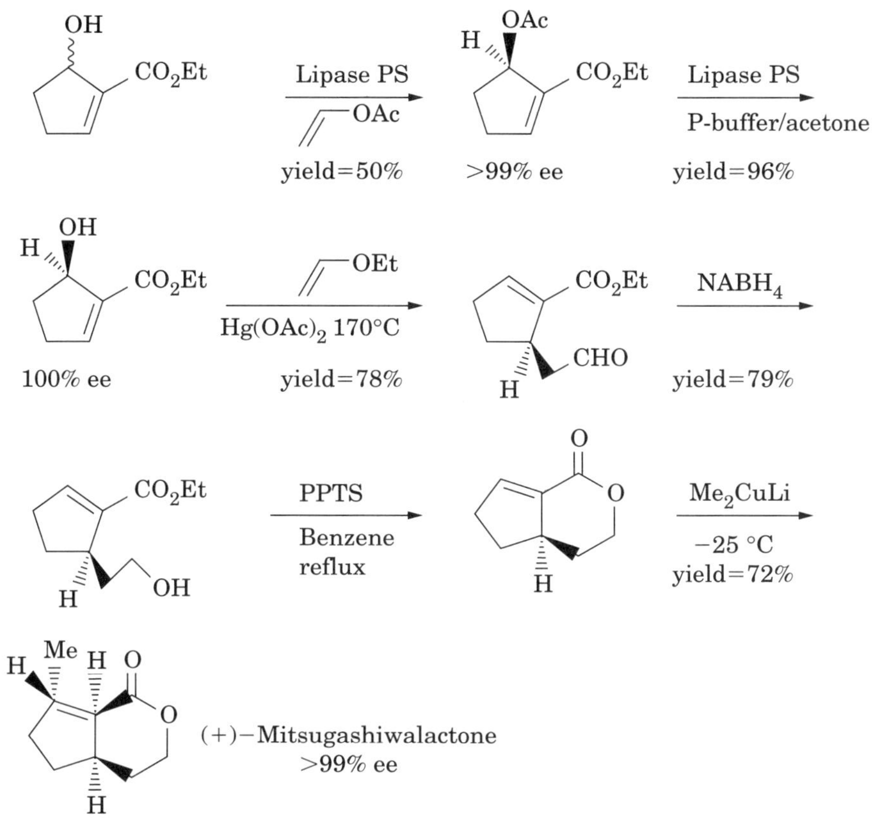

Scheme 9. Total synthesis of Mitsugashiwalactone via lipase catalyzing transesterification with vinyl acetate.

hydrolysis of a pendant acetoxy group. Screening of hydrolases for enantioselective hydrolysis of the acetate in 2-(methylsulfinyl) phenyl acetate (**1**) identified cholesterol esterase (CE) as the most enantioselective enzyme. The enantiomeric ratio, E, ranged from 10 to 25, favoring the (*R*)-configuration at sulfur. Competing chemical hydrolysis of **1** caused a large range in measured E values. A small-scale resolution of (±)-**1** yielded (*S*)-**1** with 99% ee (11% yield) after recrystallization. Changing the methyl substituent to phenyl or *n*-butyl did not significantly change enantioselectivity (E = 10 and 15, respectively), but changing it to a chloromethyl substituent did lower enantioselectivity slightly (E = 5). Changing the phenyl acetate to a naphthyl acetate (2-(phenylsufinyl) acetate versus 1-(phenylsulfinyl)-2-naphthyl acetate) increased the enantioselectivity from 10 to 19 (E value). In all cases, CE favored the (*R*)-sulfoxide. To aid the design of new resolutions with CE, Serreqi et al. proposed an empirical rule that accounts for the observed enantiopreference of CE toward these five compounds of sulfoxides and 15 compounds of other chiral aryl acetates. This empirical rule uses both the size of the substituents and their conformational preferences to predict which enantiomer reacts faster (Fig. 2).

(±)-1-(2-Thienyl)propyl acetate (±)-**1** was optically resolved by *P. capacia* lipase (PCL)–catalyzed hydrolysis to afford (*S*)-(−)-1-(2-thienyl)propyl acetate (*S*)-**1** with >99% ee (21). (*S*)-(−)-1-(2-Thienyl)propyl acetate (*S*)-**1** thus obtained was transformed to (*S*)-(+)-3-octanol (*S*)-**4**, the alarm pheromone of ants, *Crematoster castanea* and *C. liengmei* (Scheme 14).

4-Aryl-1,4-dihydropyridines are novel and potent calcium antagonists; their derivatives have been widely investigated from pharmacological points of view, and some of them have already been used therapeutically as antihypertensive drugs. These agents are known to exhibit their vasodilative activities through affinity with voltage-dependent calcium channels. Different substituents in their compounds lead to chiral derivatives possessing an asymmetric carbon at the 4-position, and the enantiomers have been reported to show different biological activities.

Ebiike et al. (22) synthesized a homochiral 4-aryl-1,4-dihydropyridine-5-phosphate (NZ105) from racemic materials. Enantioselective hydrolysis by lipase AH (*Pseudomonas* sp.) catalyst proceeded smoothly with use of propionyloxymethyl (PROM) ester derivative in organic solvents. They succeeded in converting an optically active carboxylic acid into an optically active calcium antagonist medicine (Scheme 15).

The hydrolysis of methyl D,L-3,3-difluorophenyl alanate **1a** and methyl D,L-3,3-difluoro-2-aminobutanoate **1b** and their *N*-acetyl derivatives **2a** and **2b** by subtilisin (protease) has been studied (23). All derivatives examined were enzymatically resolved to separable mixtures of the corresponding 3,3-difluoro-L-amino acids (**3a** and **3b**) or *N*-acetyl amino acids (**5a** and **5b**) and the unchanged 3,3-difluoro-D-amino acid (**4a** and **4b**) or *N*-acetyl amino acid (**6a** and **6b**). Acidic hydrolysis of methyl 3,3-difluoro-D-phenyl alanate (**4a**) or its *N*-acetyl derivative (**6a**) led to 3,3-difluoro-D-phenyl alanine (**8a**).

In the same manner, L- and D-2-amino-3,3-difluorobutanoic acids (**7b**) and (**8b**) were prepared starting from (**5b**)

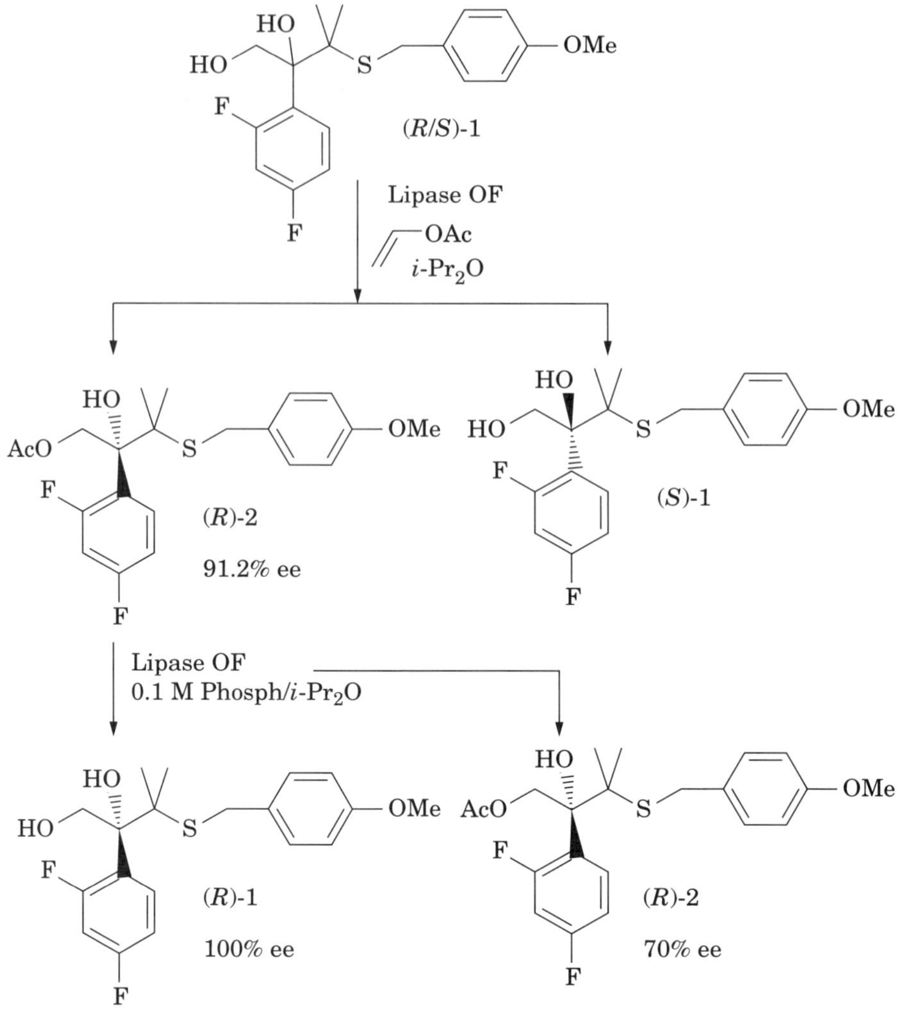

Scheme 10. Enzymatic resolution of sterically crowded diols.

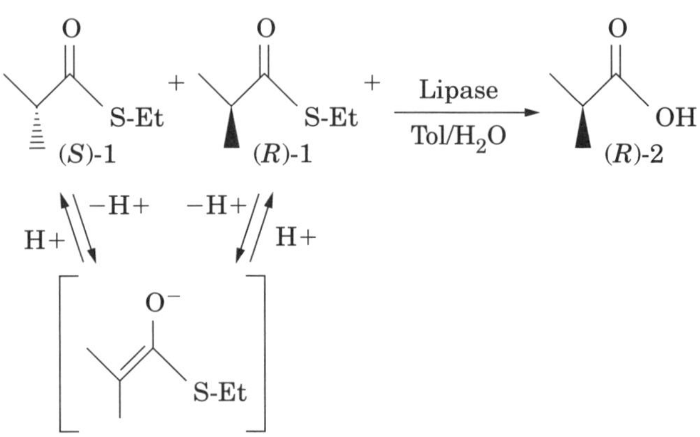

Scheme 11. Enzymatic resolution coupled with substrate racemization using a thioester substrate.

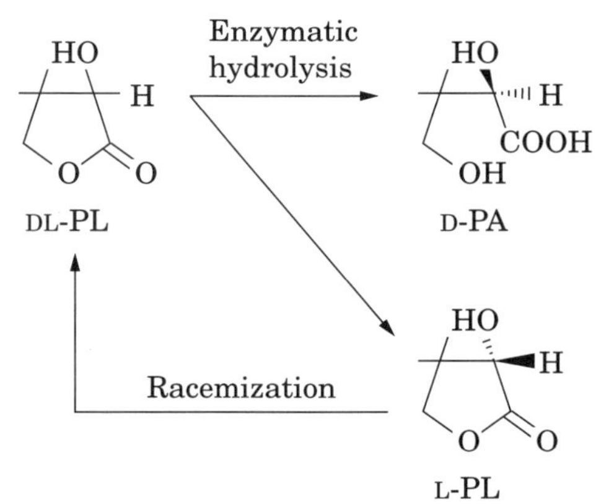

Scheme 12. Schematic presentation of the enzymatic resolution of DL-pantolactone.

and (**6b**). With these methods, unchanged methyl 3-3-difluoro-D-amino ester derivatives showed an ee of >90%, and L-amino acids were estimated to have an ee of >95% (Scheme 16).

Kinetic resolution of 2,2-disubstituted epoxides was accomplished with epoxide hydrolases from bacterial and fungal origin by using lyophilized whole microbial cells (24). In all cases investigated, biocatalytic hydrolysis was shown to occur with retention of configuration at the stereogenic center, leading to 1,2-diols and other results leading to epoxides. The selectivity of the reaction was depen-

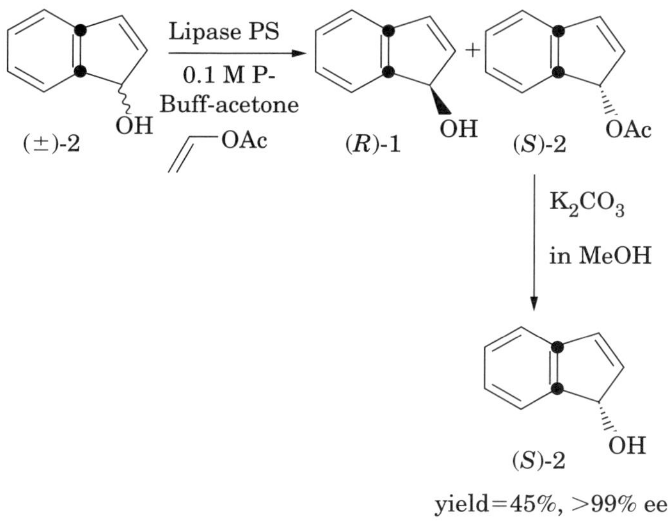

Scheme 13. Lipase-mediated resolution of inden-1-ol.

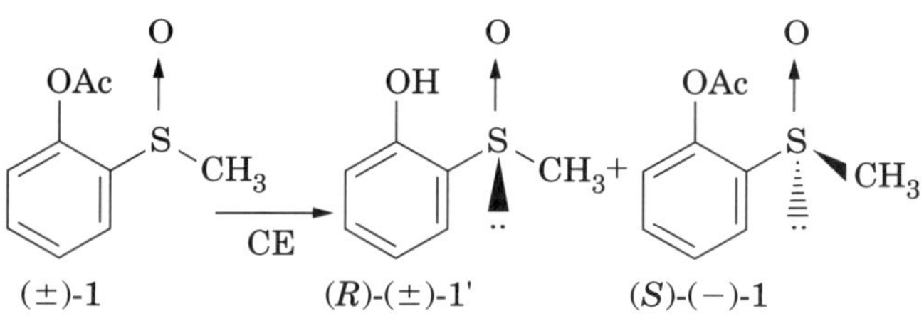

Figure 2. Kinetic resolution of sulfoxide with pendant acetoxy group using cholesterol esterase.

dent on the substrate structure and the strain used. Specifically, in the case of 2,2-disubstituted oxyrane (2-methyl-2-nonyl-1,2-oxyrane), (*S*)-(2-methyl)undecan-1,2-diol was obtained by lyophilized *Rhodococcus* sp. with high stereoselectivity (E >200) (Scheme 17).

Akita et al. (25) reported a highly stereoselective synthesis of a versatile chiral synthon possessing two stereogenic centers (2*S*,3*S*)-**1** (>99% ee) and successful synthesis of Nikkomycin B. (Scheme 18).

OPTICAL RESOLUTION BY LIPASE-CATALYZING ESTERIFICATION

The reaction for the resolution of (*R*,*S*)-ibuprofen was scaled-up to yield gram quantities of (*S*)-ibuprofen, which was accomplished through a two-stage enantioselective reaction, each stage catalyzed by Novozym 435 (26). In the first reaction, starting from 300 g of racemic ibuprofen, 88.9 g of enantio-enriched (*S*)-ibuprofen with 85% ee was produced. In the subsequent reaction, 76 g of the 85% ee materials were used to produce 38.4 g of (*S*)-ibuprofen with 97.5% ee (Scheme 19).

Synthesis of both enantiomers of fluorophoracantholide **1**, (9*S*)-10-fluorodecan-9-olide (9*S*)-(−)-**1** and (9*S*)-10-fluorodecan-9-olide (9*R*)-(+)-**1**, has been achieved through lipase-catalyzed enantioselective acetylation of methyl 10-fluoro-9-hydroxy decanoate (**2a**) with acetic anhydride in the presence of lipase amano PS (*P. capacia*) (27). The higher homologue, methyl 11-fluoro-10-hydroxyundecanoate (**2b**), has also been optically resolved. These enzymatic resolutions by acetylation of ω-fluorohydroxyalka-

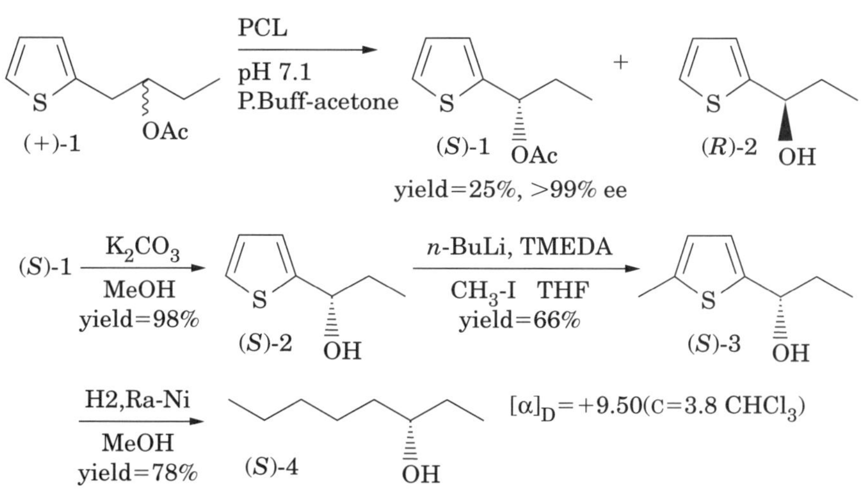

Scheme 14. Chemicoenzymatic synthesis of alarm pheromone of ants, *Crematoster* sp.

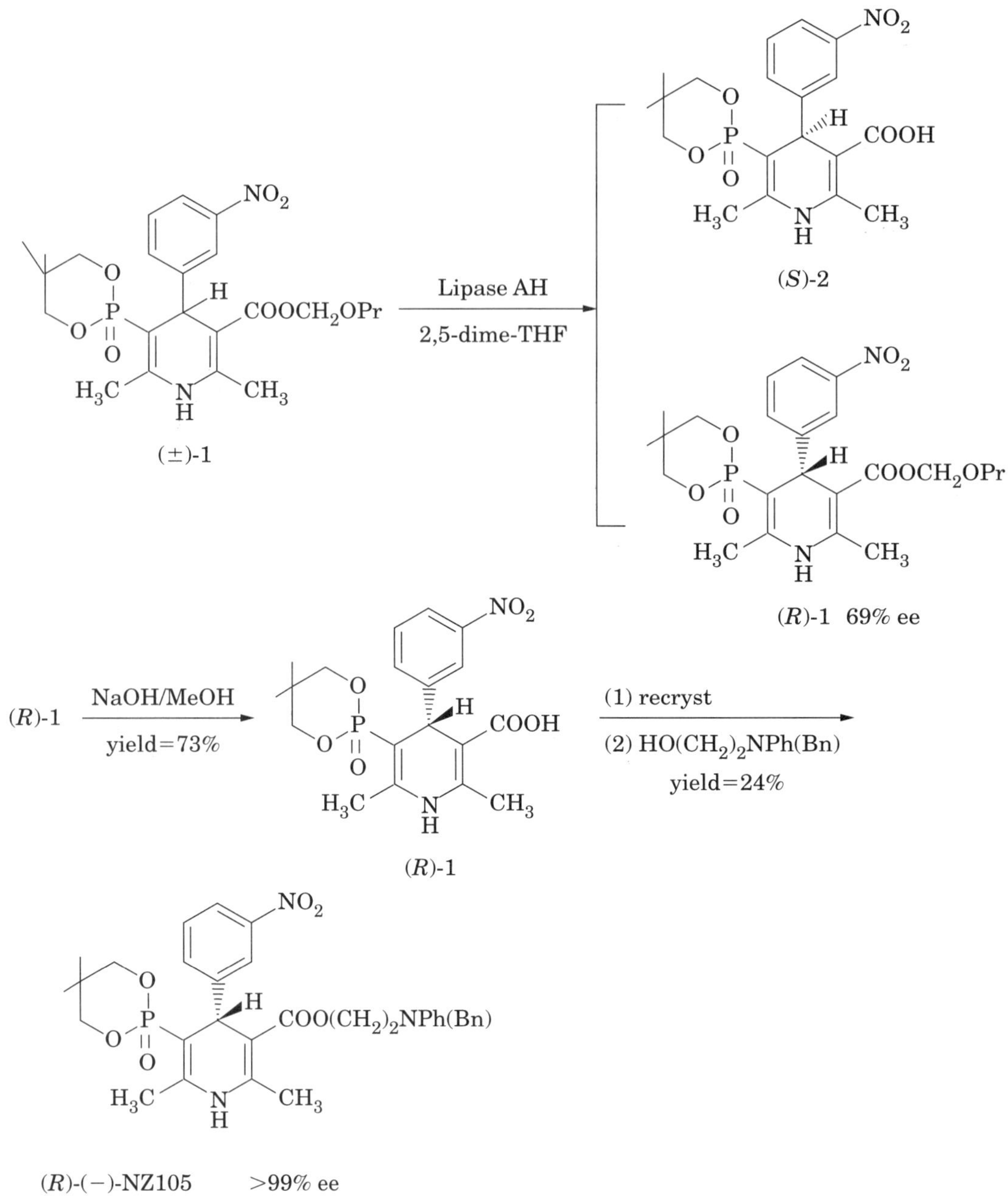

Scheme 15. Lipase-catalyzed enantiosynthesis of a calcium antagonist for medication.

noic acid esters of long-chain fatty acids represent the first lipase-catalyzed enantioselective esterification of β-fluoroalcohol (Scheme 20).

Many adrenergic drugs are 2-amino-1-arylethanols possessing a stereogenic center with a secondary alcohol function. It is well known that the two enantiomers of a chiral drug often display different biological activities. Consequently, the enantiomeric purity of such compounds is very important. The drugs of (*R*)-enantiomers of amino ethanol and (*S*)-enantiomers of amino-2-propanol are generally associated with agonist or β-blocking activity because of their structural similarity to natural catecholamines.

The enantiomers of 2-amino-1-(4-methoxy)phenylethanol were obtained at enantiomeric purity (99% ee) at 50% conversion by *P. capacia* lipase-catalyzed O-acylation of the corresponding amino alcohol (28) (Scheme 21).

The kinetic lactonization of γ-hydroxyesters by porcine pancreatic lipase (PPL) has been demonstrated by Sugai et al. (29). The reaction has been used in the synthesis of Japanese beetle pheromone with 78% ee in 29% conversion. Clearly, this chemistry has tremendous potential because of the importance of lactones in biologically active compounds. However, it is restricted to very low yields if high enantiomeric purity is desired.

Taylor et al. prepared the enantio-enriched γ-hydroxyesters from corresponding γ-ketoesters using Baker's yeast reduction (30). They attempted Baker's yeast reduction of **1** and obtained the corresponding γ-hydroxyesters **2** in a

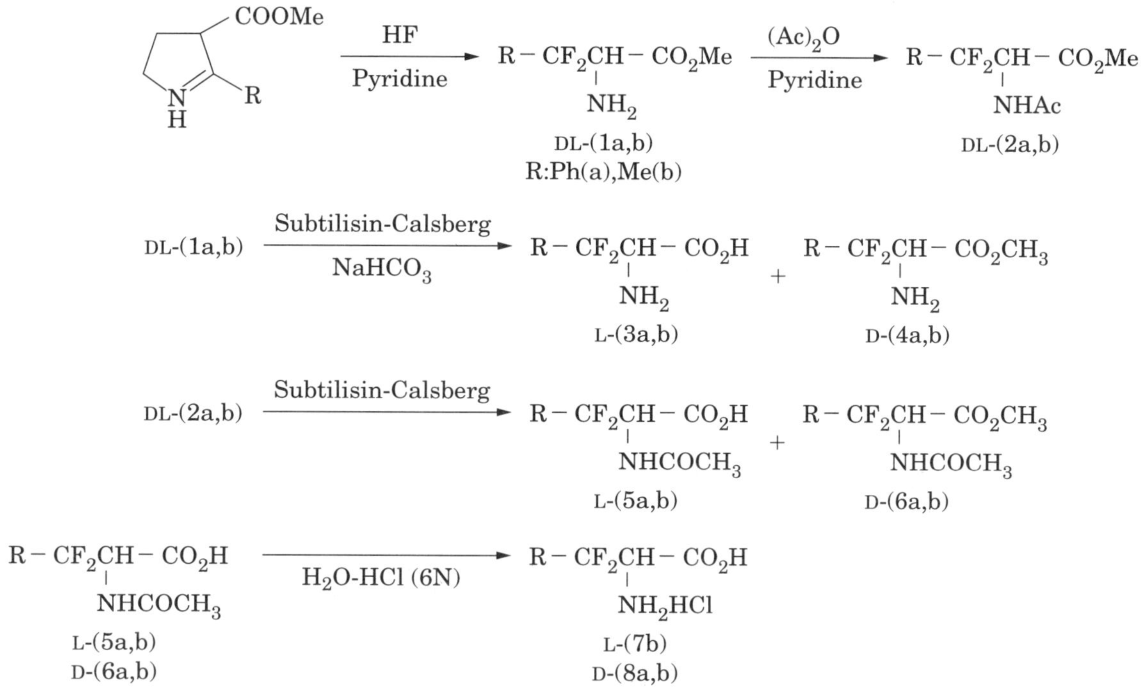

Scheme 16. Enzymatic resolution of fluoro amino acids by subtilisin.

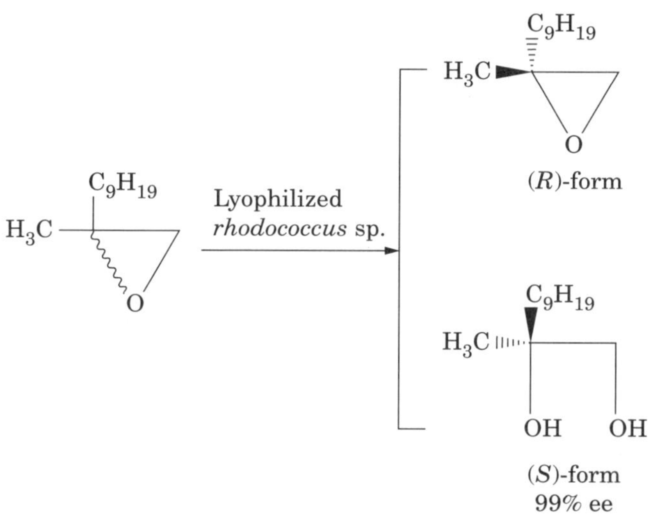

Scheme 17. Resolution of 2,2-disubstituted oxyrane using *Rhodococcus* sp. NCIMB 11216.

yield of 46% with 95% ee. This product **2** (predominantly (*S*)-**2**) was obtained by treating with PPL with 99.7% ee in 61% conversion (Scheme 22).

Taylor et al. also reported the high optical purity preparation of another two important lactones, key intermediates used in the synthesis of Japanese beetle pheromone and anti-AIDS drugs, using a complementary, two-stage enantioselective strategy. In this approach, an enzymatic kinetic resolution is performed to enrich the faster-reacting enantiomer in a starting material obtained by an earlier enantioselective step. The results show that conversions, yields, and ee can be dramatically increased by this complementary approach. The authors illustrate how modestly enantio-selective techniques can be combined with kinetic resolution procedures to yield optically pure compounds. The utility of many modestly enantio-selective techniques can be enhanced thereby.

OPTICAL RESOLUTION BY HYDANTOINASE-CATALYZING HYDROLYSIS

Racemic amino acid hydantoins can be converted into D-amino acids using a system consisting of a D-hydantoinase and a carbamoylase. D-Citrulline can be obtained from corresponding racemic hydantoin with 92% ee in 79% yield (31) (Scheme 23).

Agrobacterium tumefaciens strain 47C expresses an inducible D-hydantoinase that catalyzes the formation of optically pure *N*-carbamoyl D-amino acids from racemic hydantoin precursors. The D-hydantoinase was shown to be active and stable at higher temperature and higher pH values, thus affording favorable bioreaction conditions that result in racemization of DL-hydantoins to D-isomer. The optimal performance (reaction kinetics) was demonstrated at pH 10 and 70 °C. The enzyme exhibited a distinctive substrate specificity and was not activated further by any metal ions. *A. tumefaciens* hidantoinase was most active on 5,6-dihydro uracil and DL-5-methyl hydantoin but only slightly active on DL-benzyl hydantoin (32).

Kim and Kim (33) reported that D-hydantoinase and N-carbamoylase of the bacterium *Agrobacterium* sp I-671 were partially purified to about 90% purity on sodium dodecyl sulfate–polyacrylamide gel electrophoresis and that biochemical properties of the enzymes were characterized.

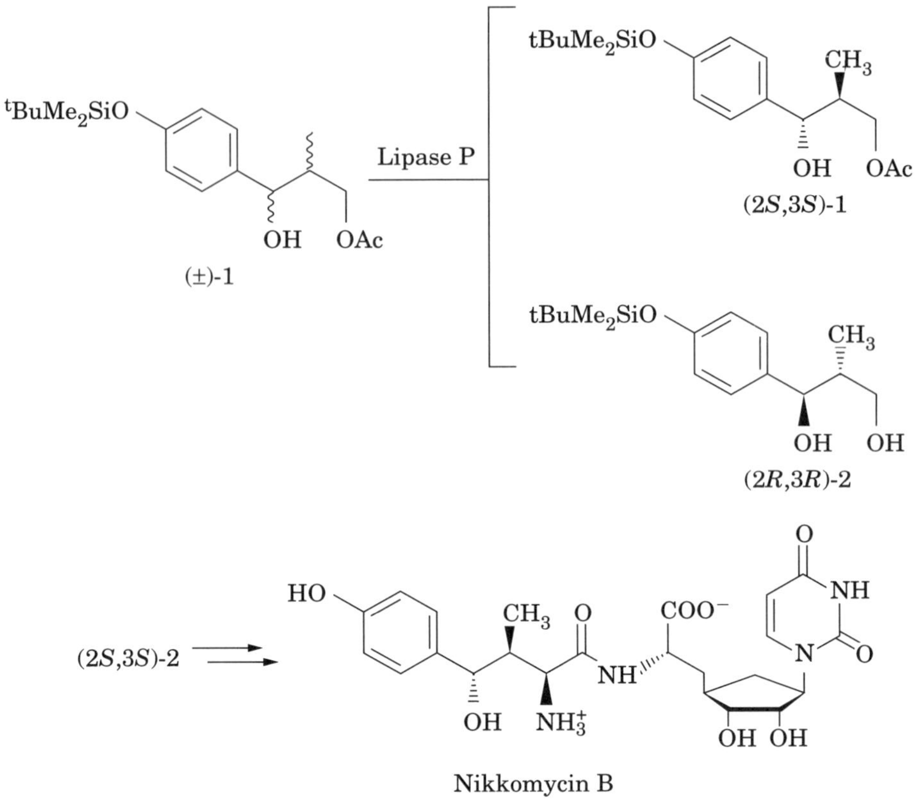

Scheme 18. A total synthesis of Nikkomycin B based on enzymatic resolution of a primary alcohol.

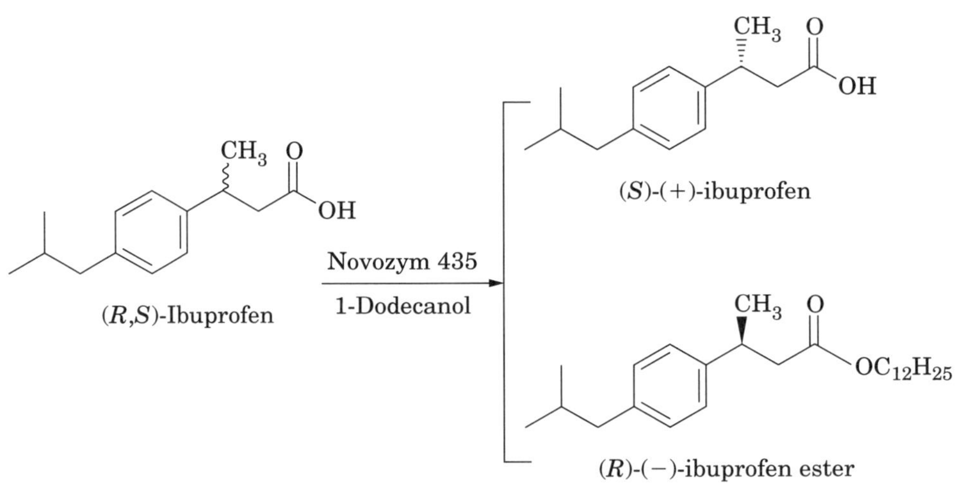

Scheme 19. The enantioselective reaction for the enzymatic resolution of (*R,S*)-ibuprofen.

The *N*-carbamoylase was found to be severely inhibited by ammonium ions coproduced with D-*p*-hydroxyphenylglycine (D-HPG). The conversion yield of D-HPG significantly increased on addition of specific adsorbents for ammonium ions. To determine the ratio of D-hydantoinase to *N*-carbamoylase, which best minimizes the accumulation of intermediate (*N*-carbamoyl-D-HPG) in the direct enzymatic production of D-HPG, a sequential reaction was numerically simulated. Simulation results coincided well with experimental data, and the optimal ratio of D-hydantoinase to *N*-carbamoylase was found to be about 1:3 by weight.

OPTICAL RESOLUTION BY OXIDOREDUCTASE-CATALYZING OXIDATION AND REDUCTION

Lipase-catalyzed kinetic resolution of secondary alcohols via hydrolysis, esterification, and transesterification has been widely used to obtain optical active compounds as described in this article.

Fantin et al. (34) succeeded in achieving the kinetic resolution of heteroaryl ethanols via oxidation both with baker's yeast and *Bacillus stearothermophilus*. These organisms were also used in the kinetic resolution of bicyclic octenol and heptenol with excellent results (35). Although

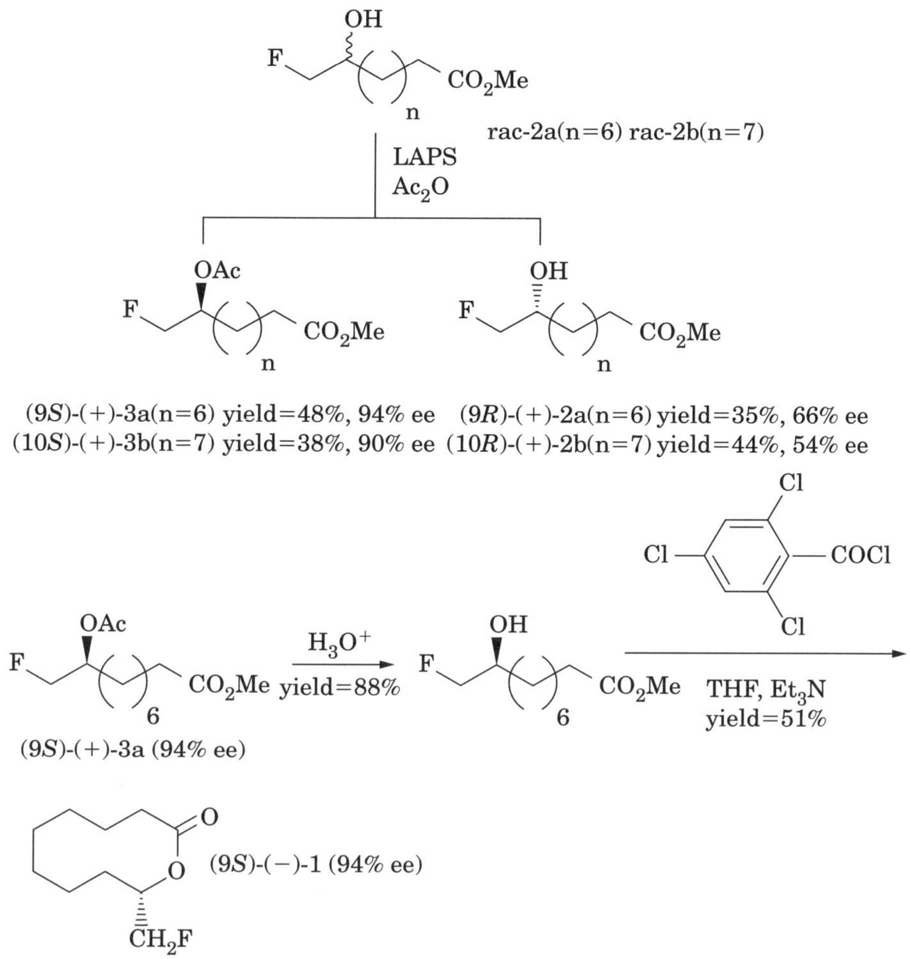

Scheme 20. Synthesis of (9*S*)-(−)-enantiomer of fluorophoracantholide.

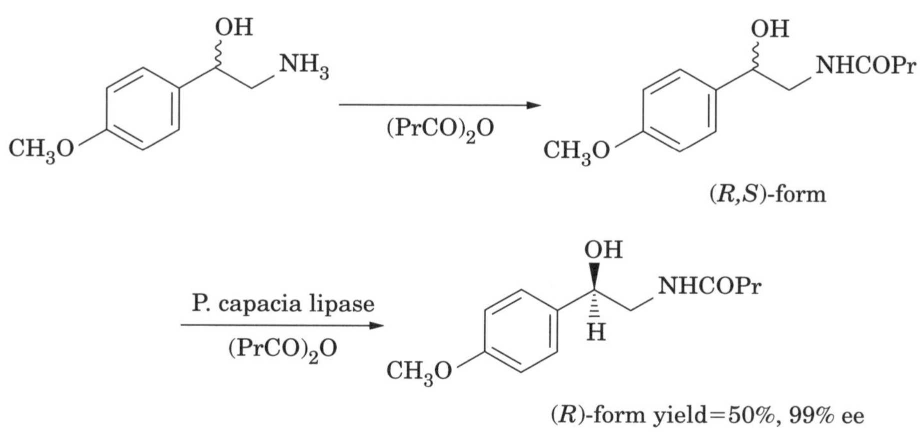

Scheme 21. Synthesis of optically pure precursor of adrenergic drugs by lipase-catalyzed acylation.

hydrolytic enzymes such as lipases and hydrolases are known to retain their activity in organic solvents, dehydrogenase and oxidoreductase are different because their catalytic reactions require the assistance of a stoichiometric amount of coenzymes, which has to be reproduced in order to elaborate the substrates catalytically in the cells. However, the organic solvents often cause serious damage on the hydrophobic cell membranes of the microbe, resulting in a leakage of the contents of the cell. To avoid this problem, the organic solvents were used with immobilized microorganisms because immobilization enhances their stability against denaturation by organic solvents.

Fantin et al. (36) reported that racemic 6-*endo*-bicyclo[3.2.0]hept-2-en-6-ol **1** was oxidized and kinetically resolved in heptane to give (1*S*,5*R*)-bicycloheptenone **3** in 49% conversion with 100% ee. Parallel experiments in water and with immobilized cells in heptane gave lower yields (41%) but equally high enantiomeric excess (99 to 100%

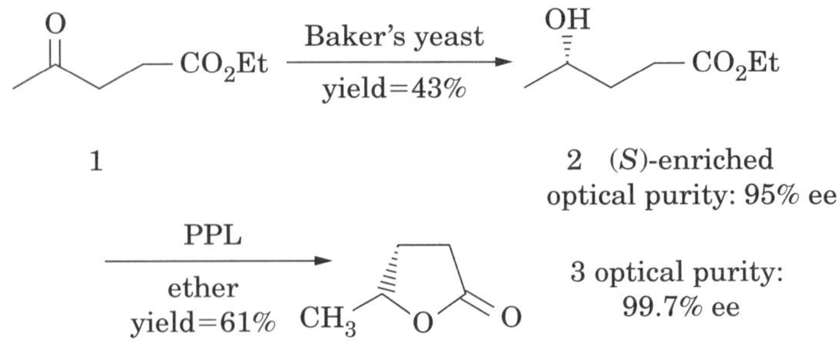

Scheme 22. The kinetic lactonization of *r*-hydroxyesters by porcine pancreatic lipase.

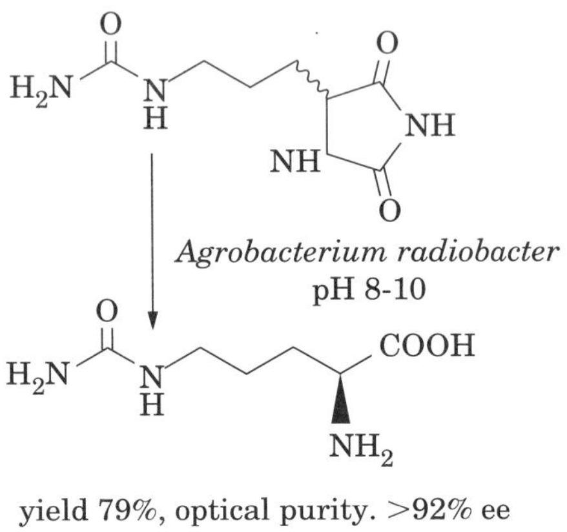

Scheme 23. D-Citrulline preparation by hydantoinase.

ee). Similar results were obtained in the oxidation of the racemic 2-*endo*-bicyclo[3.3.0]oct-7-en-2-ol **2**, but the reaction rate was higher. In fact, after only 4 h in heptane, (1*S*,5*S*)-bicyclo-octenone **4** was obtained in 48% yield and with 98% ee. In water, the yield was lower (37%). Oxidation with *B. stearothermophilus* in heptane is the first example of a reaction with bacteria carried out in organic solvents, and it is as efficient as water. The advantages of this methodology are that the procedure is simpler because the filtration of the cells and concentration of the solvents afford the crude reaction mixture. On the other hand, oxidation with the immobilized cells is very promising for use on a preparative scale (Scheme 24).

The fungus *Rhizopus arrhizus* has been used for reduction of acetophenone and its *o*-, *m*-, and *p*-methoxy derivatives **1a** to **1d** to the corresponding (*S*)-(−)-alcohols (37). To optimize the reaction time for microbial cell-catalyzed reduction, experiments were carried out with **1a** for different periods of time. Fermentation for 7 days was found to yield the best results. The 1-phenylalkanones **1a** to **1d** were therefore incubated with *R. arrhizus* in modified Czepak Dox medium for 7 days. At the end of fermentation, the transformation products 2a–d were isolated and characterized. All these alcohols were found to have negative specific rotations (Scheme 25).

The (1*R*,2*S*)- and (1*S*,2*R*)-2-methyl-2,3-epoxy-1-phenylpropanols were prepared from α-methylcinnamyl chloride by use of baker's yeast and lipase PS. Takeshita et al. (38) reported the synthesis of (1*R*,2*S*)-2-methyl-2,3-epoxy-1-phenylpropanol **3a** and (1*S*,2*R*)-2-methyl-2,3-epoxy-1-phenylpropanol **3b** starting from α-methylcinnamyl chloride **1** by use of baker's yeast and lipase PS (*Pseudomonas* sp., Amano). At first, when α-methylcinnamyl chloride **1** was fermented with baker's yeast for 36 h at 28 °C, rearrangement occurred to give 2-methyl-1-phenylallyl alcohol **2** in 68% yield. Two pairs of racemic epoxy alcohols, (**3a**, **3b**) and (**3c**, **3d**) (2:1), were separated by silica gel column chromatography (total yield 73%) after epoxidation of **2** with *m*-chloroperbenzoic acid. Enzymatic treatment of epoxy alcohols (**3a**, **3b**) with vinyl acetate in *t*-butyl methyl ether in the presence of lipase PS for 17 h or 24 h at 30 °C gave the optically active acetate **4** and alcohol **3b** in chemical yields of 31 and 46%, respectively. However, the same reaction for racemic epoxy alcohols (**3c**, **3d**) was very sluggish (under the above conditions, esterification proceeded only in 27% chemical yield in 3 weeks, and the optical yield was very low, 56% ee) (Scheme 26).

Fantin et al. (39) reported a new method for preparing optically pure secondary alcohols by combining microbial oxidation and reduction. The chiral secondary alcohols were obtained by microbial or enzymatic reductions of the

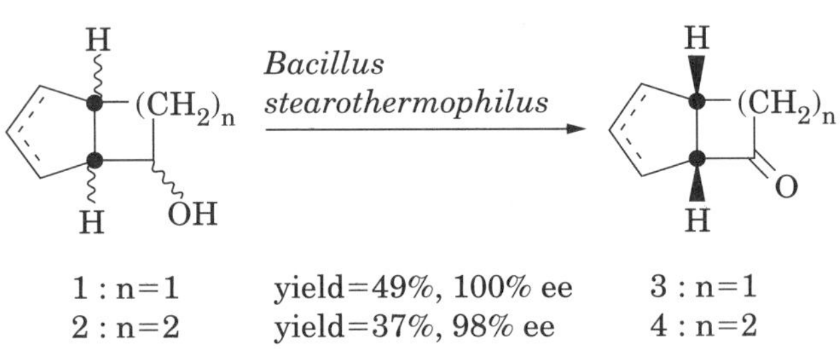

Scheme 24. Oxidation with *Bacillus stearothermophilus* in heptane.

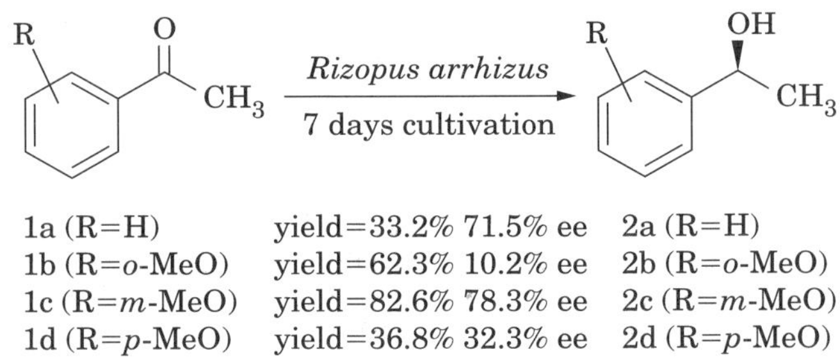

Scheme 25. Reduction of acetophenone derivatives with *Rizopus arrhizus*.

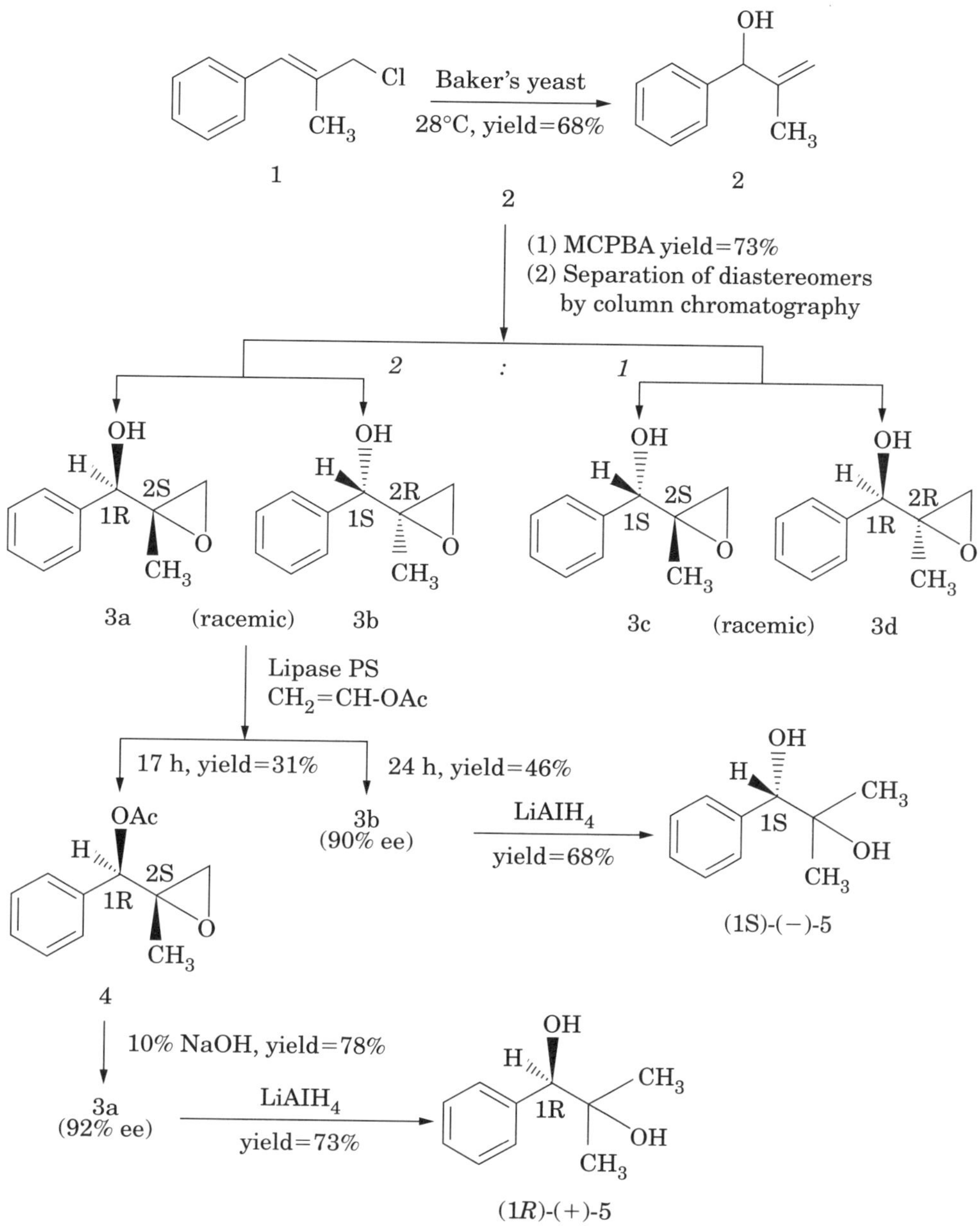

Scheme 26. Enzymatic synthesis of (1*R*,2*S*)- and (1*S*,2*R*)-2-methyl-2,3-epoxy-1-phenylpropanols with baker's yeast and lipase PS.

corresponding ketones followed by kinetic resolution of the racemic alcohols via esterification or hydrolysis of the corresponding ester with lipases. Various ketones were reduced using baker's yeast or purified alcohol dehydrogenases. The reaction with baker's yeast yielded almost exclusively the (*S*)-alcohol, whereas the enzyme produced the hydride either from the *si*-side (i.e., *Lactobacillus kefir* alcohol dehydrogenase) or the *re*-side (i.e., *Thermoanaerobium brockii* alcohol dehydrogenase) of the ketone to give (*R*)- or (*S*)-alcohols, respectively. For most cases, the stereochemical course of the reaction, which mainly depended on the steric requirements of the substrate, could be predicted from the simple model that is generally referred to as Prelog's rule. A majority of the dehydrogenases used for the stereospecific reduction of ketones and a majority of microorganisms follow Prelog's rule, giving (*S*)-alcohols. However, the enzymatic resolution of racemic alcohols with lipases, in which a high enantioselectivity is exhibited, stops at the 50% conversion stage. The authors (39) succeeded in establishing an alternative method for kinetic resolution of secondary alcohols via oxidation with baker's yeast and *B. stearothermophilus*, affording (*R*)-enantiomer in high enantiomeric excess. Moreover, they also reported an anti-Prelog behavior (formation of the (*R*)-alcohol) in microbial reductions of various carbonyl compounds with *Yarrowa lipolytica* strains, a new yeast species. They described a new approach to the high-yield synthesis of homochiral secondary (*R*)-alcohols starting from the corre-

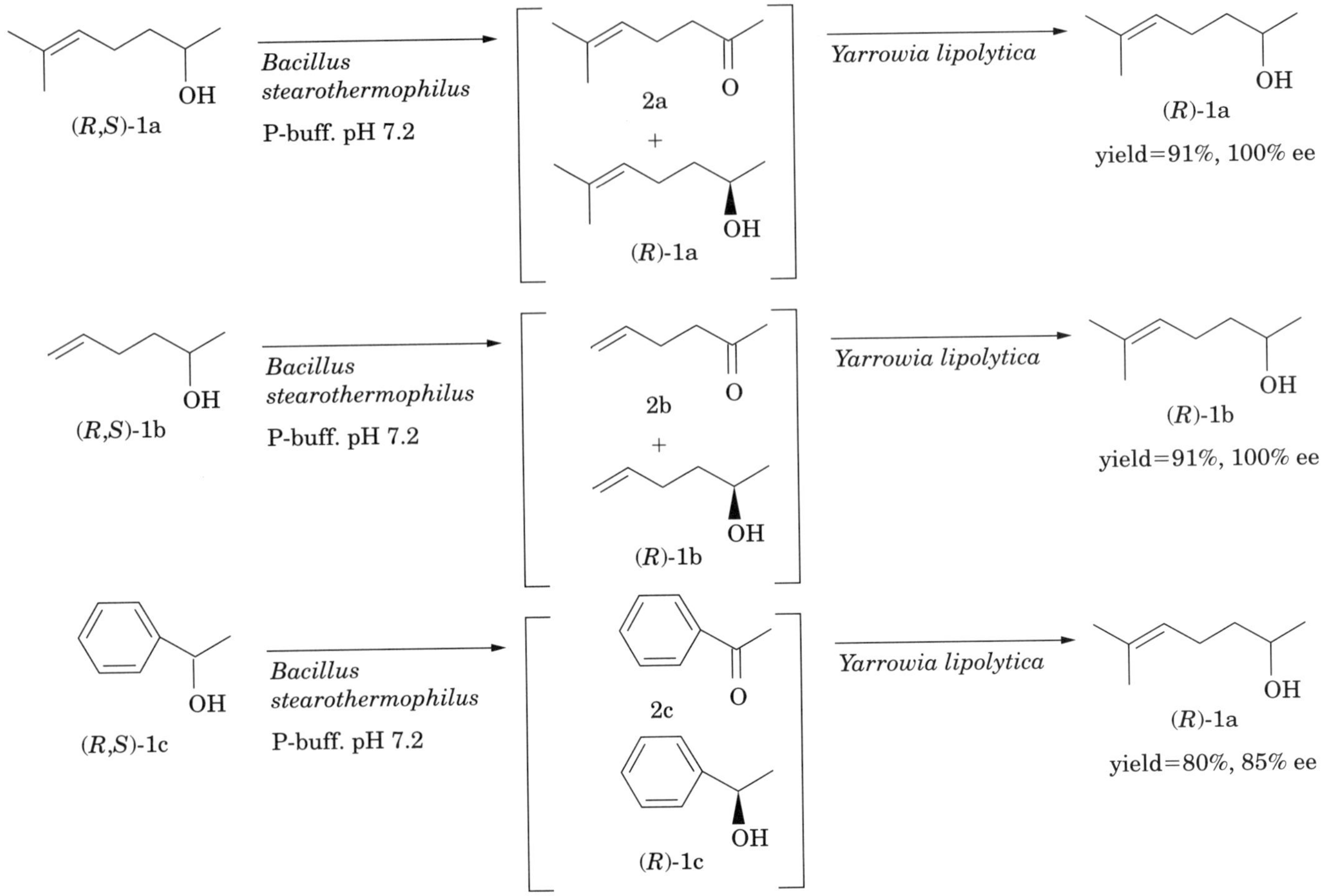

Scheme 27. Optical resolution of the racemic alcohols by a combination of microorganism-catalyzed oxydation and reduction.

sponding racemates. The approach consisted of (*1*) kinetic resolution via oxidation with *B. stearothermophilus* and (*2*) reduction of the obtained oxidation mixture with *Yarrowia lipolytica* (Scheme 27).

The asymmetric microbial degradation of (*S*)-(+)-mandelic acid was investigated in order to develop a practical process for (*R*)-(−)-mandelic acid production from racemic mandelic acid, which is commercially produced on a large scale at low cost using chemical synthesis. Among the 790 culture strains tested, microorganisms belonging to the *Brevibacterium, Pseudomonas, Rhodococcus, Rhodotorula, Rhodosporidium, Sporobolomyces*, and *Gibberela* genera exhibited high (*S*)-(+)-mandelic acid–degrading activity. *P. polycolor* IFO 3918 was determined to be the best strain and used as a biocatalyst for eliminating the (*S*)-(+)-isomer (40). The maximum rate of (*S*)-(+)-isomer degradation was obtained at 30 °C and pH 7.0. Under these optimal conditions, the (*S*)-(+)-isomer in a racemic mandelic acid 45 g/L mixture was completely degraded within 24 h, with 20 g of (*R*)-(−)-mandelic acid per liter remaining in the reaction mixture. Crystalline (*R*)-(−)-mandelic acid with a chemical purity greater than 99% and optical purity with 99.9% ee was obtained in 35% yield by acidification of the reaction mixture, extraction with ethyl acetate, and subsequent concentration.

CONCLUSION

Optical resolution by enzymes and microorganisms facilitates enantio-selective synthesis of the bioactive chiral compounds and precursors of natural compounds. Combination of organic synthesis with biotransformation should afford efficient optical-selective and regioselective preparation processes for bioactive natural products such as antibiotics, hormones, pheromones, natural fragrances, and various kinds of natural drugs.

BIBLIOGRAPHY

1. C.-S. Chen and C.J. Sih, *Angew. Chem. Int. Ed. Engl.* **28**, 695 (1989).
2. Z.-F. Xie, *Tetrahedron Asymmetry* **2**, 733 (1991).
3. R.S. Ward, *Tetrahedron Asymmetry* **6**, 1475 (1995).
4. T. Yokomatsu, N. Nakabayashi, K. Matsumoto, and S. Shibuya, *Tetrahedron Asymmetry* **6**, 3305 (1995).
5. K. Kato, M. Katayama, S. Fujii, and H. Kimoto, *Biosci. Biotech. Biochem.* **59**, 2178 (1995).
6. O. Yamada and K. Ogasawara, *Synthesis* **10**, 1291 (1995).
7. K. Tanaka and K. Ogasawara, *Synthesis* **10**, 1237 (1995).

8. S. Fernandez, M. Ferrero, and W.H. Okamura, *J. Org. Chem.* **60**, 6057 (1995).
9. T. Izumi and S. Murakami, *J. Hetrocycl. Chem.* **32**, 1125 (1995).
10. M. Lundth, O. Nordin, E. Hedenstroem, and H.-E. Hoegberg, *Tetrahedron Asymmetry* **6**, 2237 (1995).
11. M.-J. Kim, G.-B. Choi, J.-Y. Kim, and H.-J. Kim, *Tetrahedron Lett.* **35**, 6253 (1995).
12. T. Yamane, M. Takahashi, and K. Ogasawara, *Synthesis* **4**, 444 (1995).
13. O. Azzolina, S. Collina, and D. Vercesi, *Farmaco* **50**, 10, 725 (1995).
14. V. Khlebnikov, K. Mori, K. Terashima, Y. Tanaka, and M. Sato, *Chem. Pharm. Bull.* **43**, 1659 (1995).
15. D.S. Tan, M. Gunter, and G. Drueckhammer, *J. Am. Chem. Soc.* **117**, 9093 (1995).
16. M. Kataoka, K. Shimizu, K. Sakamoto, Y. Yamada, and S. Shimizu, *Appl. Microbial. Biotechnol.* **43**, 974 (1995).
17. A.S. Bommarius, M. Schwarm, K. Stingl, M. Kottenharhn, and K. Hethmature, *Tetrahedron Asymmetry* **6**, 2851 (1995).
18. M. Takahashi, R. Koike, and K. Ogasawara, *Chem. Pharm. Bull.* **43**, 1585 (1995).
19. D.-F. Tai, Y.-H. Chao, C.-Y. Huang, J.-M. Luo, Y.-T. Lin, and S.-H. Wu, *J. Chin. Chem. Soc.* **42**, 801 (1995).
20. A.N. Serreqi and R.J. Kazrauskas, *Can. J. Chem.* **73**, 1357 (1995).
21. S.-K. Kang, J.-H. Jeon, T. Yamaguchi, J.-S. Kim, and B.S. Ko, *Tetrahedron Asymmetry* **6**, 2139 (1995).
22. H. Ebiike, Y. Yamazaki, and K. Achiwa, *Chem. Pharm. Bull.* **43**, 1251 (1995).
23. A. Ayi, R. Guedj, and B. Septe, *J. Fluor. Chem.* **73**, 165 (1995).
24. M. Mishitz, W. Kroutil, and K. Faber, *Tetrahedron Asymmetry* **6**, 1261 (1995).
25. H. Akita, C.Y. Chen, and K. Uchida, *Tetrahedron Asymmetry* **6**, 2131 (1995).
26. M. Trani, A. Ducret, P. Pepin, and R. Lorti, *Biotechnology Lett.* **17**, 1095 (1995).
27. A. Sattler and G. Haufe, *Tetrahedron Asymmetry* **6**, 2841 (1995).
28. K. Lundell and L.T. Kanerva, *Tetrahedron Asymmetry* **6**, 2281 (1995).
29. T. Sugai, S. Ohsawa, H. Yamada, and H. Ohta, *Synthesis*, 1112 (1990).
30. S.K. Taylor, F. Richard, R.F. Atkinson, A.P. Almli, M.D. Carr, T.J. Van Huis, and M.R. Whittaker, *Tetrahedron Asymmetry* **6**, 157 (1995).
31. K. Draut, M. Kottenhahn, K. Macrealeas, H. Klenk, and M. Bernd, *Angew. Chem. Int. Ed. Engl.* **30**, 714 (1991).
32. D.R. Durham and J.E. Weber, *Biochem. Biophys. Res. Commun.* **216**, 1095 (1995).
33. G.-J. Kim and H.-S. Kim, *Enzyme Microbial Technol.* **17**, 63 (1995).
34. G. Fantin, M. Fogagnolo, A. Medici, P. Pedrini, S. Poli, and M. Sinigaglia, *Tetrahedron Lett.* **34**, 883 (1993).
35. G. Fantin, M. Fogagnolo, A. Medici, P. Pedrini, S. Poli, and F. Gardini, *Tetrahedron Asymmetry* **4**, 1607 (1993).
36. G. Fantin, M. Fogangnolo, P.P. Giovennini, A. Medici, P. Pedrini, and S. Poli, *Tetrahedron Lett.* **36**, 441 (1995).
37. N.A. Slavi, P.S. Patil, S.R. Udupa, and A. Banerji, *Tetrahedron Asymmetry* **6**, 2287 (1995).
38. M. Takeshita, R. Yaguchi, and Y. Unuma, *Heterocycles* **40**, 967 (1995).
39. G. Fantin, M. Fogagnolo, P. Paolo, M. Alessandro, and P. Pedrini, *Tetrahedron Asymmetry* **6**, 3047 (1995).
40. K. Takahashi, K. Nakamichi, M. Furui, and T. Mori, *J. Ferment. Bioeng.* **79**, 439 (1995).

OPTICAL SENSORS

MAHESH UTTAMLAL
Glasgow Caledonian University
Glasgow, Scotland
DAVID R. WALT
Tufts University
Medford, Massachusetts

KEY WORDS

Arrays
Biosensors
CO_2
Fermentation
Fiber optics
Monitoring
Optical sensors
pH
Sensors

OUTLINE

INTRODUCTION

Many industries such as chemicals, pharmaceuticals, and food require a means of monitoring their production processes. For example, during fermentation, pH, oxygen, carbon dioxide, amino acids, and product concentrations are only a handful of the many parameters that should be monitored to maximize the utilization of reactants and maximize product formation. Most significantly, the lack of suitable sensors in the biotechnology industry is the major obstacle to improve yields through real-time process control. Similarly, monitoring a multitude of analytes during the experimental research and development phase, such as in shaker flasks or microtiter plates, would significantly enhance the ability of biochemical engineers to optimize a process. This article presents the development and application of fiberoptic chemical sensors for bioprocess monitoring.

BIOPROCESS MONITORING

Important Analytes in Bioprocesses

The most common industrial biochemical process is batch aerobic fermentation. Most fermentation processes require control of physical, chemical, biochemical, and biomass parameters. Physical variables such as temperature, pressure, agitation speed, and feed rates are presently measured on-line with conventional sensors that are generally well-developed and reliable. The common chemical measurements of interest in all fermentation broths, from brewing beer to the most modern biotechnological process, include pH, dissolved oxygen, dissolved carbon dioxide, and redox potential. Oxygen and carbon dioxide are also monitored in the reactor vent gases using either infrared or mass spectroscopic analyzers. Next in importance are the biochemical components in the medium essential for harvest decisions, reaction optimization, and model building. These species include those of general interest in all bioreactions, such as glucose, lactate, ammonia, and phosphate, and those of specific interest, such as tetracycline, penicillin, amino acids, and ethanol. Presently these variables are analyzed off-line by removing filtered samples using sterile sampling techniques.

Requirements of Analytical Measurement Systems for Bioprocess Monitoring (1)

Chemical sensors for monitoring can be divided into four broad categories: electrochemical, optical, mass, and thermal. Bioengineers are perhaps most familiar with electrochemical sensors in the guise of pH electrodes. These sensors have been used for decades in the fermentation industry and have many of the attributes of an ideal sensor. They can be sterilized-in-place; they maintain calibration (to a reasonable degree) during the fermentation process; they are robust and can be used multiple times; and they can be interfaced to a controller that automatically adjusts the pH. The pH electrode is almost an ideal sensor from the bioengineer's perspective. Because the sensor is in direct contact with the fermentation broth, it avoids the need to acquire samples, which limits the number of samples that can be measured per unit time, and avoids errors introduced by separating, preparing, handling, transporting, and storing the samples. In addition, there is no time delay in the feedback control loop. The time-delay problem is particularly significant when intervention or corrective action is required to adjust the parameter being measured to a desired level.

By contrast, chemical sensor technology allows a device to dwell in the sampling environment. By responding rapidly and continuously to concentration changes, sensors can generate a vast amount of data. Microprocessors can store and evaluate as much data as the sensors can produce. Sensors avoid all the problems associated with sample acquisition and handling and allow for timely intervention to make appropriate adjustments.

PRINCIPLES OF CHEMICAL SENSORS AND THEIR APPLICATION TO BIOPROCESS MONITORING

In this section, a brief overview of chemical sensors, their construction, and types are given. The reader is referred to other texts (2,3) that treat the subject of chemical sensor technology in greater depth, including design, mathematical treatments, examples, and limitations.

Definition of a Chemical Sensor

A chemical sensor (4) is a device that incorporates a sensing reagent integrated with a signal transducer to afford a system that can measure a target analyte(s) (Fig. 1). The interaction of the target analyte with the sensing reagents generates a signal (electrochemical, thermal, optical) proportional to the concentration of the analyte in the bulk medium. Binding can occur by molecular recognition, such as host–guest, antibody–antigen, or protein-ligand binding, or can involve chemical reactions such as acid–base, redox, or ion exchange.

An ideal chemical sensor must possess certain characteristics for it to be a useful analytical tool in bioprocess monitoring (5). First, the sensor must be specific to the target analyte. This specificity depends on the immobilized

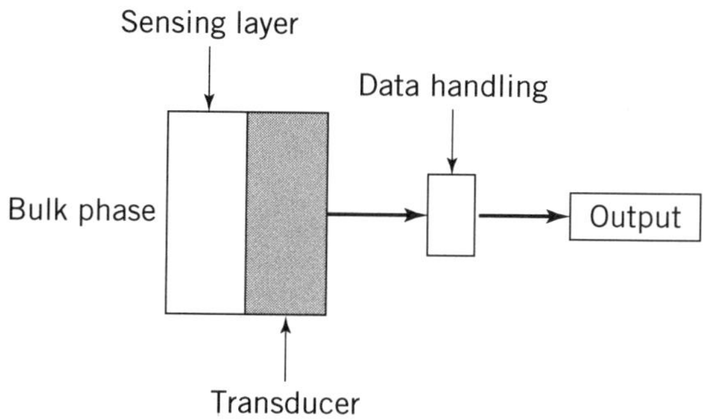

Figure 1. Components of a chemical sensor.

reagents. Biologically derived materials such as enzymes, receptor proteins, and antibodies are extremely valuable in this regard because nature has generated many analyte specific compounds. Second, the signal changes on interaction with the target analyte must be large and be sensitive in the concentration range of interest. If the latter is not possible, then sample pretreatment is required, which leads to uncertainty in the final measurement. Third, the sensor must have a response time consistent with timely intervention. For bioprocesses, response times on the order of a few seconds are sufficient. Finally, the sensor must be stable for the duration of the fermentation process; recalibration is difficult when the sensor is being used on-line or in situ.

Other important factors must be considered for application to bioprocess monitoring, such as the ability of the sensors to withstand autoclaving in situ and to avoid losing sensitivity or calibration during the process. In situ calibration is difficult and seldom practiced.

Types of Chemical Sensors

The interaction of the target analyte with the sensing layer produces a signal when it is interrogated by a suitable transducer. The transducer is usually one of four types: electrochemical, mass, thermal, or optical.

Electrochemical Sensors (6). Many chemical interactions at the sensing layer involve ionic equilibrium or electron transfer. These interactions can be interrogated using an electrochemical transducer such as a metal electrode. There are three main types of electrochemical sensors: potentiometric (symmetric), potentiometric (asymmetric), and amperometric.

Symmetric potentiometric sensors are commonly known as ion-selective electrodes (ISEs) (7) and typically consist of two electrolytes separated by a selective membrane. The potential difference between the internal reference solution and the bulk sample solution is measured with respect to a reference electrode (e.g., Ag/AgCl or saturated calomel electrode).

The classical example of an ISE is the pH electrode. First reported by Cremer (8), it is the best developed sensor and has been commercially available since the 1920s. Symmetric potentiometric devices are also available for many other single and divalent cations (e.g., Na^+, NH_4^+, Ca^{2+}). The pH electrode has been used to prepare CO_2 sensors and other potentiometric biosensors.

Asymmetric potentiometric sensors are different from ISEs in that they do not contain an internal electrolyte solution. In these devices, sometimes referred to as solid-state devices, the sensing membrane is in contact with only one electrolyte (9). This approach has led to the development of ion-selective field effect transistors (ISFETs) (10) and consists of an ion-selective membrane in contact with semiconductor surfaces. The stability of these devices has been limited and matches that of ISEs. Potentiometric sensors are equilibrium devices, and the output is governed by the Nernst equation. The potential is proportional to the log of the activity of the target analyte (11).

In amperometric devices (12), the current at an electrode is measured at a fixed electrode potential. The current generated is proportional to the concentration of the target analyte. The first report of this type of sensor was the Clark oxygen electrode (13). In this device, a layer of buffer solution is trapped on the surface of a platinum electrode by a gas permeable membrane, typically PTFE or silicone rubber. The electrode is polarized to -750 mV versus Ag/AgCl reference electrode. At this potential, any oxygen arriving at the electrode surface is reduced; thus, the magnitude of the resulting current is directly proportional to the external oxygen concentration when mass transport through the membrane is rate limiting.

Mass Sensors (14). The interaction of a target analyte with a sensing layer results in an overall increase in mass at the sensing site. Microbalances can be used to prepare chemical sensors based on the mass increases. Piezoelectric materials (e.g., quartz) are ideal for use in preparing microbalances or piezoelectric sensors. In these devices, the piezoelectric material is used to create and conduct acoustic waves. The crystals are configured into resonant structures that transmit acoustic signals in a frequency-dependent manner. A change in mass of the material on the piezoelectric crystal will cause the resonant frequency to change, which is proportional to the concentration of the analyte. A detailed mathematical treatment of the principles of this work is given elsewhere. Piezoelectric mass detectors have been used to prepare humidity gauges, thin film sensors, and gas detectors. Liquid-phase systems have also been reported for a variety of analytes.

Thermal Sensors (15). Many chemical interactions result in the evolution or absorption of heat. The change in temperature brought about by the interaction can be used to make analytical measurements. The sensing tip is usually a thermocouple or any other temperature probe such as a thermistor. The sensing layer is usually an enzyme that generates heat via catalysis of substrate to product.

Optical Sensors (16). Spectroscopic techniques are well established and are routinely used in analytical laboratories. These techniques interrogate the optical properties of molecules that vary depending on the nature of their environment or change as a consequence of a chemical interaction. Fiberoptic chemical sensors (FOCSs) consist of an indicator molecule immobilized on the distal end or along the length of an optical fiber. The fiber is used primarily for carrying light information to and from the controlling instrument. A full treatment of FOCSs is given later.

Biosensors (17). Biosensors are devices that incorporate a biological catalyst connected or integrated with a transducer. The biocatalyst catalyzes a chemical reaction, and the transducer converts the biochemical activity into an electrical, optical, or thermal signal. The biological element can be an enzyme, antibody, receptor, tissue, organism, or whole cells.

The first biosensor reported was developed for the measurement of glucose (18). The device consists of a Clark oxygen electrode with the enzyme glucose oxidase immobilized on the external side of the gas-permeable membrane. The device operates by measuring the depletion of oxygen brought about by the oxidation of glucose by glucose oxidase.

The amperometric determination of glucose has been commercialized in the form of test strips for blood glucose measurements by Medisense Inc. In their device, the oxygen dependency of the Clark device is removed by the introduction of an Fe^{2+}/Fe^{3+} redox transport system.

The literature is replete with articles on biosensors, and there are several journals dedicated to this subject.

Sampling Methods (19)

There are two main strategies for on-line fermentation analysis: either the sensor is located outside the reactor and the sample is transported to the sensor (ex situ) or it is placed directly in the fermenter (in situ).

In Situ Monitoring (20). A chemical sensor that meets all the requirements for applications monitoring can be used on-line or in situ. Several assemblies that have been developed to undertake in situ monitoring are shown in Figure 2. In situ monitoring with chemical sensors is rare and generally limited to gas sensing where the transducer surface is protected by a gas-permeable membrane. Sensors that use semipermeable membranes are subject to rapid fouling and poisoning of the sensor surface.

Ex Situ Monitoring (21). The problems associated with in situ monitoring have resulted in the development of ex situ (Fig. 3) systems. In ex situ monitoring, the sample is drawn continuously to a remote site where the analysis is carried out. The procedure removes the need for sensor sterilization, and on-line calibration is easily achieved. In many cases, sample pretreatments are required before analysis, such as filtering, adjustment of the sample pH, temperature, and dilution. The main drawbacks to ex situ monitoring are the need for the use of pumping systems, the possibility of back-contamination at the sampling port, and the effect on the analyte after sample pretreatment.

Application of Chemical Sensors to Bioprocess Monitoring (22)

In the previous section, the requirements of chemical sensors and other analytical systems to bioprocess monitoring were discussed. There are inherent problems with using sensors for bioprocesses. Some of the main difficulties that have to be overcome include the needs to provide aseptic sensors; to prevent microbial fouling, poisoning, and disruptive effects of chemical and biochemical constituents on the sensors; to avoid biochemical and chemical interferences in complex analyte mixtures; to preclude analyte chelation, binding, and buffering effects; and to perform sensor calibration, all while operating under a wide variety of aqueous physiochemical conditions.

Despite the large number of papers, books, and reviews on chemical sensors in the public domain, only a tiny fraction are dedicated to bioprocess monitoring, and very few results have been presented on the successful application of these devices to bioprocess monitoring.

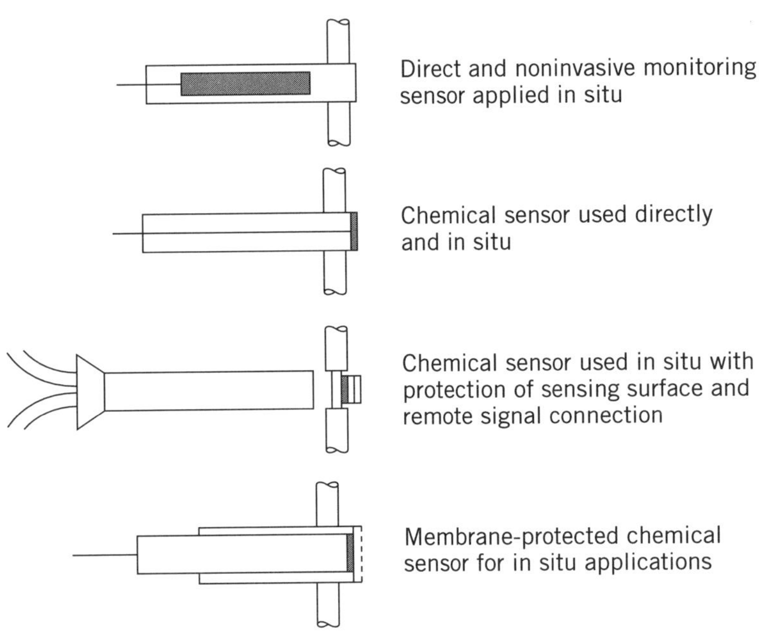

Figure 2. In situ sampling assemblies used for chemical sensors in fermentation monitoring. The sensor is in direct contact with the fermentation sample. Sample pretreatment is not possible with these configurations.

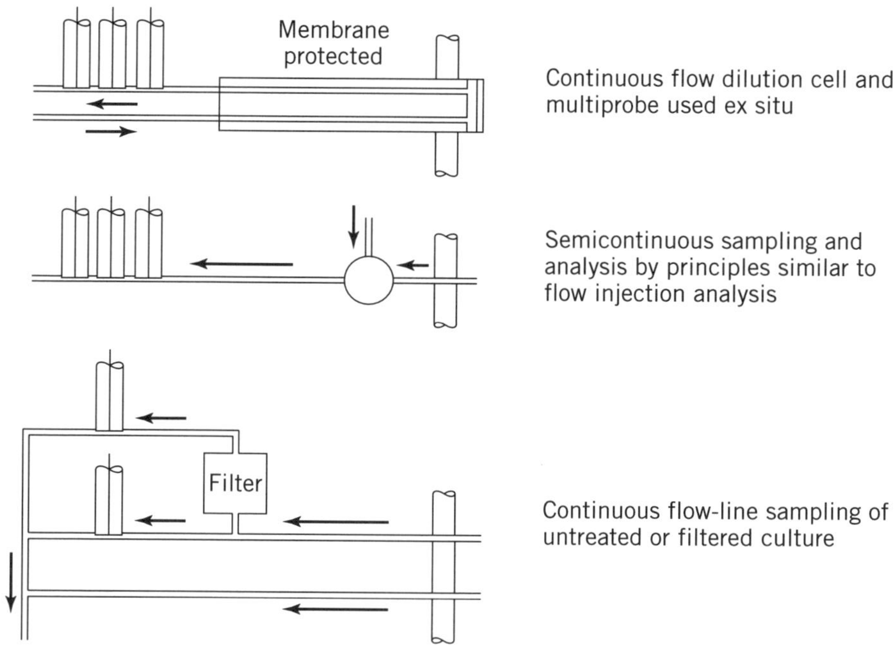

Figure 3. Ex situ sampling assemblies for using chemical sensors on-line. The sample is drawn from the fermentation vessel to the sensor(s). Sample pretreatment is possible with these configurations.

The first electrochemical sensor developed was the potentiometric pH glass electrode, and it has been modified and implemented for in-line or in situ bioprocess monitoring (23). It is by far the most widely used and the most successful chemical sensor to date. The pH sensor can be autoclaved and is generally considered biocompatible. The sensor is calibrated both before and after the fermentation, and these data are used to correct the measured pH values.

Carbon dioxide sensors based on pH measurements have also been used for in situ bioprocess monitoring (24). Carbon dioxide is measured indirectly by measuring the pH of a bicarbonate buffer solution in equilibrium with carbon dioxide. These types of sensors are based on original work by Severinghaus and Bradley (25).

The Clark oxygen sensor has also been adapted for bioprocess monitoring (26). Steam sterilizable glucose sensors based on oxygen measurements have been reported with lifetimes sufficient to last the course of the fermentation (27).

Commercial Chemical Sensor Systems for Bioprocess Monitoring

The measurement of key analytes in bioprocesses has led to the commercialization of several chemical sensors. The market is dominated by sensors for the most important analytes: pH and dissolved O_2. Table 1 summarizes a few of the commercially available sensors. Sensors for CO_2 are available but used primarily for off-line gas analysis. Most CO_2 gas analysis is performed using nondispersive FT-IR measurements. Optical sensors are not currently available commercially for bioprocess monitoring.

Table 1. Commercial Sensors for Bioprocess Monitoring

Manufacturer	Analyte	Transducer
YSI Inc. (Ohio)	pH	Potentiometric
	Glucose	Enzyme/amperometric
Russell pH Electrode Ltd (Scotland)	pH	Potentiometric
Mettler-Toledo (Greifensee, Switzerland)	pH	Potentiometric
	Dissolved O_2	Amperometric
Phoenix Electrode Inc. (Houston, Texas)	pH	Potentiometric
	Dissolved O_2	Amperometric

FIBEROPTIC CHEMICAL SENSORS

Optical Fibers

Optical fibers (28), also called fiberoptics, are used extensively in the telecommunications industry as a means for conducting information over long distances. Optical fibers are strands of glass or plastic consisting of a core, clad, and protective outer sheath. Generally, optical fibers range in diameter from 50 μm to 1 mm. The transmission of optical signals through an optical fiber is based on the phenomenon of total internal reflection. Figure 4 shows a typical optical fiber made of a core material of refractive index n_1 surrounded by a cladding material with a lower refractive index n_2. Because the index of refraction changes abruptly from the core to the clad, this type of arrangement is described as a step-index fiber. When the incident light entering the fiber strikes the core–clad interface, part of the light will be transmitted and some reflected. If light strikes the interface at an angle greater than the critical angle C as defined by equation 1,

$$\sin C = n_2/n_1 \quad (1)$$

then only light within a characteristic acceptance cone will be introduced into and carried within the fiber, as described by equation 2,

$$\sin\alpha = [(n_1^2 - n_2^2)^{1/2}]/n_0 \quad (2)$$

where sinC is the acceptance cone half angle and n_0 is the

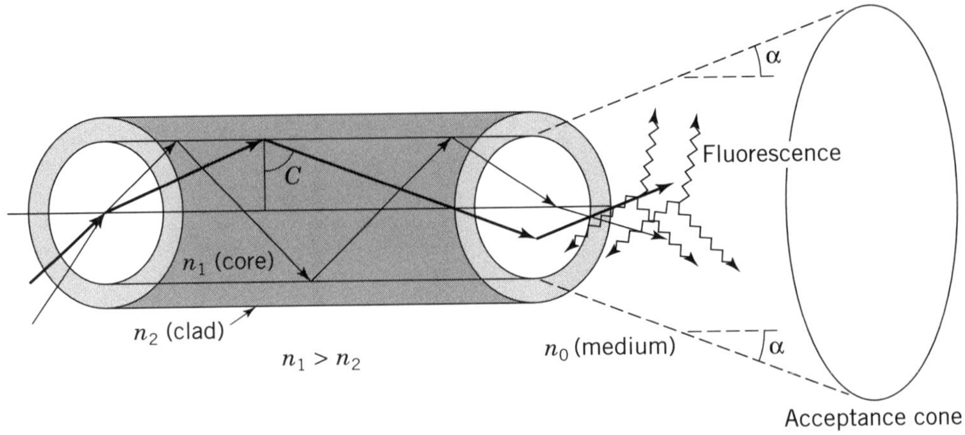

Figure 4. Cross-section of an optical fiber showing core, clad, and optical pathways.

index of refraction of the medium. By convention, collection efficiency is described by the numerical aperture (NA) or f number. As defined by equations 1 and 2, the greater the difference in refractive indices between the core and clad, the larger the acceptance half angle and therefore the more efficient the light collection of the fiber.

Preparation of FOCSs

The preparation of FOCSs requires immobilizing the indicator onto the optical fiber. The indicator is immobilized onto the core of the fiber either on the distal end or along the length of the fiber. The latter approach is used for the preparation of evanescent wave devices (16). In this article, only distal-end FOCSs will be considered because these are dominant and more useful in bioprocess monitoring.

Immobilization Techniques (29). One of the most critical aspects of optical sensors is the method of coupling the various indicating species to the optical fiber, a process known as immobilization. Immobilization techniques vary and are used according to the type of sensor being developed.

Single-Analyte Sensors. Several ways of immobilizing indicating species are shown in Figure 5. Figure 5**a** shows the most rudimentary form of an optical sensor. In this design, the optical fiber is used simply to conduct light to and from an indicator solution that is affixed to the distal end via a semipermeable membrane. Analyte diffuses across the semipermeable membrane and reacts with the indicator. Under ideal circumstances, the indicator is reversible, so that when the concentration of analyte is reduced, it dissociates from the indicator and leaves the solution in contact with the optical fiber. Many different forms of this design exist. Such designs can be extremely robust. This approach suffers from the need to assemble them by hand. Furthermore, they are not reproducible and generally possess long response times.

Figure 5**b** shows a sensor in which the sensing layer is coupled directly to the optical fiber surface. In this approach, a polymer layer containing immobilized indicator is in direct contact with the fiber. This approach is technically more demanding because it requires significant

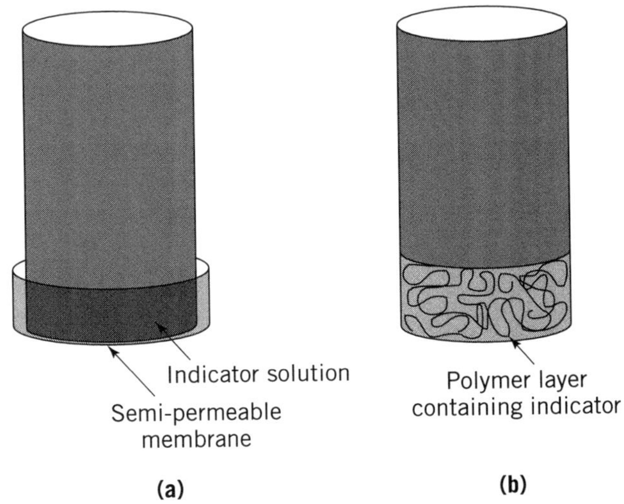

Figure 5. Immobilization methods for the preparation of FOCSs.

surface and polymer chemistry for implementation. First, the optical fiber surface is modified to possess reactive functional groups. Next, a prepolymer solution containing a modified indicator is applied to the optical fiber surface. Polymerization is initiated using either a thermal or photochemical approach. Surface reactive groups on the fiber engage in reaction with the prepolymer and functionalized indicator. In this manner, the indicator is covalently linked to the fibers' distal tip, creating a robust, resilient polymer layer to which the indicator is attached.

Multianalyte Sensors (30). In a typical measurement scenario, a multitude of analytes is often measured. Using conventional single-core optical fibers, it is necessary to use a separate optical sensor for each analyte. With additional sensors, the complexity of the readout instrumentation increases. An alternative approach uses imaging optical fibers (31). These fibers are arrays of coherent fused bundles containing thousands of individual optical fibers, each with its own clad and core, that are capable of transmitting images from one end of the array to the other. Such a transmitted image is shown in Figure 6.

As can be seen in Figure 6, the image is built up pixel by pixel similar to the way the compound eyes of insects

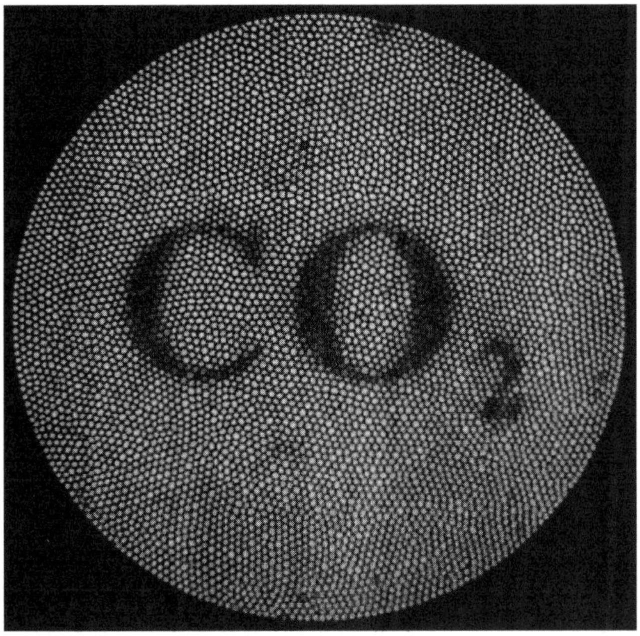

Figure 6. Imaging fiber is comprised of thousands of coherent individual fibers and is able to transmit an image.

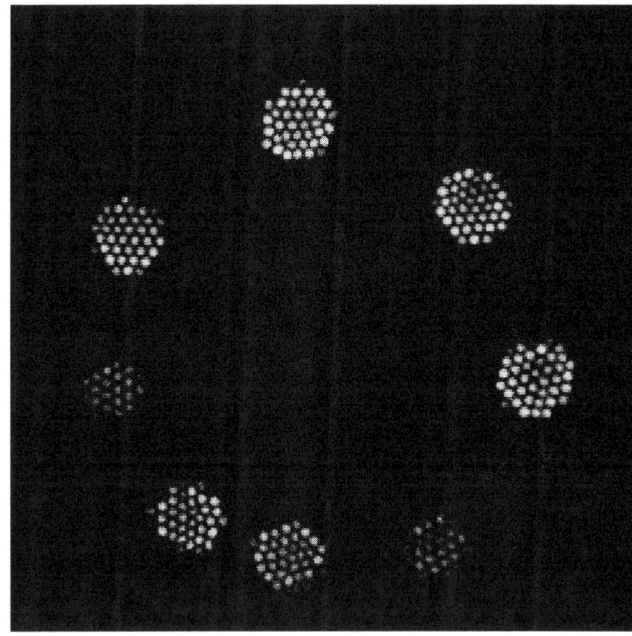

Figure 7. Collected fluoroscence image from a multianalyte imaging sensor shows sensor regions are polymerized over multiple fibers.

operate. This array architecture can be used to create a multianalyte optical sensor (32). In this approach, individual sensors are polymerized in discrete locations on the distal tip of the imaging fiber array using a photodeposition process. Each sensor element conveys an optical signal only through the optical fibers in contact with it. All sensors in the array can be addressed simultaneously. An array detector, such as a CCD camera, can be used to measure the signals from each sensor region. A example of a collected image from a multianalyte sensor array is shown in the Figure 7.

Transduction

There are a myriad of ways in which an optical fiber can be converted into a chemical sensor. In general, one must attach some type of chemical-indicating species to the optical fiber. The chemical-indicating species must provide an optical response that can be measured by the optical fiber.

Absorbance (33). In absorbance-based measurements, the presence of a particular chemical entity is detected by a change in the intensity or color of an indicator or reagent. The indicator is attached to the distal end of the optical fiber. In the presence of analyte, the absorbance increases or decreases in accordance with Beer's law:

$$A = \varepsilon bc \tag{3}$$

where ε is the extinction coefficient, b is the path length, and c is the sample concentration.

Fluorescence (34). An alternative method of detection uses fluorescence. Fluorescence sensors use fluorescent indicators that are more sensitive than absorbance-based sensors. Fluorescent indicators absorb light of one wavelength and emit light at a different wavelength. The advantage of such indicators is that they provide a black-background method because the emission signal arises from a zero background.

Theory of FOCSs

Fluorescence measurements can be made by using either the intensity of fluorescence or the fluorescence lifetime. With intensity measurements, the presence of the analyte correlates directly with the fluorescence intensity signal:

$$I_0/I = 1 + K_s[A] \tag{4}$$

where $K_s = [FA]/[F][A]$ is the binding constant, F is the fluorescence indicator, A is the analyte concentration, I_o is the fluorescence intensity in absence of analyte, and I is the fluorescence intensity in presence of analyte.

Upon binding analyte, the indicator can either increase or decrease its fluorescence intensity. An alternative intensity-based fluorescence measurement results from dynamic fluorescence quenching. This process involves collisional quenching in which the excited state indicating species collides with the analyte during the excited state lifetime. This process can be described by the Stern-Volmer equation:

$$I_0/I = 1 + K_{sv}[A] \tag{5}$$

where K_{sv} is the Stern-Volmer constant and $[A]$ is the analyte concentration.

Fluorescence Lifetime Measurements (35). A second measurement technique involves fluorescence lifetime. A variety of techniques are available for making fluorescence lifetime measurements. For most sensor applications, indicators are still used. The most common technique involves measuring the fluorescence lifetime directly. In this approach, the fluorescent indicator's lifetime is affected by analyte binding. By measuring the fluorescence decay as a function of time, the indicator lifetime can be calculated and correlated to analyte concentration. A significant advantage of this approach is that the background fluorescence does not interfere. Typical background fluorescence decays very rapidly, but most fluorescent indicators have longer lifetimes in the millisecond to nanosecond range. By watching the signal decay, one collects only signals emanating from the indicator. An increasingly popular method for performing fluorescence lifetime measurements is to use a frequency-modulated or phase-angle approach. This approach is based essentially on interferometry in which the delay between the excitation source and the emission signal is measured. The advantage of this technique is it does not require the sophisticated instrumentation necessary to measure fluorescence lifetimes. Fluorescence methods have the advantage of greater sensitivity relative to absorbance-based measurements, with fluorescence lifetime measurements requiring more sophisticated and somewhat more expensive instrumentation.

Ratiometric Measurements (36). With absorbance- or fluorescence-based measurements, it is advantageous to use ratiometric indicators. Ratiometric indicators possess either two excitation or two emission wavelengths that can be measured separately. A hypothetical example of a ratiometric indicator is shown in Figure 8. In this example, the fluorescent indicator has two distinct emission wavelengths. As the analyte concentration increases, the spectral contribution from one of the emission wavelengths increases while the other emission wavelength decreases. Replotting the data in Figure 8 shows that by plotting the ratio versus analyte concentration, the measurement becomes insensitive to the fluorescence intensity of a single peak. Furthermore, the ratiometric measurement is insensitive to changes in light intensity of the excitation source and/or the detector sensitivity. In addition, the measurement is independent of the absolute amount of dye present; dye leaching or photobleaching does not affect the measurement.

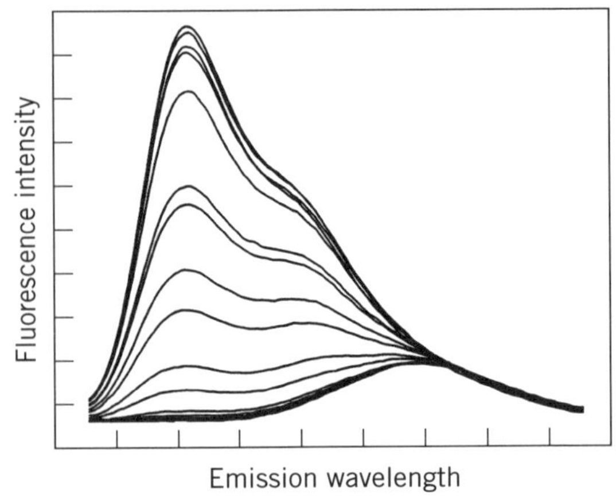

(a)

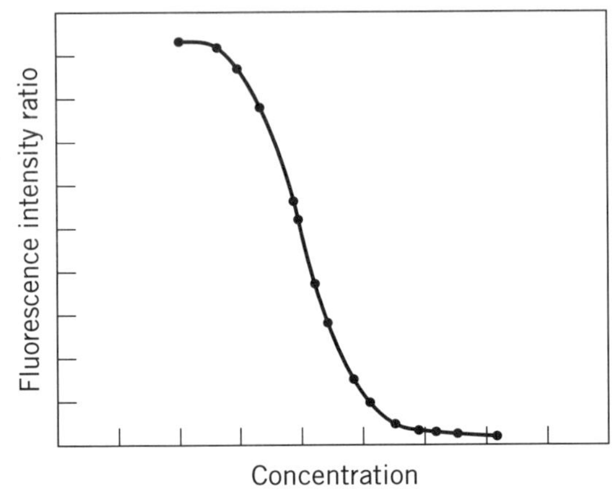

(b)

Figure 8. Ratiometric dye with two emission maxima (**a**). Replot of ratio versus analyte concentration is insensitive to instrumental variables (**b**).

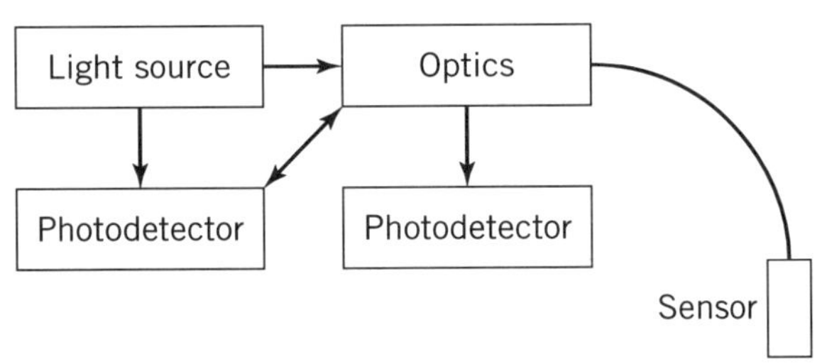

Figure 9. Typical instrumentation design for operating FOCSs.

Instrumentation (37)

The elements required to operate FOCS are shown in Figure 9. Typically, the instrument consists of a light source, optics, photodetector, and controller. The second photodetector is a reference detector that is sometimes necessary when ratiometric and lifetime measurements are not possible.

Light Source. Light sources can be either lasers or lamps, and their choice is dependent on the mode of operation. For example, for fluorescence lifetime measurements, lasers are used because they are good at producing nanosecond excitation light pulses. Table 2 provides a summary of some lasers that are used for optical sensing. In applications where a continuous light source is sufficient, then an arc lamp is often the choice. Lamps generally have relatively broad spectral emission and are much cheaper to purchase and operate than lasers. Figure 10 summarizes the available lamp types.

Optics. The optical components are responsible for focusing, splitting, directing, and selecting light from the

Table 2. Laser Light Sources

Medium	Type	Emitted wavelength (nm)
Ar	Gas	334, 351.1, 363.8, 454.5, 457.9, 465.8, 472.7, 476.5, 488.0, 496.5, 501.7, 514.5, 528.7
Kr	Gas	350.7, 356.4, 406.7, 413.1, 415.4, 468.0, 476.2, 482.5, 520.8, 530.9, 568.2, 647.1, 676.4, 752.5, 799.3
He-Ne	Gas	632.8
He-Cd	Gas	325.0, 441.6
N_2	Gas	337.1
XeF[a]	Gas	351
KrF[a]	Gas	248
ArF[a]	Gas	193
Ruby	Solid	693.4
Nd:YAG	Solid	266, 355, 532
$Pb_{1\text{-}x}Cd_xS$[b]	Solid	$2.9 \times 10^3 - 2.6 \times 10^4$
$PbS_{1\text{-}x}Se_x$[b]	Solid	$2.9 \times 10^3 - 2.6 \times 10^4$
$Pb_{1\text{-}x}Sn_xSe$[b]	Solid	$2.9 \times 10^3 - 2.6 \times 10^4$
$Pb_{1\text{-}x}Sn_xTe$[b]	Solid	$2.9 \times 10^3 - 2.6 \times 10^4$
Dye	Liquid	400–460 (Stilbene 420)
	Liquid	450–500 (Coumarin 460)
	Liquid	510–620 (Coumarin 503)
	Liquid	565–620 (Rhodamine 6G)
	Liquid	590–660 (Rhodamine 610)
	Liquid	655–720 (Oxaxine 720)
	Liquid	690–780 (Nile blue 690)

[a] Excimer media.

[b] This is a semiconductor diode laser. Lasing occurs at the *pn* junction. The emitted frequency is altered by altering the concentration of the dopant in the lead.

light source to the sensor and back to the detector. Figure 11 shows two possible configurations for operating FOCSs. Figure 11a is suitable for operating a single fluorescence-based sensor, and Figure 11b is suitable for absorbance-based systems.

Refracting, reflecting, and focusing light is achieved using lenses. Refractive optics include spherical, aspheric, and gradient index lenses, and reflection is accomplished using elliptical, spherical, and parabolic mirrors. Light splitting is achieved using optical couplers, beam splitters, or dichroic filters. Optical wavelength selection is necessary for (*1*) isolating an excitation spectral band, (*2*) selecting an optical signal from stray light, and (*3*) separating the excitation from the emission light. Even with the monochromatic light from a laser, a notch filter is required to avoid scattered or reflected laser light from reaching the detector. Wavelength selection is carried out using optical filters such as absorption and interference filters. It can also be achieved using diffraction gratings.

Photodetectors. The photodetector collects the photons emerging from the optical fiber that carries the information relating to the analytical measurement. There are three types of photodetectors: thermal, photoemissive, and quantum. The most popular types are the photoemissive or photomultiplier tubes (PMTs), which are based on the emission of electrons from a photocathode in a vacuum chamber. When photons of sufficient energy strike the photocathode, it emits an electron from the surface—the photoelectric effect. The electrons are accelerated by a voltage drop across a gap to a secondary emitting electron surface where its numbers are multiplied. The electrons are collected at the anode where they can readily be detected. Available photodetectors are summarized in Figure 12.

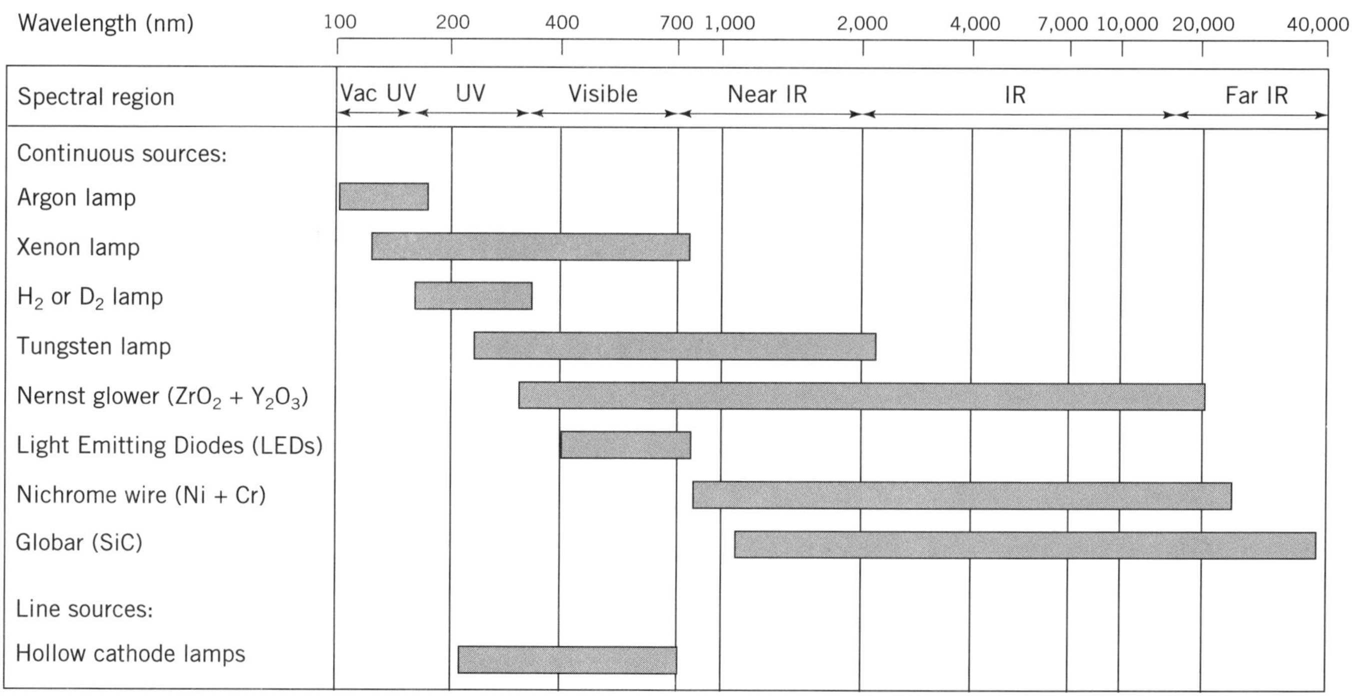

Figure 10. Light sources and their wavelengths.

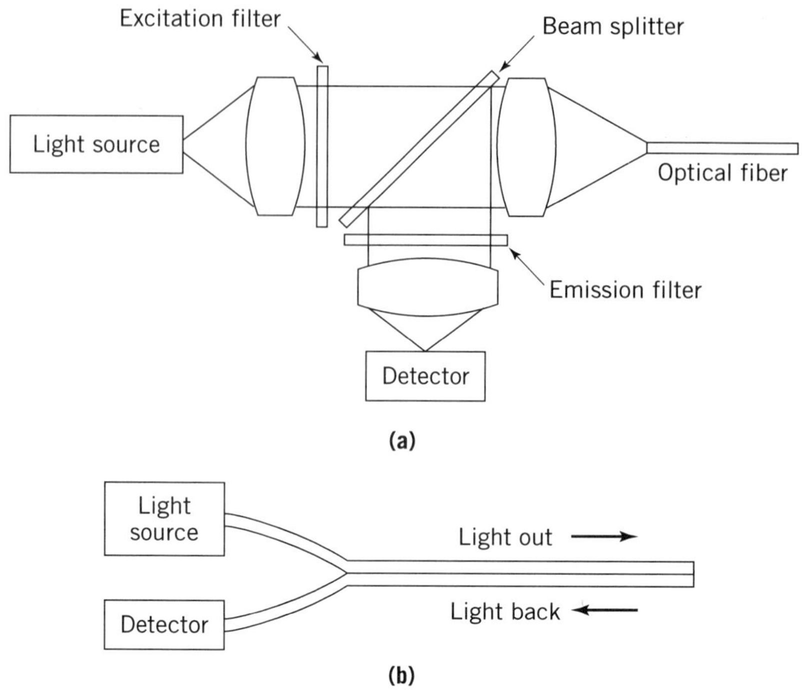

Figure 11. FOCS configurations. (**a**) is suitable for fluorescence-based devices and (**b**) is used for absorbance-based devices.

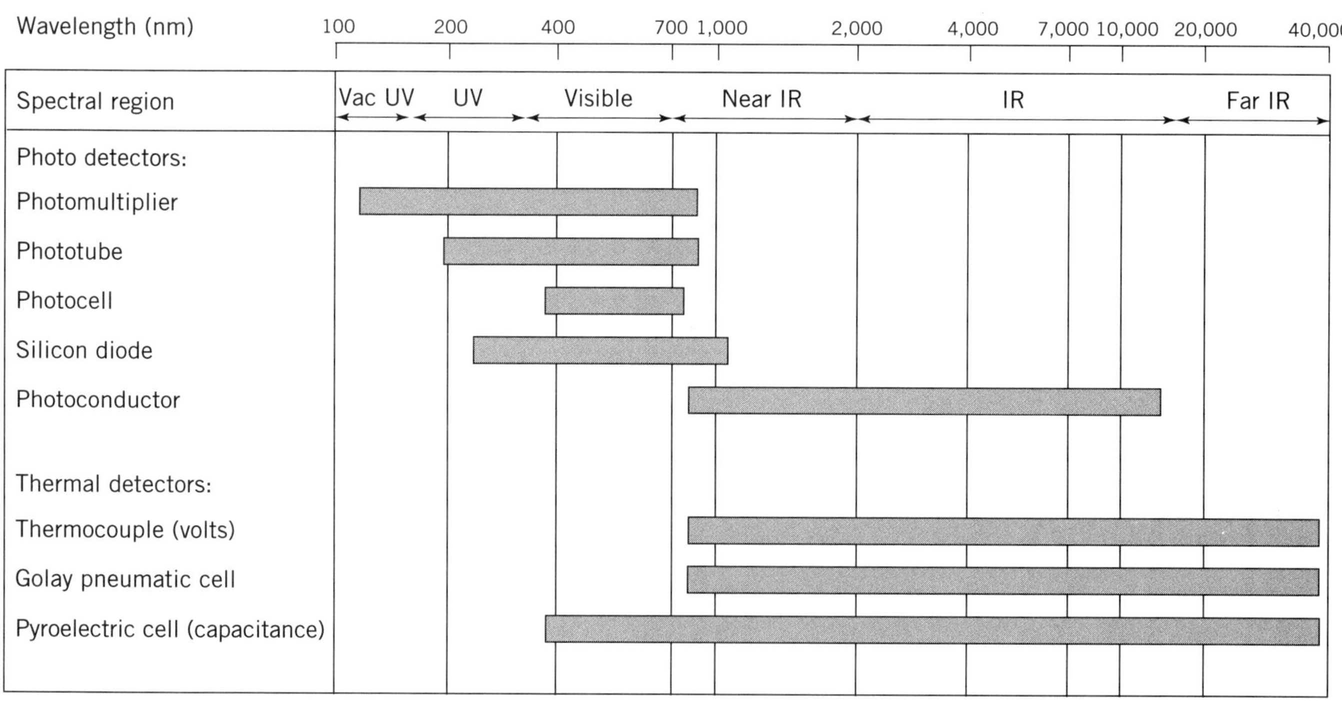

Figure 12. Light detectors and their wavelengths.

FIBEROPTIC SENSORS FOR BIOPROCESS MONITORING

The successful application of FOCSs to bioprocess monitoring has had limited success. The reasons for this lack of success will be discussed later. In this section, optical sensors that have been successfully used in fermentation analysis will be described.

pH Sensors

Measurement of pH. The measurement of pH in bioprocess is routinely carried out and is a good indicator of the state of the process. The spectroscopic determination of pH can be achieved using a variety of absorbance or fluorescent indicators. The particular indicator used depends on the pH range of interest because most indicators are only sensitive to ± 2 pH units around their pK_a. Generally, indicators are selected with a pK_a close to 7 because most fermentations are carried out between pH 5 and 9.

Theory of Spectroscopic pH Measurement (38). The spectroscopic determination solution pH is accomplished by measuring the concentration ratio of the dissociated (In^-) and undissociated (HIn) forms of the indicator. The overall equilibrium can be written $HIn \rightleftharpoons H^+ + In^-$. The acid dissociation constant, K_a, is given by $K_a = [H^+][In^-]/[HIn]$. Provided the total concentration of the indicator remains constant, then

$$\mathrm{pH} = \mathrm{p}K_a + \log([In]/[HIn]) \quad (6)$$

One form of the indicator (either associated or dissociated) must absorb or fluoresce. In the case where the absorbance of the undissociated indicator is negligible,

$$\mathrm{pH} = \mathrm{p}K_a + \log\{A/(A_0 - A)\} \quad (7)$$

where A is the absorbance of the dissociated indicator and A_0 is the absorbance when the indicator is completely dissociated. In the case of fluorescence,

$$\mathrm{pH} = \mathrm{p}K_a + \log\{I/(I_0 - I)\} \quad (8)$$

where I is the fluorescence intensity and I_0 is the intensity when the indicator is completely dissociated. A more optimal situation occurs when both forms of the indicator absorb or fluoresce. In this case, the ratio of the two indicator forms is characteristic of pH (see earlier text).

Bioprocess pH Sensors. The first measurement of fermentation pH using fiberoptics was achieved by dosing the fermentation media with a pH indicator. The optical fiber was placed in situ as a means of delivering and collecting light in order to measure the pH (39). This approach is not optimal because the introduction of a foreign species in a chemically sensitive environment can be detrimental to the outcome of the overall process.

The first example of a FOCS for fermentation pH was described by Agayn and Walt (40), and, to date, it is the only successful report of this type of sensor. The sensor consists of a fluorescein-derivatized hydroxyethyl methacrylate (HEMA) polymer covalently attached to the distal end of an optical fiber. The polymer thickness is approximately 10 μm, and the response time for a step change in pH is less than 5 s. The sensor operates in the ratiometric mode to account for system instability. The sensor can be steam sterilized in situ without any effect on the response characteristics and can be used for at least 2 months with no special storage requirements. The effects of temperature, stirring, and aeration had no significant effect on the sensor response for application to typical fermentation processes. Practical use of the device was demonstrated in the continuous measurement of pH during an *Escherichia coli* fermentation. The ability of this sensor to operate in fermentation media is attributed to the use of the ratiometric method and the use of an acrylate-based hydrogel polymer for the indicator immobilization. Acrylate polymers have been found to be compatible with fermentation and physiological media (41).

CO_2 Sensors

Measurement of CO_2. Cell activity in bioprocesses usually involves the production or uptake of CO_2. The continuous monitoring of both dissolved and off-gas CO_2 are good indicators of fermentation efficiency. Off-gas CO_2 analysis can be achieved using FT-IR analyzers (42) or mass meters, but direct measurement of dissolved CO_2 is not well established. The only chemical sensor developed for CO_2 measurements is the Severinghaus P_{CO2} electrode, as described in "Application of Chemical Sensors to Bioprocess Monitoring." This sensor has not been widely accepted because of extensive drift.

Theory of Spectroscopic CO_2 Measurement (43). The fiberoptic chemical sensor for CO_2 is the optical equivalent of the Severinghaus electrochemical sensor. In this case, the pH electrode is simply replaced by a pH-sensitive optical sensor. The CO_2 sensor consists of a pH sensor covered by a CO_2-sensitive bicarbonate solution. The solution is trapped within the fiber by a gas-permeable membrane, normally PTFE or silicone rubber, which excludes H^+ ions from the solution. The sensor measures the pH of the internal bicarbonate solution in equilibrium with the external CO_2 outside the membrane:

$$CO_{2(ext)} \rightleftharpoons CO_{2(mem)} \rightleftharpoons CO_{2(int)} + H_2O \rightleftharpoons H^+ \; HCO_3^- \rightleftharpoons 2H^+ + CO_3^{2-}$$

The relationship between the total CO_2 and H^+ is given by

$$[H^+]^3 + N[H^+]^2 - (K_1[CO_2] + K_w)[H^+] - 2K_1K_2[CO_2] = 0 \quad (9)$$

where N is the concentration of sodium bicarbonate in the solution, and K_1 and K_2 are the first and second acid dissociation constants for carbonic acid. CO_2 sensors operate in the pH and CO_2 range where the first and third terms are negligible. K_w is ignored because it is much greater than $K_1[CO_2]$. With these consideration the above equation reduces to

$$[H^+] = K_1[CO_2]/N \quad (10)$$

making the CO_2 concentration directly proportional to the H^+ concentration. A direct consequence of this analysis shows that as the bicarbonate concentration is reduced, the sensor's sensitivity is increased. Unfortunately, this increase in sensitivity is at the expense of a reduced dynamic range.

Bioprocess CO_2 Sensors. The successful measurement of dissolved CO_2 in fermentation media using a FOCS was first reported by Uttamlal and Walt (44). The sensor is based on the Severinghaus P_{CO2} electrode as described earlier. The pH indicator used in the sensor was a dual excitation fluorescent dye, hydroxypyrenetrisulfonic acid (HPTS), and allows the sensor to be operated in the ratiometric mode. The sensor has a reversible dissolved CO_2 dynamic range of 0 to 0.25 atm. The sensor can be autoclaved without affecting calibration. The response time to varying CO_2 step changes is in the range 1 to 13 minutes, in good agreement with studies carried out with the Severinghaus electrode. The temperature effect is negligible in the range 5 to 35 °C. Calibrations performed in fermentation media and clean solutions were alike, making it possible to carry out ex situ calibration. The practical suitability was tested in the monitoring of CO_2 production during the early stages of a beer fermentation.

A second system, reported by Sipior et al. (45), has been developed for off-gas CO_2 analysis and again is based on pH measurement. The sensor operates on changes in fluorescence lifetime in the frequency domain of a pH fluorophore sulforhodamine 101 (SR101) induced by an absorbance-based pH indicator, *m*-cresol purple (MCP). As the pH of the internal solution changes, the absorbance spectrum of MCP changes and alters the degree of interaction with SR101. The change in spectral overlap causes a change in the lifetime of SR101. The sensor calibration is obtained by correlating the phase change as a function of CO_2 concentration. The dynamic working range is 0 to 0.1 atm, and the response time to step changes in CO_2 is a few seconds. The sensor was used to take off-gas CO_2 measurements in an *E. coli* fermentation over a period of 10 h successfully. The lifetime of the sensor is on the order of 2 weeks.

O_2 Sensors

Measurement of O_2. The measurement of dissolved oxygen is important and, like pH and dissolved CO_2, is also an indicator of fermentation efficiency. For example, in beer fermentation, oxygen is required in the early stages for yeast to generate the lipids and sterols in the cell membrane essential for growth. The rate of oxygen uptake has also been shown to be an indicator of yeast vitality (46). Therefore, the importance of dissolved oxygen measurements in beer fermentation is twofold: as a tool to ensure the initial oxygen content is sufficient and to obtain an early indication of fermentation performance.

Theory for Spectroscopic O_2 Measurement (47). The vast majority of fiberoptic oxygen sensors are based on fluorescence quenching of the indicator. The best systems consist of a fluorescent indicator immobilized in a hydrophobic environment (e.g., siloxane). The fluorescence quantum yield is reduced by either static quenching (i.e., the indicator and oxygen associate in the ground state to give a nonfluorescent molecule), or dynamic quenching (i.e., an indicator collides with oxygen, leading to radiation-less deactivation of the excited state). The relationship between the fluorescence intensity (I), the unquenched intensity (I_o), and oxygen concentration is given by

$$I_0/I = 1 + K_{SV}[O_2] \quad (11)$$

The relationship in equation 11 is known as the Stern-Volmer equation, and it is applicable for both quenching models. If the sensor is operated using lifetime measurements, then I_o/I is replaced by τ_o/τ, where τ_o and τ are the unquenched and quenched lifetimes, respectively.

Bioprocess O_2 Sensors. The successful application of FOCS for O_2 to on-line or in situ bioprocess monitoring has not been demonstrated. However, several systems have been described where the sensors can be steam sterilized and hence are suitable for in situ applications (48,49).

Multianalyte Sensors

The measurement of a single analyte is important, but it reveals only limited information on the state of the fermentation process. A more informed analysis requires the simultaneous measurement of several analytes. Using imaging fiber bundles, Ferguson et al. (50) reported a sensor capable of simultaneously measuring O_2, CO_2, and pH in a beer fermentation.

Figure 13 shows a SEM of the individual sensors that comprise the multianalyte sensor. The working dynamic range of the individual sensors are 0 to 1 atm dissolved O_2 and 0 to 0.1 atm for dissolved CO_2; the pH range is 5.5 to 7.5.

Figure 14 shows the results obtained with the sensor during a beer fermentation. Upon pitching the yeast, O_2 in the wort decreases, pH decreases, and CO_2 increases. These results are as one would predict for such a fermentation and can be used to assess yeast vitality at very early stages of the process.

Optical Biosensors

Carbon and nitrogen sources provide fuel for cell metabolism in fermentation processes. Carbon sources generally consist of glucose and other related sugars, and nitrogen sources include a variety of amino acids. The monitoring of key carbon and nitrogen compounds indicate the state of the fermentation process. The measurement of individ-

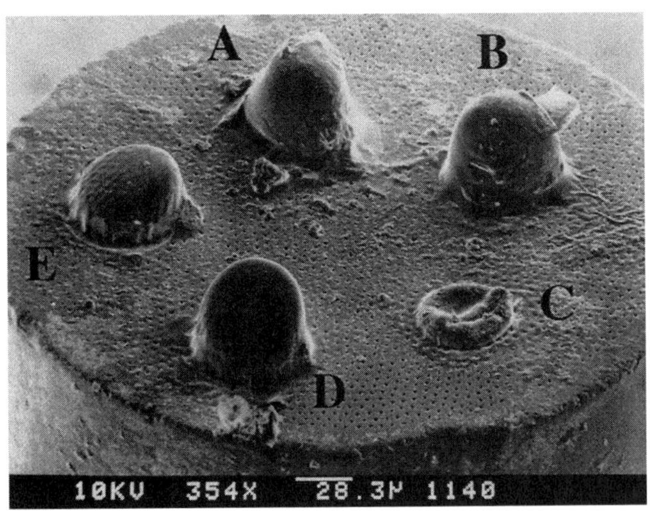

Figure 13. Multianalyte fiberoptic imaging sensor for pH, O_2, and CO_2. Sites *A* and *B* measure CO_2; *C*, pH; and *D* and *E*, O_2.

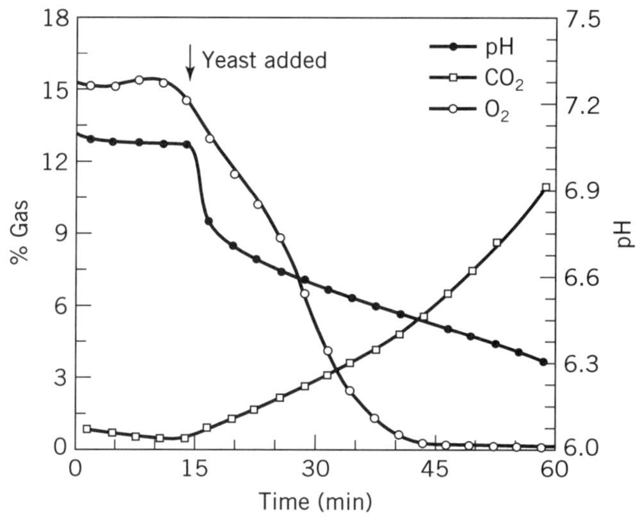

Figure 14. pH, O_2, and CO_2 profile in a beer fermentation using an imaging fiber sensor.

ual sugars and amino acids can be achieved using optical biosensors (51).

Theory of Optical Biosensors. The use of biological species in general and enzymes in particular for analysis is well established, and their incorporation into biosensors was first achieved in the 1950s by Clark and Lyons (18). In optical biosensors, the enzyme is immobilized onto the optical fiber in a suitable matrix. In most biosensors, the reaction of the biological component with the analyte is designed to generate a change in pH, O_2, or CO_2 near the sensor. The measurement of pH, O_2, and CO_2 has been described earlier. To illustrate this principle, glucose analysis can be performed using the enzyme glucose oxidase (GO/FAD):

$$\text{glucose} + \text{GO/FAD} \rightarrow \text{gluconic acid} + \text{GO/FADH}_2 \rightarrow \text{gluconate} + \text{H}^+$$

$$\text{GO/FADH}_2 + \text{O}_2 \rightarrow \text{GO/FAD} + \text{H}_2\text{O}_2$$

By either measuring the production of acid or the depletion of O_2 at the sensor, the glucose concentration can be inferred. It is also possible to measure H_2O_2 production using the indicator luminol; however, this is an irreversible measurement.

Bioprocess Biosensors. Biosensors are very difficult to use in on-line or in situ monitoring because the biological component cannot withstand autoclaving. Therefore, the use of optical biosensors is primarily for ex situ analysis. If in situ measurements are required, alternative sterilization protocols are possible, such as treating the sensor with radiation, ethanol, or antibiotics. After such treatment, the sensor can be inserted aseptically into the fermenter. There are only a few reports of optical biosensors that have been used in fermentation analysis. Short descriptions of three are given next.

Glucose. Glucose and fructose are the main carbon energy sources for cell activity in bioprocesses. Other carbohydrates such as sucrose and maltose are also used as energy sources, but they are broken down first to glucose and fructose by enzymes in the cell (52).

The development of FOCS glucose sensors using the enzyme GO/FAD has been achieved based on O_2 (53), pH (54), and H_2O_2 (55) measurements. These sensors have fast response times and have the dynamic range suitable for fermentation analysis (i.e., 0 to 50 mmol dm^{-3}); however, use in on-line bioprocess monitoring has not yet been achieved.

Amino Acids. Many bioprocesses require a nitrogen source for cell growth (52). Free amino acids, ammonium salts, urea, and gaseous nitrogen are typical of nitrogen-containing compounds used. Pure nitrogen-containing compounds are expensive, so alternatives such as corn steep liquor, yeast extract, peptones, and soy meal are commonly used. These liquors and extracts also contain a carbon source.

The measurement of nitrogen-containing compounds is therefore important in order to achieve a better understanding of the processes taking place and the state of the fermentation, with the particular nitrogen compound depending on the fermentation. An important nitrogen source are amino acids. It is possible to make optical sensors for L-amino acids using L-amino acid oxidase with transduction via the ammonia, ammonium ion, or H_2O_2 produced during biocatalysis:

$$\underset{\underset{\text{NH}_3^+}{|}}{\text{R—CH—CO}_2^-} + \text{H}_2\text{O} + \text{O}_2 \xrightarrow{\text{L-Amino acid oxidase}} \underset{\underset{\text{O}}{\|}}{\text{R—CH—CO}_2^-} + \text{NH}_4^+ + \text{H}_2\text{O}_2$$

The sensor responds to many amino acids, and the output corresponds to an aggregate amino acid concentration. Electrochemical devices based on this enzyme have been developed and used in bioprocess monitoring (56).

Another example that has been reported is the measurement of urea using the enzyme urease (57):

$$\text{Urea} \xrightarrow{\text{Urease}} 2\text{NH}_4^+ + \text{HCO}_3^-$$

The reaction generates a pH change that can be monitored using a pH sensor.

Penicillins. The discovery of penicillin by Fleming was a major breakthrough in the treatment of infectious diseases. The full-scale production was developed after World War II, and today penicillin and penicillin derivatives are produced in huge quantities (58).

Penicillin production has been monitored using both electrochemical and fiberoptic sensors. Most devices are based on the measurement of pH changes brought about by the hydrolysis of the β-lactam ring by the enzyme penicillinase (59):

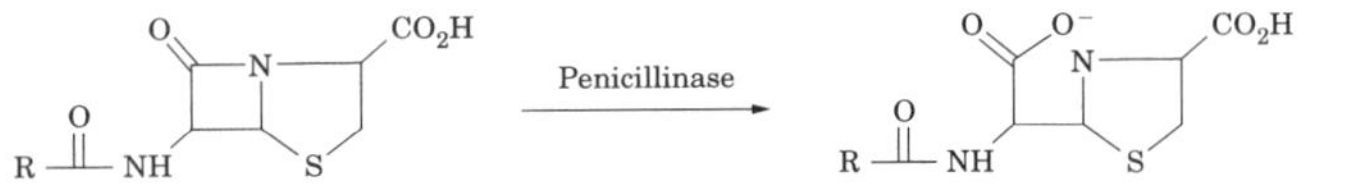

The major drawback to this method is that an independent measurement of pH must be made because the solution pH varies during the course of the fermentation. This measurement is not difficult, and pH is generally monitored anyway. A multianalyte FOCS that measures both pH and penicillin has been reported (60). The sensor was used off-line to measure penicillin production during a *Penicillium chrysogenum* fermentation (61), and results obtained from the sensor were in close agreement using a standard kit system.

DISCUSSION

The need to measure key analytes in bioprocesses is important for two reasons. First, monitoring will ensure that the process is operating correctly, and second, the information obtained will allow a better understanding of the process that regulates production. To achieve this, new methods of analysis have to be developed. FOCSs, because of their advantages over other analytical systems, are ideally suited for this type of monitoring; however, there are a number of issues that need to be solved before sensors can be implemented in bioprocess monitoring: fouling, drift, in situ calibration, and sterilization are a few of the more important ones.

Fouling

Biofouling occurs when surfaces come into contact with cells and media. When surfaces such as glass, plastic, or metal come into contact with a biological medium, protein adsorption occurs. After protein adsorption, adherent cells can attach to the protein layer. Cell and protein adhesion often result in a relatively thick layer called a biofilm that covers the end of the sensor. Such coverage compromises the measurement capability of the sensor and, in the worst cases, isolates the sensor from the medium. To avoid biofilm formation, it may be necessary to include an antifoulent, such as an antibiotic, that can be placed on the end of the sensor. This material must not interfere with the sensing capabilities of the sensor. Alternatively, it may be possible to place the sensor in a high-shear environment so that cells cannot readily adhere to the sensor surface.

Drift

In cases where internal calibration cannot be used, it may be necessary to account for sensor drift. Drift can be accounted for in several ways. If the properties of the sensor are measured for extended periods of time, it may be possible to observe a consistent change in sensor response as a function of time. For example, if dye photobleaching is a problem, one can measure the photobleaching rate as a function of time and obtain a photobleaching rate constant. This constant can then be used to correct for photobleaching effects during the fermentation process. A second method to account for sensor drift is to perform an in situ calibration in which a sample is taken from the fermentation vessel and the parameter of interest is measured using a conventional chemical analysis. The sensor is then reset to the measured value.

In situ calibration

In situ calibration for sensors being used in situ is difficult to achieve. Doping the fermentation medium with a known amount of analyte can be used, but the implications to the process can be severe. This approach can be a source of contamination, and in addition to other disadvantages, it means that in situ calibration is very rarely undertaken in nonexperimental fermentations.

For sensors that are being used ex situ, the problem is more easily solved. A pumping mechanism with valves can be used to switch between sample and a calibration solution. Sensor lifetime and reliability are increased using this method, but it is not always possible to make measurements in this way.

Sterilization

One of the most vexing issues with respect to sensor implementation is that of sterilization. The most common sterilization protocol in bioprocessing is steam sterilization. Unfortunately, many sensing chemistries are extremely sensitive to the high temperatures required during autoclaving. This problem is of particular concern when dealing with biosensors in which the active enzyme or antibody is denatured during the sterilization process. The only way to obviate this problem is to sterilize the sensor outside the fermentation vessel using a chemical approach, such as soaking the sensor in ethanol or a high concentration of antibiotic and then inserting it aseptically through the appropriate port on the fermentation vessel.

Sensor Lifetime

Another limitation of sensors, particularly biosensors, is their limited lifetime. During the long fermentation runs (e.g., months) of some mammalian cell cultures, sensors and biosensors may lose their activity or sensitivity. In this case, it may be necessary to replace a sensor during the culture. For particularly sensitive cultures in which it would be dangerous or inadvisable to replace a sensor, it may be possible to prolong the lifetime of a sensor by simply decreasing the duty cycle. For example, if photobleaching is a serious problem, one can simply calculate the sensor's total allowable light exposure and then determine how often a reading can be obtained during the fermentation run without compromising the photobleaching sensitivity.

SENSING BIOPROCESSES IN THE FUTURE

The successful monitoring and eventual control of bioprocesses in the future using optical sensors will depend on overcoming the problems discussed earlier and then on developing new molecules and chemical methodologies to undertake the analyses. The following examples of work in the early stages of development will, we hope, mature to produce ideal optical sensors for bioprocess monitoring.

Multianalyte Sensing

As sensors become more sophisticated, there will be an increasing demand to measure an ever-increasing number of analytes. In addition to the consensus analytes discussed earlier, there are a number of other species that, if monitored, would provide additional information as to the status of a particular fermentation. These analytes include trace minerals such as metal ions, complex or exotic carbon and nitrogen sources, growth regulators, enzymes or other proteins, and a wide variety of products and by-products. As information is gleaned from the Human Genome Initiative, there may be even more need to monitor additional regulators, peptides, and small organic molecules. Furthermore, there will be tremendous value placed on being able to detect specific gene sequences or, more importantly, the expression levels of messenger RNA. In principle, such gene-expression monitoring combined with the conventional and expanded set of analytes discussed above may result in the need to monitor hundreds, if not thousands, of distinct species simultaneously. Although this may seem like a daunting task, tremendous advances are being made presently in the areas of high detection sensitivity, miniaturization, and high-density arrays. Advances in all of these areas may one day lead to the ability to continuously monitor the multitude of analytes necessary to give a complete picture of the conditions and state of a particular fermentation process.

Electronic Noses

All the sensors discussed earlier are based on the conventional one-sensor–one-analyte paradigm in which a selective sensor is designed to respond to a single analyte. As the number of analytes increases, the size and complexity of the sensing system increases proportionately. A recent trend is to construct a sensing system that mimics the olfactory system. The olfactory system relies on a multitude of cross-reactive receptor cells that respond broadly to a wide variety of odorous substances (62). The complex pattern generated by the cells when an odor is presented is relayed to the olfactory bulb and then further transmitted to the higher-order cortical regions of the brain where it is recognized as a particular odor. In a similar vein, electronic noses (63,64) are based on cross-reactive arrays of sensors that differentially respond when exposed to a particular vapor. The generated response pattern is used to train a computational network that then recognizes the particular pattern upon a subsequent vapor exposure. Such a system presents a significant advantage in terms of sensor construction because additional sensors are not required to detect additional analytes; it is sufficient simply to train the network on an additional pattern. Existing manifestations of these systems are expensive and require significant sample preconditioning before sample introduction (65). A limited amount of work has been conducted on the off gases from particular fermentation processes (66). All such systems have only been demonstrated for vapor-phase measurements. The application of such an approach to liquid measurements has not been demonstrated to date. These systems can be trained in a manner more sophisticated than simply detecting the presence or absence of a particular analyte. Such a system can be trained to look at the overall quality or character of the volatile materials in the head space. In principle, such a system may allow the biochemical engineer to train for a desired outcome without necessarily knowing the specific chemical composition of the fermentation broth. For example, it may be possible to train the system on such qualities as maximum product concentration, minimum by-product concentration, off odors, and so on, without specifically training on those particular characteristics. In other words, the sensor may be able to key on surrogate components that are indicative of the desired outcome. This approach has the potential to be a very powerful tool for the biotechnologist and challenges our conventional thinking with regard to chemical monitoring.

Thermostable Enzymes

Thermostable enzymes are isolated from thermophilic microorganisms that exist at elevated temperatures in hot springs and deep ocean vents. Thermostable versions of useful enzymes are presently used in glucose sensors and as reagents for the polymerase chain reaction used in molecular biology. The ability to incorporate thermostable enzymes into biosensors may ultimately enable such sensors to be steam sterilized. Attempts to accomplish such heat stability have met with failure because existing thermostable enzymes still exhibit appreciable denaturation at elevated temperatures.

By isolating and determining the structures of thermostable proteins, researchers are beginning to understand the structural features that lead to enhanced stability. Eventually it may be possible for molecular biologists to engineer enzymes, antibodies, and receptor proteins with extremely high thermal stabilities.

Designed Receptors

An area of significant relevance to sensors is that of molecular recognition. Organic chemists are making great strides in understanding the principles governing how molecules and ions are recognized. For sensors, recognition is only half the story; it is critical that recognition be coupled to signal transaction. Rational molecular design is one approach to developing such receptors. An alternative and increasingly popular approach relies on combinatorial chemistry. In this approach, libraries of large numbers of distinct molecules are created and screened for function. Although most of the focus of combinatorial chemistry has been directed toward drug discovery (67), some recent work has used the approach to search for structures that combine binding with signal transduction (68).

SUMMARY

There are presently a limited number of analytes that can be monitored continuously during a fermentation or cell culture. The bioprocess monitoring scene is changing as new sensors become available. Optical sensors offer significant advantages because of their small size, internal calibration capability, freedom from electromagnetic interference, and multiplexing ability. As new sensing materials and sensing paradigms become available from the fields of molecular recognition and combinatorial chemistry, thermostable enzymes, microfabrication, electronic noses, and advanced sensors with powerful analytical capabilities will become available. Ultimately, such sensing systems will provide the biotechnologist and bioengineer with timely information regarding the status and composition of a particular fermentation process.

BIBLIOGRAPHY

1. M.T. Riebe and D.J. Eustace, *Anal. Chem.* **62**, 65A–71A (1990).
2. T.E. Edmonds, *Chemical Sensors*, Blackie, Glasgow, 1988.
3. J. Janata, *Principles of Chemical Sensors*, Plenum Press, New York, 1989.
4. R.W. Cattrall, *Chemical Sensors*, Oxford University Press, Oxford, 1997.
5. O.S. Wolfbeis, in J.V. Twork and A.M. Yacynych eds., *Sensors in Bioprocess Control*, Dekker, New York, 1990, pp. 95–125.
6. C. Brett and A.M. Oliveira-Brett, *Electroanalysis*, Oxford University Press, Oxford, 1998.
7. J. Koryta and K. Stulik, *Ion-Selective Electrodes*, Cambridge University Press, Cambridge, 1983.
8. M. Cremer, *Z. Biol.* **47**, 562–608 (1906) (in German).
9. M.J. Madou and S.R. Morrison, *Chemical Sensing with Solid-State Devices*, Academic Press, New York, 1989.
10. B.H. Vanderschoot, P. Bergveld, M. Bos, and L.J. Bousse, *Sens. Actuators* **4**, 267–272 (1983).
11. B.H. Vassos and G.W. Ewing, *Electroanalytical Chemistry*, Wiley, New York, 1983.
12. S.C. Chang, J.R. Stetter, and C.S. Cha, *Talanta* **40**, 461–477 (1993).
13. L.C. Clark, *Trans. Am. Soc. Artific. Internal Organs* **2**, 41–57 (1956).
14. J.F. Alder and J.J. McCallum, *Analyst* **108**, 1169–1189 1983.
15. G.C.M. Meijer, *Sens. Actuators* **10**, 103–125 (1986).
16. S.A. Momin and R. Narayanaswamy, *Optical Fibre Sensors: Current State Review*, UMIST, 1993.
17. E.A.H. Hall, *Biosensors*, Open University Press, Milton Keynes, 1990.
18. L. Clark and C. Lyons, *Ann. N.Y. Acad. Sci.* **102**, 29–45 (1962).
19. P.N. Royce, *Crit. Rev. Biotechnol.* **13**, 117–149 1993.
20. J.B. Callis, D.L. Illman, and B.R. Kowalski, *Anal. Chem.* **59**, 624A–637A (1987).
21. S.O. Enfors and N. Cleland, *Methods Enzymol.* **137**, 298–307 (1988).
22. D.J. Clarke, M.R. Calder, R.J.G. Carr, B.C. Blake-Coleman, S.C. Moody, and T.A. Collidge, *Biosensors* **1**, 213–320 (1985).
23. D.J. Clarke, B.C. Blake-Coleman, M.R. Calder, and A. Burns, *Ion Selec. Elec. Rev.* **4**, 75–133 (1982).
24. E. Puhar, A. Einsels, H. Buhler, and W. Ingold,
25. J.W. Severinghaus and A.F. Bradley, *J. Appl. Physiol.* **13**, 515–520 (1956).
26. Y.H. Lee and G.T. Tsao, *Adv. Biochem. Eng.* **13**, 35–86 (1979).
27. N. Cleland and S.-O. Enfors, *Eur. J. Appl. Microbiol. Biotechnol.* **18**, 141–147 (1983).
28. J.A. Buck, *Fundamentals of Optical Fibers*, Wiley, New York, (1995).
29. G.T. Hermanson, A.K. Mallia, and P.K. Smith, *Immobilized Affinity Ligand Techniques*, Academic Press, Sidcup, U.K., 1992.
30. P. Pantano and D.R. Walt, *Anal. Chem.* **67**, 481A–487A (1995).
31. T. Tsumanuma, K. Tanaka, S. Chigira, K. Sanada, and K. Inada, *SPIE Optical Fibers in Medicine* **906**, 92–96 (1988).
32. S.M. Barnard and D.R. Walt, *Nature* **353**, 338–340 (1991).
33. G.M. Barrow, *Physical Chemistry*, 6th ed., McGraw Hill, London, 1996, pp. 568–572.
34. J.R. Lakowicz, *Topics in Fluorescence Spectroscopy: Volume 1 Techniques*, Plenum Press, New York, 1991.
35. J.R. Lakowicz, *Principles of Fluorescence Spectroscopy*, Plenum Press, New York, 1983.
36. J.E. Jones and R.C. Spooncer, *J. Phys. Ed.* **16**, 1124–1126 (1983).
37. D.N. Modlin and F.P. Milanovich, in O.S. Wolfbeis ed., *Fiber Optic Chemical Sensors and Biosensors Vol I*, CRC Press, Florida, 1991, pp 237–301.
38. T.E. Edmonds, N.J. Flatters, C.F. Jones, and J.N. Miller, *Talanta* **35**, 103–107 (1988).
39. B.H. Junker, D.I.C. Wang, and T.A. Hatton, *Biotechnol. Bioeng.* **32**, 55–63 (1988).
40. V. Agayn and D.R. Walt, *Bio/Technology* **11**, 726–729 (1993).
41. Y. Benmakroha, S. Zhang, and P. Rolfe, *Med. Biol. Eng. Comput.* **33**, 811–821 (1995).
42. R.D. Eckles, J.M. Welles, and K. Petersen, *Measurements & Control* **161**, 83–89 (1993).
43. M.B. Tabacco, M. Uttamlal, M. McAllister, and D.R. Walt, *Anal. Chem.* **71**, 154–161 (1999).
44. M. Uttamlal and D.R. Walt, *Bio/Technology* **13**, 597–601 (1995).
45. J. Sipior, L. Randers-Eichhorn, J. Lakowicz, G.M. Carter, and G. Rao, *Biotechnol. Prog.* **12**, 266–271 (1996).
46. F.L. Peddie, W.J. Simpson, B.V. Kara, S.C. Robertson, and J.R.M. Hammond, *J. Inst. Brew.* **97**, 21–25 (1991).
47. E.R. Caraway, J.N. Demas, B.A. DeGraff, and J.R. Bacon, *Anal. Chem.* **63**, 337–342 (1991).
48. S.B. Bombat, R. Halavanahali, J.R. Lakowicz, G.M. Carter, and G. Rao, *Biotechnol. Bioeng.* **43**, 1139–1145 (1994).
49. B.H. Weigl, A. Holobar, W. Trettnak, T. Klimant, H. Kraus, P. O'Leary, and O.S. Wolfbeis, *J. Biotechnol.* **32**, 127–138 (1994).
50. J.A. Ferguson, B.G. Healey, K.S. Bronk, S.M. Barnard, and D.R. Walt, *Anal. Chim. Acta.* **340**, 123–131 (1997).
51. J.J. Ramsden, *J. Mol. Recog.* **10**, 109–120 (1997).
52. P.F. Stanbury and A. Whitikar, *Principles of Fermentation Technology*, Pergamon Press, Oxford, 1984.
53. B.P.H. Schaffar and O.S. Wolfbeis, *Biosensors Bioelectron.* **5**, 137–148 (1990).
54. W. Trettnak, M.J.P. Leiner, and O.S. Wolfbeis, *Biosensors* **4**, 1519–1523 (1989).
55. F. Schubert, F. Wang, and H. Rinneberg, *Microchim. Acta* **121**, 237–247 (1995).
56. W.J. Albery, J.E.C. Hutchins, and M. Uttamlal, *Appl. Electrochem. A.* **26**, 243–248 (1996).
57. A. Sansubrino and M. Mascini, *Biosensors Bioelectron.* **9**, 207–216 (1994).
58. D. Wilson, *Penicillin in Perspective*, Faber, London, 1976.

59. J.B.K. Nielsen, *Methods Enzymol.* **97**, 153–158 (1983).
60. B.G. Healey and D.R. Walt, *Anal. Chem.* **67**, 4471–4476 (1995).
61. L.H. Christensen, C.M. Henriksen, J. Nielsen, J. Villadsen, and M. Egelmitani, *J. Biotechnol.* **42**, 95–107 (1995).
62. J.S. Kauer, *Trends Neurosci.* **14**, 79–85 (1991).
63. K. Persand and G. Dodd, *Nature* **299**, 352–355 (1982).
64. T.A. Dickinson, J. White, J.S. Kauer, and D.R. Walt, *Nature* **382**, 697–700 (1996).
65. T.A. Dickinson, J. White, J.S. Kauer, and D.R. Walt, *Trends Biotechnol.* **16**, 250–258 (1998).
66. J.W. Gardner, T.C. Pearce, S. Fried, P.N. Bartlett, and N. Blair, *Sens. Actuat. B.* **18–19**, 240–243 (1995).
67. E.M. Gordon, M.A. Gallop, and D.V. Patel, *Acc. Chem. Res.* **29**, 144–154 (1996).
68. W.C. Still, *Acc. Chem. Res.* **29**, 155–163 (1996).

OPTICALLY ACTIVE 1,2-DIOLS, MICROBIAL PRODUCTION BY STEREOINVERSION

JUNZO HASEGAWA
MASAHIRO OGURA
SATORU TSUDA
SHUN-ICHI MAEMOTO
HIDETOSHI KUTSUKI
TAKEHISA OHASHI
Kaneka Corporation
Hyogo, Japan

KEY WORDS

Alcohol dehydrogenase
(*S*)-1,2-Butanediol
Candida parapsilosis
(*R*)-3-Methylthio-1,2-propanediol
(*S*)-4-Methyl-1,2-propanediol
(*S*)-1,2-Pentanediol
(*S*)-1-Phenyl-1,2-ethanediol
(*S*)-3-Phenyl-1,2-propanediol
(*S*)-4-Phenyl-1,2-butanediol
Stereoinversion

OUTLINE

INTRODUCTION

Optically active 1,2-diols are a versatile group as chiral starting materials for the syntheses of pharmaceuticals, agrochemicals, pheromones, liquid crystals, and so on. Several routes of preparation for optically active 1,2-diols have been reported.

1. Microbial reduction of 1-hydroxy-2-ketones (1–5)
2. Chemical reduction of optically active 2-hydroxy acids prepared from 2-keto acids by enzymatic reduction of amino acids (6–11)
3. Optical resolution of racemic 1,2-diols by lipase-catalyzed transesterification in organic media (12)
4. Chemical preparation from naturally occurring products, such as (*S*)-malic acid (2), D-mannitol (2,13,14) or (+)-pulegone (15).

We developed two efficient methods for producing various optically active (*S*)-1,2-diols with high optical purities. One is the organic synthetic method using the Grignard reaction starting from (*R*)-3-chloro-1,2-propanediol as a starting material can be produced easily from the racemate by microbial optical resolution using *Serratia marcescens* (17). The other is the microbial stereoinversion method using racemate as a substrate (18). In the course of the studies on the microbial optical resolution of racemic 1,2-pentanediol, we found a unique microbial stereoinversion consisting of two successive reactions. It is a known fact that in all enzymatic or chemical optical resolution processes, the theoretical yield of one part of enantiomer is maximally 50%. However, (*S*)-enantiomer of 1,2-pentanediol was found higher than the theoretical yield when the racemate was incubated in the culture broth of *Lodderomyces elongisporus* IFO 1676. This result indicated that this microorganism could invert (*R*)-1,2-pentanediol in the racemate to (*S*)-enantiomer. This finding led us to rescreen the microorganisms that could invert (*R*)-1,2-pentanediol to the (*S*)-enantiomer, and we found that a number of microorganisms had this ability, similar to *L. elongisporus* IFO 1676. We applied this microbial stereoinversion to various 1,2-diols by use of *Candida parapsilosis* IFO 0708, and we confirmed that nine kinds of (*S*)-1,2-diols (**1** to **9**) and (*R*)-3-methylthio-1,2-pentanediol (**10**) with high optical purities could be obtained in high yields (Scheme 1).

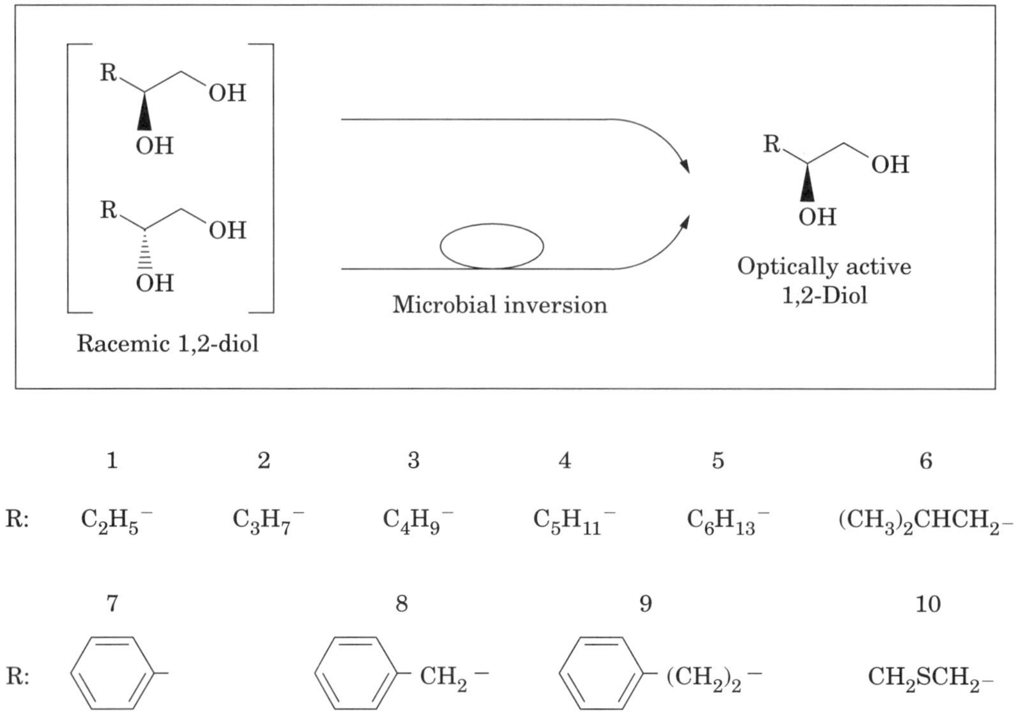

Scheme 1.

EXPERIMENTAL

Chemicals

Racemic 1,2-heptanediol (**4**), 4-methyl-1,2-pentanediol (**6**), 3-phenyl-1,2-propanediol (**8**), and 4-phenyl-1,2-butanediol (**9**) were prepared from the corresponding 1-alkenes by a method similar to that of Blau et al. (19). Other 1,2-diols used were purchased from Aldrich Chemicals Company. 1-Hydroxy-2-pentanone was prepared from 1,2-pentanediol by microbial oxidation using *Saccharomycopisis lipolytica* IFO 1551; (bp 62 °C, 6 to 7 mmHg), NMR $\delta_H(CDCl_3)$: 0.95 (3H, t, J = 7.5 Hz, CH_3), 1.68 (2H, m, J = 7.3 Hz, -$\mathbf{CH_2}$-CH_3), 2.44 (2H, t, J = 7.5 Hz, -$\mathbf{CH_2}$-(C=O)-), 3.43 (1H, broad, OH), 4.29 (2H, s, -$\mathbf{CH_2}$-OH), MS *m/z*: 102 (M^+), 71 (M^+ -CH_3O).

Microorganisms and Cultivation

All microorganisms used were stock cultures kept in our laboratory. The medium used contained 4% glucose, 0.3% yeast extract, and 10% (v/v) mineral solution, pH 7.0. The mineral solution consisted of 13% $(NH_4)_2HPO_4$, 7% KH_2PO_4, 0.8% $MgSO_4$ $7H_2O$, 0.1% NaCl, 600 ppm $ZnSO_4$ $7H_2O$, 900 ppm $FeSO_4$ $7H_2O$, 50 ppm $CuSO_4$ $5H_2O$, and 100 ppm $MnSO_4$ $4H_2O$. Microorganisms were cultured in a 500-mL Sakaguchi flask containing 50 mL of the medium at 30 °C for 1 to 2 days with shaking. When the cell grew maximally, the cells were harvested by centrifugation and were used for the reaction.

Screening of Microorganisms Producing (*S*)-1,2-Pentanediol by Stereoinversion

Cells were suspended in 50 mL of 0.1 M potassium phosphate buffer (pH 6.5) containing 250 to 500 mg of racemic 1,2-pentanediol. The reaction was conducted in 500 mL-Sakaguchi flasks that were incubated at 30 °C for 24 to 96 h with shaking. Two milliliters of the reaction mixture was used for GLC analysis of 1,2-pentanediol. The remainder was centrifuged to remove the cells, and the supernatant was concentrated to 3 to 4 mL. 1,2-Pentanediol was recovered by extraction with ethyl acetate (100 mL) and used for measurement of enantiomeric excess by HPLC.

Preparation of Cell-Free Extract and Enzyme Assay

Cells of *C. parapsilosis* IFO 0708, which grew for 24 h under the conditions described earlier, were collected and washed with 10 mM tris-HCl buffer (pH 7.0). Thirty-three grams of washed cells were suspended in 40 mL of the same buffer containing 1 mM EDTA and 0.2 mM phenylmethylsulfonylfluoride, a serine-protease inhibitor, and disrupted with a Braun homogenizer. Cell debris was removed by centrifugation (15,000 rpm × 30 min), and the supernatant was filtered through a millipore membrane (0.22 μm) and dialyzed against 10 mM tris-HCl buffer overnight. The cell-free extract was used for assay of the enzyme activities. The enzyme activities for (*R*) and (*S*)-1,2-pentanediol oxidation were assayed by measuring the increase of optical absorbances at 370 nm caused by the reduction of NAD^+ or $NADP^+$, respectively. The assay mixture in a final volume of 3.0 mL contained 3.0 μmol of $NAD(P)^+$, 0.3 mmol of (*S*) or (*R*)-1,2-pentanediol, 0.3 mmol of sodium carbonate buffer (pH 10), and a suitable amount of the cell-free extract. Incubation was performed at 25 °C in a cuvette of 1-cm light path length. The reduction of 1-hydroxy-2-pentanone was monitored by measuring the decrease of optical absorbency at 340 nm caused by the oxidation of NADH or NADPH. The assay mixture in a final volume of 3.0 mL contained 1.0 μmol of NAD(P)H, 0.3 mmol of 1-hydroxy-2-pentanone, 0.3 mmol of tris-maleate buffer (pH 6.0), and a suitable amount of the cell-free extract. The incubation was carried out under the same conditions as the method for oxidation of 1,2-pentanediol.

Preparative-Scale Reaction and Isolation of Optically Active 1,2-Diols

C. parapsilosis IFO 0708 was cultured in a 5-L jar fermenter containing 3 L of the medium at 30 °C for 23 h (aeration 0.5 vvm, agitation 450 rpm, inoculum size 1.6%). After the cultivation, racemic 1,2-diol (9 to 30g) was added to the culture broth. During the reaction, the pH of the reaction mixture was controlled at 6.5 with 4 N NaOH, and 30 g of glucose was added once a day. After the reaction, cells were removed by centrifugation, and the supernatant was concentrated to about 150 mL under reduced pressure. 1,2-Diols were extracted with 500 mL of ethyl acetate three times. The solvent was removed by evaporation, and the 1,2-diols were purified by distillation or silica gel chromatography (Wako gel G-200, solvent: *n*-hexane/ethyl acetate).

Analytical Methods

1,2-Diols were assayed by GLC using a Hitachi 163 gas chromatograph with a hydrogen flame ionization detector. Two different types of liquid phases were used. The glass column (0.3 × 100 cm) packed with Silicone OV-17 on Chromosorb WAW DMCS (Gasukuro Kogyo, Japan) was used to assay **7**,**8** and **9**. Flow rate of nitrogen gas was 20 mL/min, and column temperature was 220 °C. The other glass column (0.3 × 100 cm) packed with FAL-M 6%/TENAX GC (Wako Pure Chemical, Japan) was used to assay the other 1,2-diols and 1-hydroxy-2-ketones. Flow rate of nitrogen gas was 35 mL/min, and the column temperature was 175 °C. The samples for GLC analyses were prepared as follows: 2 mL of the reaction mixture was saturated with ammonium sulfate, 1,2-diol was extracted with 4 mL of ethyl acetate, and the organic layer was used for GLC analysis.

The enantiomeric excess of 1,2-diol was measured by HPLC analysis of the corresponding 1-tosylated derivatives except for 1-phenyl-1,2-ethanediol. Preparation and purification of 1-tosylated derivatives were done by the procedures reported (5,20). The columns and the other conditions used are shown in Table 1.

NMR and mass spectra were recorded on a Varian EM-90 spectrometer and Shimadzu QP-1000 GC-Mass spectrometer, respectively. Optical rotation was measured with a Horiba polarimeter SEPA-200. Melting points were measured using a Büchi melting point apparatus.

Table 1. Conditions of HPLC for Measurement of Enantiomeric Excess of 1,2-Diols

1,2-Diol[a]	Column[b]	Solvent[c]	Flow rate (mL/min)	Retention time (min) (R)	Retention time (min) (S)
1	OC	13:1	1.0	42	46
2	OC	13:1	1.0	36	41
3	OC	13:1	1.0	34	39
4	OC	13:1	1.0	30	36
5	OC	13:1	1.0	28	34
6[d]	OC	13:1	0.8	31	36
7	OB	30:1	0.7	33	43
8	OD	13:1	0.8	37	49
9	OD	13:1	0.8	39	51
10	OC[e]	20:1	1.0	78	72

[a]The numbers in this column were corresponding to the ones cited in Scheme 1.
[b]Daicel Chiral Cel (Ø 0.45 × 25 cm).
[c]The ratio of *n*-hexane/2-propanol.
[d]Only this compound was directly analyzed without 1-tosylation.
[e]Two columns were connected and used.

RESULTS

Screening of Microorganisms Producing (*S*)-1,2-Pentanediol by Stereoinversion

At the start of this study, we had planned the production of (*S*)-1,2-diols from racemates by stereoselective degradation of (*R*)-1,2-diols using microorganisms. We screened for suitable microorganisms for this purpose from the stock culture collection in our laboratory and found that many microorganisms degraded (*R*)-1,2-pentanediol more easily than (*S*)-enantiomer. *Lodderomyces elongisporus* IFO 1676 was selected among them and used for large-scale preparation of (*S*)-1,2-pentanediol. In this experiment, we found that the optical purity of residual 1,2-pentanediol reached approximately 100% ee as (*S*)-enantiomer when only 22% of racemic 1,2-pentanediol added in the reaction mixture was degraded. This result suggested that this microorganism could convert (*R*)-1,2-pentanediol to the (*S*)-enantiomer. This observation led us to rescreen the microorganisms that had the ability to produce (*S*)-1,2-pentanediol having high optical purity from the racemate by stereoinversion, and we found that various microorganisms had the similar ability as *L. elongisporus*. They belonged to the genera *Candida, Debaryomyces, Pichia, Stephanoascus, Endomyces, Fusarium, Giberella, Gongronella*, and others. Among them, the microorganisms that produced (*S*)-1,2-pentanediol excess of 70% molar yield from the racemate are listed in Table 2. *C. parapsilosis* IFO 0708 was used for further experiments.

Production of (*S*)-1,2-Pentanediol by *C. parapsilosis* IFO 0708

C. parapsilosis IFO 0708 was cultured in a 5-L jar fermenter containing 3 L of medium for 23 h under the conditions described earlier. After then, 75 g of racemic 1,2-pentanediol and 25 g of glucose were added to 2.5 L of the culture broth in a 5-L jar fermenter, and the pH of the reaction mixture was adjusted to 6.5 with 4 N NaOH. The reaction-accompanied cultivation was conducted for 26 h under the same conditions. Figure 1 shows the course of the reaction. Racemic 1,2-pentanediol was converted to (*S*)-1,2-pentanediol with a molar yield of 93% after incubation for 24 h. It was noteworthy that 1-hydroxy-2-pentanone was de-

Table 2. (*S*)-1,2-Pentanediol Production from the Racemate by Microbial Stereoinversion

Microorganism	Incubation period (h)	Residual 1,2-PDO[a] (A) (%)	ee of 1,2-PDO (B) (%)	(*S*)-Isomer yield[b] (C) (%)
Candida maltosa ATCC 20275	48	75	95 (*S*)	73
Candida parapsilosis IFO 0585	24	77	97 (*S*)	76
Candida parapsilosis IFO 0708	24	89	98 (*S*)	88
Candida parapsilosis IFO 1396	24	89	96 (*S*)	87
Candida tropicalis CBS 1926	24	92	91 (*S*)	88
Lodderomyces elongisporus IFO 1676	24	79	93 (*S*)	76
Pichia bovis IFO 0872	24	77	88 (*S*)	72
Gongronella butleri IFO 8080	96	84	71 (*S*)	72

[a]At the start of reaction, 10 mg/mL of racemic 1,2-pentanediol (1,2-PDO) was added except for *G. butleri* (5 mg/mL).
[b](*S*)-Isomer yield (C) = Net yield of (*S*)-isomer from the racemate = A × [1 − (100 − B)/200].

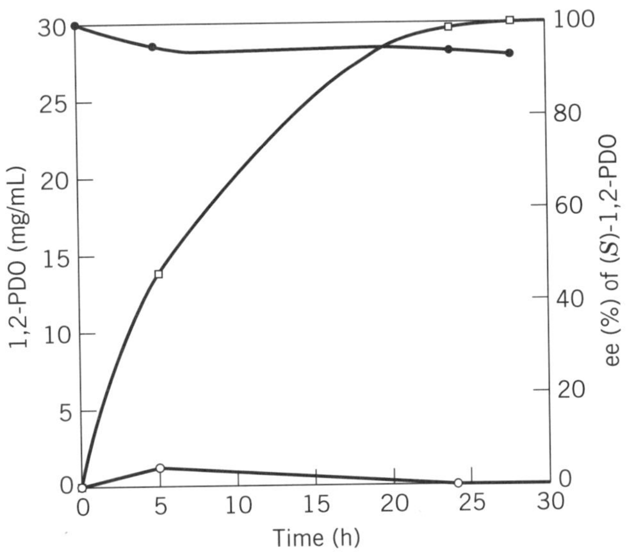

Figure 1. Production of (*S*)-1,2-pentanediol by *C. parapsilosis* IFO 0708. ● = 1,2-pentanediol; ○ = 1-hydroxy-2-pentanone; □ = enantiomeric excess of (*S*)-1,2-pentanediol.

tected in the reaction mixture. After the reaction, the cells were removed by centrifugation, and the supernatant was concentrated to 375 mL under reduced pressure. (*S*)-1,2-Pentanediol was extracted with 750 mL of ethyl acetate. The solvent was dried up by evaporation, and 66 g of yellowish oil was obtained. (*S*)-1,2-Pentanediol was purified by distillation (76 to 77 °C [3 mmHg]), and 60 g of (*S*)-1,2-pentanediol was obtained (yield 80%). GLC purity 99.9%, $[\alpha]_D^{20}$ −17.3° (c = 1, MeOH), −23.2°(c = 1, EtOH), 100% ee, (lit.(14), $[\alpha]_D^{25}$ −16.1°(c = 3, EtOH)), NMR $\delta_H(CDCl_3)$: 0.92 (3H, t, CH_3), 1.13 to 1.83 (4H, m, -$(CH_2)_2$-), 3.27 to 3.92 (3H, m, -**CH**(OH)-**CH$_2$**OH), 4.12 (2H, broad, OH), MS *m/z*: 87 (M^+-OH), 73 (M^+-CH_3O), 61 ($C_2H_5O_2^+$).

Mechanism of Stereoinversion of 1,2-Pentanediol in *C. parapsilosis* IFO 0708

To obtain the information on the stereoinversion mechanism, we used the cell-free extract of *C. parapsilosis*. Usually, alcohol dehydrogenase requires NAD^+ or $NADP^+$ as a coenzyme to catalyze the oxidation of alcohols. As the first experiment, we investigated the requirement of coenzyme for the oxidoreduction between (*R*)/(*S*)-1,2-pentanediol and 1-hydroxy-2-pentanone in this microorganism. As shown in Table 3, both NADH and NADPH served as a coenzyme to reduce 1-hydroxy-2-pentanone to 1,2-pentanediol in the cell-free extract. In the reverse reaction, NAD^+ mainly played the coenzyme role in oxidation of (*R*)-1,2-pentanediol, and the reaction rate of the oxidation of (*R*) and (*S*)-1,2-pentanediol using $NADP^+$ as a coenzyme was very slow. This NAD^+-linked (*R*)-specific 1,2-pentanediol dehydrogenase resembles the glycerol dehydrogenase isolated from *Cellulomonas* sp. (5,21) in its stereospecificity and NAD^+ requirement. Sugiura et al. reported that NAD^+-linked glycerol dehydrogenase from *Erwinia aroideae* (22) was inhibited by $CuSO_4$ and $ZnSO_4$. The effects of these metals and *p*-chloromercuribenzoate on both NADH-linked and NADPH-linked 1-hydroxy-2-pentanone reductase activities were tested. It was found that 1 mM $CuSO_4$ and 0.17 mM *p*-chloromercuribenzoate inhibited NADH-linked 1-hydroxy-2-pentanone reductase very strongly. In order to know the stereospecificity of NADPH-linked 1-hydroxy-2-pentanone reductase, 3 mM of $CuSO_4$ was added into the reaction mixture (10 mL) containing 0.50 mmol NADPH, 0.24 mmol of 1-hydroxy-2-pentanone, and the cell-free extract (2 mL). The reduction was carried out for 16 h at 25 °C. The steric configuration of 1,2-pentanediol produced under the above conditions was the

Table 3. Oxidoreduction between 1,2-Pentanediol and 1-Hydroxy-2-Pentanone by the Cell-Free Extract of *C. parapsilosis*[a]

Reaction and substrate	OD at 340 nm[a]	
Reduction (pH 6.0)	NADH	NADPH
1-Hydroxy-2-pentanone	−0.667	−0.232
Oxidation (pH 10.0)	NAD^+	$NADP^{+b}$
(*R*)-1,2-Pentanediol	0.722	0.014
(*S*)-1,2-Pentanediol	0.005	0.018

Note: The conditions of oxidation and reduction were described in the text.
[a]The change of optical density at 340 nm during the first 1 min of the reaction is listed.
[b]In the case of oxidation using $NADP^+$, the reactions were carried out at pH 8.0 using 0.1 M tris-HCl buffer instead of 0.1M sodium carbonate buffer (pH 10.0), because the reduction of $NADP^+$ was not observed at all under these conditions.

(*S*)-form, and its optical purity was very high (Table 4). The activities of NAD$^+$-linked and NADP$^+$-linked 1,2-pentanediol dehydrogenases at various pHs are shown in Figure 2. Oxidoreduction between 1,2-pentanediol and 1-hydroxy-2-pentanone by NAD$^+$-linked dehydrogenase is reversible, but reduction of 1-hydroxy-2-pentanone to (*S*)-1,2-pentanediol is practically less reversible. From these observations, the mechanism of the stereoinversion of 1,2-diol in *C. parapsilosis* may be explained as follows. Only (*R*)-1,2-pentanediol in the racemate is oxidized to 1-hydroxy-1,2-pentanone by NAD$^+$-linked (*R*)-specific dehydrogenase. Then, 1-hydroxy-2-pentanone is reduced to (*S*)-1,2-pentanediol by NADPH-linked (*S*)-specific reductase (Scheme 2).

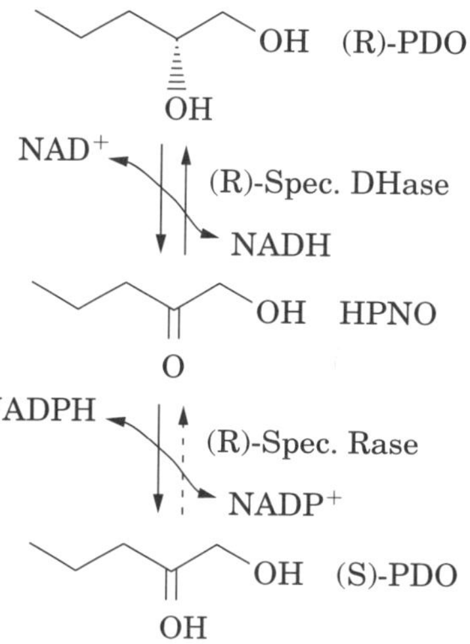

Scheme 2.

Table 4. Stereospecificity of NADPH-Linked 1-Hydroxy-2-Pentanone Reductase

Coenzyme	Inhibitor $CuSO_4$ (mM)	Formed 1,2-pentanediol Yield (%)	ee (%)
NADH	0	89	84 (*R*)
	3	0	—
NADPH	0	73	51 (*S*)
	3	31	98 (*S*)

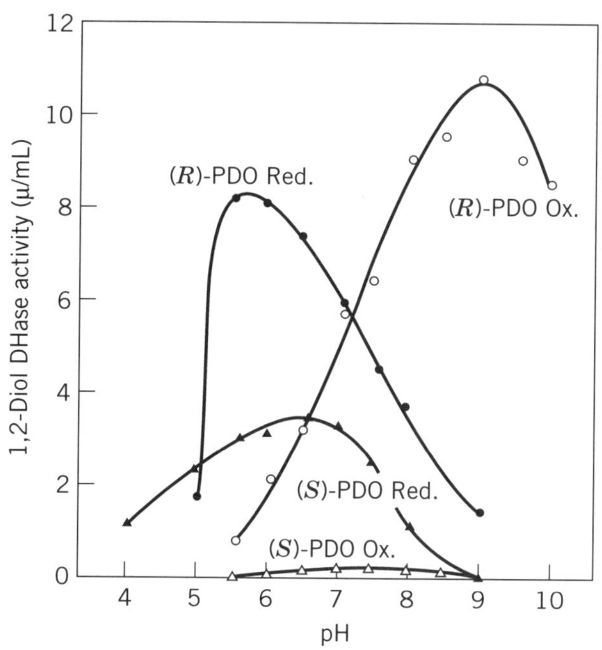

Figure 2. Effect of pH on 1,2-pentanediol oxidoreductase activities. The basic composition of the reaction mixtures was described in the text, and the following buffers were used to maintain the respective pHs: pH 4.0 to 5.5 acetate buffer, pH 5.5 to 6.5 tris-maleate buffer, pH 6.5 to 7.5 phosphate buffer, pH 7.5 to 9.0 tris-HCl buffer, pH 9.0 to 11.0 sodium carbonate buffer. ○ = NAD$^+$, (*R*)-1,2-pentanediol oxidation; ● = NADH, 1-hydroxy-2-pentanone reduction; △ = NADP$^+$, (*S*)-1,2-pentanediol oxidation; ▲ = NADPH, 1-hydroxy-2-pentanone reduction; (1 unit = 1 μmol of NAD(P)$^+$ reduction or NAD(P)H oxidation per min).

Production of Various Optically Active 1,2-Diols by Stereoinversion Using *C. parapsilosis* IFO 0708

It was very beneficial for us to apply this unique microbial reaction to production of various optically active 1,2-diols. Fourteen kinds of racemic 1,2-diols were tested: 1,2-propanediol, 1,2-butanediol (**1**), 1,2-pentanediol (**2**), 1,2-hexanediol (**3**), 1,2-heptanediol (**4**), 1,2-octanediol (**5**), 4-methyl-1,2-pentanediol (**6**), 1-phenyl-1,2-ethanediol (**7**), 3-phenyl-1,2-propanediol (**8**), 4-phenyl-1,2-butanediol (**9**), 3-methylthio-1,2-propanediol (**10**), 3-chloro-1,2-propanediol, 3-methoxy-1,2-propanediol, and 3-phenoxy-1,2-propanediol. Among them, the microbial stereoinversion was observed on 10 kinds of 1,2-diols listed in Table 5. These 1,2-diols had the same absolute configuration on the secondary alcohol. The yields and properties of optically active 1,2-diols obtained were as follows.

(*S*)-1,2-Butanediol (1). (*S*)-Isomer yield 56%, bp 65 °C (3 mmHg), (lit.(6), bp 94 to 96 °C (9 mmHg)), GLC purity 99.9%, $[\alpha]_D^{20}$ −8.57° (c = 1, MeOH), −31.9° (c = 1, EtOH), 79% ee, (lit.(6), $[\alpha]_D^{25}$ −15.35° (c = 2.6, EtOH)). NMR $\delta_H(CDCl_3)$: 0.92 (3H, t, CH_3), 1.45 (2H, m, -C**H$_2$**-CH_3), 3.23 ~ 3.83 (3H, m, -C**H**(OH)-C**H$_2$**OH), 4.03 (2H, broad, OH), MS *m/z*: 61 ($C_2H_5O_2^+$), 59 (M^+-CH_3O).

(*S*)-1,2-Hexanediol (3). (*S*)-Isomer yield 95%, bp 86 °C (3 mmHg), (lit.(1), bp 110 to 113 °C (6 mmHg)), GLC purity 99.7%, $[\alpha]_D^{20}$ −16.4° (c = 1, MeOH), −22.1° (c = 1, EtOH), 100% ee, (lit.(1), (*R*)-isomer $[\alpha]_D^{22}$ + 15.2° (c = 14, EtOH)), NMR $\delta_H(CDCl_3)$: 0.92 (3H, t, CH_3), 1.13 to 1.83 (6H, m, -$(CH_2)_3$-), 3.23 to 4.23 (3H, m, -C**H**(OH)-C**H$_2$**OH), 3.73 (2H, broad, OH), MS *m/z*: 87 (M^+-CH_3O), 61 ($C_2H_5O_2^+$).

(*S*)-1,2-Heptanediol (4). (*S*)-Isomer yield 97%, bp 101 °C (4 mmHg), (lit.(2), 67 to 70 °C (0.04 mmHg)), GLC purity 99.6%, $[\alpha]_D^{20}$ − 15.4° (c = 1, MeOH), −20.6° (c = 1, EtOH), 98% ee, (lit.(2), $[\alpha]_D^{22}$ − 16.6° (c = 11.9, EtOH)), NMR $\delta_H(CDCl_3)$: 0.87 (3H, t, CH_3), 1.10 to 1.87 (8H, m, -$(CH_2)_4$-), 3.22 to 4.13 (3H, m, -C**H**(OH)-C**H$_2$**OH), 3.75 (2H, broad, OH), MS *m/z*: 101 (M^+-CH_3O), 61 ($C_2H_5O_2^+$).

Table 5. Production of Optically Active 1,2-Diols by *C. parapsilosis* IFO 0708

1,2-Diol[a]	Concentration of substrate (g/L)	Reaction period (h)	Residual 1,2-diol (%)	e,e of 1,2-diol (%)	(*S*) or (*R*) isomer yield[b] (%)
1	10	54	62	79 (*S*)	56
2	30	26	93	100 (*S*)	93
3	10	16	95	100 (*S*)	95
4	5.0	4	98	98 (*S*)	97
5	3.0	9	96	97 (*S*)	95
6	3.0	23	97	97 (*S*)	96
7	5.0	30	100	100 (*S*)	100
8	6.3	72	72	82 (*S*)	66
9	6.3	72	100	100 (*S*)	100
10	3.0	54	66	98 (*R*)	65

[a]The numbers in this column correspond to the numbers in Scheme 1.
[b]See Table 2.

(*S*)-1,2-Octanediol (5). (*S*)-Isomer yield 95%, bp 104 °C (3 mmHg), (lit.(15), 100 to 110 °C (0.4 mmHg)), GLC purity 99.8%, $[\alpha]_D^{20}$ − 13.6° (c = 1, MeOH), −19.0° (c = 1, EtOH), 97% ee, (lit.(15), (*R*)-isomer $[\alpha]_D^{20}$ + 17.5° (c = 1.164, EtOH)). NMR $\delta_H(CDCl_3)$: 0.90 (3H, t, CH_3), 1.10 to 1.93 (10H, m, -$(CH_2)_5$-), 3.23 to 3.93 (3H, m, -**CH**(OH)-**CH$_2$**OH), 3.05 (2H, broad, OH), MS *m/z*): 115 (M^+-CH_3O), 61 ($C_2H_5O_2^+$).

(*S*)-4-Methyl-1,2-Pentanediol (6). (*S*)-Isomer yield 96%, bp 82 to 83 °C (5 mmHg), (lit.(10), 89 to 91 °C (5 mmHg)), GLC purity 99.4%, $[\alpha]_D^{20}$ − 26.3° (c = 1, MeOH, −31.5° (c = 1, EtOH), (lit.(10) $[\alpha]_D^{23}$ −24.7° (c = 1.84, EtOH), 97% ee, NMR $\delta_H(CDCl_3)$: 0.93 (6H, d, *J* = 6.6Hz, (**CH$_3$**)$_2$-CH-), 1.27 (2H, m, CH-**CH$_2$**-), 1.80 (1H, m, $(CH_3)_2$-**CH**-, 3.20 ~ 4.05 (3H, m, -**CH**(OH)-**CH$_2$**OH), 3.68 (2H, broad, OH), MS *m/z*: 117 (M^+ − 1), 87 (M^+-CH_3O), 61 ($C_2H_5O_2^+$), 57 ($(CH_3)_2C_2^+$).

(*S*)-1-Phenyl-1,2-Ethanediol (7). (*S*)-Isomer yield 100%, mp 66 to 67 °C (lit.(7), (*S*)-isomer 63 to 65 °C), GLC purity 99.9%, $[\alpha]_D^{20}$ +45.0° (c = 1, MeOH), +38.6° (c = 1, EtOH), +49.9° (c = 1, acetone), 100% ee, (lit.(4), $[\alpha]_D^{25}$ +51.2° (c = 0.25, acetone), lit(7) (*R*)-isomer $[\alpha]_D^{25}$ −39.7° (c = 4.33, 95% EtOH)), NMR $\delta_H(CDCl_3)$: 2.97 to 3.30 (1H, m, OH), 3.40 to 3.60 (1H, m, OH), 3.63 (2H, d, -CH_2-), 4.75 (1H, m, -**CH**(OH)-), 7.30 (5H, m, C_6H_5-), MS *m/z*: 138 (M^+), 120 (M^+-H_2O), 107 (M^+-CH_2O).

(*S*)-3-Phenyl-1,2-Propanediol (8). (*S*)-Isomer yield 66%, mp 47 to 48 °C, bp 140 °C (6 mmHg), (lit.(11), 101 to 102 °C (0.001 mmHg)), GLC purity 99.6%, $[\alpha]_D^{20}$ −28.2° (c = 1, MeOH), −29.5° (c = 1, EtOH), 82% ee, (lit.(11), $[\alpha]_D^{20}$ −36° (c = 1, EtOH)), NMR $\delta_H(CDCl_3)$: 2.68 (2H, d, *J* = 6.6Hz, C_6H_5-**CH$_2$**-), 3.03 (2H, broad, OH), 3.25 to 3.73 (2H, m, -**CH$_2$**OH), 3.73 ~ 4.10 (1H, m, -**CH**(OH)-), 7.22 (5H, m, C_6H_5-), MS *m/z*: 152 (M^+), 134 (M^+-OH), 133 (M^+-H_2O), 91 (C_6H_5-CH_2^+), 61 ($C_2H_5O_2^+$).

(*S*)-4-Phenyl-1,2-Butanediol (9). (*S*)-Isomer yield 100%, mp 36 to 37 °C, bp 142 °C (4 mmHg), (lit.(19), the racemate 162 to 165 °C (0.65 kPa)), GLC purity 98.8%, $[\alpha]_D^{20}$ −29.0° (c = 1, MeOH), −34.1° (c = 1, EtOH), 100% ee, NMR $\delta_H(CDCl_3)$: 1.72 (2H, m, -**CH$_2$**-CH(OH)-), 2.68 (2H, m, C_6H_5-**CH$_2$**-), 3.03 ~ 4.00 (3H, m, -**CH**(OH)-**CH$_2$**OH), 3.47 (2H, broad, OH), 7.23 (5H, m, C_6H_5-), MS *m/z*: 149 (M^+-OH), 148 (M^+-H_2O), 135 (M^+-CH_3O), 105 (C_6H_5-$CH_2CH_2^+$), 61 ($C_2H_5O_2^+$).

(*R*)-3-Methylthio-1,2-Propanediol (10). (*R*)-Isomer yield 65%, bp 101 °C (3 mmHg), GLC purity 99.7%, $[\alpha]_D^{20}$ −12.9° (c = 1, MeOH), −11.9° (c = 1, EtOH), 98% ee, MNR $\delta_H(CDCl_3)$: 2.13 (3H, s, CH_3-), 2.60 (2H, d, *J* = 6.0Hz, -S-CH_2-), 3.18 ~ 4.32 (3H, m, -**CH**(OH)-**CH$_2$**OH), 3.83 (2H, broad, OH), MS *m/z*: 122 (M^+), 104 (M^+-$H_2$0), 61 (CH_3-S-CH_2^+, $C_2H_5O_2^+$).

We tried this microbial stereoinversion for production of optically active 1,2-propanediol, 3-phenoxy-1,2-propanediol, 3-methoxy-1,2-propanediol, and 3-chloro-1,2-propanediol, but these 1,2-diols were not good substrates. In cases of 1,2-propanediol and 3-phenoxy-1,2-propanediol, their (*S*)-isomers were isolated from the reaction mixtures, but the yields and the optical purities of both (*S*)-1,2-diols were not over 50%. With the other 1,2-diols, their racemates were recovered from the reaction mixtures.

In this study, we presented a novel method for production of optically active 1,2-diols using a microbial stereoinversion. This unique bioinversion consisted of two successive reactions via 1-hydroxy-2-ketone, namely the oxidation of (*R*)-1,2-diol to 1-hydroxy-2-ketone by NAD^+-linked (*R*)-specific alcohol dehydrogenase, followed by the reduction of 1-hydroxy-2-ketone to (*S*)-1,2-diol by an NADPH-linked (*S*)-specific reductase. To the best of our knowledge, this is the most convenient procedure for producing optically active 1,2-diols among the reported methods.

BIBLIOGRAPHY

1. P.A. Levene and H.L. Haller, *J. Biol. Chem.* **74**, 343 (1927), *J. Biol. Chem.* **77**, 555 (1928); *J. Biol. Chem.* **79**, 475 (1928).
2. J. Barry and H.B. Kagan, *Synthesis*, 453 (1981).
3. D.D. Ridley and M. Stralow, *J. Chem. Soc., Chem. Commun.*, 400 (1975).

4. H. Kutsuki, I. Sawa, N. Mori, J. Ishizu, J. Hasegawa, M. Shimazaki, T. Ohashi, and K. Watanabe, *Ann. Mtg. Soc. Ferment. Technol. Japan*, 106 (1986).
5. L.G. Lee and G.M. Whitesides, *J. Org. Chem.* **51**, 25 (1986).
6. H.K. Chenault, M-J. Kim, A. Akiyama, T. Miyazawa, E.S. Simon, and G.M. Whitesides, *J. Org. Chem.* **52**, 2608 (1987).
7. J.A. Dale and H.S. Mosher, *J. Org. Chem.* **35**, 4002 (1970).
8. K. Hintzer, B. Koppenhoefer, and V. Shurig, *J. Org. Chem.* **47**, 3850 (1982).
9. N. Sakota, S. Tanaka, K. Okita, and N. Koine, *Nippon Kagaku Zasshi* **91**, 265, (1970).
10. K. Mori, *Tetrahedron Lett.* **26**, 2187 (1975).
11. W. Bergstein, A. Kleemann, and J. Martens, *Synthesis*, 76 (1981).
12. B. Cambou and A.M. Klibanov, *J. Am. Chem. Soc.* **106**, 2687 (1984).
13. T. Sugiyama, H. Sugawara, M. Watanabe, and K. Yamashita, *Agric. Biol. Chem.* **48**, 1841 (1984).
14. J. Mulzer and A. Angermann, *Tetrahedron Lett.* **24**, 2843 (1983).
15. K-Y. Ko, W.J. Frazee, and E.L. Eliel, *Tetrahedron* **40**, 1333 (1984).
16. S. Maemoto, H. Kutsuki, J. Hasegawa, and T. Ohashi, *Ann. Mtg. Agric. Biol. Chem. Soc. of Japan*, 419 (1989).
17. H. Takahashi, S. Tsuda, Y. Nakamura, M. Ogura, T. Shiraishi, and Y. Shimada, *Ann. Mtg. Agric. Biol. Chem. Soc. of Japan* 351 (1989).
18. J. Hasegawa, M. Ogura, S. Tsuda, S. Maemoto, H. Kutsuki, and T. Ohashi, *Agric. Biol. Chem.* **54**, 1819 (1990).
19. K. Blau, V. Voerckel, and L. Willecke, *J. Prakt. Chem.* **328**, 29 (1986).
20. J.P. Guett and N. Spassky, *Bull. Soc. Chim. Fr.* 4217 (1972).
21. H. Yamada, A. Nagao, H. Nishise, and Y. Tani, *Agric. Biol. Chem.* **46**, 2333 (1982).
22. M. Sugiura, T. Oikawa, K. Hirano, H. Shimizu, and F. Hirata, *Chem. Pharm. Bull.* **26**, 716 (1978).

ORGANIC COMPOUNDS, CELLULOSE CONVERSION

C.S. Gong
Ningjun Cao
G.T. Tsao
Purdue University
West Lafayette, Indiana

KEY WORDS

Cellulase
Cellulose
Ethanol
Hemicellulose
Pretreatment
Simultaneous saccharification and fermentation (SSF)
Xylose

OUTLINE

INTRODUCTION

Cellulose is the most plentiful of all of the naturally occurring organic compounds. It is a high molecular weight linear polysaccharide composed of two residues (β-1,4-linked glucose, known as cellobiose) in repeated units. Complete hydrolysis of cellulose using strong mineral acids yields glucose. The basic building blocks of cellulose are chemically reactive, making it susceptible to chemical modifications. Cellulose has great economical importance; it is processed to produce papers, fibers, and chemical derivatives to yield substances used in the manufacturing of such items as plastics, photographic films, rayon, and so on. Other cellulose derivatives are used as adhesives, explosives, and thickening agents for foods. Figure 1 shows some common uses of cellulose.

Cellulose exists in nature in two principal types: the pectocelluloses such as flax, hemp, and ramie that contain more than 80% cellulose; the lignocelluloses composed mostly of cellulose, hemicellulose, and lignin. As a basic component of plant cell walls, cellulose comprises at least 30% of all plant matters (90% of cotton and up to 50% of wood are cellulose). Table 1 shows the approximate composition of the selected lignocellulosic materials.

Because lignocellulosic materials are renewable and abundant, extensive research in recent years has been undertaken to convert cellulose, particularly agricultural residues, forest by-products, and waste cellulose, into food, chemicals, and fuels. In this article, the emphasis will be on the production of ethanol from lignocellulosic materials. Other products such as solvents (acetone and butanol) (7), 2,3-butanediol (8), and single-cell proteins (9) have been reviewed.

Relationship between Structural Features of the Cell Wall and Cellulolytic Enzymes

Native lignocellulosic materials are heterogeneous, containing cellulose, hemicellulose, lignin, ash, protein, and a

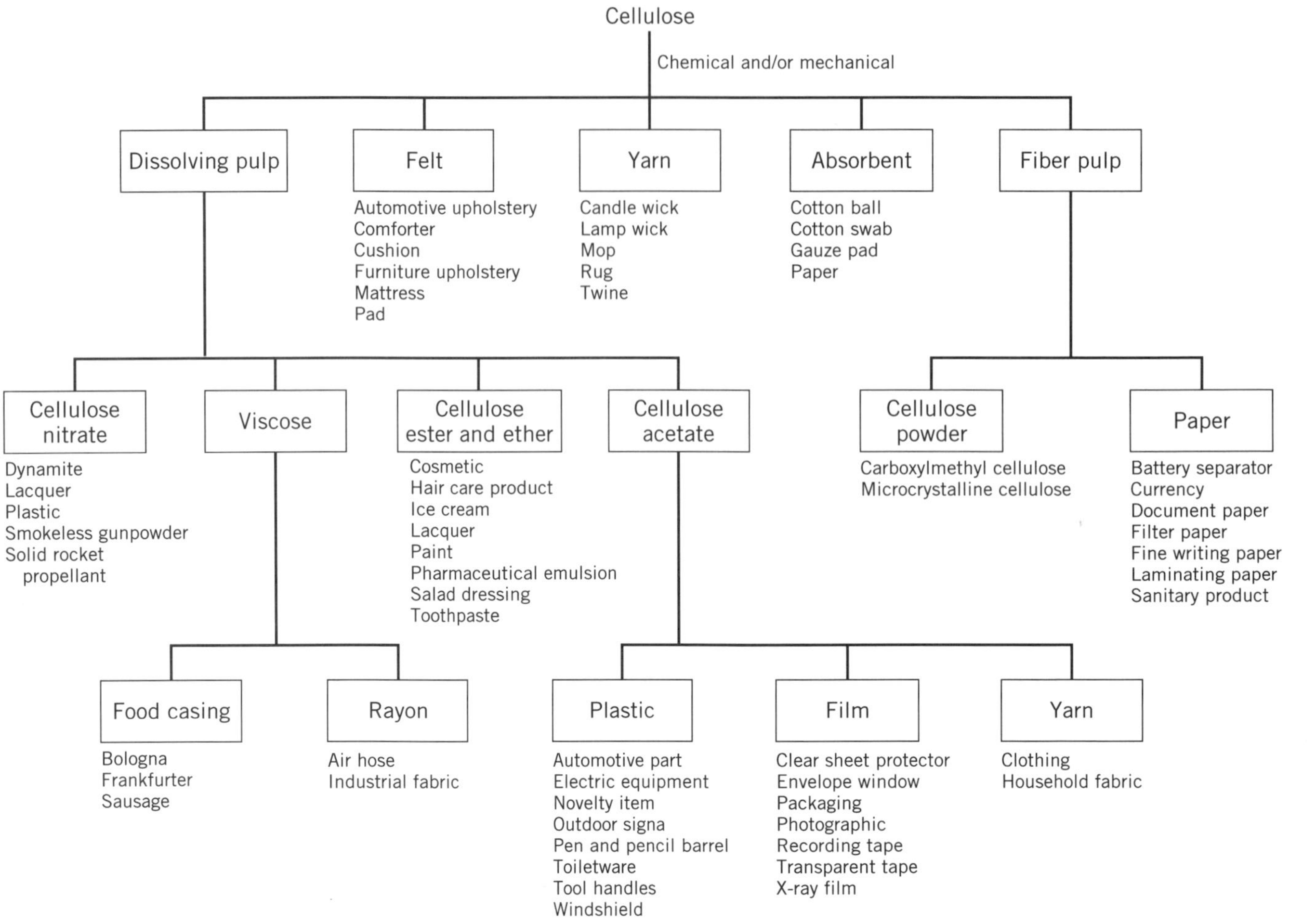

Figure 1. Common uses of cellulose. *Source*: From Ref. 1.

wide array of minor extractives. The most important structural feature in a plant cell wall is the capillary area, whose dimensions and shapes control the accessibility of lignocelluloses by hydrolytic enzymes. The lumen (Fig. 2) is the largest capillary in a plant cell, followed by pit apertures and pores in the pit-membrane, ranging from 200 Å to 10 or more μm in diameter. Most of these capillaries are filled with hemicellulose, pectic materials, and lignin, which form a protective sheath around the cellulose (10,11). Therefore, any treatment of lignocellulosic materials that solubilizes and/or removes lignin and hemicelluloses from the cell wall can enlarge the capillaries. As a result, the accessibility of cellulose to cellulase can be increased.

Molecules of cellulolytic enzymes of various microorganisms are water-soluble proteins of high molecular weight that are 13×79 Å in width and 42×252 Å in length. The relatively large enzyme molecules cannot be diffused into plant cell walls unless pretreatment is conducted to increase the size of the cell wall capillaries.

CELLULOSE PRETREATMENT

Many factors have been identified as influencing the reactivity and digestibility of the cellulose fraction of lignocellulose materials. These factors include lignin and hemicellulose contents, crystallinity of the cellulose, and the porosity of the biomass. Pretreatment of lignocellulosic materials prior to utilization is a necessary element in the biomass-to-ethanol conversion processes. The objective of the pretreatment is to render biomass materials more accessible to either chemical or enzymatic hydrolysis for efficient product generation. There are several goals of the pretreatment.

- Disrupt and remove the lignin sheath
- Remove and separate hemicellulose from cellulose
- Decrease the crystallinity of the cellulose
- Increase the accessible surface area
- Increase the pore size to facilitate the penetration of hydrolysis agents

There have been many processes developed for the treatment of various biomass. In general, agricultural residues are more easily treated than hardwood materials. The most difficult biomass to treat is that derived from softwood because it has a relatively high lignin content. Most processes exploit a variety of mechanisms to render

Table 1. Approximate Composition of Selected Plant Materials and Cellulose Waste

Materials	Cellulose (%)	Hemicellulose (%)	Lignin (%)	Ref.
Softwood				
Spruce	43	26	29	2
Pine	44	26	29	2
Hardwood				
Aspen	46	26	18	3
Black locust	49	21	22	4
Birch	38	27	23	3
Hybrid poplar	43	21	26	
Red oak	38	27	23	3
Silver maple	46	23	21	4
Sycamore	44	22	23	4
Willow	37	23	21	2
Crop residues				
Corn cobs	45	35	15	5
Corn stalks	35	25	35	5
Corn stover	41	21	17	4
Sugar cane bagasse	40	30	20	5
Wheat straw	36	28	29	2
Others				
Alfalfa hay	38	9	14	6
Nut shells	25	25	30	
Cotton	90	5	0	
Cellulose Wastes				
Paper	85	0	15	
Newsprint	61	16	21	2
Sorted refuse	60	20	20	

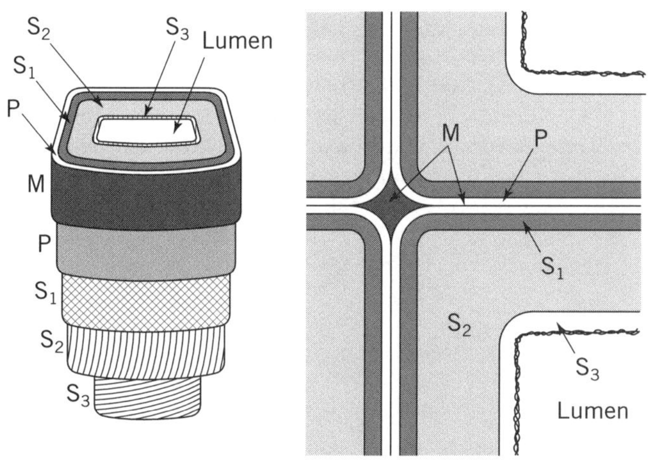

Figure 2. Diagrammatic representation of various layers of the cell wall. The intercellular materials: M, middle lamella; P, primary walls; S, secondary wall with outer (S1), middle (S2), and inner (S3) layers. *Source*: From Ref. 10.

the carbohydrate components of lignocellulosic materials more susceptible to enzymatic hydrolysis and microbial conversion (for detail see refs. 10 and 12).

The majority of the pretreatment methods involve a combination of mechanical size reduction, alkali swelling, acid hydrolysis, steam and other fiber explosion techniques, or exposure to supercritical fluids. The processes involving irradiation, strong mineral acids, cellulose solvents, and concentrated alkali chemicals are not suitable methods for the pretreatment of lignocellulosic materials because of their high cost. Biological methods, such as those applying “white rot” fungi to remove lignin from cellulose-hemicellulose, are also unsuitable due to the length of time they take, unless the lignocellulosic material is intended for single-cell protein (e.g., mushrooms) production (13). Therefore, the chosen pretreatment process must be efficient, cost effective, and environmental friendly.

Many different pretreatment approaches have been designed and tested, and some processes have also been tried in pilot scale. Ideally, the most desirable method of treatment is the dissolution of the solid materials into aqueous substrate. Many cellulose solvents at high enough concentrations can penetrate into the cellulose crystalline structure, causing the dissolution of solid cellulose. The resulting materials can be readily hydrolyzed to glucose by cellulase either in situ, or after the dissolved cellulose has been reprecipitated (14). Although cellulose solvents are powerful-solubilization agents, they are toxic, corrosive, and hazardous, and recovery costs are high. For these reasons, cellulose solvents are generally considered inappropriate for large-scale pretreatment processes.

Alkaline Treatment

Alkaline treatment causes lignocellulosic materials to swell; increased swelling leads to higher susceptibility of cellulose to saccharification. In the presence of alkaline chemicals (e.g., NaOH or NH_3), cellulose, hemicellulose, and lignin bonds can be disrupted, which permits cellulose to swell beyond normal water-swollen stages. Consequently, the pore size, intraparticle porosity, and capillary size are increased. There is also a phase change in the cellulose crystalline structure (15).

Sodium hydroxide is the most commonly used chemical in the treatment of lignocellulose to enhance digestibility for animals. It is a strong swelling agent for cellulose and may also be responsible for changing cellulose crystalline structure from cellulose I to cellulose III (decrystallization). The amount of NaOH used for treatment ranges from 2 to 20%, and the temperature for the treatment ranges from ambient to 120 °C. Under mild conditions (low concentration and low temperature) substrate components remain unchanged. Under harsher conditions, most of the lignin and hemicellulose are solubilized. Consequently, the cellulose fraction is exposed to hydrolysis agents.

Similar to NaOH, ammonia is useful in increasing the in vitro digestibility of lignocellulosic materials. It is also effective in the preparation of substrates for single-cell protein production. Ammonia is one of the most heavily used industrial chemicals. On a weight basis, aqueous ammonia is about one-third the cost of NaOH. In addition, ammonia is easily recoverable from an aqueous mixture because of its high volatility; it is also far less corrosive than mineral acids at high temperature. Increasingly, ammonia has been used in combination with other methods to treat lignocellulosic materials.

Ammonia-Recycled Percolation Process. The ammonia-recycled percolation (APR) process (16) uses ammonia as the treatment agent. In one example, the reactor was a packed-bed flow-through type (percolation reactor) and was used in a recirculation mode (Fig. 3). The intended biomass material was packed in a pressure vessel with preheated liquid ammonia passing through the substrate. The reaction temperature was 150 °C, and pressures were up to 325 psi. As the name of the process indicates, ammonia was recycled for further treatment of cellulosic materials. The scanning electron microscope examination of treated samples showed an increase in pore size and porosity. The amount of lignin removed was in the range of 23–63%, and hemicellulose removed was in the range of 20–36%.

Alkaline Peroxide Process. Chopped wheat straw was steeped in an aqueous alkaline solution (pH = 11.5) of H_2O_2 at a ratio of 0.25 g/g substrate at 25 °C. Lignin removal was 50%, and hemicellulose solubilized was almost 100%. The results of *Trichoderma* cellulase hydrolysis of cellulose indicated that glucose yield was approaching 100% with a substrate solid content of 5% (17).

Ammonia Freeze Explosion. The ammonia freeze explosion (AFEX) process is the pretreatment method that utilizes steam explosion techniques with ammonia as the chemical reagent (6,18). The intended biomass material is placed in a pressure vessel with liquid ammonia (1:1) and treated similarly as in the steam explosion process but with much lower temperatures (up to 80 °C) and pressures (up to 15 psi). Using this process, ammonia was recovered. Exposure to ammonia in the AFEX process causes swelling and partial decrystallization of the cellulose. It also caused the partial solubilization of lignin and hemicellulose resulting in the exposure of cellulose. This pretreatment method is more effective with agricultural residues (e.g., corn stover and straws) than with woody materials.

Steam Treatment with or without SO_2

Steam explosion of biomass materials has been studied by many investigators (see Ref. 19 for review). Biomass material (woodchips or wheat straw) was placed in a high-pressure stainless steel tube and was exposed to steam under pressures ranging from 250 to 650 psi at 200–240 °C for up to 20 min. The sudden pressure release caused an explosion of biomass material, thereby disrupting the lignin and hemicellulose bonding toward cellulose. The addition of SO_2 enhanced the pretreatment effect and also increased the recovery of cellulose (20–22).

Dilute Acid Pretreatment

The use of dilute acid hydrolysis pretreatment has been studied extensively (4,23). The overall goal of dilute acid pretreatment is to attain a high yield while minimizing the breakdown of hemicellulose sugars into decomposition products. The pretreatment usually involves mineral acids (e.g., HCl and H_2SO_4) at a concentration from 0.3 to 2% (w/w). The temperature for treatment ranged from 120 to 180 °C, and the length of treatment ranged from less than 1 min to 2 h or longer. The objective in dilute acid treatment is to separate hemicellulose from lignin and cellulose. It is effective in hydrolyzing hemicellulose, but a portion of lignin was also solubilized, resulting in the complication of utilization of hemicellulose carbohydrates. Depending on hydrolysis conditions, carbohydrate degradation products also formed, which interfered with microbial activities.

Because the hemicellulose fraction of biomass materials can be separated from lignin and cellulose by dilute acid treatment, cellulose becomes more reactive toward cellulase. Hemicellulose hydrolysis rates vary with acid concentration, temperature, and solid-to-liquid ratio. With most lignocellulosic materials, complete hemicellulose hydrolysis can be achieved in 5–10 min at 160 °C, or 30–60 min at 140 °C. Dilute acid hydrolysis forms the basis of many pretreatment processes; autohydrolysis and steam explosion are two examples that are based on high-temperature-dilute-acid-catalyzed hydrolysis of biomass.

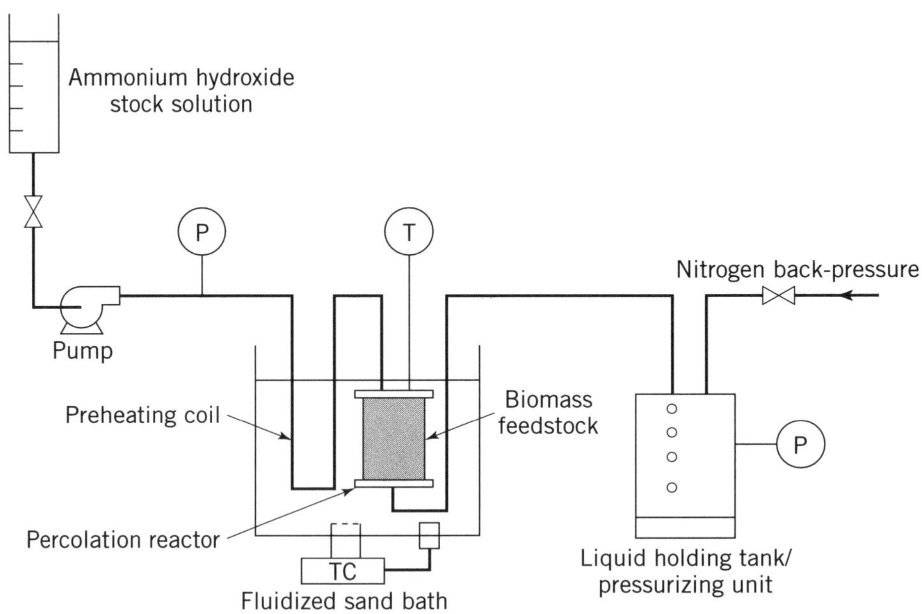

Figure 3. Schematic diagram of the ammonia-recycled percolation process. *Source*: From Ref. 16.

CELLULASE

Unlike chemical hydrolysis, enzymatic hydrolysis of cellulose has the advantage of producing only sugar as product and no sugar degradation byproducts. However, enzymatic hydrolysis of cellulose is a slow process with low yield and a long reaction time. The slow reaction is further complicated by the complex features of biomass materials. Factors affecting the rate of enzymatic hydrolysis of lignocelluloses are the following: the presence of a lignin-hemicellulose shield surrounding cellulose fibers in plant cell walls; the crystallinity of the cellulose; and the relatively small pore sizes in biomass materials that impose mass transfer limitations on the penetration of both microorganisms and the hydrolytic enzymes. Enzyme activity is also subjected to hydrolysis product feedback inhibition. Consequently, some form of pretreatment is needed to expose the cellulose matrix and increase the diffusion of cellulases into the matrix of the substrate.

There are many different microorganisms, including fungi, bacteria, and protozoa, that are capable of producing cellulase enzymes. *Trichoderma reesei* (formerly *T. viride*) is the best-known cellulase-producing microbe; it produces a highly active cellulase complex toward insoluble cellulose. A large number of fungi isolated from rotten cellulosic materials were identified, and their ability to produce cellulase was examined by Reese and Levinson (24).

The extensive research by Reese et al. resulted in the development of the multiple enzymes (C_1 and Cx) concept that could involve in the degradation of crystalline cellulose (25). It also resulted in the use of *Trichoderma reesei* as a source of most investigated active cellulase for the hydrolysis of insoluble cellulose. *Trichoderma* cellulase preparations are available in several different forms from various commercial entities.

Mode of Action of *T. reesei* Cellulase Components

It is widely recognized that true cellulolytic microorganisms produce three basic cellulase components, namely, endocellulase, exocellulase, and cellobiase. The three enzyme components act in concert to hydrolyze crystalline cellulose to its end product, glucose. The components of cellulase enzymes have been studied extensively, and the modes of action of cellulase components have been elucidated (see Refs. 26 and 27 for review). Table 2 summarizes the three major *Trichoderma* cellulase components, the substrates for measurement of enzyme activities, and the mode of action of enzymes.

Crude cellulase activity can be approximated by analyzing the soluble color intensity released from commercially available remazol brilliant blue acid-swollen cellulose (cellulose-azure, Sigma). The standard method (Procedure No. 009) for measurement of overall cellulase activity in terms of filter paper units (FPU) has been provided by National Renewable Energy Laboratory (NREL) (28).

Production of Cellulase

The basic medium for cellulase production from *T. reesei* was developed by Mandels and Reese (29). The composition of the medium is given in the following section. There have been variations in medium composition, especially in the nitrogen source and in the addition of surfactants, such as Tween 80, to reduce the cost of the medium as well as to increase the yields of cellulases. The most common substrates for cellulase production are commercially available materials such as filter paper and Solka-Floc.

Medium. The basic medium has the following composition: $(NH_4)_2SO_4$, 1.4 g/L; KH_2PO_4, 2.0 g/L; urea, 0.3 g/L; $CaCl_2$, 0.3 g/L; $MgSO_4$, 0.3 g/L; trace elements, $FeSO_4 \cdot 7H_2O$, 5.0 mg/L; $MnSO_4 \cdot H_2O$, 1.6 g/L; $ZnSO_4 \cdot 7H_2O$, 1.4 g/L; $CoCl_2$, 2.0 g/L; cellulose, 0.75–1%; protease peptone, 0.075–0.1%; Triton X-100, 0.1–0.2%; initial pH = 5.0–6.0. Rapidly metabolizable carbon sources, such as glycerol or glucose, repress cellulase formation until these carbon sources are depleted by the organism, that is, unless the hypercellulase-producing mutant Rut-C30 is used (30). The Rut-C30 produces cellulase even in the presence of glucose. The addition of small amounts (0.1%) of a readily metabolizable carbon source, such as glycerol, with peptone and cellulose reduces the lag phase for the growth of the organism, consequently increasing cellulase production. The representative cellulase production profile is presented in Figure 4 (31,32).

Supplementation of Cellulases

Commercial cellulase preparations are derived mostly from the *Trichoderma* species (*T. reesei*, and *T. koningii*). *Trichoderma* enzyme preparations are deficient in cellobiase. To enhance the saccharification of cellulose during SSF process, cellobiase derived from *Aspergillus* species is often added. The amounts of cellobiase supplementation are often at 1 unit per FPU cellulase (33).

SIMULTANEOUS SACCHARIFICATION AND FERMENTATION

In general, conversion of cellulosic materials to ethanol involves two stages: *(1)* conversion of cellulose to glucose by either chemical or enzymatic processes, and *(2)* microbial conversion of the resulting glucose to ethanol. In nature, the microbes that are rich in cellulolytic enzymes are usually those with relatively slow growth, such as wood-rotting fungi. The natural selection process has resulted in cellulolytic enzymes that are often highly regulated by the reaction end products. Cellulase activity is subjected to product feedback inhibition (34). When there is a significant concentration of glucose, the enzymatic activity of cellobiase will be affected, which results in the accumulation of cellobiose. The accumulation of cellobiose in turn inhibits cellulase activities (for details see Ref. 27). To overcome the problems caused by the glucose feedback inhibition, a process known as simultaneous saccharification and fermentation (SSF) was developed.

SSF is a process in which production of ethanol from cellulosic materials is achieved by utilizing cellulose, cellulase, ethanol producing microbes, and nutrients in the same reactor. This process is attractive because the continuous removal of sugars by fermentative organisms allevi-

Table 2. Common Names and Mode of Action of Three Cellulase Components

General category	Other Names	Substrate	Mode of action
β-1,4-Glucan glucanohydrolase (EC.3.2.1.4)	Cx Endoglucanase Endo-β-glucanase CMCase	Acid-swolen cellulose Bacterial cellulose Cellodextrins CMC	Hydrolyze cellulose in random fashion into smaller chains; create reactive site for exocellulase
β-1,4-Glucan cellobiohydrolase (EC.3.2.1.91)	C1 Exocellulase Avicelase Cellobiohydrolase	Filter paper Cellodextrins Bacterial cellulose Acid-swollen cellulose	Cellobiose producer from nonreducing end of the celulose chains
β-1,4-Glucosidase (EC.3.2.1.21)	Cellobiase	Cellobiose	Hydrolyze cellobiose into two glucose molecules

2*Source*: From Refs. 26 and 27.

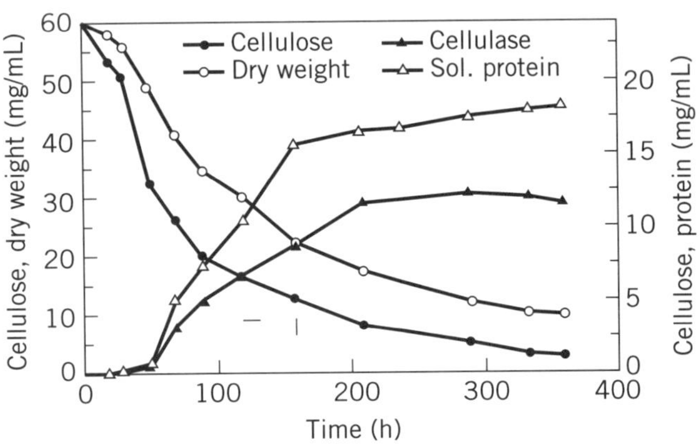

Figure 4. Typical *Trichoderma* cellulase production profile. *Source*: From Ref. 32.

ates the end-product inhibition of saccharification. The process is also simplified because only one reactor is used for SSF. The SSF process for ethanol production from cellulosic materials was reported by Blotkamp et al. (35) and was later tested in the pilot scale (for details see Ref. 36).

The advantages of SSF over the separate saccharification and fermentation of cellulose include the following:

- Reduction of contamination risk
- Lower enzyme loading requirements
- Faster hydrolysis reaction rate
- Higher product yield
- Lower operational expenses

For these reasons, SSF is the preferred process for the bioconversion of the cellulose fraction of lignocellulosic materials to ethanol and other chemicals at the present time. The SSF process involving lignocellulosic material has been reviewed in Ref. 36. There are several factors that are important for an efficient SSF process:

- The physical state of the substrate (particle size and crystallinity of the cellulose)
- Characteristics of the cellulase and cellulase concentration
- Effect of fermentation product on enzyme activity
- Compatibility of saccharification and fermentation temperature
- Characteristics of the yeast strains and yeast concentration
- Maintaining a low level of saccharification products

The nature of the cellulosic substrate is important because the degree of saccharification in the SSF system is often rate limiting with respect to ethanol production. Temperature also determines the economics of ethanol production from cellulose. In most studies, cell densities and enzyme concentration exerted little effect on the amounts of ethanol produced from cellulose under typical SSF conditions (37). In designing an efficient SSF system for the conversion of cellulose to ethanol, the fermentation temperatures should be compatible with operation temperatures, which are generally between 45 and 55 °C. The optimal temperature for the most commonly available cellulase is about 50 °C. Therefore, the use of high-temperature-tolerant microbes is desirable for the application of the SSF process for ethanol production. For the industrial production of ethanol from biomass materials, the use of yeast cultures is preferred over bacteria because the bacterial phage problem can be avoided. In addition, by-product credit can be obtained by marketing spent yeast cells as single-cell protein.

Typical industrial ethanol-producing yeast strains are mesophilic, with an optimal fermentation temperature of 30–37 °C. No true thermophilic yeast species has been reported, and only a few yeast strains that are thermotolerant as well as good ethanol fermentors have been described.

Supraoptimal Temperature Fermentation

In 1986, Anderson et al. (38) studied several thermotolerant yeast strains isolated from sugarcane factories in Australia. Most of the good ethanol-producing yeast strains belong to *Kluyeromyces marxianus* var. maxianus. Some are capable of producing over 6% ethanol (w/v) from glucose at an incubation temperature up to 43 °C after 14 h of fermentation. Over 80% of yeast cells remained viable at the end of fermentation.

Ballesteros et al. (39) studied the high-temperature SSF fermentation of pure cellulose at 42 and 45 °C at cellulase loading of 15 FPU/g substrate. The best results were achieved at 42 °C with a strain of *K. maxianus* and a strain of *K. fragilis*. Both yeast strains produced close to 38 g/L ethanol in 78 h. The results also confirmed the importance of using thermotolerant yeast in SSF processes in order to improve hydrolysis rates and achieve higher ethanol production. There are several possible benefits of using thermotolerant yeast to carry out ethanol fermentation at a supraoptimal temperature:

- Faster rate of saccharification
- Higher rates of sugar production and product (ethanol) formation
- Reduction of contamination by mesophilic yeast and bacteria
- Facilitation of product recovery
- Simplification of cooling during industrial-scale fermentation

However, yeast activities are more sensitive to inhibitory effects of ethanol with increasing temperature.

Improvement of the SSF Process

Typically, the SSF process is carried out in a continuous stirred-tank reactor (CSTR) in batch mode. Under this reaction condition, the fermentation product, ethanol, exerted its effect not only on microbes but also on cellulose hydrolysis activity. To overcome this problem and to improve the efficiency of ethanol production from cellulose, the continuous removal of end product during ethanol production should be beneficial. With this type of process application, the SSF process can be operated in a fed-batch mode. Fed-batch operation is similar to continuous operation except the fermentation broth is retained in the fermentor at all times, whereas the solid substrate is continuously fed into the fermentor.

Different methods of in situ removal of ethanol while it was being formed have been investigated (for review see Ref. 40). The processes include gas stripping, vacuum fermentation, solvent extraction, and the use of membrane bioreactors. Of these alternatives, gas stripping offers advantages in terms of its effectiveness and ease of operation. Ethanol can be recovered from the carrier gas stream by adsorbing onto activated carbon (41), or by condensation from the gas carrier (42). The experiments involving the gas (CO_2) stripping of ethanol during fermentation have been conducted in conventional stirred-tank bioreactors (41,43). Due to poor gas distribution in the stirred tanks, energy consumption for gas input into the fermentor is high. With an airlift fermentor, this drawback can be overcome. The airlift fermentor is the reactor that has a simple structure and provides better gas distribution and uniform air bubble size. The airlift fermentor is also about one-third less energy intensive than stirred tanks (44) because mechanical agitation is replaced by gas upflow to provide the required mixing. Recently, Tsao et al. (45) developed a novel airlift fermentor with a side arm (ALSA). This modified airlift fermentor improves liquid circulation velocities and minimizes the need for antifoaming agents. An ALSA fermentor coupled with gas stripping was used for the removal of ethanol from the fermentation broth during the fermentation of a substrate that contained a high glucose concentration. The ethanol concentration in the broth was maintained at under 4% (w/v) at all times due to the continuous removal of ethanol. The configuration of an ALSA fermentor integrated with a CO_2 stripping system is shown in Figure 5.

The application of this type of integrated fermentation system for high-solid SSF fermentation of biomass materials at an elevated temperature has the potential to overcome ethanol inhibition and temperature incompatibility.

XYLOSE FERMENTATION

Upon the hydrolysis of hemicellulose, xylose becomes the major monosaccharide component in the hemicellulose hydrolysate. Also, in the hydrolysates of lignocellulosic materials, mixtures of glucose and xylose will be expected. Usually, glucose in the hydrolysate can be completely fermented to ethanol within a few hours. However, the complete conversion of xylose to ethanol will take 48–72 h or longer. The nonmatching fermentation rates between glucose and xylose cause difficult design problems. The development of the SSF process has made it possible to convert lignocellulosics to ethanol with cellulase and fermenting microbes in the reactor at the same time. Again, the conversion of cellulose to glucose and finally to ethanol can be completed quickly. Nevertheless, the batch lasts a long time, until the xylose fermentation is completed.

Xylose can be metabolized by bacteria, fungi, or yeast. In bacteria, the initial step of xylose metabolism involves inducible enzymes (xylose transport enzymes, xylose isomerase, and xylulokinase). The direct isomerization of xylose to xylulose is the first step in xylose utilization by bacteria. They produce a variety of end products, including ethanol, from xylose (46). Recently, through advances in genetic engineering, a few bacterial strains have been genetically constructed to efficiently produce ethanol as the major product from xylose (47,48).

In yeast and mycelial fungi, xylose is metabolized via the coupled oxidation–reduction reactions. Xylose reductase is the enzyme involved in the reduction of xylose to xylitol, the sequential enzymatic events through the oxidation of xylitol to xylulose that lead to the utilization of xylose. Many yeast species utilize xylose readily, but ethanol production capability is very limited. Only a few yeast species such as *Pachysolen tannophilus, Candida shihatae*, and *Pichia stipitis* are relatively good ethanol producers when xylose is the substrate (49). The production of ethanol from xylose by those three yeast strains has been studied extensively in recent years. Recently, genetically engineered yeast strains have been constructed for more effective conversion of xylose to ethanol (50).

Xylose fermentation by yeast strains is subjected to regulation by dissolved oxygen (51) and by the presence of extraneous materials in the substrate (52). It is also affected by the presence of the fermentation end product,

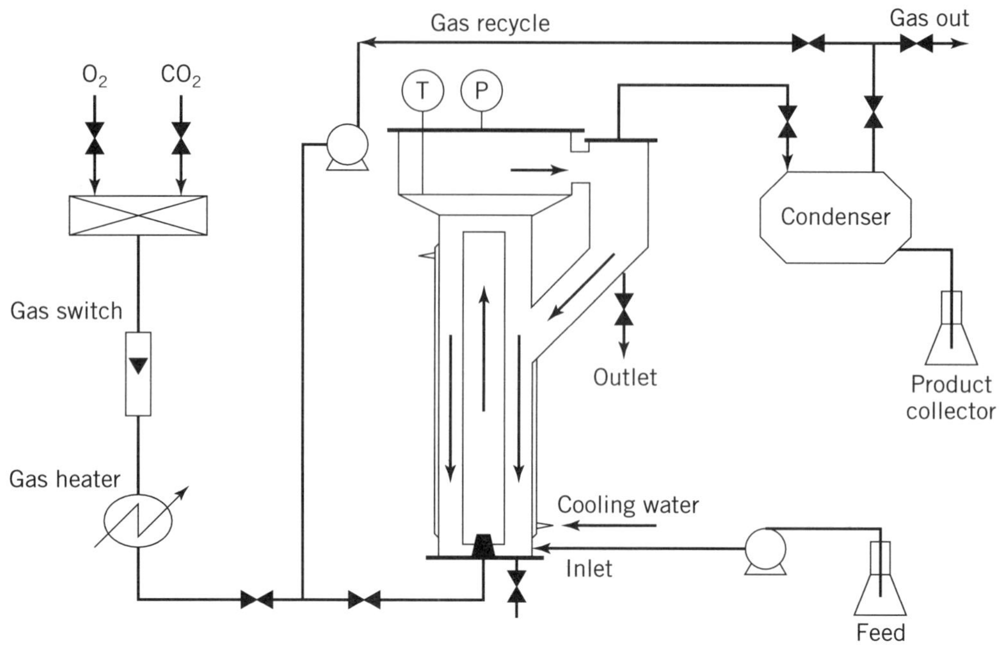

Figure 5. Schematic configuration of ALSA integrated with gas stripping. *Source*: From Ref. 45.

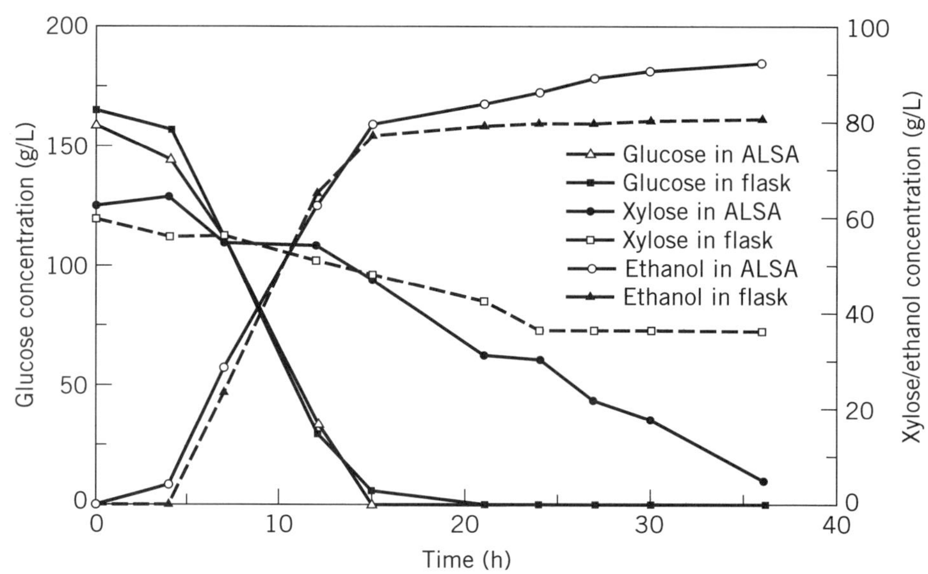

Figure 6. Fermentation of glucose–xylose mixture to ethanol by *Saccharomyces* 1400 (pLHN33) in an ALSA with gas stripping. *Source*: From Ref. 45.

ethanol. Ethanol, at about 45 g/L, will become inhibitory toward the ability of either naturally occurring or genetically modified xylose-fermenting microbes. To overcome the end-product inhibition by ethanol, Tsao et al. (45) used a modified airlift fermentor (ALSA) coupled with a gas-stripping system to remove ethanol continuously from the fermentation broth. Their results (Fig. 6) showed that xylose can be converted into ethanol without the feedback inhibition by ethanol that was encountered under typical stirred-tank fermentation conditions.

Inhibitors

In addition to producing monomer sugar residues, thermochemical processing of biomass is known to produce substances that are inhibitory to microbial growth and ethanol production. For example, furfural derived from pentose sugars, hydroxymethylfurfural, and levulinic acid from hexose sugars are known inhibitors (53). In addition, soluble phenolic compounds derived from lignin degradation and alkaline extractives are also inhibitory toward yeast in low concentration (Table 3). Another potential inhibitor is acetic acid, which can be present in relatively high concentrations (6–15 g/L); it exists as an integral part of hemicellulose. Various procedures have been investigated to minimize the inhibitory effect of the processing by-products. The methods include careful control of the pretreatment process to minimize sugar degradation, using ion exchange resins and activated charcoal to remove

Table 3. The Effects of Potential Inhibitors on Growth and Ethanol Production from Xylose

Compound	Concentration (g/L)	Inhibition (% growth)	Inhibition (% EtOH)
Furfural	2	90	90
HMF[b]	5	92	96
Acetate	10	16	38
HBA[c]	0.75	77	83
SGA[d]	0.75	55	60
Vanillin	0.5	33	53

Source: From Ref. 53.
[a]Percent inhibition of ethanol production from xylose by *Candida shehatae*.
[b]Hydroxymethylfurfural.
[c]Hydroxybenzaldehyde.
[d]Siringaldehyde.

the inhibitors, or the soaking of materials in an alkaline solution prior to hydrolysis.

BIOMASS FRACTIONATION

The rationale behind the fractionation of lignocellulose into lignin, hemicellulose, and cellulose components prior to utilization is as follows:

- Alkaline extractives, soluble lignin, and lignin-derived soluble products are inhibitory to microbial fermentation.
- Hemicellulose is much more susceptible to acid hydrolysis than cellulose.
- Cellulose carbohydrate can be fermented to ethanol much faster than can hemicellulose-derived carbohydrates (e.g., xylose).

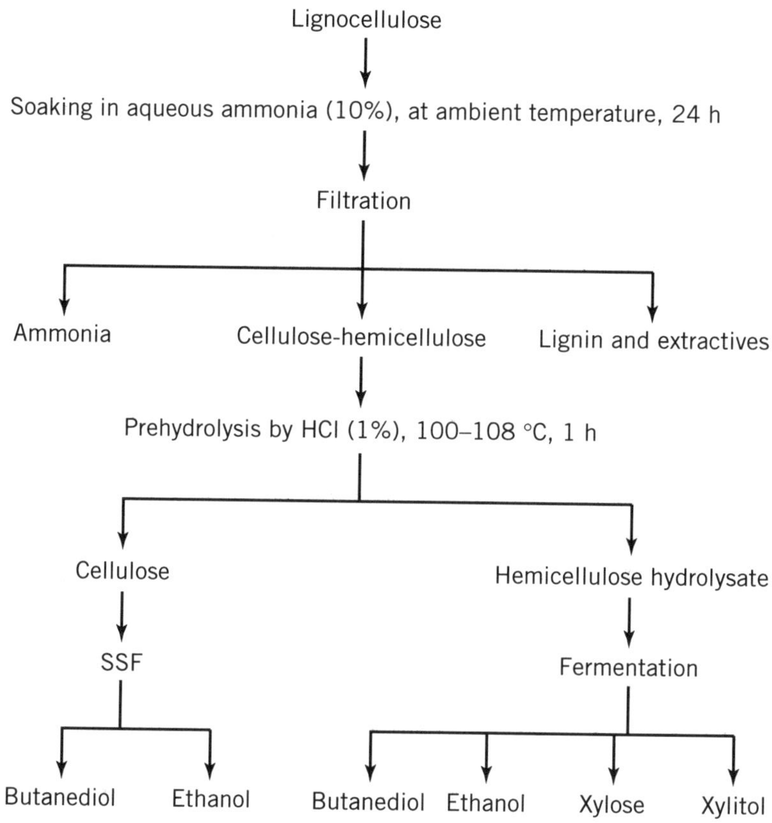

Figure 7. Flow chart of biomass fractionation and potential products. *Source*: From Ref. 55.

Table 4. Composition of Materials after Treatments

Material		Wood Chips			Corn Cobs		
Treatment		Original	After ammonia soaking	After acid hydrolysis	Original	After ammonia soaking	After acid hydrolysis
Composition (%)	Cellulose	42.9	58.09	85.37	44.88	56.2	90.4
	Xylose	22.5	30.37	4.48	32.68	38.5	5.29
	Lignin	25.0	7.11	6.97	7.41	0.85	0.91

Source: From Ref. 55.

The biomass fractionation options and their economical benefits were recently reviewed and discussed by Elander and Hsu (54). The objective of the fractionation is to separate different biomass components and subsequently convert the cellulose and the hemicellulose into separate products. To achieve this objective, mild treatment conditions such as low temperature, low concentration of reagents (or easily recoverable reagents), and low pressures are necessary. Under this biomass component fractionation scheme, cellulose sugar (glucose) and hemicellulose carbohydrates (mostly xylose) can be converted in different reactors to produce different products, if so desired.

Cao et al. (55,56) examined a fractionation option that used corn cobs and aspen woodchips as the substrates. In this biomass fractionation scheme (Fig. 7), the majority of lignin, alkaline extractives, and acetate were solubilized and separated from cellulose and hemicellulose fractions by alkaline treatment. Hemicellulose was thereafter hydrolyzed to its sugar constituents by dilute acids. Hemicellulose carbohydrates were then fermented to ethanol, 2,3-butanediol, or xylitol by xylose-fermenting microbes. The cellulose fraction, after separation from lignin and hemicellulose, was used as the substrate in the SSF process for ethanol production with a thermotolerant yeast strain as the biocatalyst, or it can be bioconverted into other products. Table 4 shows the changes of biomass components after ammonia and dilute acid treatments.

BIBLIOGRAPHY

1. E. Glade, *Industrial Uses of Agricultural Materials*, IUS-5, Economic Research Service, USDA, Washington, D.C., 1995, pp. 21–23.
2. L. Olsson and B. Hahn-Hagerdal, *Enzyme Microb. Technol.* **18**, 312–331 (1996).
3. Y.C.T. Chou, *Biotechnol. Bioeng. Symp.* **17**, 19–32 (1986).
4. R. Torget, P. Walter, M. Himmel, and K. Grohmann, *Appl. Biochem. Biotechnol.* **28/29**, 75–97 (1991).
5. F. Parisi, *Adv. Biochem. Eng. Biotechnol.* **38**, 53–87 (1989).
6. B.E. Dale and M.J. Moreira, *Biotechnol. Bioeng. Symp.* **12**, 31–43 (1983).
7. E.T. Papoutsakis, *Biomass Handbook*, Gordon and Breach Science Publishers, New York, 1989, pp. 271–286.
8. S.K. Garg and A. Jain, *Biosource Technol.* **51**, 103–109 (1995).
9. J.H. Litchfield, in M. Moo-Young ed., *Biomass Conversion Technology: Principles and Practice*, Pergamon, 1987, pp. 113–122.
10. E. Sjöström, *Wood Chemistry: Fundamentals and Applications*, Academic, New York, 1981, pp. 1–20.
11. D.S. Chahal, in D.S. Chahal ed., *Food, Feed and Fuel from Biomass*, Oxford & IBH, New Delhi, 1991, pp. 59–83.
12. J.D. McMillan, in M.E. Himmel, J.O. Baker, R.P. Overend eds., *Enzymic Conversion of Biomass for Fuels Production*, American Chemical Society, Washington, D.C., 1994, pp. 292–324.
13. J.H. Litchfield, *Microbial Technol.* **1**, 93–155 (1979).
14. N.J. Cao, Q. Xu, C.S. Chen, C.S. Gong, and L.F. Chen, *Appl. Biochem. Biotechnol.* **45/46**, 521–530 (1994).
15. M.M. Chang, Y.C.T. Chou, and G.T. Tsao, *Adv. Biochem. Eng.* **20**, 15–42 (1981).
16. H.H. Yoon, Z.W. Wu, and Y.Y. Lee, *Appl. Biochem. Biotechnol.* **51/52**, 5–19 (1995).
17. J.M. Gould, *Biotechnol. Bioeng.* **25**, 46–52 (1984).
18. M.T. Holtzapple, J.-H. Jun, G. Ashok, S.L. Patibanla, and B.E. Dale, *Appl. Biochem. Biotechnol.* **28/29**, 59–74 (1991).
19. Weil, P. Westgate, K. Kohlmann, and M.R. Ladisch, *Enzyme Microb. Technol.* **16**, 1002–1004 (1994).
20. W. Schwald, C. Breuil, H.H. Brownell, M. Chan, and J.N. Saddler, *Appl. Biochem. Biotechnol.* **20/21**, 29–44 (1989).
21. D.J. Schell, R. Torget, A. Power, P.J. Walter, K. Grohmann, and N.D. Hinman, *Appl. Biochem. Biotechnol.* **28/29**, 87–97 (1991).
22. J.E. Carrasco, M.C. Saiz, A. Navarro, P. Soriano, F. Saez, and J.M. Martinez, *Appl. Biochem. Biotechnol.* **45/46**, 23–34 (1994).
23. K. Grohmann, R. Torget, and M. Himmel, *Biotechnol. Bioeng. Symp.* **15**, 59–80 (1985).
24. E.T. Reese, and H.S. Levinson, *Physiol. Plant* **5**, 345–366 (1952).
25. E.T. Reese, S.G.H. Siu, and H.S. Levinson, *J. Bacteriol.* **59**, 485–497 (1950).
26. C.S. Gong, M.R. Ladisch, and G.T. Tsao, *Adv. Chem. Ser.* **181**, 261–287 (1979).
27. D.S. Chahal, P.S. Chahal, and G. Andre, in D.S. Chahal ed., *Food, Feed and Fuel from Biomass*, Oxford & IBH, New Delhi, 1991, pp. 281–312.
28. R. Torget, *Chemical Analysis and Testing Standard Procedures*, Procedure No. 009. NREL, Golden, Col., 1994.
29. M. Mandel and J. Weber, *Adv. Chem. Ser.* **95**, 391–414 (1969).
30. B.S. Montenecourt and D.E. Eveleigh, *Adv. Chem. Ser.* **181**, 289–301 (1979).
31. M. Mandel and D. Sternberg, *J. Ferment. Technol.* **54**, 267–286 (1976).
32. K. Nishizawa, *Biomass Handbook*, Gordon and Breach Science Publishers, New York, 1989, pp. 307–318.
33. D. Ryu and M. Mandel, *Enzyme Microb. Technol.* **2**, 91–102 (1980).
34. C.S. Gong, M.R. Ladisch, and G.T. Tsao, *Biotechnol. Bioeng.* **19**, 959–981 (1977).
35. P.J. Blotkamp, M. Tagagi, M.S. Pemberton, and G.H. Emert, *AIChE Symp. Ser.* **181**, 85–90 (1981).
36. G.H. Emert, in D.S. Chahal ed., *Food, Feed and Fuel From Biomass*, Oxford & IBH, New Delhi, 1991, pp. 313–333.
37. G.P. Philippidis, and T.K. Smith, *Appl. Biochem. Biotechnol.* **51/52**, 117–124 (1995).
38. P.J. Anderson, K. Mcneil, and K. Watson, *Appl. Environ. Microbiol.* **51**, 1314–1320 (1986).
39. I. Ballesteros, J.M. Oliva, M. Ballesteros, and J. Carrasco, *Appl. Biochem. Biotechnol.* **39/40**, 201–211 (1993).
40. P. Cen and G.T. Tsao, *Separation Technol.* **3**, 58–75 (1993).
41. P.K. Walsh, C.P. Liu, M.E. Findley, A.I. Liapis, and D.J. Siehr. *Biotechnol. Bioeng. Symp.* **13**, 629–647 (1983).
42. M.C. Dale, M.R. Okos, and P.C. Wanket, *Biotechnol. Bioeng.* **27**, 943–952 (1985).
43. H.S. Liu and H.W. Hsu. *Chem. Eng. Sci.* **45**, 1289–1299 (1990).
44. M. Trager, G.N. Qazi, U. Onfen, and C.L. Chopra. *J. Ferm. Bioeng.* **68**, 112–116 (1989).
45. G.T. Tsao, N.J. Cao, M.S. Krishnan, J.X. Du, and C.S. Gong, *Proceedings, The Seventh National Bioenergy Conference*, 1996.
46. C.S. Gong, L.F. Chen, M.C. Flickinger, and G.T. Tsao, *Adv. Biochem. Eng.* **20**, 93–118 (1981).

47. F. Alterthum, and L.O. Ingram, *Appl. Environ. Microbiol.* **55**, 1943–1948 (1989).
48. M. Zhang, C. Eddy, K. Deanda, M. Finkelstein, and S. Picataggio, *Science* 267, 240–243 (1995).
49. J.N. Nigam, R.S. Ireland, A. Margaritis, and M.A. Lachance, *Appl. Environ. Microbiol.* **50**, 1486–1489 (1985).
50. PCT Pat. WO95/13362 (Aug. 4, 1995), N.W.Y. Ho and G.T. Tsao (to Purdue University Research Foundation).
51. J.M. Laplace, J.P. Delgenes, R. Moletta, and J.M. Navarro, *Appl. Microbiol. Biotechnol.* **36**, 158–162 (1991).
52. A.V. Tran and R.P. Chambers, *Biotechnol. Lett.* **7**, 841–846 (1985).
53. J.P. Delgenes, R. Moletta, and J.M. Navarro, *Enzyme Microb. Technol.* **19**, 220–225 (1996).
54. R.T. Elander and T. Hsu, *Appl. Biochem. Biotechnol.* **45/46**, 463–478 (1994).
55. N.J. Cao, Y. Huang, C.S. Gong, and G.T. Tsao, *AICHE Series: 5th World Congress of Chemical Engineering*, 1996, pp. 16–21.
56. N.J. Cao, M.S. Krishnan, J.X. Du, C.S. Gong, N.W.Y. Ho, Z.D. Chen, and G.T. Tsao, *Biotechnol. Lett.* **18**, 1013–1018 (1996).

See also ENZYMES, PULP AND PAPER PROCESSING; HEMICELLULASES.

ORGANOSILICON COMPOUNDS

ATSUO TANAKA
TAKUO KAWAMOTO
Kyoto University
Kyoto, Japan

KEY WORDS

Bioconversion
Deacylation
Dehydrogenation
Enantioselectivity
Esterification
Hydrolysis
Oxidoreduction
Regioselectivity
Stereoselectivity
Transesterification

OUTLINE

INTRODUCTION

Silicon, which is the second most abundant element in the Earth's crust, belongs to the same group as carbon, which is one of the most fundamental elements for organisms, but is different from carbon in its characteristics (1). Silicon is less electronegative and has a bigger covalent radius than carbon (Table 1). A bond between silicon and oxygen is stronger than the corresponding bond between carbon and oxygen, whereas its bonds with carbon and hydrogen are weak.

Silicon plays an important role in the biosphere and is present mainly in living organisms, especially at a lower stage of evolutionary development. Silicate bacteria, the simplest algae, and spore plants, in fact, contain very large amounts of silicon (2,3). Recent research has indicated that silicon is also important for higher plants, animals, and humans (4), although its actual biochemical function is at present not fully defined. It seems, however, that natural biochemical processes mainly involve silicon bound to oxygen, whereas organosilicon compounds containing Si–C bonds have not been detected (1–3). Therefore, the effects of organosilicon compounds on organisms and/or enzymes are very interesting.

The explosive growth of organosilicon chemistry has created a growing awareness of its considerable utility to organic chemists and has shown that a wide spectrum of organosilicon compounds often exhibit interesting biological properties (2,3). Recently, pharmaceutical applications of organosilicon chemistry have also been gaining attention (2,3,5–7). Therefore, the production of organosilicon compounds with desired chemical and physical properties by the use of effective biocatalysts is of great importance. Several examples of bioconversion of organosilicon compounds have been summarized by Ryabov (8) and Tanaka and Kawamoto (9).

MICROBIAL REDUCTION OF ORGANOSILICON COMPOUNDS

Stereoselective microbial reduction of organosilicon compounds was first demonstrated by Tacke et al. (10). Analogous to the microbial reduction of alkyl acetoacetates to optically active 3-(*S*)-hydroxybutyrates, the sila analogues **1a** and **1b** could be transformed stereoselectively into the corresponding reduction products (*S*)-**2a** and (*S*)-**2b** by using growing cells of the yeast *Kloeckera corticis* (Fig. 1). This stereoselective reduction could be performed on a preparative scale with good yields. The optical purities obtained were about 80% ee for (*S*)-**2a** and (*S*)-**2b**.

In this case, the carbonyl groups attacked by the microorganism are relatively far away from the silicon atom. However, it was also shown that compounds containing a carbonyl group bound directly to the silicon atom can also be reduced stereoselectively by microorganisms. For example, achiral acetyldimethylphenylsilane (**3**) was transformed stereoselectively into the optically active reduction product (*R*)-**4** (86% ee) by growing and resting cells of the yeast *Trigonopsis variabilis* DSM 70714 (Fig. 2) in a yield of 70% (11). Furthermore, a wide range of microorganisms

Table 1. Physical Properties of the Silicon Atom and the Carbon Atom

Atom	Electronegativity	Bond	Bond length (nm)	Bond dissociation energy (kJ mol^{-1})
C	2.55	C—C	0.153	334
Si	1.90	C—Si	0.189	318

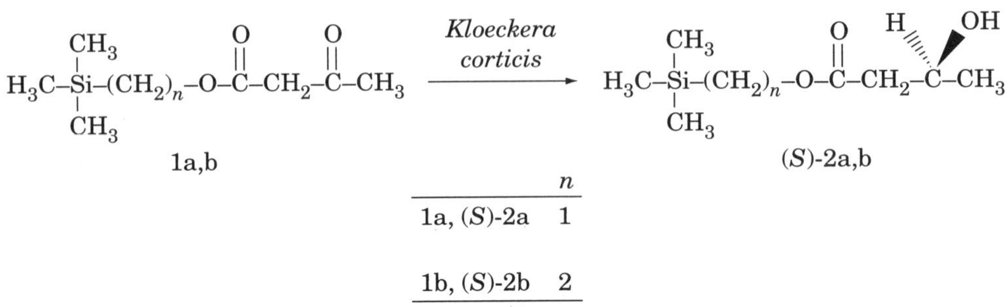

Figure 1. Stereoselective microbial reduction of sila-analogues of alkyl acetoacetates.

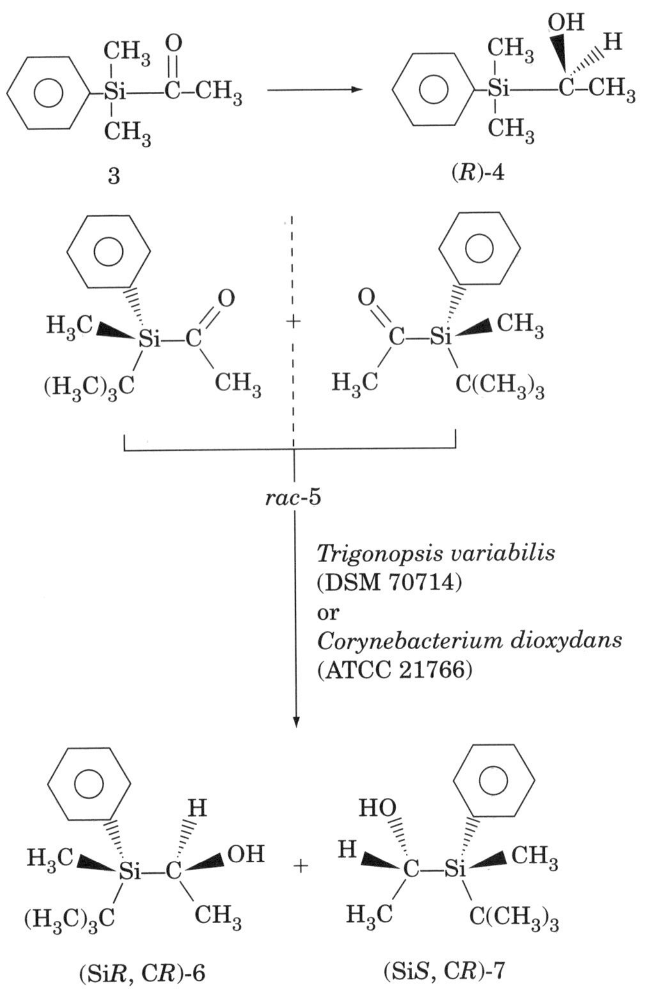

Figure 2. Microbial reduction of organosilicon compounds having a carbonyl group bound directly to the silicon atom.

such as bacteria, yeasts, fungi, and green algae showed the ability to accept **3** as a substrate for stereoselective reduction (12). Among 30 strains of microorganisms tested as resting cells, *T. variabilis* DSM 70714 was found to exhibit the highest specific activity (1.5 mg product g wet mass^{-1} min^{-1}), whereas the highest stereoselectivity (95% ee) was observed with the bacteria *Acinetobacter calcoaceticus* ATCC 31012 and *Corynebacterium dioxydans* ATCC 21766. The same bioconversion could be performed by resting cells of *Saccharomyces cerevisiae* DHW S-3 at a rate of 44–45 μmol L^{-1} min^{-1} with a yield of 40% and optical purity of more than 99% after purification (13). Plant cell suspension cultures of *Synphytum officinale* L. and *Ruta graveolens* L. served as the biocatalysts for the conversion of **3** to (*R*)-**4**, although the yields and optical purities were not satisfactory (*S. officinale*: 81% ee, yield 15%; *R. graveolens*: 60% ee, yield 9%) (14).

It is noteworthy that a carbonyl group bound to the silicon atom can be transformed by various microorganisms, as in the case of conventional ketones, although there are big differences in physical and chemical properties between these compounds. Another example of microbial reduction of organosilicon compounds is the transformation of racemic acetyl(*tert*-butyl)methylphenylsilane (*rac*-**5**) with resting cells of *T. variabilis* DSM 70714 (15) and *C. dioxydans* ATCC 21766 (16). In these cases, enantioselective reduction of both enantiomers of the racemic compounds was observed, resulting in the formation of the diastereomeric optically active silanes (Si*R*,C*R*)-**6** (*T. variabilis*: 97% ee, yield 74%; *C. dioxydans*: 99% ee, yield 20%) and (Si*S*,C*R*)-**7** (*T. variabilis*: 96% ee, yield 70%; *C. dioxydans*: 99% ee, yield 20%). Growing cells of *K. corticis* catalyzed the stereoselective reduction of 1,1-dimethyl-1-silacyclohexan-2-one (**8**) to (*R*)-1,1-dimethyl-1-silacyclohexan-2-ol ((*R*)-**9**) (92% ee, yield 60%) (Fig. 3) (17). Reduction of (+)-(benzyloxymethyl)(*tert*-butyl)methylsilyl methyl ketone ((+)-**10**) to (Si*R*,1*R*)-(+)-1-[(benzyloxymethyl)(*tert*-butyl)methylsilyl]ethanol ((+)-**11**) and its

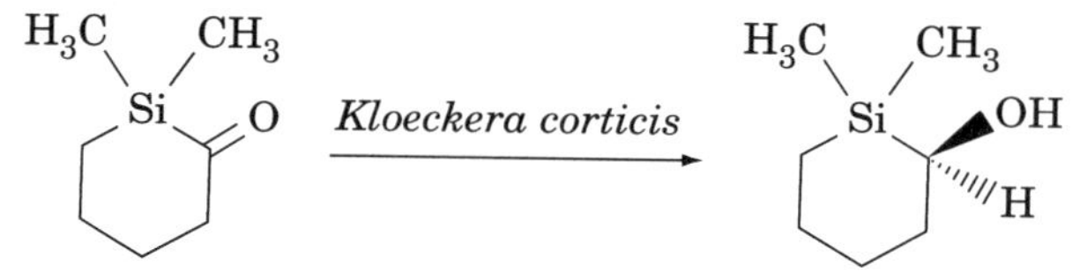

8 (R)-9

Figure 3. Microbial reduction of 1-silacyclohexan-2-one.

(Si*S*,1*R*)-enantiomer ((+)-**12**) was performed by resting free and immobilized cells of *T. variabilis* (>96% ee, yield 67–99%) (Fig. 4) to prepare (*R*)-(−)-**10** and (*S*)-(+)-**10**, the former being the intermediate in the synthesis of (*R*)-(+)-1-phenylethanol (18).

These studies clearly demonstrate that organosilicon compounds could be transformed by microorganisms and that the microbial transformation is useful for the preparation of optically active organosilicon compounds.

OXIDOREDUCTION OF ORGANOSILICON COMPOUNDS BY ALCOHOL DEHYDROGENASES

Bioconversion of organosilicon compounds by alcohol dehydrogenases (ADHs) is very attractive because of the importance of asymmetric alcohol/carbonyl oxidoreduction in organic chemistry. Although several papers are available on the reduction of organosilicon compounds by microorganisms as mentioned earlier, it is not clear whether ADH can actually catalyze the oxidoreduction of organosilicon compounds in these organisms, and it has not been possible to discuss the effect of the silicon atom on enzymatic reactions with ADH.

It was shown for the first time by Zong et al. (19) that ADH from horse liver (HLADH) could catalyze the dehydrogenation of organosilicon compounds and that the silicon atom in the substrates had unique effects on enzymatic reactions due to specific characteristics of the silicon atom. The bioconversion of three organosilicon compounds with different chain lengths between the silicon atom and the hydroxyl group ($Me_3Si(CH_2)nOH$, $n = 1–3$) by HLADH was systematically studied in comparison with the corresponding carbon compounds ($Me_3C(CH_2)nOH$, $n = 1–2$). HLADH, but not yeast alcohol dehydrogenase, catalyzed the dehydrogenation of 2-trimethylsilylethanol ($n = 2$) **(14a)** and 3-trimethylsilylpropanol ($n = 3$) **(15a)**. **14a** was a better substrate than both its carbon analogue, 3,3-dimethylbutanol **(14b)** and ethanol. The improved activity of HLADH on **14a** could be accounted for by a lower activation energy of the reaction by HLADH than that with the carbon analogue. In contrast, HLADH showed no activity on trimethylsilylmethanol ($n = 1$) **(13a)**, whereas it catalyzed the dehydrogenation of the carbon analogue, 2,2-dimethylpropanol **(13b)**. This phenomenon was explained by the β-effect of the silicon atom, that is, stabilization of a negative charge at the α-position and a positive charge at the β-position of the carbon atom from the silicon atom (1). As the silicon atom stabilizes the positive charge of the β-carbon atom, the H^- anion could be removed more easily from the β-carbon atom of **14a**, which is the first step of the dehydrogenation of alcohol catalyzed by ADH. As a result, the activation energy of the dehydrogenation of **14a** by HLADH probably became lower than that of **14b**, and thus **14a** is a better substrate than the carbon counterpart (Fig. 5). Conversely, it is more difficult for the H^- anion to be removed from the α-carbon atom of **13a** because the electron density around the α-carbon of **13a** is higher than that of **13b** owing to a lower electronegativity of the silicon atom (Fig. 5). Sonomoto et al. (20) also confirmed the same phenomena.

It was also demonstrated that HLADH could catalyze the enantioselective dehydrogenation of racemic trimethylsilylpropanols, 1-trimethylsilyl-2-propanol **(16a)** and 2-trimethylsilyl-1-propanol **(18a)** (Fig. 6) in a water–organic solvent two-layer system with coenzyme regeneration by a glutamate dehydrogenase-2-oxoglutarate system (21). Especially, the enantiomeric excess of **16a** remaining at 51% conversion was as high as 99%. In this case, 1-trimethylsilyl-1-propanol **(17a)** did not serve as a substrate. A comparative study with their carbon counterparts (**16b** and **18b**) showed results similar to those of the hydrolases to be described later, that is, substitution of the silicon atom for the carbon atom improved the enantioselectivity of HLADH.

Tsuji et al. (22) found that enantioselective dehydrogenation of primary alcohol β-hydroxyalkylsilane, 2-trimethylsilyl-1-propanol **(18a)**, could be carried out with only a catalytic amount of NAD^+ and HLADH in an aqueous system containing 5% tetrahydrofuran. When **18a** (5

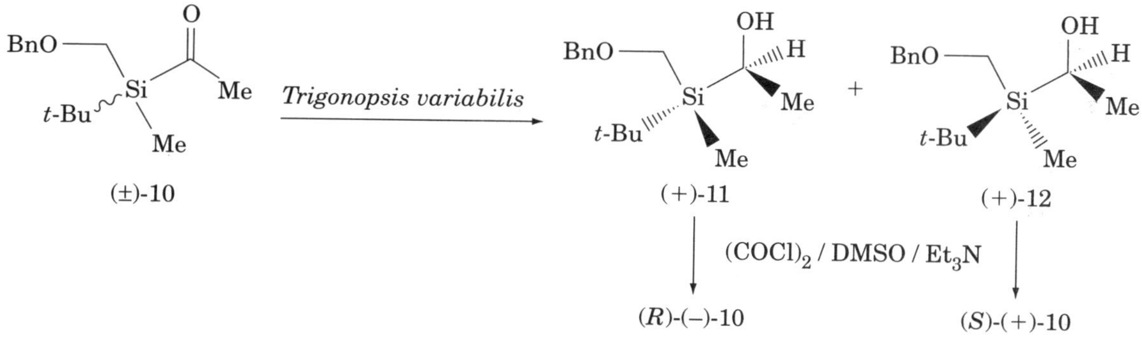

Figure 4. Microbial reduction of a silicon-containing ketone.

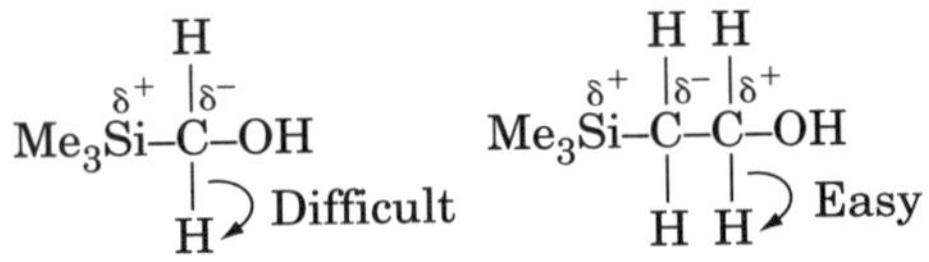

13a 14a

Figure 5. Electric effect of the silicon atom on removal of H^- from trimethylsilylalkanols.

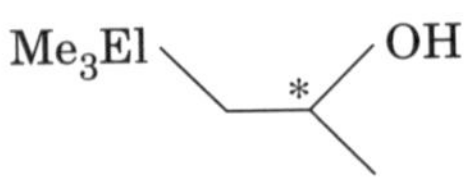

16a El = Si
16b El = C

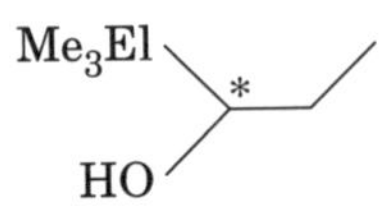

17a El = Si
17b El = C

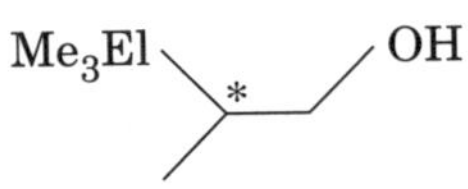

18a El = Si
18b El = C

Figure 6. Structures of three isomers of trimethylsilylpropanols and their carbon analogues.

× 10 μmol) was applied to the HLADH-catalyzed dehydrogenation with a small amount of NAD^+ (5 × 10^{-1} μmol) in the aqueous system, NAD^+ could be regenerated in situ through reduction of 1-propanal **(20)** formed by spontaneous degradation of the dehydrogenation product, 2-trimethylsilyl-1-propanal **(19)**, in the presence of water (Fig. 7). In this case, conversion ratio reached 75% after 14 h of reaction; the turnover number of NAD^+ was calculated to be about 75, where the optical purity of remaining **18a** was 90% ee. If the amount of NAD^+ was reduced to 5 × 10^{-2} or 5 × 10^{-4} μmol (**18a**, 5 × 10 μmol), the turnover number of NAD^+ reached 640 or 7,200, respectively. This novel reaction system is simple and advantageous because it does not require any other enzyme and substrate for regeneration of NAD^+, and there is no product inhibition due to the spontaneous degradation of the dehydrogenation product. On the other hand, dehydrogenation of the corresponding carbon analogue **(18b)** did not proceed under this condition because the aldehyde formed by dehydrogenation of **18b** with HLADH did not undergo the degradation to provide a substrate for NAD^+ regeneration.

These results indicate that application of organosilicon compounds makes it possible to construct an efficient and unique conversion system, which is impossible in the case of conventional carbon compounds. Other primary β-hydroxysilanes having different substituents on the silicon atom or on the stereogenic carbon atom, 2-trimethylsilyl-1-butanol, 2-trimethylsilyl-1-hexanol, and 2-dimethylphenylsilyl-1-butanol, were also found to serve as substrates in enantioselective dehydrogenation by HLADH with this novel NAD^+ regeneration system. In contrast, dehydrogenation of the secondary β-hydroxysilane, 1-trimethylsilyl-2-propanol **(16a)**, was negligible under the same conditions, although HLADH could convert **16a** with conventional coenzyme regeneration (21). The dehydrogenated product of **16a**, 1-trimethylsilyl-2-propanone, was supposed to be degraded by the addition of water into trimethylsilanol and acetone also in this case, but regeneration of NAD^+ did not occur because acetone was not recognized as the substrate by HLADH and so the dehydrogenation reaction did not proceed over the catalytic amount of NAD^+.

Recently, Wong et al. (23) have reported that NADPH-requiring *Lactobacillus kefir* ADH could catalyze the enantioselective carbonyl reduction of ketones having a large variety of side chains, including trimethylsilyl-protected terminal alkyne ketones, and NADPH regeneration in the presence of 2-propanol.

BIOCONVERSION OF ORGANOSILICON COMPOUNDS BY HYDROLASES

Hydrolases such as lipases, esterases, and proteases are the easiest enzymes to handle because hydrolases do not require coenzymes and are readily available in a large quantities. Therefore, they have been extensively used as catalysts in regio- or enantioselective conversion.

Therisod and Klibanov (24) mentioned additionally, in the regioselective acylation of secondary hydroxyl groups of C-6 protected glucose, that modification of the C-6 hydroxyl group of D-glucose with a more bulky *tert*-butyldiphenylsilyl group increased the regioselectivity of *Candida cylindracea* lipase toward the C-2 position, compared with 6-*O*-butylylglucose. It was, however, not clear how reactive the compound modified with the *tert*-butyldiphenylsilyl group was in the lipase-catalyzed acylation. Luyten et al. (25) reported in the study on enantioselective hydrolysis of several functionalized dimethyl malonates **(21a–f)** by pig liver esterase (PLE) (Fig. 8) that a *tert*-butyldimethylsilyl derivative of dimethyl 2-(hydroxymethyl)-2-methylmalo-

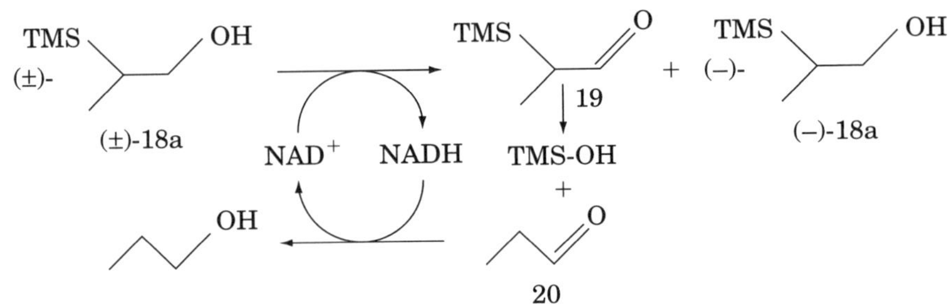

Figure 7. Horse liver alcohol dehydrogenase–catalyzed enantioselective dehydrogenation of 2-trimethylsilyl-1-propanol with in situ coenzyme regeneration. TMS, trimethylsilyl.

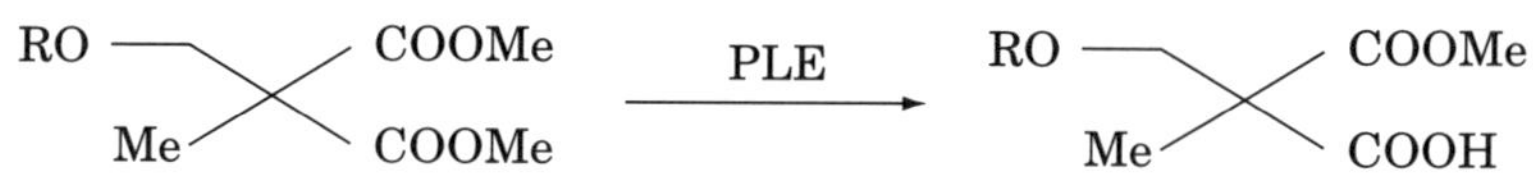

21a–f 22a–e

a, R = H
b, R = Me
c, R = (*t*-Bu)(Me)$_2$Si
d, R = $C_6H_5CH_2$
e, R = *t*-Bu
f, R = $CH_3C{=}O$

Figure 8. Enantioselective hydrolysis of functionalized dimethyl malonates by pig liver esterase.

nate **(21c)** gave a high enantioselectivity (95 % ee) but a sluggish reaction. As a result, the yield of the optically pure product ((*R*)-**22c**) was very low (49%). In this study, a (*tert*-butoxy)methyl derivative **(21e)** showed the best results, namely, a high enantioselectivity (96% ee) and a high yield (90%). Mori and Takeuchi (26) prepared enantiomerically pure (1*S*, 4*R*)-4-*tert*-butyldimethylsilyloxy-3-chloro-2-cyclopenten-1-ol **(24)** (ca. 100% ee), which is a key chiral building block for synthesis of punaglandin 4, a chlorinated marine prostanoid, through the asymmetric hydrolysis of the mixture of 3-acetoxy-5-*tert*-butyldimethylsilyloxy-1-chlorocyclopentenes **(23)** with pig pancreatic lipase (PPL) (Fig. 9) in 25% yield.

In the case of the organosilicon compounds already mentioned, functional groups attacked by hydrolases were far away from the silicon atom. Therefore, the effect of the silicon atom itself on enzymatic reactions seems very weak. It is interesting to study the bioconversion of organosilicon compounds whose silicon atom more strongly affects the functional groups to be attacked by biocatalysts. Interesting attempts were made by Therisod (27). He tried to use trimethylsilyl ethers instead of alcohols as novel nucleophiles in hydrolase-catalyzed transesterification (Fig. 10), but failed.

Kawamoto et al. (28) attempted a comparative study by using organosilicon compounds having a different chain length between the hydroxyl group and the silicon atom ($Me_3Si(CH_2)nOH$, n = 0, 1, 2, and 3) and the corresponding carbon compounds ($Me_3C(CH_2)nOH$, n = 0, 1, and 2) as an acyl acceptor in the enantioselective esterification of 2-(4-

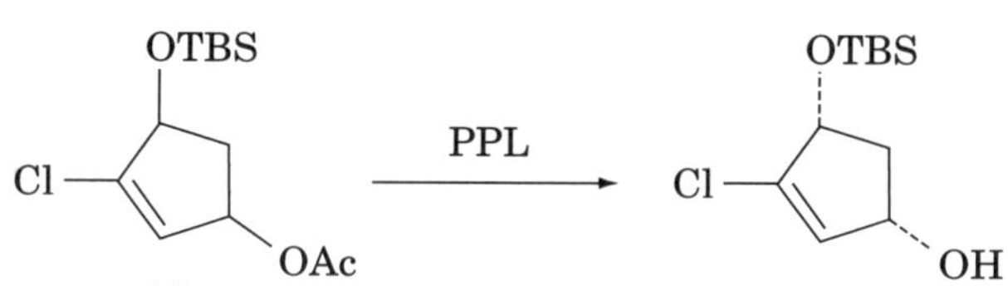

23 24

Figure 9. Enantioselective hydrolysis of a silicon-containing ester. TBS, *t*-butyldimethylsilyl.

$$CH_3(CH_2)_2COOEt + ROSiMe_3 \longrightarrow CH_3(CH_2)_2COOR + EtOSiMe_3$$

Figure 10. Attempt to use trimethylsilyl ethers as novel nucleophiles in hydrolase-catalyzed transesterification. R, hexyl or cyclohexyl.

chlorophenoxy)propanoic acid **(25)**, whose *R*-enantiomer is useful as an herbicide, with lipase OF 360 of *C. cylindracea* (Fig. 11) and discussed the effects of the silicon atom on the enzymatic reaction in connection with the distance between the hydroxyl group and the silicon atom. Trimethylsilylmethanol (n = 1) **(13a)** was found to be a particularly superior substrate, that is, the reaction rate was much higher than that with the corresponding carbon compound **(13b)**, and the enantiomeric excess of the acid remaining was above 95% at about 50% conversion. In the case of conventional substrates such as the carbon analogue and linear-chain alcohols, a high reaction rate was not consistent with a high enantioselectivity. These results indicate that organosilicon compounds may solve the unavoidable and probably inherent problems of enantioselective reactions with conventional substrates. On the other hand, no difference was observed between 2-trimethylsilylethanol (n = 2) **(14a)** and its carbon analogue, 3,3-dimethylbutanol **(14b)**, with respect to enzymatic activity and enantioselectivity. Thus, the silicon atom mimicked the carbon atom for lipase in the case of **14a** but was different from the carbon atom in enhancing the reactivity of **13a**. The phenomena were explained on the basis of the properties of the silicon atom, such as its low electronegativity and big atomic radius compared with those of the carbon atom. A lower electronegativity of the silicon atom results in a higher nucleophilicity of the oxygen atom of **13a** compared with that of the corresponding carbon analogue **(13b)**, and the hydroxyl group of **13a** is less sterically hindered due to the longer bond between Si and C than between C and C in **13b** (Fig. 12). **13a** is, therefore, more easily accessible to the acyl-enzyme intermediate and reacts as an acyl acceptor much faster than the carbon analogue. In the case of **14a**, the favorable effect of the silicon atom mentioned earlier was negligible owing to the presence of a long ethylene group between the silicon atom and the hydroxyl group. **14a**, and probably 3-trimethylsilylpropanol (n = 3) **(15a)**, were consequently regarded as similar substrates to the corresponding carbon compounds by lipase. In spite of the favorable characteristics of the silicon atom, trimethylsilanol (n = 0) did not serve as a substrate similarly to the corresponding tertiary alcohol.

To our knowledge, this work is the first study in which the effects of the silicon atom in substrates on enzymatic reaction were discussed systematically. Furthermore, this study showed for the first time the ability of organosilicon compounds to break the limit of conventional substrates, owing to their unique characteristics. According to this explanation, trimethylsilylmethylamine ($Me_3SiCH_2NH_2$) can also be expected to be a better acyl acceptor than the cor-

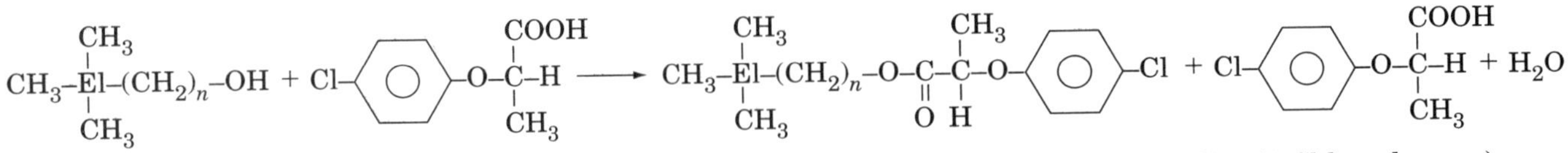

(*RS*)-2-(4-Chlorophenoxy)-propanoic acid (25)

(*S*)-2-(4-Chlorophenoxy)-propanoic acid

Figure 11. Lipase-catalyzed enantioselective esterification of 2-(4-chlorophenoxy)propanoic acid with silicon-containing alcohols and the corresponding carbon analogues. El, Si or C.

responding carbon compound, 2,2-dimethylpropylamine ($Me_3CCH_2NH_2$), in hydrolase-catalyzed amide bond formation. In fact, trimethylsilylmethylamine was a better substrate for hydrolases such as lipase OF 360 (*C. cylindracea*), lipase KLIP-100 (*Pseudomonas* sp.), lipoprotein lipase Type A (*Pseudomonas* sp.), and cholesterol esterase Type A (*Pseudomonas* sp.) than the corresponding carbon compound in amide bond formation with octanoic acid (R.S. So, Y. Masuda, T. Kawamoto, and A. Tanaka, unpublished results).

In contrast to our results, Tsai and Wei (29) reported that both **13a** and **14a** were effectively applied to the enantioselective esterification of racemic naproxen (2-(6-methoxy-2-naphthyl)propionic acid), a nonsteroidal anti-inflammatory drug, by *C. cylindracea* lipase. However, these alcohols were not good substrates for *Mucor javanicus* lipase. They have developed this esterification reaction system with **13a** (30,31).

Optical resolution of organosilicon compounds having a functional group attached to the carbon atom at the α-position from the silicon atom was also studied, and the enantioselective hydrolysis of racemic 2-acetoxy-1,1-dimethyl-1-silacyclohexane (*rac*-**26**) in an aqueous solution and enantioselective esterification of racemic 1,1-dimethyl-1-silacyclohexan-2-ol (*rac*-**27**) with triacetin in isooctane were successfully carried out by the use of a crude lipase preparation of *C. cylindracea*, leading to (*S*)-**27** (yield 71%, 95% ee) and (*S*)-**26** (yield 92%, 95% ee), respectively (Fig. 13) (32).

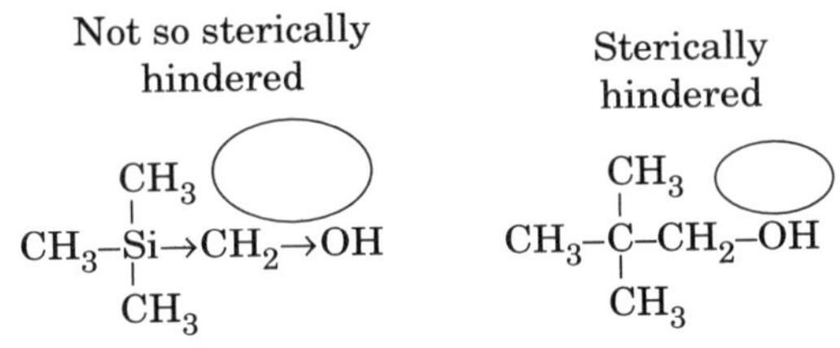

13a 13b

Figure 12. Effect of the silicon atom in trimethylsilylmethanol on esterification.

In general, it is difficult to perform the highly enantioselective conversion of primary alcohols. However, the efficient optical resolution of a primary alcohol, 2-trimethylsilyl-1-propanol **(18a)**, was achieved with 95% ee at 50% conversion by the enantioselective esterification with lipase Saiken 100 from *Rhizopus japonicus* in water-saturated isooctane (33). In the study on the enantioselective esterification of three isomers of trimethylsilylpropanols, 1-trimethylsilyl-2-propanol **(16a)**, 1-trimethylsilyl-1-propanol **(17a)**, and **18a**, and the corresponding carbon analogues (**16b, 17b**, and **18b**) (Fig. 6) with 5-phenylpentanoic acid by hydrolases in water-saturated isooctane, it was indicated that the silicon atom in the substrates generally enhanced the enzyme enantioselectivity. It is also worth noting that the β-hydroxyalkyl-

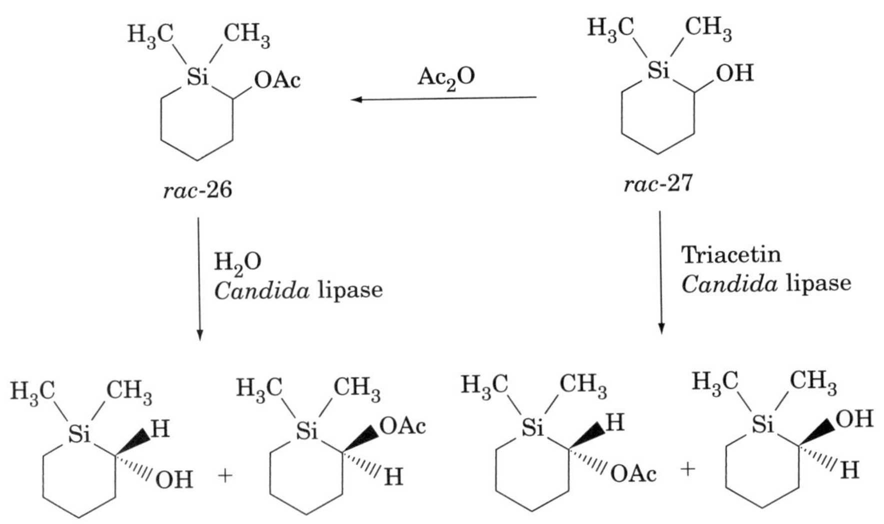

Figure 13. Lipase-catalyzed enantioselective hydrolysis and transesterification of organosilicon compounds whose functional group is attached to the carbon atom at the α-position from the silicon atom.

silanes, **16a** and **18a**, were effectively esterified by hydrolases, but they were easily converted to alkenes via β-elimination under both acidic and basic conditions (Peterson olefination) (1), and consequently, it is not possible to esterify them by chemical catalysts such as acids. Several enzymes could convert such unstable compounds with enantioselectivity. This fact is a good example showing the effectiveness of introduction of biochemical methods into organosilicon chemistry.

Optical resolution of organosilicon compounds, which have a stereogenic silicon atom, is also an important target for the application of enzymes. Furthermore, it is interesting to find out whether enzymes can recognize the chirality on the silicon atom, and such chiral silanes are expected to be new materials, synthetic reagents (34), or biologically active compounds (2). When PhSi(Me)(Et)CH_2OH **(28)** was esterified with 5-phenylpentanoic acid, various types of hydrolases exhibited a good esterification activity but low enantioselectivity. Of the enzymes examined, a commercial preparation of crude papain gave the best results, that is, a moderate reaction rate and a high enantioselectivity (35). Consequently, the enantioselective esterification of several primary alcohols, hydroxyakylsilanes, having a stereogenic silicon atom **(28–36)** was attempted (Fig. 14) with 5-phenylpentanoic acid by crude papain in water-saturated isooctane to study the effects of chain length between the silicon atom and the hydroxyl group **(28–30)** and the substitution groups on the silicon atom **(28,31–36)** (35). A short methylene chain between the silicon atom and the hydroxyl group and a phenyl substitution on the silicon atom were found to be essential for the high activity and the high enantioselectivity in the crude papain-catalyzed kinetic resolution of these organosilicon compounds. The optical purities of (+)-PhSi(Me)(Et)CH_2OH **(28)**, (+)-PhSi(Me)(Pr)CH_2OH **(31)**, (+)-*p*-Me-PhSi(Me)(Et)CH_2OH **(33)**, and (+)-*p*-F-PhSi(Me)(Et)CH_2OH **(34)**, obtained were 92 (yield 42% from the racemic substrate), 93 (yield 42%), 96 (yield 41%), and 99% ee (yield 41%), respectively.

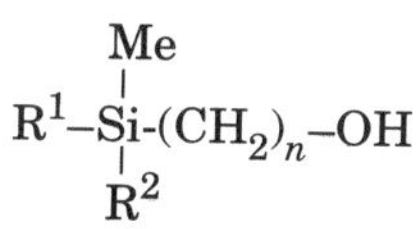

	R^1	R^2	n
28	Ph	Et	1
29	Ph	Et	2
30	Ph	Et	3
31	Ph	Pr	1
32	Ph	n–C_6H_{13}	1
33	p–Me–Ph	Et	1
34	p–F–Ph	Et	1
35	n–C_6H_{13}	Et	1
36	Pr	Et	1

Figure 14. Organosilicon compounds having asymmetric silicon atoms used in the enantioselective esterification with a crude papain.

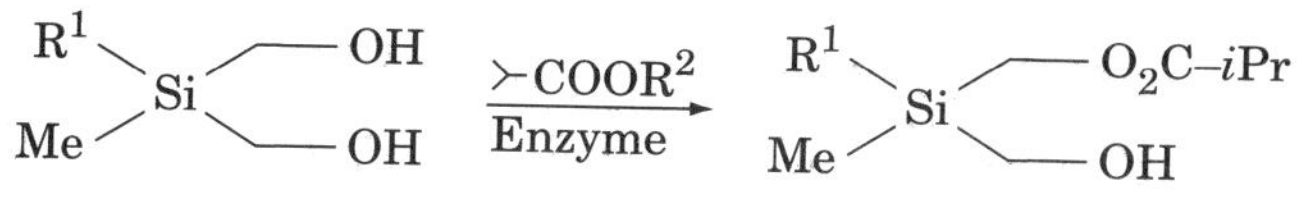

37a, b

	R^1	R^2	Enzyme
37a	Ph	Me	LCC
		∖N=≺	LCV
37b	n-C_8H_{17}	Me	LCC
		∖N=≺	LCV

Figure 15. Asymmetrization of prochiral organosilicon compounds by lipases. LCC, lipase from *Candida cylindracea*; LCV, lipase from *Chromobacterium viscosum*.

Djerourou and Blanco (36) demonstrated the preparation of silyl-chiral optically active compounds by the asymmetrization of prochiral organosilicon compounds with hydrolases. The prochiral 2-sila-1,3-propanediol derivatives **(37a,b)** were asymmetrized through transesterification with oxime ester or methylisobutyrate by lipase from *C. cylindracea* (LCC) and lipase from *Chromobacterium viscosum* (LCV) (Fig. 15), although the enantiomeric excess of the resulting esters did not exceed 70–76%.

The enantioselective hydrolysis of amide was also useful for the preparation of optically active organosilicon compounds. Optically active (*R*)-(1-aminoethyl)dimethylphenylsilane ((*R*)-**39**) was obtained in 40% yield with an enantiomeric purity of 92% ee by enantioselective hydrolysis of racemic dimethylphenyl[1-(phenylacetamido)-ethyl]silane **(38)** using immobilized penicillin G acylase from *Escherichia coli* 5K pHM 12 (Fig. 16) (37).

PREPARATION OF SILICON-CONTAINING AMINO ACIDS

Optically active nonnatural amino acids are very interesting and useful as the precursors of pharmaceuticals, agricultural chemicals, food ingredients, and so forth. Furthermore, they are also used as chiral auxiliaries for organic synthesis. Therefore, it will be very important to prepare novel nonnatural amino acids, such as silicon-containing ones. These amino acids can be expected to exhibit improved biological activities or new functions because of the characteristics of the silicon atom.

An efficient system has been developed to prepare optically active 3-trimethylsilylalanine (TMS-Ala) **(41)** by kinetic resolution with acylase I from porcine kidney from chemically synthesized *N*-acetyl-D,L-TMS-Ala **(40)** (Fig. 17) (38). Although both acylase I from porcine kidney and that from *Aspergillus melleus* showed activity on this substrate, the porcine enzyme was found to have much higher activity. The optimum pH of the reaction was 7.5, and the addition of 0.5 mM Co^{2+} accelerated the reaction. Optically pure L-**41** (>99% ee) was obtained in 72% yield. Furthermore, highly optically pure D-**41** (96% ee) could also be obtained in 76% yield by chemically hydrolyzing the residual substrate. **41** may be used as an analogue of leucine in the synthesis of bioactive peptides.

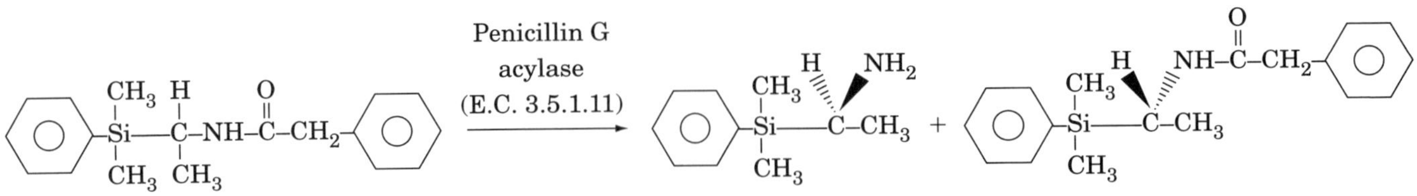

rac-38 (*R*)-39 (*S*)-38

Figure 16. Enantioselective hydrolysis of a silicon-containing amide by penicillin G acylase.

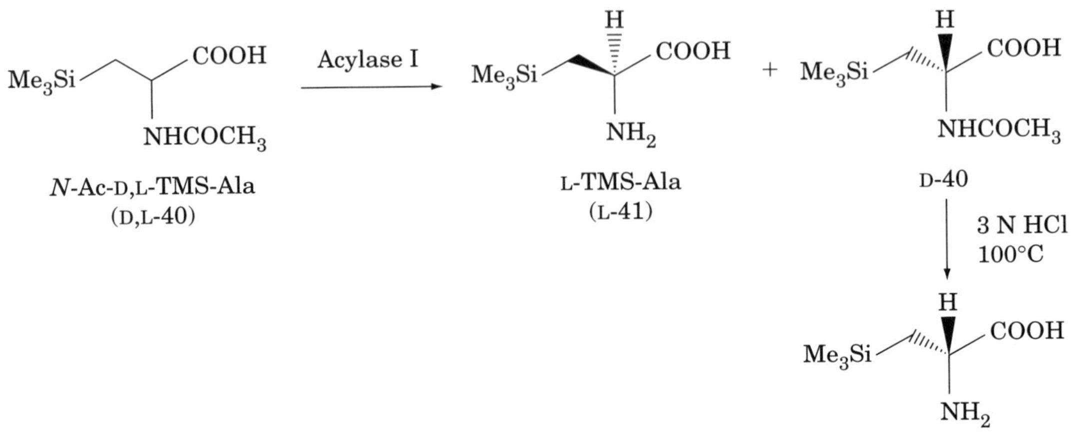

N-Ac-D,L-TMS-Ala (D,L-40)

L-TMS-Ala (L-41)

D-40

D-41

Figure 17. Preparation of optically active trimethylsilylalanine. TMS, trimethylsilyl.

As a homologue of phenylglycine and its analogues, which can be applied as side chains of penicillins and cephalosporins, optically active *p*-trimethylsilylphenylalanine **(44)** was prepared enzymatically. First, an attempt was made to obtain D-**44** from chemically synthesized D,L-5-(*p*-trimethylsilylphenylmethyl)hydantoin **(42)** via *N*-carbamoyl-D-*p*-trimethylsilylphenylalanine **(43)** through hydantoinase reaction, but none microbial strains tested exhibited the satisfactory activity to hydrolyze **42** to **43**. However, several strains, especially *Blastobacter* sp. A17p-4, showed good activity in the hydrolysis of **43** to yield **44**. Therefore, D,L-**43** prepared chemically from D,L-**42** was subjected to enantioselective hydrolysis to obtain optically active **44**. When D,L-**43** was hydrolyzed with cells or cell-free extract of *Blastobacter* sp., optical purity of D-**44** obtained was low because the cells contained not only *N*-carbamoyl-D-amino acid amidohydrolase (DCase) but also *N*-carbamoyl-L-amino acid amidohydrolase (LCase). DCase and LCase in the cell-free extract could be separated by DEAE-Sephacel-column chromatography. The optimum pH for the hydrolysis of D,L-**43** by partially purified DCase was 8.0, and addition of 2.5% *N,N*-dimethylformamide was effective to increase the substrate concentration without inactivation of the enzyme. Under the optimized conditions, highly optically pure (98% ee) D-**44** was obtained (39).

This process has been further improved by using the cell-free extract of *Blastobacter* sp. as the source of both DCase and LCase. As DCase of *Blastobacter* sp. had been known to be thermostable, the cell-free extract of this bacterium was treated at 50 °C and pH 7.0 for 40 min to inactivate LCase. After hydrolysis of D,L-**43** with the heat-treated cell-free extract, the product, D-**44**, and the residual substrate, L-**43**, were separated by silica-gel-column chromatography. L-**43** was further subjected to hydrolysis with nontreated cell-free extract as the LCase source after optimization of the reaction conditions (Fig. 18). Thus, optically pure D-**44** (99% ee) and L-**44** (99% ee) were obtained in yields of 80% and 89%, respectively, after purification (40).

CONCLUSION

This article outlines the bioconversion of various organosilicon compounds. In spite of the remarkable development of biotechnology, there have so far been only limited studies on bioconversion of organosilicon compounds. It has, however, been found that biocatalysts such as microorganisms, various types of hydrolases, and alcohol dehydrogenases are able to accept organosilicon compounds as substrates for transformation. Other types of enzymes will also be active on organosilicon compounds, although no experimental results are available at present. Introduction of biochemical methods into organosilicon chemistry has great potential because of the superior properties of biocatalysts, such as reaction selectivity, regioselectivity, enantioselectivity, stereoselectivity, and mild reaction condi-

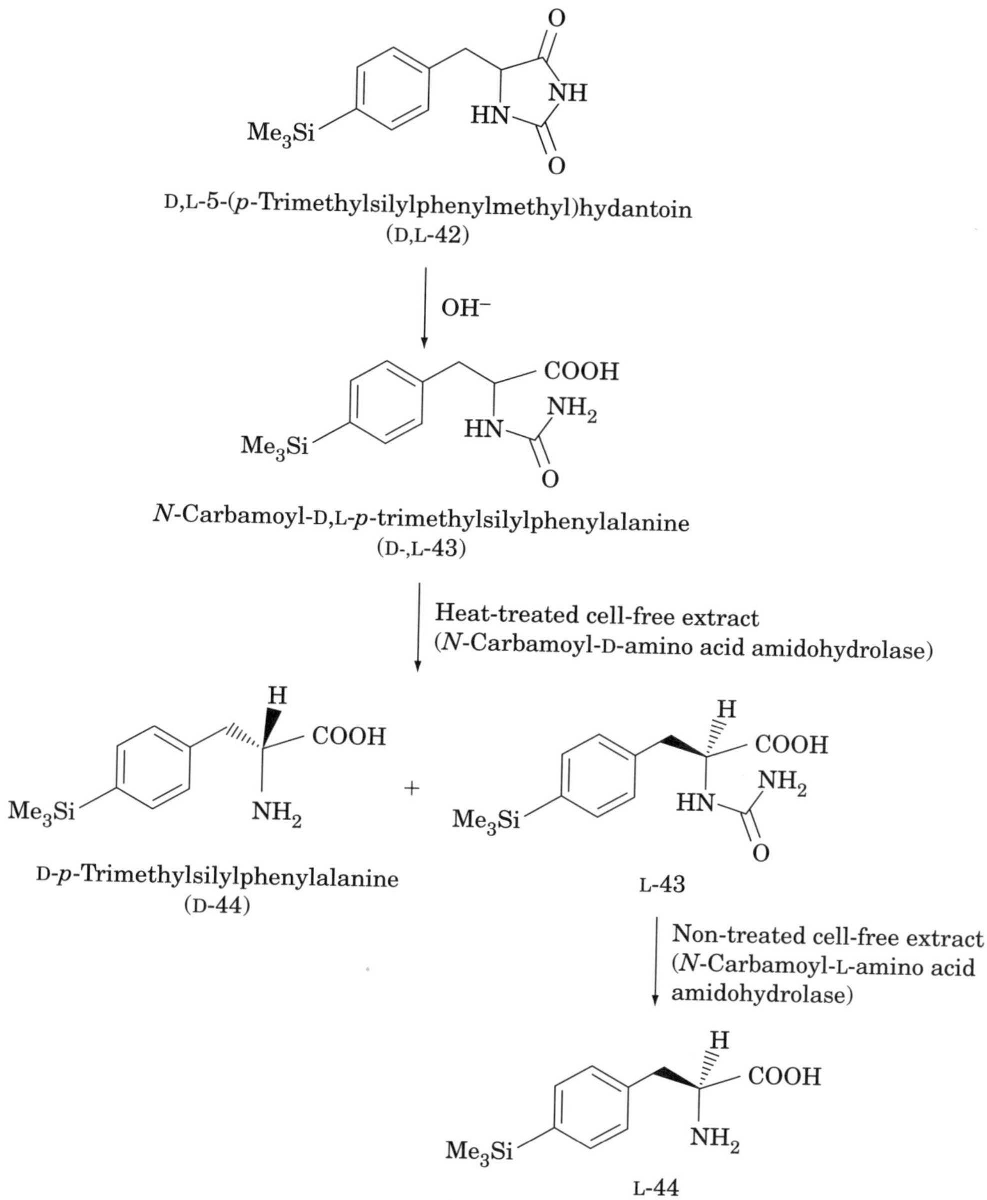

Figure 18. Preparation of optically active *p*-trimethylsilylphenylalanine.

tions. Furthermore, the application of the differences between organosilicon compounds and the corresponding carbon compounds to biochemical processes might make it possible to explore latent abilities of biocatalysts, giving fundamental information about biochemical and biological events. The field of organosilicon biochemistry, a fusion of biochemistry and organosilicon chemistry, will provide great possibilities to biotechnology in the future.

BIBLIOGRAPHY

1. E.W. Colvin, *Silicon in Organic Synthesis*, Butterworths, London, U.K., 1981.
2. R. Tacke and H. Zilch, *Endeavour New Ser.* **10**, 191–196 (1986).
3. A. Rioci, G. Seconi, and M. Taddei, *Chimicaoggi* **7**(9), 15–21 (1989).
4. R.J. Fessenden and J.S. Fessenden, *Adv. Organomet. Chem.* **18**, 275–299 (1980).
5. L.R. Garson and L.K. Kirchner, *J. Pharm. Sci.* **60**, 1113–1127 (1971).
6. C.E. Creamer, *Pharm. Technol.* March, 79–86 (1982).
7. R. Tacke and B. Becker, *Main Group Metal Chem.* **10**, 169–197 (1987).
8. A.D. Ryabov, *Angew. Chem. Int. Ed. Engl.* **30**, 931–941 (1991).
9. A. Tanaka and T. Kawamoto, *Chimicaoggi* **12**, 63–69 (1994).
10. R. Tacke, H. Linoh, B. Stumpf, W.R. Abraham, K. Kieslich, and L. Ernst, *Z. Naturforsch.* **38b**, 616–620 (1983).
11. C. Syldatk, H. Andree, A. Stoffregen, F. Wagner, B. Stumpf, L. Ernst, H. Zilch, and R. Tacke, *Appl. Microbiol. Biotechnol.* **27**, 152–158 (1987).

12. C. Syldatk, A. Stoffregen, F. Wuttke, and R. Tacke, *Biotechnol. Lett.* **10**, 731–736 (1988).
13. L. Fischer, S.A. Wagner, and R. Tacke, *Appl. Microbiol. Biotechnol.* **42**, 671–674 (1995).
14. R. Tacke, S.A. Wagner, S. Brakmann, F. Wuttke, U. Eilert, L. Fischer, and C. Syldatk, *J. Organomet. Chem.* **458**, 13–17 (1993).
15. R. Tacke, K. Fritsche, A. Tafel, and F. Wuttke, *J. Organomet. Chem.* **388**, 47–55 (1990).
16. R. Tacke, S. Brakmann, F. Wuttke, J. Fooladi, C. Syldatk, and D. Schomburg, *J. Organomet. Chem.* **403**, 29–41 (1991).
17. R. Tacke, H. Hengelsberg, H. Zilch, and B. Stumpf, *J. Organomet. Chem.* **379**, 211–216 (1989).
18. P. Huber, S. Bratovanov, S. Bienz, C. Syldatk, and M. Pietzsch, *Tetrahedron: Asymmetry* **7**, 69–78 (1996).
19. M.-H. Zong, T. Fukui, T. Kawamoto, and A. Tanaka, *Appl. Microbiol. Biotechnol.* **36**, 40–43 (1991).
20. K. Sonomoto, H. Oiki, and Y. Kato, *Enzyme Microb. Technol.* **14**, 640–643 (1992).
21. T. Fukui, M.-H. Zong, T. Kawamoto, and A. Tanaka, *Appl. Microbiol. Biotechnol.* **38**, 209–213 (1992).
22. Y. Tsuji, T. Fukui, T. Kawamoto, and A. Tanaka, *Appl. Microbiol. Biotechnol.* **41**, 219–224 (1994).
23. C.W. Bradshaw, W. Hummel, and C.-H. Wong, *J. Org. Chem.* **57**, 1532–1536 (1992).
24. M. Therisod and A.M. Klibanov, *J. Am. Chem. Soc.* **109**, 3977–3981 (1987).
25. M. Luyten, S. Muller, B. Herzog, and R. Keese, *Helv. Chim. Acta* **70**, 1250–1254 (1987).
26. K. Mori and T. Takeuchi, *Tetrahedron* **44**, 333–342 (1988).
27. M. Therisod, *J. Organomet. Chem.* **361**, C8–C10 (1989).
28. T. Kawamoto, K. Sonomoto, and A. Tanaka, *J. Biotechnol.* **18**, 85–92 (1991).
29. S.-W. Tsai and H.-J. Wei, *Enzyme Microb. Technol.* **16**, 328–333 (1994).
30. S.-W. Tsai and H.-J. Wei, *Biotechnol. Bioeng.* **43**, 64–68 (1994).
31. S.-W. Tsai and H.-J. Wei, *Biocatalysis* **11**, 33–45 (1994).
32. K. Fritsche, C. Syldatk, F. Wagner, H. Hengelsberg, and R. Tacke, *Appl. Microbiol. Biotechnol.* **31**, 107–111 (1989).
33. A. Uejima, T. Fukui, E. Fukusaki, T. Omata, T. Kawamoto, K. Sonomoto, and A. Tanaka, *Appl. Microbiol. Biotechnol.* **38**, 482–486 (1993).
34. G.L. Larson and E. Torres, *J. Organomet. Chem.* **239**, 19–27 (1985).
35. T. Fukui, T. Kawamoto, and A. Tanaka, *Tetrahedron: Asymmetry* **5**, 73–82 (1994).
36. A.-H. Djerourou and L. Blanco, *Tetrahedron Lett.* **32**, 6325–6326 (1991).
37. H. Hengelsberg, R. Tacke, K. Fritsche, C. Syldatk, and F. Wagner, *J. Organomet. Chem.* **415**, 39–45 (1991).
38. H. Yamanaka, T. Fukui, T. Kawamoto, and A. Tanaka, *Appl. Microbiol. Biotechnol.* **45**, 51–55 (1996).
39. Y. Tsuji, H. Yamanaka, T. Fukui, T. Kawamoto, and A. Tanaka, *Appl. Microbiol. Biotechnol.* **47**, 114–119 (1997).
40. H. Yamanaka, T. Kawamoto, and A. Tanaka, *J. Ferment. Bioeng.* **84**, 181–184 (1997).

OSMOTIC STRESS, SECRETION RATE

Joon Soo Ryu
Gyun Min Lee
Korea Advanced Institute of Science and Technology
Taejon, Korea

KEY WORDS

Antibody secretion rate
Hybridoma cells
Osmotic pressure

OUTLINE

INTRODUCTION

Osmolality is one of the most important physical properties of culture media. Osmolality is a measure of concentration and is defined as the molality of particles in a solution. Most cell culture media are designed to have an osmolality of 280 to 320 mOsm/kg through balancing the concentration of nutrients and osmolytes (1).

When cells are bathed in a solution of the same osmolality as the cytoplasm, they neither shrink nor swell. Osmotic pressure results when concentrations of dissolved solutes on the inside and on the outside of a cell are different. The effects of osmotic pressure on cells in culture are too complex to be discussed in detail in this article. Here, we focus on the effect of osmotic pressure on hybridoma cells in regard to cell growth and antibody secretion rates. In an effort to improve antibody productivity of hybridoma cells, the effect of osmotic pressure on hybridoma cells has been extensively studied. As a result, the following observations have been made.

1. The response of hybridoma cells to osmotic pressure varies among cell lines (2,3).
2. For most hybridomas, hyperosmotic pressure induced by either sugar or ionized salts suppresses cell growth, but it increases the specific antibody secretion rate (q_{Ab})(2–8).

3. The enhanced q_{Ab} resulting from hyperosmotic pressure is not transient and is maintained in successive passages (9).
4. Hypo-osmotic pressure also suppresses cell growth, but it does not enhance q_{Ab} (10).

Hyperosmotic pressure, which can be induced by addition of cheap salts to media, has been suggested as being an economical solution to increase q_{Ab} in hybridoma cell cultures. Despite the potential of commercial strategies based on hyperosmotic pressure, the use of hyperosmolar medium has not been popular mainly because cell growth is depressed at elevated osmolality. Hence, the enhanced q_{Ab} in a batch culture does not result in a substantial increase in the final antibody concentration (2,4–6). To overcome this drawback, several strategies, such as adaptation of cells to hyperosmotic pressure (5,8), use of osmoprotectants in media (6,7,11), and two-stage culture (12), have been applied for improved antibody production. This article deals with the responses of hybridoma cells to osmotic pressure and its application to hybridoma cell cultures for improved antibody production.

OSMOLALITY

Osmolality is a measure of concentration and is often expressed in units of osmoles. The most commonly used abbreviation, based on milliosmoles per kilogram, is mOsm/kg. One osmole of a substance is equal to 1 g mol divided by the number of particles formed by the dissociation of the molecules. If a solution is composed of a nondissociable solute, the osmolality is simply equal to the molality. For a solution of a dissociable salt, 1 mol solution is n osmole, where n is the number of ions produced per molecule.

In the literature, the terms osmolarity and osmolality are frequently used interchangeably. Osmolality is related to osmolarity in the same manner as molality is related to molarity. Thus, osmolality is a more accurate description of concentration because it is a mass per unit mass measurement and is independent of temperature.

Osmolality values of media cannot be easily calculated because media contain a wide range of compounds as well as degrees of ionization. The osmolality of the media is determined more easily by using either vapor pressure elevation or freezing-point depression. The osmometers operate based on the principle that changes in osmolalities of a solution result in changes in certain properties such as vapor pressure and freezing point. Osmolality values measured by osmometers give reliable and reproducible results with a standard deviation of less than 5%.

MEDIUM OSMOLALITY

Medium osmolality is one of the important factors that is frequently overlooked in cultivating mammalian cells. Cell culture medium is designed to have osmolality in the range of 280 to 320 mOsm/kg, basically to mimic the osmolality of serum at 290 mOsm/kg (1). The major component in maintaining osmotic pressure in cell culture media is NaCl. Other inorganic ions (K^+, Mg^{2+}, Ca^{2+}) and glucose also contribute significantly. On the other hand, larger molecules like proteins play a lesser role, and changes in their concentrations do not affect osmolality significantly.

The measurement of osmolality is a useful control step if one makes up media. It is particularly important to check osmolality if alterations are made in the constitution of the medium. During batch culture, nutrients are consumed and metabolites such as lactate and ammonia are accumulated in the media. However, medium osmolality does not change significantly (9). The osmolality of culture media can change drastically as a result of evaporation when small-scale cultures such as Petri-dish or open-plate cultures are incubated in nonhumidified incubators. In this case, a slightly hypotonic medium may be better to compensate for evaporation during incubation. Osmolality should be also carefully monitored in fed-batch culture where medium concentrates are repeatedly fed to the culture. Addition of medium concentrates results in a substantial increase in medium osmolality, although they do not contain major salts. Thus, the culture longevity, which is important for the design of high-yield processes, is often limited by elevated osmolality caused by repeated nutrient feedings (13,14).

EFFECT OF OSMOTIC PRESSURE ON HYBRIDOMA CELL GROWTH AND ANTIBODY PRODUCTION

Medium osmolality in the range of 280 to 320 mOsm/kg is known to be quite acceptable for most cells; therefore, most commercially available media are designed to have an osmolality of this range (1,15). As the osmolality goes far from this standard range, some obvious changes in cell size, cell shape, nutrient transport, cellular synthesis, specific growth rate (μ), and q_{Ab} occur. Among these changes, the most important changes for cell culturists are probably those in μ and q_{Ab}. The common osmolyte used for altering the medium osmolality is NaCl. As summarized in Table 1, elevated osmolality suppresses cell growth and increases q_{Ab}. The q_{Ab} increased up to 3.2-fold for cells grown above 300 mOsm/kg and up to 435 mOsm/kg. However, because of the depressed cell growth at higher osmolalities, similar final antibody titers were obtained in most cases.

Regardless of serum concentrations, cultivation methods (static and suspension cultures), and basal media (IMDM, DMEM, and RPMI 1640), hyperosmotic pressure substantially enhanced q_{Ab} of S3H5/γ2bA2 hybridoma in batch culture by depressing cell growth, although the degree of enhancement in q_{Ab} was influenced (2,9,10).

Ozturk and Palsson (4) altered the osmolality of the medium by addition of both ionic (NaCl- and phosphate-buffered saline) and nonionic substances (sucrose). Similar data were obtained regardless of the osmolytes used. Thus, the changes in cells in hyperosmolar media described earlier appear to be caused only by medium osmolality.

If the enhanced q_{Ab} at elevated osmolality is not transient, it can be used to maximize antibody productivity in perfusion culture systems where cell growth and production are separated. As shown in Figure 1, the enhanced q_{Ab} of hybridoma at elevated osmolality decreased while the cells were adapting to hyperosmotic pressure in long-term,

Table 1. Effect of Osmolality on Antibody Production in Batch Culture

Cell line	Osmolality (mOsm/kg)	Media	Relative q_{Ab}	Relative antibody titer	References
Hybridoma (S3H5/γ2bA2)	281	IMDM	1.0	1.00	9,15
	390	IMDM	2.5	0.86	
Hybridoma (DB9G8)	281	IMDM	1.0	1.0	2,9,15
	390	IMDM	1.1	0.9	
	440	IMDM	3.2	0.4	
Hybridoma (167.4G5.3)	290	IMDM	1.0	1.0	4
	435	IMDM	>2.0	1.0	
Hybridoma (2HG11)	300	eRDF	—	1.0	5,8
	350	eRDF	—	1.1	
Hybridoma (AB2-143.2)	352	DMEM-HamF12	1.0	—	3
	488	DMEM-HamF12	1.5	—	
Hybridoma (6H11)	330	DMEM	1.0	1.0	6
	510	DMEM	1.7–1.9	0.5	
Recombinant NSO myeloma	270	IMDM	1.0	1.0	13
	400	IMDM	1.6	0.8	

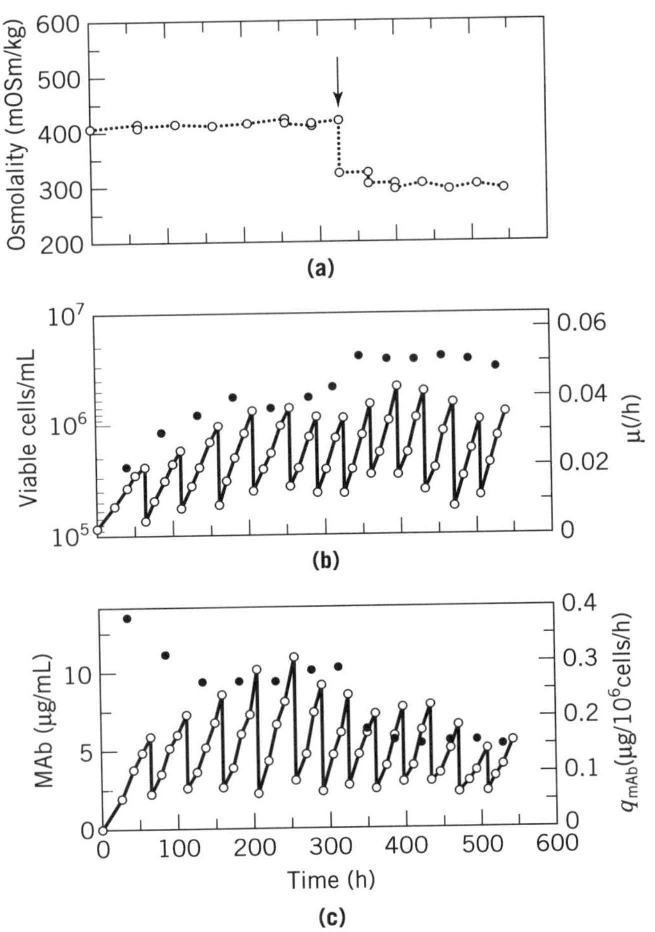

Figure 1. Repeated-fed batch culture in a spinner flask. (**a**) Osmolality, (**b**) viable cell concentration (○) and μ (●); (**c**-) antibody concentration (○) and q_{Ab} (●). The *arrow* indicates the time of switching hyperosmolar medium to standard medium.

repeated-fed-batch culture (9). However, the cells even after adaptation could still maintained the significantly enhanced q_{Ab} when compared with the q_{Ab} of cells obtained from standard medium in the batch culture, indicating the potential of using hyperosmolar medium for improved antibody production in a perfusion culture. In addition, when the hyperosmolar medium switched back to a standard medium, the physiological changes of the cells resulting from hyperosmotic pressure rapidly disappeared. Accordingly, the q_{Ab} and μ in perfusion systems may be controlled simply by changing the osmolality of medium. However, it should be noted that all the hybridomas do not display enhanced q_{Ab} at elevated osmolality. Cell-line specificity will be discussed in a later section.

Although much work has been performed on the response of hybridomas to hyperosmotic pressure, there are, to date, few reported studies on the response of hybridomas to hypoosmotic pressure.

Ryu and Lee reported the response of hybridomas (S3H5/γ2bA2 and DB9G8) to hypo-osmotic pressure resulting from NaCl subtraction (10). As shown in Figures 2 and 3, both hybridomas showed similar responses to hyperosmotic pressure in regard to cell growth and antibody production. The cell growth and antibody production at 276 mOsm/kg were comparable to those at 329 mOsm/kg (standard DMEM). Both cells grew well at 219 mOsm/kg, although their growth and antibody production were slightly decreased. When the osmolality was further decreased to 168 mOsm/kg, the cell growth did not occur. When subjected to hyperosmotic stress, both cells displayed significantly enhanced q_{Ab}. However, the cells subjected to hypoosmotic pressure did not display enhanced q_{Ab}.

Taken together, medium osmolality is one of the important factors that can be easily manipulated by addition or subtraction of inorganic salts or sugars. When subjected to hyperosmotic pressure, most hybridomas display enhanced q_{Ab} by suppressing cell growth. Hypo-osmotic pressure also suppresses cell growth but does not enhance q_{Ab}.

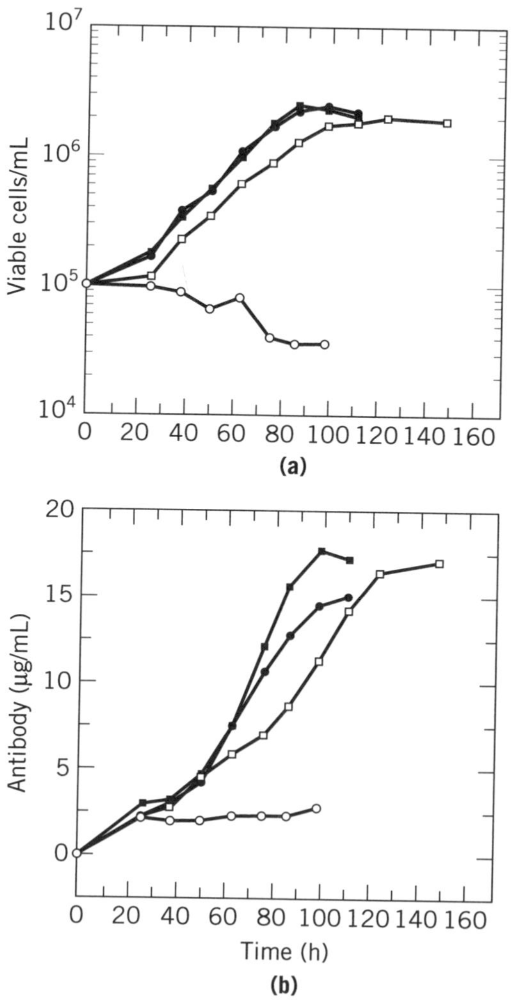

Figure 2. Cell growth (**a**) and antibody production (**b**) during a batch culture of S3H5/γ2bA2 hybridoma cells in hyperosmolar media. Data points: ○, 168; □, 219; ●, 276; ■, 329 mOsm/kg.

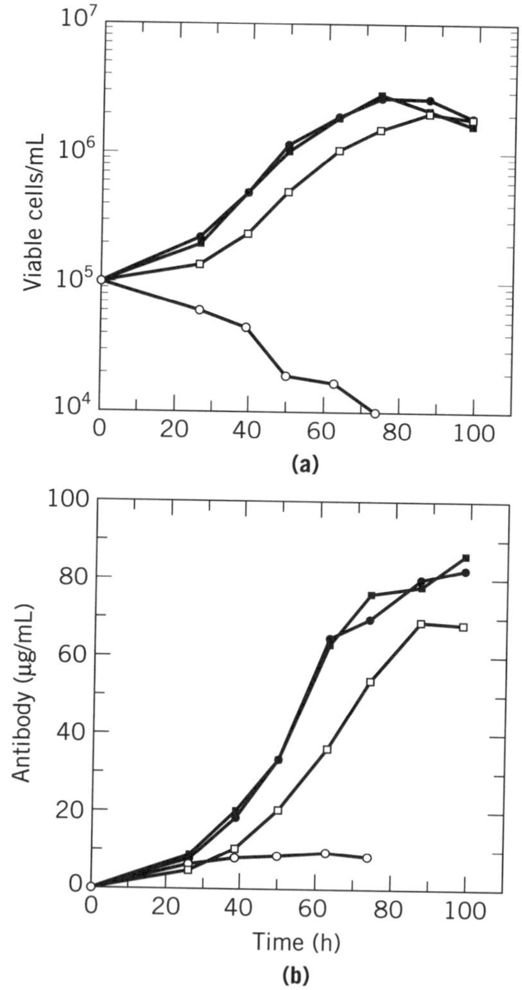

Figure 3. Cell growth (**a**) and antibody production (**b**) during a batch culture of DB9G8 hybridoma cells in hyperosmolar media. Data points: ○, 168; □, 219; ●, 276; ■, 329 mOsm/kg.

Table 2. Summary of Maximum Viable Cell Density, Maximum Antibody Titer, and q_{Ab} at Different Osmolytes

Osmolyte used	Max. viable cell density ($\times 10^6$ cells/mL)	Max. antibody titer (μg/mL)	q_{Ab} (pg/cell/h)
Control	2.92 ± 0.12	156 ± 9	0.75
NaCl	2.76 ± 0.12	299 ± 25	1.46
Sucrose	2.05 ± 0.21	257 ± 13	1.50
KCl	2.35 ± 0.21	198 ± 3	1.08

Source: Data from Oh et al. (8).

Table 3. Comparison of Maximum Viable Cell Density, Maximum Antibody Titer, and μ between Unadapted and Adapted Cells

	Unadapted cells			Adapted cells		
Osmolality (mOsm/kg)	Maximum viable cell density ($\times 10^5$/mL)	Maximum antibody titer (μg/mL)	μ (h^{-1})	Maximum viable cell density ($\times 10^5$/mL)	Maximum antibody titer (μg/mL)	μ (h^{-1})
300	34.6	140	0.046	34.6	140	0.046
350	8.4	150	0.027	31.3	265	0.036
400	7.9	75	0.038	21.3	220	0.033

Source: Data from Oh et al. (5).

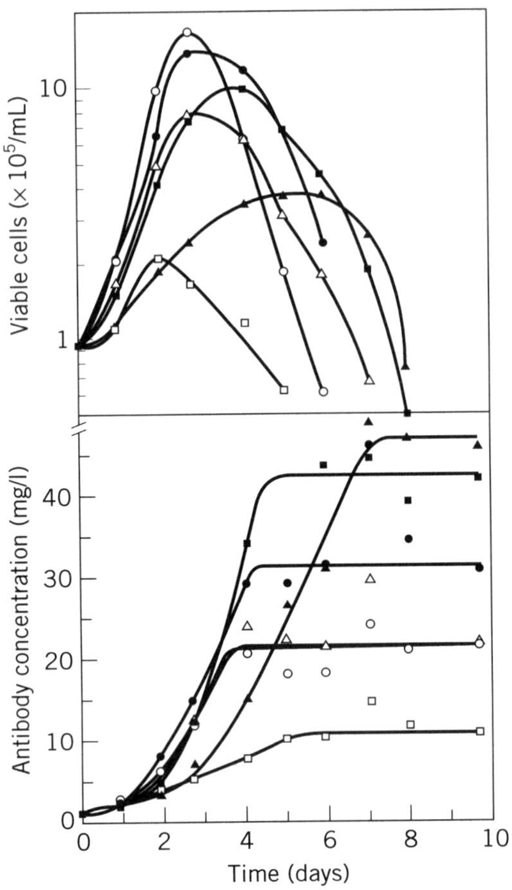

Figure 4. Hybridoma growth and antibody production in NaCl-stressed growth media in the presence and absence of glycine betaine (15 mM) (7). ○, Control, 330 mOsm/kg; △, 60 mM NaCl added, 450 mOsm/kg; ●, 60 mM NaCl and 15 mM glycine betaine added, 465 mOsm/kg; □, 100 mM NaCl added, 510 mOsm/kg; ■, 100 mM NaCl and 15 mM glycine betaine added, 525 mOsm/kg; ▲, 140 mM NaCl and 15 mM glycine betaine added, 610 mOsm/kg.

ENHANCED SPECIFIC ANTIBODY SECRETION RATE OF HYBRIDOMAS RESULTING FROM HYPEROSMOTIC PRESSURE IS CELL-LINE SPECIFIC

As mentioned earlier, some hybridomas do not display enhanced q_{Ab} at elevated osmolality.

Reddy and Miller reported the effects of osmotic pressure on antibody production in hybridoma cells that differed in production kinetics (3). AB2-143.2 cells exhibit non-growth-associated antibody production, whereas IND1 cells exhibit growth-associated production. Both hybridomas displayed enhanced q_{Ab} after abrupt batch osmotic stress. AB2-143.2 cells also increased q_{Ab} in response to gradual osmotic stress in continuous culture. In contrast, IND1 cells decreased q_{Ab} during gradual osmotic stress. It was hypothesized that this difference between cell lines in response to gradual osmotic stress in continuous culture might be related to the fact that antibody production is growth-associated in IND1, but not in AB2-143.2 cells.

Lee and Park (2) reported that two murine hybridomas exhibiting non-growth-associated antibody production (S3H5/γ2bA2 and DB9G8) showed different responses to hypersomotic pressure regarding q_{Ab}, although they showed similar depression of cell growth in hyperosmolar media. The q_{Ab} of S3H5/γ2bA2 hybridoma in a hyperosmolar medium (396 mOsm/kg) was enhanced by approximately 180% when compared with that in a standard medium (283 mOsm/kg), whereas q_{Ab} of DB9G8 hybridoma in the same hyperosmolar medium was enhanced by only 10%. Accordingly, regardless of antibody production kinetics of the cell lines, enhanced q_{Ab} of hybridomas resulting from hyperosmotic pressure appears to be cell-line specific. Thus, the strategy of using hyperosmolar media for improved antibody production cannot be applied for all cell lines.

EXPLANATION FOR ENHANCED Q_{AB} AT ELEVATED OSMOLALITY

Although the detailed mechanism of enhanced q_{Ab} is not clearly understood at the basic cellular level, several explanations based on noticeable changes induced by hyperosmotic pressure can be proposed for possible influences of NaCl on cellular functions and antibody production.

1. The increase in osmotic pressure may enhance the transport of nutrients, in particular, amino acids into the cells (5,8,11). Amino acid transport regulations in animal cells are controlled by both Na^+ ion-dependent and Na^+ ion-independent transport systems (16). Excess Na^+ ions on the outer membrane would increase the transmembrane electrochemical gradient, facilitating the simultaneous flow of ions and amino acids into the cells. Thus, the feeding of excess nutrients into the cell could increase cytoplasmic protein production, especially antibody productivity in hybridomas. The release of additional energy and precursors as a result of increased metabolism is not diverted to growth but to the synthesis and secretion of antibody (5). For hypersomotically stressed hybridomas, amino acid uptake by Na^+-dependent system A increases, whereas the activity of other Na^+-independent systems remains relatively unchanged (8). Proline, alanine, and glutamine, which are transported by this Na^+-dependent system A under hyperosmotic condition, are known to act as osmoprotective agents to compensate for osmotic imbalance of the environment (7,17). This observation would support a model for cell volume swelling and increased q_{Ab} at elevated osmolality. However, this postulate will have to be investigated further because it does not explain the workings of other osmolytes like sucrose or KCl (4,6,8). Table 2 shows that the enhancement of q_{Ab} of hybridomas subjected to hyperosmotic stress is not influenced significantly by the sources of osmotic stress.

2. The increase in osmotic pressure may induce transcriptional activation because of the change of chromatin structure, as is observed in the case of butyrate treatment (8,18–22). Acetylation of histone residue is thought to act as a transcription-stimulating event in many organisms. Unbinding of the histones from DNA results in the disper-

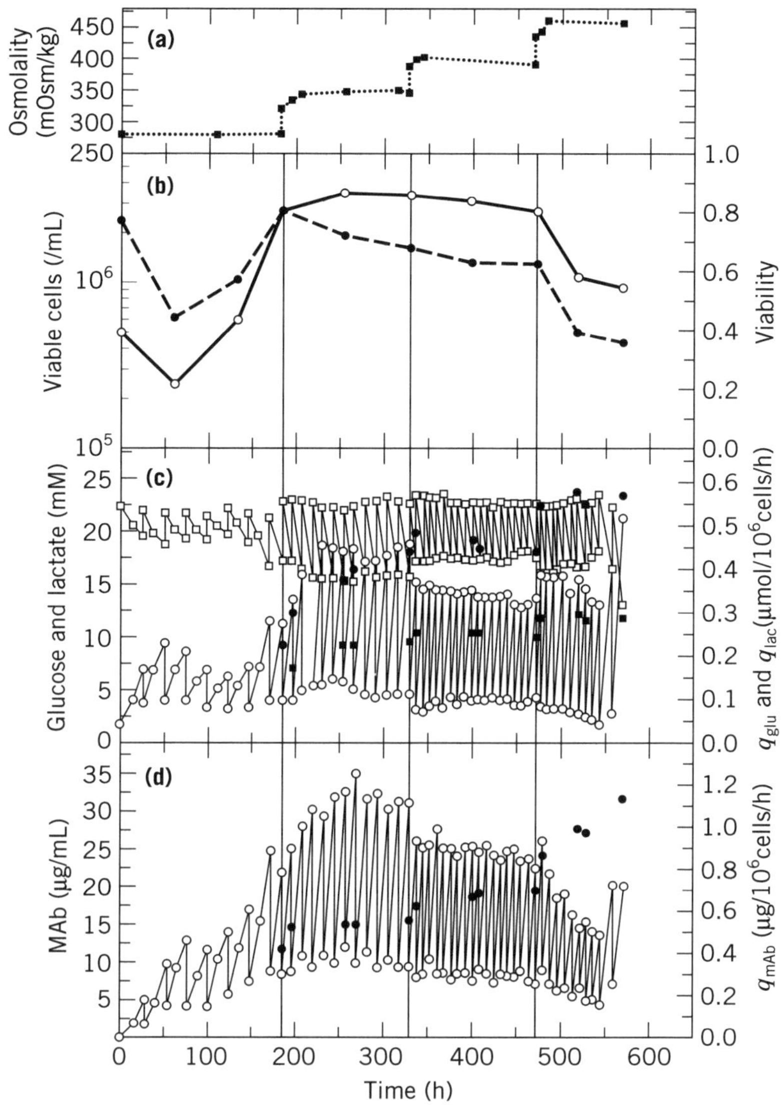

Figure 5. Immobilized cell culture with gradual hyperosmotic stress: (**a**) osmolality; (**b**) viable cell concentration of immobilized cells (○) and viability (●); (**c**) glucose (□) and lactate concentrations (○), q_{glu} (■), and q_{lac} (●); (**d**) antibody concentration (○) and q_{Ab} (●). The *vertical lines* indicate the time of elevating the osmolality of medium.

sal of the chromatin structure, which may expose segments of the chromatin, making them more accessible to RNA polymerase for mRNA transcription (21). In fact, elevated mRNA expression of certain proteins has been reported after treatment of butyrate (18,22–24).

3. The increase in osmotic pressure may stimulate the expression of chaperone-like proteins in the endoplasmic reticulum (ER) such as protein disulfide isomerase (PDI) or glucose-regulating proteins (GRPs) that are involved in protein processing and secretion (25,26).

All these postulates, however, should be confirmed by concrete experimental data, and solid, detailed mechanisms or sequential events that explain increased q_{Ab} will have to be discovered.

APPLICATION OF HYPEROSMOTIC PRESSURE TO ANTIBODY PRODUCTION PROCESSES

Hyperosmotic stress resulting from NaCl addition has been suggested as being an economical solution to increase the q_{Ab} in hybridoma cultures. However, the use of hyperosmolar medium has not been popular probably because of the following reasons. Because the enhanced q_{Ab} of hybridomas resulting from hyperosmotic stress is cell-line specific (2,3), the hyperosmolar medium cannot be used for all cell lines. Furthermore, cell growth is depressed at elevated osmolality. Hence, the enhanced q_{Ab} in batch culture does not result in a substantial increase in the final antibody titer (2,4–6). To overcome these drawbacks, the following strategies have been applied for improved antibody production.

1. Hybridoma cells are first adapted to higher osmolality media to elevate antibody levels (5,8). Before adaptation, 2HG11 hybridoma cells placed in higher osmotic pressures (350 and 400 mOsm/kg) were severely suppressed in growth down to 25% of the control (300 mOsm/kg), although total antibody titers achieved were similar to the control. After 1 week of adaptation to 350 and 400 mOsm/kg media, cell growth was not as drastically suppressed. As a result, considerably higher antibody levels were obtained at these elevated osmolalities (Table 3).

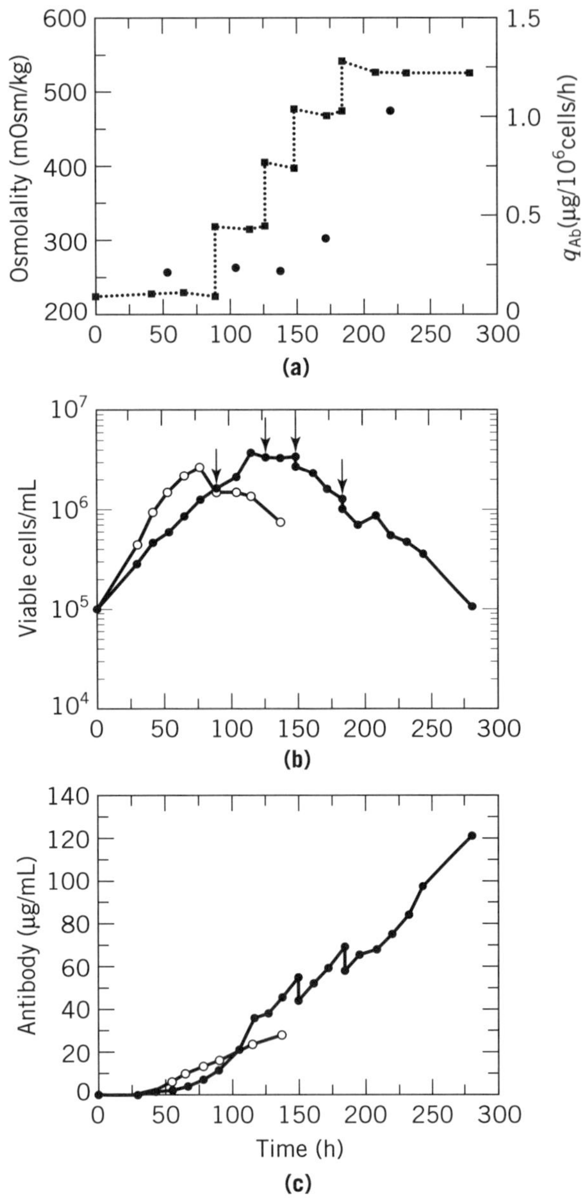

Figure 6. Fed-batch culture using a hypo-osmolar medium as an initial basal medium. (**a**) Changes in medium osmolality (■) and q_{Ab} (●). (**b**) Cell growth in batch (○) and fed-batch (●) cultures. The *arrows* indicate the addition of medium concentrates. (**c**) Antibody production in batch (○) and fed-batch (●) cultures.

2. Osmoprotective compounds, especially glycine betaine, are included in hyperosmolar media (6,7,11). Osmoprotective agents are known to exert osmoprotective effects by being accumulated inside cells to compensate for osmotic imbalance between in and out of the cells. Glycine betaine, sarcosine, proline, and glycine have all been shown to be strong osmoprotective compounds for 6H11 hybridoma cells under various hyperosmotic stress, with glycine betaine being the most effective of the four (6). When glycine betaine was included in the hyperosmolar media, q_{Ab} was increased up to 2.6-fold and maximum antibody titer up to twofold over that obtained in the control culture, as shown in Figure 4. A similar pattern of response was observed when other osmoprotective compounds (sarcosine, proline, glycine) were included in the hyperosmolar media. The improvement of cell growth by use of osmoprotective compounds did not cost increased q_{Ab} obtained under hyperosmotic condition, suggesting that medium osmolality, rather than growth rate, will determine the q_{Ab} by 6H11 hybridoma cells growing in hyperosmolar media. The use of osmoprotective compounds does not influence hyperosmotically enhanced specific metabolic rates. Accordingly, the simultaneous use of hyperosmotic stress and osmoprotective compounds can be a means to increase both q_{Ab} and maximum antibody titer in hybridoma cell culture.

3. The enhanced q_{Ab} resulting from hyperosmotic stress can be used for a higher volumetric antibody productivity in perfusion systems where cell growth and antibody production are separated. In perfusion systems such as hydrogel beads or hollow fibers, the cells maintain minimal growth rates after the cell concentrations are saturated. Thus, after achieving high cell concentration in perfusion systems, the switch of standard medium to hyperosmolar medium may significantly improve the volumetric productivity. A feasibility of using hyperosmolar medium resulting from NaCl addition for improved antibody production in a perfusion system was demonstrated by using calcium-alginate-immobilized S3H5/γ2bA2 hybridoma cells (12). First, immobilized cells were grown in the medium supplemented with 10% serum to achieve high cell concentration rapidly. Second, after the cell concentration reached more than 2×10^6 cells/mL, the medium was switched to the hyperosmolar medium supplemented with 1% serum for economical antibody production. An abrupt increase in osmolality, however, inhibited cell growth, resulting in no increase in volumetric antibody productivity. On the other hand, gradual increase in osmolality allowed further cell growth while maintaining the enhanced q_{Ab} of immobilized cells, as shown in Figure 5. Accordingly, the volumetric antibody productivity was increased by about 40% when compared with that in the control immobilized cell culture. This enhancement of volumetric antibody productivity of immobilized hybridoma cells by applying hyperosmolar stress suggests the potential of using hyperosmolar medium in other perfusion systems for improved antibody production.

APPLICATION OF HYPO-OSMOTIC PRESSURE TO ANTIBODY PRODUCTION PROCESSES

Cell culture longevity in fed-batch culture of hybridomas is often limited by elevated medium osmolality caused by repeated nutrient feedings (13,14). Shotwise feeding of 10× DMEM concentrates elevated the osmolality of medium up to 540 mOsm/kg at the end of fed-batch culture of S3H5/γ2bA2 hybridoma, which is known to be lethal to most hybridomas. S3H5/γ2bA2 hybridoma has been shown to grow without significant growth depression at 219 mOsm/kg in DMEM supplemented with 10% fetal bovine serum. In order to improve culture longevity in fed-batch culture of S3H5/γ2bA2 hybridoma, a hypo-osmolar medium (223 mOsm/kg) was used as an initial basal medium

(27). As shown in Figure 6, the use of hypo-osmolar medium delayed the onset of severe cell death resulting from elevated osmolality and allowed one more addition of 10× DMEM concentrates to the culture. As a result, a final antibody concentration obtained was 121.5 μg/mL, which is approximately 1.5-fold higher compared to fed-batch culture using a standard medium (335 mOsm/kg). When compared to batch culture, a 5.2-fold increase in the final antibody titer was achieved. Taken together, the use of hypo-osmolar medium as an initial medium in fed-batch culture improved culture longevity of S3H5/γ2bA2 hybridoma, resulting in substantial increase in the final antibody titer.

In conclusion, hyperosmotic pressure, which can be induced by addition of cheap salts to media, has been suggested as being an economical solution to increase q_{Ab} in hybridoma cell cultures. The utilization of osmotic pressure for improved antibody production in hybridoma cell cultures will be substantiated in future.

BIBLIOGRAPHY

1. R.I. Freshney, *Culture of Animal Cells: A Manual of Basic Technique*, 3rd ed., Wiley-Liss, New York, 1994, pp. 83–84.
2. G.M. Lee and S.Y. Park, *Biotechnol. Lett.* **17**, 145–150 (1995).
3. S. Reddy and W.M. Miller, *Biotechnol Prog.* **10**, 165–173 (1994).
4. S.S. Ozturk and B.O. Palsson, *Biotechnol. Bioeng.* **37**, 989–993 (1991).
5. S.K.W. Oh, P. Vig, F. Chua, W.K. Teo, and M.G.S. Yap, *Biotechnol. Bioeng.* **42**, 601–610 (1993).
6. K. Øyaas, T.E. Ellingsen, N. Dyrset, and D.W. Levine, *Biotechnol. Bioeng.* **44**, 991–998 (1994).
7. K. Øyaas, T.E. Ellingsen, N. Dyrset, and D.W. Levine, *Biotechnol. Bioeng.* **43**, 77–89 (1994).
8. S.K.W. Oh, F.K.F. Chua, and A.B.H. Choo, *Biotechnol. Bioeng.* **46**, 525–535 (1995).
9. S.Y. Park and G.M. Lee, *Bioproc. Engr.* **13**, 79–86 (1995).
10. J.S. Ryu and G.M. Lee, *Biotechnol. Bioeng.* **55**, 565–570 (1997).
11. K. Øyaas, T.E. Ellingsen, N. Dyrset, and D.W. Levine, *Cytotechnology* **17**, 143–151 (1995).
12. S.Y. Park and G.M. Lee, *Biotechnol. Bioeng.* **48**, 699–705 (1995).
13. T.A. Bibila, C.S. Ranucci, K. Glazomitsky, B.C. Buckland, and J.G. Aunins, *Biotechnol. Prog.* **10**, 87–96 (1994).
14. T.A. Bibila and D.K. Robinson, *Biotechnol. Prog.* **11**, 1–13 (1995).
15. J.S. Ryu and G.M. Lee, *Bioproc. Engr.* **16**, 305–310 (1997).
16. M.A. Shotwell, M.S. Kilberg, and D.L. Oxender, *Biochem. Biophys. Acta.* **737**, 267–284 (1983).
17. L. Silvotti, P.G. Petronini, A. Mazzini, G. Piedimonte, and A.F. Borghetti, *Exp. Cell Res.* **193**, 253–261 (1991).
18. M. Korsaka, T. Nakayama, A. Tsukamoto, and K. Kurokawa, *Cancer Res.* **50**, 3101–3105 (1987).
19. L.C. Boffa, R.J. Gruss, and V.G. Allfrey, *J. Biol. Chem.* **256**, 9612–9621 (1981).
20. A. Chabanas, E. Khoury, P. Goeltz, P. Froussard, R. Gjerset, B. Dod, H. Eisen, and J.J. Lawrence, *J. Mol. Biol.* **183**, 141–151 (1985).
21. M.J. Kruh, *Mol. Cell. Biochem.* **42**, 65–82 (1982).
22. I. Parker, J.B. deHaan, and W. Givers, *J. Biol. Chem.* **261**, 2786–2790 (1986).
23. A.J. Dorner, L.C. Wasley, and R.J. Kaufman, *J. Biol. Chem.* **264**, 20602–20607 (1989).
24. D.P. Palermo, M.E. DeGraaf, K.R. Marotti, E. Rehberg, and L.E. Post, *J. Biotechnol.* **19**, 35–48 (1990).
25. R.B. Freedman, R.E. Brockway, and N. Lambert, *Biochem. Soc. Trans.* **12**, 929–932 (1984).
26. A.S. Lee, *Curr. Opin. Biol.* **4**, 267–273 (1992).
27. J.S. Ryu and G.M. Lee, *Biotechnol. Bioeng.* **62**, 120–123 (1999).

P

PANTOTHENIC ACID AND RELATED COMPOUNDS

SAKAYU SHIMIZU
MICHIHIKO KATAOKA
Kyoto University
Kyoto, Japan

KEY WORDS

β-Alanine
Coenzyme A
Lactonohydrolase
Pantetheine
Pantoic acid
Pantothenate kinase
4-Phosphopantetheine
4-Phosphopantothenoyl-L-cysteine decarboxylase
4-Phosphopantothenoyl-L-cysteine synthetase

OUTLINE

INTRODUCTION

Pantothenic acid (*R*- or D-(+)-*N*-(2,4-dihydroxy-3,3-dimethyl-1-oxobutyl)-β-alanine; for chemical formula, see Table 1) was first isolated in the 1930s from liver and found to be an essential growth factor for yeasts (15). During this period it was also identified independently with the chick antidermatitis factor, the filtrate factor, the chick antipellagra factor, and an essential growth factor for lactic acid bacteria. Later these activities were shown to be identical with those of pantothenic acid. Because the factor could be obtained from a variety of plants and animal tissues, Williams named it pantothenic acid, meaning "from everywhere." The compound is also referred to as vitamin B_5. It was independently synthesized by two groups in 1940.

Pantothenic acid occurs in living organisms of all types, both in a free form and in conjugated forms such as coenzyme A, pantetheine (*Lactobacillus bulgaricus* factor), and 4′-phosphopantetheine (*Acetobacter suboxydans* factor) (see Table 1). The coenzyme form of the vitamin, coenzyme A, was discovered as an essential cofactor for the acetylation of sulfonamide in the liver and the acetylation of choline in the brain by Lipmann and coworkers (16). It has been since identified with "active acetate" and has been found to be essential for a variety of biochemical transacylation reactions. 4′-Phosphopantetheine is also a coenzyme form of pantothenic acid that functions as a prosthetic group of the acyl carrier protein of fatty acid synthetase, citrate-cleaving enzyme, and enzymes involved in the synthesis of peptide antibiotics. These early studies have been reviewed by several authors (15–20).

CHEMISTRY

Pantothenic acid can be obtained as a colorless, viscous oil by drying under high vacuum in P_2O_5. It is an acid and has a marked tendency to absorb water from the air. Under alkaline hydrolysis, it breaks down into β-alanine and pantoic acid. The latter readily forms a lactone, D-(−)-pantolactone, in acid solution or on heating. Acid hydrolysis of pantothenic acid gives β-alanine and pantolactone. Pantothenic acid is soluble in water, ethyl acetate, dioxane, acetic acid, ether, and amyl alcohol but is insoluble in benzene and chloroform.

The structure of pantothenic acid contains a single asymmetric center, so that it is optically active; only the natural D-(+)-isomer has vitamin activity. The absolute configuration of the vitamin has been defined as *R* (21). The conformation of the vitamin has also been reported (22).

The calcium salt of pantothenic acid, which can be obtained as needle crystals from methanol, is moderately hygroscopic and is rather more stable to heat, air, and light than is the free acid. It is soluble in water and glycerol and slightly soluble in alcohol and acetone. Reviews by Baddiley (23) and Wagner and Folkers (17) summarized early studies on the chemistry of pantothenic acid.

The naturally occurring derivatives of pantothenic acid (see Table 1) can be grouped into three types on the basis of their chemical structures: simple pantothenate derivatives, pantetheine derivatives in which cysteamine (or its analogs) attaches by an amide linkage, and coenzyme A derivatives in which the pantetheine is adenosylylated (Fig. 1).

Pantothenyl alcohol, an alcohol analog of pantothenic acid, is also a pharmaceutically important unnatural derivative. Unnatural analogs of pantothenic acid and coenzyme A have been reviewed by Shimizu (10).

Table 1. Panthothenic Acid and Its Naturally Occurring Derivatives

D-Pantothenic acid 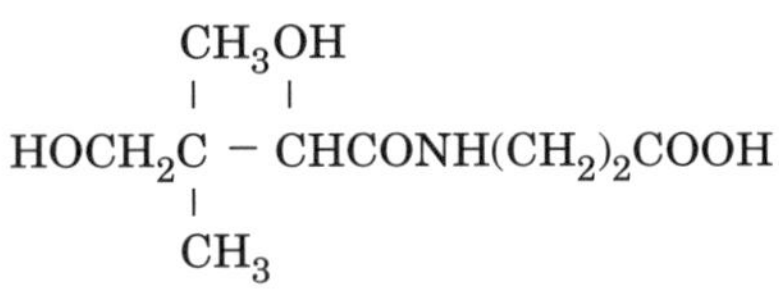 $C_9H_{17}NO_5$ MW: 219·23	Unstable, viscous oil. Extremely hydroscopic, easily decomposed by acids, bases, heat. Soluble in water, ethyl acetate, dioxane, glacial acetic acid; moderately soluble in ether, amyl alcohol; insoluble in benzene, chloroform. Solutions are stable between pH 5 and 7. $[\alpha]_D^{25}$ + 37.5°. (From Stiller et al. [1].)
Calcium D-pantothenate 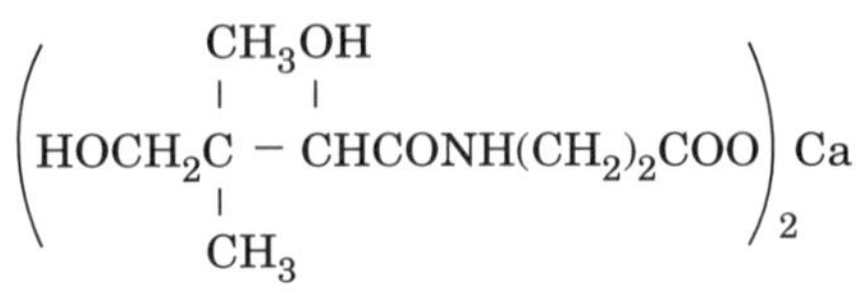 $C_{16}H_{32}CaN_2O_{10}$ MW: 476·53	White needles. Moderately hygroscopic. Soluble in water, glycerol; slightly soluble in alcohol, acetone; insoluble in ether, benzene, chloroform. Decomposed by bases. Solutions are stable between pH 5 and 7. mp 195–196° (dec); $[\alpha]_D^{25}$ + 28.2 °C (c = 5).
Sodium D-pantothenate 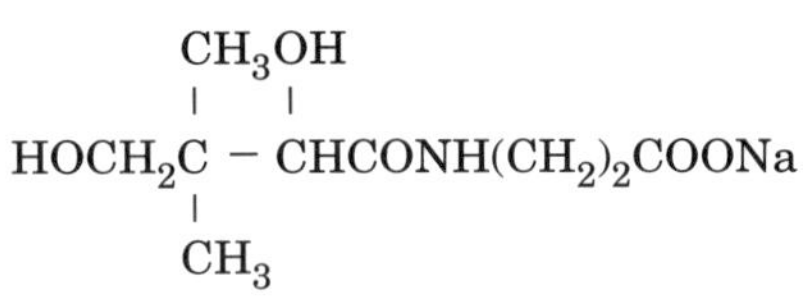 $C_9H_{16}NaNO_5$ MW: 241·21	White, hygroscopic crystals. Decomposed by acids and bases. Solutions are stable between pH 5 and 7. mp 122–124° $[\alpha]_D^{25}$ + 27.1° (c = 2). For solubility, see calcium pantothenate.
4′-Phosphopantothenic acid (Ba salt) 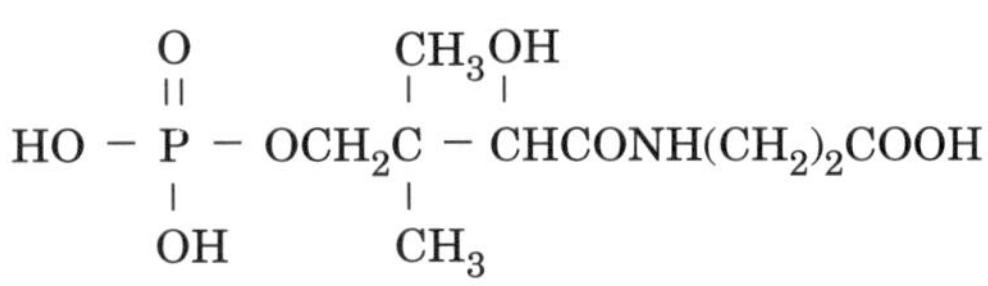 $C_9H_{16}NO_8P$ MW: 313·27	Soluble in water; insoluble in ethanol. Unstable to bases. Free acid is unstable. $[\alpha]_D^{24}$ + 9.0° (c = 3.3) (From King and Strong [2] and Okada et al. [3].)
Pantothenoyl-L-cysteine (Ba salt) 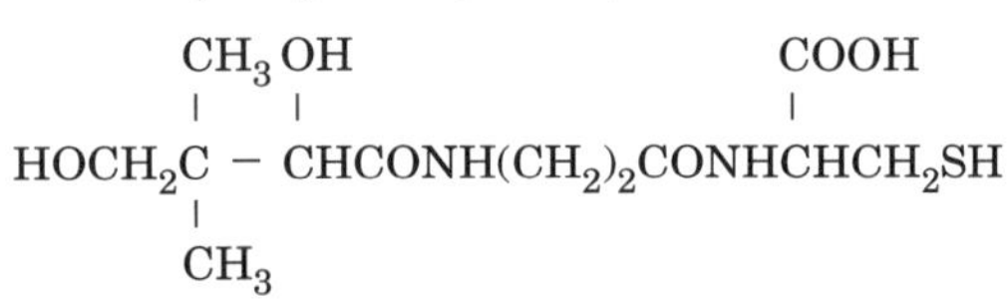 $C_{12}H_{22}N_2O_6S$ MW: 322·38	Soluble in water, methanol; moderately soluble in ethanol; insoluble in ether. Unstable to acids and bases. $[\alpha]_D$ − 14° (c = 2). (From Ohta et al. [4].)
4′-Phosphopantothenoyl-L-cysteine (Ba salt) 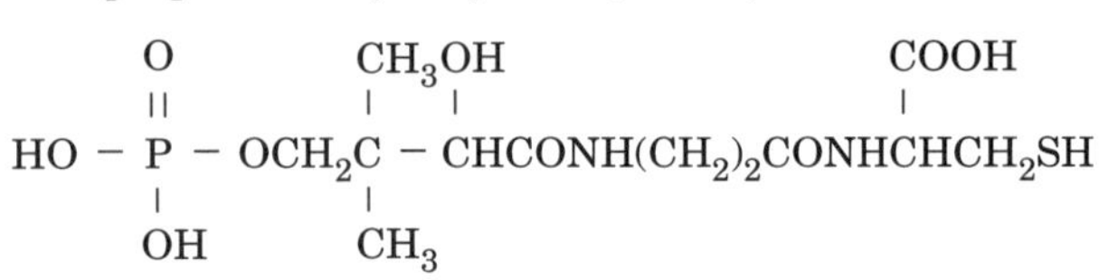 $C_{12}H_{23}N_2O_4PS$ MW: 416·42	Soluble in water; slightly soluble in alcohol. Unstable to acids and bases. Easily oxidized in air. $[\alpha]_D^{20}$ 0° (c = 2). (From Baddiley and Mathias [5] and Nagase [6].)
Pantetheine 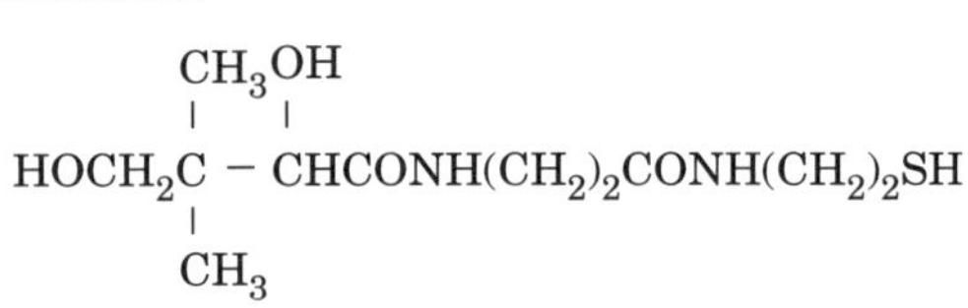 $C_{11}H_{22}N_2O_4S$ MW: 278·37	Syrup or glass. Soluble in water; slightly soluble in alcohol; insoluble in ether, benzene, chloroform, ethyl acetate. Unstable to acids and bases. Easily oxidized in air. $[\alpha]_D^{23}$ + 12.2° (c = 3.45). *Lactobacillus bulgaricus* factor. (From Shimizu et al. [7].)

Table 1. Panthothenic Acid and Its Naturally Occurring Derivatives *(continued)*

Pantetheine 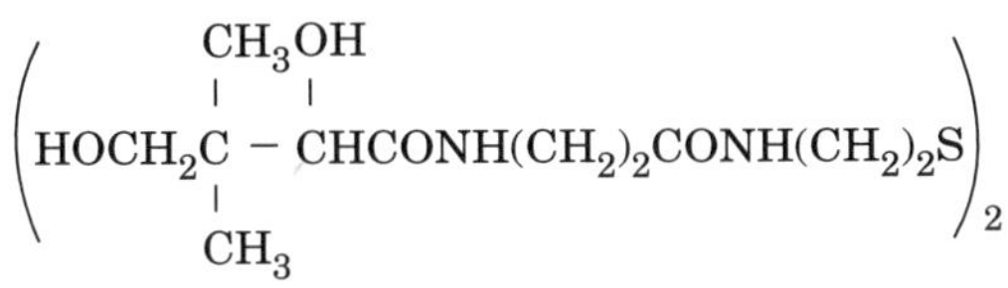 $C_{22}H_{42}N_4O_5S_2$ MW: 554·72	Disulfide form of pantetheine. Glassy, colorless to light yellow substance. Unstable to acids. $[\alpha]_D^{23}$ + 17.1° (*c* = 3.2). For solubility and references, see pantetheine.
4′-Phosphopantetheine (Ba salt) 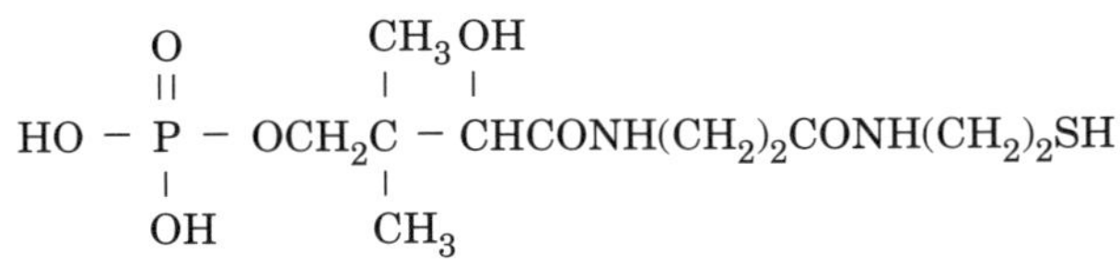 $C_{11}H_{23}N_2O_7PS$ MW: 358·35	Soluble in water; slightly soluble in ethanol; insoluble in ether. Unstable to acids and bases. Easily oxidized in air. $[\alpha]_D^{23}$ + 13.3° (*c* = 2.25). Oxidized form, $[\alpha]_D^{23}$ + 12.2°. *Acetobacter suboxidans* factor. (From Baddiley and Thain [8], Moffatt and Khorana [9], and Nagase [6].)
Dephospho-coenzyme A (Li salt) 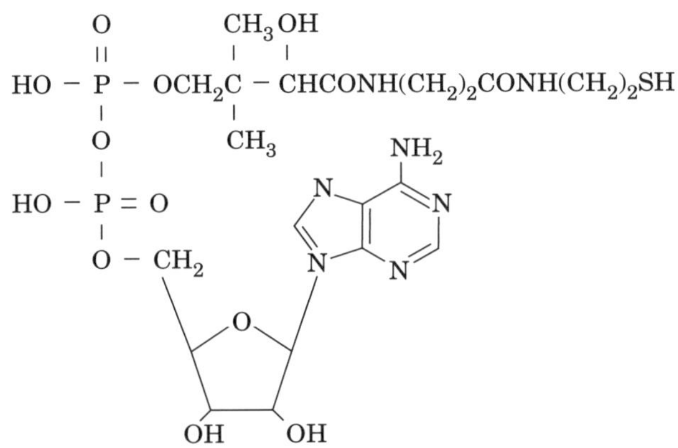 $C_{21}H_{35}N_7O_{12}P_2S$ MW: 687·56	Soluble in water, methanol; insoluble in acetone. Unstable to acids and bases. For references, see coenzyme A.
Coenzyme A 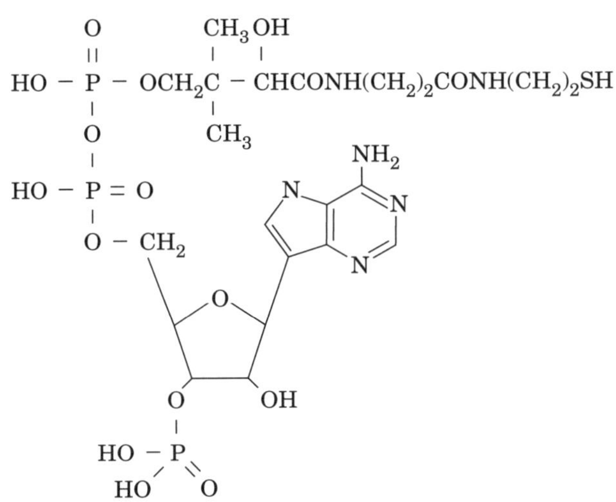 $C_{21}H_{36}N_7O_{16}P_3S$ MW: 767·55	Soluble in water; insoluble in ethanol, ether, acetone. Decomposed to pantetheine-2′,4′-cyclic phosphate and 3′,5′-ADP in 1 N NaOH (100°, 2 min). Decomposed to pantetheine-4′-phosphate and adenine in 1 N HCl (100°, 5 min). Easily oxidized in air. (From Moffatt and Khorana [9], Shimizu [10], and Shimizu et al. [11].)

Table 1. Panthothenic Acid and Its Naturally Occurring Derivatives *(continued)*

4′-Phosphopantetheine-*S*-sulfonate (Ca salt) 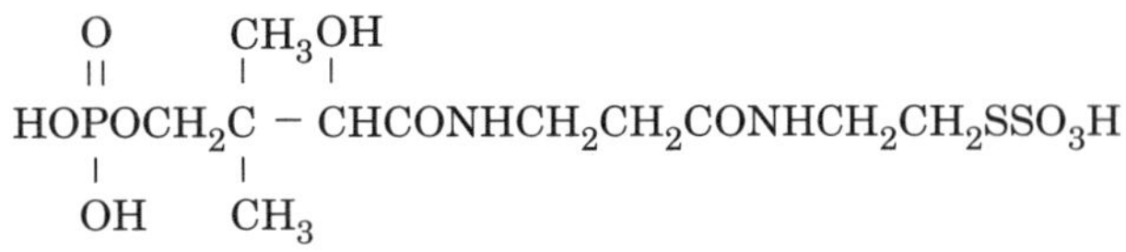 $C_{11}H_{23}O_{10}N_2S_2P$ MW: 438·41	Soluble in water. $[\alpha]_D^{22}$ + 7.2° (*c* = 1.95). *Bifidus* factor. (From Yoshioka and Tamura [12].)
4′-*O*-(*β*-glucopyranosyl)-D-pantothenic acid 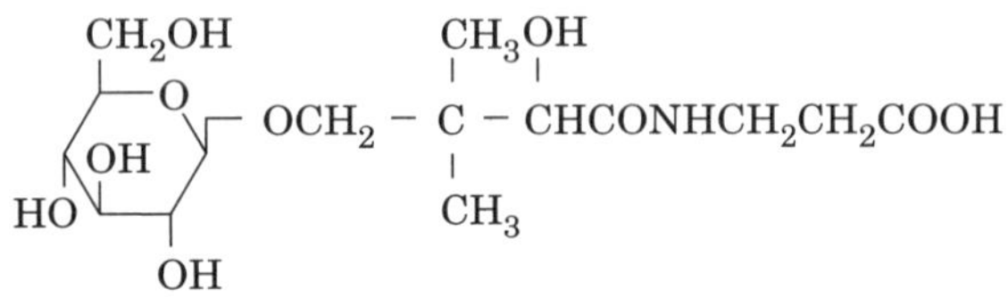 $C_{44}H_{25}O_{14}N$ MW: 791·68	Soluble in water. $[\alpha]_D^{23}$ −18.2° (*c* = 1.0). Growth factor of *Leuconostoc*. (From Amachi et al. [13].)
OA-6129A (Na salt)  $C_{20}H_{31}N_3O_7S$ MW: 457·54	Unstable to acids and bases. $[\alpha]_D^{24}$ + 11.6° (*c* = 1.0). Antibiotic produced by *Streptomyces* sp. OA-6129. (From Yoshioka et al. [14].)

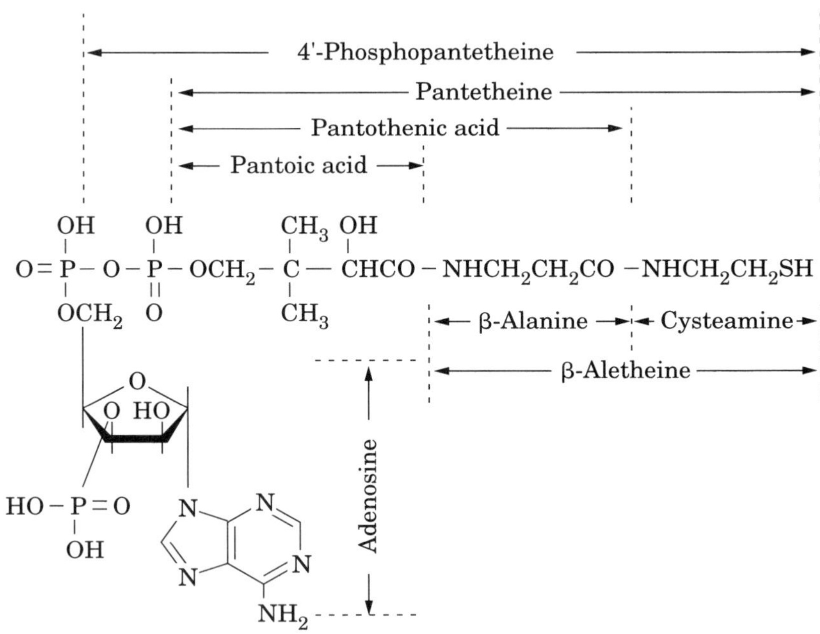

Figure 1. The structure of coenzyme A.

BIOSYNTHESIS

Although pantothenic acid cannot be synthesized by animals, microorganisms and plants are generally able to produce it from the precursors pantoic acid and *β*-alanine through catalysis of the enzyme pantothenate synthetase (EC 6.3.2.1). Animals, plants, and microorganisms can convert pantothenic acid to 4′-phosphopantetheine and coenzyme A, the metabolically active forms of the vitamin. The pathway for the biosynthesis of pantothenic acid and coenzyme A has been elucidated by several workers since the early 1950s and has been reviewed (24–27). The pathway leading to the vitamin and the coenzymes from common precursors can be summarized as shown in Figures 2 and 3.

β-Alanine

Three routes to β-alanine have been reported. Several microorganisms have been reported to form β-alanine by α-decarboxylation of L-aspartic acid (Fig. 2; reaction 1). Confirmatory evidence for this conversion was provided by Williamson and Brown (28), who purified (to apparent homogeneity), from extracts of *Escherichia coli*, an enzyme that catalyzes the α-decarboxylation of L-aspartic acid to yield β-alanine and CO_2. These investigators also reported that the enzyme is missing in a mutant of *E. coli* that requires either β-alanine or pantothenate as a nutritional factor, but is present in the wild-type strain and in a revertant strain of the mutant. It has also been suggested, on the basis of the observation that mutants of *Salmonella typhimurium* lacking the ability to degrade uracil require *N*-carbamoyl-β-alanine, β-alanine, or pantothenate, as a nutritional factor, that β-alanine is produced by decarboxylation of *N*-carbamoyl-β-alanine formed from uracil (Fig. 2; reactions 2–4) (29). β-Alanine may also be produced by transamination of malonylsemialdehyde produced from propionic acid (Fig. 2; reaction 5 or 6), because enzyme activity catalyzing this conversion has been detected in several microorganisms. However, there have been no further studies concerning this reaction.

Pantoic Acid

The route to pantoic acid from pyruvate (as shown in Fig. 2) has been elucidated mainly in *E. coli* and *Neurospora crassa*. Two enzymes catalyzing the conversion of pyruvate to α-ketoisovalerate (Fig. 2; reactions 7 and 8) in this route are shared by the route for the biosynthesis of the branched chain amino acids. In *E. coli*, two enzyme activities have been detected for the conversion of α-ketoisovalerate to ketopantoic acid (Fig. 2; reaction 9): one is dependent on tetrahydrofolate and the other is not. The physiological significance of tetrahydrofolate-independent activity seemed to be questionable because of its high K_m values for formaldehyde (10 mM) and α-ketoisovalerate (100 mM). Because a mutant lacking tetrahydrofolate-dependent activity requires pantothenate for growth although the same amount of tetrahydrofolate-independent activity is found in the same mutant, concrete evidence is provided to support the theory that the tetrahydrofolate-dependent enzyme is responsible for the ketopantoate needed for the biosynthesis of pantothenate. The tetrahydrofolate-dependent enzyme (i.e., ketopantoate hydroxymethyltransferase, EC 2.1.2.11) has been purified and characterized in some detail. The observation that pantoate, pantothenate, and coenzyme A are all allosteric inhibitors of this enzyme also supports this conclusion (30,31).

The reduction of ketopantoic acid to D-pantoic acid (Fig. 2; reaction 10) is catalyzed by an NADPH-dependent enzyme, ketopantoic acid reductase (EC 1.1.1.169). This enzyme activity has been detected in *Saccharomyces cerevisiae* and *E. coli*. The same reduction is also catalyzed by α-acetohydroxy acid isomeroreductase (EC 1.1.1.1.86), which is the enzyme responsible for the conversion of α-acetolactate to α-ketoisovalerate (Fig. 2; reaction 8) (32). Recently, Shimizu et al. (33) isolated ketopantoic acid reductase in a crystalline form from *Pseudomonas maltophilia* and characterized it in some detail. They also demonstrated that this reductase is the enzyme for D-pantoic acid formation, necessary for the biosynthesis of pantothenic acid, because mutants lacking this enzyme require either D-pantoic acid or pantothenate for growth and the revertants regain this activity.

Coenzyme A

The pathway for the biosynthesis of coenzyme A from pantothenic acid, L-cysteine, and ATP (Fig. 3) was first demonstrated by Brown (34,35) on the basis of his studies with *Proteus morganii* and early observations in the 1950s with pig liver and other organisms. Later, the validity of this pathway was confirmed by Abiko and coworkers, who separated and characterized the enzymes involved in this pathway in rat liver (26).

The first step in this pathway is the phosphorylation of pantothenic acid (Fig. 3; reaction 1) by pantothenate kinase (EC 2.7.1.33). The enzyme has been purified and characterized from rat liver (36) and *Brevibacterium ammoniagenes* (37). Both enzymes undergo allosteric inhibition by coenzyme A, the end product of the pathway. Because such feedback inhibition by coenzyme A is observed only at this step and no other regulation mechanism is known, this inhibition seems to be the most important mechanism for controlling the cellular level of coenzyme A. Pantetheine is also phosphorylated by the same enzyme to yield 4′-phosphopantetheine (Fig. 3; reaction 7), which can be converted to coenzyme A.

The condensation of 4′-phosphopantothenic acid with L-cysteine to yield 4′-phosphopantothenoyl-L-cysteine (Fig. 3; reaction 2) is catalyzed by 4′-phosphopantothenoyl-L-cysteine synthetase (EC 6.3.2.5). The mammalian enzyme requires ATP as energy for the condensation, whereas the bacterial enzyme preferably utilizes CTP (34). 4′-Phosphopantothenoyl-L-cysteine is then decarboxylated to yield 4′-phosphopantetheine (Fig. 3; reaction 3) by 4′-phosphopantothenoyl-L-cysteine decarboxylase (EC 4.1.1.36). The enzyme has been shown to be independent of pyridoxal 5′-phosphate. This step is the only one that does not require ATP among the five steps in the biosynthesis of this coenzyme.

The pyrophosphate linkage formation between 4′-phosphopantetheine and ATP to yield 3′-dephosphocoenzyme A (Fig. 3; reaction 4), and its phosphorylation (Fig. 3; reaction 5) are catalyzed, by dephosphocoenzyme A pyrophosphorylase (EC 2.7.7.3) and dephospho-coenzyme A kinase (EC 2.7.1.24), respectively. In rat liver, both enzymes are present as a complex or a bifunctional enzyme (38). The reversibility of the former reaction may be important for controlling cellular levels of coenzyme A and 4′-phosphopantetheine.

The presence of an alternate route to yield 4′-phosphopantetheine via pantetheine (Fig. 3; reactions 6 and 7) has been suggested in *Acetobacter suboxydans, Lactobacillus*

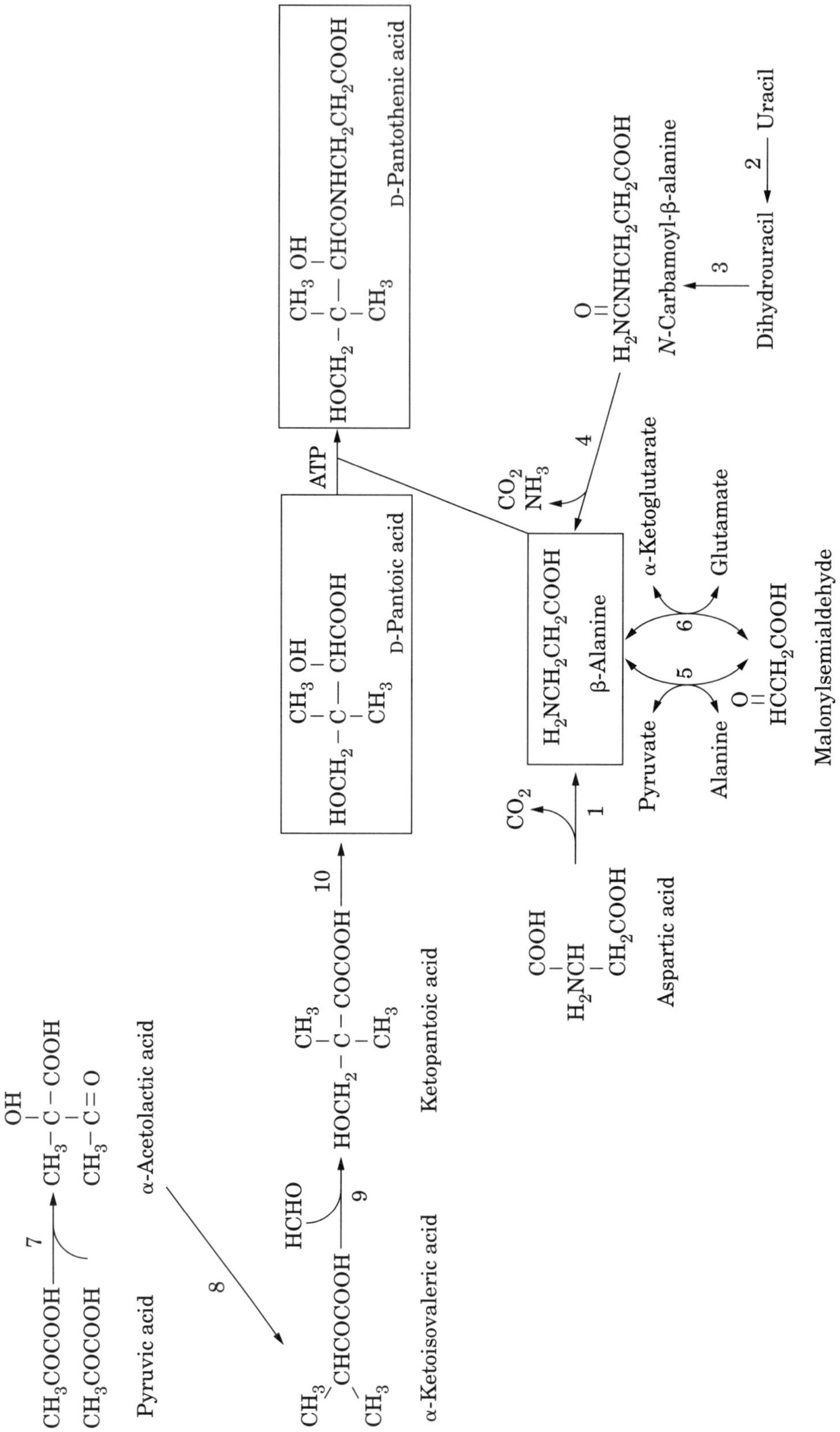

Figure 2. The pathway for the biosynthesis of pantothenic acid.

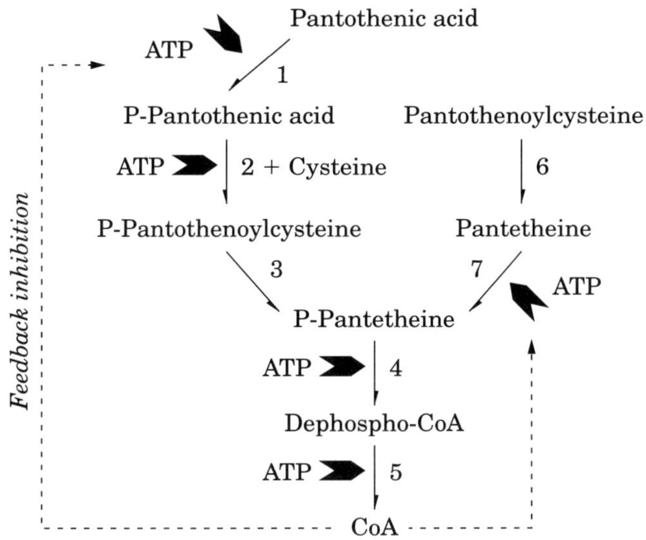

Figure 3. The pathway for the biosynthesis of coenzyme A from pantothenic acid, L-cysteine, and ATP. For chemical structures of each compound, see Table 1. CoA, coenzyme A.

helveticus, and other microorganisms (35), but confirmatory evidence is not available.

Control Mechanisms for the Biosynthesis

The allosteric inhibition of ketopantoic acid hydroxymethyltransferase of *E. coli* by D-pantoic acid, pantothenic acid, or coenzyme A may be involved as a control mechanism in pantothenate biosynthesis (30). On the other hand, such inhibition was not observed in the case of ketopantoic acid reductase of *Pseudomonas maltophilia* (33).

In the pathway to coenzyme A from pantothenic acid, the involvement of the feedback inhibition of pantothenate kinase by coenzyme A and 4′-phosphopantetheine as a control mechanism in the biosynthesis has been demonstrated (36,37,39). Because this inhibition was generally observed regardless of species and the other four steps following this reaction are not significantly inhibited by coenzyme A or 4′-phosphopantetheine, this may be one of the most important mechanisms in the control of cellular levels of coenzyme A. No other mechanism, such as repression, has been observed in either pantothenate or coenzyme A biosynthesis.

Pantetheinase, which specifically degrades pantetheine to pantothenic acid and cysteamine, may also be an important enzyme because coenzyme A can be degraded to pantetheine enzymatically and pantetheine can be reused as a precursor of coenzyme A after phosphorylation by pantothenate kinase. Cellular coenzyme A levels may be influenced by competition between pantetheinase and pantothenate kinase toward their substrate, pantetheine (40).

CHEMICAL AND MICROBIAL PRODUCTION METHODS

Pantothenic Acid

At present, commercial production of pantothenate depends exclusively on chemical synthesis. As outlined in Figure 4, the conventional chemical process involves reactions yielding racemic pantolactone from isobutyraldehyde, formaldehyde, and cyanide, optical resolution of the racemic pantolactone to D-(−)-pantolactone with quinine, quinidine, cinchonidine, brucine, and so on, and condensation of D-(−)-pantolactone with β-alanine. This is followed by isolation as the calcium salt and drying to obtain the final product. A problem associated with this chemical process apart from the use of poisonous cyanide is the troublesome resolution of the racemic pantolactone and the racemization of the remaining L-(+)-isomer. Therefore, most of the recent studies in this area have concentrated on development of an efficient method to obtain D-(−)-pantolactone.

Enzymatic resolution of racemic pantolactone can be carried out by specific fungal lactonohydrolases. Shimizu et al. found that many mold strains belonging to the genera *Fusarium, Gibberella*, and *Cylindrocarpon* stereospecifically hydrolyze D-(−)-pantolactone to D-(−)-pantoic acid (41,42). If racemic pantolactone is used as a substrate for the hydrolysis reaction by the microbial lactonohydrolase, only the D-(−)-pantolactone might be converted to D-(−)-pantoate and the L-(+)-enantiomer might remain intact. Consequently, the racemic mixture could resolved into D-(−)-pantoate and L-(+)-pantolactone as shown in Figure 5 (42,43). After the removal of L-(+)-pantolactone from the reaction mixture by solvent extraction, etc., remaining D-(−)-pantoate could be easily converted to D-(−)-pantolactone by heating under acidic conditions. The reverse reaction, that is, lactonization of D-(−)-pantoate, might also be possible for the resolution. In this case, D-(−)-pantoate in a racemic mixture of pantoate is specifically lactonized into D-(−)-pantolactone (see Fig. 5) (44). When *Fusarium oxysporum* mycelia were incubated in 700 g/L aqueous solution of racemic pantolactone for 24 h at 30 °C with automatic pH control (pH 6.8–7.2), about 90% of the D-(−)-isomer was hydrolyzed. The resultant D-(−)-pantoic acid in the reaction mixture showed a high optical purity (96% ee), and the coexisting L-(+)-isomer remained without any modification.

Practical hydrolysis of the D-(−)-isomer in a racemic mixture is carried out using immobilized mycelia of *F. oxysporum* as the catalyst. Stable catalyst with high hydrolytic activity can be prepared by entrapping the fungal mycelia in calcium alginate gels. When the immobilized mycelia were incubated in a reaction mixture containing 350 g/L racemic pantolactone for 21 h at 30 °C under automatic pH control (pH 6.8–7.2), 90–95% of the D-(−)-isomer was hydrolyzed (optical purity, 90–97% ee). After reactions repeated 180 times (i.e., for 180 days), the immobilized mycelia retained more than 90% of their initial activity (Fig. 6).

The overall process for the present enzymatic resolution is compared with the conventional chemical process in Figure 7. The enzymatic process allows skipping several tedious steps that are necessary in chemical resolution and is highly advantageous for practical purposes.

Coenzyme A

The production methods for coenzyme A roughly fall into chemical and microbial categories. The chemical methods,

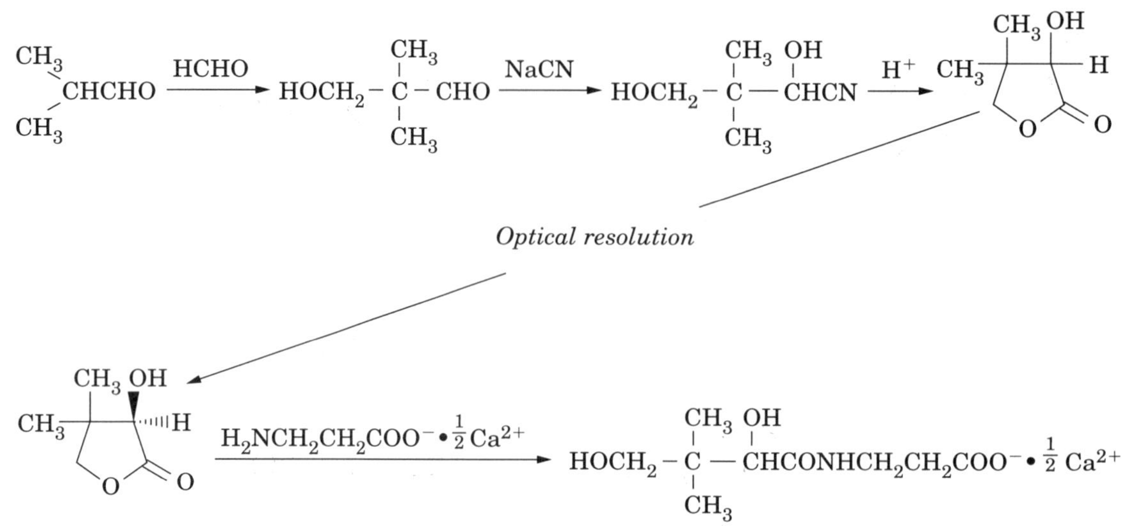

Figure 4. Outline of the chemical synthesis of D-pantothenic acid.

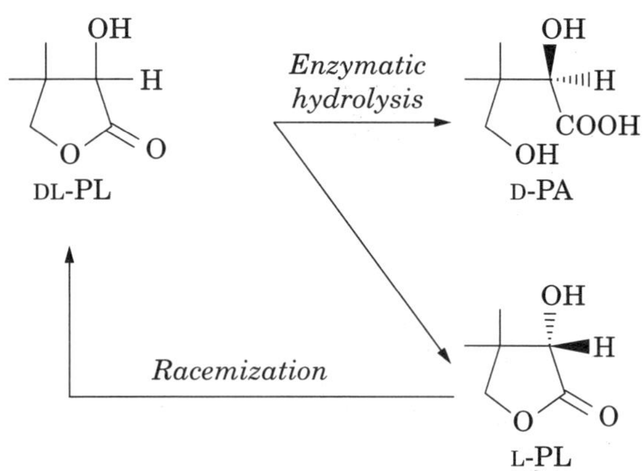

Figure 5. Schematic representation of the enzymatic resolution of racemic pantolactone. PL, pantolactone; PA, pantoate.

which have been reviewed by Shimizu (10) and Mautner (45), are too complex to be practical. Therefore, commercial production is carried out by microbiological methods. Extraction of coenzyme A from yeast cells has been performed since the early 1950s. Cells of baker's or brewer's yeasts, which are relatively rich in coenzyme A, were usually used as the source. Later, an efficient enzymatic method using *Brevibacterium ammoniagenes* cells as the catalyst was developed. These microbial methods have been reviewed by Shimizu and Yamada (46).

A successful enzymatic method using the biosynthetic route of coenzyme A from pantothenic acid, L-cysteine, and ATP was first reported by Ogata et al. (47), who found that *B. ammoniagenes* has all five enzymes necessary for the biosynthesis of coenzyme A in high activities. These three substrates, when added to a reaction mixture containing the bacterial cells, were converted to coenzyme A with a satisfactory yield (2–3 g/L). Ogata et al. also found that the same organism can accumulate coenzyme A directly in the culture medium on addition of pantothenic acid, L-cysteine, and AMP, adenosine, or adenine in the presence of a surfactant, cetylpyridinium chloride, and high levels of glucose (usually 10%), K_2HPO_4, and $MgSO_4 \cdot 7H_2O$. Under optimal conditions, the amount obtained was 5.5 g/L. Most coenzyme A in the medium was present in the disulfide form because of the vigorous shaking during the reaction. After treatment of the culture filtrate with Duolite S-30, charcoal, and Dowex 1 (Cl^-), and reduction of the disulfide, the very pure thiol form was obtained in high yield. The mechanism of this coenzyme A production has been suggested to be that shown in Figure 8 (for details, see Refs. 11 and 48).

To improve product yield further, the mechanism for regulation of biosynthesis was investigated. As described in the preceding section, the biosynthesis of coenzyme A in *B. ammoniagenes* is controlled mainly by the feedback inhibition of pantothenate kinase by coenzyme A. This was the main problem in the practical production, because the overproduced coenzyme A itself stopped the biosynthesis. Two methods to abolish this feedback inhibition have been reported.

A synthetic scheme was investigated in which the reaction is initiated by the condensation of 4′-phosphopantothenic acid and L-cysteine or the transadenosylylation of 4′-phosphopantetheine, because these routes do not involve phosphorylation of pantothenic acid or pantetheine by pantothenate kinase. Replacement of the enzymatic phosphorylation of pantothenate or pantetheine with chemical phosphorylation followed by the enzymatic reaction increased the yield of coenzyme A 10- to 20-fold. Yields from 4′-phosphopantothenic acid and 4′-phosphopantetheine were 33 g/L (molar yield based on ATP; 64.5%) and 115 g/L (100%), respectively (49). This method is applicable to coenzyme A production under ATP-generating conditions (see Fig. 8). 4′-Phosphopantothenic acid (25 g/L), L-cysteine (15 g/L), and AMP (33 g/L), when added to the culture broth of *B. ammoniagenes*, were converted to coenzyme A with a yield of 23 g/L (50).

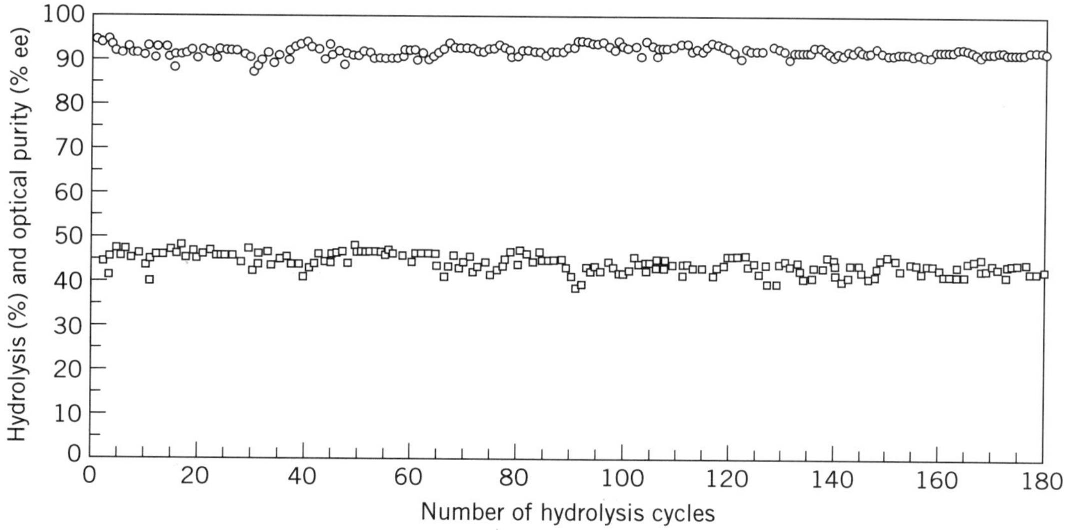

Figure 6. Stereospecific hydrolysis of pantolactone by *Fusarium oxysporum* mycelia entrapped in calcium alginate gels: 280 L immobilized mycelia (containing 15.2 kg wet cells) was incubated with 350 L aqueous racemic pantolactone solution (350 g/L) at 30 °C for 21 h. The pH of the mixture was automatically controlled at 6.8–7.2 with 15 M NH_4OH. All the immobilized mycelia filtrated from the reaction mixture were used for the subsequent reactions. The hydrolysis reactions were carried out 180 times. ○, optical purity for D-(−)-pantoate; □, hydrolysis rate.

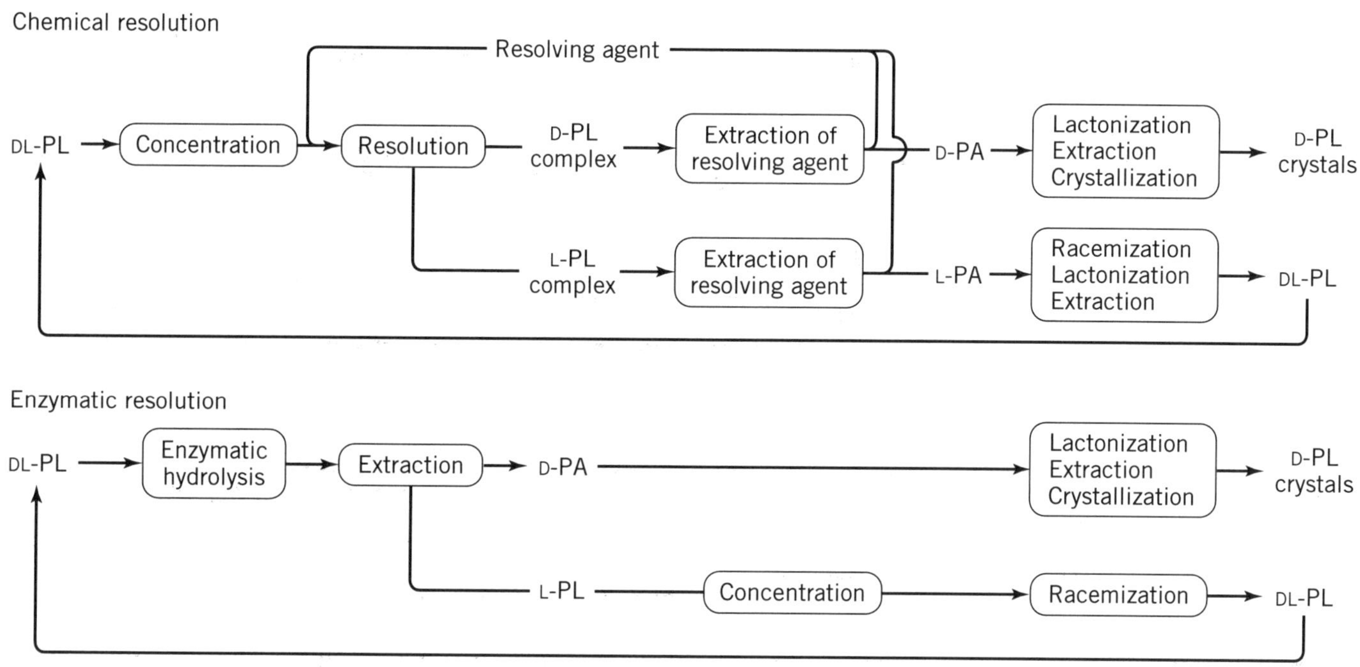

Figure 7. Comparison of enzymatic and conventional chemical resolution processes for racemic pantolactone.

Another way to improve the yield is to use mutants derepressed for the feedback inhibition or those showing elevated pantothenate kinase activity. A mutant of *B. ammoniagenes* that is resistant to oxypantetheine (the corresponding oxygen analog of pantetheine) was found to have an elevated activity of pantothenate kinase. Under ATP-generating conditions, the yields of coenzyme A from pantothenic acid (3.6 g/L), L-cysteine (1.8 g/L), and AMP (6 g/L) or from pantetheine (5 g/L) and AMP (6 g/L) were 9.3 or 11.5 g/L, respectively. These values were about threefold higher than those obtained with the parent strain, and 70–100% of the added AMP was converted to coenzyme A (51) (Fig. 9).

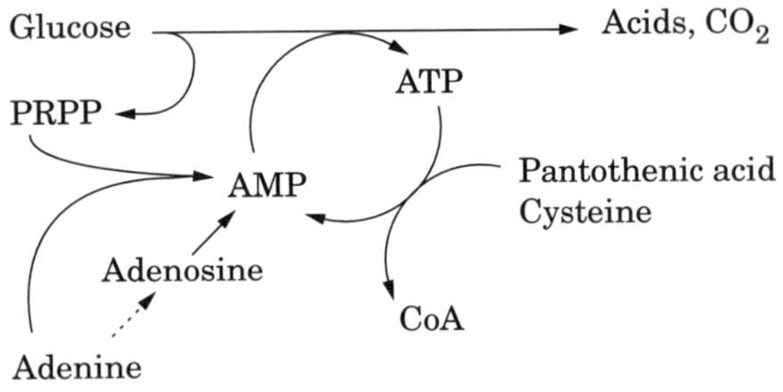

Figure 8. Reaction sequences of coenzyme A production with *Brevibacterium ammoniagenes* under ATP-generating conditions. PRPP, 5-phosphoribosyl-1-pyrophosphate; CoA, coenzyme A.

Continuous production of coenzyme A through five enzymatic steps using gel-entrapped cells of *B. ammoniagenes* has also been reported (48,52).

4′-Phosphopantetheine and Other Intermediates in Coenzyme A Biosynthesis

4′-Phosphopantetheine together with other intermediates in coenzyme A biosynthesis can be effectively synthesized by using *B. ammoniagenes* cells as the catalyst and by modifying the reaction conditions (11) as follows: 4′-phosphopantothenic acid on omission of L-cysteine from the reaction mixture, 4′-phosphopantetheine on addition of CTP or CTP and GTP in place of ATP (because pantothenic acid is phosphorylated in the presence of CTP or GTP as well as ATP and the condensation of 4′-phospho-

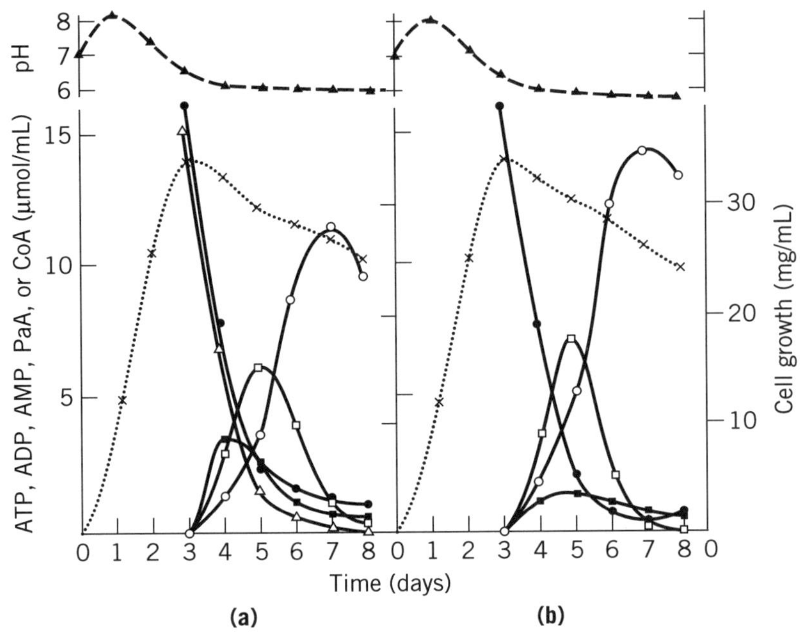

Figure 9. Time course of coenzyme A production by an oxypantetheine-resistant mutant of *Brevibacterium ammoniagenes* under ATP-generating conditions. **(a)** Production from pantothenic acid, L-cysteine, and AMP. **(b)** Production from pantetheine and AMP. ○, coenzyme A (CoA); ●, pantothenic acid (PaA); □, ATP; ■, ADP; △, AMP; ▲, pH; ×, cell growth. For details, see Shimizu et al. (51).

Table 2. Production of the Intermediates in Coenzyme A Biosynthesis by *Brevibacterium ammoniagenes*

Product	Substrates	Nucleotides	Reactions involved[a]	Dried cells	Productivity enzyme source (mg/mL) Culture broth[b]	Immobilized cells
4′-Phosphopantothenic acid	Pantothenic acid	ATP	1	3–4		1.5–2.5
4′-Phosphopantothenic acid	Pantothenic acid	AMP	1		4–5	
4′-Phosphopantetheine	4′-Phosphopantothenic acid and L-cysteine	CTP	2,3	3–4		1.8
4′-Phosphopantetheine	Pantothenic acid and L-cysteine	ITP and CTP	1–3	2–3		0.3
4′-Phosphopantetheine	Pantothenic acid and L-cysteine	GMP and CMP	1–3		3–4	
4′-Phosphopantetheine	Pantetheine	ITP	7	2–3		0.9
4′-Phosphopantetheine	Pantetheine	GMP	7		4–5	
3′-Dephosphocoenzyme A	Pantothenic acid and L-cysteine	ATP	1–5	1–2		

Note: For details, see Shimizu et al. (11,48).
[a]Numbers correspond to those given in Figure 3.
[b]In this case, the nucleoside monophosphates added are presumed to be converted to the corresponding nucleoside triphosphates and used for the reactions in a manner similar to that shown in Figure 8.

pantetheine with nucleoside triphosphate is specific for ATP), and 3′-dephosphocoenzyme A on treatment of the reaction mixture containing coenzyme A with 3′-nucleotidase. The amounts of these intermediates obtained by this method are summarized in Table 2.

ASSAY METHODS

Test microorganisms normally used for the microbiological assay of pantothenic acid are *Lactobacillus plantarum* (ATCC 8014; *L. arabinosus* 17-5), *L. casei* (ATCC 7469), and *Saccharomyces uvarum* (ATCC 9080; *S. carlsbergensis*). *Lactobacillus plantarum* is suitable for determining unconjugated pantothenate in samples. It should be noted that pantetheine, when simultaneously present in a molar ratio to pantothenate of more than 0.5, gives positive errors in the determination. *Saccharomyces uvarum* also shows almost specific growth response to free pantothenate, but β-alanine stimulates its growth. Hence, an assay procedure employing this organism is also the one chosen for determining the pantothenic acid that occurs in natural products together with other pantothenate forms. *Lactobacillus casei* responds not only to pantothenate but also several conjugated forms of pantothenate. *Lactobacillus helveticus* (ATCC 12046) and *L. bulgaricus* B1 are recommended for determination of pantetheine (or pantethine) because both these organisms require more than 100 times as much pantothenic acid as pantethine to give the same response. Treatment of samples and details of assay procedures have been described (53).

A sensitive enzymatic assay method using pantothenase has been reported (54), but the enzyme is not commercially available. Chemical and physical methods have also been reported. These are often used in determining pantothenic acid in pharmaceutical products but are not suitable for determination of natural samples because of their low sensitivity.

USE AND ECONOMIC ASPECTS

The current world capacity of calcium pantothenate production and its demand are presumed to be about 4,000 and 3,600–4,000 tons per year, respectively. It is mainly used as an additive to animal feed (about 3,000 tons per year) and as a pharmaceutical product (about 600 tons per year). Pantothenyl alcohol is used as a source of pantothenate activity for pharmaceutical vitamin products. Pantothenyl alcohol itself has no pantothenate activity; in fact, it is a competitive growth inhibitor of several pantothenate-requiring lactic acid bacteria. However, it has been demonstrated to be quantitatively converted to pantothenic acid in the animal body and to be equivalent to pantothenic acid in man.

Pantethine, the disulfide of pantetheine, and coenzyme A are also used as pharmaceutical products in several countries. They have been suggested to be effective in reducing cholesterol level, curing fatty liver, and treating related diseases.

Some sulfonate derivatives of pantetheine or coenzyme A (*Bifidus* factors), such as 4′-phosphopantetheine-*S*-sulfonate, that were originally isolated from carrot roots have been shown to be growth factors of *Bifidobacterium* (12). Addition of the *bifidus* factors to dried milk for infants has been suggested to be useful in improving the quality of the milk. A carbapenem antibiotic, OA-6129A (see Table 1) produced by *Streptomyces* sp. OA-6129, may be an interesting example suggesting a new use of the vitamin as a building block for its synthesis (14).

BIBLIOGRAPHY

1. E.T. Stiller, S.T. Harris, J. Finkelstein, J.C. Keresztesy, and K. Folkers, *J. Am. Chem. Soc.* **62**, 1785–1790 (1940).
2. T.E. King and F.M. Strong, *J. Biol. Chem.* **191**, 515–521 (1951).
3. S. Okada, O. Nagase, and M. Shimizu, *Chem. Pharm. Bull.* **15**, 713–715 (1967).
4. G. Ohta, O. Nagase, Y. Hosokawa, H. Tagawa, and M. Shimizu, *Chem. Pharm. Bull.* **15**, 644–647 (1967).
5. J. Baddiley and A.P. Mathias, *J. Chem. Soc.*, 2803–2812 (1954).
6. O. Nagase, *Chem. Pharm. Bull.* **15**, 648–654 (1967).
7. M. Shimizu, G. Ohta, O. Nagase, S. Okada, and Y. Hosokawa, *Chem. Pharm. Bull.* **13**, 180–188 (1965).
8. J. Baddiley and E.M. Thain, *J. Chem. Soc.*, 1610–1615 (1953).
9. J.G. Moffatt and H.G. Khorana, *J. Am. Chem. Soc.* **83**, 663–675 (1961).
10. M. Shimizu, *Methods Enzymol.* **18A**, 322–328 (1970).
11. S. Shimizu, Y. Tani, and K. Ogata, *Methods Enzymol.* **62**, 236–245 (1979).
12. M. Yoshioka and Z. Tamura, *Chem. Pharm. Bull.* **19**, 178–185 (1971).
13. T. Amachi, S. Iwamoto, and H. Yoshizumi, *Agric. Biol. Chem.* **35**, 1222–1230 (1971).
14. T. Yoshioka, I. Kojima, K. Isshiki, A. Watanabe, Y. Shimauchi, M. Okabe, and Y. Fukagawa, *J. Antibiot. (Tokyo)* **36**, 1473–1482 (1983).
15. R.J. Williams, *Adv. Enzymol.* **3**, 253–287 (1943).
16. F. Lipmann, *Fed. Proc.* **12**, 673–715 (1953).
17. A.F. Wagner and K. Folkers, in *Vitamins and Coenzymes*, Interscience, New York, 1964, pp. 93–137.
18. P.R. Vagelos, in P.D. Boyer ed., *The Enzymes*, 3rd ed., vol. 8, Academic Press, New York, 1973, pp. 155–199.
19. E.J. Vandamme, in A. Wiseman ed., *Topics in Enzyme and Fermentation Biotechnology*, vol. 5, Horwood-Wiley, New York, 1981, pp. 185–201.
20. H. Kleinkauf and H. von Döhren, *Trends Biochem. Sci.* **8**, 281–283 (1983).
21. R.K. Hill and T.H. Chan, *Biochem. Biophys. Res. Commun.* **38**, 181–183 (1970).
22. H. Fritz and W. Löwe, *Angew. Chem.* **74**, 751–753 (1962).
23. J. Baddiley, *Adv. Enzymol.* **16**, 1–22 (1955).
24. G.M. Brown and J.J. Reynolds, *Annu. Rev. Biochem.* **32**, 419–462 (1963).
25. G.W. Plaut, C.M. Smith, and W.L. Alworth, *Annu. Rev. Biochem.* **43**, 899–922 (1974).
26. Y. Abiko, in D.M. Greenberg ed., *Metabolic Pathways*, 3rd ed., vol. 7, *Metabolism of Sulfur Compounds*, Academic Press, New York, 1975, pp. 1–25.

27. G.M. Brown and J.M. Williamson, *Adv. Enzymol.* **53**, 345–381 (1982).
28. J.M. Williamson and G.M. Brown, *J. Biol. Chem.* **254**, 8074–8082 (1979).
29. T.P. West, T.W. Traut, M.S. Shanley, and G.A. O'Donovan, *J. Gen. Microbiol.* **131**, 1083–1090 (1985).
30. S.G. Powers and E.E. Snell, *J. Biol. Chem.* **251**, 3786–3793 (1976).
31. J.H. Teller, S.G. Powers, and E.E. Snell, *J. Biol. Chem.* **251**, 3780–3785 (1976).
32. D.A. Primerano and R.O. Burns, *J. Bacteriol.* **153**, 259–269 (1983).
33. S. Shimizu, M. Kataoka, M.C.M. Chung, and H. Yamada, *J. Biol. Chem.* **263**, 12077–12084 (1988).
34. G.M. Brown, *J. Biol. Chem.* **234**, 370–378 (1959).
35. G.M. Brown, *J. Biol. Chem.* **234**, 379–382 (1959).
36. Y. Abiko, S. Ashida, and M. Shimizu, *Biochim. Biophys. Acta* **268**, 364–372 (1972).
37. S. Shimizu, K. Kubo, Y. Tani, and K. Ogata, *Agric. Biol. Chem.* **37**, 2863–2870 (1973).
38. T. Suzuki, Y. Abiko, and M. Shimizu, *J. Biochem. (Tokyo)* **62**, 642–649 (1967).
39. D.S. Vallari, S. Jackowski, and C.O. Rock, *J. Biol. Chem.* **262**, 2468–2471 (1987).
40. C.T. Wittwer, D. Burkhard, K. Ririe, R. Rasmussent, J. Brown, B.W. Wyse, and R.G. Hansen, *J. Biol. Chem.* **258**, 9733–9738 (1983).
41. S. Shimizu, M. Kataoka, K. Shimizu, M. Hirakata, K. Sakamoto, and H. Yamada, *Eur. J. Biochem.* **209**, 383–390 (1992).
42. M. Kataoka, K. Shimizu, K. Sakamoto, H. Yamada, and S. Shimizu, *Appl. Microbiol. Biotechnol.* **43**, 974–977 (1995).
43. M. Kataoka, K. Shimizu, K. Sakamoto, H. Yamada, and S. Shimizu, *Appl. Microbiol. Biotechnol.* **44**, 333–338 (1995).
44. M. Kataoka, K. Shimizu, K. Sakamoto, H. Yamada, and S. Shimizu, *Enzyme Microb. Technol.* **19**, 307–310 (1996).
45. H.G. Mautner, *Methods Enzymol.* **18A**, 338–350 (1970).
46. S. Shimizu and H. Yamada, in H.-J. Rehm and G. Reed eds., *Biotechnology*, vol. 4, VCH, Weinheim, Germany, 1986, pp. 159–184.
47. K. Ogata, S. Shimizu, and Y. Tani, *Agric. Biol. Chem.* **34**, 1757–1759 (1970).
48. S. Shimizu, Y. Tani, and H. Yamada, in K. Venkatsubramanian ed., *Immobilized Microbial Cells*, American Chemical Soc., Washington, D.C., 1979, pp. 87–100.
49. S. Shimizu, R. Komaki, Y. Tani, and H. Yamada, *FEBS Lett.* **151**, 303–306 (1983).
50. S. Shimizu and H. Yamada, in *Third European Congress on Biotechnology*, vol. I, Verlag Chemie, Weinheim, Germany, 1984, pp. 401–405.
51. S. Shimizu, A. Esumi, R. Komaki, and H. Yamada, *Appl. Environ. Microbiol.* **48**, 1118–1122 (1984).
52. H. Yamada, S. Shimizu, and Y. Tani, in H.H. Weetall and G. Royer eds., *Enzyme Engineering*, vol. 5, Plenum, New York, 1980, pp. 405–411.
53. O.D. Bird and R.Q. Thompson, in P. Gyorgy and W.N. Pearson eds., *The Vitamins, Chemistry, Physiology, Pathology, Methods*, 2nd ed., vol. 7, Academic Press, New York, 1967, pp. 209–241.
54. R.K. Airas, *Methods Enzymol.* **122**, 33–35 (1986).

See also METABOLITES, PRIMARY AND SECONDARY.

PENICILLIN. See SECONDARY METABOLITES, ANTIBIOTICS.

PEPTIDE

HANS VON DÖHREN
HORST KLEINKAUF
Technical University Berlin
Berlin, Germany

KEY WORDS

Amino acids
Antibiotics
Biosynthesis
Hormones
Incorporation
Nonribosomal
Penicillin
Peptides
Processing
Synthetases

OUTLINE

INTRODUCTION

Peptides may be structurally defined as amino acid–derived compounds containing at least one amide (peptide) bond. Conventionally, a size limit is imposed for polypeptides of defined sequence above the range of 50 to 100 amino acids, which are termed proteins. Peptide classifications so far use either structural features, biological properties, or biosynthetic considerations (1,2). The major group of peptide antibiotics comprises structurally diverse types such as linear, cyclic and multicyclic peptides, peptidolactones, depsipeptides, and peptides modified with diverse nonpeptides moieties including acyl and aryl groups, polyketide or terpenoid-derived building blocks, or carbohydrate decorations (3). In addition, the biological properties of many compounds termed antibiotics have been shown to be diverse, thus extending the classical definition, which implies the inhibition or elimination of a certain type of organism (Table 1) (4).

From the point of view of peptide production, the bio-

Table 1. Selected Peptides of Current Interest

Compound	Structural type	Source	Properties/applications
Aspartame	Dipeptide	Synthetic	Sweetener
Glutathione	Tripeptide	Various	Detoxification and antioxidant compound
TRH	Modified tripeptide	*Homo sapiens* (hypothalamus)	Thyrotropin-releasing hormone, also stimulates prolactin release
Penicillin	Modified tripeptide	*Penicillium chrysogenum*	Antibacterial drug
Cephalosporin	Modified tripeptide	*Acremonium chrysogenum*	Antibacterial drug
Ergotpeptides	Modified tripeptide	*Claviceps* sp.	Uterine relaxant, treatment of parkinsonism, acromegaly, breast cancer
HC-toxin	Cyclotetrapeptide	*Helmintosporium carbonum*	Plant pathogenic (corn)
Enkephalins	Pentapeptides	*Homo sapiens* (brain)	Opioid activity
Actinomycin	Chromophore-attached pentapeptidolactone	*Streptomyces antibioticus*	Anticancer drug
Destruxin	Hexapeptidolactone	*Metarrhizium anisopliae*	Insecticidal peptide
Enniatin	Cyclohexadepsipeptide	*Fusarium* sp.	Antibacterial, antiviral, antihelminthic
Ferrichrome	Cyclohexapeptide	Various fungi	Siderophore
L-365,209	Cyclohexapeptide	Synthetic	Oxytocin antagonist, uterine relaxant
Vancomycins, ristocetins	Modified heptapeptides	Various Actinomycetes	Antibacterial drugs, antiviral compounds
Microcystins	Cycloheptapeptide	*Microcystis* sp. cyanobacteria	Hepatotoxic water contaminant
Lophyrotomin	Aryloctapeptide	*Lophyrotoma interrupta* (sawfly and other insects)	Hepatotoxin
Amanitins	Modified cyclooctapeptides	*Amanita phalloides*	Toxin
Surfactin	Octapeptidolactone	*Bacillus subtilis*	Surfactant, antifungal, antiviral, and antimycoplasma drug
Sandostatin	Modified octapeptide (disulfide)	Synthetic	Somatostatin agonist, treatment of acromegaly and carcinoid syndrome
PF-1022	Cyclooctadepsipeptide	*Mycelia sterilia*	Antihelminthic
Oxytocin	Nonapeptide (disulfide)	*Homo sapiens*	Smooth muscle contraction, principle birth hormone
Vasopressin	Nonapeptide amide(3 disulfides)	*Homo sapiens*	Control of diabetes insipidus
Desmopressin	Nonapeptide amide (disulfide)	Synthetic	Vasopressin analog, treatment of diabetes insipidus
Aureobasidin	Nonapeptidolactone	*Aureobasidium pullulans*	Antifungal drug
Gonadotropin-releasing hormone	Modified decapeptide Amide	*Homo sapiens* (hypothalamus)	Causes release of luteinizing hormone and follicle-stimulating hormone
Buserelin	Modified decapeptide	Synthetic	GnRH analog, treatment of prostate cancer
Tyrocidine	Cyclodecapeptide	*Bacillus brevis*	Topical antibacterial
Gramicidin S	Cyclodecapeptide	*Bacillus brevis*	Topical antibacterial, antihelminthic
Polymyxins	Heptacyclodecapeptide	*Bacillus polymyxa*	Antibacterial drug
Cyclosporin	Cycloundecapeptide	*Tolypocladium niveum*	Immunosuppressant, antifungal
Bacitracin	Modified heptacyclododecapeptide	*Bacillus licheniformis*	Antibacterial drug, proteinase inhibitor
Somatostatin	14-peptide (2 disulfides)	*Homo sapiens* (hypothalamus, gastrointestinal, tract, etc.)	Various gastrointestinal functions (antacid)
Gramicidin (linear)	Pentadecapeptide	*Bacillus brevis*	Antibacterial, ion-channel-forming
Tachyplesins	17- to 18-peptide	*Tachypleus* sp. (horseshoe crab) and related species	Antimicrobial compounds
Alamethicins	Modified nonadecapeptides	*Trichoderma viride*	Antibiotic, ion-channel-forming
Ranalexin	Modified 20-peptide (1 disulfide)	*Rana catesbeiana* (bullfrog, skin)	Antibacterial
Aborycin	Modified 21-peptide (3 disulfides)	*Streptomyces griseoflavus*	Antiviral (HIV-1)
Magainins	21- to 27-peptide	*Xenopus laevis* (skin)	Antimicrobial drugs
Melittin	26-peptide amide	Bee (venom)	Hemolytic compound
Ac-AMP	29-peptide (3 disulfides)	*Amaranthus canditus* (seeds)	Antimicrobial compound

Table 1. Selected Peptides of Current Interest *(continued)*

Compound	Structural type	Source	Properties/applications
Defensins	29- to 32-peptide	Mammalian (various sources)	Antimicrobial compounds
Human neutrophil peptide (HNP-1)	30-peptide	*Homo sapiens* (neutrophils)	Antibacterial, antifungal, cytotoxic
Cecropins	31-peptide	*Sus scrofa* (pig, small intestine)	Antibacterial (Gram negative)
Calcitonin	33-peptide amide	*Homo sapiens*	Treatment of hypercalcemia, osteoporosis
Subtilin	Modified 32-peptide	*Bacillus subtilis*	Antibacterial food preservative
Subtilosin	Modified 33-peptide	*Bacillus subtilis*	Antibacterial
Nisin	Modified 34-peptide	*Streptococcus lactis*	Antimicrobial food preservative
Mj-AMP	36-peptide	*Mirabilis jalapa* (seeds)	Antimicrobial compound
Cecropins (insect)	37- to 41-peptide amides	Various insects	Antibacterial compounds
Corticotropin (ACTH)	39-peptide	*Homo sapiens* (anterior pituitary gland)	Antibacterial compounds Control of glucocorticoid release
Thionins	34- to 47-peptide (3 or 4 disulfides)	Various plant sources	Antibacterial, antifungal
Defensins (plant)	45- to 51-peptide (4 disulfides)	Various plant sources	Antimicrobial compounds
AgaIVC	48-peptide	*Agelopsis aperta* (spider)	Venom, contains a D-Ser in position 48
Insulin	51 amino acids (2 chains, 3 disulfides)	Mammalian	Treatment of type I diabetes
Trefoil peptides	51–56 amino acids	Mammalian	Tissue repair and cell proliferation
Hirudin	65-peptide	Leeches (buccal gland)	Protease inhibitor, anticoagulant, antithrombotic agent

Note: See Ref. 11 for more extensive coverage.

synthetic origin is of primary importance. Although the ribosomal peptide biosynthetic system is an essential constituent of all self-replicating organisms, nonribosomal peptide forming systems, with few exceptions (bacterial cell wall peptides, glutathione), are not essential and are themselves gene products. So ribosomally derived peptides are directly encoded by gene transcripts, whereas nonribosmally derived compounds are the products of biocatalysts or multienzyme systems. To propagate each type of peptide in a cellular environment requires exceedingly different approaches. A decapeptide gene may be defined by a 30-bp reading frame and additional features permitting efficient transcriptional and translational processing. A biocatalyst forming a decapeptide requires approximately 40,000 bp of genetic information, and thus a considerably more complex genetic background (5,6).

Genome sequencing efforts have revealed the absence of such complex biosynthetic systems in most organisms investigated so far. These systems are especially prominent in microbial sources such as Actinomycetes, Bacilli, cyanobacteria, and filamentous fungi and are already being exploited for useful metabolites. Peptides of human origin, acting as hormones, neurotransmitters, growth factors, cytokines, and so forth are involved in the regulation of a variety of functions, including the central nervous system and cardiovascular, gastrointestinal, immunological, reproductive, and growth activities (7–10). Peptide therapeutics have been and are being developed for various treatments (Table 1).

STRUCTURAL FEATURES

The peptides compiled in Table 1 are products of different biosynthetic systems. In most cases, the current state of knowledge permits the prediction of each biosynthetic path from the structural features involved. This information permits the respective biotechnological (fermentation of microbial producers or hosts, cell culture), in vitro enzymatic, or chemical production. The new emerging approaches of combinatorial biosynthesis or synthetic peptide libraries are especially linked to biosynthetic considerations.

Biosynthesis

Peptides of biological origin differ in various structural features (Table 2) but also share the template-mediated assembly from carboxyl-activated amino acid precursors into linear precursor molecules, growing in the carboxy-terminal direction. Depending on the biosynthetic origin, the template is either nucleic acid (ribosomal) or protein (nonribosomal) (5).

Ribosomal Peptide Formation. Ribosomally formed peptides are directly gene encoded, and their mRNA template utilizes the available codons for linear assembly guided by the ribosomal architecture. Each amino acid is activated as aminoacyl-tRNA by a specific aminoacyl-tRNA synthetase; this process includes proofreading steps for structurally related compounds and achieves an accuracy of about 10^{-4} (12). Genes and their respective mRNAs encode polypeptides and proteins up to about 27,000 amino acids (titin), but the average size of translation products is about 300 amino acids.

Amino Acids. The translation process is restricted to the 20 protein amino acids and selenocysteine (13,14). Organisms may differ in their codon usage, and rare codons are thought to function in the control of both the transla-

Table 2. Structural Features of Peptides of Ribosomal and Nonribosomal Origin

	Ribosomal path	Nonribosomal path//multienzymes
Size (amino acids)	No size limitation, 2–20 residues Peptides, 20–50 residues Polypeptides, more than 50 residues Proteins, largest structures up to 30,000 residues	2 to about 50, 4 to 10 dominating; polymers up to about 1,000 residues
Amino acid constituents	21 protein amino acids (including selenocysteine) and modified amino acids	Various types of amino acids including 2-, 3- and 4-amino compounds (more than 300 known)
D-Amino acids	Not more than one, epimerized posttranslationally; in lantibiotics epimerization of several residues by dehydration and thioether formation	Often several, either incorporated directly, or epimerized during synthesis
Non–amino acid constituents in the peptide chain	Acyl residues, amines originating from decarboxylation	Various acyl residues, including aromatic acids, hydroxy acids
Cyclic structures	Rare, frequent disulfide cycles, often with several disulfide linkages; thioether cycles in lantibiotics (lanthionine)	More frequent than linear structures, various peptide bond cyclizations, but also lactones
Side-chain modifications	Hydroxylation, dehydration (Ser, Thr), side-chain cyclization (Cys to thiazoles, Thr to oxazoles, Glu to pyroGlu), frequently glycosylation	N-Methylation, hydroxylation, side-chain cyclization (Cys to thiazole), glycosylation, side-chain cross-linking (aromatic rings), glycosylation
Unusual constituents	Not known	Urea type of peptide bond, phosphoamino acids, amino-modified fatty acids–derived components (lipopeptides), mixed polyketide structures, terpenoid-derived groups
Biosynthesis	Gene can be identified; consider splicing, processing, and posttranslational modification; often prepropeptides detected	Nonribosomal enzyme systems present; peptide families are frequent in the same or related organisms; biosynthesis of rare precursors needed
Sources	Various animals and plants, sometimes bacteria	Mainly bacteria and lower fungi, occasionally plants and insects

tion rates and the folding of the primary intermediates (15,16). Heterologous expression sometimes needs codon adjustment for the intended host. Exploitation of this process to adapt synthetic nonprotein amino acids has been achieved in vitro by making use of nonutilized codons and their respective tRNAs (17). More than 100 mostly nonnatural amino acids have been introduced into peptides by biocatalytic procedures (18,19). Such procedures permit, for example, the direct incorporation of photofunctional amino acids (20). Current work focuses on the introduction of an engineered tRNA/aminoacyl-tRNA synthetase pair into *Escherichia coli*, delivering unnatural amino acids (21). D-Amino acids in ribosomally encoded peptides originate from the respective L-epimers, which are epimerized in chain by specific racemases first isolated from the spider *Agelopsis aperta* (22,23)

Peptide Processing. These primary translation products may undergo a variety of modification reactions. Besides hydrolytic removal of the initiating formyl-methionine residue (24), regions of the pre- or protranslation products may be involved in targeting and transport events, acting as signals recognized by the various proteins or nucleic acid species. Signal sequences may be removed by signal peptidases, and proteolytic processing may even continue; this is well known among hormones. Thus proopiomelanocortin gives rise to different peptides in different tissues, including ACTH, α-MSH, different forms of LPH, and β-endorphin. A rare event, now firmly established, is the joining of peptide fragments, which in analogy to the exon–intron structures of genes have been termed exteins and inteins (25). Cyclization of peptides by peptide bonds is likewise rare (26).

Side-Chain Alterations. Most common alterations besides N-terminal acylation (24) include hydroxylations of proline and lysine residues and the formation of regional specific disulfide bonds. Such processes may involve not only disulfide isomerases (27) but chaperones as well. Multiple disulfide bridges are found in peptides with up to four positionally conserved links. Such conformationally stabilized structures show antimicrobial (defensins, ranalexin, cecropins), antiviral (aborycin), and cytotoxic properties (human neutrophil peptides, HVP), or as the trefoil factor rITF from rat intestine, act as a covalently linked dimer in tissue repair and cell proliferation.

Side-chain modification of cysteine and serine residues to thiazole and oxazole derivatives are found in the gyrase inhibitor microcin and the rhizobial peptide trifolitoxin. A respective transforming enzyme complex has been characterized from the microcin B17 biosynthetic gene cluster (28). Stable thioether links between cysteine and serine or threonine side chains are prominent structural features of the lantibiotics (29–31). The name points to the unusual amino acids lanthionine and methyllanthionine formed upon hydrolysis. The respective biosynthetic clusters contain genes for enzymes catalyzing dehydration of serine or threonine to dehydroalanine and dehydrobutyrate, respectively, which undergo addition with thiols from adjacent cysteines, leading to inversion of the configuration.

In Vitro Applications of the Ribosomal System. Ribosomal peptide synthesis as a complex process involving about 150 proteins and ribonucleic acids and is being studied in detail (32). Functional in vitro systems are available from several organisms including *E. coli*, yeast, wheat germ, and rabbit reticulocytes (33). Such systems are especially useful for the production of antibacterial and cytotoxic peptides, which are not accessable by fermentation (34). Technology for mRNA production or the use of coupled transcription–translation systems introducing plasmids as informational molecules have been worked out (35). So far, peptide synthesis has been restricted to the laboratory scale, but yields up to the milligram per milliliter range have been reported. Efficiencies of such systems can be improved by cofactor regeneration (e.g., ATP). Synthesis of reporter molecules such as photoproteins permit the functional monitoring and the direct detection of, for example, translational inhibitors (36).

Nonribosomal Peptide Formation. Two basic types of peptide-forming systems account for most of the nonribosomal peptide formation structures known so far: single-step and multistep systems (Table 3). Single-step systems add amino acids to activated precursor amino acids or peptides; thus the precursors are directly activated, generally as phosphates by cleavage of ATP to ADP, and the amino acid to be added is in the free soluble state. The limit of performance appears to be the size of the substrates in relation to the enzymes involved. Covalently acting multistep systems add activated precursors to an activated amino acid, thus maintaining the activated state. In analogy to the ribosomal system, these multienzyme systems have no free intermediates, and product release involves catalysis as well (6).

Single-Step Systems. Prominent products of single-step systems are glutathione, coenzyme A, and peptides of the murein sacculus, including the uridin-diphosphomuramyl-N-acetyl precursor and various interpeptide bridges. Some steps involve aminoacyl-tRNA and are thus nucleic acid dependent (37). Enzymes involved can be identified at the gene level by characteristic motifs (38), but the class is heterogenous.

Single-step enzyme systems are obviously suited for the production of repeated sequences and polymers. Such examples include glutathion-related peptides involved in sequenstering metal ions. These have been termed phytochelatins and are induced in yeast or plant cells upon exposure to Cd^{2+}, Cu^{2+}, or Zn^{2+} (39). The compositions vary from (γGlu-Cys)nGly, (γGlu-Cys)nGlu, to (γGlu-Cys)n, with n ranging from 2 to 11 (40,41), and imply dipeptide condensations as a biosynthetic principle. Likewise, cyanophycin, a cyanobacterial storage polypeptide, is composed of the dipeptide units L-Arg-L-Asp. Synthetase and depolymerase have been identified in *Synechocystis* sp. and *Anabaena variabilis*, and the export of the polymerizing activity into *E. coli* has been demonstrated (42). Examples of amino acid polymers include γ-linked glutamate tails of folate (ranging in length from 1 to 11), β-lysine tails of nucleoside antibiotics such as nourseothricin, lysine polymers up to 25 residues in Streptomycetes, and γ-D-glutamyl capsular polymers in *Bacillus anthracis* and other spore-forming Bacilli. The molecular mechanisms of control of peptide and polypeptide chain length are unknown.

Multistep Systems. The covalent catalysis principle found in the multienzyme system closely resembles the ribosomal elongation principle. Equivalents to aminoacyl-tRNA and peptidyl-tRNA are aminoacyl and peptidyl carrier proteins. However, the protein system depends largely on fusing functional domains, which leads to multifunctional structures of impressive sizes. Grouping enzymes according to size, peptide- and polyketide-forming enzymes will occupy all top positions. On the other hand these synthetases will be at the bottom of the list if overall rates are compared. The largest known enzyme, cyclosporin synthetase, contains 41 functional domains fused into a 1.7-MDa protein. It integrates all 40 reactions, leading to a cycloundecapeptide with seven methylated peptide bonds (43). Peptide-forming systems currently under study have been compiled in Table 4.

Activation of Amino Acids and Other Carboxyls. Differing from polyketide systems, activation of amino acid carboxyl groups is an integrated function in peptide synthetases. The activation function is a preselection step in the events controlling fidelity of the peptide chain. Adenylate-forming domains share extensive homologies with acyl-CoA synthetases and the homologous insect luciferases, which likewise form acyl adenylates followed by transfer to the thiol group of CoA. Amino acid selection is controlled with varying selectivity: sometimes there is discrimination of leucine against valine or isoleucine, but occasionally all branched-chain types are accepted. Likewise phenylalanine might be discriminated against in favor of tyrosine and tryptophan, whereas some activation sites are known to accept various aromatic amino acids. Stereoselective discrimination of L- and D-residues is common, and both epimers may be direct precursors. In some cases both epimers are substrate, but the subsequent condensation reaction is stereocontrolled. Adenylates are thought to be stabilized by the interaction of the two subdomains of the adenylate domain. With low efficiency they may be subject

Table 3. Peptide-Forming Enzyme Systems

	Single-step systems	Multistep systems
Types of peptide made	Linear peptides, branched peptides	Linear and cyclic peptides, peptidolactones and depsipeptides; polymers?
Length of peptide made	2 to 5, or polymers	2 to 48 (?)
Activation of carboxyl groups	Phosphate or adenylate	Adenylate
Intermediates	Free intermediates (not clear for polymers)	Intermediates remain enzyme bound

Table 4. Nonribosomal Peptide Synthetase Systems

Peptide	Organism	Structural type	Gene(s) cloned	Enzymology
		Linear		
Bacilysin	*Bacillus subtilis*	P-2-M	(+)	(+)
Anguibactin	*Vibrio anguillarum*	R-P-2-M	+	–
ACV	*Streptomyces clavuligerus*	P-3	+	+
	Nocardia lactamdurans		+	+
	Lysobacter lactamgenus		+	+
	Aspergillus nidulans		+	+
	Penicillium chrysogenum		+	+
	Acremonium chrysogenum			
Bialaphos	*Streptomyces hygroscopicus*	P-3	+	+
Yersiniabactin	*Yersinia enterocolitica*	R-P-3-M	+	+
Phaseolotoxin	*Pseudomonas syringae* pv. ph.	P-4-M	(+)	–
Ardacin	*Kibdelosporangium aridum*	P-7-M	(+)	–
Chloroeremomycin	*Amycolatopsis orientalis*	P-7-M	+	–
Pyoverdin	*Pseudomonas aeruginosa*	R-P-8-M	+	–
Bleomycin	*Streptomyces verticillius*	P-8-M	+	–
Alamethicin	*Trichoderma viride*	R-P-19-M	–	+
		Cyclopeptides		
Enterobactin	*Escherichia coli*	P-C-E-3	+	+
HC-toxin	*Cochliobolus carbonum*	C-4	+	+
Tentoxin	*Alternaria alternata*	C-4	–	+
Echinocandin	*Aspergillus nidulans*	R-C-6	–	(+)
Microcystin	*Microcystis aeruginosa*	C-7	(+)	–
Iturin	*Bacillus subtilis*	C-8	(+)	–
Gramicidin S	*Bacillus brevis*	$C-(P-5)_2$	+	+
Tyrocidin	*Bacillus brevis*	C-10	+	+
Cyclosporin	*Tolypocladium niveum*	C-11-M	+	+
Mycobacillin	*Bacillus subtilis*	C-13	–	+
		Lactones		
Actinomycin	*Streptomyces chrysomallus*	$R-(L-5)_2-M$	+	+
Destruxin	*Metarhizium anisopliae*	L-6	+	(+)
Etamycin	*Streptomyces griseus*	R-L-7	–	(+)
Surfactin	*Bacillus subtilis*	L-8	+	+
Quinomycin	*Streptomyces echinatus*	$(R-P-4)_2$	–	+
R106	*Aureobasidium pullulans*	L-9	–	(+)
Syringomycin	*Pseudomonas syringae*	R-L-9	(+)	–
		Syringostatin		
SDZ90-215	*Septoria* sp.	L-10	–	+
		Depsipeptides		
Enniatin	*Fusarium* sp.	$C-(P_2)_3-M$	+	+
Beauvericin	*Beauveria bassiana*	$C-(P_2)_3-M$	–	+
PF1022	*Mycelia sterilia*	$C-(P_2)_4-M$	–	(+)
		Branched polypeptides		
Fengycin	*Bacillus subtilis*	P-10-C-7	+	–
Polymyxin	*Bacillus polymyxa*	P-10-C-7	–	+
Bacitracin	*Bacillus licheniformis*	P-12-C-7	+	+
Nosiheptide	*Streptomyces actuosus*	R-P-13-C-10-M	–	(+)
Thiostrepton	*Streptomyces laurentii*	R-P-17-C-10-M	–	(+)
		Branched peptidolactones		
Lysobactin	*Lysobacter* sp.	P-11-L-9	(+)	+
A21798A	*Streptomyces roseosporus*	R-P-13-L-10	(+)	+
A54145	*Streptomyces fradiae*	R-P-13-L-10	–	+
Tolaasin	*Pseudomonas tolaasii*	R-P-18-L-5	(+)	–

Note: Abbreviations: P, peptide; C, cyclopeptide; L, lactone; E, ester; R, acyl; M, modified. The structural types are defined by the number of amino, imino, or hydroxy acids in the precursor chain. The ring sizes of cyclic structures are indicated in the number following C, L, or E, defining the type of ring closure.

to aminolysis, leading to dipeptide formation without stereoselectivity (44).

Acyl and Peptidyl Carrier Domains. Adjacent to each adenylate domain is the carrier domain, homologous to the well-known acyl carrier protein of fatty acid biosynthesis. Carrier domains are posttranslationally modified with 4′-phosphopantetheine. This modification is catalyzed by specific CoA-4′-phosphopantetheine-protein transferases (45,46). The terminal cysteamine thiol of the cofactor is the acceptor of the aminoacyl moiety of the adenylate, with release of AMP. Thiol-bound intermediates can be isolated and characterized. This system property has led to the name thiotemplate mechanism.

Condensation Domains. The directed condensation of either two aminoacyl intermediates (initiation) or a peptidyl and an aminoacyl intermediate (elongation) is thought to take place at a condensation domain, interacting with the two adjacent carrier domains. This reaction is strictly stereospecific. If both epimers are present as intermediates because of the action of a reversible epimerase in the multienzyme, only the respective D-isomer is processed. The condensation domains are highly similar in structure to epimerization domains, catalyzing the conversion of thioester-attached aminoacyl or peptidylcarboxyl groups. Special types of condensation domains have been identified in systems forming thiazolidine-containing peptides such as bacitracin or yersiniabactin (47). These perform the cysteine side-chain modifications as well.

Termination Reactions. Termination of synthesis may proceed by hydrolysis (thioesterase required) or various modes of cyclization, leading to cyclic or branched cyclic peptides, peptidolactones, or cyclodepsipeptides. Flexibility of the process permits the processing of hydroxy acids as well, leading to depsipeptides, or the repeated use of templates, leading to dimers, trimers, or tetramers. Known processes are the dimerization of two pentapeptides in gramicidin S formation, the trimerization of dihydroxybenzoylseryl- or hydroxyisovaleryl-N-methylvalyl intermediates in enterobactin or enniatin synthesis, respectively, or the suspected tetramerization in the cases of bassianolide and PF1022. As in polyketide biosynthesis, repositioning of a terminating thioesterase domain may promote the release of intermediates (48). Both integrated and and associated thioesterase-like proteins are suspected to function in termination reactions (49).

Side Reactions. Two types of side reactions have been observed so far: the aminolysis of aminoacyladenylates as reactive mixed anhydrides, leading to dipeptides (44), or reactions of thioester intermediates. As in solid-phase peptide synthesis, certain steric conditions favor abortive cyclization reactions, especially of proline-containing or N-methylated dipeptides and ornithyl side chains. Less expected have been out-of-sequence dipeptide formations in ACV syntetase, actinomycin synthetase II (50), and cyclosporin synthetase. Such side reactions are promoted when lack of substrates prevents the completion of the entire catalytic cycle.

Reprogramming of Peptide Synthetases. The modular arrangement of most polycondensation pathways implies reprogramming by modular exchange. In the peptide field, functional hybrid synthetases have been constructed by module exchange between the gramicidin S synthetase system, ACV synthetase, and the surfactin synthetase system (51,52). At present the often low catalytic efficiencies of the constructs present a problem difficult to approach without the knowledge of the detailed spatial structures involved.

In Vitro Applications of Peptide Synthetases (53). The in vitro studies of peptide synthetases permitted an understanding of the catalytic principles involved. Small amounts of peptides have been produced in a variety systems (Table 3). Applying cyclosporin synthetase for preparative work, more than 100 cycosporin analogs have been prepared (43). Generally no more than two structural alterations have been found to be tolerated. Peptide synthetases obviously do not match the performance of chemical methods with regard to structural analog utilization because additional constraints are involved, including substrate selection and protein stability. However, peptide synthetases combine a number of advantages, including stereoselectivity and efficient cyclization. Another feature of special significance is the target-oriented product profile, which provides essential clues in structure–function work.

PEPTIDE PRODUCTION

Sources and Screening

Healing experiences and later ecological considerations have led to the concept of microbial warfare and eventually to the isolation of the first peptide antibiotics penicillin, tyrothricin (a mixture of linear gramicidin and tyrocidine), and gramicidin S. Although investigations of this type are continuing with the still-large reservoir of rare and difficult-to-cultivate microorganisms, new fields have been emerging that consider the defensive compounds from higher organisms. Pioneered in the fields of amphibian and insect defense strategies (54–57), antimicrobial peptides keep being discovered in mammalian and plant sources (58–61). In the field of secondary metabolites, microbial sources are generally exploited by fermentational procedures. Products of plants and animals have been investigated largely from chemical synthesis, but production can be achieved by recombinant DNA approaches (62,63).

Besides functional and chemical screening of extracts, genetic screening is now emerging as a tool to identify sources of presumably active natural products. The increasing knowledge of biosynthetic enzyme systems permits the probing of DNA isolates for the presence of peptide synthetase genes (64) or related modifying enzymes.

Fermentation. Submersed fermentation accounts for the majority of peptides produced from bacteria and filamentous fungi. Research has shifted from engineering approaches to detailed studies of the molecular mechanisms involved. Thus the productivity of cultures is increased by empirical media optimization, tight parameter control of the growth and production processes (O_2, CO_2, redox potential, cell morphology), and random and defined mutagenesis operations. The complexity of metabolite production is far from being understood.

The fermentation of β-lactam antibiotics represents the best-studied system (65). Penicillin biosynthesis is catalyzed by three enzymes: δ-(L-α-aminoadipyl)-l-cysteinyl-D-valine synthetase (ACVS), isopenicillin-N-synthase (IPNS), and acyltransferase (AAT). The respective genes are are combined in a biosynthetic cluster, which has been found to be amplified in production strains (66). Thus the gene copy number appears to be a major factor of productivity. However, production strains with a single copy of the cluster have also been selected as efficient producers. The regulation of expression is being currently studied, which affords the detailed understanding of the complex promoters involved (67). In this respect the utilization of autoregulators to induce the biosynthesis of specific metabolites is a fascinating tool already applied in peptide fermentation in Actinomycetes (68).

In addition, subcellular localization plays a significant role. Although IPNS is soluble and located in the cytosol, ACVS seems only partially soluble, and AAT is located in a peroxisome-like compartment, termed the microbody. The substrate transfer between ACVS and IPNS may involve complex formation and channeling, but the transfer of isopenicillin-N into the microbody, and the export of penicillins out of the microbody and through the cellular envelope, have not been studied so far. The role of molecular transport processes involved in biosynthetic processes is largely unknown. Export processes of bioactive compounds, especially peptides, seem also to be essential for self-protection, and thus critical for an intended overproduction. Peptide export proteins have been detected, and compounds such as cyclosporin are known to affect export systems including multi-drug-resistance proteins. Production processes utilize both product excretion and retainment in the microbial cells. Gramicidin S can accumulate up to 20% peptide within the cellular dry mass of the producer *Bacillus brevis*. Fermentation techniques to avoid feedback problems utilize solid-state approaches (69), microfiltration (70), or aqueous two-phase systems (71).

Finally, efficient substrate concentrations play a significant role in peptide fermentation. In special cases direct feeding of amino acid precursors may shift the product profile to a certain analog (72). Considering however, that the price of the penicillin precursor amino acid cystein exceeds that of the product, improvement focuses on the precursor biosynthetic systems, including uptake of basic nutrients such as glucose, ammonium, and sulfate.

Recombinant DNA Approaches in Peptide Production

The introduction of rDNA techniques permits the efficient production of various peptides independent of organ or animal sources, and thus free of viral contaminants. Current limits are related to certain posttranslational modification events, if relevant, which still have to be evaluated. An amidating enzyme system has been established (73), whereas specific glycosylation systems are still unavailable. So current production routes employ, for example, yeast or insect cell cultures that often contain similar modification systems. Prominent examples of peptides currently produced include insulin (74) and hirudin, but the vast majority of products are proteins.

Enzymatic Procedures in Peptide Production

Introduced more than 50 years ago, the reversal of proteinase reactions has been widely used in peptide synthesis (75). Strategies exist for regio- and stereospecific synthesis of peptide bonds (76,77), and new trends consider heterogeneous media (78) or even frozen aqueous solutions (79). Generally all procedures permit the construction of a single, specified peptide bond, but fragment condensations and protein modifications have been established in peptide production as well. Substrates to be converted generally need protection groups to avoid side reactions.

BIBLIOGRAPHY

1. T. Wieland and M. Bodanszky, *The World of Peptides*, Springer-Verlag, Berlin, 1991.
2. E. Gross, J. Meienhofer, and S. Udenfriend eds., *The Peptides: Analysis, Synthesis, Biology*, vols. I–IX, Academic, New York, 1979–1987.
3. H. Kleinkauf and H. von Döhren eds., *Peptide Antibiotics*, de Gruyter, Berlin, 1990.
4. W. Strohl ed., *Biotechnology of Antibiotics*, Marcel Dekker, New York, 1997.
5. H. Kleinkauf and H. von Döhren, *Eur. J. Biochem.* **236**, 335–351 (1996).
6. H. von Döhren, U. Keller, J. Vater, and R. Zocher, *Chem. Rev.* **97**, 2675–2705 (1997).
7. R.C. Hider and D. Barlow, eds., *Polypeptide and Protein Drugs: Production, Characterization and Formulation*, Ellis Horwood, Chichester, 1991.
8. A. Negro-Vilar and P.M. Conn eds., *Peptide Hormones: Effects and Mechanisms of Action*, vols. I–III, CRC Press, Boca Raton, Fla., 1988.
9. W.V. Williams and D.B. Weiner eds., *Biologically Active Peptides: Design, Synthesis and Utilization*, Technomic, Lancaster, Pa., 1993.
10. D.J. Ward, *Peptide Pharmaceuticals: Approaches to the Design of Novel Drugs*, Open University Press, Buckingham, England, 1991.
11. H. Kleinkauf and H. von Döhren, in H.-J. Rehm and G. Reed eds., *Biotechnology*, 2nd ed., vol. 7, Verlag Chemie, Weinheim, 1997, pp. 277–322.
12. J.G. Arnez and D. Moras, *Trends Biochem. Sci.* **22**, 211–216 (1997).
13. T.C. Stadtman, *Annu. Rev. Biochem.* **65**, 83–100 (1996).
14. S.C. Low and M.J. Berry, *Trends Biochem. Sci.* **21**, 203–208 (1996).
15. X. Xia, *Genetics* **144**, 1309–1320 (1996).
16. A.A. Adzubei, I.A. Adzubei, I.A. Krashennikov, and S. Neidle, *FEBS Lett.* **399**, 78–82 (1996).
17. M. Ibba, *Biotechnol. Genet. Eng. Rev.* **13**, 197–216 (1996).
18. J.D. Bain, C. Switzer, A. Chamberlin, and S. Benner, *Nature* **356**, 537–539 (1992).
19. D. Mendel, V.W. Cornish, and P.G. Schultz, *Annu. Rev. Biophys. Biomol. Struct.* **24**, 435–462 (1995).
20. T. Hohsaka, K. Sato, M. Sisido, K. Takai, and S. Yokoyama, *FEBS Lett.* **344**, 171–174 (1994).
21. D.R. Liu, T.J. Magliery, M. Pastrnak, and P.G. Schultz, *Proc. Natl. Acad. Sci. U.S.A.* **94**, 10092–10097 (1997).
22. G. Kreil, *Annu. Rev. Biochem.* **66**, 337–346 (1997).

23. S.D. Heck, W.S. Faraci, P.R. Kelbaugh, N.A. Saccomano, P.F. Tadeio, and R.A. Volkmann, *Proc. Natl. Acad. Sci.* **93**, 4036–4039 (1995).
24. R.A. Bradshaw, W.W. Brickley, and K.W. Walker, *Trends Biochem. Sci.* **23**, 263–267 (1998).
25. A.A. Cooper and T.H. Stevens, *Trends Biochem. Sci.* **20**, 351–356 (1995).
26. M. Martinez-Bueno, M. Maqueda, A. Gàlvez, B. Samyn, J. van den Beeumen, J. Coyette, and E. Valdivia, *J. Bacteriol.* **176**, 6334–6339 (1994).
27. J.M. Luz and W.J. Lennarz, *EXS 77*, 97–117 (1996).
28. Y.M. Li, J.C. Milene, L.L. Madison, R. Kolter, and C.T. Walsh, *Science* **274**, 1188–1193
29. R. Jack, F. Götz, and G. Jung, in H.-J. Rehm and G. Reed eds., *Biotechnology*, 2nd ed., Vol. 7, Verlag Chemie, Weinheim, 1997, pp. 323–370.
30. R.W. Jack, J.R. Tagg, and B. Ray, *Microbiol. Rev.* **59**, 171–200 (1995).
31. H.-G. Sahl, R.W. Jack, and G. Bierbaum, *Eur. J. Biochem.* **230**, 827–853 (1995).
32. K.H. Nierhaus, F. Fransceschi, A.R. Subramanian, V.A. Erdmann, and B. Wittmann-Liebold eds., *The Translational Apparatus*, Plenum, New York, 1993.
33. G.S. Beckler, D. Thompson, and T. Van Osbree, *Methods Mol. Biol.* **37**, 215–232 (1995).
34. W. Stiege and V.A. Erdmann, *J. Biotechnol.* **41**, 81–90 (1995).
35. L. Ryabova, E. Volianik, O. Kurnasov, A. Spirin, Y. Wu, and F.R. Kramer, *J. Biol. Chem.* **269**, 1501–1505 (1994).
36. S.V. Matveev, B.A. Illarionov, E.S. Vysotski, V.S. Bondar, S.V. Markova, and Y.B. Alakhov, *Anal. Biochem.* **231**, 34–39 (1995).
37. J.-M. Ghuysen and R. Hakenbeck eds., *Bacterial Cell Wall*, Elsevier, Amsterdam, 1994.
38. S.S. Eveland, D.L. Pompliano, and M.S. Anderson, *Biochemistry* **36**, 6223–6229 (1997).
39. W.E. Rauser, *Annu. Rev. Biochem.* **59**, 61–86 (1990).
40. J.C. Steffens, *Annu. Rev. Plant Physiol. Plant Mol. Biol.* **41**, 533–575 (1990).
41. P. Meuwly, P. Thibault, A.L. Schwan, and W.E. Rauser, *Plant J.* **7**, 391–400 (1995).
42. K. Ziegler, A. Diener, C. Herpin, R. Richter, R. Deutzmann, and W. Lockau, *Eur. J. Biochem.* **254**, 154–159 (1998).
43. H. Kleinkauf and H. von Döhren, in T. Anke ed., *Fungal Biotechnology*, Chapmann & Hall, Weinheim, 1997, pp. 147–158.
44. M.F. Byford, J.E. Baldwin, C.-Y. Shiau, and C.J. Schofield, *Chem. Rev.* **97**, 2631–2649 (1997).
45. C.T. Walsh, A.M. Gehring, P.H. Weinreb, L.E.N. Quadri, L.E.N., and R.S. Flugel, *Curr. Opin. Chem. Biol.* **1**, 309–315 (1997).
46. L.E.N. Quadri, P.H. Weinreb, M. Lei, M.M. Nakano, P. Zuber, and C.T. Walsh, *Biochemistry* **37**, 1585–1595 (1998).
47. D. Konz, A. Klens, K. Schörgendörfer, and M.A. Marahiel, *Chem. Biol.* **4**, 927–937 (1997).
48. F. de Ferra, F. Rodriguez, O. Tortora, C. Tosi, and G. Grandi, *J. Biol. Chem.* **272**, 25304–25309 (1997).
49. A. Schneider and M.A. Marahiel, *Arch. Microbiol.* **169**, 404–410 (1998).
50. F. Schauwecker, F. Pfennig, W. Schröder, and U. Keller, *J. Bacteriol.* **180**, 2468–2474 (1998).
51. T. Stachelhaus, A. Schneider, and M.A. Marahiel, *Science* **269**, 69–72 (1995).
52. A. Schneider, T. Stachelhaus, and M.A. Marahiel, *Mol. Gen. Genet.* **257**, 308–318 (1998).
53. H. Kleinkauf and H. von Döhren, *Acta Biochim. Polonica* **44**, 839–848 (1997).
54. H.G. Boman, J. Marsh, and J.A. Goodie eds., *Antimicrobial Peptides*, Ciba Foundation Symposium 186, Wiley, Chichester, 1994.
55. C.L. Bevins and M. Zasloff, *Annu. Rev. Biochem.* **59**, 395–414 (1990).
56. B.L. Kagan, T. Ganz, and R.I. Lehrer, *Toxicology* **87**, 131–149 (1994).
57. H.G. Boman, *Annu. Rev. Immunol.* **13**, 61–92 (1995).
58. R.I. Lehrer, A.K. Lichtenstein, and T. Ganz, *Annu. Rev. Immunol.* **11**, 105–128 (1993).
59. W.L. Maloy and U.P. Kari, *Biopolymers* **37**, 105–122 (1995).
60. R.E.W. Hancock, T. Falla, and M. Brown, *Adv. Microb. Physiol.* **37**, 135–175 (1995).
61. W.F. Broekaert, B.P.A. Commue, M.F.C. De Bolle, K. Thevissen, G.W. De Samblanx, and R.W. Osborn, *Crit. Rev. Plant Sci.* **16**, 297–323 (1997).
62. C. Binnie, J.D. Cossar, and D.I. Stewart, *Trends Biotechnol.* **15**, 315–320 (1997).
63. G. Gelissen, C.P. Hollenberg, and Z.A. Janowicz, *Bioprocess Technol.* **22**, 195–239 (1995).
64. K. Turgay and M.A. Marahiel, *Peptide Res.* **7**, 238–240 (1994).
65. J. Nielsen, *Physiological Engineering Aspects of Penicillium chrysogenum*, Polyteknisk Forlag, Lyngby, Denmark, 1995.
66. R.W. Newbert, B. Barton, P. Greaves, J. Harper, and G. Turner, *J. Ind. Microbiol. Biotechnol.* **19**, 18–27 (1997).
67. A. Brakhage, *FEMS Microbiol. Lett.* **148**, 1–10 (1997).
68. Y.K. Yang, M. Morikawa, H. Shimizu, S. Shioya, K.-I. Suga, T. Nihira, and Y. Yamada, *Biotechnol. Bioeng.* **49**, 437–444 (1996).
69. A. Ohno, T. Ano, and M. Shoda, *Process Biochem.* **31**, 801–806 (1996).
70. M. Taniguchi, K. Hoshino, H. Urasaki, and M. Fujii, *J. Ferment. Bioeng.* **78**, U122 (1996).
71. R. Kuboi, T. Maruki, H. Tanaka, I. Komasawa, *J. Ferment. Bioeng.* **78**, 431–436 (1994).
72. R. Thiericke, and R. Rohr, *Nat. Prod. Rep.* **10**, 263–289 (1993).
73. S.T. Prigge, A.S. Kolhekar, B.A. Eipper, R.E. Mains, and L.M. Amzel, *Science* **278**, 1300–1305 (1997).
74. M.R. Ladisch and K.L. Kohlmann, *Biotechnol. Prog.* **8**, 469–478 (1992).
75. J.S. Fruton, *Adv. Enzymol.* **53**, 239–305 (1982).
76. I. Gill, R. Lopez-Fandino, X. Jorba, and E.N. Vulfson, *Enzyme Microb. Technol.* **18**, 163–183 (1996).
77. J. Bongers and E.P. Heimer, *Peptides* **45**, 183–193 (1995).
78. I. Gill and E. Vulfson, *Trends Biotechnol.* **12**, 118–122.
79. M. Hansler and H.D. Jakubke, *J. Peptide Sci.* **2**, 279–289 (1996).

PHENYLALANINE

IAN G. FOTHERINGHAM
NSC Technologies
Mount Prospect, Illinois

KEY WORDS

Amino acid
Ammonia-lyase
Aromatic
Biosynthesis
Biocatalyst
Biotransformation
Fermentation
Phenylalanine
Resolution
Transaminase

OUTLINE

INTRODUCTION

The amino acid phenylalanine is currently manufactured worldwide at an annual scale exceeding 12,000 metric tons. The vast majority of this material is L-phenylalanine, synthesized specifically for incorporation into the dipeptide sweetener aspartame. Production of L-phenylalanine grew rapidly during the 1980s in parallel with the demand for aspartame and has grown steadily in recent years. L-Phenylalanine is also used in additional food and medical applications and is in increasing demand in pharmaceutical development as a chiral intermediate or as a precursor of chiral auxiliaries such as benzyloxazolidinone. In contrast, only a few metric tons of D-phenylalanine are used annually, also in pharmaceutical drug development, although this application is increasing. During the past two decades, a number of manufacturers have developed chemoenzymatic and purely biological routes for L-phenylalanine manufacture, as shown in Table 1. Chemoenzymatic syntheses have relied on resolution technology, applied to chemically prepared racemic phenylalanine derivatives or DL-5-monosubstituted hydantoins and biotransformation routes from chemically prepared achiral precursors of L-phenylalanine. Purely biological routes have concentrated on direct synthesis of L-phenylalanine through microbial fermentative means. Today, the largest and most commercially successful synthetic routes to L-phenylalanine production use large-scale fermentation of bacterial strains that overproduce L-phenylalanine. Such microorganisms have been isolated through a combination of classical mutagenesis selection procedures and molecular genetic manipulation, the latter deriving from a detailed understanding of the molecular biology of the pathways involved in bacterial L-phenylalanine biosynthesis. Conversely, the present commercial production of D-phenylalanine relies solely on chemoenzymatic resolution.

L-PHENYLALANINE-OVERPRODUCING MICROORGANISMS

Common Aromatic and L-Phenylalanine Biosynthetic Pathways

Efforts to develop L-phenylalanine-overproducing organisms have been vigorously pursued by Nutrasweet Company, Ajinomoto, Kyowa Hakko Kogyo, and others. The focus has centered on bacterial strains that have previously demonstrated the ability to overproduce other amino acids. Such organisms include principally the coryneform bacteria, *Brevibacterium flavum* (1,2) and *Corynebacterium glutamicum* (3,4) used in L-glutamic acid production. In addition, *Escherichia coli* (5) has been extensively studied in L-phenylalanine manufacture because of the detailed characterization of the molecular genetics and biochemistry of

Table 1. Commercial Processes for L-Phenylalanine Synthesis

Manufacturer	L-Phenylalanine processes developed
Ajinomoto	Microbial fermentation
Biotechnica	Microbial fermentation
Degussa/Rexim	Enzymatic resolution (aminoacylase)
DSM	Enzymatic resolution (amino acid amidase)
Genex	Biotransformation (phenylalanine ammonia lyase)
Kyowa Hakko	Microbial fermentation
Miwon	Microbial fermentation
Nutrasweet Co.	Microbial fermentation, biotransformation (aminotransferase)
PEI	Biotransformation (aminotransferase)
Tanabe Seiyaku	Enzymatic resolution, biotransformation (aminoacylase, aminotransferase)

its aromatic amino acid pathways and its amenability to recombinant DNA methodology. The biochemical pathway that results in the synthesis of L-phenylalanine from chorismate is identical in each of these organisms (Fig. 1**b**); the common aromatic pathway to chorismate is shown in Figure 1**a**. In each case, the L-phenylalanine biosynthetic pathway comprises three enzymatic steps from chorismic acid, the product of the common aromatic pathway. The precursors of the common aromatic pathway, phosphoenolpyruvate (PEP) and erythrose-4-phosphate, derive from the glycolytic and pentose phosphate pathways of sugar metabolism, respectively. Although the intermediate compounds in both pathways are identical in each of the organisms, there are differences in the organization and regulation of the genes and enzymes involved (2,6,7). Nevertheless, the principal points of pathway regulation are very similar in each of the three bacteria (4,6,8). Conversely, tyrosine biosynthesis, which is also carried out in three biosynthetic steps from chorismate, proceeds through different intermediates in *E. coli* than the coryneform organisms (6,8).

Regulation of phenylalanine biosynthesis occurs both in the common aromatic pathway and in the terminal phenylalanine pathway, and the vast majority of efforts to deregulate phenylalanine biosynthesis have focused on two specific rate-limiting enzymatic steps. These are the steps of the aromatic and the phenylalanine pathways carried out, respectively, by the enzymes 3-deoxy-D-arabinoheptulosonate-7-phosphate (DAHP) synthase and prephenate dehydratase. Classical mutagenesis approaches using toxic amino acid analogues and the molecular cloning of the genes encoding these enzymes have led to very significant increases in the capability of host strains to overproduce L-phenylalanine. This has resulted from the elimination of the regulatory mechanisms controlling enzyme synthesis and specific activity. The cellular mechanisms that govern the activity of these particular enzymes are complex and illustrate many of the sophisticated means by which bacteria control gene expression and enzyme activity. Chorismate mutase, the first step in the phenylalanine-specific pathway, and the shikimate kinase activity of the common aromatic pathway are subject to a lesser degree of regulation and have also been characterized in detail (9,10).

Classical Mutagenesis and Selection

Classical methods of strain improvement have been widely applied in the development of phenylalanine overproducing organisms (5,11,12,13). Tyrosine auxotrophs have frequently been used in efforts to increase phenylalanine production through mutagenesis. These strains often already overproduce phenylalanine because of the overlapping nature of tyrosine and phenylalanine biosynthetic regulation (3,14,15). Limiting tyrosine availability leads to partial genetic and allosteric deregulation of common biosynthetic steps (16). Such strains have been subjected to a variety of mutagenesis procedures to further increase the overall titer and the efficiency of phenylalanine production. In general, this has involved the identification of mutants that display resistance to toxic analogues of phenylalanine or tyrosine, such as β-2-thienyl-DL-alanine and p-fluoro-DL-phenylalanine (3,5,17). Such mutants can be readily identified on selective plates in which the analogue is present in the growth medium. The mutations responsible for the phenylalanine overproduction have frequently been located in the genes encoding the enzymatic activities of DAHP synthase, chorismate mutase, and prephenate dehydratase. In turn, this has prompted molecular genetic approaches to further increase phenylalanine production through the isolation and in vitro manipulation of these genes, as described next.

Deregulation of DAHP Synthase Activity

The activity of DAHP synthase commits carbon from intermediary metabolism to the common aromatic pathway converting equimolar amounts of PEP and E4P to DAHP (18). In *E. coli* there are three isoenzymes of DAHP synthase of comparable catalytic activity encoded by the genes *aroF*, *aroG*, and *aroH* (19). Enzyme activity is regulated by the aromatic amino acids tyrosine, phenylalanine, and tryptophan, respectively (16,20–23). In each case, regulation is mediated both by repression of gene transcription and by allosteric feedback inhibition of the enzyme, though to different degrees.

The *aroF* gene lies in an operon with *tyrA*, which encodes the bifunctional protein chorismate mutase–prephenate dehydrogenase (CMPO). Both genes are regulated by the TyrR repressor protein complexed with tyrosine. The *aroF* gene product accounts for 80% of the total DAHP synthase activity in wild-type *E. coli* cells. The *aroG* gene is repressed by the TyrR repressor protein complexed with phenylalanine and tryptophan. Repression of *aroH* is mediated by tryptophan and the TrpR repressor protein (20). The gene products of *aroF* and *aroG* are almost completely feedback inhibited respectively by low concentrations of tyrosine or phenylalanine (24), whereas the *aroH* gene product is subject to maximally 40% feedback inhibition by tryptophan (24). In *B. flavum*, DAHP synthase forms a bifunctional enzyme complex with chorismate mutase and is feedback inhibited by tyrosine and phenylalanine synergistically but not by tryptophan (25,26). Similarly, in *C. glutamicum*, DAHP synthase is inhibited most significantly by phenylalanine and tyrosine acting in concert (27), but unlike *B. flavum* reportedly does not show tyrosine-mediated repression of transcription (25,28).

Many examples of analogue-resistant mutants of these organisms display reduced sensitivity of DAHP synthase to feedback inhibition (3,5,7,25,29,30). In *E. coli*, the genes encoding the DAHP synthase isoenzymes have been characterized and sequenced (6,31,32). The mechanism of feedback inhibition of the *aroF*, *aroG*, and *aroH* isoenzymes has been studied in considerable detail (32), and variants of the *aroF* gene on plasmid vectors have been used to increase phenylalanine overproduction. Simple replacement of transcriptional control sequences with powerful constitutive or inducible promoter regions and the use of high copy number plasmids has readily enabled overproduction of the enzyme (33,35), and reduction of tyrosine-mediated feedback inhibition has been described using resistance to the aromatic amino acid analogues β-2-thienyl-DL-alanine and p-fluoro-DL-phenylalanine (33,35).

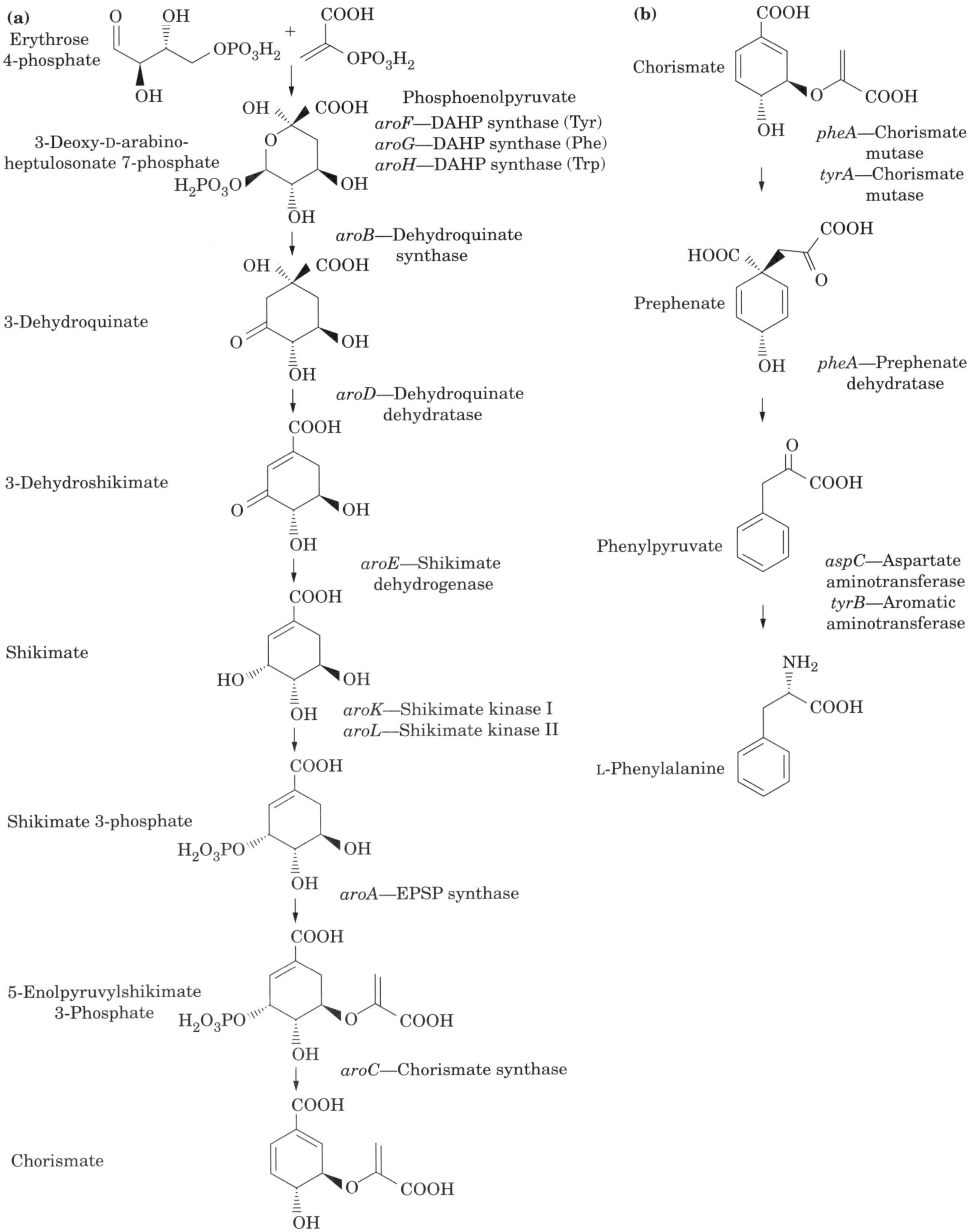

Figure 1. (**a**) The common aromatic pathway to chorismate in *Escherichia coli* K12. The enzymes responsible for each step are indicated to the right preceded by the encoding gene mnemonic. (**b**) The L-phenylalanine biosynthetic pathway from chorismate to L-phenylalanine in *Escherichia coli* K12.

Deregulation of CMPD

The three enzymatic steps by which chorismate is converted to phenylalanine appear to be identical among *C. glutamicum*, *B. flavum*, and *E. coli*, although only in *E. coli* have detailed reports appeared upon the characterization of the genes involved. In each case, the principal regulatory step is that catalyzed by prephenate dehydratase. In *E. coli*, chorismate is converted first to prephenate and then to phenyl pyruvate by the action of a bifunctional enzyme, CMPD, encoded by the pheA gene (6). The final step, in which phenyl pyruvate is converted to L-phenylalanine, is carried out predominantly by the aromatic aminotransferase encoded by *tyrB* (36). However, both the aspartate aminotransferae, encoded by *aspC* and the branched-chain aminotransferase encoded by *ilvE*, can efficiently catalyse this reaction (37). Phenylalanine biosynthesis is regulated by control of CMPD through phenylalanine-mediated attenuation of *pheA* transcription (6) and by feedback inhibition of the prephenate dehydratase and chorismate mutase activities of the enzyme. Inhibition is most pronounced upon the prephenate dehydratase activity, with almost total inhibition observed at micromolar phenylalanine concentrations (38,39). Chorismate mutase activity in contrast is maximally inhibited by only 40% (39). In *B. flavum*, prephenate dehydratase and chorismate mutase are encoded by distinct genes. Prephenate dehydratase is again the principal point of regulation, with the enzyme subject to feedback inhibition, but not transcriptional repression, by phenylalanine (40,41). Chorismate mutase, which in this organism forms a bifunctional complex with DAHP synthase, is maximally inhibited by phenylalanine and tyrosine to 65%, but this is significantly diminished by the presence of very low levels of tryptophan (42). Expression of chorismate mutase is repressed by tyrosine (29,43). The final step is carried out by at least one transaminase (44). Similarly, in *C. glutamicum*, the activities are encoded in distinct genes, with prephenate dehydratase again being more strongly feedback-inhibited by phenylalanine (45–47). The only transcriptional repression reported is that of chorismate mutase by phenylalanine. The *C. glutamicum* genes encoding prephenate dehydratase and chorismate mutase have been isolated and cloned from analogue-resistant mutants of *C. glutamicum* (8,48) and used along with the cloned DAHP synthase gene to augment L-phenylalanine biosynthesis in overproducing strains of *C. glutamicum* (48).

Many publications and patents have described successful efforts to reduce and eliminate regulation of prephenate dehydratase activity in phenylalanine overproducing organisms (5,9,34,35). As with DAHP synthase, the majority of the reported efforts have focused on the *E. coli* enzyme encoded by the *pheA* gene that is transcribed convergently with the tyrosine operon on the *E. coli* chromosome (6). The detailed characterization of the *pheA* regulatory region has facilitated expression of the gene in a variety of transcriptional configurations, leading to elevated expression of CMPD, and a number of mutations in *pheA* that affect phenylalanine-mediated feedback inhibition have been described (9,33,49). Increased expression of the gene is readily achieved by cloning *pheA* onto multicopy plasmid vectors and deletion of the nucleotide sequences comprising the transcription attenuator (Fig. 2**a**). Most mutations that affect the allosteric regulation of the enzyme by phenylalanine have been identified through resistance to phenylalanine analogues such as β-2-thienylalanine, but there are examples of feedback-resistant mutations arising through insertional mutagenesis and gene truncation (33,49). Two regions of the enzyme in particular have been shown to reduce feedback inhibition to different degrees. Mutations at position Trp338 in the peptide sequence desensitize the enzyme to levels of phenylalanine in the 2 to 5 mM range but are insufficient to confer resistance to higher concentrations of L-phenylalanine (33,49). Mutations in the region of residues 304 to 310 confer almost total resistance to feedback inhibition at L-phenylalanine concentrations of at least 200 mM (9). Feedback inhibition profiles of four such variants (JN305–JN308) are shown in Figure 2**b**, in comparison to the profile of wild-type enzyme (JN302). It is not clear if the mechanism of resistance is similar in either case, but the difference is significant to commercial application because overproducing organisms readily achieve extracellular concentrations of L-phenylalanine over 200 mM.

Precursor Supply

Rate-limiting steps in the common aromatic and phenylalanine biosynthetic pathways are obvious targets in the development of phenylalanine overproducing organisms. However, the detailed biochemical and genetic characterization of *E. coli* has enabled additional areas of its metabolic function to be specifically manipulated to determine their effect on aromatic pathway throughput. Besides efforts to eliminate additional lesser points of aromatic pathway regulation, attempts have been made to enhance phenylalanine production by increasing the supply of aromatic pathway precursors and by facilitating exodus of L-phenylalanine from the cell. The precursors of the common aromatic pathway D-erythrose 4-phosphate and PEP are the respective products of the pentose phosphate and glycolytic pathways. Precursor supply in aromatic amino acid biosynthesis has been reviewed very recently (50,51). Theoretical analyses of the pathway and the cellular roles of these metabolites suggest that the production of aromatic compounds is likely to be limited by PEP availability (52,53) because PEP is involved in a number of cellular processes, including the generation of metabolic energy through the citric acid cycle (54) and the transport of glucose into the cell by the phosphotransferase system (55). Strategies to reduce the drain of PEP by these processes have included mutation of sugar transport systems to reduce PEP-dependent glucose transport (56) and modulation of the activities of pyruvate kinase, PEP synthase, and PEP carboxylase, which regulate PEP flux to pyruvate, oxaloacetate, and the citric acid cycle (50,54,57). Similarly, the availability of E4P has been increased by altering the levels of transketolase, the enzyme responsible for E4P biosynthesis (58). In general, these efforts have been successful in directing additional flux of PEP of E4P into the aromatic pathway, although their effect has not proved to be as predictable as the deregulation of rate-limiting

(a)

−35 −10

TGTATCGCCAACGCGCCTTCGGGCGCG**TTTTTTGTTGACAGCGTGAAAACAGTACGGGTA**CTG**TACT**

AAAGTCACTTAAGGAAACAAACATGAAACACATACCGTTTTTCTTCGCATTCTTTTTTACCTTCCCC

Attenuator Region

TGAATGGGAGGCGTTTCGTCGTGTGAAACAGAATGCGAAGACGAACAATAAGGCCTCCCAAATCGGG

GGGCCTTTTTTATTGATA**ACAAAAAGGCAACACTATGACATCGGAAAACCCGTTACTGGCGCT**

−35 −10

M T S E N P L L A L

GAATTCTTTTTTGTTGACAGCGTGAAAACAGTACGGGTATAATACTAAAGTCACAAAAAGGCAACACTATGACATCGGAAAACCCGTTACTGGCGCT

Coding sequence

(b)

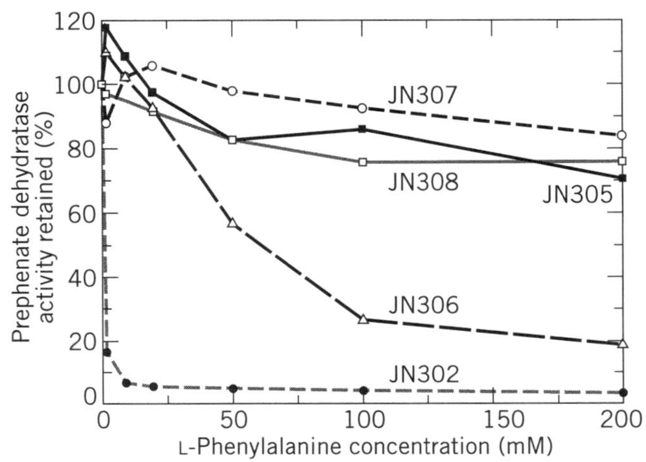

Figure 2. (**a**) The promoter region of the *E. coli* K12 *pheA* gene. The sequence of the wild-type promoter region is shown in the top part of the figure. The −35 and −10 hexamers are indicated. Bases retained in the deregulated promoter are in bold. The sequence involved in the attenuator region is shown underlined. The sequence of the deregulated *pheA* promoter region used to produce chorismate mutase/prephenate dehydratase (10) is shown in the bottom of the figure. Altered bases in the −10 region are shown underlined. (**b**) L-Phenylalanine-mediated feedback inhibition of wild-type *E. coli* K12 prephenate dehydratase (JN302) and four inhibition resistant enzyme variants (JN305-JN308). Activity is expressed as a percentage of wild-type enzyme activity.

pathway steps, and their overall impact on L-phenylalanine overproduction has not been well characterized.

Secretion of Phenylalanine from the Cell

Exodus of phenylalanine is of manifest importance in the large-scale production of phenylalanine because the amino acid is typically recovered only from the extracellular medium and washed cells. Lysis of cells to recover additional phenylalanine is not generally practical, and so L-phenylalanine remaining within the cells after washing is usually lost. Because cell biomass is extremely high in large-scale fermentations, this can represent up to 5% of total phenylalanine produced. Additionally, because L-phenylalanine is known to regulate its own biosynthesis at multiple points through feedback inhibition, attenuation, and TyrR-mediated repression, methods to reduce intracellular concentrations of phenylalanine through reduced uptake or increased exodus throughout the fermentation have been sought. Studies have addressed the means by which L-phenylalanine is both taken up and excreted by bacterial cells and whether this can be altered to increase efflux. In overproducing strains, it is likely that most phenylalanine leaves the cell by passive diffusion but specific uptake and exodus pathways also exist. In *E. coli*, L-phenylalanine is taken up by at least two permeases (59), one specific for phenylalanine encoded by *pheP* and the other encoded by *aroP* a general aromatic amino acid permease responsible for the transport of tyrosine and tryptophan in addition to phenylalanine. The genes encoding the permeases have been cloned and sequenced, and *E. coli* mutants deficient in either system have been characterized (60,61). The effect of an *aroP* mutation has been evaluated in L-phenylalanine overproduction (5). In addition, specific export systems have been identified that although not completely characterized have been successfully used to increase exodus of L-phenylalanine from strains of *E. coli* (62,63). In one such system, the Cin invertase of bacteriophage P1 has been shown to induce a metastable phenylalanine hypersecreting phenotype upon a wild-type strain of *E. coli*. The phenotype can be stabilized by transient introduction of the invertase on a temperature-sensitive plasmid replicon and is sufficient to establish L-phenylalanine overproduction in the absence of any alterations to the normal biosynthetic regulation (63).

Summary

The incremental gains made from the various levels at which phenylalanine biosynthesis has been addressed has led to the high-efficiency production strains currently in use today. Almost all large-scale L-phenylalanine manufacturing processes in present operation are fermentations using bacterial strains such as those described in this article in which classical strain development or molecular genetics have been extensively applied to bring about phenylalanine overproduction. The low substrate costs and the economics of scale associated with this approach have resulted in significant economic advantages. However, a number of alternate biotransformation and chemoenzymatic resolution strategies have been extensively developed that have also demonstrated commercially viable levels of L-phenylalanine synthesis.

BIOTRANSFORMATION ROUTES TO L-PHENYLALANINE

L-Phenylalanine Ammonia Lyase Process

The use of L-phenylalanine ammonia lyase (PAL) in various overproducing strains of the yeast *Rhodoturula* was developed by Tanabe Seiyaku Co. and by Genex Corporation in the 1980s to produce phenylalanine in the reaction shown in Figure 3. In this process, fermented cells of yeast strains *R. glutinis, R. rubra* or *R. graminis* (64–67) were recovered and washed before bioreaction under batch or immobilized conditions with *trans*-cinnamic acid and ammonia. The elevated pH and high concentration of ammonia used under the process conditions enabled the normally catabolic PAL reaction (68,69) to favor formation of L-phenylalanine. Significant development work was carried out to determine the appropriate conditions for optimal growth and maximal induction of the PAL enzyme (70) and the appropriate process conditions for retention of PAL activity (71). The principal drawbacks of the PAL process lay in substrate inhibition of the enzyme, low enzyme-specific activity, and enzyme instability. Screening regimes carried out to address these limitations resulted in the identification of an *R. graminis* strain that displayed greater enzyme specific activity and stability than the *R. rubra* and *R. glutinis* strains used previously (64). Subsequent mutagenesis and selection procedures carried out on this strain using the cinnamic acid analogue phenylpropiolic acid led to a further increase in PAL activity through a significant elevation in PAL gene expression (64). Such strain and process development strategies enabled the Genex process to achieve yields of L-phenylalanine in excess of 50 g/L, with close to 90% conversion of substrate to product (67). Nevertheless, the biotransformation could not favorably compete with the economics of fermentative production, and it is not presently used in L-phenylalanine production.

Aspartate or Aromatic Aminotransferase Process

A number of reports from PEI, Nutrasweet, and Allelix, among others, have described the use of aminotransferases (transaminases) to produce L-phenylalanine in the reaction shown in Figure 4. The reaction proceeds by a ping-pong (72,73) mechanism whereby the amino group of a donor amino acid, typically aspartic acid, is exchanged with the keto group of phenylpyruvic acid, yielding L-phenylalanine and the by-product oxaloacetate. Aminotransferases are ubiquitous in nature, and microorganisms typically possess multiple aminotransferases that are involved in the biosynthesis of a number of amino acids, often catalyzing the terminal step from the keto acid precursor (73–75). The enzymes use pyridoxal 5′-phosphate as cofactor and typically possess high catalytic rates and stereoselectivity with broad substrate specificities (76).

This is exemplified by the multiple transaminases in *E. coli* that can each catalyze the synthesis of L-tyrosine, L-aspartate, and L-valine and L-leucine in addition to L-

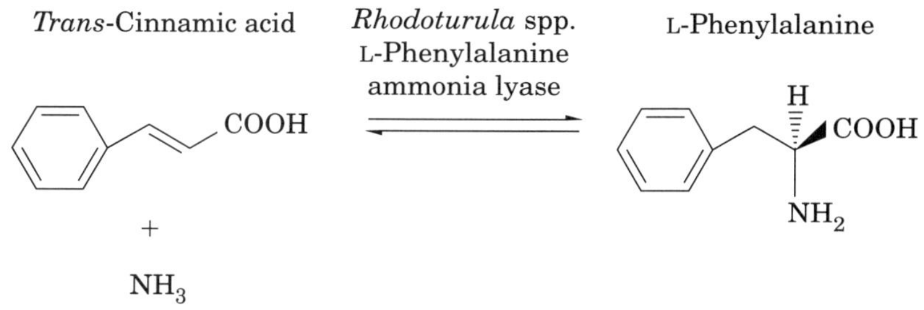

Figure 3. L-Phenylalanine-ammonia-lyase-catalyzed synthesis of L-phenylalanine from *trans*-cinnamic acid.

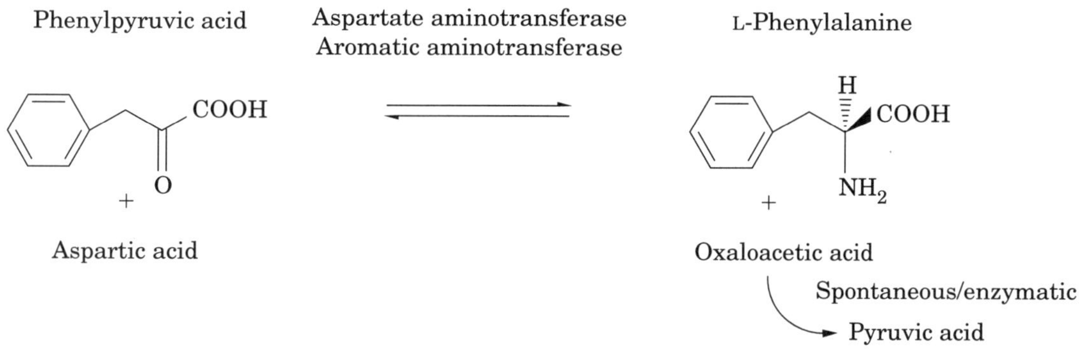

Figure 4. Aminotransferase-catalyzed synthesis of L-phenylalanine from phenylpyruvic acid using aspartic acid as the amino donor. Decarboxylation of oxaloacetic acid drives the reaction toward completion.

phenylalanine (37). Many aminotransferase encoding genes have been cloned and used to overproduce the enzymes (37,77,78). Significant sequence homology often exists among the aminotransferases of individual strains as well as between the corresponding enzymes of microbial and mammalian origin (37). In addition, the tertiary structures of a number of aminotransferases have been determined, furthering the mechanistic understanding of the enzyme and prompting mutagenesis approaches to altering enzyme properties (79). Aminotransferases from many microorganisms have been applied in L-phenylalanine biosynthesis (80–82). In the case of *E. coli*, the highly homologous aspartate and aromatic aminotransferases encoded respectively by the *aspC* and *tyrB* genes have both been developed and applied in this bioconversion (83–85). The aspartate aminotransferase has been generally preferred, having shown greater throughput and increased stability over the *tyrB* gene product (85,86). A number of L-phenylalanine biosynthetic processes involving aminotransferases have been described using either isolated enzyme or whole cells in batched or immobilized systems (82,85,87,88). In the mid 1980s, one such immobilized process was commercialized by PEI using the *E. coli* aspartate aminotransferase (85). In this case, the PPA substrate was economically derived from chemically synthesized 5-benzilidene hydantoin. The highest yields in the transaminase synthesis are generally obtained from batch processes at the expense of the greater catalyst stability of the continuous systems. The cost of catalyst regeneration in batch systems is a drawback of the aminotransferase approach, as is the need for two primary metabolites as substrates. The principal drawback, however, lies in the reaction equilibrium, which being close to 1 limits the yield of the reaction to 50% unless the keto acid by-product can be eliminated from further reaction. To overcome this limitation, commercial transamination approaches have almost invariably used aspartate as the amino donor, which upon transamination to the relatively unstable oxaloacetate will spontaneously decarboxylate to yield pyruvate. Because pyruvate is frequently not a substrate for these transaminases, this enables the reaction to proceed to a yield often exceeding 90%. The decarboxylation of oxaloacetate is more rapid at elevated temperature and has prompted the development of transaminase bioconversions carried out in excess of 50 °C (86). Aminotransferases from thermophiles have been explored in this respect as well as chemical and enzymatic steps to accelerate oxaloacetate decarboxylation and thereby optimize product yield and substrate conversion (89).

Phenylalanine Dehydrogenase Process

A similar process using the phenyl pyruvate precursor has also been developed using the enzyme phenylalanine dehydrogenase. Unlike the aminotransferases, phenylalanine dehydrogenase uses ammonia as a substrate in the reductive amination of phenyl pyruvate, thereby realizing a cost advantage. In addition, the reaction equilibrium lies heavily in favor of phenylalanine synthesis. These advantages in the reaction mechanism are diminished by the dependence of the dehydrogenase on an NADH cofactor that in turn requires enzymatic regeneration. This has been addressed through the use of a coupled enzyme system (90) in which formate dehydrogenase is used along with phenylalanine dehydrogenase in the reaction scheme shown in Figure 5. The reaction proceeds quantitatively because of the evolution of CO_2 generated in the oxidation of formate. Using partially purified enzymes in membrane reactors, this reaction has achieved phenylalanine titers and substrate conversions comparable to the aminotransferase based systems (90,91). Despite the high efficiencies (92) obtained in this bioconversion, commercial production of L-phenylalanine using this method has not been economically competitive with fermentative processes, mainly because of the costs of phenylpyruvic acid preparation. Nevertheless, in a similar process operated by Degussa using leucine dehydrogenase, this approach has been successfully applied to the synthesis of compounds of higher intrinsic value such as the unnatural amino acid tertiary leucine (93).

RESOLUTION-BASED L-PHENYLALANINE SYNTHESIS

Amino Acid Amide Resolution

In contrast to the various fermentation and bioconversion approaches that have used enzyme stereoselectivity to synthesize L-phenylalanine as a single isomer, additional methods have been developed that rely on the same stereoselectivity to resolve chemically synthesized amino acid DL-racemic mixtures. One such general approach that has been successfully commercialized in L-phenylalanine production relies on cleavage of the L-isomer of a DL-α-amino acid amide mixture by a highly enantiospecific amidase enzyme. The general procedure operated by DSM and shown in Figure 6 uses an amidase containing strain of *Pseudomonas putida* to specifically hydrolyze the L-isomer of the DL-amino acid amide mixture. The racemic amino acid mixture is prepared inexpensively from an aldehyde precursor (94,95) via the Strecker reaction, with alkaline hydrolysis of the resulting aminonitrile. Treatment with benzaldehyde after selective hydrolysis of the L-amino acid amide yields an insoluble Schiff base that forms between benzaldehyde and the residual D-amino acid amide. In this way, the residual D-amino acid amide may be readily separated (96) from the resulting L-amino acid and can then be

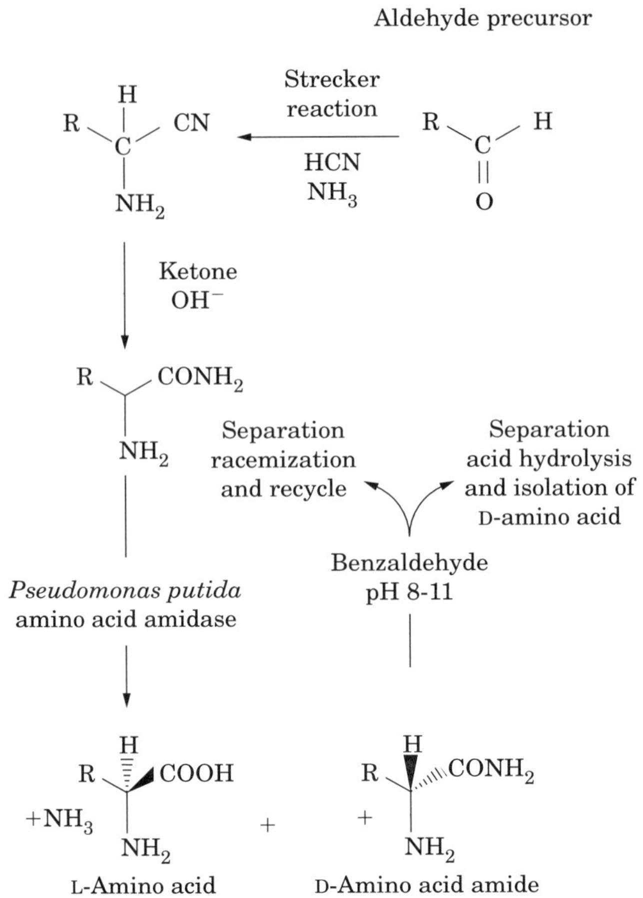

Figure 6. The DSM process for the manufacture of L- and D-α-amino acids.

chemically hydrolyzed and purified or racemized and recycled (97). The advantages of this process include general applicability to the preparation of a range of amino acids (95,98) and the opportunity to recover the D-amino acid if desired from the reaction in addition to the L-amino acid. Conversely, in the preparation of only the L-amino acid by this approach, the need to recycle the unreacted substrate is considered a drawback of the process, necessitating ad-

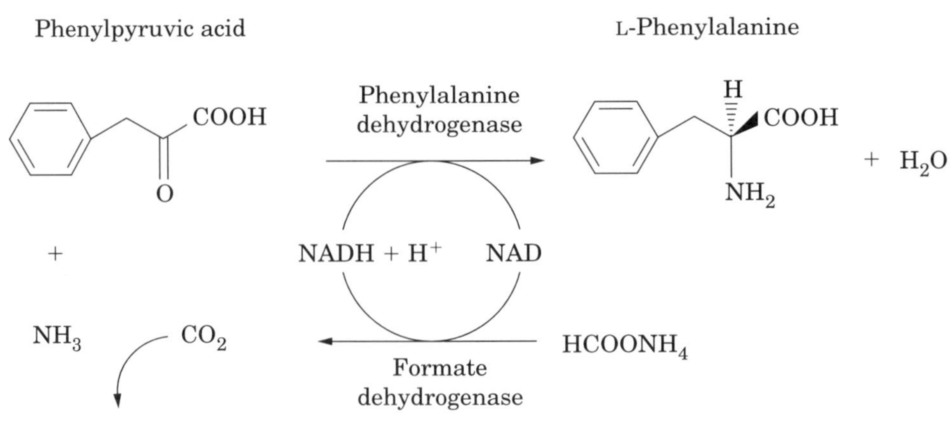

Figure 5. Reductive amination of phenylpyruvic acid to L-phenylalanine using L-phenylalanine dehydrogenase with cofactor regeneration by coupled formate dehydrogenase.

ditional complexity and cost. This aspect has been considered in a further development of this process by DSM in which *P. putida* strains possessing D- or L-amino acid amidase as well as amino acid racemase activity have been identified. The in situ coupling of either amidase with the racemase activity avoids the recycling requirement of the process and facilitates direct conversion of the DL-amino acid amide mix into the desired isomer (99).

Acylase Process

An earlier strategy involving stereoselective enzyme hydrolysis of chemically prepared racemic precursors has been operated successfully for many years by Tanabe Seiyaku. In this process, illustrated in Figure 7, a DEAE-Sephadex immobilized L-specific aminoacylase selectively hydrolyzes *N*-acetyl-L-amino acids from an *N*-acetyl-DL-amino acid mixture (100,101). After hydrolysis, the L-amino acid is isolated by selective crystallization and the remaining *N*-acetyl-D-amino acid is racemized either chemically or enzymatically (102) and recycled. The preferred aminoacylase for this reaction was obtained from the mold *Aspergillus oryzae* and is readily isolated and immobilized in the bioreactor (100). The enzyme has broad substrate specificity and has been used by Tanabe in the production of L-methionine and L-valine and addition to L-phenylalanine. More recently, this process has been further developed by Degussa/Rexim (103) to prepare bulk quantities of L-phenylalanine and a large number of L-amino acids. In addition, Tanabe has reported on the further development of the aminoacylase process through the coupling of acetamidocinnamate aminohydrolase and aminotransferase activity. In this process, L-phenylalanine is produced in two steps from acetamidocinnamic acid through a phenylpyruvic acid intermediate using immobilized cells of two independent bacterial strains (87).

DL-5-Monosubstituted Hydantoin-Based Strategies

The recycling aspect of resolution-based amino acid syntheses such as the amidase and aminoacylase processes is again considered and largely overcome in an alternative strategy using DL-5-monosubstituted hydantoins. Inexpensive chemical routes such as the Bucherer-Bergs synthesis or the condensation of aldehydes with hydantoin (104–107) are used to prepare the racemic hydantoin derivatives that are then resolved using either L- or D-enantiospecific hydantoinase and *N*-carbamoyl amino acid amidohydrolase enzymes to yield the desired amino acid isomer. The process is shown in Figure 8. In the case of phenylalanine, both L- and D-phenylalanine can be prepared from racemic DL-5-benzylhydantoin. The advantage of the hydantoinase process is that in many cases a dynamic resolution can be established whereby the unreacted isomer can be racemized in situ either chemically (108) or in the presence of a racemase (109) enzyme to enable the process to approach 100% yield. Enzymes with L- and D-specific hydantoinase activity have been described in a number of microbial species, with the D-specific activity having been identified much more frequently, often attributable to dihydropyrimidinases involved in pyrimidine catabolism (110–114). Substrate specificities vary significantly between enzyme isolates and contribute to the broad applicability of this approach (113–116). Molecular cloning has been applied in at least one case to isolate a D-hydantoinase encoding gene from a thermophilic bacillus and to subsequently overproduce the enzyme (117). The D- and L-specific *N*-carbamoyl amino acid amidohydrolases have also been shown to be widely distributed in microorganisms (113,114,116,118). The bioconversion steps are typically carried out using whole cells in immobilized systems (110,118) with racemization of the unreacted hydantoin occurring under the normal alkaline conditions of the reaction (108). In some cases, microbial racemases have been used to optimize the racemization step (109,119). The hydantoinase process offers the advantage of relatively inexpensive precursors and high conversion rates through the in situ racemization of the substrate. In addition, through the broad substrate range of hydantoinase and carbamoylase enzymes, the process is broadly applicable in L- and D-amino acid biosynthesis. However, it is in D-amino acid production that large-scale commercial application of the hydantoinase process has been more significantly applied, most notably by Recordati and Kanegafuchi in the manufacture of D-phenyl glycine and D-*p*-hydroxyphenyl glycine as components of semisynthetic penicillins and cephalosporins (120–122) but also by Degussa and Ajinomoto in the production of additional D-amino acids, including D-phenylalanine (122–124).

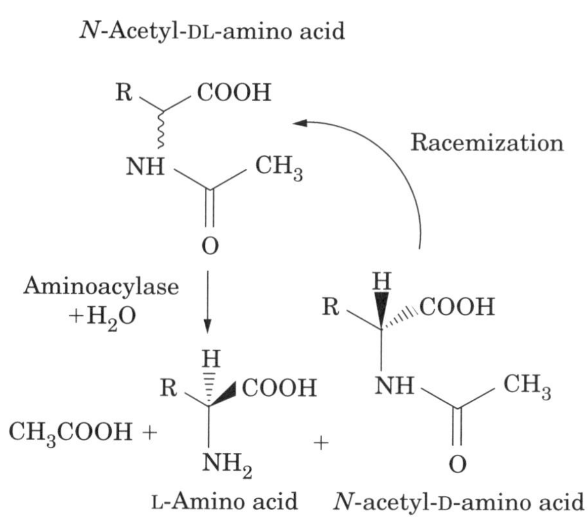

Figure 7. Resolution of *N*-acetyl-DL-amino acids using aminopeptidases yielding L-amino acid and recycling of residual *N*-acyl-D-amino acid.

CONCLUSIONS AND FUTURE PROSPECTS

Numerous commercial processes have been developed for the large-scale commercial production of L-phenylalanine since the early 1980s, fueled by the enormous increase in L-phenylalanine demand for the dipeptide sweetener aspartame. The most successful processes use microbial enzymes at some stage, either in an enantioselective hydrolysis or biotransformation step using chemically derived precursors or in whole cell fermentations using microorganisms engineered through classical and recombinant

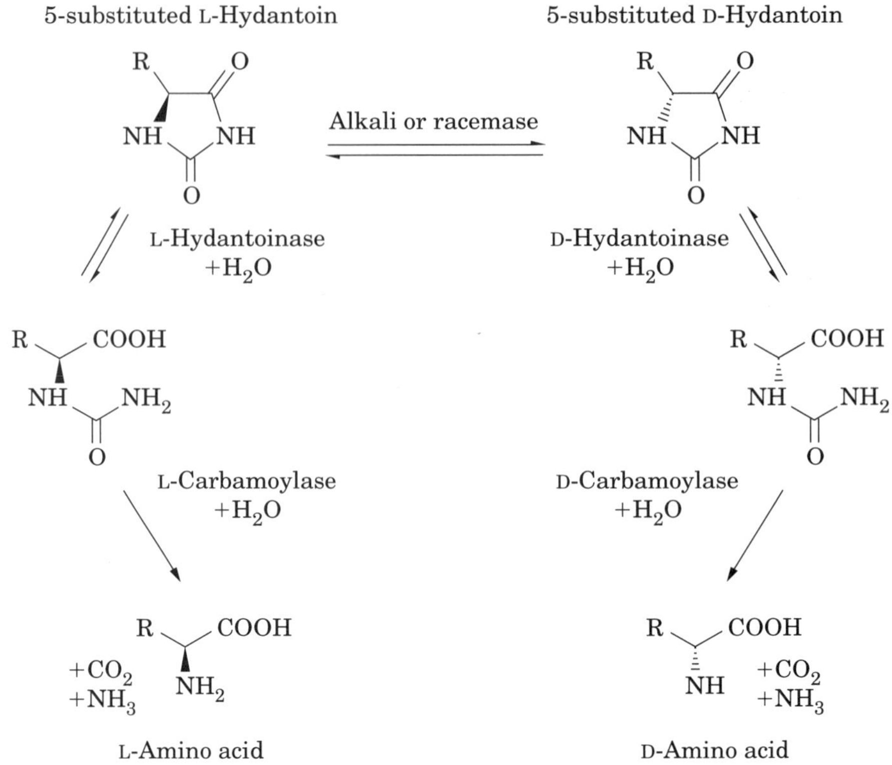

Figure 8. Hydantoinase/carbamoylase-based dynamic resolution of DL-5-monosubstituted hydantoins to either D- or L-amino acids.

strain improvement methodology. In general, the most economically competitive processes have resulted from large-scale fermentation approaches. This is mainly because of the incremental savings achieved in such processes through increased understanding of the physiology and molecular genetic regulation of L-phenylalanine biosynthesis in microbes and the low raw material costs attained through media and process development. In these processes, the concentration of L-phenylalanine can often exceed 50 g/L in crude fermentation broth with the overall incorporation of carbon into phenylalanine from a source such as glucose, equaling or exceeding the theoretical maximum calculated at around 25 to 30% (125). With this level of efficiency, it is unlikely that most major manufacturers of L-phenylalanine will continue to invest significantly in the development of fermentation processes as the return on investment is diminished.

Several elegant biotransformation approaches such as the PAL and dehydrogenase processes have been optimized through molecular genetic strategies but have been impacted by the drawbacks of greater precursor cost and in some cases limited biocatalyst stability. As a result, none of these approaches are currently used to produce L-phenylalanine. Nevertheless, a number of manufacturers, including DSM, Degussa/Rexim, and Tanabe, successfully produce L-phenylalanine through chemoenzymatic routes that are generally more versatile and are often part of more broadly applicable processes in natural and unnatural amino acid manufacture. This is apparent in the manufacture of D-phenylalanine where, in contrast to the preponderance of fermentative routes to L-phenylalanine, the synthesis is currently carried out using the hydantoinase and amidase resolution strategies mentioned earlier. Ultimately, through a greater understanding of microbial pathway engineering and the increasing availability of novel enzymatic activities, it is quite possible that the favorable economics of large-scale fermentation may also become more applicable to D-phenylalanine biosynthesis (126,127).

BIBLIOGRAPHY

1. Eur. Pat. 0264,914 (1987), R. Katsumata, and A. Ozaki.
2. I. Shiio, S. Sugimoto, and K. Kawamura, *Agric. Biol. Chem.* **52**, 2247–2253 (1988).
3. H. Hagino and K. Nakayama, *Agric. Biol. Chem.* **38**, 157–161 (1974).
4. M. Ikeda, A. Ozaki, and R. Katsumata, *Appl. Microbiol. Biotechnol.* **39**, 318–323 (1993).
5. U.S. Pat. 4,681,852, D.E. Tribe.
6. G.S. Hudson and B.E. Davidson, *J. Mol. Biol.* **180**, 1023–1051 (1984).
7. A. Ozaki, R. Katsumata, T. Oka, and A. Furuya, *Agric. Biol. Chem.* **49**, 2925–2930 (1985).
8. I. Shiio and S. Sugimoto, *Agric. Biol. Chem.* **45**, 2197–2207 (1981).
9. J. Nelms, R.M. Edwards, J. Warwick, and I. Fotheringham, *Appl. Environ. Microbiol.* **58**, 2592–2598 (1992).
10. G. Millar, A. Lewendon, M. Hunter, and J.R. Coggins, *Biochem. J.* **237**, 427–437 (1986).
11. I. Shiio, K. Ishii, K. Yokozeki, *Agric. Biol. Chem.* **37**, 1991 (1973).

12. T. Tsuchida, K. Kubota, Y. Morinaga, H. Matsui, H. Enei, and F. Yoshinaga, *Agric. Biol. Chem.* **51**, 2095–2101 (1987).
13. Y. Tokoro, K. Oshima, M. Okii, K. Yamaguchi, K. Tanaka, and S. Kinoshita, *Agric. Biol. Chem.* **34**, 1516 (1970).
14. S.O. Hwang, G.H. Gil, Y.J. Cho, K.R. Kang, J.H. Lee, and J.C. Bae, *Appl. Microbiol. Biotechnol.* **22**, 108–113 (1985).
15. U.S. Pat. 3,759,790 (1973), K. Nakayama, Sagamihara, and H. Hagino.
16. B.J. Wallace and J. Pittard, *J. Bacteriol.* **97**, 1234–1241 (1969).
17. Y.J. Choi and D.E. Tribe, *Biotechnol. Lett.* **4**, 223–228 (1982).
18. E. Haslam, in *The Shikimate Pathway*, Wiley, New York, 1974.
19. F. Gibson and J. Pittard, *Curr. Topics Cell Regul.* **2**, 29–63 (1970).
20. K.D. Brown, *Genetics* **60**, 31–48 (1968).
21. K.D. Brown and R.L. Somerville, *J. Bacteriol.* **108**, 386–399 (1971).
22. H. Camakaris and J. Pittard, *J. Bacteriol.* **115**, 1135–1144 (1973).
23. S.W.K. Im, H. Davidson, and J. Pittard, *J. Bacteriol.* **108**, 400–409 (1971).
24. K.M. Herrmann and R.L. Somerville, eds., *Amino Acids: Biosynthesis and Genetic Regulation*, Addison-Wesley, Reading, Mass., 1983.
25. I. Shiio, S. Sugimoto, and J. Miyajima, *J. Biochem.* **75**, 987 (1974).
26. S. Sugimoto and I. Shiio, *J. Biochem.* **87**, 881–890 (1980).
27. H. Hagino and K. Nakayama, *Agric. Biol. Chem.* **39**, 351 (1975).
28. H. Hagino and K. Nakayama, *Agric. Biol. Chem.* **38**, 2125–2134 (1974).
29. I. Shiio and S. Sugimoto, *J. Biochem.* **86**, 17–25 (1979).
30. S. Sugimoto, M. Nakagawa, T. Tsuchida, and I. Shiio, *Agric. Biol. Chem.* **37**, 2327–2336 (1973).
31. W.D. Davies and B.E. Davidson, *Nucleic Acids Res.* **10**, 4045–4058 (1982).
32. J.M. Ray, C. Yanofsky, and R. Bauerle, *J. Bacteriol.* **170**, 5500–5506 (1988).
33. Eur. Pat. 229,161 (1986), R.M. Edwards, P.P. Taylor, M.G. Hunter, and I.G. Fotheringham.
34. C. Förberg, T. Eliaeson, and L. Häggström, *J. Biotechnol.* **7**, 319–332 (1988).
35. S. Sugimoto, M. Yabuta, N. Kato, T. Seki, T. Yoshida, and H. Taguchi, *J. Biotechnol.* **5**, 237–253 (1987).
36. J.T. Powell and J.F. Morrison, *Eur. J. Biochem.* **87**, 391–400 (1978).
37. I.G. Fotheringham, S.A. Dacey, P.P. Taylor, T.J. Smith, M.G. Hunter, M.E. Finlay, S.B. Primrose, D.M. Parker, and R.M. Edwards, *Biochem. J.* **234**, 593–604 (1986).
38. M-J. Gething and B.E. Davidson, *Eur. J. Biochem.* **86**, 165–174 (1978).
39. T.A.A. Dopheide, P. Crewther, and B.E. Davidson, *J. Biol. Chem.* **247**, 4447–4452 (1972).
40. I. Shiio and S. Sugimoto, *J. Biochem.* **79**, 173–183 (1976).
41. S. Sugimoto and I. Shiio, *J. Biochem.* **76**, 1103 (1974).
42. I. Shiio and S. Sugimoto, *J. Biochem* **89**, 1483–1492 (1981).
43. S. Sugimoto and I. Shiio, *Agric. Biol. Chem.* **49**, 39–48 (1985).
44. I. Shiio, M. Mori, and H. Ozaki, *Agric. Biol. Chem.* **46**, 2967–2977 (1982).
45. L. de Boer and L. Dijkhuizen, *Adv. Biochem. Eng. Biotechnol.*, **41**, 1–27 (1990).
46. H. Hagino and K. Nakayama, *Agric. Biol. Chem.* **38**, 2367–2376 (1974).
47. H. Hagino and K. Nakayama, *Agric. Biol. Chem.* **39**, 331–342 (1975).
48. M. Ikeda and R. Katsumata, *Appl. Environ. Microbiol.* **58**, 781–785 (1992).
49. U.S. Pat. 4,753,883 (1988), K.C. Backman and B. Ramaswamy.
50. A. Berry, *TIBTECH* **14**, 250–256 (1996).
51. J.W. Frost and K.M. Draths, *Annu. Rev. Microbiol.* **49**, 557–579 (1995).
52. R. Patnaik and J.C. Liao, *Appl. Environ. Microbiol.* **60**, 3903–3908 (1994).
53. R. Patnaik, R.G. Spitzer, and J.C. Liao, *Biotech. Bioeng.* **46**, 361–370 (1995).
54. J.E. Miller, K.C. Backman, M.J. O, Connor, and R.T. Hatch, *J. Ind. Microbiol.* **2**, 143–149 (1987).
55. P.W. Postma, in F.C. Neidhardt ed., Escherichia coli and Salmonella typhimurium: *Cellular and Molecular Biology*. American Society for Microbiology, Washington, D.C., 1987.
56. N. Flores, J. Xiao, A. Berry, F. Bolivar, and F. Valle, *Nat. Biotechnol.* **14**, 620–623 (1996).
57. S. Bledig, Ph.D. Thesis, Univ. of Warwick, 1994.
58. K.M. Draths, D.L. Pompliano, D.L. Conley, J.W. Frost, A. Berry, G.L. Disbrow, R.J. Staversky, and J.C. Lievense, *J. Am. Chem. Soc.* **114**, 3956–3962 (1992).
59. M.J. Whipp, and A.J. Pittard, *J. Bacteriol.* **132**, 453–461 (1977).
60. J. Pi, P.J. Wookey, A.J. Pittard, *J. Bacteriol.* **173**, 3622–3629 (1991).
61. N. Honore and S.T. Cole, *Nucleic Acids Res.* **18**, 653 (1990).
62. M. Mulcahy, Ph.D. Thesis, Univ. of Dundee, 1988.
63. U.S. Pat. 5,354,672 (1994), I.G. Fotheringham, J. Ton, and C. Higgins.
64. S.A. Orndorff, N. Constantino, D. Stewart, and D.R. Durham, *Appl. Env. Microbiol.* **54**, 996–1002 (1988).
65. U.S. Pat. 4,584,269 (1986), P.J. Vollmer, J.J. Schruben, J.P. Montgomery, and Y. Huei-Hsuing.
66. S. Yamada, K. Nabe, N. Izuo, K. Nakamichi, and I. Chibata, *Appl. Environ. Microbiol.* **42**, 773–778 (1981).
67. B.K. Hamilton, H.Y. Hsiao, W.E. Swann, D.M. Anderson, and J.J. Delente, *Trends Biotechnol.* **3**, 64 (1985).
68. H.J. Gilbert and M. Tully, *J. Bacteriol.* **150**, 498–505 (1982).
69. N. Onishi, K. Yokozeki, Y. Hirose, and K. Kubot, *Agric. Biol. Chem.* **51**, 291–292 (1987).
70. C.T. Evans, K. Hanna, D. Conrad, W. Peterson, and M. Misawa, *Appl. Microbiol. Biotechnol.* **25**, 406–414 (1987).

71. C.T. Evans, D. Conrad, K. Hanna, W. Peterson, C. Choma, and M. Misawa, *Appl. Microbiol. Biotechnol.* **25**, 399–405 (1987).

72. J.F. Kirsch, G. Eichele, G.C. Ford, M.G. Vincent, and J.N. Jansonius, *J. Mol. Biol.* **174**, 497–525 (1983)

73. R.A. Jensen and D.H. Calhoun, *CRC Crit. Rev. Microbiol.* **8**, 229–266 (1981).

74. T. Tachiki and T. Tochikura, *Agric. Biol. Chem.* **37**, 1439–1448 (1973).

75. T. Tachiki, M. Moriguchi, and T. Tochikura, *Agric. Biol. Chem.* **39**, 43–50 (1973).

76. P. Christen and D.E. Metzler eds., *Transaminases*, Wiley, New York, 1984.

77. S. Kuramitsu, T. Ogawa, H. Ogawa, and H.J. Kagamiyama, *Biochemistry* **97**, 993–999 (1985).

78. M.-H. Sung, K. Tanizawa, H. Tanaka, S. Kuramitsu, H. Kagamiyama, K. Hirotsu, A. Okamoto, T. Higuchi, and K. Soda, *J. Biol. Chem.* **266**, 2567–2572 (1991).

79. E. Kohler, M. Seville, J. Jager, I. Fotheringham, M. Hunter, M. Edwards, J.N. Jansonius, and K. Kirschner, *Biochemistry* **33**, 90–97 (1994).

80. E. Bulot and C.L. Cooney, *Biotechnol. Lett.* **7**, 93 (1985).

81. C.T. Evans, W. Peterson, C. Choma, and M. Misawa, *Appl. Microbiol. Biotechnol.* **26**, 305–312 (1987).

82. K. Nakamichi, K. Nabe, Y. Nishida, and T. Tosa, *Appl. Microbiol. Biotechnol.* **30**, 243–246 (1989).

83. J. Then, A. Doherty, H. Neatherway, R. Marquardt, H.M. Deger, H. Voelskow, G. Wöhner, and P. Präve, *Biotechnol. Lett.* **9**, 680 (1987).

84. M. Robinson, R. Marquardt, and J. Then, J. McChesney, H. Neatherway, G. Wöhner, H.M. Deger, and P. Präve, *Biotechnol. Lett.* **9**, 673–679 (1987).

85. G.J. Calton, L.L. Wood, M.H. Updike, L. Lantz, J.P. Hamman, *Biotechnology* **4**, 317–320 (1986).

86. Eur. Pat. 0152,275 (1991), D. Lewis and S. Farrand.

87. K. Nakamichi, Y. Nishida, K. Nabe, and T. Tosa, *Appl. Biochem. Biotechnol.* **11**, 367–376 (1985).

88. H. Ziehr, M.R. Kula, E. Schmidt, C. Wandrey, and J. Klein, *Biotechnol. Bioeng.* **9**, 482 (1987).

89. I. Schutten, W. Harder, and L. Dijkhuizen, *Appl. Microbiol. Biotechnol.* **27**, 292 (1987).

90. W. Hummel, E. Schmidt, C. Wandrey, and M.R. Kula, *Appl. Microbiol. Biotechnol.* **25**, 175–185 (1986).

91. W. Hummel, H. Schutte, E. Schmidt, C. Wandrey, and M.R. Kula, *Appl. Microbiol. Biotechnol.* **26**, 409–416 (1987).

92. C. Wandrey, in O.M. Neijssel, R.R. van der Meer and K.Ch.A.M. Luyben eds., *Proc. of the 4th European Congress on Biotechnology, vol. 4*, Elsevier, Amsterdam, 1987, pp. 171–188.

93. U. Kragl, D. Vasic-Racki, and C. Wandrey, *Chem. Ing. Tech.* **64**, 499 (1992).

94. J. Kamphuis, W.H.J. Boesten, Q.B. Broxterman, H.F.M. Hermes, J.A.M. van Balken, E.M. Meijer, and H.E. Schoemaker, *Adv. Biochem. Eng. Biotechnol.* **42**, 134 (1990).

95. E.M. Meijer, W.H.J. Boesten, H.E. Schoemaker, and J.A.M. van Balken, in J. Tramper, H.C. van der Plas, and P. Linko eds., *Biocatalysts in Organic Synthesis*, Elsevier, Amsterdam, 1985, pp. 135–156.

96. U.S. Pat. 4,172,846 (1979), DSM/Stamicarbon.

97. Eur. Pat. 1,442,585, (1991), DSM/Stamicarbon.

98. E.M. Meijer, J. Kamphuis, J.A.M. Van Balken, H.F.M. Hermes, W.J.J. van den Tweel, M. Kloosterman, W.H.J. Boesten, H.E. Schoemaker, in V. Claassen ed., *Trends in Drug Research*, Elsevier, Amsterdam, 1990, pp. 363–382.

99. Eur. Pat. 0307023 (1989), Hermes HFM, et al.

100. T. Tosa, T. Mori, N. Fuse, and I. Chibata, *Enzymologia* **31**, 214 (1966).

101. T. Tosa, T. Mori, N. Fuse, and I. Chibata, *Agric. Biol. Chem.* **33**, 1047 (1969).

102. Eur. Pat. 304,021 (1989), Takeda Chemical Industries.

103. U.S. Pat. 4,304,858 (1981), Degussa/GBF.

104. E. Ware, *Chem. Rev.* **46**, 403 (1950).

105. Jpn. Pat. 80 136279 (1979), Ajinomoto.

106. German Pat. DE 3043259 A1 (1980), A. Kleeman and M. Samson.

107. K. Drauz, A. Kleeman, and M. Samson, *Chem.-Ztg.* **12**, 391 (1984).

108. M. Bovarnick and H.T. Clark, *J. Am. Chem. Soc.* **60**, 2426 (1938).

109. C. Gross, C. Syldatk, V. Kackowiak, and F. Wagner, *J. Biotechnol.* **14**, 363 (1990).

110. R. Olivieri, E. Fascetti, L. Angelini, and L. Degen, *Biotechnol. Bioeng.* **23**, 2173–2183 (1981).

111. K. Yokozeke, S. Nakamori, C. Eguchi, K. Yamada, and K. Mitsugi, *Agric. Biol. Chem.* **51**, 355–362 (1987).

112. A. Möller, C. Syldatk, M. Schultze, and F. Wagner, *Enzyme Microb. Technol.* **10**, 618 (1988).

113. K. Yokozeke, Y. Hirose, and K. Kubota, *Agric. Biol. Chem.* **51**, 737–746 (1987).

114. A. Yamashiro, K. Kubota, and K. Yokozeki, *Agric. Biol. Chem.* **52**, 2857–2863 (1988).

115. H. Yamada, S. Takahashi, Y. Kii, and H. Kumagai, *J. Ferment. Technol.* **56**, 484 (1978).

116. K. Yokozeke and K. Kubota, *Agric. Biol. Chem.* **51**, 721–728 (1987).

117. Y. Mukohara, T. Ishikawa, K. Watabe, and H. Nakamura, *Biosci. Biotechnol. Biochem.* **58**, 1621–1626 (1994).

118. R. Olivieri, E. Fascetti, L. Angelini, and L. Degen, *Enzyme Microbiol. Technol.* **1**, 201 (1979).

119. C. Syldatk and F. Wagner, *Food Biotechnol.* **4**, 87 (1990).

120. H. Yamada and S. Shimizu, in J. Tramper, H.C. van der Plas, P. Linko eds., *Biocatalysts in Organic Syntheses*, Elsevier, Amsterdam, 1985, pp. 19–37.

121. Eur. Pat. 0175312 (1985), Kanegafuchi.

122. K. Yokozeke, S. Nakamori, S. Yamanaka, C. Eguchi, K. Mitsugi, and F. Yoshinaga, *Agric. Biol. Chem.* **51**, 715–719 (1987).

123. German Pat. DE 3917057 (1992), Degussa.

124. Jpn. Pat. JP 86-200434 (1986), T. Ishikawa and H. Kimura.

125. C. Foerberg, T. Eliaeson, and L. Haggstrom, *J. Biotechnology*, **7** 319–332 (1988).

126. I. Fotheringham, S. Bledig, R. Senkpeil, and P. Taylor, *Abstracts of the 1995 SIM Annual Meeting*, San Jose, Calif., 1995.

127. J. Dailey, N. Grinter, R. Nelson, D. Pantaleone, R. Senkpeil, P. Taylor, J. Ton, R. Yoshida, and I. Fotheringham, *Abstracts of the IBC Enzyme Technology Symposium*, Lake Buena Vista, Fla., 1996.

PHENYLALANINE DEHYDROGENASE

YASUHISA ASANO
Toyama Prefectural University
Toyama, Japan

KEY WORDS

L-Amino acid
Bacillus badius
Bacillus sphaericus
Enantioselective synthesis
α-Keto acid
Microdetermination
NADH regeneration
Phenylalanine dehydrogenase
Phenylketonuria
Sporosarcina ureae

OUTLINE

INTRODUCTION

NAD^+-dependent phenylalanine dehydrogenase (PheDH, phenylalanine: NAD^+ oxidoreductase, deaminating (EC 1.4.1.20)) was found to occur in *Brevibacterium* sp. by Hummel et al. in 1984 (1). Asano et al. first described the enzymological properties of PheDHs with crystalline enzymes from *Soprosarcina ureae* and *Bacillus sphaericus* (2,3). Similar enzymes that catalyze a reversible oxidation-reduction of amino acids (acting on -CH-$NH_{(2)}$) with NAD^+ or $NADP^+$ as an electron acceptor are classified in EC 1.4.1., which includes glutamate dehydrogenase (GluDH, EC 1.4.1.2-4) (4), alanine dehydrogenase (AlaDH, EC 1.4.1.1), leucine dehydrogenase (LeuDH, EC 1.4.1.9), and other oxidoreductases acting on several amino acids (5).

Since its discovery (1,2,3), much attention has been paid to PheDH because it appeared to be a useful catalyst in the enantioselective synthesis of Phe and related L-amino acids from their keto analogs (6). Some microbial PheDH producers have been isolated from nature, and the enzymes have been characterized. Since 1992, PheDH has been used as a reagent in the colorimetric microdetermination of Phe to detect phenylketonuria (PKU) in the blood of neonates in Japan (7,8).

Here we deal with the occurrence, properties, and structures of bacterial PheDHs, and their applications to the synthesis and determination of amino acids.

DISCOVERY, ENZYMATIC PROPERTIES, AND FUNCTIONS OF PheDH

Hummel et al. screened for a PheDH-producing strain of bacteria from soil and isolated a *Brevibacterium* sp. by an enrichment culture technique (1). Culture conditions for PheDH formation and some reaction conditions to synthesize Phe and Tyr with the crude enzyme were optimized. Asano et al. noted PheDH production in the gram-positive spore-forming bacteria *S. ureae* SCRC-R04 (2,3), *B. sphaericus* SCRC-R79a (3), and *B. badius* IAM 11059 (7) and described its purification, crystallization, and characterization; no activity was found in yeasts. Following these studies, PheDH was given a new entry number: EC 1.4.1.20 (5). Later, *Rhodococcus maris* (9), *Nocardia* sp. (10), *Thermoactinomyces intermedius* (11), *Rhodococcus* sp. (12), *Bacillus cereus* (13), and *Microbacterium* sp. (14,15) were found to produce the enzyme. The occurrence of PheDH appears to be limited in some groups of gram-positive spore-forming bacteria and thermophilic Actinomycetes (11). The distribution of PheDH seems similar to that of LeuDH (16), although the distribution of the former is much more limited than that of the latter, which may be implicated in microbial sporulation, which shuffles carbon and nitrogen metabolism of amino acids by their reversible nature. Because PheDH is specifically distributed in *B. sphaericus* and *B. badius*, which are characterized to be poor in sugar oxidation among the genus *Bacillus* (17), PheDH seems involved in the catabolism of Phe and other amino acids in the medium.

Hummel et al. suggested that PheDH is responsible for the degradation of Phe, not its synthesis (1); however, in *S. ureae* (3), L-phenylalanine aminotransferase activity with oxaloacetic acid as an amino acceptor was detected during growth on a medium containing Phe at a range between 0.001 and 0.006 units/mg, while PheDH was highly induced (0.033 units/mg, measured in the oxidative deamination reaction). The formation of PheDH was growth associated, indicating that the enzyme is responsible for the catabolism of Phe. On the other hand, when the strain was grown in M9 medium supplemented with 0.1% yeast extract, the PheDH activity was detected only in the exponential phase at a level as high as 0.059 units/mg, which is much higher than L-phenylalanine aminotransferase activity (less than 0.001 units/mg). Because the equilibrium of PheDH favors the synthesis of the amino acid, the enzyme probably also functions in the anabolism of Phe, depending on the environment.

Enzymatic Properties

The enzymatic properties of PheDH are summarized in Table 1. The molecular weight of the PheDH subunits are in the range of 36,000 to 42,000. The higher structures vary; PheDHs from *Sporosarcina* (2,3), *Bacillus* (3), and *Microbacterium* (15) are octameric enzymes, whereas PheDH from *Rhodococcus* sp. (12) is tetrameric, that from *Rhodococcus maris* (9) is dimeric, and that from *Nocardia* (10) is monomeric. The molecular weight of PheDH ranges from 42,000 (the monomeric *Nocardia* enzyme) to 331,480 (the octameric *B. sphaericus* enzyme). The optimum pH for the reversible oxidative deamination and reductive amination varies by about 1 pH unit; the former reaction varies in the range of pH 10.4 to 11.3, whereas the latter is in the range of pH 9 to 10.3. The velocity toward the reductive amination to synthesize Phe generally is several times faster than that for the reverse reaction at optimum pH; the V_{max} values for the oxidative deamination reaction of Phe catalyzed by *Sporosarcina* and *B. sphaericus* PheDH are both 114, whereas those for the reductive amination reaction for phenylpyruvate are 598 and 416, respectively (3). The K_m values toward the preferred substrates of L-amino acids and keto acids are in the range of 0.1 to 1 mM, though much higher K_m values of around 100 mM were observed toward ammonia.

Substrate Specificity and Kinetic Properties

In the oxidative deamination reaction, Phe and L-norleucine were active as substrates for the *Sporosarcina* enzyme. The from enzymes *B. badius* and *S. ureae* show narrower substrate specificity than that from *B. sphaericus*, which has almost equal affinity toward Tyr and its keto analog (Table 2), thus *B. sphaericus* PheDH may be called *tyrosine dehydrogenase*. On the other hand, the enzymes showed wider substrate specificities in the reductive amination reaction than observed in the oxidative deamination reaction. The substrate specificity of PheDH has been extensively characterized with the *B. sphaericus* enzyme by using chemically synthesized substrates and substrate analogs (3,6). The relative rates of the reductive amination of α-keto acids and their analogs catalyzed by PheDH from *S. ureae* and *B. sphaericus* are shown in Table 3. The substrate specificity of *B. sphaericus* was examined in detail with various synthetic compounds as shown in Table 4. The enzyme was found to accommodate α-keto acids with large substituents, including compounds substituted at the β-position of pyruvic acid with a longer or bulkier group, although the relative velocity of the reductive amination reaction was low. Phenylpyruvate analogs substituted at the phenyl ring were relatively good substrates, whereas the substitution at the β-position of phenylpyruvate with a bulkier group, such as a hexyl group, greatly lowered the reaction velocity. On the other hand, pyruvate, α-ketoglutarate, α-hydroxypyruvate, benzoylformate, α-keto esters such as ethyl phenylpyruvate and ethyl α-keto-γ-phenylbutyrate, β-keto esters such as ethyl β-keto-γ-phenylbutyrate, β-keto acids such as β-keto-γ-phenylbutyrate, and α-ketoalcohols such as α-keto-β-phenylpropanol were inactive as substrates. Thus, it was shown that the enzyme acts solely on α-keto acids, and a free carboxylic acid moiety is required for a compound to be recognized as a substrate. It was also revealed that the enzyme does not differentiate the configuration of the substituent at the β-position of pyruvic acid; for example, a diastereomeric mixture of L-α-amino-β-DL-methyl-β-phenylpropionate was synthesized from α-keto-β-DL-methyl-β-phenylpropionate. Gln, Asp, methylamine, dimethylamine, trimethylamine and ethylamine (each at 200 mM) did not replace the ammonium ion in the reductive amination reaction. Thus, PheDH from *B. sphaericus* showed wide substrate specificity toward substituted pyruvic acids. The finding that the enzyme utilized α-keto-γ-phenylbutyrate and α-keto-ε-phenylvalerate, but not β-keto-γ-phenylbutyrate, shows it has a definite requirement for a distance between the carbonyl carbon and the carboxyl group of the substrates. For the *Sporosarcina* enzyme, phenylpyruvate, α-ketocaproate, α-keto-γ-methylthiobutyrate, *p*-hydroxyphenylpyruvate, and α-ketoisocaproate were active as substrates. NAD^+ was active as a cofactor for both enzymes, whereas $NADP^+$ was not. The mean value of the apparent equilibrium constant (K_{eq} = [phenylpyruvate][NADH][NH_4^+][H^+]/[Phe]-[NAD^+][H_2O]) was determined using the *Sporosarcina* enzyme to be 1.4×10^{-15} at the pH range of 7.0 to 11.5 and 25 °C, assuming the conventional concentration of water ([H_2O] = 1 M) (3). For the *B. sphaericus* enzyme, it was calculated to be 2.0×10^{-15} at pH 8.40 to 10.38 (6). The substrate specificities of PheDH from *B. badius* (7) and *R. maris* (9) are similar to that of *S. ureae* (3), which shows more limited activities than the *B. sphaericus* enzyme, although *R. maris* PheDH has a slightly higher affinity toward Tyr than *S. ureae* PheDH. It is notable that the PheDH of *Nocardia* sp. is 0.54 and 2.4 times more active toward indolepyruvate and α-ketoisocaproate, respectively, than toward phenylpyruvate (10), and *T. intermedius* PheDH is inactive toward Tyr and its keto analog (11).

STRUCTURE OF PheDH

The *pdh* genes have been cloned and sequenced, and overproduction of the *Bacillus* enzymes has been achieved (7,18,19). *Escherichia coli* JM 109/pBPDH1-DBL expresses about 120-fold higher activity of PheDH (7,200 units/L) than the wild type *B. sphaericus*. The nucleotide sequences for the *pdh* genes from *B. sphaericus* (19), *B. badius* (20), *S. ureae* (21), *T. intermedius* (22), and *Rhodococcus* sp. (12) have been reported.

Computer assessment of the data showed that the deduced primary structure of *B. badius* PheDH (20) is similar to PheDH of *B. sphaericus* (67.9% identical over 377 amino acids) (19) and *T. intermedius* (56.6% identical over 346 amino acids) (22), LeuDH from *B. stearothermophilus* (23) (49.9% identical over 353 amino acids), $NADP^+$-dependent valine dehydrogenase from *Streptomyces coelicolor* (24) (46.6% identical over 352 amino acids), PheDH from *Rhodococcus* sp. (34.7% identical over 334 amino acids) (12), and GluDH from *Clostridium symbiosum* (25.9% identical over 185 amino acids) (25). Similarities between these proteins are seen in the catalytic domains of amino acid dehydrogenases, G-G-(G or A)-K (26), and the glycine-rich

Table 1. Comparison of Properties of PheDHs

Property	Microorganism						
	B. sphaericus	*S. ureae*	*B. badius*	*R. maris*	*Nocardia* sp.	*T. intermedius*	*Microbacterium* sp.
Specific activity of final preparation (U/mg)	111	84	68	65	30	86	37
M_r by gene sequencing	331,480	330,608	330,800	–[a]	–	242,928	–
M_r gel filtration	340,000	310,000	335,000	70,000	42,000	270,000	330,000
M_r of subunit(s) by gene sequencing	41,435	41,326	41,350	36,000	42,000	40,488	41,000
Number of subunits	8	8	8	2	1	6	8
pI	4.3	5.3	3.5	–	–	–	5.8
pH optimum							
Oxidative deamination	11.3	10.5	10.4	10.8	10	11.0	12.0
Reductive amination	10.3	9.0	9.4	9.8	–	9.2	12.0
Apparent K_m (mM) value for							
L-Phenylalanine	0.22	0.096	0.088	3.80	0.75	0.22	0.10
L-Tyrosine	0.50	–	–	–	–	–	–
Phenylpyruvate	0.40	0.16	0.106	0.50	0.06	0.045	0.03
p-Hydroxyphenylpyruvate	0.34	–	–	1.30	–	–	–
NAD^+	0.17	0.14	0.15	0.25	0.23	0.07	0.20
NADH	0.025	0.072	0.21	0.043	–	0.025	0.07
Ammonia	78	85	127	70	96	106	85
Remaining activity after incubation for 10 min	100%	75%	50%	100%	50%	100%	100%
	55 °C	40 °C	55 °C	35 °C	53 °C, 2 h	70 °C, 1 h	60 °C
	pH9.0	pH9.0	pH8.0	pH7.4	pH9.5–10	pH7.2	
Reference	(3,9)	(3,21)	(7,20)	(9)	(10)	(11,12)	(15)

[a] –, data not available.

Table 2. Substrate Specificity of PheDH from *S. ureae* SCRC-R04 and *B. sphaericus* SCRC-R79a

Amino acid	*S. ureae* SCRC-R04 Relative activity (%)	*S. ureae* SCRC-R04 K_m (mM)	*S. ureae* SCRC-R04 V_{max} (units/mg)	*B. sphaericus* SCRC-R79a Relative activity (%)	*B. sphaericus* SCRC-R79a K_m (mM)	*B. sphaericus* SCRC-R79a V_{max} (units/mg)
Phe	100	0.096[a]	114[a]	100	0.22[a]	114[a]
Tyr	5.4[b]			72[c]	0.50[a]	112[a]
Trp	5.0			1.2		
His	0.1			ND[d]		
Met	4.1			3.0		
Ethionine	7.0			3.1		
Val	3.1			1.4		
Leu	2.3			1.3		
Ile	0.7			0.45		
L-*allo*-Isoleucine	4.3			0.3		
L-α-Amino-*n*-butyric acid	1.6			NT[e]		
L-Norvaline	6.3			1.3		
L-Norleucine	15	0.25[f]	20.0[f]	3.9		

Note: The oxidative deamination reaction was carried out under the standard reaction conditions. The concentration of amino acid was 10 mM unless indicated otherwise (3). Amino acids that were inactive as substrates for the *Sporosarcina* enzyme included Gly, Ala, His, Arg, Lys, Orn, Asp, Glu, Gln, Pro, Ser, Thr, Cys, L-tyrosinamide, L-tyrosine hydroxamate, D-Phe, D-Tyr, D-phenylglycine, *tert*-L-leucine, and DL-*threo*-phenylserine.
[a]Value determined from the secondary plots of intercepts against reciprocal concentration of the substrate.
[b]Measured at 1.4 mM.
[c]Measured at 0.3 mM.
[d]ND, not detected.
[e]NT, not tested.
[f]Apparent value determined from double reciprocal plot a fixed concentration of NAD^+ (2.5 mM).

Table 3. Substrate Specificity of PheDH from *S. ureae* SCRC-R04 and *B. sphaericus* SCR-R79a in the Reductive Amination Reaction

Keto acid	*S. ureae* SCRC-R04 Relative Activity (%)	*S. ureae* SCRC-R04 K_m (mM)	V_{max} (units/mg)	*B. sphaericus* SCRC-R79a Relative activity (%)	K_m (mM)	V_{max} (units/mg)
Phenylpyruvate	100	0.16[a]	598[a]	100	0.40[a]	416[a]
p-Hydroxyphenylpyruvate (5 mM)	24			136	0.34[a]	1240[a]
Indole-3-pyruvate (2 mM)	0.73			0.39		
Imidazole pyruvate	0.04			ND[b]		
α-Keto-γ-methylthiobutyrate	27			11		
α-Ketoisovalerate	2.1			5.5		
α-Ketoisocaproate	13			7.8		
DL-α-Keto-β-methylvalerate	3.2			2.9		
α-Ketovalerate	8.8			6.2		
α-Ketocaproate	32	2.44[c]	379[c]	ND[b]		

Source: From Ref. 3.
[a]Apparent value obtained from the secondary plot of intercepts against reciprocal concentration of the substrate at a fixed concentration of NH_4Cl (400 mM).
[b]ND, not detectable.
[c]Apparent value obtained from a double reciprocal plot at a fixed concentration of NADH (0.4 mM) and NH_4Cl (400 mM).

nucleotide-binding domain G-X-G-X-X-(G or A), connecting β-strand with α-helix in the region of adenine dinucleotide phosphate (ADP)-binding $\beta\alpha\beta$ folds, which is strongly conserved among $NAD(P)^+$-dependent dehydrogenases and FAD-containing oxidoreductases (27).

Reaction Mechanism

Speculation concerning the reaction mechanism of PheDH is based on the findings with respect to the substrate specificity already described, and on other experiments with GluDH and LeuDH. The formation of 2-iminoglutarate, catalyzed by GluDH, has been proved (28). The *pro-S* hydrogen at position 4 of the reduced pyridine ring of NADH is incorporated into the product Phe in the reductive amination reaction of PheDH (B-stereospecific) (3). Studies on the steady-state kinetics of the PheDH reaction revealed that the reaction proceeds sequentially (3); that is, after all three substrates (phenylpyruvate, ammonia, and NADH) bind to the enzyme, the product Phe is released. In a homology search of the deduced primary structure of PheDH, a sequence similar to the active site of GluDH, which also resembles LeuDH, has been found (29). The reaction catalyzed by PheDH would be analogous to that of GluDH from *Clostridium symbiosum* discussed on the basis of X-ray crystallographic studies by Rice et al.

Table 4. Substrate Specificity of PheDH from *B. sphaericus* SCRC-R79a (3,6)

Substrate (10 mM)	Relative activity (%)
4-Hydroxyphenylpyruvate	136
Phenylpyruvate	100
4-Vinylphenylpyruvate	52
4-Fluorophenylpyruvate	39
α-Keto-γ-methylthiobutyrate	11
α-Keto-β-DL-phenylvalerate	8.8
α-Ketoisocaproate	7.8
α-Ketobutyrate	6.3
α-Ketovalerate	6.2
α-Ketoisovalerate	5.5
α-Keto-β-DL-(4-fluorophenyl)butyrate	5.0
α-Keto-β-DL-phenylbutyrate	4.8
α-Keto-4-phenylbutyrate	3.4
α-Keto-β-DL-methylvalerate	2.9
α-Keto-ε-phenylvalerate	2.0
α-Keto-β-DL-(3-methylphenyl)butyrate	1.0
α-Ketononanoate	0.73
α-Keto-β-(2-naphthalene)propionate	0.46
Indolepyruvate (2 mM)	0.39
α-Keto-β-DL-phenyl-γ-methylvalerate	0.10

(25,26,29–32) and to that of *B. stearothermophilus* LeuDH estimated by the kinetic studies done with wild-type and mutant enzymes by Tanizawa et al. (33). The first step would be initiated by the deprotonation of the α-amino group of Phe by an acidic residue (^{126}Asp in *B. sphaericus* PheDH, which corresponds to ^{165}Asp in GluDH) acting as a general base. Subsequently, a hydride transfer from the hydrogen attached to the α-carbon to the *Si* face of NAD^+ (a hydride transfer between the *pro-S* hydrogen of NADH and the *Re* face of the imino acid) occurs (3). Next, the iminophenylpyruvate produced is hydrated by the action of a general base (^{90}Lys in PheDH, which corresponds to ^{125}Lys in GluDH and ^{80}Lys in LeuDH) in the consensus sequence (G-G-X-K, X = G or A or S). The iminophenylpyruvate breaks into phenylpyruvate, ammonia, and a proton. The orientation of phenylpyruvate is fixed, with the phenyl group accommodated in the hydrophobic pocket and the carboxyl group as an anchor, as evidenced by the finding that the enzyme does not catalyze the reductive amination of neutral α-keto-β-phenylpropanol and the α-keto esters (6). Tanizawa et al. (33) generated LeuDH mutants of *B. stearothermophilus* in which ^{80}Lys was changed to Ala, Arg, or Gln and found that the mutants showed reduced activities, but their K_m values for substrate and coenzyme did not change significantly. They analyzed the pH dependence of the reaction and observed a solvent isotope effect, suggesting that ^{80}Lys participates in catalysis as a general base and assists a nucleophilic attack of water to the substrate α-carbon atom in the oxidative deamination reaction.

PheDH of *T. intermedius* acts preferentially on Phe and Tyr (11), whereas LeuDH of *B. stearothermophilus* acts almost exclusively on Leu and some other branched-chain L-amino acids (34). The two enzymes share a similar sequence (47%). The inherent hexapeptide segment (^{124}FVHAAR) in the substrate-binding domain of a mutant PheDH of *T. intermedius* was replaced by the corresponding part of LeuDH (MDIIYQ) (35). The catalytic efficiencies (K_{cat}/K_m) of the mutant enzyme with aliphatic amino acids and aliphatic keto acids as substrates were 0.5 to 2% of those of the wild-type enzyme. However, the efficiencies for Phe and phenylpyruvate were greatly reduced to 0.008 and 0.035% of those of the wild-type enzyme, respectively. Soda et al. (36) constructed a chimeric enzyme consisting of an N-terminal domain of PheDH containing the substrate-binding region and a C-terminal domain of LeuDH containing the NAD^+-binding region. The chimeric enzyme showed 6% of that of the parental PheDH activity on Phe and had broader substrate specificity in the oxidative deamination. The substrate specificity of the chimeric enzyme in the reductive amination was an admixture of PheDH and LeuDH. The chimeric enzyme has a Gly corresponding to position 124 of PheDH and a Val corresponding to position 307.

Engel et al. showed the same effect by changing only two sites. ^{124}Gly and ^{307}Leu of PheDH from *B. sphaericus* were altered by site-specific mutagenesis to the corresponding residues in LeuDH, Ala and Val, respectively (19,37). These two residues have previously been implicated (by molecular modeling based on the X-ray crystallographic analysis of *C. symbiosum* GluDH [25,26,28–32]), as important residues in determining the degree of substrate discrimination. Of the single mutants L307V and G124A, and the double mutant G124, L307V displayed lower activities toward Phe and enhanced activity toward almost all aliphatic amino acid substrates tested compared to the wild-type; thus an amino acid dehydrogenase that shows closer resemblance to LeuDH was made from PheDH by changing only two residues.

APPLICATION OF THE ENZYME

Application of thermostable AlaDH and LeuDH is reviewed by Ohsima and Soda (34). A review on regeneration of NADH for organic synthesis was published by Chenault and Whitesides (38). By carefully choosing the differences in the substrate specificities of PheDHs, *B. sphaericus* enzyme was applied to the synthesis of various L-amino acids (6). PheDH from *B. badius* was shown be narrower in substrate specificity (7) and, therefore, was applicable to the microdetermination of Phe and phenylpyruvate in blood (8).

Synthesis of L-Amino Acids from Their α-Keto Analogs

The production of Phe as a starting material for the artificial sweetener aspartame has been a target of industrial development (39). Several enzymatic processes of Phe synthesis have been reported: L-specific hydrolysis of benzylhydantoin (40), amination of *trans*-cinnamic acid (41), transamination from an amino donor to phenylpyruvate (42), and two-step conversion starting from acetamidocinnamic acid *via* phenylpyruvate (43,44). Because the α-keto acids are generally more expensive than the L-amino acids, it has been impractical to use the α-keto acids as starting materials for the synthesis of L-amino acids. With the development of an efficient method of phenylpyruvic acid synthesis by double carbonylation of benzylchloride in the presence of a cobalt catalyst, a new enzymatic method for the synthesis of Phe from phenylpyruvate was sought (45). Previously, the enzymatic synthesis of Phe from phenylpyruvate was accomplished with the transamination reaction. However, this process requires an L-amino acid, such as Asp, as an amino donor, and after the reaction, a by-product such as oxaloacatate or its degradation product, pyruvic acid, are formed. With PheDH, ammonia can be used as a ammonium source.

The discovery of PheDH made it possible to devise a reaction scheme for the synthesis of Phe from phenylpyruvate with NADH as a reducing agent. The enzyme from *B. sphaericus* was chosen to study the application of the enzyme to the synthesis of various L-amino acids because it is very stable and shows broader substrate specificity than other PheDHs, acting on Tyr as well as Phe. Using PheDH from *B. sphaericus* and formate dehydrogenase (FDH) from *Candida* (Fig. 1), with a catalytic amount of NAD^+, Phe was synthesized in a good yield (120 g/L) (6,46).

In the study on substrate specificity of the enzyme in the reductive amination of α-keto acids, various optically pure L-amino acids were quantitatively synthesized using PheDH and FDH. Table 5 shows the yield of L-amino acids thus synthesized (6). The products from α-keto-β-DL-methylbutyrate and α-keto-β-DL-methyl-β-phenylpyruvate were identified as diastereomeric mixtures of Ile and L-*allo*-isoleucine, and L-α-amino-β-DL-methyl-β-phenylpropionic acid, respectively. L-α-Amino-γ-phenylbutyric acid (L-homophenylalanine) and other unnatural L-amino acids could be efficiently synthesized. L-Homophenylalanine is a building block of some of the angiotensin-converting-enzyme inhibitors (47). The solubility of these Phe homologs, such as Tyr, L-α-amino-γ-phenylbutyric acid, L-α-amino-ϵ-phenylvaleric acid, and so on, is so low that they are easily separated in crystalline forms from the reaction mixture by filtration. The filtered enzyme solution can be used for further repeated synthesis.

L-β-Chloroalanine is a useful intermediate for the synthesis of several L-amino acids. Conditions for the synthesis of optically pure L-β-chloroalanine from β-chloropyruvate using AlaDH, LeuDH, and PheDHs from *B. badius, S. ureae*, and *T. intermedius* with a regeneration of NADH by FDH were investigated (48). The enzymatic reaction was carried out at neutral pH because of the chemical instability of β-chloropyruvate under alkaline conditions. AlaDH from *B. stearothermophilus* IFO 12550 showed the highest activity for the production of L-β-chloroalanine at pH 7.5. L-β-Chloroalanine was produced with high chemical (>90%) and optical yields (100% enantiomeric excess) and at a high concentration (43 g/L).

Immobilized *Nocardia opaca* cells were used to synthesize Phe from phenylpyruvate and ammonia under hydrogen pressure, although no enzymological study was done (49). Thermostable L-amino acid dehydrogenases such as PheDH and formate dehydrogenase were simultaneously expressed in *E. coli*, which was utilized for the synthesis of L-amino acids from their keto analogs (50).

Continuous Synthesis of Phe

Membranes are efficiently used in the enzyme-catalyzed syntheses of optically active compounds on laboratory and industrial scales (51–53). This technique can be applied for the repeated use of the enzyme in the synthesis, especially if smaller amounts of enzyme are available. Small amounts of PheDH from *B. sphaericus* and FDH were placed in a dialysis tube to check their durability and ef-

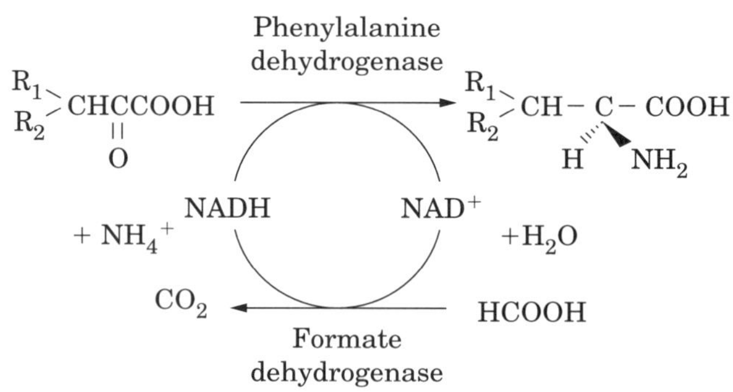

Figure 1. Synthesis of L-amino acids from their β-keto analogs by PheDH with a regeneration of NADH by formate dehydrogenase.

Table 5. Synthesis of L-Amino Acids from α-Keto Acids by Using *B. sphaericus* PheDH and FDH (6)

Substrate	Product	Yield (%)
Phenylpyruvate	L-Phe	>99
4-Hydroxyphenylpyruvate	L-Tyr	>99
4-Fluorophenylpyruvate	L-4-Fluorophenylalanine	>99
α-Keto-γ-phenylbutyrate	L-α-Amino-γ-phenylbutyric acid	99
α-Keto-ε-phenylvalerate	L-α-Amino-ε-phenylvaleric acid	98
α-Keto-β-methyl-β-phenylpropionate	L-α-Amino-β-DL-methyl-β-phenylpropionate	98
α-Ketononanoate	L-α-Aminononanoic acid	99

Note: To prevent substrate inhibition, α-keto acids have been divided into portions and added to the reaction mixture, not to exceed 50 mM.

ficiency in the continuous synthesis of Phe. The *B. sphaericus* PheDH proved to be very stable in the synthesis, with high efficiency. PheDH maintained its activity for up to 28 changes. During the 34-day operation, 10.74 g (6.51×10^{-2} mol) of Phe was synthesized, using 45 μg (5 units, 1.35×10^{-10} mol) of PheDH, ca. 7.5 mg (15 units) of FDH, and ca. 35 mg of $NAD^+ \cdot Na$ (5×10^{-5} mol). Thus, 1 mol of PheDH catalyzed the synthesis of 4.8×10^8 mol of Phe, which is 2.4×10^5 times the weight of the enzyme (6). Continuous synthesis of Phe by the use of membrane reactors has been demonstrated with PheDHs from *Brevibacterium* sp. (54) and *Rhodococcus* sp. (55,56).

To avoid the multistep purification procedure, the use of enzymes in whole cells is preferable. Acetone-dried cells of *B. sphaericus* and *C. boidinii* were also effective for Phe synthesis, providing a simple microbial method of synthesis. In a typical course of the reductive amination reaction of phenylpyruvate using the mixture of acetone-dried cells as catalysts, the concentration of Phe reached 61.5 mg/mL with a yield of more than 99% (6).

Tyr and Phe labeled with a positron-emitting radionuclide would be useful for positron emission tomography (PET) studies of tissues and neoplasms, such as measuring protein synthesis in tumor and brain tissue. ^{13}N-Labeled Phe, Tyr, and L-dopa were synthesized for PET (57). PheDH catalyzes the reductive ^{13}N amination of either phenylpyruvate or *p*-hydroxyphenylpyruvate to form [^{13}N]Phe or [^{13}N]Tyr, respectively, with NADH as a reducing agent. After short incubation of ^{13}N ammonia and either phenylpyruvate or *p*-hydroxyphenylpyruvate with the enzyme immobilized on the CNBr-activated Sepharose column, 83% of the ^{13}N ammonia was converted to [^{13}N]Phe, and 38% of the ^{13}N was converted to [^{13}N]Tyr, respectively. The labeled amino acids were purified by passage of the solution through an ion exchange column. Yields of each labeled amino acid were >30 mCi. L-Dopa was labeled by the exchange reaction between L-dopa and ^{13}N ammonia in the presence of NAD^+ and PheDH; 9% of the label was transferred to L-dopa. In a separate experiment involving the exchange reaction between Phe and ^{13}N ammonia, transfer of label from ^{13}N ammonia to Phe was 50% (Scheme 1). Thus, utilizing *B. sphaericus* PheDH is useful for preparing [^{13}N]Tyr and [^{13}N]Phe.

$$^{13}NH_3 + \text{L-dopa} \xrightarrow[\text{NADH}+H^+]{PheDH} NH_3 + \text{L-}[^{13}\text{N}]\text{dopa}$$

Scheme 1. Enzymatic synthesis of L-[^{13}N]-dopa by an exchange reaction.

Microdetermination by Cycling Assay

Quantitative determination of Phe in the plasma is important in diagnosing PKU. Several groups have been investigating the use of PheDH for the microdetermination of Phe and phenylpyruvate in blood samples (58–70). *B. badius* PheDH was found to be structurally closely related to that from *B. sphaericus*, although the former has a narrower substrate specificity suitable for the microdetermination, whereas the latter is more stable and shows a wider specificity suitable for the synthesis of various L-amino acids.

Utilizing *B. sphaericus* PheDH, enzymatic cycling assay for the determination of Phe and phenylpyruvate in deproteinized tissue extracts was devised by Cooper et al. (61). Assay 1 couples glutamine transaminase K with PheDH (Scheme 2). Assay 2 combines PheDH, L-amino acid oxidase, and catalase (Scheme 3). In both assays, Tyr and some other amino acids (or their α-keto acid analogs) can replace Phe (or phenylpyruvate) to a small extent. Thus, when measuring Phe, a correction must be made for the nonspecificity of the reaction. By removing Phe on a cation exchange column, it was possible to measure phenylpyruvate in tissue extracts. Concentrations of phenylpyruvate (μmol/kg) in normal rat liver, kidney, and brain were 2.1 $\pm$ 1.1 (n = 8), 1.8 $\pm$ 0.4 (n = 4), and 3.3 $\pm$ 0.6 (n = 4), respectively.

A spectrophotometric recycling assay for the quantitation of Phe (and phenylpyruvate) involves the coupling of PheDH with rat kidney cytosolic glutamine transaminase K (68). The latter enzyme possesses high affinity for phenylpyruvate. Recycling results in a 50-fold increase in sensitivity over that of a conventional spectrophotometric end-point analysis procedure. The spectrophotometric recycling procedure has now been adapted to the measurement of Phe in microliter quantities of human blood. The procedure is 10 times more sensitive than the end-point assay for the spectrophotometric measurement of Phe in

$$\text{L-Phe} + NAD^+ + H_2O \xrightarrow{PheDH} \text{PPA} + \text{NADH} + H^+ + NH_3$$

$$\text{L-Gln} + \text{PPA} \xrightarrow{Gln\ Transaminase\ K} \alpha\text{-Ketoglutaramate} + \text{L-Phe}$$

Net

$$\text{L-Gln} + NAD^+ + H_2O \longrightarrow \alpha\text{-Ketoglutaramate} + \text{NADH} + H^+ + NH_3$$

Scheme 2. Enzymatic cycling assay for Phe.

$$\text{L-Phe} + H_2O + O_2 \xrightarrow{L\text{-}AAO} \text{PPA} + H_2O_2 + NH_3$$

$$H_2O_2 \xrightarrow{Catalase} H_2O + 1/2\ O_2$$

$$\text{PPA} + \text{NADH} + H^+ + NH_3 \xrightarrow{PheDH} \text{L-Phe} + NAD^+ + H_2O$$

Net

$$\text{NADH} + H^+ + 1/2\ O_2 \longrightarrow NAD^+ + H_2O$$

Scheme 3. Enzymatic cycling assay for phenylpyruvate.

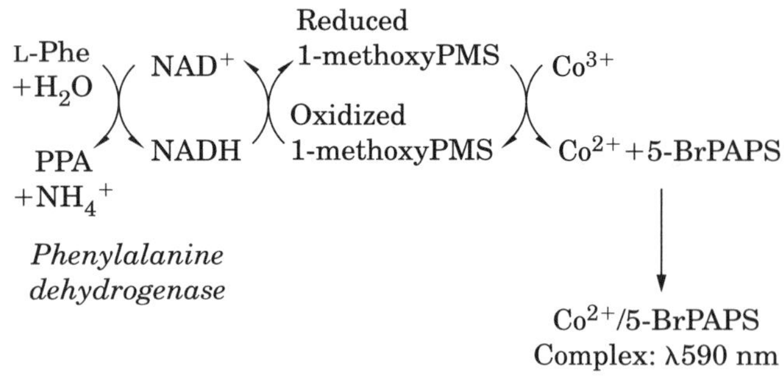

Scheme 4. Colorimetric determination of Phe in blood samples.

human blood. The findings suggest that the recycling procedure adapted for fluorometry will be even more sensitive.

Microdetermination by End-Point Assay

Naruse et al. have introduced a sensitive enzymatic microplate method using PheDH from *B. badius* for the mass screening of PKU among neonates (7,8) as an alternative method for the Guthrie test (71). The principle of this system involves two steps. First, Phe is oxidized by PheDH to phenylpyruvate, reducing NAD^+ to NADH, which then serves as the electron donor in a colorimetric reaction as follows. In the presence of NADH and the electron carrier l-methoxy-5-methylphenazium methylsulfate (l-MPMS), Co^{3+} is reduced to Co^{2+}, which produces a colored chelate with the metal indicator 2-(5-bromo-2-pyridyazo)-5-(*N*-propyl-*n*-sulfopropyl-amino)phenol (5-BrPAPS) (Scheme 4). This assay system detected Phe in filter paper blood in the concentration range of 0.02–1.50 mM. This method uses a simple TCA extraction method and a PheDH reaction with high substrate specificity that produces very sensitive colorimetric quantitation. The absorbance at 585 nm of this chelate can be measured by a colorimetric microplate reader. This highly specific and sensitive method has been used to screen neonates for PKU in Japan since 1992.

CONCLUSION

Since the recent discovery of PheDH in 1984, several microbial producers of the enzyme have been isolated and their enzymological properties and primary structures have been studied. Indeed, PheDH has been used to detect PKU in the blood of neonates in Japan since 1992. However, PheDH has many further applications to be studied. If the structure and function of the amino acid dehydrogenases are thoroughly understood, it might be possible to create a new amino acid dehydrogenase with a wider scope of application.

BIBLIOGRAPHY

1. W. Hummel, N. Weiss, and M.-R. Kula, *Arch. Microbiol.* **137**, 47–52 (1984).
2. Y. Asano and A. Nakazawa, *Agric. Biol. Chem.* **49**, 3631–3632 (1985).
3. Y. Asano, A. Nakazawa, and K. Endo, *J. Biol. Chem.* **262**, 10346–10354 (1987).
4. E.L. Smith, B.M. Austen, K.M. Blumenthal, and J.F. Nyc, *Enzymes* **11**, 293–367 (1975).
5. Nomenclature Committee of the International Union of Biochemistry, *Enzyme Nomenclature*, Academic Press, Orlando, Fla., 1992
6. Y. Asano, A. Yamada, Y. Kato, K. Yamaguchi, Y. Hibino, K. Hirai, and K. Kondo, *J. Org. Chem.* **55**, 5567–5571 (1990).
7. Y. Asano, A. Nakazawa, K. Endo, Y. Hibino, M. Ohmori, N. Numao, and K. Kondo, *Eur. J. Biochem.* **168**, 153–159 (1987).
8. H. Naruse, Y.Y. Ohashi, A. Tsuji, M. Maeda, K. Nakamura, T. Fujii, A. Yamaguchi, M. Matsumoto, and M. Shibata, *Screening* **1**, 63–66 (1992).
9. H. Misono, J. Yonezawa, S. Nagata, and S. Nagasaki, *J. Bacteriol.* **171**, 30–36 (1989).
10. L. de Boer, M. van Rijssel, G.J. Euverink, and L. Dijkhuizen, *Arch. Microbiol.* **153**, 12–18 (1989).
11. T. Ohshima, H. Takada, T. Yoshimura, N. Esaki, and K. Soda, *J. Bacteriol.* **173**, 3943–3948 (1991).
12. N.M.W. Brunhuber, A. Banerjee, W.R. Jacobs, Jr., and J. Blanchard, *J. Biol. Chem.* **269**, 16203–16211 (1994); DDBJ/EMBL/GenBank databases, accession number: RSU08381.
13. X.-L. Huang and Y. Asano, *Bull. Toyama Pref. Univ.* (in Japanese) **4**, 211–215 (1994).
14. Y. Asano, M. Tanetani, and N. Horita, *Bull. Toyama Pref. Univ.* (in Japanese) **6**, 101–106 (1996).
15. Y. Asano and M. Tanetani, *Arch. Microbiol.* **169**, 220–224 (1998).
16. S. Kuroda, K. Tanizawa, Y. Sakamoto, H. Tanaka, and K. Soda, *Biochemistry* **29**, 1009–1015 (1990).
17. J.R. Norris, in R.C.W. Berkeley and M. Goodfellow eds., *The Aerobic Endospore-Forming Bacteria: Classification and Identification* Academic Press, London 1981, pp. 337–357.
18. Y. Asano, K. Endo, A. Nakazawa, Y. Hibino, M. Ohmori, N. Numao, and K. Kondo, *Agric. Biol. Chem.* **51**, 2621–2623 (1987).
19. N. Okazaki, Y. Hibino, Y. Asano, M. Ohmori, N. Numao, and K. Kondo, *Gene* 63, 337–341 (1988); DDBJ/EMBL/GenBank databases, accession number: BACPDH.
20. A. Yamada, T. Dairi, Y. Ohno, X.-L. Huang, and Y. Asano, *Biosci. Biotech. Biochem.* **59**, 1994–1995 (1995); DDBJ/EMBL/GenBank databases, accession number: D50261.
21. DDBJ/EMBL/GenBank databases, accession number: AB001031
22. H. Takada, T. Yoshimura, T. Ohshima, N. Esaki, and K. Soda, *J. Biochem.* (Tokyo) **109**, 371–376 (1991); DDBJ/EMBL/GenBank databases, accession number: BTTPDH
23. S. Nagata, K. Tanizawa, N. Esaki, Y. Sakamoto, T. Ohshima, H. Tanaka, and K. Soda, *Biochemistry* **27**, 9056–9062 (1988).
24. L. Tang and R. Hutchinson, *J. Bacteriol.* **175**, 4176–4185 (1993).
25. K.L. Britton, P.J. Baker, D.W. Rice, and T.J. Stillman, *Eur. J. Biochem.* **209**, 851–859 (1992).
26. P.J. Baker, K.L. Britton, D.W. Rice, A. Rob, and T.J. Stillman, *J. Mol. Biol.* **228**, 662–671 (1992).
27. R.K. Wierenga, M.C.H. Demaeyer, and W.G.J. Hol, *Biochemistry* **24**, 1346–1357 (1985).
28. M.C. Hochreiter and K.A. Schellenberg, *J. Am. Chem. Soc.* **91**, 6530–6531 (1969).
29. K.L. Britton, P.J. Barker, P.C. Engel, D.W. Rice, and T.J. Stillman, *J. Mol. Biol.* **234**, 938–945 (1993).

30. T.J. Stillman, P.J. Baker, K.L. Britton, and D.W. Rice, *J. Mol. Biol.* **234**, 1131–1139 (1993).
31. P.J. Baker, A.P. Turnbull, S.E. Sedelnikova, T.J. Stillman, and D.W. Rice, *Structure* **3**, 693–705 (1995).
32. A.P. Turnbull, P.J. Baker, and D.W. Rice, *J. Biol. Chem.* **272**, 25105–25111 (1997).
33. T. Sekimoto, T. Matsuyama, T. Fukui, and K. Tanizawa, *J. Biol. Chem.* **268**, 27039–27045 (1993).
34. T. Ohshima and K. Soda, *Trends Biotechnol.* **7**, 210–214 (1989).
35. K. Kataoka, H. Takada, T. Yoshimura, S. Furuyoshi, N. Esaki, T. Ohshima, and K. Soda, *J. Biochem.* (Tokyo) **114**, 69–75 (1994).
36. K. Kataoka, H. Takada, K. Tanizawa, T. Yoshimura, N. Esaki, T. Ohshima, and K. Soda, *J. Biochem.* (Tokyo) **116**, 931–936 (1994).
37. S.Y.K. Seah, K.L. Britton, P.J. Baker, D.W. Rice, Y. Asano, and P.C. Engel, *FEBS Lett.* **370**, 93–96 (1995).
38. H.K. Chenault and G.M. Whitesides, *Appl. Biochem. Biotechnol.* **14**, 147–197 (1987).
39. A. Klausner, *Bio/technology* **3**, 301–307 (1985).
40. K. Yokozeki, K. Sano, C. Eguchi, H. Iwagami, and K. Mitsugi, *Agric. Biol. Chem.* **51**, 729–736 (1987).
41. C.T. Evans, K. Hanna, C. Payne, D. Conrad, and M. Misawa, *Enzyme Microb. Technol.* 417–421 (1987).
42. G.J. Calton, L.L. Wood, M.H. Updike, L. Lantz II, and J.P. Hammen, *Bio/technology* **4**, 317–320 (1986).
43. K. Nakamichi, K. Nabe, Y. Nishida, and T. Tosa, *Appl. Microbiol. Biotechnol.* **30**, 243–246 (1989).
44. C.T. Evans, W. Bellamy, M. Glesson, H. Aoki, K. Hanna, W. Peterson, D. Conrad, and M. Misawa, *Bio/technology* **5**, 818–823 (1987).
45. JP 60-61550 (April 9, 1985), K. Hirai and I. Ojima.
46. Y. Asano and A. Nakazawa, *Agric. Biol. Chem.* **51**, 2035–2036 (1987).
47. H. Urbach and R. Henning, *Tetrahedron Lett.* **25**, 1143–1146 (1984).
48. Y. Kato, K. Fukumoto, and Y. Asano, *Appl. Microbiol. Biotechnol.* **39**, 301–304 (1993).
49. T. Matsunaga, M. Higashijima, A. Sulaswatty, and S. Nishimura, *Biotechnol. Bioeng.* **31**, 834–840 (1988).
50. A. Galkin, L. Kulakova, T. Yoshimura, K. Soda, and N. Esaki, *Appl. Environ. Microbiol.* **63**, 4651–4656 (1997).
51. B. Nidetzky, D. Haltrich, and K.D. Kulbe, *Chemtech* **1996**, 31–36 (1996).
52. W. Hummel and M.-R. Kula, *Eur. J. Biochem.* **184**, 1–13 (1989).
53. M.-R. Kula and C. Wandrey, *Methods Enzymol.* **136**, 9–21 (1987).
54. W. Hummel, E. Schmidt, C. Wandrey, and M.-R. Kula, *Appl. Microbiol. Biotechnol.* **25**, 175–185 (1986).
55. W. Hummel, H. Schütte, E. Schmidt, C. Wandrey, and M.-R. Kula, *Appl. Microbiol. Biotechnol.* **26**, 409–416 (1987).
56. R. Campagna, and A.F. Bückmann, *Appl. Microbiol. Biotechnol.* **26**, 417–421 (1987).
57. A.S. Gelbard, A.J.L. Cooper, Y. Asano, E. Nieves, S. Filc-DeRico, and K.C. Rosenspire, *Appl. Rad. Isotopes* **41**, 229–233 (1990).
58. T. Ohshima, H. Sugimoto, and K. Soda, *Anal. Lett.* **21**, 2205–2215 (1988).
59. W. Hummel, H. Shütte, and M.-R. Kula, *Anal. Biochem.* **170**, 397–401 (1988).
60. U. Wendel, W. Hummel, and U. Langenbeck, *Anal. Biochem.* **180**, 91–94 (1989).
61. A.J.L. Cooper, L.K.H. Leung, and Y. Asano, *Anal. Biochem.* **183**, 210–214 (1989).
62. U. Wendel, M. Koppelkamm, W. Hummel, J. Sander, and U. Langenbeck, *Clin. Chim. Acta* **192**, 165–170 (1990).
63. U. Wendel, M. Koppelkamm, and W. Hummel, *Clin. Chim. Acta* **201**, 95–98 (1991).
64. K. Dooley, *Clin. Biochem.* **25**, 271–275 (1992).
65. M.A. Vilaseca, C. Farre, and F. Ramon, *Clin. Chem.* **39/1**, 129–131 (1993).
66. S. Keffler, R. Denmeade, and A. Green, *Ann. Clin. Biochem.* **31**, 134–139 (1994).
67. B.C. Cook, T.R. White, C.H. Smith, and M. Landt, *Clin. Chem.* **41**, 949–950 (1995).
68. K. Nakamura, T. Fujii, Y. Kato, Y. Asano, and A.J.L. Cooper, *Anal. Biochem.* **234**, 19–22 (1996).
69. E.W. Randell, and D.C. Lehotay, *Clin. Biochem.* **29**, 133–138 (1996).
70. T. Huang, A. Warsinke, T. Kuwana, F.W. Scheller, *Anal. Chem.* **70**, 991–997 (1998).
71. R. Guthrie and A. Susi, *Pediatrics* **32**, 338–343 (1963).

See also BIOCATAYLSIS DATABASES; PHENYLALANINE.

PHENYLGLYCINES, D-PHENYLGLYCINES

SATOMI TAKAHASHI
Kaneka Corporation
Hyogo, Japan

KEY WORDS

Asymmetric hydrolysis
Decarbamoylation
Dynamic resolution
Hydantoin
Hydantoinase
D-*p*-Hydroxyphenylglycine
D-Phenylglycine
Racemization

OUTLINE

INTRODUCTION

D-Phenylglycines is a general name of a representative group of industrially useful D-α-amino acids. They consist of 2-phenyl and substituted phenylglycine, which have D-form stereo configuration, that is, the opposite stereo configuration of L-amino acids.

D-Amino acids are often called "unnatural" amino acids, and they are not as commonly used as L-amino acids. However, they are widely distributed in various living metabolites and play important roles as physiologically active components.

In recent years, D-amino acids have become increasingly important in the pharmaceutical field as intermediates or chiral synthons for the preparation of β-lactam antibiotics, peptide hormone mimics, pyrethroids, and other compounds. Some D-phenylglycines (D-PGs) are important intermediates for well-known semisynthetic penicillins and cephalosporins, such as ampicillin or amoxicillin, as their side chains.

Among the various D-PGs, D-phenylglycine (D-PG) and D-*P*-hydroxyphenylglycine (D-HPG) are the most important intermediates and are produced and used in large amounts (several thousand metric tons per year) throughout the world. Although they had been produced by classical optical resolution methods until the 1980s, economical and efficient unique synthetic methods have been established that employ asymmetric inversions accomplished chemically (dynamic optical resolution) or enzymatically (hydantoinase process).

STRUCTURE AND NOMENCLATURE

D-PGs are substituted at the second-position proton in glycine (2-amino acetic acid) by phenyl or substituted phenyl radicals to form the D-stereo configuration, as shown in Figure 1. The absolute stereo configuration of the structure is (*R*)-form. D-PGs are enantiomers of L-phenylglycines (L-PGs), which have the reverse stereo configuration.

Representatives of D-PGs include D-phenylglycine (all X = H) and D-*p*-hydroxyphenylglycine (X = *p*-OH), which are also called D-2-amino-2-phenylacetic acid and D-2-amino-2-(4-hydroxyphenyl)acetic acid, respectively.

DERIVATIVES AND USAGE

The practical usage of D-PG and D-HPG is essentially limited to the creation of side chains for various semisynthetic penicillins (SSPs) and semisynthetic cephalosporins (SSCs), as shown Tables 1–3. Among of them, D-PG is used mainly for the side chain of ampicillin and cephalexin, and D-HPG is used mainly for the side chain of amoxicillin and cefatrizine, as shown in Figure 2.

When D-PGs are used for the production of these penicillins and cephalosporins, it is usually as active derivatives such as DANE-salt or acid chloride, which are reacted with 6-aminopenicillanic acid (6-APA, etc.) or 7-amino cephalospollins (7-ACA, 7-ADCA, etc.), as schematically shown in Figure 3.

The DANE-salts of D-PGs are prepared by the reaction of alkali metal salts (Na, K) of D-PGs with methyl acetoacetate (MAA). The acid chlorides of D-PG and D-HPG are prepared by the reaction of D-PGs with chlorinating reagents such as phosphorous pentachloride (PCl_5) or phosgene ($COCl_2$), followed by hydrogen chloride treatment, respectively.

PRODUCTION METHODS

Industrially, D-PGs have been produced by resolving the racemic DL-PGs obtained by chemical syntheses. Typical optical resolution methods are resolution by fractional crystallization of the corresponding diastereomer salt, and by predominant crystallization of the corresponding aromatic sulfuric acid salt. However, these conventional methods are disadvantageous. The former method requires many reaction steps as well as an expensive resolving agent, and the optical purity of the product is not high. The yield obtained in one cycle of the latter method is very low. Furthermore, each method requires a complicated step for racemizing the residual useless L-PGs, and a step for recyclization of the optical resolution in order to increase the overall yield of D-PGs.

To overcome these difficulties, efficient synthetic methods have recently been established by employing unique

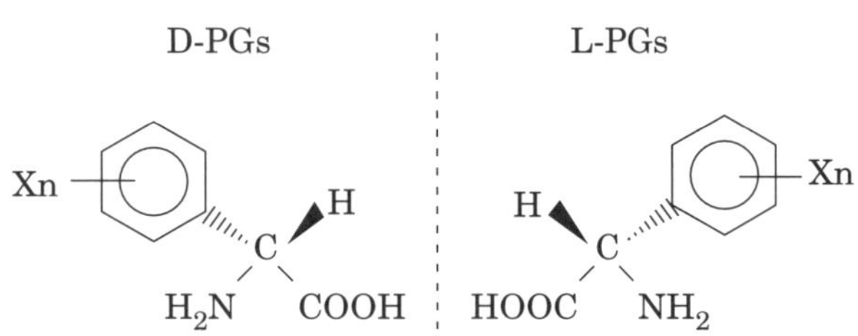

Figure 1. Stereo configurations of D-phenylglycines.

Table 1. D-PG Using Semisynthetic Penicillins (SSP) and Semisynthetic Cephalosporins (SSC)

Type	General name		
SSP	Ampicillin	Apalcillin	Azlocillin
	Bacampicillin	Helacillin	Lenampicillin
	Mezlocillin	Piperacillin	Pirbenicillin
	Pivampicillin	Sarpicillin	Sultamicillin
	Suncillin	Talampicillin	
SSC	Cefaclor	Cefpimizole	Cephalexin
	Cephaloglycin	Loracarbef	Pivcefalexin

Table 2. D-HPG Using Semisynthetic Penicillins (SSP) and Semisynthetic Cephalosporins (SSC)

Type	General name		
SSP	Amoxicillin	Aspocillin	Fumoxicillin
	Piridicillin		
SSC	Cefadroxil	Cefapadole	Cefatrizine
	Cefoperazone	Cefpiramide	Cefprozil

Table 3. Substrate Specificity of Hydantoinase from *Pseudomonas striata*

Substrate		Rerative rate
5- H-	Hydantoin	13
5- CH_3-	Hydantoin	45
5- $(CH_3)_2CH$-	Hydantoin	15
5- $(CH_3)_2CHCH_2$-	Hydantoin	48
5- $CH_3SCH_2CH_2$-	Hydantoin	48
5- 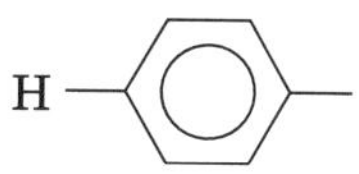	Hydantoin	25
5- 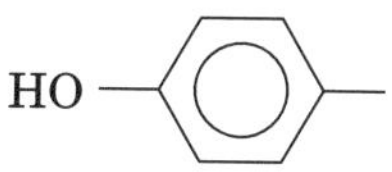	Hydantoin	16
5- 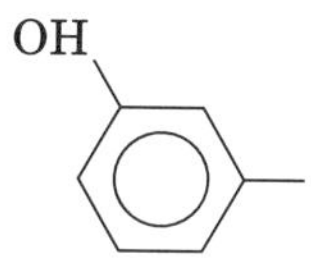	Hydantoin	5
5- 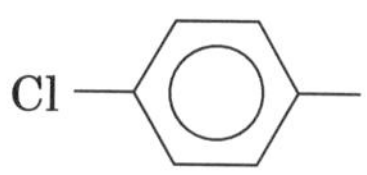	Hydantoin	19
5- 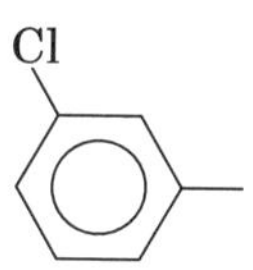	Hydantoin	10
5- 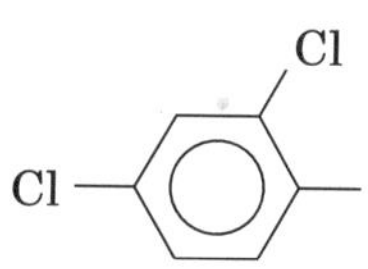	Hydantoin	7
5- 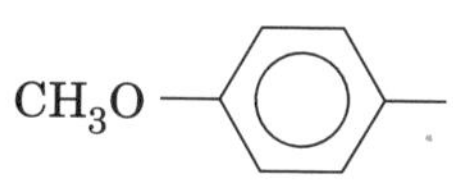	Hydantoin	4
5,5- 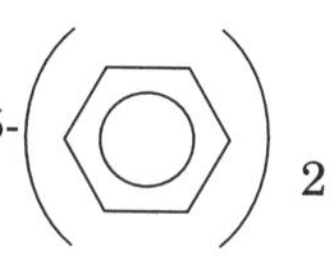	Hydantoin	0
5,5-$(CH_3)_2$	Hydantoin	0
Dihydrouracil		100

asymmetric transformations that are based on in situ racemization, also called deracemization, of the substrate. These new processes have none of the disadvantages of the optical resolution (conventional) method and are technically and economically advantageous. The overall result of these methods is an asymmetric transformation in which, in principle, 100% yield and 100% ee can be obtained.

One typical method is based on the dynamic resolution process; the other is the hydantoinase process. These methods are used for the economical production of D-PGs, including D-PG and D-HPG, with good efficiency.

Dynamic Resolution Process

In the conventional optical resolution process, the maximum yield of one cycle is not higher than 50%. However, in the dynamic resolution process, the theoretical resolution yield of D-PGs from DL-PGs can be 100% because in situ racemization of L-PGs occurs through Schiff base tautomerism with the precipitation of the diastereomer salt of the D-PGs plus the resolving agent concomitantly. The mechanism of racemization through Schiff base tautomerism is shown in Figure 4. Some aldehydes or ketones are required as a racemization catalyst to form the Schiff base of the amino acid or its derivatives in an acidic solvent such as acetic acid.

D-PG Production. The dynamic resolution process for D-PG uses DL-phenylglycin amide as the substrare, mandelic acid (MA) as the resolving agent, and benzaldehyde as the racimezation calalyst, followed by acid hydrolysis (1,2). This process is shown schematically in Figure 5.

D-HPG Production. The starting material, DL-HPG, is basically produced from glyoxylic acid, phenol, and ammonia. In DL-HPG preparation, there are two processes: the "one-pot" process and the via-HMA process. In the one-pot process, glyoxylic acid, phenol, and ammonia are reacted at the same time. But the yield is not high (not more than 40%) because of the complex formation of DL-HPG produced with glyoxylic acid under the reaction conditions. So usually, DL-HPG is produced via the reaction of 4-hydroxymandelic acid (HMA), which is produced from glyoxylic acid and phenol, with ammonia or its salts, in good yield (60 to 70%).

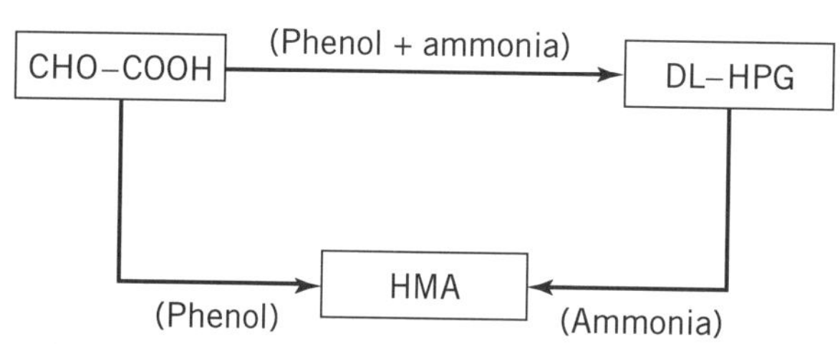

DL-HPG can be efficiently resolved into D-HPG by two dynamic resolution processes: (*1*) a dynamic preferential resolution process using arylsulfonic acid (Ar · SO_3H) to precipitate D-HPG · arylsulfonate salt preferentially by the salt seeding (3,4), and (*2*) a dynamic diastereomer resolution process using D-bromocamphorsulfonic acid (BCSA) as the resolving agent (5), as shown in Figure 6. Because of the high reproducibility and high yield, the later process is generally used for commercial D-HPG production. The yield of D-HPG is approximately 90%, and there is almost 100% ee.

Hydantoinase Process

The hydantoinase process, which is an attractive chemicoenzymatic process from both the economical and technical standpoints, was initially developed for the production of various D-amino acids by employing microbial hydantoinase (6). In the process, racemic DL-hydantoins,

General name	Structure
Ampicillin	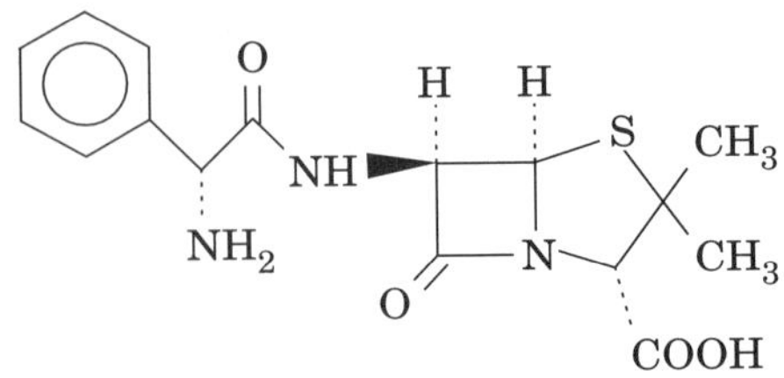
Cephalexin	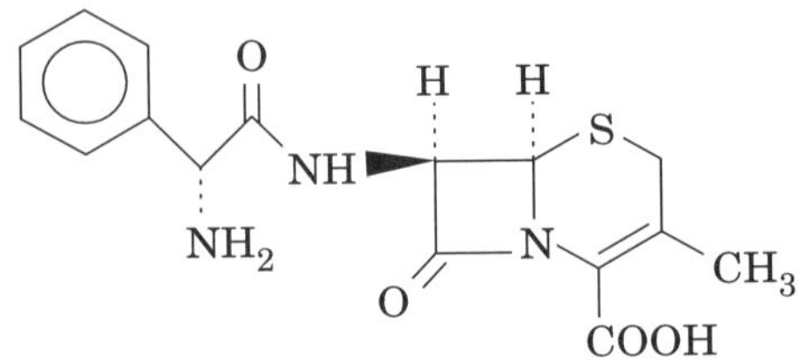
Amoxicillin	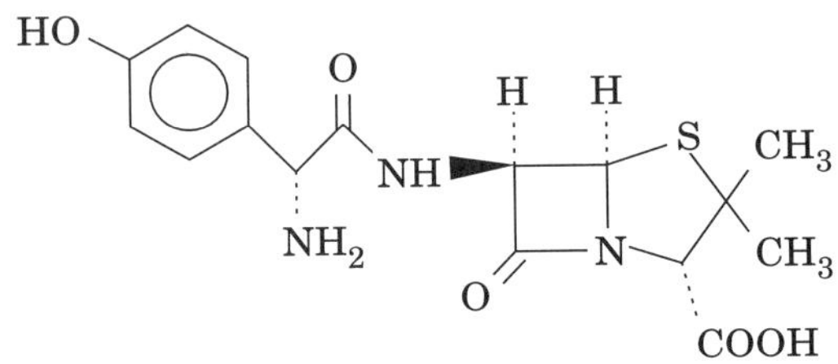
Cefatrizine	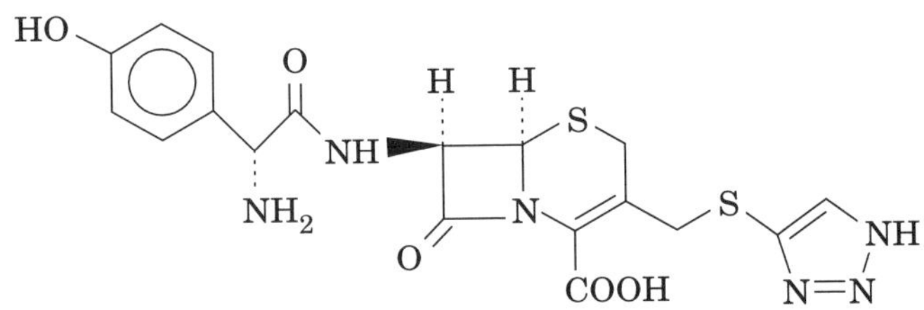

Figure 2. The representative semisynthetic penicillins and semisynthetic cephalosporins.

prepared by a Bucherer-Berg reaction from aldehyde, can be completely transformed into the corresponding optically active *N*-carbamoyl-D-amino acids, because in situ racemization of L-hydantoins occur mainly through keto–enole tautomerism of the hydantoins under the enzymatic reaction conditions. The *N*-carbamoyl-D-amino acids thus obtained can be readily transformed into D-amino acids in good yield by two methods: the chemical method using nitrous acid, or the enzymatic method using *N*-carbamoyl-D-amino acid amidehydrolase.

D-PG can be efficiently produced by the hydantoinase process from benzaldehyde, but the chemical resolution method is usually used for this purpose owing to its easy racemization of the unuseful L-PG.

After the establishment of a new synthetic method for DL-5-(4-hydroxyphenyl)hydantoin (p-HPH), the substrate for D-HPG production, the hydantoinase process was developed mainly for D-HPG production; a practical hydantoinase process was established and began to be used commercially. At present, most commercial D-HPG is produced by this process.

Hydantoinase and D-amino Acid Production. Despite many attempts at L-amino acid production from DL-hydantoins, D-amino acid production from hydantoins was not achieved until the end of the 1970s.

Dudley et al., in the course of studies on the metabolic fate of the anticonvulsants ethotoin and dilantin, found that only the D-form of 5-phenylhydantoin was hydrolyzed into *N*-carbamoyl-D-phenylglycine under the catalysis of animal dihydropyrimidinase (E.C.3,5,2,2), which catalyzed the hydrolysis of dihydropyrimidines to *N*-carbamoyl-*β*-amino acids (7). On the basis of these findings, Cecere et al., studied the specificities of the enzyme and showed that the D-form of several 5-substituted hydantoins were hydrolyzed to provide the corresponding *N*-carbamoyl-D-amino acids (8).

The distribution of hydantoin-hydrolyzing activity in microorganisms was investigated by Yamada and Takahashi (9), and high activity was found in various bacteria belonging to the genera *Pseudomonas, Aerobacter, Agrobacterium, Bacillus, Corynebacterium*, and *Actinomycetes*. A hydantoin-hydrolyzing enzyme (hydantoinase) was purified and crystallized from the cell extract of *Pseudomonas striata*, and its properties were established in some detail (10). The bacterial enzyme showed its highest activity and affinity toward dihydrouracil, suggesting its identity with dihydropyrimidinase. The enzyme also catalyzed the

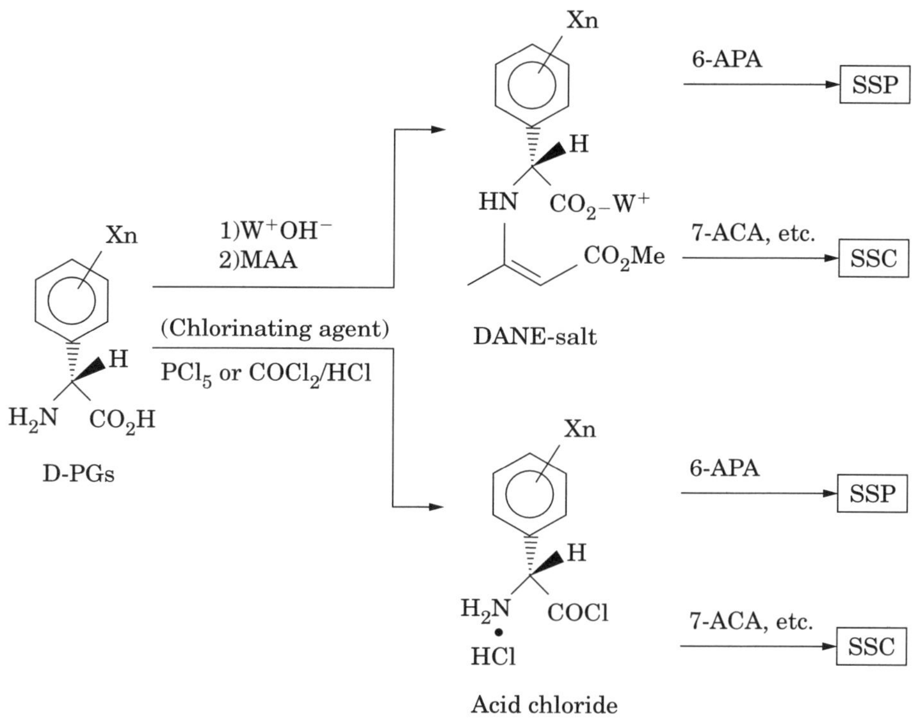

Figure 3. Derivatization of D-PGs and the synthesis of semisynthetic penicillins (SSP) and semisynthetic cephalosporins (SSC).

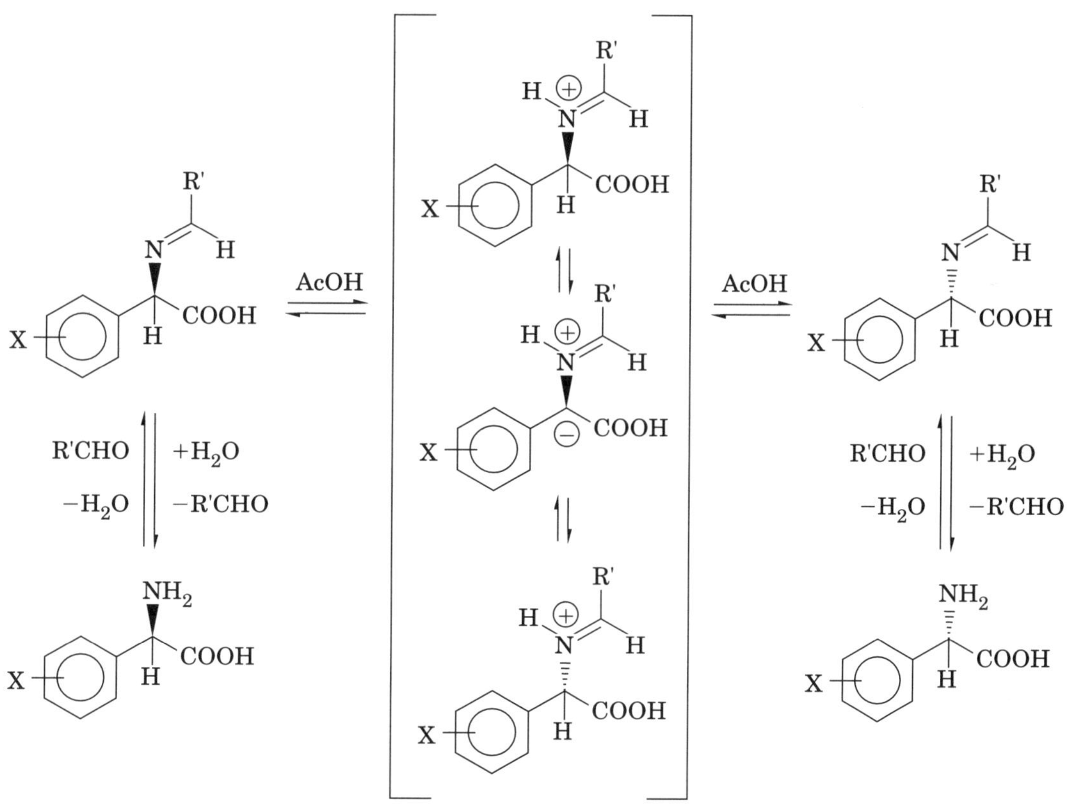

Figure 4. Racemization mechanism.

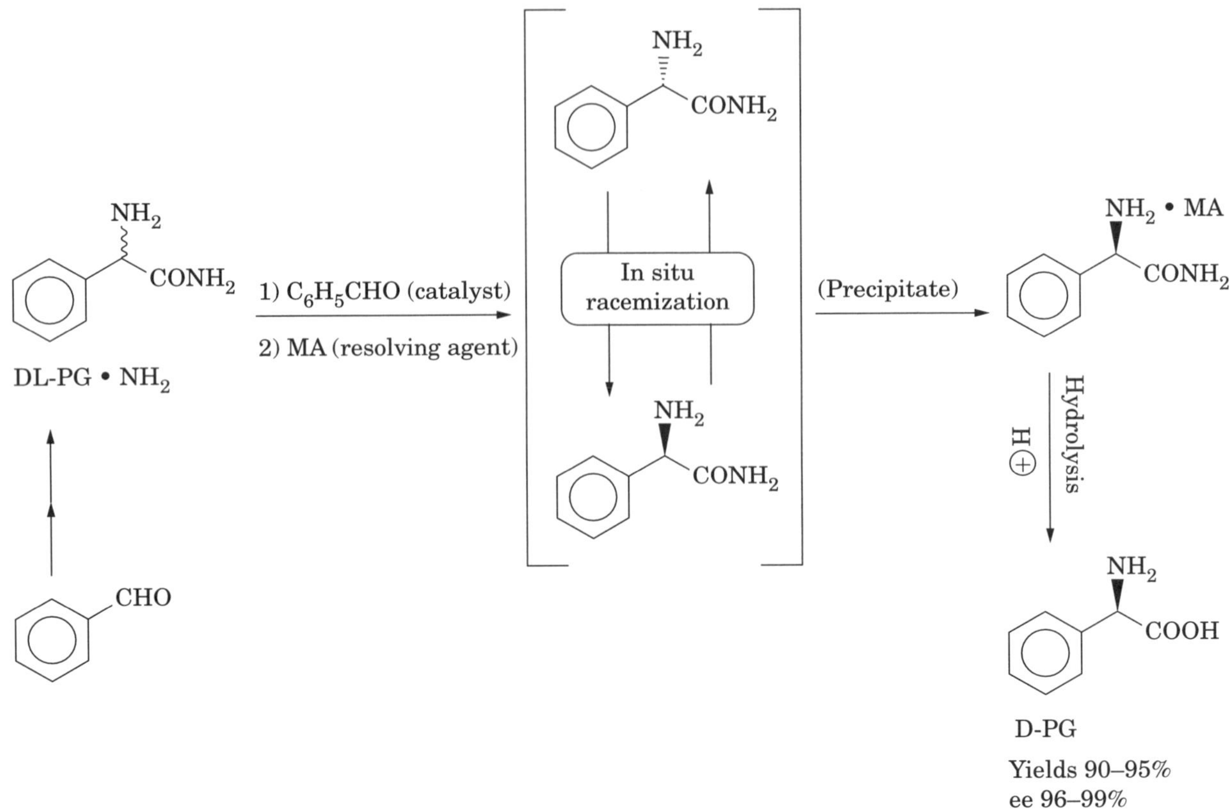

Figure 5. D-PG production.

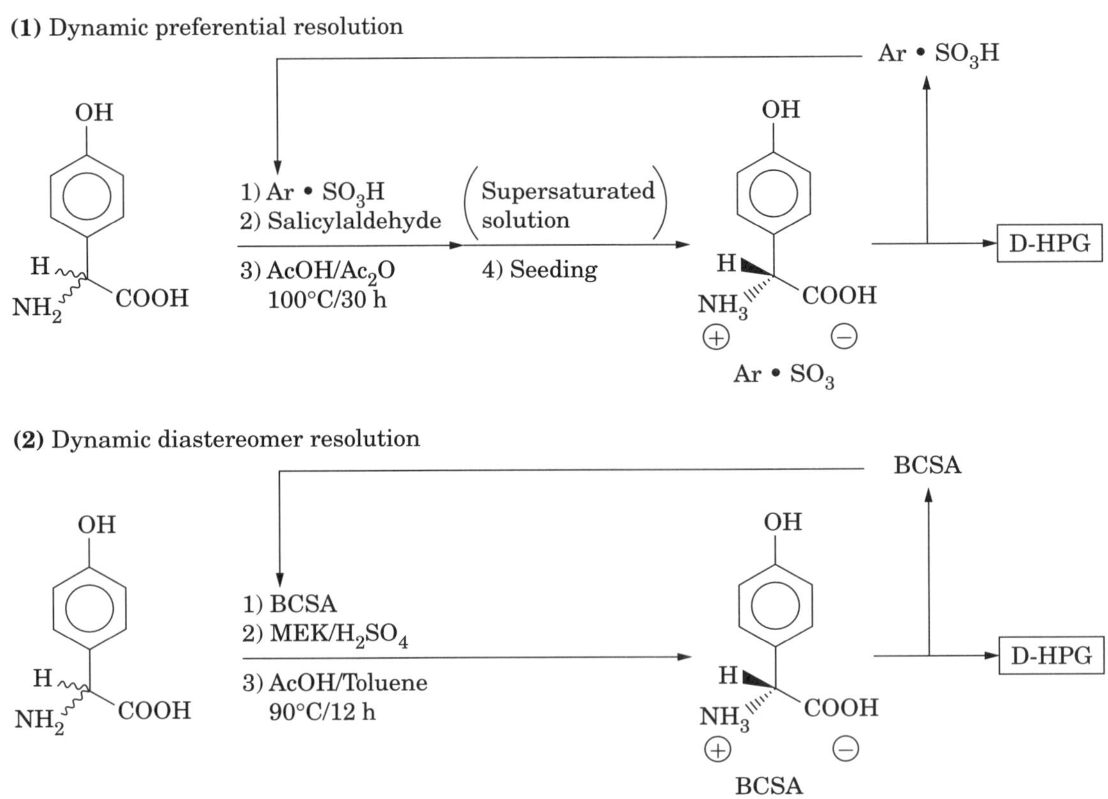

Figure 6. Resolution of DL-HPG into D-HPG.

hydrolysis of a variety of 5-substituted-D-hydantoins, as shown in Table 3. The D-form of 5-aliphatic and aromatic hydantoin such as 6-phenyl, 5-(4-hydroxyphenyl), and 5-thienylhydantoin were hydrolyzed easily to the corresponding *N*-carbamoyl-D-amino acids. However, other hydantoins having a charged group in the amino acid moieties were generally resistant to hydrolysis.

The *N*-carbamoyl-D-amino acids thus obtained were quantitatively transformed to D-amino acids by treating them with nitrous acid under acidic conditions. These results suggested that the combination of the D-stereoselective hydantoinase reaction and the decarbamoylation by nitrous acid was useful as a simple method of D-amino acid production (6).

This chemicoenzymatic process consists of two chemical steps (steps 1 and 3) and one enzymatic step (step 2), as follows:

1. Hydantoin synthesis (generally according to Bucherer–Berg synthesis)
2. Transformation of hydantoin to *N*-carbamoyl-D-amino acid (microbial hydantoinase)
3. Decarbamoylation of *N*-carbamoyl-D-amino acid (nitrous acid treatment)

D-HPG Production. The hydantoinase process for D-HPG production consists of three steps (11), as follows:

1. Hydantoin (5-(4-hydroxyphenyl)hydantoin:p-HPH) synthesis from glyoxylic acid (GOA)
2. Transformation of the hydantoin into carbamoyl-D-HPG using hydantoinase
3. Decarbamoylation of carbamoyl-D-HPG into D-HPG

However, three sub-types are differentiated in the step 3 decarbamoylation:

1. Chemical decarbamoylation using sodium nitrite
2. Enzymatic decarbamoylation using immobilized *N*-carbamoyl-D-amino acid amidehydrolase
3. Enzymatic decarbamoylation using bacterial cells that contain hydantoinase and *N*-carbamoyl-D-amino acid amidohydrolase

A summary is shown in Figure 7.

Synthesis of p-HPH. p-HPH is an important intermediate in the hydantoinase process for D-HPG production. It is known that p-HPH can be synthesized by the reaction of 4-hydroxybenzaldehyde, ammonium carbonate, and sodium cyanide, according to Bucherer–Berg's method. However, this method requires the use of dangerous sodium cyanide and expensive aldehyde. Furthermore the crude hydantoin obtained is contaminated by impurities caused by the oxidative side reaction of the phenol nucleus in an alkaline condition, or it may take on a brownish color.

A novel and unique method for preparing p-HPH involving amidoalkylation of phenol with glyoxylic acid derivatives was established (12). According to this method, p-HPH can be readily prepared in high purity and good

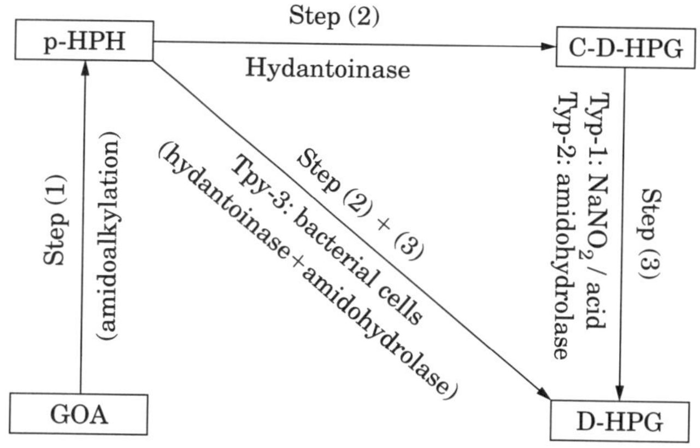

Figure 7. Hydantoinase process of HPG production.

yield from relatively inexpensive raw materials such as phenol, glyoxylic acid, and urea.

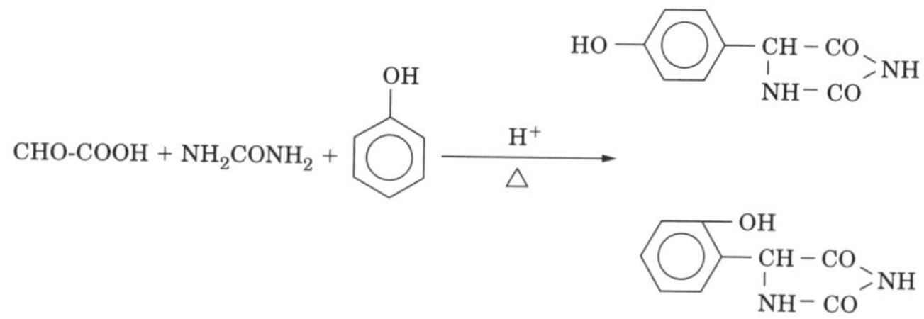

The main product of this reaction is p-HPH, but its isomer, 5-(2-hydroxyphenyl)hydantoin (o-HPH), is produced concomitantly. For the efficient synthesis of p-HPH, the ratio of p-HPH to o-HPH and total HPH yield should be increased concurrently. However, practically speaking, because o-HPH is hard to crystallize and more soluble, p-HPH can be easily obtained in a high state of purity by merely precipitating and separating it from the reaction mixture.

More recently, various methods for p-HPH synthesis have been investigated using glyoxilic acid derivatives. Among them, the method via *p*-hydroxymandelic acid, which is prepared from glyoxylic acid and phenol, seems promising because by-product o-HPH preparation can be minimized.

Transformation of p-HPH into N-Carbamoyl-D-HPG. *Enzyme (hydantoinase) preparation.* Hydantoinase-producing microorganisms can be found in a wide range of genera; high activity is found particularly in bacteria. The stable hydantoinase from thermophilic microorganisms is useful for high productivity.

Cells with high hydantoinase activity can be obtained by culturing the microorganism in a medium supplemented by pyrimidine base or a metabolite such as uracil, thymine, or β-alanine as the inducer. The accumulation of hydantoinase in the cells is further increased when a metal ion such as manganese is added to the medium together with the inducer.

Recently, to drastically increase hydantoinase levels, genetic engineering methods were attempted and resulted in reports of 4- to 40-fold higher enzyme activity with an overexpression plasmid vector. Hydantoinase from thermophilic microorganisms such as thermophilic *Bacillus* seem to be very useful because their stability is rather higher than that of mesophilic microorganisms.

Asymmetric hydrosis of p-HPH. In the asymmetric hydrosis of p-HPH, racemic p-HPH can be completely transformed into *N*-carbamoyl-D-HPH by the action of hydantoinase. Technically and economically, this is a great advantage for the industrial production of D-HPG. The mechanism of this interesting reaction seems to be as follows.

It is well known that hydantoins are readily racemized in dilute alkaline solution through base-catalyzed tautomerism. In practice, p-HPH undergoes spontaneous racemization very easily under mild conditions such as those of the enzymatic reaction. In the reaction system for the asymmetric hydrolysis of p-HPH, only the D-form of p-HPH is susceptible to enzymatic hydrolysis. Unreactive L-p-HPH undergoes rapid spontaneous racemization in the same system. However, the *N*-carbamoyl-D-HPG formed is never racemized under these conditions. Consequently, in this system the enzymatic hydrosis of the hydantoin ring and chemical racemization of the substrate proceed simultaneously, so that DL-p-HPH can be completely transformed into the D-form of *N*-carbamoyl-HPG (Figure 8).

Hydantoinase is employed via intact whole cells or immobilized enzyme resin. In the immobilized form, the stable hydantoinase from a thermophilic microorganism such as thermophilic *Bacillus* is very useful for good performance even after repetitive use.

The concentration of the substrate DL-p-HPH that is available to the reaction depends on the enzyme activity used. A large portion of p-HPH is present in suspended form because the solubility of p-HPH in water is very low (ca. 50 mM). However, this is not an obstacle to the reaction, because the substrate is successively dissolved during the progress of the reaction in alkaline pH (ca. pH9). It is preferable to maintain the pH by adding alkaline solution successively, because the pH lowers in the course of hydrolysis; a drop in pH will result in a lowered reaction rate. It is also effective to cover the reaction mixture with an inert gas such as nitrogen to avoid the oxidative side reaction of the phenol nucleus. Under these optimum conditions, the yield of *N*-carbamoyl-D-HPG formed is almost quantitative.

Decarbamoylation of N-Carbamoyl-D-HPG *N*-Carbamoyl-D-HPG produced by enzymatic hydrolysis can be readily converted into D-HPG by decarbamoylation with nitrous acid under acidic conditions. The principle of this oxidative reaction is based on the Van Slyke determination, and the reaction seems to be a consecutive reaction. With respect to the stereochemistry of the reaction, the retention of the configuration is achieved completely. Therefore, optically pure D-HPG can be readily obtained in good yield. The decarbamoylation is preferably carried out by reacting *N*-carbamoyl-D-HPG with approximately equimolar nitrous acid in an aqueous medium in the presence of a strong mineral acid such as sulfuric or hydrochloric acid. It is convenient to employ a water-soluble salt of nitrous acid such as sodium nitrite or potassium nitrite. Because this decarbamoylation is an exothermic reaction and generates large quantities of gas (N_2 and CO_2), an aqueous solution of nitrous acid is added gradually to the reaction medium with agitation to facilitate a smooth reaction. The reaction temperature is usually kept below 20

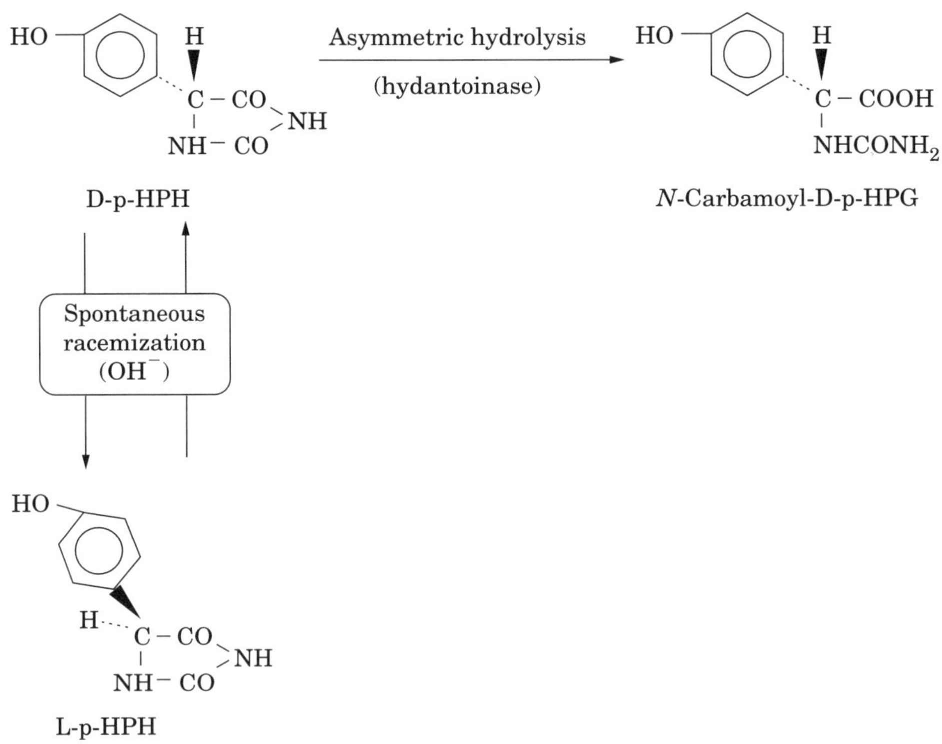

Figure 8. The mechanism of asymmetric hydrolysis of DL-p-HPH.

°C to avoid a side reaction such as the further degradation of D-HPG into 4-hydroxymandelic acid and other compounds. Under these optimum conditions, the yield of D-HPG from *N*-carbamoyl-D-HPG is almost quantitative.

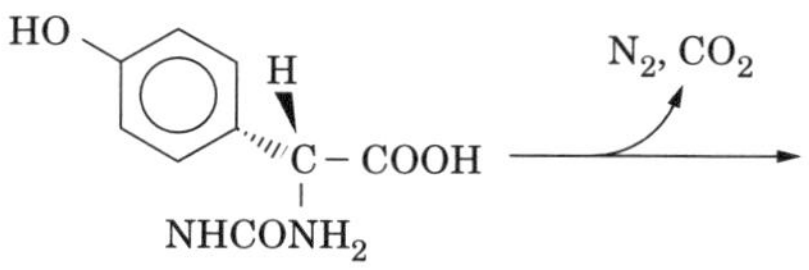

N-Carbamoyl-D-HPG

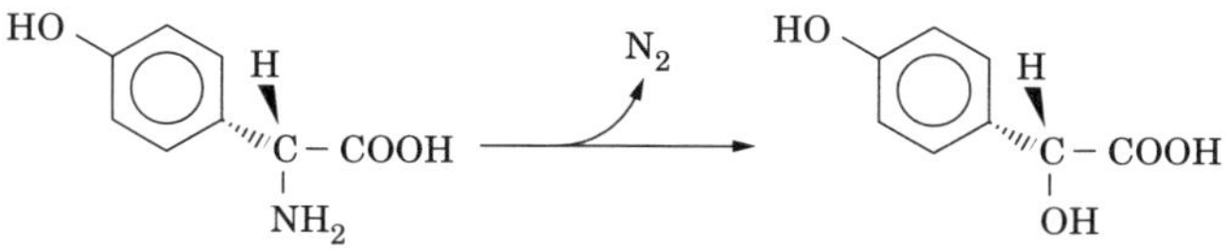

D-HPG 4-Hydroxybenzaldehyde

BIBLIOGRAPHY

1. Dutch Pat. 900.0386 (September 16, 1991), W.H. Boesten and W.H.J. Boesten (to Stamicarbon BV and DSM NV).
2. Dutch Pat. 900.0387 (September 16, 1991), W.H. Boesten and W.H.J. Boesten (to Stamicarbon BV and DSM NV).
3. C. Hongo, R. Yoshioka, M. Tohyama, S. Yamada, and I. Chibata, *Bull. Chem. Soc. Jpn.* **56**, 3744–3747 (1983).
4. C. Hongo, R. Yoshioka, M. Tohoyama, S. Yamada, and I. Chibata, *Bull. Chem. Soc. Jpn.* **85**, 433–436 (1985).
5. Eur. Pat. 173921 (March 12, 1986), V.W. Jacewixz (to Beecham Group PLC).
6. S. Takahashi, T. Ohashi, Y. Kii, H. Kumagai, and H. Yamada, *J. Ferment. Technol.* **57**, 328–332 (1979).
7. K.H. Dudley, T.C. Butler, and D.L. Bius, *Drug Metab. Disp.* **2**, 103–112 (1974).
8. F. Cecere, G. Galli, and F. Morisi, *FEBS Lett.* **57**, 192–194 (1975).
9. H. Yamada, S. Takahashi, Y. Kii, and H. Kumagai, *J. Ferment. Technol.* **56**, 484–491 (1978).
10. S. Takahashi, Y. Kii, H. Kumagai, and H. Yamada, *J. Ferment. Technol.* **56**, 492–498 (1978).
11. K. Aida, I. Chibata, K. Nakayama, K. Takinami, and H. Yamada, *Progress Industrial Microbiology, Vol. 24, Biotechnology of Amino Acid Production*, Kodansha, Tokyo, 1986, pp. 269–279.
12. T. Ohashi, S. Takahashi, T. Nagamachi, K. Yoneda, and H. Yamada, *Agric. Biol. Chem.* **45**, 831–838 (1981).

PICHIA, OPTIMIZATION OF PROTEIN EXPRESSION

Koti Sreekrishna
Procter and Gamble Co.
Ross, Ohio

KEY WORDS

Alcohol (methanol) oxidase promoter
Clonal variation
Expression plasmid
Fermentation
Methylotrophic yeast
mRNA secondary structure
Protease deficient strain
Secretion
Synthetic gene
Transcriptional barrier

OUTLINE

INTRODUCTION

Since the advent of the recombinant DNA technology, a wide variety of expression systems for protein production have become available. These include systems based on bacteria, lower eukaryotes (yeast and fungi), invertebrates (cells or larvae), vertebrates (cells or whole animal), and plants (cells or whole plant). Of these, only the yeast and fungal expression systems have many of the attributes of both the bacterial and higher eucaryotic expression systems. Yeasts have the robust growth characteristics and ease of manipulation of a bacterial system and at the same time perform many posttranslational modifications found in higher eukaryotes. The purpose of this article is to highlight the strategies one can apply to optimize protein expression in the *Pichia* yeast expression system.

BACKGROUND

Pichia pastoris is a methylotrophic yeast. It is able to use methanol for energy as well as for growth. It possesses a highly regulated pathway for the utilization of methanol (Fig. 1). Synthesis of the enzymes of methanol metabolism increase rapidly once the cells are placed in the methanol medium. The most dramatic effect is seen for methanol oxidase, which accounts for 35% of the cellular protein in cells adapted to grow on methanol. An extensive proliferation of the peroxisomes, known to sequester methanol oxidase and dihydroxy acetone synthase, is also observed in methanol grown cells (1).

Pichia was initially investigated at Phillips Petroleum Company as a potential source of single cell protein. Thus, a very efficient fermentation process (cell density > 130 g dry cell weight per L and biomass productivity > 10 g/L-h) was developed (2). Though impressive, this process could not compete with the economics of production of soy protein. After this setback, Phillips Petroleum Company directed its efforts in developing *Pichia* as an expression system for the production of recombinant proteins. This has turned out to be a worthwhile endeavor.

STRATEGIES FOR OPTIMIZATION OF PROTEIN EXPRESSION

Typical *Pichia* expression vectors (Fig. 2), for example, pHIL-A1 (3), pHIL-D2 (4), pHIL-D7 (5) and pPIC9 (6) are based on the strong methanol oxidase gene (*AOX1*) promoter (7). A wide variety of proteins have been produced in this system (Table 1). The final yield of a protein expressed in *Pichia* is largely influenced by its inherent properties. Nevertheless, the production yield of a protein can be significantly enhanced by using a strategy that takes into account the multiple factors that influence protein expression.

Cellular State of the Expression Cassette

The expression cassette can be introduced into *Pichia* cells by way of chromosomal integration or autonomous replication. However, the integrative approach is preferable because it has the following advantages: (*1*) expression cassette stability; (*2*) ability to generate clones that contain multiple copies of the expression cassette (see section on gene dosage); (*3*) control over the sites of integration (*HIS4* or *AOX1* loci); and (*4*) ability to engineer different modes of integration (with or without the eviction of the *AOX1* coding sequences) (see section on methanol utilization phenotype).

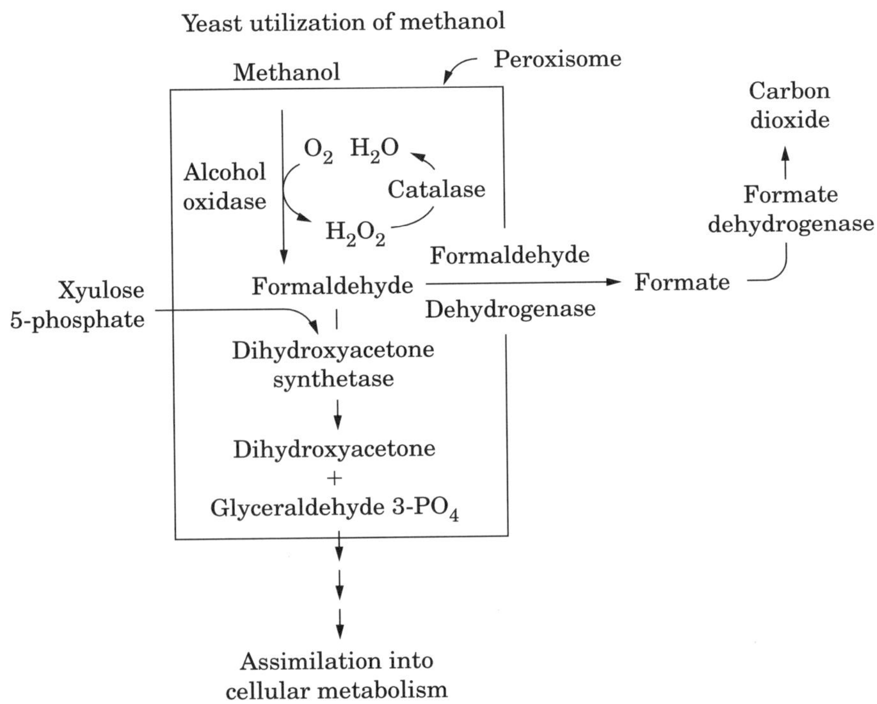

Figure 1. Methanol utilization pathway in methylotrophic yeasts.

Plasmids based on autonomous vectors such as pHIL-A1 (Fig. 2) are maintained in *Pichia* cells at low copies and are rapidly lost from the population of dividing cells. In a small number of cells, autonomous plasmids do eventually integrate. Thus, to arrive at a stable clone, a secondary selection screen should be implemented. This makes autonomous plasmids less attractive for protein expression. However, owing to their ability to transform *P. pastoris* spheroplasts at a very high frequency, the autonomous plasmids are useful for cloning genes in *Pichia* by functional complementation.

Site of Integration of the Expression Cassette

Both *AOX1* and *HIS4* sites have been used successfully for expression. Occasional loss of the expression cassette at the *HIS4* locus because of intrachromosomal crossover between mutant *his4* and the good *HIS4* has been observed. Thus, the *AOX1* locus is the preferred site for stable expression.

Methanol Utilization Phenotype, Mut$^+$ or Mut$^-$ of the Host

Transformation of a *P. pastoris his4* strain using linear DNA with the ends bearing homology to the 5′ and 3′ regions of the *AOX1* chromosomal locus (e.g., *Bgl*II digested pHIL-D7, or pPIC9 or *Not*I-digested pHIL-D2 [Fig. 2]) results in the site-specific eviction of the *AOX1* structural gene, as illustrated in Figure 3. The eviction of the *AOX1* occurs at a frequency of 1 to 5 per 20 His$^+$ transformants. *AOX1*-deleted clones show a slower growth phenotype (Mut$^-$) on minimal methanol medium, as compared to *AOX1* intact clones, which have phenotypically normal growth characteristics (Mut$^+$) on minimal methanol medium. The Mut$^+$ transformants arise presumably because of circularization of the linear DNA inside the yeast cell and subsequent integration into one or more of the *AOX1* or *HIS4* chromosomal loci. Thus, in a single transformation experiment, both Mut$^+$ and Mut$^-$ transformants can be obtained.

For intracellular expression, it is preferable to use Mut$^-$ cells because they will have a lower level of alcohol oxidase protein and the expressed protein can be more readily purified. For secretion, either one of Mut$^+$ or Mut$^-$ constructs can be used. There is no significant difference between these two types of host cells in the secretion of human serum albumin (74,75).

Gene Dosage: Exploiting the Clonal Variation of Expression

In many instances of *Pichia* expression of heterologous proteins, namely, expression of LACZ (8), hepatitis B surface antigen (9), invertase (33), and human serum albumin (HSA) (4), a single copy of the expression cassette was sufficient for optimal production. Deliberately increasing the copy number had no significant effect on production in these instances. Yet, in other cases, multiple copies of the expression cassette (> 10) are a must for high-level expression. A dramatic effect of copy number on protein production in *P. pastoris* is seen in the production of human tumor necrosis factor (TNF), tetanus toxin fragment C, *Bordetella pertussus* pertactin P69, and mouse epidermal growth factor (EGF) (6,13,14,74,76). In some rare instances, an increase in copy number has a negative effect on the production level (16).

Because the effect of gene copy number on expression is unpredictable, it is necessary to examine the production level as a function of gene dosage. The spheroplast method of transformation of *P. pastoris* results in transformants with a wide range of copy numbers (14,74,76). Analysis of as few as 100 individual clones for protein production is generally adequate to arrive at a good producer. If other methods of transformation (e.g., LiCl method or electroporation) that do not give rise to multicopy transformants at a high frequency are used, then more efficient screens can be used, such as colony hybridization (or dot blot analysis) with DNA probes or selection for multicopy integrants based on increased level of resistance to antibiotic G418 (77) or Zeocin (Invitrogen, San Diego, Calif.).

Alternatively, multicopy constructs of the expression cassette can be obtained by transformation with DNA concatamers or by using multicopy construction vectors such as pAO815 (3,16). This is not a method of choice because it permits production analysis over only a narrow range of gene dosage.

Translational Optimization: 5′ Untranslated Region

The nucleotide (nt) sequence and the length of the 5′ untranslated region (5′ UTR) can be detrimental to optimal protein translation. The leader length of the highly expressed *AOX1* mRNA is 114 nt long, and the sequence is A + U rich (7). For optimal synthesis of heterologous proteins, it is essential that the 5′ UTR of the expression cassette should closely resemble that of the *AOX1* mRNA. Ideally, it is preferable to make it identical to that of the *AOX1* mRNA. The expression level of HSA is increased more than 50-fold by adjusting the 5′ UTR to be closer to that of the *AOX1* mRNA. Expression plasmid such as pHIL-D7 can be used to readily make 5′ exact constructs.

Translational Optimization: Initiation Codon AUG Context

The translation initiation codon AUG should be avoided in the 5′ UTR to ensure efficient translation of mRNA from the actual translation start site. The mRNA secondary structure around the initiator AUG may be adjusted, so that AUG is relatively free of secondary structure (74). This can be accomplished by redesigning the initial portion of the coding sequences with alternate codons (3,74).

Transcriptional Optimization

Genes with high A + T nucleotide clusters are inefficiently transcribed because of premature termination. One such sequence, ATTATTTATAAA, present in HIV-gp120 has been identified to block transcription in *P. pastoris*, and when this stretch is altered to TTTCTTCTACAAG, the premature termination is abolished (17). Because there are many yet unidentified AT-rich stretches that act as transcriptional terminators, a general strategy would be to redesign the genes using *P. pastoris* preferred codons (3,74) so as to have an A + T content in the range of 30 to 55%. By using this approach, it has been possible to con-

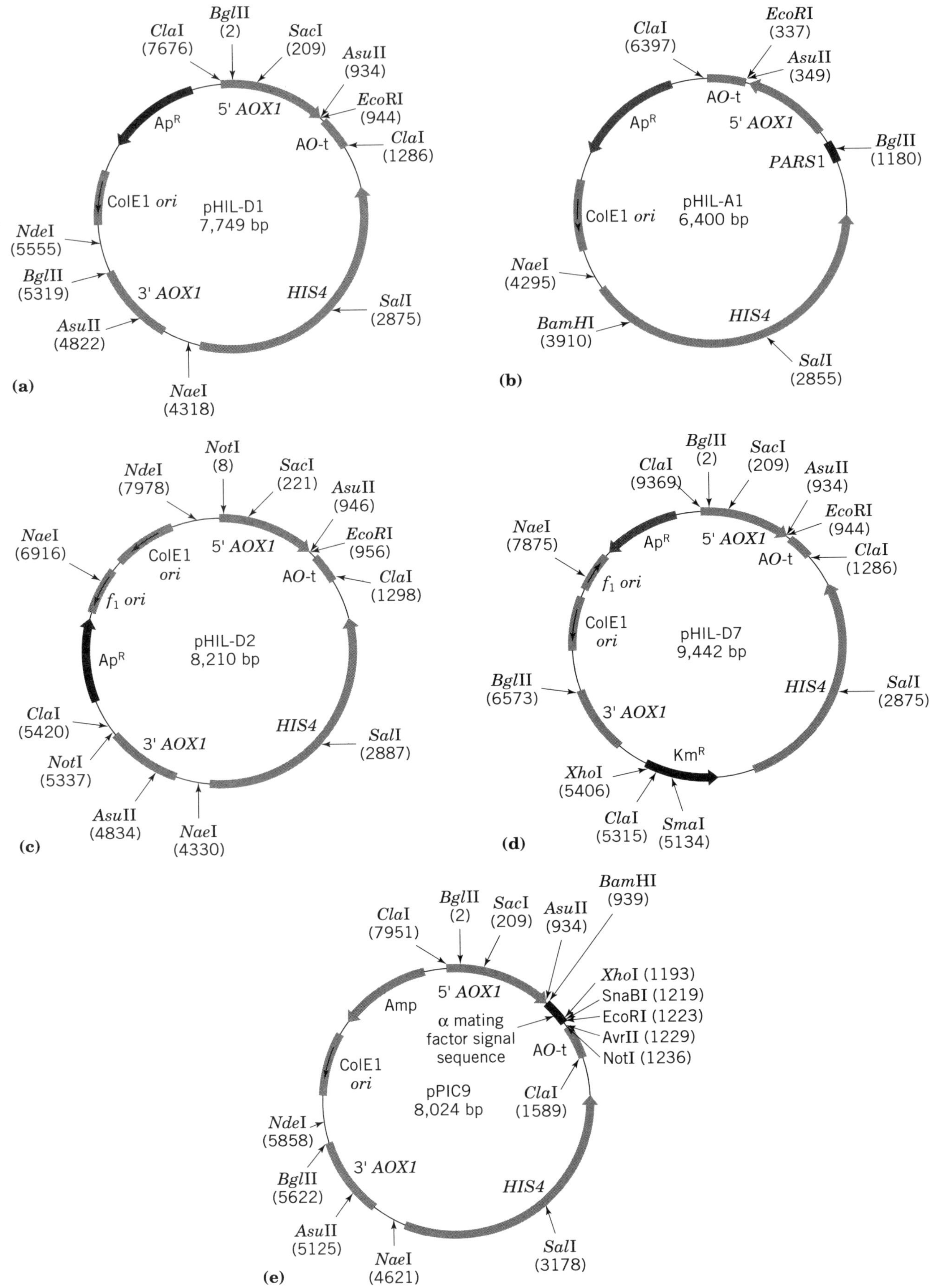

BglII (2)
SacI (209)
ClaI (7676)
AsuII (934)
EcoRI (944)
ClaI (1286)
5' AOX1
AO-t
Ap^R
ColE1 ori
pHIL-D1
7,749 bp
NdeI (5555)
BglII (5319)
3' AOX1
HIS4
SalI (2875)
AsuII (4822)
NaeI (4318)
(a)
ClaI (6397)
EcoRI (337)
AsuII (349)
AO-t
5' AOX1
Ap^R
BglII (1180)
PARS1
ColE1 ori
pHIL-A1
6,400 bp
NaeI (4295)
BamHI (3910)
HIS4
SalI (2855)
(b)
NotI (8)
NdeI (7978)
SacI (221)
AsuII (946)
EcoRI (956)
NaeI (6916)
ColE1 ori
5' AOX1
AO-t
ClaI (1298)
f1 ori
Ap^R
pHIL-D2
8,210 bp
ClaI (5420)
NotI (5337)
3' AOX1
HIS4
SalI (2887)
AsuII (4834)
NaeI (4330)
(c)
BglII (2)
ClaI (9369)
SacI (209)
AsuII (934)
EcoRI (944)
NaeI (7875)
Ap^R
5' AOX1
AO-t
ClaI (1286)
f1 ori
ColE1 ori
pHIL-D7
9,442 bp
BglII (6573)
3' AOX1
HIS4
SalI (2875)
Km^R
XhoI (5406)
ClaI (5315)
SmaI (5134)
(d)
BamHI (939)
BglII (2)
ClaI (7951)
SacI (209)
AsuII (934)
XhoI (1193)
SnaBI (1219)
EcoRI (1223)
AvrII (1229)
NotI (1236)
5' AOX1
Amp
α mating factor signal sequence
AO-t
ColE1 ori
pPIC9
8,024 bp
ClaI (1589)
NdeI (5858)
BglII (5622)
3' AOX1
HIS4
AsuII (5125)
NaeI (4621)
SalI (3178)
(e)

struct strains for efficient production of tetanus toxin fragment C (13) and *Bacillus sphaericus* mosquitocidal toxins (BSP1 and BSP2) (3,15).

Product Secretion

If a protein can be secreted, it is the desired mode of protein production because of the ease of product recovery. Furthermore, certain normally secreted proteins, such as HSA and growth hormones, remain insoluble when expressed intracellularly. For a protein that is not normally secreted, it may be difficult to cause secretion. However, there are some encouraging results with respect to yeast secretion of cytochrome P-450 (78) and rat liver elongation factor (55) that suggest that at least in some instances it may be possible to secrete a normally intracellular protein.

The choice of secretion signal is rather arbitrary (Table 1). In several cases, (HSA, invertase, bovine lysozyme, etc.), the native signal sequence is adequate. We have found that the native secretion signals present in matrix metalloproteinases 1,2,3,9 (MMP-1,2,3,9) and tissue inhibitor of matrix metalloproteinases 1 (TIMP-1) are functional in the *P. pastoris* system. If a native secretion signal sequence is not available, then a signal sequence based on the *S. cerevisiae* invertase (Suc2p), the *S. cerevisisae* α-mating factor (αMF), or the *P. pastoris* acid phosphatase (Pho1p) can be used.

The pre-pro αMF signal works very efficiently, especially for the secretion of a large variety of proteins, including the smaller-sized products such as aprotinin, EGF, thrombomodulin fragment, blood factor XII, a fragment of amyloid β-protein, antibody single-chain Fv fragment, and ghilanten (Table 1). In making protein fusions to the αMF signal, it is prudent to retain the Glu-Ala spacers adjacent to the Kex2-like protease cleavage site: (... V-S-S-L-E-Lys-Arg-$^{\text{vKex2p}}$Glu-Ala-^{vDapp}Glu-Ala-^{vDapp}fused protein). The presence of the Glu-Ala spacers help to alleviate the steric interference by the fused protein, resulting in an efficient cleavage of the pro sequence by the *P. pastoris* Kex2-like protease (16). The Glu-Ala spacer sequence is subsequently cleared by a diamino peptidase (Dapp) to yield the protein of interest free of additional N-terminal amino acid residues. In spite of taking this precaution, we have noticed that the pro sequence is not cleaved when the αMF signal is fused with human brain-derived neurotrophic factor (BDNF) (mol. wt. ~ 14 kDa). In this situation, BDNF is secreted into medium as a 26 to 30 kDa product, which is the expected result if the pro portion of the αMF signal is not cleaved and is heterogeneously glycosylated at the multiple glycosylation sites present in the αMF pro peptide (5).

Acid phosphatase (Pho1p) secretion signal sequence, invertase secretion signal, and a hybrid sequence consisting of Pho1p secretion signal containing a Kex2 protease recognition sequence has also been used successfully for high level (>g/L) secretion of certain proteins (Table 1).

Choice of the Secretion Signal

In most of the examples listed in Table 1, the relative efficacies of the different secretion signals for a given product have not been examined. Only in a limited number of cases has a rough comparison been made between native signal versus pre-pro αMF. For example, in the case of invertase secretion, both the extent of glycosylation and secretion kinetics are enhanced when the invertase signal sequence

Figure 2. Typical *Pichia* expression vectors. (**a**) pHIL-D1(3) is an *E. coli–P. pastoris* shuttle vector, with sequences required for selection in each host. The left half of the plasmid is a portion of pBR322, from *Cla*I (at position 7676 in the map) to *Pvu*II site (modified to *Bgl*II [at nt 5319 in the map]). This segment of pBR322 contains the ampicillin-resistance encoding gene (Ap^{R}) and the origin of replication (ColE1 *ori*). The *Eco*RI site in this segment has been eliminated. The DNA elements comprising the rest of the plasmid are derived from the genome of *P. pastoris*, except for short regions of pBR322 used to link the *Pichia* sequences. The *P. pastoris* elements in the plasmid are as follows: (*1*) 5′ *AOX1*, approximately a 1,000 bp of the alcohol oxidase promoter. The *AOX1* sequences following the nt A of the ATG start codon have been removed by BAL 31 treatment and the synthetic linker 5′-GGAATTC added to generate a unique *Eco*RI cloning site. (*2*) AO-t, approximately a 300 bp of the *AOX1* terminating sequence. (*3*) *P. pastoris* histidinol dehydrogenase gene, *HIS4*, contained on a 2,800-bp fragment to complement the defective *his4* gene in the *P. pastoris* strains such as GS115, SMD1168, etc. (*4*) Region of 3′*AOX1* DNA approximately 650 bp, which together with the 5′*AOX1* region is necessary for the site-directed displacement of *AOX1* coding sequences (see Fig. 3). The unique *Sac*I (at nt 209 in the map) and *Sal*I (at nt 2875 in the map) sites present in the plasmid can be used for site-directed integration of the entire plasmid into the *AOX1* and *HIS4* loci of *P. pastoris*, respectively. (**b**) Plasmid pHIL-A1 is an *E. coli–P. pastoris* shuttle vector, with sequences required for selection and autonomous replication in each host. The left half of the plasmid is a modified portion of pBR322 containing the Ap^{R} gene and the ColE1 *ori*. The DNA elements comprising the rest of the plasmid are derived from the genome of *P. pastoris*, except for short segments of pBR322 used to link the yeast sequences. The various *P. pastoris* sequences are as follows: (*1*) 5′*AOX1*, approximately 750-bp portion of the *AOX1* promoter with the *Eco*RI (at position 337 in the map) engineered as previously described (Fig. 1). (*2*) AO-t and *HIS4* are same as in Figure 1. (*3*) *PARS1*, approximately 190-bp segment of a *P. pastoris* autonomous replication sequence. The unique *Sal*I (at nt 2885 in the map) site can be used to direct the integration of pHIL-A1 into the *HIS4* loci of *P. pastoris*. (c) pHIL-D2 (4) has a 458-bp DNA containing the f_1-*ori* inserted between the original *Dra*I sites of pBR322. The *Not*I sites at nt 8 and 5337 are used for site-directed displacement of *AOX1* coding sequences. The other details are the same as described for pHIL-D1 (**a**). (**d**) pHIL-D7 is a shuttle plasmid containing the kanamycin-resistance-encoding gene (Km^{R}) and is useful for selection of multicopy transformants based on increased level of resistance to G418. Unlike the commonly used vectors such as pHIL-D1 and pHIL-D2, we have designed pHIL-D7 so that it has a unique *Asu*II site at nt position 934 (the second *Asu*II site present in the 3′*AOX1* has been eliminated). Therefore, the sequence TTCGAAACG can be added immediately upstream of the start codon (ATG) of the gene of interest, and an *Eco*RI site can be added downstream of the stop codon. The modified gene can then be inserted between the *Asu*II (at nt 934 in the map) and *Eco*RI (at nt 944 in the map) region TTCGAAACGAGGAATTC of pHIL-D7. Thus, the modified gene will have identical 5′ UTR to that present in *AOX1*. The other details are same as described for pHIL-D1 (**a**). (**e**) pPIC9 (6), a secretion vector with αMF secretion signal and multicloning sites. The other details are same as for pHIL-D1 (**a**).

Table 1. Partial List of Heterologous Proteins Expressed in *P. pastoris*

Intracellular expression

E. coli β-galactosidase (LACZ) (8)
Hepatitis B surface antigen (HBsAg) (9)
Human tumor necrosis factor-α (TNF-α) (10)
Human TNF-α analogues (11)
Salmon growth hormone (SGH) (K. Sreekrishna and K. A. Parker, unpublished observations, 1987)
Streptokinase (12)
Tetanus toxin fragment C (13)
Bordetella pertussus pertactin P69 (14)
Human inteleukin 2 (W. R. McCombie, personal communication, 1988)
B. sphaericus mosquitocidal components, BSP1, BSP2, and BSP1 + BSP2 (15)
Superoxide dismutase (SOD) (16)
Human serum albumin (HSA) (4)
HIV gp120 (17)
Porcine leukocyte lipoxygenase (18)
$p110^{RB}$ retinoblastoma tumor suppressor protein 919
Hamster prion protein (20)
Spinach phosphoribulokinase (21)
Spinach cytochrome C oxidase (22)
Spinach glycolate oxidase (23)
Dogfish shark cytochrome P450c17 (hydroxylase and lyase activities) (24)
Algal phytochrome (25)
Green fluorescent protein (GFP) (26)
Arabidopsis chaperone protein ATJ2 (27)
Corn cytochrome *b* and cytochrome *c* reductases (28)
Mouse thioether methyltransferase (29)
Polyomavirus large T antigen (30)
Rat liver mitochondrial carnitine palmitoyltransferase I and II (31)
Mouse $5HT_{5A}$ serotonin receptor (32)

Secretion using native secretion signal sequence

S. cerevisiae invertase (*SUC2*) (33)
Human serum albumin (HSA) (4)
Bovine lysozyme (34)
Matrix metalloproteinases MMP-1,2,3, and 9 (5)
Tissue inhibitor of matrix metalloproteinases-1 (TIMP-1) (5)
MMP-1,2,3, or 9 coexpressed with TIMP-1 (5)
Barley α-amylases 1 and 2 (35)
Cathepsin E (36)
D-alanine carboxypeptidase (dacA) of *Bacillus stearothermophilus* (37)
Pectate lyase pel B from *Fusarium solani f.* sp. *pisi* (38)
IgE receptor (α subunit) (39)

Secretion using pre–pro α mating factor (αMF) secretion signal sequence

Aprotinin (16)
S. cerevisiae invertase (*SUC2*) (16)
Human epidermal growth factor (EGF) (16)
Mouse epidermal growth factor (EGF) (6)
Insulin-like growth factor-1 (IGF-1) (40)
Kunitz protease inhibitor (nexin-2) (41)
Human vascular endothelial factor (VEGEF) (42)
κ-Bungaratoxin (43)
Cathepsin L pro-peptide (44)
Human fibroblast collagenase, matrix metalloproteinase-1 (MMP-1) (45)
N-Lobe of human serum transferin (46)

Table 1. Partial List of Heterologous Proteins Expressed in *P. pastoris* (*continued*)

C. albican hyphal surface protein HWP1 (47)
Coffee bean α-galactosidase (48)
Major histocompatibility (MHC) class II human leukocyte antigens DRα-Fos and DRβ-Jun (49)
Human immunodeficiency virus envelope glycoprotein gp120 (17)
Ghilanten (50)
Human factor XII (51)
Cellulose binding domain (CBD)–factor X variant fusion protein (52)
Cellulose binding domain (CBD)–steel factor fusion protein (1996 Invitrogen) (52)
Human gelatinase B (matrix metalloporiteinase-9, MMP-9) fragment (53)
Measles viral glycoprotein (54)
Rat liver elongation factor-2 (eEF-2) (55)
Human growth hormone (56)
Drosophila notch protein EGF-like domains (57)
Type II fish antifreeze protein (58)
Follicle stimulating hormone, FSH α and β chains (59)

Secretion using acid phosphate secretion signal sequence

Human vascular endothelial factor (VEGEF) (60)
Pectate lyase pelC from *Fusarium solani* f. sp. *pisi* (61)
Catalytic domain of 92-kDa gelatinase (matrix metalloproteinase-9, MMP-9) (62)
Angiotensin-converting enzyme (ACE) and its mutants (63)
Rabbit plasma cholesteryl ester transfer protein (CETP) (64)
β-Cryptogein elicitor protein secreted by *Phytophthora cryptogea* (65)
Candida albicans aspartic proteinases (66)
Angiostatin (an integral fragment of plasminogen) (67)
Rabbit reticulocyte 15-lipoxygenase (68)
A single chain antibody (69)
Human monocyte chemotactic protein-3 (MCP-3) (70)

Secretion using a hybrid of acid phosphatase signal sequence with a synthetic pro sequence and a KEX2 protease cleavage site

Tick anticoagulant peptide (TAP) (71)

Secretion using invertase (SUC2) secretion signal sequence

A single-chain antibody (P. Mezes, personal communication, 1992)
Porcine tumor necrosis factor-α (TNF-α) (72)
Boophilus microplus Bm 86 antigen (73)

Note: As of November 1998, the number of proteins expressed in the *Pichia* system is 172.

is substituted with the pre-pro αMF sequence (secretion $t_{1/2}$ of 10 min for pre-pro αMF signal, compared to 90 min for the native invertase signal) (16). Thus, in this case, one would avoid using αMF signal if hyperglycosylation is not desired. However, if higher productivity (product yield per liter per hour) is desired, then it is preferable to use the pre-pro αMF signal. The secretion kinetics can also impact on the recovery of the intact protein, if the expressed protein is susceptible to proteolysis in the *Pichia* broth. For example, the yield of full length MMP-1 from the culture medium is significantly improved by using the αMF signal instead of the native signal (5,45,47).

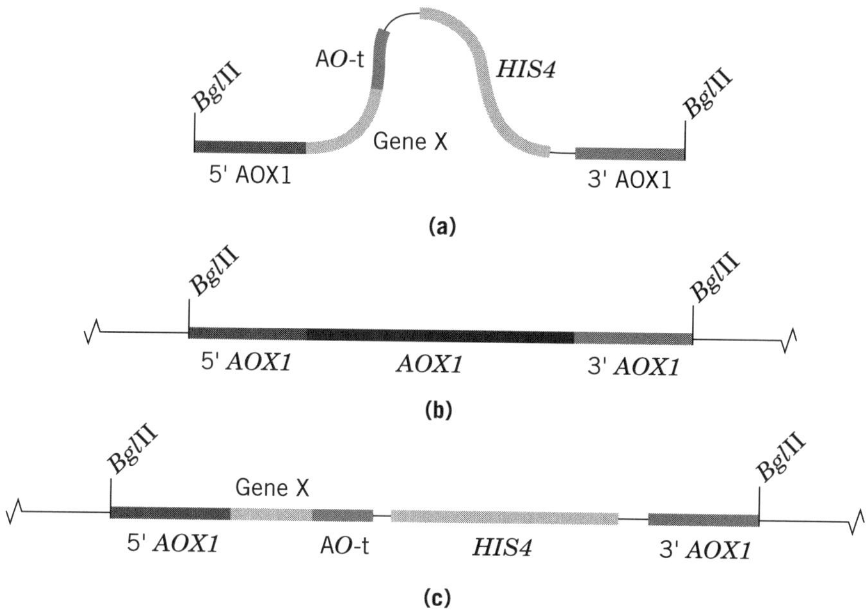

Figure 3. Site-specific eviction of *AOX1* by gene replacement. (**a**) *Bgl*II-digested DNA derived from pHIL-D1 expression plasmid used for transformation. Gene X is the gene of interest cloned at the *Eco*RI site of pHIL-D1. (**b**) *P. pastoris* chromosomal *AOX1* locus. (**c**) Chromosomal structure resulting from the replacement of the entire *AOX1* locus by the transforming DNA (shown in **a**).

Product Stabilization by Media Manipulations

In some instances, a secreted protein is rapidly degraded in the *Pichia* broth. The stability of a protein in the *Pichia* broth can be improved by manipulating the pH of the induction medium. The recommended pH range for experimentation is between 2.8 to 8.0. The pH control is best maintained in a fermenter, although it is possible to have a reasonable pH control even in shake flask cultures with the use of appropriate buffers (phosphate buffer for pH 5.7 to 8.0; alanine-HCl buffer for pH 2.5 to 3.6; citrate buffer for pH 3.0 to 6.0). The product stability is further enhanced by addition of casamino acids (1 to 4%) or yeast extract (1%) plus peptone (2%) to the medium (6,16,40,74,75). For example, secretion of HSA was significantly improved by raising the pH of the medium from 5.2 to 6.0, and the yield was further enhanced by the addition of yeast extract and peptone (4,74). Production of mouse EGF was favored at pH 3.0 in the presence of casamino acids (6). Casamino acids are preferable to yeast extract plus peptone, because the peptide components of peptone (such as bovine collagen fragments) can interfere in product analysis and recovery (5). Addition of 5 mM EDTA to the medium also improves product stability. It should be noted that media manipulation can significantly alter the profile of protein components, so that previously unnoticed proteins can become evident (5). The susceptibility of a protein to proteolysis in the *Pichia* culture medium can be determined beforehand if sufficient quantity of protein is available and proper adjustments can be made to the culture medium.

Protease-Deficient Strains

In addition to the media optimizations described in the previous section, the product yield can be further improved by using a protease-deficient *Pichia* strain. Two generally applicable techniques have been used to generate protease deficient strains. In one approach, a specific *Pichia* protease is knocked out by site-specific gene disruption, and in the other approach, a *Pichia* strain is engineered to express a specific protease inhibitor. Both types of protease-deficient strains have been found to be superior for expression of certain products. Even while using the protease deficient strains, it is essential to optimize the culture medium as discussed in the previous section.

By using the gene disruption approach, three protease-deficient strains, SMD1168 (*his4, pep4*), SMD1165 (*his4, prb1*), and SMD1163 (*his4, pep4, prb1*), that are defective in one or more of protease A (PEP4) or protease B (PRB1) have been constructed. Because a functional PEP4p is necessary for the maturation of PRB1p, the use of the single knockout strain SMD1168 alone should be adequate. The production yield of insulin-like growth factor-1 (40) and ghilanten (50) are improved twofold to threefold by using SMD1168 as compared to the strain GS115, which has intact PEP4p.

Co-expression of MMP-9 along with its inhibitor, TIMP-1, significantly improves the expression level of MMP-9 (5). Furthermore, TIMP-1 expression results in overall stabilization of several *Pichia* secreted proteins. Thus, strains engineered to express protease inhibitors such as aprotinin (16) and TIMP-1 (5) should be useful for high-level expression of proteins that are susceptible to rapid degradation.

Strategies for Coexpression of Proteins

In many instances it is required or desirable to express two or more heterologous proteins at the same time. The fact that *Pichia* can stably integrate multiple copies of an expression cassette implies that it should be possible, at least in principle, to express numerous heterologous proteins at the same time in *Pichia*. The first example of simultaneous expression of two proteins is that of the construction of *Pichia* strains expressing high levels of BSP1 (43 kDa) and BSP2 (52 kDa) components of *Bacillus sphaericus* mosquitocidal toxin (15). Since then, there have been few other reports in literature for coexpression of two proteins in *Pichia* (5,49,51).

Two generally applicable approaches have been used for coexpression. In one approach, *Pichia* is cotransformed with two expression cassettes, the resulting His^+ transformants are screened for the presence of both the expression cassettes by PCR and subsequently examined for expression (49). In another approach, a *Pichia* expression plasmid is engineered with two expression cassettes (Fig. 4) so that the DNA used for transformation has both the expression cassettes present on the same fragment. In this situation, every His^+ transformant is expected to contain both the expression cassettes. Plasmids such as pAO815 or pAO856 (3,16) can also be used to readily construct expression plasmids for simultaneous expression of two or more products.

One can also use sequential transformation to construct *Pichia* strains that coexpress two or more different proteins, because numerous transformation markers, namely His^+, Arg^+, Suc^+, $G418^R$, and $Zeocin^R$ markers are available for *Pichia*.

Leucine Zipper Dimerization Motif Mediated Protein Assembly

Several membrane receptors, such as major histocompatibility complex (MHC) class I molecules, MHC class II molecules, T-cell receptors, are multiprotein complexes. These receptors are comprised of α- and β-chains, each of which have a large NH_2-terminal extracellular portion, a transmembrane region, and a short C-terminal cytoplasmic tail. Efforts to express soluble $\alpha\beta$-heterodimers of MHC class II molecules by coexpressing just the extracellular portions of α and β chains have proved impossible. The reason for this is that the transmembrane region of MHC class II α- and β-chains facilitate the proper assembly of $\alpha\beta$ heterodimer, and in their absence, assembly is hampered. Interestingly, this problem has been overcome (49) by substituting the transmembrane regions of α- and β-chains with the leucine zipper dimerization motifs from the transcription factors Fos and Jun, which are known to assemble as stable, soluble heterodimers. The leucine zippers are characterized by five leucines that are spaced periodically at every seventh residue (heptad repeat); each heptad repeat contributes two turns of the α-helix (3.5 residues/turn). The leucine residues play a special role in dimerization, and they form the interface between the two α-helices in the coiled coil. The Fos-Jun dimer is soluble because of the charged amino acid residues on the outer surface of the coiled coil. The Fos/Jun leucine zipper dimerization domains (Table 2) may have general applicability in the expression of extracellular domains of multiprotein membrane receptors.

Product Toxicity

In some instances, the product being expressed may be toxic to *Pichia*. In such cases, a stepwise induction is suggested. First, the cell mass is built up using glucose as sole carbon source, taking care to see that glucose is not limiting. Under these conditions, very little product is made because of strong repression of the *AOX1* promoter by glucose. Subsequent to this, the product production phase is initiated with methanol feed. Whether a particular product is toxic to *P. pastoris* can be evaluated by comparing the growth of transformants in 1% sorbitol (or 100 mM alanine) media to that on sorbitol plus methanol or alanine plus methanol media. If the expressed protein is toxic, then the growth in the presence of methanol will be drastically impaired.

Strategies for Induction of Protein Expression

Expression vectors for heterologous expression in *P. pastoris* are typically derived from *AOX1* regulatory sequences (7). One or more of the following induction strategies can be used for optimal protein expression. Refer to the "Fermentation Process" section for induction of expression in the fermenter. Refer to the "Media Composition" section for details of media composition.

Continuous Induction for Intracellular Expression. The *AOX1* promoter can be maximally activated (derepression or induction) by growing cells on methanol as the sole source of carbon and energy in shake flasks, shake tubes, or plates at 30 °C. The doubling time on methanol is ≈6 h as compared to <2 h on glucose or glycerol medium. Even the cells in which the *AOX1* gene has been evicted (Mut^-) can grow on methanol. However, the growth rate is slow (doubling time, 18 to 24 h) because it is solely dependent on a secondary alcohol oxidase, *AOX2*, which is expressed only at low levels (79). In such cases the growth rate can be enhanced by supplementing the minimal methanol medium with 1% sorbitol or 100 mM alanine. The advantage of using sorbitol or alanine lies in the fact that these compounds do not interfere with the induction of *AOX1* promoter. Induction of expression on plates is carried out by streaking or plating cells on a methanol medium (MM) plate.

Stepwise Induction for Intracellular Expression. For liquid cultures, cells are initially grown to an OD_{600nm} of 2 to 10 on a nonactivating carbon source such as glycerol or glucose (minimal glycerol yeast extract medium [MGY] or minimal dextrose medium [MD]) in shake flasks or tubes at 30 °C. The cells are then harvested by centrifugation and shifted to MM and incubated with shaking at 30 °C for 1 to 4 days. Induction of expression on plates is carried out by streaking or plating cells on a nitrocellulose or cellulose acetate membrane placed on an MD or MGY plate. Once colonies appear, the filter is lifted and placed on an MM plate.

Efficient Secretion of Proteins. The conditions described thus far work well for the intracellular accumulation of heterologous proteins, but are rather inadequate for secreted proteins. The following protocols work well for secreted proteins for both Mut^+ and Mut^- cells (3,75).

Shake Tube Cultures. First cells are grown to saturation in 10 mL buffered minimal glycerol yeast extract medium (BMGY) placed in a 50 mL tube (2 to 3 days). The OD_{600nm} of culture will be in the range of 10 to 20. The cells are harvested by centrifugation, and the cell pellet is

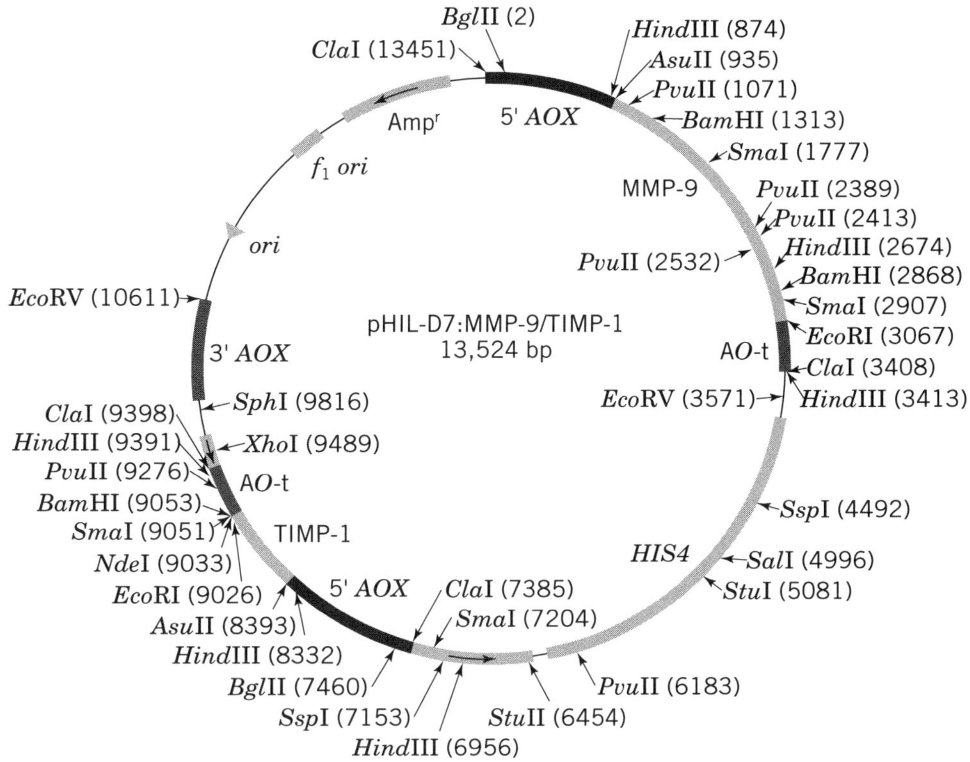

Figure 4. Expression plasmid for Coexpression of MMP-9 and TIMP-1 (5). This plasmid is derived from pHL-D7 (Fig. 2**d**) *Source*: Courtesy of Keith E. Kropp.

Table 2. Linker (7 Amino Acid) and Fos/Jun Leucine Zipper Dimerization Domains (40 Amino Acid)

VD GGGGG LTDT*L*QAETDQ*L*EDEKSA *L*QTEIAN *L*LKEKEK *L* EFILAAH
VD GGGGG RIAR*L*EEKVKT*L*KAQNSE*L*ASTANM *L*REQVAQ *L*KQKVMNH

Source: Data from Kalandadze et al. (49).

resuspended with 2 mL of buffered minimal methanol–yeast extract medium (BMMY). The tube is covered with a sterile gauze (four layers of cheesecloth; USP type VII gauze, 20 × 12 mesh from Ultimed International Inc., Glendale Heights, Ill.) and incubated in the shaker for 4 days. Aliquots of supernatant can be analyzed for product secretion at desired time intervals. Fresh media should be added to account for loss of media caused by evaporation.

Shake Flask Cultures. Cells are grown in 1 L of BMGY in a 2-L flask. Cells are harvested and resuspended in 50 mL of BMMY in a 300- to 400-mL low baffle flask, covered with four layers of cheesecloth as described in the previous section, and incubated in a 30 °C shaker for 2 to 4 days. The media supernatant is analyzed for product at different time intervals. Fresh BMMY media is added as necessary to prevent loss of liquid caused by evaporation.

Plates. Cells are patched or plated on nitrocellulose filter (sterile), placed on a square or circular BMGY plate, and incubated at 30 °C for 2 to 3 days. Once colonies have grown to ≈2 mm in size, the filter is removed and placed over a BMMY plate. The filter containing colonies is covered with a sterile nitrocellulose filter. If desired, a cellulose acetate filter can be interfaced between the two nitrocellulose filters and incubated at 30 °C for 2 to 3 days. The secreted protein on the nitrocellulose filter can be analyzed by ELISA or another suitable procedure.

FERMENTATION PROCESS

The general methods for production of biomass as well as for the production of heterologous proteins using Mut^+ and Mut^- cells in both continuous and batch modes of fermentation are described here. The process can be scaled up (>1,000 L) or scaled down (0.2 L) as desired. Fermentation can be conducted over a wide pH range (3.0 to 6.5) at 30 °C. (see "Media Composition" section for details of media.) For example, HSA secretion yield is improved more than threefold by using pH 5.85 compared to the generally used pH of 5.0 (4). In the case of secretion of the V1 domain of CD4 (amino acid residues number 1 to number 106 of mature CD4), intact product was seen only at acidic pH (2.5 to 3.5) (R.G. Buckholz, personal communication, 1990). As high as a twofold to fourfold increase in secreted yields of human EGF and human insulin-like growth factor-1 are observed at pH 3.0 (40).

Fermentation Equipment

A typical benchtop fermenter has a 2- to 20-L capacity, a 1- to 10-L operation volume, and monitors and controls for pH, dissolved oxygen (DO), agitator speed, temperature, air flow, pressure, foam, and weight.

Fermenters can be custom built or purchased from one of the numerous commercial sources such as B. Braun Biotech Inc., Allentown, Penn.; Biolaffitte, SA, France; New

Brunswick Scientific, Edison, N.J.; Porton Instruments Inc., Hayward, Calif.; Infors, UK Ltd., Crewe, Cheshire, Great Britain; Caltec Scientific Ltd., Edmonton, Alberta, Canada; Hotpack Corp., Philadelphia, Penn.; J & H Berge Inc., S. Plainfield, N.J.; Setec Inc., Livermore, Calif.; Techne, Inc., Princeton, N.J.; and VWR Scientific Products, W. Chester, Penn.

Sixfors, offered by Infors, UK Ltd., can bridge the gap between shake flasks and fermenters. They can be used for small-scale (typically 0.3 L) continuous, batch, or cascade processes where the growth from one vessel is passed into another.

Methods for Monitoring the Fermentation Process

Bio-Rad's portable fermentation monitoring analyzer can be used on the spot to provide fast HPLC analysis of the concentration of methanol, glycerol, ethanol, glucose, acetic acid, lactic acid, fructose, maltotriose, and maltose. The equipment is compatible with automatic sampling and computerized data analysis. Analysis takes only 10 min. The conditions used are as follows:

1. *Instrument*: Bio-Rad's fermentation monitoring analyzer
2. *Column*: Fermentation monitoring column (150 × 7.8 mm)
3. *Sample*: Extracellular broth, prefiltered through a 0.22-μm syringe filter; sample volume: 10 to 20 μL
4. *Injector*: Rheodyne injector valve with a 10-μL injection loop
5. *Solvent*: 0.002 N H_2SO_4 in water at a flow rate of 0.8 mL/min (isocratic, so only one pump is required)
6. *Temperature*: 65 °C
7. *Detector*: Refractive index detector

To determine the wet cell weight of the culture, 1 mL of the fermenter broth is centrifuged in a microfuge, and the cell pellet is weighed after carefully decanting the cell-free supernatant. To determine the washed dry cell weight of the culture, cells are harvested by centrifugation (3000 × *g* for 10 min at room temperature) from a known volume of the fermenter culture (typically 10 to 50 mL). The cell pellet is washed twice with water (10 to 50 mL), dried overnight at 100 °C, and weighed. Alternatively, the dry cell weight can be readily determined in less than 10 min with the use of a moisture balance such as the Mettler LP16 Infrared Dryer (Mettler Instrument Corporation, Hightstown, N.J.).

Mass transfer is determined by measuring the air flow to the fermenter and the composition of both the inlet and outlet air with a Perkin-Elmer gas analyzer. The O_2 transfer rate (OTR) is calculated as follows:

$$\text{OTR} = f(C_{\text{in}} - C_{\text{out}})/V$$

where f is flow rate L/h, C_{in} and C_{out} are the concentration in mmol of O_2 in the inlet and outlet gases, and V is the ungassed broth volume. OTR can also be established from yields using the equation

$$\text{OTR} = \mu X/Y \times O_2$$

where μ is the specific growth rate or dilution rate (h^{-1}), X is the cell density (g/L), and $Y \times O_2$ is the yield on O_2 (g cells/mmol O_2).

Heat transfer rate and heat load can be determined by measuring the temperature of the circulating water in and out of the fermenter heat exchanger.

Productivity (P) for the continuous culture is determined using $P = DX$, where D is the dilution rate (h^{-1}), and X is the cell density. An increase in D and/or X will result in higher productivity. Because D also equals the specific growth rate μ, the dilution rate is limited by the microorganism's intrinsic characteristics. Once μ_{max} or D_{max} is reached, P can only be increased by increasing the cell density. A set of parameters determined for the high-productivity process for *P. pastoris* on methanol in a 1,500-L fermenter are given in Table 3.

Extracellular protein concentration (for secreted proteins) can be estimated by using a Lowry-type protein analysis on TCA-precipitated material and analyzed for specific product by SDS-PAGE and immunological methods. Aliquots of cells are lysed to prepare cell extracts for protein analysis.

Continuous Culture of Mut$^+$ and Mut$^-$ *Pichia* on Methanol

Fermentation is carried out in two steps, first in the batch mode on glycerol as the carbon source, followed by continuous mode on methanol-containing medium.

Inoculum for the fermenter is prepared by growing cells in 1 L of MMD, YMPD, YMPGy, MGyB, MMGY, or YPD grown to an OD_{600nm} of 2 to 10 in a 2-L shake flask. This volume of inoculum is used for inoculating a 20-L fermenter containing 9 L of basal salt FM21 medium containing 5% glycerol (higher levels may be toxic to the cells) or 5 to 10% glucose and pH adjusted to 5.0 (by using 50% NH_4OH solution or NH_3 gas). Biotin stock (4 mL) and PTM1 trace mineral mix (11 mL) are also added. Fermentation is conducted until the glycerol or glucose (carbon source) is exhausted. During the run, dissolved O_2 is maintained at >20%, with the agitator speed set between 500 to 1,500 rpm and a vessel pressure of 2 to 3 psi. Foaming is con-

Table 3. Parameters Determined for a High Cell-Density Process in a 1,500-L Fermenter

Substrate	Methanol
Substrate concentration (g/L)	263
Dilution rate (h^{-1})	0.11
pH	3.5
Temperature (°C)	30
Cell mass (g of cells/L)	105
Productivity (g/L h)	11.6
Yield (g of cells/g of methanol)	0.4
OTR (mmol O_2/L h)	880
Oxygen consumption (g O_2/g cells)	2.42
Heat release	
kcal/L h	109
kcal/mol O_2	123.9

Source: Adapted from Shay et al. (80).

trolled through the addition of a 5% Struktol J673 (Strucktol Company of America, Stow, Ohio) or Mazu DF 37C (Mazer Chemicals, Inc., Gurnee, Ill.). The cell yield expected for the batch phase on 5% glycerol under these conditions is 20 to 25 g of washed dry cell weight per liter.

Continuous fermentation is established by feeding FM21-methanol (15% v/v) for Mut^+ cells or FM21-methanol (1% v/v) plus 15% glycerol, sorbitol, or alanine for Mut^- cells. The feed for continuous culture is also supplemented with PTM1 (1.1 mL of stock solution per liter) and biotin (0.4 mL of the stock solution per liter). Feed sterilization is carried out by filtration (Pall Ultipor disposable filter assembly DFA 4001 AR, 0.2 μm). Feed addition is achieved with a Milton-Roy positive-displacement metering pump (Model 2396 Duplex). Continuous culture is performed as a chemostat under steady-state conditions where the dilution rate (D) is equal to the growth rate of the population. The growth rate of cells can be controlled by adjusting the flow rate of fresh medium into the fermenter. Typical D values range from 0.056 h to 0.11 h. Dissolved oxygen is held in the 45 to 75% air saturation range by varying air flow, vessel pressure, or agitator speed. The maximum cell mass achievable under these conditions is around 80 g of washed dry cell weight per liter with a cell productivity of approximately 10 g/L h. If a higher cell mass is desired, the amount of carbon source in the feed can be raised (e.g., 25% methanol for Mut^+ and 1% methanol plus 25% glycerol, sorbitol, or alanine for Mut). This would also necessitate a proportional increase in the amount of minerals, trace elements, and biotin. As high as 110 g/L washed dry cell weight with 11.6 g/L h. productivity are achievable in this process. Highest induction of the *AOX1* promoter occurs using methanol (Mut^+ cells) or methanol plus sorbitol or alanine (Mut^-) as the carbon source in the feed. Intermediate levels of induction are seen with methanol plus glycerol feed.

Fed-Batch Fermentation of Mut^+ and Mut^- *Pichia* on Methanol

Fermentation is carried out in three steps: glycerol batch phase, glycerol fed-batch phase, and methanol-fed batch phase.

One liter of inoculum (prepared as previously described for the continuous fermentation protocol) is used for inoculating a 20-L fermenter containing 5 L of basal salt BSM medium with 5% v/v glycerol (higher levels may be toxic to the cells) or 5 to 10% glucose as carbon source; the pH of which is adjusted to 5.0 by using 50% NH_4OH solution or NH_3 gas. Biotin stock (40 mL) and PTM1 trace mineral mix (40 mL) are also added. Fermentation is conducted as previously described until all the carbon source (glucose or glycerol) is completely consumed. This phase should take 18 to 24 h. After this, a fed-batch phase on glycerol is initiated with a 50% w/v glycerol feed (500 mL of 100% glycerol + 480 mL water + 10 mL each of PTM1 and biotin stock solution) at a feed rate of 18 mL/h/L initial fermentation volume with the aid of a peristalic pump. Glycerol feeding is carried out for 4 h. The cell yield at this point will be in the range of 180 to 220 g/L of wet cells (equivalent to approximately 30 to 55 g washed dry cell weight per liter). This phase can be manipulated by varying the concentration of glycerol or the duration of fermentation to achieve optimal heterologous protein production in the methanol-fed batch phase. Some suggested cell yield ranges that should be tested for are listed in Table 4.

The fed-batch phase on methanol is initiated with a methanol feed (980 mL of 100% methanol + 10 mL of PTM1 + 10 mL of biotin stock solution) at a rate of 1 to 4 mL/L/h of the initial culture volume. The methanol-feed flow rate is adjusted such that with Mut^- strains, the methanol concentration in the fermenter is maintained in the 0.2 to 0.5% v/v level, whereas with the Mut^+ strains, methanol levels in the fermenter approach zero and the fermenter is run in a methanol-limited fashion. Dissolved oxygen in both cases is maintained at the 20 to 70% range. If dissolved oxygen falls below 20%, the methanol feed is stopped, and nothing should be done to increase oxygen rates until an upward spike in the dissolved oxygen level is seen. At this point, adjustments (rpm, aeration, vessel pressure, oxygen feed) can be made. Maintaining the dissolved oxygen above 20% may be difficult depending on the OTR of the fermenter. With stainless steel vessels, the system can be pressurized up to 15 to 30 psi to increase OTR. Also, the oxygen feed (air plus oxygen mixture) at 0.1 to 0.3 vvm can be used to maintain adequate levels of dissolved oxygen. The methanol fed-batch phase generally lasts for 70 h. However, it may prolong to more than 200 h if the feed rate is slowed down because of the fall in dissolved oxygen levels. Another factor that may also influence the fermentation time is the secretion rate of the heterologous protein. Longer times may be necessary to allow for accumulation of high levels of a slowly secreted protein in the broth. Depending on the product being produced, adjustments in the media (addition of casamino acids) and pH will have to be made to increase product yield. One technique that can be used to lower the pH is by setting the pH to the desired lower value (e.g., pH 3.0) and then letting the culture pH to decrease to the new set point of pH 3.0 as a result of cellular metabolism (40).

Strategies for Multicycle Fermentation

Considerable time is expounded on setting up a fermentation run. It would make sense to reap as many cycles of product production as possible for each start-up. Continuous fermentation on methanol, which works well for biomass production (1,2,80), is largely not applicable for recombinant protein production, especially with Mut^- strains, because they grow poorly on methanol. Use of methanol plus glycerol mixed feed restores the cell productivity, with considerable level of protein expression; optimal level is not achievable because of partial repression of the *AOX1* promoter by glycerol (11). This problem can be overcome by using a sorbitol plus methanol mixed feed, because sorbitol is a nonrepressing carbon source. However, sorbitol is a poorer carbon source than glycerol and also needs supplementation with 1% each of malt extract, peptone, and yeast extract for improved growth. Nevertheless, this strategy has been used successfully to accomplish several cycles of MMP-2 production in a Biostat-B (Braun) benchtop fermenter (5). In this approach, after recovery of 90% of the fermenter sample from the first fermentation

Table 4. Cell Yield Ranges

Mut^- intracellular expression	50–100 g/L dry cell weight or 200–400 g/L wet cell weight
Mut^+ intracellular expression	30–80 g/L dry cell weight or 140–320 g/L wet cell weight
Mut^- secretion	20–40 g/L dry cell weigth or 80–160 g/L wet cell weight
Mut^+ secretion	12–80 g/L dry cell weight or 50–300 g/L wet cell weight

Note: It is obvious from the suggested ranges for cell density that in several instances the glycerol fed-batch phase is not even necessary.

cycle, the fermenter is rebooted with an appropriate volume of fermenter growth media containing sorbitol as carbon source and supplemented with 0.1% yeast extract and 0.2% peptone. After growth for 16 to 20 h, a fed-batch fermentation with sorbitol plus methanol nutrient feed is continued for another 72 h for MMP-2 production. The whole process of product recovery and rebooting the fermenter can be repeated as many times as one desires. By this approach, using a 4-L fermenter with one start-up, we were able to generate 25 L of MMP-2 broth in less than 4 weeks. This approach is also useful for Mut^+ cells; it reduces overall methanol consumption because most of the growth is supported by sorbitol. Methanol can be added to the sorbitol feed at any desired point, to initiate induction of expression.

CONCLUSIONS AND FUTURE PERSPECTIVES

Problems generally encountered in protein expression may be overcome in most instances by taking due consideration of the factors that influence protein expression. Nearly every expression project has given new insight into the intricacies of the *Pichia* system. In spite of due consideration and careful planning, certain proteins are difficult to express in this system. Now that the expression system is readily available from Invitrogen (San Diego, Calif.), hundreds of investigators around the world are currently exploring the system, and their results will undoubtedly provide new insights as well as expand our knowledge base about the utility and limitations of the system.

Within the near future, it is anticipated that *Pichia*-produced HSA will reach the market. It will make one think of what to do with the tens of thousands of tons of *Pichia* yeast that would become available as a by-product of HSA production plants. It is tempting to speculate that the *Pichia* HSA strain would also be engineered to produce an intracellular insecticidal toxin, and thus the *Pichia* cells resulting from the HSA production can be used as a biopesticide.

Several modifications have recently been made to the *P. pastoris* expression vectors by Invitrogen (San Diego, Calif.). These newer vectors (pPICZ A,B,C series, and p-PICZα A,B,C series) are smaller in size and have $Zeocin^R$ marker. Furthermore, they allow in-frame fusion to *myc* epitope and His_6 tags. It has been previously noticed that the addition of eight amino acid carboxy terminal tags to TNF rendered the tagged proteins completely insoluble under nondenaturing conditions, whereas the untagged TNF was completely soluble (11). Thus, a tag enthusiast should bear in mind that in some instances, a tag can bring about an undesirable change in the property of a protein.

Overexpression of ubiquitin (Ubi) seems to enhance the secretion of a human leukocyte protease inhibitor in *S. cerevisiae* (81). Ubi is recognized to be a normal secreted component of *P. pastoris* (5). Thus, it is presumable that the overexpression of Ubi in *P. pastoris* may also enhance heterologous protein secretion. A *P. pastoris* host strain that overexpresses Ubi can be engineered using one of the available Ubi cDNAs. Also, a Ubi expression plasmid can be introduced into any *P. pastoris* strain by using one of the dominant selection markers for transformation, such as invertase, *SUC2* (82), $G418^R$ (77,83) or $Zeocin^R$ (available from Invitrogen, San Diego, Calif.). Thus, the effect of Ubi overexpression on product secretion can be tested in any of the existing production strains. A note of concern may be warranted while using $zeocin^R$ for selecting *P. pastoris* transformants. $Zeocin^R$ is a strong mutagen, and thus cells challenged with Zeocin*R* on a complex medium may acquire some undesirable change that may interfere with growth and thus reduce the overall productivity under the growth conditions in a fermenter.

A major concern for large-scale use of *Pichia* in the production of biologicals has been the substrate methanol, which is toxic, inflammable, and volatile. With the existing technology in a typical production plant, one will have to use several tons of methanol for the production phase of the operation. It will certainly be an added advantage if the amount of methanol required can be reduced or even eliminated. Two recent developments in this direction have been made. In one case, a mutant strain of *Pichia* that is able to turn on the *AOX1* promoter in the absence of methanol has been generated (A.A. Sibirny, personal communication, 1992). This mutant still responds to repression by glucose and ethanol. In another instance, a *Pichia* strain in which both the alcohol oxidase structural genes *AOX1* and *AOX2* have been deleted has been constructed and shown to express heterologous protein LACZ in response to methanol induction (V. Chiruvolu, personal communication, 1995). This novel alcohol oxidase null strain (*aox1, aox2*) uses considerably lower amounts of methanol for production of hetrologous protein.

Another development that one can speculate is that *Pichia* will be engineered to express human post-translational modification enzymes (e.g., glycosyl transferases, amidating enzyme, vitamin K-γ carboxylase, etc.), so as to obtain *Pichia* strains that can be used to produce more authentic human proteins.

MEDIA COMPOSITIONS

Stock Solutions

Note: For filter sterilization of various solutions and liquids, filter wares (disposable or reusable types) equipped with cellulose acetate or cellulose nitrate membranes (pore size 0.2 to 0.22 μm) from one of the several manufacturers (Nalgene Company, Rochester, N.Y.; Costar Corporation, Cambridge, Mass.; Corning Glass Works, Corning, N.Y.) can be used. For filter sterilization of methanol and methanol-containing media, only cellulose acetate membranes (0.2 to 0.22 μm) are suitable, because methanol does not filter through cellulose nitrate membranes of pore size 0.2 μm.

10× YNB. Dissolve 13.4 g of yeast nitrogen base without amino acids (YNB, Difco Labs., Detroit, Mich.) in 100 mL of water (heat if necessary) and filter sterilize. This solution can be stored for over a year at 4 °C.

500× B. Dissolve 20 mg of D-biotin (Sigma Chemicals, St. Louis, Mo.) in 100 mL of water and filter sterilize.

100× H. Dissolve 400 mg L-histidine in 100 mL of water (heat if necessary) and filter sterilize.

10× D. Dissolve 20 g of D-glucose in 100 mL water. Autoclave for 15 min or filter sterilize. Stores well for years at room temperature.

10× GY. Mix 100 mL of glycerol with 90 mL of water. Filter sterilize. Stores well for years at room temperature.

10× M. Mix 5 mL of methanol (100%) with 95 mL of water. Filter sterilize and store at 4 °C.

100% methanol. Filter sterilize pure methanol (100%). Store at room temperature in a fireproof cabinet.

Minimal Media Compositions

MD. Mix 100 mL of 10× YNB, 2 mL of 500× B, and 100 mL of 10× D with 800 mL of autoclaved water (include 15 g bacto agar for plates).

MM. Mix 100 mL of 10× YNB, 2 mL of 500× B, and 100 mL of 10× M with 800 mL of autoclaved water (include 15 g bacto agar for plates).

MGY. Mix 100 mL of 10× YNB, 2 mL of 500× B, and 100 mL of 10× GY with 800 mL autoclaved water (include 15 g bacto agar for plates).

All these liquid media and plates store well for several weeks at 4 °C.

Minimal media with other carbon sources (such as D-sorbitol, D,L-alanine) are prepared by using the desired carbon source at 10 g/L in place of glucose in MD. Minimal media containing a mixture of carbon sources can also be prepared by combining two or more desired substrates in the growth medium.

Supplemental Minimal Media Compositions

Minimal media are supplemented with necessary supplemental nutrients such as amino acids depending on the specific requirement of a given strain. For example, *P. pastoris* strains GS115 and KM71, commonly used in molecular genetic manipulations, are auxotrophic for histidine. Such strains will grow in minimal media only in the presence of supplemental histidine. However, once transformed with *HIS4* (histidinol dehydrogenase gene) they readily grow in the absence of histidine.

The composition of supplemental minimal histidine media (suitable for histidine auxotrophic strains such as GS115) is as follows. Other supplemental media can be prepared depending on the need of a particular strain in use.

MDH. Mix 100 mL of 10× YNB, 2 mL of 500× B, 100 mL of 10× D, and 10 mL of 100× H with 790 mL of autoclaved water (include 15 g agar for plates).

MMH. Mix 100 mL of 10× YNB, 2 mL of 500× B, 100 mL of 10× M, and 10 mL of 100× H with 790 mL of autoclaved water (include 15 g agar for plates).

MGyH. Mix 100 mL of 10× YNB, 2 mL of 500× B, 100 mL of 10× GY, 10 mL of 100× H with 790 mL of autoclaved water (include 15 g agar for plates).

MGyB. Dissolve 11.5 g KH_2PO_4, 2.66 g K_2HPO_4, 6.7 g YNB, pH 6.0, and 20 mL glycerol in 1 L water and autoclave.

All these liquid media and plates will store well for several weeks at 4 °C.

Supplemental minimal histidine media with other carbon sources is prepared by adding a similar amount of histidine as above to the minimal media with the desired carbon source.

Complex Media Composition

YPD. Dissolve 10 g of bacto yeast extract, 20 g of peptone, and 20 g of glucose in 1,000 mL of water (also include 15 g bacto agar for slants and plates) and autoclave for 20 min.

YMPD. Dissolve 3 g of yeast extract, 3 g of malt extract, 5 g of peptone, and 10 g of glucose in 1 L of water and autoclave.

YMPGy. Same as YMPD with the exception that 10 mL of 100% glycerol is used instead of 10 g of glucose.

Secretion Media Composition

BMGY. Mix 100 mL of 1 M potassium phosphate buffer, pH 6.0, 100 mL of 10x YNB, 2 mL of 500x biotin (refer to growth and storage section for composition of stock solutions), and 10 mL of glycerol. Filter sterilize and add to an autoclaved solution of 10 g yeast extract and 20 g peptone in 788 mL water (15 g bacto agar is included for plates).

BMMY. Same as BMGY, with the exception that 5 mL of methanol is added in the place of 10 mL glycerol.

Note: Yeast extract and peptone in the above media can be replaced by 1% casamino acids. The pH 6 suggested here may not be optimal for every secreted product. Experimentation with pH values in the range 2.5 to 8 (by using appropriate buffers) is suggested to determine the optimal pH for a particular product. Some suggested buffers are as follows:

Phosphate buffer for pH range 5.7 to 8.0
Alanine-HCl buffer for pH range 2.5 to 3.6
Aconitic acid-NaOH buffer for pH range 2.5 to 5.7
Citrate buffer for pH range 3.0 to 6.2

Avoid buffers such as succinate buffer, because succinate can serve as a carbon source and will repress activation of the *AOX1* promoter.

Fermentation Media Composition

FM21 basal salt media

Composition is for 1 L final volume in water:

Phosphoric acid, H_3PO_4 (85%)	3.5 mL
Calcium sulfate, $CaSO_4 \cdot 2H_2O$	0.15 g
Potassium sulfate, K_2SO_4	2.4 g
Magnesium sulfate, $MgSO_4 \cdot 7H_2O$	1.95 g
Potassium hydroxide, KOH	0.65 g
Biotin stock solution	
Biotin	0.2 g/L

PTM1 trace salts

Composition is for 1 L final volume in water:

Cupric sulfate ($CuSO_4 \cdot 5H_2O$)	6.0 g
Manganese sulfate ($MnSO_4 \cdot H_2O$)	3.0 g
Ferrous sulfate ($FeSO_4 \cdot 7H_2O$)	65.0 g
Zinc sulfate ($ZnSO_4 \cdot 7H_2O$)	20.0 g
Silfuric acid (H_2SO_4)	5.0 mL
Cobalt chloride ($CoCl_2 \cdot 6H_2O$)	0.5 g
Boric acid (H_3BO_3)	0.02 g
Sodium molybdate ($NaMoO_4 \cdot 2H_2O$)	0.2 g
Potassium iodide (KI)	0.1 g

BSM medium composition

Composition is for 1 L final volume in water:

Phosphoric acid, H_3PO_4 (85%)	26.0 mL
Calcium sulfate, $CaSO_4 \cdot 2H_2O$	0.9 g
Potassium sulfate, K_2SO_4	18.0 g
Magnesium sulfate, $MgSO_4 \cdot 7H_2O$	14.0 g
Potassium hydroxide, KOH	4.0 g

GLOSSARY OF *P. pastoris* VECTORS

pHIL-A1	Autonomously replicating vector (3)
pHIL-D1	Integration vector with or without deletion of AOX1 structural gene (3)
pPIC3	pHILD1 type vector with multiple cloning sites (3)
pAO815	pHILD1 type vectors for making multicopy expression units (3)
pAO856	pAO815 with a unique *Bgl*II site (3)
pHIL-D2	Modified pHILD1 with *Not*I site and fl ori (3)
PHIL-D3	Derived from pHIL-D2 for making constructs with exact 5′-UTR (3)
pHIL-D4	pHILD1 with kanamycin-resistance marker (3)
pPIC3K	pPIC3 with kanamycin-resistance gene (3)
pHIL-D5	pHILD2 with kanamycin-resistance gene (3)
pHIL-D7	pHILD4 with a unique *Csp45*I (*Asu*II) site for making exact 5′-UTR constructs (75)
pHIL-S1	Secretion vector with *P. pastoris* acid phosphatase secretion signal (3)
pPIC9	Secretion vector with *S. cerevisiae* αMF pre-pro signal (6)
pPIC9K	pPIC9 with kanamycin-resistance gene (77)
pPICZ A, B and C	Zeocin^R marker and *myc* epitope–His_6 tag (Invitrogen Inc.)
pPICZα A, B, and C	Zeocin^R marker, *myc* epitope–His_6 tag, and αMF secretion sequence (Invitrogen Inc.)

GLOSSARY OF *P. pastoris* STRAINS

NRRL Y-11430-SC5 (wild type) (3)

GS115 (*his4*)—this strain is also known as GTS115 (3)

KM71 (*his4, aox1::ARG4*) (3)

PPF1 (*his4, arg4*) (3)

Mc100-3 (*aox1::ARG4, aox2::his4, his4, arg4*) (79)

Protease deficient strains (derived by protease A (*PEP4*) and/or protease B (*PRB*) gene disruption (3):

- SMD1163 (*his4, pep4, prB1*)
- SMD1165 (*his4, prB1*)
- SMD1168 (*his4, pep4*)

BIBLIOGRAPHY

1. G.H. Wegner and W. Harder, in H.W. Van Verseveld and J.A. Duine eds., *Microbial Growth on C_1 Compounds*, Martinus Nijhoff, Boston, 1986, pp. 139–149.
2. G.H. Wegner, *FEMS Microbiol. Rev.* **87**, 279–284 (1990).
3. K. Sreekrishna and K.E. Kropp, *Non Conventional Yeasts in Biotechnology*, Springer, Berlin, 1996, pp. 203–252.
4. K. Sreekrishna, K.A. Barr, S.A. Hoard, W.D. Prevatt, R.E. Torregrosa, R.E. Levingston, J.A. Cruze and G.H. Wegner, *Fifteenth International Congress of Yeast Genetics and Molecular Biology*, Hague, The Netherlands, topic number 09-37B.
5. K. Sreekrishna, R.G. Brankamp, K.E. Kropp, D.T. Blankenship, J.T. Tsay, P.L. Smith, J.D. Wierschke, A. Subramaniam, and L. Birkenberger, *Gene* **190**, 55–62 (1997).
6. J.J. Clare, M.A. Romanos, F.B. Rayment, J.E. Rowedder, M.A. Smith, M.M. Payne, K. Sreekrishna, and C.A. Henwood, *Gene*, **105**, 205–212 (1991).
7. S.B. Ellis, P.F. Brust, P.J. Koutz, A.F. Waters, M.M. Harpold, and T.R. Gingeras, *Mol. Cell. Biol.* **5**, 1111–1121 (1985).
8. J.M. Cregg and K.R. Madden, *Dev. Ind. Microbiol.* **29**, 33–41 (1988).
9. J.M. Cregg, J.F. Tschopp, C. Stilman, R. Siegel, M. Akong, W.S. Craig, R.G. Buckholz, K.R. Madden, P.A. Kellaris, G.R. Davis, B.L. Smiley, J. Cruze, R. Toregrossa, G. Velicelebi, and G.P. Thill, *Bio / Technology* **5**, 479–485 (1987).
10. K. Sreekrishna, L. Nelles, R. Potenz, J. Cruze, P. Mazzaferro, W. Fish, M. Fuke, K. Holden, D. Phelps, P. Wood, and K. Parker, *Biochemistry* **28**, 4117–4125 (1989).
11. K. Sreekrishna, R. Potenz, J.A. Cruze, W.R. McCombie, K.A. Parker, L. Nelles, P.K. Mazzaferro, K.A. Holden, R.G. Harrison, P.J. Wood, D.A. Phelps, C.E. Hubbard, and M. Fuke, *J. Basic Microbiol.* **28**, 265–278 (1988).
12. M.J. Hagenson, K.A. Holden, K.A. Parker, P.J. Wood, J.A. Cruze, M. Fuke, T.R. Hopkins, and D.W. Stroman, *Enzyme Microb. Technol.* **11**, 650–656 (1989).
13. J.J. Clare, F.B. Rayment, S.P. Ballantine, K. Sreekrishna, and M.A. Romanos, *Bio / Technology* **9**, 455–460 (1991).
14. M.A. Romanos, J.J. Clare, K.M. Beesley, F.B. Rayment, S.P. Ballantine, A.J. Makoff, G. Dougan, N.F. Fairweather, and I.G. Charles, *Vaccine* **9**, 901–906 (1991).

15. Eur. Pat. 0586892A1 (March 16, 1994), K. Sreekrishna, W.D. Prevatt, G.T. Thill, G.R. Davis, P. Koutz, K.A. Barr, and S.A. Hopkins (to Research Corporation Technologies, Inc.).
16. G.P. Thill, G.R. Davis, C. Stillman, C. Holtz, R. Brierly, M. Engel, R. Buckholz, J. Kinney, S. Provo, T. Vedvick, and R.S. Siegel, *Proceedings of the Sixth International Symposium on Genetics of Industrial Microorganisms*, Societe Francaise de Microbiologie, Paris, 1990, pp. 477–490.
17. C.A. Scorer, R.G. Buckholz, J.J. Clare, and M.A. Romanos, *Gene* **136**, 111–119 (1993).
18. R.G. Reddy, T. Yoshimoto, S. Yamamoto, and L.J. Marnett, *Biochem. Biophys. Res. Commun.* **205**, 381–388 (1994).
19. D. Antelman, D. Giroux, J. Christian, P. Shabram, and R. Gregory, *Current Topics in Gene Expression Systems*, Invitrogen Inc., San Diego, 1994, p. 25.
20. S. Weiss, M. Famulok, F. Edenhofer, Y.H. Wang, I.M. Jones, M. Groschup, and E.L. Winnacker, *J. Virol.* **69**, 4776–4783 (1995).
21. H.K. Brandes, F.C. Hartman, T.Y. Lu, and F.W. Larimer, *J. Biol. Chem.* **271**, 6490–6496 (1996).
22. N. Shiraishi, *Plant Physiol.* **108**, 72–81 (1995).
23. M.S. Payne, K.L. Petrillo, J.E. Gavagan, L.W. Wagner, R. Dicosimo, and D.L. Anton, *Gene* **167**, 215–219 (1995).
24. J.M. Trant, *Arch. Biochem. Biophys.* **326**, 8–14 (1996).
25. S.H. Wu and J.C. Lagarias, *Proc. Natl. Acad. Sci. USA* **93**, 8989–8994 (1996).
26. E.Z. Monosov, T.J. Wenzel, G.H. Luers, J.A. Heyman, and S. Subramani, *J. Histochem. Cytochem.* **44**, 581–589 (1996).
27. R. Zhou, B. Kroczynska, and J.A. Miernyk, *Current Topics in Gene Expression Systems*, Invitrogen Inc., San Diego, 1996, p. 26.
28. N. Shiraishi and W.H. Campbell, *Current Topics in Gene Expression Systems*, Invitrogen Inc., San Diego, 1996, p. 38.
29. J. Hoffman, and M. Huff, *Current Topics in Gene Expression Systems*, Invitrogen Inc., San Diego, 1996, p. 46.
30. Y-C. Peng and N.H. Acheson, *Current Topics in Gene Expression Systems*, Invitrogen Inc., San Diego, 1996, p. 54.
31. Y. de Vries, D.N. Arvidson, H.R. Waterham, J.M. Cregg, and G. Woldegiorgis, *Current Topics in Gene Expression Systems*, Invitrogen Inc., San Diego, 1996, p. 59.
32. H.M. Weiss, W. Haase, H. Michel, and H. Reilander, *FEBS Lett.* **377**, 451–456 (1995).
33. J.F. Tschopp, G. Svelow, R. Kosson, W. Craig, and L. Grinna, *Bio / Technology* **5**, 1305–1308 (1987).
34. M.E. Digan, S.V. Lair, R.A. Brierly, R.S. Siegel, M.E. Williams, S.B. Ellis, P.A. Kellaris, S.A. Provow, W.S. Craig, G.V. Velicelebi, M.M. Harpold, and G.P. Thill, *Bio / Technology* **7**, 160–164 (1989).
35. N. Juge, J.S. Anderson, D. Tull, P. Roepstroff, and B. Svensson, *Protein Exp. Purif.* **8**, 204–214 (1996).
36. M. Yamada, T. Azuma, T. Matsuba, H. Iida, H. Suzuki, K. Yamamoto, Y. Kohli, and H. Hori, *Biochim. Biophys. Acta* **1206**, 279–285 (1994).
37. C.W. Despreaux and R.F. Manning, *Gene*, **131**, 35–41 (1993).
38. W. Guo, L.G. Candelas, and P.E. Kolattukudy, *J. Bacteriol.* **177**, 7070–7077 (1995).
39. M. Basu, M.Y. Nettleton, E. Dharm, R.F. Manning, C.W. Despreaux, J. Hakimi, and J.P. Kochan, *American Society for Biochemistry and Molecular Biology / Biophysical Society Joint meeting*, Houston, Texas, Feb. 9–13, 1992, Abstract #1507.
40. U.S. Pat. 5,324,639 (1994), R.A. Brierley, G.R. Davis, and G.C. Holtz.
41. S.L. Wagner, R.S. Siegel, T.S. Vedvick, W.C. Raschke, and W.E. van Nostrand, *Biochem. Biophys. Res. Commun.* **186**, 1138–1145 (1992).
42. D. Mohanraj, T. Olson, and S. Ramakrishnan, *Biophys. Res. Commun.* **215**, 750–756 (1995).
43. J.J. Fiordalisi, P.L. James, Y. Zhang, and G.A. Grant, *Toxicon* **34**, 213–224 (1996).
44. E. Carmona, E. Dufour, C. Plouffe, S. Takebe, P. Mason, J.S. Mort, and R. Menard, *Biochemistry* **35**, 8147–8157 (1996).
45. S.A. Rosenfeld, O.H. Ross, M.C. Hillman, J.I. Corman, and R.L. Dowling, *Protein Exp. Purif.* **7**, 423–430 (1996).
46. A.B. Mason, R.C. Woodworth, R.W.A. Oliver, B.N. Green, L-N. Lin, J.F. Brandts, B.M. Tam, A. Maxwell, and R.T.A. MacGillivary, *Protein Exp. Purif.* **8**, 119–125 (1996).
47. J.F. Staab, C.A. Ferrer, and P. Sundstrom, *J. Biol. Chem.* **271**, 6298–6305 (1996).
48. A. Zhu, C. Monahan, Z. Zhang, R. Hurst, L. Leng, and J. Goldstein, *Arch. Biochem. Biophys.* **324**, 65–70 (1995).
49. A. Kalandadze, M. Galleno, L. Foncerrada, J.L. Strominger, and K.W. Wucherpfennig, *J. Biol. Chem.* **271**, 20156–20162 (1996).
50. R.G. Brankamp, K. Sreekrishna, P.L. Smith, D.T. Blankenship, and A.D. Cardin, *Protein Exp. Purif.* **6**, 813–820 (1995).
51. C.E. White, N.M. Kempi, and E.A. Komives, *Structure* **2**, 1003–1005 (1994).
52. M.M. Guarna, A. Boraston, E.A. Amandoron, H.C.F. Cote, R.T.A. MacGillivary, R.A.J. Warren, and D.G. Kilburn, *Current Topics in Gene Expression Systems*, Invitrogen Inc., San Diego, 1996, p. 28.
53. N. Roy, M.M. Smith, M. Navre, and G. Das, *Current Topics in Gene Expression Systems*, Invitrogen Inc., San Diego, 1996, p. 56.
54. I. Duffy and B.K. Rima, *Current Topics in Gene Expression Systems*, Invitrogen Inc., San Diego, 1996, p. 41.
55. P. Bargis, A. Dumont-Miscopein, J.P. Lavergne, and J.P. Reboud, *Current Topics in Gene Expression Systems*, Invitrogen Inc., San Diego, 1996, p. 42.
56. L.L. Escamilla-Trevino, M. Guerrero-Olazaran, J.M. Viader-Salvado, and H.A. Barrera-Saldana, *Current Topics in Gene Expression Systems*, Invitrogen Inc., San Diego, 1996, p. 43.
57. H.A. Gonzalez and E.A. Komives, *Current Topics in Gene Expression Systems*, Invitrogen Inc., San Diego, 1996, p. 44.
58. M.C. Loewen, A.J. Daugulis, and P.L. Davies, *Current Topics in Gene Expression Systems*, Invitrogen Inc., San Diego, 1996, p. 47.
59. C.A. Seid, *Current Topics in Gene Expression Systems*, Invitrogen Inc., San Diego, 1996, p. 58.
60. D. Mohanraj, T. Olson, and S. Ramakrishnan, *Growth Factors* **12**, 17–27 (1995).
61. W. Guo, L. Gonzalez-Candelas, and P.E. Kolattukudy, *Arch. Biochem. Biophys.* **323**, 352–360 (1995).
62. J.M. Shipley, G.A.R. Doyle, C.J. Fliszar, Q-Z. Ye. L.L. Johnson, S.D. Shapiro, H.G. Welgus, and R.M. Senior, *J. Biol. Chem.* **271**, 4335–4341 (1996).
63. R. Sadhukan, G.C. Sen, and I. Sen, *J. Biol. Chem.* **271**, 18310–18313 (1996).
64. H. Kotake, Q. Li, T. Ohnishi, K.W.S. Ko, L.B. Agellon, and S. Yokoyama, *J. Lipid Res.* **37**, 599–605 (1996).
65. M.J. O'Donohue, G. Boissy, J-C. Huet, C. Nespoulous, S. Brunie, and J-C. Pernollet, *Protein Exp. Purif.* **8**, 254–261 (1996).
66. K. Jaton-Ogay, J. Wyniger, D. Sanglars, S. Foundling, and M. Monod, *Current Topics in Gene Expression Systems*, Invitrogen Inc., San Diego, 1996, p. 32.

67. B.K.L. Sim, H. Liang, R. Lapcevich, J.W. Madsen, W. He, A.H. Fortier, S. Plum, A. Ruiz, and M.S. O'Reilly, *Current Topics in Gene Expression Systems*, Invitrogen Inc., San Diego, 1996, p. 33.
68. K. Schwarz, S. Borngraber, and H. Kuhn, *Current Topics in Gene Expression Systems*, Invitrogen Inc., San Diego, 1996, p. 57.
69. J.A. Whyte and C. Cunningham, *Current Topics in Gene Expression Systems*, Invitrogen Inc., San Diego, 1996, p. 62.
70. S. Masure, L. Paemen, P. Proost, J. Van-Damme, and G. Opdenakker, *J. Interferon Cytokine Res.* **15**, 955–963 (1995).
71. Y. Laroche, V. Storme, J.D. Meutter, J. Messens, and M. Lu-wereys, *Bio / Technology* **12**, 1119–1124 (1994).
72. M.D. Moody, J. Ray, K. Murphy, and D. Tannatt, *Current Topics in Gene Expression Systems*, Invitrogen Inc., San Diego, 1996, p. 51.
73. M. Rodriguez, R. Rubiera, M. Penichet, R. Montesinos, J. Cremata, V. Falcon, G. Sanchez, R. Bringas, C. Cordoves, M. Valdes, R. Lieonart, L. Herrera, and J. de la Fuente, *J. Biotechnol* **33**, 135–146 (1994).
74. K. Sreekrishna, *Industrial Microorganisms: Basic and Applied Molecular Genetics*, American Society of Microbiology, Washington D.C., 1993, pp. 119–126.
75. K.A. Barr, S.A. Hopkins, and K. Sreekrishna, *Pharm. Eng.* **12**, 48–51 (1992).
76. K. Sreekrishna, W.R. McCombie, R. Potenz, K.A. Parker, P.K. Mazzaferro, G.M. Maine, J.L. Lopez, D.K. Divelbiss, K.A. Holden, R.D. Barr, and M. Fuke, *Fifth Annual Biotech USA Industry Conference and Exhibition*, San Francisco, 1988.
77. C.A. Scorer, J.J. Clare, W.R. McCombie, M.A. Romanos, and K. Sreekrishna, *Bio / Technology* **12**, 181–184 (1994).
78. A.C. Swart, P. Swart, S.P. Roux, K.J. van der Merwe, I.S. Pretorius, and A.J.C. Steyn, *Endocrine Res.* **21**, 289–295 (1995).
79. J.M. Cregg, K.R. Madden, K.J. Barringer, G.P. Thill, and C.A. Stilton, *Mol. Cell. Biol.* **9**, 1316–1323 (1989).
80. L.K. Shay, H.R. Hunt, and G.H. Wegner, *J. Ind. Microbiol.* **2**, 79–85 (1987).
81. Y. Chen, D. Pioli, and P.W. Piper, *Bio / Technology* **12**, 819–823 (1994).
82. K. Sreekrishna, J.F. Tschopp, and M. Fuke, *Gene* **59**, 115–125 (1987).
83. K. Sreekrishna, T.D. Webster, and R.C. Dickson, *Gene* **28**, 73–81 (1984).

See also CHINESE HAMSTER OVARY CELLS, RECOMBINANT PROTEIN PRODUCTION; EXPRESSION SYSTEMS, E. COLI; EXPRESSION SYSTEMS, MAMMALIAN CELLS; INSECT CELL CULTURE, PROTEIN EXPRESSION; INSECT CELLS AND LARVAE, GENE EXPRESSION SYSTEMS.

PILOT PLANTS, DESIGN AND OPERATION

BETH H. JUNKER
Merck Research Laboratories
Rahway, New Jersey

KEY WORDS

Biopharmaceutical
Bioprocessing
Pilot plant
Process development
Scale-up

OUTLINE

INTRODUCTION

Several diverse roles, objectives, and purposes are associated with a bioprocessing, biochemical, or biopharmaceutical pilot plant. These include, but are not limited to, manufacture of clinical supplies, process development for new products in the company pipeline, process improvement including technical support for existing products on the market, and production of biologically produced materials (which are not intended for the clinic) to support basic research efforts such as screening, assays, and disease mechanism elucidation. Unlike chemical pilot plants, pharmaceutical pilot plants are used not only to scale up laboratory processes but also to produce developmental quantities of chemicals (and biological substances) for safety, toxicological, and clinical studies (1). Thus, balancing the demands for scale-up and development research with those for clinical production is a key dilemma for pilot plant operation (2). The timing of clinical production is critical because expensive and time-consuming clinical programs can be delayed, thus postponing product application filings, approvals, and launches (2). Although traditionally pilot plants have not manufactured material for sale, accelerated launch dates, deferred manufacturing facility capital commitments, and other areas of project uncertainty have prompted consideration and use of pilot plants for manufacturing in recent years.

The types of products manufactured using biologically based methodologies include primary metabolites, secondary metabolites (ranging from pharmaceuticals to biopolymers), constitutive proteins, recombinant proteins, intracellular or extracellular enzymes and biocatalysts, antigens, viruses, polysaccharides, and DNA itself. These products are made using a wide range of cell cultivation and isolation techniques suitable for bacteria, yeast, fungi, actinomycetes, suspension animal cells, anchorage-dependent animal cells, insect cells, and plant cells. Processing equipment varies considerably among these diverse systems.

Designing for extensive processing flexibility can be expensive, potentially in excess of $100 million dollars, depending on the specific process requirements, size, and intended use of the pilot plant. Processing requirements can change dramatically as products are deemphasized and reemphasized in development. These decisions on priorities are likely not to result from the robustness of the process but rather from the state of the clinical data or potential market share. The ability to process multiple products in an acceptable manner typically is desirable to maximize facility use. Although designers may need to predict future types of processes and products, they should be cautious about overdesigning the facility in terms of containment and flexibility (3).

The general areas required in a pilot plant might include support labs for bench-scale development and quality control, production suites for scale-up, utility areas, and personnel offices, as well as ample storage areas for raw materials, supplies, product, samples, and documentation (2). There should be a clear division between controlled and noncontrolled areas, by utilizing signs at a minimum and perhaps by physically restricting access. General design criteria might focus on modularity, simplicity, future growth provisions, redundancy, construction costs, operating costs, and air and water waste containment (4). Future operating costs are an important consideration for pilot plants because ample funds are necessary for raw materials, staffing, maintenance, and additional capital modifications to permit improvements.

OPERATIONAL CONCEPTS AND DESIGN REQUIREMENTS

Several operational concepts influencing design requirements need to be resolved before initiation of design activities. Agreement should be sought among end users, operations staff, compliance auditors, environmental consultants, and safety officers as well as from appropriate levels of management directly or indirectly influencing and controlling capital expenditures. A written document summarizing consensus on these issues may then become the basis for conceptual design activities.

Major decisions influencing pilot plant capabilities include whether the facility will produce bulk drugs (thus falling under the regulation of Center for Drug Evaluation and Research (CDER) and following guidelines from 21CFR 210/211) or biologics (thus falling under the regulation of Center for Biologics Evaluation and Research (CBER) and following guidelines from 21 CFR 210/211 and 600) (5). The level of validation and the validation philosophy are directly related to the end use of the resulting pilot plant products, specifically whether they are used for process development, clinical supplies (presafety/basic research, safety, phase I, phase IIa, IIB, phase III), or material for sale. Validation strategies, as well as other operational and design issues, are also influenced by whether the pilot plant is administratively located within the manufacturing or research division. Individual company preferences and guidelines for similarity between manufacturing and pilot plant areas need to be identified and evaluated for their impact on the proposed pilot plant facility. Typical areas targeted for standardization within a company are vessel designs, computer systems and interfaces, equipment vendors, and construction specifications.

Design requirements center around the desired initial and end product stages of processing, specifically the forms of the pilot plant inputs and outputs. Example stages include master/working seed preparation, fermentation broth, initial captured product (e.g., microfiltration retentate, ultrafiltration concentrate, cell paste), crude product (e.g., rich ion exchange fraction), finished isolated or sterile bulk, and filled vials. Ambient, refrigerated, and frozen storage requirements based on these inputs and outputs then need to be addressed in the areas of seed, raw materials, intermediates, clinical supplies, and process samples.

The scale of the pilot plant relative to current or proposed manufacturing scale and relative to laboratory scale needs to be established. Generally, a 1:5 or 1:10 scale-up is common, but smaller ratios can be advantageous in terms of minimizing risk during factory start-up. The appropriate level of biosafety also must be considered, with good large scale practice (GLSP) being sufficient for most recombinant work. Some pilot plant areas may require biosafety level 1–large scale (BL1-LS) and/or biosafety level 2-large scale (BL2-LS), and careful consideration of multiple-use/decontamination issues for BL1- and BL2-containing facilities is necessary. Solvent use areas need to be identified and minimized if possible because cost implications for explosion-proof design can be substantial.

Primary motivations to design a multiuse pilot plant are the conservation of capital and improved response time for clinical manufacturing needs for new development products. The strategy for multiuse should be identified and considered throughout the design process. Common strategies (6) include using totally dedicated equipment and limiting production to one product at one particular production stage, using campaigned equipment in which multiple products are sequentially processed in the same equipment "campaign style," with documented product changeover procedures completed to minimize cross-contamination potential, and conducting concurrent manufacturing in dedicated equipment in which several products are simultaneously produced in segregated areas with segregation achieved via physical separation or closed systems. These strategies result in minimization of cross-contamination potential through engineering controls, standard operating procedures (SOPs), or temporal segregation (7). Typical pilot plants combine the second and third strategies by providing segregated and dedicated areas for certain types of processing (e.g., virus and nonvirus areas, GLSP versus BL1/BL2 areas, beta-lactam antibiotic processing, heat-resistant spore-former cultivation) (8). Example layouts for product-specific and common areas depending on the strategy selected have been described (9).

Guidelines for the use of open versus closed systems for various processing stages need to be devised and justified based on available and cost-effective technologies as well as quality requirements. In a closed system, the product is not exposed to the immediate environment. The quality of materials entering a closed system (such as water, steam, or air entering a bioreactor) is controlled by its quality as well as the manner in which these materials are entering the closed system (such as filter sterilization, autoclaving, or steam sterilization of system connections) (6). In an open system, the product is exposed directly to the surrounding environment, thus necessitating the creation of a controlled or sometimes even a closed system around the open system. One example of this strategy is the transfer of seed vial contents in a biosafety cabinet. Other examples might include the supply of higher-quality filtered air to areas for the cooldown of autoclaved materials and assembly of clean equipment. Thus, levels of containment can be constructed until all open systems are contained within controlled or closed systems to meet quality requirements for the specific processing stage.

The specific issue of potential exposure of product to spore-forming organisms has been debated extensively. The closed system concept appears to minimize the concern for airborne spores typically found in the environment. Cultivation guidelines for spore-forming organisms within a multiuse facility appear to have been redefined to evaluate the robustness of the spore type formed. Careful consideration of segregation has been given when cultivating organisms such as *Bacillus* that form difficult-to-kill spores. A separate, self-contained area with no shared equipment, air systems, or entrances minimizes exposure concerns for other products. The relatively heat-sensitive, substantially larger, fungal spores have not been highlighted as a substantial cross-contamination concern. Thus, segregation for these types of processes is not as critical, although it still might be implemented due to company preferences for separate bulk drug and biologics processing areas.

DESIGN

The initial generation of flow charts for a bioprocessing, fermentation, or isolation pilot plant is best conducted based on model processes that are developed and critiqued by several end users of the equipment and facility. These model processes should be carefully selected as to give adequate and appropriate representation to current and reasonable future needs of the pilot plant. As they form a preliminary but sound basis for equipment and utility sizing, they should include unit operations as well as expected cycle times. In some cases, simulation (10–12) or actual laboratory- or pilot-scale testing (13) of critical equipment under typical processing conditions or with expected process streams may be warranted to further define design specifications. Example model processes have been published for vaccines (14) and intracellular and extracellular products (15–18). Resulting flow charts can focus on raw materials for fermentation or isolation, seed development, fermentation, recovery, crude product isolation, final product isolation, product finishing, buffers, utilities, and waste (including chemical waste, biological waste, and any necessary offgas treatment) (15).

The requirements for these model processes, coupled with an estimate of the required facility output, can be used to construct a preliminary architectural layout that divides the facility into sections (termed modules, cubes, or suites). These sections are surrounded by controlled corridors and accessed via airlocks (3). Based on this initial layout, flow patterns are developed to permit the logical, usually unidirectional, flow of personnel, raw materials, product, waste, and clean or used equipment throughout the facility. Flows are designed primarily to minimize cross-contamination from multiple products and from clean and used equipment while maintaining access and flexibility (3). This separation often is accomplished by using clean and return corridors located on either side of the processing area as well as by temporal segregation using appropriate SOPs. Evaluation of these flow patterns should be conducted carefully and should incorporate end users as well as quality auditors. In some cases, companies may solicit outside good manufacturing practice (GMP) reviews of the project at this time from external consultants, a second design firm, or even FDA representatives. Gathering and organizing up-to-date accurate information and comments about the impact on model process needs are critical factors in forming a successful basis of design (19).

Example layouts for process flows for biotech facilities have been published (6,20,21). An extensive reference (including (piping and instrument diagrams (P&IDs), layouts, and flow diagrams) for the design of biopharmaceutical plant equipment, utilities, heating, ventilation, and air-conditioning (HVAC), and waste treatment also has been published (22). Although it was based on acceptable practices in 1991, the strategies and concepts for design presented are still quite relevant. Adequate space around the equipment needs to be reserved for operations and maintenance access as well as for related portable equipment. The movement, segregation, storage, and cleaning of portable equipment such as vessels, skids, hoses, and carts should be fully examined.

Various reports of successful and problematic aspects of project design, construction, start-up, and validation have been published. Cost control issues and the benefits of formulating a master schedule (23) need to be reviewed in terms of their potential validation impact. Construction concerns and "lessons learned" from multiple biotech installations also have been summarized (24). One case study of a laboratory facility highlights the need for teamwork among client, design firm, and construction manager (25). This trio can be extended to include the validation contractor in an effort to minimize "finger pointing" and "blame storming" when unforeseen problems are uncovered. Regardless of who may actually be perceived as being "at fault," the company owning the facility ultimately is adversely affected by substantial start-up and validation delays. A comprehensive overview of biochemical pilot plant design and operational guidelines appeared in the mid-1980s (26).

Process Equipment

The major decision regarding process equipment is that of similarity versus diversity as it applies to specific process steps as well as multiple products. The extent of commonality required in equipment specifications is an important operational factor to be considered because equipment specifications are based on inputs from model processes. For a pilot plant to accept a manufacturing interface role, at least a portion of the equipment design and selection should match that of the factory which the pilot plant is intended to support. For example, if the factory intends to use continuously sterilized medium for large-scale fermentation, then there should be provisions in the pilot plant to perform continuous sterilization as products get closer to the transfer to production. Alternatively, depending on individual company preferences for the pilot plant role, later clinical batches (phase II and beyond) might be manufactured in the factory to facilitate eventual process transfer and ensure consistency of equipment.

Skid-mounted equipment has been often selected in recent installations because it puts the design burden entirely on one vendor, who specifies the appropriate quality, supply rate, and pressure of the required utilities. Skid-mounted equipment can be fixed or portable, which permits different types of processing in the same space as necessary. Consistency in skid designs with respect to components and spare parts may be a drawback as vendors might not be willing to build skids to individual company specifications without added costs. This situation can be particularly problematic if the skid manufacturer has arranged a low-cost deal with a vendor (e.g., diaphragm valves) or programmed software on a programmable logic controller (PLC) that may not be the vendor of choice for the rest of the plant. There also may be incompatibilities in piping, utility requirements, or software when the skid is connected (27) if miscommunications occurred between the skid vendor and the facility designer.

The traditional alternative to skid-mounted designs is to purchase the equipment components separately and request the facility designer to develop the system. One advantage to this approach is that total flexibility in equipment selection now rests with the pilot plant designers, who may be quite confident in their ability to design the equipment based on their extensive prior experience. Drawbacks of this approach include placement of the design burden on the facility as well as added design costs for one-of-a-kind systems. Specific references are available for flow charts and detailed designs for fermentation and harvest equipment (28–31) as well as for filtration units (32), chromatographic equipment (20,33,34), and centrifuges (35,36). A list of typical process, utility, and support equipment that might be required in a bioprocessing pilot plant has been compiled (Table 1).

For a bioprocessing facility, certain aspects of equipment and piping specifications should be uniform throughout the installation where possible. These specifications might include requirements for orbital welding of lines, documented weld inspections, passivation, pressure testing, self-draining (sloped) piping, minimal lengths for dead legs and pockets, restricted use of flanged connections, level of polish on product contact surfaces, stainless steel quality (typically 316L for product contact surfaces), and use of FDA-approved polymers in specific applications (gasket, o-rings, valve seats, distribution piping). Proper storage of fabricated piping awaiting installation as well as prominent identification of valves and lines also must be considered. Valve specifications can be important to determine early in the design estimation phase. Valve capital cost contribution is high because large numbers are required and any installed valves (either diaphragm or ball) need subsequent maintenance. Consideration should be given as to whether live steam, set up as a steam block, should be used to continually purge the back side of product contact valves and whether live steam should be used for tracing in between double o-rings on a port or manway (28). While such designs traditionally have increased sterility assurance for lengthy secondary metabolite fermentations, they may create detrimental localized hot spots during animal cell cultivation. Neither the equipment itself nor its disposable or replaceable parts should release any extractable substances (after the initial postinstallation rinsing and cleaning steps) into the product (28). An overview of sanitary piping design, installation, and validation concerns is available (37).

Transfer piping strategy is another major design decision. One common strategy is for several transfer lines to meet at a transfer panel or process manifold, typically located at a high point between two or more areas, and be connected using spool or "jumper" pieces to move material in the desired fashion (3). Steam entry is at the high point and condensate drainage at the low point for steam-in-place (SIP) during sterile transfers. Clean-in-place (CIP) solutions are passed along the direction of the transfer. Advantages to this approach include reduced piping distribution costs and space requirements, which can multiply rapidly when multiple tanks and multiple types of process transfers are required. Disadvantages center on operational restrictions concerning the elapsed time between successive transfers through a common manifold system as well as cross-contamination associated with transfers of nonsterile media/buffer, sterile media/buffer, inoculum, harvested broth, CIP solutions, and water through common lines. A listing of product and nonproduct transfer line/utility piping that might be required in a bioprocessing facility has been compiled in Table 2.

Both sterilization and cleaning sequences need to be defined and documented during the equipment and transfer piping specifications because retrofits after installation can be time consuming and expensive. P&IDs should be traced through these sequences before approval by both operational and validation personnel. Requirements for validation testing and in-use monitoring of equipment operation also need to be identified during the equipment specification phase. Examples include sizing condensate lines to permit adequate drainage of condensate when spore strips are present, specification of additional validation ports for thermocouples on vessels and condensate

Table 1. Types of Processing Equipment and Utilities Possibly Required for a Bioprocessing Pilot Plant

Fermentation and harvest	Isolation	Utilities	Support
Culture storage freezer	Low-pressure chromatography columns	Clean steam generator	Sterilizing autoclave
Incubator/shaker	High-pressure chromatography columns	USP purified water system	Decontamination autoclave
Biosafety cabinet	Solvent blending and delivery system	WFI system	Depyrogenation ovens
Laminar flow hood	Sterile filtration assembly	Deionized water	Laminar flow hoods for cooldown and equipment assembly
Seed fermentor(s)	Laminar flow hood	Process (city) water	Glasswasher(s) (glassware, vials, stoppers)
Production fermentor(s)	Liquid transfer pumps	Compressed clean air	Buffer preparation tanks
Nutrient feed tank(s)	Dispensing machines	Compressed instrument air	Media preparation tanks
Harvest tank(s)	Chemical fume hood	Compressed plant air	In-line mixer
Microfiltration skid	Portable tanks	Compressed gases (O_2, N_2, CO_2, NH_3)	In-process sample refrigerator/freezer
Ultrafiltration skid	Laboratory-scale centrifuge	Electricity	Raw material, cold/frozen storage
Pilot-scale centrifuge	Ultracentrifuge	Chilled water	Floor and bench scales
Homogenizer	In-process refrigerator/cold room/freezer	Cooling tower water	Process and equipment monitoring/alarm systems
Continuous sterilizer	Product refrigerator/freezer	Plant steam	Uninterrupted power source (UPS)
Filter integrity tester	Syringe/vial filling equipment	Glycol	Offgas analyzer (mass spectrometer)
Sterile tubing welder	Lyophilizer	HVAC and building automation system	Bar-code system
		CIP skid(s)	
		Biowaste inactivation system	
		Environmentally potent chemical destruction system	
		Dust collection system	

Table 2. Transfer Lines and Utility Piping Possibly Required in a Bioprocessing Pilot Plant

Product contact	Nonproduct contact
Utilities	Utilities
Clean water	Chilled/cooling water and/or glycol
Clean air	Instrument/plant air
Clean steam	Plant steam
CIP solutions	Contained sewer
Other compressed gasses (CO_2, O_2, N_2, NH_3)	Chemical (noncontained sewer)
Process	Sanitary water
Sterilized medium or buffer	Vacuum system
Sterilized nutrient/acid/base/antifoam feeding	Natural gas
Nonsterilized medium or buffer	
Inoculum	
Harvest	

lines, installation of product contact utility sampling stations (especially at worst-case locations), and installation of temperature sensors on critical lines to monitor sterilization and proper trap operation.

Utilities

Product contact utilities might include water, clean steam, compressed air and other gases, and CIP systems. Nonproduct contact utilities generally include chilled water, instrument air, plant steam, and plant air, although in some cases plant steam may be appropriate for product contact (Table 3). A listing of product and nonproduct contact utilities is compiled in Table 2. Multiple utility use points are located within and among suites for both product and nonproduct utilities; these are organized into utility stations for each equipment skid. To minimize piping, it might seem convenient to use product contact utilities in nonproduct contact applications such as compressed clean air on a vessel jacket, relying on a check valve for backflow prevention. This approach can risk disaster because the operation and quality of utilities can affect processing throughout the entire facility. Cross-contamination for product contact utilities among suites is rare because of positive pressures of utilities piping into each module.

Peak utility loads should be based on carefully investigated assumptions that are transferred from design directly to validation testing. Although it may be possible to run several large fermentors simultaneously, each at maximum airflow rate, it may be less likely that several large fermentors are cooled simultaneously after completion of sterilization. Thus, air compressors might be sized assuming maximum load while process chillers might be sized on the basis of an operationally realistic fractional load. Assumptions about processing cycle times and tank volumes are important particularly when sizing CIP systems and

Table 3. Example Plant Steam and Clean Steam Achievable Quality Comparisons

Quality attribute	Clean steam	Plant steam
Bioburden	<0.1 cfu/mL	<0.1 cfu/mL
Total organic carbon	<0.5 ppm	<2.0 ppm
Endotoxin	<0.25 EU/mL	<0.5 EU/mL
pH	5–7	5–8
Conductivity	Meets USP 23 Suppl. 5 standards	N/A

liquid waste decon systems. Shorter (20-h) *Escherichia coli* fermentations might require a smaller tank volume [1,000 L] but are run more frequently, compared with longer (400-h) fungal fermentations, which might require a larger tank volume (20,000 L) but are run less often.

Product Contact. The most critical product contact utility is often water because of the sheer volume needed during typical processing steps. It is well known that the quality of water needs to be appropriate for the intended application, but water system design decisions often are conservative. In many installations, purified water is used for upstream processes while WFI is reserved for downstream isolation and for cultures sensitive to endotoxin such as animal cells (38). A brief overview of various methods to produce pure water is available (39).

Water-for-injection (WFI) is often favored for the entire facility for uniformity. Specifications are set for parameters such as bioburden (typically 0.1 cfu/mL) and endotoxin (typically 0.25 enzyme units, EU/mL) as well as USP chemical tests that include conductivity, pH, and total organic carbon (TOC). WFI systems containing stills generally are designed as a hot 80 °C loop with point-of-use coolers that can provide water at a user-selected temperature. These coolers can form deadlegs and thus require substantial flushing before use. It is also possible to design a cold or ambient WFI loop with periodic (typically daily) heating to 80 °C or ozonization for sanitization purposes. In some applications, reverse osmosis (RO) systems are used, but care of the membranes is critical to consistent operation.

USP purified water is an attractive alternative to WFI for various upstream applications. Specifications include bioburden (typically 100 cfu/mL) and USP chemical tests. Although such systems can be operated hot or cold, recent success with ambient systems sanitized with ozone and ambient systems sanitized chemically (40) has been reported. Other noncompendial grades of water may be required. In these cases, quality attributes typically are established by the facility to address the needs of specific processing steps (Table 4). Examples might be deionized water with bioburden (typically <500 cfu/mL) and resistivity (>1 Mohm) specifications or process water (municipal water passing through a break tank).

General design elements for water systems include the desired flow rate at each use point and the desired number of simultaneous use points in operation. Use-point positioning should consider the height of the take-off valve, automation of take-off valves, and sink or sewer location for flushing, as well as cleaning, sanitization, and acceptable noncoiled storage for associated use-point hoses. These values are combined to obtain the total "take-off" flow rate; the system and piping are then sized such that velocity of the remaining flow rate is sufficient to minimize bioburden and biofilm growth. Suggested methods for obtaining this residual velocity have been outlined (41). In addition, water systems should be designed with specific sanitization procedures identified at the outset. Steam, ozone, or chemical sanitization methods are most common, although steam sanitization generally has not been recommended for plastic loop systems because of concerns about distribution pipe sagging. Plastic systems utilizing polyvinylidene fluoride (PVDF) instead of stainless steel are becoming more common and effective at achieving even WFI conditions. A direct comparison of plastic piping with stainless steel with respect to ion leaching, smoothness, and cfu counts has been made (42).

Clean steam can be generated from a purified water feed using plant steam for supplying heat of vaporization. Increased heat exchanger fouling may occur if a controlled feed water supply is not used. Clean steam distribution piping is designed with regulators to reduce the pressure from 5 atm down to about 2–2.5 atm as needed for typical SIP operations. Sanitary pressure gauges, placed near or on equipment skids, can assist in determining dynamic pressures during SIP cycles. Steam sampling stations generally take the form of a removable sanitary trap to which a sample cooling device (typically a heat exchanger) is attached. Consequently, a source of cooling water is necessary, either piped from local lines or transported by gravity via a portable holding tank about 20 L in volume. Traps should be placed in header branches and equipment supply lines such that condensate buildup does not occur when skid equipment is isolated for repair. Periodic cleaning of the clean steam system assists in cleaning fouled heat exchange surfaces and in minimization of rouging.

Compressed air in product contact generally is used for fermentation sparger air supply and pressurized transfers. Oil-free compressors are favored, with intakes positioned away from other building effluents. Heating up to and holding at 200 °F (93 °C), typically accomplished by design of the compression ratio and residence time in a receiver, is desirable to minimize airborne biological contaminants such as bacteriophage, although there is some evidence that submicrometer (0.1 μm) filtration also can be effective. Humidity can then be reduced from the compressed air using refrigeration and desiccant dryers. Moisture removal is desirable to a dew point of −40 °C. Air is then microfiltered (0.2 μm) at the point of use to remove microbes and particulates. Compressed air may be sampled for any or all of the following depending on the application: identity, hydrocarbon level, microbial content, and particulates. Other gases may be required for the facility such as oxygen enrichment for microbial cells, carbon dioxide for pH control of bicarbonate-buffered media, nitrogen and oxygen supplementation for animal or insect cells, and appropriate gas blending and control systems. Planning should consider storage of bottled gas cylinders outside the facility or in segregated maintenance areas to minimize the transport of unclean cylinders in controlled locations.

Table 4. Comparison of Example USP Purified Water, Deionized Water, and Process Water Achievable Quality Attributes

Quality attribute	USP purified water	Deionized water	Process (city) water
Bioburden	<100 cfu/mL	<500 cfu/mL	<500 cfu/mL
Coliforms	Absent	Absent	Absent
Total organic carbon	<0.5 ppm	<2.5 ppm	<5 ppm
Endotoxin	<0.25 EU/mL (for information only)	N/A	N/A
pH	5–7	5–8	5–8
Conductivity/resistivity	Meets USP 23 Suppl. 5 standards	>1 Mohm	N/A

CIP skids can be portable and moved into place in the suite at a designated utility location or one or more fixed main skids can be located permanently in the utility area. CIP systems are composed of supply and return pumps, one or more cleaning agent dilution tanks, provisions for heating cleaning agents, and a PLC for sequence control and associated data acquisition. Because multiple tankage and pumps are required (20), these skids often are located in utility areas where acid- or caustic-based cleaning agents arriving in drums can be suitably diluted. More than one CIP skid may be necessary to provide product segregation and minimize cross-contamination. Multiple skid requirements depend on design decisions concerning the level of containment and product segregation (3) as well as capacity.

Nonproduct Contact. The type of cooling/heating system desired, while not in product contact, has widespread design implications. Recirculating temperature control loops containing high surface area/unit volume heating and cooling heat exchangers, an expansion tank, and a circulation pump are common. These systems rely on indirectly heating or cooling of jacket fluid so their response can be slow, particularly if heat exchangers were not sized according to user expectations. The involvement of heat exchangers means that practical application may be limited to vessels a few thousand liters in size. Jacket fluids can be water, glycol, or novel fluids with wide temperature ranges. Operation of the circulation pump and internal (as well as external) integrity of the heat exchangers are also critical. Cooling heat exchangers can use either chilled water or glycol, whereas heating heat exchangers generally use plant steam. Dual parallel filter housings, containing 10 to 50-μm elements, are advantageous to reduce heat exchanger fouling. Sediment buildup on filters can be monitored using measured pressure differentials across the filters, and the assembly can be designed so that one filter may be replaced while maintaining flow to the second filter.

Direct application of heating or cooling fluids to process jackets is the traditional alternative. Consecutive application of chilled water and steam on the jacket is acceptable if the design evacuates the jacket using higher-pressure compressed air such that banging is minimized. Use of glycol and steam on the same jacket can have environmental implications from glycol losses. Regardless of the indirect or direct mode of application, jacket design can greatly affect heat transfer effectiveness. Dimpled or straight jackets are preferred for smaller vessels, with external half-pipe cooling coils used in larger vessels. The largest manufacturing scale vessels utilize internal cooling coils but these present additional internals for cleaning and increased sterility risks, particularly at welds, because cooling water can be at a higher pressure than the fermentor contents.

Filtered dried instrument air, typically at a pressure near 100 psi, is required for solenoids. Pressure specifications for control valve operation vary considerably among vendors. Plastic hosing, although convenient and cost effective, should be evaluated carefully for instrument air lines, especially on hot equipment where heat-induced leaks can develop and in solenoid cabinets where crimping can slice the tubing. A backup instrument air compressor can minimize operational consequences from supply pressure dips. A separate compressed air source, at a higher pressure than chilled/cooling water and jacket steam, is desirable for evacuation of vessel jackets to minimize risk of backflow into the instrument air system.

Natural gas usage, if required, might best be addressed using bottled gas for safety reasons, which discourage distribution piping throughout the facility. Favored for flaming of openings in many microbiological procedures, the use of Bunsen burners in a biosafety cabinet or laminar flow hood should be examined carefully with company safety department representatives. Care should be taken to minimize the potential for reaching the lower explosive limit near the motor of the cabinet or hood. Compatibility of plastic laboratory ware with flames also should be ensured.

HVAC. HVAC is a key and costly component of the facility. Its design and operation needs to be understood by operating, maintenance, and quality assurance personnel. Requirements for air cleanliness, as well as the ability to achieve them, are determined by HEPA filtration, airflow volume, and pressurization room design as well as the type of operation being conducted (Table 5) (3). HEPA filters are often used for inlet air but are only used for exit air when required by biosafety containment. The specific classification of processing areas depends on processing step quality concerns. Classification level, representing the maximum number of particles permitted per cubic foot of air sampled, of an area increases from fermentation (class 10,000–100,000; 20–40 room changes per hour) to purification (class 1,000–10,000; up to 100 room changes per hour) (3). Within a room there are cleaner areas surrounding open transfers (e.g., biosafety cabinets or laminar flow hoods) that are class 100 (540 changes per hour) (43). For viable particles, the range is from class 100,000, which permits 2.5 cfu/ft^3, down to class 100, which permits 0.1 cfu/ft^3 with

Table 5. Example Controlled and Noncontrolled Area Achievable Quality Comparisons Based on Room Design

Type of area	Pressurization	Airflow changes (#/h)	Inlet filtration	Nonviable (#/ft)	Viable (cfu/ft^3)
Bulk fermentation processing	None	60	Coarse	<250,000	<25
Laboratory	Positive	60	Coarse	<100,000	<2.5
Laboratory	Positive	10	HEPA	<100,000	<2.5
Laboratory, clean construction	Positive	25	HEPA	<100,000	<2.5
Laboratory, clean construction	Positive	20	HEPA	<10,000	<0.5
Laboratory, contains autoclave	Positive	100	HEPA	<100,000	<2.5
Laboratory, contains glasswasher	Negative	50	HEPA	<100,000	<2.5

intervening levels generally determined by individual application (43). Sufficient air velocities also are necessary to dilute particulate contaminants. In some cases, separate air handlers are considered for critical sterile areas, inoculum preparation areas, high particle or dust areas, and segregated product areas. In other instances, airflow is designed in a once-through pathway, which may increase heating and cooling utility costs substantially.

Inlet air enters from the top of the room, flows downward, and exits from ducts near the floor. Any recirculation, which may be desirable to minimize capital costs for heating and cooling equipment as well as operational costs, should be only from the room where the air originated, limited to areas that do not generate substantial amounts of particles, and pass through inlet HEPA filters again. This strategy is beneficial to the life span of the HEPA filters and the ease of maintaining constant temperature and humidity (3). Exit ducts should be placed near equipment within a processing area that generates the most particulates or aerosols so as not to spread them throughout the area. Specific examples include autoclaves, glass washers, and cell-disrupting homogenizers.

Room pressurizations should be higher for cleaner areas and lower for areas that generate particles (3). Higher pressurization levels should be assigned to airlocks between adjacent areas for which it is desired to minimize cross-contamination. Relative pressurizations between rooms sharing a common wall should be considered to supplement or minimize reliance on completely sealing rooms during their construction. The use of door sweeps and gasketted ceiling tiles to achieve and maintain pressurization should be evaluated relative to floor cleanability and ceiling maintenance access, respectively. Pressure differentials might be as high as 0.03 in of water between classified areas and as high as 0.05 in between classified and nonclassified areas. Lower pressure differentials might be acceptable depending on the specific product so long as the direction of airflow remains consistent. Interlocks, control, monitoring, and alarm building automation systems may be designed to reduce pressure losses when doors are opened and to alert users to unsatisfactory conditions.

Facility temperature is generally maintained on the cool side because gowning tends to make workers warm. Inaccurate assumptions concerning the impact of radiant heating may cause temperatures to be higher than desired on certain days of operation. Humidity should be comfortable to avoid drying of workers' respiratory membranes during extended hours. Care should be taken to specify tolerance limits that have quality implications relevant to the processing steps expected to be conducted in that area to avoid unnecessary facility costs. While specific example guidelines are 23 ± 3 °C for temperature and 40–60% relative humidity (44), substantially larger ranges may be acceptable for many processing applications. An extensive P&ID design of HVAC for clean spaces is available as a published reference (43). In addition, consideration should be given to Federal Standard 209E, Airborne Particulate Cleanliness Classes in Clean Rooms and Clean Zones, as well as European Economic Community (EEC) regulations that establish parallel categories of Grade D (100,000), Grade C (10,000), and Grades A + B (different designs of class 100) (43).

Containment

Similar issues exist for product, environmental, and personnel protection. Often the resulting design specifications overlap with those items necessary for sterile or sanitary operation. This concept is addressed in the literature (45) in which solutions to prevent microbial transport for the purposes of containment were compared with those to maintain sterility for fermentors. A unified approach is recommended with the goal of minimizing unnecessarily stringent specifications that can substantially augment capital and operating costs. Environmental and personnel protection issues should be evaluated for each new process undertaken by the facility, along with product protection quality concerns. Samples of broth, and other product, intermediate, and waste streams, should be submitted for aquatic and human toxicity testing in advance of processing in the pilot plant when appropriate. Preliminary laboratory data then might be available to evaluate potential hazards in conjunction with published documentation concerning the relationships of the questionable compounds, raw materials, or organism with known hazards.

Product Protection. The major concept for product protection is the closed system in which all material entering or leaving the process must be controlled in a specified and documented manner. This requirement extends to product contact utilities such as purified water, clean steam, compressed air, and controlled/monitored room air so that when equipment is opened for cleaning the quality of the surrounding environment is consistently at a known and documented level. The equipment is cleaned to an acceptable quality level. After sterilization, material enters the closed system via a sterilizing filter, steamed connection to an autoclaved bottle assembly, or sterilized transfer line.

Offgas passes through a 0.2 μm steam-sterilized vent filter, often contained in a low-pressure steam-jacketed housing preceded by a coalescing filter, condenser, or vent heater to avoid pluggage from moisture. Double mechanical seals are utilized with a barrier fluid of a controlled quality (typically condensed clean steam) at a higher pressure than the vessel. A closed system also can be used to surround an open operation such as the use of isolator technology (i.e., glove box) as an alternative to clean rooms or biosafety cabinets (46,47).

Environmental Protection. Environmental protection concerns focus on emissions in air, liquid, and solid waste. State air permits typically are required for new vessels and need to be altered if additional air treatment devices such as vent filters, incinerators, or scrubbers are added. Hydrophobic vent filters can provide emergency foam control and prevent external product release. These filters can be monitored for pluggage over cultivation time using pressure differentials with plugged filters able to be replaced during a run only if a bypass line has been installed. Liquid emissions include accidental spills/sewerings, which can be minimized by locking vessel bottom valves, blocking open sewer connections, collecting sample line flush waste, and providing clear written instructions to personnel regarding batch disposition using batch sheets and appropriate signs. Dikes, troughs, and vaults can be used to contain large spills or divert inadvertently sewered material. Sample waste can be minimized using low-volume sampling devices. Thermal or chemical (less common) batch or continuous kill tanks need to be considered for biological hazards. For chemical hazards, other appropriate waste destruction and minimization systems may include hot caustic hydrolysis, reverse osmosis, and evaporative condensers. Batch leakage into cooling water utility systems can be minimized by pressurizing utility systems higher than the batch, but this might introduce contamination if jacket cooling coils develop leaks. Alternatively, the utility system return headers can be isolated from general use until their contents are treated and/or tested. This solution might be attractive for cooling water as well as condensate, especially when it is returned to river sources. Care must be taken to avoid sequestering large quantities of potential waste for long periods of time. Solid waste can be controlled procedurally using clearly labeled containers.

Personnel Protection. Personnel protection is not as great a concern now that most recombinant organisms can be grown under typical operating conditions of GLSP. For other pilot plant processes requiring personnel protection, either the culture is rated at or above BL1-LS or a compound produced by the culture adversely affects human heath (e.g., a genotoxin, mutagen, teratogen, or carcinogen). Environmental monitoring of the operating area is necessary for the compound or culture of interest using air or swab sampling techniques. For optimal recovery, swabs should be soaked in a solvent known to dissolve the compound of interest. Screening and monitoring of personnel potentially in contact with the process (operator, mechanics, janitors, engineers, visitors) must be arranged and notification given to those personnel at unacceptable levels of risk relative to their individual health histories. A key element of personnel protection is personnel protective equipment (PPE). This equipment can include gowns, uniforms, safety glasses, lab coats, and gloves at a minimum but can be extended to include face shields and respirators as required. Spill response equipment, procedures, and training are also necessary.

Reduction to practice techniques are presented in the literature for general biocontainment issues (48) and biocontainment regulations and requirements (49). Specific examples are available for the design of waste inactivation systems (50,51), contained facility design and operation (52,53), design details for cleanliness and containment for floors, walls, ceilings, and penetrations (54), and contained sampling systems (55).

Instrumentation

The main purpose of a pilot plant is to collect data in as many ways as possible. Thus, a large fraction of the pilot plant capital allocation might be devoted to instrumentation. Traditional instruments for on-line monitoring, particularly for fermentation, include pH (56), dissolved oxygen (DO) (57–59), temperature, agitator speed, agitator power draw, backpressure, foam detection, batch level or weight, airflow rate, and vent gas analysis via mass spectrometry (60). Redundant instrumentation is desired for critical parameters such as pH and DO. Instrumentation for isolation monitoring includes monitoring of conductivity, pH, ultraviolet wavelength intensity, and refractive index detection. Useful utility instrumentation includes on-line total organic carbon (TOC) and resistivity/conductivity meters. Local readouts of transmitter values can be useful, particularly for pressure and possibly temperature, although the potential for differences between field and control room values may have to be addressed.

Many novel sensors have been developed including dissolved carbon dioxide probes (56,61), redox sensors (56), glucose sensors (62), near-infrared detectors (63,64), fiber optic biosensors (65), fluorometry of intracellular NADPH (66), and cell density probes (67–69). Another novel measurement device example incorporates fermentor side streams to measure viscosity on-line (70). A description of useful sensors may be found in the literature (71). Characteristics of on-line and in situ devices include cleanability, minimal drift, minimal interferences, versatility, sterilizability, ease of calibration, and several other factors. Utilization of any sensor should be evaluated for its effect on bioreactor sterility as well as product quality.

Automation and Control

The need for flexibility and change within the pilot plant (as compared with production) needs to be apparent in control schemes, event sequencing, data collection and presentation, and alarming (72). Increased amounts of automation restrict abilities of the end user to implement changes, but do reduce the number of manual operations and thus possibly headcount. Specifically the control system needs to readily permit set-point and alarm changes. Control systems can be distributed or centralized, with distributed control systems (DCS) favored because there are individual controllers/computers for each unit, often connected to a central computer via a high-speed network. Thus, the loss of one centralized computer is not catastrophic (72).

While there can be "islands of automation" within a facility, connectivity in a centralized location for data monitoring, set-point/alarm changes, and data archiving might be desirable. Substantial advantages exist for localized controls on skids when many field-based inputs are required based on an operator's evaluation of field conditions. A local readout of monitoring data versus a centralized "control room" monitoring area (or even both) is dependent on the degree of operator supervision versus intervention required for the processing. This decision is related directly to the flexibility of operators to go in and out of the field in terms of gowning procedures and ease of access. The ability to remotely input set-point or alarm changes can be useful when containment or product protection levels are high. Multiple areas for setting parameter values need to be considered carefully from the aspect of operational control to avoid duplication of desired changes and to minimize batch record log omissions. Verbal communications between the field and the control room using radios or intercoms also need to be considered.

Distributed control systems are favored for process management. System architecture requirements vary depending on the processing requirements but examples for a non-PC based system have been published (73). Distributed control systems also can be PC-based using commercial PC control/data acquisition software (74). In either case, some type of process-level controller is required such as a PLC, loop controller (LC), personal computer–based controller) (PCBC) (32). The ability for those without computer expertise to easily modify and understand control sequences using newly developed commercial control software is attractive but requires effective change control and documentation procedures. The ability to readily configure the data acquisition system to download and archive data is a consideration important for later operation and process development.

There are several types of computer inputs and outputs (72), which include *(1)* digital inputs in the form of an on–off signal used for foam probe, pressure, flow, or temperature switches; *(2)* digital outputs in the form of an on–off signal used for the flow of antifoam, open–close valves, or pump/motor start/stops; *(3)* analog inputs in the form of an amplified output in the range of 4–20 mA used for temperature, pressure, pH, dissolved oxygen, flow, and power draw; and *(4)* analog outputs in the form of a proportional signal that controls the valve opening via an I/P conversion, typically used for back pressure, cooling, and airflow valves or for signals such as motor speed.

The type of control most frequently utilized is the proportional integral differential (PID) control loop. It is tuned using three parameters to minimize oscillations, which can change depending on the specific process (e.g., for dissolved oxygen cascade control of a fermentor cultivating faster-growing *E. coli* versus slower-growing yeast cells) or over the course of the same process (e.g., for glucose and ethanol consumption phases of a baker's yeast fermentation) (75). These three parameters are the size of the error relative to the proportional band with smaller bands resulting in more sensitive systems, integral of the error over time with longer error durations resulting in greater corrections, and rate of error change with time with larger increases in correction with time calculated the faster the error is growing with time (71). Typically the third parameter is at a minimal value near zero for the optimally tuned loop.

A conscious decision needs to be made concerning the desired level of automation for CIP, SIP, and vessel-to-vessel transfers. Higher levels of automation may ensure reproducibility of operation, reduce reliance on operational staff, and simplify SOPs. Raising the level of automation increases capital, validation, and maintenance costs while reducing flexibility. Operating personnel who routinely observe automated processing steps may be less familiar with the details of automated versus manual equipment operational sequences. A domino effect may exist in that automation of one valve may lead to automation of several related valves, severely increasing design complexity and cost. Limit switches often are installed that confirm to the control system that the desired automatic valve has opened or shut as requested. Automatic pulsed diaphragm valves have been successful for nutrient feeding as well as for sterilization of transfer lines in automatic plants (76). One prominent example of a highly automated but apparently reliable fermentation facility is operated for the manufacture of cephamycins as well as for pilot use (77). Typical diagrams for monitoring and control of a fermentor (20,72) and for monitoring and control of an ultrafiltration unit (32) have been published.

Data Acquisition and Archiving

The collection and management of data during pilot plant operations are critical, with the collection interval dependent on the nature of the processing step. For fermentations, intervals might be every 15–20 min during secondary metabolite or animal cell cultivations but might shrink to 5–10 min for faster-growing bacterial or yeast cultures, especially when a key process addition or change is planned to be made on the basis of offgas or dissolved oxygen data. Typically, 5- to 10-min intervals are appropriate for microfiltration or centrifugation operations. Smaller data collection intervals may create substantial data archiving demands that may be alleviated by only saving those data substantially different from previous values, based on some user-defined value. Continuous data collection in the form of chromatograms is typically required for chromatographic operations.

In-process trending of data is attractive to permit timely evaluation of process status and troubleshooting of processing problems. Automatic recording of alarms and changes to the control system, preferably sorted according to the equipment and/or batch on which they occurred, also can help trace back processing problems and actions taken. On-line data might be archived on a common server by some unique batch delimiter, according to the established facility SOP for archiving data. One formidable challenge is to incorporate off-line data in the same or a closely related reference database; this off-line data can get fragmented among the notebooks and files of individual researchers. Novel customized systems for bar-coding samples taken from the field and creating a database for results can be helpful and can also create possibilities for the use of robotic systems for sample analysis. Remote access is an attractive option for monitoring processes from off-site locations. Remote alarming can be triggered to a satellite pager, thus removing the necessity for round-the-clock coverage. Remote control of operations can be administratively challenging to regulate and document as well as a potential safety issue. Security issues must always be considered for any established remote capability.

Warehousing

Warehousing needs for the facility need to be determined based on expected production capacity. Common items that might be stored include ambient, cold, or frozen solid or liquid hazardous and nonhazardous raw materials (controlled temperature and/or relative humidity storage), disposables, equipment, and spare parts as well as sample and product storage. Computerized inventory systems, organized by process or equipment, can be useful for most of these items with reminders to reorder when supplies decline to a critical level. Marker equipment also might be purchased for long-lead replacement parts for which redundancy was not desired during the original facility design. Sizing of these storage areas should consider typical lot volumes of prepared culture media and solids to minimize repetitive release testing for multiple lots. For a pilot plant, in which several diverse processes may be run, additional process-specific storage often is required. The relative amount of storage necessary in the building should be compared with other on-site and off-site options. Vendor manufacturing schedules often require purchasing larger quantities of rare raw materials to avoid future shortages. In this instance, if on-site storage is not available, vendor storage or a third-party warehouse alternative may be required. Warehouse access may need to be restricted depending upon the nature of its contents. There should be segregation for released, unreleased, and quarantined materials either by physical separation or by designated labels.

While the entire facility should have a comprehensive pest monitoring program, as well as an SOP, there may be a need to directly address pest control, particularly for flours and meals used in certain secondary metabolite fermentations. These food ingredients when purchased for pilot-scale fermentation purposes may not pass through fumigation unless specifically requested on the purchase order. Typically, larger customers in the food industry have invested capital for on-site fumigators, but commercial nutrient vendors may not have this type of equipment because of their high rate of inventory turnover.

Staging areas should be available for collecting raw materials and supplies for a specific batch. Some method of inventory assessment and control should be available; one convenient method is to use computerized bar-coding to check inventory as well as to confirm use of specified materials when charging occurs. In addition, controlled sampling areas (to provide raw material protection, environmental, and/or personnel protection) for sampling raw materials may be necessary.

Backup Systems and Redundancy

A critical design decision centers on the level of backup or redundancy required in the pilot plant. While typically none of the material created in the pilot plant is "for sale," mechanical problems can delay processing and clinical material production, which may interrupt clinical supplies and compromise the success of ongoing studies. Even if no clinical material is at risk, process development efforts are hampered when experiments are frequently interrupted for mechanical support-type equipment failures (4). Scenarios that might be protected against include loss of clean steam lubrication to agitator seals, loss of positive air pressure on a sterile tank, and loss of electrical power to a controller. Corresponding equipment needs might include redundant clean air compressors, clean steam generators, and purified or WFI water generation systems, with the peak plant capacity needs split between two systems rather than relying on a single system sized for peak usage. Such redundancy usually is achievable because utility costs are low compared to equipment costs (4). An uninterrupted power source (UPS) system sized to include the most critical equipment (typically controllers and data acquisition devices) might also be installed. Backup power for large electrical loads such as motors might not be economical; however, consideration might be given to installing connections for a gas-powered generator if future evaluations altered the economics.

Future Expansion and Modification

Because future processing needs may change, the pilot plant must adapt to maintain its usefulness to the company. Large initially underutilized spaces might be outfitted with utility headers to minimize processing disruption during future installation. Oversizing of utilities for anticipated future loads (4) as well as unforeseen procedural modifications warranted on the basis of validation testing results such as additional rinsing cycles assures subsequent flexibility. Selection of portable equipment permits exchanges for new processes, provided there are procedures developed for storage and start-up of this equipment.

One aggressive approach to increase versatility is the availability of a modular mobile validated pilot plant (78), composed of a system of interconnected and integrated modules. Modules are selected according to processing needs and can be assembled rapidly to form a processing plant capable of undergoing validation. Production modules include fermentation, recovery, and purification; the utility module includes clean water and clean steam; the personnel support module includes lockers and changing areas; and the HVAC module houses the air handling and filtration equipment. Additional modules can then be added or deleted as required.

OPERATION

Each pilot plant facility must decide for itself whether it is required to operate in a continuous mode of GMP compliance or whether it can selectively apply GMP practices to certain batches [but not to other concurrently run batches] or during certain specific processing periods. A conservative solution might be to operate continually in GMP compliance, thus insuring facility readiness and preventing omission of a required procedure for an actual batch destined for the clinic. Dual GMP and non-GMP operation serves to create two standards and may cause confusion among operating and maintenance staff. If dual operating modes are considered, then a clear distinction between GMP and non-GMP operation must be established, documented, and maintained.

Maintenance

Preventative. Preventive maintenance (PM), performed during an annual facility shutdown, is probably the most important factor influencing smooth operation throughout the upcoming year. The scheduling of an annual facility shutdown is desirable for several reasons including the ease of lockout and tagging of hazardous energy sources, the time economies of concentrating several maintenance workers in a single area, and the ability to collect and document completion of a large fraction of maintenance work at one time. This yearly PM binder can then serve as the model for maintenance personnel for the following year's shutdown. Individual PM work orders, spread over the course of a year, result in substantial time and effort expended in scheduling, preparing, and documenting the work done. Interspersing of annual PM work among batch processing inevitably results in PM delays due to equipment and/or manpower unavailability. Guidelines for planning and executing an annual plant shutdown are available (79).

All PM work orders should be listed on a single spreadsheet to be reviewed before the facility shutdown and augmented with any modifications or one-time replacement work orders that are requested. As PM work is completed, documentation might be signed by the tradesperson as applicable and reviewed by mechanical and operational staff for completeness. Problems identified with equipment or instrumentation should be highlighted as they occur and then entered into change control work order systems so that they can be tracked for their potential effect on validation.

Before the annual shutdown, transfer line manifolds and utility supply lines, which typically cannot be shut down entirely for maintenance throughout the year, might be leak tested to identify faulty valves in need of replacement. After the annual shutdown, equipment start-up procedures should be documented, preferably through the use of equipment start-up batch sheets. These procedures might include equipment cleaning, testing of instrument calibrations such as level, temperature, and pressure (80), and vessel integrity testing for leaky valves. Instrument calibration accuracy should be evaluated for random or systematic errors, with random error within accepted tolerances being the expected outcome.

Some general PM procedures have been outlined that can apply to several areas in a pilot plant, both product and nonproduct contact. Sanitary fittings, which tend to loosen during repeated heating and cooling cycles, should be checked, with gaskets replaced and clamps tightened as required. Any flanges in product contact should be tightened to a specified torque. PM inspections of vessel internals should be conducted with lock washers on nuts that may loosen due from repeated heating and cooling. A PM should be developed for inspection of vessel cooling coil leaks that can result in inadvertent cooling water contact with product (81). For agitated systems, there should be a regular examination of foot bearings, motor/drive vibration, oil analysis, and shaft runout. Double mechanical seals, the failure of which can lead to condensate buildup or batch contamination, should be tested regularly and replaced preventatively based on a minimum life expectancy (82). An outline of specific PM procedures for laboratory fermentors can be applied to larger-scale fermentors (83). An example of documentation for the maintenance of horizontal laminar flow hoods also is available (84) that can be readily adapted to other HEPA filter units. PMs should be conducted on all utility systems, which may include rebuild of air compressor heads, descaling of heat exchange surfaces, and replacing of air dryer desiccant, among other items recommended by the equipment manufacturer. Finally, novel predictive maintenance evaluation techniques should be explored and implemented as appropriate to identify imminent equipment failures (85).

It is most appropriate to perform any annual validation work (e.g., fermenter SIP, autoclave load pattern, controlled temperature unit) at the conclusion of the annual PM shutdown. In this manner, proper equipment operation can be documented and the effectiveness of PM procedures can be evaluated. Often inadvertent mistakes or omissions during maintenance can be identified before they have an adverse, and possibly widespread, quality impact.

Ongoing. A system of tracking outstanding maintenance by equipment or by process should be designed. Each piece of equipment should have a clearly distinguished tag name. Work orders need to be initiated, evaluated for their effect on validation, marked as complete when mechanical work is done, and finalized when process personnel have checked the job. A computerized database can readily track maintenance and identify repetitive maintenance problems. Such a system permits easy review of outstanding work orders for each piece of equipment by the operation group before it is readied for service. All maintenance work orders need to be tracked within one system; care should be taken not to bypass the operations area when maintenance needs are identified by those directly controlling the mechanical staff.

Interim maintenance work, such as calibrations and subannual PM work, should be documented and filed in a manner equivalent to the yearly PM results. PM schedules should be reviewed based on the frequency and results of interim maintenance work. Unusually common equipment failures can be identified through maintenance database searches so that appropriate longer-term repair efforts or replacement can be justified and undertaken.

Calibrations. Additional concerns are associated with calibrations because important processing decisions may be based on the accuracy of the instrument readings. In some cases, specific instruments are designated as "critical" for a processing step; in other cases, desired specification ranges for the processing step are targeted based on certain identified instruments. These two methods provide assurance that product quality is not compromised if any of the other instruments fail in the equipment being used.

Although some facilities distinguish between calibration procedures for critical and noncritical instrumentation, it may be prudent for a pilot plant to calibrate all instruments in a reliable and documented manner because their accuracy may be important for useful process development data. Calibrations should be performed using National Institute of Standards and Technology (NIST) traceable instruments and should include full loop calibration to the process monitor as appropriate. A specific calibration and maintenance SOP for the instrument is most desirable, but a general SOP referencing vendor manual might also be acceptable, particularly in the interim period while instrument calibration SOPs are being developed. The as-found and as-left conditions should be documented along with the serial number of the standard used. Establishment of an acceptable tolerance for each instrument ensures consistency. When an instrument is found to be outside the tolerance specified, the impact of this calibration error on previously executed batches may need to be evaluated. Thus, a confirmation of the error might be warranted before correcting the calibration. Records should be evaluated for the amount of instrument drift, which may indicate the need to either recalibrate on a more frequent basis or else replace or upgrade aging equipment. An example calibration sheet is available (86).

Modifications and Change Control. Any validated facility needs to devise procedures for tracking changes on equipment as well as on processes and computers. Incomplete documentation of changes can obviate extensive time and money spent on validation testing. Although the exact contents may change, such a procedure starts during the initial design phase as equipment specifications are altered, continues through the facility start-up and qualification phases, then moves directly into the facility operational phase. Adherence to change control procedures may be difficult during facility start-up as many unexpected items are discovered to be incorrect and management pressure to have the facility operational may mount. Depending on the philosophy of the individual facility, the scope of this change control procedure may include product and nonproduct contact equipment to varying degrees. The procedure identifies the individual or individuals that need to approve changes and determine before or in concert with the change implementation what additional validation testing is required. Often, processing personnel, being most familiar with the equipment being modified and the reason for its modification, can provide valuable leadership in this area and initiate contact with quality and validation personnel as required for input.

Generally replacements in kind are accepted without additional testing unless the item being replaced was specifically tested or calibrated during the initial validation effort. Example items and associated testing might include a replacement or rebuilt motor or associated drive (measure motor amperage draw at various agitator speed settings) or replacement instrumentation (calibrate transmitter). Replacement items in this category need to be tagged appropriately in the field with the same tag number as the original item and serial numbers updated in any centralized maintenance data bank as well as recorded on installation documentation.

Modifications or obvious changes in equipment or instrumentation need to be evaluated for their potential impact on the validated state of the equipment. This impact is particularly important for SIP and CIP cycles. Such modifications may be minor, such as a change in gasket or o-ring material of construction, or major, such as a change in piping. An example procedure is available that can ensure proposed changes are reviewed by operations, maintenance, and quality for their potential impact (87). Specific attention should be given to alterations needed in P&IDs, SOPs, PMs, batch sheets, or training programs. Equipment modifications, necessary to accommodate a new process, need to be identified early and their impact evaluated (2). It is critical to recognize the need to balance the request for change with an evaluation of its validation impact; changes cannot be discouraged in a multipurpose pilot plant but need to be controlled and managed in a logical manner.

Staffing

Personnel needs are best defined based on the production process (88), but in a pilot plant setting production processes change rapidly. A decision needs to be made early in the design process concerning the level of off-shift activity reasonably expected in the facility because this can severely impact items such as utility capacity and operation. Round-the-clock coverage, while demanding for personnel involved, may be necessary a large fraction of the days to ensure success and timeliness of production. The use of unionized personnel, technicians, or process development staff should be incorporated into the educational and manual workload expectations for the operating staff. A core operations group, responsible for the facility, can ensure continuity in areas such as SOP updates, maintenance work order tracking, batch sheet content, and equipment operation. Such a group might be supervised by an area manager responsible for coordination and scheduling who might interface with those involved directly in process development. Processing groups might then rotate through the pilot plant as their projects demanded, accompanied by a member of the operations group. Alternatively, a dedicated pilot production process development staff might be instituted, responsible for translation of laboratory processes into GMP batches. A shift report or log should be established and continually updated as to the mechanical and operational status of individual pieces of equipment and utilities.

Laboratory Support

To provide adequate and accurate sample analyses, laboratory support needs should be identified and addressed through either on-site or contract laboratories. Targets for testing might include final product, in-process samples, raw materials, sterility, culture purity, environmental quality of controlled environments, and utility quality (Table 6). While the goal of these laboratories might be to work toward using validated methods and approved SOPs for all assays, this goal might not be attainable for project-specific assays until the later stages of a successful development project. Initially, in the early project stages, only notebook procedures might be available to guide laboratory testing. Information directly impacting quality should be communicated directly to operating personnel and appropriate investigations conducted both in the field and in the laboratory for failing results.

A microbiological laboratory is critical for fermentation-based operations to prepare seed cultures and to test new culture seed vials for culture purity and productivity in lab-scale processes. An examination of up-to-date sterility/culture purity results before vessel-to-vessel transfers should be undertaken as close to possible to the time of transfer to minimize adverse impacts of contaminated cultures. In a pilot plant setting, tension may exist between the time required to adequately test new cultures and the development pressure to use the latest mutant or transformant in a pilot-scale cultivation to evaluate its capabilities. Reasonable procedures need to be established in advance for the introduction of new processes or new cultures into the pilot plant. Laboratory-prepared nutrients, trace ingredients, and sterilized additions need to be evaluated for their quality impact, and the procedures used for their preparation should be documented where possible.

Standard Operating Procedures

SOPs are essential for the planning and standardization of various procedures conducted by operational staff and maintenance personnel that need to be completed before, during, and after pilot plant batches (2). Typical categories include SOPs for equipment operation, facility and equipment cleaning, sampling, sterilization/sanitization, maintenance, and documentation procedures for tasks such as product changeover, change control, and batch sheets (Table 7). When referencing SOPs as part of the batch record, working versions of SOPs can be helpful to track progress in the field. An SOP on the requirements for writing, issuing, and inactivating an SOP is also useful. SOPs focus operation, guide operations staff, document the order and scope of activities, standardize approaches, define regulatory needs and constraints, and increase the speed of new personnel training (2). They should be written by a person familiar with the facility or equipment operation that is the subject of the SOP, reviewed by another person who might be asked to follow the SOP, and approved by a quality group. Each SOP should have an effective date of issue, be available to those who need it, and be archived when a future revision is created.

The need for structure and consistency within an SOP must be balanced by the need for flexibility in a pilot plant when unexpected process requirements arise. Procedures must be defined for what to do when an SOP is altered intentionally to accommodate a new process or a surprise development in an existing process. Consideration to SOPs should be given during the equipment design phase so that tasks can be performed in a logical clockwise or counter-clockwise fashion (88). SOPs should be clear so that the expected outcome is assured. Specific examples of details often omitted include statements regarding settings for pressure switches and acceptable calibration error tolerances, insertion diagrams, part numbers, and manual ref-

Table 6. Types of Testing (Sampling and Analysis) Possibly Required by On-Site or Off-Site Support Laboratories for a Bioprocessing Pilot Plant

Facility and Utility monitoring	Sterilization cycle validation	Cleaning cycle validation	Process monitoring
Total organic carbon	Indicator spore strip incubation	Total organic carbon (swab and rinse water)	Spectrophotometer
pH	Indicator water spore suspension incubation	pH	pH meter
Conductivity	Indicator media spore suspension incubation	Conductivity	HPLC for organic acids, amino acids, sugars
Endotoxin (LAL)	Sterility	Endotoxin (LAL)	HPLC for product/impurity analysis
Bioburden and coliforms	Culture purity (bacterial and fungal)	Bioburden (swab and rinse water)	LC-mass spectrometry
Airborne viable particles	Filter integrity testing	Ultraviolet light absorbance	Protein gel analysis
Airborne total particles	Microbial identification	Visible light absorbance	Osmometer
Surface, viable particles	Gram stains	Filtered solids	Microscope
Compressed gas, viable particles		Dissolved solids	Coulter counter
Compressed gas, nonviable particles			Conductivity
Compressed gas, identity			Metabolite analyzer
Noncondensable gases in steam condensate			Plate readers
			Blood gas analyzer
			Laboratory-scale control/troubleshooting experiments

Table 7. Example Types of Procedures Possibly Required in a Bioprocessing Pilot Plant

Administrative	Process equipment	Laboratory equipment	Maintenance	Operations
Processing abnormality	Sterilization-in-place	Autoclave operation: sterilization and decontamination	Instrument calibration	Gowning
Investigation of contamination	Cleaning-in-place	Autoclave load patterns	Equipment preventative maintenance	Water system flushing/ sampling
Product changeover	Sanitization	Glasswasher, operation	Utility preventative maintenance	Steam system sampling
Writing of an SOP	Vessel setup and operation	Glasswasher, load patterns	HVAC preventative maintenance	Spill containment
Personnel training on equipment/operations	Transfer line/panel operations	Working in biosafety cabinet	Equipment change control	Facility housekeeping and cleaning
Personnel training on assays	Batch harvest (noncontained)	Working in laminar flow hood	Equipment start-up after processing hiatus	Facility start-up after processing hiatus
Personnel training on computers	Batch harvest (contained)	Incubator/shaker operation	Equipment start-up after major mechanical work	Daily utility checkout
Issuing and review of batch sheets	Equipment integrity test	Freezer	Equipment storage during processing hiatus	In-process sampling and analysis
Raw material release and storage	Utility systems operation	Refrigerator/cold room	Computer system change control	Data archiving
Investigation of monitoring excursions	HVAC systems operation	Tubing welder use and cleaning	Computer system failure recovery	Computer system operation
Forklift certification	Media/buffer preparation	Manual washing	Shutdown/interim maintenance documentation	Monitoring-controlled environments
Pest control	Specific equipment SOPs	Usage of scales and checkweights	Equipment isolation for maintenance	Tracking of process/ equipment status
Ongoing validation testing		Calibration of pH meter	Issuing and tracking of work orders	
Ongoing facility monitoring		Facility monitoring assays		
Introduction of new products/processes		In-process monitoring assays		
		Fume hood use		
		Seed preparation		

erences to aid in following the procedure, and reliance solely on valve numbers, not valve purposes, to describe the procedure. While SOPs should be reviewed periodically every few years to determine if they need clarification, care should be taken to minimize the number of similar versions of the same SOP issued to the field. Operational clarification memos, issued to all operational staff and signed off when read, might be used to provide additional SOP clarification without reissuing the SOP during the interim period between SOP reviews. All major SOP changes should be documented, evaluated for their effect on validation, and communicated to staff in a documented manner.

Safety

Because concerns regarding product exposure to personnel are similar to concerns for personnel exposure to product, GMP guidelines overlap with safety regulations in many respects such as training, SOPs, preventative maintenance including calibrations, and validation/change control (2). Safety captions can be required for certain raw materials based on review of material safety data sheet (MSDS) documentation. Broths, as well as isolated compounds, might be submitted for toxicity testing as soon as they are cultivated at a larger scale. Safety is a component of installation qualification as pressure vessel documentation and utility connections are reviewed. Hazard operability (HAZOP) reviews for pressure vessels (e.g., fermentors, autoclaves) and utilities (e.g., clean steam generators) often are required by company policy and summaries included within equipment qualification reports. HAZOP action items need to be addressed promptly and modifications to equipment or procedure made to minimize risk. Equipment design needs to be reviewed for the ease of lockout and removal of hazardous energy sources (e.g., bleed valves for the evacuation of process and utility lines, field-mounted pressure gauges to confirm depressurization) as well as the individual isolation of skids/units for maintenance. In addition, specific SOPs for preparing equipment for maintenance might be required.

Regular safety inspections with documented action item lists communicated to all affected personnel (i.e., operating staff, technicians, engineers, management) can identify and correct nonoptimal working conditions and cluttered processing areas. Exposure of personnel to excessive sound, heat, or dust should also be considered during facility design as well as monitored during facility operation. Although personal protective equipment such as ear plugs, heat-resistant gloves, and respirators is available, when possible engineering controls such as sound dampeners, insulation, and dust collectors may be better solutions. Safety during validation testing should always be considered, particularly when groups outside of operations are involved in setting up or operating equipment and when sources of hazardous energy are involved. Whether during validation testing or operational use, fittings of any sort should be tightly secured and should not be tightened when equipment is under pressure or hot, specifically during SIP and CIP procedures. Summaries of generally accepted laboratory and pilot plant safety procedures to consider are available (89).

According to ASME code, overpressurization protection using relief devices is required for vessels more than 6 in in diameter. The background for this regulation as well as details of pressure relief valves and rupture disks is available (90). Sanitary rupture disks are most desirable for product contact; once blown, processing must stop until they are replaced. Sometimes a rupture disk is installed, followed by a pop safety valve rated at a slightly higher pressure with a pressure gauge between to indicate rupture of the disk (48). If this arrangement is used, then the cleanliness of the area between the rupture disk and pop safety valve must be demonstrated for the batch to not be adversely impacted if the disk ruptures but the pop safety valve does not release. Removal of the entire rupture disk assembly (not the removal of the disk from the assembly) for cleaning should be checked with company policy but may be acceptable in most cases.

Another often-overlooked matter is that the outlets of safety relief devices (which might be unexpectedly spewing steam, hot water, or broth after relief) should be oriented such that personnel safety is considered for those who may be standing near them. One method is to pipe outlets to outside containment troughs, but it may not be desirable to have a potential connection to the outside. A second alternative is to create a designated area within the pilot plant processing area. While outlets to the floor beside the equipment should be minimized, if they are found to be necessary they should be carefully situated outside the normal path of personnel.

Training

Initial and ongoing training programs need to be developed and documented for each individual associated with the facility, based on the facility needs. A training SOP should be developed, which might include a training manual containing useful items such as key SOPs, facility layouts, equipment diagrams, and policies. Initial training might include items from vendor-supplied training sessions to validated load pattern arrangements. Often this training falls at a peak workload period during the facility start-up and validation when attention can be diverted into seemingly more urgent areas. In other cases, more individuals require training than can reasonably be accommodated in the area surrounding the equipment. When permitted, such training might be recorded for later review by the staff when they have fully assumed responsibility for operation. Another alternative is for those just trained to provide training to others, which may be suitable for certain operations. For best results, training should occur with written documentation available to those being trained for reference.

Ongoing training should be done on a regular basis; once every 1 or 2 months seems most appropriate for an active pilot plant facility. Training should include supervisors, operators, engineers, and maintenance personnel as well as validation and GMP staff. Training might include a process review that highlights critical steps and equipment, SOP reviews, identification of personnel safety and environmental containment issues, and review of PM and ongoing maintenance needs, as well as what operational personnel should look for with respect to proper equipment function (88). Training sessions are one method to communicate recent SOP changes as well as to explain why certain procedures are necessary. These might also be used for periodic SOP reviews to obtain feedback from those using the SOP frequently. Attendance should be required and sessions recorded when possible; those new to the facility should review prior training sessions for the past few years as applicable.

Validation

The recent guideline issued by the FDA (91) makes pilot plants eligible for product licensure should company needs require a product launch from a pilot plant facility. When manufacturing clinical materials, pilot plants are evaluated under the guidelines for investigational new drug products, among other regulations which clearly state that GMPs apply for drug products approved for clinical trials in humans and animals (92). Individual company philosophies vary on the suitability or desirability of their pilot plants for this task because using a pilot facility for manufacturing reduces its availability for the production of clinical material. The level of validation effort required is best considered during the initial facility design. Validation can be done retrospectively, although the collection of the required documentation may be more challenging and modifications necessary for testing may be costly.

The major steps for validation ensure that installed equipment and utilities operate in a manner acceptable for the process to be run in that equipment. Because multiple processes, many of which are not necessarily identified at the time of validation, may be conducted in a pilot plant, operational ranges need to be set to accommodate a wide range of reasonably expected processes. For example, an incubator might be validated for temperatures between 25 and 37 °C to accommodate common bacterial, yeast, and fungal cultures, but not necessarily at 45 °C to accommodate thermophiles. The selection of worst-case scenarios can be difficult for a pilot plant if the associated worst-case conditions are unduly burdensome for normal operation. Specifically, the cleaning procedures required after a 3-week fungal production cultivation are significantly more complex than those required after a 2-day fungal seed cultivation of the same culture. Similarly, the sterilization cycle hold time for a concentrated sugar solution is substantially longer than that required for a heat-sensitive waterlike growth medium. Approaches to resolving these multiproduct validation issues can be complicated by desires to follow manufacturing validation guidelines, which most likely were established based on a single product.

Often validation, qualification, verification, commissioning, and start-up are used with specific definitions in mind for individual facilities and organizations. Generally, validation implies the highest rigor of direct reproducible testing of a specific procedure. Qualification encompasses a wider range of documentation collection and testing of capabilities. Verification might imply a single test against an identified standard. Commissioning can refer to documented troubleshooting of newly installed equipment, whereas start-up might refer to initial equipment checkouts directly after installation. The major validation phases are divided to focus on qualification followed by validation. Installation qualification (IQ) focuses of the verification of proper installation, appropriate utility connections, and adherence to manufacturer and purchase order specifications (93). Operational qualification (OQ) includes testing of equipment function and an acceptance criteria for performance (93). Performance qualification (PQ) centers on process-specific testing for SIP, CIP, and autoclave sterilizing and decontamination load patterns. For computers, a system life cycle methodology incorporating specifications, program design, coding, testing, start-up, operation, and maintenance has been adopted (93).

Process validation centers on the reproducibility of the process in the manufacturing area. Parameters influencing this reproducibility might be best studied in the pilot plant to obtain acceptable operating ranges for those processing parameters, such as cultivation temperature, pH, and dissolved oxygen, found to influence product quality. The set of criteria for proceeding to the next processing step, specifically volumetric or specific productivity, yield, and purity, must be validated in a documented fashion. The critical processing steps important to process quality should be identified and validated (93). Other noncritical process steps may be studied for their impact on manufacturing productivity. Process validation may require multiple identical pieces of equipment, such as fermentors, at the laboratory and process scale to define acceptable processing ranges in a parallel rather than sequential manner.

Batch and Facility Records

Documentation associated with a new facility can be extensive initially and can increase substantially during the time that the facility operates. While centralizing all documentation in a single area can be attractive, duplicate copies of some items often are needed in other locations as well as for archival backup purposes. For example, equipment manuals and facility drawings may be needed near the actual location of the equipment, in the maintenance area, and in the centralized documentation area. Validation documentation, consisting of IQ, OQ, PQ, and load pattern and SIP/CIP reports, is generated both during the initial validation as well as during annual retesting and ongoing environmental monitoring programs. Storage of change control documentation might be accomplished through notebooks and file folders for individual pieces of equipment or by individual suites. SOPs might be archived in a centralized location but also posted near equipment for ready reference. Access and sign-out procedures for the control of documentation need to be developed.

A robust system for batch records needs to be created, based on individual batch sheets for processing steps or for individual pieces of equipment or some appropriate combination of the two approaches. Issued batch sheets should be tracked by a unique sequential number with the product, processing step, and/or equipment number clearly identified. There is a need to track the use of equipment by product and batch, particularly for multipurpose facilities. While this might be done through individual equipment use logs, it is also necessary to include such tracking within the batch record itself.

Newly authored or completed batch sheets should be reviewed by the operations area and may be required to be approved by a quality group before execution. Batch sheets should have clear written instructions as well as a signature, date, and time, as required, for each step. Handwritten instructions, likely to be added during the processing of pilot-scale batches, may need to be reviewed by representatives from operation or quality groups. Brief justifications for alterations in batch sheet instructions also should be considered. An written explanation documenting that the material from the current step is acceptable for continued processing in a subsequent step is recommended. Care should be taken to use common templates wherever possible to simplify successive batch sheet differences and minimize the review effort. Executed batch sheets should be reviewed for completeness by the operations area, by those in charge of the batch, and by a quality group as required. A tracking system should be instituted to monitor the review of completed batch sheets as well as to ensure that all issued batch numbers are accounted for in the batch record filing cabinets. Photocopying, microfilming, or scanning of completed batch sheets serves to provide a backup in case of loss or damage. Paperless batch records, as well as electronic signatures, have also been considered to minimize documentation volume, and guidelines for implementation and validation have been published (94).

BIBLIOGRAPHY

1. P. Basu, J. Quaadgras, J. Holleman, R. Mack, and A. Noren, *Chem. Eng. Prog.* **96**, 66–75 (1997).
2. P. Basu, J. Quaadgras, R. Mack, and A. Noren, *Chem. Eng. Prog.* **94**, 67–74 (1998).
3. M. Burnett, V. Santamarina, and D. Omstead, *Ann. N.Y. Acad. Sci.* **646**, 357–66 (1991).
4. P. Barrer, *Biotechnology* **1**, 661–666 (1983).
5. F. Bader, in M. Ladisch and A. Bose eds., *Harnessing Biotechnology for the 21st Century*, American Chemical Society, Washington, D.C., 1992, pp. 228–231.
6. J. Odum, *Pharm. Eng.* **15**, 8, 10–18, 20 (1995)
7. F. Bader, A. Blum, B. Garfinkle, D. MacFarlane, T. Massa, and T. Copmann, *BioPharm* **5**, 32–40 (1992).
8. M. Hamers, *Biotechnology* **11**, 561–570 (1993).
9. S. Fitzpatrick, A. Ma'ayan, and J. Wagget, *Chem. Eng. Prog.* **86**, 26–31 (1990).
10. M. Narodoslawsky, *Chem. Biochem. Eng. Q.* **5**, 183–187 (1991).
11. N. Samsatli and N. Shah, *Trans. Inst. Chem. Eng. Part C* **74**, 221–231 (1996).

12. N. Samsatli and N. Shah, *Trans. Inst. Chem. Eng. Part C* **74**, 232–242 (1996).
13. D. Schell, *Appl. Biochem. Biotechnol.* **51/52**, 549–557 (1995).
14. M. Cannales, A. Enriquez, E. Ramos, D. Cabrera, H. Dandie, A. Soto, V. Falcon, M. Rodriguez, and J. de la Fuente, *Vaccine* **15**, 414–422 (1997).
15. E. Eliezer, *BioPharm* **6**, 24–29 (1993).
16. R. Sato and M. da Costa, *Biotechnol. Lett.* **18**, 275–280 (1996).
17. D. Petrides, E. Sapidou, and J. Calandranis, *Biotechnol. Bioeng.* **48**, 529–541 (1995).
18. N. Titchener-Hooker, D. Gritsis, K. Mannweiler, R. Olbrich, S. Gardiner, N. Fish, and M. Hoare, *BioPharm* **4**, 34–38 (1991).
19. J. Zawistowski and J. Rago, *Pharm. Eng.* **14**, 24, 26, 28–30, 32 (1994).
20. A. Shahidi, R. Torregrossa, and Y. Zelmanovich, *Pharm. Eng.* **15**, 72–83 (1995).
21. O. Doblhoff-Dier, S. Huss, R. Litos, R. Plail, F. Unterluggauer, M. Reiter, and H. Katinger, *Process Biochem.* **26**, 201–207 (1991).
22. K. Nelson, in A. Prokop, R. Bajpai, and C. Ho eds., *Recombinant DNA Technology and Applications*, McGraw-Hill, New York, 1991, pp. 509–565.
23. K. Leach, *Pharm. Eng.* **13**, 54–62 (1993).
24. J. Odum, *BioPharm* **5**, 36–38 (1992).
25. L. Scutti and S. Stark, *Pharm. Eng.* **13**, 47–52 (1993).
26. B.K. Hamilton, J.J. Schruben, and J.P. Montgomery, in A. Demain and N. Solomon eds., *Manual of Industrial Microbiology and Biotechnology*, American Society for Microbiology, Washington, D.C., 1986, pp. 321–344.
27. P. James, *Pharm. Eng.* **18**, 72–82 (1998).
28. R. Wisniewski and C. Burman, in A. Lubiniecki and S. Vargo eds., *Regulatory Practice in Biopharmaceutical Production*, Wiley-Liss, New York, 1994, pp. 407–445.
29. M. Charles and J. Wilson, in B. Lydersen, N. D'Elia, and K. Nelson eds., *Bioprocess Engineering: Systems, Equipment and Facilities*, Wiley, New York, 1994, pp. 3–67.
30. Y. Chisti, *Chem. Eng. Prog.* **88**, 55–58 (1992).
31. J. Alford, B. Allen, S. Bisch, L. Brill, D. Clapp, and L. Danaher, *Ann. N.Y. Acad. Sci.* **721**, 326–336 (1994).
32. T. Ransohoff, M. Murphy, and H. Levine, *BioPharm* **3**, 20–26 (1990).
33. A. Barry and R. Chojnacki, *BioPharm* **7**, 43–47 (1994).
34. S. Fulton, A. Shahidi, N. Gordon, and N. Afeyan, *Biotechnology* **10**, 635–639 (1992).
35. G. Aronsson and P. Zadorecki, *Aust. J. Biotechnol.* **1**, 17–20 (1987).
36. J. Mahar, *BioPharm* **6**, 42–51 (1993).
37. J. Odum, *Pharm. Eng.* **12**, 8–12 (1992).
38. K. Nelson, *BioPharm* **1**, 34 (1988).
39. J. Williams, *Biotechnology* **7**, 75–76 (1989).
40. B. Junker, M. Stanik, J. Adamca, K. LaRiviere, M. Abbatiello, and P. Salmon, *Bioprocess Eng.* **17**, 277–286 (1997).
41. G. Gray, *Pharm. Eng.*, **17**, 28–33 (1997).
42. M. Burkhart, J. Wermelinger, W. Setz, and D. Muller, *PDA J. Pharm. Sci. Technol.* **50**, 246–251 (1996).
43. M. Del Valle, *Pharm. Eng.* **15**, 14–22 (1995).
44. A. Petrossian, N. Smart, and R. Projetto, *BioPharm* **6**, 40–45 (1993).
45. C.H. Brooks and P.D. Russell, *Process Biochem.* **21**, 77–80 (1986).
46. J. Akers, *BioPharm* **7**, 43–47 (1994).
47. J. Akers, *Pharm. Technol.* **19**, 28 (1995).
48. S. Miller and D. Bergmann, *J. Ind. Microbiol.* **11**, 223–234 (1993).
49. J. Van Houten and D.O. Fleming, *J. Ind. Microbiol.* **11**, 209–215 (1993).
50. D. Janssen, P. Lovejoy, M. Simpson, and L. Kennedy, in P. Yu ed., *Fermentation Technologies: Industrial Applications*, Elsevier, New York, 1990, pp. 388–393.
51. J. Kossik *Gen. Eng. News* **18**, 21 (1998).
52. L. Kennedy, M. Boland, D. Janssen, and M. Frude, in P. Yu ed., *Fermentation Technologies: Industrial Applications*, Elsevier, New York, 1990, pp. 383–387.
53. R. Maigetter, F. Bailey, and B. Miller, *BioPharm* **3**, 22–29 (1990).
54. J. Odum, *BioPharm* **6**, 42 (1993).
55. J. Hodgson, *Biotechnology* **12**, 983–987 (1994).
56. K. Gary, *American Biotechnology Laboratory* **7**, 26–33 (1989).
57. Y. Lee and G. Tsao, in T. Ghose, A. Fiechter, and N. Blakebrough eds., *Advances in Biochemical Engineering*, vol. 13, Springer-Verlag, New York, 1979, pp. 35–86.
58. M. Johnson, J. Borkowski, and C. Engblom, *Biotechnol. Bioeng.* **6**, 457–468 (1964).
59. R. Elsworth, *Chem. Eng.* 63–71 (1972).
60. P. Salmon and B. Buckland, in M. Rhodes and P. Stanbury eds., *Appl. Microb. Physiol.* , 131–163 (1997).
61. M. Uttamlal and D. Walt, *Biotechnology* **13**, 597–601 (1995).
62. J. Bradley, A. Kidd, P. Anderson, A. Dear, R. Ashby, and A. Turner, *Analyst* **114**, 375–379 (1989).
63. G. Macaloney, J. Hall, M. Rollins, I. Draper, B. Thompson, and B. McNeil, *Biotechnol. Tech.* **8**, 281–286 (1994).
64. G. Macaloney, J. Hall, M. Rollins, I. Draper, K. Anderson, J. Preston, B. Thompson, and B. McNeil, *Bioprocess Eng.* **17**, 157–167 (1997).
65. V. Agayn and D. Walt, *Biotechnology* **11**, 726–729 (1993).
66. A. Humphrey, K. Brown, J. Horvath, and H. Semerjian, in A. Fiechter, H. Okada, and R. Tanner eds., *Bioproducts and Bioprocesses*, Springer-Verlag, Berlin, 1989, pp. 309–320.
67. D. Kell, G. Markx, C. Davey, and R. Todd, *Trends Anal. Chem.* **9**, 190–194 (1990).
68. B. Junker, J. Reddy, K. Gbewonyo, and R. Greasham, *Bioprocess Eng.* **10**, 195–207 (1994).
69. L. Olsson and J. Nielsen, *Trends Biotechnol.* **15**, 517–522 (1997).
70. W. Groot, R. Mijnhart, R. van der Lans, and K. Luyben, *Biotechnol. Tech.* **5**, 371–376 (1991).
71. S. Collins, *Appl. Microb. Physiol.*, 75–101 (1997).
72. J. Petersen, *Pharm. Technol.* **12**, 42 (1988).
73. O. Doblhoff-Dier, S. Huss, R. Litos, R. Plail, F. Unterluggauer, M. Reiter, and H. Katinger, *Process Biochem.* **26**, 201–207 (1991).
74. H. Blachere, *Gen. Eng. News* **18**, 23 (1998).
75. K. Jones, D. Williams, D. Phipps, and P. Montgomery, *IFAC Symp. Ser. (Adv. Control Chem. Processes)* **8**, 35–40 (1992).
76. F. Wilde, *Gen. Eng. News* **18**, 24 (1998).
77. H. Eiki, I. Kishi, T. Gomi, and M. Ogawa, in M. Ladisch and A. Bose eds., *Harnessing Biotechnology for the 21st Century*, American Chemical Society, Washington, D.C., 1992, pp. 223–227.
78. E. Bardone, B. Lani, and G. Cassani, *Pharm. Eng.* **14**, 34–38 (1994).
79. G. Williams, *Plant Services* **19**, 115–118 (1998).

80. B. Junker, J. Lynch, J. Leporati, J. Schmitt, J. Gieger, T. Garah, M. Stober, and P. Salmon, *Bioprocess Eng.* **17**, 279–287 (1995).
81. C. Perkowski, *Biotechnol. Bioeng.* **26**, 857–859 (1984).
82. D. Todhunter, *Bioprocess Eng. Symp.*, 97–103 (1989).
83. F. Kleppinger, *Int. Biotechnol. Lab.* **5**, 28 (1987).
84. F. Kleppinger, *Int. J. Pharm. Compd.* **1**, 344–345 (1997).
85. K. Mobley, *Plant Services* **19**, 147 (1998).
86. D. Wade, *Maint. Technol.* 44–46 (1989).
87. C. Kennedy, *BioPharm* **8**, 34–37 (1995).
88. C. Perkowski, *BioPharm*, 62–65 (1987).
89. V. Singh, *Chem. Age India* **39**, 387–388 (1988).
90. J. Kossik, *Gen. Eng. News* **18**, 25 (1998).
91. Center for Biologics Evaluation and Research, *Fed. Reg.* **60(132)**, 35750–35753 (1995).
92. Center for Drug Evaluation and Research, Food and Drug Administration, Washington, D.C., March 1991.
93. R.G. Werner, H. Langlouis-Gau, F. Walz, H. Allgaier, and H. Hoffman, *Arzneim.-Forsch.* **38**, 855–862 (1988).
94. 21 CFR Part 11, *Fed. Reg.* **62(54)**, 13429–13466 (1997).

See also CHROMATOGRAPHY, COMPUTER-AIDED DESIGN; CLEANING, CLEANING VALIDATION; METABOLITES, PRIMARY AND SECONDARY; SECONDARY METABOLITE PRODUCTION, ACTINOMYCETES, OTHER THAN STREPTOMYCES.

PLASMID DNA REPLICATION

MARCELO E. TOLMASKY
California State University, Fullerton
Fullerton, California

LUIS A. ACTIS
Miami University
Oxford, Ohio

JORGE H. CROSA
Oregon Health Sciences University
Portland, Oregon

KEY WORDS

ColE1
Covalently closed circular DNA
Extrachromosomal elements
Incompatibility
Linear plasmids
pT181
R6K

OUTLINE

INTRODUCTION

Plasmids are extrachromosomal replicons found in Gram-negative and Gram-positive bacteria as well as in some lower eukaryote organisms. They are present in bacterial cells replicating at a specific number of copies per cell. Their size varies from a few to several hundred kilobase pairs, and bacterial cells can harbor more than one plasmid species (1).

The term *plasmid* originally was used by Lederberg to describe all extrachromosomal hereditary determinants. Currently, the term is restricted to the autonomously replicating extrachromosomal DNA of bacteria. Although plasmids replicate autonomously, they generally rely on some host-encoded factors for their replication. Although not essential for the survival of bacteria, plasmids may encode a wide variety of genetic determinants that permit their bacterial hosts to survive better in an adverse environment or to compete better with other microorganisms occupying the same ecological niche. Plasmids are found in a wide variety of microorganisms, and it is as difficult to generalize about plasmids as it is to generalize about the microorganisms that harbor them. Plasmid DNA is mostly isolated as covalently closed, circular double-stranded DNA molecules, although recently linear plasmids have been described in certain bacterial species (2). Furthermore, single-stranded DNA plasmids were identified as intermediates during the process of replication in certain Gram-positive bacteria (3). Plasmids also include the replicative forms of filamentous coliphages and the prophage state of phages such as P1 (4).

The medical importance of plasmids that encode for antibiotic resistance (R plasmids) and those that contribute directly to microbial pathogenicity, such as for instance iron transport in several pathogens or the presence of adhesions, invasins, or antiphagocytic proteins, is well documented, as is the role played by plasmids in bacteria of importance in agriculture and industry (5–8). Some plasmid genes encoding antibiotic resistances and other traits are frequently located in transposable elements or integrons (9–11). This produces great variation and flexibility in the constitution of plasmids. These extrachromosomal elements are of equal importance, however, for the study of the structure and function of DNA. Plasmids have taken on paramount importance in recombinant DNA technology (12). In this section we describe the mechanisms of the control of plasmid DNA replication and examine the replication of plasmids ColE1 and R6K, found in Gram-negative organisms, and pT181, found in Gram-positive bacteria. In addition, we also discuss linear plasmids. For the replication mechanisms of other plasmids we refer the reader to other reviews (13–23).

ColE1-TYPE PLASMIDS

The ColE1-type family encompasses both a large number of naturally occurring plasmids and many of the most commonly used cloning vehicles (16,18,24–30). The mechanism controlling initiation of replication of these plasmids is among the best understood, perhaps reflecting the importance of this family of plasmids in recombinant DNA applications (11,16–18,31–36).

Replication of plasmids containing the ColE1 origin is initiated at a unique startsite *(ori)* (37,38) (Fig. 1). Unlike in other plasmid families, initiation of replication of ColE1 is not mediated by a plasmid-encoded protein; instead ColE1 requires the host DNA polymerase I (Pol I) (39). In addition, initiation of replication of ColE1 requires the presence of a host-encoded RNA polymerase and ribonuclease H (RNase H) (39–41). Replication of the plasmid is initiated by the synthesis of an RNA molecule by the host RNA polymerase 555 bases upstream of *ori* (Fig. 1, panel a). This RNA species, called RNA II, extends about 700 nucleotides from its initiation (Fig. 1, panels b and c) (40,42). The 3′ of this RNA molecule subsequently forms a duplex with the DNA of the plasmid at a region close to the origin of replication, which is located at nucleotide 555 of RNA II (Fig. 1, panel c) (40). The formation of the duplex structure (known as coupling) begins following the synthesis of a portion of the RNA II molecule. The coupling event is strongly dependent on the formation of a specific secondary structure at the 5′ end of the RNA II molecule that results in an interaction between the RNA portion on the *ori* and an upstream region of this molecule with the template DNA (Fig. 1, panel c) (43–46). The appropriate conformation of RNA II necessary to form the hybrid with DNA is shown in Figure 2 (46,47). The duplex RNA II–DNA structure acts as a substrate for the subsequent action of RNase H, which cuts the RNA at the replication origin, leaving a 3′-hydroxyl group that serves as the primer for DNA synthesis catalyzed by Pol I (Fig. 1, panels c and d) (40). It has been shown that certain point mutations in RNA II prevent RNA II from adopting the right conformation, which affects hybridization with the template DNA (44,45,47). Once Pol I begins the addition of deoxynucleotides, the remaining RNA hybridized to the DNA strand is digested by RNase H at other sites on the molecule and by the 5′-3′ exonuclease activity of Pol I. RNase H also functions to remove the RNA primer from the growing DNA strand (Fig. 1, panel d) (48). Replication of ColE1 then proceeds unidirectionally with the initiation of the lagging strand synthesis at specific ColE1 sites.

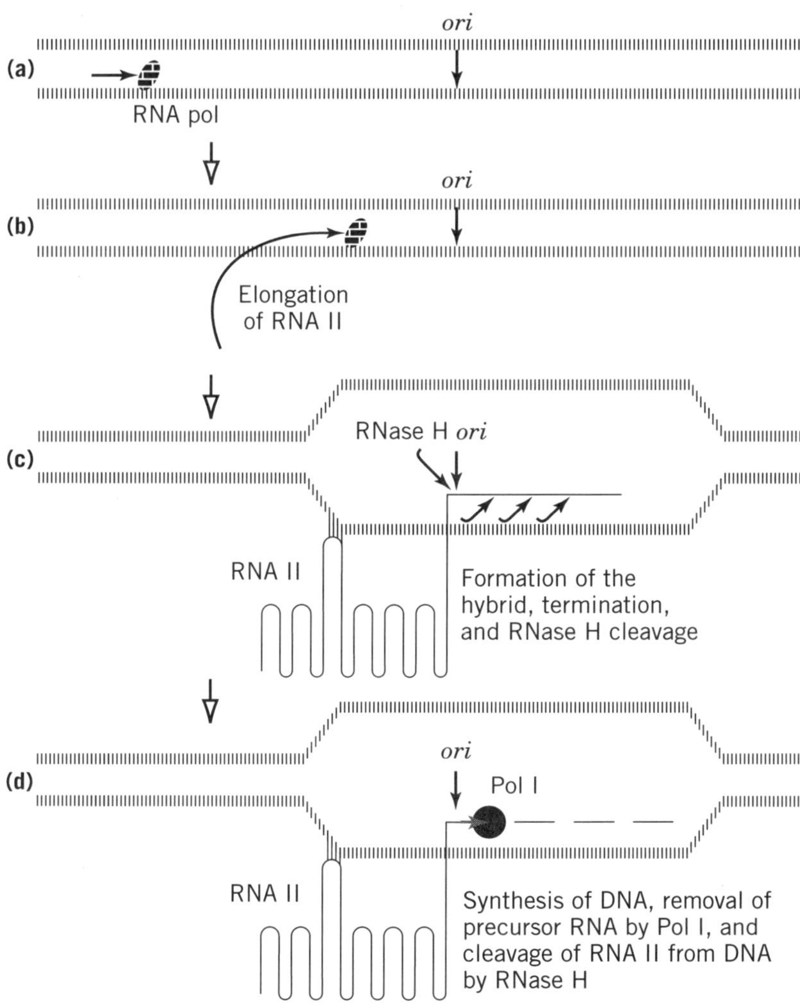

Figure 1. Mechanism of initiation of ColE1 replication in wild-type *E. coli*. The horizontal black arrow in (**a**) indicates the transcription start site (555 nucleotides upstream of *ori*) of RNA II. The origin of replication is indicated by a vertical arrow. ↘, indicates sites of action of RNase H; IIIIII, DNA; ■, RNA; , RNA polymerase I. See the text for details.

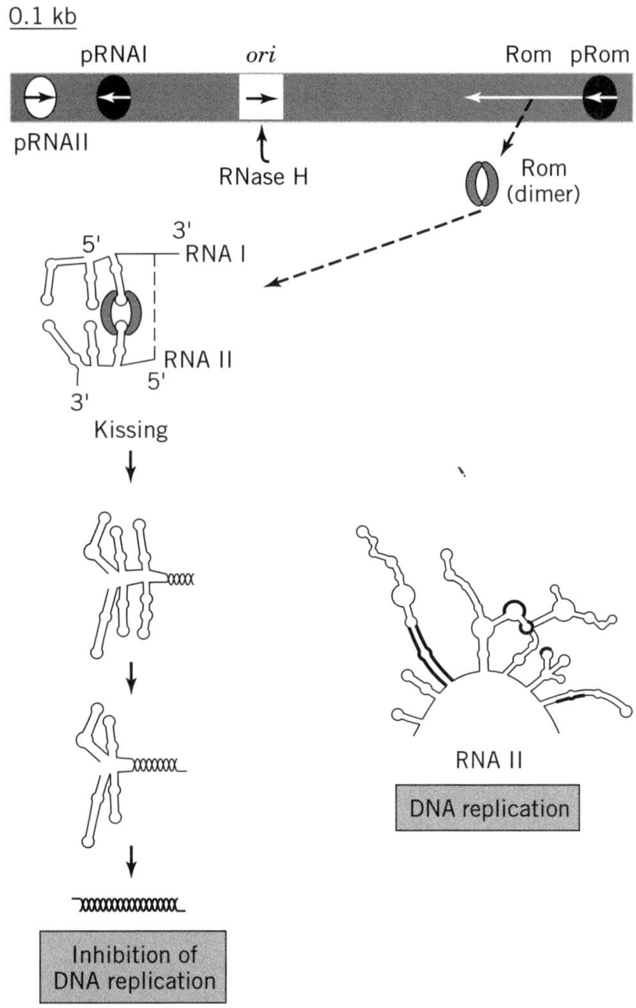

Figure 2. Proposed mechanism of the succession of events involved in the regulation of the initiation of replication of ColE1 by RNA I and Rom. The upper portion shows a genetic map of the origin of replication of ColE1. ||||| represents ColE1 DNA, with the promoters indicated by arrows inside circles. The open arrow indicates the *rom* gene. In the left panel, a schematic of the mechanism of RNA I and RNA II interaction that leads to the inhibition of DNA synthesis is shown. The first interaction between RNA I and RNA II (kissing) is reversible and stabilized by the Rom protein. The dimeric form of Rom is shown both below its gene and in the interaction with RNA I and RNA II. The right panel of the figure indicates the secondary structure that leads to DNA replication. For this to occur, an RNA II molecule that is not coupled with RNA I binds to DNA and acts as a primer after RNase H digestion. The structures reported by Fitzwater et al. as well as Masukata and Tomizawa are almost identical (46,47). A more detailed scheme of the possible interactions between the RNA and Rom species has been reported elsewhere (18). *Source*: Adapted from Ref. 18.

It has been further shown that replication of plasmids containing the ColE1 origin of replication can proceed using two alternative modes of replication that allow for replication in the absence of either RNase H alone or RNase H and Pol I (49–51) (Fig. 3). These two mechanisms appear to not be utilized in wild-type cells, however, and probably represent adaptations to specific host mutations (49,51–53). One of these mechanisms is independent of the presence of Pol I and involves the elongation of RNA II coupled to the template DNA. The lack of RNase H allows the formation of a single-stranded DNA region that can extend to a length that is adequate for assembly of a replisome and initiation of synthesis on the opposite DNA strand (lagging strand) (Fig. 3, panel a) (18,50).

Another initiation mechanism has been demonstrated in RNase H–deficient mutants that is dependent of Pol I. In this case, the extended RNA II species can be recognized by Pol I and used as a primer (Fig. 3, panel b) (49). However, this mechanism of initiation has proved to be inefficient in in vitro experiments (40). Regardless of the replication mode, however, RNA II must be produced in the correct structural configuration to act as the primer for DNA replication (49,51,54).

Control of initiation of replication of ColE1 is mediated by the interaction of RNA II with a 108-nucleotide antisense RNA transcript, called RNA I, that is encoded in a region that overlaps the coding region for RNA II (Fig. 2) (40,55). RNA I binds to nascent RNA II by complementary base pairing, triggering a conformational change in RNA II that prevents coupling with DNA, thereby inhibiting replication initiation (42,44,55–57). Within the secondary structures of both RNA species there are several stem-loops (Fig. 2). Interaction between both RNA species starts at complementary loops in a reversible process known as "kissing," which results in the generation of an unstable initial complex (Fig. 2) (58–62). After this initial contact takes place, an irreversible process leads to the formation of an RNA duplex between RNA I and RNA II (Fig. 2) (56,60,62,63). RNA I achieves its inhibitory effect only if it is present during a short, specific interval in the synthesis of RNA II (56,64). This situation occurs most frequently because RNA I transcription occurs as much as five times more often than transcription of RNA II. This results in the formation of a productive primer only in about 1 out of every 20 RNA II molecules that are started (65). The available amount of RNA I in the cell is influenced by polyadenylation (66,67) and the bacterial RNase E, which cleaves the 5′ end of RNA I, rendering it inactive (68).

A second factor involved in the control of initiation of replication of ColE1 is the plasmid-encoded 63–amino acid protein Rom (RNA one modulator), also known as Rop (Repressor of primer) (Fig. 2) (56,60,62,69,70). The gene encoding Rom is located downstream of *ori* (Fig. 2). In the early stages of plasmid replication, Rom binds to the stem-loop portions of the unstable initial complex (58,59), reducing the dissociation constant of the complex, thus creating a pathway for stable binding of RNA I and RNA II (Fig. 2) (60,71). As a consequence, ColE1 deletion derivatives that lack *rom* present a higher copy number than plasmids that have the complete ColE1 replication region (62,72). The presence of Rom, however, is not needed for viability of the plasmid, and in some RNA-regulated replicons is not present (16).

Incompatibility is the failure of two plasmids to be stably inherited in the same cell line. This phenomenon is a consequence of sharing elements of plasmid inheritance functions such as replication or partition (1,17,73). RNA I

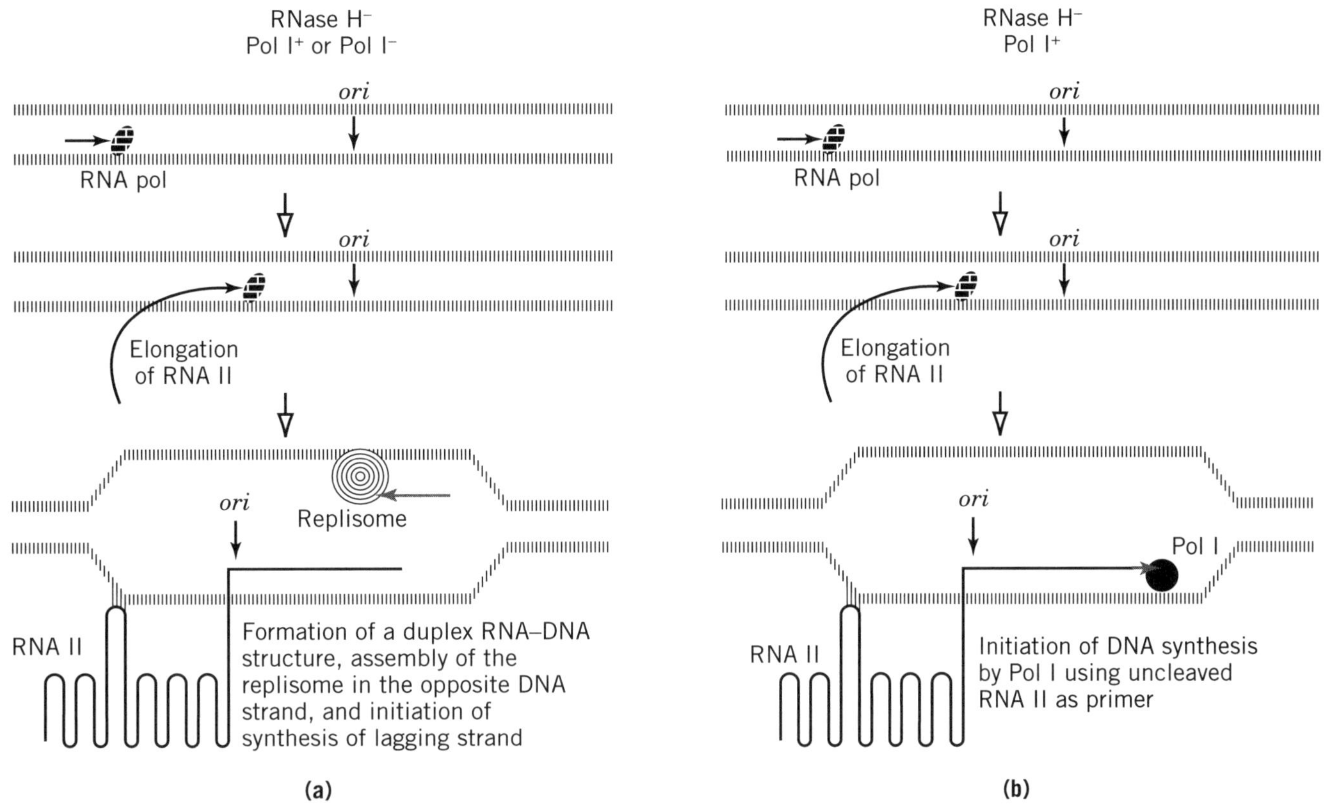

Figure 3. Alternative mechanisms of initiation of replication of ColE1 in RNase H–deficient mutants. (**a**) Proposed mechanism of initiation in RNase H–deficient *(rnh)* and Pol I positive *(polA$^+$)* or negative *(polA)* mutants (52). (**B**) Mechanism proposed to occur in *rnh, polA$^+$* mutants. The black circle represents Pol I. The horizontal black arrow at the top indicates the transcription start site; concentric circles represent the replisome; and *ori* shows the origin of replication in wild-type cells. IIIIII, DNA; ■, RNA. *Source*: Adapted from Ref. 18.

is the main incompatibility determinant in ColE1-type plasmids. Two plasmids that depend on the same RNA I species for regulation of initiation of replication can not coexist in the same cell (1,17,73). It has been shown that even single nucleotide changes can have profound effects in the incompatibility properties of ColE1-type plasmids (55,57).

THE R6K PLASMID

R6K is a naturally occurring conjugative plasmid that encodes resistance to the antibiotics ampicillin and streptomycin (74). This plasmid is an example of an extrachromosomal element of intermediate size and copy number because it is about 38 kb in size and has a copy number of 13 to 40 per cell (75). These features, together with a unique mode of replication, make R6K an attractive system to study the genetic and molecular mechanisms involved in plasmid DNA replication. In addition, this plasmid and its replication components were among the first used to generate gene fusions, transcription enhancement, protein tagging, and site-specific proteolysis (76–79); it is currently utilized for high expression and rapid isolation by affinity chromatography of a large variety of proteins. Furthermore, R6K derivatives were instrumental in designing a series of suicide vectors successfully used to generate mutants by allelic exchange or transposition mutagenesis in Gram-positive bacteria (80,81).

Plasmid R6K is a member of a growing group of replicons encoding an initiator protein that binds to nucleotide sequence repeats, called iterons, located within *ori*. This group of iteron replicons includes the *Escherichia coli oriC* (82) and the plasmids F (83), pSC101 (84), P1 (85), PMJ101 (86), Rts1 (87), the REPI replicon of pColV-K30 (88), and the RK2- and RP4-related plasmids (89). R6K has three origins of replication, named *ori*α, *ori*β, and *ori*γ (90,91), all clustered within a 4-kb DNA fragment (Fig. 4). In addition, the *pir* and *bis* genes encoding the π and Bis replication proteins, respectively, were found within this replication region. Near this region is located the replication termination site containing the τ sequence that binds the host-encoded ter protein and arrests the movement of the replication fork (90,92,93). The main active origins in vivo are α and β, while *ori*γ tends to remain inactive (90,91) due to the synthesis of a silencer RNA encoded immediately downstream from the *ori*γ sequence (94). The silencing activity of this RNA was explained by the formation of a silencer RNA-activator RNA hybrid, driven to completion of the R6K-encoded replication protein, that leads to the in-

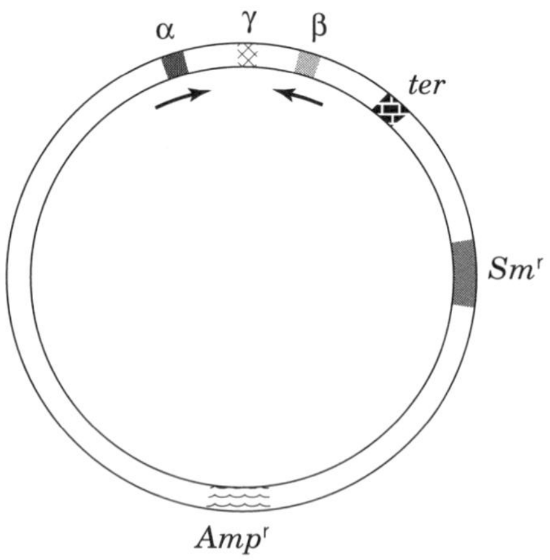

Figure 4. Diagram of R6K showing the 4-kb replication region containing the α-, β-, and γ-origins. The location of the termination site (*ter*) and the genes encoding resistance to streptomycin *(Sm)* and ampicillin *(Amp)* are also indicated. The arrows mark the in vivo direction of the initial replication from the α- and β-origins.

activation of *ori*γ (95,96). However, *ori*γ can replicate autonomously when the other two origins are deleted and the *pir* gene is provided in *cis* or in *trans* (97–101). Thus, *ori*γ has served as the simplest model system derived from R6K to study the replication of an iteron-containing DNA molecule.

The γ origin is the major initiator binding site and is required in cis for initiation of plasmid replication from the other two sites (102–104). Therefore, the γ origin behaves as a prokaryotic enhancer-type element in the sense that DNA–protein interactions at this site induce significant changes in DNA structure that facilitate initiation of DNA replication from distantly located origins (105). Figure 5a shows a diagram of the complete *ori*γ replication region. The molecular organization of this origin is similar to other plasmid origins, although it contains two functionally distinct segments (104,106). The 277-bp core segment, common to all three origins, is essential for replication, and consists of three distinct regions: (*1*) the AT-rich region, bound by π and the integration host factor (IHF) protein, (*2*) seven 22-bp repeats bound by π, and (*3*) a multiprotein-binding region interacting with DnaA, IHF, and RNA polymerase. Immediately to the left of the core lies the 106-bp origin enhancer, which includes a DnaA box and a small segment containing the *stb* locus (107). Even though the enhancer is not absolutely required for replication at low level of wild-type π-protein, it was found that the enhancer region is required for stable maintenance of γ-origin-containing plasmids. Although the *stb* locus has some similarities with the *par* locus of pSC101, the partition systems of these two plasmids differ from each other in several aspects (107). It has been proposed recently by Wu et al. (107) that a host-encoded protein may be binding to the *stb* repeats. The finding that R6K derivatives carrying the intact three origins can bind in vitro to both inner and outer membrane fractions of *E. coli* (108) appears to support this hypothesis. An alternative explanation is that *stb* mediates plasmid partition by altering the structure of γ-origin-containing plasmids (107). The host-encoded protein Fis also binds to 10 sites in the γ-origin that overlap all the previously identified binding sites for the R6K-encoded π-protein and the host-encoded DNA binding proteins DnaA, IHF, and RNA polymerase (109). However, it appears that the Fis protein is required for plasmid replication only when π-copy-up variants and the penicillin-resistant marker are simultaneously used, suggesting that plasmid sequences such as those encoding antibiotic resistance play a role in plasmid DNA replication.

The analysis of the functional α-replicon revealed that it is composed of two elements that must be present in *cis* and oriented as in the intact R6K: (*1*) a 580-bp fragment containing the α-origin, and (*2*) the 277-bp core segment located within the γ-origin (110) (Fig. 5b). These two elements are separated by approximately 3 kbp, which are not required for the α-origin activity. The 580-bp fragment contains a long 98-bp palindrome, and it was suggested that this sequence serves as the recognition signal for initiation of DNA replication in the α-origin. DNA homology analysis also revealed the presence of a 23-bp sequence that resembles the seven 22-bp iterons found in the core segment (110) and plays a role in the π-protein-mediated looping between the γ- and α-sequences (105). In addition, this fragment contains a DnaG binding site that can serve as DnaB loading site by DnaB–DnaG interaction (111).

The minimal β-replicon was defined as a 2-kbp fragment that encompasses the following elements: (*1*) the γ-core region, (*2*) the *pir* gene encoding the π-initiator protein, (*3*) the *bis* gene encoding the 17.2-kDa Bis protein, and (*4*) the β-origin (102,112–114) (Fig. 5c). The Bis protein is required only by the β-origin, and its synthesis is coupled in cis to the expression of π-protein from an unaltered *pir* gene (113). The β-origin also contains a half 22-bp iteron and a 98-bp palindrome that has 96% homology with the α-origin hairpin (110). The presence of these pallindromic sequences at the *ori*α and *ori*β regions was previously assessed by electron microscopy. The half iteron and the long palindromic sequence are required for the π-mediated looping between the γ- and β-sequences (105) and the initiation of DNA replication at the *ori*β, respectively.

The π-initiator protein is essential for replication from each R6K origin and, although it can be provided in cis or trans to regulate the activity of the γ- and α-origins, it is required in cis for activation of the β-origin (97,102, 110,114,115). This protein is a homodimer with a 36-kDa molecular weight for the monomeric form, which is lysine rich and weakly basic (98). In its native dimeric form (116), or as a hybrid with β-galactosidase (78,117) or collagen (77), it has been shown to bind to the seven iterons in the γ-origin as well as to an eighth iteron and a smaller inverted pair of repeats located in the operator-promoter region of the *pir* gene (78,116–118). In addition, this initiator protein interacts with the iterons located in the α- and β-iterons; however, these contacts are weak and require the enhancing effect of π already bound to the seven γ-iterons (105,119).

The π-protein has both positive and negative roles in the replication of R6K (120). The positive role of π is displayed by its ability to enhance replication of *ori*α and *ori*β. This activity is mediated by the promotion of conforma-

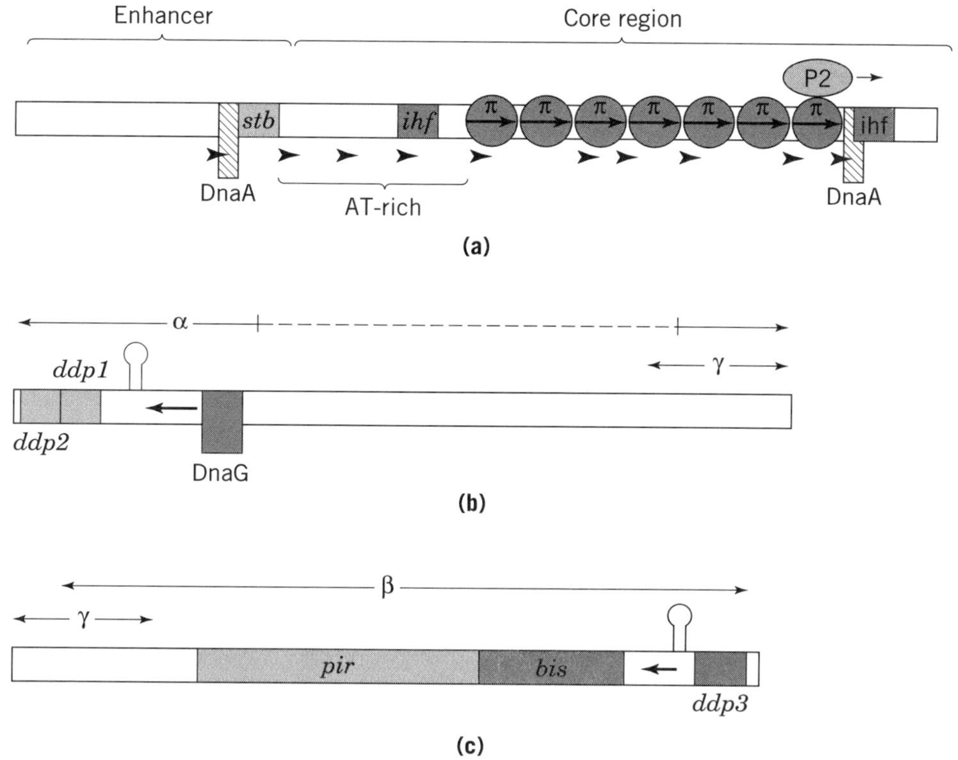

Figure 5. Diagram of the three R6K origins of DNA replication. (**A**) Components of the enhancer and core regions of the γ-origin. The long rectangle represents the DNA region containing binding sites for the host-encoded proteins DnaA (tall rectangles), IHF (short rectangles), and RNA polymerase (oval). P2 and the open arrow represent the promoter site and the direction of transcription of the *pir* gene, respectively. The thick arrows represent the seven iterons that bind the π-replication protein (circles). *stb* represents the locus involved in plasmid maintenance. The 10 arrowheads represent the location of the Fis binding sites. (**B**) Components of the active α-origin. The double-headed arrows indicate the DNA fragment containing the α-origin and the γ-core region required in cis for active replication. The dashed line represents the intervening sequences not involved in DNA replication. The hairpin marks the location of the long inverted repeat located in the α-origin, followed by the genes encoding the DDP1 and DDP2 proteins (small rectangles). The thick arrow and the large rectangle indicate the position of an iteron and a DnaG binding site, respectively. (**C**) Components of the active β-origin. The double-headed arrows indicate the DNA fragment containing the γ-core region and β-origin required in cis for active replication. The β-origin includes the *pir* and *bis* genes (long rectangles), a half iteron (half thick arrow), the *oriβ* long inverted repeat (hairpin), and the gene encoding the DDP3 protein (short rectangle).

tional changes, such as DNA unwinding and bending, in the replication region that cause activation of these two origins by looping out intervening sequences located between the *oriγ* core region and the α- and β-origins (119,121). Site-directed mutagenesis showed that π-protein is necessary but not sufficient for activation of *oriβ* and probably *oriα* (105). This observation suggested that the DNA looping process must also be required for the transfer of a multiprotein complex capable of initiating DNA replication. Supporting this hypothesis, it was recently reported that π specifically interacts with the host-encoded helicase DnaB replication protein (111). This result indicates that DnaB is initially recruited by π, already bound to the γ-origin, and then delivered to the two other origins by the DNA looping induced by the R6K-encoded replicator protein.

The π-protein has been shown to possess two negative activities on the replication of R6K. One of them is as an autorepressor at the level of transcription and involves its binding to iteron sequences located within the operator-promoter site of the *pir* gene. These interactions act by either preventing the binding of the RNA polymerase to the promoter or displacing the RNA polymerase from promoter–enzyme complexes (99,118,122). The other negative role of the π-protein is its negative control of R6K replication. Biochemical and genetic experiments showed that increased intracellular levels of initiator protein will either lower the plasmid copy number or completely prevent its replication (120). The copy-up phenotype of some of these mutants correlates with base changes within a 100-bp fragment of the *pir* gene and can be suppressed by the presence of the wild-type initiator protein at concentrations lower than that of the mutant protein (123). This negative regulatory activity of π was explained by molecular models involving either direct π–DNA interactions or association of π with either other π molecules or with host-encoded proteins (120,124). One of these models, termed *handcuffing*, is based on the observation that π has the ability to associate two DNA molecules containing γ-origin sequences and to enhance the DNA ligase–catalyzed multimerization of a single DNA fragment carrying this R6K origin of replication (124). In addition, it was shown that the negative domain of π is located in the N-terminal region of this protein and that the π-mediated inhibition of R6K replication does not require direct binding to DNA (125). Consequently, the origins located within these DNA–protein complexes are unable to initiate replication, most likely because π-induced DNA structure alterations in the origins or potential π–host protein interactions are prevented. In summary, all these observations suggest that the dual activity of π and the regulation of R6K replication is the result of a competition between positive γ–α and γ–β interactions and the inhibitory aggregation of γ-containing molecules, all mediated by the π-replication protein.

Recently, three novel R6K-encoded proteins, designated DDP1, DDP2, and DDP3, were identified (126). DDP1 and DDP2 are encoded by two tandem genes located at the 5′ end of the long inverted repeat of the α-origin, while the gene encoding DDP3 was mapped at the 3′ end of the β-origin long inverted repeat. These three proteins are required for the distortion of the DNA structure of the two R6K active origins. Although the distortions caused by these proteins are potentially linked to R6K replication, they are not equivalent to those described in other replication regions previously characterized (82). It was also suggested (126) that this distortion system serves to synchronize the initiation of replication and establishes the direction of replication from the α- and β-origins by generating a "locked" preinitiation protein–DNA complex.

PT181-TYPE PLASMIDS

The rolling circle type of replication is most often utilized by a number of small, high-copy-number plasmids from Gram-positive bacteria (13,15,21,127). However, plasmids that replicate using this mechanism have recently been identified in a number of Gram-negative bacteria (128–132).

Gram-positive plasmids that replicate using rolling circle mechanisms are classified into five families based on sequence comparisons and genetic organization of the replication regions (13,16,21,133). All plasmids in these families replicate by an asymmetric rolling circle pathway that resembles that of the single-stranded filamentous bacteriophages (16,21,134). One of the best-understood replication mechanisms is that of the pT181-type family. We describe the replication of pT181 as a paradigm of rolling circle replicating plasmids. A more comprehensive description of replication of plasmids in other families can be found elsewhere (13,15,21,127).

A schematic representation of the pT181 replication region and the mechanism of initiation of replication is shown in Figure 6. This plasmid, derived from *Staphylococcus aureus*, encodes a 38-kDa initiator protein, RepC, that has sequence-specific endonuclease and topoisomerase I–like activities (137). To initiate replication, RepC nicks one of the pT181 DNA strands (leading strand) at a specific site, generating a free 3′-OH end that is used as primer for subsequent DNA synthesis (137). The functional structure of the RepC protein is a homodimer (RepC/RepC) (135,138) that recognizes and binds to a site (Rep binding site) that encompasses an inverted repeat (IR III in Fig. 6b) (136). A domain of six amino acids has been identified in RepC that is important in the recognition and interaction of the Rep binding site (Fig. 6) (139). The binding efficiency of RepC is increased by the presence of *cmp*, a 100-bp cis-acting replication enhancer located about 1 kb from the nicking site (140,141). It has been recently demonstrated that an *S. aureus* protein, CBF1, binds *cmp* and increases distortion of the already bent *cmp* locus (142). Whether this binding is associated with the enhancing activity of *cmp* is still not known. Upon binding of the RepC homodimer to IR III, DNA in this region is induced to bend (143) (Fig. 6b). This is closely followed by a change in structure of RepC, DNA melting, and formation of a cruciform structure at the IR II region (Fig. 6b) (139). The melting step is facilitated by the presence of an AT-rich inverted repeat region (IR I) located upstream of IR II (Fig. 6b) (16). The formation of a cruciform structure may result in an approximation of the nicking site of the leading strand and the active site of RepC. This process involves a tyrosine residue that appears to facilitate the generation of the nick. After the endonuclease attack, the RepC protein remains covalently bound to the 5′ end of the DNA by a phosphotyrosine bond (93), probably remaining attached throughout replication of the leading strand. This is not true for all plasmids replicating through a rolling circle mechanism, however, for it was recently shown that in a derivative of pMV158 the initiator protein does not remain covalently bound to the DNA after nicking (144). After generation of the 3′-OH terminal end by RepC, an initiation complex is formed with the probable participation of DNA polymerase III, the helicase PcrA, and single-strand binding protein (Fig. 6b) (16,145,146).

Following (or during) replication, the initiator protein RepC becomes modified (RepC*) by the addition of a 10- to 12-mer oligodeoxynucleotide identical to the sequence located immediately 3′ to the origin of the leading strand. In addition, the two enzymatic activities of RepC are lost during this time (138,147). Further analysis has demonstrated that the active initiator homodimer RepC/RepC becomes the inactive heterodimer RepC/RepC* after it has been used for replication of pT181 (135).

Replication of the lagging strand starts at a different origin than replication of the leading strand. Fig. 6a shows the location of the lagging strand origin in pT181 (148). This origin, also known as SSO (single-strand origin) or *palA*, comprises about a 160-bp palindromic DNA sequence (16,21). Synthesis of the leading strand generates a displaced single-stranded DNA that allows the origin of replication of the lagging strand to adopt the correct conformation to serve as a priming site. Replication of this strand does not require any protein encoded by the plasmid (149). In addition, some rolling circle replicating plasmids have more than one origin of replication for its lagging strand that is functional in different hosts (16). Taken to-

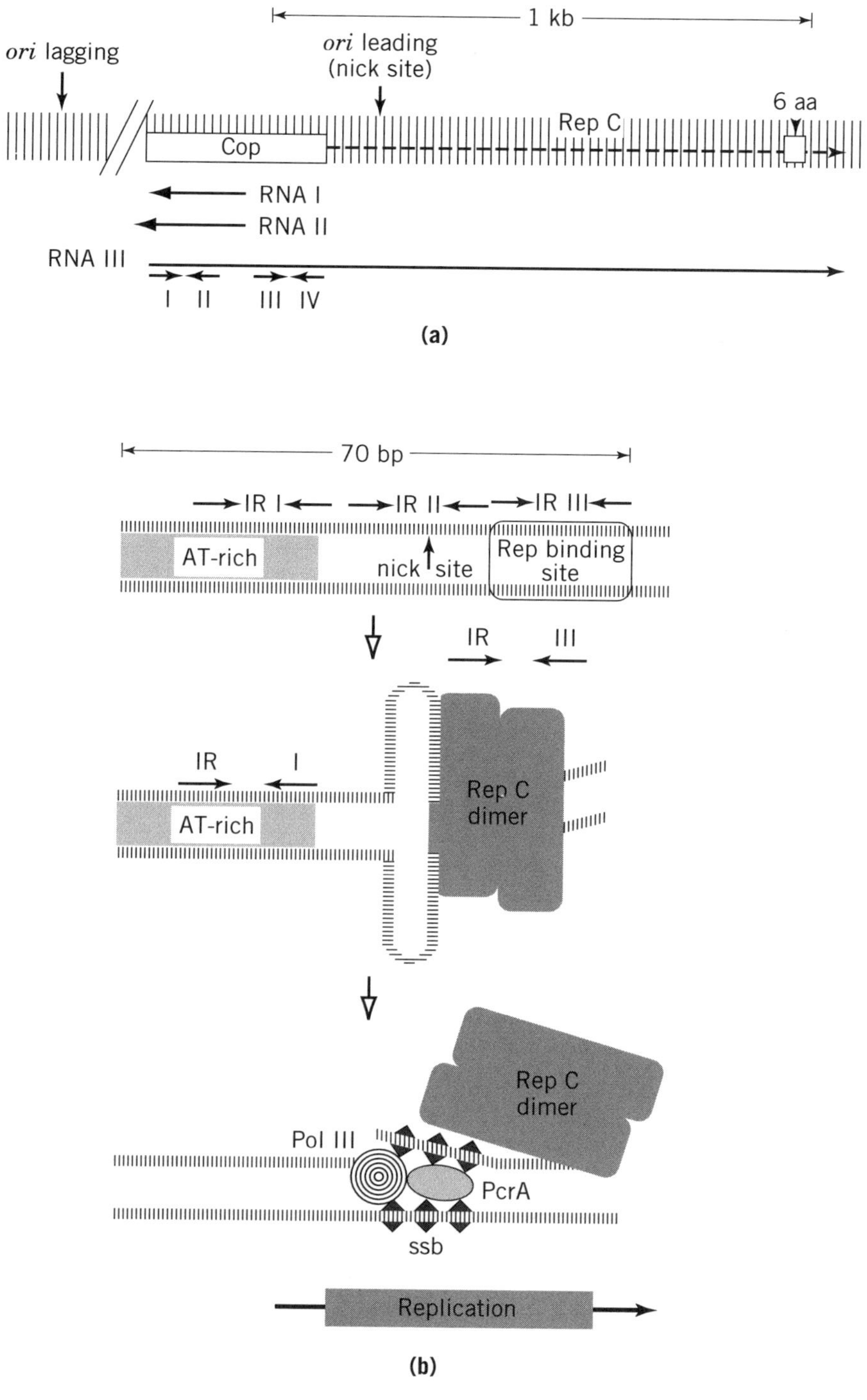

Figure 6. Replication region and model for initiation of replication of the plasmid pT181. (**a**) Diagram of the pT181 replication region. The bar represents pT181 DNA. The two origins of replication (leading and lagging strands) are indicated by different vertical arrows. The broken-line arrow represents the RepC coding region. The six–amino acid region of RepC recognized by *ori* is indicated as a white square. The two diagonal lines indicate that there is a larger distance than shown between the *ori* of the lagging strand and the *ori* of the leading strand. The black arrows represent RNA molecules. The location of the *cop* region is indicated by a box. The small arrows under RNAIII represent the inverted repeats that participate in attenuation. For the sake of clarity, these inverted repeats are not at scale. (**b**) Model for initiation of replication of pT181. The first diagram shows the region encompassing the origin of replication of the leading strand and the three inverted repeats. This region is also referred to as the DSO (double-stranded origin). The RepC binding site located at IR III is boxed. The second diagram shows the formation of the cruciform structure after binding of the RepC homodimer, the bending of the DNA at the binding region, and the change in structure of RepC. The melting process is facilitated by the presence of an AT-rich region that includes IR I. The third diagram shows that after nicking, a replisome is assembled at *ori*, and presumably, DNA polymerase III initiates replication in the presence of the helicase PcrA and single-strand binding protein (ssb). *Source*: Adapted from Refs. 16, 135, and 136.

gether, these facts suggest that lagging strand origins of replication may be important in determining the host range of the plasmid (16).

Termination of synthesis of the leading strand occurs at the nick site by a strand transfer mechanism that is mediated by the replication initiator protein RepC (150). In pT181, once the nick site has been replicated and extended a few nucleotides beyond this site, one of the subunits of the RepC dimer contacts the growing strand. This interaction initiates a strand transfer reaction, resulting in the formation of a single-stranded monomer (old strand), a double-stranded molecule where one of the strands is newly synthesized, and a dimer in which one of the monomers is attached to the oligonucleotide resulting from the extension of replication beyond the nick site (16,21). Although the presence of the nicking site is required, the initiator protein recognition site is not essential for termination (21).

Regulation of initiation of replication of pT181 is achieved through the control of the synthesis of RepC (16,21,151,152). The organization of the pT181 replication regulation region *(cop)* is shown in Fig. 6a. The mRNA

(RNA III) encoding RepC contains a leader sequence typical of genes that are regulated through attenuation (153). This leader sequence has four interacting sequences (I–IV). Sequence I, a nine-nucleotide sequence called *preemptor* can form duplexes with sequences II and III (Fig. 7). In addition, there are two antisense RNA species (RNA I and RNA II) complementary to the leader sequence of RNA III. Analysis of the secondary structure of the leader portion of RNA III in the presence or absence of antisense RNA indicated that upon interaction of the two complementary RNA species, a stem-loop structure between inverted repeats III and IV is formed followed by an AUUUUUU sequence that acts as a transcription termination signal (Fig. 7) (154). On the other hand, if antisense RNA is not present, the interaction between inverted repeats I (the preemptor) and III generates a structure that prevents the formation of the III-IV stem-loop (Fig. 7) (154). Under these conditions, transcription of *repC* takes place. Mutations in the preemptor have been shown to lead to an absolute requirement for RepC in trans (154,155). For example, Figure 7 shows the effect on the secondary structure of RNA III caused by a three-nucleotide replacement mutation in the preemptor. In absence of antisense RNA, the stem-loop structure involving inverted repeats III and IV is formed, no RepC synthesis occurs, leading to a plasmid that has an absolute requirement for RepC in trans (154,155) (Fig. 7).

The main incompatibility elements in pT181-type plasmids are the antisense RNA and the origin of replication of the leading strand (16,154,155). The origin of replication of the leading strand directly binds RepC, titrating the protein, which is present at limiting levels in the cytosol due to the controlling action of the antisense RNAs.

LINEAR PLASMIDS IN BACTERIA

A Unique Plasmid Anatomy

Bacterial genomes have long been considered to contain only circular DNA molecules. However, the development of orthogonal field-pulsed agarose gel electrophoresis in the mid-1980s allowed the discovery of the first linear bacterial chromosome in *Borrelia burgdorferi* (156,157), the causative agent of Lyme disease. Since then, linear replicons have been found in many bacterial genera including *Agrobacterium, Streptomyces, Thiobacillus, Nocardia, Rhodococcus*, and even *Escherichia* (158). The first linear plasmid was found in *Streptomyces rochei* in 1979 (159),

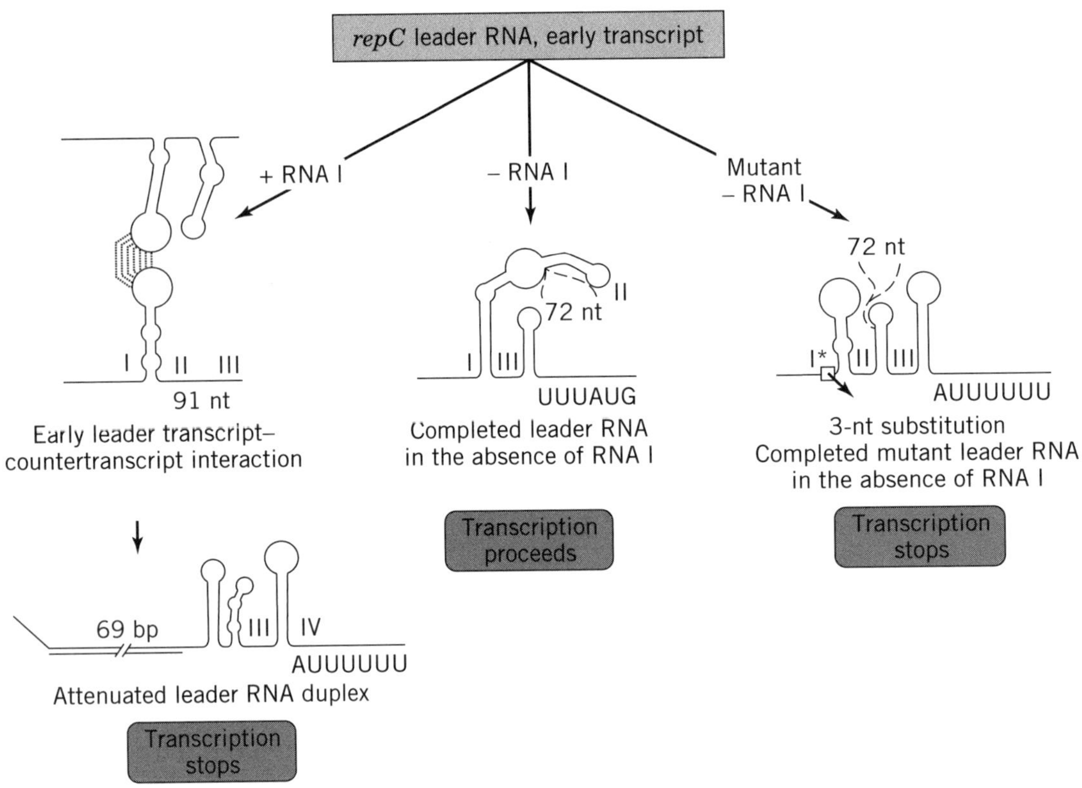

Figure 7. Mechanism of regulation of expression of RepC by antisense RNA. In this model, the early RNA III transcript can interact with antisense RNA (RNA I), allowing the formation of a stem-loop between inverted repeats III and IV that results in a transcription termination signal. In the absence of antisense RNA, the inverted repeat III is sequestered by inverted repeat I, leading to completion of the *repC* mRNA. The three-nucleotide-change mutation in inverted repeat I is also shown. The secondary structure of this mutated RNA III results in formation of the stem loop III–IV even in the absence of antisense RNA, thus leading to a complete dependence of RepC in trans for replication. *Source*: Adapted from Refs. 21 and 154.

and others have been detected in the Gram-positive bacteria *Rhodococcus fascians* and *Nocardia opaca* as well as in the Gram-negative *Thiobacillus versutus* (158). The *Streptomyces* linear plasmids range in size from nine to several hundred kilobases, and all of them contain terminal inverted repeats of different lengths according to the size of the plasmids where they are present. The linear plasmids in *Borrelia* also show a wide range of sizes, typically ranging from 15 to 200 kb and are distributed widely among the members of this species. The *Borrelia* plasmids are unique among extrachromosonal elements because some of them carry critical genes such the *guaA* and *guaB* biosynthetic genes (160), as well as genes encoding the essential major outer surface Osp or Vmp lipoproteins (161–164). In addition, it was determined that some of these linear plasmids are present stably in low copy number, about the equivalent of one per chromosome, and can be cured by exposing *Borrelia* cultures to the DNA gyrase inhibitor novobiocinin (165). These characteristics suggest that the replication and partition processes of linear plasmids are well controlled during cell division. All the facts, together with the observation that the *Borrelia* chromosome has attributes of a linear DNA molecule, has led to the idea that these linear plasmids might be regarded more properly as minichromosomes (166).

The presence of these linear DNA molecules brings an interesting biological problem associated with the protection of the DNA ends from exonuclease degradation, and the complete replication of these plasmids. These problems are solved in eukaryotes by the addition of telomere repeats by reverse transcription of a short RNA incorporated in the telomerase (167,168). On the other hand, prokaryotes can protect and replicate completely these linear plasmids using two alternative mechanisms.

The ends of linear *Streptomyces* plasmids have telomeric structures also found in adenoviruses and some prokaryotic phages as well as in almost all eukaryotic plasmids. Sequence analysis of pSL2, a linear plasmid present in *S. rochei*, revealed the presence of an inverted terminal repetition of 614 bp and 11 sites of interrupted homology (169,170). These terminal DNA structures, together with specific DNA binding proteins, are thought to be involved in the juxtaposition of the two plasmid termini containing identical or very similar nucleotide sequences. This structure, known as the racket frame–like DNA model and shown in Figure 8a, also includes a terminal protein covalently attached to the 5′ DNA termini, which is required to protect the DNA from degradation and to complete the replication of the 3′ overhanging ends, as described later.

The structure of the 16- and 49-kb linear plasmids of *Borrelia burgdorferi* consists of a double-stranded DNA chain connected at each end by a perfect palindromic AT-rich hairpin loop (Fig. 8b) (2,171). In addition, each end contains a conserved 19-bp inverted repeat sequence, a telomeric structural feature found in all linear plasmids analyzed in *Borrelia* to date (158,162,163). These features have similarities to the telomeres of other linear double-stranded replicons, including among them the genomes of vaccinia and other animal poxviruses (172), the mitochondria of the yeast *Pichia* (173), and the iridovirus, which causes African swine fever (174). The fact that the latter and at least one *Borrelia* species share a common tick vector suggests that the *Borrelia* linear plasmids may have originated by horizontal transfer between kingdoms (2). These telomeric structures indeed play a role in preserving the integrity of these plasmids, and in their replication before cell division.

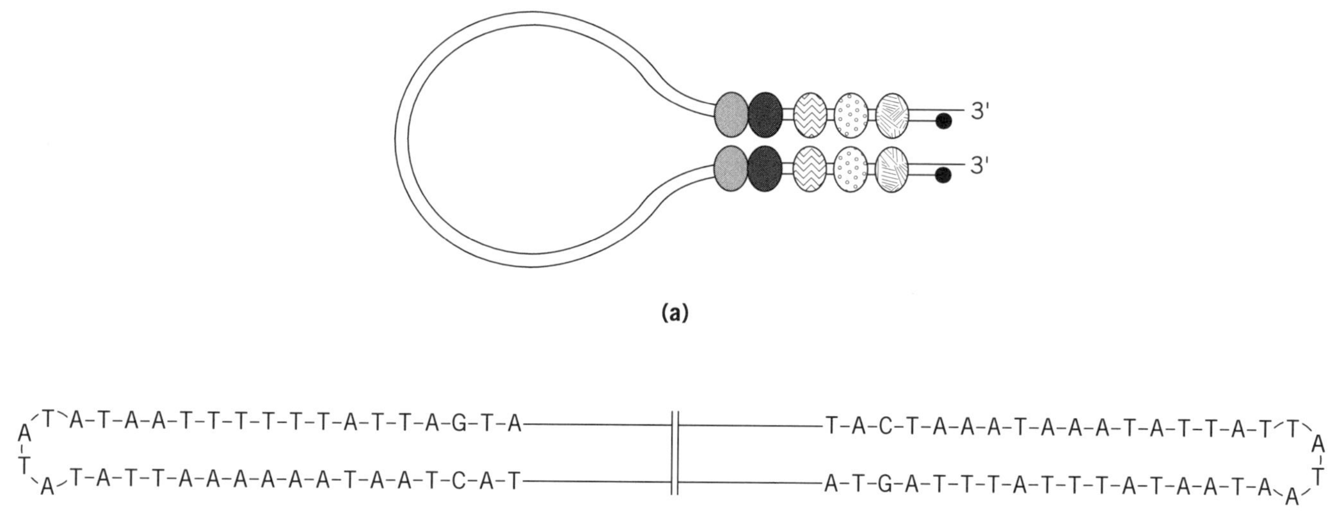

Figure 8. Telomeric structure of linear plasmids. (**a**) Racket frame structure proposed for linear *Streptomyces* plasmids. The ovals represent juxtaposition proteins that recognize specific regions of palindromic symmetry, bringing together the plasmid termini. The black circles represent the terminal protein attached to the 5′ ends, which is required to complete the replication of a 280-nucleotide segment at both plasmid termini. (**b**) Structure and partial nucleotide sequence of the telomeric termini of linear plasmids isolated from *Borrelia*.

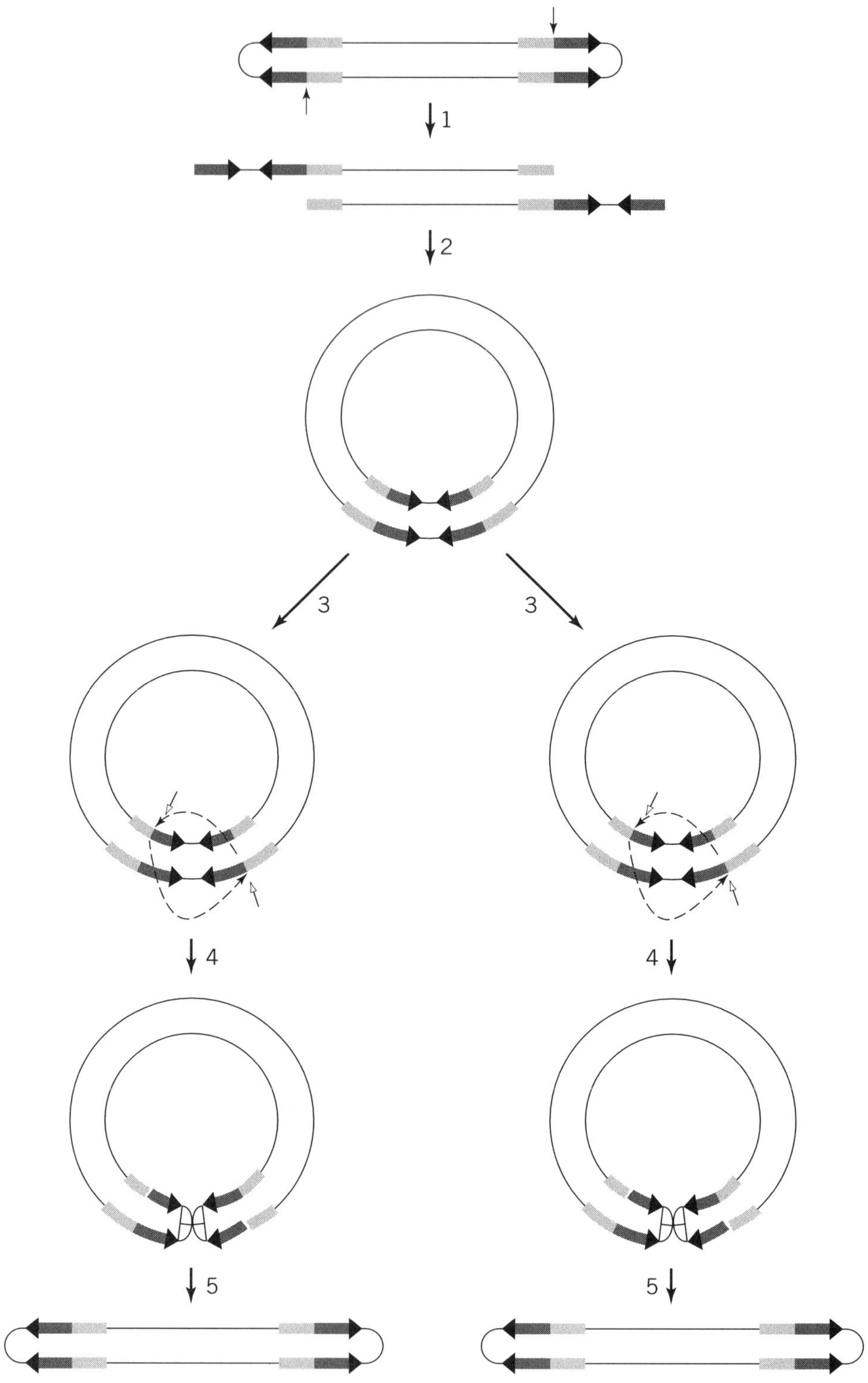

Figure 9. Model depicting the different stages leading to the formation of circular intermediates and replication of linear plasmids in *Borrelia*. According to this model, in step 1 the ends of the plasmid are open by a nick (indicated by the arrows with open arrowheads) within the inverted repeats (represented by the thick arrows) located at the plasmid termini. The open linear plasmid circularizes due to the presence of complementary sequences at its termini (step 2) and replicates as a circular replicon (step 3). A second nick is introduced within the repeated sequences, and the free single-stranded ends pair back (indicated by the dashed arrows) with their complementary copies located in the same monomer and, thus, reconstitute the hairpins at each plasmid end (step 4). After DNA ligation (step 5), two copies of the original linear plasmid are generated. *Source*: Adapted from Ref. 157.

Replication of Linear Plasmids

It was proposed initially that the linear plasmid of *Streptomyces* replicates by a protein-primed replication mechanism (158,175) that has been well characterized for adenovirus, phage ø29, and *Bacillus subtilis* (168). In this mechanism of DNA replication, the telomere is the origin of replication, and it is recognized by specific DNA binding proteins that promote the unwinding of the double helix and serve as the primer for a specific DNA polymerase. However, it was later observed that the *S. clavuligerus* linear plasmid pSCL can replicate as a circular DNA molecule when the telomeres are removed and the ends are ligated (176). This observation was further confirmed by the de-

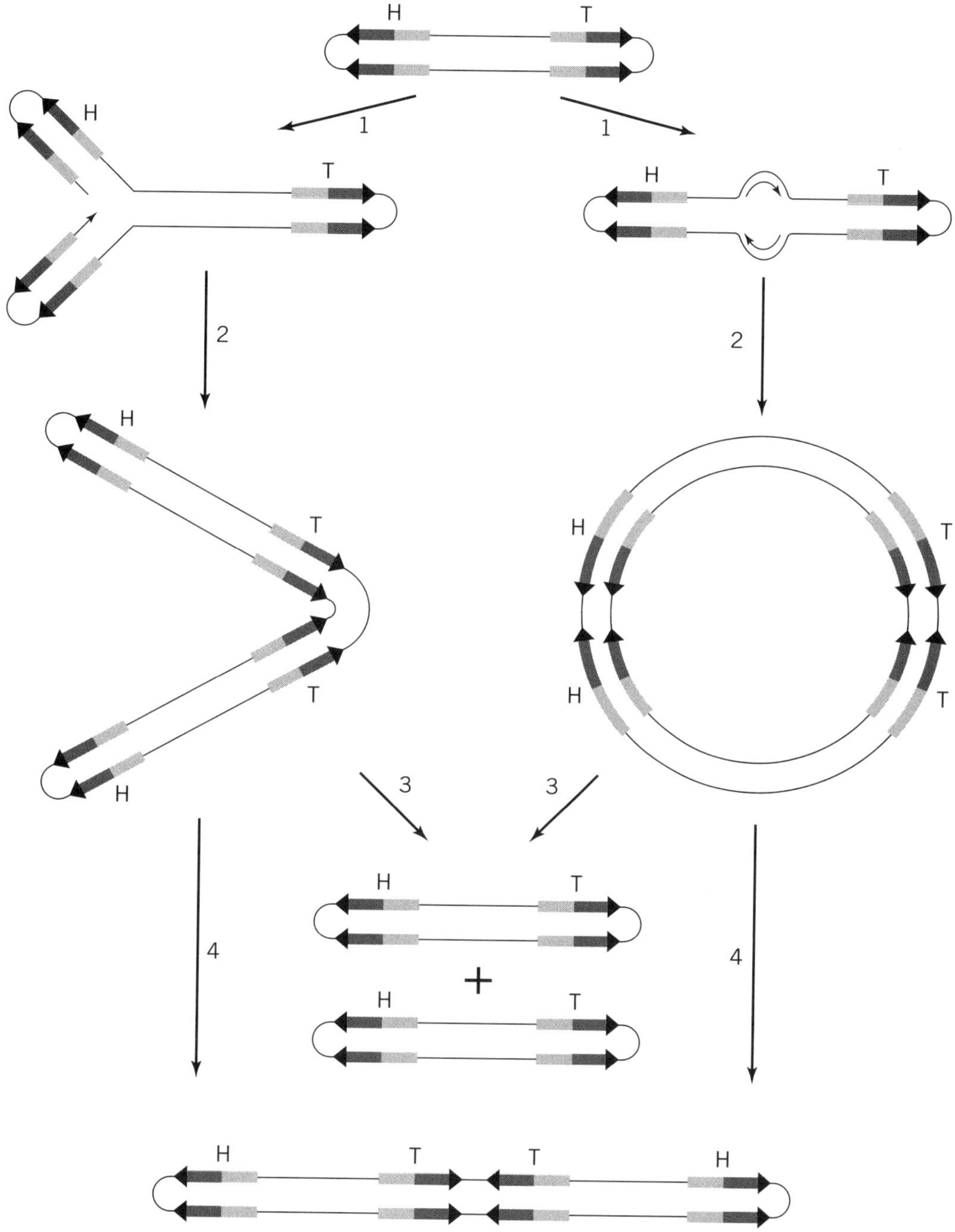

Figure 10. Model showing an alternative replication mechanism and the formation of tail (T)-to-tail dimers of linear plasmids in *Borrelia*. The model depicted in the left half starts with the cleavage of the head (H) telomere and the initiation of DNA replication (step 1). The replication of the entire linear replicon results in a tail-to-tail dimer (step 2) that can result in either two copies of the original monomers (step 3), after resolution of the tail telomeres, or remain as a tail-to-tail dimer due to a failure in the plasmid segregation mechanism (step 4). The model shown in the right half predicts the initiation of DNA replication from an internal origin (step 1) that results in the formation of a circular intermediate (step 2). Segregation of this intermediate by independent DNA cleavage events at each telomere junction leads to the generation of two copies of the original linear replicon (step 3). Conversely, a failure of the cleavage of the tail telomere results in the formation of a tail-to-tail dimer (step 4). *Source*: Adapted from Ref. 178.

tailed analysis of the mechanism of replication of pSLA2, another *Streptomyces* plasmid that contains protein covalently attached to the 5′ end (175). Ligation of pSLA2 deletion derivatives lacking the telomeric termini to a selectable marker resulted in a high-copy-number circular extrachromosomal element. This observation suggested that this linear plasmid must contain an internal site capable of promoting DNA replication. Two-dimensional agarose gel electrophoresis analysis showed that this plasmid is indeed replicated primarily bidirectionally from an internal origin located near the center of the plasmid, toward the ends, rather than by full-length strand displacement initiated at the telomeres. However, pSLA2 still requires a protein-primed strand-displacing mechanism to complete the replication of a 280-nucleotide segment at both plasmid termini. It was suggested (175) that the synthesis of the 5′ terminal segment of the lagging strand of this linear plasmid uses as a template the 3′ overhang of the leading strand and is primed by the protein covalently attached to the 5′ end of the mature plasmid.

The replication mechanisms of linear plasmids in *Borrelia* are poorly understood; however, our knowledge of these novel replicons encoding essential genes has been extended by a series of very recent publications (166,177,178). The replication of some of the *Borrelia* linear plasmids involves circular intermediates. In addition, it was recently reported that the 180-kb plasmid of *Borrelia hermsii* can exist and replicate either as a linear or a monomeric circular element (166). According to these observations, it was proposed that the monomeric circular intermediate is formed by a head-to-tail junction, after a nick introduced in the inverted repeat opens the plasmid termini (Fig. 9). Because the termini are complementary, the plasmid circularizes and can replicate as any other well-characterized circular replicon. This circular intermediate is then resolved into a linear double-stranded DNA structure by a second nick within the terminal inverted repeats. Because the terminal repeats are complementary, a linear molecule should be reconstituted.

More recently, it was found that *B. burgdorferi* sensu lato contains atypical large linear plasmids ranging from 92 to 105 kb (178). These plasmids carry *p27* and *ospAB*, genes that were also detected in other isolates on the 50-kb linear plasmid pAB50. A more detailed analysis of the larger plasmids demonstrated that they are formed by tail-to-tail dimerization of pAB50. The presence of such dimers can explain the unusual plasmid variability observed among different isolates and may provide new information regarding the replication mechanism of these linear replicons. It was postulated (178) that these linear dimers are the result of failed segregation after DNA replication by a mechanism similar to that described for vaccinia viruses. In this model proposed by Marconi et al. (178) and shown in Figure 10, initiation of plasmid replication proceeds from one of the termini, after the head hairpin loop in this particular case is nicked, allowing the formation of tail-to-tail dimer intermediates. Alternatively, these dimers can arise by DNA replication initiated from an origin located within a linear monomer that results in a circular replication intermediate. In the normal replication process, the circular intermediate is resolved into two linear monomers by independent cleavage events at each telomere junction. Cleavage failure at the tail hairpin by a specific DNA cleavage system results in two monomers linked tail to tail (Fig. 10), a possibility that was confirmed experimentally by restriction analysis and Southern blot DNA hybridization experiments (178). However, this same analysis demonstrated that not all *Borrelia* large plasmids were originated via dimer formation, and alternative DNA replication systems that are still uncharacterized must exist.

Some of the proteins involved in the replication of the linear plasmids in *Borrelia* were identified by determining the complete nucleotide sequence of the 16-kb plasmid lp 16.9 isolated from *B. burgdorferi* B31 (177). This study revealed the presence of 15 open reading frames named A to O. The predicted proteins encoded by OrfM and OrfN showed homology to proteins involved in plasmid and cell replication. OrfM is homologous to MinD, a cytoplasmic membrane protein with ATPase activity required for the correct placement of the division site (179,180). OrfM also has homologies with plasmid partition proteins such as ParA or SopA (181,182), and the RepB protein encoded by the pheromone-responsive plasmid pAD1 in *Enterococcus faecalis* and involved in the control of the copy number of this extrachromosomal element (183). The predicted features and the primary sequence of the OrfN hypothetical polypeptide are similar to those of the CopB DNA binding proteins, also known as RepB or RepA2, involved in the control of plasmid copy number in Gram-negative bacteria (184). This sequence analysis also led to the hypothesis that OrfN may interact with the short repeat sequence located within the promoter region of OrfM and, thus, control the expression of the latter (177).

In summary, sequence analysis of *Borrelia* plasmids showed that the few proteins identified are similar to prokaryotic proteins involved in plasmid replication and maintenance in Gram-positive and Gram-negative bacteria. In addition, these results do not support the previous conclusion that the linear plasmids in *Borrelia* are indeed derived from animal viruses. Furthermore, it seems that circular and linear plasmids of this bacterial genus have a common origin and share DNA replication mechanisms.

BIBLIOGRAPHY

1. M.E. Tolmasky, L.A. Actis, and J.H. Crosa, in J. Lederberg ed., *Encyclopedia of Microbiology*, vol. 3, Academic Press, San Diego, 1992, pp. 431–442.
2. J. Hinnebusch and A.G. Barbour, *J. Bacteriol.* **173**, 7233–7239 (1991).
3. H. te Riele, B. Michel, and D. Ehrlich, *Proc. Natl. Acad. Sci. U.S.A.* **83**, 2541–2545 (1985).
4. A. Abeles, L. Reaves, B. Youngren-Grimes, and S. Austin, *Mol. Microbiol.* **18**, 903–912 (1995).
5. R. Brubaker, *Clin. Microbiol. Rev.* **4**, 309–324 (1991).
6. J.H. Crosa, *Microbiol. Rev.* **53**, 517–530 (1989).
7. A. Jones, E. Lai, K. Shirasu, and C. Kado, *J. Bacteriol.* **178**, 5706–5711 (1996).
8. Y. Lee, S. Jin, W. Sim, and E. Nester, *Gene* **179**, 83–88 (1996).
9. R. Hall and C. Collis, *Mol. Microbiol.* **15**, 593–600 (1995).
10. N. Kleckner, *Annu. Rev. Genet.* **15**, 341–404 (1981).
11. M. Tolmasky and J. Crosa, *Plasmid* **29**, 31–40 (1993).

12. S. Cohen, *Gene* **135**, 67–76 (1993).
13. M. Espinosa, G. del Solar, F. Rojo, and J. Alonso, *FEMS Microbiol. Lett.* **130**, 111–120 (1995).
14. M. Filutowicz, S. Dellis, I. Levchenko, M. Urh, F. Wu, and D. York, *Prog. Nucleic Acids Res. Mol. Biol.* **48**, 239–273 (1994).
15. A. Gruss and S. Ehrlich, *Microbiol. Rev.* **53**, 231–241 (1989).
16. D. Helinski, A. Toukdarian, and R. Novick, in F. Neidhardt ed., Escherichia coli *and* Salmonella, vol. 2, ASM Press, Washington, D.C., 1996, pp. 2295–2324.
17. B. Kittel and D. Helinski, in D. Clewel ed., *Bacterial Conjugation*, Plenum Press, New York, (1993).
18. U. Kues and U. Stahl, *Microbiol. Rev.* **53**, 491–516 (1989).
19. D. Manen and L. Caro, *Mol. Microbiol.* **5**, 233–237 (1991).
20. K. Nordstrom, *Cell* **63**, 1121–1124 (1990).
21. R. Novick, *Annu. Rev. Microbiol.* **43**, 537–565 (1989).
22. J. Viret, A. Bravo, and J. Alonso, *Microbiol. Rev.* **55**, 675–683 (1991).
23. E. Wagner and R. Simons, *Annu. Rev. Microbiol.* **48**, 713–748 (1994).
24. D. Astill, P. Manning, and M. Heuzenroeder, *Plasmid* **30**, 258–267 (1993).
25. M. Bagdasarian, and M. Bagdasarian, in P. Gerhardt ed., *Methods for General and Molecular Bacteriology*, American Society for Microbiology, Washington, D.C., (1993), pp. 406–417.
26. K. Dery, R. Chavideh, V. Waters, R. Chamorro, L. Tolmasky, and M.E. Tolmasky, *Plasmid*, **38**, 97–105 (1997).
27. J. Fu, H. Chang, Y. chen, Y. Chang, and S. Liu, *Plasmid* **34**, 75–84 (1995).
28. N. Nomura and Y. Murooka, *J. Ferment. Bioeng.* **78**, 250–254 (1994).
29. R. Old and S. Primrose, Principles of Gene Manipulation, Blackwell Scientific, Oxford, 1985.
30. J. Sambrook, E. Fritsch, and T. Maniatis, *Molecular Cloning: A Laboratory Manual*, 2nd ed., Cold Spring Harbor Laboratory, Cold Spring Harbor, N.Y., 1989.
31. G. Cesareni, C. Helmer, and L. Castagnoli, *Trends Genet.* **7**, 230–235 (1991).
32. Y. Eguchi, T. Itoh, and J. Tomizawa, *Annu. Rev. Biochem.* **60**, 631–652 (1991).
33. E. Gerhardt, H. Wagner, and R. Simmons, *Annu. Rev. Microbiol.* **48**, 713–742 (1994).
34. S. Merlin and B. Polisky, *J. Mol. Biol.* **248**, 211–219 (1995).
35. K. Nordstrom and E. Wagner, *Trends Biochem. Sci.* **19**, 294–300 (1994).
36. B. Polisky *Cell* **55**, 929–932 (1988).
37. J. Inselburg, *Proc. Natl. Acad. Sci. U.S.A.* **71**, 2256–2259 (1974).
38. J. Tomizawa, M. Ohmori, and R. Bird, *Proc. Natl. Acad. Sci. U.S.A.* **74**, 1865–1869 (1977).
39. T. Itoh and J. Tomizawa, *Cold Spring Harbor Symp. Quant. Biol.* **43**, 409–418 (1978).
40. T. Itoh and J. Tomizawa, *Proc. Natl. Acad. Sci. U.S.A.* **77**, 2450–2454 (1980).
41. T. Itoh and J. Tomizawa, *Nucleic Acids Res.* **10**, 5949–5965 (1982).
42. J. Tomizawa, T. Itoh, G. Selzer, and T. Som, *Proc. Natl. Acad. Sci. U.S.A.* **78**, 1421–1425 (1981).
43. L. Castagnoli, R. Lacatena, and G. Cesareni, *Nucleic Acids Res.* **13**, 5353–5367 (1985).
44. H. Masukata and J. Tomizawa, *Cell* **44**, 125–136 (1986).
45. H. Masukata and J. Tomizawa, *Cell* **36**, 513–522 (1984).
46. H. Masukata and J. Tomizawa, *Cell* **62**, 331–338 (1990).
47. T. Fitzwater, Y. Yang, X. Zhang, and B. Polisky, *J. Mol. Biol.* **226**, 997–1008 (1992).
48. G. Selzer, and J. Tomizawa, *Proc. Natl. Acad. Sci. U.S.A.* **79**, 7082–7086 (1982).
49. S. Dasgupta, H. Masuakata, and J. Tomizawa, *Cell* **51**, 1113–1122 (1987).
50. H. Masukata, S. DasGupta, and J. Tomizawa, *Cell* **51**, 1123–1130 (1987).
51. H. Ohmori, Y. Murakami, and T. Nagata, *J. Mol. Biol.* **198**, 223–235 (1987).
52. A. Kornberg and T. Baker, *DNA Replication*, 2nd ed, W.H. Freeman, New York, 1991.
53. D. Kingsbury and D. Helinski, *Biochem. Biophys. Res. Commun.* **41**, 1538–1544 (1970).
54. S. Naito and H. Uchida, *J. Bacteriol.* **166**, 143–147 (1986).
55. R. Lacatena and G. Cesareni, *Nature* **294**, 623–626 (1981).
56. J. Tomizawa, *Cell* **47**, 89–97 (1986).
57. J. Tomizawa and T. Itoh, *Proc. Natl. Acad. Sci. U.S.A.* **78**, 6096–6100 (1981).
58. Y. Eguchi and J. Tomizawa, *Cell* **60**, 199–209 (1990).
59. Y. Eguchi, and J. Tomizawa, *J. Mol. Biol.* **220**, 831–842 (1991).
60. J. Tomizawa, *Cell* **40**, 527–535 (1985).
61. J. Tomizawa, *J. Mol. Biol.* **212**, 683–694 (1990).
62. J. Tomizawa and T. Som, *Cell* **38**, 871–878 (1984).
63. J. Tomizawa, *Cell* **38**, 861–870 (1984).
64. E. Wong, and B. Polisky, *Cell* **42**, 959–966 (1985).
65. S. Lin-Chao and H. Bremer, *Bacteriol.* **169**, 1217–1222 (1987).
66. L. He, F. Soderbom, E. Wagner, U. Binnie, N. Binns, and M. Masters, *Mol. Microbiol.* **9**, 1131–1142 (1993).
67. F. Xu, S. Lin-Chao, and S. Cohen, *Proc. Natl. Acad. Sci. U.S.A.* **90**, 6756–6760 (1993).
68. T. Tomcsanyi and D. Apirion, *J. Mol. Biol.* **185**, 713–720 (1985).
69. G. Cesareni, M. Cornelissen, M. Lacatena, and L. Castagnoli, *EMBO J.* **3**, 1365–1369 (1984).
70. M.I. Lacatena, D. Banner, L. Castagnoli, and G. Cesareni, *Cell* **37**, 1009–1014 (1984).
71. J. Tomizawa, *J. Mol. Biol.* **212**, 695–708 (1990).
72. A. Twigg and D. Sherrat, *Nature* **283**, 216–218 (1980).
73. R. Novick, *Microbiol. Rev.* **51**, 381–395 (1987).
74. N. Datta and P. Kontomichalou, *Nature (Lond.)* **208**, 239–241 (1965).
75. P. Kontomichalou, M. Mitani, and R.C. Clowes, *J. Bacteriol.* **104**, 34–44 (1970).
76. D. Bastia and J. Germino, *Cell* **32**, 131–140 (1983).
77. J. Germino and D. Bastia, *Proc. Natl. Acad. Sci. U.S.A.* **81**, 4692–4696 (1984).
78. J. Germino and D. Bastia, *Cell* **32**, 131–140 (1983).
79. J. Germino, J.G. Gray, H. Charbonneau, T. Vanaman, and D. Bastia, *Proc. Natl. Acad. Sci. U.S.A.* **80**, 6848–6852 (1983).
80. V. De Lorenzo and K.N. Timmis, *Methods Enzymol.* **235**, 386–405 (1994).
81. V.L. Miller and J.J. Mekalanos, *J. Bacteriol.* **170**, 2575–2583 (1988).

82. D. Bramhill and A. Korngerg, *Cell* **54**, 915–918 (1988).
83. T. Morotsu, K. Matsubara, H. Sugusaki, and M. Takanami, *Gene* **15**, 257–271 (1981).
84. C. Vocke and D. Bastia, *Proc. Natl. Acad. Sci. U.S.A.* **80**, 6557–6561 (1983).
85. A.L. Abeles, K.M. Snyder, and D.K. Chattoraj, *J. Mol. Biol.* **173**, 307–324 (1984).
86. C. Bidinost, J.H. Crosa, and L.A. Actis, *Plasmid* **31**, 242–250 (1994).
87. Y. Kamio, Y. Itoh, and Y. Terawaki, *J. Bacteriol.* **170**, 4411–4414 (1988).
88. A.E. Gammie and J.H. Crosa, *Mol. Microbiol.* **5**, 3015–3023 (1991).
89. B.L. Kittell and D.R. Helinski, *Proc. Natl. Acad. Sci. U.S.A.* **88**, 1389–1393 (1991).
90. J.H. Crosa, L. Luttropp, and S. Falkow, *J. Bacteriol.* **126**, 454–466 (1976).
91. J.H. Crosa, *J. Biol. Chem.* **255**, 11075–11077 (1980).
92. T. Horiuchi and M. Hidaka, *Cell* **54**, 515–523 (1988).
93. C. Thomas, D. Balson, and W. Shaw, *J. Biol. Chem.* **265**, 5519–5530 (1990).
94. I. Patel and D. Bastia, *Cell* **47**, 785–792 (1986).
95. I. Patel and D. Bastia, *Cell* **51**, 455–462 (1987).
96. P.R. Sista, C.A. Hutchinson III, and D. Bastia, *Genes Dev.* **5**, 74–82 (1991).
97. R. Kolter, M. Inuzuka, and D.R. Helinski, *Cell* **15**, 1199–1208 (1978).
98. J. Germino and D. Bastia, *Proc. Natl. Acad. Sci. U.S.A.* **79**, 5475–5479 (1982).
99. A. Shafferman, R. Kolter, D.M. Stalker, and D.R. Helinski, *J. Mol. Biol.* **161**, 57–76 (1982).
100. M. Inuzuka and D.R. Helinski, *Biochemistry* **17**, 2567–2573 (1978).
101. A. Shafferman and D.R. Helinski, *J. Biol. Chem.* **258**, 4083–4090 (1983).
102. D.M. Stalker, R. Kolter, and D.R. Helinski, *J. Mol. Biol.* **161**, 33–43 (1982).
103. J.H. Crosa, L.K. Luttropp, and S. Falkow, *J. Mol. Biol.* **124**, 443–468 (1978).
104. W.L. Kelly, I. Patel, and D. Bastia, *Proc. Natl. Acad. Sci. U.S.A.* **89**, 5078–5082 (1992).
105. A. Miron, S. Mukherjee, and D. Bastia, *EMBO J.* **11**, 1205–1216 (1992).
106. F. Wu, I. Goldberg, and M. Filutowicz, *Nucleic Acids Res.* **20**, 811–817 (1992).
107. F. Wu, I. Levchenko, and M. Filutowicz, *J. Bacteriol.* **177**, 6338–6345 (1995).
108. P.F. Jemilohun, C.W. Clark, and E.R. Archibold, *Biochem. Biophys. Res. Commun.* **221**, 186–192 (1996).
109. F. Wu, J. Wu, J. Ehley, and M. Filutowics, *J. Bacteriol.* **178**, 4965–4974 (1996).
110. A. Shafferman, Y. Flashner, I. Hertman, and M. Lion, *Gen. Genet.* **208**, 263–270 (1987).
111. P.V.A.L. Ratnakar, B.K. Mohanty, M. Lobert, and D. Bastia, *Proc. Natl. Acad. Sci. U.S.A.* **93**, 5522–5526 (1996).
112. J.H. Crosa, L.K. Luttropp, and S. Falkow, *Cold Spring Harbor Symp. Quant. Biol.* **43**, 111–120 (1979).
113. P. Mukhopadhyay, M. Filutowicz, and D.R. Helinski, *J. Biol. Chem.* **261**, 9534–9539 (1986).
114. M. Shon, J. Germino, D. Bastia, *J. Biol. Chem.* **257**, 13823–13827 (1982).
115. R. Kolter and D.R. Helinski, *J. Mol. Biol.* **161**, 45–56 (1982).
116. M. Filutowicz, E. Uhlenhopp, and D.R. Helinski, *J. Mol. Biol.* **187**, 225–239 (1985).
117. J. Germino and D. Bastia, *Cell* **34**, 125–134 (1983).
118. W. Kelly and D. Bastia, *Proc. Natl. Acad. Sci. U.S.A.* **82**, 2574–2578 (1985).
119. S. Mukherjee, H. Ericson, and D. Bastia, *Cell* **52**, 375–383 (1988).
120. M. Filutowicz, M.J. McEachern, and D.R. Helinski, *Proc. Natl. Acad. Sci. U.S.A.* **83**, 9645–9649 (1986).
121. S. Mukherjee, I. Patel, and D. Bastia, *Cell* **43**, 189–197 (1985).
122. M. Filutowicz, G. Davis, A. Greener, and D.R. Helinski, *Nucleic Acids Res.* **13**, 103–114 (1985).
123. D.M. Stalker, M. Filitowicz, and D.R. Helinski, *Proc. Natl. Acad. Sci.* **80**, 5500–5504 (1983).
124. M.J. McEachern, M.A. Bott, P.A. Tooker, and D.R. Helinski, *Proc. Natl. Acad. Sci. U.S.A.* **86**, 7942–7946 (1989).
125. A. Greener, M.S. Filutowicz, M.J. McEachern, and D.R. Helinski, *Mol. Gen. Genet.* **224**, 24–32 (1990).
126. Y. Flashner, J. Shlomai, and A. Shafferman, *Mol. Microbiol.* **19**, 986–996 (1996).
127. G. del Solar, M. Moscoso, and M. Espinosa, *Mol. Microbiol.* **8**, 789–796 (1993).
128. D. Galli and D. Leblanc, *J. Bacteriol.* **177**, 4474–4480 (1995).
129. A. Gielow, L. Diederich, and W. Messer, *J. Bacteriol.* **173**, 73–79 (1991).
130. H. Kleanthous, C. Clayton, and S. Tabaqchali, *Mol. Microbiol.* **5**, 2377–2389 (1991).
131. H. Yasukawa, T. Hase, A. Sakai, and Y. Masamune, *Proc. Natl. Acad. Sci. U.S.A.* **88**, 10282–10286 (1991).
132. N. Zhang and J. Brooker, *Plasmid* **29**, 125–134 (1993).
133. J. Alonso and M. Espinosa, in K. Hardy ed., *Plasmids: A Practical Approach*, 2nd ed., vol. 138, IRL Press, New York, 1993, pp. 39–63.
134. P. Baas and H. Jansz, *Curr. Topics Microbiol. Immunol.* **136**, 31–70 (1988).
135. A. Rasooly, P. Wang, and R. Novick, *EMBO J.* **13**, 5245–5251 (1994).
136. F. Wang, S. Projan, V. Herniquez, and R. Novick, *EMBO J.* **12**, 45–52 (1993).
137. R. Koepsel, R. Murray, W. Rosenblum, and S. Khan, *Proc. Natl. Acad. Sci. U.S.A.* **82**, 6845–6849 (1985).
138. A. Rasooly and R. Novick, *Science* **262**, 1048–1050 (1993).
139. F. Wang, S. Projan, V. Herniquez, and R. Novick, *J. Mol. Biol.* **223**, 145–158 (1992).
140. M. Gennaro and R. Novick, *J. Bacteriol.* **168**, 160–166 (1986).
141. V. Henriquez, V. Milisavljevic, J. Kahn, and M. Gennaro, *Gene* **134**, 93–98 (1993).
142. Q. Zhang, S. Soares de Oliveira, R. Colangeli, and M. Gennaro, *J. Bacteriol.* **179**, 684–688 (1997).
143. R. Koepsel and S. Khan, *Science* **233**, 1316–1318 (1986).
144. M. Moscoso, G. del Solar, and M. Espinosa, *J. Biol. Chem.* **270**, 3772–3779 (1995).
145. S. Iordanescu, *J. Bacteriol.* **175**, 3916–3917 (1993).
146. S. Iordanescu, *J. Mol. Biol.* **221**, 1183–1189 (1991).
147. A. Rasooly, S. Projan, and R. Novick, *J. Bacteriol.* **176**, 2450–2453 (1994).
148. A. Gruss, H. Ross, and R. Novick, *Proc. Natl. Acad. Sci. U.S.A.* **84**, 2165–2169 (1987).

149. P. Birch and S. Khan, *Proc. Nat. Acad. Sci. U.S.A.* **89**, 290–294 (1992).
150. S. Iordanescu and S. Projan, *J. Bacteriology* **170**, 3427–3434 (1988).
151. I. Carleton, S. Projan, S. Highlander, and R. Novick, *EMBO J.* **3**, 2407–2414 (1984).
152. R. Novick, S. Projan, S. Kumar, C. Carleton, S. Gruss, S. Highlander, and J. Kornblum, in D. Helinski, S. Cohen, D. Clewell, D. Jackson, and A. Hollaender ed., *Plasmids in Bacteria*, Plenum Press, New York, 1985, pp. 299–320.
153. R. Landick, C. Turnbough, and C. Yanofsky, in F. Neidhardt, R. Curtis III, J. Ingraham, C. Lin, K. Low, B. Magasanik, W. Reznikoff, M. Riley, M. Schachter, and H. Umbarger eds., Escherichia coli *and* Salmonella: *Cellular and Molecular Biology*, 2nd ed., American Society for Microbiology, Washington, D.C., 1996, pp. 1263–1286.
154. R. Novick, S. Iordanescu, S. Projan, J. Kornblum, and I. Edelman, *Cell* **59**, 395–404 (1989).
155. R. Novick, G. Adler, S. Projan, S. Carleton, S. Highlander, A. Gruss, S. Khan, and S. Iordanescu, *EMBO J.* **10**, 2399–2405 (1984).
156. C. Baril, C. Ricahud, G. Baranton, and I. Saint Girons, *Res. Micriobiol.* **140**, 507–516 (1989).
157. M.S. Ferdows and A.G. Barbour, *Proc. Nat. Acad. Sci. U.S.A.* **86**, 5969–5973 (1989).
158. J. Hinnebusch and H. Tilly, *Mol. Microbiol.* **10**, 917–922 (1993).
159. T. Hayakawa, T. Tanaka, K. Sakaguchi, N. Otake, and H. Yonehara, *J. Gen. Appl. Microbiol.* **25**, 255–260 (1979).
160. N. Margolis, D. Hogan, K. Tilly, and P.A. Rosa, *Bacteriol.* **176**, 6427–6432 (1994).
161. A.G. Barbour, *Trends Microbiol.* **1**, 236–239 (1993).
162. A.G. Barbour and C.F. Garon, *Science* **237**, 409–411 (1987).
163. T. Kitten and A.G. Barbour, *Proc. Natl. Acad. Sci. U.S.A.* **87**, 6077–6081 (1990).
164. R.H. Plasterk, M.I. Simon, and A.G. Barbour, *Nature (Lond.)* **318**, 257–263 (1985).
165. J. Hinnebusch and A.G. Barbour, *Bacteriol.* **174**, 5251–5257 (1992).
166. M.S. Ferdows, P. Serwer, G.A. Griess, S.J. Norris, and A.G. Barbour, *J. Bacteriol.* **178**, 793–800 (1996).
167. F. Meinhardt, F. Kempken, J. Kamper, and K. Esser, *Curr. Genet.* **17**, 89–95 (1990).
168. M. Salas, *Annu. Rev. Biochem.* **60**, 39–71 (1991).
169. H. Hirochika, K. Nakamura, and K. Sakaguchi, *EMBO J.* **3**, 761–766 (1984).
170. K. Sakaguchi, H. Hirochika, and N. Gunge, in S.N.C.D.R. Helinski, D.B. Clewell, D.A. Jackson, A. Hollaender eds., *Plasmids in Bacteria*, vol. 30, Plenum Press, New York, 1985, pp. 433–451.
171. J. Hinnebusch, S. Bergstrom, and A.G. Barbour, *Mol. Microbiol.* **4**, 811–820 (1990).
172. B.M. Baroudy, S. Venkatesan, and B. Moss, *Cell* **28**, 315–324 (1982).
173. N. Dinouel, R. Drissi, I. Miyakawa, F. Sor, S. Rousset, and H. Fukuhara, *Mol. Cell. Biol.* **13**, 2315–2323 (1993).
174. A. Gonzalez, A. Talavera, J.M. Almendral, and E. Vinuela, *Nucleic Acid Res.* **14**, 6835–6844 (1986).
175. P.-C. Chang, and S.N. Cohen, *Science* **265**, 952–954 (1994).
176. D. Shiffman and S.N. Cohen, *Proc. Natl. Acad. Sci. U.S.A.* **89**, 6129–6133 (1992).
177. A.G. Barbour, C.J. Crater, V. Bundoc, and J. Hinnebusch, *J. Bacteriol.* **178**, 6635–6639 (1996).
178. R.T. Marconi, S. Casjens, U.G. Munderloloh, and D.S. Samuels, *J. Bacteriol.* **178**, 3357–3361 (1996).
179. P.A.J. de Boer, R.E. Crossley, A.R. Hand, and L.I. Rothfield, *EMBO J.* **10**, 4371–4380 (1991).
180. A.W. Varley and G.C. Stewart, *J. Bacteriol.* **174**, 6729–6742 (1992).
181. A.L. Abeles, S.A. Friedman, and S.J. Austin, *J. Mol. Biol.* **185**, 261–272 (1985).
182. H. Mori, A. Kondo, A. Ohshima, T. Oruga, and S. Hiraga, *J. Mol. Biol.* **192**, 1–15 (1986).
183. K.E. Weaver, D.B. Clewell, and F. An, *J. Bacteriol.* **175**, 1900–1909 (1993).
184. J. Lopez, I. Andres, J.M. Ortiz, and J.C. Rodriguez, *Nucleic Acid Res.* **18**, 7177 (1990).

PLURONIC POLYOLS, CELL PROTECTION

David W. Murhammer
University of Iowa
Iowa City, Iowa

KEY WORDS

Cell culture scale-up
Pluronic polyols
Sparging protection

OUTLINE

INTRODUCTION

In recent years, it has become apparent that there is a need for animal cell cultures to produce some products (e.g., complex recombinant DNA proteins) because of the inability of other expression systems (e.g., bacteria) to properly perform post-translational modifications (e.g., glycosylation and proper folding). Unfortunately, animal cells lack a protective cell wall and thus are subject to damage from sparging and other environmental stresses. In large-scale bioreactors, however, aerating the cultures through direct sparging is the most practical method of supplying oxygen to the culture medium. Therefore, in order to realize the full potential of using animal cells to produce recombinant DNA proteins on an industrial scale, there is a critical need for large-scale sparged bioreactors capable of supporting animal cell growth. Currently, the generally accepted approach to solving this problem is to supplement the medium with a protective agent, such as the commercially available surfactant Pluronic F-68.

PROPERTIES OF PLURONIC POLYOLS

A wide range of Pluronic polyols are available that are capable of protecting animal cells in suspension cultures from the adverse effects of sparging. Of these, Pluronic F-68 is by far the most widely used. Properties of Pluronic polyols that are relevant to the selection of a protective agent are discussed in this section.

Synthesis and Structure

Pluronic polyols, commercially available from the BASF Corporation (Wyandotte, Mich.), are block copolymer nonionic surfactants consisting of a center block of poly(oxypropylene) (hydrophobe) and end blocks of poly(oxyethylene) (hydrophile). They are synthesized by the two-step process shown in Figure 1 (1). In the first step, a poly(oxypropylene) block is synthesized by the addition of propylene oxide to propylene glycol. The poly(oxyethylene) blocks are then added to both ends by reacting the resulting poly(oxypropylene) with ethylene oxide. The oxyalkylation steps are carried out in the presence of an alkaline catalyst (e.g., potassium hydroxide) that is subsequently neutralized and removed from the final product. The properties of the Pluronic polyols are altered by varying the relative sizes of the poly(oxypropylene) and poly(oxyethylene) blocks (note that the sizes of the two poly(oxyethylene) blocks are statistically identical). The properties of a variety of Pluronic polyols are given in Table 1.

Nomenclature

The letters L, P, and F in the Pluronic polyol nomenclature represent liquid, paste, and flake, respectively (Table 1) (1). The first two digits of the three-digit polyol number, or the first digit of a two-digit polyol number, are indicative of the hydrophobe [poly(oxypropylene)] molecular weight, for example, 6, 8, and 12 represent average hydrophobe molecular weights of 1,750, 2,250, and 4,000, respectively. The last digit in the polyol number represents approximately one-tenth the percentage of hydrophile [poly(oxyethylene)] in the molecule. For example, Pluronic F-68 is a flake (i.e., a white powder), has a hydrophobe molecular weight of approximately 1,750, and contains approximately 80% hydrophile.

Impurities

Commercial-grade Pluronic polyols have been demonstrated to contain low molecular weight impurities, including aldehydes, formic acid, and acetic acid (2,3). In addition, peroxide derivatives of the Pluronic polyols can be formed during steam sterilization. The formation of the peroxide derivatives can be prevented by filter sterilizing the Pluronic polyol solutions instead of using steam sterilization. The low molecular weight impurities can be removed from aqueous solutions of Pluronic polyols by silica-gel column absorption (2).

Hydrophilic–Lipophilic Balance

The hydrophilic–lipophilic balance HLB is an empirical measure of the emulsifying capabilities of surfactant mol-

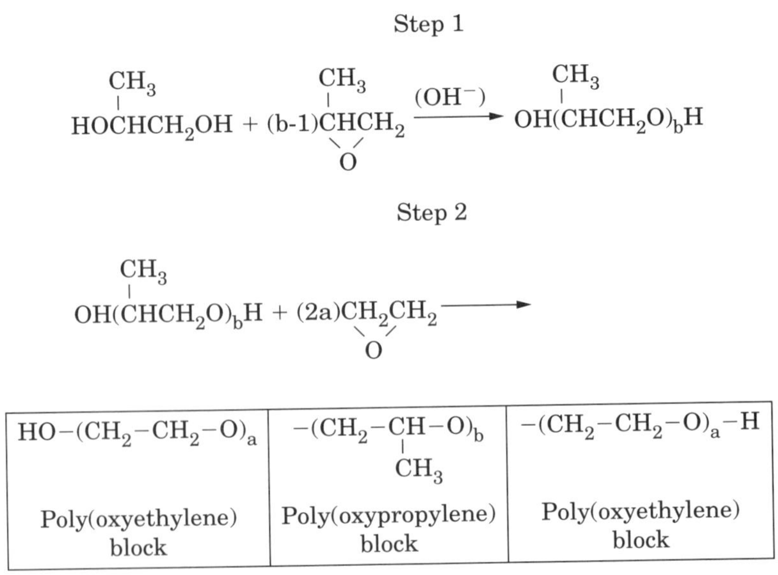

Figure 1. Pluronic polyol synthesis.

Table 1. Pluronic Polyols Tested for Their Effect on *Spodoptera frugiperda* Sf-9 Insect Cell Growth in the Absence and Presence of Sparging

Pluronic polyol	Average MW	wt % hydrophobe	HLB range	Effect on cell growth in the absence of sparging	Protects from sparging?
L-61	2,000	90	1–7	Cell lysis	N/A
L-121	4,400	90	1–7	Cell lysis	N/A
P-103	4,950	70	7–12	Cell lysis	N/A
P-123	5,750	70	7–12	Cell lysis	N/A
P-84	4,200	60	12–18	Cell lysis	N/A
P-104	5,900	60	12–18	Cell lysis	N/A
P-65	1,900	50	12–18	Inhibition	N/A
P-105	6,600	50	12–18	Inhibition	N/A
L-35	1,900	50	18–23	None	Yes
F-127	12,600	30	18–23	None	Yes
F-38	4,700	20	>24	None	Yes
F-68	8,400	20	>24	None	Yes
F-108	14,600	20	>24	None	Yes

Source: Adapted from Murhammer and Goochee (6).

ecules (4). Lower values of the HLB are representative of molecules with a higher degree of hydrophobicity. In general, molecules with high HLB values tend to be water soluble, and those with low HLB values tend to be oil soluble. The HLB values given in Table 1 are given in terms of ranges to indicate that the effective HLB of the Pluronic polyols varies with the emulsion system (5). From the results given in Table 1, it is clear that Pluronic polyols with low HLB values are unsuitable for use as protective agents because medium supplemented with 0.2% (w/v) of these Pluronic polyols either lysed *Spodoptera frugiperda* Sf-9 cells or significantly inhibited their growth in the absence of sparging, specifically in 50-mL spinner flasks oxygenation by surface diffusion (6). It was further demonstrated that at higher HLB values that Sf-9 cell growth was not inhibited in the absence of sparging and that protection from the adverse effects of sparging was provided in airlift bioreactor studies. Thus, the key to finding a Pluronic polyol that served as a protective agent for the Sf-9 cells in a sparged environment corresponded to finding one that did not inhibit cell growth in the absence of sparging. It should be noted that the results may be different with other cell lines.

Foaming Characteristics

Excessive foaming is generally undesirable in animal cell cultures. Therefore, if Pluronic polyols are used that exhibit unacceptable levels of foaming, then an antifoam should also be added to the medium (6). Of the Pluronic polyols given in Table 1, F-68, P-105, and P-65 are the best for foam generation (5). Therefore, a number of other Pluronic polyols may be preferable if foaming is problematic (e.g., L-35 or F-127).

Surface Tension

Because the Pluronic polyols are surfactants, it would be expected that they would reduce the air–water static surface tension. Indeed, according to BASF literature (5), the addition of 0.1% (w/v) Pluronic polyols to aqueous solutions reduces the static surface tension from 72 dyne/cm for pure water to a range of 33 (L-121) to 52 dyn/cm (F-38). In addition, the static surface tension of a 0.1% (w/v) aqueous solution of Pluronic F-68 is 50 dyn/cm. Michaels et al. (7), however, have given a convincing argument that the dynamic surface tension is a more suitable measure for sparged systems since interfacial properties are sensitive to the physical state of the interface. They found that the reduced dynamic surface tension afforded by Pluronic F-68 addition correlated well with reduced cell–bubble interactions and thereby provided cell protection.

Oxygen Transport

Zhang et al. (8) found that the addition of Pluronic F-68 to growth medium had minimal effect on K_La values for large bubbles (~5 mm diameter), but significantly enhanced K_La for intermediate (~1 mm diameter) and micron-sized (~100 μm diameter) bubbles. It is generally accepted that surfactants can affect oxygen transport from bubbles in two different ways: affecting the diffusional resistance and affecting the interfacial area. The first affects the K_L, and the second affects the a in K_La. The addition of surfactants usually results in a decreased K_L (9) and an increased a (10). Consistent with this expected behavior, Zhang et al. (8) found that there was a significant reduction in the Sauter mean bubble size for the intermediate and micron-sized bubbles, thereby resulting in an increased a. Pluronic F-68 addition, however, had minimal effect on the size of the large bubbles. No direct measurements directly related to the K_L term were taken in this study, although the results clearly demonstrate that the presence of the Pluronic F-68 resulted in a greater increase in a than decrease in K_L.

PROTECTIVE EFFECTS OF PLURONIC POLYOLS

It has been clearly demonstrated that Pluronic polyols can protect insect and mammalian cells from the adverse effects of sparging. There is also some evidence, although not as conclusive, that Pluronic polyols can provide protection from shear forces resulting from agitation. Evidence supporting these protective effects and the corresponding mechanisms are discussed in this section.

Sparging Damage

In 1968, Kilburn and Webb (11) were the first to report the use of Pluronic polyols to protect animal cells from sparging. Specifically, they found that supplementing the medium with 0.02% (w/v) Pluronic F-68 protected mouse LS cells from the adverse effects of sparging in short-term studies. They did not, however, show any results from long-term tests, such as an entire bioreactor run. No comprehensive studies regarding the use of Pluronic polyols to protect animal cells from the adverse effects of sparging were conducted until the late 1980s, when Handa-Corrigan et al. (12,13) and Murhammer and Goochee (6,14,15) demonstrated the ability of Pluronic F-68 to protect mammalian and insect cells, respectively. Handa-Corrigan et al. (12,13) demonstrated that the survival of a variety of mammalian cells (hybridomas, NS1 myeloma, BHK-21, and lymphoblastoid [RAJI]) in the presence of sparging depends on cell type and bubble size and frequency (i.e., small bubbles and high flow rates are more detrimental). They further demonstrated that the protective effect of Pluronic F-68 was concentration dependent, with 0.1% (w/v) being sufficient to protect cells under the most severe conditions reported. Murhammer and Goochee (14) investigated the protective effect of Pluronic F-68 in both airlift and sparged agitated bioreactors. It was found that supplementing the medium with 0.1% (w/v) Pluronic F-68 resulted in a major protective effect and that 0.2% (w/v) provided full protection to Sf-9 cells grown in both types of bioreactors. It was also demonstrated that recombinant protein synthesis (produced after infection of the cells with a recombinant baculovirus) in the airlift bioreactor was significantly reduced compared to the unsparged control even after increasing the Pluronic F-68 concentration to 0.5% (w/v). This was hypothesized to result from the increased susceptibility of virally infected cells to lysis in a sparged environment. Murhammer and Goochee (6) further demonstrated that many other Pluronic polyols could serve as protective agents in airlift bioreactors (Table 1).

After the development of conclusive evidence regarding the ability of Pluronic F-68 to protect mammalian and insect cell cultures from the adverse effects of sparging, the research focus in this field shifted to the mechanisms of this protective effect and of cell damage in the absence of protective agents. In regards to bubbles in sparged bioreactors, there are three regions in which cell damage could potentially occur: (*1*) the bubble formation region at the gas distributor, (*2*) the rising bubble region between the gas distributor and medium surface, and (*3*) the bursting bubble region at the medium surface. The results of Handa-Corrigan et al. (13) and Tramper et al. (16) clearly demonstrated that the primary source of cell damage is not in the region of rising bubbles. Specifically, they demonstrated that cell damage decreased as the height of the bioreactors increased. In addition, Michaels et al. (17) demonstrated that cell-to-bubble interactions in the bulk of the bioreactor are not responsible for cell damage in studies in which the headspace above the medium where bubbles can burst was eliminated. Murhammer and Goochee (15) demonstrated that it is possible to have cell damage in the sparger region and that this damage correlates with a high pressure drop across the sparger. In most practical applications (i.e., when the sparger is properly designed), however, significant cell damage only occurs at the medium surface in the bursting bubble region, as first proposed by Handa et al. (12). It has also been demonstrated that cells tend to adhere to air–medium interfaces and that fatal damage occurs to these attached cells as the bubbles burst at the medium surface, that is, the energy released by the bursting bubbles results in extensive cell lysis (18–21).

Michaels et al. (22) suggested that the protective effect of Pluronic polyols could result either from a biological mechanism (i.e., through changes in the cells' ability to resist shear) or from a physical mechanism (i.e., through changes in the medium properties that affect the level or frequency of forces experienced by cells). They concluded that the protective effect of Pluronic F-68 in sparged bioreactors is caused by a physical effect. Consistent with this hypothesis, it was later demonstrated that the protective effect of Pluronic F-68 resulted from its ability to cover the medium–bubble surface and thereby prevent cell adhesion to the bubbles (21,23). Thus, cells are not in the immediate vicinity of bursting bubbles at the medium surface.

It has also been demonstrated that Vero cells grown on microcarriers are damaged in sparged bioreactors (24). It was suggested that the addition of Pluronic F-68 to the medium may be able to protect cells grown on microcarriers from the adverse effects of sparging. Corresponding studies, however, were not conducted. This is certainly an issue that needs to be addressed in future research.

Agitation Damage

Although a few studies have reported that supplementing the medium with Pluronic F-68 can protect attachment-independent cells growing in suspension from agitation damage (25), this is generally not a relevant issue because insect (15) and mammalian (17) cells are commonly tolerant to relatively high agitation rates. For example, it was found that Sf-9 cell damage did not occur unless the agitation rate was high enough to induce bubble incorporation via cavitation or vortexing (15). Further, supplementing the medium with Pluronic F-68 protected the cells from damage caused by bubble incorporation by either of these mechanisms. It is assumed that the corresponding mechanisms of cell damage without Pluronic F-68 and protection by Pluronic F-68 are similar to those reported above for sparged bioreactors.

In spite of the lack of evidence demonstrating that Pluronic F-68 can protect cells from agitation damage, there is some evidence that Pluronic F-68 can make cells more resistant to shear damage. For example, Goldblum et al. (26) demonstrated that the addition of Pluronic F-68 (0.2 and 0.3% [w/v]) to the medium protected Sf-9 and *Tricho-*

plusia ni Tn-368 insect cells from shear damage in viscometer studies in a concentration-dependent manner. In contrast to these results, Michaels et al. (22) found that Pluronic F-68 did not make CRL-8018 cells more shear tolerant. These results suggest that the ability of Pluronic F-68 to increase a cell's shear tolerance may be cell dependent. Direct interaction with the cell (e.g., incorporation of the hydrophobic portion of Pluronic F-68 with the cell membrane) is a possible mechanism through which Pluronic F-68 may increase the shear tolerance of cells. Several lines of evidence are consistent with an interaction of Pluronic F-68 with the cell membrane. First, Pluronic polyols of low HLB values lysed cells (Table 1). Second, Pluronic F-68 inhibits the uptake of trypan blue dye that would normally be taken up in the absence of Pluronic F-68 (14). Third, treating sickled erythrocytes with Pluronic F-68 reduced their rigidity (27). Fourth, Pluronic F-68 increases the fluidity of hybridoma membranes, which correlated with increased shear sensitivity (28).

APPLICATION

If attachment-independent insect or mammalian cells are to be grown in sparged bioreactors, then it is recommended that the medium be supplemented with a Pluronic polyol. Many commercially available serum-free media already contain Pluronic F-68; therefore, additional supplementation may not be necessary. If supplementation is necessary, then a 10% (w/v) aqueous solution of Pluronic F-68 (obtained from BASF Corp.) can be prepared and filter sterilized. This solution can then be used to supplement the medium at 0.1 to 0.2% (w/v) Pluronic F-68. Ready-made solutions of Pluronic F-68 are also commercially available (e.g., Sigma Chemical Company, St. Louis, Mo.). The resulting medium should first be tested in the absence of sparging (e.g., in spinner or shaker flasks) to detect any potential inhibitory effects. Note that purification of the Pluronic F-68 (as mentioned above) may be necessary to remove inhibitory impurities. The next step is to test the ability of the medium to protect cells in small-scale sparged bioreactors. Success in protecting the cells at this scale is a good indicator of similar success in larger-scale sparged bioreactors. The use of Pluronic F-68 allows sparging with micron-sized bubbles that significantly enhance culture oxygenation. If Pluronic F-68 is not successful, many other Pluronic polyols are available for testing (Table 1). Another concern is that Pluronic polyols may interact with the protein of interest and interfere with downstream purification (29). Winzerling et al. (30) have demonstrated, however, that recombinant insect transferrin can be purified from Sf-9 cell supernatant containing 0.1% (w/v) Pluronic F-68 by using high metal ion capacity gel chromatography.

BIBLIOGRAPHY

1. I.R. Schmolka, *J. Am. Oil Chemists' Soc.* **54**, 110–116 (1977).
2. P.K. Bentley, K.C. Lowe, O.L. Johnson, S.K. Sharma, C. Washington, and S.S. Davis, *J. Pharm. Pharmacol.* **40** (suppl), 143P (1988).
3. P.K. Bentley, R.M.C. Gates, K.C. Lowe, D.I. de Pomerai, and J.A.L. Walker, *Biotechnol. Lett.* **11**, 111–114 (1989).
4. P. Becher and R.L. Birkmeier, *J. Am. Oil Chem. Soc.* **41**, 169–172 (1964).
5. BASF Publication, *Pluronic and Tetronic Surfactants*, BASF, Ludwigschafen, Germany, 1987.
6. D.W. Murhammer and C.F. Goochee, *Biotechnol. Prog.* **6**, 142–148 (1990).
7. J.D. Michaels, J.E. Nowak, A.K. Mallik, K. Koczo, D.T. Wasan, and E.T. Papoutsakis, *Biotechnol. Bioeng.* **47**, 420–430 (1995).
8. S. Zhang, A. Handa-Corrigan, and R.E. Spier, *Biotechnol. Bioeng.* **40**, 252–259 1992.
9. J.D. Sheppard and D.G. Cooper, *J. Chem. Tech. Biotechnol.* **48**, 325–336 (1990).
10. J.J. Heijnen and K. Van't Riet, *Chem. Eng. J.* **28**, B21–B42 (1984).
11. D.G. Kilburn and F.C. Webb, *Biotechnol. Bioeng.* **10**, 801–814 (1968).
12. A. Handa, A.N. Emery, and R.E. Spier, *Dev. Biol. Standard.* **66**, 241–253 (1987).
13. A. Handa-Corrigan, A.N. Emery, and R.E. Spier, *Enzyme Microb. Technol.* **11**, 230–235 (1989).
14. D.W. Murhammer and C.F. Goochee, *Bio / Technology* **6**, 1411–1418 (1988).
15. D.W. Murhammer and C.F. Goochee, *Biotechnol. Prog.* **6**, 391–397 (1990).
16. J. Tramper, D. Joustra, and J.M. Vlak, in C. Webb and F. Mavituna, eds., *Plant and Animal Cells*, Ellis Horwood Ltd., Chichester, England, 1987, pp. 125–136.
17. D. Michaels, A.K. Mallik, and E.T. Papoutsakis, *Biotechnol. Bioeng.* **51**, 399–409 (1996).
18. F. Bavarian, L.S. Fan, and J.J. Chalmers, *Biotechnol. Prog.* **7**, 140–150 (1991).
19. J.J. Chalmers and F. Bavarian, *Biotechnol. Prog.* **7**, 151–158 (1991).
20. M. Garcia-Briones and J.J. Chalmers, *Ann. N.Y. Acad. Sci.* **665**, 219–229 (1992).
21. D. Chattopadhyay, J.F. Rathman, J.J. Chalmers, *Biotechnol. Bioeng.* **45**, 473–480 (1995).
22. J.D. Michaels, J.F. Petersen, L.V. McIntire, and E.T. Papoutsakis, *Biotechnol. Bioeng.* **38**, 169–180 (1991).
23. M. Jordan, H. Sucker, F. Widmer, and H.M. Eppenberger, in R.E. Spier, J.B. Griffiths, and J.B. Berthold eds., *Animal Cell Technology: Products of Today, Prospects for Tomorrow*, Butterworth-Heinemann, Oxford, U.K., 1994, pp. 302–304.
24. D.E. Martens, E.A.A. Nollen, M. Hardeveld, C.A.M. van der Velden-de Groot, C.D. de Gooijer, E.C. Beuvery, and J. Tramper, *Cytotechnology* **21**, 45–59 (1996).
25. A. Mizrahi, *J. Clin. Microbiol.* **2**, 11–13 (1975).
26. S. Goldblum, Y.-K. Bae, W.F. Hink, and J. Chalmers, *Biotechnol. Prog.* **6**, 383–390 (1990).
27. C.M. Smith, R.P. Hebbel, D.P. Tukey, C.C. Clawson, J.G. White, and G.M. Vercellotti, *Blood* **69**, 1631–1636 (1987).
28. O.T. Ramirez and R. Mutharasan, *Biotechnol. Bioeng.* **36**, 911–920 (1990).
29. N. Kioukia, A.W. Nienow, M. Al-Rubeai, and A.N. Emery, *Biotechnol. Prog.* **12**, 779–785 (1996).
30. J.J. Winzerling, D.Q.-D. Pham, S. Kunz, P. Samaraweera, J.H. Law, and J. Porath, *Protein Exp. Purif.* **7**, 137–142 (1996).

POLY(3-HYDROXYALKANOATES)

Birgit Kessler
Bernard Witholt
Institute of Biotechnology
Zürich, Switzerland

KEY WORDS

Bacteria
Bioplastic
Biopolymer
Polyhydroxyalkanoates (PHA)
Polyhydroxybutyrate (PHB)

OUTLINE

INTRODUCTION AND HISTORY

Plastics have been produced by the chemical industry since the early 1930s. They constitute a very important group of materials, because their high molecular weight and low reactivity make them especially suited for applications for which durable, inert materials are required. However, these very same properties have recently been recognized as major drawbacks of plastic materials. Whenever plastic is disposed of, it remains at the site of disposal, and its resistance to biological breakdown causes it to accumulate in the environment. The recognition of this undesired property has led to a search for new plastic materials that disappear after their disposal, the so-called biodegradable plastics. Several biological polymers that could fulfill this requirement are currently under investigation. Most of these biopolymers belong to the class of polysaccharides or polyhydroxyalkanoates (PHAs). In this review, we focus on the second class, the PHAs.

The first example of PHAs to be discovered was polyhydroxybutyrate (PHB). In 1926, Lemoigne isolated and characterized PHB from *Bacillus megaterium* (1). Since this first report, PHB accumulation has been found in many microorganisms: representatives of Gram-negative and Gram-positive species (i.e., autotrophs, heterotrophs, phototrophs, aerobes, anaerobes) and archaebacteria (as reviewed elsewhere [2–4]). Bacteria synthesize and accumulate PHAs as carbon and energy storage materials or as a sink for redundant reducing power under conditions of limiting nutrients in the presence of excess carbon source. When the supply of the limiting nutrient is restored, the PHA can be degraded by intracellular depolymerases and subsequently metabolized as a carbon and energy source.

In 1976, Imperial Chemical Industries (ICI Ltd., UK) recognized the potential applicability of PHB to replace some of the oil-derived commodity plastics. Although the bacterially produced PHB was relatively expensive compared to petrochemical plastics, it was expected that the price of crude oils would rise to high levels, making the production of PHB economically feasible. Because no rise in oil prices occurred, PHB came to be seen as a biodegradable and biocompatible plastic material. Mainly because of their high price, PHB and PHB/HV (a copolymer consisting of hydroxybutyrate and hydroxyvalerate) known under the tradename Biopol™ have only recently found their way to the market. In 1990, the German company Wella released a new shampoo packaged in bottles made of Biopol™. Early in 1996, Monsanto (USA) bought the Biopol™ business from ZENECA BioProducts, a daughter company of ICI. Monsanto is now in the process of generating transgenic plants (i.e., soybean and rapeseed) for future agricultural production of PHB and PHB/HV, which is expected to result in a significantly lower price of the final product.

The discovery of a polyester that consists mainly of hydroxyoctanoate monomers by de Smet et al. (5) in 1983 was the first example of a new group of polymers that can contain a wide variety of different monomers. Thus far, more than 90 different monomers have been found in the polymers (6). Among these are 3-hydroxy acids of 6 to 14 carbon atoms with saturated, unsaturated, straight, or branched aliphatic side chains. Furthermore, monomers with various different functional groups in the side chain, such as halogen atoms, hydroxy-, epoxy-, cyano-, carboxyl-, and esterified carboxyl groups, have been introduced into medium-chain-length (MCL) PHAs (4,6,7).

PHAs can be classified into two groups depending on the number of carbon atoms in the monomer units: short-chain-length (SCL) PHAs, which consist of 3–5 carbon atoms, and MCL PHAs, which consist of 6–14 carbon atoms (8). The general structure of PHAs is shown in Figure 1.

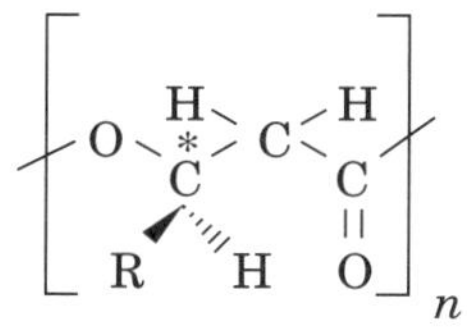

$R = CH_3\text{-}C_2H_5$ Short-chain-length PHAs
$R = C_3H_7\text{-}C_{14}H_{29}$ Medium-chain-length PHAs

Figure 1. General structure of polyhydroxyalkanoates (PHAs). PHAs are polyesters with (*R*)-3-hydroxy acids as monomer unit. The monomers of short-chain-length (SCL) PHAs consist of 3–5 carbon atoms, and those of medium-chain-length (MCL) PHAs consist of 6–14 carbon atoms. The molecular weights of these biopolymers range from 2×10^5 to 3×10^6, and the number of monomers (n) varies between 1,000 and 30,000.

The monomer units in these microbial polyesters are all in the *R*-configuration because of the stereospecificity of the biosynthetic enzymes. The molecular weights of PHAs range from 2×10^5 to 3×10^6, depending on the specific PHA, the microorganism from which the PHA was obtained, and the growth conditions. In this review, the term *PHA* refers to polyhydroxyalkanoates generally. PHB (or PHV) refers to specific SCL PHAs, and MCL PHA refers to all PHAs that contain monomers with six or more carbon atoms.

SCL PHA

PHB, the best-known of the SCL PHAs, not only appears in microorganisms as storage material but is also ubiquitous in nature. A fascinating development in recent years has been the discovery of the very wide distribution of PHB as a low molecular weight oligomer (120–200 monomers) in microorganisms, plants, and animals, including humans (9,10). In many cases this form of PHB is found as a PHB–calcium polyphosphate complex in membranes that seems to function as an ion channel through cell membranes (11,12). In *Escherichia coli*, the complex has been found to occur in large amounts in the membrane of cells made competent for genetic transformations (13–15). In addition, PHB of similar molecular weight has been found in eukaryotes, for example, in various plant tissues, in organs of animals (such as beef heart, chicken liver, porcine kidney), and in human blood plasma (16). In this review, we focus on the high molecular weight polymer that has considerable potential as an industrial biodegradable plastic.

Physiology and Biosynthesis of PHB

The biochemistry of PHB biosynthesis is very well studied. In most bacteria, such as *Alcaligenes eutrophus*, PHB is synthesized in a three-step reaction starting with acetyl-CoA when cultivated on carbohydrates, pyruvate, or acetate (Fig. 2) (17,18). Two acetyl-CoA molecules are coupled to form acetoacetyl-CoA in a condensation reaction catalyzed by 3-ketothiolase. The product is subsequently stereoselectively reduced to (*R*)-3-hydroxybutyryl-CoA in a re-

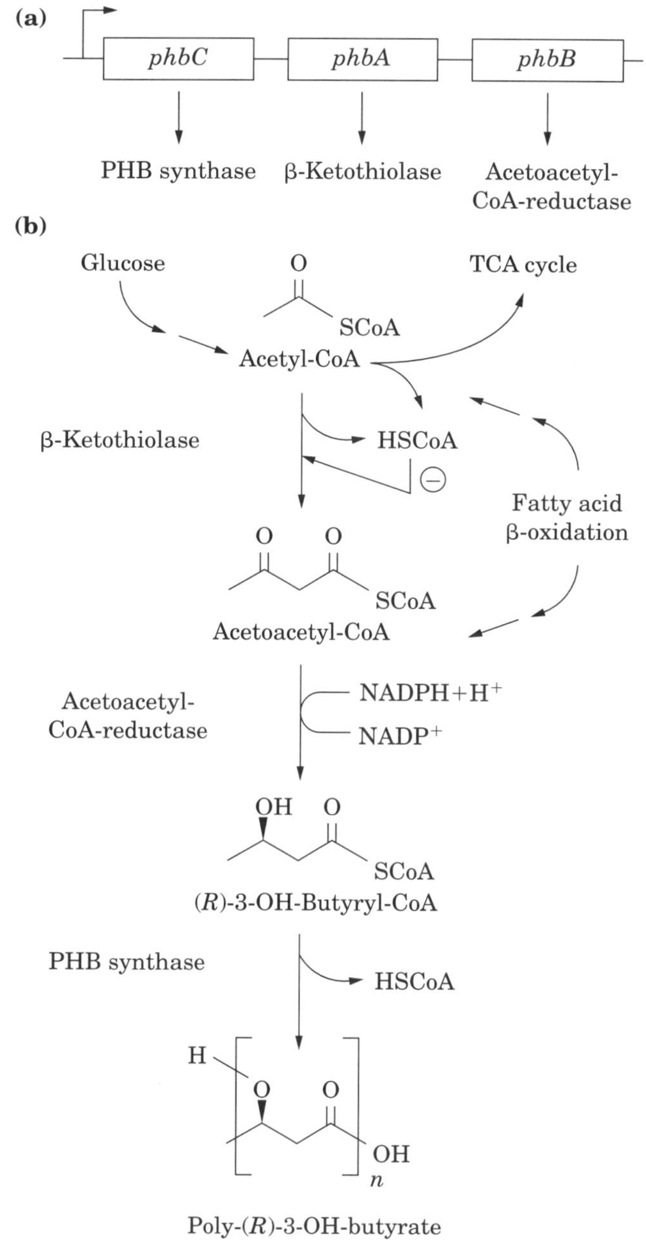

Figure 2. Polyhydroxybutyrate (PHB) synthetic genes and PHB synthesis pathway of *Alcaligenes eutrophus*. (**a**) The genes *phbA*, *phbB*, and *phbC* of *Alcaligenes eutrophus* are located in one operon and code for the β-ketothiolase, acetoacetyl-CoA-reductase, and PHB synthase, respectively. (**b**) In the three-step PHB synthesis pathway, two acetyl-CoA molecules are coupled to form acetoacetyl-CoA in a condensation reaction catalyzed by β-ketothiolase. The β-ketothiolase is negatively regulated by the product coenzymeA (HSCoA), which is also a product when acetyl-CoA enters the TCA cycle under nonlimited conditions. The product is subsequently stereoselectively reduced to (*R*)-3-hydroxybutyryl-CoA in a reaction catalyzed by NADPH-dependent acetoacetyl-CoA reductase. Finally, PHB is synthesized by polymerization of (*R*)-3-hydroxybutyryl-CoA molecules by the PHB synthase. *Source*: From van der Leij et al. (103).

action catalyzed by NADPH-dependent acetoacetyl-CoA reductase. Finally, PHB is synthesized by polymerization of *(R)*-3-hydroxybutyryl-CoA molecules by the PHB synthase (19). A few exceptions to this general pathway are known, such as *Rhodospirillum rubrum*, where (*S*)-3-hydroxybutyryl-CoA is generated and converted to *(R)*-3-hydroxybutyryl-CoA by two stereospecific enoyl-CoA hydratases (20).

As pointed out in the introduction, PHB biosynthesis and accumulation occur in most bacteria if a carbon source is provided in excess and if another nutrient, such as nitrogen, sulfur, phosphate, iron, magnesium, potassium, or oxygen, is limiting (21,22). PHB synthesis is regulated at the enzymatic level, and the availability of suitable substrates for PHB synthesis routes results from an imbalanced supply of nutrients to the cell via physiological regulation pathways than have been studied in *Azotobacter beijerinckii* (23) and *Alcaligenes eutrophus* (24,25). It was found that the intracellular concentration of acetyl-CoA and free coenzyme A play a central role in the regulation of polymer synthesis (26). Under balanced growth conditions acetyl-CoA is oxidized via the tricarboxylic acid (TCA) cycle. In this reaction sequence, NADH is generated and is used for biosynthetic purposes. When growth ceases the NADH concentration increases, which reduces the activity of the TCA cycle enzymes citrate synthase and isocitrate dehydrogenase. As a result acetyl-CoA cannot be oxidized via the TCA cycle and enters the PHB synthetic pathway. The 3-ketothiolase enzyme of this pathway is inhibited by free CoA, which is generated by oxidation of acetyl-CoA via the TCA cycle during normal growth.

PHB/HV Copolymer Biosynthesis

A copolymer of PHB and PHV can be synthesized by *Alcaligenes eutrophus* and many other microorganisms from either a mixed substrate of glucose and propionic acid or a direct precursor of 3-hydroxyvalerate (e.g., valeric acid) (27). If propionate is fed, essentially the same biochemical pathway as for PHB synthesis is used, but propionyl-CoA and acetyl-CoA are condensed by the ketothiolase to give 3-ketovaleryl-CoA, which leads to the incorporation of 3-hydroxyvalerate monomers in the polymer. 3-Hydroxybutyrate arises in the normal way. Alkanoic acids of odd-number carbon chain length can also serve as carbon source (28,29). In this case, the 3-hydroxyvalerate in the polymer arises directly from the β-oxidation of these fatty acids. Recently, it was shown that low dissolved oxygen concentrations enhance 3-hydroxyvalerate incorporation into the polymer produced by *A. eutrophus* (30).

Several microorganisms have been isolated that are capable of accumulating PHB/HV from substrates apparently unrelated to, and not immediate precursors of, 3-hydroxyvalerate. These organisms include *Corynebacterium, Nocardia*, and *Rhodococcus* strains (31) and mutants of *A. eutrophus* (32). When supplied with hexanoic acid as substrate, *Rhodococcus* sp. was able to accumulate a terpolymer containing 3-hydroxybutyrate, 3-hydroxyvalerate, and 3-hydroxyhexanoate monomers. In addition, *Rhodococcus* sp. NCIMB40126 and *Chromobacterium violaceum* are capable of producing a homopolymer consisting of 3-hydroxyvalerate when grown on valeric acid as sole carbon source (31,33).

Copolymers of 3-hydroxybutyrate and 4-hydroxybutyrate can be synthesized by *A. eutrophus* from 4-hydroxybutyric acid, 1-4-butanediol, butyrolactone, and 4-chlorobutyric acid (34). Terpolymers of 3-hydroxybutyrate, 3-hydroxyvalerate, and 5-hydroxyvalerate can be produced by *A. eutrophus* fed with either 5-chlorovalerate alone or 5-chlorovalerate or 5-hydroxyvalerate in combination with valeric acid (35). Polymers that contain 4-hydroxyvalerate and 3-hydroxypropionate have been synthesized by a variety of organisms, including *A. eutrophus* (36). The substrates used were 4-hydroxyvalerate, 4-valerolactone, and 3-hydroxypropionic acid. Most *Alcaligenes* strains tested and *Pseudomonas oxalaticus* accumulated a terpolymer of 3-hydroxybutyrate, 3-hydroxyvalerate, and 4-hydroxyvalerate from 4-valerolactone or 4-hydroxyvalerate as sole carbon source (36). *Alcaligenes eutrophus* and *Burkholderia cepacia* are able to synthesize terpolyesters that consist of 3-hydroxybutyric acid, 3-hydroxyvaleric acid, and 2-methyl-3-hydroxybutyric acid when tiglic acid is used as sole carbon source or in combination with gluconic acid (37).

Molecular Biology and Enzymology of SCL PHA Synthesis. In *A. eutrophus*, the structural genes for PHA synthesis are organized in the *phbCAB* operon (see Fig. 2), coding for PHB synthase, β-ketothiolase, and NADPH-dependent acetoacetyl-CoA reductase, respectively (38–40). The transcription start of *phbC* was identified 307 base pairs upstream from the translation initiation codon by S1 nuclease protection assay, and a sequence that exhibits a striking homology to the *E. coli* σ^{70} promoter consensus sequence was found (39).

The PHB synthase structural genes have been cloned and sequenced from various bacteria (8). Comparison of the protein primary structures revealed that PHB synthases represent a group of highly homologous enzymes. The PHB synthase of *A. eutrophus* has been studied in detail in several laboratories. It is a large enzyme with identical subunits that exhibit a relative molecular mass of 63.9 kDa (38) and specificity for 3-hydroxyacyl-CoA of C_3 to C_5 units (41). Site-directed mutagenesis demonstrated that the cysteine$_{319}$ residue is required for PHB synthase activity in *A. eutrophus* (42). Furthermore, evidence has been obtained that the enzyme requires posttranslational modification by phosphopantetheine to provide a second thiol group per PHB synthase subunit (42), which could support the mechanistic model for the polymerization process that depends on the presence of two thiol groups (43).

A putative model for the polymerization process based on de novo fatty acid synthesis is presented in Figure 3. This model illustrates only one possible reaction mechanism, and other models are also possible. In an initiation or priming process (Fig. 3a), the 3-hydroxybutyric acid moiety from 3-hydroxybutyryl-CoA is transferred to the thiol group of the phosphopantotheine-modified serine, coupled with the release of coenzyme A. This 3-hydroxybutyryl moiety is then translocated to the thiol of the cysteine. It has been shown recently that acylation of cysteine$_{319}$ causes a shift of the monomeric form of the synthase to its dimeric form, and this shift is accompanied by an increase in its specific activity and a decrease in the lag

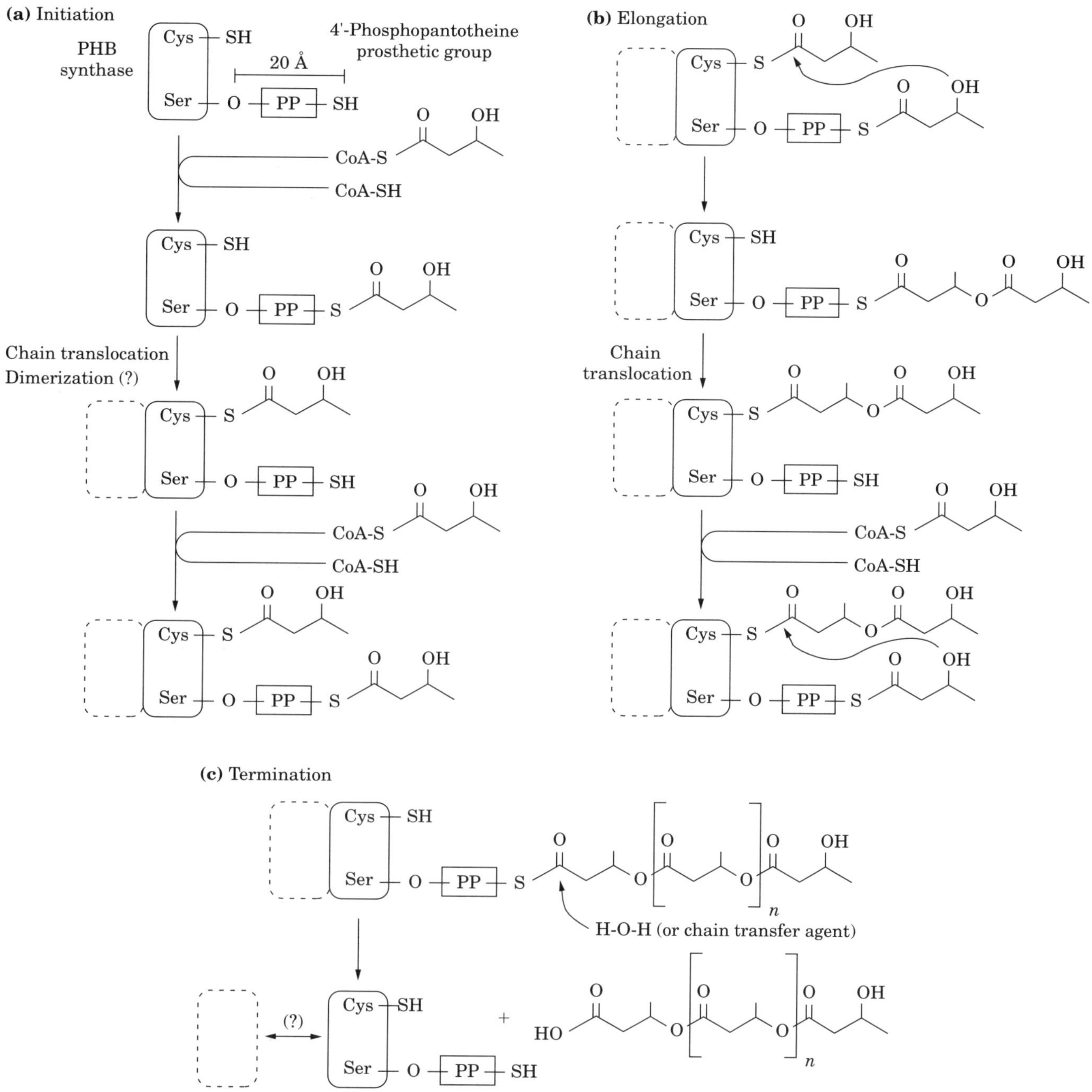

Figure 3. The hypothetical model of the catalytic PHB synthase mechanism is based on analogy to the de novo fatty acid synthesis and on information available about PHB synthases. The conserved cysteine$_{319}$ and the posttranslationally modified serine$_{259}$ of the PHB synthases that are likely to be involved in the reaction mechanism are indicated. In the initiation step (**a**) the 3-hydroxybutyric acid moiety from 3-hydroxybutyryl-CoA is transferred to the thiol group of the phosphopantotheine modified serine, coupled with the release of coenzyme A. It is supposed that this 3-hydroxybutyl moiety is translocated to the thiol group of the cysteine and that a shift of the monomeric form of the synthase to its dimeric form takes place. A second 3-hydroxyacyl moiety is added to the thiol group of the phosphopantotheine modified serine. During chain elongation (**b**) the 3′-hydroxy group of the entering monomer attacks the thiol ester of the previously entered monomer to form a covalently bound dimer/oligomer. Termination takes place when a chain transfer with water occurs, and the polymer chain is released from the enzyme (**c**). A chain transfer agent might be involved in this process. This model does not show that the polymer chain might be translocated from the posttranslationally modified serine to the thiol group of the cysteine of the other subunit as is the case in de novo fatty acid synthesis.

phase of polymer formation (44). A second monomer is then again added to the thiol group of the phosphopantotheine-modified serine, and its 3′-hydroxy group attacks the thiol ester of the first monomer to form a covalently bound dimer (Fig. 3b). After translocation of the formed dimer to the thiol group of the cysteine (either of the same subunit as shown in Fig. 3b or of the other enzyme subunit), the thiol group of the posttranslationally modified serine can accept a new 3-hydroxybutyric acid moiety, and by repeated propagation steps the covalently bound oligomer or polymer grows by one unit during each cycle. It is assumed that the polymer is released from the enzyme if a chain transfer with water occurs that terminates polyester biosynthesis (Fig. 3c). Doi and coworkers have postulated that a chain transfer agent, which might be an enzyme with a water molecule in its active site, could be involved in the termination process (45–47).

A combined chemical and enzymatic procedure has been developed to synthesize PHB granules in vitro (48). Purified PHB synthase from *A. eutrophus* was exposed to synthetically prepared *(R)*-3-hydroxybutyryl-CoA, and PHB was formed, thereby establishing the minimal requirements for PHB formation. The in vitro polymerization system yielded PHB with a molecular mass higher than 10×10^6 Da, exceeding by an order of magnitude the mass of PHB typically extracted from microorganisms. Furthermore, the molecular mass of the polymer could be controlled by the initial PHB synthase concentration.

The PHB synthases of *Chromatium vinosum* and *Thiocystis violacea* represent a different class of synthases that consists of two different types of subunits encoded by *phaE* and *phaC* (49,50). Together, these genes constitute one single operon. The C-terminus of PhaC contains high homology to the central region of the PHB synthase of *A. eutrophus*. However, the PhaC protein lacks the N-terminal part and is much smaller than the other synthases. The PhaE protein may take over essential functions of the larger PHB synthases in the polymerization process (8). The PHA synthase of *Thiocapsa pfennigii* is also composed of two different subunits, similar to those of *Chromatium vinosum* and *Thiocystis violacea*. From the substrate range, however, this synthase seems to be clearly distinguished from all other PHA synthases because it can synthesize copolyesters that consist of 3-hydroxyalkanoates ranging from C_3 to C_{14}, or even 4- and 5-hydroxyalkanoic acids (51).

Granule Structure

The PHA polymers are accumulated in the bacterial cells as defined granules (Fig. 4). Because the granule structure of SCL and MCL PHAs produced by different organisms seems to be very similar, both types of polymer granules are discussed together. Nuclear magnetic resonance (NMR) spectroscopy of various bacteria has clearly demonstrated that the polyester in the cells occurs in a metastable amorphous state and that it is stored as a mobile elastomer (52–54). The bouyant density of PHB granules is about 1.18–1.24 g cm^{-3} (55–57), in contrast to MCL PHA granules, which have a bouyant density of approximately 1.05 g cm^{-3} (58). The isolated granules consist of polyester, proteins, and lipids. The composition was first determined for PHB granules of *Bacillus megaterium*, which consist of 97.7% polyester, 1.87% proteins, and 0.46% lipids or phospholipids (43).

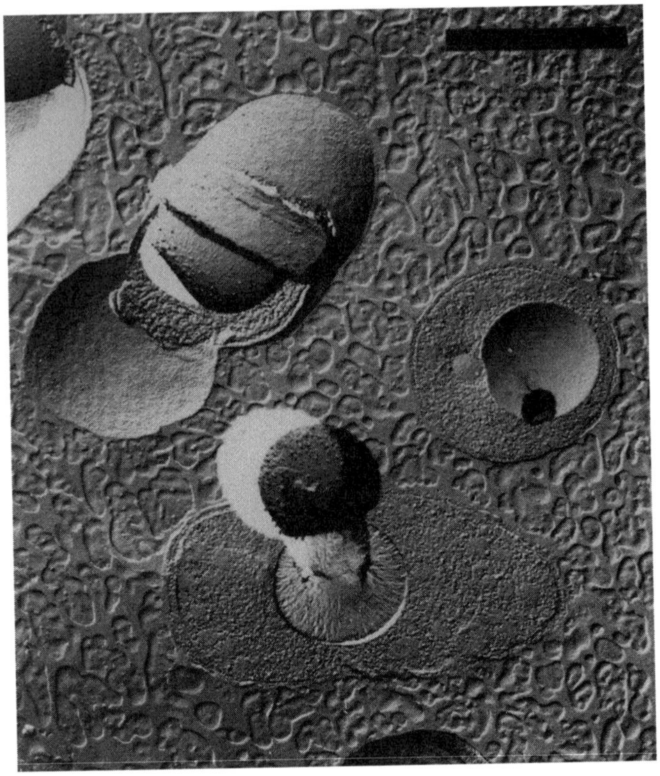

Figure 4. Polyhydroxyalkanoate (PHA) granules in *Pseudomonas oleovorans*. Freeze-fracture electron micrograph of octane-grown *P. oleovorans* cells containing PHA granules that were deformed during the fracture process and show the typical mushroom-type deformation of granules consisting of MCL PHA. The bar represents 0.5 μm.

Several different models of granule structure have been proposed. First indications that the native PHA granules in *Pseudomonas oleovorans* might be surrounded by a phospholipid monolayer in which the polymerase enzymes are embedded have been found by Witholt and coworkers (59,60). In contrast to this model, Stuart et al. (61) presented a model in which the hydrophobic PHA core is surrounded by two protein layers with a phospholipid layer in between. In this model, the polymerase is supposed to be embedded only in the inner layer. Recently, determination of the thickness of the boundary layer surrounding the polymer inclusions of different bacteria confirmed that a lipid monolayer is more likely than two protein layers with a lipid layer between them (62). In addition to the polymerases, other proteins such as depolymerases, granule-associated proteins (GAPs, also called phasins), and proteins of unknown function seem to be embedded in the phospholipid layer (63,64). Phasins are in general low molecular weight proteins and are assumed to form a close protein layer at the surface of the granules, providing the interface between the hydrophilic cytoplasm and the hydrophobic core of the PHA inclusion (63). In *A. eutrophus*, the first evidence was found by mutant studies for the structural importance of a 24-kDa GAP (called GA24) (65). Mutants lacking the GA24 protein synthesized only one large PHB granule per cell in contrast to 10–30 granules as found in the wild type. The amino acid sequence of the GA24 protein revealed two closely related stretches consisting exclusively of nonhydrophilic amino acids at the C-terminal region, which are presumably involved in the binding of GA24 to the granules. Similar granule-forming proteins were identified and isolated from various other strains, such as *Acinetobacter* sp., *Rhodococcus ruber*, *Chromatium vinosum*, *Methylobacterium extorquens*, and *Pseudomonas oleovorans* (64,66–73).

Industrial Production of PHB and PHB/HV

Many bacteria have been screened to produce PHB or PHB/HV. However, only a few have been used for biotechnological production of these polyesters on a large scale. Here, we focus on the two examples that have found application in an industrial process.

The suitability of a bacterium for PHA production depends on many different factors such as stability and safety of the organism, growth and accumulation rates, achievable cell densities and PHA contents, extractability of the polymer, molecular weights of accumulated PHA, range of utilizable carbon sources, costs of the carbon source and the other components of the medium, and occurrence of by-products (74). Recently, the theoretical yield (the yield based on the reaction stoichiometry) of PHB has been estimated for several substrates (75). Regeneration of nicotinamide nucleotides, which are used as cofactors for PHA synthesis, has been taken into account in this analysis.

In the production plant of ZENECA BioProducts at Billingham (UK), the PHB homopolymer and the copolyester PHB/HV have been produced at scales up to 200,000 L (74). The process utilized a mutant of *A. eutrophus* (76), and the fermentation was carried out in a two-step fed-batch process. During the first step, the cells were grown in a mineral salt medium with glucose as sole carbon and energy source and with a calculated amount of phosphate based on the known requirements of the organism to allow production of a given amount of biomass. As the culture grew, phosphate was depleted from the medium, and during the second step when phosphate was limiting, the culture started to store polymer. Glucose was fed to the culture, and the fermentation was continued until the required polymer content was reached. Each phase lasted for approximately 48 h, and final biomass dry weights of 100 g/L were achieved (77). The copolymer of PHB/HV was made by substituting a mixed feed of glucose and propionic acid in the polymer accumulation phase. The 3-hydroxyvalerate content of the copolymer was controlled by adjusting the ratio of the two substrates in the feed (77).

A different industrial process has been developed at the Biotechnologische Forschungsgesellschaft in Linz (Austria) for the production of PHB homopolymer (78,79). In this process strain *Alcaligenes latus* DSM1124 was employed, which can accumulate PHB during balanced cell growth up to 80% of the cellular dry matter. Therefore, a one-step fed-batch fermentation process was carried out using a mineral salts medium with sucrose as sole carbon source. A biomass of 60 g/L was achieved (79). In 1993, however, the company stopped the production of PHB/HV.

Recovery

The cells that contain PHB/HV must be separated from the broth by conventional procedures such as centrifugation, filtration, or flocculation–centrifugation. After the biomass is harvested, cells must be disrupted to recover the polymer. Several methods have been developed for the recovery of PHAs. The method most often used involves extraction of the polymer from the biomass with solvents (e.g., chloroform, methylene chloride, propylene carbonate, dichloroethane). Large amounts of solvent are required in this process. Several other methods have been developed, such as sodium hypochloride treatments for the differential digestion of non-PHA cellular materials (80). Although this method is effective, it causes severe degradation of PHB, resulting in a reduction in the molecular weight (81).

ZENECA has developed a nonsolvent-based recovery process as an alternative to solvent extraction for the commercial production of PHB and PHB/HV by *A. eutrophus* (77,82). In this process the cells were first exposed to 80 °C and subsequently treated with a cocktail of various hydrolytic enzymes consisting of lysozyme, phospholipase, lecithinase, the proteinase alcalase, and others. Most of the cellular components were hydrolyzed by these enzymes, whereas the polymer remained intact. After washing, flocculation, and drying, the polymer was recovered as a white powder that was melted, extruded, and converted into chips, the form in which the polymer was supplied to the fabricators.

In contrast, the Biotechnologische Forschungsgesellschaft has applied a solvent-based process for the recovery of PHB from *A. latus*. The cells were harvested by centrifugation, and subsequently the PHB was extracted with methylene chloride from the suspended cells and precipitated out of the solvent by the addition of water. After drying of the polymer, a polyester with 99% purity was obtained (79). The process also included recovery of the solvent.

Properties of PHB and Its Copolymers

Comprehensive overviews on the physical properties of PHB and its copolymers, or on the material properties of the manufactured objects from those polymers, have been published previously (27,74,83–85) and are summarized next.

The PHB homopolymer is a completely stereoregular polyester, with all asymmetric carbon atoms in the (*R*)-configuration, and it is therefore highly crystalline. The high crystallinity (typically 55–80%) makes it relatively stiff and brittle. The glass transition temperature (T_g of PHB lies between 5 and 9 °C, and the melting point (T_m) lies between 173 and 180 °C (Table 1). PHB polymer decomposes at approximately 200 °C, which is close to its melting temperature. In its solid, crystalline state, the polymer is a right-handed helix, a condition that is retained in chloroform solution. The mechanical properties

Table 1. Comparison of Polymer Properties of Polyhydroxybutyrate and Its Copolymers with Polypropylene

Physical properties	P(3HB)	P(3HB/3HV) 90:10	P(3HB/3HV) 80:20	P(4HB)	P(3HB/4HB) 90:10	P(3HB/4HB) 10:90	PP
Melting point (°C)	179	150	135	53	159	50	170
Tensile strength (MPa)	40	25	20	104	24	65	34.5
Young's modulus (GPa)	3.5	1.2	0.8	149	—	100	1.7
Elongation to break (%)	3.0	20	100	1000	242	1080	400

Source: Data taken from Lee (2) and Byrom (74).
P(3HB), poly(3-hydroxybutyrate); P(3HB/3HV), poly(3-hydroxybutyrate-*co*-3-hydroxyvalerate); P(4HB), poly(4-hydroxybutyrate); P(3HB/4HB), poly(3-hydroxybutyrate-*co*-4-hydroxybutyrate); PP, polypropylene.

Table 2. Enzymes Involved in the Pathways Leading to MCL PHA Formation and Their Respective Genes in *Escherichia coli*

Enzyme	Gene in *E. coli*
1. Acyl-CoA synthetase	*fadD*
2. Acyl-CoA dehydrogenase	*fadE*
3. Enoyl-CoA hydratase or crotonase	*fadB*
4. (*S*)-3-OH-Acyl-CoA dehydrogenase (NAD dependent)	*fadB*
5. 3-Ketoacyl thiolase	*fadA*
6. 3-ketoacyl synthase I; II; III	*fabB, F, H*
7. Ketoacyl-ACP reductase	*fabG*
8. 3-OH-Acyl-ACP dehydratase	*
8a. 3-OH-Decanoyl-ACP dehydratase	*fabA*
9. Enoyl-ACP reductase	*
10. Fatty acid thioesterase I; II	*tesA, B*
11. Ketoacyl-CoA reductase (NADPH dependent)	*
12. (*R*)-3-OH-Acyl-CoA dehydrogenase (NAD dependent)	*
13. 3-OH-Acyl-CoA epimerase	*fadB*
14. Enoyl-CoA hydratase II	*
15. Enoyl-CoA isomerase	*fadB*
16. (*R*)-3-OH-Acyl (ACP → CoA) transferase	*
17. Several acyl transferases or acylases	*
18. PHA polymerase	None

Source: From van der Leij et al. (103).
Note: Additional genes in *E. coli* have the following functions: *fadH* encodes a 2,4-dienoyl-CoA reductase; *fadL* encodes a protein that is needed for fatty acid transport across the membrane; and *fadR* codes for a regulatory protein.
*The respective gene is not known.

of PHB including Young's modulus (3.5 GPa) and the tensile strength (40 MPa) are similar to those of polypropylene (Table 1). However, the elongation to break for PHB is about 3%, which is significantly lower than that of polypropylene (400%).

The flexibility of the material is greatly improved when 3-hydroxyvalerate units are incorporated into the polymer, resulting in a decrease of Young's modulus below 0.7 and a decrease of tensile strength below 30 MPa. The elongation to break also increases as the comonomer fraction increases. The melt temperature is greatly depressed, down to 130 °C, dependent on the 3-hydroxyvalerate content (Table 1), but the degradation temperature is little affected. This property generates a relatively wide window of conditions that allow thermal processing as a melt without degradation of the material. PHB/HV copolymers with different 3-hydroxyvalerate content, ranging from 0 to 30%, are commercially available as Biopol™ resins.

Copolymers of 3-hydroxybutyrate and 4-hydroxybutyrate monomers cannot form isomorphic crystals because of the extent of the structural differences between the monomer units. The glass transition temperature decreases from 5 to −50 °C, and the melting temperature decreases from 180 to 54 °C as the mol% of 4-hydroxybutyrate increases from 0 to 100%. Young's modulus, tensile strength, and % elongation at break (for 94 mol% 4-hydroxybutyrate) are reported as 55 MPa, 39 MPa, and 500%, respectively. The T_g and T_m of copolymers consisting of 3-hydroxybutyrate and 3-hydroxypropionate decrease with an increase in the 3-hydroxypropionate (from 0 to 26 mol%) from 4 to −8.1 °C and 180 to 85 °C, respectively. The degree of crystallinity also decreases with an increase of 3-hydroxypropionate.

Recombinant Organisms

Successful biosynthesis of PHB in *E. coli* was independently achieved by two different groups (39,40). Polymer formation was only observed when all three PHB biosynthesis genes of *A. eutrophus* were expressed in *E. coli*; expression of *phbC* alone did not result in PHB production. These studies were extended by the production of the copolymer PHB/HV in recombinant *E. coli* (*fadR atoC*) mutants containing the PHB biosynthetic genes and using glucose and propionic acid as carbon source. The *E. coli* (*fadR atoC*) mutants produce the enzymes required for the catabolism of SCL fatty acid constitutively. 3-Hydroxyvalerate contributed up to 39 mol% to the accumulated polyester (86). Synthesis of PHB by recombinant *E. coli* does not require limitation of a specific nutrient, but is dependent on the amount of acetyl-CoA available and is significantly enhanced by the addition of various complex nitrogen sources, amino acids, or oleic acid (87). Application of recombinant *E. coli* in fed-batch culture resulted in cell densities of 101 g L^{-1}, containing approximately 80% PHB per cell dry weight (88) after 39 h of cultivation.

Under specific conditions, such as pH 6.0, *E. coli* produced PHB with an extremely high molecular mass of 20 MDa (47). Furthermore, it was shown that the amount of PHB synthase in recombinant *E. coli* plays a key role in controlling the molecular weight and the polydispersity of the polymer (89). With an increased amount of PHB biosynthetic enzymes, a decrease in polyester molecular weight was observed.

For the first time, PHB biosynthesis was successfully conferred to *Arabidopsis thaliana*, a small oil seed plant, by introducing the *phbB* and *phbC* gene from *Alcaligenes eutrophus* (90). When the enzymes necessary to catalyze the synthesis of PHB from acetyl-CoA were produced in the cytoplasm, only small amounts of PHB were synthesized. Plants were viable but growth was very poor, possibly because of a reduction of acetyl-CoA needed for growth. Targeting of all three required proteins to the plastids of *Arabidopsis thaliana* resulted in normal growth, active enzymes, and PHB accumulation up to 14% of total cell dry weight (91). However, the molecular weight distribution of the plant-produced PHB was much broader than that of typical bacterial PHB (92).

The *phbC* gene of *Alcaligenes eutrophus* has also been introduced into *Saccharomyces cerevisiae*, which enabled this yeast to accumulate PHB to up to 0.5% of cell dry weight (93). A novel pathway for PHB synthesis has been engineered in insect cells (*Spodoptera frugiperda* (94,95). In these insect cells, a heterologous dehydrase-deficient fatty acid synthase from rats is produced in addition to the PHB synthase of *A. eutrophus*. The multifunctional fatty acid synthase effectively replaces the β-ketothiolase and acetoacetyl-CoA reductase activities of the natural prokaryotic PHB pathway and, therefore, provides the precursor 3-hydroxybutyryl-CoA for conversion to PHB (Fig. 5). This strategy suggests a new method for generating structurally diverse polyhydroxyalkanoates by metabolic engineering.

MCL PHA

Although many species of bacteria are reported to synthesize PHB and PHV, including some *Pseudomonas* strains, the ability to store polymers that contain monomers of 3-hydroxy acids of carbon chain length C_6 and longer, here referred to as MCL PHAs, is unique to the group of pseudomonads.

Physiology and Biosynthesis of PHA

Three main pathways are involved in the synthesis of PHA precursors in *Pseudomonas putida* (96,97). These were revealed by labeling studies, NMR spectroscopy, and gas chromatography–mass spectroscopy (GC-MS) analyses of PHA from cells that were grown on simple carbon substrates such as glucose and/or fatty acids. *P. putida* derives PHA from precursors made via the following routes. *(1)* Chain elongation, in which acyl-CoA is extended with acetyl-CoA, is comparable with the major route for PHB monomer synthesis, where the acyl-CoA is one of the two acetyl-CoA molecules (Fig. 6). It has been found that about 3–10% of the monomeric units of PHA are one C_2 unit longer than the fatty acid used as substrate (98–100). However, this route is less important for PHA synthesis than it is for PHB synthesis. *(2)* Fatty acid degradation by β-oxidation is the main pathway when fatty acids are used

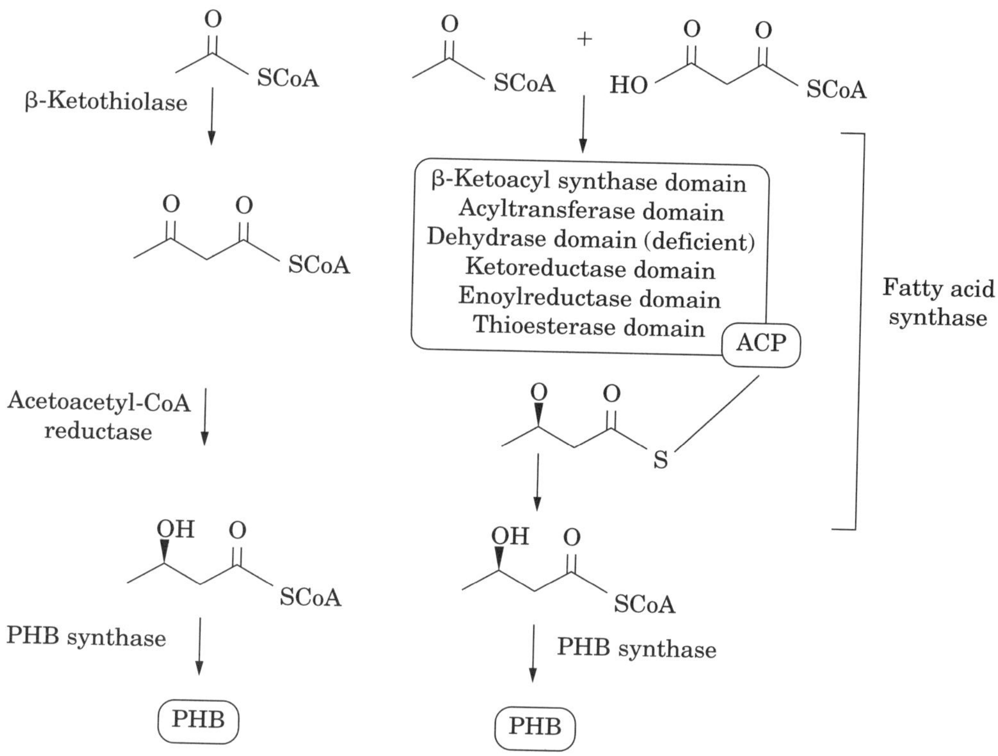

Figure 5. Engineered polyhydroxybutyrate (PHB) synthesis pathway in insect cells. The bacterial PHB synthesis pathway described in Figure 2 is shown on the left, and the engineered PHB synthesis pathway in insect cells on the right. The condensation of acetyl-CoA and malonyl-CoA followed by ketoreduction of acetoacetyl-acyl carrier protein (ACP) are performed by the fatty acid synthase multifunctional protein. The different catalytic domains of the enzyme are listed. The dehydrase domain is not functional. *Source*: From data of Williams et al. (94).

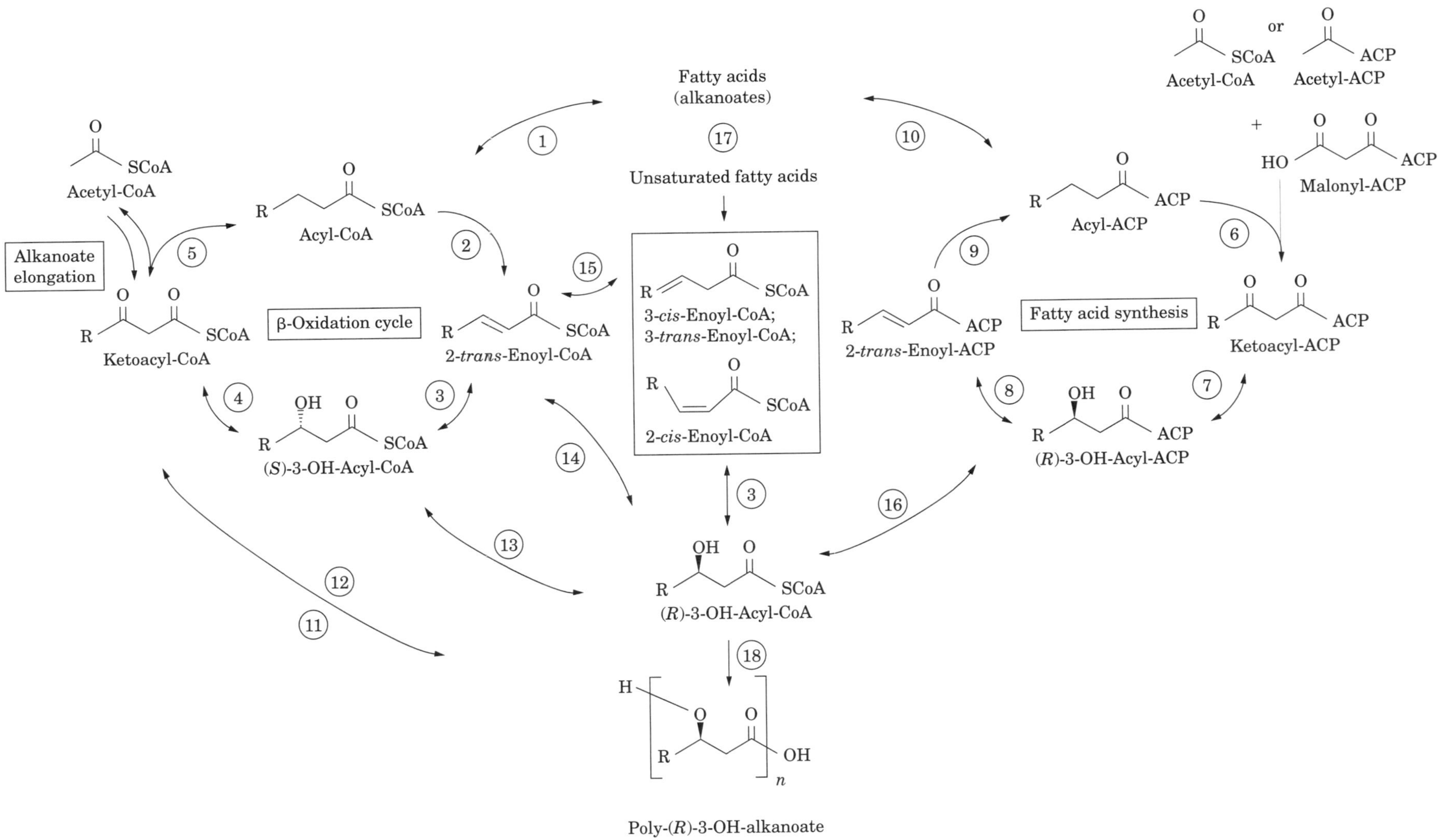

Figure 6. Major pathways involved in MCL polyhydroxyalkanoate (PHA) synthesis in *Pseudomonas*. Alkanoate elongation, β-oxidation, and fatty acid biosynthesis are involved in MCL PHA synthesis in *Pseudomonas*. The numbering of the enzymatic activities is as given in Table 2. *Source*: From van der Leij et al. (103).

as substrate. *(3)* De novo fatty acid biosynthesis is the main route during growth on simple carbon compounds. The enzymes involved in the pathways depicted in Figure 6 do not have the same activity for different chain length substrates. Thus, the PHA polymerase seems to have a preference for the incorporation of C_8 to C_{10} monomers. Such preferences might have substantial influences on the fluxes and the contributions of each pathway.

The intermediate precursors of PHA are either ketoacyl-CoA, (*S*)-3-hydroxyacyl-CoA, enoyl-CoA, or (*R*)-3-hydroxyacyl-acyl carrier protein (ACP). Regarding the similarity between the PHB polymerase and the PHA polymerases (see following), it seems likely that the ultimate substrate for polymerization is the (*R*)-form of the CoA-activated 3-hydroxy fatty acid intermediates. Because the monomeric units in PHA, like those in PHB, are in the (*R*)-form, and the PHB and PHA polymerases are homologous, the intermediates in fatty acid metabolism must presumably be converted to the (*R*)-form of 3-hydroxyacyl-CoA before transesterification (Fig. 6). The PHA precursors are generally considered to be available in a much lower in situ concentration than the PHB precursors because they are intermediates of ongoing fatty acid metabolic cycles.

Molecular Biology and Enzymology

Knowledge of the genetic and physiological regulation of PHA biosynthesis is relatively limited compared to what is known about PHB biosynthesis because PHA synthesis has only been studied since 1983 (5). The *pha* loci of *Pseudomonas oleovorans* (101) and *Pseudomonas aeruginosa* (102) consist of four ORFs: the ORFs *phaC1* and *phaC2* code for polymerases and are separated by an ORF called *phaZ*, which encodes a depolymerase; downstream of these genes is an ORF called *phaD*, which encodes a putative granule surface protein (103) (Fig. 7). Subsequently, more ORFs have been identified downstream and in opposite orientation to the *pha* locus of *P. putida* and *P. oleovorans*, which presumably encode granule-associated proteins (H. Valentin, unpublished results, 1996). The sequence of this region has not yet been published.

At least two promoters are involved in the transcription of the *pha* genes in *P. aeruginosa* (102). Two promoters are

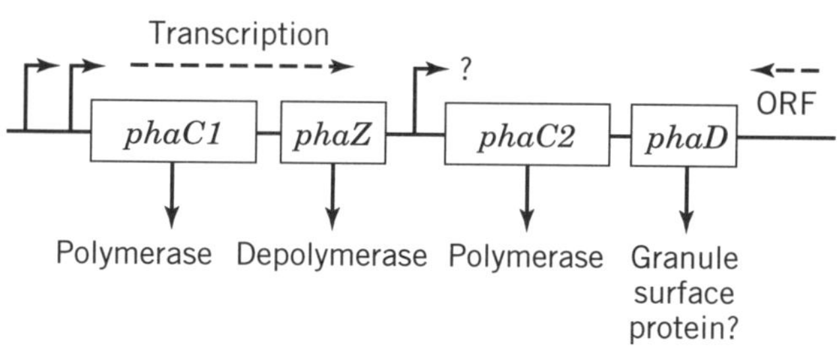

Figure 7. Organization of *pha* genes in *P. oleovorans* and *P. aeroginosa*. Two polymerases and one depolymerase are encoded by *phaC1, phaC2*, and *phaZ*, respectively. The *phaD* gene may encode a granule-associated protein. More open reading frames (ORF) in opposite orientation were found downstream of these genes. Putative promoters are indicated with open arrows (see text for details). *Source*: From Steinbüchel et al. (8) and van der Leij et al. (103).

upstream of the *phaC1* gene and resemble the consensus sequences for σ^{54}-dependent and σ^{70}-dependent promoters. It is not clear whether a third promoter upstream of the *phaC2* gene in *P. aeruginosa* is active (102). Northern blot experiments indicated the presence of two mRNAs, one carrying *phaC1* and *phaZ*, the other *phaC2* and *phaD*. However, a transcriptional start site accompanying the putative promoter upstream of *phaC2* could not be experimentally shown, suggesting that this promoter may not be functional.

The two PHA polymerases show identities of 54–59% at the protein sequence level when compared within each species. When corresponding PHA polymerases from different species are compared, they show 69–80% identity at the amino acid level. The identities between these PHA polymerases and the *A. eutrophus* PHB polymerase are nearly 40% (101,102). According to the gene sequences, the *Pseudomonas* PHA polymerases have molecular weights of 62–64 kDa. The enzymes show a small difference in substrate specificity (104).

The *phaZ* gene product, a PHA depolymerase with a predicted molecular weight of 31.5 kDa, shows homology with a range of hydrolytic enzymes (59,101). It contains the common lipase box Gly-X-Ser-X-Gly, which is also found in extracellular PHA depolymerases (105).

Production of MCL PHA

Pseudomonas strains such as *P. oleovorans, P. putida, P. aeroginosa*, and *P. resinovorans* are the major strains used for MCL PHA production (100,106,107). Batch, fed-batch, and continuous cultivation techniques have been extensively investigated at the laboratory scale (60,108,109), but MCL PHA production up to a 200-L scale has also been carried out (110). Efficient PHA production requires rapid formation of biomass and product. In batch and fed-batch fermentations, similar to the PHB production process, cells are first grown to a desired concentration without nutrient limitation, after which an essential nutrient is limited to trigger efficient PHA synthesis. During this nutrient limitation stage the residual cell concentration (defined as the cell concentration minus the PHA concentration) remains almost constant, and the cell concentration increases only because of the intracellular accumulation of PHA.

P. oleovorans has been used to produce MCL PHAs in two-liquid-phase cultivations (111,112). The organism was cultured in media consisting of an aqueous phase containing all minerals necessary for growth and a second liquid phase containing *n*-alkanes as sole carbon and energy source. Two-liquid-phase, single-stage chemostat, and fed-batch cultivations of *P. oleovorans* using *n*-octane as the carbon source resulted in highest volumetric productivities of 0.58 g L^{-1} h^{-1} (108) and 0.25 g L^{-1} h^{-1} (109), respectively. In a two-liquid-phase, two-stage continuous cultivation a volumetric productivity of 1.06 g L^{-1} h^{-1} with a final cellular PHA content of 66% per total cell dry weight has recently been reported (113). The monomer composition of the synthesized polymer is mainly dependent on the strain used and the carbon source applied. The cultivation conditions have only a minor influence on the composition of the polymer. However, it was shown that temperature and concentration of the substrate has an influence on the molecular weight of the polymer (45). The monomer composition of PHAs produced from different alkanoates using *P. oleovorans* is shown in Table 3 (100). Recently, it was reported that *Pseudomonas putida* U grown on octanoic acid as the sole carbon source accumulated a homopolymer of poly-3-hydroxyoctanoate (114).

Different *Pseudomonas* strains have been used to produce MCL PHAs from renewable resources, such as fatty acids (29,100) and carbohydrates (96,107,115). In fed-batch cultivations, *P. putida* KT2442 was cultivated to high cell densities (92 g/L) with oleic acid as a substrate. With this process, an overall volumetric productivity of 1.6 g L^{-1} h^{-1} was achieved with a final cellular PHA content of 40% per total cell dry weight (116). Furthermore, MCL PHAs could be produced from glucose, fructose, and glycerol using *P. putida* KT2442 (96). In addition to the major constituent 3-hydroxydecanoate, six other monomers were found to be present: 3-hydroxyhexanoate, 3-hydroxyoctanoate, 3-hydroxydodecanoate, 3-hydroxydodecenoate, 3-hydroxytetradecanoate, and 3-hydroxytetradecenoate (Table 4). The degree of unsaturation of PHA was slightly influenced by the cultivation temperature (96).

Various functional groups can be introduced into the polymer by applying different substrates in the growth medium. Olefin groups were introduced by adding *n*-alkenes (111,112) or alkenoates (29,117,118) to the medium. Furthermore, different halogen atoms were incorporated in the polymer when halogenated alkanoates were used as substrates (119–122). A homopolymer of 99% purity consisting of 3-hydroxy-5-phenylpentanoate is synthesized by *P. oleovorans* when grown on 5-phenylvalerate (123). Polymers containing different mol% of phenyl groups (124,125), or other aromatic compounds such as phenoxy groups (126,127) or cyanophenoxy groups (128,129), were produced by adding phenyl alkanoates, phenoxy alkanoates, or cyanophenoxy alkanoates, respectively, as a cocarbon source.

Table 3. Composition of MCL Polyhdroxyalkanoates Produced by *Pseudomonas oleovorans* Grown on Different Alkonoates as Sole Carbon Source

	Composition (mol% of total 3-hydroxy fatty acids)						
Carbon source	C6	C7	C8	C9	C10	C11	C12
Hexanoate	95		5				
Heptanoate		100					
Octanoate	8		91		1		
Nonanoate		35		65			
Decanoate	8		75		17		
Undecanoate		28		59		13	
Dodecanoate	6		57		32		5
Tridecanoate		32		48	5	14	
Tetradecanoate	7		59		30		4
Pentadecanoate		32		47	8	13	
Hexadecanoate	8		50		30		12

Source: Huisman et al. (100).
HC, 3-hydroxyhexanoate; HH, 3-hydroxyheptanoate; HO, 3-hydroxyoctanoate; HN, 3-hydroxynonanoate; HD, 3-hydroxydecanoate; HUD, 3-hydroxyundecanoate; HDD, 3-hydroxydodecanoate.

Table 4. Influence of Cultivation Temperature on MCL Polyhydroxyalkanoate Composition in *Pseudomonas putida*

			Composition (mol% of total PHA)						
Substrate	T (°C)	% PHA	C6	C8	C10	C12:1	C12	C14:1	C14
Glucose	15	20.5	0.4	8.4	64.8	16.7	6.0	4.0	<0.1
	30	17.0	<0.1	6.9	74.3	8.8	7.7	1.6	<0.1
Decanoate	15	48.5	5.9	50.9	43.2	—	—	—	—
	30	38.1	5.3	52.3	42.4	—	—	—	—

Source: Huijberts et al. (96).
C6, 3-hydroxyhexanoate; C8, 3-hydroxyoctanoate; C10, 3-hydroxydecanoate; C12:1, 3-hydroxy-5-*cis*-dodecenoate; C12, 3-hydroxydodecanoate; C14:1, 3-hydroxy-7-*cis*-tetradecenoate; C14, 3-hydroxytetradecanoate; —, not detectable.

Recently, it was shown that MCL PHAs can be produced by specific recombinant *E. coli* strains that are deficient in fatty acid *β*-oxidation and in addition contain the PHA polymerase-encoding gene of *P. aeroginosa* or *P. oleovorans* (130,131). The recombinant strains were cultured on LB medium or mineral salt medium supplemented with yeast extract and alkanoates as additional carbon source. Up to 21% PHA per cell dry weight could be achieved by these recombinant *E. coli* strains.

Recovery

Recovery procedures for MCL PHA resemble those originally developed for the production of PHB. A number of solvent extraction processes have been assessed to separate MCL PHAs from biomass. These usually involve the use of a chlorinated solvent such as chloroform (111) or methylene chloride. An alternative, nonsolvent-based extraction process was developed as well to make the overall production process more attractive (132,133). The biomass is separated from the medium by centrifugation, and treated with a protease cocktail and a detergent to solubilize all cell components. Removal of the solubilized cell material and concentration of the resulting PHA suspension is achieved by cross-flow microfiltration. The submicrometer MCL PHA granules display a density close to that of water (58), as a result of which a MCL PHA suspension does not settle (134); in fact, it forms a highly stable polymer latex. The overall purity of the latex amounts to 95%. MCL PHAs produced for applications that require higher purity, like those in biomedicine, must be further purified by extraction of PHA from the latex using methylene chloride and subsequent precipitation in a methanol–water mixture.

The applicability of MCL PHA as a latex product has been evaluated by Marchessault et al. (135). A typical latex product would be a coating. Unlike that reported for PHB and its derivatives (135), additional treatments (e.g., heat, solvent vapor) to enhance coalescence into a film are not necessary, because MCL PHA does not readily crystallize. Heat, UV, or electron beam sources have been applied to cross-link polymer chains and fix the film. Mcl-PHA coatings are transparent with good clarity and exhibit excellent barrier properties to water.

Properties of MCL PHAs

Owing to their enzymatic synthesis, as is the case for PHB, MCL PHAs have an exceptional stereochemical regularity and therefore are fully isotactic. This allows MCL PHAs to achieve some crystallinity, which can be as high as 25% (136). The molecular weight is of the order of 2×10^5 g/mol (Table 5) (45,112). The glass transition temperature is usually well below room temperature, ranging from −43 °C to −25 °C for MCL PHAs (112). Melting temperatures depend on both the thermal history and the nature of the pendant chain. Values reported vary between 39 °C and 61 °C with MCL PHA, based on *n*-octane or *n*-octanoate (PHO) (112,136,137). The glass transition temperature of PHO is −35 °C (112,138,139). Depending on the thermal history and the test conditions, the tensile modulus varies between 2.5 and 9 MPa, the tensile strength between 6 and 10 MPa, and the maximum elongation between 300 and 450% (138,139). These data classify PHO as a thermoplastic elastomer (136).

Because the low melting temperature and crystallization rate hamper the applicability of MCL PHAs as thermoplasts, cross-linking was anticipated to overcome both limitations (138). Cross-linking was achieved by conventional techniques, namely electron-beam irradiation (138), peroxide cross-linking (140), or vulcanization (141). In all three studies, the presence of olefin groups was shown to enhance chemical cross-linking.

Sufficient cross-linking appeared to prevent polymer crystallization, so a true rubber was produced. Properties were therefore constant over a wide temperature range from the glass transition temperature (T_g, − 30 °C) up to the decomposition temperature (~180 °C) (138). The rubber modulus could be controlled up to 10 MPa by varying the cross-link density (138). Similar to PHO, maximum elongations amounted to more than 300% for the irradiated material, but other cross-linking techniques produced lower values. Peroxide cross-linking even resulted in a loss of material integrity (140). The tensile strength of the cross-linked material never exceeded 1 MPa and tear resistance was poor (140,141).

BIODEGRADATION

PHAs can be degraded either intracellularly by intracellular depolymerases of the accumulating strain, to supply accumulated carbon and energy in the absence of a suitable exogenous sources (43), or extracellularly by extracellular depolymerases for consumption of oligomers or monomers by fungi, yeast, and bacteria. Relatively little is known about intracellular degradation of PHAs. It was shown that intracellular PHA depolymerases are unable to hydrolyze extracellular PHAs, and extracellular depolymerases cannot hydrolyze intracellular granules (142), apparently as the result of differences in the physical structures of the completely amorphous intracellular granules and the partially crystallized extracellular PHA. Under sterile or aseptic conditions, PHA is degraded by a hydrolytic mechanism, especially at high pH and high temperature (143,144). This type of degradation is important for medical applications, such as the use of PHA in drug release carriers or surgical sutures.

Because PHAs are so abundant, a wide range of bacteria and fungi have developed the capability to utilize PHAs (145). It has been shown that PHA degradation occurs in a large variety of complex ecosystems, including oxic and anoxic environments such as soil, seawater, or sludge. Microbiological parameters of the ecosystem, like population density, microbial diversity, microbial activity, and spatial distribution of microorganisms, can influence the degradation of the polymers. Furthermore, physicochemical parameters of the ecosystem, such as temperature, pH, water content, oxygen content, redox potential, and nutrient supply, can all affect biodegradation. The polymer itself influences biodegradability as well. The most important factors are (*1*) the stereoregularity, (*2*) the crystallinity, (*3*) the molecular mass, and (*4*) the monomeric composition of the polymer. Finally, material processing can also affect the extent of biodegradability.

The effect of different environments on the degradation rate of PHB and PHB/HV has been studied by different groups (146–154). It was shown that the polymer (1-mm molding) was completely degraded after 6 weeks in anaerobic sewage, after 75 weeks in soil, and after 350 weeks in seawater. Concerning functional MCL PHAs, it could be shown that even after cross-linking of unsaturated side chains the polymer remained biodegradable (138). However, it should be noted that at higher cross-link densities the rate of degradation was reduced.

Eleven groups of PHA-degrading bacteria can be distinguished by their substrate (PHA) specificities (155). In addition, 95 genera of fungi capable of degrading PHAs have been identified (155). The synthesis of PHA depolymerases is highly regulated. Most PHA-degrading bacteria repress depolymerase synthesis in the presence of a soluble carbon source that permits high growth rates. However, after exhaustion of the nutrients, the synthesis of depolymerase is derepressed (156,157).

Table 5. Physical Characteristics of Saturated and Unsaturated MCL Polyhydroxyalkanoates Isolated from *Pseudomonas* Strains

Strain	Substrate	Mw	Mn	Mw/Mn	T_g (°C)	T_m (°C)	Reference
P. oleovorans	*n*-Octane	178,000	99,000	1.8	−36.5	58.5	112
P. oleovorans	*n*-Octene	242,000	101,000	2.4	−36.6	—	112
P. putida KT2442	Oleic acid	140,000	70,000	1.9	ND	ND	96
Pseudomonas sp. NCIMB40135	Glucose	143,000	61,900	2.32	< −60	54.8	115

Note: Mw, weight-average molecular weight; Mn, absolute number of average molecular weight; ND, not determined.

The extracellular PHA depolymerases of many bacteria have been purified and characterized. Depending on the depolymerase, the hydrolysis products are monomers [e.g., *Comamonas* sp. (156)] or monomers and dimers [e.g., *Pseudomonas fluorescens* (157)] or a mixture of oligomers (monomers to trimers) as in the case of *Alcaligenes faecalis* and *Pseudomonas lemoignei* depolymerases (158). In a second step, the oligomers are hydrolyzed to monomers by oligomer hydrolases (159,160). All PHA depolymerases analyzed so far are specific for polymers that consist of monomers in the (*R*)-configuration. Chemically synthesized poly-(*S*)-hydroxybutyrate is not degraded by PHB depolymerases (161).

Nine bacterial PHA depolymerase genes that encode depolymerases with specificities for SCL PHAs have been cloned and analyzed (Ref. 155 and references cited therein). The enzymes contain a signal peptide (which is cleaved off during passage across the plasma membrane), a large catalytic domain at the N-terminus, a C-terminal substrate-binding domain, and a linking region in between. Three strictly conserved amino acids, namely serine, aspartate, and histidine, constitute the active center of the catalytic domain. The serine is part of the lipase-box motif Gly-X-Ser-X-Gly, which has been found in all known serine hydrolases such as lipases, esterases, amd serine proteases (162). The oxygen atom of the serine side chain is the nucleophile that attacks the ester bond and is supported by the imidazole ring of the histidine. The positive charge of the latter is stabilized by the carboxyl group of the aspartate.

Only one gene that encodes an extracellular MCL PHA depolymerase has been cloned and analyzed so far (157,163). In contrast to most SCL PHA depolymerases, the MCL PHA depolymerase of *P. fluorescens* GK13 consists of two identical subunits, and except for small regions adjacent to the lipase-box motif there are no significant homologies to SCL PHA depolymerases.

APPLICATIONS

PHAs display an unique combination of features. First, in contrast to synthetic polymers, the natural PHAs have the fundamental advantage of being renewable resources not dependent on the supply of petroleum. Second, as described in the previous section, they are genuinely biodegradable, that is, they can be completely metabolized into harmless, naturally occurring molecules. Although complete degradation can easily be achieved, the shelf life of the PHA products is virtually unlimited. Colonization of the hydrophobic surface is an arduous task for the degrading organisms, and thus a high biological activity is a prerequisite for biodegradation. Cheaper biodegradables, such as the starch-based plastics, are available however. Therefore, the most important feature of these polyesters may well be their hydrophobicity, which allows them to surpass carbohydrate-based biodegradable competitors in moisture resistance.

Applications of PHB

As mentioned previously, PHB and its derivates are all highly crystalline materials and therefore possess a high modulus, making them suitable for polypropylene-type applications including injection molding, extrusion blow molding, and fiber-spray gun molding. The polymers can be manufactured into many different materials with a wide range of applications (164). The first commercial application of PHB or PHB/HV, sold under the trade name Biopol™, was in packaging for personal hygiene products, for example, as biodegradable shampoo bottles. Later applications included packing containers, bottles, tubes, wrappings, bags, tubes, and disposable razors. In these cases the rather costly Biopol™ was used for its environmentally friendly image to increase product sales. In addition, applications that require biodegradability for functional reasons were developed. Among those, garbage bags, which should ideally be composted together with their contents, fishing nets, which can cause severe damage once they are lost in the marine environment, and agricultural delivery devices for nutrients, fertilizers, herbicides, and insecticides are important examples.

Applications of MCL PHA

No applications for MCL PHAs have been reported yet. Complementing the intrinsically rigid and brittle properties of PHB and its derivates, MCL PHAs are expected to support applications that require flexibility and elastomeric properties. MCL PHAs will have very limited applicability as thermoplasts. Amorphous MCL PHAs, cross-linked and non-cross-linked, are more likely to penetrate the plastics market (110). The capability of pseudomonads to incorporate many different monomer units into MCL PHAs suggests various possibilities to produce tailor-made polymers. Incorporation of unsaturated pendant groups to enhance cross-linking is one example (138). A recent study reported that nitrophenoxy groups have been introduced, aiming to create chiral polymers for nonlinear optical applications (128). Specific functionalities for surface modification could be introduced as well. In the long run, it should be feasible to custom design new polymers and then program the bacteria to produce them.

PHA Application in Blends and Composites

Polymer blending offers interesting possibilities of preparing inexpensive (biodegradable) materials with useful mechanical properties. In this context, PHAs could form an important source of biodegradable additives. Several blending studies have been performed on PHB or PHB/HV copolymers with other dissimilar compounds. For example, the miscibility and compatibility of these polyesters with chlorinated polymers, polyamides, polycarbonates, cellulose derivatives, and other functional polymers are well documented (85,165). At present, applications of PHAs as commodities are primarily limited by their high cost, and thus attention is being focused on composite products with plastics constituting only a minor part of the product, such as paper coatings like the plastic film moisture barrier in food or drink cartons and in sanitary napkins. In these cases, total product biodegradability can be attained for little extra cost.

Medical Applications

A notable feature of PHB is that it is biocompatible, producing a mild foreign-body response to any implant. Degradation of PHB in vivo is significantly faster than the in vitro hydrolysis in neutral buffer at body temperature, indicating that enzymes present in vivo catalyze the degradation process (166). The only degradation product is (*R*)-3-hydroxybutyric acid, which is a normal mammalian metabolite and is resorbed without metabolic or histological disturbance. Like PHB, MCL PHA might be biocompatible and useful in biomedical applications. The rigid PHB might be useful in bone fracture fixation (167), whereas the flexible elastomeric MCL PHAs might find uses in vascular grafts, surgical swabs, wound dressings, breast prostheses, and other implants (110).

Chiral Synthons

Hydrolysis of PHAs yields chiral (*R*)-3-hydroxy fatty acids, which could be a feedstock for stereospecific synthesis and the associated pharmaceutical industry (27,168). Concerning the enormous number of monomers that have been incorporated into various polymers, many different synthon precursors could be provided by hydrolyzing the polymers.

ECONOMICS

The market price of Biopol when produced by ZENECA BioProducts was about \$16 kg^{-1} (169). Current advances in fermentation and purification technology, as well as the application of metabolic engineering techniques for expanding the spectrum of utilizable substrates and for improving polymer concentration in the cells, are likely to lower the price of PHB to \$4 kg^{-1} (169). A preliminary economic study for the production of mcl-PHAs showed that the price for octanoate-based PHAs would be of the order of \$20 kg^{-1} for an annual production of 1,000 tons (110,132). Optimizing the polymer content of the cells, using cheaper carbon sources as substrates, reducing the chemicals used during downstream processing, and reducing energy consumption is expected to lower the MCL PHA price to a minimum of \$5 kg^{-1} (110,132).

It can be assumed that production of PHAs by using native or recombinant bacteria is in the long term only required for development purposes and for the production of specific functionalized PHAs. It is more likely that efforts in generating transgenic plants for PHA production purposes will succeed in ultimately reducing the price of the polymer to that of other large-scale agromaterials such as starch or conventional commodity plastics, or less than \$1 kg^{-1} (103,170). Therefore, an approach to insert the PHB biosynthesis genes into crop plants (e.g., soybean and rapeseed), which have the agronomic properties to provide high yields of PHB per hectare, has been undertaken by CEREGEN/Monsanto and is currently under way.

OUTLOOK

The development of PHAs as a commercial product remains limited to PHB and its copolymer, PHB/HV. As yet, no applications in the fabrication of articles made of MCL PHAs have been reported. There are probably at least two reasons for this. There has been little material available for development work, and the lower polymer concentration in the cells and the different recovery process means that the material cost of MCL PHA will remain even higher than for PHB for the foreseeable future.

Alternative strategies for PHA production are shown in Figure 8. Future large-scale production of PHB (and MCL PHA) in transgenic plants should ultimately provide a cheaper and sustainable technology for the production of these materials. This advance is essential if PHAs are to be used for commodity purposes and must compete on the plastic market. Improvements in the competitiveness of PHAs on the market could also derive from environmental legislation and from decreasing availability of oil and concomitant rising prices of synthetic plastics.

The development of polymers containing exotic monomers into commercial products will depend on the identification of niche applications that will support the high cost of production arising from both the cost of the monomers used and the small scale of the operation to make these specialty materials. Such applications are likely to include nonlinear optical devices or medical implants of the isotactic polymers, and the use of the stereospecific monomers as chiral synthons for the production of fine chemicals. Furthermore, capsules made of PHA for controlled release of bioactive agents, such as drugs or fertilizers, could find broad application in medicine and agriculture.

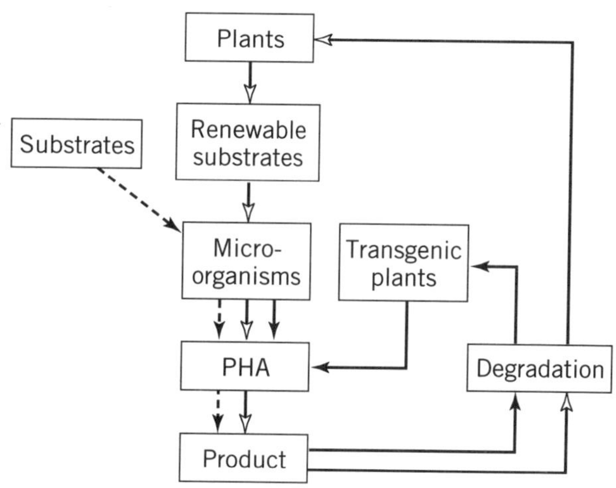

Figure 8. Overview of alternative polyhydroxyalkanoate (PHA) production strategies for different applications. Dependent on the application purpose, three different PHA production strategies can be considered. Bacterial fermentations can be used mainly for production of smaller amounts of functionalized, high-cost PHAs for specialties (--▶). Application of renewable sources as sole carbon source in bacterial fermentations or in addition to cosubstrates containing a specific functional group will reduce the price of the product (——▷). Low-cost products with specific properties could be produced by this strategy. Production of inexpensive and biodegradable PHAs for commodities (——▶) will be best achieved by large-scale cultivation of transgenic plants.

BIBLIOGRAPHY

1. M. Lemoigne, *Bull. Soc. Chem. Biol.* **8**, 770–782 (1926).
2. S.Y. Lee, *Biotechnol. Bioeng.* **49**, 1–14 (1996).
3. A. Steinbüchel, in D. Byrom ed., *Biomaterials. Novel Materials from Biological Sources*, Macmillan, Basingstoke, 1991, pp. 123–213.
4. C. Sasikala and C.V. Ramana, in S.L. Neidleman and A.I. Laskin eds., *Advances in Applied Microbiology*, vol. 42, Academic Press, San Diego, Calif., 1996, pp. 97–218.
5. M.J. de Smet, G. Eggink, B. Witholt, J. Kingma, and H. Wynberg, *J. Bacteriol.* **154**, 870–878 (1983).
6. A. Steinbüchel and H.E. Valentin, *FEMS Microbiol. Lett.* **128**, 219–228 (1995).
7. R.W. Lenz, Y.B. Kim, and R.C. Fuller, *FEMS Microbiol. Rev.* **103**, 207–214 (1992).
8. A. Steinbüchel, E. Hustede, M. Liebergesell, U. Pieper, A. Timm, and H. Valentin, *FEMS Microbiol. Rev.* **103**, 217–230 (1992).
9. R.N. Reusch and H.L. Sadoff, *Proc. Natl. Acad. Sci. U.S.A.* **85**, 4176–4180 (1988).
10. R.N. Reusch, *Proc. Soc. Exp. Biol. Med.* **191**, 377–381 (1989).
11. R.N. Reusch, R.P. Huang, and L.L. Bramble, *Biophys. J.* **69**, 754–766 (1995).
12. D. Seebach, A. Brunner, H.M. Burger, R.N. Reusch, and L.L. Bramble, *Helv. Chim. Acta* **79**, 507–517 (1996).
13. R.N. Reusch and H.L. Sandoff, *J. Bacteriol.* **156**, 778–788 (1983).
14. R.P. Huang, and R.N. Reusch, *J. Bacteriol.* **177**, 486–490 (1995).
15. C.E. Castuma, R.P. Huang, A. Kornberg, and R.N. Reusch, *J. Biol. Chem.* **270**, 12980–12983 (1995).
16. R.N. Reusch, A.W. Sparrow, and J. Gardiner, *Biochim. Biophys. Acta* **1123**, 33–40 (1992).
17. G. Gottschalk, *Arch. Mikrobiol.* **47**, 225–229 (1964).
18. G. Gottschalk, *Arch. Mikrobiol.* **47**, 230–235 (1964).
19. G.W. Haywood, A.J. Anderson, and E.A. Dawes, *FEMS Microbiol. Lett.* **57**, 1–6 (1989).
20. G.J. Moskowitz and J.M. Merrik, *Biochemistry* **8**, 2748–2755 (1969).
21. H.G. Schlegel, G. Gottschalk, and R. von Bartha, *Nature* **191**, 463–465 (1961).
22. A. Steinbüchel and H.G. Schlegel, *Appl. Microbiol. Biotechnol.* **31**, 168–175 (1989).
23. P.J. Senior and E.A. Dawes, *Biochem. J.* **134**, 225–238 (1971).
24. V. Oeding and H.G. Schlegel, *Biochem. J.* **134**, 239–248 (1973).
25. G.W. Haywood, A.J. Anderson, L. Chu, and E.A. Dawes, *FEMS Microbiol. Lett.* **52**, 91–96 (1988).
26. G.W. Haywood, A.J. Anderson, L. Chu, and E.A. Dawes, *FEMS Microbiol. Lett.* **52**, 259–264 (1988).
27. P.A. Holmes, in D.C. Bassett ed., *Development in Crystalline Polymers*, vol. 2, Elsevier, London, 1988, pp. 1–65.
28. M. Akiyama, Y. Taima, and Y. Doi, *Appl. Microbiol. Biotechnol.* **37**, 698–701 (1992).
29. G. Eggink, H. van der Wal, G.N.M. Huijberts, and P. de Waard, *Ind. Crops Prod.* **1**, 157–163 (1993).
30. G. Lefebvre, M. Rocher, and G. Braunegg, *Appl. Environ. Microbiol.* **63**, 827–833 (1997).
31. A.J. Anderson and E.A. Dawes, *Microbiol. Rev.* **54**, 450–472 (1990).
32. A. Steinbüchel and U. Pieper, *Appl. Microbiol. Biotechnol.* **37**, 1–6 (1992).
33. A. Steinbüchel, E.M. Debzi, R.H. Marchessault, and A. Timm, *Appl. Microbiol. Biotechnol.* **39**, 443–449 (1993).
34. Y. Doi, A. Segawa, and M. Kunioka, *Int. J. Biol. Macromol.* **12**, 106–111 (1990).
35. Y. Doi, A. Tamaki, M. Kunioka, and K. Soga, *Macromol. Chem. Rapid Commun.* **8**, 631–635 (1987).
36. H.E. Valentin, A. Schönebaum, and A. Steinbüchel, *Appl. Microbiol. Biotechnol.* **36**, 507–514 (1992).
37. B. Füchtenbusch, D. Fabritius, and A. Steinbüchel, *FEMS Microbiol. Lett.* **138**, 153–160 (1996).
38. O.P. Peoples and A.J. Sinskey, *J. Biol. Chem.* **264**, 15298–15303 (1989).
39. P. Schubert, E.F. Steinbüchel, and H.G. Schlegel, *J. Bacteriol.* **170**, 5837–5847 (1988).
40. S.C. Slater, W.H. Voige, and D.E. Dennis, *J. Bacteriol.* **170**, 4431–4436 (1988).
41. G.W. Haywood, A.J. Anderson, L. Chu, and E.A. Dawes, *Biochem. Soc. Trans.* 1046–1047 (1988).
42. T.U. Gerngross, K.D. Snell, O.P. Peoples, A.J. Sinskey, E. Csuhai, S. Masamune, and J. Stubbe, *Biochemistry* **33**, 9311–9320 (1994).
43. R. Griebel, Z. Smith, and J.M. Merrick, *Biochemistry* **7**, 3676–3681 (1968).
44. J. Wodzinska, K.D. Snell, A. Rhomberg, A.J. Sinskey, K. Biemann, and J. Stubbe, *J. Am. Chem. Soc.* **118**, 6319–6320 (1996).
45. K. Hori, K. Soga, and Y. Doi, *Biotechnol. Lett.* **16**, 709–714 (1994).
46. F. Koizumi, H. Abe, and Y. Doi, *J. Macromol. Sci. Pure Appl. Chem.* **A32**, 759–774 (1995).
47. S. Kusaka, H. Abe, S.Y. Lee, and Y. Doi, *Appl. Microbiol. Biotechnol.* **47**, 140–143 (1997).
48. T.U. Gerngross and D.P. Martin, *Proc. Natl. Acad. Sci. U.S.A.* **92**, 6279–6283 (1995).
49. M. Liebergesell and A. Steinbüchel, *Appl. Microbiol. Biotechnol.* **38**, 493–501 (1993).
50. M. Liebergesell and A. Steinbüchel, *Eur. J. Biochem.* **209**, 153–150 (1992).
51. A. Steinbüchel, in H.-J. Rehm and G. Reed eds., *Biotechnology*, vol. 6, VCH, Weinheim, 1996, pp. 405–464.
52. Y. Kawaguchi and Y. Doi, *FEMS Microbiol. Lett.* **70**, 151–156 (1990).
53. G.N. Barnard and J.K.M. Sanders, *J. Biol. Chem.* **264**, 3286–3291 (1988).
54. S.R. Amor, T. Rayment, and J.K.M. Sanders, *Macromolecules* **24**, 4583–4588 (1991).
55. K.W. Nickerson, *Appl. Environ. Microbiol.* **43**, 1208–1209 (1982).
56. H. Bauer and A.J. Owen, *Colloid Polym. Sci.* **266**, 241–247 (1988).
57. D.M. Horowitz, J. Clauss, B.K. Hunter, and J.K.M. Sanders, *Nature* **363**, 23 (1993).
58. H. Preusting, J. Kingma, G. Huisman, A. Steinbüchel, and B. Witholt, *J. Environ. Polym. Degrad.* **1**, 11–21 (1993).
59. G.W. Huisman, Ph.D. Thesis, Rijksuniversiteit Groningen, The Netherlands, 1991.

60. H. Preusting, J. Kingma, and B. Witholt, *Enzyme Microb. Technol.* **13**, 770–780 (1991).
61. E.S. Stuart, R.W. Lenz, and R.C. Fuller, *Can. J. Microbiol.* **41**, 84–93 (1995).
62. F. Mayer and M. Hoppert, *J. Basic Microbiol.* **37**, 45–52 (1997).
63. A. Steinbüchel, K. Aerts, W. Babel, C. Follner, M. Liebergesell, M.H. Madkour, F. Mayer, U. Pieper-Fürst, A. Pries, H.E. Valentin, and R. Wieczorek, *Can. J. Microbiol.* **41**, 94–105 (1995).
64. R.C. Fuller, J.P. O'Donnel, J. Saulnier, T.E. Redlinger, J. Foster, and R.W. Lenz, *FEMS Microbiol. Rev.* **103**, 279–288 (1992).
65. R. Wieczorek, A. Pries, A. Steinbüchel, and F. Mayer, *J. Bacteriol.* **177**, 2425–2435 (1995).
66. U. Pieper-Fürst, M.H. Madkour, F. Mayer, and A. Steinbüchel, *J. Bacteriol.* **176**, 4328–4337 (1994).
67. U. Pieper-Fürst, M.H. Madkour, F. Mayer, and A. Steinbuchel, *J. Bacteriol.* **177**, 2513–2523 (1995).
68. M. Liebergesell, B. Schmidt, and A. Steinbüchel, *FEMS Microbiol. Lett.* **99**, 227–232 (1992).
69. M. Liebergesell and A. Steinbüchel, *Biotechnol. Lett.* **18**, 719–724 (1996).
70. R. Wieczorek, A. Steinbüchel, and B. Schmidt, *FEMS Microbiol. Lett.* **135**, 23–30 (1996).
71. M.A. Schembri, A.A. Woods, R.C. Bayly, and J.K. Davies, *FEMS Microbiol. Lett.* **133**, 277–283 (1995).
72. C.G. Follner, W. Babel, and A. Steinbüchel, *Can. J. Microbiol.* **41**, 124–130 (1995).
73. C.G. Follner, M. Madkour, F. Mayer, W. Babel, and A. Steinbüchel, *J. Basic Microbiol.* **37**, 11–21 (1997).
74. D. Byrom, *FEMS Microbiol. Rev.* **103**, 247–250 (1992).
75. T. Yamane, *FEMS Microbiol. Rev.* **103**, 257–264 (1992).
76. H.G. Schlegel and G. Gottschalk, *Biochem. Z.* **342**, 249–259 (1965).
77. D. Byrom, in E.A. Dawes ed., *Novel Biodegradable Microbial Polymers*, Kluwer, Dordrecht, 1990, pp. 113–117.
78. U.J. Hänggi, in E.A. Dawes ed., *Novel Biodegradable Microbial Polymers*, Kluwer, Dordrecht, 1990, pp. 65–70.
79. O. Hrabak, *FEMS Microbiol. Rev.* **103**, 251–256 (1992).
80. D.R. Williamson and J.F. Wilkinson, *J. Gen. Microbiol.* **19**, 198–209 (1958).
81. E. Berger, B.A. Ramsay, J.A. Ramsay, C. Chaverie, and G. Braunegg, *Biotechnol. Tech.* **3**, 227–232 (1989).
82. D. Byrom, *Trends Biotechnol.* **5**, 246–250 (1987).
83. Y. Doi, *Microbial Polyesters*, VCH, New York, 1990.
84. P.J. Hocking and R.H. Marchessault, in G.J.L. Griffin ed., *Chemistry and Technology of Biodegradable Polymers*, Blackie A&P, Glasgow, 1994, pp. 48–96.
85. R. Sharma and A.R. Ray, *J. Macromol. Sci. Rev. Macromol.* **C35**, 327–359 (1995).
86. S. Slater, T. Gallaher, and D. Dennis, *Appl. Environ. Microbiol.* **58**, 1089–1094 (1992).
87. S.Y. Lee and H.N. Chang, *Can. J. Microbiol.* **41**, 207–215 (1995).
88. S.Y. Lee, K.S. Yim, H.N. Chang, and Y.K. Chang, *J. Biotechnol.* **32**, 203–211 (1994).
89. S.J. Sim, K.D. Snell, S.A. Hogan, J. Stubbe, C.K. Rha, and A.J. Sinskey, *Nat. Biotechnol.* **15**, 63–67 (1997).
90. Y. Poirier, D.E. Dennis, K. Klomparens, and C. Somerville, *Science* **256**, 520–523 (1992).
91. C. Nawrath, Y. Poirier, and C. Somerville, *Proc. Natl. Acad. Sci. U.S.A.* **91**, 12760–12764 (1994).
92. Y. Poirier, C. Somerville, L.A. Schechtman, M.M. Satkowski, and I. Noda, *Int. J. Biol. Macromol.* **17**, 7–12 (1995).
93. T.A. Leaf, M.S. Peterson, S.K. Stoup, D. Somers, and F. Srienc, *Microbiology UK* **142**, 1169–1180 (1996).
94. M.D. Williams, J.A. Rahn, and D.H. Sherman, *Appl. Environ. Microbiol.* **62**, 2540–2546 (1996).
95. M.D. Williams, A.M. Fieno, R.A. Grant, and D.H. Sherman, *Protein Expr. Purif.* **7**, 203–211 (1996).
96. G.N.M. Huijberts, G. Eggink, P. de Waard, G.W. Huisman, and B. Witholt, *Appl. Environ. Microbiol.* **58**, 536–544 (1992).
97. G.N.M. Huijberts, T.C. de Rijk, P. de Waard, and G. Eggink, *J. Bacteriol.* **176**, 1661–1666 (1995).
98. H. Brandl, R.A. Gross, R.W. Lenz, and R.C. Fuller, *Appl. Environ. Microbiol.* **54**, 1977–1982 (1988).
99. G.W. Haywood, A.J. Anderson, and E.A. Dawes, *Biotechnol. Lett.* **11**, 471–476 (1989).
100. G.W. Huisman, O. de Leeuw, G. Eggink, and B. Witholt, *Appl. Environ. Microbiol.* **55**, 1949–1954 (1989).
101. G.W. Huisman, E. Wonink, R. Meima, B. Kazemier, P. Terpstra, and B. Witholt, *J. Biol. Chem.* **266**, 2191–2198 (1991).
102. A. Timm and A. Steinbüchel, *Eur. J. Biochem.* **209**, 15–30 (1992).
103. F.R. van der Leij and B. Witholt, *Can. J. Microbiol.* **41**, 222–238 (1995).
104. G.W. Huisman, E. Wonink, G.J.M. de Koning, H. Preusting, and B. Witholt, *Appl. Microbiol. Biotechnol.* **38**, 1–5 (1992).
105. D. Jendrossek, B. Müller, and H.G. Schlegel, *Eur. J. Biochem.* **218**, 701–710 (1993).
106. B.A. Ramsay, I. Saracovan, J.A. Ramsay, and R.H. Marchessault, *Appl. Environ. Microbiol.* **58**, 744–746 (1992).
107. A. Timm and A. Steinbüchel, *Appl. Environ. Microbiol.* **56**, 3360–3367 (1990).
108. H. Preusting, W. Hazenberg, and B. Witholt, *Enzyme Microb. Technol.* **15**, 311–316 (1993).
109. H. Preusting, R. van Houten, A. Hoefs, E. Kool van Langenberghe, O. Favre-Bulle, and B. Witholt, *Biotechnol. Bioeng.* **41**, 550–556 (1993).
110. G.J.M. de Koning, M.B. Kellerhals, C. van Meurs, and B. Witholt, *J. Environ. Polym. Degrad.* **4**, 243–252 (1996).
111. R.G. Lageveen, G.W. Huisman, H. Preusting, P. Ketelaar, G. Eggink, and B. Witholt, *Appl. Environ. Microbiol.* **54**, 2924–2932 (1988).
112. H. Preusting, A. Nijenhuis, and B. Witholt, *Macromolecules* **23**, 4220–4224 (1990).
113. W.M. Hazenberg, Ph.D. Thesis, ETH Zürich, Switzerland, 1997.
114. D. Carnicero, M. Fernandez Valverde, L.M. Canedo, C. Schleissner, and J.M. Luengo, *FEMS Microbiol. Lett.* **149**, 51–58 (1997).
115. G.W. Haywood, A.J. Anderson, D.F. Ewing, and E.A. Dawes, *Appl. Environ. Microbiol.* **56**, 3354–3359 (1990).
116. G.N.M. Huijberts, Ph.D. Thesis, Rijksuniversiteit Groningen, The Netherlands (1996).
117. K. Fritzsche, R.W. Lenz, and R.C. Fuller, *Int. J. Biol. Macromol.* **12**, 85–91 (1990).
118. Y.B. Kim, R.W. Lenz, and R.C. Fuller, *J. Polym. Sci. Appl. Polym. Chem.* **33**, 1367–1374 (1995).
119. Y.B. Kim, R.W. Lenz, and R.C. Fuller, *Macromolecules* **25**, 1852–1857 (1992).

120. C. Abe, Y. Taima, Y. Nakamura, and Y. Doi, *Polym. Commun.* **31**, 404–406 (1990).
121. O. Kim, R.A. Gross, W.J. Hammar, and R.A. Newmark, *Macromolecules* **29**, 4572–4581 (1996).
122. Y. Doi and C. Abe, *Macromolecules* **23**, 3705–3707 (1990).
123. K. Fritzsche, R.W. Lenz, and R.C. Fuller, *Makromol. Chem.* **191**, 1957–1965 (1990).
124. J.M. Curley, B. Hazer, R.W. Lenz, and R.C. Fuller, *Macromolecules* **29**, 1762–1766 (1996).
125. Y.B. Kim, R.W. Lenz, and R.C. Fuller, *Macromolecules* **24**, 5256–5260 (1991).
126. J.J. Song and S.C. Yoon, *Appl. Environ. Microbiol.* **62**, 536–544 (1996).
127. H. Ritter and A.G. Vonspee, *Macromol. Chem. Phys.* **195**, 1665–1672 (1994).
128. O.Y. Kim, R.A. Gross, and D.R. Rutherford, *Can. J. Microbiol.* **41**, 32–43 (1995).
129. R.A. Gross, O. Kim, D.R. Rutherford, and R.A. Newmark, *Polym. Int.* **39**, 205–213 (1996).
130. S. Langenbach, B.H.A. Rehm, and A. Steinbüchel, *FEMS Microbiol. Lett.* **150**, 303–309 (1997).
131. Q. Ren, Ph.D. Thesis, ETH, Zurich, Switzerland, 1997.
132. G.J.M. de Koning, M. Kellerhals, C. van Meurs, and B. Witholt, *Bioprocess Eng.* **17**, 15–21 (1997).
133. G.J.M. de Koning and B. Witholt, *Bioprocess Eng.* **17**, 7–13 (1997).
134. R.H. Marchessault, F.G. Morin, S. Wong, and I. Saracovan, *Can. J. Microbiol.* **41**, 138–142 (1995).
135. R.H. Marchessault, C.J. Monasterios, and P. Lepoutre, in E.A. Dawes ed., *Novel Biodegradable Microbial Polymers*, Kluwer, Dordrecht, 1990, pp. 97–112.
136. R.H. Marchessault, C.J. Monasterios, F.G. Morin, and P.R. Sundararajan, *Int. J. Biol. Macromol.* **12**, 158–165 (1990).
137. R.A. Gross, C. DeMello, R.W. Lenz, H. Brandl, and R.C. Fuller, *Macromolecules* **22**, 1106–1115 (1989).
138. G.J.M. de Koning, H.H.M. van Bilsen, P.J. Lemstra, W. Hazenberg, B. Witholt, H. Preusting, J.G. van der Galiën, A. Schirmer, and D. Jendrossek, *Polymer* **35**, 2090–2097 (1994).
139. K.D. Gagnon, R.W. Lenz, R.J. Farris, and R.C. Fuller, *Macromolecules* **25**, 3723–3728 (1992).
140. K.D. Gagnon, R.W. Lenz, R.J. Farris, and R.C. Fuller, *Polymer* **35**, 4358–4367 (1994).
141. K.D. Gagnon, R.W. Lenz, R.J. Farris, and R.C. Fuller, *Polymer* **35**, 4368–4375 (1994).
142. J.M. Merrick and M. Doudoroff, *J. Bacteriol.* **88**, 60–71 (1964).
143. S.J. Holland, A.M. Jolly, M. Yasin, and B.J. Tighe, *Biomaterials* **8**, 289–295 (1987).
144. N.D. Miller and D.F. Williams, *Biomaterials* **8**, 129 (1987).
145. H. Brandl, R. Bachofen, J. Mayer, and E. Wintermantel, *Can. J. Microbiol.* **41**, 143–153 (1995).
146. J. Mergaert, G. Glorieux, L. Hauben, V. Storms, M. Mau, and J. Swings, *Syst. Appl. Microbiol.* **19**, 407–413 (1996).
147. J. Mergaert, A. Wouters, C. Anderson, and J. Swings, *Can. J. Microbiol.* **41**, 154–159 (1995).
148. J. Mergaert, A. Webb, C. Anderson, A. Wouters, and J. Swings, *Appl. Environ. Microbiol.* **59**, 3236–3238 (1993).
149. D.F. Gilmore, S. Antoun, R.W. Lenz., S. Goodwin, R. Austin, and R.C. Fuller, *J. Ind. Microbiol.* **10**, 199–206 (1992).
150. D.F. Gilmore, S. Antoun, R.W. Lenz, and R.C. Fuller, *J. Environ. Polym. Degrad.* **1**, 269–274 (1993).
151. B.H. Briese, D. Jendrossek, and H.G. Schlegel, *FEMS Microbiol. Lett.* **119**, 396 (1994).
152. L.R. Krupp and W.J. Jewell, *Environ. Sci. Technol.* **26**, 193–198 (1992).
153. C.L. Yue, R.A. Gross, and S.P. McCarthy, *Polym. Degrad. Stabil.* **51**, 205–210 (1996).
154. H. Abe and Y. Doi, *Macromolecules* **29**, 8683–8688 (1996).
155. D. Jendrossek, A. Schirmer, and H.G. Schlegel, *Appl. Microbiol. Biotechnol.* **46**, 451–463 (1996).
156. D. Jendrossek, I. Knoke, R.B. Habibian, A. Steinbüchel, and H.G. Schlegel, *J. Environ. Polym. Degrad.* **1**, 53–61 (1993).
157. A. Schirmer, D. Jendrossek, and H.G. Schlegel, *Appl. Environ. Microbiol.* **59**, 1220 (1993).
158. K. Nakayama, T. Saito, T. Fukui, Y. Shirakura, and K. Tomita, *Biochim. Biophys. Acta* **827**, 63–72 (1985).
159. F.P. Delafield, K.E. Cooksey, and M. Doudoroff, *J. Biol. Chem.* **240**, 4023–4028 (1965).
160. Y. Shirakura, T. Fukui, T. Tanio, K. Nakayama, R. Matsuno, and K. Tomita, *Biochim. Biophys. Acta* **748**, 331–339 (1983).
161. Y. Doi, Y. Kumagai, N. Tanahashi, and K. Mukai, in M. Vert ed., *Biodegradable Polymers and Plastics*, Royal Soc. of Chemistry, London, 1992, pp. 139–148.
162. K.E. Jäger, A. Steinbüchel, and D. Jendrossek, *Appl. Environ. Microbiol.* **61**, 3113–3118 (1995).
163. A. Schirmer and D. Jendrossek, *J. Bacteriol.* **176**, 7065–7073 (1994).
164. P.A. Holmes, *Phys. Technol.* **16**, 32–36 (1985).
165. H. Verhoogt, B.A. Ramsay, and B.D. Favis, *Polymer* **35**, 5155–5169 (1994).
166. E. Piskin, *J. Biomater. Sci. Polym. Ed.* **6**, 775–795 (1995).
167. C. Doyle, E.T. Tanner, and W. Bonfield, *Biomaterials* **12**, 841 (1991).
168. R.H. Marchessault, T.L. Bluhm, Y. Deslandes, G.K. Hamer, W.J. Orts, P.R. Sundararajan, and M.G. Taylor, *Macromol. Chem.* **19**, 235–254 (1988).
169. S.Y. Lee, *Trends Biotechnol.* **14**, 431–438 (1996).
170. Y. Poirier, C. Nawrath, and C. Somerville, *Biotechnology* **13**, 142–150 (1995).

POLYHYDROXYALKANOATES, SEPARATION, PURIFICATION, AND MANUFACTURING METHODS

JULIANA RAMSAY
Queen's University
Kingston, Ontario
Canada

BRUCE RAMSAY
Queen's University
Kingston, Ontario
Canada

KEY WORDS

Polyhydroxyalkanoates (PHA)
Polyhydroxybutyrate (PHB)
Processing
Production
Separation

OUTLINE

POLY-3-HYDROXYALKANOATES AS BIODEGRADABLE PLASTICS

Microbially produced biodegradable plastics can be divided into four classes: polysaccharides, polymers of amino acids, naturally occurring polyesters, and semisynthetic polyesters (Table 1). All possess linkages that are susceptible to nonspecific, extracellular, microbial enzymes that exist in nature and thus are inherently environmentally degradable. Their structures can be used as the basis for the design of synthetic biodegradable polymers. Unlike many common synthetic plastics, those microbially produced are often at least partially water soluble or tend to be easily hydrolyzed, especially at extremes of temperature and pH. These characteristics aid in their biodegradability but greatly limit their use as plastics. Although their properties as plastics may be greatly improved through chemical modification, care must be taken to maintain environmental degradability.

The ideal plastic is durable, biodegradable, recyclable, and produced from renewable resources such as carbohydrates. Poly-3-hydroxyalkanoates (PHAs) are a family of intracellular polyesters produced by a wide variety of microorganisms. They are water insoluble, relatively resistant to aqueous hydrolysis, and readily environmentally degradable. They can be produced from renewable resources, and, although temperature sensitive, they are potentially recyclable. PHAs have mechanical properties similar to those of some common synthetic plastics. Like conventional thermoplastics, PHAs can be recycled with some loss of molecular weight and mechanical properties. Unlike many conventional polymers, PHAs are completely environmentally degradable and will decompose rapidly to CO_2 and H_2O during composting. Degradation has also been demonstrated in soil burial studies and in the marine environment.

PHAs whose polymeric units are predominantly five carbons or fewer are classified as short-chain-length (SCL) PHAs, and all others are referred to as medium-chain-length (MCL) PHAs. Homopolymeric poly-3-hydroxybutyrate (PHB) has properties that are frequently compared to those of polypropylene, except that it is more brittle and its melting temperature is high enough that there may be significant thermal degradation on melt extrusion. Lower-melting, more-flexible PHB can be obtained by annealing or addition of plasticizers. However, copolymers of 3-hydroxybutyrate and 3-hydroxyvalerate (P[HB-co-HV]) and of 3-hydroxybutyrate and 4-hydroxybutyrate (P[3HB-co-4HB]) are more flexible and have lower melting temperatures than PHB, making them more suitable for the manufacture of extruded products. MCL PHAs have even lower melting points and glass transition temperatures; they are tough, rubberlike thermoelastomers, especially after cross-links between the polymer chains have been formed by heat or chemical induction.

PHAs would seem to be ideal biodegradable plastics, but their production is much different from that of conventional plastics. The following sections describe how PHAs may be produced by widely varying methodologies, with details of diverse approaches to PHA fermentation processes and the separation of the polymer from the biomass. Existing commercial processes and future possibilities for PHA production with an emphasis on cost reduction are also discussed.

Table 1. Classes of Microbially Produced Biodegradable Plastics

Class	Linkage	Example	Effects of H_2O at different pH and temperature	Fermentation characteristics
Polysaccharides	Glycosidic	Pullulan	Often soluble or hydrolyzed	High yield, simple separation but cost increases with purity; production limited by O_2 transfer
Polyaminoacids	Peptide	Poly-γ-glutamate	Often soluble or hydrolyzed	Still experimental; separation simple; yield should be high
Polyesters	Ester	Polyhydroxyalkaoates	Not soluble; resistant to hydrolysis	Yield depends on type of polymer and on the process; separation can be expensive
Semisynthetic polymers of organic acids	Ester	Polylactides	Generally not soluble but are hydrolyzed	High yield

THREE ROUTES TO PHA PRODUCTION

Chemical Synthesis

PHB can be produced synthetically from racemic β-butyrolactone, and P(HB-co-HV) is made from a similar mixture of β-butyrolactone and β-valerolactone with triethylaluminum and water as catalyst (1). Although the stereospecificity of the resulting polymer can be controlled to a certain extent by adjusting the ratio of triethylaluminum and water, synthetic PHAs are usually optically inactive and hence only partially biodegradable.

Synthesis in Transgenic Plants

SCL PHA. PHB can be synthesized in plants in which the bacterial reductase and polymerase genes have been introduced. Although virtually all organisms possess the remaining gene in the SCL PHA pathway, coding for 3-ketothiolase, its activity and specificity vary and it may not be present in all eukaryotic organelles. Production is usually increased by cloning the 3-ketothiolase gene with the other two. Although transgenic plants are easily created, productivity remains problematic. The SCL PHA pathway was first cloned into *Arabidopsis thaliana*, the *Escherichia coli* of plant biotechnology (2). The yield was low, 0.1% dry weight, and the plants were stunted. However, granules similar to those found in bacteria were produced. The initial cloning strategy targeted the cytoplasm, and granules were even found in the cell nucleus (Fig. 1). Subsequent work has targeted the plastid, which is also the site of starch and oil synthesis in plants, resulting in increased yields of up to 14% dry weight in the leaves of *A. thaliana*. These plants grew well but were chlorotic, possibly because of large amounts of PHB in the chloroplasts. Present research centers on oil- and starch-producing plants because these are already adapted to have large amounts of storage material in the plastids (3), but other strategies are also being assessed.

MCL PHA. Because of its high melting point and brittle nature, the PHB homopolymer might have limited application even if its production cost in transgenic plants was low. Although SCL PHA copolymers such as P(HV-co-HB) have been synthesized in plants (H. Valentin, personal communication), it is not obvious how the metabolic pathways of plants can be manipulated to maximize their production. It may be easier to produce MCL PHA in plants. Unfortunately progress is limited by a lack of understanding of the link between fatty acid synthesis and MCL PHA synthesis. Nevertheless, small quantities of MCL PHA have recently been produced in recombinant *A. thaliana* (Y. Poirier, personal communication).

Microbial Synthesis

SCL PHA. Many bacterial species accumulate PHB, but PHA copolymers are more commercially valuable. Most PHB-accumulating bacteria will produce P(HB-co-HV) under the appropriate conditions. When propionic acid is supplied during a fermentation process, some of the acid is converted to propionyl-CoA. A portion of this propionyl-CoA is condensed with acetyl-CoA via β-ketothiolase to produce five-carbon monomers; this is the most common pathway of P(HB-co-HV) production. Because acetate is needed for this synthetic route and some propionic acid is decarboxylated before monomer synthesis, usually less than 50% of the monomers will be HV even if a feed of pure propionic is added (i.e., no glucose or acetate). This yield can be increased in *Ralstonia eutropha* (previously *Alcaligenes eutrophus*) by decreasing the dissolved oxygen concentration in the bioreactor during the accumulation phase (4).

High yields of HV are obtained when pentanoic acid is used because pentanoic acid is metabolized directly to L-(3)-hydroxyvaleryl-CoA, which is subsequently racemized (at least in *R. eutropha*) to D-(3)-hydroxyvaleryl-CoA (5). Although some acetate is produced leading to HB monomer formation, PHAs with greater than 90% HV content can be obtained. SCL PHA producers have not been reported to accumulate monomers with side chains longer than C_2 or C_3 but can produce P(3HB-co-4HB) when fed 4-hydroxybutyric acid or 1,4-butanediol. *R. eutrophus* accumulates terpolyesters of 3HB-, 3HV-, and 5HV-monomers when fed a mixture of pentanoic and 5-chloropentanoic acids. Even among SCL-accumulating microorganisms, there may be differences in the PHA synthetic pathway. For example, in *Rhodospirillum rubrum*, L-(+)-3-hydroxybutyrl-CoA is racemized to D-(−)-3-hydroxybutyrl-CoA using enoyl-CoA hydrases that have not been found in *Azotobacter beijerinckii*.

In most SCL PHA–accumulating bacteria, high intracellular levels of reduced nicotinamide adenine dinucleotide (NADH) and acetyl-CoA greatly stimulate the rate of PHA synthesis. Thus, if the growth rate is limited by a nutrient other than the source of carbon and energy, PHA is accumulated in the cell. For this reason, PHA production is usually divided into two phases. In the first phase, balanced growth proceeds with relatively little PHA accumulation. When the concentration of the limiting nutrient in batch culture falls below a critical value, the rate of PHA synthesis increases while synthesis of other cellular components decreases during the second phase. Because cosubstrates such as propionic acid are expensive and can be used for the synthesis of biomass other than PHA, these chemicals are only added during the accumulation phase. This limitation probably results in nonhomogeneity of polymer composition in different PHA granules with some (produced during the growth phase) containing almost exclusively PHB. There is still debate about the existence and importance of monomer turnover in PHA granules. In contrast, carbon and energy limitations increase depolymerase activity, leading to a decrease in cellular PHA content. The control of these biochemical pathways varies greatly among different bacterial species. For example, *Alcaligenes latus* does not require nitrogen or phosphorous limitation; it can accumulate large amounts of PHB during balanced growth (6).

The most important economic considerations are substrate costs, reactor productivity, and separation costs. SCL PHA can be produced from hydrolyzed starch (glucose), lactose (using recombinant *E. coli* on cheese whey), cane or beet molasses (using *Azotobacter vinelandii* or *Al-*

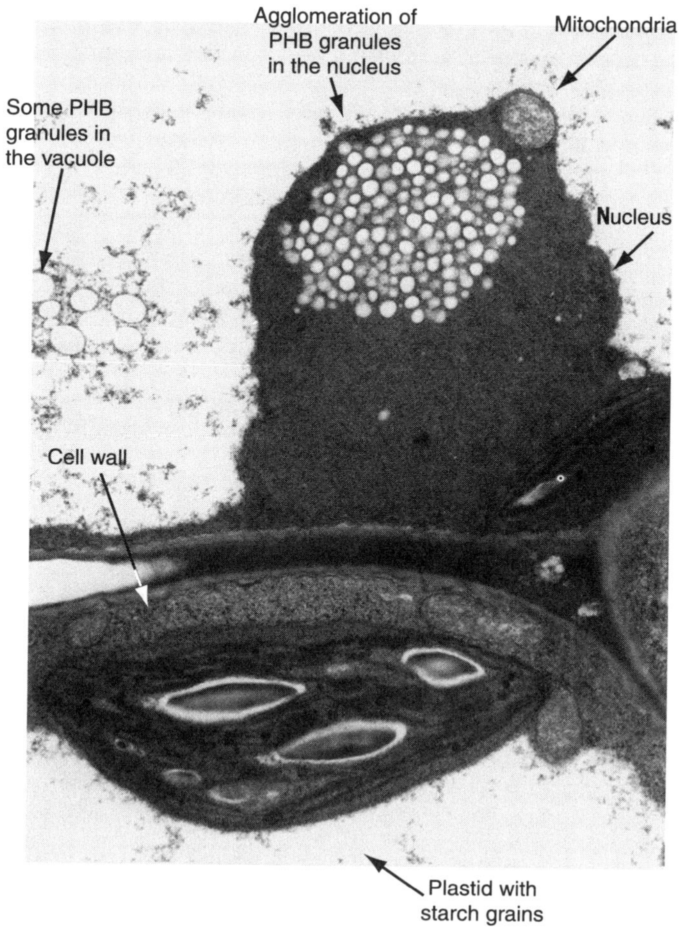

Figure 1. Polyhydroxybutyrate (PHB) granules in *Arabidopsis thaliana* (electron micrograph courtesy of Yves Poirier, Université de Lausanne; Chris Somerville, Carnegie Institute; and Karen Klomparens, Michigan State University).

caligenes latus on sucrose), waste wood (fermented, hydrolyzed hemicellulose), methanol, and even mixtures of hydrogen and carbon dioxide (using *R. eutropha*). Addition of the appropriate cosubstrate such as propionic acid can result in P(HB-co-HV) synthesis in any of these processes although the %HV can vary considerably depending on the bacterial strain. High productivity have been reported for several of these substrates. Using *R. eutropha* growing on glucose, 164 g L^{-1} of biomass containing 76% PHB has been produced in a fed-batch process in only 50 h (7). Densities well in excess of 100 g L^{-1} have been reported for fed-batch processes using *E. coli* strains with multicopy plasmids containing the key SCL PHA synthetic genes (8) as well as for methylotrophs growing on methanol (9). In recombinant *E. coli* strains, PHB content is frequently reported as being in excess of 90% of the biomass dry weight, making the separation process much simpler.

MCL PHA. The highest productivity results from using octanoic or nonanoic acids, but these are toxic at concentrations of several grams per liter and their low solubility limits the growth rate as the result of poor mass transfer. The salts of these acids are much more soluble, but the final product concentration in fed-batch culture is limited by accumulation of toxic concentrations of sodium ions in the medium. In any event, the present cost of both the salts and the acids prohibits their use in commercial production. Most microorganisms that make MCL PHA produce lipases and grow well on vegetable oils. If the lipase concentration limits the production rate, the oils can be chemically hydrolyzed to liberate their fatty acids. A *Pseudomonas putida* strain has been grown in fed-batch culture from oleic acid (a fatty acid commonly found in plant oils) to a biomass concentration of 92 g L^{-1} containing 45% PHA in 26 h (10). If of sufficient low density, these PHAs may float to the surface after breakage of the bacterial cell and can thus be easily and cheaply separated from the rest of the biomass.

SEPARATION OF PHA FROM BIOMASS

General Considerations

All separation processes to date have been developed for microbial rather than plant biomass, but many of these approaches could also be applied to recovery of PHA from plants. As a first step in the recovery from microbial biomass, the cells are separated from the fermentation broth, usually by centrifugation or flocculation. In the latter case, the pH of the fermentation broth is adjusted to 9; it is heated, then the pH is readjusted to 5 (11) and the biomass recovered by centrifugation or filtration. The wet biomass may be freeze- or spray-dried to remove water. Processes to separate the intracellular PHA from the rest of the microbial biomass may be divided into two categories: *(1)* selective solubilization of PHA in an organic solvent, leaving the majority of the other biomass components in suspension, and *(2)* destruction (i.e., enzymatic solubilization) of cellular components other than PHA, leaving the PHA granules in suspension.

Physical treatments such as drying, heating, freezing and thawing, repeated centrifugation, and exposure to acids and organic solvents change the amorphous nature of native PHA to a semicrystalline state.

Selective Solubilization of PHAs: Solvent Extraction

The majority of the patented separation processes describe the extraction of PHB from microbial biomass using organic solvents such as chlorinated hydrocarbons (e.g., chloroform or 1,2-dichloroethane), azeotropic mixtures (e.g., 1,1,2-trichloroethane with water [12]), chloroform with either methanol, ethanol, acetone, or hexane (13), and cyclic carbonates (e.g., hot [120–150 °C] ethylene carbonate or 1,2-propylene carbonate [14]) in which the polymer is soluble. With the wider variety of PHAs that can be produced today, the choice of solvents should be carefully considered. In general, solvents that are suitable for PHB should be equally good for any SCL PHA and MCL PHA. The reverse may not be true. For example, while semicrystalline PHB is insoluble in acetone, MCL PHA will dissolve in it.

Cells may be pretreated to make the polymer more accessible to the extraction solvent. This may be done by washing in a hydrophilic solvent in which the polymer is insoluble such as methanol or acetone for SCL PHA and methanol for MCL PHA or by means of physical disruption such as grinding, wet-milling, French press, or freezing and thawing cycles. Pretreatment with acetone has been shown to have the added advantage of removing certain polar lipid impurities and increasing the amount of PHB recovered (15). After solvent extraction, the polymer solution is separated from the cellular residue by filtration or centrifugation. PHB is then precipitated from the chlorinated hydrocarbons or azeotropic mixtures with a cold nonsolvent such as methanol. In the case of ethylene carbonate and 1,2-propylene carbonate, most of the PHB precipitates on cooling (14).

Difficulties are encountered at polymer solution concentrations greater than 5% by weight because these concentrations tend to be very viscous. As a result, large amounts of solvent are required to obtain sufficiently dilute solutions to enable easy separation. Even with efficient solvent recovery procedures, some solvent loss is expected. This method is not considered economical as solvent recovery and raw material costs are high. However, solvent extraction is useful when the PHA content of the biomass is low or when high polymer purity is required. A significant portion of the PHB may remain in the biomass after solvent extraction (15).

Although thermal degradation of PHB in air or a nitrogen atmosphere has been well studied (16), few examples show that PHB degrades in organic solvents. The molecular weight of PHB decreased in propylene carbonate and ethylene carbonate at 110–140 °C (14), in a chloroform–methanol mixture at room temperature (17), and in methylene chloride, chloroform, and 1,2-dichloroethane at their boiling points (15).

Although solvent extraction is commercially uneconomical, it is a commonly used laboratory technique for PHA recovery, especially for molecular weight determination. Usually, methanol-pretreated (or ethanol-pretreated), lyophilized biomass is heated in chloroform under reflux conditions for about 1 h, the polymer solution is recovered by filtration or centrifugation, and the polymer is precipitated by cold ethanol or methanol. The polymer may be further purified by redissolving in chloroform and reprecipitating with cold ethanol. It should be noted that about 25% of the original molecular weight (determined by chloroform extraction under ambient conditions) was lost when refluxed in chloroform for 1 h (15).

Solubilization of Cellular Components Other than PHA

Enzymatic Methods. Classical enzymatic treatment (involving lysozyme, proteinases, DNAses, etc.) with or without a surfactant step can be used to solubilize the cellular biomass (18). Biologists have used these methods to prepare native PHB granules to study intracellular enzymatic degradation. On a commercial level, Holmes and Lim (19) described a process in which the fermentation broth may be first heated to 80 °C to denature DNA and some proteins. A series of enzymes (such as lysozyme, phospholipase, lecithinase, or alcalase [proteinase]) are then used to obtain a product that is 90–95% pure with 6–7% proteins and some peptidoglycan as major impurities. There may be a final wash with hydrogen peroxide to make a whiter product. This is the reported method of choice for the current production of P(HB-co-HV) (20), which is sold under the tradename Biopol by Monsanto plc. When PHA granules recovered by enzymatic treatment are freeze-dried, they have a hard, crystalline shell and a noncrystalline core (21).

Chemical Methods. Strong oxidizing agents such as sodium hypochlorite have been used to dissolve the cellular biomass under conditions that result in significant loss of PHA molecular weight. However, by optimizing the operating conditions, PHB with an average molecular weight of 600,000 g/mol and 95% purity has been obtained from *R. eutropha* biomass containing 50% PHB with an initial molecular weight of 1,200,000 g/mol. When the biomass is pretreated with a surfactant, a reduced contact time with hypochlorite is needed to obtain PHB with the same purity. This method allows an even higher molecular weight (800,000 g/mol) to be obtained (22). Never-dried granules obtained by surfactant–hypochlorite treatment also have an external crystalline shell and an amorphous core (23).

Physical Methods. Although there are many methods (e.g., reciprocating pump, ball mill) to physically disrupt microbial cells, few processes that are based on these principles have been proposed for the large-scale recovery of PHAs. Holmes and Jones (24) patented a process in which a biomass suspension was heated to 220 °C under nitrogen in an autoclave, then forced through a fire-jet into a tank of cold water. Cell breakage occurred, releasing PHB granules, which were suspended in 50% aqueous acetone to remove lipids and to cause flocculation.

COMMERCIAL PRODUCTION

History

PHB remained an academic curiosity until the 1950s and 1960s when W.R. Grace and Co. in the United States produced small quantities for commercial evaluation. They obtained patents for production processes and manufactured articles such as sutures and prosthetic devices (25) but eventually abandoned the project because of low PHB yields and difficulties in the separation process. About 10 years later, Imperial Chemical Industries (ICI) began a program in PHA production and product development. They patented the first commercial process to produce P(HB-co-HV) (26).

Current Production Methods

Tredegar Industries has used recombinant *E. coli* to produce more than 100 g L^{-1} of biomass containing at least 80% PHB by dry weight at the pilot-scale level. Mild detergent treatment and washing steps followed by spray-drying result in a white product that is 98% pure (27).

Fermentation and separation processes were developed for the pilot-scale production of PHB using an *Alcaligenes latus* strain. As much as 1 ton per week of PHB has been produced in Austria in a 15,000-L bioreactor (28). This technology is currently owned by the German company, Biomer. The Biomer fermentation process is relatively simple with multistaging from the petri dish to a shaker flask to a small fermenter, which is then used to inoculate the production reactor (Fig. 2a). If sufficient carbon source (molasses) is added during the fed-batch process, as much as 60 g L^{-1} of PHB may be obtained. The resulting biomass is washed in water. The concentrated cell suspension is then subjected to methylene chloride extraction (Fig. 2b). The resulting slurry is separated in a decanter. The solvent phase is then injected into hot water in a precipitation vessel where the methylene chloride evaporates and the PHB precipitates. After water removal (drying), the PHB can then be processed into a marketable product.

Biopol has been produced by Zeneca from glucose and propionic acid in 35,000-L airlift reactors and in stirred tank reactors up to 200,000 L (29). Monsanto uses conventional stirred-tank fermenters to produce 1,000 tons of Biopol per year. Phosphate limitation (after about 60 h of growth) is used to trigger the accumulation phase in a strain of *R. eutropha*. Phosphate limitation was found to give better results than nitrogen limitation (30). Although solvents (chloroform or methylene chloride extraction after a hot methanol wash) may be used when high purity is required, the P(HB-co-HV) is usually obtained by treatment of the biomass with a series of surfactants, enzymes, and possibly oxidizing agents to dissolve the non-PHA biomass. The product is essentially a latex that may then be washed, spray-dried, compounded with nucleating agents and plasticizers, and extruded in the form of pellets. Several grades of Biopol are available at a cost of about $8 per pound. Based on process development at IPT (Sao Paulo), Copersucar, the Brazilian ethanol cooperative is now using hydrolyzed sucrose and NH_4^+ limitation to produce SCL PHA. these are purified by solvent extraction. Pilot-scale production capacity is currently 30 to 60 tons per annum. Cost savings from the burning of bagasse to supply energy and from cheap cane sugar allow Copersucar to sell PHA for much less than $8 per pound.

PROCESSING TO FINAL PRODUCTS

Conventional equipment used to process polyethylene can be used for Biopol to form blow-molded, injection-molded, and film products. A key requirement is temperature control. It is recommended to operate at the lowest possible temperature and to keep the residence time at a minimum because the melting point of PHB is about 178 °C and its degradation temperature begins at 180 °C. The melting point of the copolymers decreases as the HV content increases, which allows HV copolymers to be processed at lower temperatures with less risk of thermal degradation. Plasticized high-HV Biopol can have a melting point as low as 136 °C. For most applications, Monsanto suggests processing at a maximum temperature of 170 °C and a maximum residence time of 3 min, and that molds and blow-pin temperatures be kept at 60 ± 5 °C (31). Mechanical properties deteriorate rapidly if the average molecular weight falls below 400,000 g mol^{-1}.

PHAs are moisture resistant and have gas barrier properties similar to the best coated films, so that they have potential use in products which require a plastic moisture barrier. Possible markets include mulch films, food and agricultural packaging, drink cartons, feminine hygiene products, disposable diapers, and disposable kitchen items such as plastic films that are difficult to separate for recycling. Coatings may be applied with PHA as an amorphous latex, which can be reconstituted from the crystalline form of the polymer or obtained directly from a bacterial suspension by dissolving the other cellular macromolecules (cell wall components, DNA, etc.). In the latter case, it is important that the granules are not dried before use as they would become crystalline. The latex may be applied as a spray to a material such as paper, dried, and cross-linked using heat or ultraviolet light. The MCL PHAs are promising as coatings because they are less crystalline than the SCL PHAs (in their crystalline state), and the microorganisms that synthesize them can introduce different functional groups on the molecule to make specific polymers (32,33).

The mechanical properties, morphology, biodegradability, and thermal and crystallization behavior of PHAs blended with nonbiodegradable polymers (such as polyvinyl acetate) and with biodegradable materials (such as wood cellulose fibers and starch) have been reviewed (34). In general, immiscible blends or blends with other biodegradable polymers are more biodegradable than miscible blends or blends with a nondegradable polymer.

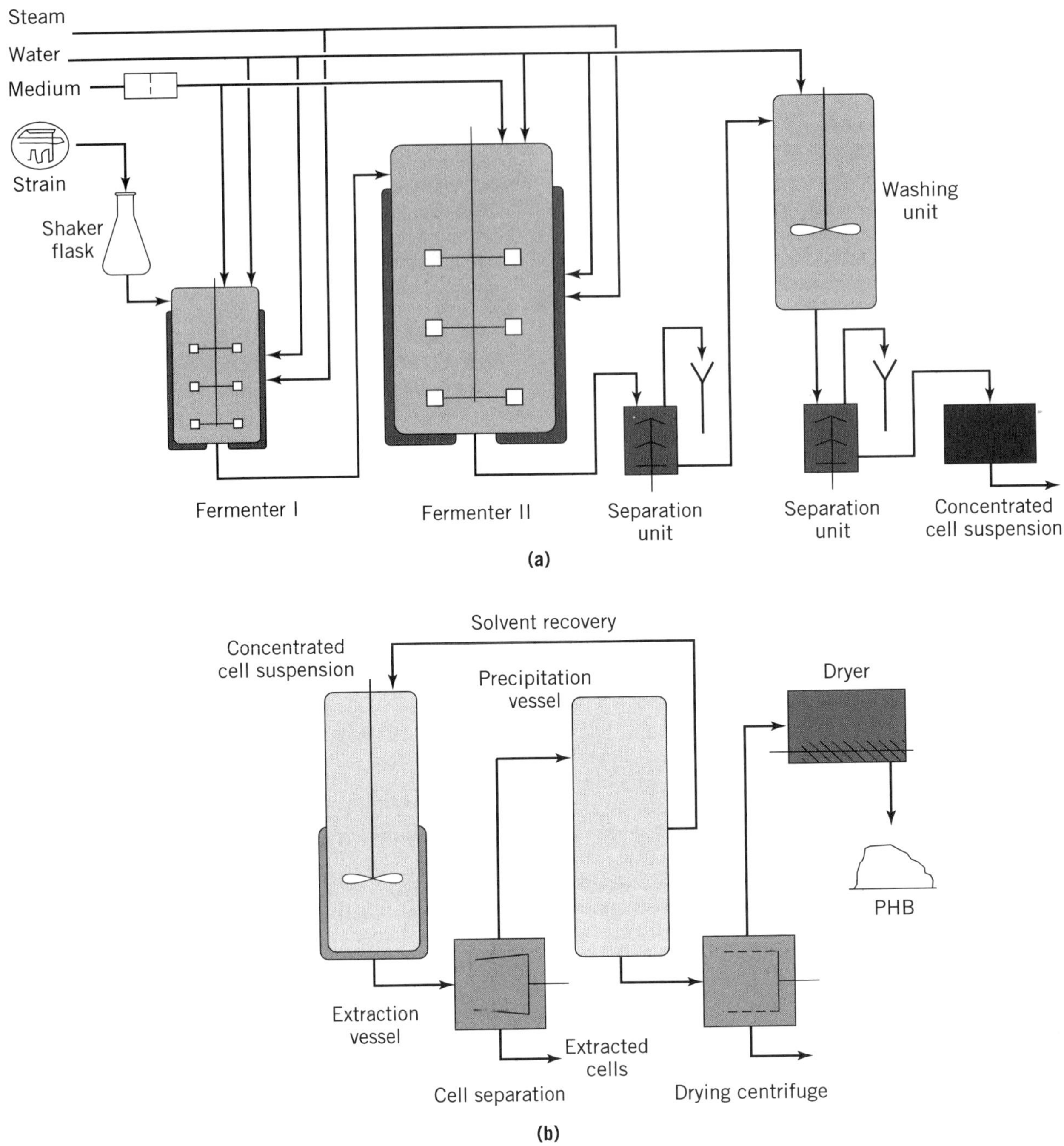

Figure 2. (**a**) Biomer polyhydroxybutyrate (PHB) fermentation process. (**b**) Biomer PHB separation process.

FUTURE PROSPECTS

Economic Considerations

Cost of production is not a major consideration if PHA is to be used exclusively for specialty applications. However, if PHA is to be economically competitive with synthetic plastics, it must be produced on a far larger scale than any other aerobically produced microbial product. An additional problem arises from the fact that PHAs are intracellular products. Thus, the amount that can be produced per cell has physical limitations.

In a typical PHA fermentation process, the cost is divided fairly evenly between substrate cost, capital and operating expenditures in the fermentation process, and the cost of separating the PHA from the microbial biomass. Typically the yield of dry biomass from a hexose sugar such as glucose is about 42% for *R. eutropha*. If the final biomass contains 75% by weight of recoverable PHA, then the yield of PHA from the total amount of substrate supplied is about 32%. Thus, if carbon substrate was available at $500 per metric ton, the cost of the carbon source alone would be $1,560 per metric ton, making the product twice as expensive as bulk synthetic plastics.

New methods such as the incorporation of temperature-sensitive plasmids for the production of autolytic enzymes should reduce separation costs. The direct use of PHA-containing bacteria in melt processing (i.e., no separation process apart from water removal) has also been proposed. However, even assuming a negligible cost for the carbon source by using a waste material such as cheese whey permeate, and neglecting the separation costs, studies have shown that PHA produced in bioreactors could not be economically competitive with synthetic bulk plastics in the foreseeable future. The volumetric production rate cannot compete with chemical processes, and this rate is limited by the oxygen transfer rate of present-day bioreactors. New technologies must be developed for commercial PHA production.

New Technologies for the Production of PHA

In vitro synthesis of PHA granules may be commercially feasible. In *R. eutropha*, the maximum rate of PHA synthesis occurs early in the accumulation phase and deteriorates slowly thereafter. One would therefore expect that the maximum amount of PHA accumulation would depend on enzymatic activity or on NADPH or substrate supply. However, the most detailed study on this process points to a physical limitation (30) where polymer synthesis slows to a virtual stop simply because there is no more space available in the cell. If this is indeed the case, then a solution is to produce the polymer outside of the cells as has been achieved previously (35,36). Economic exploitation of this concept would require a source of the biosynthetic enzymes, substrates, NADP, and a method for NADP reduction. This would eliminate separation costs because the product could be used directly as a latex.

It is likely that, in the near future, starch, oilseed, or leguminous plants will be used to produce PHAs. Other plant materials such as cellulose and starch are already used to make plastics or plasticlike materials. Hemicelluloses have been used directly or converted to organic acids for PHA production at a laboratory scale. Thus, any waste from the separation of PHA from plants could be hydrolyzed, fermented to lactic, acetic, and propionic acids, and then fed into bioreactors for the production of specialty PHAs. Polylactides and other polymers of organic acids could be produced in the same plant as could polysaccharides. Plant materials may become a major source of plastic materials in the next century, with PHAs leading the way.

ACKNOWLEDGMENT

We acknowledge Y. Matteau for the preparation of Figure 2.

BIBLIOGRAPHY

1. D.A. Holden and R.H. Marchessault, *Polymer Preprints* **29**, 594–595 (1988).
2. Y. Poirier, D.E. Dennis, K. Klomparens, and C.R. Somerville, *Science* **256**, 520–523 (1993).
3. C. Nawrath, Y. Poirier, and C. Somerville, *Mol. Breed.* **1**, 105–122 (1995).
4. G. Lefebvre, M. Rocher, and G. Braunegg, *Appl. Environ. Microbiol.* **63**, 827–833 (1997).
5. Y. Doi, A. Tamaki, M. Kunioka, and K. Soga, *J. Chem. Soc. Chem. Commun.*, 1635–1636 (1987).
6. G. Braunegg and B. Bogensberger, *Acta Biotechnol.* **5**, 339–345 (1985).
7. B.S. Kim, S.C. Lee, H.N. Chang, Y.K. Chang, and S.I. Woo, *Biotechnol. Bioeng.* **43**, 892–898 (1994).
8. B.S. Kim, S.Y. Lee, and H.N. Chang, *Biotechnol. Lett.* **14**, 811–816 (1992).
9. D. Bourque, Y. Pomerleau, and D. Groleau, *J. Microbiol. Biotechnol.* **44**, 367–376 (1993).
10. G.N.M. Huijberts, H. van der Wal, R.A. Weusthuis, and G. Eggink, Abstract, *5th Int. Symp. on Bacterial Polyhydroxyalkanoates*, Davos, Switzerland, August 1996.
11. U.S. Pat. 4,358,583 (September 11, 1982), J. Walker and J.R. Whitton (to Imperial Chemical Industries, PLC, Great Britain).
12. U.S. Pat. 4,705,604 (November 10, 1987), N. Vanlautem and J. Gilain (to Solvay and Company, Belgium).
13. U.S. Pat. 4,562,245 (December 31, 1985), R.F. Stageman (to Imperial Chemical Industries, PLC, Great Britain).
14. U.S. Pat 4,101,533 (July 18, 1978), R.M. Lafferty and E. Heinzle (to Agroferm AG, Chur, Switzerland).
15. J.A. Ramsay, E. Berger, R. Voyer, C. Chavarie, and B.A. Ramsay, *Biotechnol. Tech.* **8**, 589–594 (1994).
16. N. Grassie, E.J. Murray, and P.A. Holmes, *Polym. Degrad. Stabil.* **6**, 95–103 (1984).
17. A.H. de Mola, M. Marx-Figini, and R.V. Figini, *Makromol. Chem.* **176**, 2655–2667 (1975).
18. R. Greibel, Z. Smith, and J.M. Merrick, *Biochemistry* **7**, 3767–3681 (1968).
19. U.S. Pat. 4,910,145 (March 20, 1990), P.A. Holmes and G.B. Lim (to Imperial Chemical Industries, PLC, Great Britain).
20. D. Byrom, in E.A. Dawes ed., *Novel Biodegradable Microbial Polymers*, Kluwer, Dordrecht, Netherlands, 1990, pp. 113–117.
21. C. Lauzier, J.-F. Revol, and R.H. Marchessault, *FEMS Microbiol. Rev.* **103**, 299–310 (1992).
22. J.A. Ramsay, E. Berger, B.A. Ramsay, and C. Chavarie, *Biotechnol. Tech.* **4**, 221–226 (1990).
23. C.A. Lauzier, C.J. Monasterios, I. Saracovan, R.H. Marchessault, and B.A. Ramsay, *Tappi J.* **76**, 71–77 (1993).
24. Eur. Pat. Appl. 46,335 (February 24, 1980), P.A. Holmes and E. Jones (to Imperial Chemical Industries, PLC, Great Britain).
25. U.S. Pat. 3,225,766 (December 28, 1965), J.N. Baptist and J.B. Ziegler (to W.R. Grace and Co., New York).
26. Eur. Pat. 69,497 (April 15, 1987), P.A. Holmes, L.F. Wright, and S.H. Collins (to Imperial Chemical Industries, PLC, Great Britain).
27. F. Merchant, D.E. Dennis, R. Merchant, and C.F. Batrell, Abstract, *4th Int. Symp. on Bacterial Polyhydroxyalkanoates*, Montreal, Canada, August 1994.
28. U.J. Hänggi, in E.A. Dawes ed., *Novel Biodegradable Microbial Polymers*, Kluwer, Dordrecht, 1990, pp. 65–70.
29. D. Byrom, *Int. Biodeterior. Biodegrad.* **31**, 199–203 (1993).
30. D.G.H. Ballard, P.A. Holmes, and P.J. Senior, *Recent Adv. Mech. Synthet. Aspects Polymer.* **215**, 293–314 (1987).
31. *Biopol Nature's plastic: Properties and Processing*, Monsanto, St. Louis, Mo., 1996, pp. 13–15.

32. K. Fritzsche, R.W. Lenz, and R.C. Fuller, *Int. J. Biol. Macromol.* **12**, 85–91 (1990).
33. R.G. Lageveen, G.W. Huisman, H. Preusting, P. Ketelaar, G. Eggink, and B. Witholt, *Appl. Environ. Microbiol.* **54**, 2924–2932 (1988).
34. H. Verhoogt, B. Ramsay, and B. Favis, *Polymer* **35**, 5155–5169 (1994).
35. R.J. Griebel and J.M. Merrick, *J. Bacteriol.* **108**, 782–789 (1971).
36. T.U. Gerngross and D.P. Martin, *Proc. Natl. Acad. Sci. U.S.A.* **92**, 6279–6283 (1995).

PROCESS CONTROL, STRATEGY AND OPTIMIZATION

D. Dochain
Université Catholique de Louvain
Louvain-La-Neuve, Belgium

KEY WORDS

Automatic control
Bioprocess
CSTR
Fed-batch reactor
Nonlinear control
Optimal control
Optimization
Parameter estimation
Software sensor
State estimation

OUTLINE

INTRODUCTION

Industrial-scale biotechnological processes have progressed vigorously over the past decades. Generally speaking, the problems arising from the implementation of these processes are similar to those of more classical industrial processes, and the need for monitoring systems and automatic control to optimize production efficiency, improve product quality, or detect disturbances in process operation is obvious. Nevertheless, automatic control of industrial biotechnological processes is clearly developing very slowly. There are two main reasons for this slow development:

1. The internal working and dynamics of these processes are as yet poorly grasped, and many problems of methodology in modeling remain to be solved. It is difficult to develop models that take into account the numerous factors that can influence the specific bacterial growth rate. The modeling effort is often tedious and requires many experiments to produce a reliable model. Reproducibility of experiments is often uncertain because of the difficulty in obtaining the same environmental conditions. Moreover, because these processes involve living organisms, their dynamic behavior is strongly nonlinear and nonstationary. Model parameters cannot remain constant over a long period; they vary, for example, due to metabolic variations of biomass or to random and unobservable physiological or genetic modifications. It should also be noted that the lack of accuracy of the measurements often leads to identifiability problems.
2. Another essential difficulty lies in the absence, in most cases, of cheap and reliable instrumentation suited to real-time monitoring. To date, the market offers very few sensors capable of providing reliable on-line measurements of the biological and biochemical parameters required to implement high-performance automatic control strategies. The main variables (i.e., biomass, substrate, and synthesis product concentrations) generally need to be determined through laboratory analyses. The cost and duration of the analyses obviously limit the frequency of the measurements.

Figure 1 shows a schematic view of a computer-controlled bioreactor. In the illustrated situation, the influent flow rate is modulated via a control algorithm implemented in the computer. As explained later, the control algorithm combines different informations about the process provided by on-line measurements and some knowledge about the process dynamics. The on-line measurements may undertake some mathematical treatment via software sensors to provide extra information about some key process parameters (such as specific growth rates) and unmeasured variables.

The article is organized as follows. The basic concepts of automatic control of bioprocesses, with a particular emphasis on the notions of optimization, stability, and disturbances, are discussed first. Then I introduce the basic in-

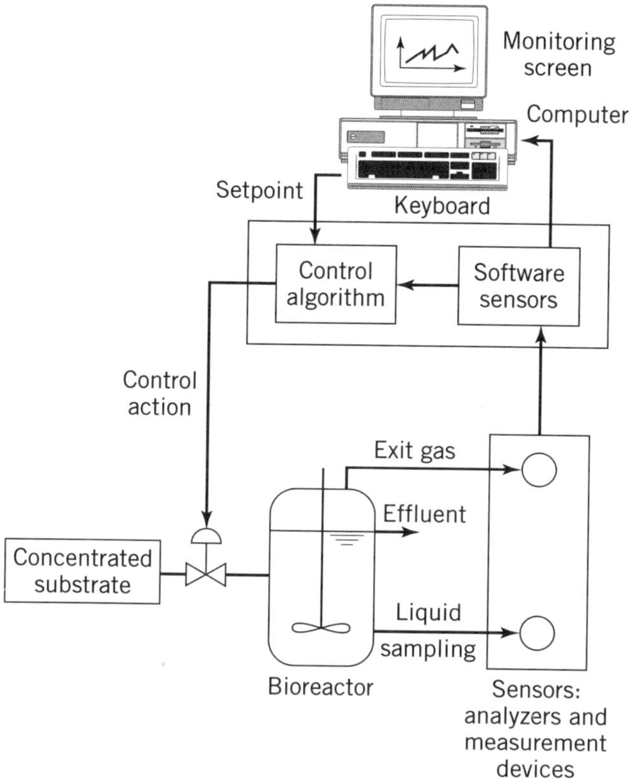

Figure 1. Schematic view of a computer-controlled bioreactor.

gredients of automatic control (dynamical model; feedback; proportional, integral, and feedforward actions), and briefly compare linear and nonlinear control and adaptive and nonadaptive control.

BIOPROCESS CONTROL: BASIC CONCEPTS

The objective of automatic control when applied to any process (including bioprocesses) is to run and maintain the process in stable, optimal operating conditions in spite of disturbances.

Let us first give some more insights about the concepts that are covered by the words *stable, optimal*, and *disturbances*. And let us consider the following simple example to illustrate these concepts: a simple microbial growth reaction in a stirred tank reactor. The microbial growth reaction is characterized by the following reaction scheme:

$$S \rightarrow X \tag{1}$$

with S the limiting substrate and X the biomass. Mass balances for both components S and X lead to the following set of dynamic equations:

$$\frac{dS}{dt} = -DS - \frac{1}{Y_{X/S}}\mu X + DS_{in} \tag{2}$$

$$\frac{dX}{dt} = -DX + \mu X \tag{3}$$

where D is the dilution rate (h^{-1}) (i.e., the ratio of the influent flow rate F_{in}[L/h] over the volume V[L] occupied by the reacting medium in the reactor), $Y_{X/S}$ is the yield coefficient, μ is the specific growth rate (h^{-1}), and S_{in} is the influent substrate concentration (g/L). Note the following:

1. The model is a dynamic model in order to capture the time evolution of the process: this is an essential ingredient of automatic control (discussed in more detail later).
2. The model is valid whether the operation mode is continuous (then the reactor is sometimes called a chemostat or a continuous stirred tank reactor [CSTR]), batch (then D is equal to zero), fed batch (then the volume is not constant and its time variation is given by the following equation: $dV/dt = F_{in}$).
3. We use the same notation for the components (in reaction scheme 1) and their concentration (in equations 2 and 3).

Let us now consider that the specific growth rate here is of the Haldane type:

$$\mu = \frac{\mu^* S}{K_S + S + S^2/K_I} \tag{4}$$

with K_S is the Monod constant, K_I is the inhibitor constant, and μ^* is a constant connected to the maximum specific growth rate μ_{max} as follows:

$$\mu^* = \mu_{max}\left(1 + 2\sqrt{\frac{K_S}{K_I}}\right) \tag{5}$$

Let us also consider that the objective is to control the process by using the dilution rate D (or, more precisely, the influent flow rate F_{in}) or the influent substrate concentration S_{in} as the control action.

Disturbances

Let us illustrate the notion of disturbances with the preceding example. If the dynamic model (equations 2–4) is assumed to be perfect, and if the dilution rate D is the only manipulated input, then the influent substrate concentration S_{in} is a disturbance: variations of S_{in} will disturb the process by inducing variations in the time evolution of the variables S and X.

Assume now that the dynamic model is not perfect, for example, that some of the parameters μ^*, K_S, and K_I of the Haldane model (equation 4) are changing with time (e.g., due to microorganism adaptation): these variations are also disturbances with regard to the considered process model. Another similar disturbance may come from the presence of a side reaction (e.g., maintenance or microorganism death) that has been neglected in the earlier dynamic model but that may not be negligible under some operating conditions. Then, in case of maintenance, for instance, the true dynamic model of the process is described by equation 3 and the following modified mass balance equation for S:

$$\frac{dS}{dt} = -DS - \frac{1}{Y_{X/S}}\mu X - k_m X + DS_{in} \tag{6}$$

where k_m is the maintenance coefficient.

Finally, typical disturbances may be due to process failures (e.g., actuator, sensor, or even stirring system). Even if those disturbances are inherently different, it is important to consider all the disturbances that are likely to take place during the process operation when analyzing and testing the performances of a designed controller that has to be implemented on a real bioprocess.

Stability

The notion of stability has been largely considered and studied in dynamic systems and automatic control and has led to different mathematical definitions and tests (see Ref. 1). Here we shall concentrate on Lyapunov stability. Qualitatively the notion of stability can be roughly explained as follows. Let us consider that the process is at some equilibrium point (or steady state). The equilibrium point will be stable if the process does not go far from this state for small deviations. It will be unstable if the variables of the process move away from the equilibrium point and if their variations become larger and larger as time increases. Stability is obviously of vital concern in automatic control: it is essential to design controllers that keep processes in stable conditions on the one hand and that are capable of stabilizing unstable processes on the other hand.

Let us illustrate the concept of stability on the microbial growth example and how to test it. The Haldane model is represented in Figure 2. What we shall now show is that values of the substrate concentration on the left-hand side of the maximum specific growth rate

$$\mu^* \Big/ \left(1 + 2\sqrt{\frac{K_S}{K_I}}\right)$$

correspond to stable equilibrium points, whereas the ones on the right-hand side correspond to unstable equilibrium points.

Equilibrium Points. An equilibrium point is, by definition, a constant state (denoted here $[\bar{S}, \bar{X}]$) that satisfies the equation of the dynamic model, the following algebraic equations in our example:

$$\frac{dS}{dt} = 0 \Rightarrow -\bar{D}\bar{S} - \frac{1}{Y_{X/S}}\mu(\bar{S})\bar{X} + \bar{D}\bar{S}_{in} = 0 \tag{7}$$

$$\frac{dX}{dt} = 0 \Rightarrow -\bar{D}\bar{X} + \mu(\bar{S})\bar{X} = 0 \tag{8}$$

for given constant values of $\bar{D}$ and $\bar{S}_{in}$. If we consider the Haldane model (Fig. 2), it is straightforward to check that there are three possible equilibrium points:

1. Operational equilibrium point 1: $\bar{S}_1 \leq \sqrt{K_S K_I}$, $\bar{X}_1 = Y_{X/S}(\bar{S}_{in} - \bar{S}_1)$
2. Operational equilibrium point 2: $\bar{S}_2 > \sqrt{K_S K_I}$, $\bar{X}_2 = Y_{X/S}(\bar{S}_{in} - \bar{S}_2)$
3. Washout equilibrium point: $\bar{X}_3 = 0, \bar{S}_3 = \bar{S}_{in}$.

The washout may occur for any value of $\bar{D}$ and $\bar{S}_{in}$. It is also the only possible equilibrium point for $\bar{D} > \mu_{max}$. It is obviously an undesirable steady state. The values of $\bar{S}$ of the last two equilibrium points correspond indeed to the two solutions of the second-order algebraic equation obtained by introducing the Haldane equation (equation 4) into the equilibrium point equation of X (equation 8):

$$\frac{\bar{D}}{K_I}\bar{S}^2 + (\bar{D} - \mu^*)\bar{S} + \bar{D}K_S = 0 \tag{9}$$

Stability Analysis. For the stability analysis, let us consider the linear approximation of the model (equations 2–4) around the equilibrium points (i.e., the linearized tangent model). Whatever the input u (D, S_{in}, or both), the linearized tangent model around each operational equilibrium point is written as follows:

$$\frac{dx}{dt} = Ax + Bu \tag{10}$$

where x is the vector of the deviations of S and X from the equilibrium values

$$x = \begin{bmatrix} S - \bar{S} \\ X - \bar{X} \end{bmatrix} \tag{11}$$

and the matrix A is given by the following expression:

$$A = \begin{bmatrix} 0 & \Omega \\ -\frac{1}{Y_{X/S}}\bar{D} & -\frac{1}{Y_{X/S}}\Omega - \bar{D} \end{bmatrix} \tag{12}$$

with

$$\Omega = \frac{\mu^*\bar{X}\left(K_S - \frac{\bar{S}^2}{K_I}\right)}{\left(K_S + \bar{S} + \frac{\bar{S}^2}{K_I}\right)^2} \tag{13}$$

The stability of the equilibrium points is determined by the

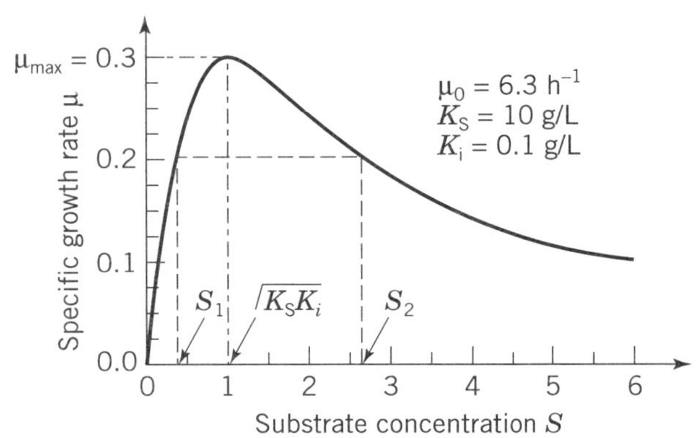

Figure 2. The Haldane model.

eigenvalues of the matrix A: the equilibrium point will be stable if the real parts of all the eigenvalues are negative, it will be unstable if one eigenvalue has a positive real part. In our example, the eigenvalues are equal to $(-\bar{D})$ and $(-1/Y_{X/S}\Omega)$. If the first eigenvalue is always negative, the second one is negative for the first equilibrium point $(\bar{S}_1, \bar{X}_1)$ and is positive for the second one $(\bar{S}_2, \bar{X}_2)$. This means that $(\bar{S}_1, \bar{X}_1)$ is a stable equilibrium point and that $(\bar{S}_2, \bar{X}_2)$ is an unstable one.

Regulation versus Tracking

Two typical situations are usually considered in automatic control: constant setpoint control (regulation) and trajectory tracking. Regulation is typical of the control of continuous reactors, where the objective is usually to maintain the process in an a priori chosen steady state despite the disturbances. But in many instances, such as in batch and fed-batch reactors but also for start-up of reactors and in case of production grade changes, regulation is not appropriate: the objective is then to drive the process from some initial state to some final desired state by following some predetermined trajectory. In the simple microbial growth example, if we consider a fed-batch reactor and if the objective is to optimize the production of biomass, then an optimal trajectory is an "exponential" profile for X: the controller must then be designed so as to maintain the process as close as possible to the optimal profile.

Optimization

The optimal conditions in which the controller is supposed to run the bioprocess are typically defined so as to optimize production efficiency, improve product quality, or improve the economic profit of the plant. Optimization is obviously a helpful tool for determining the optimal operating conditions. Static optimization techniques can be used for the determination of optimal constant setpoints, whereas dynamic optimal control techniques will be useful for determining the best operating profile.

It must be stressed, however, that the results obtained via optimization have to be handled with great care. The reason lies in the high sensitivity of the optimization techniques on the quality of the process model (including the model structure, i.e., its equations, and the precision on the values of the parameters of the chosen model). Indeed the models used for describing the dynamics of bioprocesses (often mainly more specifically the kinetics) are typically characterized by a large uncertainty (as already mentioned).

Let us illustrate these ideas with our simple microbial growth example with Haldane kinetics. Assume that only the substrate concentration S is accessible for on-line measurement (and will therefore be used for feedback control) and that the objective is to maximize the production of biomass X.

In a CSTR, the objective is to operate the process in a steady state that maximizes the productivity (i.e., the quantity of biomass per unit time). This means that the objective function is

$$J = \bar{D}\bar{X} \tag{14}$$

The solution of the optimization requires the knowledge of the process model in steady state (equations 4, 7, and 8) (equality constraints), as well as (physical) inequality constraints on the process variables

$$0 \le \bar{S} \le \bar{S}_{in},\ 0 \le \bar{X} \le Y_{X/S}\bar{S}_{in} \tag{15}$$

The value of the substrate concentration that maximizes J

$$J = \bar{D}\bar{X} = \frac{\mu^*\bar{S}}{K_S + \bar{S} + \bar{S}^2/K_I} Y_{X/S}(\bar{S}_{in} - \bar{S}) \tag{16}$$

is obtained by calculating the derivative of J with respect to $\bar{S}$ and by setting its value to zero

$$\frac{dJ}{d\bar{S}} = 0 \rightarrow \bar{S}_{optimal} = \frac{-K_SK_I + K_I\sqrt{K_S^2 + \bar{S}_{in}(\bar{S}_{in} + K_I)}}{\bar{S}_{in} + K_I} \tag{17}$$

and then by checking that the sign of the second-order derivative is negative (to have a maximum):

$$\frac{d^2J}{d\bar{S}^2}(\bar{S}_{optimal}) < 0 \Rightarrow \bar{S}_{optimal} \text{ is a maximum} \tag{18}$$

and that the inequality constraints are fulfilled. This optimization problem was originally treated by D'ans et al. (2). (Indeed there is another value of $\bar{S}$ for which dJ/dS is equal to zero, but it is negative and therefore obviously rejected.)

In a fed-batch reactor, the optimization problem is somewhat different. Assume that the time of the fed-batch reactor operation is fixed. Then the objective is to obtain the highest value possible at the end of the fermentation. The objective function is then

$$J = X(t_f)V(t_f) \tag{19}$$

Intuitively, it is obvious that the optimum will be obtained when the biomass is produced at its maximum production rate, that is, if the value of substrate concentration is kept at the value that maximizes the specific growth rate

$$S_{optimal} = \sqrt{K_SK_I}$$

This result can be formally obtained as follows by considering optimal control theory.

The general problem of optimal control is to find an optimal trajectory of the control input $u(t)$ that drives the process from its initial state to its final state and at the same time minimizes a cost function

$$J = g[x(t_f), t_f] + \int_{t_0}^{t_f} h[x(t), u(t), t]dt \tag{20}$$

where x is the state vector, t_0 and t_f are the initial and final time, respectively, and the process dynamics are described by the following ordinary differential equation:

$$\frac{dx}{dt} = f[x(t), u(t), y] \tag{21}$$

and the input $u(t)$ obeys the following practical constraints:

$$U_{\min} \leq u(t) \leq U_{\max} \tag{22}$$

The optimal control solution is based on the Hamiltonian H:

$$H \triangleq \lambda(t)^{\mathrm{T}}(f_1(x) + b(x)u) = \varphi + \psi u \tag{23}$$

where λ represents the adjoint variables, or costates. They are the solution of the following differential equation:

$$\frac{d\lambda}{dt} = \frac{\partial H}{\partial x} \tag{24}$$

For time-invariant systems linear in the input

$$\frac{dx}{dt} = f_1(x) + b(x)u \tag{25}$$

and the cost function J without integral term ($J = g[x(t_f), t_f]$), it can be shown that the optimal control has the following form:

$$u(t) = U_{\min} \quad \text{if } \psi < 0 \tag{26}$$

$$= U_{\max} \quad \text{if } \psi > 0 \tag{27}$$

$$= u_{\text{sing}} \quad \text{if } \psi = 0 \tag{28}$$

The optimal control solution consisting of applying only $U_{\min}$ and $U_{\max}$ is called bang-bang control. The so-called singular control input u_{sing} is obtained by computing the first- and second-order time derivatives of ψ and equating them to zero.

In our fed-batch reactor example, assume that the control input is F_{in}. The dynamics of the process are given by equations 2 and 3 plus the volume equation ($dV/dt = F_{\text{in}}$) and a Haldane specific growth rate model (equation 4). Then the Hamiltonian H is given by the following expression:

$$H = \lambda_1\left(-\frac{F_{\text{in}}}{V} S - \frac{1}{Y_{X/S}}\mu X + \frac{F_{\text{in}}}{V} S_{\text{in}}\right) + \lambda_2\left(-\frac{F_{\text{in}}}{V} X + \mu X\right) + \lambda_3 F_{\text{in}} \tag{29}$$

It is routine to check that $d\psi/dt = 0$ if $\partial\mu/\partial S = 0$ (i.e., u_{sing} corresponds to the value that maximizes μ). If we choose the initial value of the substrate concentration $S(t_0)$ corresponding to the one that maximizes μ

$$(S(t_0) = \sqrt{K_S K_I})$$

the optimal trajectory is obtained by applying the singular control input from t_0 to t_f; that is,

$$F_{\text{in,sing}} = \frac{\mu^* X V \sqrt{K_S K_I}}{Y_{X/S}(2K_S + \sqrt{K_S K_I})(S_{\text{in}} - \sqrt{K_S K_I})} \tag{30}$$

with X and V computed via the mass balance equations:

$$\frac{dX}{dt} = \frac{\mu_{\max}\sqrt{K_S K_I}}{2K_S + \sqrt{K_S K_I}} X - \frac{F_{\text{in}}}{V} X \tag{31}$$

$$\frac{dV}{dt} = F_{\text{in}} \tag{32}$$

Figure 3 illustrates the high sensitivity of optimal control to modeling errors in numerical simulation. We have considered the fed-batch example with the following model parameter and initial values:

$$\mu^* = 5\text{h}^{-1},\ K_S = 10 \text{ g/L},\ K_I = 0.1 \text{ g/L},\ Y_{X/S} = 1 \tag{33}$$

$$S(0) = 0.9 \text{ g/L},\ X(0) = 0.1 \text{ g/L},\ V(0) = 10 \text{ L},\ V(t_f) = 20 \text{ L},\ S_{\text{in}} = 10 \text{ g/L} \tag{34}$$

We have also considered the optimal control when 10% error is made on the value that maximizes μ; that is, the operator believes that the optimal substrate concentration is equal to 9 (instead of 10) g/L. This has been done by setting the saturation constant K_S to 8.1 (instead of 10) g/L in the optimal control law (equation 30). Note that even in the presence of a reasonable model error, the optimal control is moving far away from the optimal substrate concentration.

Let us mention among the large body of scientific literature on optimization and optimal control the following references: Ref. 3, which the treats the optimization of chemical processes, and Ref. 4, which covers the essential items of the optimal control of bioprocesses and treats it then in an original manner.

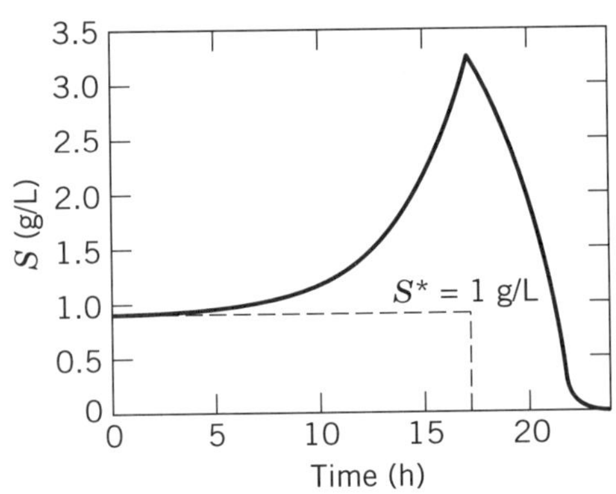

Figure 3. Sensitivity of the optimal control of a fed-batch reactor to modeling errors.

BIOPROCESS CONTROL: BASIC INGREDIENTS

Dynamic Model

The first essential basic ingredient for the design of automatic control algorithms is a dynamic model of the process. The need of dynamic models for the control design is motivated by the intrinsic need to handle transients for transferring the process from one state to another or from one setpoint to another, for following a desired trajectory, or for rejecting or reducing the effect of disturbances. In the latter, disturbances influence the dynamic behavior of the process: knowledge of the process dynamics, and in particular of the dynamic effect of the disturbance, is helpful in selecting the appropriate strategy to reduce and possibly reject their negative influence on the process operation.

The dynamic models to be considered for control design can take different forms. The preceding sections suggest that the most obvious model forms are the mass balance models. But many other forms are possible, including linear models deduced, for example, by linearization of the mass balance models or by identification from input–output data (e.g., Ref. 5) (such models are also called black-box models). The input–output models may be nonlinear; neural network models are presently a quite popular version of nonlinear black-box models (e.g., Ref. 6). A possibly attractive form of dynamic models of bioprocesses are hybrid models, which are an intermediate form between mass balance and neural network models (e.g., Ref. 7). In the following, mainly for conciseness and pedagogical reasons, we concentrate on mass balance models.

Feedback

Another essential basic ingredient of automatic control is the introduction of a feedback loop: On-line (real-time) information about the process (typically on-line measurement of the variable to be controlled) is provided to the controller, which then provides the control action to the process. The concept of feedback loop is illustrated in Figure 4.

Proportional Action

Most feedback control algorithms (except very simple controllers such as on-off controllers) modulate the amplitude of the control action proportionally to the control error (i.e., the difference between the desired value of the controlled variable and its measured value), which is known as the proportional action.

Integral Action

The proportional action is usually not sufficient to guarantee a control error equal to zero in steady state: this motivates the introduction in the control scheme of an integral action (i.e., a term proportional to the integral of the control error).

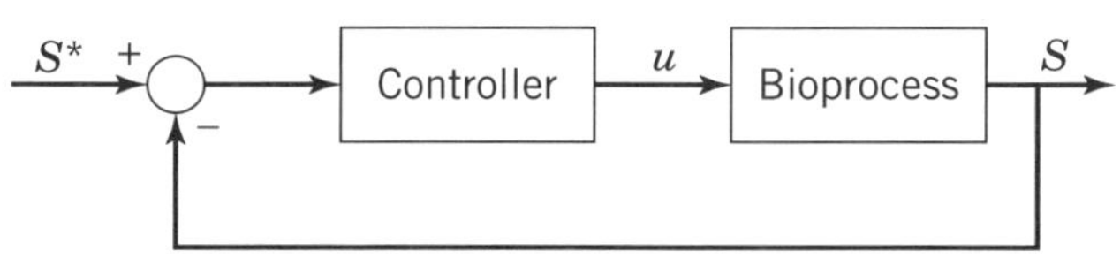

Figure 4. Feedback loop for the control of S in a bioprocess.

Feedforward Action

If the effects of a disturbance on the bioprocess dynamics are known, the controller may be designed by also incorporating a feedforward action to anticipate the possibly negative effects of the disturbance. We shall illustrate in the following how to incorporate feedforward action when the influent substrate concentration S_{in} is a measured disturbance.

Linear versus Nonlinear Control

The controller can be designed on the basis of either a linear or nonlinear model of the process. As earlier, let us illustrate the design with a simple example: the control of the substrate concentration S by acting on the dilution rate D (or more precisely the influent flow rate F_{in}) in a stirred tank reactor with a simple microbial growth reaction.

Let us start with the first option: linear control. The control design is based on a linear model of the process (e.g., the linearized tangent model [equations 10 and 11] around some steady state [$\bar{S}, \bar{X}, \bar{D}, \bar{S}_{in}$]. If we consider the control of S at a prescribed value S^* (e.g., equal to $\bar{S}$), a linear proportional integral (PI) regulator will then typically have the following form (see, e.g., Ref. 8):

$$D = \bar{D} + K_p\left[(S^* - S) + \frac{1}{\tau_i}\int_0^t (S^* - S(\tau))d\tau\right], \quad K_P > 0,\ \tau_i > 0 \qquad (35)$$

where K_P is the controller gain and τ_i is the integral time, or reset time.

Indeed a similar PI regulator can be computed on the basis of a nonlinear model of the process (e.g., the mass balance equation of the substrate concentration S) (2). The control algorithm will then be written as follows:

$$D = \frac{C_1(S^* - S) + C_2\int_0^t (S^* - S(\tau))d\tau - k_1\mu X}{S_{in} - S}, \quad C_1 > 0,\ C_2 > 0 \qquad (36)$$

The preceding controller equation could have been obtained by combining the mass balance equation, equation 2, with the following desired closed-loop second-order dynamic equation:

$$\frac{d(S^* - S)}{dt} = -C_1(S^* - S) - C_2\int_0^t (S^* - S(\tau))d\tau \qquad (37)$$

for S^* constant. (Indeed by differentiating equation 37 once with respect to time, one obtains the usual second-order linear equation $d^2x/dt^2 + C_1dx/dt + C_2x = 0$ with $x = S^* - S$.) In simple terms, equation 37 imposes that if at some time instant the substrate concentration S is different from its desired value S^*, the controller will force S

to converge to S^* with a rate determined by the design parameters C_1 and C_2.

Let us present at this point the advantages and drawbacks of the approaches introduced. Before doing so, it is important to note that the second algorithm (equation 36) is more sophisticated than the first one (equation 35). In particular, the second controller requires the knowledge (on-line measurement) of the biomass concentration X and of the influent substrate concentration S_{in} and the knowledge of the specific growth rate model. When comparing both approaches, it is important to keep these differences in mind.

Linear Control. The linear controller, if it is correctly tuned (see, e.g., Ref. 8 for tuning rules), will most probably do a good job if it is used for regulation around a steady state. It also has the advantage to be easily implemented via standard industrial PI regulators. However, its performance (in particular, its transient performance) might be degrading if the process is moving away from the steady state for which it has been calibrated and tuned. This typically happens for start-up and grade changes in CSTRs and might be even more critical for batch and fed-batch reactors in which the process state typically follows large variations (just think of the "exponential" growth of the biomass in this type of reactor). This is illustrated in Figure 5, where the PI regulation of S has been numerically simulated with the same model as in Figure 3 and under the following conditions:

$$S^* = 1 \text{ g/L}, K_p = 2, \tau_i = 0.5, S(0) = 0 \text{ g/L} \quad (38)$$

The values of the design parameters of the PI have been chosen so as to correspond to a closed-loop dynamics close to the mean open-loop dynamics. We note that the controller gain is too high at the beginning (and induce large variations of the controlled output S) and then too small (the controller is not able to maintain the controlled output S to its desired value during the second part of the "exponential" growth phase).

Nonlinear Control. The main advantage of the nonlinear controllers is directly related to the mentioned drawback of the linear controllers: if correctly designed and tuned, it will be able to handle a larger spectrum of the operating domain of the bioprocess. In the example, the main source of nonlinearity is the term $k_1\mu X$. The presence of this term in the controller is particularly important in the control of fed-batch reactors; as pointed out by Axelsson (9), the integration of a term that characterizes the "exponential" biomass growth is a key factor for the efficiency of the control action. The other term, $S_{in} - S$, also plays an important role; if S_{in} is a disturbance in the context of the control of the bioprocess, then this term is indeed a feedforward term that allows anticipation of the effect of variations of S_{in} on the performance of the closed-loop system (process plus controller).

However, first the implementation of the control algorithm requires a computer instead of a standard industrial PI (which today can hardly be considered a drawback). Secondly (as for any controller), the performance depends on the quality and reliability of the model on which the design has been based. In other words, in the given example, this means that we assume that the hydrodynamics in the reactor are fairly represented by completely mixed conditions and that the simple growth reaction is largely dominant with regard to possible other side reactions. Besides the implementation as such of the controller, equation 36 requires the values of X and also of S_{in}. In many cases one may expect that if S is measured on-line, it should not be difficult to also measure S_{in} (in many instances, you do not even need to measure it because the influent substrate concentration is known by user's choice). But the situation is usually different for the biomass concentration: very often X is not measured on-line, or, to say it differently, S and X are not often available at the same time for on-line measurement. Finally, the nonlinear controller (equation 36) relies on the knowledge of (the model of) the specific growth rate μ; as suggested previously, this is a major source of uncertainty in bioprocess models.

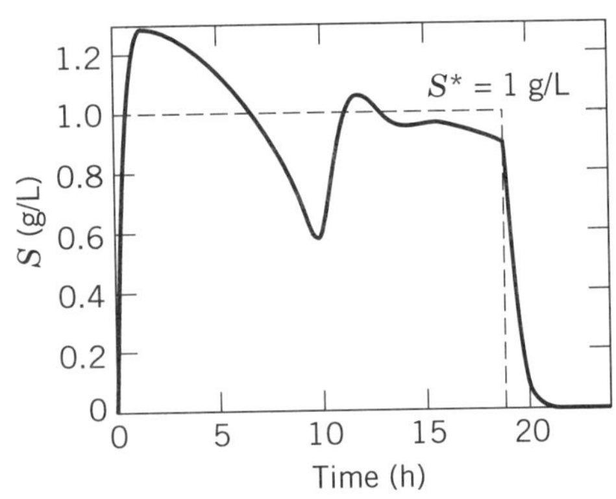

Figure 5. Numerical simulation of a linear proportional integral control of a fed-batch reactor.

Adaptive versus Nonadaptive Control

One possible way to handle the previously mentioned difficulties (lack of on-line measurements, uncertainties on the kinetics models) is to use adaptive (linearizing) control (see Ref. 10). In the example, this means that the control law will be implemented by considering on-line estimates of X and μ. An on-line estimate of the biomass concentration can be provided by an asymptotic observer (for a systematic design, see Ref. 10). In our example, it will take the following form. We first define an auxiliary variable Z:

$$Z = S + \frac{1}{Y_{X/S}} X \quad (39)$$

The dynamics of Z are readily derived from the mass balance equations of X and S:

$$\frac{dZ}{dt} = DS_{in} - DZ \quad (40)$$

The on-line estimate $\hat{X}$ of X is then given by computing Z from equation 40 and by rewriting equation 39:

$$\hat{X} = Y_{X/S}(Z - S) \quad (41)$$

or equivalently, one can replace the term $X/Y_{X/S}$ in the controller equation with $Z - S$.

Let us now consider the on-line estimation of μ. Different methods can be used to give an on-line estimate $\hat{\mu}$ of μ. One possible estimator is the observer-based estimator (see Ref. 10), which specializes in our example as follows:

$$\frac{d\hat{S}}{dt} = -\hat{\mu}(Z - S) + DS_{\text{in}} - DS + \gamma_1(S - \hat{S}),\ \gamma_1 > 0 \quad (42)$$

$$\frac{d\hat{\mu}}{dt} = -\gamma_2(S - \hat{S}),\ \gamma_2 > 0 \quad (43)$$

$\hat{S}$ is an estimate of S whose role is to drive the estimation of the unknown value of the specific growth rate μ on-line; this is achieved by comparing $\hat{S}$ with the measured value S (this is the role of the correction terms $S - \hat{S}$ in equations 42 and 43). γ_1 and γ_2 are design parameters that have to be calibrated to obtain the best on-line estimates (10).

Because the parameter estimation introduces an integral action in the control loop (10,11), the term $C_2\int_0^t(S^* - S(\tau))d\tau$ in the control law (equation 36) is not necessary anymore. Then, in our example, the adaptive linearizing controller is given by the following equation:

$$D = \frac{C_1(S^* - S) - \hat{\mu}(Z - S)}{S_{\text{in}} - S},\ C_1 > 0 \quad (44)$$

with Z given by equation 40 and $\hat{\mu}$ given by 42 and 43.

Different applications of adaptive linearizing control to bioprocesses have been described (12–15). One of the key features of adaptive linearizing can be illustrated on the basis of the example; it allows incorporation of the well-known characteristics of the process dynamics while keeping the usual features of classical controllers:

- Proportional action via the term $C_1(S^* - S)$
- Integral action via the adaptation mechanism (equations 42 and 43)
- Feedforward action via the presence of S_{in}; the controller is capable of anticipating the effect of a variation of the influent substrate concentration; for instance, an increase of S_{in} results in a decrease of the control input D inversely proportional to this increase.

As for the nonadaptive, nonlinear PI, the controller contains a state estimate via the term $k_1\mu X$ (or more precisely $\hat{\mu}[Z - S]$). If the process is in good working condition, in particular if the biomass is in a good state, then it is possible to treat large amounts of substrate, because the control action D is proportional to X.

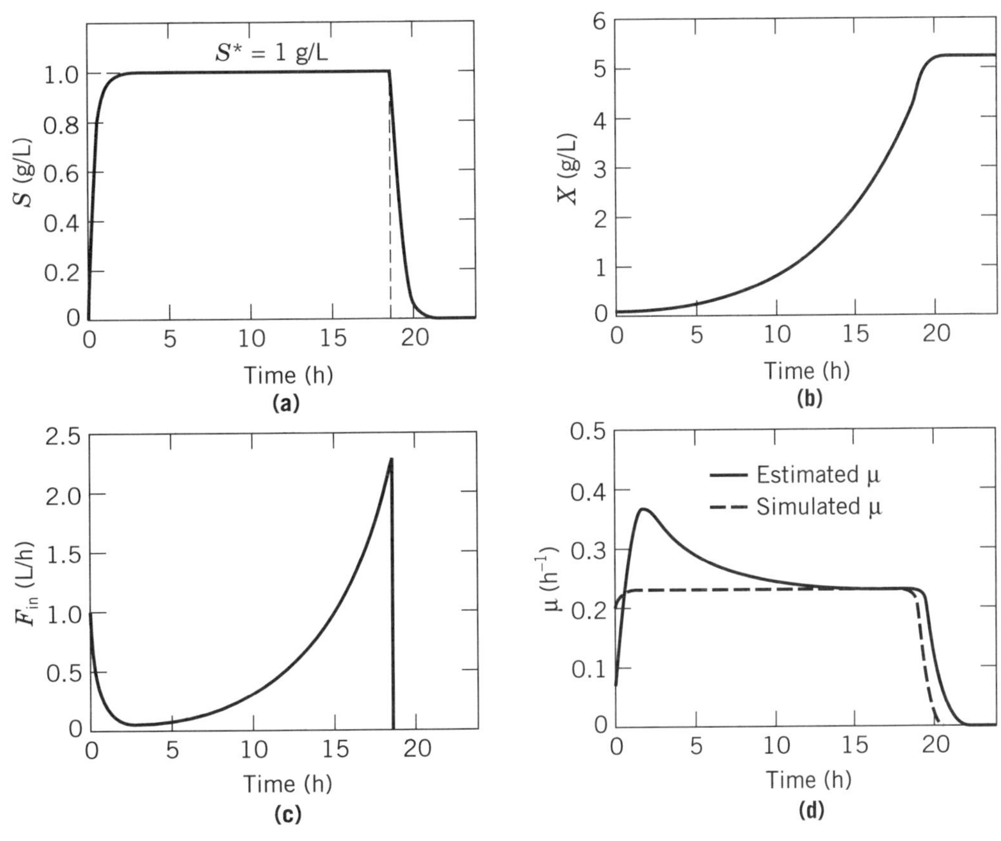

Figure 6. Adaptive linearizing controller of a fed-batch reactor. (**a**) Controlled output S; (**b**) biomass concentration X; (**c**) control input F_{in}; (**d**) specific growth rate μ.

Finally, note that besides the integral action, the estimation of physical parameters (here μ) has the further advantage of giving useful information that can be used for monitoring the process and possibly also for analyzing the internal working of the process.

Figure 6 illustrates the performance in numerical simulation of the adaptive linearizing controller. The same model parameters as in Figures 3 and 5 have been used. The following design parameters and initial values have been considered:

$$S^* = 1 \text{ g/L}, C_1 = 2, \gamma_1 = 5, \gamma_2 = \frac{\lambda_1^2}{4(Z - S)} \quad (45)$$

$$\hat{\mu}(0) = 0.1 \text{ h}^{-1}, \hat{S}(0) = S(0), Z(0) = S(0) + 0.5X(0), S(0) = 0.5 \text{ g/L} \quad (46)$$

The initial value of S has been set to 0.5 g/L, which allows the opportunity to see that the controller has no problem forcing S to converge to its desired value (see Fig. 6a). Figures 6b and 6c exhibit the exponential profile of the biomass X and the control input F_{in}. Finally, Figure 6d gives the on-line estimation of the specific growth rate μ.

Other Approaches

The (adaptive) nonlinear controller presented in the preceding section belongs to a large class of control strategies, often labeled model-based control. This includes control approaches such as model predictive control, in which constraints (such as input constraints) may be handled explicitely in the design (8). We have restricted the presentation to mass balance models, but obviously any dynamic model of the bioprocess (e.g., hybrid model), as long as it is reliable, can be considered in the control design.

CONCLUSIONS

In this article we have first concentrated on the basic concepts of automatic control of bioprocesses, with a particular emphasis on the notions of optimization, stability, and disturbances. In a second step the basic ingredients of automatic control (dynamic model, feedback, proportional, integral, and feedforward actions) have been introduced, and linear and nonlinear control and adaptive and nonadaptive control have been compared.

An appealing avenue for future applications in bioprocesses is an integrated process design approach, which is already largely used in other areas such as aeronautics and that implies the integration of the control strategies in the design of the process from the very beginning (16).

ACKNOWLEDGMENTS

This article presents research results of the Belgian Programme on Inter-University Poles of Attraction initiated by the Belgian State, Prime Minister's office for Science, Technology, and Culture. The scientific responsibility rests with its author.

BIBLIOGRAPHY

1. J.L. Willems, *Stability Theory of Dynamical Systems*, Nelson, London, 1970.
2. G. D'Ans, D. Gottlieb, and P. Kokotovic, *Automatica* **8**, 729–736 (1972).
3. T.F. Edgar and D.M. Himmelblau, *Optimization of Chemical Processes*, McGraw-Hill, New York, 1988.
4. J. Van Impe, Ph.D. Thesis, Katholieke Universiteit, Leuven, Belgium, 1993.
5. L. Ljung, *System Identification: Theory for the User*, Prentice-Hall, Englewood Cliffs, N.J., 1987.
6. G. Montague and J. Morris, *TIBTECH* **12**, 312–323 (1994).
7. F. Feyo De Azevedo, B. Dahm, and F.R. Oliviera, *Comput. Chem. Eng.* **21**, S751–S756 (1997).
8. D.E. Seborg, T.F. Edgar, and D.A. Mellicamp, *Process Dynamics and Control*, Wiley, New York, 1989.
9. J.P. Axelsson, Ph.D. Thesis, Lund Institute of Technology, Lund, Sweden, 1989.
10. G. Bastin and D. Dochain, *On-Line Estimation and Adaptive Control of Bioreactors*, Elsevier, Amsterdam, 1990.
11. M. Perrier and D. Dochain, *International Journal of Adoptive Control and Signal Processing* **7**, 309–321 (1993).
12. L. Chen, G. Bastin, and V. Van Breusegem, *Automatica* **31**, 55–65 (1995).
13. Y. Pomerleau, M. Perrier, and D. Bourque, in A. Monack and K. Schogerl eds., *Proc. 6th Int. Conf. Computer Appl. Biotechnol.*, Pergamon Press, Oxford, 1995, pp. 107–112.
14. Y. Pomerleau and G. Viel, in N.M. Karim and G. Stephanopoulos eds., *Proc. 5th Int. Conf. Computer Appl. Biotechnol.* Pergamon Press, Oxford, 1992, pp. 315–318.
15. P. Renard, D. Dochain, G. Bastin, H. Naveau, and E.J. Nyns, *Biotechnol. Bioeng.* **31**, 287–294 (1988).
16. J. Harmand, Ph.D. Thesis, Université de Perpignan, Perpignan, France, 1997.

See also AMMONIA TOXICITY, ANIMAL CELLS; MEDIUM FORMULATION AND DESIGN; PILOT PLANTS: DESIGN AND OPERATION; PROCESS MONITORING; SECONDARY METABOLITE PRODUCTION, ACTINOMYCETES, OTHER THAN STREPTOMYCES.

PROCESS MONITORING

A.P.F. TURNER
S.F. WHITE
Cranfield Biotechnology Centre
Bedfordshire, United Kingdom

KEY WORDS

Biomass
Biosensors
Carbon dioxide
Chromatography
Oxygen
pH
Pressure
Temperature
Viscosity

OUTLINE

INTRODUCTION

For most modern industrial production methods, process monitoring plays a key role; bioprocess systems are no exception to this rule. Effective methods of monitoring are required in order to develop, optimize, and maintain biological reactors at maximum efficiently. An optimized process should lead to streamlined performance, reductions in running and material costs, and improvements in quality control.

A number of bioprocess monitoring instruments have been regular features of the industry for many years (Fig. 1), in particular, sensors capable of measuring physical parameters. The challenge is to develop cost-effective devices capable of measuring a much wider range of parameters, providing process operators with a deeper insight into the process. Eventually effective monitoring will lead to better control systems, providing cost and quality benefits.

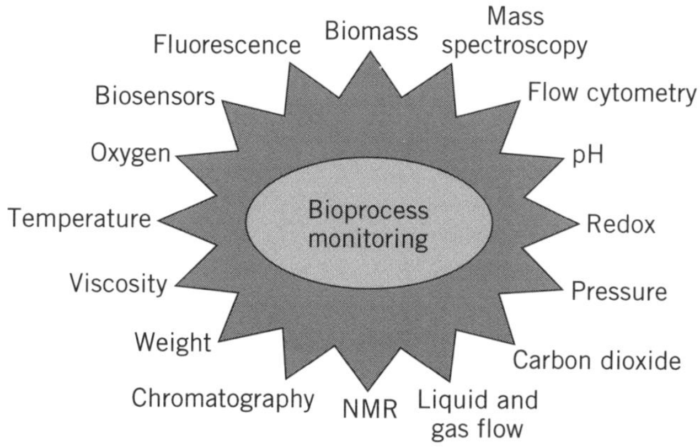

Figure 1. Major areas of bioprocess monitoring.

OPERATING FEATURES

For any measuring technique, there are a number of criteria that the device must satisfy if it is to accepted by commercial bioprocess operators (Fig. 2). The following list of desirable characteristic is not exhaustive, but does cover the main areas of concern.

Reliability

Reliability is a key issue for bioprocess monitoring equipment. Confidence in the ability of an instrument to maintain its credibility in terms of performance is a fundamental parameter. This encompasses a number of features relating to the operation of the instrument, for example, ease of use, maintenance, repair, and replacement. A successful monitoring instrument will have a low failure rate; again, this is related to reliability.

Accuracy

Accuracy can be described as the relationship between the measured value (by the instrument) and the actual value of a bioprocess variable. An accurate instrument will achieve a low percentage error between these two values. Generally this can be stated as

$$\% \text{ error} = (\text{measured value} - \text{true value})/\text{true value} \times 100\%$$

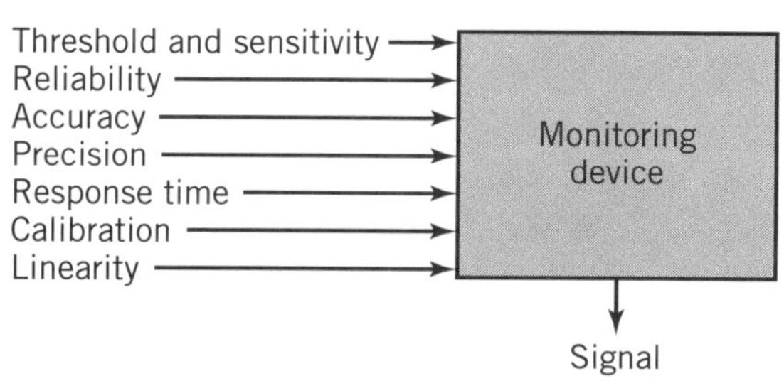

Figure 2. Main criteria for a bioprocess monitoring instrument.

Precision

Precision is a measure of instrument reproducibility, that is, the ability to obtain the same value with repeated measurements of a process variable (at a constant level). It follows that a precise instrument may not, necessarily, be accurate. Therefore it is important to distinguish between these two parameters.

Response time

The measurement of any process variable will entail a time delay between change in the parameter and display of the measured value. This response time should not be detrimental to the progress of the bioprocess, particularly if the measurement is linked to a control action. Response times will be affected not only by the type of instrument but also by the method of measurement. Off-line sampling and measuring can involve a number of time-consuming steps. In contrast, in situ devices can provide a real-time measurement.

Calibration

In order to maintain the accuracy of a sensor it is (generally) desirable to carry out a calibration step. This is carried out using set standards of comparison. The effects of calibration can impinge on the process, in other words, it may be difficult to carry out calibration of an in situ sensor, whereas an on-line device could easily incorporate a calibration step(s) during routine running.

Linearity

Under ideal conditions, the output signal from a sensor would be directly proportional to the analyte concentration. However, this is not always the case, and other models have to be used to reach a true value. Despite this drawback, linearity is not an overriding prerequisite for a successful monitoring instrument. Developments in modern software can be adapted to compensate for nonlinearity.

Threshold and Sensitivity

Sensitivity is the magnitude of the output signal per unit change in the target analyte concentration. The lowest level of detection is related to the sensitivity and the signal-to-noise ratio. A number of factors influence the sensitivity of a monitoring device, including sensor design, the operating environment, periods of maintenance, and interfering noise levels. The sensitivity will influence the dynamic range above which the device becomes saturated; the highest level of detection will be the threshold limit for the device.

METHODS OF MONITORING

In addition to the variety of monitoring devices available, a number of methods (for carrying out the measurement exist (Fig. 3). Sampling and sample handling is a vital issue, affecting both the accuracy and frequency of measurements (1). Broadly, the main approaches to sampling can be described in one of several ways.

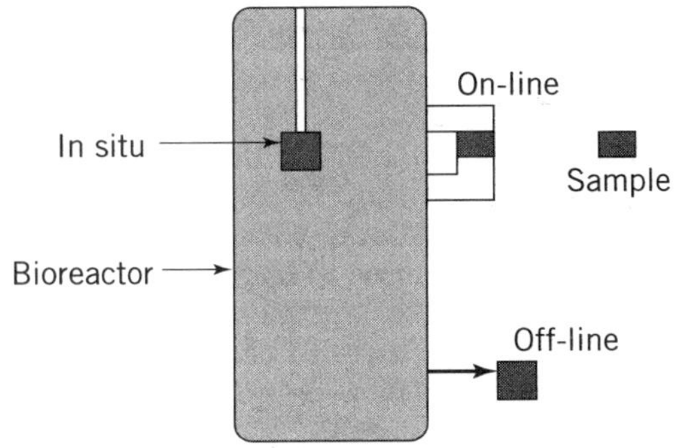

Figure 3. Methods of monitoring.

Off-Line or At-Line Monitoring

Off-line monitoring involves taking a sample from the bioreactor and carrying out the measurement at a different site, usually under laboratory conditions. Off-line sampling can be detrimental in terms of cost, efficiency, and threat to asepsis. Manually removing and measuring a sample requires technical labor. The received signal will not be "real time" because of the delays. In order to obtain the sample, the sterility of the bioreactor must be maintained. Hence, provision must be made to achieve this. In some cases, decentralized equipment has become available that allows measurements to be made simply close to the production process (at-line), reducing some of the delays associated with laboratory analysis.

In Situ or In-Line Monitoring

In situ sensors are placed directly in the bioprocess vessel. The use of in situ sensors has long been established in the bioprocess industry (2), where "dip-in" devices are used to monitor a number of parameters such as pH and dissolved gas concentrations. A number of advantages are gained by operating sensors in this fashion, including real-time monitoring (sampling-related time delays are eliminated) with the sensors operating continuously (any rapid change in the analyte concentration can be readily observed), and labor requirements and problems of contamination are both significantly reduced. However, in situ sensors can have a number of drawbacks, for example, the sensor system must be amenable to sterilization; if the lifetime is short, replacement may be difficult (during a process run); the surface of the sensor could become fouled by components of the growth medium, affecting the signal output.

On-Line Monitoring

Methods of on-line monitoring involve the automatic removal and measurement of sample or sample stream from the bioreactor. One example of this has been the development (since the early 1980s) of flow injection analysis (FIA). This is a liquid handling technique that has proved flexible in adapting to most chemical and biochemical reaction procedures (3), representing an effective compromise between the desirability of in situ monitoring and the technical ease of off-line measurements. The main advantages of on-line methods include the following: sensor sterilization can be readily accomplished, sample pretreatment (e.g., gassing, dilution, removal of interferents), and sensor calibration can be built into the system. The main disadvantages are a need for an effective and reliable sampling system and the fact that the signal is discontinuous; frequency measuring rate is determined by the limitation of the overall FIA arrangement. All of these methods have their advantages and disadvantages. The choice of approach adopted depends on a number of factors, not least of which will be the availability of both sensor and system.

MEASURING DEVICES

pH Measurement

Monitoring the pH of a number of industrial biological reactions is an extremely effective way of determining their course, providing the operator with valuable information. Measuring pH, along with temperature, is one of the most common practices in bioprocess monitoring (4). Correcting action can be taken, either manually or automatically, to prevent unwanted increases or decreases in pH.

The most common form of pH sensor used for fermentation monitoring is based on the electrode design introduced by Ingold in 1947. Under operating conditions, the probe must always be overpressurized (with respect to the fermentation broth) to keep a positive flow of electrolyte out of the porous plug. A combined glass and reference electrode that is capable of withstanding repeated sterilization at temperatures of 121 °C and pressures of 138 kN/m^2 is required. In this case, the term *combined* refers to the single unit containing both electrodes. Figure 4 illustrates the main features of the pH electrode.

Typically, the reference electrodes used are silver/silver chloride (Ag/AgCl) or calomel (Hg/HgCl) and maintained

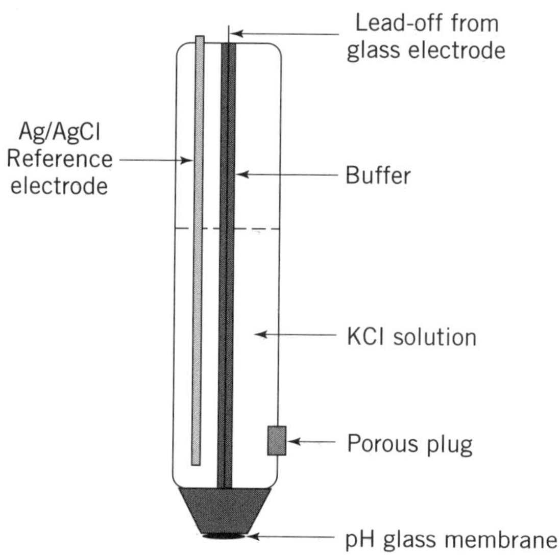

Figure 4. pH electrode.

in a concentrated solution of potassium chloride in a buffer solution within the body of the sensor. Under these conditions an equilibrium is established with the electrolyte:

$$\mathrm{Hg/HgCl_2/KCL} \rightarrow \mathrm{2Hg} + \mathrm{2Cl^-} \rightarrow \mathrm{Hg_2Cl_2} + \mathrm{2e^-} \quad (1)$$

This leads to a situation whereby the reference electrode potential is dependent on the activity of the chloride ions.

Generally the sensor format is that of a single unit glass-reference design. It has been known since the early years of this century that certain forms of glass are electrically conducting. Furthermore, the electrical potential at the glass-liquid interface changes with activity of hydrogen ions. This potential is directly related to the activity of the H^+ via the Nernst equation,

$$E = E^0 + (RT/\eta F) \ln aH^+ \quad (2)$$

where E is the measured potential, R the gas constant, T the absolute temperature, F the Faraday constant, and ln aH^+ the activity. With this type of pH electrode, the potential determining process is determined by interactions between H^+ activity (in the sample solution) and sianol groups on the glass.

Autoclavable pH probes are available (for sterilization purposes), but this can lead to a reduction in useful probe life. Generally the sensor's response does degenerate over time with repeated use. Interactions with the medium constituents can lead to problems. However, proprietary cleaning agents are available to regenerate the probe. A recent advance in the field of pH monitoring has been the introduction of sensors using ion-selective field effect transistors (ISFETs). These are solid-state devices based on the use of semiconductor structures (5,6). The pH is determined by measuring the change in the voltage required to maintain a zero current between the gate and drain electrode as the transmembrane potential alters. Several commercial companies have marketed ISFET-based pH meters with potential applications in a number of areas. These sensors are robust, virtually shatterproof, and are easily cleaned. In addition, they require no filling solution and are easy to operate and maintain.

Temperature

Maintaining optimum conditions for any bioprocess will invariably involve monitoring and controlling the temperature of the broth. Generally, bioprocesses are monitored over a temperature range of between 0 °C and 100 °C (excluding the sterilization cycle). Furthermore, this may require a control regime operating within a narrow range of temperatures. A number of devices are available for obtaining an accurate measurement of temperature conditions during a bioprocess operation, based on a range of techniques.

Thermistors. Thermistors are semiconductor devices composed of agglomerates of polycrystalline metallic oxides (e.g., magnesium oxide, cobalt oxide, ferric oxide). The operating characteristic of these devices relies on a sensitive change in resistance, resulting from temperature fluctuations (7). This change is a function of the absolute temperature. The output response from thermistors is of a nonlinear temperature versus resistance curve, with the resistance decreasing as the temperature increases. Generally the response can be described by the following equation:

$$R(T) = R_0 \exp[B(1/T - 1/T_0)] \quad (3)$$

Thermistors tend to be cheap and stable, with rapid response times.

Resistance Thermometers. Resistance thermometers are based on the changes in the electrical resistance of metallic conductors, with changing temperature (8). This relationship can be described by the following equation:

$$R = R_0(1 + \alpha T) \quad (4)$$

R is resistance. T is the temperature, and α the temperature coefficient of linear expansion.

Platinum is the preferred metal for these devices, but other metals can be used (e.g., nickel, tungsten, or copper). The advantage of platinum is that it can be obtained in a highly pure form, hence the electrical properties of the metal can be well characterized. A platinum wire of 100-Ω resistance at 0 °C is typically used. Generally, calibration of the sensor is carried out at this temperature. Readings can be obtained using a Wheatstone bridge circuit. Overall, these sensors tend to be precise, accurate (±0.25%), reliable, and sensitive. Platinum sensors are stable in both sterilization and fermentation conditions.

Thermocouples. Thermocouples are based on the Seebeck effect, first postulated in 1821. It was noted that a circuit consisting of two dissimilar metals joined together at junctions maintained at different temperatures will generate an electrical current. The electromotive force generated will be determined by the choice of metals and the respective temperatures. For practical applications, one of the junctions (termed the *cold junction*) is maintained at a constant temperature and used as a reference point. The second junction is in contact with the media to be monitored; fluctuations in the temperature can be studied by monitoring the current response (7).

The response (EmV versus T[K]) approximates to a parabolic curve. Generally it is preferable to measure over the linear range of this response. One advantage of thermocouples is that they do not require an external electrical supply. However, they are susceptible to cold junction problems, with the need to maintain a constant temperature.

Mercury-in-Glass Thermometers. Mercury-in-glass thermometers can be used directly in small batch fermenters. However, the fragility and potential toxicity of these thermometers restricts their use. In order to overcome the problem of using mercury, alcohol can be substituted. Generally these sensors are not very accurate (especially over a wide range) and provide only an indication of the temperature. However, these thermometers have the advantage of being both portable and providing a direct reading.

Bimetallic Thermometers. Bimetallic thermometers usually consist of a bimetallic helical coil surrounded by a protective tube or wall. The coil winds or unwinds with changes in temperature and causes movement of a fixed pointer (8). By affixing a pen to the fixed point, a temperature profile of the bioprocess can be obtained. Despite being more robust than the mercury thermometer, they are less accurate (the range determined by the metals used) and, again, are limited to giving only an indication of temperature.

Pressure

Many bioprocesses operate under conditions of overpressure. Monitoring the magnitude of this pressure is an important factor, both in terms of safety and optimization of the process. Industrial and laboratory fermenters are designed to operate up to a safe working pressure. Increasing the applied pressure above the upper limit can be dangerous. Furthermore, maintaining a positive reactor head pressure can prevent contamination of the bioreactor. In order to facilitate effective sterilization, fermenters need an accurate pressure monitor. Pressure will also affect the solubility of gases (such as O_2 and CO_2). Several different approaches have been used in the development of pressure gauges.

Bourdon Tube Pressure Gauge. The Bourdon tube pressure gauge is composed of an arched or spiral-shaped tube. Part of the tube is of an elliptical shape, made of either metal or quartz. Under pressure, the elliptical section tends to become circular. The interior of the tube is influenced by the process pressure, whereas the exterior is under atmospheric pressure. The overall effect of the change in cross-sectional area is to straighten out the curved shape of the tube. One end of the tube is fixed to an inlet from the reactor. The other end is sealed and allowed to move; this is connected (via a geared section and pinion) to an indicator needle (9). Pressure can be read directly from the gauge. These devices tend to be robust and economical.

Strain Gauges. Strain gauges are based on the phenomenon that when a wire is placed under strain, the electrical resistance changes (7). This is due to both the alteration in geometry and resistivity changes. The electrical output can be related to changes in pressure. Typically the sensors are constructed using either a metallic or semiconductor element, usually in a grid structure composed of a number of conductor threads. Generally the change in resistance to pressure can be related by the following equation:

$$\Delta R/R = \Delta L/L + \Delta\rho/\rho - \Delta S/S \quad (5)$$

R is resistance, L is length of thread (a multiple of the number of threads), ρ is the resistivity of the material, and S is the cross-sectional area of the material.

Piezoelectric Manometers. Certain materials, such as quartz crystal, posses an asymmetrical electrical charge distribution. When an alternating electric field is applied across a piezoelectric material, small mechanical deformations (to the crystal) are generated (10). Alternatively, if the crystal is subjected to pressure on one face, an electrical charge will be generated on the opposite face. This effect (the piezoelectric effect) forms the basis of operation for piezoelectric manometers. By attaching electrodes to one face of the crystal, changes in pressure can be relayed as an electrical output.

Diaphragm-Type and Pressure Bellows Sensors. Diaphragm-type sensors incorporate a membrane that deforms as a result of pressure from either a liquid or gas, forming a pressure difference across the membrane. This movement can be monitored and a corresponding readout obtained, indicating the pressure of the matter under investigation. Generally, the membranes used can be made from materials such as stainless steel, palladium, or silicon. When in operation, these sensors can suffer from the deposition of material on the surface, leading to erroneous measurements. In order to prevent this, periodic checks should be made.

Pressure bellow sensors operate, as the name implies, by monitoring the movement of a bellows when subject to a pressure. Changes in the bellows extension length can be related to pressure. The bellows can be linked directly to a variable transformer that is capable of recording the physical movement and displaying a readout.

Viscosity

Information on the rheology, or viscosity, can help in ensuring the efficiency of a biological process (11). An example of this would be the effectiveness of oxygen transfer throughout a medium and determining the degree of branching of filamentous microorganisms. In addition, viscosity can effect pumping, mass transfer, and mixing. Viscosity is the apparent shear resistance between adjacent layers of liquids or gases moving at different speeds. Typically, in fluids this is the result of molecular cohesion; rising temperatures lead to a decrease in viscosity. In contrast, viscosity increases for gases under conditions of rising temperature. This results from an increased measured molecular activity.

Newtonian Fluids. Consider two parallel surfaces with area A and distance Y apart. If a shearing force F is applied to one side, it will move with a velocity V:

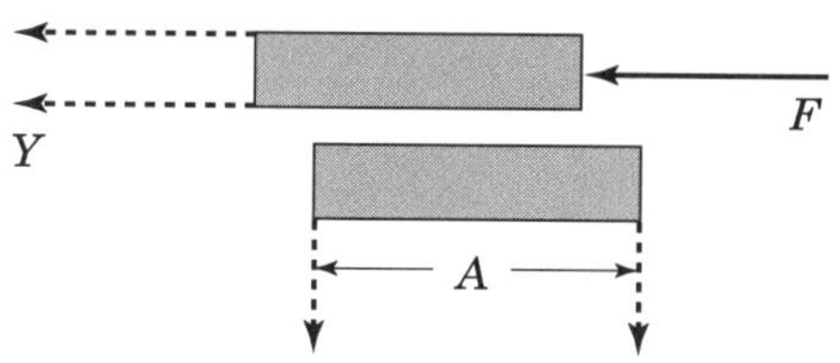

It is found that for ideal or Newtonian fluids the shear stress $F/A(\tau) = dV/dY$ multiplied by a constant (μ). This constant is called the coefficient of absolute (or dynamic) viscosity. The term dV/dY is called the *shear rate* (γ). Plotting τ against γ gives the flow curve characteristics of a particular fluid. Newtonian fluids display a straight-line response, passing through the intersection of the axis. The slope of this gradient is equal to μ. Typical units of viscosity are Newtons per second per square meter (NS m^{-2}) and centipoise. Examples of Newtonian fluids include all gases, liquids with low molecular weight (notable exceptions are the colloidal suspensions and polymeric solutions with long molecules, e.g., specially developed mineral oils), water, and some oils.

Non-Newtonian Fluids. Non-Newtonian fluids are classed according to their flow curves, which may be of various shapes according to the nature of the fluid (12). For each fluid the ratio between shear stress and shear rate is not constant.

Bingham Plastic. For these fluids a certain minimum shear (τ^0) is required before viscous flow takes place, after which the relationship is linear. Examples include some fats and plastic melts, some greases, and detergent slurries. Perhaps the best everyday example is toothpaste, a certain amount of pressure must be applied to the tube before the paste begins to flow.

Pseudoplastic Fluids. Apparent viscosity decreases with shear rate, tending to a constant value at high shear rates. This behavior is typical of suspensions of long-shaped particles of high-polymer solutions such as cellulose derivatives. The process has been explained as the result of progressive alignment of particulars or molecules as shear rate increases. Examples include starch suspensions, biological fluids, and rubber solutions.

Dilatant Fluids. In contrast to the pseudoplastic fluids, dilatant fluids display an increasing apparent viscosity with increased shear rate. Examples include sugar solutions, starch, and wet cement aggregates.

Casson Plastic. These fluids display a decreasing apparent viscosity with increasing shear rate after a yield stress (τ^0) value has been exceeded. Examples include blood, printing ink, and some fruit juices.

Generally, bioprocess media display non-Newtonian behavior, that is, a nonlinear response to shear rate. These fluids do not have a constant viscosity, it varies according to the shear rate. In practice, because of the problems of maintaining sterility, most viscosity measurements are made off-line.

There are a number of commercially available instruments (viscometers), applicable to the measurement of bioprocess rheology properties (Fig. 5). Generally, these instruments can be classified into three different types:

1. *Cone and plate viscometer*. This instrument consists of a flat plate in contact with an inverted cone, the fluid to be measured fills the gap between the cone and the plate. In operation, the cone is rotated in the fluid, and the angular velocity and torque are measured. These values can be related to the shear stress and rate.
2. *Coaxial cylinder rotary viscometer*. This instrument consists of two cylinders, one located inside the other. The fluid to be measured fills the gap between the two cylinders. Again, the angular velocity and torque are measured when one of the cylinders is held stationary and the other rotated.

Figure 5. The CVO rheometer (Bohlin Instruments).

3. *Impeller viscometer*. This device is similar to the coaxial cylinder viscometer. However, the fluid measurements are made using a rotating impeller inside the stationary cylinder.

Oxygen

Generally, oxygen measurement falls into two main categories: dissolved oxygen and exit gas analysis. For most aerobic fermentations, an accurate determination of the dissolved oxygen concentration is of particular importance, for example, in maintaining the concentration above a specified minimal level. In addition, oxygen transport processes can be determined only by using an accurate method of measurement.

The exit gas analysis for oxygen, usually in conjunction with CO_2 measurement, can be used to determine the metabolic state of aerobic microorganisms and their oxygen uptake.

Dissolved Oxygen Probes. A wide range of dissolved oxygen probes based on the use of electrochemical determination are available (13). Typically these fall into two main types (14): galvanic (potentiometric) or polarographic (amperometric or Clark). In essence these electrodes measure the activity (or partial pressure) of oxygen present, rather than the concentration. The range covered by these devices is approximately 0.001 to 1000 ppm (15). Both the galvanic and polarographic electrodes use an oxygen-permeable membrane (e.g., Teflon, polyethylene, and polypropylene) to separate the electrode components from the fermentation media. Similarly, the measurement for both types is based on the electrochemical reduction of oxygen at a cathode.

Polarographic Electrodes. The principal components of the polarographic oxygen sensor (16) include a precious metal cathode (e.g., gold or platinum), a silver/silver chloride (Ag/AgCl) anode, and an electrolyte solution (e.g., sodium chloride). During operation the following reactions take place at the respective electrodes:

$$\begin{aligned} \text{Cathode:} \quad & O_2 + 2H_2O + 2e^- \rightarrow H_2O_2 \\ & + 2OH^- \rightarrow H_2O_2 + 2e^- \rightarrow 2OH^- \\ \text{Anode:} \quad & Ag + Cl \rightarrow AgCl + 2OH^- \end{aligned} \tag{6}$$

Typically, polarographic (or amperometric) sensors operate at a fixed voltage, 0.6 to 0.8V, between the two electrodes. This voltage range represents the diffusion-controlled plateau of oxygen reduction. In effect, the current recorded can be directly related to the oxygen concentration.

Galvanic Electrodes. In contrast to the polargraphic oxygen electrode, the galvanic sensor (17) employs a non–noble metal (zinc, lead, or cadmium) anode with a noble metal cathode (silver or gold). The reactions occurring at the electrodes are

$$\begin{aligned} \text{Cathode:} \quad & O_2 + 2H_2O + 4e \rightarrow 4OH^- \\ \text{Anode:} \quad & 2Pb(e.g.) \rightarrow 2Pb^{2+} + 2e^- \end{aligned} \tag{7}$$

In contrast to the polargraphic sensors, galvanic electrodes require no external voltage source. The combination of a sacrificial anode and noble metal cathode is sufficient to provide an adequate electromotive force, resulting in a spontaneous reduction of oxygen at the electrode.

Despite their differences in operation, both types of electrodes are subject to similar constraints. During steady state, oxygen flux to the cathode is dependent on a number of transport-related steps including movement of oxygen through the bulk of the fermentation media and diffusion across the membrane and through the supporting electrolyte. In relation to the fermentation media, the viscosity and local hydrodynamic characteristics (in the vicinity of the electrode face) will play an important role. An example of this would be the incorrect placing of the electrode in a highly viscous media displaying non-Newtonian characteristics. If the electrode is situated in an area of quiescent fluid, the received signal may not be an accurate representation of oxygen partial pressure throughout the bioreactor.

The design of the sensor will determine its operating characteristics. These include the membrane thickness, the membrane diffusion coefficient, and the surface area of the cathode. Temperature can also have a significant influence on the sensor response. Generally a change in temperature will affect both the solubility and diffusion of oxygen through the membrane. In order to compensate for these effects a built-in thermistor is generally included in commercial devices.

There are several problems encountered by both types of sensor during bioprocess operations. Fouling of the sensor membrane surface by components of the bioprocess media can lead to erroneous signals. With continued use, this could include the growth of microorganisms on the membrane surface. The effect of a buildup on the membrane would be to change the diffusion characteristics of the sensor (see equation 8). Dissolved oxygen electrodes are also susceptible to signal drift during operation. In addition, gas bubbles and restricted flow toward the sensor can affect the received signal. Other important considerations are the effect of sterilization on the sensor, and calibration. Typically, calibration is carried out after sterilization, using medium gassing to obtain the full range of sensor values. In order to avoid sterilization problems, the sensor can be withdrawn into an isolating lock where it can be sterilized using milder conditions. The actual measurements can be expressed in one of several ways. One approach is to use percent of saturation on a scale of 0 to 100% oxygen saturation. Alternatively partial pressure can be used,

$$pO_2 = (p_{atm} + \Delta p - pH_2O)0.2095 \qquad (8)$$

where $(p_{atm} + \Delta p)$ is the absolute pressure and pH_2O is the partial water pressure. Finally a direct concentration can be obtained using Henry's law coefficient:

$$c = pO_2/H \qquad (9)$$

The coefficient value is subject to alteration through interactions with media components such as sugars, products, salts, and so forth (18). Hence, these factors must be taken into consideration when determining oxygen concentration.

O_2 Optodes. A recent development for measuring oxygen, and indeed a range of other analytes, has been the introduction of sensors based on the use of photometric transducers: optodes (19). The driving force for the development of these devices was linked to the evolution of immobilization technologies and fiber optics. The optical fibers (used to transmit information over long distances: 2–100 m) can be sited in harsh environments (e.g., high temperature and extremes of pH), allowing in situ real-time monitoring. A fiber is constructed using an optically transparent material called the *core* (this is the guide path for the light) and a cladding, surrounding the core. Light is totally reflected at the cladding-core interface, allowing it to be transported over long distances without significant loss. The light interacts with a sensitive element or layer, generating a signal.

Several spectroscopic techniques have been used in the construction of optodes; of these, fluorescence has been the most widely used in the construction of O_2 sensors. Fluorescence is the emission of light from an absorbing species that has been excited by light of a different wavelength. Optical sensors for oxygen are constructed using an immobilized flurophore that undergoes dynamic quenching. Increasing concentrations of the quencher lead to a decrease in the fluorescence. The relationship between fluorescence intensity and extent of quenching can be summarized as

$$I_0/I = 1 + K_{sv} \times [Q] \qquad (10)$$

where I_O and I represent the measured fluorescence intensity in the presence and absence of the quenching agent, K_{sv} the Stern-Volmer proportionality constant, and Q the concentration of the quenching agent (20).

For a number of applications, dyes, whose fluorescent properties depend on the presence of a target analyte (e.g., oxygen), can be used. The dye can be attached directly to the surface of the fiber or onto a membrane. With the most common configuration, where sensing takes place at the end of the fiber light is returned after acting with the sensitive layer. This can occur through the same fiber or through a parallel fiber in a bifurcated device. The magnitude of the interaction is determined at the other end of the fiber using a detector system (21).

These sensors have a number of advantages compared with electrochemically based systems including that they do not require a reference signal, they can be miniaturized, multisignaling from a range of analytes measured at the same time can be achieved, and the fibers can be steam sterilized. Despite these advantages there are a number of drawbacks to these devices, including interferences from ambient light, photobleaching and washout of the sensitive indicator, and the limited range of some devices. A number of fiber-optic-based O_2 optodes have been described, using a range of indicator molecules (e.g., 22,23,24). Their use in measuring oxygen concentrations during bioprocess applications will indubitably become more widespread as the technology develops.

Tubing Methods. An alternative approach to measuring dissolved oxygen content is to use the tubing method described by Philips and Johnson (25). This method is based on the use of a polymeric tubing (e.g., permeable Teflon) immersed in the fermenter media. A stream of helium or nitrogen gas is passed through the coil. Oxygen in the fermenter medium diffuses through the tubing into the carrier gas stream. The oxygen concentration in the inert gas is then determined using a suitable detector, for example, a paramagnetic gas analyser. Despite the longer response times for this method (2 to 10 minutes for a reading) compared with the electrode approach, it is a reliable technique. In addition the tubing is robust and can withstand repeated sterilization cycles.

Exit Gas Analysis. There a number of methods available for determining the oxygen concentration of exit gas from a bioprocess reactor. Several of these are based on exploiting the strong affinity shown by oxygen to a magnetic field. The paramagnetic gases, such as oxygen, display a positive magnetic susceptibility. There are two types of detectors based on this phenomenon: the magnetopneumatic and thermomagnetic analysers.

Magnetodynamical Analyzer. The magnetodynamical analyzer is based on a dumbbell-shaped di-magnet held in a nonuniform magnetic field (15). The dumbbell structure is free to rotate about an axis. In the presence of oxygen, the dumbbell is rotated (under the action of the increased magnetic force) in proportion to the oxygen partial pressure. By applying a known corrective electrostatic force to return the dumbbell to its original position, the current required can be directly related to the partial pressure.

Thermomagnetic Analyzers. Thermomagnetic analysers (26) are based on the observation that changes in temperature affect magnetic susceptibility (χ). Generally this is described by the Curie law,

$$\chi T = \text{constant} \tag{11}$$

whereby χ is inversely proportional to the absolute temperature. The instrument consists of a flow-through ring element divided up into two measuring chambers. The detection site is composed of a thin-walled glass tube containing resistors that heat the incoming gases. These resistors form part of a Wheatstone bridge circuit and are used to detect changes in resistance resulting from flowrate variations. When the gas is heated, oxygen loses a high proportion of its magnetic susceptibility. This gas, in turn, is displaced by incoming cool oxygen, generating a convection current or "magnetic wind." The flow of current (or heat transport) causes a change in resistance, imbalancing the Wheatstone bridge. In effect the convection current can be related to the oxygen concentration.

Carbon Dioxide

For many bioprocesses, the measurement of CO_2 is an important feature. Increased levels of dissolved carbon dioxide can inhibit growth and reduce the production of secondary metabolites.

Electrochemical Probe. Steam-sterilizable electrochemical probes for determining dissolved CO_2 partial pressure are commercially available. These function in a similar way to oxygen probes and are based on a pH sensor immersed in a saturated bicarbonate buffer, separated from the bioreactor fluid by a hydrophobic membrane. Dissolved CO_2 gas molecules diffuse from the bioreactor media, through the hydrophobic membrane, into the carbonate buffer. The bicarbonate buffer equilibrates with the dissolved CO_2 gas molecules:

$$CO_2 + H_2O \rightarrow HCO_3^- + H^+$$
$$a\text{H}^+ = K(a\text{CO}_2)/(a\text{HCO}_3) \tag{12}$$

Here, a is the activity. Hence, increases or decreases in pH can be directly related to the concentration of dissolved CO_2. This logarithmic relationship requires a two-point calibration, which is usually carried out by measuring the pH response, following substitution of the carbonate buffer with a reference solution.

CO_2 Optrodes. Optical sensors for the detection of CO_2 have also been described. Generally, these sensors are analogous to CO_2 electrochemical systems, whereby changes in pH resulting from the formation of bicarbonate ions are detected. One example (27) described the use of a fluorescein dye as a pH indicator. All of the advantages and disadvantages already outlined for the O_2 sensor can be equally applied to these devices. Future developments in this field should lead to a more widespread use of these sensors.

Exit Gas Analysis. With a well-mixed aerobic fermentation, the concentration of dissolved CO_2 can be estimated from analysis of the exit gas. One particularly important approach used for determining the exit gas CO_2 concentration is infrared (IR) analysis. Particular molecules absorb IR radiation at characteristic wavelengths. This absorption is related to the chemical structure of the molecule, that is, the type and number of function groups present. When these molecules absorb IR radiation, the vibrational and rotational states are altered. Observing this absorbance (in either the gas, solid, or liquid phase, or combinations of these), the concentration of the target analyte can be determined (28). In the gas phase, CO_2 absorbs strongly at wavelengths $<15\ \mu$m. The main components of an IR analyzer (29) consist of a suitable radiation source, a method of selecting the range of wavelengths required, a sample and reference cell, a detector, and a means of displaying the information in terms of CO_2 concentration. Generally, CO_2 is monitored using a positive filtering method. With this method, the optical system detector is filled with CO_2. The system detects the reduced radiation energy of the measuring beam reaching the detector. Comparing the reference and sample responses provides a value for the CO_2 concentration present. One of the main problems encountered with this approach is conflicting absorbance patterns from "interfering" molecules (i.e., those absorbing over the same wavelength range).

Redox Potential

Monitoring the redox potential of a bioprocess medium can provide information about the equilibrium between oxidizing and reducing species (electron acceptors and donors, respectively) present. Measurement of redox potential is achieved using a combined metal-reference electrode system. Typically the metal electrode can be gold, iridium, or platinum; with platinum being the usual choice. The reference electrode is either Ag/AgCl or calomel. The redox sensor is linked to a pH meter that is fitted to provide a readout in millivolts. During operation, the redox potential (measured between the metal and reference electrodes) varies as the logarithm of the ratio of oxidizable and reducible components in the media. The redox value varies linearly with the pH of the media and with the logarithm of dissolved oxygen tension. This can lead to complications when trying to interpret the readout from the sensor. A full description of the use and problems associated with the redox sensor was outlined by Kjaergard (28).

Chromatography

Chromatography has proved to be a valuable tool for numerous applications in diverse areas of analysis. The principle is based the separation of a mixture of compounds by means of a suitable matrix. Target analytes are separated by their characteristic interaction with the matrix. Following separation, the compounds can be detected using a range of detectors (7) including flame ionization (used for gas chromatography applications), UV detectors, electrochemical detectors, infrared and refractive index detectors, conductivity detectors, and mass spectrometers. A number of chromatographic methods are available, several of which can be used for bioprocess monitoring. Overall chromatographic analysis is carried out off-line. However, recent developments have shown that on-line methods of analysis can be developed.

Liquid Chromatography. For the bioprocess industry, liquid chromatography is used in sample analysis as a preparative method for the large-scale purification of biomolecules and as recovery technique for high-value products such as therapeutics and pharmaceuticals. One important development in this field has been the introduction of high-pressure liquid chromatography (HPLC). This approach to liquid chromatography involves the use of well-characterized homogeneous columns that contain small (3–5 μm-diameter) particles. In operation, these columns can withstand high pressures (up to 6,000 psi). The overall effect is to achieve a fast, accurate separation of the components in a sample. Liquid chromatography permits the separation of a sample solution into consistent components through the interaction of two phases. One type incorporates a solid stationary phase in conjunction with a fluid mobile phase containing the sample of interest. During their passage over the stationary phase, sample components are separated by their degree of solubility or, alternatively, their affinity for the solid matrix. The result is a separation profile indicative of the component mixture of the sample. There are a numerous examples of HPLC systems being used on-line to monitor bioprocess production (e.g., 30,31,32). Typically, on-line monitoring is carried out using a sampling system based on membrane filtration (33,34).

Liquid–Liquid Chromatography. With the liquid–liquid chromatography method, the stationary phase is coated with a suitable fluid. The sample is carried in the liquid phase under pressure. Partitioning between the two liquids leads to a separation of the sample components. Detection of the separated components can be carried out using detectors such as UV absorption, conductivity, or refractive index. Liquid–liquid chromatography can be further divided into normal-phase type, in which the carrier liquid is less polar than the stationary fluid phase, and reverse-phase type, in which the carrier liquid is more polar. The reverse-phase method is particularly useful for separating organic compounds in an aqueous carrier solvent.

Ion Exchange Chromatography. Ion exchange chromatography is used to separate ionic compounds; typically from an aqueous carrier. This approach is particularly suited to measuring low concentrations of ions. The method is based on the ionic interaction between cations or anions in the sample solution, with the corresponding (oppositely charged) functional group on the column matrix.

Gas Chromatography. Gas chromatography (GC) employs gas as the mobile phase and is principally used for separating and measuring the composition of volatile compounds such as aldehydes and alcohols, for example, in a bioreactor headspace. Again, this is a chromatographic approach that has been widely used for bioprocess monitoring. Connected to a suitable sampling system, GC can be used in an on-line format (35,36). An important development in the field of analysis has been the introduction of Fourier transform infrared spectroscopy (FT-IR) instrumentation (37). Combined with chromatography (e.g., GC), this approach can provide a powerful tool for the accurate, sensitive analysis of gas mixtures. This technique will undoubtedly become more important over time.

Gel Filtration. Gel filtration is a method of separation based on molecular size and is carried out using an appropriate polymeric matrix. By selecting the correct pore-size exclusion limit, molecules in a sample solution can be separated. Small molecules, relative to the gel "sieve," pass rapidly through the chromatography column. However, larger molecules diffuse into the polymer matrix, where their passage is impeded. Hence, elution time from the column is determined by molecular size.

Mass Spectrometry

The use of on-line mass spectroscopy (MS) has developed since the first report of Reus et al. (38). A wide range of gasses, both free and dissolved (e.g., CO_2, O_2, CH_4), can be measured. In addition, volatile organic compounds such as methanol, ethanol, acetone, and simple organic acids can be monitored. The technique is based on the rupture of molecules by a high-energy source into corresponding ions. Magnetic separation of the charged ions may then be achieved, based on their mass: charge ratio. Each compound produces a characteristic display pattern. Typically an instrument consists of an electron bombardment source, the magnetic ion separation system, and the detector.

There are several types of mass spectrometer available. The quadrapole MS is based on the use of a radio frequency system to separate the ions. A combination of applied voltage and frequency range produces a change in the trajectory of the ions. The term *quadrapole* comes from the four rods used to produce the radio frequencies. There is an alternative MS instrument that is potentially more precise and stable. This instrument is based on the use of a magnetic field to separate ions. Two types of magnetic instruments are available. One system uses a fixed magnetic field and several detectors. The other uses a variable controlled magnetic field with one detector system.

Mass spectrometers offer a number of advantages for process monitoring including speed, specificity, precision, and stability. The major drawback to the use of mass spectrometers is cost; in addition some peaks can be difficult to resolve using the less elaborate instruments. Despite these drawbacks, MS monitoring of volatile compounds has been widely reported (39,40,41).

Weight

Monitoring fluctuations in the weight of a vessel used for bioprocesses is an effective way of measuring the contents and flow rate of additions. There are a number of load cells available for this task. The principle of a load cell (8) is based on measuring the compressive strain that is placed on the device (e.g., a solid or tubular steel cylinder) when under an axial load. Electrical resistance strain gauges can be included in the device structure. A proportional electrical resistance to the applied load can then be measured; this varies in relation to changes in load. Obviously, a temperature compensator must be included in the instrument to account for heat effects on the resistors. These devices need to be rugged and have long-term stability. For reasons of safety, it is also desirable for the device to be accurate. A load cell designed to detect tensile forces can be used for weight measurements of suspended vessels.

Liquid Level Measurement

Liquid levels in a vessel can be determined using a number of techniques (42). Determining the level can assist in formulating mass balances and measuring nutrient additions. Hydrostatic pressure can be measured using a single sensor positioned at the bottom of the vessel. Alternatively, two sensors spaced at the top and bottom of the vessel (differential design) can be employed. These sensors are electromechanical devices based on the deformation of a spring; the resulting signal is displayed as an electrical signal (capacitance or reactance).

Conductivity Sensors. Conductivity sensors are particularly suited to monitoring levels of foam (in a number of fermentations), where in excess it can cause problems. The sensor consists of a stainless steel probe, insulated except for the tip. The device can be used with noncorrosive conducting fluids. If the liquid or foam level rises to contact with the tip, an electric current is passed through the sensor; the foam acts as an electrolyte and the vessel as a ground. The sensor can be coupled to a control device that dispenses antifoam to counter the increased level.

Capacitance Sensors. Capacitance sensors operate by detecting changes in the relative dielectric constant of the media, compared with air. Measurements are made by detecting variations in electrical capacity brought about by changes in the liquid level.

Acoustic Sensors. Acoustic sensors operate via transducers that generate and detect ultrasonic waves. Two formats are available: a single device that both transmits and receives the signal, and two separate transducers with one function. By monitoring the time it takes for the sound wave to travel, the liquid level can be determined. With the single device the sound wave is directed onto the liquid surface. When the level rises, the time delay is shortened.

Temperature Probes. Liquid levels can be measured using a series of thermistors sited vertically through the vessel. During operation, an electrical current is applied to the sensors, raising their temperature above the ambient. When the liquid (or foam) level reaches a sensor positioned lower down the vessel, the sensor cools. This causes a temperature difference and, hence, a change in electrical resistance. The resistance can be displayed as an output signal indicating the liquid level.

Flow Measurement and Control

Flow measurement is an important feature for many bioprocesses (7), for both gases and liquids. These measurements are carried out on both the influent and effluent streams.

Gas Flow Measurement. The most common form of gas measurement is carried out using a variable area flow meter, or rotameter (7). This device consists of vertically mounted (usually conical) tube enclosing a free-floating body that is able to move up (floating in the gas flow of interest) and down a tapered bore running through the tube body. Typically, the tube is constructed of glass or metal and the float is a ball or hollow thimble shape. Because the position adopted by the flow rate is dependant on both the gas flow and the viscosity of the medium, the instrument requires careful calibration.

Another important instrument is the thermal mass flow meter. The measuring principle is based on the use of a heating device situated in the gas flow, which is used to heat the gas. The mass flow rate can derived from the heat balance,

$$Q = H/C_p(T_2 - T_1) \tag{13}$$

where Q is the mass flow of the gas, H is the enthalpy, C_p is the specific heat of the gas, T_1 is the temperature of the gas before heat is transferred to it, and T_2 is the gas temperature after the transfer of heat. The mass flow detected using this approach is independent of pressure and temperature.

Liquids. A number of techniques are available to measure liquid flow rates (Pons 1991). The choice of method will be determined by the nature of the liquid (e.g., viscosity, conductivity, etc.). For nonsterile liquids, instruments such as the turbine flow meter can be used. Fluid flow over a turbine blade array is proportional to the rotation speed of the turbine, hence a direct correlation can be made.

A common approach to measuring fluid flow in ancillary pipework is to use a differential pressure device, which utilizes the pressure drop caused by a constriction in the pipeline. This approach is based on the general Bernoulli equation for fluid flow. The constriction causes an increase in the velocity of the fluid, which in turn results in a corresponding pressure drop across the constriction. The differential pressure is a function of the flow velocity and density of the fluid. Measurement of the pressure difference can be used to determine fluid velocity.

Measurement of fluid flow under sterile conditions can be carried out using a magnetic-inductive flow meter. This device can operate with a fluid that includes particulate matter in suspension. The outside of the device comprises two windings that, in operation, are supplied with an alternating current. This induces a magnetic field. A pair of electrodes is placed in conjunction with the device. The average flow velocity is proportional to the electrical potential induced in the field. Potential differences in the field are detected by the electrode arrangement. Hence, accurate flow measurements can be made.

Biomass

The monitoring of biomass concentration can be carried out using a number of techniques (43,44). Conventionally, biomass concentration is measured off-line using labor-intensive, time-consuming methods such as dry weight cell, plate or microscopic cell count, and measuring the optical density of diluted samples (45). However, a number of more rapid methods have been developed. Of particular interest is the development of real-time on-line methods. Generally, methods for determining biomass concentration can be divided into two classifications: direct and indirect. The former is based on determining the physical properties of the cell and its components. In contrast, indirect methods measure factors related to the cell and its activity (e.g., respiration, electrochemical behavior, and nutrient fluctuation).

Bioluminescence and Chemiluminescence. The use of bioluminescent techniques is based on determining levels of adenosine triphosphate (ATP) concentration. Generally, levels of ATP remain constant for living cells, decreasing when the cells die. ATP concentration can, therefore, be related to biomass. A particular enzyme that has proved useful for this method is luciferase, which catalyzes the following reaction:

$$\text{ATP} + \text{luciferin} + O_2 \xrightarrow{\text{luciferase}} \text{oxyluciferin} + \text{AMP} + \text{light} \quad (14)$$

By detecting the light produced by this reaction, ATP concentrations, and hence, biomass, can be determined. This method can detect cell numbers a low as 10^5 cells ml^{-1} (46); it can be automated, and the assay time is fairly rapid (45). However, the method does suffer from several drawbacks, for example, the extraction of ATP may be incomplete, there may be free ATP present from other sources, and there maybe degradation of ATP by the extraction reagents.

Chemiluminescence is based on the detection of light produced by the protein-catalyzed oxidation of luminol in the presence of hydrogen peroxide.

Acoustic Resonance Densitometry. Acoustic resonance density is based on determining the change in a resonant frequency that results from changes in cell density (47). This is a noninvasive method that does not require contact between the instrument and the culture medium. The method incorporates an oscillatory circuit, amplifier, and test cell. Theoretically, the fluid density of the sample can be calculated from the square of its oscillation. Recent reports have described using this approach to monitor cultures of hybridomas and human lymphoma cells (48). Further advantages for this method include independence from flow rate and viscosity. However, there are disadvantages including the need for a filtration system (to enable the resonance density of the medium to be measured in the absence of cells), with the inherent problems associated with such systems, including poor sensitivity and problems caused by bubbles and particulate matter.

Capacitance, Conductivity, and Electrochemical Methods. The electrical properties of cells have been exploited in the development of a number of techniques. Changes in media capacitance have been related to biomass concentration (e.g., a decreasing capacitance coupled with an increasing biomass concentration) (49). In addition, it has been suggested that cell viability is linked to capacitance measurements (50).

Several in situ commercial instruments are currently available for determining biomass concentration based on capacitance measurements. The Biomass Monitor (Aber Instruments, Aberystwyth, U.K.) measures the radio frequency conductance and capacitance of a cell suspension using a constant voltage, four-terminal, phase-sensitive detector system (Figure 6). Furthermore, the probes can be inserted directly into bioreactors using a standard 25-mm-diameter (Ingold-type) port. The probes are fully sterilizable and can be cleaned in situ during operation.

Fluorescence. Fluorescence is a characteristic possessed by a number of important biological compounds including proteins, enzymes, and coenzymes. Simply defined, it is the absorbance of light energy at a particular wavelength, followed by reemission at a longer wavelength. After the passage of light energy, the compound returns to its ground-state level. In the field of bioprocess monitoring, fluorescence has been used, in particular, to monitor reduced nicotinamide adenine dinucleotide (NADH) and reduced nicotinamide adenine dinucleotide phosphate (NADPH) concentrations. These compounds are irradiated at 340 nm and emit at 450 nm. Both compounds are present in living organisms and are vital components of metabolic processes. By monitoring the intensity of fluo-

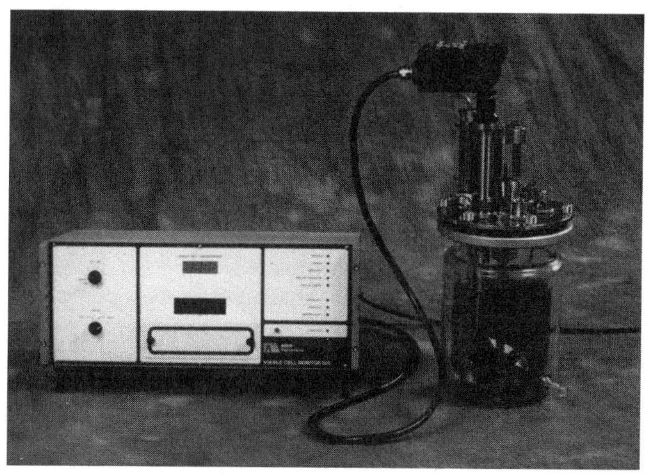

Figure 6. The Aber Instrument Viable Cell Monitor S20, connected to an Applikon fermenter.

rescence at the characteristic emission wavelength, it should be possible to calculate the total biomass concentration.

Unfortunately, the use of fluorescence as a marker for cell concentration does suffer from a number of drawbacks including that the fluorescence signal can originate from changes in metabolic state and not from cell concentration levels, sensitivity may be affected by the presence of compounds that can quench both the excitation and emission wavelengths, and other compounds may fluoresce at the same wavelength. Nevertheless, the use of this approach has been reported in a number of papers (51,52,53).

Flow Cytometry. Flow cytometry is a measuring technique based on the irradiation of a sample solution (containing a cell population) with a suitable light source, followed by monitoring of the scattered or absorbed light. In addition, fluorescence can be used as the measuring parameter. This technique can be used to ascertain a number of cellular features, such as the accumulation of cellular components (e.g., DNA, RNA, and proteins), and cell dynamics (e.g., cell size distribution). Furthermore, flow cytometry can be used to differentiate and quantify a range of species populations present in a mixed medium.

Light Scattering and Turbidity. Optical techniques based on either the scattering of light (nephelometry) or the degree of transmitted light or optical density (turbidity) have been developed for determining biomass (44). By monitoring the degree of light scattering using nephelometry, both cell numbers and mass can be determined. This approach is particularly suited for bioprocesses that involve low cell concentrations, where the background (compared) level is near zero. Turbidity measurement can be used for both on-line and off-line determination of biomass (54). Light scattering by a turbid medium results from a number of factors including particle size, shape, and number. Quantification is based on the Lambert-Beer law (55).

Biosensors

One possible answer to the problem of monitoring metabolites, both in situ and on-line, may be solved in the near future by the use of biosensors. Since their conception in the 1960s, these devices have generated considerable interest. This has spread to a diverse range of fields including clinical diagnostics, environmental protection, bioprocess monitoring, and defence applications (56). In general the operation of a biosensor is characterized by three functional steps: recognition, physiochemical signal generation, and signal processing. The biological component (e.g., enzyme, whole cell, antibody, and cell receptor) imparts a high degree of selectivity on the biosensor. Coupled to the biological component; the transducer is designed to respond to the changing physicochemical parameters caused by the specific interaction of the biological component with the substrate. A range of transducers have been used in the development of biosensors and include the electrochemical (potentiometric and amperometric), optical, calorimetric, piezoelectric, and thermometric (57). Most of these types have been adopted in the development of biosensors intended for bioprocess monitoring. For a more detailed insight into both the fundamentals and potential for biosensors, a number of specialized publications are available (58). Enzyme-modified field effect transistors (FETs), whereby the biologically active layer is positioned on top of the ion electrode membrane, have also been used for bioprocess monitoring (59).

An off-line biosensor system (Yellow Spring, Inc.) has been available for a number of years (Fig. 7). The instrument can be used to determine a range of metabolites (e.g., glucose, lactate, and ethanol). Recently the company has introduced an on-line instrument. Indeed, on-line biosensor systems have been demonstrated using a variety of biological-transducer systems (60,61). A wide range of metabolites have been monitored using various bioprocess regimes (62). Commercially, a significant number (63) of biosensor devices are available for measuring a range of analytes. However, at present, the majority of these devices are aimed at the medical diagnostic market. The goal is to develop cheap, reliable sensors that can operate under a range of bioprocess conditions for extended periods with a minimum of maintenance.

Nuclear Magnetic Resonance

Nuclear magnetic resonance is based on the detection of the response from particular nuclei when exposed to a

Figure 7. The YSl 2700 biochemistry analyzer.

magnetic field and electromagnetic radiation. Following absorption, the resonance frequency is shifted in a characteristic response pattern, according to the environment of the sample under detection. This approach can be used to determine a range of intracellular factors such as ATP, ADP, sugar phosphate, and polyphosphate, as well as pH. In these examples, concentrations of ^{31}P are used to characterize the compounds. This is an off-line monitoring system that is currently expensive; hence, its use is primarily in the research field, not in routine production environments.

Artificial Intelligence

The data obtained using the various measuring instruments already discussed are invariably used to control and optimize the bioprocess being carried out. Recent developments in the field of artificial intelligence have led to investigations into the use of such systems for improving bioprocess control, based on the received measurement output signals. This has included the use of both knowledge-based expert systems (64,65) and neural networks (e.g., 66,67) during bioprocess operation. A recent report (68) described the successful use of a neural network as a tool for evaluating the received measurement from an enzyme (penicillin-G amidase) pH-FET sensor linked to a flow injection system.

Undoubtedly, the adaptation of such "intelligent" systems will develop over the coming years and will play an important role in the precise control of bioprocess applications.

CONCLUSION

Although a diverse range of monitoring equipment is available, only a relatively narrow range of the more reliable instruments is used routinely in practice. As culture techniques become more elaborate and high-value-added products are produced, conventional methods will either prove insufficient or require supplementation with a range of sensors able to directly monitor key process parameters. Despite a clear demand for new sensors (e.g., lactate, glutamine, and glutamate for animal cell culture), cell cultivation represents only a modest market for analytical instrument manufacturers. Hence, while progress is to be expected, it may be slower than might be wished.

BIBLIOGRAPHY

1. B. Mattiasson, H. Hakanson, *TIBTECH* 11, 136–142 (1993).
2. S.L. Brooks, I.J. Higgins, J.D. Newman, and A.P.F. Turner, *Enzyme Microb. Tech.* **13**, 946–955 (1991).
3. J. Ruzicka and E.H. Hansen, *Flow Injection Analysis*, 2nd ed., Wiley & Sons, New York, 1988.
4. C.C. Westcott, *pH Measurements*, Academic Press, New York, 1978.
5. P. Bergveld, *IEEE Trans. Biomed. Eng.* BME **19**(5), 342–353 (1972).
6. J.M. Thompson, *Med. Biol. Eng. Comput.* **28**, B29–B33 (1990).
7. M.N. Pons, in M.N. Pons ed., *Bioprocess Monitoring and Control*, Carl Hanser Verlag, Munich, 1991, pp. 4–28.
8. P.F. Stanbury and A. Whitaker, *Principles of Fermentation Technology*, Pergamon Press, Oxford, (1984).
9. B.G. Liptak, in B.G. Liptak ed., *Instrumentation in the Process Industries*, Chilton, Philadelphia, 1969, pp. 38–94.
10. H. Schafer, V. Graeger, and R. Kobs, *Sensors Actuators* **17**, 521–527 (1989).
11. A.H. Scragg, *Bioreactors in Biotechnology: A Practical Approach*, Ellis Horwood, Chichester U.K., 1991.
12. P.M. Doran, *Bioprocess Engineering Principles*, Academic Press, London, 1995.
13. Y.H. Lee and G.T. Tsao, *Adv. Biochem. Eng.* **13**, 35–86 (1979).
14. B. Atkinson and F. Mavituna, *Biochemical Engineering and Biotechnology Handbook*, 2nd ed., Macmillan, Basingstoke, U.K., 1991, Chapter 18.
15. K.G. Carr-Brion, in K.C. Carr-Brion ed., *Measurement and Control in Bioprocessing*, Elsevier Science, New York, 1991, pp. 37–66.
16. L.C. Clark and C. Lyons, *Ann. N.Y. Acad. Sci.* **102**, 29–45 (1962).
17. P. Van Hemert, D.G. Kilbrun, R.C. Righelato, and A.L. Van Wezei, *Biotechnol. Bioeng.* **11**, 549–560 (1969).
18. G. Quicker, A. Schumpe, B. Konig, and W.-D. Deckwer, *Biotechnol. Bioeng.* **23**, 635–650 (1981).
19. D.W. Lubbers and N. Opitz, in J. Grote, D. Reneau, and G. Thews eds., *Oxygen Transport to Tissue, II*, Plenum, New York, 1976, pp. 65–68.
20. M.A. Arnold, *Anal. Chem.* **64**, 1015A–1025A (1992).
21. J. Janata, in J. Janata ed., *Principles of Chemical Sensors*, Plenum Press, New York, 1989, pp. 241–283.
22. M.R. Surgi, in D.L. Wise ed., *Applied Biosensors*, Butterworths, Boston, 1989, pp. 249–290.
23. R.S.D. Wolthuis, J.C. McRae, E. Hartl, G.L. Saaski, K. Mitchell, and R. Willard, *IEEE Trans. Biomed. Eng.* **39**, 185–193 (1992).
24. K.W. Berndt and J.R. Lakowicz, *Anal. Biochem.* **201**, 319–325 (1992).
25. D.H. Philips and M.J. Johnson, *J. Biochem. Microbiol. Technol. Eng.* **3**, 261–275 (1961).
26. J.E. Brown, R.K. Kaminski, A.C. Blake, and A. Brodgesell, in B.G. Liptak ed., *Instrument Engineer's Handbook*, vol. 1, Chilton, Philadelphia, 1969, pp. 713–918.
27. Y. Kawabata, T. Kamichika, T. Imasaka, and N. Ishibashi. *Anal. Chim. Acta* **219**, 223–229 (1989).
28. L. Kjaergard, in T.K. Ghose ed., *Advanced Biochemical Engineering*, Springer-Verlag, New York, 1977, pp. 131–145.
29. M.H. Gordon and R. Macrae, in *Instrumental Analysis in the Biological Sciences*, Blackie, Glasgow, 1987, pp. 133–145.
30. K. Holzhauer-Rieger, W. Zhou, and K. Schugerl, *J. Chromatogr.* **499**, 609–615 (1990).
31. W. Smolenski and J.M. Suflita, *J. Microbiol. Methods* **6**, 71–79 (1987).
32. H. Kurokawa, Y.S. Park, S. Iijima, and T. Kobyashi. *Biotechnol. Bioeng.* **44**, 95–103 (1994).
33. K. Schugerl, in C. Christiansen, L. Munck, and J. Villadsen eds., *Proceedings of the 5th European Congress on Biotechnology*, vol. II, Munksgaad International, Copenhagen, 1990, pp. 1188–1193.
34. K. Schugerl, in K. Schurgerl ed., *Biotechnology*, vol. 4, VCH, New York, 1991, pp. 149–180.

35. D.M. Comberbach, J.M. Scharer, and M. Moo-Young, *Biotechnol. Lett.* **6**, 91–96 (1984).
36. J.K. McLaughlin, C.L. Mayer, and E.T. Papoutsakis, *Biotechnol. Bioeng.* **27**, 1246–1257 (1985).
37. C. Fujimoto and K. Jinno, *Anal. Chem.* **64**, 476A–481A (1992).
38. M. Reuss, H. Piehl, and F. Wagner, *Eur. J. Appl. Microbiol.* 1, 323–325 (1975).
39. S.J. Coppella and P. Dhurjati, *Biotechnol. Bioeng.* **29**, 679–689 (1987).
40. E. Heinzle, *Adv. Biochem. Eng. Biotechnol.* **35**, 1–45 (1987).
41. S. Chauvatcharin, K.B. Konstantinov, K. Fujiyama, T. Seki, and T. Yoshida, *J. Fermen. Bioeng.* **79**, 465–472 (1995).
42. M. Reuss, in R.K. Finn and P. Prave eds., *Biotechnology Focus I: Fundamentals.Applications.Information*, Oxford University Press, New York, 1988, pp. 153–190.
43. B. Sonnleitner, G. Locher, and A. Fiechter, *J. Biotechnol.* **25**, 5–22 (1992).
44. K. Konstantinov, S. Chuppa, E. Sajan, S. Yoon, and F. Golini, *TIBTECH* **12**, 324–333 (1994).
45. C.M. Harris and D.B. Kell, *Biosensors* **1**, 17–84 (1985).
46. E.W. Chappelle, G.L. Piccioli, and J.W. Deming, *Methods Enzymol.* **57**, 656–672 (1978).
47. B.C. Blake-Coleman, D.J. Clarke, M.R. Calder, and S.C. Moody, *Biotechnol. Bioeng.* **28**, 229–235 (1986).
48. D.G. Kilburn, P. Fitzpatrick, B.C. Blake-Coleman, D.J. Clarke, and J.B. Griffiths, *Biotechnol. Bioeng.* **33**, 1379–1484 (1989).
49. K. Mishima, A. Mimura, Y. Takahara, K. Asami, and T. Hanai, *J. Ferment. Bioeng.* **72**, 291–295 (1991).
50. R.M. Matanguihan, K.B. Konstantinov, and T. Yoshida, *Bioprocess Eng.* **11**, pp. 213–222 (1994).
51. G. McMichael, W.B. Armiger, J.F. Lee, and R. Matharasan, *Biotechnol. Techn.* **1**, 213–218 (1987).
52. J.K. Li and A. Humphrey, *J. Ferment. Bioeng.* **74**, 104–111 (1992).
53. K.B. Konstantinov, P. Dhurjati, T. Van Dyke, W. Majarian, and R. LaRossa, *Biotechnol. Bioeng.* **42**, 1190–1198 (1993).
54. B. Sonnleitner, G. Locher, and A. Fiechter, *J. Biotechnol.* **25**, 5–22 (1992).
55. K.F. Reardon and T. Scheper, in K. Schugerl ed., *Biotechnology*, vol. 4, VCH, New York, 1991, pp. 179–223.
56. E. Kress-Rogers, *Handbook of Biosensors and Electronic Noses*, CRC Press, Boca Raton, Fla., 1997.
57. R.F. Taylor and J.S. Schultz, *Chemical and Biological Sensors*, IOP Publishing, Bristol, U.K., 1996.
58. A.P.F. Turner, *Advances in Biosensors*, suppl. 1, 2, 3, JAI Press, London, 1991, 1992, 1993, 1994.
59. T. Kullick and R. Ulber, in R. Freitag ed., *Biosensors in Analytical Biotechnology*, Academic Press, San Diego, 1996, pp. 23–33.
60. B. Mattiasson and B. Danielsson, in R.F. Taylor and J.S. Schultz eds., *Chemical and Biological Sensors*, IOP Publishing, Bristol, 1996, pp. 533–549.
61. U. Bilitewski and I. Rohm, in E. Kress-Rogers ed., *Handbook of Biosensors and Electronic Noses*, CRC Press, Boca Raton, Fla., 19, 435–468 (1997).
62. U. Bilitewski and I. Rohm, in R. Freitag ed., *Biosensors in Analytical Biotechnology*, Academic Press, San Diego, 1996, pp. 163–191.
63. J.D. Newman and A.P.F. Turner, *Chemistry & Industry*, May 1994, pp. 374–378.
64. M.N. Karim and A. Halme, in N.M. Fish, R.I. Fox, and N.F. Thornhill, eds. *Computer Applications in Fermentation Technology: Modelling and Control of Biotechnological Process*, Elsevier Applied Science, Barking, U.K., 1989, pp. 37–46.
65. K.B. Konstantinov and T. Yoshida, *Biotechnol. Bioeng.* **39**, 479–486 (1992).
66. J. Thibault, V. van Breusegem, and A. Cheruy, *Biotechnol. Bioeng.* **36**, pp. 1041–1048 (1990).
67. Q. Zhang, J.F. Reid, J.B. Litchfield, J. Ren, and S.-W. Chang, *Biotechnol. Bioeng.* **43**, 483–489 (1994).
68. B. Hitzman, T. Kullick, R. Quack, and M. Venschott, in R. Freitag ed., *Biosensors in Analytical Biotechnology*, Academic Press, San Diego, 1996, pp. 129–144.

See also CONDUCTIVITY; PILOT PLANTS, DESIGN AND OPERATION; PROCESS CONTROL, STRATEGY AND OPTIMIZATION; SECONDARY METABOLITE PRODUCTION, ACTINOMYCETES, OTHER THAN STREPTOMYCES; YEAST, BAKER'S.

PROCESS VALIDATION

DANE W. ZABRISKIE
Biogen, Inc.
Cambridge, Massachusetts
THOMAS M. SMITH
ALAN R. GARDNER
SmithKline Beecham
King of Prussia, Pennsylvania

KEY WORDS

Cell culture
Clearance of contaminants
Consistency
Fermentation
Good manufacturing practices (GMP)
Operating parameters
Operating ranges
Proven acceptable ranges
Purification
Validation
Viral clearance

OUTLINE

INTRODUCTION

Process validation for pharmaceuticals has been defined by the U.S. Food and Drug Administration (FDA) as "establishing documented evidence which provides a high degree of assurance that a specific process will consistently produce a product meeting its pre-determined specifications and quality attributes" (1). In spite of the apparent clarity of this definition, pharmaceutical companies vary widely in their approaches to process validation. Operational definitions of process validation are often developed by firms to suit their own validation philosophies. Subcategories of process validation such as formal process validation, process characterization, process development, process evaluation, and the like can result, and the meaning of these phrases is highly specific to a particular company. For the purposes of this article, the term *process validation* refers to any activity consistent with the FDA definition. The discussion that follows is necessarily subjective, and it does not reflect the view of any particular company or the industry at large. Rather, it describes concepts used by those who practice this craft in industry.

This article is concerned with processes that produce the active ingredient in a biopharmaceutical product for use in humans as therapeutics or vaccines. Biopharmaceuticals, protein agents derived from recombinant gene expression or hybridoma production systems, include various biotechnology products such as monoclonal antibodies, enzymes, hormones, soluble receptors, growth factors, and immunogens. Pharmaceutical manufacturing processes are divided into two subprocesses. The primary process produces the bulk biological substance (BBS), also referred to as the active pharmaceutical ingredient (API). The secondary process transforms the BBS into the final product, which is what is actually sold. Secondary processing steps include formulation, sterile filtration, filling, lyophilization (in some cases), capping, and packaging. This article is limited to the validation of the primary processes which produce biopharmaceutical BBS.

Pharmaceutical manufacturers typically classify validation activities into the following broad categories: process validation, equipment and facility validation, cleaning validation, analytical methods validation, and computer validation. Equipment and facility validation is further subdivided into design qualification, installation qualification (IQ), operational qualification (OQ), and performance qualification (PQ). Process validation is viewed by some experts to be an element of PQ.

Process validation is a regulatory requirement to enable product licensure. Limited process validation is also critical for clinical trial materials to ensure product safety. Process validation regulations originated within the FDA, which was created in 1906, initially to prevent the adulteration of products. The sulfanilamide disaster led the Food, Drug, and Cosmetics (FD&C) Act in 1938 to ensure product safety. The thalidomide disaster in 1962 led to the Kefauver–Harris Amendment to the FD&C Act, which required demonstration of product safety and efficacy. It was in this revision that the concept of good manufacturing practices (GMP) originated. In 1978 the act was further revised to include the concepts underlying current GMPs (cGMPs) and good laboratory practices (GLPs). The origins for process validation derive from FDA draft guidelines and workshops that occurred in the late 1970s and early 1980s (2). *Guideline on General Principles of Process Validation* (1), published in 1987, represented the first (and still only) regulatory document focusing on this topic. The definition for *process validation* quoted earlier was taken from this FDA document and has gained wide acceptance by other regulatory agencies around the world. This guidance was specifically intended for the production of finished drug products. Today, the regulatory basis in the United States for process validation is contained within cGMPs for finished pharmaceuticals (drugs) (3). The *Code of Federal Regulations* (21 CFR 211.110 states that "control procedures shall be established to monitor the output and validate the performance of those manufacturing processes that may be responsible for causing variability in the characteristics of in-process material and the drug product." It (21 CFR 211.113) also states that "appropriate written procedures, designed to prevent microbiological contamination of drug products purporting to be sterile, shall be established and followed. Such procedures shall include validation of any sterilization process."

Until recently, these FDA validation requirements and guidelines only applied explicitly to drug products and did not apply to bulk biologics. Nevertheless, there was a trend for these regulatory principles to apply implicitly to areas more broadly than originally defined. There are numerous publications by industrial scientists, but currently no umbrella regulatory document exists on the validation of bulk production processes for biopharmaceuticals. Important aspects to this subject are captured within numerous other regulatory documents that focus on other aspects of manufacturing and control. Table 1 summarizes some of these references. Regulatory requirements that pertain to process validation are changing and are likely to continue to change for the foreseeable future. The most current information on process validation and many other regulatory subjects may be accessed on the Internet (Table 2). Proposals published recently by the FDA pertain to the cGMPs for finished pharmaceuticals and include content on process validation (18). How these proposals will be extended to bulk biological products has become clearer with the recent publication by the FDA of a draft guidance for the manufacturing of APIs (19) and the guidance for submission of the chemistry, manufacturing, and controls (CMC) section of a biologics license application (BLA) for most biopharmaceutical products (20).

In addition to being a regulatory requirement, process validation provides important benefits to the manufacturer. Biopharmaceuticals are among the most expensive products to manufacture, costing hundreds to thousands of dollars per gram to produce. The assurance of process consistency therefore enables a manufacturer to maximize plant productivity and obtain favorable cost efficiency by reducing the incidence of process failures, rejection of lots, and reworking of rejected but salvageable lots. In today's manufacturing environment where plant capacity is closely matched with product demand, unexpected process failures can lead to product shortages, with negative financial and company product image impacts, particularly

Table 1. Selected References on Process Validation

Validation topic	Comment	Reference
Viral and contaminant clearance, cell line stability	A good general discussion of these topics with examples from Genentech's Activase® experience	4
Purification and general topics	A special issue of BioPharm that compiles 12 papers published in this journal in 1994 and 1995	5
Viral clearance	ICH guideline common to the United States, Europe, and Japan	6
Cell line characterization	ICH guideline common to the United States, Europe, and Japan	7
General	The first and still only regulatory document in the world dedicated to this subject; some content is becoming obsolete	1
Prospective, concurrent, and retrospective studies	Focuses on sterile drug products but contains useful general guidance	8
Operating ranges, proven acceptable ranges	An early discussion of this subject	9
Cell age	Describes scientific approaches to demonstrating consistency of cell line characteristics used at Genentech for tPA	10
Chromatography	Industrial perspective on the validation of chromatography purification steps	11
General	Inclusive discussion for biotech products beginning with master cell bank and ending with the final product; includes example from the tPA process	12
Experimental design	Application of design of experiment (DOE) techniques	13
Tangential flow filtration	Industrial perspective on the validation of tangential flow filtration steps	14
Cell line characterization and viral clearance	FDA Points to Consider in the Characterization of Cell Lines Used to Produce Biologicals (1993)	15
Cell line characterization, viral clearance, and manufacturing changes	FDA Points to Consider in the Manufacture and Testing of Monoclonal Antibody Products for Human Use (1997)	16
Viral clearance	Note for guidance published by the CPMP for use in the European Union (1996)	17

Note: ICH, International Conference on Harmonization; CPMP, Committee for Proprietary Medicine.

Table 2. Internet Sites with Current Information on Process Validation and Other Regulatory Information

For Searchable GMPs for the FDA	http://www.access.gpo.gov/cgi-bin/cfrassemble.cgi
ICH Documents	http://www.ifpma.org/ich5.html
FDA Documents	http://www.fda.gov
European Union (European Union Drug Regulatory Authorities)	http://www.eudra.org/frame/frametest1.html

for life-saving products. Process validation is a key means to minimize the occurrence of unexpected process failures. Process validation can also provide a manufacturer with operational flexibility to respond to inevitable production problems (e.g., power failures, equipment breakdown, and scheduling conflicts). To the extent that they can be anticipated in the process validation program, ranges can be established for operational parameters that, if maintained, will ensure that the products meet specifications.

These important benefits only accrue after considerable investment because process validation is costly and time consuming. It occupies the full period from producing supplies for the first clinical trial through the start of production in the commercial plant and continues throughout the commercial lifetime of the product. During product development, for example, more resources from process development, operations, quality control, and quality assurance groups can be consumed in validating the process, equipment, and facility than were used to develop the process and to produce clinical supplies. Nevertheless, careful planning and sound scientific strategies make process validation cost effective and meet regulatory requirements.

There are several different classes of process validation. *Clearance validations* demonstrate the capability of a process to remove or inactivate contaminants to and beyond the degree necessary to ensure product safety. Of greatest concern are contaminants that have toxic, pharmacological, or infectious potential in humans. Typical contaminants in biopharmaceutical processes with these potential liabilities include DNA and proteins derived from the recombinant host cell line, certain components of culture media (e.g., methotrexate and other drugs used to maintain selection pressure, insulin, bovine serum albumin, growth factors, and animal-derived components of culture media), viruses derived from the host cell line or raw materials, additives for viral inactivation (e.g., guanidine), and affinity chromatography ligands (e.g., monoclonal antibodies, protein A). *Validation of operating ranges* involves the determination of a range for critical operating parameters through which excursions can occur without affecting the quality of the product. The usage of the term *operating parameters* here is taken very broadly to include not only compositions, temperature, pH, and related factors but also holding times of in-process intermediates, variations in raw materials, cycle limits on multiuse resins, age of raw materials, and age limit of cells used in the process. *Validation of process consistency* must be established through the characterization of multiple lots of product produced by separate production runs operated and controlled under the same conditions. Each of these classes of validation are described in greater detail in this article.

PROCESS VALIDATION STRATEGY AND DOCUMENTATION

Process validation generally starts after a process is defined in development and evolves with the progression of clinical trials. Process validation for phase I and II trials is limited to addressing potential product safety concerns. Validations for the clearance of model viruses and retroviruses and host cell DNA are sufficient. More extensive validation activities occur in parallel with production of phase II and III clinical supplies and the conducting of the phase II and III clinical studies. At this stage, the virus clearance validations are expanded to include a greater number of model viruses spanning a greater range of virus characteristics. The full range of clearance and operating range validations are undertaken in a pilot plant or development laboratories. Process consistency validation, however, is derived from the pilot and commercial plants. License applications can be filed with consistency data derived from a pilot plant. However, the awarding of a license is contingent on the filing of process and product consistency data derived from a pilot plant or commercial manufacturing facility. In the United States, three qualification lots are usually sufficient; the European Union usually requires five lots. Process validation continues after product launch for the length of time the product is manufactured. Validation maintenance programs often call for revalidation of specific aspects of the process at regular intervals. Trending or control charting of process variables can lead to refinements of operating range limits. Revalidation is also required when the process is modified because of improvement, equipment alterations, raw material changes, process scale-up, substitution of a processing step with an alternate, resequencing process step, and so forth.

Validation is subdivided into prospective, concurrent, and retrospective according to the timing relationship between the validation work and production (1). Prospective validation studies "are conducted prior to the [commercial] distribution of a new product, or product made under a revised manufacturing process, where the revisions may affect the product's characteristics" (1). Concurrent validation studies employ data derived from production runs for product in distribution. How prospective and concurrent validations are conducted is often defined in preapproved validation protocols. Retrospective validation studies demonstrate, "in some measure, the adequacy of the process by examination of accumulated test data on the product and records of the manufacturing procedures used. Retrospective validation can also be useful to augment initial premarket prospective validation for new products or changed processes" (1). Retrospective validation is generally not conducted according to a preapproved validation protocol, but this practice may change in the future.

In reality, most comprehensive process validation programs involve all three types of validation throughout the lifetime of a product. Consistency validations are most often run prospectively in full-scale production equipment for clinical supplies or commercial purposes. Concurrent and retrospective validations can be used to set new operating ranges not covered by previous validation work or to modify (usually narrow) operating ranges based on manufacturing experience using statistical process control, control charting, or related techniques. A final precaution should be taken regarding recent regulatory concerns on retrospective validation. Retrospective validations are acceptable as an adjunct to a well-defined and comprehensive prospective validation program, but they cannot substitute for it.

Another technical consideration to be defined is the scale at which process validation studies are to be conducted. One could assume that the most authoritative results would be obtained by conducting validation experiments at the scale intended for commercial manufacturing. This assumption normally leads manufacturers to validate process consistency using the equipment to be licensed. It is not unusual, however, to identify 50 operating parameters for which low and high operating range limits are sought. This could involve 100 experiments (if interactions among parameters were disregarded) and up to 1×10^{15} experiments (if interactions are included), which is clearly not feasible in full-scale equipment. The infectivity and limited quantity of viral reagents used in clearance validations prevent these studies from being executed at full scale or in a GMP facility. GMP prohibits the intentional introduction of infectious agents into processing areas (21). In other clearance validations, limitations on the quantity of reagents used for spiking (e.g., host cell proteins, host cell DNA, and affinity ligands) prevent these studies from being conducted in full-scale equipment.

One approach to these problems is to scale down processing steps. Scale-down factors from 10- to 10,000-fold have been employed in industry. In some cases, validation experiments can be run in parallel using multiple identical scaled-down units (e.g., fermentors), which enable statistical design of experimental techniques and a significant shortening in the overall time expended for validation experiments. In other cases a hybrid approach can be taken in which samples of in-process streams taken from the full-scale process are used as feed streams to scaled-down unit operations to assess performance. Virus clearance studies that employ live viruses and retroviruses are conducted in this manner. The use of scaled-down systems for virus clearance validations also facilitates adequate biological containment.

If scaled-down models are to be employed, then it is necessary to establish that the models are predictive of full-scale performance. A useful scaled-down model may be developed by attending to three considerations: performance, design, and quality (PDQ). The performance of the scaled-down model should be identical to or at least predictive of the full-scale process. Typical performance characteristics are titer, cell density, viability, processing rates, and step yields. In cases where there is a performance difference between the scales, it should be quantified and the basis for the difference should be understood from technical and scientific principles whenever possible. The degree of variation in the model should not be greater than variation at full scale to ensure that process changes are distinguishable from natural variation in the model. During the design of the scaled-down model, it is important to observe the same technical principles used to scale up a unit operation. Typical examples are maintaining constant culture split ratios in cell culture and to preserve chromatography bed heights and linear flow velocities in purification (22,23) during scale-up or scale-down. The quality of the intermediate produced by each scaled-down unit operation should be identical or similar to its full-scale counterpart in its analytical attributes. Finally, it is important to document the qualification of scaled-down models used for process validation.

As the definition of *process validation* states, documentation of validation work is essential. The validation master plan defines the overall validation philosophy, the systems to be validated, the general testing scheme to be employed, the responsibilities of various groups, and the approval mechanisms of the program. It may also define the evolution of validation throughout the life cycle of the process, from its origins in process development through validation maintenance during commercial manufacturing. Another class of document is the validation protocol, which prospectively defines the details for testing of each validation stage. The validation protocol is defined by the FDA as "a written plan stating how validation will be conducted, including test parameters, product characteristics, product equipment, and decision points on what constitutes acceptable test results" (1). The final document is the process validation report, which includes the validation protocol, deviations from the protocol, the results from the validation experiment, and the conclusion regarding the validation status for the processing stage. An important phrase in the definition of process validation is establishing documented evidence. Validation experiments must be documented in reports to be acceptable to regulatory agencies. Copies of the most critical validation reports and summaries of the others are included as part of the license application (20). All other process validation reports must be available to the regulators for inspections as part of the licensing process or during routine postlicense inspections. Failure to submit essential process validation reports can lead to a refusal to file determination by the FDA in the United States. A failure to produce appropriate validation reports during regulatory inspections can lead to a rejected license application or other regulatory actions such as the issuance of an FDA 483 form.

CLEARANCE VALIDATIONS

Clearance validations are generally concerned with the capacity of the recovery and purification process to remove process contaminants. Of particular concern are contaminants with toxic, pharmacological, or infectious potential in humans; studies of these risks are thus among the first to be undertaken in a validation program. The overall approach to this form of validation can be illustrated using the clearance of DNA.

DNA contaminants could contain genes that encode harmful proteins (e.g., toxins and oncogene products), viruses, or retroviruses endogenous to the host. If these gene contaminants could transfect patient cells (resulting in retention and duplication of the gene), then expression of the gene product is a theoretical possibility. The World Health Organization (WHO) has established a 100 pg/dose limit to DNA contaminants in biopharmaceutical products. Proposals are under consideration to increase this limit to 10 ng/dose. One approach to validating adequate clearance of DNA by the process is to measure the DNA level in the BBS over the full range of process operating conditions. If these measurements fall below the 100 pg/dose limit, then the process is validated for this clearance capability. This approach is taken for contaminants whose assays are sufficiently sensitive to quantify levels in the BBS.

Often the assays for DNA and many other process contaminants are not sufficiently sensitive (24) to quantify the contaminant in the bulk product, and an alternate validation approach must be used. Clearance factors are established for upstream unit operations for which DNA levels can be measured in the in-process streams. For downstream steps for which DNA levels are immeasurable, host cell DNA is added (spiked) to the feed stream for each unit operation at a level sufficient to detect it in the output stream. Clearance factors for the downstream steps can then be calculated. The clearance factors are then summed together to give a total clearance factor for the process. This total may be adjusted downward to eliminate double counting steps that have similar mechanisms of clearance. If, for example, a process includes two anion exchange steps, only one would be included in the analysis of clearance. This is based on the notion that virus particles capable of being removed by this mechanism (i.e., anion exchange) would only be removed by the first step. No further clearance could be expected in the second or additional anion exchange steps. The total clearance factor is then applied to the range of DNA levels present in unprocessed bulk harvests. If the calculated DNA level in the BBS is below the 100 pg/dose level, then the process is validated with respect to DNA clearance.

Spiking studies frequently use a hybrid of the full-scale process and scaled-down process models during validation. During full-scale operation, samples of in-process streams are collected and stored. Portions of these samples are spiked with the appropriate contaminant(s) and are then processed over a scaled-down model. In some cases, the spiked species may be labeled with a radioactive label to facilitate quantitative detection in in-process streams. This approach is often used for measuring DNA clearance.

When animal cells are employed during production, clearance of putative viruses constitutes the largest group of clearance validations. The general approach is similar to that described for DNA clearance (25). The product safety concern is to minimize the risk of viruses or retroviruses contaminating the product. The viruses can originate as active or latent viruses associated with the host cells (i.e., endogenous viruses) or can be introduced with raw materials (i.e., adventitious viruses), particularly those derived from animals. In most cases, if viruses exist at all, they are present at levels below detection in in-process streams, making spiking–clearance studies necessary for all steps in purification. For infectious contaminants such as viruses, there are two modes of clearance. The first involves removal by each purification step, similar to the manner in which DNA is removed. The second involves inactivation, where additives or conditions are used to disable the virus without removing it.

The determination of the viral load to the process depends on the specifics of the host cell system. For Chinese hamster ovary (CHO) cells, the load can be estimated by measuring the level of retrovirus-like particles (RVLPs) present in the unprocessed bulk harvest. CHO cells are known to produce RVLPs, which resemble intracytoplasmic type A retroviruses and cell membrane budding type C retroviruses (26). Even though these particles are defective in their ability to infect other cells, they nevertheless provide a basis for estimating a retroviral load in the process. Levels of 10^5 to 10^7 RVLPs per milliliter of unprocessed bulk are typical. The level of viral clearance to target depends on the perceived risk to the patient. One target for retroviral clearance in processes involving CHO cells is to allow no more than 1 RVLP per million doses of final product (4). The total process retroviral clearance and the level of RVLPs in the unprocessed bulk are used to calculate an estimate of the RVLPs in the final product. The process is considered to be validated for retroviral clearance when the estimated level of RVLPs in the final product is less than or equal to the target. Process clearance factors of 15–20 logs are typical for murine retroviruses. The FDA has suggested that a minimum of 3 logs of clearance is necessary based on the initial level of contamination entering the purification process (16).

For most viruses other than endogenous retroviruses, the validation exercise becomes more academic; there is no reasonable basis for establishing the virus load in the process because there should not be any present. In these cases, measured clearance factors are reported to provide some assurance that if these types of viruses were inadvertently introduced into the process, there would be a defined capacity to clear them.

The spiking studies use model viruses that are representative of the most likely contaminants and when grouped together cover a broad spectrum of virus characteristics. Clearance studies are often performed for a number of viruses to test the ability of the process to clear viruses in general. The choices of model viruses depend on the cell line employed and the viruses could contain. For CHO cells, typical model viruses include a small, nonenveloped virus (e.g., simian virus [SV40], human polio virus 1, animal parvovirus), a medium-to-large enveloped RNA virus (parainfluenza virus, influenza virus, Sinbis virus), a medium-to-large DNA virus (e.g., herpes simplex virus 1 and pseudorabies virus), and a murine retrovirus (e.g., murine leukemia virus) (6). The first three viruses are chosen because they cover a range of virus characteristics, and the murine retrovirus is chosen to model one that could infect the CHO host cell line.

OPERATING RANGES

The purpose of validating process operating ranges is to demonstrate the robustness of the process to excursions in operating parameters that are likely to be encountered in a practical manufacturing environment. The process can be subdivided into a series of validation stages or modules. In some cases, a stage corresponds to a specific processing step (e.g., production fermentation or a chromatography step). In other cases, a stage may be a series of steps (e.g., media preparation or seed expansion). The choice of defining validation stages depends on the specifics of the process and the manner in which the process is controlled. One validation staging strategy for a typical biopharmaceutical process is shown in Figure 1.

Once the stages are identified, inputs and outputs to each stage are defined. Typical inputs include the operating parameters, properties and storage duration of the feedstock derived from the previous upstream step, and quality attributes of critical raw materials used in the step. Typical outputs include the performance attributes for the stage (e.g., titer and yield) and the properties of the product stream (product homogeneity, purity, contaminant levels, and chromatogram shape). For chromatographic steps, the shape of the chromatogram has been used to indicate step performance. Suggested quantitative measures of chromatogram shape have been proposed, such as SPEW, for the retention times (or chromatogram distance) when

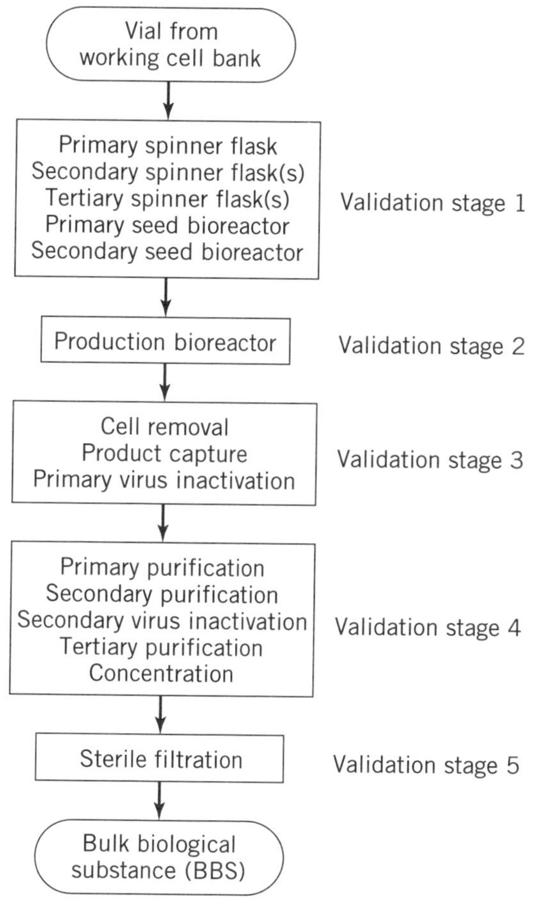

Figure 1. Validation of a process can be simplified by dividing the process into a sequence of validation stages, some of which include multiple processing steps.

the product collection starts, peaks, and ends and the width of the elution peak (27). In general, all in-process controls must be included within the list of process step outputs. Representative inputs and outputs for a fermentation stage and a purification stage are illustrated in Figure 2.

It is usual to prioritize the inputs to be validated based on the *(1)* sensitivity of process outputs to variations in each input, *(2)* proximity of the validation stage to the final product, and *(3)* technical difficulty in controlling an input to within a specified range. Assessment may be based on an analysis of all data that are available and documented, including process development, monitoring of clinical manufacturing, and any experiments specifically designed to gauge the sensitivity of the process to variations. The most important stage inputs must be included in the operating range validation program.

Having identified the parameters to be validated, the next step is to define the criteria on the outputs that will be used to determine whether a validation experiment is successful. These criteria are typically a mixture of process performance results (e.g., titer, yield, and concentration) and quality of the product intermediate (e.g., protein purity, product homogeneity, and contaminant levels). The need for both types of criteria is based on the assumption that no amount of testing of the end product can ensure consistency without also showing the process to be controlled. These criteria are based on a variety of considerations and experience with the process. In some cases, these criteria are based on the output range results from operational history of the process during development or clinical supply production. In other cases, they may be based on equipment limitations. For example, a minimum titer from the production fermentation tank may be established by the resin bed size and lower capacity limit for the first chromatography step in a process. In still other cases, output criteria may be derived from product specifications. If, for example, product purity is primarily determined by the performance of the first purification step of a five-step process producing a monoclonal antibody, then the product purity specification can be used to define a purity output criterion for the first step.

It is then necessary to define an overall experimental paradigm. Usually, the experimental program begins with a set of proposed operating ranges for the inputs to be evaluated for each validation stage. These ranges may be based on equipment capability, theoretical considerations, process development data, operating history, the need for flexibility in manufacturing, or other considerations. Each validation stage is then run using inputs at the limits of the proposed operating ranges. These experiments can be run to investigate one input at a time or in parallel using statistical design of experiment (DOE) techniques. The outputs from these experiments are obtained, and the proposed operating ranges become proven acceptable ranges (PARs) (9) if the output criteria are met. If the output cri-

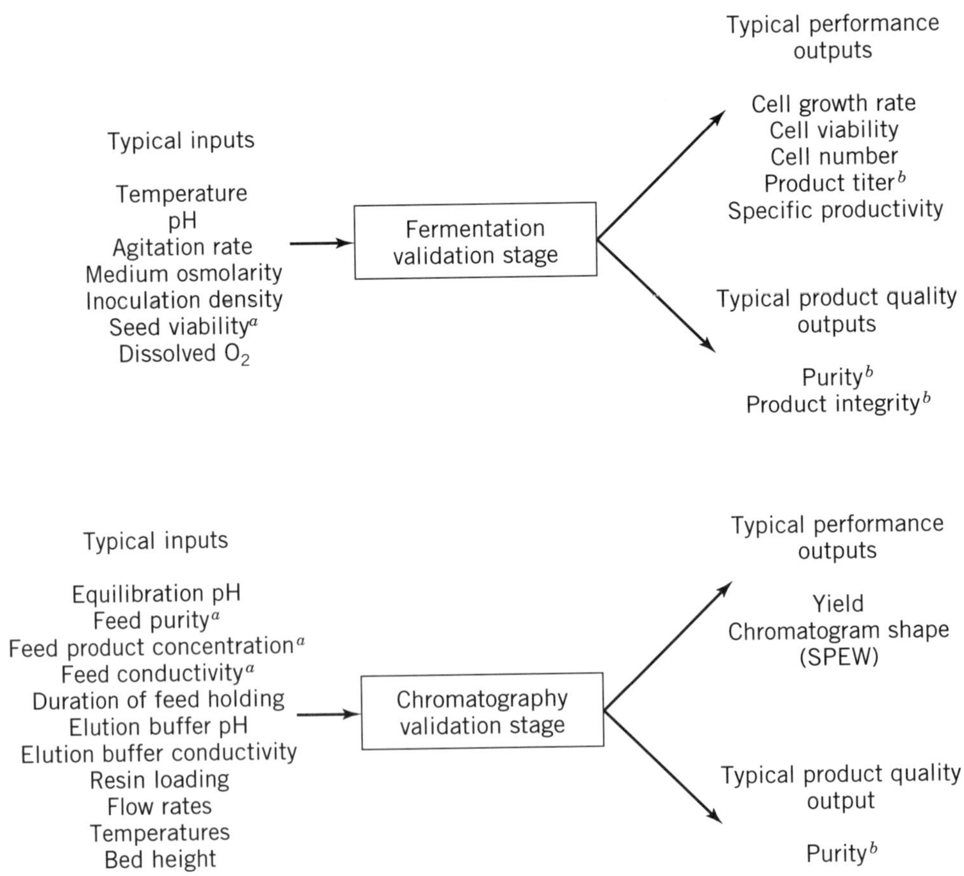

Figure 2. Each validation stage is driven by a set of inputs and produces a set of outputs. The outputs may be related to process stage performance or product quality. The general goal is to validate that the outputs are maintained within satisfactory ranges when the inputs are permitted to vary within defined ranges. [a]Forward linkage input–output from previous stage; [b]forward linkage output–input to next stage.

teria are not met, then the input ranges are narrowed and the experiments are repeated until the criteria are met.

This approach is taken for each validation stage. However, it may be necessary to explore interactions among the validation stages. For example, the performance of a purification step or the quality of the product it produces often depends on the quality of the input feed stream. Because the input feed stream is an output from the previous purification step, one way to handle these interactions is to include the input stream properties as other inputs to the validation stage. Input streams of high and low quality are then required. One approach to obtain these streams is to operate the upstream purification step under best- and worst-case conditions. Alternatively, by knowing the function of the purification step to be validated (e.g., endotoxin removal, host cell protein removal, DNA removal, or removal of product isoforms), one can prepare input streams of high and low quality by spiking in the appropriate species to be removed. The pair-wise validation of processing steps has been referred to as forward linkage validation. *Worst case* and *most appropriate challenge* are terms that describe the validation of a process run with its critical operating parameters set at the limits of their operating ranges (11).

PARs are generally used to establish maximum operating ranges (MORs) bounded by action limits for the important process variables. When a process variable exceeds the outer limits of an MOR, then the impact on the process and the product may be unknown. This event often requires corrective action to be taken by operations staff and various QA systems. Depending on the importance of the variable, the duration and extent of the excursion beyond the MOR, and other operating data from process operations, a variety of follow-up actions may be appropriate. These actions can include relatively minor responses such as requiring a more detailed batch record review or more analytical testing than is normally required for product release when appropriate analytical systems and databases are available. Other QA responses include launching an investigation to look for the cause of the problem, requiring the demonstration of lot stability before release, requiring reprocessing of the lot, or rejecting the lot.

VALIDATION OF PROCESS CONSISTENCY

From the standpoint of process control, it is necessary to determine the extent to which the process and product vary naturally under normal operating conditions. This knowledge is important in defining the normal operating range (NOR) bounded by alert limits for each important process variable. When a NOR is exceeded, operations staff may have to take corrective action. NORs must fall within the MORs established in most cases by the PAR studies. When NORs exceed MORs, the process is not sufficiently controlled to provide "a high degree of assurance that a specific process will consistently produce a product meeting its pre-determined specifications and quality attributes" (1). NORs are typically calculated from replicate runs of the process that produces product that meets specifications. The calculation of a 99% or 95% confidence interval is a typical means of establishing NORs.

The relationship among ranges for any process variable is shown in Figure 3. Process development or optimization often establishes a desired value known as a target or setpoint (e.g., 37 °C) for a process variable. Under normal operating conditions, however, a process variable varies within an NOR (e.g., 37 ± 0.5 °C). Validation studies can determine a PAR (e.g., 37 ± 2.0 °C), which can be used to set the MOR (e.g., 37 ± 2.0 °C). The MOR falls within a broader range bounded by the edge of failure (e.g., 32–40 °C). A process variable that exceeds the edges of failure will cause the process or product to fail, but the edges of failure are not typically determined during process development or validation except by accident. The ranges over which the variable can be controlled (e.g., 20–80 °C) and for which it can exist (e.g., ≥ −273 °C) are still broader (9).

POSTLAUNCH PROCESS VALIDATION

Process validation is a requirement for product licensing. However, it continues throughout the lifetime of the product. Experience with a process in the commercial manufacturing facility is often very limited when the facility files license applications. Experience may mean production of as few as three or five lots to meet the qualification lot requirements for filing in the United States or Europe, respectively. These numbers of lots are insufficient to generate meaningful process or product specifications. As a consequence, initial process validation data are generally derived from pilot plant experiences obtained during the production of supplies for clinical trials or the development of the process. Regulatory authorities accept this information as sufficient for licensing, but many of the MORs, NORs, and product specifications may be refined during actual commercial production experience. In most cases, changes to these ranges are reported to regulatory authorities in annual updates, particularly when the change re-

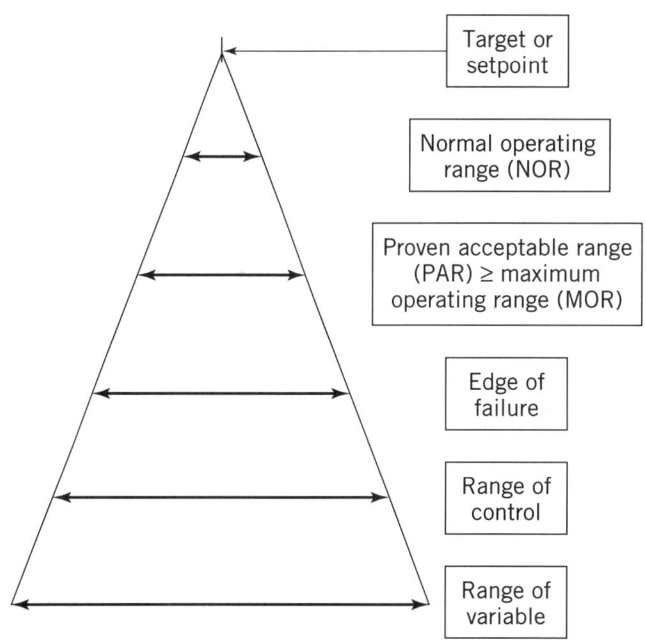

Figure 3. Relationships among ranges for process variables.

sults in a narrowing of a control range or an improvement in product quality. In other cases, however, the change may need to be proposed to the FDA and regulatory approval gained before implementation.

FUTURE PROSPECTS

The significance of process validation has grown from its origins in the mid 1970s to the present time. Of major importance to the biopharmaceutical industry has been the total absence of medical incidents traceable to process control problems, largely because of the application of validation principles. More recently, regulatory authorities have reconsidered their former positions that most changes to the manufacturing process, including relatively minor ones, led to a product that might be different. In the past, clinical studies were required to establish the safety and efficacy of the new product. Today, many changes can be justified using analytical data and process validation studies without the need to undertake a clinical trial. More recently, the FDA has dropped the requirement for filing an establishment license application (ELA) for "therapeutic recombinant DNA-derived products and/or monoclonal antibodies" (20). The FDA has also dropped the requirement for the FDA Center for Biologics Evaluation and Review (CBER) to release test each lot of product with this classification. These changes are due largely to the increase in powerful analytical methods that are available to test these types of products and the greater sophistication of validation to characterize the process.

Because of the growing importance of process validation and the absence of all but the most general guidance from regulatory agencies, it is likely that process validation will become the subject of future regulatory focus by the FDA and the European Union. Most cGMP regulatory requirements are specified for final product, and efforts in the United States to define cGMPs specifically for the production of bulk pharmaceutical chemicals (also known as APIs) are in progress (19). It is in these guidelines that we can expect more practical directives on process validation to be detailed. Given the expense of process validation work, guidance that leads to improved definition of requirements and elimination of process validation studies of marginal significance will be welcome. A reexamination of the balance among prospective, concurrent, and retrospective validation approaches will also be useful. Industry will need to contribute to these debates to ensure that validation costs and their impact on the overall health care system cost are directed toward areas with the greatest impact on the safety and efficacy of biopharmaceutical products.

ACKNOWLEDGMENTS

We thank Tony Lubiniecki, Peter Calcott, and John Bennan for their thoughtful reviews of this manuscript.

BIBLIOGRAPHY

1. U.S. Food and Drug Administration Center for Drugs and Biologics and Center for Devices and Radiological Health, *Guideline on General Principles of Process Validation*, May 1987.
2. K.G. Chapman, *Pharmaceutical Technology* **15**, 82–96 (1991).
3. Current Good Manufacturing Practice in Manufacturing, Processing, Packing, or Holding of Drugs, *Code of Federal Regulations*, Part 210, Title 21, 1998; Current Good Manufacturing Practice for Finished Pharmaceuticals, *Code of Federal Regulations*, Part 211, Title 21, 1998.
4. A.L. Lubiniecki, M.E. Wiebe, and S.E. Builder, in A.S. Lubiniecki ed., *Large-Scale Mammalian Cell Culture Technology*, Marcel Dekker, New York, 1990, pp. 515–541.
5. Biotech Product Validation, *BioPharm Validation Supplement* (1995).
6. International Conference on Harmonization of Technical Requirements for Registration of Pharmaceuticals for Human Use (ICH), *Viral Safety Evaluation of Biotechnology Products Derived from Cell Lines of Human and Animal Origin*, March 5, 1997.
7. International Conference on Harmonization of Technical Requirements for Registration of Pharmaceuticals for Human Use (ICH), *Quality of Biotechnological Products: Analysis of the Expression Construct in Cells Used for Production of r-DNA Derived Protein Products*, November 1995.
8. PMA's Validation Advisory Subcommittee, *Pharmaceutical Technology* **9**, 78–82 (1995).
9. K.G. Chapman, *Pharmaceutical Technology* **8**, 22–36 (1984).
10. M.E. Wiebe and S.E. Builder, in F. Brown and A.L. Lubiniecki eds., *Genetic Stability and Recombinant Product Consistency*, Karger, Basel, 1994, pp. 45–54.
11. Parenteral Drug Association, *Journal of Parenteral Science and Technology* **46**, 87 (1992).
12. R.G. Werner, H. Langlouis-Gau, F. Walz, H. Allgaier, and H. Hoffmann, *Arzneim.-Forsch.* **38**, 855–862 (1988).
13. L.D. Torbeck and R.C. Branning, *Pharmaceutical Technology* **20**, 108 (1996).
14. Parenteral Drug Association Biotechnology Task Force on Purification and Scale-Up, *Journal of Parenteral Science and Technology* **46**, s3–s13 (1992).
15. U.S. Food and Drug Administration, Center for Biologics Evaluation and Research (CBER), *Points to Consider in the Characterization of Cell Lines Used to Produce Biologics*, 1993.
16. U.S. Food and Drug Administration, Center for Biologics Evaluation and Research (CBER), *Points to Consider in the Manufacture and Testing of Monoclonal Antibody Products for Human Use*, April 28, 1997.
17. CPMP Biotechnology Working Party, *EU—Note for Guidance on Virus Validation Studies: The Design, Contribution, and Interpretation of Studies Validating the Inactivation and Removal of Viruses*, 1996.
18. Current Good Manufacturing Practice: Proposed Amendment of Certain Requirements for Finished Pharmaceuticals, *Fed. Regist.*, 61(87), 20104 (1996).
19. U.S. Food and Drug Administration, *Guidance for Industry: Manufacturing, Processing, or Holding Active Pharmaceutical Ingredients*, Discussion Draft, March 1998.
20. U.S. Food and Drug Administration, Center for Biologics Evaluation and Research (CBER) and Center for Drug Evaluation and Research (CDER), *Guidance for Industry for the Submission of Chemistry, Manufacturing, and Controls Information for a Therapeutic Recombinant DNA-Derived Product or a Monoclonal Antibody Product for In-Vivo Use*, August 1996.
21. Biological Products: General; Physical Establishment, Equipment, Animals, and Care, *Code of Federal Regulations*, Part 600.11a, Title 21, 1998.

22. International Conference on Harmonization of Technical Requirements for Registration of Pharmaceuticals for Human Use (ICH), *Guideline on Quality of Biotechnology Products: Viral Safety Evaluation of Biotechnology Products Derived from Cell Lines of Human or Animal Origin*, December 1, 1995.
23. G. Sofer, *BioPharm* **9**, 51–54 (1996).
24. A. Riggen, G.C. Davis, and T.L. Copmann, *BioPharm* **9**, 36–41 (1996).
25. A.J. Darling and J.J. Spaltro, *BioPharm* **9**, 42–50 (1996).
26. K.P. Anderson, M.-A. L. Low, Y.S. Lie, G.-A. Keller, and M. Dinowitz, *Virology* **181**, 305–311 (1991).
27. R.J. Seely, H.D. Wight, H.F. Fry, S.R. Rudge, and G.F. Slaff, *BioPharm* **7**, 41–48 (1994).

See also ASTM STANDARDS FOR BIOTECHNOLOGY; CLEANING, CLEANING VALIDATION; GOOD MANUFACTURING PRACTICE (GMP) AND GOOD INDUSTRIAL LARGE SCALE PRACTICE (GLSP); STERILIZATION-IN-PLACE.

PRODUCTION OF L-AMINO ACIDS BY AMINOACYLASE

TADASHI SATO
TETSUYA TOSA
Tanabe Seiyaku Company, Ltd.
Osaka, Japan

KEY WORDS

Aminoacylase
L-Amino acid
Optical resolution
Racemization

OUTLINE

Bibliography

At present, fermentative and chemical synthetic methods are used for the industrial production of L-amino acids instead of conventional isolation from protein hydrolysates. However, chemically synthesized amino acids are optically inactive racemic mixtures of L- and D-isomers. The L form is the physiologically active natural form and is therefore the required form for medicine and food. To obtain L-amino acid from the chemically synthesized DL form, optical resolution is necessary.

Generally, optical resolution of racemic amino acids can be carried out by physicochemical, chemical, enzymatic, and biological methods. Among these methods, an enzymatic method developed by Chibata et al. (1) using mold aminoacylase is one of the most advantageous procedures, yielding optically pure L-amino acids. The reaction catalyzed by the enzyme is as follows.

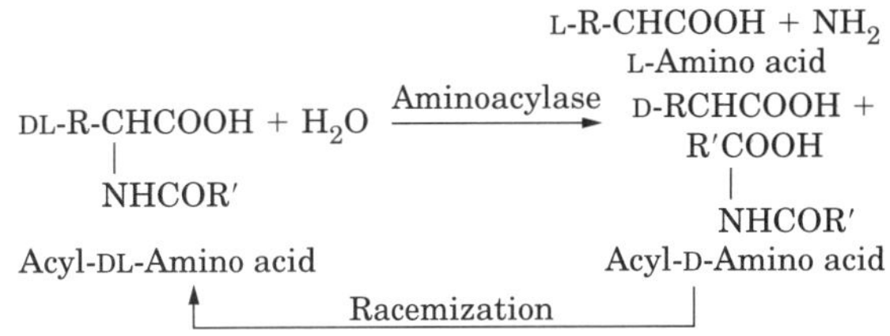

Chemically synthesized acyl-DL-amino acid is asymmetrically hydrolyzed by aminoacylase to give L-amino acid and the unhydrolyzed acyl D-amino acid. After concentration, both materials are easily separated on the basis of the difference in their solubilities. Acyl-D-amino acid is racemized and reused for the resolution procedure.

Chibata et al. (1) studied enzymes catalyzing the reaction and found that aminoacylase produced by *Aspergillus oryzae* has high activity and broad substrate specificity, that is, it can asymmetrically hydrolyze many kinds of acyl-DL-amino acids.

From 1953 to 1969, this mold aminoacylase was used for the industrial production of several L-amino acids. This method was one of the most advantageous procedures for the industrial production of L-amino acids. However, because the enzyme reaction was carried out batchwise by incubating a mixture containing the substrate and soluble enzyme, the procedure had some disadvantages for industrial purposes. For instance, to isolate L-amino acid from the enzyme reaction mixture, it was necessary to remove enzyme protein by pH and/or heat treatment. Thus, even if enzyme activity remained in the reaction mixture, the enzyme had to be discarded because there was no suitable procedure for isolating the active enzyme from the mixture. In addition, a complicated purification procedure was necessary for the removal of proteins and coloring materials contaminating the crude enzyme preparations usually used for industrial purposes.

To overcome these disadvantages and to improve this enzymatic method, we extensively studied the continuous optical resolution of DL-amino acids using a column packed with immobilized aminoacylase (2). An efficient and automatically controlled enzyme reactor system was achieved, and since 1969, we have been industrially operating a series of these reactors in our plants.

Various immobilization methods were investigated, and relatively active and stable immobilized aminoacylases were obtained by ionic binding to DEAE-Sephadex (3), covalent binding to iodoacetylcellulose (4), and entrapping into polyacrylamide gel (5). However, for the industrial operation of a continuous enzyme reaction using an immobilized enzyme, it was necessary to satisfy many conditions. Thus, the characteristics of these three immobilized aminoacylases were compared, and aminoacylase bound to DEAE-Sephadex was chosen as the most advantageous enzyme preparation for the industrial production of L-amino acids, because (*1*) preparation is easy, (*2*) the activity is high, (*3*) the stability is high, and (*4*) regeneration of deteriorated immobilized enzyme is possible.

To design the most efficient enzyme column, the following are important factors: (*1*) the effect of flow rate of substrate on the reaction rate, (*2*) the flow of the substrate solution, (*3*) the effect of column dimension on the reaction rate, and (*4*) the pressure drop in the column.

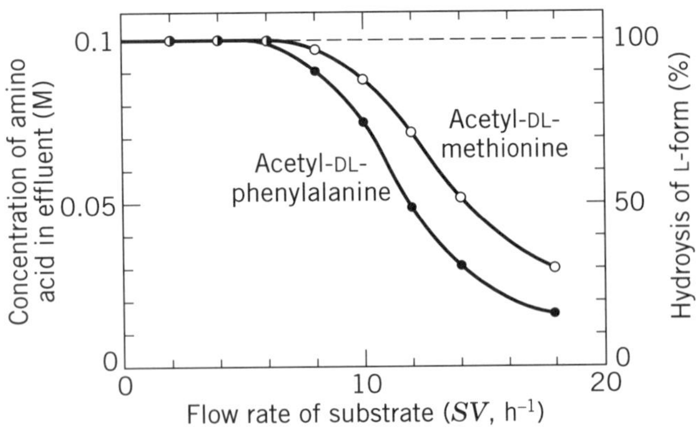

Figure 1. Continuous hydrolysis of acetyl-DL-amino acids by an immobilized aminoacylase column. Relationship between flow rate of substrate and extent of hydrolysis. Aminoacylase was immobilized by ionic binding to DEAE-Sephadex. Concentration of substrate, 0.2 M (containing 5×10^{-4} M Co^{2+}); reaction temperature, 50 °C.

Because it is one of the factors considered in the design of an enzyme reactor, the relationship between the flow rate of a substrate solution passing through the enzyme column and the extent of the reaction must be investigated in advance. We investigated the relationship between the flow rate of a solution of acetyl-DL-methionine or acetyl-DL-phenylalanine and the extent of the reaction, and the results are shown in Figure 1 (2). An aqueous solution of acetyl-DL-amino acid (0.2M, pH 7.0 to 7.5, containing 5×10^{-4} M Co^{2+}) was passed through a DEAE-Sephadex-aminoacylase column at various flow rates at 50 °C. In Figure 1, flow rate is expressed in terms of space velocity. The space velocity is the volume of liquid passing through a given volume of immobilized enzyme in 1 h divided by the latter volume. If the space velocity is reduced, that is, the flow rate is slowed, the reaction proceeds more effectively and the relationship between the flow rate and the extent of the reaction is sigmoidal. As shown in Figure 1, when the concentration of L-amino acid in the effluent reached 0.1 M, the reaction stopped. This result indicates that the L-form of 0.2 M acetyl-DL-amino acid was completely hydrolyzed and its D-form was not hydrolyzed, that is, asymmetric hydrolysis occurred.

The pressure drop of the enzyme column is also important factor in the design of an enzyme reactor. Thus, the relationship between the flow rate of substrate solution and the pressure drop of the column was investigated using columns of various lengths. The pressure drop of the DEAE-Sephadex-aminoacylase column was found to be proportional to the flow rate of substrate solution and the column length.

Table 1. Production of L-Amino Acids on 1,000-L DEAE-Sephadex Aminoacylase Column

		Yield (theory) of L-amino acids per	
L-Amino acid	Space velocity	24 h (kg)	20 days (kg)
L-Alanine	1.0	214	6,420
L-Methionine	2.0	715	21,450
L-Phenylalanine	1.5	594	17,820
L-Tryptophan	0.9	441	13,230
L-Valine	1.8	505	15,150

In addition, by considering some other properties of the immobilized enzyme, such as optimum pH, the effect of temperatures on the reaction rate and stability, and so on, we designed an enzyme reactor system for continuous production of L-amino acids, as shown in Figure 2. In this system, the flow rate and pH of substrate solution and the operating temperature can be automatically controlled and recorded.

Since 1969, continuous optical resolution of acetyl-DL-amino acids has been carried out using this system, and several kinds of optically active amino acids such as methionine, phenylalanine, valine, and others have been industrially produced at Tanabe Seiyaku Company, Japan. Several examples of the production of L-amino acids are summarized in Table 1, which shows the space velocity and the theoretical yield for each amino acid produced in a 1,000-L aminoacylase column. That was the first industrial application of immobilized enzymes.

The immobilized aminoacylase was very stable, as shown in Figure 3, and maintained 60 to 70% of its initial activity after continuous operation for 30 days at 50 °C. The enzyme column can be completely regenerated after prolonged operation by the addition of an amount of aminoacylase corresponding to the lost activity. The carrier, DEAE-Sephadex, is also very stable and has been used for more than 5 years in our column process without any change in adsorption capacity for the enzyme, shape, or pressure drop. A continuous enzyme reaction using such a stable immobilized enzyme gives not only high productivity per enzyme unit, but also high product yield, because contamination of the effluent by such impurities as proteins and coloring substances does not occur, so the product can be obtained by simple purification and the amount of substrate required is reduced. Further, automatic operation reduces labor costs, and a significant reduction in production cost can be expected in comparison with the conventional batch process using soluble enzyme. A comparison of the production costs of L-amino acids by the conventional batch process using soluble enzyme and by the continuous process using immobilized enzyme in our plant is shown in Figure 4. In the immobilized enzyme process, the overall production cost is more than 40% lower than that of the conventional batch process using soluble enzyme. Savings of enzyme and labor costs are the main contributors, as are the increase in product yield from the easy isolation of L-amino acids from the reaction mixture. Although DEAE-Sephadex is a relatively expensive carrier, production costs are not greatly affected by the cost of the carrier, because it can be used for a very long period, as mentioned.

Besides immobilization by ionic binding, mold aminoacylase from *Aspergillus* sp. was also immobilized by covalent binding to alkylaminosilanized porous glass with glutaraldehyde or to the diazonium derivative of acylaminosilanized porous glass, and these immobilized aminoacylases were used for continuous preparation of L-amino acids from acetyl-DL-amino acids (6).

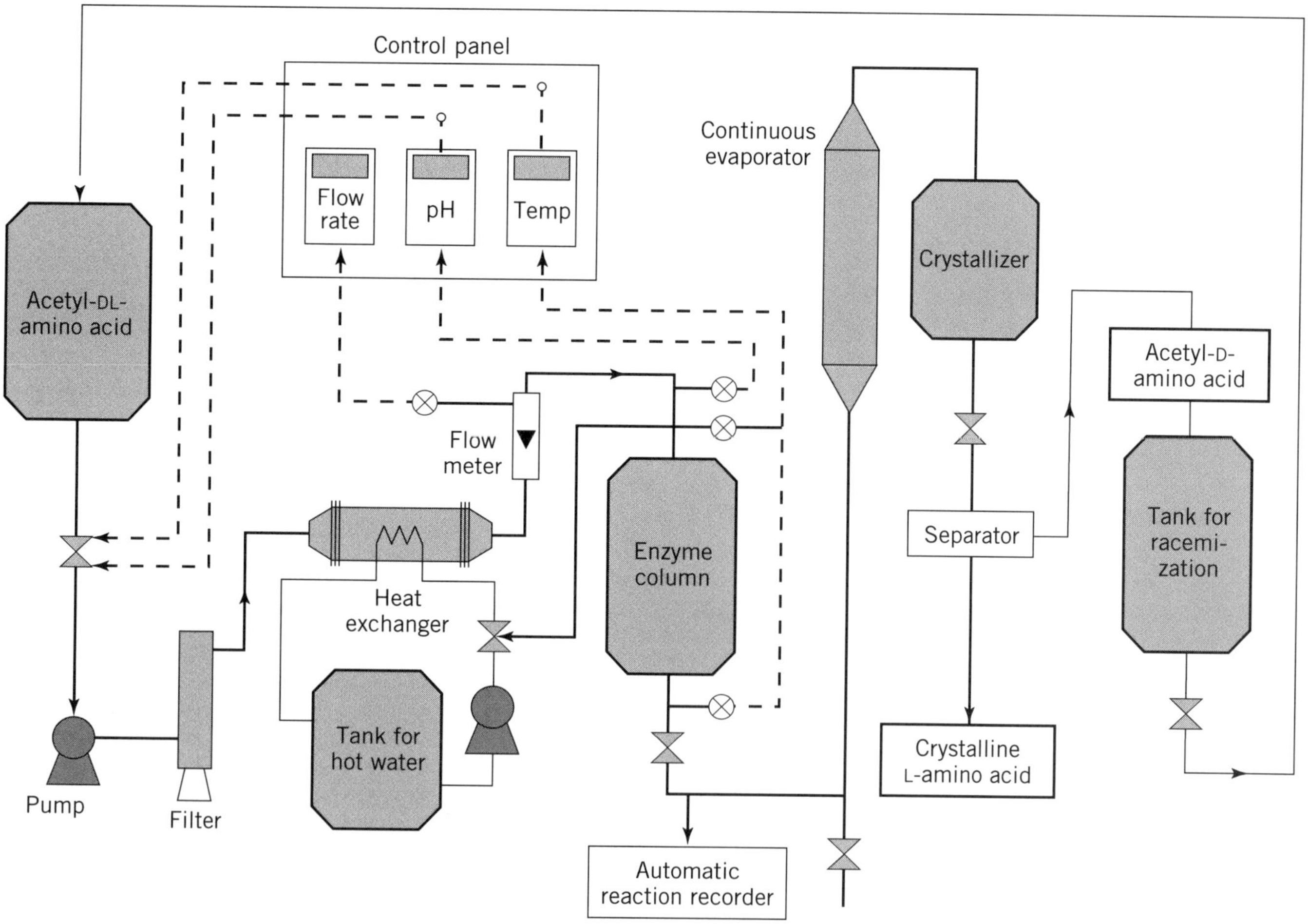

Figure 2. Flow diagram for the continuous production of L-amino acids by immobilized aminoacylase.

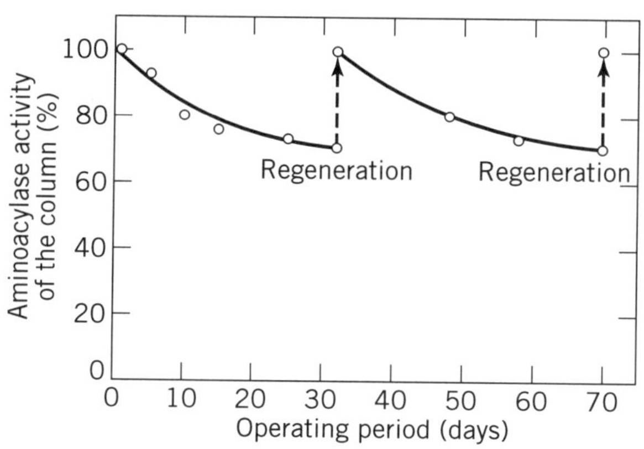

Figure 3. Stability and regeneration of an immobilized aminoacylase column. Aminoacylase was immobilized by ionic binding to DEAE-Sephadex, and 0.2 M acetyl-DL-methionine (containing 5×10^{-4} M Co^{2+}) was continuously passed through a column packed with the immobilized aminoacylase at a flow rate of SV = 2 at 50 °C.

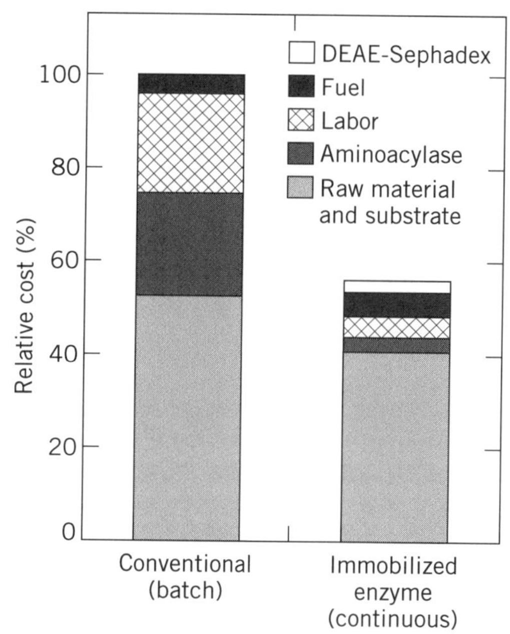

Figure 4. Comparison of production costs of L-amino acids by batch and continuous processes.

Bacterial and animal aminoacylases have been also immobilized. Bacterial aminoacylase has been covalently bound to diazotized polyaminostyrene (7,8). The aminoacylase from pig kidney was immobilized by ionic binding to DEAE-cellulose (9) and by covalent binding to the azide derivative of Enzacryl AH (10) or diazotized Enzacryl AA (11). However, these preparations were used batchwise and on only a laboratory scale for the preparation of L-amino acids from acyl-DL-amino acids.

BIBLIOGRAPHY

1. I. Chibata, T. Ishikawa, and S. Yamada, *Bull. Agric. Chem. Soc. Jpn.* **21**, 300–303 (1957).
2. I. Chibata, T. Tosa, T. Sato, T. Mori, and Y. Matuo, *Proceedings of the 4th International Fermentation Symposium: Fermentation Technology Today*, 1972, p. 383–389.
3. T. Tosa, T. Mori, N. Fuse, and I. Chibata, *Biotechnol. Bioeng.* **9**, 603–615 (1967).
4. T. Sato, T. Mori, T. Tosa, and I. Chibata, *Arch. Biochem. Biophys.* **147**, 788–796 (1971).
5. T. Mori, T. Sato, T. Tosa, and I. Chibata, *Enzymologia* **43**, 213–226 (1972).
6. H.H. Weetall and C.S. Detar, *Biotechnol. Bioeng.* **16**, 1537–1544 (1974).
7. M.A. Mitz and L.J. Summaria, *Nature* **189**, 576–577 (1961).
8. M.D. Lilly, C. Money, W.E. Hornby, and E.M. Crook, *Biochem. J.* **95**, 45p (1965).
9. T. Barth and H. Maskova, *Collect. Czech. Chem. Commun.* **36**, 2398–2402 (1971).
10. Y. Ohno and M. Stahmann, *Macromolecules* **4**, 350–352 (1971).
11. H. Maskova, T. Barth, B. Jirovsky, and I. Rychlik, *Collect. Czech. Chem. Commun.* **38**, 943–947 (1973).

See also AMINO ACIDS, PRODUCTION PROCESSES; BIOCATALYSIS DATABASES; FOOD PROCESS ENGINEERING.

PRODUCTION OF L-GLUTAMIC ACID

SHUKUO KINUSHITA
Kyowa Hakko Kogyo Co., Ltd.
Tokyo, Japan

KEY WORDS

Amino acid fermentation
Biotin
Corynebacterium
Fermentation process
Food taste enhancer
Glutamic acid
Penicillin

OUTLINE

INTRODUCTION

L-Glutamic acid monosodium salt was first discovered by Dr. Kikunae Ikeda as a taste component of *konbu*, a sea tangle, which is a food taste-enhancing material traditionally used in Japan. Based on this discovery, L-glutamic acid monosodium salt was industrially produced from wheat gluten by hydrolysis with hydrochloric acid (1). Later, the raw material was replaced by soybean protein, and a synthetic method was studied and finally industrialized. In addition attempts were made to produce α-ketoglutaric acid, which is a precursor of L-glutamic acid, by organic acid fermentation. These attempts were aimed at producing L-glutamic acid by transamination from the resulting α-ketoglutaric acid.

In 1957, Kinoshita et al. succeeded in establishing a process to produce L-glutamic acid by direct fermentation from a carbohydrate and ammonia using a microorganism, *Corynebacterium glutamicum*, which they had isolated from nature (2). Compared to the then conventional hydrolytic process comprising the hydrolysis of natural protein, this direct fermentation process made it easy to secure the production raw material, and at the same time facilitated a drastic reduction of cost.

With establishment of the fermentative process, the monosodium glutamate production industry showed rapid growth. In recent years, the production worldwide has been estimated to reach to about 600,000 tons per year as L-glutamic acid (Table 1). As the production cost was cut

Table 1. Production of Monosodium Glutamate

Years	Production of MSG (t/year)
1954	7,652
1960	22,177
1970	92,935
1988	340,000
1994	592,000

Note: MSG, monosodium glutamate.

owing to improvement in production technologies, the production plants, initially operated only in Japan, have been extended worldwide, particularly to South East Asia where there are abundant supplies of raw materials and also consumption of the product (Table 2).

SYNTHETIC PATHWAY OF L-GLUTAMIC ACID

In the biosynthesis of L-glutamic acid (Fig. 1), a key step is a reductive ammonia-addition reaction of α-ketoglutaric acid, which constitutes the TCA cycle:

$$\alpha\text{-ketoglutaric acid} + \text{NADPH} + \text{H}^+ + \text{NH}_4 \xleftrightarrow{\textit{glutamate dehydrogenase}} \text{L-glutamic acid} + \text{NADPH}^+ + \text{H}_2\text{O}$$

NADPH, the coenzyme of this reaction, is supplied through coupling in the reaction to form α-ketoglutaric acid:

$$\text{isocitric acid} + \text{NADPH}^+ \xleftrightarrow{\textit{isocitrate dehydrogenase}} \alpha\text{-ketoglutaric acid} + \text{CO}_2 + \text{NADPH} + \text{H}^+$$

L-GLUTAMIC ACID FERMENTATION

Initially, a glutamic acid–producing bacterium was isolated as a biotin-requiring mutant of *C. glutamicum*, and a fermentative production process in which the strain is cultured in a medium containing a limited concentration of biotin was industrialized. Afterward, use of cheap cane molasses attracted attention as a raw material carbohydrate; however, cane molasses contains a large amount of biotin. To enable use of cane molasses, the penicillin shot method was developed, which made it possible to accumulate L-glutamic acid in the presence of excess biotin, and the industrial technology had been further advanced. Subsequently, techniques to use oleic acid–requiring strains or to add saturated higher fatty acid esters (or saturated higher fatty acid) such as polyoxyethyleneglycol parmityl ester were developed. Also, use of various carbon source materials other than carbohydrate was examined. L-Glutamic acid fermentation from *n*-paraffin (3,4) has been much studied, and the production process has become almost as well established as the use of carbohydrate. However, the production from *n*-paraffin has never been commercialized because of the high cost of the main raw material, as well as other reasons. Currently, the penicillin shot method using cane molasses, beet molasses, glucose, and so forth as raw materials is most frequently employed in industry because of cost, stable production, and other reasons.

Table 2. Worldwide Capacity for Monosodium Glutamate, 1994

Area	Capacity (estimated) (t/year)
U.S.	60,000
Japan	279,000
Far East (except Japan)	392,700
Europe	72,500
South and Central America	53,000
Total	857,200

METABOLIC MECHANISMS OF L-GLUTAMIC ACID PRODUCTION

Change of the Membrane-Permeation System

As already mentioned, the L-glutamic acid production process was initially established using a biotin-requiring bacterial strain and strict control of biotin concentration. It was later found that production could be accomplished even under excess biotin conditions by adding penicillin, using oleic acid-requiring microorganisms, and adding saturated higher fatty acid esters such as polyoxyethyleneglycol parmityl ester or fatty acid itself. It was also found that, under high oleic acid concentration, although L-glutamic acid was not produced by the addition of higher fatty acids or their esters, the penicillin shot method was still effective. Today, these phenomena are explained as attributable to dependency of L-glutamic acid accumulation on the changes of the membrane-permeation system (Fig. 2). More particularly, biotin plays a role in the biosynthesis of oleic acid by glutamic acid–producing bacteria, and under the biotin-controlled condition, oleic acid formation is repressed. This causes a change in fatty acid composition and phospholipids that constitute the cell membrane and, consequently, a change in permeability of L-glutamic acid through the membrane. Oleic acid–requiring mutants and the direct addition of saturated higher fatty acid bring about a similar change in composition whether or not biotin is present, and produce a similar effect. Hence these conditions do not give rise to L-glutamic acid accumulation at a high concentration of oleic acid. On the other hand, it is believed that the addition of penicillin inhibits synthesis of cell walls rather than changes in the composition of fatty acid that constitutes the membrane, thereby causing a difference in the osmotic pressure between the outside and the inside of the membrane. It is this pressure differential that causes the membrane to undergo a change that leads to L-glutamic acid accumulation.

When *n*-paraffin is used as a raw material, the mechanism of L-glutamic acid production is explained by a similar change of membrane permeability, although a threonine-requiring strain is used in place of the biotin-requiring strain used in carbohydrate fermentation; and the presence of a copper ion additionally affects L-glutamic acid formation (4).

Other Factors for Excess Formation of L-Glutamic Acid

There has been a report on the change in the activity of the L-glutamic acid–producing enzyme system in its L-glutamic acid formation phase (6). Furthermore, in some cases, L-glutamic acid–producing strains with an enhanced biosynthetic enzyme system have been obtained by recombinant technology. Thus, L-glutamic acid production may not be totally explained by "leakage" of the amino acid. This may be a problem to be studied in the future.

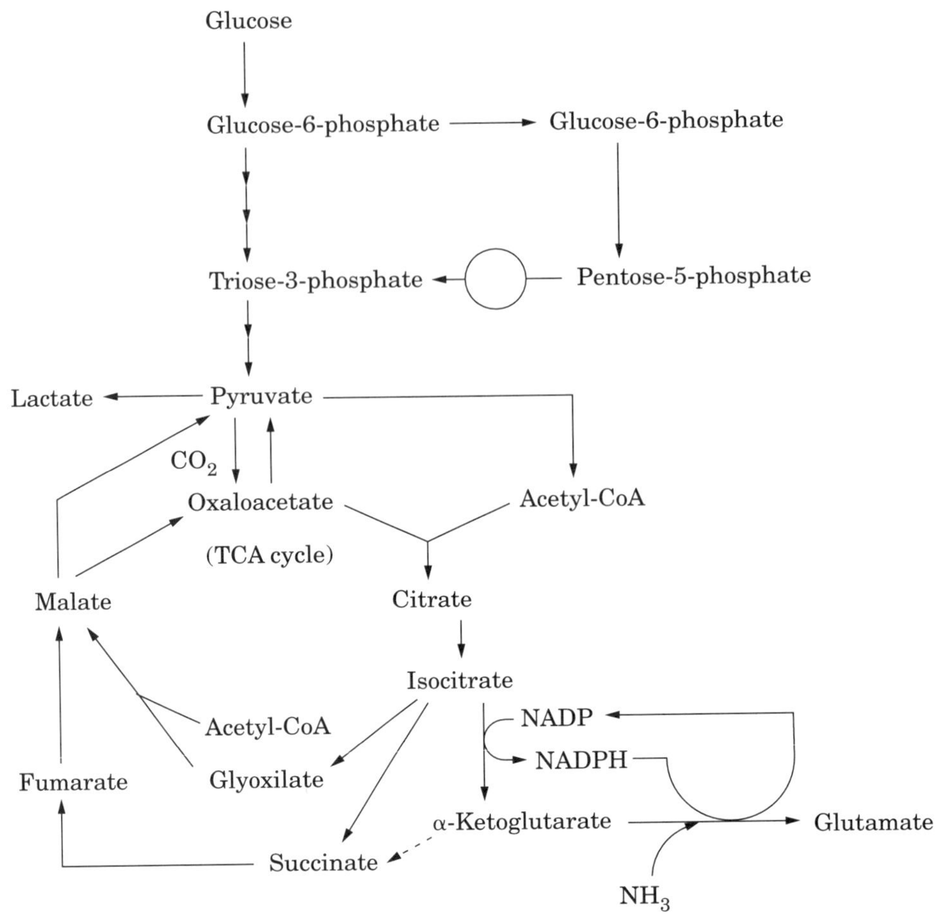

Figure 1. The biosynthetic pathway of glutamate.

In a recent report (7) a gene for detergent-sensitivity (*dtsR*) carried by a wild-type strain was identified using a polyoxyethylene sorbitan fatty acid ester–sensitive strain of *Brevibacterium lactofermentum*, and a mutant lacking this gene was found to accumulate L-glutamic acid in a medium containing excess biotin and in the presence of polyoxyethylene sorbitan oleate ester. Under these conditions the wild-type strain did not accumulate the amino acid. This strain had its specific activity of 2-oxoglutarate dehydrogenase complex (an enzyme involved in the TCA cycle that converts α-ketoglutaric acid to succinyl-CoA) reduced to one-third of that possessed by the wild-type strain. From these findings, Kimura et al. (8) proposed a mechanism of L-glutamic acid fermentation: "The target of controlling biotin concentration and addition of surfactants is biotin enzyme complex containing DtsR, and the reduction of the activity of this biotin enzyme complex may possibly cause glutamic acid formation." According to this report, the sensitive strain propagates in response to the concentrations of oleic acid and polyoxyethylene sorbitan oleate ester in the medium. Protein, a product of the DtsR gene, had homology with a biotin enzyme derived from other organisms and was involved in the biosynthesis of higher fatty acid. Furthermore, accumulation of L-glutamic acid has been reported by a mutant strain having membrane permeability to L-glutamic acid. The mutation was induced from a wild-type methanol-assimilating strain that does not accumulate L-glutamic acid in the medium (9).

These findings suggest some ideas for the improvement of techniques to construct L-glutamic acid–producing strains by the recombinant technology. In the future, more efficient fermentation production of L-glutamic acid may be possible through diverse analyses of production processes and mechanisms, and construction of mutant strains.

INDUSTRIAL PRODUCTION OF MONOSODIUM GLUTAMATE

Fermentation Processes

The typical L-glutamic acid–producing wild strain of *C. glutamicum* isolated from nature requires biotin for growth and accumulates L-glutamic acid in the medium when cultured under conditions that include sufficient ammonium ions and controlled biotin concentration. This glutamic acid–producing strain shows different forms of fermentation, depending upon fermentation conditions other than biotin concentration. Basic fermentation conditions and the forms of fermentation corresponding to them are shown in Table 3. Glutamic acid production requires sufficient oxygen, appropriate amounts of phosphoric acid, and a nitrogen source, as well as controlled biotin concentration. This strain grows vigorously in a medium containing excess biotin and does not accumulate L-glutamic acid. However, fermentation production under

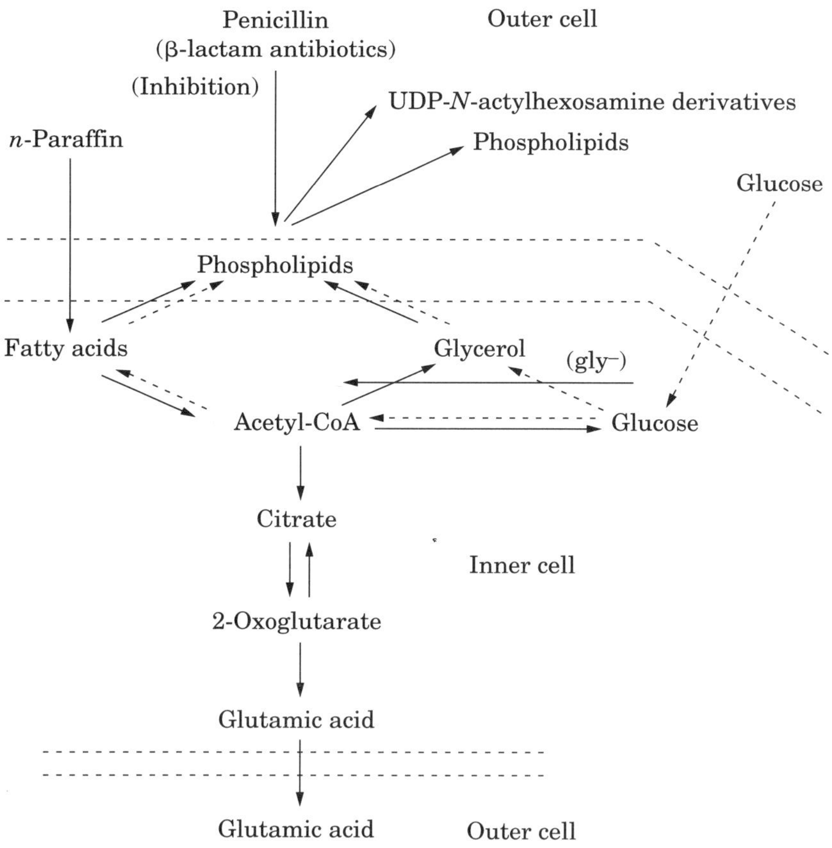

Figure 2. The excretion model of L-glutamic acid in glutamate-producing bacteria. *Source*: From Ref. 5.

Table 3. Coversion of Fermentation by Glutamic Acid–Producing Bacteria with Nutrient and Atmospheric Regulation

Regulatory factors	Coversion of fermentation
Oxygen	Lactate or succinate ↔ Glutamate (anaerobic) (aerobic)
NH_4^+	α-ketoglutarate ↔ Glutamate ↔ Glutamate (unsatisfy) (optimal) (oversatisfy)
pH	*N*-Acetylglutamine ↔ Glutaminate (or Glutamine) (acidic) (basic)
Phosphate	Valine ↔ Glutaminate (high)
Biotin	Lactate or succinate ↔ Glutamate (satisfy) (deficiency)

Source: From Ref. 10.

controlled biotin concentration involved two difficulties: *(1)* available raw materials were restricted (particularly, it was difficult to use cane molasses due to its high biotin content) and *(2)* fermentation production lacked stability because cell growth was not stable. Hence in order to solve this problem, various processes were attempted to accumulate L-glutamic acid under excess biotin conditions. Among these, there are two methods that have become employed as typical methods for industrial production. One is the penicillin shot method in which penicillin is added to cells cultured with sufficient biotin, and the other is the addition of saturated fatty acids or their esters to accumulate L-glutamic acid. The latter method was developed with the elucidation of the physiological effect of the penicillin shot. At present, the penicillin shot method is considered to be advantageous because it is stable in industrial operation. This method is discussed in detail later.

Raw Materials

From the viewpoints of stable supply and low cost, cane molasses (hereinafter referred to simply as "molasses"), beet molasses, and glucose are popular as the main raw material carbon sources. When molasses is used, it is necessary to employ a technique for excess biotin conditions described in the article GLUTAMIC ACID PRODUCING MICROORGANISMS because molasses contains excess biotin. On the other hand, when beet molasses and glucose are used, it is necessary to add biotin to the medium, whether using conditions of excess biotin or limited biotin. The following discusses the penicillin shot method using molasses.

When molasses is used as a main raw material, necessary components such as carbon, nitrogen, phosphoric acid, potassium, sodium, sulfuric acid radical, various metal ions, vitamins, and so on can be derived from this raw material. Any components revealed to be lacking by preliminary tests are supplemented to the medium.

In the following description of the penicillin shot method of L-glutamic acid fermentation, *C. glutamicum* and molasses are used, respectively, as the strain and the raw material, and culturing is carried out by a fed-batch system.

The fermentation tank is a basic aeration/agitation tank, shown in Figure 3. Generally, jacket-type (small fermenter) and coil-type (large fermenter) cooling devices are used for controlling the temperature, but cooling may also be effected by a system that cools the fermentation liquor with an external heat exchanger. In the L-glutamic acid fermentation from carbohydrate, optimization of oxygen supply is an essential condition, and equipment needs be selected to satisfy this condition. A seed medium prepared in a seed tank is transferred to a main fermenter containing a sterilized (using the equipment mentioned later) initial medium, and culturing is initiated after the culturing conditions are adjusted as predetermined by earlier tests. Penicillin G potassium salt is added (8–10 units/L) during the logarithmic growth phase. By adding penicillin, the strain undergoes changes, mainly in membrane permeability, to produce L-glutamic acid, and L-glutamic acid accumulates in the medium. Culturing is then carried out by continuously adding sterilized liquid molasses and controlling the pH around neutrality with ammonia. Liquid molasses is added until the strain no longer produces L-glutamic acid. As shown in Figure 4, which gives one example of a time course of fermentation, generally, the L-glutamic acid production phase continues only 30 or so hours after the addition of penicillin. L-Glutamic acid is finally produced in a concentration of 90–105 g/L. Unlike the culturing process with controlled biotin concentration, it is a characteristic feature of the penicillin shot method that culturing can be carried out under optimum conditions for microbial growth during the early phase of culturing. The advantages of this process are that culturing can be controlled easily, cell concentration is maintained almost at a constant level, and process control (i.e., the rate of molasses addition) is relatively easy. As shown in Table 3, glutamic acid fermentation undergoes different patterns as a result of certain factors. For example, an excessively high rate of carbohydrate addition leads to increased con-

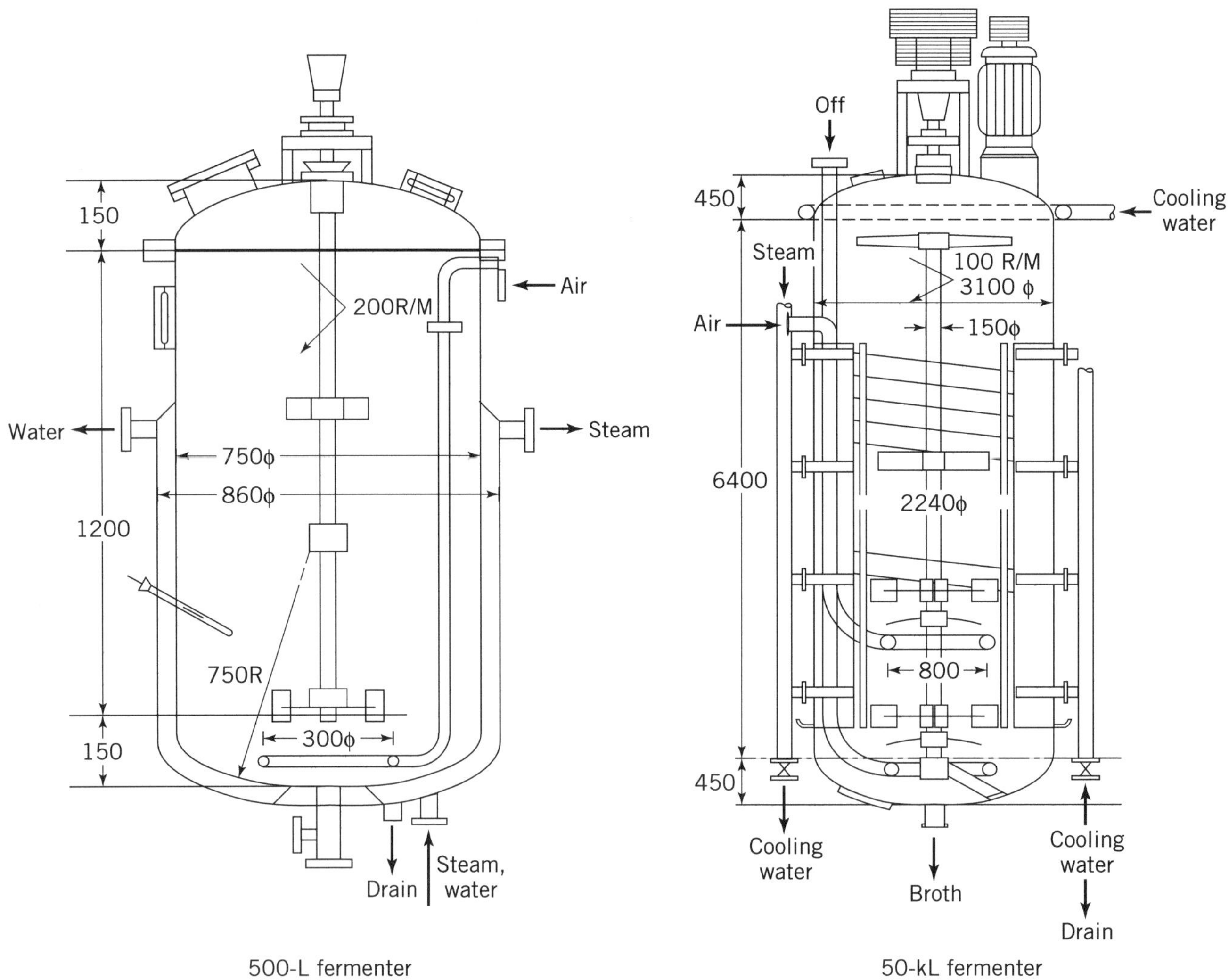

Figure 3. Example of a fermenter. *Source*: From Ref. 11.

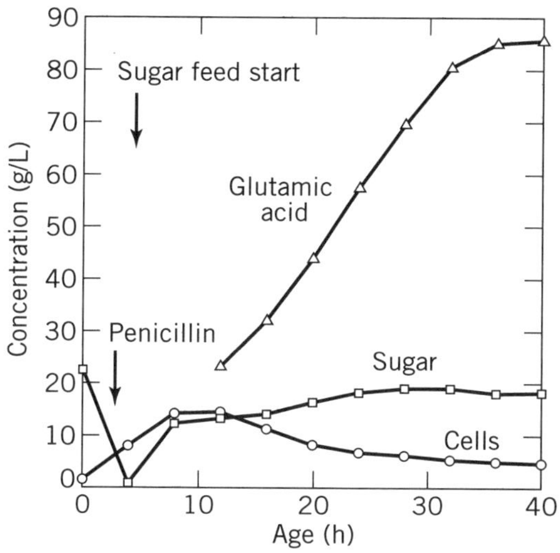

Figure 4. A fed-batch culture producing L-glutamate (*C. glutamicum*).

sumption of oxygen, resulting in oxygen shortage and, consequently, lactic acid and other organic acid formation. The control of cell growth is delicate when accomplished by the control of required substances, and the penicillin shot method is superior in ensuring stable productivity. Furthermore, the penicillin treatment concomitantly reduces the fraction of insoluble cells to a remarkable extent, which makes the recovery of L-glutamic acid from the fermentation broth much easier. This makes it possible to omit the mechanical separation of cells, resulting in simplification of the whole process.

It is, however, a problem that under the conditions currently used, the L-glutamic acid productivity lasts only about 30 h, as already mentioned. Generally, continuous processes are used because they improve productivity, but there is no continuous process that can practically operated due to the limited period of L-glutamic acid formation.

Purification: Direct Concentration and Crystallization

The outline of the purification step is shown in Figure 5. This step consists of two sub steps: recovering the slurry of L-glutamic acid from the fermentation broth (GA step), and producing the final product, monosodium L-glutamate, from L-glutamic acid (MSG step).

The GA step is directed to efficient recovery of high-purity GA crystals from the fermentation broth and the MSG-crystallized mother liquor resulting from the MSG step, and discharges the depleted GA-crystallized mother liquor out of the system. According to the penicillin shot method, in which most of the cells in the fermentation broth are lysed and solubilized, physical separation of the cells, such as with centrifugation, can be excluded, and the cell components are discharged into the GA-crystallized mother liquor. Precipitation is effected at the isoelectric point. GA crystals are known to take either the α- or the β-form. The α-crystal, which can be readily isolated, needs to be selectively precipitated because removal of impurities is accomplished by washing the crystals with water.

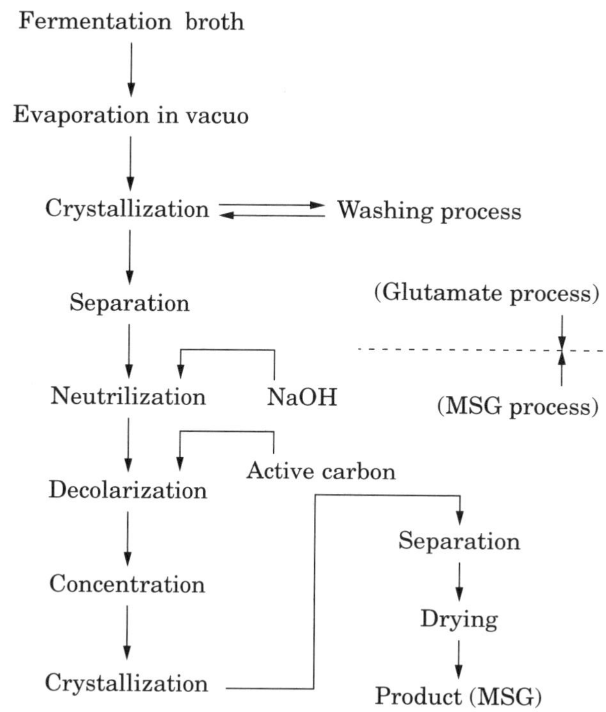

Figure 5. Process flow of purification of monosodium glutamate (MSG).

In the MSG step, ensuring high quality of the final product is essential, and improvement of the yield is required. To this end, both high purity of the GA crystals and efficient use of activated carbon must be attained. The resulting MSG-crystallized mother liquor, which still contains a high concentration of dissolved MSG, is sent back to the GA step to enhance the overall purification yield. In the whole process, from fermentation to final product, the GA-crystallized mother liquor and a decolorized cake of activated carbon are the main wastes discharged from the system. If the high quality of the product is to be ensured, it is apparent that the properties of the fermentation broth strongly affect the yield as a whole. Also, efficient use of the GA-crystallized mother liquor in view of utility and cost is an important factor in evaluating the process. Thus, in MSG production by the direct concentration and crystallization method, the process as a whole, from the selection of raw materials to the final product production, consists of linked steps that cannot be simply compared with other methods such as the resin methods. The purification yield of MSG from L-glutamic acid contained in the fermentation broth is estimated to be nearly 100% (78% in terms of ratio by weight and L-glutamic acid recovery). If the GA-crystallized mother liquor, which is rich in nutrient sources of plant origin and is used as an organic fertilizer, is efficiently used, this process as a whole constitutes an efficient, recycleable production system.

As is already mentioned, yield of L-glutamic acid is strongly affected by the ratio of this substance to the total solids in the fermentation broth. This ratio varies depending upon the sugar purity of the raw material used (ratio of sugar to total solids) and the conversion to L-glutamic acid based on sugar. Because sugar purity varies greatly, depending on the different raw materials used, selection of appropriate raw materials is important.

FUTURE PROSPECTS OF COMMERCIAL MONOSODIUM L-GLUTAMATE PRODUCTION

Genetic Recombination Technology

There have been many reports concerning the application of current genetic recombination technology to the construction of amino acid–producing bacterial strains. Among them are studies on *Escherichia coli*, for which there has been advanced analysis on the genetic level, and those strains of coryneform bacteria that are most widely used for amino acid production. As shown in Figure 1, the biosynthetic pathway of L-glutamic acid is relatively simple compared with that of other amino acids and requires strong assimilability of ammonia, which indicates that coryneform bacteria are preferred producers of this substance. Figure 6 shows one example of improving L-glutamic acid productivity using recombinants of coryneform bacteria. In this example, productivity was improved by introducing a variant-type gene of L-glutamic acid biosynthetase into *C. glutamicum*. Although the productivity did not reach a practical level, this suggests the possibility of constructing an industrially feasible strain by making further improvements to both the donor gene and receptor microorganism. Currently, success in the application of recombinant technology to construction of amino acid–producing strains has been reported in connection with amino acids that have a relatively complicated mechanism of metabolic regulation, such as the aromatic amino acids histidine and threonine. With respect to amino acids that are produced in good yields as a result of conventional construction method, such as L-glutamic acid, so far there have been no reports published on remarkable productivity improvements achieved by recombinant technology. However, there is still high technical potential, and drastic improvement in production may be expected in the future using this technology.

Strain		L-Glutamic acid produced (mg/dL)
No. 5707	(parent)	600
No. 12	(donor, glu^-)	0
No. 26	(donor, glu^-)	0
AJ 11566	(No. 12/pAM 286)	1,010
AJ 11567	(No. 26/pAM 286)	1,000

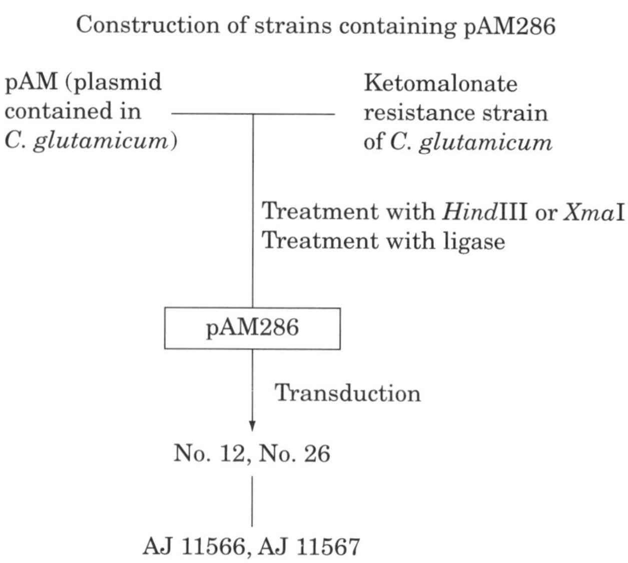

Figure 6. Glutamic acid production by recombinant *C. glutamicum*. *Source*: From Ref. 12.

Continuous Fermentation

The penicillin method batch-type L-glutamic acid fermentation gives only a short period of high L-glutamic acid productivity. Though continuous fermentation is desired to improve productivity, it is difficult to establish a continuous process with the currently available technologies. Studies on measures to stabilize the productivity, and establishment of the process with techniques other than the penicillin method are desired. Further improvement in productivity may be expected in combination with genetic recombination technology.

Replacement of Raw Materials: L-Glutamic Acid Fermentation with Petrochemicals

In the fermentation processes, the main raw materials are carbohydrate derived from molasses, corn, beet molasses, and so forth, with ammonia as a nitrogen source. Although industrial production has been carried out by fermentation with carbohydrate, the possibility of using petrochemicals including normal paraffin, ethanol, methanol, and acetic acid as fermentation raw material has been extensively studied (13–15). However, fermentation production using petrochemicals as a raw material has not been industrialized, except for partial use of acetic acid, because of the cost (16).

Acetic acid, derived from petrochemicals, may partially replace molasses. Glutamic acid–producing microorganisms from carbohydrate can easily utilize acetic acid if it is in low concentration, and acetic acid may be used in industrial processes if its concentration in the culture medium is kept at a low level. Acetic acid is supplied during the fermentation. With this method, although the cost of raw materials for culturing increases, the purification yield can be enhanced by increasing the ratio of L-glutamic acid to the total solid of fermentation broth. The feasibility must be evaluated with respect to the process as a whole.

Due to the rising cost of petroleum, L-glutamic acid production by fermentation from petrochemicals, particularly normal paraffin, is presently not practical, and the prospects for the future are not optimistic. However, this technology is discussed in this section because L-glutamic acid fermentation (and, similarly, fermentation to produce other amino acids) from hydrocarbons may provide one option for replacing raw materials. Naturally isolated, hydrocarbon-assimilating L-glutamic acid–producing microorganisms did not show a requirement for biotin, unlike those producing L-glutamic acid from a carbohydrate such as molasses; but they did require other vitamins for growth. Typical L-glutamic acid producers of the genera *Corynebacterium* and *Arthrobacter* had a requirement for thiamin. As in the case with fermentation using carbohydrate, L-glutamic acid was accumulated in the medium under the controlled concentration of the required vitamin, and the membrane permeability underwent a change when the microorganism was in the L-glutamic acid production phase. Also, the addition of penicillin or saturated fatty acids or their esters to the cells growing in the presence of a sufficient concentration of the required substance converted the cell from the growing phase to the L-glutamic acid production phase. Whereas carbohydrates are metabolized through glycosylation, hydrocarbons are metabolized through β-oxidization. However, subsequently, L-glutamic acid formation undergoes similar metabolic regulation in both processes.

With respect to the reaction processes, carbohydrates are used basically as an aqueous solution, whereas hydrophobic hydrocarbons separate into a layer of oil on water. Furthermore, oxidization of hydrocarbons requires much more oxygen compared with the oxidation of carbohydrates, although an advantage of the hydrocarbon process is that continuous feeding of the raw material, as required with carbohydrates, is not necessary, and, therefore, high-concentration culturing is possible in a batch culturing manner. The equipment for glutamic acid production must provide efficient dispersion of the oil to enlarge the surface area of the oil layer and also must provide efficient aeration and agitation to supply the large volume of oxygen required. Under optimum conditions, glutamic acid production from hydrocarbons is very similar to production from carbohydrate. There is no major difference between the two processes in the purification step, except that fermentation from hydrocarbons does involve steps for removing trace residual oil and cells.

Environmental Problems

The main raw materials, molasses, beet molasses, and glucose, are agricultural products. It is necessary to establish a production process that takes into consideration the possibility of recycling the wastes produced. The current industrial process uses the remaining fractions to produce organic fertilizers, and the process as a whole is designed to be a closed system.

BIBLIOGRAPHY

1. S. Kinoshita ed., *Hakko to Kogyo*, Dainihon Tosho, Tokyo, 1975, p. 125.
2. S. Kinoshita, S. Udaka, and M. Shimono, *J. Gen. Appl. Microbiol.* **3**, 193–205 (1957).
3. K. Yamada, J. Takahashi, and K. Kobayashi, *Agric. Biol. Chem.* **27**, 773–783 (1963).
4. K. Tanaka, K. Yamaguchi, and S. Kinoshita, *Amino Acid Nucleic Acid* **17**, 136–142 (1968).
5. H. Soda, K. Takinami, I. Chibata, K. Nakayama, and S. Yamada eds., *Amino Acid Fermentation*, Gakkai Shuppan Center, Tokyo, 1986, p. 201.
6. K. Sonntag, J. Schwinde, A.A. de Graaf, A. Marx, B.J. Eikmanns, W. Wiechert, and H. Sahm, *Appl. Microbiol. Biotechnol.* **44**, 489–495 (1995).
7. E. Kimura, C. Abe, Y. Kawahara, T. Nakamatsu, and H. Takuda, *Biochem. Biophysic. Research Communication* **234**, 157–161 (1997).
8. E. Kimura, C. Abe, Y. Kawahara, and N. Nakamatsu, *Protein Nucleic Acid Enzyme* **42**, 2633–2640 (1997).
9. H. Motoyama and H. Anazawa, *Kagaku to Seibutsu* **35**, 123 (1997).
10. T. Uemura and H. Soda eds., *Fermentation and Microorganisms*, Asakura Shoten, Tokyo, 1971, p. 53.
11. Y.C. Su and K. Yamada, *Bull. Agric. Chem. Soc. Jpn.* **24**, 69–74 (1960).
12. Tsuchida, K. Miwa, S. Nakamori, and H. Momose, Japanese Unexamined Patent Publication No. 148296/81, November 17, 1981.
13. K. Miyai, I. Tsuruo, R. Ogawa, S. Hayakawa, Y. Akimoto, and K. Goto, *Nippon Nogieikagaku Kaishi* **37**, 32–36 (1963).
14. S. Okumura, R. Tokawa, T. Tsunoda, K. Kohno, T. Matsui, and N. Miyaji, *Nippon Nogeikagaku Kaishi* **36**, 141–159 (1962).
15. Association of Amino Acid and Nucleic Acid ed., *Amino Acid Fermentation*, Kyoritsu Shuppan, Tokyo, 1972, p. 15.
16. T. Tsunoda, I. Shiio, and K. Mitsugi, *J. Gen. Appl. Microbiol.* **7**, 18 (1961).

PROFESSIONAL SOCIETIES, ASSOCIATION OF BIOMOLECULAR RESOURCE FACILITIES

RONALD L. NIECE
University of California–Irvine
Irvine, California

KEN WILLIAMS
Yale University
New Haven, Connecticut

CLAYTON NAEVE
St. Jude Children's Research Hospital
Memphis, Tennessee

OUTLINE

INTRODUCTION

The primary purpose of this article is to benefit the tens of thousands of individuals who either utilize biomolecular analyses and syntheses provided by resource laboratories or help to oversee, administer, and regulate the hundreds of biopolymer core laboratories that are dedicated to support research, quality control, and quality assurance programs. The core laboratories that provide these biopolymer analyses and syntheses are in turn supported by the Association of Biomolecular Resource Facilities (ABRF), an international, nonprofit organization of more than 300 core laboratories that is truly remarkable in terms of the ever-widening expanse of its research and outreach activities; the dedication, volunteerism, and extent of participation of its members; and the laudable goals to which it aspires. These are well summarized in its mission statement:

> A. To promote and support resource facilities, research laboratories, and individual researchers regarding operation, research, and development in the areas of methods, techniques, and instrumentation relevant to the analysis and synthesis of biomolecules.
> B. To provide mechanisms for the self-evaluation and improvement of procedural and operational accuracy, precision, and efficiency in resource facilities and research laboratories.
> C. To provide a mechanism for the education of resource facility and research laboratory staff, users, administrators, and interested members of the scientific community.

Following a brief history, which describes the genesis of bioanalytical core laboratories and the organization they engendered, the ABRF, are nine sections that chronicle the work of ABRF's research committees. These sections are followed by a section on the effective use of biomolecular resource facilities, and the article concludes with a prognostic look into the expanding role that core laboratories are likely to play in the postgenome era.

HISTORY AND ACTIVITIES OF THE ABRF

Ronald L. Niece
University Of California–Irvine
Irvine, California

The formation of shared biotechnological instrumentation laboratories in the 1980s provided a major and highly efficient mechanism to bring state-of-the-art biomolecular syntheses, analyses, and advice to tens of thousands of investigators at minimal cost. Growth in the number of biotechnological core laboratories and the parallel growth in the diversity of services they offer reflect the increasing sophistication of the protocols and instrumentation needed to conduct research in the life sciences. The ever-increasing complexity of the equipment increases the cost of conducting research in multiple ways. In addition to higher capital costs, operating expenses are higher due to high-purity custom reagents and increased service contract and maintenance costs. Specialized parts and highly trained instrument service representatives contribute to the cost of high-performance instrumentation. Laboratory staff costs are higher because of the need for experts in operation of the instrumentation and interpretation of the resulting data. The establishment of biomolecular resource laboratories has provided expensive and sophisticated technology at minimal cost.

Core laboratories began to evolve in the early 1980s from individual research laboratories and programmatic centers that were often equipped with an automated amino acid analyzer and/or spinning cup protein/peptide sequencer and that provided analyses on an informal or collaborative basis. This evolution was spurred on by the introduction and commercialization of the gas phase protein/peptide sequencer (1) and automated oligonucleotide synthesizers (2) and by the inauguration of the Shared Instrumentation Grant Program of the National Institutes of Health in 1981.

An awareness that many laboratories were providing services for other researchers (i.e., they were carrying out noncollaborative analyses on samples not prepared in their own laboratories) emerged at the Symposium of American Protein Chemists in San Diego, California, in 1985. This led to the informal association of Research Resource Facilities (RRF) (3), which met annually to exchange views on sample handling in the core facility setting, improvements in protocols, and advances in technology. A significant innovation of RRF was the inauguration of research committees devoted to analyzing instrumentation capabilities and to providing a realistic definition of state of the art in terms of what could be accomplished by the majority of the practitioners of the art. The first "unknown" sample presented to affiliates was a peptide designed to assess the performance of automated protein sequencers and amino acid analyzers (4). In 1989 the informal RRF group incorporated as the Association of Biomolecular Resource Facilities (ABRF). ABRF research committees based on other themes including peptide synthesis, mass spectrometry, nucleic acids chemistry, DNA sequencing, laboratory quality and compliance, and carbohydrate analysis were established subsequently (see following sections).

As it has been from the beginning, a major focus of ABRF is to establish realistic expectations of what can be accomplished with currently available technology in resource and research laboratories. The laboratories using the technology provide the best definition of requirements for sample preparation or design, instrument performance, and personnel proficiency. Since 1988 ABRF research committees have designed experiments and distributed test samples and syntheses for membership participation (see later sections) following the model described (4) for the 1988 sequence and amino acid analysis sample. An "unknown" sample or synthetic project that reflects a typical task is distributed to members with the recommendation that it receive no more nor less care than would any other request made to the laboratory.

The research committees analyze the resulting data and synthesized products as described in the sections that follow. The results help establish a useful assessment of the range of performance and capabilities on instruments in member laboratories. Because the data and products are submitted anonymously, individual core facilities cannot be identified. The anonymity helps ensure that samples and data returned are in fact representative of what happens on a day-to-day basis and allows member laboratories to compare their performance relative to that of their peers and relative to the level appropriate to their constituency. Unless a laboratory knows that it is performing below the average level defined by peers using similar instruments, reagents, and protocols, there is little incentive to improve. The best analyses and highest-quality synthetic products highlight the level of performance possible under actual operating conditions and identify the conditions needed to attain this high level of performance. Surveys of the composition, capabilities, and performance of core laboratories provide a complementary means of defining realistic expectations (5–7). In addition, survey data on operations and financial support and expenses of core facilities is collected anonymously as with the "unknown" samples.

Comparison of performance in terms of sensitivity, throughput, turnaround time, and so forth with the average provides a performance baseline that can serve as an incentive for performing as effectively as one's peers. The detailed financial data provided by the surveys have documented the economics of core facilities. Although the centralized laboratory with shared instrumentation provides considerable economies of scale and pooled expertise, the survey data demonstrate that only a small percentage are able to recover 100% of operating expenses from user fee income.

Results of the research studies are distributed directly to the members who participated and are also published (3–50). Originally, the research results were presented at an annual satellite meeting held in conjunction with the Symposium of the Protein Society. Today symposia, workshops, and tutorials are presented by the ABRF. Symposia on leading-edge technologies presented at national and international scientific meetings introduce the scientific community to the range of capabilities available in core facilities while highlighting the constraints on sample preparation or design required for success. Because ABRF consists of the users of the technology, the workshops emphasize the practical tools needed for improved and high performance. The tutorials are popular and effective tools for helping core facility staff expand their repertoire of capabilities and increase their level of expertise.

The research undertaken by the ABRF has been funded by federal grants, corporate sponsors, institutional contributions, and membership dues. These funds are leveraged by voluntary contributions of individual research committee members, their laboratories, and their institutions. Research committees consist of volunteer ABRF members with a designated liaison from the executive board. ABRF is an incorporated, nonprofit organization led by an elected, volunteer executive board. Business activities are handled by the Federation of American Societies for Experimental Biology (FASEB) business office (ABRF, 9650 Rockville Pike, Bethesda, MD 20814-3998; 301-571-8300; FAX: 301-530-7049; E-mail: *abrf@faseb.org*). Official ABRF business is conducted at an annual business meeting.

The ABRF Corporate Sponsorship program provides a unique opportunity to enhance communication between the manufacturers and suppliers of biotechnological instrumentation and the laboratories that depend on this equipment. ABRF research activities help define how effective the instruments and protocols are in the field, the level of training required for their optimum use, and what happens to this performance as the instruments age. Identifying performance levels, new applications, and unforeseen problems that accompany the running of "real" samples that were not anticipated by manufacturers promotes improvements and corrective actions. ABRF surveys of satisfaction with instrument performance, service support, reagent and supply costs, and application support offer unbiased group feedback from the day-to-day users who best understand the instrumentation and protocols.

Since 1994 the ABRF has presented an annual award for outstanding contributions to the development of technologies used in core facilities. An independent, anonymous panel of respected scientists presents the ABRF executive board with candidates for this award. This ABRF award recognizes contributions to science through technological advances and is accompanied by $10,000 and an opportunity to address the ABRF members at their annual meeting. In 1994 and 1995 Beckman Instruments funded the awards; in 1996 Hewlett-Packard began sponsoring of the award. The first award was presented to Dr. Frederic Sanger, the winner of two Nobel prizes, for his development of methods for sequencing proteins (chemistry, 1958) and for sequencing DNA (chemistry, 1980). In 1995 Dr. Klaus Biemann was honored for his contributions in the development of mass spectrometry for analysis of biomolecular polymers and for his training of mass spectrometrists. Dr. David Lipman received the 1996 award for developments of computational methods for analysis of protein and DNA sequences. The 1997 award was presented to Dr. Lloyd Smith for his contributions to the advancement of automation in DNA sequencing. The 1998 award went to Nobel laureate (chemistry, 1984) Dr. Bruce Merrifield for developing the concept and strategies for solid-phase synthesis of peptides, which opened the way for peptide and other organic syntheses to become routine.

ABRF educational initiatives target member laboratories, the scientific community, which represents its user base, the manufacturers of instrumentation and reagents used in core facilities, and the administrators and decision makers who deal with science, scientists, and the delivery of research resources. To fulfill the educational needs, ABRF promotes several activities beyond conducting and publishing research reports. Since 1986 symposia have been organized in conjunction with the Protein Society and the International Congress of Biochemistry and Molecular Biology and joint symposia have been presented at the FASEB and American Society of Biochemistry and Molecular Biology (ASBMB) annual meeting. Workshops and symposia have been presented at meetings of the American Peptide Society, American Society of Cell Biology, American Society of Mass Spectrometry, European Protein Society, International Genome Sequencing and Analysis Conference, and the Methods in Protein Structure Analysis. From 1986 to 1998 the ABRF organized workshops and symposia at 32 different meetings and held three independent meetings, each of which were attended by 500 to 1,000 individuals.

Another significant medium for information exchange and education of members has been the newsletter, *ABRFnews*. The newsletter contains peer-reviewed articles including summaries of meetings, methods and review papers, and laboratory tips. A new dimension in publication of methods in biomolecular research began in 1997 with the launch of the electronic *Journal of Biomolecular Techniques: The Official Methods Journal of the ABRF (http://www.abrf.org/JBT/JBT.html)* edited by John Shively. Beginning in 1998, *ABRFnews* and the electronic journal were merged into one under the title *Journal of Biomolecular Techniques*, with Clayton Naeve as the editor in chief. To maintain the advantages of the electronic journal, accepted rapid communications are immediately posted on the Web site followed by reprinting in the next issue of the journal. Electronic communications using an E-mail bulletin board (*abrf@aecom.yu.edu*) connect more than 900 members around the world; the novice and expert can almost instantaneously assess instrument and reagent problems, discover if other laboratories have similar concerns, and help one another design experimental protocols. The ABRF Web site (*www.abrf.org*) contains written protocols distributed at workshops and tutorials, the results of many research studies, and comprehensive archives of the discussions on the E-mail bulletin board. The Web site averages more than 100 inquiries per hour to one or more documents.

The concept that developed into ABRF began with six scientists in 1986. In 1998 there were more than 900 members listed in the ABRF directory. Members reside in 26 different countries on six continents. Three-quarters (76%) are in the United States, from 40 different states, half (55%) are in academic institutions, 13% in research institutions, 28% in industry, and the balance in government laboratories. Although these numbers are illustrative, they fail to adequately convey the underlying reason for the success of ABRF: The ABRF is uncommon in terms of the unrelenting volunteer spirit and commitment of its members and its single-minded purpose of helping to bring state-of-the-art biotechnology to bear on the myriad array of biochemical and biomedical challenges that are its reason for being. It is unique in its multipronged, targeted research by its research committees, which study technical capabilities, current protocols, and emerging technologies.

AMINO ACID ANALYSIS

DANIEL J. STRYDOM
Bio Nebraska, Inc.
Lincoln, Nebraska

JOHN W. CRABB
Cleveland Clinic Foundation
Cleveland, Ohio

Amino acid analysis (AAA) is one of the major analytical techniques used in the biochemical and biotechnical environments. It is the method of choice for reliable quantification of peptides and proteins and also provides useful qualitative information in the evaluation of synthetic peptides and recombinant proteins, the identification of proteins based on composition, and the detection of unusual amino acids. AAA also has applications in protein sequence studies, such as in the design of fragmentation strategies, the identification of amino- and carboxy-peptidase reaction products, and the quantification of samples subjected to Edman degradation to detect those with blocked N-termini. In addition, AAA is used for the characterization of feed and food samples and physiological fluids.

Two basic approaches to quantitative AAA receive widespread use, namely (*1*) classical postcolumn derivatization followed by ion exchange chromatography and (*2*) precolumn derivatization followed by reverse-phase HPLC. Excellent instrumentation and more than one method is available for these basic approaches. Because AAA applications are of such varied character, no single methodology can be shown to be superior in all respects. The technology is deceptively simple but actually is one of the more complex and difficult in protein chemistry. The greatest benefit of AAA is its quantitative capability; however, measuring many different amino acids with quantitative accuracy is precisely what makes the analysis so difficult. AAA is a statistical type of measurement; amino acids vary widely in their stability/lability to acid hydrolysis, oxidation, and modification. Trp and Cys are destroyed by HCl hydrolysis, Ser and Thr are partially destroyed, Ile and Val are slow to cleave, Met is subject to oxidation, and Asn and Gln are completely deamidated. Variable, but ubiquitous, background (most frequently Gly, Ala, and Ser) plus variable instrument and analyst performance can confound the analysis, particularly at and below the low picomole level. As a consequence, realistic expectations concerning the accuracy and precision of AAA must be tempered with an appreciation for the difficulties of the technology.

The annual ABRF AAA studies between 1988 and 1997 provide a broad performance comparison of AAA methods and instruments (Table 1). In addition to the reports in Techniques in Protein Chemistry (Table 1), the ABRF AAA studies have been highlighted in the *ABRFnews* and reported elsewhere (12,13,49). The number of participating

Table 1. Summary of ABRF Amino Acid Analysis Studies, 1988–1996

Year	Sample description	Number of participants	Sample amount[a] (μg)	Average % error[b]	Study focus	Reference
1988	Synthetic peptide (STD-2)	45	5	19.3	Accuracy and sensitivity	4
1989	β-Lactoglobulin (89AAA)	17	0.2	25.3	Accuracy, precision, sensitivity	10
		26	5.0	13.9	Hydrolysis methods; and Cys quantification	
1990	Myoglobin control (90AAA1)	41	1.3	13.5	AAA in the presence of buffer salts and on PVDF membrane	15
	Myoglobin + salt (90AAA2)	35	1.3	19.7		
	Myoglobin on PVDF (90AAA3)	30	0.9	26.9		
1991	Bovine serum albumin (91AAA)	51	150	9.5	Accuracy with unlimited sample; Trp and Cys quantification	18
1992	Chymotrypsin (92AAA)	59	56	10.5	Trp and Cys quantification	21
1993	Synthetic peptide (93AAA)	50	23	12.5	Phosphoamino acids and hydroxyproline analyses	26
1994	RNase + angiotensin (94AAA1)	58	30	10.9	Hydrolysis and chromatography methods; cys and glucosamine analyses	30
	Prehydrolyzed 94AAA1 + glucosamine (94AAA2)	60	30	6.5		
1995	Myoglobin on PVDF (95AAA)	77	30	21.4	AAA on PVDF membrane	35
1996	Triose phosphate isomerase (96AAA)	71	10	11.9	Protein identification from AAA databases	48
1997	Bovine serum albumin, lysozyme on PVDF (97AAA)	40	1 5 10	30 19 19	AAA on electroblotted PVDF Protein identification from AAA	49

[a]Sample amounts provided since 1991 were meant to be adequate for all methods and not to restrict performance.
[b]Average percentage compositional error for each study is shown for all methods combined without results for Cys, Trp, or unusual residues.

laboratories per year has varied from about 40 to 77 (Table 1); approximately 38% of the participants since 1989 have used postcolumn ninhydrin methodology, approximately 48% have used precolumn phenylthiocarbamyl (PTC) methods, and 14% have used other AAA methods. The ABRF AAA studies are crafted to examine some aspect of the AAA technology to help participating laboratories evaluate and improve their capabilities; participation is organized in an anonymous and noncompetitive manner. In general, these studies demonstrate that more than one AAA method works well and that the skill of the analyst is probably the most important determinant of success. More specifically, these collaborative trials have evaluated AAA performance with high and low amounts of sample, in the presence of salt, on PVDF membrane, with emphasis on difficult residues (e.g., cysteine/cystine, tryptophan, and phosphoamino acids), with special focus on hydrolysis and chromatography conditions, or with attention to protein identification methods based on amino acid composition data. Each year, these studies have published the average levels of AAA performance and also typically spotlighted the highest-quality analyses attained with different AAA methods.

The overall AAA accuracy in the past 10 years of ABRF studies has ranged from 6.5% to 30% average compositional error per study (Table 1). Accuracy is strongly dependent on the type and amount of sample provided, hydrolysis conditions, and level of background contamination during the analysis. The ninhydrin-based postcolumn analyses require more sample (typically 5–10 μg per analysis) than precolumn derivatization–based techniques (0.5–1 μg per analysis). Each year a broad range of accuracy is apparent, including some analyses that are grossly wrong and others that are exceptionally good (e.g., 3–5% average error). The studies indicate that knowledgeable and skillful analysts can expect high-quality results with approximately 10% average compositional error from more than one AAA method.

Instrument calibration and peptide/protein hydrolysis to individual amino acids are critically important for high-quality AAA. General parameters for sample hydrolysis have been considered frequently in the ABRF studies (e.g., vacuum, temperature, acid additives, liquid acid versus acid vapor), and two trials have specifically evaluated the hydrolysis process by providing both a hydrolyzed and an unhydrolyzed sample for analysis (10,30). Study results clearly demonstrate the variability of the hydrolysis process from laboratory to laboratory (30) and emphasize the importance of controlling this step. Vapor-phase acid hydrolysis has become the most popular methodology over the course of these trials. Instrument calibration has been examined also, and results demonstrate that replicate analysis of amino acid standards such as Pierce standard H provided superior results, whereas hydrolyzing the standard appears to be disadvantageous (18).

Quantification of cysteine is difficult because hydrolysis in constant-boiling HCl destroys cysteine and partially destroys the disulfide-linked, dimeric form of the residue, cystine. Tryptophan is destroyed by routine HCl hydrolysis and is even more challenging to quantify than cysteine. Cys is best analyzed by various pretreatments of the sample, such as oxidation, reduction and alkylation, or perhaps optimally by adding disulfide exchange reagents during acid hydrolysis (21). Trp appears best analyzed by adding dodecanethiol to the hydrochloric acid used for hydrolysis (21).

Rare and modified residues are of continual interest to many investigators. Phosphoamino acids and hydroxylated and glycosylated residues have been examined in the ABRF collaborative trials; such analyses are not routine for most laboratories. No current AAA methodology allows quantitation of phosphoamino acids in a single analysis in conjunction with all the common amino acids (26). Optimal hydrolysis conditions must be determined for each sample; however, precolumn analyses such as the phenylthiocarbamyl (PTC) method were more effective in the chromatographic resolution of phosphoamino acids than the postcolumn ninhydrin method (26). In analyses of the 93AAA sample, 41% of the facilities detected an unknown amino acid and 75% of these correctly identified it as hydroxyproline. Glucosamine in the 94AAA2 sample was detected by only 13% of the participants, although all had been alerted to its possible presence.

Comparison of ninhydrin and PTC methodologies in the ABRF AAA studies reveals statistically significant differences. For example, the interlaboratory variability of Arg, His, Leu, Lys, Met, Pro, and Val is dramatic between these two techniques, perhaps due to chromatography issues and reagent-related peaks. Pro is the only of these amino acids that appears to be analyzed more consistently and accurately by PTC than by ninhydrin methods, primarily due to its low color yield with ninhydrin. Met is underestimated by ninhydrin sites, probably due to oxidation and to chromatographic disturbances, and it is overestimated by PTC sites. The latter can be due to overlap of the PTC Met peak with reagent-related peaks. The oxidation of Met leads to high Arg values in PTC analyses due to coelution of the oxidation product with Arg, whereas ninhydrin methods frequently underestimate Arg, most likely due to the very broad chromatography peak and consequent integration problems. Other residue-specific problems were addressed in the 1991 study (18), including low values for PTC Asp and PTC Glu caused by metal contamination from glass hydrolysis tubes that often can be remedied by drying the samples with EDTA prior to hydrolysis. Although more than one ABRF AAA study has shown precolumn derivatization techniques more sensitive than ninhydrin methods, the 1990 study demonstrated that postcolumn derivatization methods generally provided better AAA results in the presence of buffer salts.

Two ABRF studies have evaluated AAA on PVDF membranes (15,35), and both resulted in compromised accuracy (21–27% average error) relative to nonimmobilized samples. The 1995 study further demonstrated significantly high backgrounds on blank PVDF membranes (35), suggesting that perhaps the higher error from PVDF immobilized samples is due in part to background contamination that is difficult to extract from the membranes. The 1997 ABRF AAA study readdressed AAA on PVDF by probing whether PVDF electroblotted samples exhibit less background contamination. Results in 1997 indicated that the average error for samples blotted to PVDF was still about twofold higher than for solution samples (49).

The 1996 ABRF AAA study (48) demonstrated that 90% of participating laboratories could correctly identify an unknown sample protein (triose phosphate isomerase) from their AAA data (11.9% average compositional error) using the World Wide Web and either ExPASy *(http://expasy.hcuge.ch/ch2d/aacompi.html)* or PROPSEARCH *(http://www.EMBL-heidelberg.de/aaa.html)* database computer query programs. Results from the 1997 AAA study on PVDF membrane suggest that correct protein identifications can be made with AAA data containing up to 15% average error. In the years to come, AAA will likely become a more widely recognized and valuable tool for the identification of proteins.

PROTEIN SEQUENCING

JOSEPH FERNANDEZ
Rockefeller University
New York, New York

KATHRYN STONE
Yale University
New Haven, Connecticut

The ABRF Protein Sequence Research Committee focuses primarily on providing member laboratories with "unknown" protein/peptide samples that are representative of samples routinely submitted for protein sequence analysis. Participating laboratories characterize the primary structure of the protein/peptide, or the sequence, typically from the amino-terminal end of the sample utilizing automated Edman degradation chemistry.

Chemistry and Sample Preparation

The chemistry used during Edman degradation (51) consists of consecutive cycles; one cycle contains the following steps: (*1*) coupling of phenylisothiocyanate (PITC) with the α-amino group or amino-terminal amino acid of the protein/peptide at pH 9–10 to form a phenylthiocarbamyl (PTC) group, (*2*) removal of chemistry by-products using organic washes, (*3*) cleavage with anhydrous trifluoroacetic acid (TFA) to generate a protein/peptide minus the first amino acid (or N-1 protein/peptide, which contains a free α-amino group for a subsequent cycle of Edman sequencing), along with an anilinothiazolinone (ATZ) derivative of the released amino-terminal amino acid, (*4*) extraction of the ATZ derivative from the N-1 protein/peptide, and (*5*) conversion of the ATZ derivative to the more stable phenylthiohydantoin (PTH) form. The PTH amino acid derivative can then be identified by its retention time on reverse phase HPLC, during which amino acids, chemistry artifacts, and derivatized amino acids are resolved from each other. Figure 1 shows chromatograms distributed to ABRF members as a sequence assignment exercise (46) and represent an example of a PTH amino acid standard HPLC profile, along with Edman cycles showing residues 1–7 of the peptide. Cysteine is destroyed during the Edman chemistry and thus is usually reduced and alkylated before automated sequence analysis (52).

The performance and capabilities of automated Edman degradation chemistry are generally characterized by initial yields, repetitive yields, and lag (carryover). Initial yield refers to the recovery of the first amino acid, typically in picomoles, and is generally expressed as a percentage of the total picomole amount of protein/peptide analyzed (typically 50–80%). Repetitive yield provides a measure of the average amino acid recovery after each cycle. Although it usually ranges from 90 to 99%, repetitive yield also varies with sample and instrumentation. Lag or carryover means the yield, expressed as a percentage, of the amino acid that appears in the subsequent cycle ($n + 1$) relative to the actual cycle (n). Repetitive yield and lag indicate the degree of completion of the Edman chemistry and generally determine how far a protein can be sequenced before the signal becomes indistinguishable from the noise.

Edman degradation is only possible when the protein/peptide has a free amino (N) terminus. If a naturally occurring or other chemical group is present at the N-terminus, no data can be obtained and the sample is considered blocked. Some of the more common naturally occurring blocking groups are acetylated amino acids, formylated methionine, and pyroglutamic acid. Generally, the percentage of naturally blocked proteins isolated from an organism increases along with the complexity of the organism. Hence, while only a few bacterial proteins are blocked, approximately 80% of soluble proteins isolated from a mammalian tumor cell line appear to be blocked (53). Artificial blockages can occur during electrophoresis, but certain precautions, such as using high-purity reagents, can be taken to help decrease this type of blocking (54). A lack of sequence or PTH amino acid signal can also be due to an insufficient amount of protein/peptide used during sequencing. The reason for a lack of sequence (a blocked amino terminus or an insufficient amount of protein/peptide) can often be inferred by performing hydrolysis and an AAA (see previous section) of the sample and its sequencing support, after sequencing.

Samples for sequence analysis by Edman degradation are generally prepared in one of three forms: (*1*) in solutions that often contain nonvolatile buffers (phosphate, Tris, SDS, etc.), (*2*) in volatile buffers, such as the trifluoroacetic acid/acetonitrile solutions used to elute reverse-phase HPLC columns, and (*3*) noncovalently bound (via electrophoretic transfer) to PVDF membranes (55). All samples in these forms can be directly sequenced, although certain instruments are better designed for one sample form over another.

ABRF Protein Sequence Studies

Over the past 10 years, the ABRF Protein Sequence Research Committee has distributed annually an unknown amino acid sequencing sample, along with a survey, to member laboratories. The purpose of these studies is to provide participating laboratories with a mechanism for evaluating their own sequencing capabilities, particularly as new instruments and techniques are introduced. Because many laboratories are involved in these studies (the average number of participating laboratories is 69; see Table 2), they provide a reasonable approach to determine realistic expectations regarding the average amount of protein/peptide that can be routinely sequenced. In addition, the quality or accuracy of the resulting data can be determined because the identity of the sample was unknown to the participants. Each study has been designed with a particular emphasis: sensitivity of sequencing (STD-1, ABRF-93SEQ, and 97SEQ), posttranslational modification (ABRF-92SEQ), heterogeneity (ABRF-89SEQ, ABRF-95SEQ, ABRF-96SEQ, and ABRF-97SEQ), protein bound peptides (which mimic large proteins) on PVDF or in solution (ABRF-90SEQ and ABRF-91SEQ), identification of problematic residues such as cysteine and tryptophan (ABRF-94SEQ, ABRF-95SEQ, and ABRF-96SEQ), and distance sequenced (ABRF-95SEQ). In addition, emphasis is placed on positive correct calls or positive accuracy, which is defined as the number of correct positive calls per total number of positive calls. A summary of these studies is found in Table 2.

The amount of sample supplied in these studies has ranged from a high of 500 pmol (ABRF-92SEQ) to a low of 10 pmol (ABRF-97SEQ). In the first study (STD-1, 1988 [4]), 100 pmol of a unique 40-residue synthetic peptide was supplied. The positive accuracy for this study was quite good, with ~95% correct. Because real samples often contain some heterogeneity, the following study (ABRF-89SEQ [8]) contained a mixture of two unique synthetic peptides at a 5:1 molar ratio (248:48 pmol). The lengths of these peptides were similar to the STD-1 sample, and the positive sequence calling accuracy for the major sequence was again ~95%. This indicated that the participating laboratories were quite good at differentiating a major sequence from a minor sequence when the abundance of the minor sequence was ~20% of the major. Furthermore, it provided a guideline of about 80% for the minimal requirement of purity for samples destined for N-terminal sequencing. A sequence mixture was also provided as a sequence assignment exercise in the ABRF-96SEQB (46) study, where participating laboratories received the HPLC chromatograms of a sequence mixture in place of an actual sample. The positive accuracy for the major, 10-pmol, 22-mer sequence was still 95%. In this study, the accuracy for the minor 14-mer sequence, which was at the 2-pmol level, was also determined and found to be 85%. This lower sequence calling positive accuracy most likely resulted from either the very low amount of sample sequenced or the fact that it was a minor component.

The ABRF-93SEQ (25) sample had the same composition and length as the STD-1 sample. However, only 50 pmol of ABRF-93SEQ was supplied as compared with the 100 pmol of the STD-1 sample that was distributed for the 1988 study (Table 2). A comparison of several sequencing performance parameters (positive sequence calling accuracy, number of positive correct assignments, and average

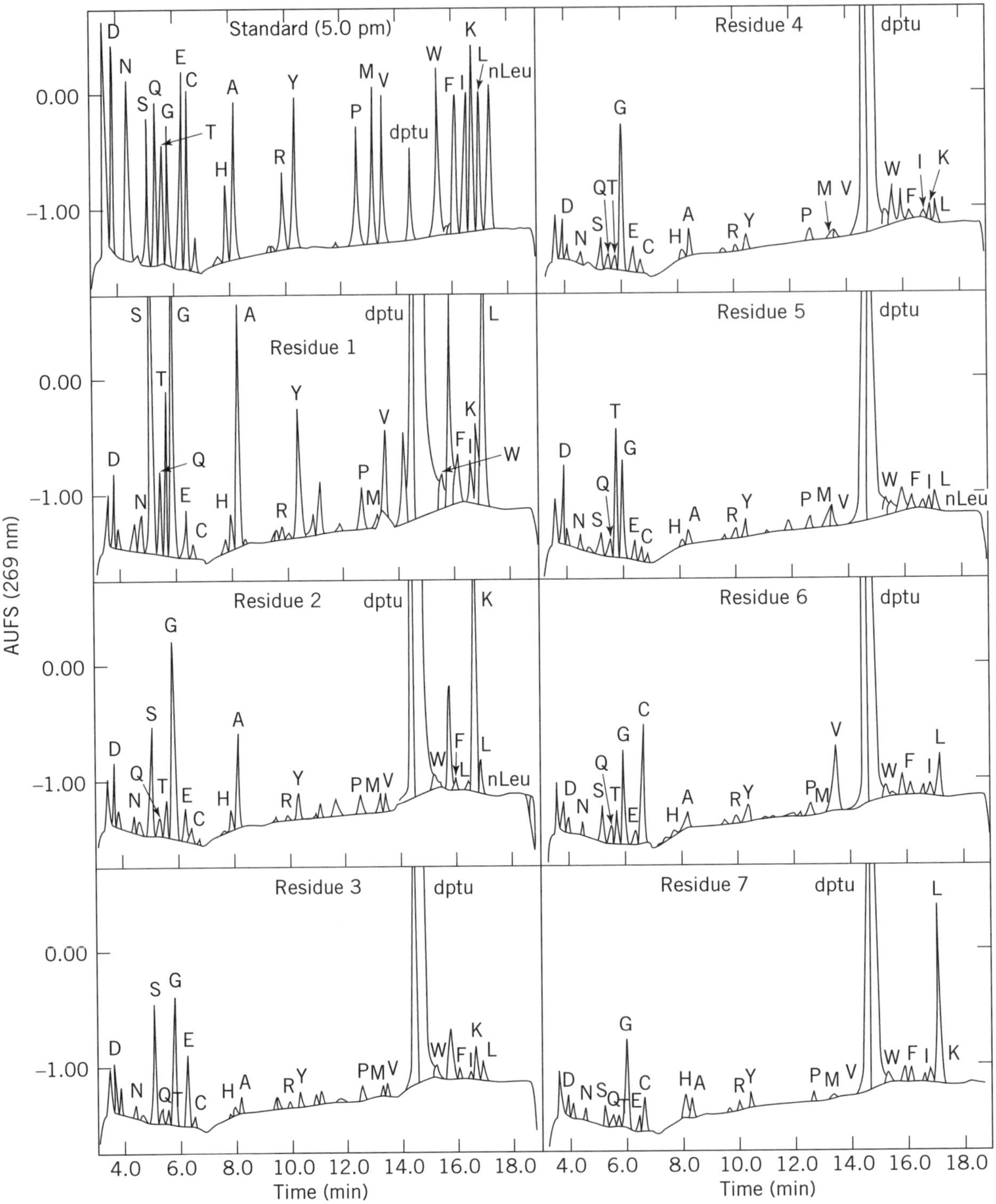

Figure 1. PTH chromatograms from ABRF-96SEQ Dataset B that was distributed as a sequence assignment exercise. Cysteine was present in both the standard and sample as the carboxamidomethylated derivative. The derivative nLeu represents PTH norleucine and is present only in the standard, and diphenylthourea (dptu) is an Edman chemistry by-product. The sequence of the major peptide indicated was leucine-lysine-serine-tryptophan-threonine-cysteine-leucine (LKSWTCL) and the minor was tyrosine-alanine-glutamic acid-glycine-aspartic acid-valine-histidine (YAEGDVH).

Table 2. Summary of ABRF Protein Sequencing Studies, 1988–1997

Sample	Sample description	Number mailed	Number responses[a]	Amount sent (pmol)	Positive accuracy (%)	Reference
STD-1	40-residue synthetic peptide	103	47 (47%)	100	~95	4
ABRF-89SEQ	40-residue synthetic peptide (major)	123	49 (40%)	240	~95	8
	43-residue synthetic peptide (minor)			48		
ABRF-90SEQ	29-residue synthetic peptide covalently attached to acetylated transferrin (90%) bound to PVDF	132	56 (42%)	30	83	14
ABRF-91SEQ	29-residue synthetic peptide dried with acetylated transferrin (90%)	198	90 (45%)	80	83	16
ABRF-92SEQ	37-residue synthetic peptide with two unusual amino acids	225	74 (32%)	500	94	22
ABRF-93SEQ	40-residue synthetic peptide with same composition as STD-1	274	80 (29%)	50	91	25
ABRF-94SEQ	Lactoferrin to assist sequencing labs in identification of cysteine and tryptophan	258	78 (30%)	50	95	31
ABRF-95SEQ	Recombinant protein to determine maximum length of sequences that can be obtained; sample also contained four heterogeneous positions	252[b]	71 (28%)	45	78	37
ABRF-96SEQA	Sequence calling exercise (single sequence)	210[b]	95 (45%)	40	99.8	46
ABRF-96SEQB	Sequence calling exercise (major)	210[b]	95 (45%)	10	96	46
	Sequence calling exercise (minor)			2	85	46
ABRF-97SEQ	21-residue synthetic peptide (major)	215	50 (23%)	10	92	
	14-residue synthetic peptide (minor)			2	72	

[a]Percentage based on number of responses divided by number mailed.
[b]Samples for ABRF-95SEQ, ABRF-96SEQ, and ABRF-97SEQ were mailed only to facilities that perform protein sequencing.

repetitive yield) between the STD-1 and ABRF-93SEQ studies indicated there was no significant improvement in sequencing performance over this 5-year period. Again, however, the amount of peptide provided in the ABRF-93SEQ study was half that provided in the STD-1 study.

To better study the factors that affect sequence calling accuracy, the ABRF-96SEQA and ABRF-96SEQB (46) studies were sequence calling exercises. In each case, the PTH chromatograms were supplied to participants rather than an actual sample to be sequenced. The ABRF-96SEQA sample was at a relatively high level (40 pmol) and had the highest degree of positive correct calls for any ABRF sequence study, with an accuracy of 99.8%. This study indicated that PTH chromatograms generated from straightforward, high-level samples with no heterogeneity can be used to assign a positively called sequence with a very high level of confidence. ABRF-96SEQB was a sequence mixture as described previously, and here the positive sequence calling accuracy was 96%. ABRF-97SEQ also contained a mixture of two peptides and was similar to ABRF-96SEQB in quantity (~10:2 pmol ratio) and peptide composition. The ABRF-97SEQ study, which involved member laboratories running the sample and interpreting the resulting data, was designed to be comparable with the ABRF-96SEQB sequence calling exercise. The positive sequence calling accuracy of ABRF-97SEQ major was 92%, which was slightly lower than the ABRF-96SEQB major (96%). However, a comparison of the positive sequence calling accuracy obtained from ABRF-97SEQ samples run by the committee using the same instrumentation as the ABRF-96SEQB analysis shows the same positive sequence calling accuracy (96%). Thus, the positive sequence calling accuracy appears to be more dependent on instrumentation (model and upkeep) and sample handling and loading than on data interpretation.

ABRF-90SEQ (14) differed from all of the other studies in that this sample was adsorbed to a PVDF membrane. A major advancement in Edman sequencing was realized when Matsudaira (55) discovered that proteins could be sequenced from sodium dodecyl sulfate polyacrylamide gel electrophoresis (SDS PAGE) after electroblotting to a PVDF membrane. Sequencing from PVDF membranes propelled SDS PAGE from an analytical tool to a preparative procedure; it is still the method of choice for the final purification of proteins destined for N-terminal and internal (see next section) sequencing. The goal of the ABRF-90SEQ sample was to mimic a large electroblotted protein. Hence, a 29-residue synthetic peptide was covalently attached to an acetylated transferrin, mixed with yeast glucose-6-phosphate isomerase, and adsorbed onto a PVDF membrane. The sequence identified for this sample achieved only an 83% accuracy with respect to positive calls. This decrease from ~95% positive accuracy for previous studies may have been due to the difficulty of the sequence, the PVDF support, or background from the acetylated protein.

Therefore, the following study, ABRF-91SEQ (16), used a sample similar to ABRF-90SEQ. In this study, the 29-residue synthetic peptide was not covalently attached to the transferrin, and the sample (also containing the glucose-6-phosphate isomerase) was supplied in a dried form, which was dissolved before loading onto the sequencer. The positive sequence calling accuracy was also 83% for this study, and thus the PVDF appeared to not have influenced the ability to call this sequence. The lower sequence calling accuracy in these two studies was probably due to the sample itself, which demonstrates nicely the dependence of the quality of the sequence data on the sample. In this instance, the lower sequence calling accuracy for the ABRF-91SEQ (83%) as compared, for instance, with the STD-1 sample (95%) may well be attributed to the increasing background that would be expected to accompany sequencing of the relatively high molecular weight (~80,000) transferrin protein that was in the ABRF-91SEQ sample. That is, the small fraction of nonspecific cleavage that occurs during each cycle of Edman degradation becomes more significant as the size of the protein (and thus the probability that it will contain an increased number of acid labile peptide bonds) is increased.

Two other challenges faced in amino acid sequencing are identification of posttranslational modifications and identification of difficult residues. ABRF-92SEQ (22) contained two posttranslationally modified amino acids—hydroxyproline and phosphoserine. Hydroxyproline was identified by 45% of the study participants using Edman sequencing alone, whereas only 13.9% were able to identify phosphoserine. Participants were encouraged to use amino acid composition, mass spectrometry, capillary zone electrophoresis (CZE), or other techniques to properly identify these residues. However, very few participants used techniques other than Edman degradation for sequence identification.

ABRF-94SEQ (31) focused on tryptophan and cysteine identification, which tend to be problematic because PTH cysteine is destroyed during Edman chemistry, and PTH tryptophan, recovered in low yields, often elutes on or near an Edman chemistry by-product, diphenylurea. Tryptophan identification in this study was somewhat improved (86%) over several previous studies (~70%), possibly because of the addition of isopropanol to solvent B. The isopropanol helps to separate the diphenylurea from tryptophan. Because cysteine degrades during Edman sequencing to dehydroalanine, it is usually necessary to reduce and alkylate the cysteines in the sample before sequencing. The ABRF-94SEQ study provided participants with seven different techniques that could be used to modify cysteines, with the hope of improving cysteine (Cys) identification. These techniques included cysteine modification to carboxymethyl-Cys, carboxamidomethyl-Cys, pyridylethyl-Cys, Cys-S-propionamide, Cys-S-dimethylpropionamide, *N*-isopropylcarboxyamidomethyl-Cys, and aminopropyl-Cys. Cysteine identification improved from ~53% in the ABRF-93SEQ sample to 82% in the ABRF-94SEQ sample.

In the ABRF-96SEQB (46) study, the cysteines were carboxamidomethylated with iodoacetamide by the committee, the sample was sequenced by the committee, and the PTH HPLC chromatograms were sent out to participating laboratories. In this study, 99% of the cysteines were identified correctly. This suggested that some difficulty was being encountered by facilities in alkylating the sample, willingness to alkylate the sample, or separating the PTH cysteine derivative from the Edman chemistry by-products or other PTH amino acids.

The ABRF-97SEQ study attempted to address this issue by alkylating the cysteines to Cys-S-propionamide using acrylamide before distributing the sample. This test sample contained a 10:2 pmol mix of two synthetic peptides and was the lowest amount of sample provided thus far in these studies. Two controls were also included with the sample, a PTH-Cys-S-propionamide standard and an internal sequencing standard. The PTH-Cys-S-propionamide standard was included so that separation of this derivative could be obtained before sequencing the sample. The cysteine calling accuracy for the ABRF-97SEQ sample was 88% for the cysteine in cycle 4 and 97% for the cysteine in cycle 13. Thus, although the positive sequence calling accuracy for cysteine was not quite as high in the ABRF-97SEQ study (Cys-13 = 97%) as in the ABRF-96SEQ study (Cys-13 = 99%), there was definite improvement over the ABRF-95SEQ study (Cys-15 = 63%). The 17-residue internal sequencing standard peptide (56) was composed of norleucine in cycles 1, 6, 11, and 16 and succinylated lysine in the other positions. Because this peptide is composed of nonnaturally occurring amino acids, it can be mixed with "unknown" sequencing samples. This allows independent and continuous monitoring of the sequencer's performance during all sequencing runs and rapid detection of instrument problems. The low amount of the ABRF-97SEQ sample was expected to be a challenge for the participating laboratories. However, based on the positive sequence calling accuracy for the 50 participating laboratories, the accuracy (92%) was not significantly decreased even for this low amount of sample (see Table 2).

In the ABRF-95SEQ (37) study, the sample was a recombinant 20K protein that contained microheterogeneity (2:1 molar ratio) in positions 6–9. The purpose of this sample was to assess participating laboratories' abilities to recognize microheterogeneity and to determine the length of sequence assignment that is possible. The longest correct sequence observed was 62 residues, including 1 cysteine and 2 tryptophans. The best responses (based on number of residues sequenced and sequence calling accuracy) were obtained on new-generation sequencers, specifically the Hewlett Packard G1005A and Perkin Elmer/Applied Biosystems Division 494 sequencers. Hence, this study suggests that these new instruments may result in improved performance. The overall average positive sequence calling accuracy on this sample was only 78%, which increased to 80% if the four cycles of microheterogeneity were excluded. The majority of the errors occurred in the later cycles and could have been minimized by terminating the sequence earlier, which illustrates another common problem, overcalling of sequences. Basically, the confidence of sequence calling decreases as the sequencing run progresses due to both loss of signal and (especially with relatively high molecular weight proteins) increased background.

Instrument Technological Improvement/Design over the Course of ABRF Studies

The first automated protein sequencer, based on Edman degradation chemistry, was designed by Edman and Begg in 1967 (57) and was manufactured by Beckman. This instrument, a Series 890 Beckman Automated Protein/

Peptide Sequenator (a liquid-phase sequencer), used a spinning cup to perform the coupling and cleavage reactions and was a major step forward in protein sequencing. A solid-phase automated sequencer was developed by Laursen in 1971 (manufactured by Sequemat) in which a polypeptide was covalently attached to derivatized controlled-pore glass beads to minimize sample washout. The PTH conversion and PTH amino acid analysis for both the liquid-phase sequencer and the solid-phase sequencer were done off-line and required several nanomoles of starting material.

A major advance in instrumentation occurred in 1981 with the introduction of the gas-phase sequencer (1), which was made commercially available by Applied Biosystems, Inc. The commercial instrument performed the entire Edman chemistry (including conversion), delivered the coupling base and TFA as gases (greatly reducing background), miniaturized the chemistry plumbing, and, most importantly, increased sensitivity to ~500 pmol of starting material. Samples were analyzed by application to a glass fiber disk coated with a polymer support (polybrene), which helps retain the sample. Shortly after the introduction of the gas-phase sequencer, Applied Biosystems introduced an on-line HPLC to serve as a dedicated PTH analyzer. This change automated the sequencing process, increased the quality of the data, and decreased the amount of starting material required to ~100 pmol. Minor improvements were made by Applied Biosystems such as incorporation of computer control of the instrument, data collection, and the introduction of pulsed-liquid delivery of TFA, which increases repetitive yields. Another sequencer, manufactured by Porton, was one of the first instruments to run multiple consecutive samples in series on a single instrument.

The next major change in Edman chemistry involved the introduction of biphasic columns (half reverse-phase-type material and half hydrophilic-silica-type material) by Hewlett-Packard to analyze samples. This allows samples to remain in the sequencer without use of a polymer support and also allows loading of samples in various salt-containing solutions. In the past few years, Perkin Elmer/Applied Biosystems Division has further miniaturized the instrument design and incorporated in-line sensors to regulate exactly how much reagent is delivered, thus optimizing the chemistry. Perkin Elmer/Applied Biosystems Division recently introduced a sequencer that uses a capillary HPLC, which increases sensitivity to below 1 pmol of starting material, so that PTH amino acid signals of ~50 fmol can be identified. Hence, the overall amount of protein/peptide required for Edman sequencing has decreased over the years from nanomoles (Beckman spinning cup) to subpicomole amounts (Procise cLC) and represents a greater than 10,000-fold increase in sensitivity in 17 years.

Due to these developments, it is currently relatively routine for a laboratory that specializes in high-sensitivity sequencing to obtain sequence from ~5 pmol of sample and, if the sequencer and PTH analyzer are carefully optimized, from 1–1.5 pmol of sample. In this regard, it is important to note that although reverse-phase HPLC-purified peptides typically have initial sequencing yields of ~50%, initial yields for SDS-PAGE purified samples are often less than 25% if electroblotting is not optimized or artificial N-terminal blockage occurs. Assuming 50% electroblotting efficiency from an SDS polyacrylamide gel to a PVDF membrane and an initial sequencing yield of ~15% for a sample that has been subjected to SDS-PAGE, 700 femtomoles of protein would need to be subjected to SDS PAGE to provide the minimum PTH amino acid signal of ~50 fmol required for the cLC instrument. It is important to note that although such a sample might yield 5–10 residues of sequence, a 10-fold greater amount would almost certainly result in higher-quality data and more positively called residues. Thus, it is difficult to respond to the often asked question of how little material is required because the answer depends on the quality and quantity of the sequencing data, the extent of N-terminal blocking that may have occurred during sample preparation, the molecular weight of the protein, the sequence itself, the instrumentation available, and the level of experience of the instrument operator. Current trends in amino acid sequencing include the use of mass spectrometry for obtaining a primary sequence from low and in some instances subpicomole amounts of peptides generated from gel-separated samples (58). In addition, much greater emphasis is being placed on using amino acid composition and mass spectrometric data to identify proteins via protein database searches without resorting to Edman degradation.

INTERNAL PROTEIN SEQUENCING

SHEENAH MISCHE
Rockefeller University
New York, New York

DAVE SPEICHER
Wistar Institute
Philadelphia, Pennsylvania

ULF HELLMAN
Ludwig Institute for Cancer Research
Uppsala, Sweden

KEN WILLIAMS
Yale University
New Haven, Connecticut

One- and two-dimensional PAGE offer powerful methods for the isolation of microgram quantities of individual proteins (59,60). Two-dimensional gels (in which the first dimension is usually isoelectric focusing and the second is SDS PAGE) are generally considered the method of choice for highest-resolution separation of complex protein mixtures at the microgram level and are often used to study the phenotypic-dependent alterations of protein expression in total cellular extracts or enriched cell fractions.

Either one- or two-dimensional PAGE is also often used for final purification of proteins destined for N-terminal sequence analysis. In either case the separated proteins are then electroblotted onto a PVDF membrane, first described as a support for protein microsequencing 10 years ago (55), stained with a dye such as Amido Black or Coomassie Blue, and the band of interest excised and placed in an automated peptide/protein sequencer. However, if the protein has a blocked (e.g., acetylated) amino terminus (estimated occurrence of this modification is approximately 80% in soluble proteins from Ehrlich Ascites cells [53]), no useful sequence data will result. In addition, there are other reasons that may eliminate the direct N-terminal sequencing approach, including the need for noncontiguous stretches of primary sequence data for DNA cloning and/or the need to generate peptides that could be used for rapid protein identification via an increasing array of mass spectrometric approaches. All these considerations inspired the development of protein digestion methods that can be carried out in gel slices and on PVDF membranes. Although proteolytic digestion and peptide purification have been essential methods for the protein chemist in the study of protein structure and function, novel approaches were needed that conserved the limited quantities of available protein.

In 1987 Aebersold et al. (61) described a method for generating internal sequence from tryptic peptides recovered from proteins electroblotted to nitrocellulose. In rapid succession, methods were reported for in situ digestion of protein bands from Coomassie Blue–stained gels (62,63), proteins bound to PVDF (64), and cationic derivatized PVDF (65). It soon became apparent that microscale generation and isolation of internal peptides were extremely useful in obtaining primary sequence information from N-terminally blocked proteins, and internal sequence analysis radically changed the approach taken in protein sequencing laboratories. Additional refinements that improved each procedure (66–71) resulted in the rapid emergence of internal sequence analysis as a technique for obtaining sequence data from low abundance proteins and has created demands for these analyses that often are associated with unrealistic expectations.

In keeping with its role of education and implementation of new techniques and methodologies, the ABRF created the Internal Protein Sequence Research Committee in 1995 with the purpose of helping member laboratories evaluate or implement internal protein sequence analysis in their laboratories. The study goals that the Internal Protein Sequence Research Committee seeks to achieve are fivefold: (*1*) provide a mechanism for ABRF laboratories to anonymously compare their internal sequencing capabilities with those of other core laboratories; (*2*) provide a reasonable sample and well-proven protocols to facilitate introduction of this technology into laboratories that do not offer these services; (*3*) obtain data that can help determine the relative efficacy of internal sequencing from PVDF blots versus gel samples; (*4*) determine if there are any significant commonalities among the best in-gel and PVDF digestion protocols to help optimize these protocols, and (*5*) compile data obtained by multiple laboratories on the same "unknown" sample, to define more realistic expectations for internal sequencing.

The results of the two ABRF studies completed provide a basis for estimating the amounts of protein that are likely to be required for internal sequence analysis as well as the protein- and methodological-dependent limitations inherent to this approach and their probable impact on the results. Refer to the work of Williams and coworkers (47) for the details of the 1996 study. In brief, each ABRF study involved a microscale enzymatic digestion of protein bound to a membrane support or within an SDS PAGE, Coomassie Blue–stained band, followed by elution of the resultant peptides into a volume of buffer that often ranges from 50 to 200 μL. This peptide mixture can then be subjected directly to mass spectrometry and/or the peptides can be fractionated and a limited number individually sequenced by traditional chemical (Edman) degradation (see "Protein Sequencing" for a brief description of Edman degradation).

Although it is possible to use mass spectrometry to directly sequence individual peptides from complex protein digests, the expense (often more than $400,000) of the necessary equipment and the considerable expertise needed to successfully carry out this approach and to accurately interpret the data has so far limited its availability to a very small fraction of core laboratories. Increasingly, however, a much larger fraction of core laboratories are being equipped with the instrumentation needed to determine the masses (but not the sequences) of many of the peptides present in proteolytic digests. The resulting peptide masses are then searched against peptide mass databases that contain theoretical masses for all peptides expected from digesting known proteins with trypsin and other proteases. Although peptide mass searching provides an extremely rapid approach to identifying known proteins (72–75), which usually requires the use of only a small fraction of the digest, we believe most laboratories recommend that such an identification be confirmed by mass spectrometry or Edman sequencing of at least one peptide.

Microanalytical preparation of peptides from in situ proteolysis for Edman sequence analysis usually employs reverse-phase HPLC using a C18 [$CH_3(CH_2)_{17}$-] stationary phase that is covalently linked to a silica-based resin packed into a 2.1-mm or smaller internal diameter column that is most often equilibrated with 0.05–0.1% trifluoroacetic acid in water. Peptides are eluted by linear acetonitrile gradients that extend from 0–5% (v:v) in the beginning of the gradient to 80–100% (v:v) at the end of the gradient. One or more peptides are selected for Edman degradation, with or without prescreening by mass spectrometric analysis (76), and the primary sequence is used to search protein and DNA sequence databases (GenBank, Protein Identification Resource [PIR], European Molecular Biology Laboratory [EMBL], Swiss Protein) to determine whether the protein has been previously identified.

As mentioned, two studies of these analytical strategies have been carried out by the ABRF Internal Protein Sequence Research Committee. The 1996 study was a comparative HPLC mapping study in which the participants received 70 pmol each of two recombinant proteins, 28 kDa and 30 kDa, which were identical except for an insert in the 30-kDa protein that generated a 15-residue tryptic peptide unique to the 30-kDa protein. A mixture of these two proteins was separated by one-dimensional SDS PAGE and either provided as Coomassie Blue–stained in-gel samples or electrophoretically transferred to PVDF membranes and stained with Amido Black. A synthetic peptide in which the sequence of the 15-residue unique peptide had been randomized served as an external standard calibrant. This standard was used to normalize data from various laboratories for use of different flow cell path lengths, flow rates, absorbance wavelengths, and so forth. In-gel and PVDF blanks and representative protocols for each type of sample were provided. Participants who requested samples were asked to (*1*) perform a tryptic digestion of the 28-kDa, 30-kDa, and blank samples and subject all three digests plus the external standard to comparative HPLC; (*2*) identify the target peptide by its presence in the 30-kDa chromatogram, absence in the 28-kDa chromatogram, and elution near that of the external standard peptide; and (*3*) determine the amino acid sequence of the target peptide by traditional Edman degradation and/or mass spectrometric analysis.

For the 1997 Internal Sequencing Study, participating laboratories were asked to perform a tryptic digest on 50 pmol of a known 41-kDa recombinant protein, either in-gel or on PVDF, and to identify the target peptide contained within. A predigested sample of a recombinant protein containing a 23.5-kDa portion of the 41-kDa protein was supplied to assess the ability of the participating laboratory's instrumentation to perform the chromatography required to separate tryptic digestion mixtures. An external peptide standard was supplied that had similar, but not identical, chromatographic behavior to the target peptide. This peptide standard was used by the participants to help identify the target peptide and by the Internal Protein Sequence Research Committee to calculate the relative recoveries of tryptic peptides from each participating laboratory. In-gel and PVDF blanks and representative protocols for each type of sample were provided. In addition to providing a predigested sample, the 1997 study differed further from the 1996 study in that whereas the 1996 target peptide sequence was not in any commonly used database and was thus unique, the 1997 target peptide corresponded to an expected tryptic peptide from a protein whose sequence is included in all commonly used databases. Thus, the 1997 target peptide sequence could be deduced by utilizing observed peptide masses and/or partial Edman or mass spectrometric fragmentation data to search sequence databases.

Figure 2 shows one of the data sets that were returned in the 1997 study. In this instance, the participating laboratory carried out both in situ gel and on-PVDF membrane digests; as evident from this figure, both approaches provided excellent results. Although almost all significant absorbance peaks are unique to the samples, the digests carried out on "blank" gel and PVDF pieces reveal a few absorbance peaks that can be readily eliminated from further study based on their appearance in these control digests. These peaks may result from trypsin autolysis products, buffer-related contaminants, reagents used in SDS PAGE, or residual Coomassie Blue or Amido Black stain.

Several important conclusions follow from the data that resulted from the 1996 and 1997 studies, as summarized in Table 3. First, because hydrolysis and amino acid analysis of representative gel and PVDF bands from the 1997 study indicated that almost threefold more protein was present in the gel bands, the 41-kDa protein used in this study probably has a blotting efficiency of only 19/50 pmol (Table 3), or 38%, which is much lower than the >80% efficiency reported previously (77). Second, although the apparent recovery of the 41-kDa gel band following SDS

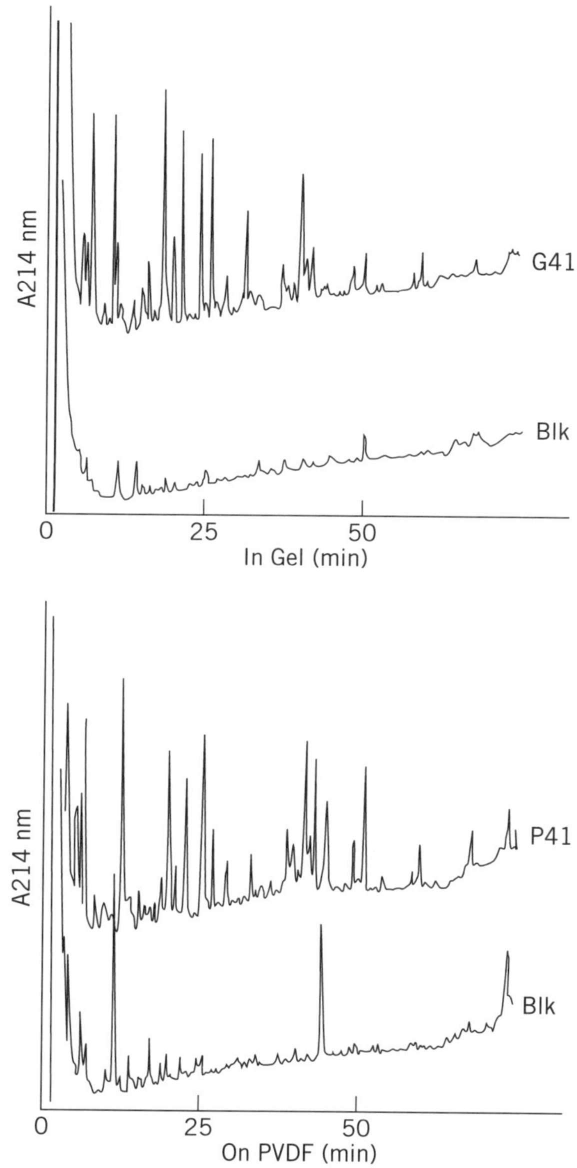

Figure 2. HPLC chromatograms submitted by participant 5 in the 1997 ABRF Internal Sequencing Study. The upper two chromatograms were obtained from tryptic digests of 50 pmol of the 41-kDa protein that were carried out either in the Coomassie Blue–stained gel band (upper panels) or on the Amido Black–stained PVDF membrane (lower panels). In both cases the corresponding digests carried out on blank sections of gels or PVDF membranes are shown in the lower chromatograms. HPLC was carried out at 50 μL/min on a 1- × 100-mm C18 column that was equilibrated with buffer A (2% acetonitrile, 0.057% trifluoroacetic acid in water) with detection at 214 nm. Peptides were eluted with linear gradients to 40% B (60 min), 60% B (65 min), and finally to 100% B (100 min). Although hydrolysis and amino acid analysis of several gel and PVDF protein bands indicated almost three-fold more protein in the gel samples (Table 3), the similar peak heights suggest a similar final yield of peptides from both the PVDF and gel samples. This suggests the facility that submitted the data may have obtained higher relative peptide yields from its PVDF as opposed to their in-gel digests.

Table 3. Summary of ABRF Internal Sequencing Studies

Parameter	In-Gel		On PVDF	
	1996[a]	1997	1996[a]	1997
Amount of protein subjected to SDS PAGE	70 pmol	50 pmol	70 pmol	50 mol
Amount of protein in stained gel or PVDF band[b]	ND	52 pmol	ND	19 pmol
Median sequencing yield from target peptide[c]	2.3 pmol	8.9 pmol	2.5 pmol	2.9 pmol
Overall target peptide yield[d]	6.4%	24%	7.6%	8.0%
Target peptide recovery from stained gel or PVDF band[e]	ND	23%	ND	21%

ND, no data.
[a]See Ref. 47 for further information regarding assumptions that were made to calculate overall peptide yield and other data from the 1996 study.
[b]As estimated by acid hydrolysis of the gel or PVDF band followed by amino acid analysis.
[c]Yield of valine (4) and of phenylalanine (3) in the target peptides used in the 1996 and 1997 studies, respectively.
[d]Calculated from the estimated initial peptide sequencing yield in cycle 1 (which was based on the median sequencing yield from the target peptide corrected to cycle 1 based on an estimated repetitive yield of 90%) divided by the estimated amount of protein loaded onto the SDS polyacrylamide gel.
[e]Calculated from the estimated initial peptide sequencing yield divided by the amount of protein in the stained gel or PVDF band.

PAGE and staining with Coomassie Blue was high, a previous study (70) demonstrated recoveries as low as 44% following SDS PAGE, Coomassie Blue staining, and washing of gel bands that contained relatively low molecular weight proteins. Hence, in the case of the six proteins (which had a median size of 15 kDa) used in this study (70), the median recovery of protein after washing of the Coomassie Blue–stained gel bands before in-gel digestion was 68% of the amount loaded onto the gel. Taken together, these data demonstrate that some proteins exhibit unusually high losses during electrotransfer to PVDF membranes, whereas other proteins exhibit unusually high losses using in-gel digestion methods.

Comparison of the overall target peptide yields for the two ABRF studies (Table 3) suggests that even when similar in-gel approaches are used, there can be substantial differences in the recovery of an individual peptide from two different proteins. As calculated from the estimated initial amount of peptide that actually sequenced versus the amount of protein subjected to SDS PAGE, the overall target peptide recovery from the 1996 in-gel samples was 6.4% as compared with 24% in 1997. This four-fold difference in peptide recovery makes it difficult to provide a precise answer to the often asked question of how much protein is needed for internal sequencing. Clearly, the answer can vary by at least fourfold, depending on the protein of interest, assuming, of course, that the amount of the protein of interest that was subjected to SDS PAGE is accurately known. Because we believe the most common means of estimating the amount of protein submitted for internal sequencing (comparative staining intensity) is also subject to at least two- to fourfold variability (depending again on the protein of interest), it is quite possible that a state-of-the-art core laboratory might correctly provide a fourfold or greater range in response to the question of how much protein is needed for internal sequencing. Alternatively, they may suggest an amount of protein that is sufficiently large that the internal sequencing project will likely succeed even if the protein in question stains more intensely with Coomassie Blue or Amido Black than an average protein and/or if peptide yields from this protein are below average. In view of the apparent fourfold or greater range in peptide yields following in situ enzymatic cleavage of different proteins, it is also apparent that studies and publications based on results from only a few nonrepresentative proteins may underestimate the amount of an average protein that may be needed to succeed with the same procedure. In this regard, it appears that neither of the two proteins used in the studies summarized in Table 3 are average. That is, the average recovery of 409 peptides that were sequenced from in-gel digests of 191 "unknown" proteins was 12.2% (70). In contrast, the 6.4% target peptide recovery for the 30-kDa protein used in the 1996 ABRF study (Table 3) suggests that this protein may be more difficult than an average protein to digest, and the 24% average in-gel target peptide recovery in the 1997 ABRF study (Table 3) suggests the 41-kDa protein used in this study may digest in a gel matrix much better than does an average protein.

As one means of approaching the question of how much protein is generally needed for internal sequencing, this question was asked of the participants in the 1997 study, which determined that the median response was 50 pmol protein. The results from the 1997 study indeed suggest that a core laboratory that routinely offers internal sequencing will indeed usually succeed if 50 pmol or more protein is subjected to SDS PAGE. Hence, 83% of the core laboratories that attempted to sequence a single peptide derived from the 41-kDa protein distributed in the 1997 ABRF study succeeded in identifying the protein based on database searches of the resulting sequence. Obviously, the percentage of laboratories that could have succeeded in this study is higher than 83%, because many laboratories isolated numerous other peptides from this protein in addition to the target peptide and many of these were probably suitable for sequencing. It appears from the 1997 ABRF Internal Sequencing Study that the 50-pmol estimate is sufficiently robust that even when one step in the procedure does not proceed optimally (such as the much lower than average blotting efficiency of the 41-kDa protein used in the 1997 study), usable internal sequences are nonetheless likely to be derived from the sample.

Another important response in the survey portion of the 1997 ABRF study was the answer to the question of what percentage of received samples are already in the databases. Because the median response was 75%, it is clear that the overall goals of internal protein sequencing will increasingly shift to protein identification (i.e., proteome research) and to the elucidation of posttranslational modifications that usually cannot be predicted a priori from the ever-increasing wealth of cDNA sequences in nonredundant databases and that mass spectrometry will play an ever-increasing role in this endeavor.

Finally, the answer to the question of which approach (in-gel or on-PVDF digestion) is best for internal protein sequencing is that the two methods appear to be quite comparable. That is, when peptide recoveries are calculated from the Edman sequencing yields divided by the amount of protein in the stained gel or PVDF band (as opposed to the amount that was loaded onto the gel), the target peptide recoveries for both PVDF and gel samples were just above 20% in the 1997 study (Table 3). The optimal method for a specific project depends on the protein of interest and the personal preferences and level of expertise that the core laboratory workers to whom the sample will be submitted has with each of these approaches. Based on two ABRF studies, each of which involved about 40 core laboratories, we conclude that if a 50-pmol protein sample is subjected to SDS PAGE and then prepared according to the protocol recommended by the core laboratory to which the sample will be submitted, the probability of obtaining internal peptide sequences is likely to be much higher than 83%.

PEPTIDE SYNTHESIS

LYNDA F. BONEWALD
University of Texas Health Science Center
San Antonio, Texas
RUTH HOGUE ANGELETTI
Albert Einstein College of Medicine
Bronx, New York

The goal of peptide resource laboratories is not only to supply the biomedical researcher with the high-quality peptides critical for scientific research at an affordable cost but also to actively research, test, and design new investigative peptides. To meet this challenge, the peptide chemist must be aware of, be able to learn, and be capable of instituting the latest techniques of peptide chemistry and evaluation. The ABRF Peptide Synthesis Research Committee was established to assist personnel of its member facilities in attaining this goal.

Use of peptides is diverse, ranging from use as antigens for the production of antibodies and as enzyme substrates or inhibitors to use for structure–function studies, for screening purposes, or as agonists or antagonists in biological assays and systems. For these applications, either large amounts of a small number of peptides or small amounts of a large number of peptides are usually necessary. More peptides with modifications are required to address important scientific questions such as the roles of phosphorylation and dephosphorylation in cell signaling. Peptide chemists are in the process of improving and developing new strategies for synthesis and evaluation of modified peptides. Examples of other modifications include biotinylated and fluoresceinated peptides for detection and quantitative assays, modifications that allow cross-linking on exposure to ultraviolet light, cyclized peptides for enhanced stability leading to enhanced biological activity, and a variety of other modifications of both the amino- and carboxy-termini and modifications of amino acid side chains.

To achieve the goal of supplying necessary, critical peptides for research, the following two components are essential: skilled, knowledgeable personnel and state-of-the-art instrumentation. Peptide synthesis laboratory personnel must be flexible and learn and institute new technologies quickly. Automated instrumentation belies the actual skill and experience required for peptide laboratory personnel. Peptide synthesis is not just "button pushing." Each peptide can vary considerably in the amount of design and planning required for synthesis. Some peptides require only routine synthesis, cleavage, desalting and/or purification, and analysis, whereas others may generate unique challenges at any or all phases of production. Many of these challenges lead to research projects involving collaboration between the peptide chemist and the biomedical researcher.

The laboratory must contain or have access to state-of-the-art instrumentation. The standard peptide synthesis instrument that produces a single peptide at one time is becoming inefficient except in cases of producing longer peptides. New instruments are being designed for multiple arrays and for more stringent chemistries and synthesis conditions. New instrumentation under development that can produce multiple peptides at one time with on-line cleavage is becoming state of the art. Though not perfected, these instruments can meet the demand for the production of large numbers of peptides. With the ability to produce large numbers of peptides comes the need to desalt and purify peptides efficiently and in a timely manner. HPLC systems are being developed to provide autoinjection and autocollection of either large numbers or large amounts of sample. As discussed later, the optimal means to ensure that the correct peptide has been made is by mass spectrometric analysis.

The Peptide Synthesis Research Committee's raison d'être has been to evaluate the quality of peptides produced by its member laboratories and to introduce and analyze new chemistries and technologies in solid-phase peptide synthesis. Since its inception in 1990 the Peptide Synthesis Research Committee has addressed several aspects of peptide production. The first two studies (17,20) were designed to determine and address potential problems in synthesis, cleavage, and side-chain deprotection. For peptide analysis, mass spectrometry with an accuracy of one atomic mass unit was found to be essential. HPLC and capillary electrophoresis are useful for quantitation but can be misleading. A single peak on a chromatogram can represent an incorrect peptide. HPLC is most useful for purification and collection. AAA is useful for quantitation but cannot identify specific problems, only that a problem may or does exist. Amino acid sequence analysis is costly, time-consuming, and cannot aid in identifying protected amino acids. Mass spectrometry proved to be most useful in identifying problems in synthesis, cleavage and deprotection, and identification of correct peptides.

Because cleavage from the resin was shown to be more problematic than synthesis or side-chain deprotection in the first and second study, this step was specifically addressed in the third study (27). However, no correlations could be found between common side reactions occurring during cleavage and specific cleavage methods. The human element may play a critical role in the quality of peptide production.

The fourth study (32) was undertaken to determine if core laboratories had improved from the first study; therefore, the exact same peptide was requested for analysis as was requested in the first study. The submitted peptides were found to have increased considerably in quality from the first to the fourth study (in the first study, 78% of submitted samples had less than 25% of the correct product in contrast to the fourth study, in which 93% contained greater than 25% of the correct product). The sixth study (45) showed even greater improvement (almost 95% of the submitted samples contained greater than 70% correct product).

The fifth study (36) addressed problems in disulfide bond formation and introduced cyclization to the member laboratories. It was found that peptide cyclization can be performed on a routine basis with high yield. The sixth study (45) was undertaken to examine potential problems in racemization. It was found that histidine racemization may not be a major problem, but researchers should evaluate their peptide application beforehand to determine whether racemization is important. If even low levels of racemization could pose a potential problem, the racemized peptide can be separated from the correct peptide by HPLC techniques.

In the seventh study, the Peptide Synthesis Research Committee introduced and analyzed phosphorylated peptides requested from its member laboratories. This study was initiated because peptide modifications are becoming necessary for a number of applications. One very important modification is the role of phosphorylation in cell signaling and transduction. This study should determine whether it is feasible to synthesize posttranslational peptide modifications, such as tyrosine phosphorylations.

Other important observations and conclusions drawn from the studies conducted by the Peptide Synthesis Research Committee include the following: A shift in the type of chemistry used in solid-phase peptide synthesis has occurred from Boc to Fmoc chemistry (from 50% in 1991 [17] to 98% in 1996 [45]). This is probably due to Fmoc chemistry being safer, more flexible, and less problematic. One observation that has been repeated with each year's study is the unexpected occurrence of human errors in peptide synthesis. Each year's study contains a few surprises. For example, one year a submitted test peptide was synthesized in the wrong direction. It is not known whether these incorrect peptides are mainly due to inexperience or to human error, but they should convince all concerned that postsynthesis peptide analysis is critical. The best combination of techniques to ensure the correct product is HPLC (see Fig. 3) and mass spectrometric analysis (see Refs. 17, 20, 27, 32, 36, 38, 44, and 45).

The peptide laboratory must be flexible and capable of constantly changing to meet the needs of biomedical research. Academic peptide laboratories cannot compete with commercial peptide facilities without support from their institutions. The rationale for institutional support is the availability of advice, consultation, potential collaboration, potential for modifications and chemistries not readily available through commercial ventures, and reduced cost to the investigator. The common goal of the peptide chemist and biomedical researcher should be scientific endeavor and not profit or monetary gain. The future of

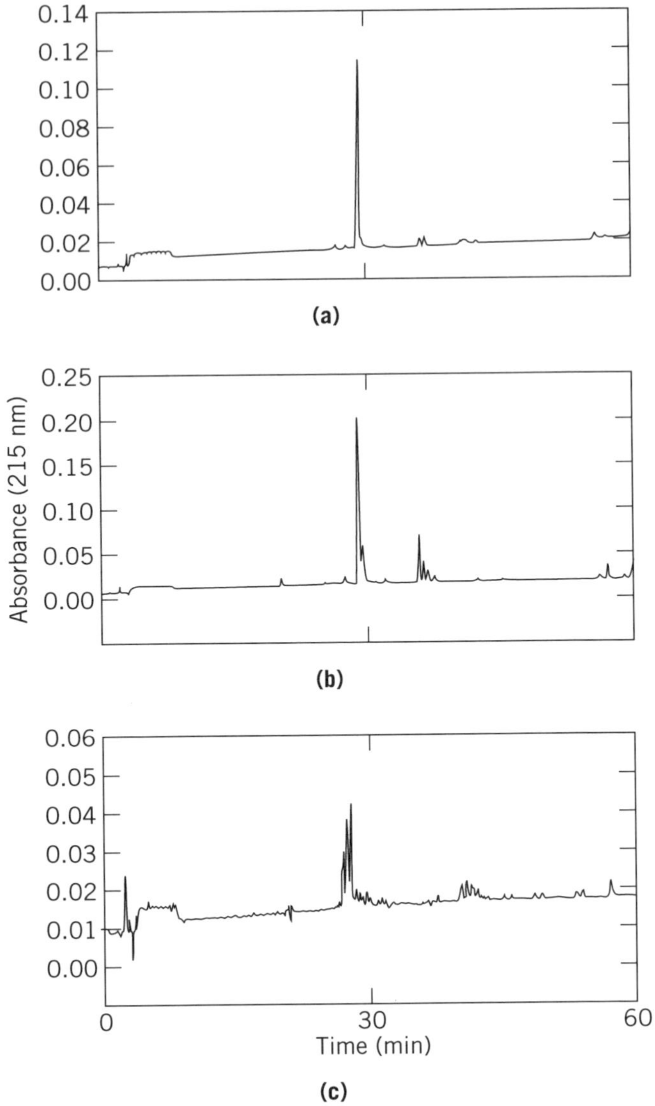

Figure 3. Analytical reverse-phase HPLC chromatograms from three peptide samples submitted in the ABRF 1995 Peptide Synthesis Study. Biomedical researchers should determine whether their studies require purified (greater than 70% correct peptide) or highly purified (greater than 95% correct) peptides. If the study does not require highly purified material, then considerable time and cost can be saved. Usually 70% correct peptide can be obtained with a good synthesis followed by a desalting step. To obtain highly purified peptides, HPLC is usually necessary. Examples of the linear peptide used from the ABRF 1995 study are shown (36). **(a)** Greater than 96% pure peptide; **(b)** good peptide, greater than 76% pure; **(c)** poor synthesis. The major peak by HPLC from any synthesis should always be analyzed by mass spectrometry to ensure that the peak represents the correct peptide.

peptide laboratories will definitely include new chemistries, new peptide and amino acid modifications, the synthesis of longer, larger peptides, and new instrumentation. Perhaps in the future, combinatorial libraries may be generated for basic research or such exotic chemistries as peptide nucleic acids will be utilized. The ABRF Peptide Synthesis Research Committee will continue to assist its members with these challenging and changing goals.

CARBOHYDRATE ANALYSIS

R. REID TOWNSEND
University of California–San Francisco
San Francisco, California

Determining the monosaccharide content of glycoproteins (*1*) provides the basis for further structural studies, (*2*) suggests classes of oligosaccharide chains, (*3*) provides unequivocal evidence that the polypeptide chain is glycosylated, and (*4*) may serve as a measure of production consistency for therapeutic glycoproteins. The presence of GalNAc (as $GalNH_2$ after de-*N*-acetylation from acid hydrolysis) suggests mucin-type *o*-linked structures. Only Man and GlcNAc suggest oligomannosidic-type structures; GlcNAc, Man, Gal, and Fuc suggest lactosamine-type oligosaccharides; and Glc and Gal suggest glycosylation of hydroxylysine and hydroxyproline residues. Historically, monosaccharide analysis has been a three step process of (*1*) release of residues from oligosaccharide structures and the polypeptide backbone (*2*) derivatization for gas chromatography (GC) or HPLC, and (*3*) chromatography with quantitative detection. It is well recognized that breakage of glycosidic and protein–sugar bonds by acid hydrolysis or methanolysis is often incomplete and results in some destruction of monosaccharide residues, although many studies have been performed to optimize the yield of released monosaccharides. For GC, volatile monosaccharide derivatives must be prepared (78,79). For HPLC, derivatization is used to increase hydrophobicity for reverse-phase chromatography and to increase the sensitivity of detection (for review, see Ref. 80). The chemical diversity and lability among neutral, amino, and anionic classes of monosaccharides necessitates different release and derivatization methods to analyze all of the sugars from a glycoprotein. The detection of carbohydrates on gold electrodes in a low-volume flow cell after separation using high-pH anion-exchange chromatography (HPAEC) obviates the need for derivatization of neutral, amino, or anionic monosaccharides in samples that contain ~100 pmol of individual monosaccharides (for review, see Ref. 81). The commercialization of HPAEC with pulsed amperometric detection (PAD) has resulted in a wider availability of monosaccharide analysis using a single method. Most recently, fluorophore-based methods, which are available in a kit format from vendors, should further increase the general availability of monosaccharide analysis. Finally, the oligosaccharides on a single protein occur as an array of structures whose proportions are difficult to quantify, and the type and proportions of structures can vary from lot to lot, even after isolation from the same source.

In 1995 the ABRF Carbohydrate Analysis Research Committee (CARC) initiated a study to determine the ability of research laboratories and core facilities to identify and quantify carbohydrates (43), which are commonly found in N- and O-linked oligosaccharides of glycoproteins from serum, cell membranes, and recombinant expression systems. Test sample A was ultrapure grade bovine serum albumin from Calbiochem (San Diego, California), and test sample B was bovine fetuin, purified by the Spiro method from Gibco (Grand Island, New York). The two test samples were lyophilized from water and quantified by hydrolysis followed by amino acid analysis. Based on the average of the values for alanine and valine, test sample B contained 25 nmol fetuin, which was chosen because it is one of the most well-characterized glycoproteins (82–88). All participating laboratories ($n = 29$) correctly identified which sample was glycosylated. In 14 cases the glycosylated sample was identified by AAA or colorimetric assays. The amino acid compositions were analyzed further using PROPSEARCH (89), which can be accessed via the Web site *(www.embl-heidelberg.de/aaa.html)* to identify the test samples as bovine serum albumin and bovine serum fetuin. Monosaccharides were quantified using HPAEC PAD (14 laboratories), HPLC after precolumn derivatization with 1-phenyl-3-methyl-5-pyrazolone (90) (one laboratory), gas chromatography and mass spectrometry (four laboratories), or gel electrophoresis after derivatization with 2-aminoacridone (one laboratory) (91).

Analysis of the glycosylated test sample by 4 CARC laboratories and 10 participating laboratories using HPAEC PAD gave the following results (in mol/mol of protein): GalN, 2.3 ± 39%; GlcN, 16 ± 23%; Gal, 12 ± 44%; and Man, 9 ± 55%. (The error is expressed as percent standard

Table 4. Monosaccharide Composition of Bovine Fetuin as Determined by HPLC (Mole Sugar per Mole Protein)

Monosaccharides	Literature[a] (1984–1991)	CARC[b] (1995–1996) (%)	External laboratories (%)
GalNAc (3)[c]	1.2–3.5	2.7 ± 20[d]	2.2 ± 41
Man (9)	6.4–8.3	9.6 ± 15	8.0 ± 50
GlcNAc (15)	11.5–16.4	17.2 ± 9	15.2 ± 27
Gal (12)	9.3–13.0	12.4 ± 24	11.6 ± 43
Neu5Ac (12)	7.2	14.8 ± 9.5	15.2 ± 27

[a]Reviewed in Ref. 78.
[b]Values after iterative analyses by CARC.
[c]Values in parentheses are theoretical and estimated.
[d]Based on the amino acid analysis of the bovine fetuin test sample (25 nmol from an average of alanine and valine content). The percent standard deviation is given in terms of σ_{n-1}.

deviation.) The results from GC analysis (alditol acetate and trimethylsilyl derivatives) from four laboratories also showed large errors: GalNAc, 1.7 ± 43%; GlcNAc, 9.2 ± 36%; Gal, 13.6 ± 35%; and Man, 9.6 ± 26%. The analysis of *N*-acetylneuraminic acid by both the CARC and participating laboratories gave a lower percent standard deviation by both HPLC (15.2 ± 17%) and GC (13.2 ± 23%). Finding these errors too large, CARC undertook a systematic internal study to identify sources of error in monosaccharide analysis by HPAEC PAD. First, it was established that interlaboratory variability was significantly greater than the analysis of the same bovine fetuin sample when performed on a single instrument over a 6-month period: GalN, 2.4 ± 8.8%; GlcN, 16.4 ± 8.6%; Gal, 12.8 ± 15% and Man, 7.8 ± 4.6%. Because all studies were performed with external standards, monosaccharide standards from four CARC laboratories were analyzed in the same laboratory. The percent standard deviation was no greater than 15% and for most monosaccharides was less than 5%, indicating that standard preparation was not a major contributor to the error among CARC laboratories. Based on literature reports and experience within CARC laboratories, standard hydrolysis conditions were adopted for neutral and amino sugars (2 M trifluoroacetic acid for 3 h at 100 °C) and sialic acids (2 M acetic acid for 3 h at 80 °C). Because monosaccharides can be destroyed under acid hydrolysis conditions used for their release, CARC used external standards that had been hydrolyzed under the mentioned conditions. Untreated external standards and standards that had been subjected to the same hydrolysis conditions as the glycoprotein were then compared. Unexpectedly, the amino sugar electrochemical responses in the hydrolyzed standards were increased as much as twofold, suggesting that the higher values observed for amino sugar determination, both in some CARC and participating laboratories, were due to this phenomenon. The lower peak areas were attributed to either selective loss of amino sugars through the injection path or reaction of free amino groups on the standard nonacetylated sugars with the aldehyde functions of the monosaccharides in both the standard mixture and hydrolyzates.

To control further for variability associated with hydrolysis of samples, a single hydrolyzate of the bovine fetuin sample was prepared in one laboratory (in a single vessel) and then distributed to the other CARC laboratories for analysis by HPAEC PAD. Hydrolyzed, external monosaccharide standards were also used. The following reduction in percent standard deviation was observed from four CARC laboratories: GalN, 2.7 ± 20%; GlcN, 17.2 ± 9.1%; Gal, 12.4 ± 22%; and Man, 9.6 ± 15%. In summary, by adopting standard hydrolysis conditions, ensuring that the external standards prepared in different laboratories gave similar responses and using hydrolyzed external standards, the interlaboratory error of HPAEC PAD analysis was significantly improved; however, interlaboratory error remained greater than those observed using a single instrument.

Table 4 compares the monosaccharide compositions of bovine fetuin from literature values with those obtained by CARC after the iterative analyses discussed earlier and with those from participating laboratories, all using HPLC. Overall, the mean values obtained by both the CARC laboratories and the participating laboratories vary by no more than two residues; however, the error range associated with the external study was significantly greater. The theoretical values (indicated in parentheses in Table 4) were estimated from the oligosaccharide structures and the number of modified sites on the protein. Bovine fetuin has three N-linked sites that contain biantennary (84,86) and triantennary structures (82,87,88; Fig. 4). In addition, at least three O-linked sites, occupied with sialylated di- and tetrasaccharides (83,85; Fig. 4), are present (83). The value of three GalNAc residues assumes complete site occupancy of three O-linked sites; however, it is well recognized that glycosylation at serine and threonine residues is frequently not stoichiometric. An approximately threefold range of values has been reported in the literature and by the external laboratories participating in the 1995–1996 study for the GalNAc composition of fetuin (Table 4). The CARC laboratories found an average value of 2.7, with a range of 2–3 residues.

If all three N-linked sites are fully occupied, nine Man residues would be present in fetuin. The value obtained by CARC approached this value, with an average of eight reported in the external study and a larger error. The theoretical value of 15 for GlcNAc accounts only for the composition of three triantennary structures; however, the O-linked tetrasaccharide also contains GlcNAc (Fig. 4). Whether the value of 17 residues from the CARC analyses reflects a significant amount of the tetrasaccharide in fetuin will require further studies. The 12 theoretical Gal residues are estimated from three triantennary structures and three O-linked disaccharides, and the 1995–1996 results approach this value (Table 4). The determination does not indicate the presence of biantennary structures which have one less Gal residue, nor the tetrasaccharide, which has one more Gal residue. Complete sialylation of 12 Gal residues would result in 12 Neu5Ac residues. The greater value for Neu5Ac indicates complete sialylation and is also consistent with the presence of the known tetrasialylated triantennary structures in bovine fetuin (87,88).

Overall, these results show that quantitative monosaccharide analysis in combination with amino acid analysis, to determine the amount of protein, gives molar ratios of sugars to protein that are in agreement (within one to two residues) with the oligosaccharide content. However, these studies have also shown that there is a significantly larger error associated with analyzing the same hydrolyzate on different instruments (43). Future studies will focus on the reanalysis of the same test sample to identify factors that are related to differences in instrument configurations and interlaboratory analyses.

NUCLEIC ACIDS

C.W. Naeve
St. Jude Children's Research Hospital
Memphis, Tennessee

P.S. Adams
Trudeau Institute
Saranac Lake, New York

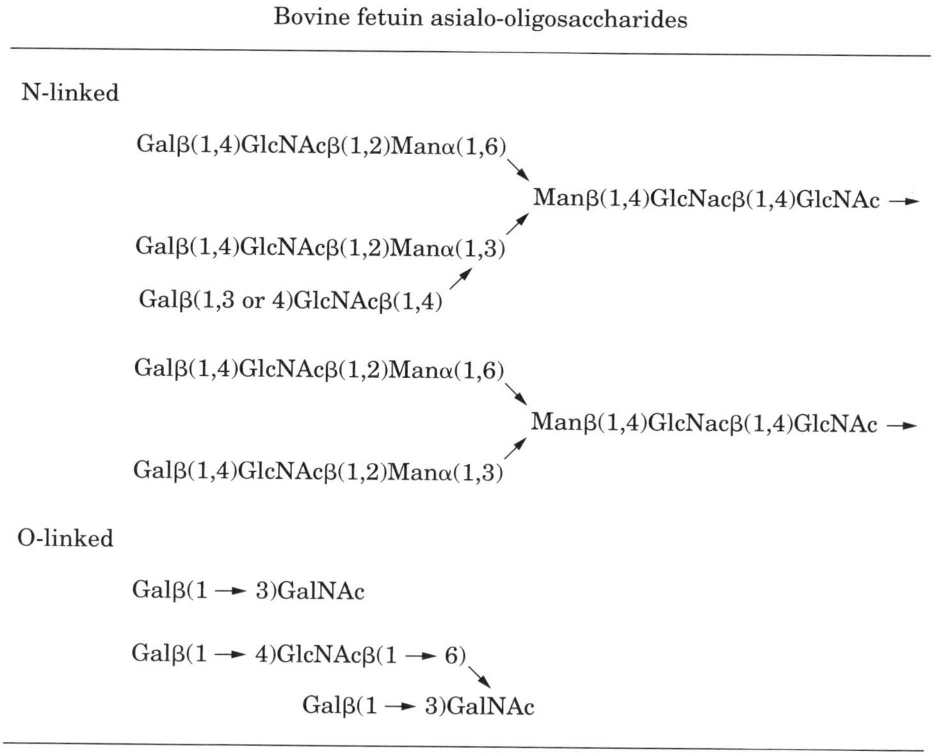

Figure 4. Structures of bovine fetuin asialo-oligosaccharides.

The dramatic growth in the use of molecular biology methods in biomedical research over the past 15 years has seen a corresponding increase in the number of core laboratories engaged in providing essential DNA chemistry services, especially automated DNA synthesis and DNA sequencing. To effectively utilize these resources, it is important for the investigator to have realistic expectations regarding the current capabilities of this technology. To determine the level of technology that can be routinely reached, provide a means of self-evaluation, and gain valuable comparative data beneficial to the field, the ABRF regularly surveys member laboratories and tests facility capabilities through distribution of "unknown" samples for analysis. Two ABRF committees have been established to evaluate nucleic acid chemistries. The Nucleic Acids Committee has published three studies since its inception in 1991, the first designed to assess the type of nucleic acid services provided by core facilities (24), the second designed to assess the accuracy of automated DNA sequencing services (29), and the most recent designed to determine the quality of synthetic oligonucleotides produced by member laboratories (33). The tremendous increase in use of automated DNA sequencing led to the establishment of a second committee in 1995 whose focus is specifically DNA sequencing. The DNA Sequence Research Committee recently completed a general survey and a study designed to assess member laboratories' capabilities in sequencing a moderately difficult GC-rich template (39; *http://www.dfci.harvard.edu/abrfdnaseq*). The results of the most recent of these studies are summarized in the following and provide an overview of the current capabilities of core laboratories that provide DNA synthesis and DNA sequencing services.

ABRF DNA Synthesis Studies

The ABRF has conducted two studies regarding DNA synthesis in core laboratories; the first in late 1992 was a survey designed to obtain information on instrumentation used, favored chemistries, modifications, charge-back fees, and so forth (24). The most recent study was conducted in April 1995 when member laboratories were again asked to respond to a brief survey but were also asked to provide the crude trityl-off, deprotected, and lyophilized products from a 25-base and 50-base 0.2-mmol-scale DNA synthesis for analysis by the Nucleic Acids Committee (33). Seventy-one laboratories participated in the survey and submitted 208 oligonucleotides prepared on a distribution of instruments from various vendors that was essentially unchanged from the 1992 survey. New developments in instrumentation (higher capacity or throughput, lower reagent consumption), lower reagent prices, and increasing competition from commercial providers has led to substantial reductions in charge-back fees; the median rates for oligonucleotide synthesis decreased 20–30% from the previous study to $1.90/base for the 40-nmol-scale synthesis (the actual total cost of synthesis is estimated to be $3.80/base, assuming an institutional subsidy of 50%). Median annual output from these facilities had increased more than 20% to 1,440 oligonucleotides per year. Although new developments in DNA synthesis chemistry have occurred (e.g., fast deblocking reagents), most respondents (80%) continued to use phosphoramidites with the conventional *N*-benzoyl and *N*-isobutyryl protection groups and controlled pore glass (CPG) supports, which suggests that these traditional reagents continue to provide the highest overall satisfaction.

The synthetic oligonucleotides submitted to the committee were analyzed by capillary electrophoresis (CE) to estimate overall yield and purity and were used as primers in DNA sequencing reactions as a functional measure of quality. Overall yield by CE of most of the oligonucleotides was very good. Purity ranged from 14 to 95%, with an average of 74% for 25-mers (Fig. 5) and from 1 to 92% with an average of 50% for 50-mer (data not shown). Average median coupling efficiencies of 98.9% for 25-mers and 98.7% for 50-mers were calculated. Coupling efficiencies of >98% are usually guaranteed by reagent and instrument manufacturers, and this number has been used previously as a minimum standard. Thus, the vast majority (approximately 85%) of crude oligonucleotides submitted for analysis exceeded the minimum standard. To put coupling efficiencies into perspective, it should be noted that with a coupling efficiency of 98%, the maximum yield of desired product for a 25-mer, 50-mer, and 100-mer synthesis would be 61%, 37%, and 14%, respectively, with the remaining material in each preparation resulting from $n - 1, n - 2, n - 3, \ldots$ species. Because most facilities perform quality assurance measures and offer purification services, many of the substandard oligonucleotides would have been detected or eliminated before distribution.

As an additional measure of quality, the crude 25-mers were tested for their ability to function as primers in automated DNA sequencing reactions. A total of 103 of 104 25 mer submitted generated sequence data, and most of the data were remarkably good. Four primers produced perfect sequences over a 400-base window downstream from the primer target; half generated fewer than four errors (including insertions, deletions, substitutions, and ambiguities), and only those in the bottom quartile generated more than seven errors. The number of sequencing errors generated by each 25-base primer was compared with the CE-measured purity of that primer (see Fig. 5). Surprisingly, primer purity does not strictly correlate with sequencing fidelity. Primers that were >70% pure generally produced sequence data with an accuracy >97.5% over the 400-base window. Primers with a purity <70% yielded mixed sequencing results, although 18 of 25 primers that were only 54–70% pure still generated good (<10 errors) sequence data. Overall, more than 85% of the crude 25-mer oligonucleotides yielded unedited sequence data that were >97.5% error free over the 400-base test window. Because unprocessed postsynthesis samples were requested, these results also suggest that purification or desalting provides only marginal benefit for the synthesis in which the average coupling efficiency exceeds 98.5%.

ABRF DNA Sequencing Studies

The ABRF has conducted two formal studies that focus exclusively on DNA sequencing services. A 1995 survey, conducted by the Nucleic Acids Committee (29), was designed to determine the accuracy of sequences obtained using dye-primer or dye-terminator chemistry and *Taq* polymerase. A standard double-strand DNA template was mailed to 80 member laboratories; 44 responded by submitting 83 DNA sequence data sets for analysis. These were grouped by chemistry utilized (dye terminator or dye primer) and whether postrun editing was employed (edited or unedited). The results showed that, using *Taq* polymerase, dye-primer chemistry provided the longest and most

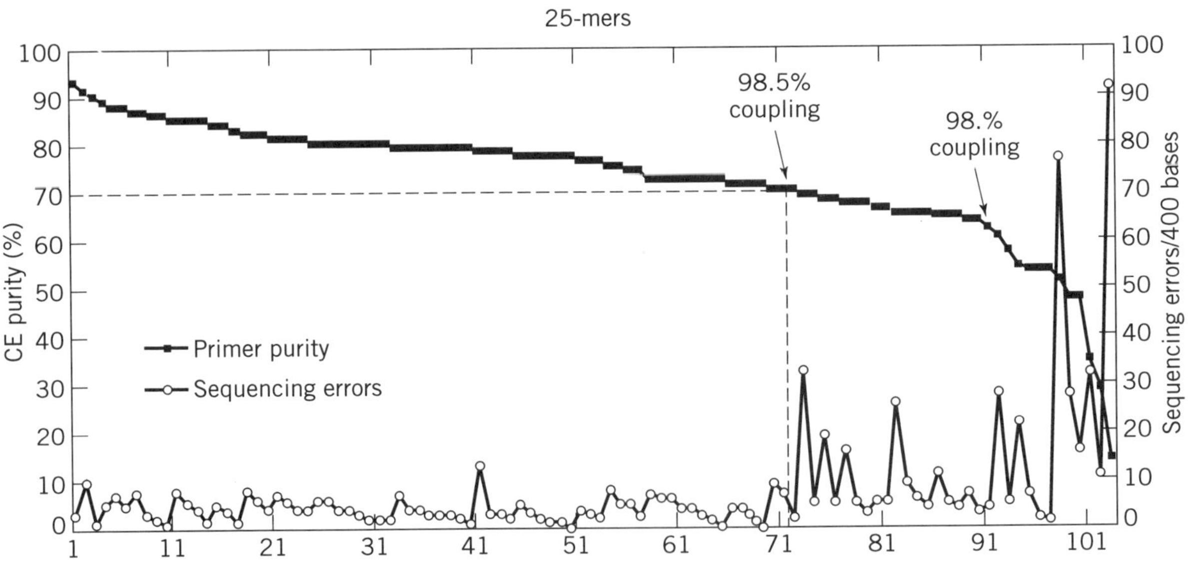

Figure 5. Oligonucleotide purity and performance as sequencing primers. Purity and sequence performance of 25 mer. The capillary electrophoresis (CE) determined purity (■) of the 25-mer oligonucleotides is illustrated in descending order from highest (95%) to lowest (14%) purity. Also plotted is the number of sequencing errors (○) (deletions, insertions, miscalls, and ambiguities) found in the first 400 bases of unedited, automatically called sequence when the 25 mer were used as sequencing primers. CE purity of 70%, representing approximately 98.5% average coupling efficiency, is indicated. The industry standard 98% average coupling efficiency is indicated.

accurate sequence. Edited dye-primer data were >95% accurate out to 400–450 base pairs, whereas edited dye-terminator data were accurate at this level only to 300–350 base pairs. However, most laboratories in this sampling (75%) prefer the dye-terminator protocol, presumably because of its versatility and convenience. Laboratories that manually edit dye-primer data were able to obtain an additional 100 base pairs of data. Surprisingly, editing of dye-terminator data did not increase the length of read for this template.

The DNA Sequence Research Committee's 1996 survey was designed to assess member laboratories' ability to sequence a moderately difficult double-stranded plasmid DNA sample containing a GC-rich insert (69% GC over 926 base pairs) while studying the effectiveness of protocols developed for GC-rich samples, changes in sequencing chemistry, performance according to sequencing hardware, the type of products used, and the effect of editing. In addition, a questionnaire was included to allow comparisons with previous studies (6,7,24). Responses to the questionnaire revealed that the number of facilities performing automated DNA sequencing and the number of samples sequenced per year continues to grow. In 1989, 12 laboratories participated in the survey and reported sequencing an average of 240 templates per year. In comparison, in 1993, 37 laboratories reported sequencing an average of 2,100 templates per year, and the 1996 survey showed 53 laboratories sequencing an average of 4,560 templates per year, of which 76% were double-stranded plasmids and 21% were polymerase chain reaction (PCR) products. Fluorescent techniques were used by all contributing facilities in 1996, dye-terminator chemistry was used by 82%, and the new TaqFS (Perkin Elmer/Applied Biosystems Division [PE/ABD]) enzyme was used by two-thirds. The 53 laboratories that participated in the 1996 study generated 78 sequencing data sets for analysis. The majority of respondents submitted data obtained using dye-terminator sequencing chemistry and the enzyme TaqFS. The sequences were ranked according to length of entirely correct sequence. Keeping in mind that this was a difficult sequence and a routine run was requested, the average length of correct unedited sequence for all machines was 284 bases and the longest was 615 bases. Gel length, manual data review, and sequencing chemistry all influenced sequence accuracy and length. The average correct unedited length of read for PE/ABD Stretch instruments equipped with 36- or 48-cm plates was 362 bases compared with an average of 232 bases for all other instruments. In several direct comparisons, it was shown that an average of 157 more bases of correct sequence were obtained by manual editing for this difficult template. Some facilities submitted data plus or minus dimethylsulfoxide (DMSO) in the sequencing reactions. In most cases DMSO increased the length of correct sequence; however, many of the top-ranked responses did not use DMSO. The 1997 study focused on elucidating the benefits of DMSO, altered thermocycling procedures, and editing when sequencing difficult GC rich templates. A summary of the average error-free length of read depicted by machine, well to read (length of the gel), and a comparison of edited versus unedited sequence can be seen in Figure 6.

In summary, studies conducted by the ABRF reveal that the vast majority of DNA core facilities that offer DNA synthesis services produce synthetic oligonucleotides of high quality. The typical facility in 1996 utilized cyanoethyl phosphoramidite chemistry to produce more than 1,400 synthetic oligonucleotides per year. The average chargeback fee was $1.90/base for 40-nmol scale (an estimated total cost of $3.80/base assuming an institutional subsidy of 50%). Surprisingly, even crude primers of relatively low purity were found to function well when used in automated DNA sequencing. Primers with a purity of >70% were found to function well as DNA sequencing primers without any further purification or even desalting, whereas primers with purity <70% were less likely to function well in DNA sequencing reactions. The average DNA core laboratory that offered automated DNA sequencing services in 1996 typically used dye-terminator chemistry and TaqFS polymerase. A median of 4,560 templates per year were sequenced for an average fee of $29.50/template (estimated total cost $59/template assuming an institutional subsidy of 50%). Minimum template requirements were 0.5–1.0 μg DNA, from which the user could expect to obtain 400–800 base pairs depending on the length of the gel. A recent study (39) has shown that postrun editing of the raw data can substantially increase the sequence accuracy from difficult GC-rich templates when using the most popular chemistry and enzyme.

MASS SPECTROMETRY

Gary Hathaway
California Institute of Technology
Pasadena, California

Kristine M. Swiderek
ZymoGenetics
Seattle, Washington

Often regarded as a supportive or adjunct technique by biomolecular resource laboratories, and initially employed by them largely as a procedure for the quality control of synthetic peptides (17), mass spectrometry is rapidly becoming a major analytical tool for biopolymer analysis (in several instances it has become the only available tool for their structural analysis). With the advent of electrospray ionization (ESI) and matrix-assisted laser desorption ionization (MALDI) methods, mass spectrometry has become a popular and powerful method for analyzing large biopolymers such as peptides, proteins, and both native and modified oligonucleotides (92–97). No longer an exercise reserved for specialty laboratories, the use of mass spectrometry for sequencing picomole quantities of peptides and proteins is becoming widespread (see test sample results later in this article), with a few laboratories able to achieve results in the femtomole range (98,99). Used in conjunction with automation and databases, such sequence and fragmentation mass data can provide identification of biologically interesting proteins isolated by two-dimensional gel electrophoresis (100–102). This technique, when applied to the study of cellular response to a given stimulus as reflected by its entire complement of proteins

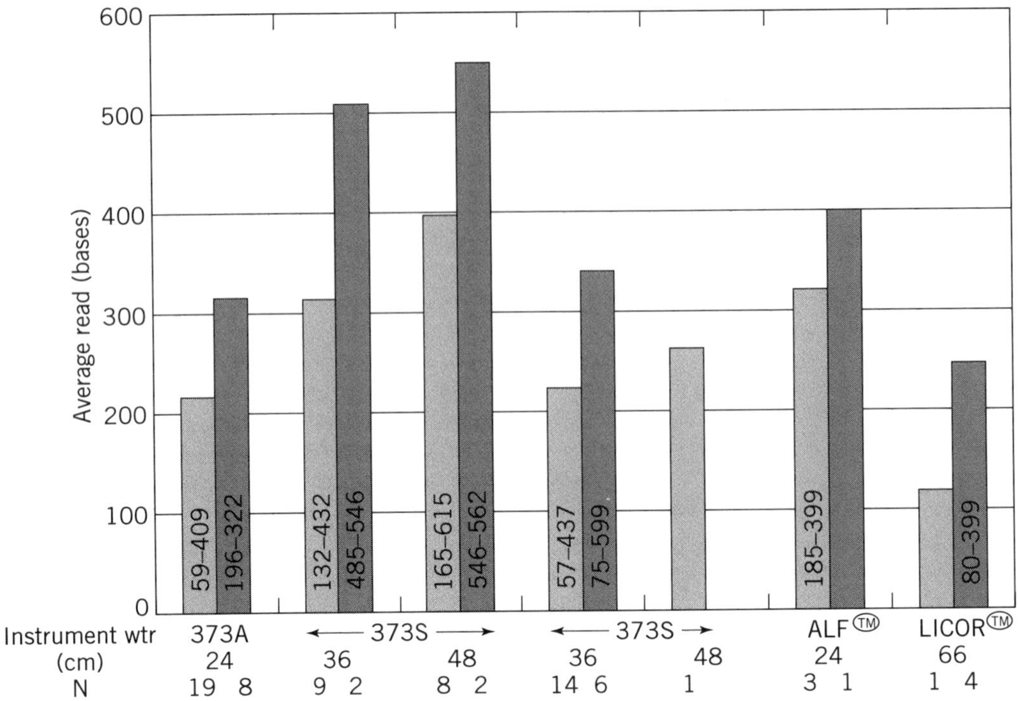

Figure 6. A summary of the average error-free length of read depicted by machine (instrument) and well to read (WTR; length of the gel in centimeters). Values for unedited sequences are open bars, and values for edited sequences are shaded bars. Ranges and the number of samples per group *(N)* are shown.

at a particular point in time, has come to be known as proteome science (103) and is an area of intense effort. If compared with the conventional N-terminal sequence analysis by Edman degradation, the direct generation of peptide fragments and their sequence analysis within coupled mass analyzers (MS/MS) is faster and more sensitive (although data interpretation can sometimes take longer) (104). It has the advantage that modified amino acids can be identified more easily and that a blocked N-terminus does not present a problem for a successful analysis. On the other hand, amino acids of equal mass (such as isoleucine and leucine) generally cannot be distinguished, and glutamine and lysine, which differ by only 0.036 mass units, are determined only with difficulty or after chemical modification of lysine. Another disadvantage is that sequence analysis utilizing mass spectrometric techniques is limited to relatively short peptides. Peptides longer than about 25 amino acids can be sequenced with mass spectrometry, but only if special techniques (which may not be generally available in core laboratories) such as ladder formation through cyclic chemical degradation are employed or by the use of exoproteases such as carboxypeptidase and aminopeptidase (105,106). This allows readout of the sequence by a single mass determination. Automated sequencing by Edman degradation is often less ambiguous and easier to interpret than de novo sequence analysis of peptides by mass spectrometry, which remains a somewhat laborious technique that requires great expertise. Nevertheless, for the identification of peptides and proteins of known sequence, mass spectrometry has become the method of choice.

The goals of the ABRF Mass Spectrometry Committee include determining current state-of-the-art mass spectrometry capabilities in resource laboratories, providing test samples to permit anonymous self-evaluation, and fostering mass spectrometric techniques relevant for the biochemist. As part of an initial research study, the committee conducted a survey designed to obtain information about current mass spectrometric capabilities in 284 ABRF laboratories. The survey revealed that close to 90% of the respondents were engaged in mass spectrometry, and the respondents expressed a desire to receive a test sample designed to address basic issues of mass accuracy and resolution. In addition, the test sample would measure how well the laboratories identified peptides in mixtures that utilize mass spectrometric data with database analysis. To assist them in the identification of the unknown sample, participants were allowed to use adjunct methods such as HPLC separation of components, amino acid analysis, and enzymatic digestion by endo- and exoproteases. The more advanced laboratories were invited to perform collisional activated dissociation (CAD) and/or postsource decay (PSD) fragmentation to achieve direct mass information on fragments and to use this information for the process of identification.

Initial Results of the Survey

As mentioned, 95 out of 107 (88.8%) laboratories that responded reported using some form of mass spectrometry. The majority of the laboratories had only recently added the technique; 19 had less than 1 year experience, 60 reported between 1 and 10 years experience, and only 8 reported more than 10 years experience in mass spectrometry. More than 98% of the mass spectrometers were reported to have been obtained commercially, and although most laboratories had only one, surprisingly, six laboratories reported having more than three. The choice of hardware showed about half (52.5%) that used MALDI or time of flight (TOF) and close to 37% used ESI/quadrupole instrumentation. A few used the newer technology of ESI quadrupole ion trap analysis. Only two laboratories reported using fourier transform ion cyclotron resonance mass analyzers (FTICR). According to the results, the most frequently analyzed samples were peptides and proteins, but 36 laboratories reported doing carbohydrate or glycoprotein sample analysis and 24 reported analyzing oligonucleotides, and other samples were lipids, peptide nucleic acids (oligonucleotide analogs that have a peptidelike backbone), and small organic molecules. Most sites ask for 1 to 10 pmol of sample for a mass determination. Surprisingly, and despite recently reported and dramatic results using mass spectrometry as the sole means for the identification of proteins (98,99), a majority surveyed said they perceived mass spectrometry as a supportive technique.

Finally, to gauge current interest, respondents were asked what they felt was most important in terms of addressing the future needs of the core laboratory in the field of mass spectrometry. The highest responses were increased testing, more education in mass spectrometry, mass spectrometric techniques, evaluation of instrumentation, and methods development.

Preliminary Results of the Analysis of ABRF MS1 and MS2 Test

Participants were asked to analyze two samples. MS1 was *N*-acetylated porcine angiotensinogen with average mass 1,801.085. MS2 was the same peptide mixed in equal portions with the *N*-acetylated peptide from human, mass 1,802.075. The two peptides differed only in four residues at their C-termini. Laboratories were asked to determine these values with highest accuracy and to report both mass values and resolution (calculated by the full width, half mass [FWHM] method) as a test of ability to separate closely spaced ions. Participants were told that the second sample contained two peptides and that HPLC might be required to separate them. Also, it was suggested that certain other procedures such as AAA, exoprotease or endoprotease digestion, and CAD or PSD fragmentation might be helpful in identifying the peptides.

Preliminary results based on 26 sites reporting show an average error of 387 ppm (this is equivalent to 99.96% accuracy and gives an uncertainty of ± 0.8 Da for a 1.8-kDa peptide). This amount of error is close to the difference in mass of the two peptides in MS2. Routine introduction of error at this level would prevent the distinction of an amidated from a free carboxylic acid at the C-terminus of a 2-kDa peptide. Error ranged from a high of 1,873 ppm (99.81% accuracy) to a low of 3 ppm (99.9997% accuracy). Interestingly, the error spread for two samples reported by a single site ranged from a low of 5 ppm to a high of 1,327 ppm, with an average spread of 227 ppm. Because it is assumed that the same technique and hardware were used to determine both samples, the explanation for this difference in accuracy may lie with the particular methodology employed or operator technique. The overall dependence of accuracy on hardware was similar for the two most prominent types, with values ranging from 8 to 541 ppm for ESI or quadrupole equipment and 3 to 1,435 ppm for MALDI or TOF. Resolution varied widely, with a high of 10,606 measured with a quadrupole ion trap. Values ranged from 514 to 3,001 FWHM for ESI or quadrupole and 120 to 5,457 for MALDI or TOF instruments. Five laboratories correctly identified both peptides, and 13 identified at least one of the peptides. Thus nearly 70% (18 of 26) of the respondents were able to identify one or both peptides. This result demonstrates that a significant number of resource laboratories possess a high degree of development both in the ability to use mass spectrometry and in the sophisticated adjunct methods such as database analysis for the identification of peptides and proteins.

Conclusions

The results of the survey and test sample analyses showed that although the average resource laboratory is relatively young in the field of mass spectrometry it already possesses a high degree of expertise not only for the accurate measurement of peptides but also in the ability to analyze a diverse set of sample types, to employ advanced techniques, and in general bring mass spectrometry to bear in a concerted way with other established methodologies in solving difficult analytical problems in biology. This was exemplified by the many laboratories that correctly identified test samples not only from mass measurements but also by combining data from AAA enzymatic digestion, and advanced mass spectrometric techniques. Results clearly demonstrated the high general expertise and awareness of many laboratories of a wide variety of database search algorithms and the ability to access them via the Internet.

This initial research study better defines the work of the mass spectrometry research committee. Based on the results of the survey and test samples, future studies may now be devised to focus exactly on the interests of laboratories engaged in the application of mass spectrometry to problems of biological significance.

SURVEYS OF CORE LABORATORIES

Ronald L. Niece
University of California–Irvine
Irvine, California

Kathryn M. Ivanetich
University of California–San Francisco
San Francisco, California

The typical, or average, biotechnology core laboratory can be characterized by a description of the instruments, space, staff, budget, and performance parameters for the resources provided. This information has been periodically obtained from general surveys of ABRF member laboratories. The first survey to provide a formal description of biomolecular core laboratories was conducted in 1987 (5);

it identified several services including protein/peptide sequencing, AAA, HPLC purification of proteins and peptides, peptide synthesis, and DNA synthesis, which are now considered traditional services. The most recent published data from a survey of facility operations (7) have shown that core facilities have been quick to incorporate new instrumentation into their repertoire and now offer technologies as diverse as capillary electrophoresis, DNA sequencing, mass spectrometry, and carbohydrate analysis (see Fig. 7). The survey data also show that many core facilities are resources not only for their home institutions but for outside investigators as well. These core facilities offer a broad range of services (Fig. 7), with outside charges usually being higher than in-house charges (7).

Combined data from several surveys of ABRF member core laboratories (7,34) have shown that academic institutions and research institutes house more than three-quarters of member laboratories, and industry has most of the remaining 25%, with only a small percentage of such core facilities in governmental settings (universities excluded). Core facilities require instrumentation, suitable laboratory and office space, appropriately trained personnel, and financial resources. A typical core facility is staffed with four full-time staff members, including one member with a doctoral degree, one with a master's degree, and two with bachelor's degrees. Directors have a median of 13 years experience in the analysis and synthesis of biological molecules, and staff have a median of 6.9 years experience (34). A typical facility houses six major instrument systems (50) in 1,000 sq. ft. of laboratory space. An average, non-capital, annual operating budget is $250,000, half of which comes from charge-back fees for services and reagents. Nearly all core facilities require substantial subsidization, averaging 50% of total operating budget, to cover the difference between user fee income and total expenses.

Performance with respect to the quantity of sample needed or scale of synthesis, turnaround time, and charge for selected services is listed in Table 5. The level of subsidy an institution provides its core facility is largely determined by the portion of operating budget that the facility obtains from its users. For highly subsidized facilities the user charges are relatively low, reflecting the high subsidy; for the 1–2% of facilities that operate without subsidy, the charge is the actual cost of performing the service, which very likely is much higher than the corresponding charge from a heavily subsidized facility. The spread in charges as shown by the large standard deviations may reflect both differences in subsidy and specifics of what is included in the standard charge. Both lower subsidy from institutional sources and higher levels of performance by the facility (including sensitivity, scale, extent of service, and turnaround time) are reflected in higher charges for users. Services for outside users generally incur an additional charge.

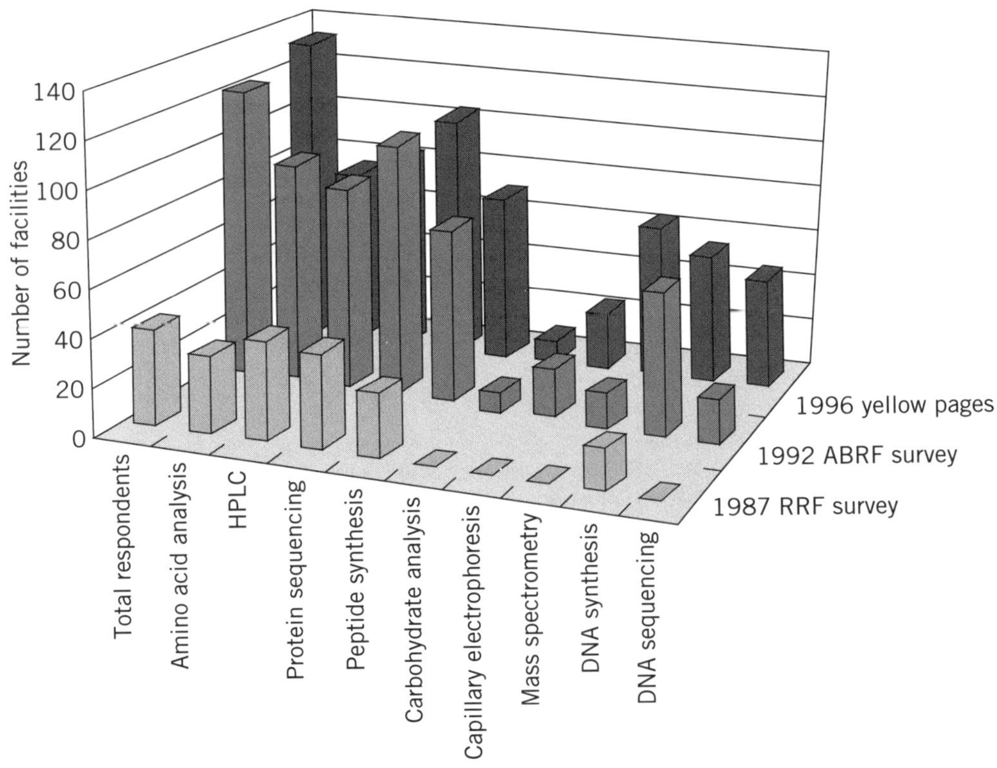

Figure 7. The number of core facilities offering selected services. Data from survey of laboratories interested in the informal organization, Research Resource Facilities (RRF) (5), respondents to survey conducted by the ABRF Survey Committee in 1992 (7), and the 1996 ABRF Membership Directory, Section II (Yellow Pages): Facilities Offering Services to Outside Investigators (available from ABRF business office or ABRF Web site).

Table 5. Services, Turnaround, Charges, and Costs Determined from ABRF Surveys

	Scale of service		Turnaround (days)	In-house charge ($)		Subsidy (% of operating costs)	Calculated cost without subsidy ($)
	Mean ± SD	Units	mean ± SD	Setup mean ± SD	Per cycle mean ± SD	mean ± SD	
Amino acid analysis	32 ± 220	μg	6 ± 4	35 ± 19		45 ± 45	64
HPLC	0.51 ± 1.4	nmol	7 ± 6	106 ± 74		47 ± 39	200
Protein sequencing	73 ± 92	pmol	9 ± 7	117 ± 79	17 ± 10	42 ± 41	788[a]
Peptide synthesis	0.56 ± 1.3	mmol	20 ± 12	194 ± 156	35 ± 12	35 ± 36	1,214[a]
Carbohydrate analysis	81 ± 74	μg	6 ± 4	20 ± 14			
Capillary electrophoresis	17 ± 18	pmol	3 ± 2				
Mass spectrometry	0.15 ± 0.2	nmol	6 ± 4	74 ± 48		30 ± 33	106
DNA synthesis	0.2 ± 0.1	μmol	4 ± 2	22 ± 13	3 ± 1	21 ± 33	123[a]
DNA sequencing	2 ± 2	μg	6 ± 4	41 ± 32		30 ± 37	59

Note: Mean ± standard deviation (SD) for scale of service with units noted, anticipated turnaround time for completion of the synthesis or analysis, in-house charge with setup and per-cycle charges (where appropriate), subsidy, and calculated charge for full cost recovery.
[a]Protein sequence (25 residues), synthetic peptide (15 residues), oligodeoxyribonucleotide (25 bases).
Source: Data from Ivanetich et al. (7).

The self-reported data on performance in the member laboratories collected by the surveys describe analytical capabilities in operating laboratories. This assessment of performance level is corroborated by the analysis of results from the "unknown" samples distributed by the ABRF research committees (see other sections of this article). In addition, members of ABRF have critiqued instrumentation by rating performance, service, technical support, and manufacturer-provided reagents. These evaluations, provided by many instrument users, have helped guide manufacturers in meeting the needs of the research and quality control communities by identifying deficiencies and useful features (11,50). A continuing concern of core facilities and their users has been the cost of consumables (7,11) because these expenses most frequently are included in the portion of operating costs that is not subsidized.

A number of trends can be discerned from the responses to these surveys conducted over a period of nearly 10 years. First, the number of core facilities continues to increase, from 146 ABRF member facilities in 1988 to well over 300 today. The average operating budgets of these facilities have increased from $158,000 to $279,000 per year; the level of institutional subsidization has remained fairly constant at approximately 50%. Although the majority of ABRF member facilities are rooted in protein chemistry, the number of core facilities that provide DNA chemistry services has increased substantially. For example, the number of ABRF core laboratories that offer DNA sequencing has doubled in the most recent (1996) survey. Although costs for protein chemistry services have either remained steady (e.g., amino acid analysis) or increased (e.g., peptide synthesis and protein sequencing), the cost of DNA chemistry services has declined dramatically. DNA synthesis costs decreased ~74% between 1988 (5) and 1993 (7). Core facilities are often the providers of advanced technology, as evidenced by the rapid implementation of new instrumentation and methodologies. Mass spectrometry was not listed as a service by any of the respondents in 1987, but in 1992 more than 10% offered the service and in 1998 nearly half of the respondents offered mass spectrometry (K. Ivanetich, unpublished data).

In summary, the average core facility characterized by survey data includes six instrument systems operated by four staff in about 1,000 sq. ft. Services offered include the traditional protein/peptide sequencing, AAA HPLC purification of proteins and peptides, peptide synthesis, and DNA synthesis and newer technologies such as capillary electrophoresis, DNA sequencing, mass spectrometry, and carbohydrate analysis. Most core facilities receive about 50% of their operating budgets from institutional funds to supplement their cost-recovery income. Core facilities evaluate instrument performance and service and technical proficiency through anonymous surveys. These data identify instrument performance levels and define the state of the art in real-world settings.

LABORATORY QUALITY AND COMPLIANCE

NADINE RITTER
Abbott Laboratories
Abbott Park, Illinois

BETH FOWLER
Autoimmune, Inc.
Lexington, Massachusetts

JOHN DOUGHERTY
Lilly Research Laboratories
Indianapolis, Indiana

AUGUSTINE SMITH
Abbott Laboratories
Abbott Park, Illinois

ELEANOR CANOVA-DAVIS
Genentech
San Francisco, California

BARBARA GHRIST
Genentech
San Francisco, California

The roles of analytical resource facilities in biotechnology range from supporting early research and discovery efforts to conducting validated testing for product development. Facilities that comply with the specific regulations governing product manufacturing and testing can also test drug products manufactured for commercial purposes. There are significant differences in the documentation and compliance requirements placed on analytical laboratories depending on how a client intends to use the data generated. Discussions between the client and the facility should identify these considerations, preferably before the analyses are conducted. If the facility can meet the client's quality and compliance needs, using a contract analytical facility can provide numerous advantages in biotechnology product development.

Most resource facilities have the instrumentation and expertise needed to investigate and physicochemically characterize many biomolecules of interest. In the research and discovery phase, most investigations are aimed at identifying active moieties. A company's research and development team could contract with a resource facility to perform selected analytical techniques that may yield critical clues about the nature of the biomolecules being studied. This strategy leverages the research and development resources across a greater breadth of technologies than may be available in-house.

In later developmental phases, an analytical profile or physicochemical "fingerprint" of the product in its final purified form may be needed. The objectives include determination of the composition of the biological material, its impurities or degradation products, and its stability under various conditions. These data may be used to establish the product's identity, purity, potency, and strength. As in discovery, the analytical methods employed may go beyond a company's in-house capabilities.

Some analytical methods, such as protein or DNA sequence analysis, are needed for only short periods in the product's development cycle. Final product characterization studies may utilize numerous methods but only on a few lots of material. Methods used in stability studies may be required only at periodic intervals over the study's lifetime. Rather than establishing and maintaining a complex analytical laboratory, clients can choose a facility with the exact instrumentation needed to perform a given analysis. A client can stay abreast of advances in technology each time a new study is contracted by choosing laboratories with the desired analytical capabilities. In addition, highly skilled resource facility scientists can assist in planning the appropriate experiments needed to obtain the best results. Finally, outsourcing selected analytical services allows a company to focus more of its internal resources on controlling the critical quality parameters of the product itself.

The level of quality assurance in a resource facility can be matched to the needs of the client base. For any laboratory, clients (and staff) are principally concerned with data integrity. Properly functioning instrumentation, adequate documentation of experiments, and traceability of results to the samples tested are minimal expectations in all laboratories. If the objective of the analysis is to provide information used in product discovery, additional quality measures may be employed. Calibration of equipment, controlled temperature storage of materials, expiration dating of reagents and materials, and documentation of these and other routine laboratory activities demonstrate the quality control of facility operations. In addition, when sample storage, handling and preparation, general and specific experimental protocols, instrument operation parameters, and data analysis methods are well documented according to established facility procedures, the client may consider such a laboratory well suited to provide reliable, reproducible analyses on which to base product design decisions.

A sometimes misused designation for general laboratory quality measures is the term good laboratory practices (GLPs). GLPs are a specific set of U.S. Food and Drug Administration (FDA) requirements for nonclinical (i.e., animal-based) laboratory studies conducted to support product development in the preclinical safety phase (107). The GLPs apply to the procedures used to characterize the test article (the substance used for dosing in the animal and toxicology experiments) to determine its stability, homogeneity, and concentration (108).

The GLP regulations include requirements that must be met by contract facilities that perform specified parts of the total study. At this phase, product's dosing formulation—usually a mixture—is being analyzed rather than the purified active biomolecule. Although a biomolecular resource facility may support the study with selected biochemical analyses, it typically does not perform the entire GLP study. The extent to which the GLP regulations apply to a laboratory depends on the role it will have in designing, conducting, documenting, archiving, and auditing the quality assurance of the study.

One element of setting up a GLP study is the designation of two key parties: a study director and a quality reviewer. The responsibilities of these entities are outlined in the regulations. Also included in the regulations are laboratory operational measures aimed at ensuring the quality of the testing results, such as those listed earlier. These measures are sometimes called good laboratory practices in casual use. When selecting a facility that claims to be a GLP laboratory, a client should clarify the meaning of this designation, discuss the role of the laboratory in the overall study plan, and define the directing and reviewing activities to prevent any misunderstandings.

Analytical procedures to support the chemical, manufacturing, and controls (CMC) sections of pharmaceutical product applications are more likely performed by contract resource facilities. Typically, this function is termed *characterization* of the product. Analytical tests such as protein sequence analysis, mass spectrometry, and AAA may be among these procedures. For these types of analyses, GLP conditions are not formally required, particularly the establishment of the study director and quality reviewer. However, facilities that implement some of the operational elements of quality described in the GLP regulations are better able to provide traceable, auditable proof of the analyses performed. As such, clients can have more confidence in reporting these results in the product application for regulatory approval.

Analytical resource facilities may be contracted to perform lot release testing on commercial products. The U.S.

regulations governing such testing are called current good manufacturing practices (cGMPs) and are published in the *Code of Federal Regulations* section 610 for biological products (109) and sections 210 and 211 for pharmaceuticals (110). There are significant regulatory requirements for virtually every aspect of laboratory operations, including the physical environment in which the laboratory is housed. An analytical laboratory that is designated cGMP must meet stringent quality measures including instrument and equipment standard operating procedures, established method testing procedures, personnel training procedures, data review and approval, and archiving of all files associated with laboratory activities and results. In addition, most process and testing methods are expected to be validated. Modifications to any of these operating procedures or validations must be controlled via an established change control system. Before contracting with (or designing) a laboratory for cGMP testing, the facility should be audited by individuals knowledgeable in both the federal regulations and analytical laboratory operations to ensure the facility meets or exceeds the level of quality expected.

Another type of quality designation that some analytical laboratories may obtain is the ISO 9000 series of measures (American National Standards Institute, 11 West 42nd St., 13th Floor, New York, NY [111]). These measures were established to provide defined levels of quality metrics in businesses ranging from law firms to lamp manufacturing and can be applied to resource facilities. The basis of these measures is to define and document the quality system of an organization: namely, to designate the responsible individual and the method of data management. ISO 9000 registration is obtained when the organization seeks and passes inspection from an accredited third party (112,113). Certification is conferred for a specified period of time, after which reinspection and recertification are required. ISO 9000 certification adds to the quality of an analytical facility in direct proportion to the level of operations already present. Usually, such laboratories are relatively small organizations with clearly defined roles and responsibilities. Interviews with facility directors and audits of laboratories allow the client to assess the quality system in place and to evaluate its integrity.

An additional global regulatory initiative is taking shape in the European Union (European Standard EN 46001 [see Ref. 114]). Similar to the ISO 9000 requirements, EN 46000 is currently expected to apply to a specific classification of medical products, although the nature of the regulations may expand in the future. Likewise, quality principles are being codified in many other countries around the world. This trend reflects a movement toward global harmonization of quality practices and product regulation, as seen in activities by the International Committee on Harmonization (see Ref. 115).

It is important to note that these quality and compliance designations are not mutually inclusive or exclusive. A laboratory may be ISO 9000 registered without complying with the GLP or cGMP regulations. Samples designated for a GLP study may be handled in the same facility that performs analyses for other, nonregulated studies, as long as the GLP samples are handled in accordance with applicable FDA regulations. Both the client and the resource facility must understand the differences among the various quality measures to have confidence in the compliance of laboratory operations and the appropriate use of the data generated.

In general, analytical resource facilities are ideally equipped to provide the various types of biomolecular analyses required for biotechnology products. As analytical methods and techniques evolve, these facilities are positioned to incorporate technical advances as ongoing improvements to their services. In addition, many of these laboratories may be able to provide the level of quality needed to support specified regulatory uses of the data obtained. As partners in product development, these facilities provide highly skilled, technically advanced teams of scientists with the instruments and expertise needed to perform complex, ever-changing methods. Additionally, as regulatory agencies begin to accept specified biotechnology products for licensing (116), adopting a strategy of outsourcing special bioanalytical services may prove to be an effective way to manage product development costs while ensuring the quality of biotechnology products.

EFFECTIVE USE OF BIOMOLECULAR RESOURCE FACILITIES

CLAYTON NAEVE
St. Jude Children's Research Hospital
Memphis, Tennessee

Progress in biomedical research is becoming increasingly dependent on the utilization of shared resources or core facilities. These facilities typically provide biotechnology reagents and services that require skilled staff and expensive instruments that are most effectively and economically provided by an institution in the form of shared resources. Such core facilities span a broad range in the size and spectrum of services and reagents offered and may include protein or DNA sequencing; DNA, RNA, or peptide synthesis; mass spectroscopy; carbohydrate analysis; and many other possible services. Because the use of sophisticated instrumentation is high in these core facilities, the investigator often assumes that these are turnkey operations (i.e., submit a request, get a reagent or result). In reality, these resources play a critical partnership role in the research endeavor and should be viewed as such to succeed. If you have a core facility available that provides the services you require, the following suggestions may help optimize your use of their expertise.

Before Submission of a Request

Familiarize Yourself with the Core Facilities Available to You and the Range of Expertise, Services, and Reagents They Provide. Knowing that the protein core facility at your institution can or cannot sequence proteins at the femtomole level or knowing that the DNA core facility at your institution will only accept DNA sequencing templates cloned into specific vectors may influence your experimental design. If cost is a deterrent, communicate with the facility to understand the basis for its service charges; some facilities and services are highly subsidized by their host institutions and some are not (see the "Surveys of Core Laboratories" section of this article). One of the reasons for the seemingly high charges at your facility may be that its degree of subsidization via nonuser fee income is much less than the ~50% average level (7); hence, its charges may cover much more than just 50% of the cost of providing the requested service.

Talk to Your Core Facility Director and Staff before You Write the Grant, Design the Protocol, Plan the Experiment, or Run the Sample That Will Utilize Their Facility. Communicating with facility staff will ensure that funds can be budgeted for the needed analyses and syntheses, and the core laboratory can verify that the biotechnological aspects of the proposal are indeed reasonable. Proposing to load an SDS PAGE gel slice directly onto a protein/peptide sequencer or proposing a postsource decay (sequencing) study on a laser desorption or TOF mass spectrometer that is not equipped with a reflectron would not enhance the probability of your grant application being funded. Take advantage of the resource of expertise and services available.

Understand the Limits of the Technology. For instance, although protein hydrolysis and amino acid analysis provide the amount of individual amino acids in a protein or peptide sample, they do not provide the number of amino acid residues unless an accurate mass is available, the sample is pure, and corrections are made for peptide bonds that are difficult to hydrolyze and for loss of less-stable amino acids.

Follow Your Core Facility's Sample Preparation Guidelines. Proper sample preparation is probably the most important step in ensuring success with most biotechnology chemistries. Preparation of clean DNA templates for automated sequencing is a classic example; follow the guidelines given to you explicitly and you will greatly increase your probability of success. Consult with facility staff in planning the appropriate sample preparation strategy before sample submission.

After Receipt of Your Data or Reagents

Understand the Significance of the Data Produced. If your DNA sequencing facility provides you with four-color electropherograms and unedited sequence data, you must be aware that miscalls occur and that it is your responsibility to evaluate the quality of the data and correct errors. Suppose you have submitted a sample for MALDI or TOF mass analysis and the observed mass differs from the predicted by 0.01% using reflectron mode analysis. Is this within the accuracy range of the instrument (thus helping to confirm the identity of the peptide) or does this suggest the presence of a modification? To assess significance, you must familiarize yourself with the technology in use to judge the results obtained. Obviously, the core facility staff will help you with this assessment, but the ultimate responsibility is yours.

Work with the Facility to Correct Any Real or Perceived Deficiencies. If you are unhappy with the results obtained, talk to the facility's staff or director. If problems do occur, your input and experience can often be critical in diagnosing and correcting difficulties. On the other hand, the problems that you believe are present (e.g., absence of cysteine during the amino acid sequencing of a peptide known to contain this amino acid) may reflect an oversight or lack of knowledge on your part (in this case the failure to appropriately modify all cysteine residues before submitting the sample for amino acid sequencing).

What Should You Do If Your Core Laboratory Does Not Provide the Service You Need? First, ask why this particular service is not available. There are many reasons why a given service may not be available, including insufficient demand, staff, expertise, equipment, or space to justify offering the service. For instance, many core laboratories do not offer automated C-terminal protein sequencing because of the relatively meager demand, the large amounts of protein required, and the relatively few amino acid residues that can be sequenced. It is thus important to first determine why the service is not available. Second, if there is sufficient demand to justify the addition of this service, help the core laboratory bring this new service to fruition. Third, ask for a referral, because your core laboratory may well know an outside core laboratory that is particularly well suited to providing the requested service. In this instance, it is important to tell staff at your core laboratory what happened at the outside core laboratory so they may better advise others who follow.

Support Your Core Laboratory. A close partnership between the core laboratory and its users is critical for success. Just as the user needs access to state-of-the-art biotechnologies, the core laboratory can only succeed with the support of its users. Support includes acknowledging all services provided by the core laboratory and forwarding reprints of these publications to the core laboratory, responding to requests made by the core laboratory for research proposals and other support of their shared instrumentation grant applications, and helping the core laboratory gain institutional support, which may range from subsidization of operating costs (to help lower user fees) to equipment acquisitions to gaining additional space. Finally, because much of the work carried out in core laboratories goes without reward in terms of publications, be appreciative of the efforts of core laboratory staff and do not hesitate to coauthor publications with them when their efforts and input go beyond that which justifies a simple acknowledgment.

FUTURE OF CORE LABORATORIES

John Stults
Genentech
San Francisco, California

What does the future hold for core laboratories? The needs of the biological community are changing rapidly, as are the economics and technology of science. As a result, the twenty-first century will present challenges and exciting new possibilities for core laboratories.

Paradigm Shift in the Postgenome Era

The single most important transforming factor in the immediate future is the availability of enormous amounts of genomic sequence. Partial nucleotide sequences from human expressed sequence tags (ESTs) (117,118) already number in excess of 1,000,000 (119). Complete genomes have been sequenced for a number of organisms, including *Saccharomyces cerevisiae* (120). The imminent completion of a number of other genomes, including human, and the advent of the postgenome era will provide an information resource previously unimaginable. This information glut will trigger a paradigm shift in the ways that core laboratories operate and in the services they provide. The emphasis on sequence determination for new proteins, particularly by DNA sequencing and cloning, an overwhelming and perhaps inordinate goal of biology for the past 20 years, will give way to renewed emphasis on biological principles and protein function and its relationship to structure (121). For core laboratories, this information will likely mean that much of traditional protein sequence determination will give way to protein identification via an array of rapidly evolving approaches and technologies, many of which utilize mass spectrometry, and that traditional DNA sequencing will give way to powerful genetic analysis techniques based on hybridization to high-density DNA oligonucleotide arrays (122).

Automation

Automation of many tasks will continue to be a key factor in further reducing operating costs and increasing productivity and reproducibility. Proper, efficient sample preparation for analysis has been and will continue to be a bottleneck for rapid and reliable analyses, making it a primary target for automation. DNA synthesis chemistries and instrumentation have already reached a state of extremely high throughput and reliable operation that is perhaps manifested best by the 10-fold reduction in the in-house charge for a 25-mer DNA oligonucleotide that has occurred over the past 10 years (5,33). Instrumentation is now available that can synthesize and, in some cases, automatically purify 48 or more DNA oligonucleotides. DNA sequencing is rapidly approaching high-throughput operation such that some instruments can analyze as many as 96 samples in parallel. To keep pace with this instrumentation, increased attention is being given to the use of robotics to automate DNA template preparation. Peptide synthesis is moving more and more toward automation and high-throughput operation with the availability of instruments capable of synthesizing 12 or more peptides followed by the cleavage of the newly synthesized peptides from their solid supports. In the case of both DNA and peptide synthesis, increased automation accompanied by decreased reagent costs due to the smaller synthesis scales adopted has resulted in further cost reductions. Amino acid analysis has been a reasonably high-throughput method for years, with a typical instrument able to complete an analysis in 1 h. Protein sequencing remains slow by comparison, yet its operation continues to be simplified and its sensitivity of detection increased, and automated operation for analyzing multiple samples in tandem is now available. In contrast, mass spectrometry is inherently a rapid technique. Automation of mass spectrometry, although done for years with GC and MS, has only recently become available for protein and peptide analysis. The capabilities of the instrumentation are sure to keep advancing; in general, the higher the throughput of the instrumentation, the more suited it is for use in core laboratories.

Future Opportunities

Undoubtedly, the availability of the sequences of the approximately 100,000 human proteins will pose many more questions than it answers and will spawn a new area of investigation called proteome research. This availability will also result in a renaissance in studies directed toward understanding how protein structure relates to function. In their simplest form, the questions are as follows: What are the functions of these proteins? How do they accomplish their tasks? How can we take advantage of this knowledge to intervene in a bewildering array of diseases? It is not difficult to illustrate our ignorance in this regard. Currently, the only reliable means for predicting three-dimensional structure and function from a newly discovered protein sequence is to carry out a database search and hopefully identify homologous proteins in the same family or in other organisms. If the sequence is truly unique, no laboratory currently is capable of accurately predicting the function of this protein, its three-dimensional structure, or how the intricate folding of its one-dimensional amino acid sequence gives genesis to function. Compared with the 100,000 human proteins, the current availability of only a few dozen structures for protein–ligand complexes surely pales in comparison and bodes extremely well for increasing the need for identifying sites of protein–protein and protein–ligand cross-linking and for the use of synthetic peptide analogs and other approaches to identify active sites in proteins. Similarly, opportunities for core laboratories abound in terms of the biophysical instrumentation (e.g., spectrofluorometers, scanning and isothermal calorimeters, analytical ultracentrifuges, surface plasmon resonance, and circular dichroism instruments) needed to quantitatively characterize these interactions.

All these factors suggest that the future of biotechnology core laboratories is indeed bright and far more interesting than in the past, with the tedium of determining primary sequences being replaced rapidly by the more interesting and pertinent task of determining the reason for that order. Core laboratories are a tremendous intellectual and biotechnological resource whose importance can only be increased by the increasing financial constraints that helped promote their rapid emergence and by the ever-increasing complexity that seems to lie beneath each biochemical and biomedical discovery. Although core laboratories are synonymous with sophisticated instrumentation, in the end it is the efforts of people that lead to advancements in science. Ultimately, it is the interdisciplinary knowledge, research experience, dedication, and ingenuity of core laboratory personnel that will lead to their continued importance and success.

ACKNOWLEDGMENTS

We especially thank all past and present members of all of the ABRF Research Committees that compiled the data on which much of this article is based and all of the member laboratories who participated in this research. Special thanks are given to all the individuals who reviewed sections of this article for their critical insights and to the National Science Foundation (DIR9003100), the Department of Energy (DE-FG02-95ER61839), and all past and present ABRF members and corporate sponsors who have provided support for the ABRF.

BIBLIOGRAPHY

1. R.M. Hewick, M.W. Hunkapiller, L.E. Hood, and W.J. Dreyer, *J. Biol. Chem.* **256**, 7990–7997 (1981).
2. M.J. Gait, in M.J. Gait ed., *Oligonucleotide Synthesis: A Practical Approach*, IRL Press, Oxford, 1984, pp. 1–22.
3. R.L. Niece, D. Atherton, A.V. Fowler, R. Kutny, and A.J. Smith, in K.A. Walsh ed., *Methods in Protein Sequence Analysis*, Humana Press, Clifton, New Jersey, 1987, pp. 317–319.
4. R.L. Niece, K.R. Williams, C.L. Wadsworth, J. Elliott, K.L. Stone, W.J. McMurray, A.V. Fowler, D. Atherton, R. Kutny, and A.J. Smith, in T. Hugli ed., *Techniques in Protein Chemistry*, Academic Press, San Diego, Calif., 1989, pp. 89–110.
5. K.R. Williams, R.L. Niece, D. Atherton, A.V. Fowler, R. Kutny, and A.J. Smith, *FASEB J.* **2**, 3124–3130 (1988).
6. R.L. Niece, C.M. Beach, R.F. Cook, G.M. Hathaway, and K.R. Williams, *FASEB J.* **5**, 2756–2760 (1991); **6**, 792 (1992).
7. K.M. Ivanetich, R.L. Niece, M.F. Rohde, E. Fowler, and T.K. Hayes, *FASEB J.* **7**, 1109–1114 (1993).
8. D.W. Speicher, G.A. Grant, R.L. Niece, R.W. Blacher, A.V. Fowler, and K.R. Williams, in J.J. Villafranca ed., *Current Research in Protein Chemistry*, Academic Press, San Diego, Cal., 1990, pp. 159–166.
9. R.L. Niece, D. Atherton, C.M. Beach, R.F. Cook, A.V. Fowler, G.M. Hathaway, R. Kutny, A.J. Smith, and K.R. Williams, *J. Protein Chem.* **9**, 262–263 (1990).
10. J.W. Crabb, L.H. Ericsson, D. Atherton, A.J. Smith, and R. Kutny, in J.J. Villafranca ed., *Current Research in Protein Chemistry*, Academic Press, San Diego, Cal., 1990, pp. 49–61.
11. C.M. Beach, G.M. Hathaway, T.K. Hayes, A.J. Smith, and R.L. Niece, *ABRF News* **2**, Supplement 1–7 (1991).
12. R.L. Niece, L.H. Ericsson, A.V. Fowler, A.J. Smith, D.W. Speicher, J.W. Crabb, and K.R. Williams, in H. Jörnvall, J.-O. Höög, and A.-M. Gustavsson eds., *Methods in Protein Sequence Analysis*, Birkhäuser Verlag, Basel, 1991, pp. 133–141.
13. L.H. Ericsson, D. Atherton, R. Kutny, A.J. Smith, and J.W. Crabb, in H. Jörnvall, J.-O. Höög, and A.-M. Gustavsson eds., *Methods of Protein Sequence Analysis*, Birkhauser Verlag, Basel, 1991, pp. 143–150.
14. K.Ü. Yüksel, G.A. Grant, L.M. Mende-Mueller, R.L. Niece, K.R. Williams, and D.W. Speicher, in J.J. Villafranca ed., *Techniques in Protein Chemistry II*, Academic Press, San Diego, Calif., 1991, pp. 151–162.
15. G.E. Tarr, R.J. Paxton, Y.-C.E. Pan, L.H. Ericsson, and J.W. Crabb, in J.J. Villafranca ed., *Techniques in Protein Chemistry II*, Academic Press, San Diego, Calif., 1991, pp. 139–150.
16. D.L. Crimmins, G.A. Grant, L.M. Mende-Mueller, R.L. Niece, C.A. Slaughter, D.W. Speicher, and K.Ü. Yüksel, in R. Hogue-Angeletti ed., *Techniques in Protein Chemistry III*, Academic Press, San Diego, Calif., 1992, pp. 35–43.
17. A.J. Smith, J.D. Young, S.A. Carr, D.R. Marshak, L.C. Williams, and K.R. Williams, in R. Hogue-Angeletti ed., *Techniques in Protein Chemistry III*, Academic Press, San Diego, Calif., 1992, pp. 219–229.
18. D.J. Strydom, G.E. Tarr, Y.-C.E. Pan, and R.J. Paxton, in R. Hogue-Angeletti ed., *Techniques in Protein Chemistry III*, Academic Press, San Diego, Calif., 1992, pp. 261–274.
19. R.L. Niece and R.H. Angeletti, *FASEB J.* **7**, A1056 (1993).
20. G.B. Fields, S.A. Carr, D.R. Marshak, A.J. Smith, J.T. Stults, L.C. Williams, K.R. Williams, and J.D. Young, in R. Hogue-Angeletti ed., *Techniques in Protein Chemistry IV*, Academic Press, San Diego, Calif., 1993, pp. 229–238.
21. D.J. Strydom, T.T. Andersen, I. Apostol, J.W. Fox, R.J. Paxton, and J.W. Crabb, in R. Hogue-Angeletti ed., *Techniques in Protein Chemistry IV*, Academic Press, San Diego, Calif., 1993, pp. 279–288.
22. S.M. Mische, K.Ü. Yüksel, L.M. Mende-Mueller, P. Matsudaira, D.L. Crimmins, and P.C. Andrews, in R. Hogue-Angeletti ed., *Techniques in Protein Chemistry IV*, Academic Press, San Diego, Calif., 1993, pp. 453–461.
23. R.L. Niece and L.H. Ericsson, *FASEB J.* **8**, A1263 (1994).
24. R.T. Pon, G.A. Buck, R.L. Niece, M. Robertson, A.J. Smith, and E. Spicer, *BioTechniques* **17**, 526–534 (1994).
25. J. Rush, P.C. Andrews, D.L. Crimmins, J.E. Gambee, G.A. Grant, S.M. Mische, and D.W. Speicher, in J.W. Crabb ed., *Techniques in Protein Chemistry V*, Academic Press, San Diego, Calif., 1994, pp. 133–141.
26. K.Ü. Yüksel, T.T. Andersen, I. Apostol, J.W. Fox, J.W. Crabb, R.J. Paxton, and D.J. Strydom, in J.W. Crabb ed., *Techniques in Protein Chemistry V*, Academic Press, San Diego, Calif., 1994, pp. 231–240.
27. G.B. Fields, R.H. Angeletti, S.A. Carr, A.J. Smith, J.T. Stults, L.C. Williams, and J.D. Young, in J.W. Crabb ed., *Techniques in Protein Chemistry V*, Academic Press, San Diego, Calif., 1994, pp. 501–507.
28. C.W. Naeve and R.L. Niece, *FASEB J.* **9**, A1260 (1995).
29. C.W. Naeve, G.A. Buck, R.L. Niece, R.T. Pon, M. Robertson, and A.J. Smith, *BioTechniques* **19**, 448–453 (1995).
30. K.Ü. Yüksel, T.T. Andersen, I. Apostol, J.W. Fox, R.J. Paxton, and D.J. Strydom, in J.W. Crabb ed., *Techniques in Protein Chemistry VI*, Academic Press, San Diego, Calif., 1995, pp. 185–192.
31. J.E. Gambee, P.C. Andrews, K. DeJongh, G. Grant, B. Merrill, S. Mische, and J. Rush, in J.W. Crabb ed., *Techniques in Protein Chemistry VI*, Academic Press, San Diego, Calif., 1995, pp. 209–217.
32. G.B. Fields, R.H. Angeletti, L.F. Bonewald, W.T. Moore, A.J. Smith, J.T. Stults, and L.D. Williams, in J.W. Crabb ed., *Techniques in Protein Chemistry VI*, Academic Press, San Diego, Calif., 1995, pp. 539–546.
33. R.T. Pon, G.A. Buck, K.M. Hager, C.W. Naeve, R.L. Niece, M. Robertson, and A.J. Smith, *BioTechniques* **21**, 680–685 (1996).
34. L. Bibbs, N.D. Denslow, C. Glabe, K. Ivanetich, C.W. Naeve, and R.L. Niece, *ABRF News* **7**, 18–21 (1996).
35. A.M. Mahrenholz, N.D. Denslow, T.T. Andersen, K.M. Schegg, K. Mann, S.A. Cohen, J.W. Fox, and K.Ü. Yüksel, in D.R. Marshak ed., *Techniques in Protein Chemistry VII*, Academic Press, San Diego, Calif., 1996, pp. 323–330.

36. R.H. Angeletti, L. Bibbs, L.F. Bonewald, G.B. Fields, J.S. McMurray, W.T. Moore, and J.T. Stults, in D.R. Marshak ed., *Techniques in Protein Chemistry VII*, Academic Press, San Diego, Calif., 1996, pp. 261–274.
37. K. De Jongh, J. Fernandez, J.E. Gambee, G.A. Grant, B. Merrill, K.L. Stone, and J. Rush, in D.R. Marshak ed., *Techniques in Protein Chemistry VII*, Academic Press, San Diego, Calif., 1996, pp. 347–358.
38. G.B. Fields, L. Bibbs, L.F. Bonewald, J.S. McMurray, W.T. Moore, A.J. Smith, J.T. Stults, L.C. Williams, and R.H. Angeletti, in P.T.P. Kaumaya and R.S. Hodges eds., *Peptides: Chemistry, Structure, and Biology*, Mayflower Scientific, 1996, pp. 52–54.
39. P.S. Adams, M.K. Dolejsi, S. Hardin, S. Mische, B. Nanthakamur, H. Riethman, J. Rush, and P. Morrison, *BioTechniques* **21**, 678 (1996).
40. P.S. Adams, M.K. Dolejsi, S. Hardin, S.M. Mische, B. Nanthakamur, H. Riethman, J. Rush, and P. Morrison, *Eighth Annual Genome Sequencing and Analysis Conference*, 1996, B-26.
41. P.S. Adams, M.K. Dolejsi, S. Hardin, S.M. Mische, B. Nanthakamur, H. Riethman, J. Rush, and P. Morrison, *http://mbcf.dfci.harvard.edu/abrfdnaseq* (last updated February 8, 1999).
42. G.A. Grant and M.W. Crankshaw, in B.J. Smith ed., *Methods in Molecular Biology*, vol. 64: *Protein Sequencing Protocols*. Humana Press, Clifton, 1997, New Jersey, pp. 197–215.
43. R.R. Townsend, A. Manzi, R.K. Merkle, M.F. Rohde, M. Spellman, A.J. Smith, and S. Carr, *ABRF News* **8**, 14–21 (1997).
44. R.H. Angeletti, L.F. Bonewald, and G.B. Fields, in G.B. Fields ed., *Methods in Enzymology*, Academic Press, San Diego, Calif., 1997, pp. 697–717.
45. R.H. Angeletti, L. Bibbs, L.F. Bonewald, G.B. Fields, J.W. Kelly, J.S. McMurray, W.T. Moore, and S.T. Weintraub, in D.R. Marshak ed., *Techniques in Protein Chemistry VIII*, Academic Press, San Diego, Calif., 1997, pp. 875–890.
46. J. Fernandez, A. Admon, K. DeJongh, G.A. Grant, W. Henzel, W.S. Lane, K.L. Stone, and B. Merrill, in D.R. Marshak ed., *Techniques in Protein Chemistry VIII*, Academic Press, San Diego, Calif., 1997, pp. 69–78.
47. K.R. Williams, U. Hellman, R. Kobayashi, W.W. Lane, S.M. Mische, and D.W. Speicher, in D.R. Marshak ed., *Techniques in Protein Chemistry VIII*, Academic Press, San Diego, Calif., 1997, pp. 99–109.
48. K.M. Schegg, N.D. Denslow, T.T. Andersen, Y.A. Bao, S.A. Cohen, A.M. Mahrenholz, and K. Mann, in D.R. Marshak ed., *Techniques in Protein Chemistry VIII*, Academic Press, San Diego, Calif., 1997, pp. 207–216.
49. A. Mahrenholz, T.T. Andersen, Y. Bao, S.A. Cohen, N.D. Denslow, J. Hulmes, P.E. Hunzijer, K. Mann, K.M. Schegg, and K. West, ABRF Amino Acid Analysis Survey: Identification of Proteins Electroblotted to PVDF. *http://www.abrf.org/ABRF/ResearchCommittees/aaarescomm.html* (last updated February 9, 1999).
50. K.M. Ivanetich, L. Bibbs, R.L. Niece, N.D. Denslow, C.W. Naeve, M. Rohde, and L.H. Ericsson, *Genetic Engineering News* **17**, 17, 19, 47 (1997).
51. P. Edman, *Acta Chem. Scand.* **4**, 283 (1950).
52. M.W. Hunkapiller, in A.S. Bhown ed., *Protein/Peptide Sequence Analysis: Current Methodologies*, CRC Press, Boca Raton, Fla., 1988, pp. 87–117.
53. J.L. Brown and W.K. Roberts, *J. Biol. Chem.* **251**, 1009–1014 (1976).
54. M.W. Hunkapiller, E. Lujan, F. Ostrander, and L.E. Hood, *Methods Enzymol.* **91**, 227–236 (1983).
55. P. Matsudaira, *J. Biol. Chem.* **262**, 10035–10038 (1987).
56. J.I. Elliott, K.L. Stone, and K.R. Williams, *Anal. Biochem.* **211**, 94–101 (1993).
57. P. Edman and G. Begg, *Eur. J. Biochem.* **1**, 80 (1967).
58. A. Shevchenko, M. Wilm, O. Vorm, and M. Mann, *Anal. Chem.* **68**, 850–858 (1996).
59. U.K. Laemmli, *Nature* **227**, 680–685 (1970).
60. P.H. O'Farrell, *J. Biol. Chem.* **250**, 4007–4021 (1975).
61. R.H. Aebersold, J. Leavitt, R.A. Saavedra, L.E. Hood, and S.B.H. Kent, *Proc. Natl. Acad. Sci. U.S.A.* **84**, 6970–6974 (1987).
62. H. Kawasaki, Y. Emori, and K. Suzuki, *Anal. Biochem.* **191**, 332–336 (1990).
63. J. Rosenfeld, J. Capdevielle, J.C. Guillemot, and P. Ferrara, *Anal. Biochem.* **203**, 173–179 (1992).
64. J. Fernandez, M. DeMott, D. Atherton, and S.M. Mische, *Anal. Biochem.* **201**, 255–264 (1992).
65. S.D. Patterson, D. Hess, T. Yungwirth, and R. Aebersold, *Anal. Biochem.* **202**, 193–203 (1992).
66. U. Hellman, C. Wernstedt, J. Góñez, and C.-H. Heldin, *Anal. Biochem.* **224**, 451–455 (1995).
67. S. Best, D.F. Reim, J. Mozdzanowski, and D.W. Speicher, in J.W. Crabb ed., *Techniques in Protein Chemistry V*, Academic Press, New York, 1994, pp. 205–213.
68. J. Fernandez, L. Andrews, and S.M. Mische, *Anal. Biochem.* **218**, 112–117 (1994).
69. H. Erjument-Bromage, M. Lui, D. Sabatini, S.H. Snyder, and P. Tempst, *Protein Sci.* **3**, 2425–2446 (1994).
70. K. Williams, M. LoPresti, and K. Stone, in D. Marshak ed., *Techniques in Protein Chemistry VIII*, Academic Press, San Diego, Calif., 1997, pp. 79–90.
71. U. Hellman, in R.M. Kamp, T. Choli-Papadoupoulou, and B. Whittmann-Liepold eds., *Protein Structure Analysis: Preparation, Characterization, and Microsequencing*, Springer-Verlag, Berlin, 1997, pp. 97–104.
72. D.J.C. Pappin, P. Hojrup, and A.J. Bleasby, *Curr. Biol.* **3**, 327–332 (1993).
73. J.R. Yates, III, S. Speicher, P.R. Griffin, and T. Hunkapiller, *Anal. Biochem.* **214**, 397–408 (1993).
74. W.J. Henzel, T.M. Billeci, J.T. Stults, S.C. Wong, C. Grimley, and C. Watanabe, *Proc. Natl. Acad. Sci. U.S.A.* **90**, 5011–5015 (1993).
75. M. Mann, P. Hajrup, and P. Roepstorff, *Biol. Mass Spectrom.* **22**, 338–345 (1993).
76. K.R. Williams, S.M. Samandar, K.L. Stone, M. Saylor, and J. Rush, in J.M. Walker ed., *The Protein Protocols Handbook*, Humana Press, Totowa, New Jersey, 1996, pp. 541–555.
77. J. Mozdanowski and D.W. Speicher, in J.J. Villafranca ed., *Current Research in Protein Chemistry: Techniques, Structure, and Function*, Academic Press, San Diego, Calif., 1990, pp. 87–93.
78. B. Lindberg and J. Lonngren, *Methods Enzymol.* **50**, 3–33 (1978).
79. R.K. Merkle and I. Poppe, *Methods Enzymol.* **230**, 1–15 (1994).
80. R.R. Townsend, in C. Horvath and L.S. Ettre eds., *Chromatography in Biotechnology*, American Chemical Society, Washington, D.C., 1993, pp. 86–101.
81. R.R. Townsend, in Z. El Rassi ed., *Carbohydrate Analysis: High Performance Liquid Chromatography and Capillary Electrophoresis*, Elsevier, New York, 1995, pp. 181–209.

82. R.G. Spiro, *J. Biol. Chem.* **237**, 382–388 (1962).
83. R.G. Spiro and V.D. Bhoyroo, *J. Biol. Chem.* **249**, 5704–5717 (1974).
84. R.R. Townsend, M.R. Hardy, T.C. Wong, and Y.C. Lee, *Biochemistry* **25**, 5725–5731 (1986).
85. A.S.B. Edge and R.G. Spiro, *J. Biol. Chem.* **262**, 16135–16141 (1987).
86. E.D. Green, G. Adelt, J.U. Baenziger, S. Wilson, and H. van Halbeek, *J. Biol. Chem.* **263**, 18253–18268 (1988).
87. B. Bendiak, M. Harris-Brandts, S.W. Michnick, J.P. Carver, and D. Cumming, *Biochemistry* **28**, 6491–6499 (1989).
88. D.A. Cumming, C.G. Hellerqvist, M. Harris-Brandts, S.W. Michnick, J.P. Carver, and B. Bendiak, *Biochemistry* **28**, 6500–6512 (1989).
89. U. Hobohm, T. Houthhaeve, and C. Sander, *Anal. Biochem.* **222**, 202–209 (1994).
90. D.J. Strydom, *J. Chromatogr.* **678**, 17–23 (1994).
91. C.M. Starr, R.I. Masada, C. Hague, E. Skop, and J.C. Klock, *J. Chromatogr.* **720**, 295–321 (1996).
92. M. Dole, L.L. Mack, R.L. Hines, R.C. Mobley, L.D. Ferguson, and M.B. Alice, *J. Chem. Phys.* **49**, 2240–2249 (1968).
93. J.B. Fenn, M. Mann, C.K. Meng, S.F., Wong, and C.M. Whitehouse, *Science* **246**, 64–71 (1989).
94. M. Karas and F. Hillenkamp, *Anal. Chem.* **60**, 2299–2301 (1988).
95. R. Beavis and B.T. Chait, *Rapid Commun. Mass Spectrom.* **3**, 432–435 (1989).
96. G.J. Currie and J.R. Yates, III, *J. Am. Soc. Mass Spectrom.* **4**, 955–963 (1993).
97. U. Pieles, W. Zurcher, M. Schar, and H.E. Moser, *Nucleic Acids Res.* **21**, 3191–3196 (1993).
98. D.C. Stahl, K.M. Swiderek, M.T. Davis, and T.D. Lee, *J. Am. Soc. Mass Spectrom.* **7**, 532–540 (1996).
99. M. Wilm, A. Shevchenko, T. Houthaeve, S. Breit, L. Schweigerer, T. Fotsis, and M. Mann, *Nature* **379**, 466–469 (1996).
100. P.R. Griffin, M.J. MacCoss, J.K. Eng, R.A. Blevins, J.S. Aaronson, and J.R. Yates, III, *Rapid Commun. Mass Spectrom.* **9**, 1546–1551 (1995).
101. S.D. Patterson and R. Aebersold, *Electrophoresis* **16**, 1791–1814 (1995).
102. C.H. Wheeler, S.L. Berry, M.R. Wilkins, J.M. Corbett, K. Ou, A.A. Gooley, I. Humphery-Smith, K.L. Williams, and M.J. Dunn, *Electrophoresis* **17**, 580–587 (1996).
103. P. Kahn, *Science* **270**, 369–370 (1995).
104. K. Bieman, *Protein Sci.* **4**, 1920–1927 (1995).
105. M. Bartlett-Jones, W.A. Jeffery, H.F. Hansen, and D.J.C. Pappin, in J.W. Crabb ed., *Techniques in Protein Chemistry VI*, Academic Press, San Diego, Calif., 1995, pp. 3–11.
106. D.H. Patterson, G.E. Tarr, F.E. Regnier, and S.A. Martin, *Anal. Chem.* **67**, 3971–3978 (1995).
107. *Fed. Regist.* Part VI, 21 *Code of Federal Regulations*, 58.
108. U.S. FDA, *Digest of Questions and Answers*, June 20, 1981.
109. U.S. FDA, 21 *Code of Federal Regulations* 610.
110. U.S. FDA, 21 *Code of Federal Regulations* 210 and 211.
111. A. Thayer, *Chem. Eng. News* 12–17 (1993).
112. M. Breitenberg, NISTIR No. 4721, U.S. Department of Commerce, National Institute of Standards and Technology, Standards Code and Information Program, Office of Standard Service, November 1991.
113. J.H. Stratton, *Quality Progress* 67–69 (1992).
114. European Standard EN 46001, CEN/CENELEC, Brussels, Belgium, August 1996.
115. Discussions of the Activities of the International Committee on Harmonization of Technical Requirements for the Registration of Pharmaceuticals for Human Use, *Fed. Regist.* 59 (1994).
116. Elimination of the Establishment Licensing Application for Specified Biotechnology and Specified Synthetic Biological Products, *Fed. Regist.* **61**, 94 (1996).
117. M.D. Adams, J.M. Kelley, J.D. Gocayne, M. Dubnick, M.H. Polymeropoulos, H. Xiao, C.R. Merril, A. Wu, B. Olde, R.F. Moreno, A.R. Kerlavage, W.R. McCombie, and J.C. Venter, *Science* **252**, 1651–1656 (1991).
118. L. Hillier, G. Lennon, M. Becker, M.F. Bonldo, B. Chiapelli, S. Chissoe, N. Dietrich, T. DuBuque, A. Favello, W. Gish, M. Hawkins, M. Hultman, T. Kucaba, M. Lacy, M. Le, N. Le, E. Mardis, B. Moore, M. Morris, J. Parsons, C. Prange, L. Rifkin, T. Rohlfing, K. Schellenberg, M.B. Soares, F. Tan, J. Thierry-Meg, E. Trevaskis, K. Underwood, P. Wohldman, R. Waterston, R. Wilson, and M. Marra, *Genome Res.* **6**, 807–828 (1996).
119. National Center for Biotechnology Information dbEST, *http://www.ncbi.nlm.nih.gov/dbEST/* (released February 9, 1999).
120. A. Goffeau, B.G. Barrell, H. Bussey, R.W. Davis, B. Dujon, H. Feldmann, F. Galibert, J.D. Hoheisel, C. Jacq, M. Johnston, E.J. Louis, H.W. Mewes, Y. Murakami, P. Philippsen, H. Tettelin, and S.G. Oliver, *Science* **274**, 546–549 (1996).
121. S.M. Tilghman, *Genome Res.* **6**, 773–780 (1996).
122. M. Chee, R. Yang, E. Hubbell, A. Berno, X.C. Huang, D. Stern, J. Winkler, D.J. Lockhart, M.S. Morris, and S.P. Fodor, *Science* **274**, 610–614 (1996).

PROFESSIONAL SOCIETIES, SOCIETY FOR INDUSTRIAL MICROBIOLOGY (SIM)

CHRISTINE L. CASE
Skyline College
San Bruno, California

ROBERT SCHWARTZ
Abbott Laboratories
North Chicago, Illinois

OUTLINE

BACKGROUND OF THE ORGANIZATION

The Society for Industrial Microbiology (SIM) is a nonprofit professional association dedicated to the advancement of applied microbiological sciences, especially as they apply to industrial products, biotechnology, materials, and processes. A primary objective of SIM is to serve as a liaison among the various specialized fields of applied microbiology. It promotes the exchange of scientific information through meetings and publications.

Industrial microbiology is the application of scientific and engineering principles to the processing of materials by microorganisms or other biological agents to create useful products for the benefit of humankind. These microorganisms may be natural isolates, laboratory-selected mutants, or genetically engineered microbes. Members of SIM include bacteriologists, molecular biologists, mycologists, chemists, engineers, geologists, marine biologists, and plant pathologists. These scientists, engineers, technicians, administrators, and educators represent many organizations in industrial, educational, federal, and state laboratories, and are involved in research, process development, production, and quality control for a wide variety of products and services including antibiotics and other pharmaceuticals, vaccines, diagnostic tests, foods, and beverages; biocides; microbial products such as enzymes, amino acids, and carbohydrates; agricultural biotechnology; oil recovery and mining; and bioremediation. SIM sponsors an annual scientific meeting as well as special topic conferences.

In 1995, then-president-elect of SIM, Steve Orndorff, commissioned the Strategic Planning Committee with responsibility for charting a course for SIM for the next 5 years. The committee developed a statement of the mission, visions, and values of SIM as well as a strategic plan. These concepts were published for comment by the membership, and after due consideration were formally adopted by the board to become part of the society's formal charter.

> *Mission*: To enhance and strengthen the professional activities of all those who practice, teach, or study microbiological and related sciences to solve practical or industrial problems for the benefit of humanity. We do this by providing high-quality publications, informative and congenial scientific meetings, synergistic career development interactions, practical education of our successors, and effective advocacy of relevant issues.
>
> *Vision*: To be the most valued representative of professionals in our field. To be the most productive catalyst of career development interactions in our field. To be the most widely acclaimed organizer of scientific conferences in our field. To be the most respected publisher of scientific work in our field.
>
> *Values*: We perform to the highest standards of professional integrity and scientific ethics. We direct our intellectual resources to benefit and protect humankind. We respect and treat fairly our employees, colleagues, employers, and suppliers. We inform and educate the public within the framework of our nonprofit status. We rely for our effectiveness on the productive volunteerism of our members. We value the benefit of interactions between scientists with common interests. We hold ourselves accountable for being responsible public citizens and scientists. We take responsibility for educating and nurturing our successors.

In 1996, the board of directors recommended to the membership to change the society's name to the Society for Industrial Microbiology and Biotechnology to reflect the careers and technologies that have developed since the 1980s. In 1998, the society membership rejected the name change (390 to 369 votes) in favor of remaining the Society for Industrial Microbiology.

The common denominator of applied or industrial microbiology is microbial science in the service of humans. In fact, the science of microbiology developed from problems associated with some microbial processes that had immediate application to humans. Microorganisms have been used for the production of food, clothing, and drink for centuries but leavening bread, fermenting wine, producing vinegar, and retting flax had been practiced in ignorance. It was not until 1864 that Louis Pasteur defined the role of microorganisms in fermentations. Industrial microbiology was slow to develop after Pasteur's initial experiments, probably because of the lack of good culture methods and selective media. In the late 1800s, Jokichi Takamine brought the Koji process for producing amylase to the United States. He received the first U.S. patent on an enzyme production process and established the first fermentation company in the United States (1). Except for a few outstanding contributions mainly involving lactic acid, acetone–butanol fermentations, and production of yeast and yeast products, industrial microbiology did not progress very rapidly until after 1900. Samuel C. Prescott initiated the first course in industrial biology in 1896 at the Massachusetts Institute of Technology. The first textbook in industrial microbiology was not published until 1940, however (2).

Industrial microbiologists in the 1940s worked with deterioration of military material, solvent fermentations, and penicillin production. As these scientists began to accumulate experimental results, there was a natural desire to share their findings with scientific colleagues at a national society meeting and in scientific journals. However, there was no central organization in which to share their findings because these scientists came from a variety of scientific specialties including plant pathology, medical laboratories, materials engineering, and academic mycology and bacteriology. Moreover, it was difficult for a purely practical paper in microbiology to secure a place on the program of existing societies or to find an opportunity for publication in the established journals. These scientists became unhappy, and a conflict developed between basic and applied microbiology practitioners. In 1951, the Society of American Bacteriologists (SAB) authorized a new journal called *Applied Microbiology*. However, the first issue was not published until January 1953.

The decision to publish *Applied Microbiology* took too long in the minds of the industrial microbiologists of the late 1940s. By that time, industrial microbiologists felt that applied and industrial aspects of microbiology were being neglected by the SAB. Walter Ezekiel, Bureau of Ordnance, Department of the Navy (Washington, D.C.) wrote to and spoke with a number of individuals during the summer and autumn of 1949 and suggested formation of a new Society for Industrial Microbiology. Ezekiel arranged with Raymond L. Taylor, associate administrative secretary of the American Association for the Advancement of Science (AAAS), to schedule and advertise a meeting of interested persons during the AAAS meeting in New York (3). Ezekiel published a short announcement of the meeting in *Science*, in which he stated that "a new society of industrial microbiologists is to be organized to cover the borderline work in application of microbiologic knowledge or processes to work with industrial materials. The society plans to handle . . . microbiologic manufacturing processes and microbiologic assay. The organization meeting will be held December 29 at 4:00 P.M. at the Hotel McAlpine Ballroom" (4).

On that date, more than 250 interested persons met in the ballroom of the Hotel McAlpine in New York City. There, it was decided to form a new society to affiliate with the newly formed American Institute of Biological Sciences (AIBS). Charles Thom, Northern Regional Research Laboratory (Peoria, Ill.) was elected acting chair and Charles L. Porter, Purdue University (West Lafayette, Ind.), acting secretary. Ezekiel's motion to establish a scientific organization known as the Society for Industrial Microbiology, with its general scope the field of microbiology as applied to industrial materials and processes, was passed. It was decided at that meeting, over the objections of a few, that the group would meet the following year, again with AIBS. Thom appointed the following members of the Organizing Committee: Maynard M. Baldwin (Battelle Memorial Institute, Columbus, Ohio), F. G. Walter Smith (University of Miami Marine Laboratory, Coral Gables, Fla.), Earnest A. Walker (U.S. Department of Agriculture, Washington, D.C.), William L. White (Farlow Herbarium, Harvard University), Ezekiel, and Porter (5).

The committee met several times during the first year to prepare a program and write a constitution for the new society in preparation for the first annual meeting to be held with AIBS at Ohio State University, September 11–13, 1950. In Ohio, Thom was elected the first president and Porter the secretary-treasurer. In 1951, the new SIM met with AIBS at the University of Minnesota and Thom was elected president for a second term. In 1952, the meeting was held at Cornell University.

The need for an interdisciplinary society first conceived by Ezekiel became apparent, and membership grew at a rapid pace. In 10 years, membership in the new society had grown from 75 to 550. SIM was incorporated as a nonprofit organization in 1960 in Washington, D.C. Membership reached 2,000 in 1989. In 1989, Paula Myers-Keith was the first woman elected president of SIM. For many years, the annual meetings were held along with AIBS on university campuses. In 1967, SIM held its first independent annual meeting, at the University of Western Ontario. Since 1985, all annual meetings have been held at commercial convention centers rather than universities. The annual meeting is a weeklong event featuring plenary sessions, roundtables, poster sessions, workshops, and extensive technical exhibits.

Between 1960 and 1970, SIM sponsored summer institutes on microbiological problems of current interest with topics including industrial microbial genetics and federal regulations for disinfectants, drugs, and cosmetics. In 1960, SIM sponsored the initial conference on antimicrobial agents. This was a mammoth undertaking for the small organization, and sponsorship was subsequently released to the American Society for Microbiology (Interscience Conference on Antimicrobial Agents and Chemotherapy, ICAAC) (6). In 1978, the Alma Dietz actinomycete roundtable discussions were included and have since become a part of all subsequent annual meetings.

In 1987, reaffirming its commitment of service to the microbial biotechnologist, SIM approved the organization of the Special-Topic Conference Series. These conferences run 2 to 3 days and usually result in the production of a monograph. Initiated by George Somkuti (president, 1985–1986), the first event in the series was the International Conference on the Biotechnology of Microbial Products: Novel Pharmacological and Agrobiological Activities (BMP), held March 13–16, 1988. BMP is held every 3 years. Since then, SIM has sponsored or cosponsored special conferences on a regular basis. Other conference series include Recent Advances in Fermentation Technology (RAFT), initiated in 1995. RAFT is held biannually and is cosponsored with the Biochemical Technology Division of the American Chemical Society. The aim is the free exchange of newly developed fermentation technologies related to current highly productive, consistent, scaleable, and economic processes. Fermentation microbiology, biochemistry, biocatalysis, and biochemical engineering aspects are explored. The Sixth Genetics and Molecular Biology of Industrial Microorganisms (GMBIM) Conference, cosponsored with Indiana University Institute of Molecular and Cellular Biology, was held in 1996 at Indiana University. This continuing series provides a forum for academic and industrial scientists to exchange information on the latest developments in the genetics and biochemistry of microbes of industrial interest.

The business of SIM is conducted by the board of directors (which is elected by the membership), 12 standing committees, and the executive secretary. The board consists of the president, president-elect, secretary, treasurer, past president, and four directors. In 1959, as president of SIM, Porter said, "There are a number of organizations including the Society of American Bacteriologists and the American Chemical Society which have grown so large that they have become unwieldy. The administration of these groups has lost personal contact with the membership . . . They must enforce rules and regulations which hamper individual initiative, incentive, and recognition. At their meetings the sessions devoted to volunteer papers are so clogged that each presentation is limited to 10 or 12 minutes. . . . Membership beyond 1200–1500 should be discouraged, for then, we like other groups mentioned, would lose that personal contact which is one of the principal attractions" (7).

At present, most of the membership is in the United States, but interest in industrial microbiology is international and SIM has more than 200 members in 38 countries outside the United States. Employment opportunities in industrial microbiology and biotechnology have increased dramatically since 1959; consequently, SIM now has a membership of more than 1,900, in addition to 75 corporate members. However, SIM continues to foster an environment in which individuals are recognized and have opportunities to participate. Modern communications including the Internet are available so that members have fast and easy access to society resources via electronic mail and a World Wide Web home page. At the annual meetings, speakers give 30-minute presentations of their work. Additionally, there are numerous opportunities for members to serve on committees and become officers. Local sections provide more opportunities for scientists, engineers, technicians, and students to meet and form personal contacts.

AWARDS

SIM has established the Committee on Awards and Honors to provide monetary and honorary awards to its members and others eminent in the field. Awards are made at the annual meeting. In 1967, the first Charles Thom Award was presented to Kenneth B. Raper (president, 1952–1953). This award is bestowed annually to persons making outstanding contributions in the field of industrial microbiology. The second Thom Award was presented to Arthur M. Kaplan (president, 1960–1961) at the 1970 annual meeting held at the University of Rhode Island. The Charles Porter Award of Merit is given in recognition of outstanding contributions to the growth and success of SIM. The first Porter Award was given to Earnest A. Walker in 1960.

A fellowship status was approved by the SIM membership at the 1984 Annual Business Meeting. Fellowship is a membership grade of distinction in the Society acknowledging significant research and/or service contributions to the professions of applied microbiology. A sustained record of such contributions while a member of SIM is the main criterion for consideration to fellowship. No more than 10% of the members are eligible for fellowship status at any given time. The first group elected as SIM fellows in August 1985 were Arnold L. Demain (president, 1990–1991), Arthur E. Humphrey, Richard P. Elander (president, 1973–1974), Warren P. Iverson, Clifford W. Hesseltine (president, 1957–1958), H. Boyd Woodruff (president, 1954–1956), Leland A. Underkofler (president, 1967–1968), Kaplan, and Raper.

The Selman A. Waksman Teaching Award was established in 1989, and the first recipient was Douglas E. Eveleigh. The award is presented to a recipient every other year at the annual meeting. The nominee must be an active, full-time professor at a recognized institution of higher education for a minimum of 10 years or have attained emeritus status. The nominee must be actively involved in research in an area of industrial or applied microbiology or biotechnology. Annually, SIM present the Student Abstract Award for the best abstract submitted for the annual meeting.

The Panlabs Award Lecture is sponsored by Panlabs and is presented by individuals of world renown in the field of microbiology. Each Panlabs Lecturer presents an address at the annual meeting. The Schering–Plough Research Institute Young Investigator Award is made to a society member who is under 36 years of age at the time of nomination. The nominee must have made a significant contribution in industrial microbiology or biotechnology that indicates promise of a professional career of merit. This award is made on alternate years with the Waksman Teaching Award.

LOCAL SECTIONS

Regional groups called local sections are governed by their own constitutions and elected officers. Local sections provide an opportunity for individuals involved in biotechnology to meet, discuss topics of interest, attend talks given by noted industrial and academic speakers, and attend local symposia on important contemporary subjects. Membership includes all interested microbiologists, chemists, and engineers from students to advanced industrial, academic, and government professionals.

PUBLICATIONS

SIM News is the official publication of the society. It was first published in 1951 and entitled *SIM News Letter*; the name was changed in 1972. *SIM News* is distributed bimonthly to all members. It contains technical review articles, news of special significance in the field of applied microbiology, and reports of many activities of the society including local sections reports, placement opportunities, and meeting notices.

The first volume of *Developments in Industrial Microbiology (DIM)* was published in 1960. This peer-reviewed annual publication contained papers from the annual meeting. *DIM* was published annually for 30 years (31 volumes).

The *Journal of Industrial Microbiology and Biotechnology (JIMB)* is an international, peer-reviewed journal that seeks to further scientific knowledge and to disseminate information in biotechnology, fermentation and industrial microbiology, biodegradation, biodeterioration, quality control, and other areas of applied microbiology. The concept for a new technical publication for the society was presented by SIM President C. Herb Ward (1983–1984), and the first issue of *JIMB* was published in March 1986 as the *Journal of Industrial Microbiology (JIM)*. The first editor in chief was George E. Pierce. The name *JIM* was used continuously through 1996, when the board of directors adopted the new name, *JIMB*. *JIMB* will consider publication of original research papers, critical reviews, short communications, and letters to the editor. All papers submitted are subject to peer review.

The *Developments in Industrial Microbiology Series* complements *JIMB* and *SIM News* by providing in-depth reviews of specific subject areas. The monographs are focused on topics of interest to scientists in industrial and environmental microbiology. Authors for the individual series are identified from presentations at special conferences sponsored by SIM and selected symposia at the annual meeting. The address for SIM is 3929 Old Lee Highway, Suite 92A, Fairfax, VA 22030-2421. Internet: ⟨*info@simhq.org*⟩⟨*www.simhq.org*⟩.

BIBLIOGRAPHY

1. J.W. Bennett, *Takamine: Documents from the Dawn of Industrial Biotechnology*. Miles, Berkeley, Calif., 1988.
2. G. Reed, *Prescott and Dunn's Industrial Microbiology*, 4th ed., AVI Publishing, Westport, Conn., 1983.
3. W.N. Ezekiel, in *Developments in Industrial Microbiology*, vol. 8, Society for Industrial Microbiology, Washington, D.C., 1967, pp. 431–434.
4. Announcements, *Science* **110**, 674 (1949).
5. R.M. Rogers, *ASM News* **41**, 642–647 (1975).
6. L.A. Underkofler and R.D. Schwartz, *SIM News* **42**, 260–262 (1992).
7. C.L. Porter, in *Developments in Industrial Microbiology*, vol. 1, Society for Industrial Microbiology, Washington, D.C., 1960, pp. 261–267.

PROFESSIONAL SOCIETIES, THE PROTEIN SOCIETY

ROBERT NEWBURGH
The Protein Society
Bethesda, Maryland

Protein science is the study of proteins. The development of new techniques, including those from molecular biology, analytical chemistry, immunology, spectroscopy, diffraction, and computations, has provided opportunities to probe structural and functional relationships of proteins in detail undreamed of until recent times. These opportunities underscore the importance of broad interactions among investigators of diverse backgrounds to resolve fundamental questions. The Protein Society was established in 1986 to promote these interactions by providing regular opportunities for protein scientists to exchange ideas, findings, and techniques via both scientific meetings and the society's journal, *Protein Science*.

The goal of the society is to promote the discipline of protein science by these means:

1. Promoting research and scientific endeavor in the field of protein science
2. Promoting education in protein science to a wide variety of constituencies including established investigators, young scientists, students, others interested in science, and physicians interested in diseases pertinent to the field
3. Promoting and extending current understanding of the importance of protein science, particularly in regard to the methods, concepts, and problems of structural biology, cell and molecular biology, and biotechnology, as they relate to the behavior and properties of proteins.

The society attempts to foster effective interactions between investigators who share these interests. To further this objective, the society seeks to expand its membership and activities in ways that will enhance interpersonal communication among scientists.

To achieve these goals, the society invests its resources in

1. Management of the Society
2. Conduct of scientific meetings
3. Publication of scientific research
4. Educational activities
5. Fostering relations with scientists in other countries interested in protein science

Further information regarding the society can be obtained from the executive officer, Dr. Robert Newburgh, The Protein Society, 9650 Rockville Pike, Bethesda, MD 20814-3998 (phone 301-571-0662; Fax 301-571-0666; e-mail newburgh@protein.faseb.org).

PROTEIN ADSORPTION, EXPANDED BED

SIDDARTHA GHOSE
Aston University
Birmingham, United Kingdom

KEY WORDS

Adsorbent
Chromatography
Expanded bed chromatography
Expanded beds
Fluidized beds
Whole-broth adsorption

OUTLINE

INTRODUCTION

Traditional biochemical processes can be characterized into three major operational divisions: *(1)* the upstream process, which deals primarily with preparation and pretreatment of the raw material from which the specific target product is to be directly or indirectly obtained, *(2)* the actual production phase, in which the chemical transformation of the raw material takes place to produce the end product in which the desired target product is obtained along with associated by-products, and *(3)* the downstream processing, which is specifically involved with the recovery and purification of the target product. Progress in the area of recombinant DNA technology has made it commonplace to clone and express specific products for diagnostic and therapeutic applications within a variety of host microorganisms. This advancement has resulted in a pace of development of production phase technology that has far surpassed that of the recovery and purification areas. Recent years have seen significant advances in this latter area, and, with the incorporation of affinity-based interactions, it is now possible to include more powerful and specific operations for improving the overall efficiency of the recovery process (1).

Downstream processing refers to the collective set of operations involved in obtaining the target product in its required end form. The name derives from its relative position to the actual production stage in the process flow sheet. Factors such as the starting concentration of the target product, its final required purity, and the type of impurities to be removed dictate the exact choice of operations, and the sequence is necessarily adapted for each particular process. The use of microorganisms in biochemical processes makes cell harvesting, disruption (for the product residing within the cytoplasmic spaces as an intracellular product), and solid–liquid separation essential initial stages in the sequence of operations. Processes in which the product is expressed as an extracellular component require no cell disruption stage, although clarification of the feedstock from whole cells or cellular debris is always essential before subsequent treatment. The common solid–liquid separation methods traditionally used are centrifugation (2,3) and cross-flow microfiltration (4). The use of aqueous two-phase systems for primary treatment of feedstock (5) and cell debris removal has also been reported (6–9).

Biochemical product recovery nearly always involves one or more packed bed chromatographic processes as an often essential step of the separation protocol, usually in the secondary and final purification stages (10). It has now developed into a powerful tool for product separation at the analytical and preparative (11) scale of operation. The wide variety of modes of interaction between the solubilized product and ligands on the solid adsorbent (11–13), ease of scale-up and cleaning-in-place procedures, commercial availability of equipment and consumables, and the well-documented operation have given this technology wide acceptance in academic and commercial research and development. There is now a significant amount of information on the optimization of chromatographic separations (14). However, full utilization of its potential implies clarification of the biological feedstock before application onto a packed bed, failing which bed blockage is inevitable.

This article describes in some detail the use of expanded bed chromatography, which has been described as the only new unit operation that the bioprocess industry has seen for decades (15). This technique offers the possibility of directly applying the unclarified feedstock onto the bed, bypassing the conventional solid–liquid separation techniques while simultaneously capturing the desired product from the liquid phase. A typical sequence of separation operations is shown schematically in Figure 1, in which the relative position of expanded bed chromatography is also indicated. As will become clear from the description of its operation, the incorporation of this unit will enable facilities to replace cost-intensive and often problematic solid–liquid separation techniques such as centrifugation and cross-flow microfiltration. In addition, there is the added benefit of a chromatographic step early in the separation protocol. One must bear in mind that this technology is still in its development phase, and it is too early yet to conclude that it will become an integral part of a downstream processing protocol. However, as witnessed from the Second International Conference on Expanded Bed Adsorption recently held in the United States, there has been a growing awareness of the potential of the technology and it may indeed prove to have a promising future in bioseparations.

Before a detailed description of various aspects of expanded bed systems, a point of explanation should be made

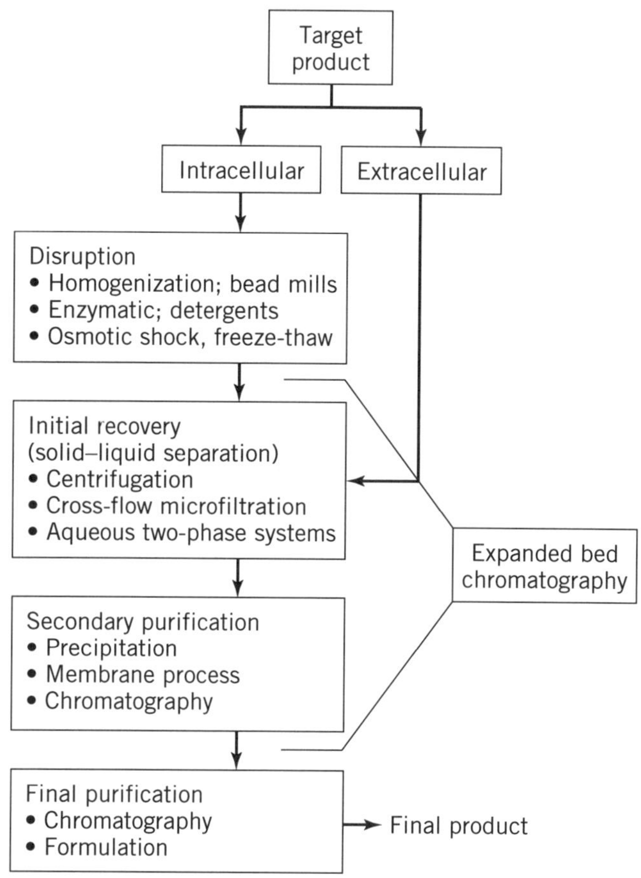

Figure 1. Typical sequence of operation in the recovery and purification of proteins from biological feedstock. The relative position of expanded bed as a viable alternative operation is also indicated.

regarding the use of the terms *fluidized bed* and *expanded bed*. During the initial development of this technology, the terms were often used exclusively of one another. In classical chemical engineering operations, fluidized beds were associated with catalytic reactors in which a high degree of solid–fluid mixing was maintained to ensure good heat and mass transfer. Expanded beds, on the other hand, would benefit product capture if good plug-flow-type hydrodynamics could be obtained. Clearly mixing needed to be minimized and bed stability ensured. The two terms were clearly in contradiction with each other; to maintain the existence of the critical difference, the use of the term *fluidized beds* was avoided in the context of biological separations by *expanded beds*. Over the years, however, this distinction has eroded, and now both are interchangeably used to mean essentially the same thing. The same applies in this article.

THEORY

To understand the physical process by which an expanded bed is obtained and some of the basic expressions describing its various states, we begin with a settled bed of adsorbent packed in a column with no top flow adapter. When an upward flow of liquid is introduced at the lower end, there is a tendency for the adsorbent particles to move in the direction of flow. This movement occurs only when the liquid velocity exceeds the minimum fluidization velocity of the particles. The bed then undergoes a transient expansion phase, at the end of which the adsorbent particles attain a steady position when the upward force due to the liquid flow exactly counterbalances the downward gravitational force on the particles. The bed is then said to be fluidized and will remain in this state as long as the two forces counterbalance each other.

When the fluid velocity equals the minimum fluidization velocity of the particles, the bed is said to be in a state of incipient fluidization. At this point the bed can be considered packed for the purpose of estimating the pressure drop across it and fluidized for calculating the buoyant weight of the particles. By using the Ergun equation for pressure drop and equating the two forces, we obtain an expression for the minimum fluidization velocity:

$$150\,\frac{(1-\varepsilon_{\mathrm{mf}})^2}{\varepsilon_{\mathrm{mf}}^2}\cdot\frac{\mu\cdot u_{\mathrm{mf}}}{d_{\mathrm{p}}^2}+1.75\,\frac{1-\varepsilon_{\mathrm{mf}}}{\varepsilon_{\mathrm{mf}}^2}\cdot\frac{\rho u_{\mathrm{mf}}^2}{d_{\mathrm{p}}} = (1-\varepsilon_{\mathrm{mf}})(\rho_{\mathrm{p}}-\rho)g \quad (1)$$

The pressure drop term on the left-hand side of equation 1 has two parts; the first term denotes the contribution due to laminar flow, and the second is due to turbulent flow. Depending on the flow regime, one could simplify the expression by neglecting the appropriate term. In the laminar flow region, the minimum fluidization velocity can be derived from equation 1 as

$$u_{\mathrm{mf}} = \frac{\varepsilon_{\mathrm{mf}}^2}{150(1-e_{\mathrm{mf}})}\cdot\frac{1}{\mu}\,d_{\mathrm{p}}^2(\rho_{\mathrm{p}}-\rho)g \quad (2)$$

The physical parameters that affect the velocity at which a settled bed of particles begins to fluidize can be seen from equation 2 and are the viscosity of the fluid, the particle size of the adsorbent beads, and the density difference between the adsorbent and the liquid. An accurate estimate of the bed voidage at the point of minimum fluidization is also important for estimating u_{mf}. For spherical particles, 0.4 is always a good first approximate valid when the packing is not overly compact. Wen and Yu (16) have found a value of 0.42 to be satisfactory.

Once the bed is in a fluidized state, a simple relation exists between the bed voidage and the fluid velocity. This relation is valid over a given range of the latter, and the most commonly used relationship is that described by Richardson and Zaki (17), which relates the superficial fluid velocity, the terminal settling velocity of the particle, and the bed voidage. The relationship is expressed as

$$\frac{u}{u_{\mathrm{t}}} = \varepsilon^n \quad (3)$$

where n is the Richardson–Zaki coefficient. The determination of n and u_{t} is usually obtained from a logarithmic plot of experimentally determined values of the bed voidage and the corresponding flow velocity and forms a fundamental part of the study of these systems.

However, bed voidage is not an easily measured parameter and would in turn require further experimentation. A more convenient approach is to use the relationship between the bed voidage at a given bed expansion, ϵ, in terms of the expanded bed height, H_{exp}, the initial (or settled) bed height, H_{o}, and the voidage of the settled bed ϵ_{o}. Thus we have

$$\varepsilon = 1 - \frac{H_{\mathrm{o}}}{H_{\mathrm{exp}}}\cdot(1-\varepsilon_{\mathrm{o}}) \quad (4)$$

The ratio $H_{\mathrm{exp}}/H_{\mathrm{o}}$ is referred to as the degree of bed expansion. Experimentally, one would measure the stable expanded bed height, H_{exp}, at a given flow velocity through the column (18–20). The data can then be plotted to obtain n and u_{t} for the bed. Analysis of equations 3 and 4 shows us that a high u_{t} for a particle results in a lower bed expansion and vice versa.

The terminal velocity is the upper limit of validity of the Richardson–Zaki expression, indeed of fluidization in general. At a fluid velocity equal to the terminal velocity of the particles, the particles are no longer suspended in the fluid but travel along with it. In the context of expanded bed chromatography, this leads to what is known as elutriation, where the adsorbent particles are physically removed from the column. Clearly, the latter phenomenon is highly undesirable, and it is therefore essential that for effective operation of expanded beds, the velocity range lies well within the limits of the minimum fluidization velocity at the lower end and the terminal velocity at the higher. The value of u_{t} can also be determined theoretically from Stokes law; thus

$$u_{\mathrm{t}} = \frac{d_{\mathrm{p}}^2(\rho_{\mathrm{p}}-\rho)g}{18\mu} \quad (5)$$

The coefficient n has been estimated by Richardson and Zaki (17) to be a function of both the particle to column diameter and the Reynolds number. The correlations are valid for uniform spherical particles $>100\ \mu$m in size. Garside and Al Dibouni showed that a different set of correlations are required when the fluidized bed voidage is greater than 0.85. Two expressions were proposed by them—one general type and a second similar to the Richardson–Zaki type of formulation. Both are valid over the entire range of Reynolds number. Some common expressions developed by different workers for estimating the values of n and u_t are shown in Table 1. Values determined experimentally or through the correlations are used for comparative evaluation of different adsorbents for their potential use as fluidized bed adsorbents. The equations are a simple and effective starting point for evaluating some of the fundamental characteristics of an expanded bed.

PRINCIPLES OF OPERATION

The operational procedure for protein recovery using an expanded bed is similar in principle to that of a conventional packed bed column and is schematically depicted in Figure 2. As mentioned in the previous section, the initial step involves formation of the stable expanded bed from its settled or packed state. This is done by first moving the top flow adapter to a sufficient height to permit free movement of the top of the bed in the upward direction of liquid flow. When flow of the equilibrating buffer is initiated, bed expansion occurs and the top of the bed rises. The bed height increase is usually characterized by a linear phase followed by a progressively reducing nonlinear rate of change until a final stable height is achieved. This expanded height depends on a number of physical factors of the system, which are described in the next section. Once a stable bed has been obtained, the top flow adapter is usually repositioned to lie anywhere between 1 and 15 cm above the top of the bed. The actual position depends on the mode of operation employed (see later section). With the top adapter in position, the bed is essentially ready for initiation of feedstock load.

The flow is then switched from the buffer to the unclarified feedstock, usually by an appropriate valve mechanism. During the feedstock load, the solid matter, consisting of cell debris or sometimes whole cells, travels directly through the column while some of the solubilized products are adsorbed onto the column. Ideally there will be no retention of the particulate debris within the column, but in practice one could well observe a degree of nonspecific binding to the adsorbent particles. This issue becomes of particular relevance during cleaning-in-place and column regeneration. Feed load is usually terminated when the outlet concentration of the product being recovered is approximately 5–10% of its concentration in the feed. At this point the inlet flow is then switched back to the equilibrating buffer to wash out the residual particulate matter from the expanded bed.

When the outlet liquid is completely clear of insoluble material, the product can be eluted from the column, most commonly in the packed mode and after the bed is first allowed to settle by terminating the flow of liquid. The top flow adapter is then lowered onto the settled bed, bringing it to a packed mode. The bound products are then eluted by the appropriate elution buffers, usually in a single step because specific separation is not the primary objective at this stage of the recovery process. When elution is complete, the column is regenerated and where required cleaned. The bed is then expanded with the equilibration buffer and made ready for the next cycle of feedstock load.

Bed Stability

The sequence of steps in the operation of an expanded bed is relatively straightforward and simple to incorporate. As

Table 1. Fluidization Parameters for Uniform-Sized Spherical Particles

Reference	Validity range	Equations
Richardson and Zaki (17)	100–6,350 d_p (μm) 1.06–11.25 ρ_p (g/mL) 0.81–2.89 ρ_1 (g/mL) 10^{-3}–113.10^{-3} μ (Ns/m^2)	$\frac{u}{u_t} = \varepsilon^n$ where $n = 4.65 + 20\frac{d_p}{d_c}$ (Re < 0.2) $n = \left(4.4 + 18\frac{d_p}{d_c}\right)\mathrm{Re}^{-0.03}$ (0.2 < Re < 1) $n = \left(4.4 + 18\frac{d_p}{d_c}\right)\mathrm{Re}^{-0.01}$ (1 < Re < 200) $n = 4.4\mathrm{Re}^{-0.1}$ (200 < Re < 500) $n = 2.4$ (Re > 500)
Garside and Al Dibouni (xx)	167–3,070 d_p (μm) 2.8 ρ_p (g/mL) 1.0 ρ_1 (g/mL) 10^{-3} μ (Ns/m^2)	$\frac{(u/\varepsilon \cdot u_t) - A}{B - (u/\varepsilon u_t)} = 0.06\mathrm{Re}^{(\varepsilon+0.2)}$ where $A = \varepsilon^{4.14}$ $B = 0.8\varepsilon^{1.28}$ for $\varepsilon \le 0.85$ $B = \varepsilon^{2.65}$ for $\varepsilon > 0.85$
Wen and Yu (16)	200–350 d_p (μm) 2.37–7.84 ρ_p (g/mL) 1.0 ρ_1 (g/mL) 10^{-3} μ (Ns/m^2)	$\varepsilon^{4.7}\left(\frac{d_p \cdot g \cdot (\rho_p - \rho_1)\rho_1}{\mu^2}\right) = 18\mathrm{Re} + 2.7\mathrm{Re}^{1.687}$ $0.01 < \mathrm{Re} < 10^4$

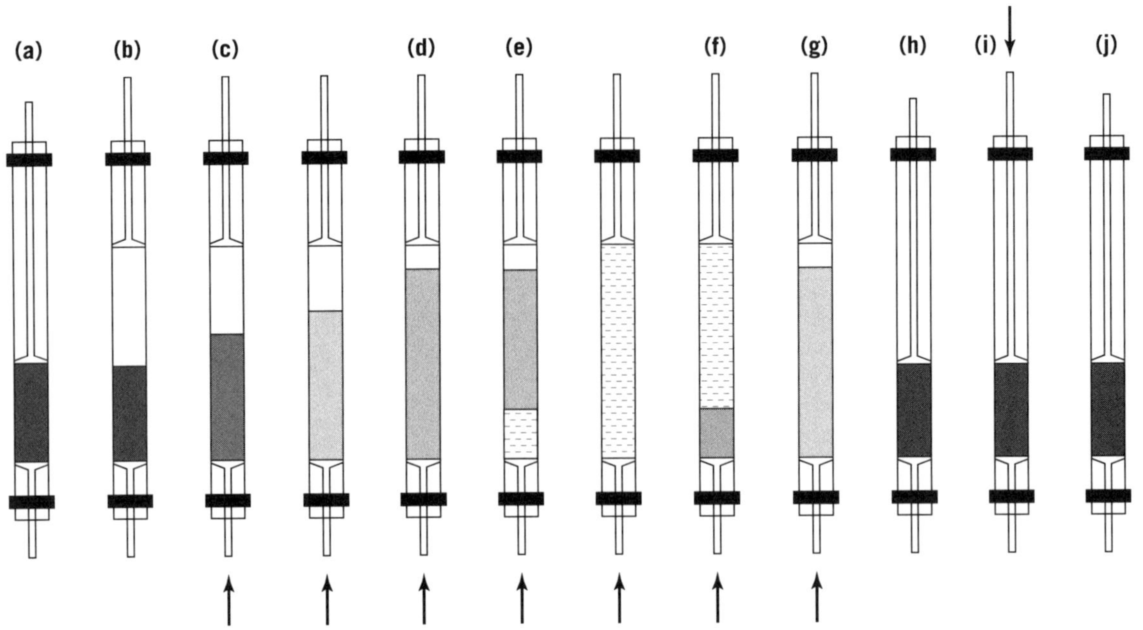

Figure 2. The various stages of operation of expanded bed chromatography from start-up to column regeneration. **(a)** Starting with a settled bed of adsorbent, **(b)** the adapter is moved up to remove any limitations on the movement of the bed. **(c)** Liquid flow through the column is then initiated from the bottom in the upward direction, during which time the bed undergoes a transient phase until **(d)** it finally reaches a stable steady-state height. The bed is equilibrated with buffer and **(e)** feedstock load initiated until the column is saturated, as seen from the breakthrough of bound product from the column. **(f)** The load is terminated and buffer wash started until **(g)** all the homogenate is washed out from the column. **(h)** The flow is then terminated, the bed allowed to settle, and the flow adapter lowered to attain a packed bed configuration, after which **(i)** the products are eluted, usually in the reverse direction. **(j)** The column is then cleaned and regenerated and made ready for the next run.

described earlier, the first step is obtaining a stable expanded bed; although not immediately obvious, this is a critical requirement with implications for overall process efficiency. A stable bed can be defined as one in which the liquid–liquid and solid–liquid mixing is low. Bed stability becomes a necessary requirement for good plug-flow-type hydrodynamics, which in turn affects the product binding characteristics. This is best quantified through residence time distribution (RTD) studies by analyzing the outlet profile of a tracer pulse in exactly the same manner in which it would be done for a packed bed system. Tracers such as acetone are effectively used in the absence of any particulate material. However, RTD experiments conducted in the absence of such particulates only indicate the state of the bed before feedstock load is initiated. Unlike packed bed chromatography, in which the hydrodynamics of the bed can be assumed to be more or less the same throughout the load phase, expanded beds are more than likely affected to a certain degree by the presence of the particulate matter. Although the influence need not necessarily be significant, it is nevertheless an important point to remember when conducting stability tests in these systems.

The tracer outlet profile can be evaluated by the moments rule to obtain an estimate of the Peclet number or the vessel dispersion number (21), defined as

$$\mathrm{Pe} = \frac{u}{D_{ax} \cdot L} \tag{6}$$

where D_{ax} is the liquid axial dispersion coefficient. Values of Pe greater than 40 are considered acceptable in terms of the flow hydrodynamics. The Peclet number is used interchangeably with the Bodenstein number, defined in the same manner and used extensively by Karau et al. (20) and Thommes et al. (22).

For the experienced user of expanded bed chromatography columns, the power of visual observation, although highly qualitative and impossible to validate, can often prove to be sufficient in ascertaining the extent of bed stability. Liquid–liquid (and thus solid–liquid) mixing is often a result of radial flow inequalities. Some of the common causes are poor distribution at the inlet to the column (which often transmits through the full length), inaccurate vertical alignment of the column (which is particularly important in scale-down systems), and aggregation of the adsorbent particles (which leads to significant variation in the bed voidage along the length of the column and thus affects the flow pattern). The consequential gross flow inadequacies can be more than sufficiently detected by the naked eye, and although an accurate determination of the Peclet number is not possible, visual observation often gives sufficient indication when a bed has yet to reach a stable configuration. Experience clearly plays a significant role in the use of visual observation as a tool for determining bed stability, and it has only limited applicability. Nevertheless, it may often prove sufficient when a rigorous residence time distribution experiment is not essential.

Stability of expanded beds is achieved through the inherent particle size distribution of the adsorbent. From the theory of fluidization outlined in the previous section, smaller particles fluidize more readily than larger ones in a given liquid stream, assuming that the particles are uniformly dense. For fluid velocity ranging between the minimum fluidization velocity of the smaller particle to the terminal velocity of the larger one, different situations arise, as depicted in Figure 3. As the flow velocity is increased such that $u_{\mathrm{mf,large}}, u_{\mathrm{t,large}}, u_{\mathrm{t,small}} > u > u_{\mathrm{mf,small}}$ the smaller particles fluidize and attain a steady-state position somewhere along the length of the column. Continual increase in the flow until $u_{\mathrm{t,large}}, u_{\mathrm{t,small}} > u > u_{\mathrm{mf,small}}, u_{\mathrm{mf,large}}$ results in both the small and large particles being fluidized. However, the increased flow velocity requires a longer column length for the inertial forces to drop to the level to be sufficiently counterbalanced by the buoyant force of the smaller particles. It therefore attains its steady state higher up along the column relative to both its initial position and that of the larger particle. A further increase such that $u > u_{\mathrm{mf,large}}, u_{\mathrm{t,small}}$ would wash the small particles out from the column and move the steady-state position of the larger particles higher up. Finally, when $u > u_{\mathrm{t,large}}$, the larger particles are also elutriated.

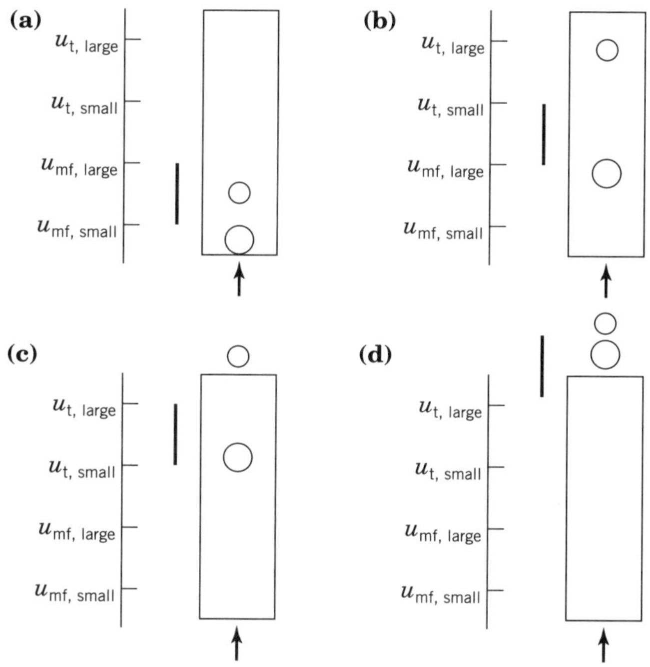

Figure 3. Schematic diagram showing the effect of flow velocity on the fluidization of two particles of different sizes. **(a)** As the flow velocity is increased such that $u_{\mathrm{t,large}}, u_{\mathrm{t,small}}, u_{\mathrm{mf,large}} > u > u_{\mathrm{mf,small}}$ the smaller particles fluidize and attain a steady-state position somewhere along the length of the column. **(b)** Continual increase in the flow until $u_{\mathrm{t,large}}, u_{\mathrm{t,small}} > u > u_{\mathrm{mf,large}}, u_{\mathrm{mf,small}}$ results in both the small and large particles being fluidized. The increased flow velocity, however, requires a longer column length for the inertial forces to drop to the level to be sufficiently counterbalanced by the buoyant force of the smaller particle. It therefore attains its steady state higher up along the column relative to both its initial position and that of the larger particle. **(c)** A further increase such that $u_{\mathrm{t,large}} > u > u_{\mathrm{t,small}}, > u_{\mathrm{mf,large}}, u_{\mathrm{mf,small}}$ would wash the small particles out from the column and move the steady-state position of the larger particles higher up. Finally, when $u > u_{\mathrm{t,large}}$ the larger particles are also elutriated. The vertical line shows the flow velocity range on a relative scale between the limits.

Extending the preceding discussion to an expanded bed of adsorbent with a defined particle size distribution means that the particles will be stratified along the length of the column. This is schematically depicted in Figure 4 for a bed of adsorbent in the settled and stable expanded states. In the absence of gross flow inequalities, each particle remains in its steady-state position, with the larger ones at the bottom or inlet to the column and the smaller ones near the top (23). This configuration can be maintained indefinitely. In reality there is localized movement of the particles, but it contributes little to liquid–liquid mixing within the system.

The other most notable method for stabilizing fluidized beds is through the use of magnetic fields (24–27). A magnetic material is incorporated within the adsorbent beads, and the superimposition of the magnetic field acts to limit the movement of the adsorbent particles. The flow of particulate-laden feedstock can often be conducted at relatively higher flow velocities, enhancing the efficiency of its wash through the column.

Modes of Operation

Although stability has been discussed in terms of liquid–liquid and solid–liquid mixing, bed height movement must be considered. Because the top of the bed is not restricted in any way, changes in the flow or physical properties of

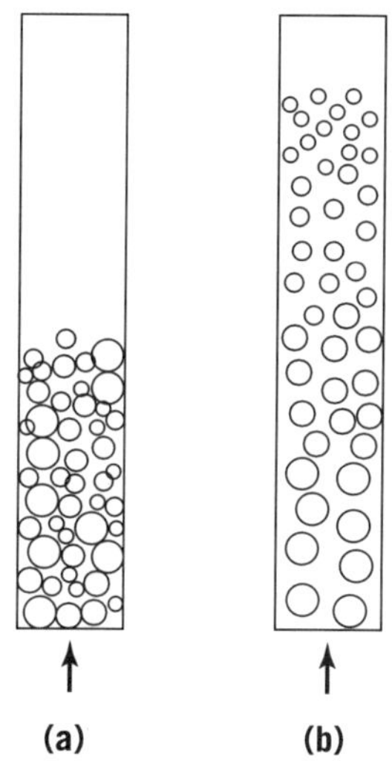

Figure 4. Schematic diagram showing the stratification of particles based on size along the length of the column going from **(a)** the randomly distributed settled bed state to **(b)** the stable expanded one.

the inlet liquid directly, although not always instantly, influence the bed height. Thus, when switching between buffers of different viscosities, densities, or both, the bed exhibits a concomitant movement to achieve a stable height corresponding to the new physical conditions to which it has been subjected. This movement is also seen when feedstock load onto a stable expanded bed is initiated. The significantly different physical properties of the particulate containing feedstock from the equilibrating buffer implies that on switching to feedstock, the bed height will change to a new stable position, assuming that the flow rate through the entire operation is kept constant. Consequently, there are two distinct possibilities for modes of operation of the bed:

1. In a *constant flow rate* mode, the bed is operated with the flow rate maintained at its initial value, allowing the bed height to vary if necessary. Because biological feedstocks are in general more viscous and dense than the standard buffers commonly used in biochemical processes, switching from buffer to feedstock load is manifested in terms of the expanded bed height increasing through an initial transient phase to a new position. This increase is further influenced by the presence of particulate matter.Two operational implications arise as a result. Firstly, the position of the top flow adapter needs to be managed appropriately, either through continuous repositioning during the course of the run or by selecting an initial position significantly higher than the top of the bed before the feedstock load. In this mode the gradually increasing bed voidage during the course of feedstock load would be beneficial in the context of solids removal. Protein adsorption could be adversely affected, however, because of increasing mass transport distances, likely making it a critical rate limiting step in the process. The latter effect would be more pronounced for high molecular weight proteins.
2. In a *constant bed height* mode the bed is maintained at its initial expansion, and any tendency for the bed height to change is corrected by appropriate adjustment in the flow rate. An obvious advantage is that there is no need to adjust the position of the top flow adapter, simplifying equipment design. It is imperative, however, to be able to monitor the top of the expanded bed and any changes that it undergoes to effectively control flow rate.Protein adsorption in this case would perhaps be more efficient but would be offset by slower solids removal because of the progressive reduction in the flow rate to maintain the bed height at its initial level. The extent to which flow rate adjustment is required depends on the percentage of solids in the feedstock, with their removal being hindered at the higher solid content.

EQUIPMENT

Because of the nature of the feedstock typically used in expanded bed chromatography applications, the equipment varies from that used in conventional packed bed chromatography, although in principle the equipment undergoes only minor modifications. Certain design considerations with respect to the column and flow distributor would need to be satisfied, as discussed in the following sections.

Column

Expanded bed chromatography is a low-pressure operation that requires minimal modification of conventional packed bed columns. A principle design consideration is the inlet flow distributor, which must be able to handle particulate material in the feedstock being loaded onto the column. This aspect is discussed in detail later. Another point of practical relevance is that the column must be of sufficient length to handle the desired degree of bed expansion. Two- to threefold bed expansions are typical; for a settled bed height of 20 cm, this translates to a minimum of 60-cm column length. In practice, columns are usually 1 m long so that the top flow distributor can be suitably positioned to allow for an increase in bed expansion when unclarified feedstock is being processed in the column.

The commercially available columns now range from 2.0 to 1,200 cm in diameter. The first column to be marketed was the STREAMLINE 50™, which is extensively used by researchers today. Given the potential for commercial application, the subsequent generation of columns was scaled up to 20, 60, and 120 cm in diameter. Although useful for process scale operations, the size clearly limited their use for methods development and process validation work. Scale-down of the expanded bed columns followed, and column sizes of 2.0 and 2.5 cm are now available. I conducted the first scaled-down validation, and data from columns sizes of 0.5, 1.0, 5.0, and 25.0 cm in diameter showed excellent correlation in expansion properties and protein adsorption breakthrough.

Table 2 shows the major suppliers of the expanded bed chromatography columns and available specifications. Included in the table is the small glassware company that tailor-made the small-diameter columns I used in the scaled-down studies described earlier. These columns are fitted with a sintered glass disk as the inlet and outlet flow distributor. Columns up to 20 cm in diameter are constructed of glass, and larger ones are made from stainless steel. Glass columns have the advantage of allowing visualization of the process. This observation is particularly useful when column blockage results in excessive back pressure, reducing the flow through the column and causing the bed to rapidly settle.

Flow Distributor

The importance of the flow distributor at the inlet has been a neglected issue in column design. There are several parallel requirements, beginning with the need for a uniform distribution of the liquid at the column inlet. A degree of variation from this ideal situation is always encountered, and this particular parameter is not considered to be critical in determining the overall performance of the bed. One of the reasons is the higher ratio of length to diameter, which offers a dampening effect for poor flow profiles that

Table 2. Commercial Suppliers of Columns for Expanded Bed Chromatography

Supplier	Column diameters (cm)	Materials	Distributor type
Amersham Pharmacia Biotech AB.	2.5, 5.0, and 20.0	Glass	Stainless steel disk and mesh arrangement
Amersham Pharmacia Biotech AB.	60 and 120	Stainless steel	Stainless steel disk and mesh arrangement
UpFront Chromatography A/S.			Distribution through stirred baffles in the column base

result from maldistribution at the inlet. A more relevant requirement of flow distributors in the context of expanded bed chromatography is that it should allow free passage of the particulate material present in the feedstock. The high voidage of the bed itself permits almost unhindered flow, and bed blockage is therefore more likely to be a problem at the flow distributor. Although larger pore sizes are clearly beneficial in overcoming this limitation, it is also important to remember that the adsorbent particles themselves must be retained within the column.

The most common distributor type is the perforated stainless steel plate fitted with a stainless steel wire mesh on top. This type is found in all of the Amersham Pharmacia Biotech (APB) columns available commercially. The number of holes in the perforated plate depends on the column diameter it fits. The wire mesh serves to prevent the liquid from jetting into the bed at the column inlet by offering a distribution surface. It also has a dual purpose of preventing the adsorbent particles from leaking out from the bottom of the column. De Luca et al. (19) have investigated bed expansion characteristics using distributors with various numbers of orifices and arrangements from the center. The settled bed height and bed expansion were found to be important parameters that influence mixing within the bed.

Although all expanded bed work using APB columns relies on these distributors, little has been published about the problems of blockage when using unclarified feedstock. This is not to say that the problem is nonexistent. Column back pressure buildup occurs because of gradual accumulation of the particulate material in either the pores of the perforated disk or the space between the disk and the mesh. In new columns and those seeing their first feedstock challenge subsequent to cleaning and regeneration, the buildup is relatively slow, and a much longer load time is possible before the column needs to be back flushed to clear the appropriate blockage. Back flushing is usually done by reversing flow direction for a short period, often at an enhanced rate. When the loading is resumed, the back pressure will be at its initial value; however, the buildup to the next back flush condition often occurs more quickly. The procedure can in principle be repeated until the end of the load cycle. The exact length of the initial and subsequent load periods depends on the parameters of the systems in question, including the solids concentration in the feedstock, the type of flow distributor, and the flow rate. Thus no reports on investigations of the influence of different parameters on the load efficiency of flow distributors have been published.

An interesting approach to flow distributor design has been adopted by UpFront Chromatography (Denmark). Its novel method essentially bypasses the conventional type of distributor and mesh from the flow path of the feedstock. In contrast to the commonly used method for load described earlier, the UpFront system introduces the feedstock from a side inlet and above the bottom support mesh. Distribution of the material is ensured through the use of a stirring device located immediately above the bottom support mesh. Although distributor blockage is clearly avoided in this system, the stirring speed needs to be carefully selected, balancing the requirement for uniform distribution over the inlet area with the need to avoid vortex formation and increased liquid–liquid mixing in the column itself. The scale-up potential of this distributor type needs to be addressed. The company manufactures 2-, 5-, and 20-cm-diameter columns using this technology.

Bioprocessing (England) has another approach to obtaining uniform flow distribution. Standard chromatography columns in which beds of glass ballotini are placed directly above the bottom support mesh are used. The expanded bed adsorbent is then added on top. When flow of buffer is initiated through the column, the bed of adsorbent fluidizes but the glass ballotini do not because of their significantly higher density. Uniform distribution is obtained, and the mixing of adsorbent and ballotini beads is avoided. Proper selection of ballotini size is important. The smaller ones have been shown to offer better distribution, equivalent to the use of a perforated disk with many orifices. However, one must remember that smaller particles have a lower minimum fluidization velocity, and the selection needs to balance the need for small-sized ballotini with maintaining a larger difference in the minimum fluidization velocity between it and the adsorbent beads.

Adsorbent

As with packed bed chromatography, the adsorbent must offer a basis for interaction with the solubilized components of the feedstock to enable product capture. In addition, expanded bed adsorbents must also possess good fluidization characteristics. During the flow of particulate bearing feedstock through an expanded bed, the adsorbent should, under the flow conditions employed, remain in a fluidized state and not tend to elutriate from the column. This can be achieved if the terminal settling velocity of the adsorbent is significantly higher than any of the other solid matter present in the system. The three physical properties of the system that most significantly influence the terminal velocity of the adsorbent—particle diameter, density difference between the particle and the liquid, and liquid viscosity—can be seen from the Stokes equation, described earlier.

Liquid viscosity is essentially dependent on the feedstock and is for the most part not an adjustable parameter, although some feed pretreatment is certainly possible. However, regarding the adsorbent, one can essentially experiment with the particle diameter and the density difference between the adsorbent and the liquid. The problem of suitably dense adsorbents can be offset to some extent by using size as a parameter for increasing the terminal velocity of an adsorbent particle. Thus a material of lower density and higher particle size can be used in expanded bed chromatography. However, a significant disadvantage lies in the resulting increase in mass transport resistance. The diffusion time is related to the square of the particle diameter; pore diffusion thus becomes the rate-limiting step in larger particles. Another disadvantage of expanded beds is increased interparticle distances, which may lead to slower transport of solute from the bulk to the surface of the adsorbent. The presence of particulate debris may cause further mass transport limitations and pose problems of adsorbent fouling and pore blockage.

The optimal adsorbent for use in expanded bed chromatography should therefore be small and very dense, and, as with any chromatography adsorbent, the dynamic bed capacity (DBC) should be acceptably high. There is an intrinsic link between these different parameters for effective expanded bed operation. DBC can be improved by using smaller particle sizes; however, efficient removal of the particulate debris present in the feedstock is an essential process requirement, and the two conditions can be simultaneously satisfied through the use of high-density adsorbents. Although good adsorbent design is critical in expanded bed operation, only limited work has been undertaken on its improvement, and most applications of this technology thus far have centered on the use of commercially available material. Only a few such materials exist currently. Table 3 summarizes some of the more common adsorbents currently used in expanded bed applications.

APPLICATIONS

The growing realization of the potential of expanded bed chromatography is borne out by the rapidly increasing number of reported applications of this technology in biological product recovery. The growth of this subject area over the last decade can be judged by following the literature for reported work during that period. The pioneering work in this field by Chase and coworkers relied on the use of conventional packed bed chromatographic adsorbents (28,29). Yeast cell homogenates were also commonly used in model systems. They are a suitable starting material given the many proteins present, any one of which can be targeted for recovery by expanded bed operation and offering the scope for use with ion exchange and affinity adsorbents (30,31).

Applications of expanded bed chromatography for the recovery of product directly from feedstock has moved on from the yeast homogenate systems that were studied a few years ago. There are now examples of applications for the recovery of both native and recombinant proteins from yeast (32) and bacterial cells (33–35), milk (36), and monoclonal antibodies from hybridoma cells.

The examples mentioned are but a few of the growing list of new applications that are being reported. The recently held Second International Conference on Expanded Bed Chromatography gives further evidence of the growing awareness of the potential of this technology in biological product recovery (see the book of abstracts for further information). In following the work in this area, it is obvious that the applications are for the most part based on conventional operational procedures and differ only in the system for which the recovery is being studied. Although the theory-oriented investigations have strived to tackle interesting issues of this technology, the applications mostly been bound by conventional operating protocols, similar to what has been described in the previous sections.

Two applications warrant specific mention because of their novel and interesting approach and are briefly highlighted. The work by Chase and coworkers on the use of a continuous countercurrent extractor (37–39) extends the

Table 3. Commercial Suppliers of Tailor-Made Adsorbents for Expanded Bed Chromatography

Supplier	Adsorbent trade name	Ligands	Density (g/mL)	Size	Matrix type
Amersham Pharmacia Biotech AB.	STREAMLINE™	DEAE SP Phenyl Recombinant protein A	1.2	100–300 μm, mean size ~200 μm	Cross-linked agarose with crystalline quartz core
Bioprocessing Ltd.	PROSEP™	Protein A Thiosorb	1.3	100 μm; pore size ~70–1,000 Å	Porous, dense glass beads
UpFront Chromatography A/S.	Mimo™	Mixed mode	1.0–1.6		Cross-linked agarose with core of one glass particle
	FastMabs™	Antibody	1.0–1.5		
	UFC	Ion exchange Hydrophobic Metal chelate	1.0–1.6 1.0–1.6 1.0–1.6	40–300 μm	
Biosepra	HyperD				Mineral oxide based

principles of expanded bed chromatography to continuous protein separation. A perfluorocarbon-based adsorbent (40) was pumped between four different contactors, each representing in sequence one of the four stages (i.e., load, wash, elution, and regeneration). Crude unclarified feedstock was pumped into the first contactor and product obtained from the third. The use of a specific affinity ligand permitted highly purified product to be collected at the outlet of the elution stage.

Another interesting methodology has been demonstrated by Lyddiatt and coworkers in which a fluidized bed was coupled with batch fermentation for direct product removal through either continuous or intermittent recycle of the reactor contents through the fluidized bed. Inherent problems of holding the fermentation material outside the controlled reactor environment were addressed through the use of a multibed manifold comprising smaller beds. Each bed was operated until the adsorbent was saturated and then removed from the flow loop for product recovery, cleaning, and regeneration.

Expanded bed protein adsorption has now reached a stage in its development where a good understanding of the potential and drawbacks can be effectively assessed. Issues related to the bed design, performance, and application have been addressed in the literature. Although the potential of this technology is well accepted, it is dangerous to assume that it will offer the solution to many recovery and separation problems. Many aspects are still either not considered important or not understood. The worth of the technology will be seen only when they become an integral part of bioprocesses. Toward this goal, the first step may already have been taken: There is now at least one report of a biotechnology company having patented a process for plasmid DNA production in which expanded bed adsorption is incorporated in the purification protocol.

BIBLIOGRAPHY

1. J.E. Ramirez-Vick and A.A. Garcia, *Separation Purif. Methods* **25**, (1996).
2. A.I. Clarkson, M. Bulmer and N.J. Titchener-Hooker, *Bioprocess Eng.* **14**, 81–89 (1996).
3. J.P. Maybury, K. Mannweiler, N.J. Titchener-Hooker, M. Hoare, and P. Dunnill, *Bioprocess Eng.* **18**, 191–199 (1998).
4. A. Pessoa and M. Vitolo, *Appl. Biochem. Biotechnol.* **70**, 505–511 (1998).
5. D.P. Harris, A.T. Andrews, G. Wright, D.L. Pyle, and J.A. Asenjo, *Bioseparations* **7**, 31–37 (1997).
6. S.G. Walker and A. Lyddiatt, *J. Chromatogr., B* **711**, 185–194 (1998).
7. G.M. Zijlstra, M.J.F. Michielsen, C.D. de Gooijer, and L.A. van der Pol, *Bioseparations* **6**, 201–210 (1996).
8. M. Ritopalomares and A. Lyddiatt, *J. Chromatogr., B* **680**, 81–89 (1996).
9. G.M. Zijlstra, M.J.F. Michielsen, C.D. de Gooijer, and L.A. van der Pol, and J. Tramper, *Biotechnol. Bioeng.* **12**, 363–370 (1996).
10. I.J. Isaacs, *Australasion Biotechnol.* **6**, 88–92 (1996).
11. S.R. Narayanan, *J. Chromatogr., A* **658**, 237–258 (1994).
12. M. Leonard, *J. Chromatogr.* **699**, 3–27 (1997).
13. E. Boschetti, *J. Chromatogr., A*, **658**, 207–236 (1994).
14. Q.M. Mao and M.T.W. Hearn, *Biotechnol. Bioeng.* **52**, 204–222 (1996).
15. D.K. McCormick, *Biotechnology* **11**, 1059 (1993).
16. Y.U. Wen and Y.H. Yu, *Chem. Eng. Prog.* **62**, 100–111 (1966).
17. J.F. Richardson and W.N. Zaki, *Trans. Inst. Chem. Eng.* **32**, 35–53 (1954).
18. G.S. Finette, Q.M. Mao, and M.T.W. Hearn, *J. Chromatogr., A*, **743**, 57–73 (1996).
19. L. De Luca, D. Hellenbroich, N.J. Titchener-Hooker, and H.A. Chase, *Bioseparations* **4**, 311–318 (1994).
20. A. Karau, C. Benken, J. Thömmes, and M.-R. Kula, *Biotechnol. Bioeng.* **55**, 54–64 (1997).
21. O. Levenspiel, *Chemical Reaction Engineering*, 2nd ed., Wiley, New York, 1972.
22. J. Thommes, M. Weiher, A. Karau, and M.-R. Kula, *Biotechnol. Bioeng.* **48**, 367–374 (1995).
23. H.A. Chase, *TIBTECH* **12**, 296–303 (1994).
24. M. Goto, T. Imamura, and T. Hirose, *J. Chromatogr.* **690**, 1–8 (1995).
25. L. Nixon, C.A. Koval, R.D. Noble, and G.S. Slaff, *Chem. Mater.* **4**, 117–121 (1992).
26. B.E. Terranova and M.A. Burns, *Biotechnol. Prog.* **5**, 98–104 (1989).
27. A.S. Chetty and M.A. Burns, *Biotechnol. Bioeng.* **38**, 963–971 (1991).
28. H.A. Chase and N.M. Draeger, *J. Chromatogr.* **597**, 129–145 (1992).
29. Y.K. Chang, G.M. McCreath, N.M. Draeger, and H.A. Chase, *Trans. Inst. Chem. Eng.* **71**, 299–303 (1993).
30. Y.K. Chang, G.E. McCreath, and H.A. Chase, *Biotechnol. Bioeng.* **48**, 355–366 (1995).
31. Y.K. Chang and H.A. Chase, *Biotechnol. Bioeng.* **49**, 204–216 (1996).
32. F. Raymond, D. Rolland, M. Gauthier, and M. Jolivet, *J. Chromatogr.* **706**, 113–121 (1998).
33. M. Hansson, S. Ståhl, R. Hjorth, M. Uhlén, and T. Moks, *Biotechnology* **12**, 285–288 (1994).
34. G. Maurizi, V. DiCioccio, G. Macchai, P. Bossu, C. Bizzarri, U. Visconti, D. Boraschi, A. Tagliabue and P. Ruggiero, *Prot. Expression Purif.* **9**, (1997).
35. H.J. Johansson, C. Jagersten, and J. Shiloach, *J. Biotechnol.* **48**, 9–14 (1996).
36. A. Degener, M. Belew, and W.H. Velander, *J. Chromatogr.* **799**, 125–137 (1998).
37. R.O. Owen, G.E. McCreath, and H.A. Chase, *J. Mol. Recognit.* **9**, 575–584 (1996).
38. R.O. Owen, G.E. McCreath, and H.A. Chase, *Biotechnol. Bioeng.* **53**, 427–441 (1997).
39. R.O. Owen and H.A. Chase, *J. Chromatogr., A* **757**, 41–49 (1997).
40. G.E. McCreath and H.A. Chase, *J. Mol. Recognit.* **9**, 607–616 (1996).

PROTEIN AGGREGATION, DENATURATION

Catherine H. Schein
University of Texas Medical Branch
Galveston, Texas

KEY WORDS

Aggregation
Biophysical techniques
Co-solvent effects
Homology modeling
Kinetics
Measuring
Protein folding
Surface force
Three-dimensional structure

OUTLINE

INTRODUCTION

Because some protein components, such as the tyrosine side chain, have negligible solubility in water, proteins in solution exist in a precarious equilibrium. Most proteins form oligomers or aggregate at higher densities, a phenomenon that can to some extent be prevented by the addition of solvent additives (1) and mutations targeted to surface areas involved in aggregate formation (2). Solvent additives have proved invaluable in areas ranging from protein formulation for drug purposes to sample preparation for high-resolution structure determination techniques. Using cosolvents is, however, a matter of trial and error, with little framework for predicting their effects or even optimizing concentration. This is partly because of the inherent problems in following the early stages of protein aggregation.

This article summarizes biophysical techniques useful for measuring solubility and following precipitation. Maintaining solubility depends on subtle interactions of side chains and backbone atoms to balance the cavitation energy needed to maintain a solvent pocket for the bulk of the protein by the energy of interaction of hydrophilic groups with the solvent. Traditionally, proteins are presented as unique but ill-characterized globules remarkable only with respect to size and net surface charge. Direct methods to determine the characteristics of the protein surface may allow a better accounting for protein–solvent interactions in the presence of cosolvents. Energy functions that account for the various forces on the protein have been derived; their application to different solvents is now being tested. Interpreting empirical data to derive atomic solvation parameters or deriving bulk solvent parameters may lead to a thermodynamic or kinetic accounting for the effect of cosolutes on protein stability and solubility.

Designing mutations that prevent aggregation can be greatly helped by a high resolution three-dimensional structure of the protein–protein complex. A relatively unsophisticated home computer or work station now allows easy assess to the steadily increasing data banks of detailed 3-D protein structures, which were previously accessible only with expensive viewing systems. Faster and more efficient computational methods have greatly expanded our abilities to model protein structures by homology to proteins of known structure or ab initio using protein threading and other algorithms from the primary sequence. Thus, a section is also dedicated to methods for obtaining and using structural information to predict areas of the protein that may be involved in oligomerization and aggregation.

METHODS FOR MEASURING SOLUBILITY AND PROTEIN ASSOCIATION

For details about the molecular interactions involved in complex formation, many methods measure the absorption or scattering of some form of electromagnetic radiation by the sample (UV/vis and IR/Ramen spectroscopy, light scattering, laser light scattering, neutron diffraction and X-ray crystallography) (3). A beam of radiation sent through the sample can:

1. Pass through the sample unchanged
2. Be absorbed by the sample and emitted at the same wavelength
3. Be absorbed by the sample and emitted at a different wavelength
4. Be scattered by the sample

Various methods can be used to detect the effects of the sample on the beam (4). Figure 1 is a summary of methods aligned according to the electromagnetic radiation spectrum. At opposite ends of this spectrum lie the two highest resolution methods, X-ray crystallography and NMR spectroscopy, which yield complete structural information at the atomic level. However, combining results from lower resolution methods can be used to develop a working model of complexes and to obtain information about the aggregation process. In the simplest and most commonly used method to measure protein concentration and aggregation, a coherent beam of visible or UV light is passed through the sample. The light emerging on the other side of a cuvette of known path length is measured and related to the concentration of the sample. Some information about the nature of the sample can be obtained by observing how much light is absorbed when the wavelength is varied over the whole spectrum (a so-called UV/vis scan). In large proteins, the tryptophan side chains within the protein may be shielded, and as the protein unfolds, this tryptophan absorption at 280 nm becomes more pronounced. Absorption in the visible range is usually due to the presence of a chromophore (e.g., Haem absorption) or particulates. Fluorescence spectroscopy measures the energy of the emitted light, and laser light scattering measures the angle of the reflected light to determine particulate size (5). Many other, more advanced methods, summarized in Table 1, have been used to measure the quaternary state of proteins in solution.

Combining Methods to Follow Aggregation and Precipitation and Determine the Structure of Complexes

The size of small complexes in solution can be accurately determined by equilibrium centrifugation, native gel electrophoresis, or gel sizing chromatography (using for example Sephadex beads or HPLC). However, these methods are not particularly useful for studying aggregated protein. In vitro aggregation can have many forms, extending from a drastic, sudden precipitation of large flocculent masses of protein, often visible to the naked eye, to the gradual formation of tiny, invisible particles. The most valuable experiments look at the initial phases. The simplest direct quantitation of aggregation is thus to measure the absorption of visible light by the sample, if the particles are small enough to remain in solution, or to follow the precipitation by the drop in UV absorption of a protein solution where the particles are allowed to settle or are removed by centrifugation. If aggregation is related to unfolding, it may be possible to follow the increase in OD_{280} (as the tryptophan residues become exposed upon unfolding) and then its decrease as the protein aggregates.

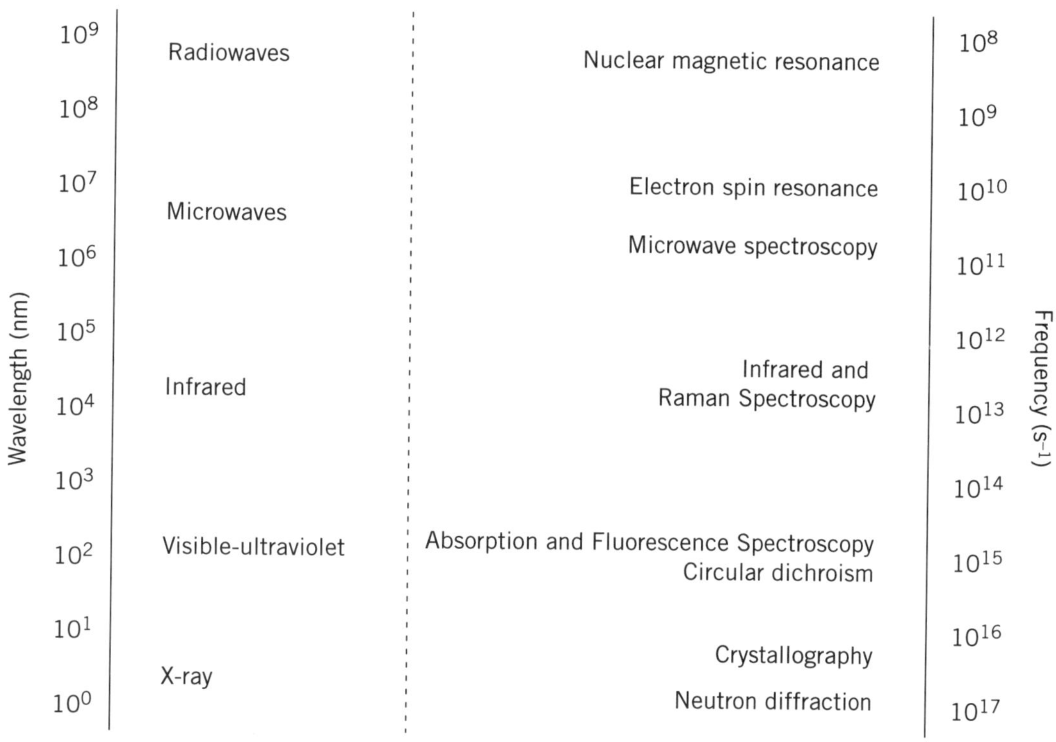

Figure 1. Methods used for analyzing the structure of biological molecules and complexes thereof as a function of position in the electromagnetic spectrum.

Table 1. Physical Methods for Monitoring Changes in Protein Structure and Interprotein Interactions Involved in Aggregation

Method (abbreviation)	Brief description	Protein needed[a]	Sample preparation/ buffer interference[b]	Best used for
Equilibrium (ultra) centrifugation (EC)	Protein complex is centrifuged at high speed in a density gradient, and its equilibrium position used to determine the size of the molecule or complex.	High	Gradient components may affect protein association.	Determining the oligomeric or aggregation state of soluble complexes. Insoluble aggregates settle at the bottom of the tube or float on the surface of the gradient.
Gel sizing chromatography (GSC)	Soluble proteins are passed over a column of porous polymer. Molecules small enough to enter the pores are held back and elute according to size, larger ones pass through unimpeded with the void volume. Can be done with capillary HPLC columns on the microscale.	Moderate to high	Sample should be concentrated for application, a step that may induce aggregation or oligomerization.	Sizing small, soluble complexes of proteins; large aggregates either elute with the front or precipitated at the top of the column (latter may slowly redissolve, giving unreliable data).
Polyacrylamide gel electrophoresis (PAGE)	Proteins are separated according to size by electrophoresis through acrylamide gels of varying density. Proteins may be detected in a variety of ways: by staining, autofluorescence, or after transfer to a membrane, by reaction with a specific antibody or nucleic acid fragment.	Concentrated sample required, but small amounts of protein may be visualized by silver or gold staining.	If the sample is not to be eluted for further analysis, apply less protein and use more sensitive dyes for accurate size determination.	Sizing small, soluble complexes of proteins; large aggregates stick in the slot.
UV/Vis absorption (UV-abs)	A beam of coherent light is passed through the sample, and the absorption by the protein constituents is measured. Proteins absorb between 210 and 220 nm because of orbital transitions of the peptide backbone, which are overwhelmed by that of the side chains of D, N, E, Q and R, and at 280 nm because of aromatic side chains of W, Y, F; chromophores in the protein may absorb throughout the visible range as well.	Low	Only for protein in solution; samples should be diluted to be in the linear range of the spectrophotometer	The denaturing effect of cosolvents on proteins; following aggregation by decrease in the fluorescence absorption and/or increase in absorption in the visible range.
Fluorescence transfer (FT)	A fluorescence acceptor and donor group are coupled to a protein. When excited by light, the quenching of donor fluorescence by the acceptor serves as a measure of the distance in the complex between them.	High	Only for protein in solution, buffer interference; requires coupling of donor and acceptor molecules usually to free cysteine groups in the proteins.	Distance between labeled side chains in a single molecule or oligomer; most useful with protein complexes with a well-established 3-D structure.

Optical rotatory dispersion (ORD)/ Circular dichroism (CD)	In ORD, the rotation of a beam of polarized light by the sample is measured by a coupled polarizer/photomultiplier on the exit side of the cuvette. The angle of rotation of the second polarizer to obtain the highest intensity of emitted light is plotted as a function of the wavelength. In CD, molar ellipticity is measured as the difference in absorbence by the sample of an incident beam rotated to the left or right (I_L and I_R) by a timed electric field change. A photomultiplier on the exit side measures absorbence as a function of time, and a microprocessor interprets the data to determine the mean residue ellipticity (which is proportional to $I_L - I_R$) as a function of wavelength.	Moderate	Only for protein in solution, buffer interference possible.	Secondary structure of proteins in various media; also to follow changes in the secondary structure as a function of denaturation or aggregation. Computer-aided spectral interpretation can be used to obtain more complicated data about the protein configuration.
Electron microscopy (EM)	In vacuo, a beam of electrons is passed through a target with the (usually stained or coated) object affixed to it or imbedded in a matrix to obtain a 2-D image. In essence, one measures the electron density of the sample after coating to obtain a 2-D image. Cryo-EM techniques use unstained samples embedded in vitreous ice. High-speed calculations can be used to phase many images of particles (6,7).	Low	Fixation/staining may be needed. Consistent sample preparation, crucial to the cryo-EM technique, can be quite difficult.	Determining the ultrastructure of large, electron-dense, complexes; particularly useful to determine the shape of particles and their size distribution. Current resolution is in the range of 10 Å.
Electron diffraction (ED)	By tilting the target during EM and measuring reflection of the electron beam at various angles from the incident, a 3-D image can be reconstructed.	Low to high	2-D crystals or membrane patches; viral particles.	Obtaining ultrastructural data about regularly structured monolayers such as the membranes of purple bacteria.
Scanning tunneling microscopy (STM)	The electron stream from the object is detected across a vacuum barrier by a sensitive probe. The microscope itself is in reality the probe, and the name comes from the tunnel current, which flows when there is a small electric current between the tip of the probe and the conducting surface to be scanned (which should be less than 1 nm away).	Low	Resolution is higher for conductive samples; fixation/staining may be needed for proteins and nucleic acids.	Determining the shape and size of fixed and coated large complexes or protein-DNA complexes; resolution is theoretically at the atomic level; suited to the study of larger protein aggregates because it gives a 3-D image directly without the image reconstruction methods used for 3-D transmission EM images.
Atomic force microscopy (AFM)	Technique (also called scanning probe microscopy) related to STM and using similar apparatus, more suited to biological samples because it is done under a solvent phase rather than in vacuo (8). Very little different from hydration STM, based on the electrical conductivity of molecularly thin water layers that adsorb to the sample surfaces in a humid atmosphere.	Low	Attachment to a surface necessary, can be done in a variety of solvents.	Direct viewing of the ultrastructure of large, not necessarily regular biomolecular complexes without staining. Methods are being developed for viewing protein-RNA complexes.

Table 1. Physical Methods for Monitoring Changes in Protein Structure and Interprotein Interactions Involved in Aggregation *(continued)*

Method (abbreviation)	Brief description	Protein needed[a]	Sample preparation/ buffer interference[b]	Best used for
Fourier transform-infrared absorption (FTIR)	Absorption of light in the infrared region (2.5–250 μm) by the sample is measured and the output beam resolved by Fourier transformation. *IR difference spectra*, deconvoluted in the amide I band region (between 1620 and 1700 cm^{-1}) can show secondary structure in terms of % α-helix, β-sheet, etc. (9,10).	High	Can also be used for precipitates.	Determining secondary structure elements and alterations thereof under different conditions.
Infrared attenuated total reflection spectroscopy (IR-ATR)	Infrared beam is reflected through an optically transparent germanium plate coated with sample; internal reflections along the plate increases sensitivity.	Low to moderate	Membrane patches attached to a special plate.	Membrane protein interactions with each other and external molecules.
Scanning fluctuation correlation spectroscopy (FCS)	Measures particle number concentrations by monitoring spontaneous equilibrium fluctuations in the local concentration of fluorescent species in a small (femtoliter) subvolume of a sample.	Low	Little problem with buffer interference	Detect molecular aggregation for dilute, submicromolar samples by directly counting particles.
Elastic (classical, static) laser light scattering (SLLS)	For elastic (classical, static) light scattering, the scattering of a beam of polarized laser light as it passes through the sample is measured and the obtained scatter factor can be used to calculate both the weight average molecular weight of the particles in solution as well as the radius of gyration.	Moderate	Scattering methods are usually not greatly affected by the buffer.	Determining the size of aggregates and precipitates. To obtain the scattering weight of a particle, one measures the scattering factor at several different scattering angles (q) and then plots the scattering factor as a function of q. Extrapolating back to $q = 0$ will yield the scattering weight.
Quasielastic (dynamic) laser light scattering (DLLS)	Similar to SLLS; but changes in the frequency from the translational (Brownian) movement of the scattering particles are also measured. The broadness of the intensity distribution of the emitted light for frequencies around the primary monochromatic beam frequency is directly related to the diffusion coefficient of the particles, which can then be related to the hydrodynamic radius if a model for the particle shape is available.	Moderate	The beam should not be strong enough to damage the sensitive biological specimen.	Kinetics of particle coagulation by following the decrease in diffusion coefficient as the particle size increases (11); determine particle molecular weight as a function of time; to follow course of aggregation in any solvent; compare samples in different buffers if viscosity is accurately measured.

Raman, Raman resonance (RRS)	Light scattered at the incoming frequency, so-called Raleigh scattering, is measured in classical (elastic) and quasielastic (dynamic) light scattering. A small fraction of the laser light is scattered at frequencies higher (anti-Stokes) and lower (Stokes) than the incident beam. These scattered beams are measured in Raman scattering, which has many variations.	High	Sample fluorescence may interfere with Raman resonance measurement of precipitates.	Follow quaternary interactions of soluble complexes and to compare the secondary structures of proteins under various conditions; to compare the effects on secondary structure of precipitating agents.
Neutron scattering (NS)	Scattering of a neutron beam by the sample is measured.	High	Beam time is limited.	Determining the size of larger proteins and complexes.
Mass spectrometry (MS)	The magnetic field induced deflection of a vaporized, ionized molecule is used to determine its mass. Types are characterized by the method used to ionize and detect the sample. Advanced computational methods can be used to interpret structural data from the cracking pattern of larger molecules.	Varies	Usually protein is bound to a solid (e.g., nitrocellulose).	Very accurate molecular weight of monomers; use in combination with chemical cleavage methods may aid in determining tertiary and quaternary structure. Electrospray ionization (ESI-MS) has been used to characterize complexes of proteins and protein/nucleic acids (12).
Electron spin resonance (ESR)	Microwaves are absorbed by paramagnetic substances, and the change in the energy level of the electron spins measured.	Moderate		Structure and dynamics of lipid membranes; free radical detection in proteins; state of transition metals in proteins.
Nuclear magnetic resonance (NMR)	A concentrated solution of sample, possibly isotopically enriched for nuclei with a magnetic moment different from 0 (^{1}H, ^{13}C, ^{15}N, ^{31}P), is subjected to a high magnetic field. In the most common method, after the sample is pulsed, the decay in amplitude of the emitted radiowaves with time is measured and converted to a frequency spectrum by Fourier transformation. The magnitude and direction of the pulse can be altered to detect certain types of interactions between nuclei. The strength of the local magnetic field around an individual atom is decreased by the presence of covalently bound or spatially close neighboring groups, causing changes in the position of the resonance lines (chemical shifts) in the output frequency spectrum. Peak intensities in NOESY spectra reflect the distances between individual spins. Distance geometry methods and restrained molecular dynamics calculations are used to calculate the 3-D structure of proteins or protein–protein complexes.	High	Most solvents interfere with measurements. Labeling of the sample with stable isotopes may be necessary. Oligomers may be artificially generated by the high protein concentration required for measurement.	Determining the exact tertiary structure of small proteins, nucleic acids, and complexes. Combining results from different NMR experiments can be used to monitor changes in the resonance frequencies upon complex formation, and thus give specific information on binding sites. Various techniques (e.g., resonance transfer) can be used to measure interactions between two molecules even without establishing a structure. See Ref. 13 for a discussion of NMR techniques to study protein complexes.

Table 1. Physical Methods for Monitoring Changes in Protein Structure and Interprotein Interactions Involved in Aggregation *(continued)*

Method (abbreviation)	Brief description	Protein needed[a]	Sample preparation/ buffer interference[b]	Best used for
Protease mapping (PM)	Fragments left after cleaving with a site-specific protease are analyzed (by gel chromatography and sequencing or MS/MS sequencing) to determine surface accessible areas	High	Need to have specific sites for proteases in the protein; buffers should not interfere with techniques used to identify protein fragments.	Tertiary structure of an isolated protein or protected areas in complexes.
Chemical-based cleavage (CC)	Like PM, but an activated metal ion cleavage agent is incorporated at a free sulfhydryl group, and the cleavage is triggered by chemical or physical stimuli.	High	Buffers used cannot stimulate cleavage or interfere with subsequent methods to determine the size and constitution of fragments; need free sulfhydryl groups in the protein	Tertiary structure of an isolated protein or areas of contact in complexes.
Surface plasmon resonance biosensors (SPR)	A surface plasmon is an oscillation of free electrons that propagates across the surface of a conductor, typically a thin film of gold or silver. The sample to be measured is immobilized on one side of the gold film, e.g., in a hydrogel, and plasmon excitation is started with a beam of light. The intensity of plasmon generation (measured as the reflected light reaching the detector) is related to the refractive index of the sample. This can be altered by flowing a binding substance over the surface where it is immobilized.	Low to moderate	No labeling of the biomolecules is required, but the two media must have different refractive indices. To avoid rebinding artifacts, the bound protein should not be overloaded.	Real time analysis of (unlabeled) protein binding reactions (14).
Total internal reflection fluorescence (TIRF)	Fluorescently labeled proteins in aqueous solution interact with a solid phase (e.g., fused silica) coated with bilayers of synthetic phospholipids, unlabeled proteins, or even whole cells.	Moderate	Two media must have different refractive indices.	Determining interactions between membrane-bound and free proteins.
Small-angle X-ray scattering (SAXS)	The scattering of X-rays at small angles can be used to obtain greater resolution of X-ray crystal structures.	Moderate	Crystals required and special detection/computation systems.	Resolving the structure of large complexes to higher resolution than possible with diffraction data alone.
X-ray diffraction (XRD)	A protein crystal is bombarded with a high intensity X-ray beam of known wavelength and polarity. The diffraction pattern of the emitted energy, measured as spots on X-ray film or using an area detector, can be used to determine the position of individual atoms in the 3-D structure of the protein or complex.	High	Crystals required and extensive experimentation must be done to solve the phase problem.	The only method now in use that can give the 3-D structure of large, regular protein complexes at the atomic level. Adaptations are now being developed to use it to determine dynamics as well.

Note: Data presented with respect to the protein requirements, impedance by common buffers, and the type of information obtainable about the protein in solution or precipitate. For more details, see related references in Table 2; for general information about the methods, see Galla (4), cited specific references, or other books on protein characterization techniques.

[a]Protein concentration required for measurement. Low is in the nM–μM range, high means 0.1 mM or higher. Note that highly concentrated protein solutions are usually required to obtain crystals and that a lot of protein is wasted during the process of crystallization. Thus, one needs to start with about 100 mg of purified protein for X-ray analysis, although only a few μgs may find their way to the beam.

[b]Buffers used for sample preparation for all methods should be sterile and ultrafiltered (0.45 μ or smaller pore diameter filter) to remove particulate matter. Buffer requirements for various methods are discussed in the references given.

Although rapid onset suggests aggregation is a noncontrolled process, there are reports of sequence specificity based on the pattern of higher molecular weight aggregates seen on gels (15). An approximately similar pattern of bands was seen, whether a protein formed inclusion bodies alone or in the presence of another aggregating protein. Although this would indicate that each protein was only aggregating with itself, the authors could only study aggregates small enough to enter a nondenaturing gel, and a substantial portion of the protein in all the experiments remained at the top of the polyacrylamide gel (PAG). More recent data using computer simulations and light scattering suggest a cluster–cluster mechanism rather than a stepwise monomer addition polymerization (16). The aggregation of polypeptide chains during refolding occurred by multimeric polymerization, in which two multimers of any size could associate to form a larger aggregate and did not require a sequential addition of monomeric subunits.

Generally, combinations of methods are applied to proteins, such as the HIV-Rev, where it is important to determine the 3-D structure, but problems arise in obtaining this by high resolution methods. The Rev protein, whose function is to move viral unspliced RNA from the nucleus to the cytoplasm, is essential for the establishment of infection. Although establishing its 3-D structure would be very useful for designing inhibitors that could be used therapeutically, severe aggregation problems have hindered its characterization by X-ray or NMR spectroscopy. Data from STEM, cryo-EM of lysine cross-linked filaments, and Raman spectroscopy have been used to develop a low resolution (21 Å) model of Rev filament structure that may be useful in designing inhibitors (17). Structural modeling tools possibly useful for this protein are discussed later. To further illustrate how the methods of Table 1 can be applied to study the interaction of proteins, Table 2 is a list of some of the protein complexes that have been studied by combining biophysical methods.

CHARACTERIZING PROTEIN–SOLVENT INTERACTIONS

Protein Stability in a Practical Sense

As Privalov has pointed out (85), if one assumes that all deviations from the X-ray crystal structure represent a new and separate state of a protein, the net thermodynamic stability of a protein at any temperature above a few degrees Kelvin is essentially zero. The thermodynamic framework works well for small molecules that can assume a limited number of stable conformations. However, proteins change conformation continuously; phenylalanine side chains are often free to rotate around the Cβ-phenyl bond and as many as 40% of the amide protons of a protein exchange with the solvent in a few seconds (86). Solvent effects make these dynamic considerations even more complicated; the rate of hydrogen exchange of a protein changes when even the isotopic nature of the solvent is changed (87).

To simplify the problem, thermodynamic stability has been defined as the energy needed to switch the protein structure from a folded, active native state (in reality a collection of substates around a mean native structure) to a completely unfolded, denatured state. Formation of the denatured state in a dilute protein solution can be induced by slowly increasing temperature until the transition is indicated by extreme changes in the protein's activity or protease susceptibility, or in the protein's characteristic UV absorption, circular dichroism, or more recently, NMR spectrum. The temperature at which this change occurs is referred to as the thermal denaturation temperature, or T_d. Thermodynamic stability data can be collected in a few hours, but the measurements cannot be done at high-protein density because aggregation usually precedes total denaturation. The T_d varies considerably from one protein to another and is a function of the pH and other properties of the solution in which the measurement is made. The free energy of unfolding ($\Delta G^{\circ}_{unfolding}$), which can be calculated from the T_d, is approximately 12 ± 5 kcal/mol at 30 °C for all proteins. This number is equated with the free energy of stabilization of proteins in aqueous solution according to the two-state model (88). As native, properly folded structures aggregate less than unfolded, denatured ones, there is an intimate relationship between solubility and thermodynamic stability. However, although extreme differences in the T_d can be used as a direct indication of the stability of the protein structure (peptides may lose their secondary structure at $T < 12$ °C, whereas thermophilic proteins retain it in a boiling water bath), the T_d of typical proteins is in too narrow a range to serve as an indicator of structure or purity. We know that proteins vary greatly in their stability in solution or the dry state. It has been shown that the $\Delta G_{unfolding}$ for RNase T1 can be almost doubled by adding 0.2 M Na_2HPO_4 or certain other salts, but the increase amounts to only 5 kcal/mol (86). Many additives that stabilize proteins against aggregation do not affect the T_d. Aggregation occurs at temperatures well below the T_d for proteins, indicating that a protein need not unfold completely to aggregate.

Thus, other methods must be used to determine protein stability in solution under nonextreme conditions. A two-state model ignores the less abrupt (and therefore less easy to measure) changes in the protein structure that precede complete unfolding. Such changes, coupled with an increased rate of aggregate-producing encounters, probably lead to temperature-dependent protein precipitation. Stability for practical purposes is the empirically determined resistance of a protein to oligomerization or to inactivation by other mechanisms. Stability can then be equated with the persistence of solubility or activity and determined kinetically as the protein half-life (measured by protein concentration or activity after centrifugation) in a solution of specified pH, solute concentration, and temperature (1).

The kinetic method of measuring stability requires several days and moderate amounts of protein. To shorten the time required for kinetic measurements of solution stability, an additional stress can be put on the system by agitating it or adding a protease. For example, resistance to mechanical shaking was found to be a useful indicator of solution half-life for 10 enzymes present in liver extracts (89). Use of more sensitive methods to detect protein aggregation should reduce the protein required for stability measurements. Osmolytes useful for formulating recombinant human growth hormone and interferon-γ were de-

Table 2. Model Systems for the Study of Protein Aggregation

Model system	Studied with	Examples and key results
		Multiprotein complexes
Ribosomes	XRCrys, SANS, EM, neutron diffraction	Early crystals of ribosomes from thermophilic organisms diffracted to >20 Å were used for comparing the exterior structure with EM photos. Crystals diffracting to <5 Å were used, combined with image reconstruction, to determine structural details (18). Neutron diffraction of ribosome crystals was used to localize the RNA within the core of the particles using difference maps prepared at different D_2O/H_2O (altering the relative concentrations of D_2O and H_2O in the solvent changes the relative diffraction caused by RNA or protein in the crystals) (19). Freezing crystals stabilized them in the X-ray beam. A 37-Å resolution 3-D structure of ribosomes from rabbit reticulocytes was calculated from electron micrographs of uranyl-acetate stained single-particle specimens (20); whereas cryo-EM and neutron scattering data can be used to model the protein/nucleic acid interactions (21).
Nucleosomes	EM, XRCrys	The crystal structure of the nucleosome core particle was developed to high resolution using a variety of isolation and protein techniques and modern image analysis (22).
Photosynthetic reaction center and other membranes	EM, XRCrys, NMR, electron diffraction	Crystals diffracting to ~3 Å resolution required the development of small amphiphilic molecules (23). Other techniques include NMR of ^{19}F-labeled bacteriorhodopsin fragments in a membrane-like milieu (24) and IR difference spectra to monitor structural changes in ^{15}N-labeled or deuterated bacteriorhodopsin (25). EM of purple membranes from bacteria are a natural 2-D crystalline array of lipids and bacteriorhodopsin. As the array is only one unit cell thick, it is not suitable for X-ray analysis, but yields electron diffraction patterns with spots to beyond 4 Å resolution. By tilting the sample, a series of patterns are obtained that can be converted to a 3-D model of the structure of bacteriorhodopsin in the membranes (26, 27). EM images the hexagonally packed regular protein monolayer from the cell wall of the bacterium *Deinococcus radiodurans* (28).
Virus assembly	CD, EM, XRCrys, SAXS, STM	CD showed the membrane-associated coat protein of bacteriophage M13 (5.2 kD) changed from an α-helical conformation when it was in a membrane environment (i.e., in solution with cholate or in the capsid) to a β-sheet form when aggregated (aggregation followed with PAGE) (29). NMR studies of detergent-solubilized M13 coat protein, in combination with the results of sedimentation equilibrium, Raman and CD studies, indicated two conformers that represent the nonequivalent monomers of an asymmetric dimer (30) STM was used to view bacteriophage T4 polyheads (31). Small-angle X-ray scattering was used to follow the rapid assembly of tobacco mosaic virus protein induced by a temperature jump. The minimum counting time was 7.5 seconds (32). Assembly of the icosahedral shells of the dsDNA bacteriophages, herpesviruses, and acid adenoviruses requires proteins not found in the mature virion, termed scaffolding proteins. The bacteriophage P22 precursor procapsid contains approximately 300 scaffolding molecules within a shell composed of 420 coat protein subunits (33). Cryo-EM has been used to characterize the scaffolding/assembly of herpes simplex virus particles (34).
Antibody–antigen interactions	XRCrys, NMR, SLLS, DLLS, STM	Crystal structures of antibodies or bacterially produced antibody fragments with peptide antigen (35) or proteins can diffract to <2.5Å (36) and can be used to identify the atomic features of the binding site. NMR studies of binding of an antibody fragment (Fab) to a peptide (37). As the on/off rate of the peptide is short compared to the spin-lattice relaxation time of the Fab and peptide protons, a NOESY spectrum in the presence of a vast excess of peptide has extra cross-peaks (transfer NOEs) that are due to magnetic exchange between the bound and free peptide fractions and are not present when the peptide and Fab are present in a 1:1 ratio (38). Dynamic light-scattering measurements of antibody–antibody complex aggregate formation indicated fractal dimensions rather than a defined geometric shape (39). STM ultrastructural images of unstained immunoglobulin G, air dried on the surface of highly oriented pyrolytic graphite (HOPG), required only 1 μl of a 0.5 mg/mL solution of protein in 50 mM of tris buffer (40).
Casein micelles	SANS, light scattering, EM	Neutron scattering from suspensions of casein micelles measurements were combined with EM data to characterize the structure as packed micelle monomer subunits (41).

Table 2. Model Systems for the Study of Protein Aggregation *(continued)*

Model system	Studied with	Examples and key results
Inclusion body formation	EM, SLLS	The traditional method to determine the presence of aggregation after differential centrifugation is gel electrophoresis (also coupled with immunoblotting or isoelectric focusing [42]). Inclusion bodies contain protein polymers (43,44) resistant to SDS and β-mercaptoethanol solubilization. Inclusion bodies (IBs) are seen as electron dense particles in EM thin sections of whole *E. coli* cells; the morphology of isolated IBs was studied by scanning and freeze-etch EM (45). IBs (at least the polymeric forms that will enter a PAG) appear to consist of only one protein, not nonspecific coaggregates with other proteins (15). More recent data using computer simulations and light scattering suggest a cluster–cluster mechanism rather than a stepwise monomer addition polymerization (16).
β-amyloid fibril formation	CD, static and dynamic light scattering	β-amyloid peptide (A-β) is the primary protein component of senile plaques in Alzheimer's disease patients. Assembly of synthetic A-β into amyloid fibrils in phosphate-buffered saline was followed by static and dynamic light scattering. CD was used to show A-β's structure varied when dissolved in organic solvents commonly used in neurotoxicity assays. In pure DMSO, A-β had no detectable β-sheet content; in 0.1% trifluoroacetate, the peptide contained one third β-sheet; and in 35% acetonitrile/0.1% trifluoroacetate, A-β was two-thirds β-sheet, equivalent to the fibrillar peptide in physiological buffer (46).
Bacterial flagella assembly, microtubules	EM and cryo-EM	EM demonstrated in vitro reconstruction of organelles such as *Salmonella* flagellar filaments (phosphotungstic acid stained preparations) (47) and microtubules (frozen hydrated or glutaraldehyde/tannic acid fixed and stained) (48).
Lipoxygenase substrate binding	EMR, X-ray absorption spectroscopy, infrared CD magnetic CD	Spectroscopic studies, including electron magnetic resonance, X-ray absorption spectroscopy, infrared circular dichroism, and magnetic circular dichroism, have been applied to compare lipoxygenases from varied sources and with different substrate positional specificity. Soybean lipoxygenase-1 contains a novel three-turn π-helix near the iron center (49).
Nucleotide binding	NMR	The conformation of MgATP bound at the active site of *Salmonella typhimurium* 5-phospho-α-D-ribose 1-diphosphate synthetase (PRibPP synthetase) (50) and conformations of the adenosine moiety of MgADP and MgATP bound to rabbit muscle creatine kinase (51) have been investigated by 2-D transferred-NOE spectroscopy (TRNOESY).
Fibrinolase/Zn binding	NMR and CD	The Zn binding enzyme fibrolase (pI is 6.7) shows a minimum in its solubility profile (which can be followed with CD or NMR spectroscopy as a structural change) below pH 5. Zn loss, induced by rising temperature or lowered pH or adding chelators (EDTA, DTT), leads to a rapid loss of α-helical structure, exposure of hydrophobic residues, and aggregation (52).
Calmodulin/Ca	XRCrys, SANS	Small angle X-ray and neutron scattering of calmodulin in complex with two peptides (53).
		Specific protein interactions
Profilin/actin	XRCrys	Crystals of profilin/actin diffracting to 1.8-Å resolution required ATP hydrolysis to form and a 3.2 M ammonium sulfate/ATP bathing solution for stability (54).
RecA protein/DNA	EM, SANS, STM	RecA protein complexes with itself and with DNA were characterized under various conditions by combining EM and small-angle neutron scattering (55) STM images to a resolution of a few nM has been obtained for coated specimens of RecA bound to DNA (56).
Other protein DNA complexes	EM, STM	A combination of electron microscopy, image analysis, 3-D reconstruction (57) and STM has been used to show that the active form of the DnaB protein, the primary replicative helicase in *E. coli*, is a hexamer. Depending on the presence of different nucleotide cofactors (ATP, ATP γ-S, AMP-PNP or ADP), two different hexameric forms of the DnaB ring were found, one with a three-fold symmetry and one with six-fold symmetry. EM and image analysis was used to establish the relative orientation *E. coli* RuvA and RuvB bound to model Holliday structures (58,59).
Clotting factors with inhibitors	XRCrys	Crystals structures of the complex of thermitase (279 amino acids) with the inhibitor Eglin C (70 amino acids), resolved to <2 Å resolution (60) and human α-thrombin with its inhibitor hirudin to 2.3 Å resolution (61).
Cyclosporin/cyclophilin	XRCrys, NMR	NMR of the binding of cyclophilin to cyclosporin indicate that cyclosporin A and FK506 (but not rapamycin) undergo drastic conformational changes when bound to their respective binding proteins (62).

Table 2. Model Systems for the Study of Protein Aggregation *(continued)*

Model system	Studied with	Examples and key results
α-Crystallins	EM, SLLS, DLLS	Dynamic light scattering data (determine particle molecular weight as a function of time) combined with EM to determine shape of aggregates of bovine α-crystallin. Static light-scattering data indicated that the sample preparation methods used for (negative staining) EM had induced some particle breakdown (63).
Making lac repressor tetramer into a dimer or monomer; mechanism of isopropyl 1 thio-β-D-galactoside (IPTG) binding	SAXS, SANS, XRCrys, SDM, equilibrium sedimentation, fluorescence emission quenching	Tetrameric (4 × 360 amino acid) wild-type lac repressor was modeled based on scatter data (size and shape of tetramer) initially until a crystal structure was available. Targeted SDM-generated monomers formed preferentially dimers or did not associate, while retaining DNA binding function. A point mutation (Y282D) was sufficient to disrupt tetramer formation and generate monomeric protein (64). IPTG (inducer) binding reduced the fluorescence emission intensity of the repressor in solution. Low concentrations of IPTG reduced the fluorescence quenching effect of KI on the emission intensity (65).
Making a dimer from monomeric staphylococcal nuclease	XRCrys, sedimentation equilibrium	A serendipitous mutant of staphylococcal nuclease (6 amino acids in a surface loop, Δ114–119) forms a stable dimer (Kd $< 10^{-8}$ M). The carboxyterminal helices of each monomer swap places in the dimer (66).
		In vitro aggregation
	Light scattering, laser light transmission fluctuation	For monoclonal antibody aggregation in the presence of antigen, the estimates of the radius of gyration in solution, as measured by both classical and dynamic light scattering, agreed very well with those estimated from electron micrographs of the same aggregates (67). The aggregation of low-density lipoprotein (LDL) induced by acetylation, carbamylation, maleylation, or oxidation was evaluated by laser light transmission fluctuation. Aggregated LDL, but not unmodified monomer, stimulates the uptake of cholesteryl ester in arteriosclerotic cells (68).
Protein denaturation	CD, FTIR, IR-ATR, RS, SLLS, NMR	Raman spectroscopy has been used extensively to study the quaternary interactions of hemoglobin and to compare the secondary structures of proteins under various conditions (69,70). For example, 10–20% solutions of α-chymotrypsin and chymotrypsinogen were used for a Raman spectroscopic analysis of secondary changes induced by pH and pressure (71). A relationship between the salt type and concentration used for precipitation of α-chymotrypsin and the disordering of the secondary structure was determined from Raman spectroscopy (72). When the method was extended to another 11 proteins, the general result was that the precipitated protein contained on the average a higher percentage of β-sheet structure as determined by the amide I band intensity and location; the effect was more pronounced when the chaotrophic salt KSCN was used compared to the structure stabilizing salt, Na_2SO_4 (73).
Salt precipitation of protein	RS, SANS, fluorescence scatter peak, SLLS, DLLS	SANS in combination with ultracentrifugation has been used to determine the particle weight of malate dehydrogenase from a halophilic bacteria in different salt solutions. In 1–2 M NaCl solution, the protein is a stable dimer and binds much more water and NaCl than in low salt solution, where the salt-loving protein unfolds (74). Stopped-flow turbidimetry has been used to derive a kinetic model for the precipitation of α-chymotrypsin by different salts (75).
Crystallization pathways	Fluorescent light scattering, EM, SLLS, DLLS	Quasielastic light scattering study of the effect of salts on lysozyme and concanavalin A showed precipitating salts decrease translational diffusion coefficient, whereas those furthering crystallization do not change it and solutions remain monodisperse with increasing salt (76). Light scattering studies of solutions of canavalin at a fixed salt concentration (where the protein was known to form rhombohedral crystals) (77). Stereo-electron microscopy of frozen specimens used to characterize crystallization methods for lysozyme and tomato bushy stunt virus particles (330 Å). Crystallization intermediates had a more compact structure than aggregates that lead to amorphous precipitates (78). A symmetric dimer of a polypeptide in an organic solvent suggested as a model for the effects of crystal packing and solution dynamics on structure determination by X-ray crystallography and solution NMR methods (79).

Table 2. Model Systems for the Study of Protein Aggregation *(continued)*

Model system	Studied with	Examples and key results
Complexes with phospholipids and surfactants	Solid state NMR, NMR, ATR/IR, ESR	The complete structure determination of a polypeptide in a lipid bilayer derived from solid-state NMR observations (80). IR has also been used to study the binding of CO_2 to carbonic anhydrase and mutants thereof (81), and combinations of ^{31}P-NMR and ATR-IR have been used to study pulmonary surfactants (82,83). The interaction of phospholipid with (*R*)-3-hydroxybutyrate dehydrogenase, a phosphatidylcholine-requiring membrane enzyme, has been studied using ESR spectroscopy of spin-labeled lipids, both as ordered multibilayers and in lipid vesicle suspensions (liposomes) (84).

Note: The systems described in this chapter are amenable to characterization by several methods. See the section under the appropriate method for details and references. CD = circular dichroism; EM = electron microscopy; FTIR = Fourier-transformed infrared spectroscopy; IR-ATR = infrared attenuated total reflection spectroscopy; RS = Raman spectroscopy; SE = sedimentation equilibrium; SANS = small-angle neutron scattering; SAXS, small-angle X-ray scattering; SLLS/DLLS = static or dynamic laser light scattering; STM = scanning tunneling microscopy; XRCrys = X-ray crystallography.

tected by their ability to protect the native state of the protein against the stress of emulsification in organic solvent and re-extraction into buffer. Recovery of native protein was determined by native size exclusion HPLC followed by circular dichroism (CD), both rapid techniques that require relatively little protein (90).

Determining the Surface Charge and Hydrophobicity of a Protein

Again, practicality dictates the use of chromatographic methods to determine the approximate surface charge and hydrophobicity of a protein in solution. Because aggregation of integral membrane proteins after transfer to an aqueous environment is a well-described phenomenon (91), there is a tendency to regard precipitation in aqueous solvents as an indication that a protein is hydrophobic. In reality, all proteins are to some extent hydrophobic, with tightly packed cores that exclude water (92,93), and even proteins with many polar residues at their surfaces, such as interferon-γ and RNase A, will aggregate at high concentrations.

As with stability, the surface charge of a protein is not easily defined. An isoelectric point (pI) measured by isoelectric focusing or chromatofocusing indicates the pH at which the protein has no net charge at very low salt concentrations (high ionic strength prevents proper electrophoresis and interferes with the establishment of the ampholyte-pH gradient). These methods cannot be used for proteins (e.g., T7RP) that precipitate in low salt buffers. Further, the effective pI of a protein is greatly affected by buffer conditions. Due to the binding of salt, one cannot assume that a protein in solution will be negatively charged at a pH above its pI in a solution containing salts or metal ions (e.g., acidic caseins bind Ca^{2+} and appear positively charged at pH 7 (16). Oligomerization can alter the apparent isoelectric point of a protein by more than a pH unit, as has been shown for insulin (94).

At pH 7.5 and 50 mM salt, most proteins will bind to DEAE-coupled (positively charged) resins if they are negatively charged and to phosphate and other negatively charged resins if they are positively charged. The degree of surface charge can be estimated from the salt concentration required for elution from DEAE or CM-conjugated resin columns; this can also give indirect evidence about the degree of surface exposure of altered residues. For example, when positively charged residues (Arg10 or Lys7) were changed to uncharged glutamine by site-directed mutagenesis from the N-terminus of RNase A, the mutants eluted at lower salt from CM-Sepaharose at pH 8 than did the wild type; the double mutant did not bind to the column at this pH (95). Similarly, C-terminal deletions of human interferon-γ eluted at progressively lower salt concentrations from CM-Sephadex as more positively charged residues (Lys [K] and Arg [R]) were deleted (Table 3).

Later in the purification, after the solubility of the protein has been better defined, chromatofocusing in an appropriately stabilizing buffer can be used to more accurately estimate the protein's pI. Once one has enough purified protein, the pI can be determined by measuring the amount of KOH necessary to increase the pH of a solution of protein in buffer from 2.5 to 11 compared to that needed to titrate the buffer alone. Gel methods for following the changes in surface charge during protein folding and aggregation have also been developed (42).

Similarly, binding to resins coupled with hydrophobic groups, such as Phenyl- or Octyl-Sepharose® (Pharmacia) indicates the presence of hydrophobic residues at the protein surface. Proteins are applied in high salt (0.7 to 1 M ammonium sulfate), which increases hydrophobic interactions, and then eluted with a decreasing salt gradient. Most proteins elute between 0.5 and 0.1 M salt; very hydrophobic proteins will not elute into low salt buffer unless the polarity is decreased by adding ethylene glycol. If a protein does not bind to Phenylsepharose, it either has a very hydrophilic surface (e.g., RNase A) or is aggregated.

One can determine the hydrophobicity of a purified protein or follow changes in exposure of hydrophobic groups during folding by measuring interaction with a hydrophobic dye or radioactive tracer [e.g., 1-anilino-8-napthalene-sulfonate (98), or ^{125}I-TID, 3-(trifluoromethyl)-3-(*m*-^{125}iodophenyl)diazirine (99)].

Empirical Models for Salting-in and Precipitation with Various Agents

Empirical Models to Account for Salt/pH Effects. Generally, charged proteins can be salted-in by counterions. This

Table 3. Proteolytically Cleaved Human IFN-γ Elutes from CM-Sephadex at Progressively Lower Salt Concentrations Depending on the Number of Positively Charged Residues Removed from the C-Terminal

Human IFN-γ	C-Terminal	[NACl] eluted
Full length	– PAAKTGKRKRSQMLFRGRRASQ	300–500 mM
IFN-$\gamma\Delta$12	– PAAKTGKRKR	200–300 mM
IFN-$\gamma\Delta$14	– PAAKTGKR	80–100 mM

Note: After overnight cleavage at room temperature with one unit of either the protease Factor Xa (Biolabs) (for the Δ12 protein) or Endopeptidase Arg-C (Sigma) for the Δ14 protein, samples were diluted fivefold with 10 mM $NaHPO_4$ buffer, pH 5.7, and applied to small columns of CM-Sephadex C25 (Pharmacia) (96,97). The indicated eluates were then concentrated by centrifuging in Centricon-3 microconcentrators (Amicon). After automated Edman degradation, which indicated that the N-terminus of the proteins was unaltered, the number of amino acids deleted was determined by electrospray ionization mass spectroscopy.

is caused by the ion binding to the protein and neutralizing the net charge at its surface (see Table 4 for dielectric constants of common ions). The solubility of lysozyme, a positively charged protein, was shown to vary more with the nature of the anion added than the cation; the anion dependence followed the Hofmeister series (100). On the other hand, the solubility of caseins with pI between pH 3 and 5 varies with the nature of the cation: sodium, potassium, and ammonium caseinates are soluble; calcium caseinates form micelles (101). Casein solubility varies inversely with the atomic radius of the divalent cations added; Zn^{2+} and Cu^{2+} are more effective precipitants than Co^{2+} (102).

The interaction of salts as a function of pH with very soluble proteins can be accounted for mathematically, on a case-by-case basis. Detailed studies have been reported for a few proteins, including porcine pancreatic amylase (103) and ovalbumin (104). The latter study found that at 30 °C, the solubility of ovalbumin (C_o) in ammonium sulfate solutions of varying concentration C_A could be described by the simple quadratic equation:

$$\log_{10}(C_o) = 5.06 - 0.006t\text{-}0.205\ C_A + 0.5(\text{pH-}4.58) + 1.1(\text{pH-}4.58)^2$$

where concentrations of both ovalbumin and ammonium sulfate are expressed in g/100g H_2O, and 4.58 is the isoelectric point of ovalbumin. The data agreed quite well with earlier measurements made at 18 °C if the temperature correction was added.

Precipitation with Polyethylene Glycol and Other High MW Polymers. Precipitation as a means of purifying or storing a protein in a dry state must be done in such a way that after removal of the precipitant, a soluble and active product is obtained. Organic solvents that perturb the aqueous shell around the protein (ethanol, acetone, and even trichloroacetic acid [105]) are used industrially for large-scale protein precipitation, where maintaining a low temperature during the procedure is usually required, but the most useful general precipitants are those that do not denature the protein, such as polyethylene glycol (PEG). Models for precipitation with such compounds are usually based on excluded volume, meaning that the protein is forced into a progressively smaller area as the polymer occupies more of the solvent until the critical solubility limit of the protein is reached.

One recent study analyzed precipitation of common proteins (bovine serum albumin, human serum albumin, and ovalbumin) with various weights of PEG, and assumed:

1. the protein is globular and has homogenous surface properties
2. the precipitating polymer is linear, nonionic, and homogenous

With regard to the first assumption, the model is only useful for charged proteins if the ionic strength of the solvent is adjusted to minimize electrostatic protein–protein interactions (106). To predict the phase diagram, the Gibbs free energies (evaluated using a second-order perturbation

Table 4. Electronegativity Values for Protein Atoms and Common Ions Associated with Proteins

Electronegativity of protein atom constituents		Ionic character of bond and (percent ionic)		Electronegativity of ions	
H	2.1	C-H, S-H	0.4 (4%)	Li	1.0
C	2.5	C-O	1.0 (22%)	Na	0.9
N	3.0	N-C	0.5 (7%)	K, Rb	0.8
O	3.5	OH	1.4 (39%)	Ca	1.0
S	2.5	NH	0.9 (19%)	Cu	1.9
P	2.1	P-O	1.4 (39%)	Zn	1.6
				Fe	1.8
				Cl	3.0
				F	4.0

theory around a hard sphere as the reference) of protein in the solid and liquid phases were equated. Protein solubility was lowest (concentration of protein in the solid phase increased) for higher polymer concentrations, larger protein sizes, higher molecular weight of polymer, weaker protein–polymer interactions, poorer solvent, higher ionic strengths, and pH values close to the isoelectric point (107).

Because precipitation models are typically based on data for large, quite soluble proteins with the polymer used for precipitation of the same size and at about the same density as the protein being precipitated, it is difficult to apply the results directly to most laboratory-scale processes. Here, typically, the polymer greatly outweighs the protein in solution, and the intrinsic solubility of the protein is much less than the standard proteins.

Models for Determining the Effect of Co-solvents on Protein Folding

Accounting for the interaction between parts of the protein and the surrounding solvent is one of the most rapidly developing areas in protein folding. There are several theories to explain the effects of denaturants and osmolytes on protein solubility. The eventual outcome of this work is to understand how a protein folds, with an eye to predicting tertiary from primary structure. Determining the effects of osmolytes and denaturants on protein stability in solution should enable us to predict a priori what cosolutes are best suited to a protein of a given composition.

Osmolytes have strikingly different effects on different proteins. One cannot simply add any osmolyte to a protein solution and assume it will increase solubility; indeed, addition of carbohydrates or other osmolyte to a concentrated protein solution will often induce precipitation. The osmolytic amino acids betaine and proline, which protect *Escherichia coli* against urea and salt stress (108), can destabilize DNA helices (109). Adding side chains to betaine generated forms that are protective against high salt but not urea and vice versa (108). Curiously, although osmolytes have the general property of increasing the net solubility of proteins in solution, that is, one can achieve a higher concentration of the protein before aggregation occurs, none of the individual amino acids are more soluble in the osmolytic solutions for which there are data (0.5 M sucrose [110]; 1 M trimethyl amine *N*-oxide [TMAO] [111]). There is also no really significant increase, regardless of side chain, in amino acid solubility in the presence of 2 M urea, although a cyclic glycine dipeptide designed to mimic the backbone unit is about 20% more soluble (111).

The Thermodynamic Approach. The thermodynamic approach is based on the relative ability of molecules to stabilize the native state or the denatured state, thus accounting for their effects in a thermodynamic framework. As one cannot explain the specificity of osmolytes by difference in solubilities of the amino acid side chains, the preferential hydration theory suggests that osmolytes are repelled from the surface of the protein, whereas denaturants bind. The basic theory resembles the excluded volume theory of protein precipitation and deals with the protein as a hydrated ion. The binding of water molecules to the surface of the protein serves to shield charges and protect the protein from interaction with other proteins. According to the model, the concentration of the cosolvent in the immediate vicinity of the protein is either less than (for osmolytes) or greater than (for denaturants) that in the bulk solution. Thus, the water shell around a protein in osmolytic solution will be relatively richer in water molecules than the bulk solution. The theory assumes first that such a concentration gradient surrounding the protein exists and that this can serve as a driving force for stabilization.

The concentration of cosolvent near the protein surface is not easily measured. Experimental results to demonstrate that this is different from that in the bulk solution are based on measurements of the partial molar volume of the protein in various solutions or more recently, infrared spectroscopy to measure O-H stretching rates in the vicinity of the protein (112). Barrier effects caused by the presence of high concentrations of cosolvents in the binding of CO to myoglobin have been analyzed by frequency resolved calorimetry (113). To develop more general ways to account for the effects of solvent at the protein surface, computer models have been developed. Explicit modeling of solvent, by inserting the protein into a shell of solvent of varying thickness, is computationally inefficient. Drastic perturbations have to be imposed in the unfolding simulation, such as high temperature or pressure or by applying special algorithmic means to unfold a protein (114–116), which are likely to induce pathways not found in nature. Alternatively, one can implicitly model solvent molecules, in an approach based on the continuum approximation and the calculation of the solvent accessible surface area (SASA).

This approach was experimentally validated by a recent statistical data analysis that showed that for 45 proteins, the measured free energy and heat capacity of unfolding correlates directly with the change in SASA during the process (117). A recent method was introduced to efficiently calculate the SASA (118). By changing the protein–solvent interaction via the SASA and extrapolating to zero cosolvent concentration, a folding pathway that should have more physiological meaning can be induced and monitored.

In this case, the protein–solvent interaction term E_{hyd} is computed in the continuum approximation by

$$E_{hyd} = \sum_{i=1,atoms} \sigma_i A_i$$

where A_i is the SASA of atom i and σ_i is a solvation-parameter (119–121) depending on the atom type and solvent composition. A variety of data are available for the determination of σ_i values for individual protein atoms. Several such atomic solvation parameter (ASP; σ_i) sets have been derived (119,120,122–126) based on deriving a transfer free energy (ΔG_{tr}) for moving organic molecules (usually amino acids or their analogues) from one medium to another. Determination of the ΔG_{tr} can be calculated from the relative solubility of the groups in the two media. In the latter case, solubilities in the two solutions (C_w and C_s for maximal solubility in water or alternative solvent, respectively) yield ΔG_{tr}, by assuming that the ratio of the activity coefficients is ~1, according to the equation (127):

$$\Delta G_{tr} = RT \ln \frac{C_w}{C_s}$$

These ΔG_{tr} values can be directly related to the accessible surface area of each amino acid (A_i) by the equation: $\Delta G_{tr} = \Delta\sigma_i A_i$, and it is assumed for parameter extraction that the ΔG_{tr} of a residue R is the sum of the free energies of each atom:

$$\Delta G^{R}_{tr} = \sum_{\text{atoms}j} \Delta\sigma_j A_j$$

Eisenberg and McLachlan (119), in one of the earliest systematic studies, used data from Fauchère and Pliska on the energy of transfer of amino acid side chains from n-octanol to water (128) to derive a set of five ASP values ($\Delta\sigma$(C), $\Delta\sigma$(N/O), $\Delta\sigma$(O−), $\Delta\sigma$(N+), $\Delta\sigma$(S)). Their calculated ΔG_{tr} for the side chains using this set was linearly related to the observed values, and others have shown that the parameter sets, with slight modification, can drive disturbed structures back to their native state (118,129–131). Although one can also expand the parameter list by distinguishing aliphatic, methylene or carbonyl-C, and backbone or guanidyl nitrogen atoms, too many parameters may be detrimental in determining accurately the effect of the interaction. Other data well suited to the derivation of ASPs are those for the transfer potential of side chain analogues to a vapor phase (132) or based on measurement of the hydration enthalpy of N-acetyl amino acid amides (133), or the isoentropic pressure coefficients of tripeptides (134). Data are also available to derive ASP sets for solvents containing denaturants (135) or osmolytes (110).

There are data available for the solubility of protein components in ethanol, octane, and other solvents that could be used to derive ASPs for calculation purposes and direct data for purifying proteins with aqueous ethanol extraction, or following the unfolding of proteins or peptides in various solvent conditions (136,137). As the ability of water-miscible organic solvents to solubilize naphthalene is logarithmically related to the concentration at which they denature well-known proteins (trypsin, cytochrome c, chymotrypsinogen, chymotrypsin, laccase, and myoglobin) (138), ASP sets derived for one solvent may be related to other solvents according to simple correlation constants.

In support of using such ASPs, the hydrations seen around atom types in protein crystals (Table 5) show specific chemical interactions of the individual atoms with water that are unaffected by the crystal lattice. Neutral and negatively charged oxygen atoms are the most hydrophilic, followed by positively charged nitrogen atoms, neutral nitrogens, carbons, and sulfurs ($O^- \sim O > N+ > N \gg C \sim S$, where ~ means approximately the same).

Interactions between protein and bound water are fundamentally similar to those between protein and bulk fluid water, measured in partition experiments (139). The linear correlation coefficient of the data from Table 5 with side-chain free energy of transfer between cyclohexane and water (132,140) was $r = -0.95$ and with residue retention on reversed-phase HPLC using a pH 7.5 average scaler = -0.93 and $r = -0.86$ for the data of Fauchère (128).

The Kinetic Approach. A kinetic-based accounting for the effects of cosolvents on aggregation is based on their alteration of bulk solution properties, primarily the viscosity, macroscopic dielectric constant, and the chemical potential, which affect also the unfolding of the protein and its thermodynamic equilibrium energy. These changes can also have pronounced effects on the kinetics of unfolding, refolding, and transitions between substrates, which must be dealt with to account for the effects of osmolytes on aggregation and solubility. Aggregation, the major barrier to obtaining clear thermodynamic data, accelerates with increasing protein concentration. This strongly suggests that protein precipitation is directly related to the rate of protein–protein interaction. Kinetically, the role of osmolytes in maintaining proteins in equilibrium solution can be described by equations initially based on Newton's law of viscosity and Archimedes' principle. A protein in solution is in buoyant equilibrium with the water structure and viscosity-based forces accelerating its precipitation. To remain in aqueous solution, the protein must inhabit a cavity in the solvent structure somewhat larger than its surface area. The energy to create such a cavity has been estimated, based on the surface tension of water being 72 dyn/cm, to be 104 cal/Å/mol (141). This energy must be offset by the energy of binding of water molecules to the charged groups on the protein molecule. This may explain why more highly charged proteins, such as RNase A, are quite stable in solution in the presence of moderate amounts of salt, in contrast to proteins with large hydrophobic surface areas. A compact globular shape has the lowest surface area; unfolding the protein increases its surface area (and area for interaction with other molecules) and alters its specific density. For example, for a protein in a temperature gradient, the change in velocity of the molecule can be summarized as a function of the (temperature-dependent) viscous forces $[\nabla \cdot \tau]$ and buoyant forces $[\rho\beta g\Delta T]$:

$$\rho \frac{Dv}{Dt} = -[\nabla \cdot \tau] - \rho\beta g(T - T)$$

where ρ is the specific density (mass/unit volume), g is the gravitational constant, and β is the coefficient of volume expansion.

The second term in kinetic effects in protein unfolding states simply that the viscosity of the solution will affect the time required for folding, an effect summarized (142) as:

$$\bar{t}(n) = t_0 e^{(\Delta E(n)/k_b T)^2}$$

where $t(n)$ is the lifetime of a typical microstate, and t_0 is the time scale for a typical motion of a large segment of the chain, which is a factor of the local barriers and the solvent viscosity. For the case of viscosity (friction) effects on geminate binding reactions, please see Ansari et al. (143) for a restatement of Kramer's law in order to apply it to proteins.

In addition, one must account for changes in the dielectric constant of the medium and for the polarizability (α) of the protein and its cosolvents on the interaction energy between protein particles as a function of distance (R). Classic thermodynamics says that this energy, for an oscillating system with no limits on the sphere of motion, should be:

$$\Delta E(R)_{\mathrm{dis}} = -3h\nu/4(\alpha^2/R^6)$$

where the oscillator energy $h\nu$ can be equated with the ionization potential (IP). Thus, the energy of interaction decreases as the distance between particles decreases to the sixth power and the polarizability increases. As the distance between particles decreases directly with the concentration of the protein, small increases in concentration will greatly accelerate the aggregation rate.

With respect to the ionization potential, the constituent atoms of proteins, with the exception of hydrogen, have very similar electronegativity values (Table 4), and most of the bonds within a protein are basically covalent. Note that hydration is seen in crystal structures at precisely the atoms in proteins where the ionic character of the bonds is most pronounced (Table 4). Obviously, substituting a hydrogen atom for any of the alkali metals will increase the local dipole moment within the protein. The polarizing effect of substituting a larger ion is intermediate. These considerations may be used as the basis for analysis of salt interactions with proteins, if the surface charge is known. Although such models will still require experimentation, they may be used as an optimization framework that may reduce the data needed.

COMPUTER DESIGN OF MORE SOLUBLE PROTEINS

Combinations of the methods discussed in the first section have been used to generate mutants of insulin with altered oligomerization and solubility and to change the lac repressor from a tetramer to a dimer or monomer, to name but a few examples (see Table 2). Residues involved in aggregation were identified with the help of a detailed 3-D structure, generated by high-resolution methods, and site-directed mutagenesis was used to test which residues could be altered without affecting function or the global fold of the protein (144). All these studies required years of work. Fortunately, computational advances now allow us to have a reasonable view of the protein structure for many proteins based on their sequence homology to other proteins. This sort of modeling can be implemented to solve problems even during the initial isolation of a protein of known sequence. As discussed elsewhere (2), there are a few empirical rules for identifying areas of proteins that cause aggregation. The few good examples of controlling aggregation of proteins through alteration of primary structure started from high resolution X-ray crystallographic or NMR structures of protein complexes or oligomers. It is now possible to view the 3-D structure of proteins whose coordinates have been deposited in the protein data base (PDB) (for downloading from the Internet, see the notes to the Internet site "Pedro's Macromolecular Tools") on a regular Macintosh or PC terminal; stereo viewing software can also be had for a reasonable price. Good viewing systems such as the Silicon Graphics Indigo series or the Evans and Sutherland Molecular Graphics systems allow more detail and sophisticated, real-time rotation of the molecule. Even when the 3-D fold of the protein is known, it is still not easy to identify aggregation sites; one can, however, identify surface-exposed amino acid stretches that might be involved in aggregation or intermolecular contacts rapidly and test the effects of site-directed mutagenesis at these positions on solubility. New methods for mimicking protein folding by computer allow hope that we may soon be able to identify problem areas in the primary sequence. Two basic assumptions may greatly simplify calculations of the protein aggregation problem:

1. Protein folding proceeds first by folding into a series of domains, which then interact with each other specifically to form the final stable global structure.
2. During this process, certain domains may form inter- rather than intramolecular contacts. Once two molecules are so intertwined, their residual domains may be forced to associate with other molecules, leading rapidly to a mass of partially folded, interlocked protein monomers. This idea, based on the domain swapping hypothesis of Bennett et al. (145), is from structural analysis of oligomeric proteins, including *Diphtheria* toxin, staphylococcal nuclease, α-spectrin, and antisialidase, with swapped domains ranging from 14 to 252 amino acids.

The problem would then become identifying probable domains within the protein that are sufficiently separated during folding so as to allow another molecule to intercalate during the folding process. One can view surface groups on proteins, searching for areas of exposed hydrophobic surface, for example, that may be involved in aggregation. There are also specific programs that allow one to view the dynamics of interaction between molecules (DOCK, from Kuntz's group at UCSF, can be obtained over the Internet). DOCK is most suitable for viewing the interaction between small sections of the molecule and small

Table 5. Hydrations Seen in Crystal Structures

Protein atom group	Atom types included in group	Water ions/occurrence
Neutral O	Q, N, D, T, Y and all carbonyl O atoms	0.53
Negative O	D and E side chain O	0.51
Positive N	H, K, R side chain N except Arg Nε, which is fairly nerutral	0.44
Neutral N	W, N and Q side chain N and Arg Nε, plus all main chain N	0.35
Carbon/sulfur	All carbon and sulfur atoms	0.08

ligands. For example, the 3-D structure of neurotrophin (also called p75NGFR or LANR) was modeled by its homology to regions in the receptor of tumor necrosis factor. Hypothetical complexes of this neurotrophin model structure with nerve growth factor were then modeled with DOCK 3.5. Viewing of hundreds of predicted orientations with MIDAS revealed similarities between motifs in ligand-binding domains of the neurotrophin and tumor necrosis factor receptor and pinpointed differences that could account for specificity (146).

However, experimental structural data are not available for most proteins, and structures of complexes of proteins are even rarer. For a significant percentage of newly identified protein sequences, which have homology to a protein of known sequence homology, one can predict with some degree of confidence their overall topology and stretches of amino acids likely to be surface exposed. This does not mean that such models can replace experimental data. The protein-folding problem is mathematically described as a system with frustration, because the number of computing steps required to solve the problem increases faster than any power of the size of the system (147). This section covers some of the possibilities for using sequence data and molecular graphics tools to generate model 3-D structures. Some programs from the Internet are presented in addition to the self-correcting distance geometry (SECODG) approach to the modeling being developed in the computational biology group of the UTMB. Several different pathways, starting from an initial sequence alignment, can be used to obtain such constraints (Fig. 2).

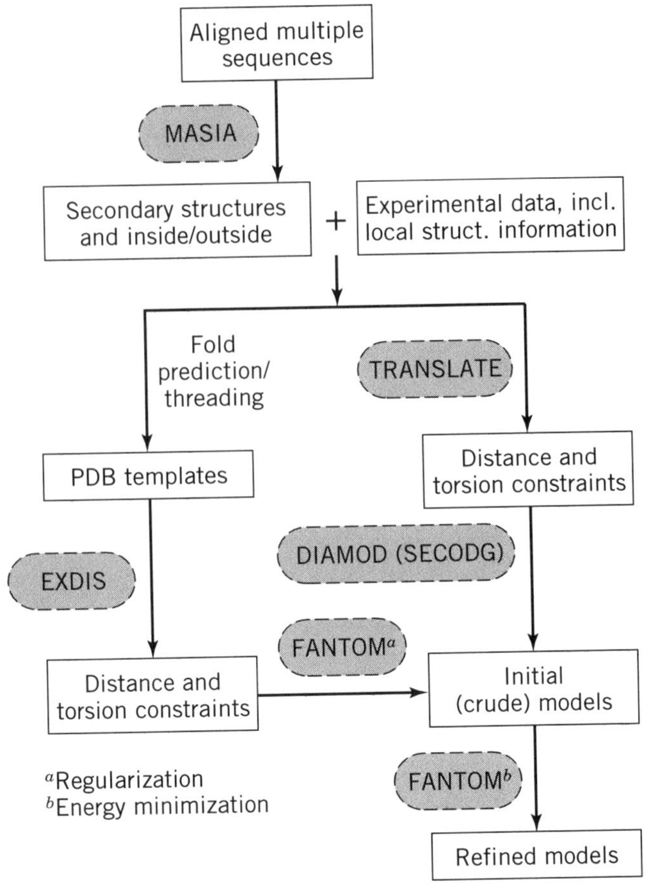

Figure 2. Flow chart describing protein modeling strategies depending on the degree of homology of the protein sequence to those of known structure. The *shaded boxes* enclose names of computational biology software that is being developed in the computational biology group at UTMB. The left and right sides of the flow chart represent procedures for generating models by independent routes. See text for details. *Source*: Figure kindly supplied by Dr. Kizhake Soman of UTMB, who retains copyright.

Automatic Homology Modeling

If a structure for a protein with significant (>30 to 40%) sequence identity exists, a 3-D-structure model can be obtained over the Internet. For example, we tested SWISSMODEL using a 62-amino-acid section that aligns with other homeobox proteins from the Pem protein (involved in control of hormone generation in the testes). The murine Pem homeodomain homologous area

RQMPLQGSRFAQHRLRELESILQRTNSFDVPREDL
DRLMDACVSRVQNWFKIRRAAARRNRR

sequence was cut out of a text file and pasted into the appropriate area of the form for submission to SWISSMODEL, a program on the Internet that is underwritten by Glaxo Research and constructed by Peitsche. An e-mail response with coordinates for a 3-D structure based on the structure of the Fujitaratsu homeodomain (PDB files 21FJL and 11 FJL) was returned quickly. Although for proteins with sufficiently high homology to one in the PDB the template selection is automatic, it is always a good idea to choose an appropriate model oneself by extensive homology (BLAST or equivalent) searches before requesting the modeling.

SWISSMODEL uses the PDB to obtain structural information and automatically generates a model for sequences that share significant similarities with at least one protein of known 3-D structure. Given an unknown sequence, SWISSMODEL uses a FastA and BLAST search to determine what protein in the PDB could be a possible structural homologue based on the degree of primary sequence similarity. A sequence with a FastA score 10 SD above the mean of the random scores and a Poisson unlikelihood probability P(N) of 10–5 will be considered. The program as of 1998 requires at least 35% residue identity with 40% of the target sequence (or overall 25% identity). The automatic alignment tool used within the program suite is SIM, but the user on a second level can choose to submit other alignments or structures for use in the modeling (148). Hyperlinks between SWISSMODEL and the 2-D gel electrophoresis data bank of SWISS-2DPAGE (see *http://expasy.hcuge.ch*) can be used to map *E. coli*, *Haemophilus influenzae*, *Mycoplasma genitalium*, *M. tuberculosis*, and *Bacillus subtilis* sequences on the basis of charge and size to proteins seen on the gel grid. A similar gel database is being accumulated for yeast and human proteins (149).

The ease, speed, and impressive level of prediction of SWISSMODEL is certainly a tribute to the Glaxo team. A significant amount can be learned from visualizing the structure using Midas software on a Silicon graphics terminal, and residues involved in DNA binding, on the basis of the homologous protein used to develop the model, can be predicted. However, the program only works for proteins where a high degree of sequence homology exists to protein of known structure. By some estimates, only 10 to 15% of *E. coli* proteins have a suitable modeling template in the PDB using the above cutoff levels of residue identity; one would anticipate an even lower probability for an unknown mammalian sequence. For our example, when the whole PEM protein sequence, or even the 63-amino-acid sequence mentioned earlier that aligns with many other homeodomain proteins, was sent in the same fashion to SWISSMODEL and submitted without naming a target protein for modeling, the program was unable to generate even a model of the area that contained the homeobox. A second problem with using such simple models to design site-directed mutagenesis experiments is that although the program will be able to predict a global fold for the protein, local conformations that can account for specific properties are less easy to predict. For this reason, developing a working model (as of the date of this article) means using manually interactive programs that require more work on the part of the investigator.

Other methods are available from the Web and should also be investigated. For example, one can use the TOPITS program, which threads protein sequences onto model structures (150). Here, sequence residue identity is less important than the stability of the final structure. An example of using this program is shown in Figure 3.

Modeling Using Self-Correcting Distance Geometry with the Programs CLUSTAL, MASIA, NOAH, DIAMOD, and FANTOM to Develop a 3-D Model of a Protein

Figure 2 summarizes methods used by our group for structure prediction based on a unique algorithm based on self-correcting distance geometry (SECODG) from Dr. Werner Braun and coworkers. The SECODG-based program suite has the advantage that experimental results can be directly incorporated into the modeling process by manual insertion of distance constraints (DCs), limits on the distance between two atoms in a 3-D structure. Experimental DCs can be derived experimentally from NMR or crystal data or from other types of biophysical data, for example, from knowing where secondary structure elements lie, which two cysteine residues form a disulfide bond, or which residues lie within the active site (which means they must lie within a certain distance from each other). The method starts with BLAST searches and multiple sequence alignment to determine whether a structure exists for a protein with reasonable sequence homology. If a homologous structure is available, the PDB coordinates can be regularized and DCs extracted automatically by related software programs.

Step 1: Proteins with a significant identity pattern to the test protein are searched for in a protein sequence database (for example, SWISSPROT).

Step 2: CLUSTALW (downloaded from the Internet) is used to align the best-matched proteins in a block to identify areas of high homology with other proteins. These sequences are then used to search for known 3-D structures in the PDB, a composite of tertiary structure details.

Step 3: The program MASIA (formerly MULTAN [130]) analyzes the aligned sequences and identifies areas of probable secondary structure.

Step 4: The results from steps 2 and 3 are combined to generate a list of distance and dihedral angle constraints. These constraints can be combined with results from other biophysical methods, such as knowledge of where disulfide links and active site residues are in proteins.

Step 5: The SECODG-based package of NOAH/DIAMOD then uses these constraints to generate a family of structures that best fit these constraints. The NOAH program is a structure-based filter, which means in simple terms that it uses a rough 3-D structure to determine whether a list of distance constraints is internally consistent.

Step 6: The local structure is refined using the energy minimization program FANTOM, to determine structures with lowest energy.

For examples of how this method can be applied to real situations, see the discussion by Mumenthaler and Braun (130). Figure 4 is an example of the structure generated by this method, before energy minimization, for the rat and murine Pem homeodomains by the author and Hong Yao Zhu (Univ. of Texas Medical Branch).

Modeling Complexes

Modeling the interaction of several molecules is important for our understanding of aggregation. Extensive modeling studies for protein complexes has been done, for example, to analyze the allosteric mechanism of the tetrameric lac repressor. Here, years of site-directed mutagenesis studies were done before a crystal structure of the repressor was available for study (152). If crystal structures of the proteins involved are available, and there is some biophysical data about the binding site, one can make relatively exact models using graphics programs such as MIDAS.

Of course, a model must be validated experimentally and theoretically. Tests for structure validity can be based on a binary code where all residues are considered polar except for seven nonpolars (CILMFWV). The arrangement of hydrophobic and polar residues alone as evaluated by a scoring scheme (hydrophobic fitness score) (153) recognized the native fold out of 2,000 near-native structures generated for each of five small monomeric proteins. Other tests for structural validity based on identifying structures with minimum thermodynamic energy are being developed.

Protein Design

The field of protein design (designing proteins that fold in a certain manner) is proving to have nearly as much impact on our understanding of protein folding as the direct measurement of natural proteins. Proteins can be designed to have a specific overall structure with defined secondary elements and selected for solubility (154,155). There have been several recent protein design projects that have illus-

RANK	EALI	LALI	IDEL	NDEL	ZALI	PIDE	LEN2	PDB-ID	NAME
1	76.87	124	22	4	1.76	10	625	1nfk	Nuclear factor κB
2	76.25	117	47	6	1.72	10	378	7api	Modified α1 antitrypsin
3	75.98	120	16	6	1.71	14	576	1vnc	Chloroperoxidase
4	75.40	124	48	8	1.68	13	272	1dih	Dihydrodipicolinate reductase
5	75.20	123	33	4	1.67	11	373	2ach	α1 Antichymotrypsin
6	75.15	124	19	3	1.66	15	1189	1prc	Photosynthetic reaction center
7	74.43	122	20	4	1.62	15	285	2ebn	Endo-β-*N*-acetylglucosaminidase
8	74.38	120	42	6	1.62	15	1078	1nir	Nitrite reductase
9	74.15	110	22	5	1.61	14	280	1pyp	Inorganic pyrophosphatase
10	73.68	123	33	6	1.58	20	403	1atl	Atrolysin C
11	73.62	117	25	4	1.58	14	228	1udh	Uracil DNA-glycosylase
12	73.57	112	15	6	1.58	19	314	1tag	Transducin α + GDP + Mg
13	73.22	117	26	5	1.56	12	449	1rtm	Mannose-binding protein a
14	73.18	120	25	5	1.56	17	421	1ses	Seryl-tRna Synthetase
15	73.15	123	41	4	1.55	10	279	1thm	Thermitase
16	72.38	124	38	3	1.51	6	338	1cse	Subtilisin carlsberg

Figure 3. Output files from the TOPITS program (151) for template proteins when an alignment of HIV-Rev protein sequences was submitted as input. Listed are the best 16 hits of protein structures whose folds Rev is predicted to adopt. EALI = alignment score; LALI = length of the alignment (between rev and each of these proteins); IDEL = number of inserted residues; NDEL = number of insertion points; ZALI = alignment score based on which the list has been ranked; PIDE = % pairwise sequence identity; and PDB-ID = the 4-letter PDB code for each protein. For reliable predictions, ZALI > 3.0. Thus, the fold prediction only reflects remote homology with Rev. *Source*: Unpublished data used with the kind permission of Kizhake Soman and Werner Braun, Structural Biology, UTMB Galveston, who retain copyright.

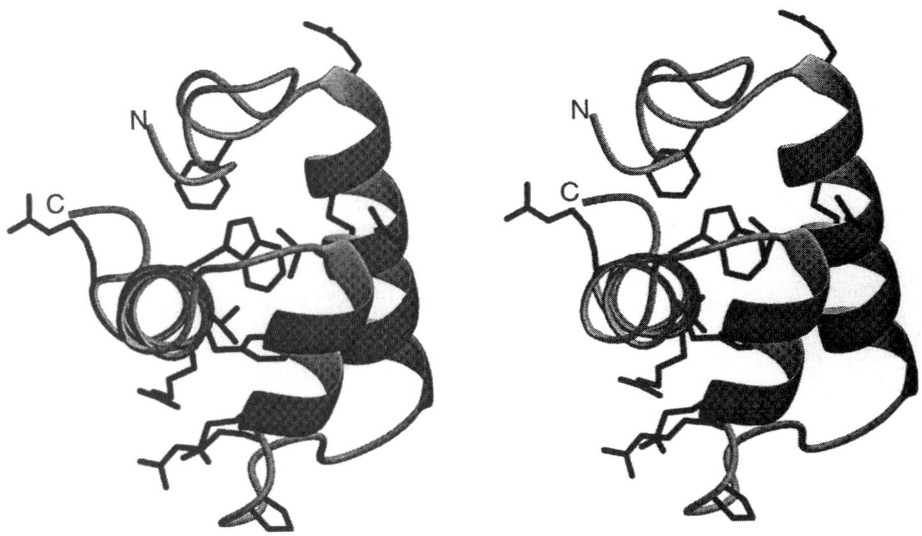

Figure 4. Stereo view of a model structure for the murine PEM homeodomain region, derived by homology modeling based on the 1TTF structure in the PDB using the DIAMOD program. The side chains shown are identical in both the modeled sequence and the template structure. *Source*: Unpublished data from the author and Dr. Hong Yao Zhu, who retain the copyright.

trated how well we can plan a primary sequence that will form a desired structure. Although the design of α-helical proteins is fairly advanced, β-sheet design is still hindered by solubility problems.

For this reason, the Paracelsus prize (small in cash value but honorable in the extreme) was claimed many years after it was offered to any group who could change the structure of a protein from helical to a sheet (or v.v.) without altering more than 50% of the residues. The winning protein was a composite of an area of protein G that formed a β-sheet, which was altered to a four-helix bundle by substituting residues at discrete positions from the ROP protein. Helix-forming residues were inserted, and residues known to favor sheet formation were removed (156). Despite the apparent difficulty humans have in changing the structure of a peptide from α-helical to β-sheet, such a mechanism has been proposed for natural puzzles. For example, aggregation of the proteins associated with scrapies is suggested to be a conversion from a largely helical conformation to a β-sheet-based structure in the aggregate (157,158).

Computational methods have also been introduced into the area of protein design. For example, a protein that contained no histidine, cysteine, or chelated metal ion was designed that exactly fit the X-ray crystal structure of the zinc finger protein. The structure of the synthesized model protein, which required no metal ion for maintenance, was determined by NMR to be identical to the original Zn finger region (159). With time, similar modeling algorithms will presumably allow the design of proteins that match in both structure and their functional abilities, for example, DNA binding (160,161).

CONCLUSION

The study of protein oligomerization with various methods is leading to a better understanding of the atomic basis for aggregation. Many different forms of biophysical data can be used to develop better models of the processes. Understanding how the interactions between solvents and the structure of proteins affects interprotein contact will help design better methods for controlling aggregation and processes dependent on resolubilizing proteins after precipitation.

ACKNOWLEDGMENTS

I wish to thank all my coworkers at UTMB, particularly Kizhake Soman, Hong Yao Zhu, Yuan Xu, Robert Frackiewicz, and Werner Braun, Wayne Bolen for sharing unpublished work and critical thoughts, and David Volk for critically reading the manuscript. The work related to Figures 2 to 4 was supported by grants DE-FG03-96ER62267 from the Department of Energy and #454940 from the John Sealy Memorial Endowment Fund. I also thank those who sent reprints during the preparation of this work and apologize in advance to any authors whose work, as a result of my oversight, is not adequately cited here.

BIBLIOGRAPHY

1. C.H. Schein, *Bio / Technology* **8**, 308–317 (1990).
2. C.H. Schein, *Pharm. Acta Helv.* **69**, 119–126 (1994).
3. C.H. Schein, in G. Georgiou and E. de Bernardez-Clark eds., *Protein Refolding* ACS Books, Washington, D.C., 1991, pp. 21–34.
4. H.-J. Galla, *Spektroskopische Methoden in der Biochemie*, Georg Thieme Verlag, Stuttgart, 1988.
5. B.J. Berne and R. Pecora, *Dynamic Light Scattering*, Wiley, New York, 1976.
6. E. Mancini, F. de Haas, and S. Fuller, *Structure* **5**, 741–750 (1997).
7. Z. Zhou, W. Chiu, K. Haskell, H. Spears, J. Jakana, F. Rixon, and L. Scott, *Biophys. J.* **74**, 576–588 (1998).
8. L. Haggerty and A.M. Lenhoff, *Biotechnol. Prog.* **9**, 1–11 (1993).
9. H.L. Casal, U. Köhler, and H.H. Mantsch, *Biochim. Biophys. Acta* **957**, 11–20 (1988).
10. R.C. Hester and J.C. Austin, in E.D. Schmid, F.W. Schneider, and F. Siebert, eds., *Spectroscopy of Biological Molecules: New Advances*, Wiley, Chichester, England, 1988, pp. 3–10.
11. H. Versmold and W. Härtl, *J. Chem. Phys.* **79**, 4006–4009 (1983).
12. T.D. Veenstra, L.M. Benson, T.A. Craig, A.J. Tomlinson, R. Kumar, and S. Naylor, *Nat. Biotechnol.* **16**, 262–266 (1998).
13. A.J. Wand and S.W. Englander, *Curr. Opin. Biotechnol.* **7**, 403–408 (1996).
14. D.G. Myszka, *Curr. Opin. Biotechnol.* **8**, 50–57 (1997).
15. M. Speed, D. Wang, and J. King, *Nature Biotechnol.* **14**, 1283–1287 (1996).
16. M. Speed, J. King, and D. Wang, *Biotechnol. Bioeng.* **54**, 333–343 (1997).
17. N.R. Watts, M. Misra, P.T. Wingfield, S.J. Stahl, N. Cheng, B.L. Trus, A.C. Steven, and R.W. Williams, *J. Struct. Biol.* **121**, 41–52 (1998).
18. A. Yonath and H.G. Wittmann, *TIBS* **14**, 329–335 (1989).
19. M. Eisenstein, R. Sharon, Z. Berkovitch-Yellin, H.S. Gewitz, S. Weinstein, E. Pebay-Peyroula, M. Roth, and A. Yonath, *Biochemie* **73**, 879–886 (1991).
20. A. Vershoor and J. Frank, *J. Mol. Biol.* **214**, 737–749 (1990).
21. F. Mueller and R. Brimacombe, *J. Mol. Biol.* **271**, 545–565 (1997).
22. K. Luger, A. Mader, R. Richmond, D. Sargent, and T. Richmond, *Nature* **389**, 251–260 (1997).
23. J. Deisenhofer and H. Michel, *EMBO J.* **8**, 2149–2170 (1989).
24. V.F. Bystrov, A.S. Arseniev, A.L. Barsukov, A.L. Lomize, G.V. Abdulaeva, A.G. Sobol, I.V. Maslennikov, and A.P. Golovanov, in O. Jardetsky ed., *Protein Structure and Engineering*, Plenum Press, New York, 1989, pp. 111–138.
25. K.J. Rothchild, Y.W. He, D. Gray, P.D. Roepe, S. Pelletier, R.S. Brown, and J. Herzfeld, *Proc. Natl. Acad. U.S.A.* **86**, 9832–9835 (1989).
26. T.A. Ceska and R. Henderson, *J. Mol. Biol.* **213**, 539–560 (1990).
27. R. Henderson, J.M. Baldwin, T.A. Ceska, F. Zemlin, E. Beckmann, and K.H. Downing, *J. Mol. Biol.* **213**, 899–929 (1990).
28. M. Amrein, Z. Wang, and R. Guckenberger, *J. Vac. Sci. Technol.* **9**, 1276–1281 (1991).
29. R.B. Spruijt, C.J.A.M. Wolfs, and M.A. Hemminga, *Biochemistry* **28**, 9158–9165 (1989).
30. G.D. Henry and B.D. Sykes, *J. Mol. Biol.* **212**, 11–14 (1990).
31. M. Amrein, R. Dürr, H. Winkler, G. Travaglini, R. Wepf, and H. Gross, *J. Ultrastruct. Mol. Struct. Res.* **102**, 170–177 (1989).
32. Y. Hiragi, H. Inoue, Y. Sano, K. Kajiwara, T. Ueki, and H. Nakatani, *J. Mol. Biol.* **213**, 495–502 (1990).

33. B. Greene and J. King, *Virology* **225**, 82–96 (1996).
34. Z. Zhou, S. MacNab, J. Jakana, L. Scott, W. Chiu, and F. Rixon, *Proc. Natl. Acad. Sci. U.S.A.* **95**, 2778–2783 (1998).
35. R.L. Stanfield, T.M. Fieser, R.A. Lerner, and I.A. Wilson, *Science* **248**, 712–719 (1990).
36. G. Boulot, J.-L. Eisele, G.A. Bentley, T.N. Bhat, E.S. Ward, G. Winter, and R.J. Poljak, *J. Mol. Biol.* **213**, 617–619 (1990).
37. R. Levy, T. Assulin, T. Scherf, M. Levitt, and J. Anglister, *Biochemistry* **28**, 7168–7175 (1989).
38. J. Anglister, *Q. Rev. Biophys.* **23**, 175–203 (1990).
39. J.G. Rarity, R.N. Seabrook, and R.J.G. Carr, *Proc. R. Soc. London, Sec. A.* **423**, 89–102 (1989).
40. R.J. Leatherbarrow, M. Stedman, and T.N.C. Wells, *J. Mol. Biol.* **221**, 361–365 (1991).
41. P. Stothart, *J. Mol. Biol.* **208**, 635–638 (1989).
42. P.J.M. van den Oetelaar, B.M. de Man, and H.J. Hoenders, *Biochim. Biophys. Acta* **995**, 82–90 (1989).
43. C.H. Schein and M.H.M. Noteborn, *Bio / Technology* **6**, 291–294 (1988).
44. C. Albiges-Rizo and J. Chroboczek, *J. Mol. Biol.* **212**, 247–252 (1990).
45. G.A. Bowden, A.M. Paredes, and G. Georgiou, *Bio / Technology* **9**, 725–730 (1991).
46. C. Shen and R. Murphy, *Biophys. J.* **69**, 640–651 (1995).
47. T. Ikeda, S. Asakura, and R. Kamiya, *J. Mol. Biol.* **209**, 109–114 (1989).
48. R.H. Wade, D. Chrétien, and D. Job, *J. Mol. Biol.* **212**, 775–786 (1990).
49. B. Gaffney, *Ann. Rev. Biophys. Biomol. Struct.* **25**, 431–459 (1996).
50. G.M. Jarori, N. Murali, R.I. Switzer, and B.D. Rao, *Eur. J. Biochem.* **230**, 517–524 (1995).
51. N.J. Murali, G.K. Jaori, S.B. Landy, and B.D. Rao, *Biochemistry* **32**, 12941–12948 (1993).
52. D. Pretzer, B. Schulteis, C.D. Smith, D.G. Vander Velde, J.W. Mitchell, and M.C. Manning, *Pharm. Biotechnol.* **5**, 287–314 (1993).
53. J. Trewhalla, D.K. Blumenthal, S.E. Rokop, and P.A. Seeger, *Biochemistry* **29**, 9316–9324 (1990).
54. C.E. Schutt, U. Lindberg, J. Myslik, and N. Strauss, *J. Mol. Biol.* **209**, 735–746 (1989).
55. E. DiCapua, M. Schnarr, R.W.H. Ruigrok, P. Lindner, and P.A. Timmins, *J. Mol. Biol.* **219**, 557–570 (1990).
56. M. Amrein, A. Stasiak, H. Gross, E. Stoll, and G. Travaglini, *Science* **240**, 514–516 (1988).
57. A. Stasiak, I.R. Tsaneva, S.C. West, C.J. Benson, X. Ya, and E.H. Egelman, *Proc. Natl. Acad. Sci. U.S.A.* **91**, 7618–7622 (1994).
58. X. Yu, S.C. West, and E.H. Egelman, *J. Mol. Biol.* **266**, 217–222 (1997).
59. X. Yu, M.J. Jezewska, W. Bujalowski, and E.H. Egelman, *J. Mol. Biol.* **259**, 7–14 (1996).
60. P. Gros, C. Betzel, Z. Dauter, K.S. Wilson, and W.G.J. Hol, *J. Mol. Biol.* **210**, 347–367 (1989).
61. T.J. Rydel, K.G. Ravichandran, A. Tulinsky, W. Bode, R. Huber, C. Roitsch, and J.W. Fenton, *Science* **249**, 277–280 (1990).
62. W. Braun, W. Kallen, V. Mikol, M.D. Walkinshaw, and K. Wüthrich, *FASEB J.* **9**, 63–72 (1995).
63. P. Schurtenberger and R.C. Augusteyn, *Biopolymers* **31**, 1229–1240 (1991).
64. J.M. Chen and K.S. Matthews, *Biochemistry* **33**, 8728–8735 (1994).
65. W.I. Chang and K.S. Matthews, *Biochemistry* **34**, 9227–9234 (1995).
66. S.M. Green, A.G. Gittis, A.K. Meeker, and E.E. Lattman, *Nat. Struct. Biol.* **2**, 746–751 (1995).
67. R.M. Murphy, H. Slayter, P. Schurtenberger, R.A. Chamberlin, C. Colton, and M.L. Yarmush, *Biophys. J.* **54**, 45–56 (1988).
68. V.V. Tertov, I.A. Sobenin, Z.A. Gabbasov, E.G. Popov, and A.N. Orekhov, *Biochem. Biophys. Res. Commun.* **163**, 489–494 (1989).
69. D.L. Rousseau and M.R. Ondrias, in D.L. Rousseau ed., *Optical Techniques in Biological Research*, Academic Press, Orlando, Fla., 1984, pp. 65–132.
70. R.W. Williams, *Methods Enzymol.* **130**, 311–331 (1986).
71. L. Heremans and K. Heremans, *Biochim. Biophys. Acta* **999**, 192–197 (1989).
72. T.M. Przybycien and J.E. Bailey, *Biochim. Biophys. Acta* **995**, 231–245 (1989).
73. T.M. Przybycien and J.E. Bailey, *Biochim. Biophys. Acta* **1076**, 103–111 (1991).
74. G. Zaccai and H. Eisenberg, *Trends Biol. Sci.* **15**, 333–337 (1990).
75. T.M. Przybycien and J.E. Bailey, *AIChE J.* **35**, 1779–1790 (1989).
76. V. Mikol, E. Hirsch, and R. Giege, *J. Mol. Biol.* **213**, 187–195 (1990).
77. W. Kadima, A. McPherson, M.F. Dunn, and F.A. Jurnak, *Biophys. J.* **57**, 125–132 (1990).
78. S.D. Durbin and G. Feher, *J. Mol. Biol.* **212**, 763–774 (1990).
79. S. Pascal and T. Cross, *J. Mol. Biol.* **241**, 431–439 (1994).
80. R. Ketchem, K. Lee, S. Huo, and T. Cross, *J. Biomol. NMR* **8**, 1–14 (1996).
81. J.F. Krebs, F. Rana, R.A. Dluhy, and C.A. Fierke, *Biochemistry* **32**, 4496–4505 (1993).
82. F.R. Rana, A.J. Mautone, and R.A. Dluhy, *Biochemistry* **32**, 3169–3177 (1993).
83. F.H. Rana, J.S. Mautone, A.J. Dluhy, *Biochemistry* **32**, 27–31 (1993).
84. K.R. Klein, B. McIntyre, J.O. Fleischer, S. Trommer, *Biochemistry* **35**, 3044–3049 (1996).
85. P.L. Privalov, *Adv. Protein Chem.* **33**, 167–241 (1979).
86. G. Wagner, *Q. Rev. Biophys.* **16**, 1–57 (1983).
87. M.J. Parker and A.R. Clarke, *Biochemistry* **36**, 5786–5794 (1997).
88. R. Jaenicke, *Naturwissenschaften* **75**, 604–610 (1988).
89. D.F. Mann, K. Shah, D. Stein, and G.A. Snead, *Biochim. Biophys. Acta* **788**, 17–22 (1984).
90. J.L. Cleland and A.J. Jones, *Pharm. Res.* **13**, 1464–1475 (1996).
91. M. McCloskey and M. Poo, *Int. Rev. Cytol.* **87**, 19–81 (1984).
92. G.D. Rose, A.R. Geselowitz, G.J. Lesser, R.H. Lee, and M.H. Zehfus, *Science* **229**, 834–838 (1985).
93. P.E. Wright, H.J. Dyson, and R.A. Lerner, *Biochemistry* **27**, 7167–7175 (1988).
94. N.C. Kaarsholm, S. Havelund, and P. Hougaard, *Arch. Biochem. Biophys.* **263**, 496–502 (1990).
95. E. Boix, M.V. Nogués, C.H. Schein, S.A. Benner, and C.M. Cuchillo, *J. Biol. Chem.* **269**, 2529–2534 (1994).
96. C.H. Schein and M. Haugg, *Biochem. J.* **307**, 123–127 (1995).

97. C.H. Schein, *In Vitro Toxicology* **10**, 275–285 (1997).
98. Y. Goto and A.L. Fink, *Biochemistry* **28**, 945–952 (1989).
99. R.D. Mitchell, H.K.B. Simmerman, and L.R. Jones, *J. Biol. Chem.* **263**, 1376–1381 (1988).
100. M.M. Ries-Kautt and A.F. Ducruix, *J. Biol. Chem.* **264**, 745–748 (1989).
101. J.E. Kinsella, *CRC Crit. Rev. Food Sci. Nutr.* **21**, 197–262 (1984).
102. H.M. Farrell, T.F. Kumosinski, P. Pulaski and M.P. Thompson, *Arch. Biochem. Biophys.* **265**, 146–158 (1988).
103. R. Boistelle, *J. Cryst. Growth* **123**, 109–120 (1992).
104. R. Judge, M. Johns, and E. White, *J. Chem. Eng. Data* **41**, 422–424 (1996).
105. Eur. Pat. Appl. 61,250 (1982), C. Weissmann and C.H. Schein (to Biogen N.V.).
106. M. Guo and G. Narsimhan, *Sep. Sci. Technol.* **31**, 1777–1804 (1995).
107. R. Guo, M. Guo, and G. Narsimhan, *Ind. Eng. Chem. Res.* **35**, 3015–3026 (1996).
108. K. Randall, M. Lever, B.A. Peddie, and S.T. Chambers, *Biochim. Biophys. Acta* **1291**, 189–194 (1996).
109. C.S.V. Rajendrakumar, T. Suryanarayana, and A.R. Reddy, *FEBS Lett.* **410**, 201–205 (1997).
110. Y. Liu and D.W. Bolen, *Biochemistry* **34**, 12884–12891 (1995).
111. A.J. Wang and D.W. Bolen, *Biophys. J.* **71**, 2117–2122 (1996).
112. F. Demmel, W. Doster, W. Petry, and A. Schulte, *Eur. Biophys. J* **26**, 327–335 (1997).
113. T. Kleinert, W. Doster, H. Leyser, W. Petry, V. Schwartz, and M. Settles, *Biochemistry* **37**, 717–733 (1998).
114. V. Daggett and M. Levitt, *Proc. Natl. Acad. Sci. U.S.A.* **89**, 5142–5146 (1992).
115. M. Hao, M. Pincus, S. Rackovsky, and H. Scheraga, *Biochemistry* **32**, 9614–9631 (1993).
116. P. Hunenberger, A. Mark, and W. van Gunsteren, *Proteins: Struct., Funct., Genet.* **21**, 169–213 (1995).
117. J.K. Myers, C.N. Pace, and J.M. Scholtz, *Protein Sci.* **4**, 2138–2148 (1995).
118. R. Fraczkiewicz and W. Braun, *J. Comp. Chem.* **19**, 319 (1998).
119. D. Eisenberg and A.D. McLachlan, *Nature* **316**, 199–203 (1986).
120. T. Ooi, M. Oobatake, G. Némethy, and H.A. Scheraga, *Proc. Natl. Acad. Sci. U.S.A.* **84**, 3084–3090 (1987).
121. J. Vila, R.L. Williams, M. Vasquez, and H.A. Scheraga, *Proteins* **10**, 199–218 (1991).
122. L. Wesson and D. Eisenberg, *Protein Sci.* **1**, 227–235 (1992).
123. G.I. Makhatadze and P.L. Privalov, *J. Mol. Biol.* **232**, 639–659 (1993).
124. M. Delarue and P. Koehl, *J. Mol. Biol.* **249**, 675–690 (1995).
125. J.S. Evans, S.I. Chan, and W.A. Goddard III, *Protein Sci.* **4**, 2019–2031 (1995).
126. M.D. Cummings, T.N. Hart, and R.J. Read, *Protein Sci.* **4**, 2087–2099 (1995).
127. A.J. Wang and D.W. Bolen, *Biochemistry* **36**, 9101–9108 (1997).
128. J.L. Fauchère and V. Pliska, *Eur. J. Med. Chem. Chim. Ther.* **18**, 369–375 (1983).
129. B. von Freyberg, T.J. Richmond, and W. Braun, *J. Mol. Biol.* **233**, 275–292 (1993).
130. C. Mumenthaler and W. Braun, *J. Mol. Model.* **1**, 1–10 (1995).
131. B. von Freyberg and W. Braun, *J. Comput. Chem.* **12**, 1065–1076 (1991).
132. A. Radzicka and R. Wolfenden, *Biochemistry* **27**, 1664–1670 (1988).
133. G. Barone, G.G. Della, V.P. Del, C. Giancola, and G. Graziano, *Biophys. Chem.* **51**, 193–199 (1994).
134. G.R. Hedwig and H. Hoiland, *Biophys. Chem.* **49**, 175–181 (1994).
135. J.K. Myers, C.N. Pace and J.M. Scholtz, *Protein Sci.* **4**, 2138–2148 (1995).
136. I. Karle, J. Flippen-Anderson, K. Uma, and P. Balaram, *Proteins: Struct., Funct., Genet.* **7**, 62–73 (1990).
137. I. Karle, J. Flippen-Anderson, K. Uma, and P. Balaram, *Biopolymers* **33**, 827–837 (1993).
138. C.M. Rosell, A.M. Vaidya, and P.J. Halling, *Biochim. Biophys. Acta* **1252**, 158–164 (1995).
139. N. Thanki, J.M. Thornton, and JM. Goodfellow, *J. Mol. Biol.* **202**, 637–657 (1988).
140. P.R. Gibbs, A. Radzicka, and R. Wolfenden, *J. Am. Chem. Soc.* **113**, 4714–4715 (1991).
141. A. Fersht, *Enzyme Structure and Mechanism*, W.H. Freeman, New York, 1984.
142. J. Bryngelson, J. Onuchic, N. Socci, and P. Wolynes, *Proteins: Struct., Func., Genet.* **21**, 167–195 (1995).
143. A. Ansari, C.M. Jones, E.R. Henry, J. Hofrichter, and W.A. Eaton, *Biochemistry* **33**, 5128–5145 (1994).
144. C.H. Schein, *Curr. Opin. Biotechnol.* **4**, 456–461 (1993).
145. M. Bennett, M. Schlunegger, and D. Eisenberg, *Protein Sci.* **4**, 2455–2468 (1995).
146. B.S. Chapman and I.O. Kuntz, *Protein Sci.* **4**, 1696–1707 (1995).
147. U.H.E. Hansmann and Y. Okamoto, *Preprint SC-94-20*, Konrad Zuse Zentrum, Berlin, available via anonymous-ftp: *ftpserv01.zib-berlin.deSC94-20cd / pub get README* (1994).
148. M.C. Peitsch, *Biochem. Soc. Trans.* **24**, 274–279 (1996).
149. M.C. Peitsch, M.R. Wilkins, L. Tonella, J.C. Sanchez, R.D. Appel, and D.F. Hochstrasser, *Electrophoresis* **18**, 498–501 (1997).
150. B. Rost, *Methods Enzymol.* **266**, 525–539 (1996).
151. B. Rost, in C. Rawlings ed., *Third International Conference on Intelligent Systems for Molecular Biology* AAAI Press, Menlo Park, Calif., 1995, pp. 314–321.
152. L. Li and K.S. Matthews, *J. Biol. Chem.* **270**, 10640–10649 (1995).
153. E. Huang, S. Subbiah, J. Tsai, and M. Levitt, *J. Mol. Biol.* **257**, 716–725 (1996).
154. M.W. West and M.H. Hecht, *Protein Sci.* **4**, 2032–2039 (1995).
155. S. Roy, K. Helmer, and M.H. Hecht, *Fold. Design* **2**, 89–92 (1997).
156. S. Dalal, S. Balasubramanian, and L. Regan, *Nature Struct. Biol.* **4**, 548 (1997).
157. S.B. Prusiner, G. Telling, F.E. Cohen, and S.J. Dearmond, *Semin. Virol.* **7**, 159–173 (1996).
158. J. Safar, P.P. Roller, D.C. Gajdusek, and C.J. Gibbs Jr., *Protein Sci.* **2**, 2206–2216 (1993).
159. B. Dahiyat, C. Sarisky, and S. Mayo, *J. Mol. Biol.* **273**, 789–796 (1997).
160. J.L. Pomerantz, S.A. Wolfe, and C.O. Pabo, *Biochemistry* **37**, 965–970 (1998).
161. J.S. Kim and C.O. Pabo, *Proc. Natl. Acad. Sci. U.S.A.* **95**, 2812–2817 (1998).

PROTEIN EXPRESSION, SOLUBLE

CATHERINE H. SCHEIN
University of Texas Medical Branch
Galveston, Texas

KEY WORDS

Bacillus subtilis
Escherichia coli
High-density cultivation
Proteolysis
Secretion
Temperature effect

OUTLINE

INTRODUCTION

My first publications on techniques to avoid the formation of inclusion bodies in *Escherichia coli* (1,2) were met with many positive responses. Many others had experienced the problems summarized in the second column of Table 1 and were tired of dealing with "solutions of renatured proteins" that were in reality particulate microdispersions of inactive protein. They also shared my concern that proteins treated with strong denaturants were not the best starting point for biophysical studies, especially in cases where the specific activity had not been determined previously. Techniques such as protein NMR revealed structural heterogeneity in proteins purified from inclusion bodies (IBs), and few proteins purified from IBs have been useful therapeutically.

However, a recent review of the literature would indicate that renaturation from IBs, probably because it was the earliest-described method for producing large quantities of protein in *E. coli*, is still the predominant method. Generally, if isolation from IBs does not work, usually for the reasons in the second column of Table 1, the next step is to express the protein in yeast, insect cells, or transfected mammalian cells. Although expression in yeast is a fine alternative (see Refs. 3 and 4 for a summary of methods), both of the latter methods are significantly more expensive than bacterial cell culture. Insect cell cultivation is an excellent method for production of larger proteins, but stably transformed lines are given to variability, tests for expression are time-consuming, and the scale-up is more difficult than with bacteria or yeast.

Isolation from inclusion bodies is frequently used to purify proteins for characterizing the results of site-directed mutagenesis, precisely the area where accurate determination of specific activity and physical properties is most needed. University groups, who often appear to be racing with competitors to study the greatest number of mutants in the shortest time, frequently resort to IB purification schemes. Having little access to bioreactors and fermentors, they rely on shaker flask cultivation. The medium during growth in shaker flasks rapidly becomes anaerobic and acidic, meaning that the bacteria are struggling to grow and not able to produce optimal levels of product. Under these conditions, significant production requires completely subverting the bacterial metabolism into recombinant protein production, which usually means the use of induction methods such as temperature shock that favor IB production.

Obtaining an apparently pure protein in solution after renaturing from IBs is no guarantee of activity (5). For example, human tyrosine kinase Lck was completely inactive if expressed insolubly in *E. coli* and renatured from inclusion bodies. The protein was active, however, as measured by autophosphorylation, if solubly expressed by coproduction with either gro ESL or TRX (6).

Further, a mutant of a protein that renatures well from IBs may differ considerably from the wild type in solubility as well as specific activity. Ribonuclease A is probably the most common model for protein refolding experiments. However, a graduate student with experience in expressing RNase A mutants in *E. coli* and isolating them from IBs was unable to obtain any of a mutant that represented the ancestral protein of the hooved animals (the gene contains 16 amino acid changes chosen by evolutionary considerations). Undergraduate students in a lab course recloned the gene into an *E. coli* secretion system (7), expressed the mutant protein in a soluble form, and were able to determine its specific activity with respect to that of the modern protein (author's unpublished results).

As illustrated by the this example, expressing recombinant proteins solubly does not necessarily require great expertise. However, unlike expression in IBs, which can be done under relatively uncontrolled conditions, high-level expression of proteins in a soluble form requires control over many variables during growth and induction. These include temperature, medium pH, oxygenation, and concentrations of essential nutrients such as glucose (or glyc-

Table 1. Perceived Advantages and Actual Disadvantages of Expressing Recombinant Proteins as Inclusion Bodies in *Escherichia coli*

Advantages of IBs	Disadvantages
Many mutants can be tested because a single flask (or multiplate well) per mutant suffices for comparative studies.	Mutants of proteins can differ greatly in their specific activity and ability to refold.
High expression level means reduced fermentation costs.	Refolding shifts problems and costs downstream.
Production can be monitored with PAGE or immunoblotting.	Production cannot be monitored directly by activity.
Cytoplasmic proteins are washed away, simplifying purification. 95% purity can be quickly achieved.	The other 5% contaminants are hydrophobic, insoluble membrane proteins and cell wall fragments
Removal of cytoplasmic proteins early means less proteolysis.	Significant proteolysis accompanies refolding in dilute solution
The major contamininants are oligomers and misfolded or proteolyzed forms of the desired protein.	Separation of multiple forms of the same protein is the most difficult purification step.
The pL promotor induced by raising temperature often yields where other systems fail.	If the protein does not refold well, another expression system will still be needed.

erol) and toxic by-products of the bacteria, including acetate and carbon dioxide. Several industrial biotechnology groups have developed elegant solutions for all the problems encountered in soluble expression (8,9). Presumably because of stability problems with respect to both the host strain and plasmid, continuous culture methods are generally not used. The highest-yielding processes to date are based on fed-batch technology, which is also the standard for antibiotic production.

Most researchers wish to obtain their protein in a soluble form with a minimum amount of work. I thus begin by suggesting alterations in growth conditions that can be tested with most expression systems. The second section then discusses alternative host strains and plasmids that are better for directing soluble protein expression. The third section is devoted to discussing ways of altering the protein itself to avoid IB formation.

Today, bacterial culture and cloning techniques are used by many whose specialties are far removed from the area of microbiology. For this reason, the fourth section contains detailed suggestions for preparing and maintaining stock cultures and inocula, as well as a phage-free environment. One must stress the importance of good sterile technique, a topic that is never presented in other areas of the physical sciences. The last section includes tips on maintaining solubility during harvesting and the preliminary washing/lysis steps. Cosolvents for retaining solubility of proteins in solution are discussed elsewhere (10).

ALTERING GROWTH CONDITIONS TO INCREASE SOLUBLE PROTEIN EXPRESSION

Lowering Growth Temperature

For intracellular production, simply lowering the growth temperature (T) (1,11) can greatly increase the soluble fraction of protein. Table 2 lists many recombinant proteins from many different sources that are more soluble when expressed in *E. coli* at a lower growth temperature. The list includes several fusion proteins, for example, mouse DNA polymerase delta expressed as a glutathione-*S*-transferase conjugate (19). Although most of the examples in Table 2 used 25–30 °C for production, there is evidence that one can lower the temperature much further to increase the yield of soluble protein. The vaccinia virus large subunit of ribonucleotide reductase (vvR1, 87 kDa) was completely insoluble in *E. coli* when expressed at 37 °C, but mostly soluble if induced at 15 °C (29). The quantity of soluble and insoluble human androgen receptor (AR) fusion protein produced at different times following induction at 12, 22, and 37 °C indicated that while much more insoluble protein was expressed at higher temperature, soluble receptor expression was much higher at the lower temperatures. The amount of soluble protein did not correlate well with the ligand-binding capability of the receptor, indicating there was some bacterial factor needed for folding to a binding-competent state whose capacity was exhausted at higher expression levels (there is little evidence of proteolytic degradation of the cytosolic fraction from their gels) (15). Six different areas of the protein were converted to insoluble protein at the higher temperatures.

There is a lower limit of about 8–10 °C below which *E. coli* will not grow at all. In general, K12 strains grow very slowly below 23 °C; at 10 °C only a few dozen proteins are produced (42).

Point mutations may make a protein more or less prone to temperature-induced aggregation. For example, human triose phosphate isomerase is soluble at 37 °C during production in *E. coli*, as are two mutants, Met14Gln and Arg98Gln, which are partially and completely inactive, respectively. The (inactive) protein containing both mutations, however, was produced in a soluble form only at a growth temperature of 28 °C (43).

Although in some cases the production of soluble protein may appear to be the same at different growth temperatures (44), lowering the growth temperature has two additional great advantages. The first is that at lower temperature the O_2 solubility in the growth medium is much higher. Thus the cell density can be increased before the culture grows anaerobically. Anaerobic growth, and particularly the production of acetate, has been shown to limit cell growth and the production of protein (45).

The second advantage is that several plasmid expression systems have been shown to be more stable when bacteria are grown at lower temperatures. This may be due to the fact that the half-life of the antibiotics used to select plasmid-bearing bacteria during the cultivation also have a longer half-life at the lower temperature.

Table 2. A Few Proteins That Are More Soluble When Produced in *Escherichia coli* Grown at 30 °C or Less

Protein	Ref.
Animal / mammalian	
Bovine heart mitochondrial ubiquinol–cytochrome *c* reductase (fusion protein)	12
Chick brain β-tubulins, 33 kDa	
Chicken riboflavin carrier protein	13
Cystic fibrosis transmembrane conductance regulator (fused to glutathione-*S*-transferase)	14
Human androgen receptor (AR) fusion protein	15
Human initiation factor-4E	16
Human Interferon-α2	
Human Interferon-β	17
Human Interferon-γ	
Human transforming growth factor-β	18
Human triose phosphate isomerase double mutant (M14Q/R98Q)	
Mouse DNA polymerase delta	19
Murine cAMP-dependent protein kinase (catalytic subunit)	20
Murine Mx	
Rabbit muscle glycogen phosphorylase	21
Rat phosphatidylinositol transfer protein	22
Regulator of chromosome condensation (RCC1)	23
Xenopus Cu, Zn superoxide dismutase	24
Plant	
Radish 3-*OH*-3-methyl glutarylcoenzyme A reductase	25
Rice lipoxygenase L-2	26
Yeast α-glucosidase P1	27
Viral	
Herpes simplex thymidine kinase fusion protein	28
Vaccinia virus large subunit of ribonucleotide reductase (vvR1, 87-kDa)	29
Yeast / fungal	
Cephalosporin-C acylase from Pseudomonas N176	30
Isopenicillin-*N*-synthases	31
Bacterial	
Bacteroides fragilis metallo-β-lactamase (truncated protein)	32
Diphtheria toxin and fusion proteins thereof	33
Heparinase II and III from *Flavobacterium heparium*	34
Lycopene cyclase from *Erwinia uredovora*	35
P22 tailspike	36
Ricin A-chain	37
Salmonella typhimurium CheY mutant	38
Salmonella typhimurium tryptophan synthetase C-terminal deletions	39
Shigella flexneri invasion plasmid antigens his-fusion	40
T4 DNA polymerase	41

Note: Where no reference is cited, the data are from Ref. 11. Proteins are not listed that are produced in a soluble form at 30 °C but for which no comparative data is provided in the reference concerning solubility at a higher growth temperature.

Higher growth temperature can also inhibit processing. When chicken riboflavin carrier protein, a phosphoglycoprotein, was expressed in the T7 system, at 37 °C, two products were seen in whole cells lysates for protein with and without the pelB leader sequence encoded in the pET20b vector. Protein solubilized primarily from IBs did not contain the signal sequence, which was also absent in soluble protein expressed at 22 °C (13).

Possible Reasons for the Effect of Temperature. The reasons for the effect of lowering growth temperature are hard to pinpoint. The temperature effect is not simply due to slowing the production rate of the protein or to a decreased growth rate. The rate of synthesis of a particular protein does not necessarily increase with temperature. The protein Mx was sequestered into IBs at 37 °C, regardless of the absolute production rate (which was controlled by varying the distance between the Shine–Dalgarno sequence and the initial ATG in otherwise similar expression plasmids) (11). Further, decreasing the growth rate in other ways does not increase soluble expression; to the contrary, substrate limitations lower recombinant protein expression (9).

It is possible that compartmentalization of protein production is more efficient at the lower growth temperature. For example, RNase A was insoluble at 37 °C when produced cytoplasmically, but soluble at this temperature when directed to the periplasm/medium in a secretion system (7). In mammalian cells, proteins expressed in the nucleus were more susceptible to thermal denaturation in situ compared with those expressed in the cytoplasm (46).

Another possibility is the presence or lack of some proteins important for folding or aggregation at the different temperatures. Phosphorylation may occur at the higher temperature, thus altering the characteristics of the protein, as has been shown for Gro EL (8). Proteolysis increases at higher temperatures, as does the catabolism of amino acids involved in central metabolic pathways, such as serine (47). Studies of the expression of various heat shock proteins, which are induced by growth at higher temperatures, have indicated that their presence alone cannot mimic the effects on cell growth of increasing temperature. Indeed, as we will see in the section "Altering the Host Strain and Vector to Direct Soluble Expression", coexpression of some of these proteins can selectively increase the amounts of soluble recombinant protein. A possible explanation for the effect of primary sequence on IB formation is discussed in the section "Altering the Protein for Solubility."

Medium Composition

Use of a defined medium (i.e., one in which poorly defined nitrogen sources such as yeast extract or casein hydrolysate are not used) is preferable. Several defined media have been developed to allow high-density bacterial growth (48). Glycerol may be a better carbon source than glucose because less acetate is produced (45). In general, growth limitations should be avoided, with the exception of expression with negatively regulated induction systems, such as those based on the tryptophan promotor. Batch feeding during induction has been shown to increase protein yield considerably. Mathematical models have also been developed to aid in increasing cell density (49).

Before eliminating casein hydrolysates from the growth medium, as one does for example in producing labeled proteins for NMR experiments, one should insure that their presence is not essential for controlling protease activity. Addition of protease inhibitors to the growth medium, in the form of an ill-defined extract of potato, was essential for production of interferon in *Bacillus subtilis* (50). Specific inhibitors of toxic enzymes may also be added to the growth medium. Yields of vvR1, for example, were improved by adding the ribonucleotide reductase inhibitor hydroxyurea (29).

Adding compatible cosolvents during the culture may also help maintain solubility, whereas in some cases their positive effect may actually be during harvesting (see section "Controlling Other Variables That Affect Solubility"). Increasing medium osmolarity by the addition of nonmetabolizable sugars may also increase protein solubility at higher temperatures: 0.4 M sucrose in the growth medium increased the production of soluble β-lactamase protein (51). There seem to be relatively few examples, however, where compatible cosolvents have been tried, perhaps because the levels of the compounds must be optimized and their use can greatly increase medium costs. The solubility of bovine heart mitochondrial ubiquinol–cytochrome *c* reductase, expressed as a fusion protein with glutathione, was improved by adding 2.5 mM betaine and 440 mM sorbitol to the growth medium. Presumably, the increased osmolarity of the growth medium forced the uptake of betaine and a corresponding increase of glycyl-betaine within the cells (12).

Good aeration, even beyond the O_2 uptake requirements of the culture, can increase production of proteins in *E. coli*. Rapid aeration reduces the concentration of CO_2 and acetate, both of which can be toxic. High acetate production has been implicated in reducing yields, especially in shaker flasks, even when the pH is controlled by using highly buffered media. Strains that produce less acetate have been created by metabolic pathway engineering; however, these may grow too slowly to be useful. Controlling the growth conditions of a prototrophic strain with few mutational markers so that it did not experience anerobiosis or stress has been shown to be a more useful way to reduce acetate production in large-scale, high-density culture (9).

Proteins induced under stress conditions may serve to control the production of acetate. The universal stress protein (USPA), a 13.2-kDa protein, is induced by a variety of stress conditions. This protein's role during stress may be to control the pathways of glycolysis, because mutants in the USPA gene secrete large amounts of acetate (52).

Other methods to increase the oxygen availability may be preferable to heavy aeration, which can lower the yields of secreted proteins by increasing the shear force (increased agitation) and denaturation at the air–liquid interface of bubbles. For example, surface aeration with supplemental O_2 was used to produce interferon-α in *Bacillus subtilis* because large activity losses were found when air or even plain nitrogen was bubbled through a solution of medium and the purified protein (53). The antifoam agent needed to control foaming at high volumetric airflow rates (VVM) can also seriously lower yields, especially for secreted proteins, and should be used sparingly.

Induction Conditions

The amount and time of inducer addition should be strictly controlled. Many have found it advantageous to induce at low cell density in shaker flask culture to insure that the cells are growing actively. However, in a bioreactor with efficient batch feeding and good aeration, much higher densities can be used. A higher cell density may be preferable with induction at very low temperatures, as for example when using a cold shock promotor, because the cells will grow quite slowly after that point. Most industrial processes use very high cell density; for example, production of the fusion protein of fibroblast growth factor–Saponin (a mitotoxin designed to treat cancer and restenosis) relied on fed-batch cultivation with a rich medium feed to reach $A_{600} = 85$ before inducing with isopropyl-D-thiogalactopyranoside (IPTG) at 30 °C (pET11a vector/BL21 [DE3] host). The yield of soluble mitotoxin, which could be isolated in good yield using heparin affinity chromatography, was 2.2 mg/L/A_{600} (54). The concentration of inducer may also be important; several authors have reported better yields of soluble protein if the IPTG inducer concentration was reduced from the usual 1 mM (22,23,55). Yields of vvR1, for example, were improved by inducing at a low cell density and using a low concentration (0.05 mM) of IPTG (29).

ALTERING THE HOST STRAIN AND VECTOR TO DIRECT SOLUBLE EXPRESSION

Choice of Host Strain

Expression versus Transformation. Correct host strain selection is also essential. Table 3 lists the markers carried by a few commonly used *E. coli* strains. Although some of their markers may determine how well a strain produces protein, several of them are potentially dangerous leftovers selected for the use for which the strain was originally designed. For example, *supE* is an amber suppressor that inserts a glutamine at UAG codons; it was added to strains of *E. coli* to allow the replication of early M13 vectors. These vectors were required to contain amber mutations that would control their propagation on accidental release, according to NIH guidelines that no longer apply (62). For this reason, older bacterial strains with fewer markers may be preferable for production purposes (9).

Careful cultivation and preservation of host strains is essential for optimal production. Because host strains for protein production often do not transform well, strains such as JM101 or HB101 are used for transformations after ligation or for vector propagation. Choose a host strain that is suitable for the problem at hand.

Low Protease Strains and Those with Better Redox Characteristics. Use of protease minus host strains (63) is often an advantage. *E. coli* BL21 (DE3), a protease-deficient strain, was used to express bacterial glutamyl tRNA re-

Table 3. Several Common *Escherichia coli* Cloning or Host Strains and Their Genotypes

Strain	Genotype	Ref.
C600	*supE44, thi-1, hsdR, thr-1, leuB6, lacY1, tonA21, mcrA*	56
HB101	*supE44, hsdS20*(r-Bm-B)*recA13ara-14proA2lacY1gal / K2rpsL20xyl-5mtl-1*	57
JM101	F′, *traD36, proAB, lacZ,* ΔM15/Δ(*lac,pro*), *supE44, thi-1*	58
JM109(DE3)	[F′, *traD36, proAB, lacI*^{q}, *lacZ,* ΔM15] λ(DE3), Δ(*lac,pro*), *endA1,hsdR17*(Rk$^-$Mk$^+$)*supE44, thi-1, recA1, gyrA96,relA1*	45
RV308	*lac74-galISII*::OP308-*strA*	59
XL1-Blue	*endA1, hsdR17*(Rk$^-$Mk$^+$)*supE44, thi-1, λ-, recA1, gyrA96, relA1, lac*[F′, *proAB,lacI*^{q}ΔM15,TN10(*tet*^{R})]	60

Note: See tables by Sambrook et al. (61) for more strains and an explanation of the abbreviations.

ductase. The synthesis of 5-aminolevulinic acid (ALA) and porphyrins, which were secreted into the medium, could be used to follow the production of the protein, which was soluble when produced at 37 °C (64).

Protease-minus strains of *E. coli* are particularly valuable for production of soluble proteins and those expressed from secretion vectors. A *lon*$^-$ *htpr*$^-$ strain (lacking the lon protease of *E. coli* [65]) significantly increased the production of RNases in a secretion system (7), and the strain SG21173 (*lon*$^-$, *htpR*$^-$, *clp*) was used to express the human T-cell receptor V beta 5.3 under the control of the hybrid trc promoter containing the *E. coli* heat-stable enterotoxin II (STII) signal sequence for protein secretion (66). There are also strains that facilitate the formation of disulfide bonds in the reducing environment of the cytoplasm (67). These should be considered for expression of proteins with multiple cysteine disulfides.

Coexpressing Other Proteins. Cocloning of molecular chaperones, which can include such diverse molecules as thioredoxin reductase (68) and, for membrane proteins, phospholipids (69) can be of help, but they must be selectively chosen (Table 4). Coexpression of the *E. coli* heat shock protein GroESL has solubilizing effects only on certain mammalian proteins. For example, coexpression of GroESL increased the solubility of four of eight mammalian proteins (c-myb, CRE-BP1, p53, and the *Xenopus* protooncogene product Mos), while coexpression of thioredoxin increased the solubility of these and the other four (SnoN, c-Myc, adenovirus E1A, and the human tyrosine kinase Lck) (6). Thioredoxin also increased the production of antibodies (68).

There is also a report that association with GroEL actually decreased the solubility of the cystic fibrosis transmembrane conductance regulator (expressed as a glutathione-*S*-transferase fusion protein). A 60-kDa protein coeluted from affinity columns with the protein and was shown by sequencing to be GroEL. The authors were able to break up the complex and purify the protein (which was induced at 25–28 °C) by adding 5 mM MgATP to the bacterial extract and incubating at 37 °C (14).

Other reports indicate that GroESL may have a solubilizing effect only at certain temperatures. The aggregation of preS2-*S*′-*β*-galactosidase, a three-domain fusion protein in the cytoplasm of *E. coli*, was reduced three- to sixfold by cooverexpression of the DnaK and DnaJ molecular chaperones over a wide range of growth temperatures (30–42 °C), while an increase in the concentration of the GroEL and GroES chaperonins had a significant effect only at 30 °C. Coimmunoprecipitation experiments confirmed that preS2-*S*′-*β*-galactosidase formed a stable complex with DnaK, but not with GroEL, at 42 °C. When the intracellular concentration of chromosomal heat shock proteins was increased by overproduction of the heat shock transcription factor σ (33), or by addition of 3% ethanol (v/v) to the growth medium, a two- to threefold higher recovery of active enzyme was observed at 30 and 42 °C, but not at 37 °C. The overexpression of all heat shock proteins or specific chaperone operons did not significantly affect the synthesis rates or stability of preS2-*S*′-*β*-galactosidase and did not lead to the disaggregation of preformed inclusion bodies. Rather, the improvements in the recovery of soluble and active fusion protein resulted primarily from facilitated folding and assembly. The authors suggested that titration of the DnaK–DnaJ early folding factors may trigger the formation of preS2-*S*′-*β*-galactosidase inclusion bodies (70).

Coexpression of proteins that improve solubility or the overall growth of the bacteria can also be useful. For example, coexpression of *Vitreoscilla* hemoglobin can increase the total cellular protein of various common *E. coli* cloning strains (JM101, pRed2 and W3110) and *B. subtilis* (71). Their production characteristics should also be improved because onset of oxygen limitation is delayed (72). Cocloning demonstrated that the function of at least one of the six homologous insulin-like growth factor binding proteins is to assist in the formation and maintenance of the native disulfides of IGF-I (73).

Cocloning of specific antibodies to the protein may increase yields. For example, the heat-induced aggregation of carboxypeptidase A could be greatly reduced in vitro in the presence of monoclonal antibodies that did not inhibit its activity (74). In addition, coexpression of tRNA genes for the preferred mammalian codons may increase the expression of mammalian genes in *E. coli*. For example, the

Table 4. Useful Expression Partners for Recombinant Proteins

DNA-k and DNA-J
E. coli GroESL
Phospholipids
Thioredoxin
Antibodies to the protein
Vitreoscilla hemoglobin

expression of argU tRNA enhances expression of interferon (coded from the mammalian sequence) in *E. coli* (75).

Alternative Hosts to *E. coli*. Despite the millions of different bacteria in nature, very few are used currently as hosts for protein production. *E. coli* itself, isolated originally from infant diarrhea, was chosen as an experimental strain by Jacques Monod soon after World War II because it was relatively nontoxic and grew quickly. Other bacteria that have been used as hosts for recombinant protein production were selected because they were known for medical reasons (*Straphylococci, Bacillus calmette-guerin*), were approved for use in some food-related applications (*Lactobacillus* [76]), or were used industrially for enzyme or antibiotic production (*Bacillus subtilis* and *Bacillus amyloliquefaciens*).

B. subtilis systems, based on previous industrial processes for enzyme production, are mostly suitable for the production of stable proteins that are resistant to proteases (77). If the basic problems of protease degradation and processing problems that may lead to insoluble product (53,78) can be avoided, there are several good expression vectors, including one based on the α-amylase signal sequence and promotor (50). *Bacillus licheniformis*, useful for producing bacterial enzymes on the industrial scale (79), has been used to produce active human IL-3 of therapeutic quality (80). *B. calmette-guerin* has been used to produce a *Schistosoma mansoni* antigen (81). *Lactobacillus lactis*, found in cheese cultures, is also a useful expression host (82) and may have promise for vaccines (83).

Recently, *Staphylococcus carnosus* was used for surface display of a malaria antigen (84), and a secretion/expression system in *Salmonella* has been suggested for vaccine production (85). There are "shuttle vectors" available that can function in both *E. coli* and various alternative host bacteria. For example, expression of *Trympanosoma cruzi* polyantigens using the vector pRIT21, which contains the origins of replication for both *E. coli* and *Staphylococcus aureus*, allowed expression in both strains. Seven different antigens showed less proteolysis when secreted from *S. aureus*; with purification yields of 80%, between 4 and 16 mg/L of the different antigens for vaccine production were obtained (86).

Various *Vibrio* species were used to express the nontoxic B-subunit of *E. coli* heat-labile enterotoxin linked to a 27-residue C-terminal fragment of Pol, the catalytic subunit of DNA polymerase of herpes simplex virus type 1. The highest yields were reported to be 3.5 mg/L; 0.3 mM EDTA added to the growth medium blocked proteolysis of the secreted protein (87).

Altering the Vector

Choice of Promotor. Using a prokaryotic promoter that is tightly regulated and can be induced to express large amounts of protein at low temperatures (23–30 °C) is essential for soluble expression. For example, expression levels in the T7 system can be increased by tighter transcriptional control before induction (88). In view of the effect of temperature on solubility, it is interesting that a whole group of *E. coli* proteins is induced when the culture temperature is lowered to below 20 °C. The promotor for one of these genes, cold shock promotor CspA, is the basis of a vector developed specifically for low-temperature induction of recombinant proteins (89). The expression levels possible with this promotor at the optimal induction temperature compare favorably with those obtained with the tac promotor (90). The Csp promotor has the advantage that temperature shift alone induces production. It should certainly be considered for expression of highly aggregation-prone or unstable gene products.

Older promotors, such as those based on the tryptophan operon, are useful for low-temperature expression, as is a modified tryptophanase promotor (91). Secretion of specific proteins from *E. coli* is already possible using a variety of vectors (7,92). Secretion expression systems are particularly suited to mutant analysis (93) because the culture supernatant can be directly analyzed. Further study of the factors affecting transport of proteins is needed.

Selecting Signal Sequence or Fusion Partners. The sequence of the protein itself can be altered by adding linkers to improve protein folding, solubility, protease resistance, yield, and secretion into the culture medium (Table 5). Expression of small proteins or fragments of a larger protein may be possible only if the desired sequence is expressed as a fusion protein (98). The *E. coli* maltose binding protein (MalE) is particularly useful because fusion molecules with it, besides being easily purified, are often secreted. For example, CD4, the human T-lymphocyte receptor for the AIDS virus, HIV, as well as many different functional deletions that retained their ability to bind monoclonal antibodies and HIV gp160 and inhibit HIV binding to $CD4^+$ cells, were fused as multiple repeats to both ends of a single MalE molecule. The hybrid molecules (expressed in L-broth at 30 °C) were exported into the periplasm (95). Pollen major allergen Bet v1 from *Betula verrucosa* (White Birch) has been cloned and expressed in *E. coli* as a fusion with maltose-binding protein and a factor Xa proteolytic cleavage site. The cloning strategy, based on a polymerase chain reaction, positioned the factor Xa proteolytic site so that the authentic amino terminus of Bet v1 was generated after cleavage. Fusion protein (expressed solubly at 37 °C) was isolated by amylose affinity chromatography and enzymatically cleaved by incubation with factor Xa (99). The kringle V domain of the lipoprotein apolipoprotein(a) from human liver was solubly incubation with factor Xa (99). The kringle V domain of the lipoprotein apolipoprotein(a) from human liver was solubly expressed in *E. coli* as a MalE fusion protein. As with the other MalE fusions, this protein was purified by amylose–agarose affinity chromatography by eluting with 10 mM maltose (100).

Several linkers, such as hexahistidine, can aid in the detection and purification of recombinant proteins. Most vectors also include a protease site between the linker and the protein. The 54-kDa subunit of the canine signal recognition particle (SRP) was expressed *E. coli* using a T7 expression system with an amino-terminal extension encoding an initiating methionine and 10 histidine residues followed by an enterokinase cleavage site. His-SRP (54 kDa) was purified to give a single band on SDS-PAGE in 20% yield (0.3 mg from 500 mL culture) (101). The hexo-

Table 5. Linkers That Improve the Solubility of Fusion Proteins or Allow More Specific Purification

Linker	Comments
Calmodulin (94)	Calmodulin binds to phenol sepharose tightly.
E. coli maltose binding protein (MBP or MalE)	Purify fusion protein by binding to amylose beads and eluting with maltose. Secretion into the periplasm of soluble fusion protein is often possible even with 37 °C growth. Proteins and deletions thereof may be cloned at either or both ends (95).
Glutathione-*S*-transferase (GST)	Purify fusion protein by binding to glutathione beads. Fusions may be more soluble.
Hexa- (or more) polyhistidine linker	Allows purification with metal chelate columns.
Thioredoxin (TRX)	Very commonly used vector system. Antibodies to thioredoxin may be useful in purifying the fusion proteins (96).
Ubiquitin (97)	Easily cleavable fusion protein.

kinase of *Schistosoma mansoni* has been expressed in *E. coli* as a fusion protein including an N-terminal polyhistidine tag and enterokinase cleavage site. The enzyme was purified by metal chelate chromatography to > 95% homogeneity, based on analysis by SDS gel electrophoresis and isoelectric focusing (102). This sort of expression system has the advantage that the protein can often be expressed without an N-terminal methionine, but the disadvantage that yields are reduced due to the additional processing step. Even specific proteases may cleave the protein itself. Thus initially high expression levels may turn into disappointing bottom-line figures.

Thioredoxin fusions of many proteins are soluble, especially if produced at a lower growth temperature, active in some cases before removal of the thioredoxin portion, and may be purified by heating lysates to 80 °C. A-DDDDK-linker, the recognition site for mammalian intestinal protease enteropeptidase, allows removal of the thioredoxin moiety (103). IGF-1 was more than 100 times more soluble when fused to one or two Z-domains (from Staph protein A), and the solubilizing effect could also be seen during in vitro refolding (104).

There is a fine interplay between the temperature effect on solubility and the linker. For example, the solubility of the his fusions with the IpaC invasion plasmid antigens of *Shigella flexneri* (antibodies to which may be useful in treating Shigellosis dysentery) was improved by production at 30 °C. However, no protein was found when the IpaB antigen was similarly cloned. Both histidine-linked proteins were partially soluble at 37 °C if expressed as fusions with Trx (40).

Human osteocalcin (hOC), a 49–amino acid peptide produced mainly by bone osteoblasts, was fused to the glutathione S-transferase (GST) gene, and the fusion protein was produced mainly in a soluble form. The affinity-purified fusion protein was cleaved with activated protease factor X, releasing the rhOC portion. The structure of rhOC was confirmed by mass spectrometry and amino acid sequencing (105).

Changing Codon Use without Altering Sequence. Bacterial and mammalian codon usage differ significantly. Genes for several proteins have been synthesized to reflect the bacterially preferred codons for degenerate amino acids and to eliminate elements in the noncoding region of genes that lower productivity. For example, RNase A was expressed at a much higher level when synthesized from a synthetic gene that reflected codon usage than was its murine counterpart expressed from the natural mammalian gene (7). Alternatively, one can look for specific mammalian codon usage in the cDNA that may inhibit translation. For example, 4.9% of the adult isoform of human cardiac troponin (288 amino acids) is arginine coded by the (rarely used in *E. coli*) codons AGG and AGA. Changing two areas where these codons occurred consecutively (AGG [165] and AGG [166], and AGG [215] and AGG [216]) to the bacterially preferred codon CGT increased expression 10-fold (one pair); there was a 40-fold increase when both pairs of the rare arginine codons were replaced (106). In a similar example, electrospray ionization mass spectrometry of recombinant link module from human tumor necrosis factor–stimulated gene 6 protein revealed that much of the protein was a mixture of forms where between one and four of the arginine residues had a lysine substitution. When the AGA codons for the arginine residues were substituted to the bacterially preferred codon CGT, the substitution problem was eliminated, and the level of expression was greatly increased (107). One can also try using strains that express tRNAs for the mammalian preferred codons, as already discussed.

A combination of methods may be necessary. For example, the aggregation of the vaccinia virus large subunit of ribonucleotide reductase (vvR1, 87 kDa) (29) was prevented by lowering the induction temperature to 15 °C, the cell density during induction, and the concentration of the inducer (to 0.05 mM IPTG). Addition of hydroxyurea, an inhibitor of ribonucleotide reductase, further increased production of soluble vvR1 in a dose-dependent manner.

ALTERING THE PROTEIN FOR SOLUBILITY

Why Do IBs Form?

The reason for the formation of IBs has been the subject of much debate since the phenomenon was first observed. An answer to this question is vital to achieving maximum soluble expression in *E. coli*. Lowering growth temperature to achieve more soluble protein does not work for all proteins; further, we do not understand the physical characteristics that account for why some proteins are perfectly soluble even at 42 °C, but point mutations thereof can render the proteins insoluble at all temperatures. Size is not a variable because even very large proteins (e.g., heparinases II and III, see Table 2, are over 700 amino acids long) can be expressed in soluble form.

Although even point mutations can render proteins more or less soluble (for reviews see Refs. 2 and 108), specific effects cannot be directly related to hydrophobicity (Gravy scores [31]) or content of a specific amino acid such as cysteine or proline. A multifactorial analysis (109) also did not yield a satisfactory explanation for the effect of sequence on solubility.

The only general observation is that there is a good deal of misfolded protein and proteolytic fragments in IBs. IB formation in *E. coli* was probably first observed when cells were treated with amino acid analogues (110); this relationship between abnormal protein and aggregation has been born out by other studies. In general, proteolysis accompanies IB formation (11); indeed several authors attributed the effects of a lower growth temperature on solubility to reduced proteolytic cleavage of their proteins (17,55). Certain intermediate states in protein aggregation are detectable by some physical techniques but difficult to define at the sequence level (111).

Two recent observations may shed some light on the process of aggregation. The first is that protein aggregates, as viewed by PAGE, appeared identical in the presence or absence of another protein. Thus the protein appeared to aggregate only with itself and not to coaggregate with a potential partner (112). Of course, the authors could not analyze the more insoluble, larger molecular weight aggregates that did not penetrate into the gel, but the results strongly suggest that aggregation follows a very specific pathway.

The second observation is that the oligomers seen in many crystal structures are not due to the monomers aligning themselves at surface-exposed hydrophobic sites. Rather, the monomers actually exchange a whole section of the protein to form a compact structure. Requirements for the swapped domain are that it have both a hydrophilic and hydrophobic face and that it be joined to the rest of the protein by a flexible linker. "Domain swapping" is one possible mechanism for aggregation that accounts for many of the observations about IB formation (113):

- *Specificity*. The swapped domain is connected to the oligomeric partner in much the same way as to the original monomer.
- *Speed*. Thermodynamically speaking, the major barrier to overcome during the initial formation of the dimer molecule is the barrier toward simultaneously opening two monomers that lie within a permissible distance of one another to permit establishing a bond. Once a dimer has formed, however, addition of another monomer requires only the opening up of one more molecule at a time. Thus aggregation will proceed quite quickly once the initial template exists.
- *Proteolysis*. IB formation is usually accompanied by proteolysis, and proteolytic fragments are often seen in solubilized preparations from IBs. If one assumes that a long, flexible linker joins the swapped domains, this essentially unstructured area would present a good site for the attack of proteases. The domains liberated by hydrolysis of the linker might serve as fragments to further participate in the aggregate formation.
- *Association with synthesis*. Pulse labeling studies with *E. coli* overexpressing the *Salmonella typhimurium* gene *CheY* have shown that newly formed protein aggregates more quickly than existing protein (38). However, preexisting protein can also become involved in IBs if incubation is continued for a sufficiently long time (11). This makes sense if one assumes that protein immediately after synthesis is not yet completely folded. Possibly the stable domains fold first, followed by the folding of the whole protein. A protein in solution exhibits a certain amount of folding-out (so-called breathing) that can be observed by techniques such as hydrogen exchange. Presumably, new protein whose secondary structure is formed but is still unfolded at the tertiary level can exchange domains easily, whereas existing protein can join the aggregate only at the appropriate stage in its breathing state. The movement increases with temperature, which can also account for the observed temperature dependence of aggregation. Aggregation would thus be one aspect of the protein solution half-life, which can be related to temperature by an Arrhenius plot (ln k vs. $1/T$, where k is an intrinsic instability constant and T is the temperature in °K) (114).
- *Irreversibility*. Because the area of the domain that is swapped is substantial, the energy required to form (and hence to separate) oligomers is very large. The domain-swapping model can also explain to some extent the influence of heat shock proteins and chaperonins on solubility.
- *Other Miscellaneous Observations*. It has also been noted that there is active protein in IBs (115), which suggests strongly that the protein is not completely misfolded. Some IBs have the ability to be separated by ion exchange. Also there is a lack of a general relationship between protein sequence and aggregation. Domain swapping accounts quite neatly for the effects of specific point mutations (116) and their corresponding supressors on the solubility of proteins (117).

If one accepts that domain swapping is involved in oligomer formation, designing peptides to prevent aggregation may be a greatly simplified process, especially if a crystal structure is available or one has knowledge of interacting faces withing the protein. One can direct changes in the protein itself to inhibit its aggregation, or design peptides that will bind reversibly to these sites and prevent the initiation of aggregation (106).

These peptides need not be very large; for example, a simple tetramer, Arg-Gly-Asp-Ser, from fibrinogen was sufficient to inhibit the aggregation of platelets (118). A few discrete amino acid changes at the end of the B-chain of insulin considerably reduced the oligomerization state (119).

Two examples show that this mechanism can be exploited if one can detect amino acid regions in turns. In the first, a monomeric λ-cro protein was made from the homodimer by duplicating the dimer interface region (six amino acids) at the C-terminus of the molecule. The monomeric mutant was more thermostable and appeared from NMR studies to have a solution structure similar to that of the dimer (120). *E. coli* maltose-binding protein is soluble, but a double point mutant (Gly32Asp/Ile33Pro) is insoluble whether it is targeted to the cytoplasm or periplasm by expression without or with its signal sequence, respectively. These mutations are located in a turn/loop region between two helical regions in the crystal structure (121).

Mutation to Increase Solubility

Single point mutations may greatly improve solubility, whereby the observed increase in soluble protein may also be a secondary effect of increased stability to proteolysis or improved interaction with a chaperone (122). Single point mutations may alter the effect of temperature on solubility. The solubility and stability to proteolysis during production in *E. coli* of a 101–amino acid fragment from the human respiratory syncytial virus (RSV) major glycoprotein (G-protein) was significantly improved by altering several phenylalanine residues. Two cysteine-to-serine changes, did not affect the solubility or stability of the gene product. The mutant G-proteins appeared antigenically similar to the parent and had similar secondary structures (123).

Destruction due to proteases can be minimized by identifying degradation "hot spots" in proteins (124). Some of these sites have been discussed elsewhere (2,108). It should be noted that such sites in mammalian proteins may be one mechanism for controlling their activity. For example, proteolytic cleavage at the C-terminal of human interferon-γ, which also occurs in blood, can increase activity several fold. Efficient production of this cytokine in *E. coli* was possible only if the N-terminal residues Cys-Tyr-Cys, which are also not found in serum isolates, were removed from the coding sequence (125).

CONTROLLING OTHER VARIABLES THAT AFFECT SOLUBILITY

Inoculum Preparation and Storage

One of the greatest causes of variation in the production of proteins in bacteria is plasmid and host strain instability. The plasmid should be checked periodically during large-scale cultivation. Loss of plasmid, or alterations, including changes in the sequence of the desired protein or even deletion of the whole segment and insertion of unrelated DNA into the plasmid, can occur. Several different insertion sequences are known in *E. coli*. Correct preparation of inocula to minimize variation is essential and is thus dealt with here in some detail.

Most strain stability problems are not discussed openly in papers dedicated to studying the proteins themselves (with notable exceptions [126]). Because many expression problems can be ameliorated by care in preparing the inoculum, a detailed protocol is included here for several stages.

The following general recommendations are made for the start of culture:

1. Larger inocula are preferred.
2. Glycerol cultures should not be reused, and they should never be stored at temperatures above −70 °C.
3. Glycerol cultures should be made from the first culture after a transformation.

Glycerol Culture Preparation and Storage. The following protocol proved reliable for culture of various *E. coli* strains producing interferon and other proteins in differing vector backgrounds:

1. A transformation was done by standard protocols (62), and the bacteria were grown up on selective solid medium.
2. Five to six colonies from the transformation were picked individually into numbered flasks containing 50 mL of selective media (L-broth plus antibiotic) and grown up on a shaker to midexponential phase. Then between 50 and 100 glycerol cultures were made from each flask by combining 0.4–0.5 mL of the culture with 0.5 mL 50% glycerol in L-broth. (In practice, sterile Eppendorf tubes—with a freezer-safe lock closure—or cryotubes were filled with the glycerol solution, closed, and autoclaved. The culture was added sterilly and the tube was reclosed.) The cultures were stored on ice for 2 h, frozen overnight at −20 °C, and then moved to −70 °C or liquid nitrogen storage. In addition, 2× 1 mL of each culture was centrifuged (5 min centrifugation in a cold room in an Eppendorf centrifuge) and the pellet frozen for plasmid analysis.
3. The remaining culture was induced in an appropriate fashion, the cells were lysed, and the amount of protein or its activity was determined. A plasmid prep was made of the selected colony, and the restriction map compared to that expected. The culture with the highest yield was chosen for further analysis (assuming its restriction map was in order), and the other cultures and the glycerol cultures made from them were discarded.*Note*: It is unwise to ignore oddities in the restriction map. Because 1 μL of a commercial *Eco*RI or *Bam*HI preparation will cut an entire miniprep of plasmid completely in a matter of minutes, unexplainable high molecular weight bands are probably not residual host DNA. Unexpected bands may indicate either latent bacteriophage contamination or insertion sequences in the plasmid. When in doubt, throw it out (the culture that is) and do another transformation.
4. To each of the chosen glycerol cultures, afix a freezer-stable label with the plasmid name, host strain, resistance marker, date, and initials of the experimenter. Cultures should be stored in numbered boxes, and a file kept of their location in the −70 °C

or liquid nitrogen storage. Excess glycerol cultures of strains no longer in use should be weeded at regular intervals (it is best not to wait until the compressor needs replacing or the contents of the freezer must be moved).

Cultures prepared and selected in this fashion are relatively stable; the author recently prepared Mx protein from a culture that had been frozen in a −70 °C freezer (with all the temperature variations that can represent) for more than 10 years. Nonetheless, one should maintain a good plasmid library of minipreps of vectors as the ultimate backup.

Trouble-Shooting Protein-Producing Recombinant Bacteria

Most host *E. coli* strains carry a portfolio of mutations that may be essential to their success in producing protein. Several cloning and expression strains and their genotypes are listed in Table 5. The requirements for a cloning strain (basically, high efficiency of transformation) are different from those for a good expression host. The latter should grow vigorously on minimal media and be low in proteases, whereas the former, such as HB101, should have few host restriction systems and high sensitivity to the selection marker. Do not store host strains on agar plates in the cold room; the best method is to prepare them as glycerol cultures and store at −70 °C or liquid nitrogen temperature.

It is vital to continuously monitor the quality of these host strains, especially when several researchers share stock cultures. For best results, one should prepare several hundred glycerol cultures of common host strains when they are obtained in the lab. One of these should be used per experiment, preferably by adding the whole 1 mL to 50 mL of growth media used to induce transformation phenotype. Although not suitable for long-term storage or in cases where a high yield is necessary from the transformation, competent cells can be stored frozen and used for several weeks after. When performing a transformation, a control plate of competent cells with no DNA added should be used to test for development of antibiotic resistance. Look for differences in the characteristic color, size, and odor of host strain colonies streaked on a nonselective agar medium.

If necessary, one can test the host strain for the presence of markers. Reversion to a protease-producing phenotype can be detected by plating on agar containing 1% casein. A complete listing of the meaning of various markers for *E. coli*, with alternate notation possibilities, can be found on the internet (61); the more common mutations are described by Sambrook et al. (62).

Protecting Your Cultures from Contamination. When bacteriophages were first introduced, in the 1960s, several labs were sufficiently afraid of their invasiveness as to limit work with them to discrete areas. This procedure should be adhered to today. A wide-spread M13 phage contamination can be caused from one shaker accident that is not adequately cleaned up, and months and even years of subsequent work can be wasted if the problem is ignored. Because switching to an F^- *E. coli* strain will limit possible hosts, the best method is to prevent the original contamination.

Frequent hand washing with antiseptic hand soap and even wearing disposable gloves is essential to avoid cross-contamination of cultures when handling plasmids and other replication-competent nucleic acids. Surfaces should be covered with absorbant paper that is changed regularly, preferably at the end of every experiment. The counter underneath and all cold room surfaces should be washed down with 2% chlorine bleach or Javell water, or 0.5% hydrogen peroxide, frequently. Isolation precautions such as those used throughout the medical field and the food preparation industry should be automatic during work with recombinant organisms. One should avoid touching hair or face during work, and long hair should be tied back. Protective coats should be removed at the door to the lab and never worn in public areas, especially lunchrooms.

Discourage other groups from using your shakers unless they are willing to conform to the following rules, which seem to be obvious:

1. All cultures must be stoppered properly (preferably with a cotton wad wrapped in cheese cloth or some other thick filtering substance) and the stopper fixed in position with tape so that it will not fly off during culture. Aluminum foil is not adequate. Flasks should fit the holder, and one should avoid wrapping paper towels and such around smaller flasks to accommodate them to a larger holder. Multiple test tube cultures must be in a properly secured rack, not taped to the side of a flask holder.
2. Flasks should be removed from the shaker area or room if additions are made after inoculation.
3. The shaker should be cleaned often with an antiseptic solution.
4. UV sterilization using a fluorescent light (one with a time switch can be installed for little expense in any area) is an excellent method to halt the growth of microorganisms, if the light has not been used for more than 300 h of operation and is not left on for more than 15 min at a time. Improperly used, UV irradiation is worse than no sterilization at all because a weak light can select for mutant bacteria that will be resistant to most killing methods. An easy way to determine if the lamp is still in order is to inoculate an agar plate with 1 mL of a dense culture of bacteria (preferably an aerobic spore former such as *Bacillus*, but *E. coli* will do) and leave it open under the lamp for 15 s (time accurately!). If any colonies grow on the plate after 24 h at 30–37 °C, install a new fluorescent bulb.

Fermentor Culture Starting from Glycerol Cultures. Even a 30,000-L fermentor can be inoculated from 1 mL of culture if the plasmid is stable. Do not reuse a glycerol culture; use it once and dispose of it.

Several methods can be used to determine plasmid stability as a function of culture time. One can compare the specific plasmid content of the bacteria at the end of a plasmid miniprep and determine the colony-forming units of a transformation into HB101 or a high frequency of transformation strain.

Stability of Selection Stress. Most antibiotics (in particular, penicillin derivatives) should not be stored in frozen aqueous solution, nor should antibiotic-containing agar plates be stored before use. The half-life of the closed β-lactam ring is extremely short in aqueous solution, particularly at neutral pH. Ampicillin should be dissolved in sterile water, filter sterilized, and added to the growth medium as close to time of inoculation as humanly or mechanically possible.

Before ignoring this caveat, note that those frozen ampicillin solutions inherited from the previous lab generation, while most certainly lacking their ability to lyse bacteria, may inhibit growth of nontransformed bacteria. Week-old refrigerated antibiotic-containing plates used for transformation selection will show no colonies for the control Ca^{2+} cells. As the transformed bacteria grow, they secrete β-lactamase that, coupled with the instability of the antibiotic itself at 37 °C, will rapidly eliminate any selective pressure in that area of the plate (or in the vessel during a large-scale cultivation). The invisible nontransformed bacteria will start to grow. Colonies picked from the transformation plates will perforce contain nontransformed but not necessarily nonviable bacteria. The initially inhibited nonproducers will be able to grow, diluting the plasmid bearers and lowering yield. If the plasmid confers any negative growth characteristics, the plasmid-containing bacteria will rapidly become the minority in the culture vessel. Additional proteins from all those nonproducing bacteria will render the purification even more difficult.

Maintaining Solubility during Harvesting and Cell Lysis

If one is fortunate, the protein to be isolated has distinct characteristics that separate it from the pack. Unfortunately, the majority of naturally occurring proteins are moderately hydrophobic, with a pK_a of between 6 and 7. This means they will elute from hydrophobic interaction columns at salt concentrations between 0.1 and 0.5 N, and from ion exchange resins with the vast majority of bacterial proteins. Thus a typical protein, if expressed at low concentration in the bacterial cytoplasm, will be difficult to separate by general methods of purification and may be destroyed by proteases before ever being seen on a gel.

Here, IB formation would appear to confer an enormous advantage. The protein is clearly visible from the earliest stages of expression on a polyacrylamide gel, and an apparently purified preparation can be quickly obtained. The denaturants used at the early stages of preparation inhibit most proteases, and the overall yield, estimated visually, would seem to be very high.

The apparent ability to wash away contaminants from IBs is often a Trojan horse. Membrane fragment and other insoluble proteins persist through many detergent washes (indeed, there is a report that some proteins are rendered insoluble by these agents, particularly Triton X-100). These high molecular weight, often highly modified proteins are not always visible on gel electrophoresis because they "stick in the slot" or do not stain with silver stain reagents or Coomassie blue. An even greater problem arises: the protein is contaminated with proteolytic fragments, alternate folded forms, and oligomers of itself. If solubly expressed protein requires isolation from bacterial proteins with similar characteristics, the problem in IB schemes comes down to isolating the protein from what is essentially the same protein. This is the hardest of all purification steps.

Even proteins that are initially soluble in the bacteria may be induced to aggregate, sometimes irreversibly, during cell breakage and washing. Aggregation and proteolysis can be minimized by adding a proteolysis inhibitor to the cell culture at the end of growth and maintaining a low temperature during centrifugation. Especially on the large scale, this may require special cooling equipment for centrifuges, and small batch processing for producing cell lysates using high-pressure equipment such as a Manton/Gaulin or French press.

As noted earlier, detergents such as Triton X-100 and Nonidet P-40 can render proteins insoluble when added during cell washing/breakage steps. However, others have noted that extraction with 1% Triton X-100 was necessary for obtaining high yields of bovine heart mitochondrial ubiquinol–cytochrome *c* reductase (fusion protein with glutathione). They solved later aggregation problems by resolubilizing the protein in the detergent dodecyl maltoside in 300 mM NaCl; the resultant protein appeared active because it gave a blue shift in the visible spectrum when incubated with ubiquinone (12). The preliminary breakage steps to maintain the solubility of T7 RNA polymerase had to be done in buffer containing large amounts of glycerol and detergent to prevent the protein from precipitating. Once its extremely narrow solubility range was determined (10), the purification was straightforward.

Even the pH of the extraction buffer can have a profound effect on solubility. For example, UMP-kinase from *E. coli* was insoluble if extracted from cells with a pH 7.4 Tris buffer, but was soluble if 1 mM UTP was added to this buffer or if 100 mM borate buffer pH 9 was used for the extraction. Reversible aggregation could be induced by adding Mg to the UTP-containing solutions or lowering the pH. A point mutant, D159N, was soluble at up to 5 mg/mL at pH 7.4 (127).

As the last example demonstrates, the effects of point mutations on solubility do not fit easily into a clear set of empirical or theoretical rules. The reader is referred to Refs. 92 and 108 for further examples and discussion.

CONCLUSIONS

Proteins produced in a soluble form in bacteria are easier to purify and work with than those refolded from inclusion bodies. Expression of soluble proteins can be improved by judicious choice of the vector and host strain, and manipulation of the inoculation, growth, and induction conditions. Very high per liter yields of purified product have been achieved, and the overall yield for some secretion systems is higher than those possible with inclusion bodies.

BIBLIOGRAPHY

1. C.H. Schein, *Bio / Technology* **7**, 1141–1149 (1989).
2. C.H. Schein, *Curr. Opin. Biotechnol.* **2**, 746–750 (1991).

3. M.C. Hensing, R.J. Rouwenhorst, J.J. Heijnen, J.P. van Dijken, and J.T. Pronk, *Antonie van Leeuwenhoek* **67**, 261–279 (1995).
4. A.R. Mistry, L. Falciola, L. Monaco, R. Tagliabue, G. Acerbis, A. Knight, R.P. Harbottle, M. Soria, M.E. Bianchi, C. Coutelle, and S.L. Hart, *BioTechniques* **22**, 718–729 (1997).
5. J.W. Harper and N.J. Logsdon, *Biochemistry* **30**, 8060–8066 (1991).
6. T. Yasukawa, C. Kanei-Ishii, T. Maekawa, J. Fujimoto, T. Yamamoto, and S. Ishii, *J. Biol. Chem.* **270**, 25328–25331 (1995).
7. C.H. Schein, E. Boix, M. Haugg, K.P. Holliger, S. Hemmi, G. Frank, and H. Schwalbe, *Biochem. J.* **283**, 137–144 (1992).
8. P. Carter, R.F. Kelley, M.L. Rodrigues, B. Snedecor, M. Cavarrubias, M.D. Velligan, W.L.T. Wong, A.M. Rowland, C.E. Kotts, M.E. Carver, M. Yang, J.H. Bourell, H.M. Shepard, and D. Henner, *Bio/technology* **10**, 163–167 (1992).
9. U. Horn, W. Strittmayer, A. Krebber, U. Knupfer, M. Kujau, R. Wenderoth, K. Muller, S. Matzku, A. Plückthun, D. Riesenberg, *Appl. Microb. Biotechnol.* **46**, 524–532 (1996).
10. C.H. Scein, *Bio/Technology* **8**, 308–317 (1990).
11. C.H. Schein and M.H.M. Noteborn, *Bio/Technology* **6**, 291–294 (1988).
12. L. Yu, K.P. Deng, and C.A. Yu, *J. Biol. Chem.* **270**, 25634–25638 (1995).
13. Sooryanarayana, P.R. Adiga, and S.S. Visweswariah, *Protein Express. Purif.* **7**, 147–154 (1996).
14. C. Randak, A.A. Roscher, H.B. Hadorn, I. Assfalgmachleidt, E.A. Auerswald, and W. Machleidt, *FEBS Lett.* **363**, 189–194 (1995).
15. B. Cooper, J.A. Gruber, and M.J. McPhaul, *J. Steroid Biochem. Mol. Biol.* **57**, 251–257 (1996).
16. S. Morino, H. Hazama, M. Ozaki, Y. Teraoka, S. Shibaata, M. Doi, H. Ueda, T. Ishida, and S. Uesugi, *Eur. J. Biochem.* **239**, 597–601 (1996).
17. T. Mizukami, Y. Komatsu, N. Hosoi, S. Itoh, and T. Oka, *Biotechnol. Lett.* **8**, 605–610 (1986).
18. H.F. Seow, C.R. Goh, A.G. Porter, and L. Krischnan, *Gene* **83**, 117–129 (1989).
19. R. Hindges and U. Hubscher, Gene **158**, 241–246 (1995).
20. L.W. Slice, and S.S. Taylor, *J. Biol. Chem.* **264**, 20940–20946 (1989).
21. M.F. Browner, P. Rasor, S. Tugendreich, and R.J. Fletterick, *Protein Eng.* **4**, 351–357 1991.
22. J.M. Tremblay, G.M. Helmkamp, and L.R. Yarbrough, *J. Biol. Chem.* **271**, 16139–16143 (1996).
23. A.M. Tsai, M.J. Betenbaugh, and J. Shiloach, *Biotechnol. Bioeng.* **48**, 715–718 (1995).
24. A. Battistoni, M.T. Carri, A.P. Mazzetti, and G. Rotilo, *Biochem. Biophys. Res. Commun.* **186**, 1339–1342 (1992).
25. A. Ferrer, C. Aparicia, N. Nogués, A. Wettstein, T.J. Bach, and A. Boronat, *FEBS Lett.* **266**, 67–71 (1990).
26. Y. Shirano, and D. Shibata, *FEBS Lett.* **271**, 128–130 (1990).
27. E. Kopetzki, G. Schumacher, and P. Buckel, *Mol. Gen. Genet.* **216**, 149–155 (1989).
28. J. Fetzer, and G. Folkers, *Pharm. Pharmacol. Lett.* **2**, 112–114 (1992).
29. M.B. Slabaugh, R.E. Davis, N.A. Roseman, and C.K. Mathews, *J. Biol. Chem.* **268**, 17803–17810 (1993).
30. Y. Saito, Y. Ishii, T. Fijimura, H. Sasaki, Y. Noguchi, H. Yamada, M. Niwa, and K. Shiomura, *Ann. N.Y. Acad. Sci.* **782**, 79–86 (1996).
31. S. Tiow-suan, and D.S.H. Tan, *Biochem. Mol. Biol. Int.* **35**, 1069–1078 (1995).
32. J.H. Toney, J.K. Wu, K.M. Overbye, C.M. Thompson, and D.L. Pompliano, *Protein Express. Purif.* **9**, 355–362 (1997).
33. W.R. Bishai, R. Rappuoli, and J.R. Murphy, *J. Bacteriol.* **169**, 5140–5151 (1987).
34. H.S. Su, F. Blain, R.A. Musil, J.J.F. Zimmermann, K.F. Gu, and D.C. Bennett, *Appl. Environ. Microbiol.* **62**, 2723–2734 (1996).
35. G. Schnurr, N. Misawa, and G. Sandmann, *Biochem. J.* **315**, 869–874 (1996).
36. A. Mitraki, and J. King, *FEBS Lett.* **307**, 20–25 (1992).
37. M. Piatak, J.A. Lane, W. Laird, M.J. Bjorn, A. Wang, and M. Williams, *J. Biol. Chem.* **263**, 4837–4843 1988.
38. J. Klein and P. Dhurjati, *Appl. Environ. Microbiol.* **61**, 1220–1225 (1995).
39. L.H. Yang, S. Ahmed, and E.W. Miles, *Protein Express. Purif.* **8**, 126–136 (1996).
40. W.L. Picking, J.A. Mertz, M.E. Marquart, and W.D. Picking, *Protein Express. Purif.* **8**, 401–408 (1996).
41. T.C. Lin, J. Rush, E.K. Spicer, and W.H. Konigsberg, *Proc. Natl. Acad. Sci.* **84**, 7000–7004 (1987).
42. P.G. Jones, R.A. VanBogelen, and F.C. Neidhardt, *J. Bacteriol.* **169**, 2092–2095 (1987).
43. V. Mainfroid, P. Terpstra, M. Beauregard, J.M. Frere, S.C. Mande, W.G.J. Hol, J.A. Martial, and K. Goraj, *J. Mol. Biol.* **257**, 441–456 (1996).
44. M.-C. McKenna, C. Muchardt, R. Gaynor, and D. Eisenberg, *Protein Express. Purification* **5**, 105–111 (1994).
45. S. Sakamoto, I. Terada, Y.C. Lee, K. Uehara, H. Matsuzara, and M. Iijima, *Appl. Microbiol. Biotechnol.* **45**, 94–101 (1996).
46. A.A. Michels, V.-T. Nguyen, A.W.T. Konings, H.H. Kampinga, and O. Bensaude, *Eur. J. Biochem.* **234**, 382–389 (1995).
47. R.G. Matthews, and F.C. Neidhardt, *J. Bacteriol.* **171**(5), 2619–2625 (1989).
48. D. Riesenberg, *Curr. Opin. Biotechnol.* **2**, 380–384 (1991).
49. L. Andersson, L. Strandberg, L. Haggstrom, and S.O. Enfors, *FEMS Microbiol. Rev.* **14**(1), 39–44 (1994).
50. I. Palva, P. Lehtovaara, L. Kèèrièinen, M. Sibakov, K. Cantell, C.H. Schein, K. Kashiwagi, and C. Weissmann, *Gene* **22**, 229–235 (1983).
51. G.A. Bowden, A.M. Paredes, and G. Georgiou, *Bio/Technology* **9**, 725–730 (1991).
52. T. Nystrom and F.C. Neidhardt, *J. Bacteriol.* **175**, 3949–3956 (1993).
53. C.H. Schein, K. Kashiwagi, A. Fujisawa, and C. Weissmann, *Bio/Technology* **4**, 719–725 (1986).
54. J.R. McDonald, M. Org, C. Shen, Z. Parandoosh, B. Sosnowski, S. Bussell, and L.L. Houston, *Protein Purif. Express.* **8**, 97–108 (1996).
55. G.R. Smith, R. Ford, J.D. Bryant, R.L. Gambley, T.K. Meghia, R.M. Harding, and J.L. Dale, *Arch. Virol.* **140**, 1817–1831 (1995).
56. T.V. Huynh, R.A. Young, and R.A. Davis, D. Glover ed., *DNA Cloning*, Oxford University Press, Oxford, 1985, pp., 56–110.
57. F. Bolivar and K. Backman, *Methods Enzymol.* **68**, 245 (1979).
58. C. Yanish Perron, J. Viera, and J. Messing, *Gene* **33**, 103–199 (1985).
59. R. Maurer, B.J. Meyer, and M. Ptashne, *J. Mol. Biol.* **139**, 147–161 (1980).
60. W.O. Bullock, J.M. Fernandez, and J.M. Short, *Biotechniques* **5**, 376–379 (1987).

61. M. Kroger, and R. Wahl, *Nucleic Acids Res.* **25**, 39–42 (1997).
62. J. Sambrook, E.F. Fritsch, and T. Maniatis, *Molecular Cloning: A Laboratory Manual*, Cold Spring Harbor Laboratory Press, Cold Spring Harbor, New York, 1989.
63. G.T. Yarranton and A. Mountain, in R.R. Rees, M.J.E. Sternberg, R. Wetzel, eds., *Protein Engineering: A Practical Approach*, IRL Press, New York, 1992, pp. 303–325.
64. W. Chen, L. Wright, S. Lee, S.D. Cosloy, and C.S. Russell, *Biochim. Biophys. Acta* **1309**, 109–121 (1996).
65. S.A. Goff, L.P. Casson, and A.L. Goldberg, *Proc. Natl. Acad. Sci. U.S.A.* **81**, 6647–6651 (1984).
66. B. Andrews, H. Adari, G. Hannig, E. Lahue, M. Gosselin, S. Martin, A. Ahmed, P.J. Ford, E.G. Hayman, and S.C. Makrides, *Gene* **182**, 101–109 (1996).
67. A.J. Derman, W.A. Prinz, D. Belin, and J. Beckwith, *Science* **262**, 1744–1747 (1993).
68. K. Proba, L. Ge, and A. Plückthun, *Gene* **159**, 203–207 (1995).
69. M. Bogdanov, J. Sun, H.R. Kaback, and W. Dowhan, *J. Biol. Chem.* **271**, 11615–11618 (1996).
70. J.G. Thomas and F. Baneyx, *J. Biol. Chem.* **271**, 11141–11147 (1996).
71. P.T. Lallio and J.E. Bailey, *Biotechnol. Prog.* **12**, 31–39 (1996).
72. J.E. Bailey, A. Sburlati, V. Hatzimanikatis, K. Lee, W.A. Renner, and P.S. Tsai, *Biotechnol. Bioeng.* **52**, 109–121 (1996).
73. S. Hober, A. Hansson, M. Uhlen, and B. Nilsson, *Biochemistry* **33**, 6758–6761 (1994).
74. T. Katzav-Gozansky, E. Hanan, and B. Solomon, *Biotechnol. Appl. Biochem.* **23**, 227–230 (1996).
75. O.L. Garcia, B. Gonzalez, A. Menendez, A.E. Sosa, J.R. Fernandez, H. Santana, and N. Meneses, *Ann. N.Y. Acad. Sci.* **782**, 79–86 (1996).
76. H. Hashiba, R. Takiguchi, K. Jyoho, and K. Aoyama, *Biosci. Biotechnol. Biochem.* **56**, 190–194 (1992).
77. M. Takagi, S.P. Lee, and T. Imanaka, *J. Ferment. Bioeng.* **81**, 557–559 (1996).
78. D. Dahan, R. Srikumar, R. Laprade, and J.W. Coulton, *FEBS Lett.* **392**, 304–308 (1996).
79. D.M. Rothstein, P.E. Devlin, and R.L. Cate, *J. Bacteriol.* **168**, 839–842 (1986).
80. H. Burger, R.W. van Leen, L.C. Dorssers, N.L. Persoon, P.J. Lemson, and G. Wagemaker, *Blood* **76**, 2229–2234 (1990).
81. L. Kremer, G. Riveau, A. Baulard, A. Capron, and C. Locht, *J. Immunol.* **156**, 4309–4317 (1996).
82. C. Rush, L. Hafner, and P. Timms, *Appl. Microbiol. Biotechnol.* **47**, 537–542 (1997).
83. P.M. Norton, R.W.F. Le Page, and J.M. Wells, *Folia Microbiol.* **40**, 225–230 (1995).
84. P. Samuelson, M. Hansson, N. Ahlborg, C. Andreoni, F. Gotz, T. Bachi, T.N. Nguyen, H. Binz, M. Uhlen, and S. Stahl, *J. Bacteriol.* **177**, 1470–1476 (1995).
85. S. Liljeqvist, D. Haddad, K. Berzins, M. Uhlen, and S. Stahl, *Biochem. Biophys. Res. Commun.* **218**, 356–359 (1996).
86. J.I. Moreno, *Protein Express. Purif.* **8**, 332–340 (1996).
87. A. Loregian, T.R. Hirst, H.S. Marsden, and G. Palu, *Protein Express. Purif.* **8**, 381–389 (1996).
88. N. Mertens, E. Remaut, and W. Fiers, *Biotechnology* **13**, 175–179 (1995).
89. J.A. Vasina and F. Baneyx, *Appl. Environ. Microbiol.* **62**, 1444–1447 (1996).
90. J.A. Vasina and F. Baneyx, *Protein Express. Purific.* **9**, 211–218 (1997).
91. K.C. Sitney, M.B. Mann, G.W. Stearns, A.D. Menjares, J.L. Stevenson, M.D. Snavely, J.C. Fieschko, C. Curless, and L.B. Tsai, *Ann. N.Y. Acad. Sci.* **782**, 297–310 (1996).
92. C.H. Schein, *Curr. Opin. Biotechnol.* **4**, 456–461 (1993).
93. E. Boix, M.V. Nogués, C.H. Schein, S.A. Benner, and C.M. Cuchillo, *J. Biol. Chem.* **269**, 2529–2534 (1994).
94. Y. Ishii, T. Nakano, N. Honma, N. Yuyama, Y. Yamada, H. Watarai, T. Tomura, M. Sato, H. Tsumura, and T. Ozawa, *J. Immunol. Methods* **186**, 27–36 (1995).
95. J.M. Clement, M. Jehanno, O. Popescu, W. Saurin, and M. Hofnung, *Protein Express. Purif.* **8**, 319–331 (1996).
96. R.R. Dickason, R.A. Edwards, J. Bryan, and D.P. Huston, *J. Immunol. Methods* **185**, 237–244 (1995).
97. M.H.M. Koken, H.H. Odijk, M. van Duin, M. Fornerod, and J.H. Hoeijmakers, *Biochem. Biophys. Res. Comm.* **195**, 643–653 (1993).
98. M. Larsson, E. Brundell, L. Nordfors, C. Hoog, M. Uhlen, and S. Stahl, *Protein Express. Purif.* **7**, 447–457 (1996).
99. M.D. Spangfort, H. Ipsen, S.H. Sparholt, S. Aasmul-Olsen, M.R. Larsen, E. Mortz, P. Roepstorff, and J.N. Larsen, *Protein Express. Purif.* **8**, 365–373 (1996).
100. X. Chenivesse, T. Huby, J. Chapman, D. Franco, and J. Thillet, *Protein Express. Purif.* **8**, 145–150 (1996).
101. S. Patel and B.M. Austen, *Protein Express. Purif.* **8**, 283–294 (1996).
102. R.L. Armstrong, J.E. Wilson, and C.B. Shoemaker, *Protein Express. Purif.* **8**, 374–380 (1996).
103. E.R. LaVallie, E.A. DiBlasio, S. Kovacic, K.L. Grant, P.F. Schendal, and J.M. McCoy, *Bio/Technology* **11**, 187–193 (1993).
104. E. Samuelsson and M. Uhlen, *Ann. N.Y. Acad. Sci.* **782**, 486–494 (1996).
105. S.M. Kakonen, J. Hellman, K. Pettersson, T. Lovgren, and M. Karp, *Protein Express. Purif.* **8**, 137–144 (1996).
106. X. Hu, Q. Shi, T. Yang, and G. Jackowski, *Protein Express. Purif.* **7**, 289–293 (1996).
107. A.J. Day, R.T. Aplin, and A.C. Willis, *Protein Express. Purif.* **8**, 1–16 (1996).
108. C.H. Schein, *Pharm. Acta Helv.* **69**, 119–126 (1994).
109. D.L. Wilkinson and R.G. Harrison, *Bio/Technology* **9**, 443–448 (1991).
110. W.F. Prouty, M.J. Karnovsky, and A.L. Goldberg, *J. Biol. Chem.* **250**, 1112–1122 (1975).
111. J. King, C. Haase-Pettingell, A.G. Robinson, M. Speed, and A. Mitraki, *FASEB J.* **10**, 57–66 (1996).
112. M.A. Speed, D.I.C. Wang, and J. King, *Nat. Biotechnol.* **14**, 1283–1287 (1996).
113. M.J. Bennett, M.P. Schlunegger, and D. Eisenberg, *Protein Sci.* **4**, 2455–2468 (1995).
114. C.Ó. Fágáin, *Biochim. Biophys. Acta* **1252**, 1–14 (1995).
115. K. Tokatlidis, P. Dhurjati, J. Millet, P. Béguin, and J.-P. Aubert, *FEBS Lett.* **282**, 205–208 (1991).
116. R. Wetzel, J.L. Perry, and C. Veilleux, *Bio/Technology* **9**, 731–737 (1991).
117. A. Mitraki, B. Fane, C. Haase-Pettingell, J. Sturtevant, and J. King, *Science* **253**, 54–58 (1991).
118. S. Krishnamurthi, T.A. Dickens, Y. Patel, C.P. Wheeler-Jones, and V.V. Kakkar, *Biochem. Biophys. Res. Comm.* **163**, 1256–1264 (1989).
119. D.N. Brems, L.A. Alter, M.J. Beckage, R.E. Chance, R.D. DiMarchi, L.K. Green, H.B. Long, A.H. Pekar, J.E. Shields, and B.H. Frank, *Protein Eng.* **5**, 527–533 (1992).

120. M.C. Mossing and R.T. Sauer, *Science* **250**, 1712–1715 (1990).
121. J.M. Betton and M. Hofnung, *J. Biol. Chem.* **271**, 8046–8052 (1996).
122. P.A. Nygren, S. Stahl, and M. Uhlen, *Trends Biotechnol.* **12**, 184–188 (1994).
123. M. Murby, E. Samuelsson, T.N. Nguyen, L. Mignard, U. Power, H. Binz, M. Uhlen, and S. Stahl, *Eur. J. Biochem.* **230**, 38–44 (1995).
124. M. Murby, M. Uhlen, and S. Stahl, *Protein Express. Purif.* **7**, 129–136 (1996).
125. C.H. Schein and M. Haugg, *Biochem. J.* **307**, 123–127 (1995).
126. M.-C. McKenna, C. Muchardt, R. Gaynor, and D. Eisenberg, *Protein Express. Purif.* **5**, 105–111 (1994).
127. L. Serina, N. Bucurenci, A.M. Gilles, W.K. Surewicz, H. Fabian, H.H. Mantsch, M. Takahashi, I. Petrescu, G. Batelier, and O. Bârzu, *Biochemistry* **35**, 7003–7011 (1996).

See also β-GALACTOSIDASE, ENZYMOLOGY; ENZYMES, PULP AND PAPER PROCESSING; GENE TRANSFER, GRAM-POSITIVE BACTERIA; INSECT CELLS AND LARVAE, GENE EXPRESSION; PROTEIN SECRETION, *SACCHAROMYCES CEREVISIAE*; *PICHIA*, OPTIMIZATION OF PROTEIN EXPRESSION.

PROTEIN GLYCOSYLATION

NIGEL JENKINS
Eli Lilly, Inc.
Indianapolis, Indiana
PIR M. SHAH
De Montfort University
Leicester, U.K.

KEY WORDS

Carbohydrates
Glycosylation
Recombinant proteins
Review

OUTLINE

INTRODUCTION

Protein glycosylation, the addition of sugar residues to the polypeptide backbone, is the most extensive of all the cellular posttranslational modifications made to proteins in the eukaryotic cell. It entails attachment and remodeling of oligosaccharides in the endoplasmic reticulum (ER) and Golgi apparatus prior to secretion of the protein or its appearance at the cell surface (1,2). Almost all proteins on the mammalian cell surface (including receptors) are glycoproteins (3), and the majority of secreted proteins are also glycosylated (with notable exceptions such as insulin, growth hormone, and some interleukins). Glycosylation has been shown to influence the secretion, solubility, stability, antigenicity, and blood clearance of glycoproteins. In some cases, carbohydrates form part of the ligand binding site. In cases where this does not occur, the influence of carbohydrates may be overlooked until the protein enters animal and human clinical trials. There have been four major areas of advance in recent years:

1. Development of methods for the accurate structural analysis of carbohydrates attached to glycoproteins (glycans)
2. Cloning of key glycosyltransferase enzymes in the glycan biosynthetic pathways.
3. Understanding the cellular influences on protein glycosylation.
4. Understanding the varied biological functions of carbohydrates in glycoproteins.

In this article the methods and findings that underpin the science of glycobiology are reviewed, with particular focus on their application in the biotechnology industry. Differences in carbohydrate structures that may arise from choosing alternative gene expression systems and cell culture conditions are discussed, together with methods of carbohydrate analysis that are appropriate at each stage of product development. For reviews of other aspects of glycobiology the reader should consult reviews on biosynthetic pathways (4,5), the biological properties of carbohydrates (6), and the cellular influences on the glycosylation process (2,7).

Oligosaccharide Structures

Three distinct types of glycosylation are known to occur in higher plants and animals, two of which are shown in Figure 1:

1. N-linked glycosylation of proteins; oligosaccharides are attached to the amino group of asparagine residues, usually within the sequence Asn-X-Ser/Thr.
2. O-linked glycosylation; smaller oligosaccharides are linked to serine or threonine residues.
3. Glucosamine and mannose residues are incorporated into the glycosyl–phosphatidylinositol (GPI) anchor of some membrane proteins.

In addition, single fucose residues (8) or small glycans linked via Asn-X-Cys have been found on some blood proteins (9).

N-Linked Glycosylation

N-linked glycosylation is characterized by a glycosidic bond to the R-group of an asparagine residue of the peptide backbone. A tripeptide core or the consensus peptide sequence Asn-X-Ser/Thr (where X is any amino acid apart from proline) must be present for this particular form of glycosylation. The presence of this amino acid sequence by no means guarantees its glycosylation, because other factors such as protein secondary structure and its folding status within the ER are important influences (10,11). For example, the immunomodulator interferon-γ (IFN-γ) has two potential N-glycosylation sites (Asn_{25} and Asn_{97}), but exists in three glycoforms: doubly glycosylated (at both Asn_{25} and Asn_{97}), singly glycosylated (at Asn_{25} only), and nonglycosylated (12). Similarly, in the blood protease tissue plasminogen activator (tPA), only two N-glycosylation sites (Asn_{117} and Asn_{448}) are fully occupied: Asn_{184} is only 50% glycosylated due to steric hindrance (11).

N-linked oligosaccharides fall into three structural categories: complex, high-mannose (or oligomannose), and hybrid, dependent on the extent of glycan remodeling in the later stages of the glycosylation process (Fig. 1). All three categories contain the same chitobiose core structure of two *N*-acetylglucosamine (GlcNAc) units attached to the asparagine containing sequon, which is extended by three mannose (Man) units, with a single fucose (Fuc) residue sometimes attached to the GlcNAc of the reducing end (next to its peptide attachment site).

1. Complex oligosaccharides contain two or more branches attached to the core, and the terms *biantennary, triantennary*, and *tetraantennary* reflect the number of branches. Sugar residues such as GlcNAc, galactose (Gal), *N*-acetylgalactosamine (GalNAc), Fuc, and sialic acids (*N*-acetyl [NeuAc] and *N*-glycolyl [NeuGc] neuraminic acids) extend the structures.
2. Oligo (5–9) and high-mannose (>9 Man) structures contain only mannose residues attached to the core structure.

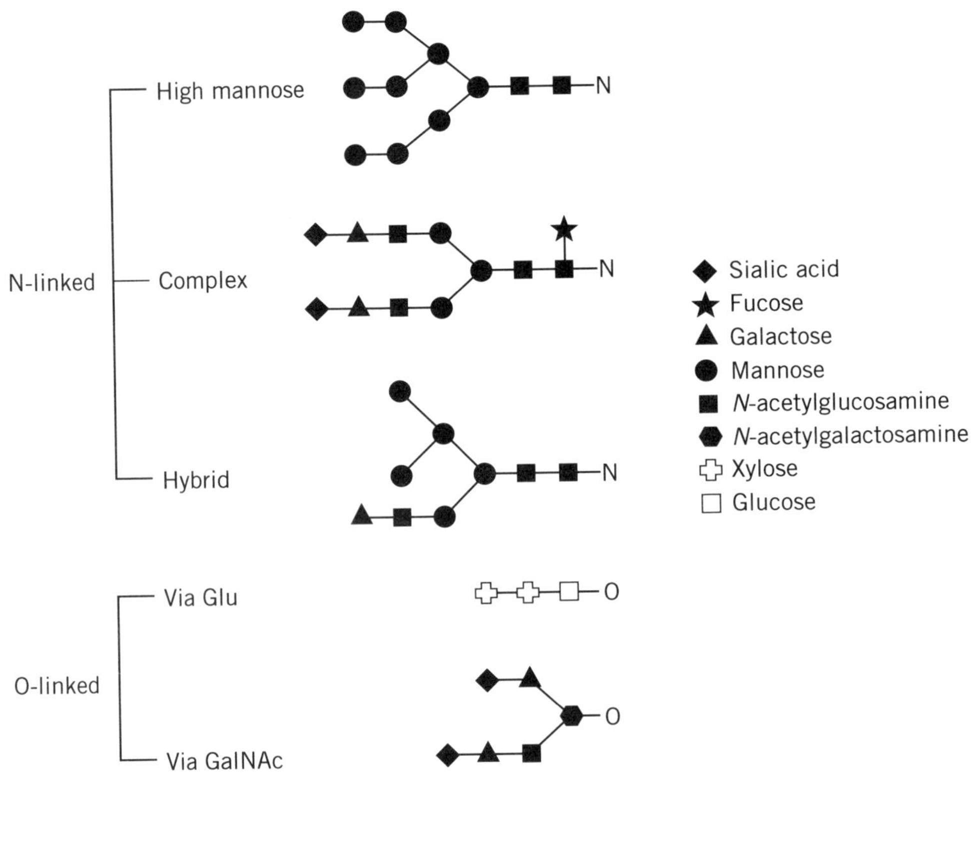

Figure 1. Common oligosaccharide structures found on glycoproteins.

3. Hybrid structures contain at least complex sugar-containing antennae, with other α1-3 branches containing only mannose residues.

N-glycosylation occurs in both the ER and Golgi apparatus and is controlled by a large number of glycosyltransferases and glycosidases. Glycosyltransferases are grouped into families, based on the type of sugar they transfer to the protein (e.g., galactosyltransferases and sialyltransferases) (5). Theoretically, if all enzymes in the pathway are fully active, the required donor sugar nucleotides are present, and no structural hindrances are present, all glycans will be fully processed into complex structures terminating in sialic acids. But in practice, this is rarely observed and a mixture of partially and fully processed glycans is more typical of N-linked glycoproteins.

Figure 2 shows a simplified pathway in the biosynthesis of *N*-glycans. A common core glycan ($Glc_3Man_9GlcNAc_2$) is built on a dolichol pyrophosphate lipid carrier, first in the cytoplasm and then in the ER. This oligosaccharide core structure is attached to the target asparagine of the nascent polypeptide chain as it enters the ER. The enzyme responsible for catalyzing this reaction, oligosaccharyl transferase, is closely associated with the ribosome (step 1). Within the ER lumen the three terminal glucose units are removed by α-glucosidases to give $Man_9GlcNAc_2$ (steps 2 and 3) that is further modified to $Man_8GlcNAc_2$ by α2-mannosidase (step 4). This trimmed structure is no longer recognized by the ER chaperone calnexin and is transported to the *cis*-Golgi via an intermediate compartment. Here it is further trimmed by additional mannosidases until a $Man_5GlcNAc_2$ structure remains (step 5), which is a substrate for GlcNAc transferase I to add GlcNAc to the 1-3 antenna (step 6). The presence of the GlcNAc on this arm of the glycan acts as a signal for possible further modifications. These include the removal of further Man residues (mannosidase II, step 7); addition of antennary GlcNAc (via GlcNAc transferase II, step 8), and bisecting GlcNAc residues (via GlcNAc-transferase III); creation of further branches (via GlcNAc-transferase IV and V); and core α6-fucosylation (step 9).

Therefore, the absence GlcNAc transferase I activity, or the subsequent removal of GlcNAc, as has been suggested for some insect cell lines (13), may result in simple oligomannose or hybrid structures being present on the mature glycoprotein. There appears to be no requirement for more complex glycan processing to precede secretion of glycoproteins from the cell or their appearance on the plasma membrane. Certain pathways are mutually exclusive, for example, the addition of a bisecting GlcNAc residue via GlcNAc-transferase III can prevent further branching of the antennae by GlcNAc-transferase V. The final stages of *N*-glycan processing take place in the *trans*-Golgi and involve the addition of galactose and sialic acids (steps 10 and 11), and α1-6- or α1-3-linked fucose residues. It is the presence and activity of these Golgi enzymes that gives rise to differences in glycan structures (microheterogeneity) produced by alternative cell hosts; these factors vary with culture conditions. Certain cell types posses unusual glycosyltransferases that facilitate the addition of cell-specific glycan structures. For example, up to 20% of IgG molecules produced by human B-lymphocytes possess a bisecting GlcNAc residue β1-4-linked to the central β-linked mannose of the core glycan, which can play a role in antibody-dependent cell-mediated cytotoxicity (ADCC). The GlcNAc transferase III enzyme required for this modification is not active in all cells, indeed only certain rodent cell lines such as the rat Y0 myeloma are able to produce recombinant antibodies containing this bisecting residue (14). Another example is sulfation of antennary GalNAc residues in certain pituitary glycoprotein hormones such as luteinizing hormone (LH), thyrotrophin, and proopiomelanocortin (15). The specific GalNAc transferase and a terminal GalNAc sulfotransferase recognize protein motifs in the nascent peptide (e.g., Pro-Leu-Arg) and are mainly restricted to the anterior pituitary gland. Therefore, only cell lines derived from the pituitary gland or endothelium (such as At20 and 293 cells) are able to perform this sulfation (16).

O-Linked Glycosylation

O-linked glycosylation is a simpler process involving a smaller number of sugar residues (typically 1–6) and occurs exclusively in the Golgi apparatus, where a core GalNAc residue attaches via an O-glycosidic bond to the R-group of serine or threonine (Fig. 1). There is no obvious and invariant peptide sequon surrounding the serine or threonine attachment site, although a computer algorithm has been devised to predict the likely occupation of potential sites, based on sequences in other O-linked glycoproteins (17). Large structural proteins such as mucins have multiple O-linked oligosaccharides, which may dominate the mass of the mature glycoprotein. Also, regions that contain high proportions of serine, threonine, and proline have been shown to be extensively glycosylated. For example, immunoglobulin class A_1 contains a proline-rich hinge region between the Fab and Fc regions. Of the 21 amino acids within this hinge, there are 10 Pro, 5 Ser, and 4 Thr; and the 5 Ser are all O-glycosylated (18). Several core O-linked oligosaccharides have been described attached to serine or threonine, which can be elongated by addition of further Gal, GlcNAc, Fuc, and sialic acid residues.

Glycophosphatidylinositol Residues

Carbohydrates are also components of the glycophosphatidylinositol (GPI) anchor used to secure certain proteins to cell membranes (19). All have the common core structure ethanolamine-PO_4-Man-Man-Man-$GlcNH_2$-myoinositol-PO_4-lipid. Ethanolamine forms a bridge to the membrane protein, and the inositol phospholipid is inserted onto the lipid bilayer of the cell's plasma membrane. There are certain protein-specific and species-specific modifications to this basic structure (for further information see Ref. 19).

INFLUENCES ON THE GLYCOSYLATION PROCESS

Ultimately, the process of glycosylation is controlled by the activity of glycosyltransferases and glycosidases, the ac-

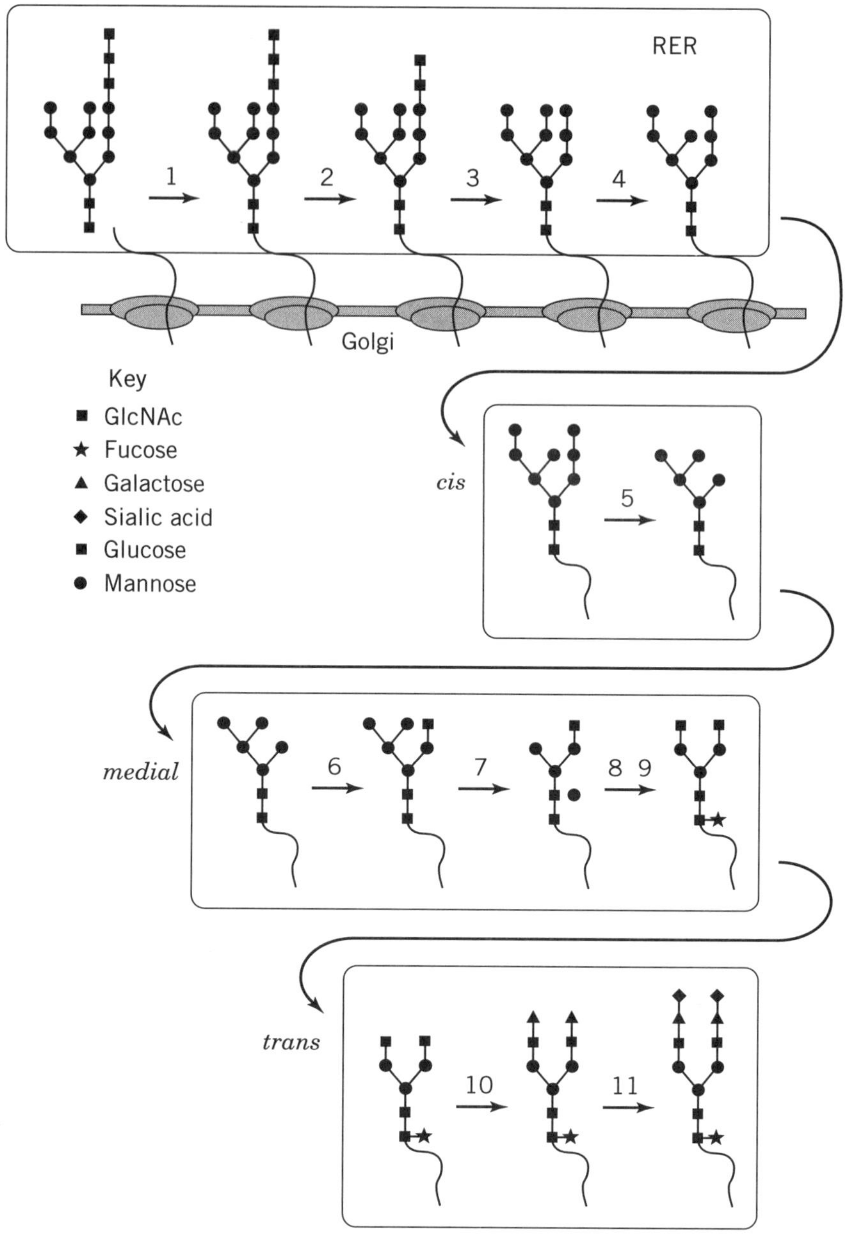

Figure 2. Steps in the N-glycosylation pathway. RER, rough endoplasmic reticulum.

cessibility of their glycan substrates on the glycoprotein, and the availability of cofactors and nucleotide sugar donors (e.g., CMP-NeuAc, GDP-Man, and UDP-GlcNAc). At the cellular level, these control points are influenced by the secondary and tertiary structure of the protein in the ER and Golgi, and the host cell type chosen for the recombinant protein expression and its physiological status.

As more recombinant glycoproteins are used therapeutically and for vaccines, the importance of correct glycosylation has become evident. The availability of alternative expression systems for producing recombinant glycoproteins, together with advances in analytical methods for glycans, has enabled comparisons to be made between glycoproteins from different sources. Crucially, the authenticity of recombinant glycoproteins compared to their natural human counterparts can be assessed, and choices made, concerning the most appropriate expression system.

Host Cell Line

Microbial cells are often the first choice for recombinant protein expression because of their high yields and simple media components. However, common bacterial expression systems such as *Escherichia coli* are incapable of expressing large proteins with authentic mammalian-type glycosylation. Hypermannosylation (the addition of a large number of mannose residues to the core oligosaccharide) is a common property of most yeast strains and can compromise the efficacy of recombinant protein vaccines (20). There is also some evidence to suggest that different O-glycosylation sites are used by yeast and mammalian cells (21).

The few studies reporting the production of therapeutic proteins in plants have suggested smaller *N*-glycan structures are present compared with mammals and lack outer residues such as sialic acids, which may compromise activity. Another obstacle may be the presence of potentially allergenic residues such as core α1-3-linked fucose and xylose (22) and substantial activity of endoglycosidases causing glycan cleavage.

The baculovirus-infected insect cell expression system has been used to express complex proteins and its short production time gives it an advantage over stable mammalian expression systems. Several cell lines have been developed using this technique, including *Spodoptera frugiperda*–derived cell lines Sf9 and Sf21 and *Estigmena acrea*–derived cell line Ea4. Some studies have shown that this system can attach complex oligosaccharides (23) using Sf9 cells, although this result is controversial because other researchers have shown that only simple oligomannose structures can be observed with Sf9. Other insect cell lines differ in their glycosylation capacity; for example, the Ea4 and *Trichoplusia ni* (TN-368) cell lines have been shown to add some complex antennary (GlcNAc and Gal) residues to glycoproteins (24,25).

Although most mammalian cell lines studied possess the intracellular machinery to perform complex N- and O-linked glycosylation to produce proteins that are not dissimilar to those of humans, there are some notable exceptions. Mouse and pig cells have a tendency to add NeuGc instead of NeuAc as the terminal sialic acid of glycoproteins. In contrast, NeuGc is absent from adult humans (due to inactivation of the CMP-NeuAc hydroxylase gene), and the NeuGc residue can be immunogenic (26). Rodents and pigs, like most mammals, express the enzyme α1-3-galactosyltransferase, which generates Galα1,3-Galβ1,4-GlcNAc residues on membranes and secreted glycoproteins. In humans, apes, and Old World monkeys the gene has become inactivated through frameshift mutations. For further details of specific glycosylation machinery in different cell types the reader is referred to another review (2).

Relatively few studies have been reported on the glycosylation of recombinant proteins expressed in the milk of transgenic animals. Those published from experiments in transgenic goats (27) and mice suggest that a lower level of sialylation and higher fucosylation may be achieved. It is now feasible to remodel glycoproteins in situ by the transgenic expression of extra glycosyltransferases in the mouse mammary gland (28). The use of human host cell lines poses a significant risk of retroviral transmission and is not generally favored by the regulatory authorities. Additionally, the transformation event required in most cases to produce a stable cell line may itself result in altered glycosylation profiles (29).

Cell Culture Conditions

It is well documented that the cell culture environment can influence both the macroheterogeneity and microheterogeneity of oligosaccharides in recombinant glycoproteins by affecting both the glycan biosynthetic process and the secretion or release of degradative enzymes, such as exoglycosidases (30). The process of adaptation from serum-containing to serum-free medium may lead to more heterogeneous glycans production in recombinant BHK-21 cells (31) and may improve galactosylation in recombinant CHO cells (14).

The ambient glucose concentration affects the amount of glycosylation of monoclonal antibody produced by human hybridomas in batch culture (32), and of IFN-γ produced by CHO cells in continuous culture (33). However, both lipids and nucleotides are also required for glycan biosynthesis. Lipid supplements alone or in combination with lipoprotein carriers can improve the N-glycosylation site occupancy of IFN-γ (34), and provision of cytidine and uridine can also alter protein glycosylation capacity of hepatocytes by increasing the availability of nucleotide sugars (35). Substances that are commonly used to increase productivity, such as sodium butyrate, should be checked for their influence on glycosylation (36).

Over time, an increase in the percentage of oligomannose (predominantly $Man_5GlcNAc_2$) and truncated structures has been observed in batch and fed-batch cultures of NSO myelomas producing recombinant IgG (37) and in interferon-γ made by recombinant CHO cells (38). Perfusion or chemostat culture of BHK-21 and CHO cells leads to a more consistent glycosylation pattern over time, and perturbations of nutrient availability (glucose or amino acids) can lead to changes in macroheterogeneity (33). Mild hypoxia has minimal effects on the glycosylation of tPA produced by recombinant CHO cells (39) but influences the level of sialylation of recombinant FSH (40). Similarly, pH changes within the range 6.9 to 8.2 in the cell culture medium do not have a dramatic effect on the glycosylation profile of recombinant placental lactogen expressed in CHO cells, however there is some evidence for underglycosylation outside this range (41). Increases in the concentration of ammonium ion in the culture medium above 2 mM may compromise sialylation (42). Degradative glycosidase enzymes have been detected in CHO cell lysates and culture supernatants, and sialidase is most active at neutral pH (43). This sialidase has been purified from culture supernatant of CHO cells (44) and can degrade glycans from proteins such as recombinant gp120. Although CHO cells also produce an α-L-fucosidase, this enzyme is incapable of releasing core fucose from intact recombinant glycoproteins (45). Sialidase, β-galactosidase, β-hexosaminidase, and fucosidase can also be detected at low levels in supernatants from mouse 293, NSO, and hybridoma cells, but the sialidase activity is much lower than that found in CHO cells (46). For more information, see GLYCOSYLATION OF RECOMBINANT PROTEINS.

EFFECTS OF GLYCOSYLATION

The presence, extent, and type of glycosylation influence the functional and structural properties of glycoproteins. Therefore, data acquired on the structure–function relationships of glycans present on one recombinant glycoprotein cannot always be applied to another. A complete listing of the structure–function relationships of glycans found on glycoproteins is beyond the scope of this article (see Refs. 6 and 47); however, the main processes on which protein glycosylation has been shown to exert influence are the following:

1. Pharmacokinetics and clearance in vivo (particularly influenced by the extent of sialylation)
2. Immunogenicity, which is carbohydrate residue specific
3. Protein solubility and protease resistance (6,7)

REGULATORY CONSIDERATIONS

Because of the efficacy and safety considerations, both the Food and Drug Administration in the United States and the European Medicines Agency now demand more comprehensive carbohydrate analysis before licensing recombinant glycoproteins. It is important to understand that some degree of heterogeneity exists in most natural as well as recombinant glycoproteins. Therefore, current guidelines accept glycan heterogeneity, but it must be shown to be consistent within prescribed boundaries between lots of the finished product. The inherent glycan heterogeneity and its manipulation by genetic or enzymatic engineering poses interesting challenges to the patenting of glycoprotein drugs, and several patents have already been filed that claim to alter the glycosylation process for biotechnological purposes.

STRATEGIES FOR ANALYSING GLYCOPROTEINS

See also GLYCOSYLATION OF RECOMBINANT PROTEINS.
Much of the increased knowledge concerning the influences on the glycosylation pathways and the biological significance of glycan variations has derived from recent advances in complex carbohydrate analysis. Using the appropriate equipment and techniques, it is now possible to unequivocally recognize most glycan structures and deduce their relative proportions in a complex glycoprotein. Current efforts are focused on improving the sensitivity of such techniques and making them more user friendly.

Sample Preparation

The glycoprotein must first be isolated and purified, and it is essential to know if it is secreted or membrane bound. The two classes of glycoproteins have different physicochemical properties; hence, the isolation protocol has to be adjusted for each type, with detergents usually required for membrane-bound proteins. Common precipitation and chromatographic techniques used for protein purification can generally be applied to glycoproteins. Chromatographic techniques tend to be based on charge, size, or affinity, and it is worth noting that of all the common sugar residues found in glycoproteins only the sialic acids are charged (negative). Also, protein A from *Staphylococcus aureus* Cowan I strain or proteins G and C from *Streptococcus*, which are commonly used to purify IgG, are both glycosylation independent. The purity of the glycoprotein can be assessed using SDS-polyacrylamide gel electrophoresis (expect multiple bands for complex glycoproteins and/or oligomers). Its yield can be quantified using protein stains (Bradford or Lowry method), A_{280}, HPLC, or immunoassays. We recommend performing at least two different assays to estimate protein yield because errors at this stage may have disastrous consequences for further analysis.

Preliminary Analysis

The simple SDS-PAGE gel analysis can be extended to yield data on the extent of protein glycosylation. For example, a portion of the sample can be digested with PNGase F (Oxford Glycosciences, U.K.) (to remove any *N*-glycans) and a combination of sialidase and *O*-glycanase (to remove *O*-glycans), and its mass shift compared on gels to that of undigested glycoprotein. Alternatively, kits are available, such as the FACE system from Dextra Laboratories (Reading, U.K.), which use a combination of isoelectric focusing gels and carbohydrate stains. Bear in mind that there are many other causes of protein heterogeneity than glycosylation: modifications such as proteolysis are common and will result in multiple bands. These preliminary tests are useful in estimating the degree and type of glycosylation and will help determine the final glycosylation analysis strategy.

Lectin blots may also be used to detect certain sugar residues or linkages (48). Lectins are carbohydrate-binding proteins that have been used successfully to identify and even isolate some glycoproteins. Lectin blots are a convenient way of identifying glycoproteins run on SDS-PAGE gels and then transferred to nitrocellulose sheets by techniques analogous to Western blotting. The blots are washed, blocked, and then incubated with the desired lectin that is already conjugated to a convenient enzyme (e.g., alkaline phosphatase or peroxidase) or light-generating compound. Control experiments using an appropriate competing saccharide should be run to verify that specific binding is being observed. Even using these controls, there are incidents of false identification, and we are reluctant to rely on these methods entirely. Similarly, colorometric assays are available to approximate the percentage of total sugars in each sample, but these assays are not very sensitive (5–10 nmol).

Glycopeptide Preparation

Glycopeptide preparation is often a prerequisite to establishing the location and structures of individual glycans within a complex glycoprotein. Controlled proteolytic cleavage may be carried out with enzymes such as trypsin and chymotrypsin (49). The exact protease to be chosen for depends on the size of the glycopeptides and its peptide substrate specificity; chemical cleavage, such as with cyanogen bromide, may also be employed. Care should be taken to prevent incomplete or nonspecific digestion because this can compromise the subsequent resolution of individual glycopeptides by reverse-phase HPLC (see later). Before enzyme or chemical cleavage it is desirable to modify any disulfide bonds present in the glycoprotein (50). This can be accomplished by reduction and carboxymethylation in the presence of a denaturant (8 M urea).

Releasing Glycans from Glycoproteins

There are a plethora of chemical and enzymatic methods that have been used to release carbohydrates from glycoproteins, with varying degrees of success (Table 1). The enzymatic methods (such as peptide-*N*-glycosidase F [PNGase F]) (see Table 2) used to cleave the glycosidic linkage between GlcNAc and the asparagine of N-linked glycans) generally preserve intact the peptides and glycans, but sometimes there are concerns over their efficiency (51). Most chemical cleaving methods such as hydrazinolysis (a nonselective method for releasing glycans using anhydrous hydrazine to cleave both N- and O-glycosidic bonds [Fig. 3]), tend to destroy the peptide at the glycosylation site(s). There is also a risk of further breakdown to monosaccharides and their degradation products; therefore the reaction must be carefully controlled (52). A clean-up step is often required before progression to further analysis. Following these preliminary steps (not all may be required), the analysis should progress to a combination of one or more of the techniques listed next. The strategy routinely used by us is described in Figure 4.

Gas–Liquid Chromatography

Gas–liquid chromatography GLC methods are highly efficient for the analysis of carbohydrates released from glycoproteins, especially volatile monosaccharide derivatives. A mass spectrometer can be connected to the GLC and used to characterize these resolved peaks, in comparison with known standards. Identification of each peak is achieved by the combination of its mass and the relative retention times (53). However, these GLC methods have largely been superseded for more complex oligosaccharides by HPLC alternatives, due to increased sensitivity, which enables smaller sample sizes to be analyzed.

High-Performance Liquid Chromatography

Analysis of glycopeptides or carbohydrate chains released from glycoproteins often requires high-resolution methods such as HPLC. The most common mode of HPLC for glycopeptide purification is reverse-phase HPLC (RP-HPLC), which separates on the basis of peptide, glycopeptide, or glycan hydrophobicity. Other modes employed are normal-phase partition and ligand exchange (49). RP-HPLC requires the presence of hydrophobic groups such as N-Ac-, deoxy-, -OMe, or OAc groups for longer retention time and better separation. Oligomannose- or high-mannose-type oligosaccharides (which have only two N-Ac groups) are retained on the RP-HPLC column and are not well resolved compared with complex-type oligosaccharides. Glycopeptides, released sugars (chemically or enzymically), or derivatized glycans can be resolved by HPLC and can be analyzed on a mass spectrometer. A common use for this technique is the separation of glycopeptides corresponding to individual glycosylation sites within a glycoprotein, following controlled proteolysis (50). The resultant glycopeptides are then analyzed by mass spectrometry and exoglycosidase arrays.

Table 1. Glycosidase Enzymes Used for Selective Cleavage of Oligosaccharides for Structural Analysis

Methods	Comments
Mild acid hydrolysis	Quantitative release of terminal nonreducing fucose
Strong acid hydrolysis	Quantitative release of hexose, deoxyribose, pentoses, and hexosamines
Methanolysis	Quantitative release of neutral monosaccharides and hexosamines
Hydrazinolysis	Nonselective release of glycans by cleaving both N-linked and O-linked bonds; can be optimized for specific N-linked or O-linked cleaving
Smith degradation	Quantitative release of monosaccharides
Acetolysis	Cleaves 1-6-linkages in oligosaccharides

High-Performance Anion Exchange Chromatography with Pulsed Amperometric Detection

Another procedure for the direct analysis of carbohydrates uses the high-performance anion-exchange mode (utilizing pellicular resin columns) of HPLC with pulsed amperometric detection (HPAE-PAD, pioneered by Dionex [54]). This is suitable for the analysis of released N- and O-linked glycans in quantities >0.2 nmol (55). Oligomannose-type glycans are eluted at low alkali concentrations, and an increase in the content of neutral sugar residues results in higher retention times. The negatively charged glycans that contain sialic acids are retained most. The method can also be used for analyzing released monosaccharides, such as resolving different forms of sialic acids, with a detection limit of around 20 pmol.

Capillary Electrophoresis

Capillary electrophoresis (CE) is an emerging technology that may be used for direct analysis of oligosaccharide derivatives (56). The main advantages of CE are its simplicity and high resolving power. There are a number of modes of operation that have been developed for CE, for example, free-zone capillary electrophoresis (CZE), micellar electrokinetic capillary chromatography (MECC) (57), affinity electrophoresis, gel electrophoresis, isoelectric focusing (IEF), and isotacophoresis. The method can be used for resolving released glycans, glycopeptides, or glycolipids (58), and the sensitivity of CE can be improved by employing a laser-induced fluorescence detector (59).

Mass Spectrometry

Mass spectrometry has proved to be an invaluable tool for glycosylation analysis. There are a number formats that can be used, such as electrospray ionisation mass spectrometry (ESI-MS) (60), liquid chromatography mass spectrometry (LC-MS) (61), and matrix-assisted laser desorption/ionization time-of-flight mass spectrometry (MALDI-TOF-MS) (49,50) (Fig. 4). In MALDI-TOF, which is more tolerant to common salts, a variety of matrices have been developed to maximize ionization of glycopeptides or released glycans (62).

Table 2. Glycosidase Enzymes Used for Selective Cleavage of Oligosaccharides for Structural Analysis

Enzyme	Source	Specificity	pH
PNGase F	*Flavobacterium Meningosepticum*	N-Glycosidic linkage between GlcNAc and asparagine	7.0
Endoglycosidase H	*Streptomyces plicatus*	Oligomannose and hybrid-type *N*-glycans within the chitobiose core	5.0
α-l-Fucosidase	*Aspergillus niger*	Fucose α1-2-galactose	3.8
	Bacillus fulminans	Fucose α1-2-galactose	6.6
	Clostridium perfringens	Fucose α1-2-galactose	6.0
	Corynebacterium	Fucose α1-2-galactose	8.5
	Almond emulsin	Fucose α1-3 and 4-GlcNAc	5.5
	Bovine epididimis	Nonspecific	6.5
	Charonia lampas	Nonspecific	3.3
	Turbo cornutus	Nonspecific	4.0
α-d-Galactosidase	*Aspergillus niger*	Galα1-3Gal	4.5
	Coffee bean	Nonspecific	6.0
	Fig	Nonspecific	6.0
β-d-Galactosidase	Jack bean	Galβ1-6GlcNAc > Galβ1-4GlcNac > Galβ1-3GlcNAc	3.5
	Streptococcal 6646K	Galβ1-4GlcNAc > Galβ1-3GlcNAc	5.5
	Aspergillus niger	Galβ1-4GlcNAc > Galβ1-6GlcNac > Galβ1-3GlcNAc	4.0
α-d-Mannosidase	Jack bean	Manα1-2 > Manα1-6 > Manα1-3	4.0
	Aspergillus niger	Manα1-2 and Manα1-4 > Manα1-6 > Manα1-3	4.5
	Aspergillus saoti	Manα1-2	4.5
			5.0
β-d-Mannosidase	*Helix pomitia*	ManGlcNAc$_2$Asn	4.5
β-*N*-Acetylgalactosaminidase	*Aspergillus niger*	GalNAc (terminal)-Ser	3.5
β-*N*-Acetylhexosaminidase	Jack bean	GlcNAcβ1-2,3,4,6	6.0
	Aspergillus niger	GlcNAβ1-3,4	5.0

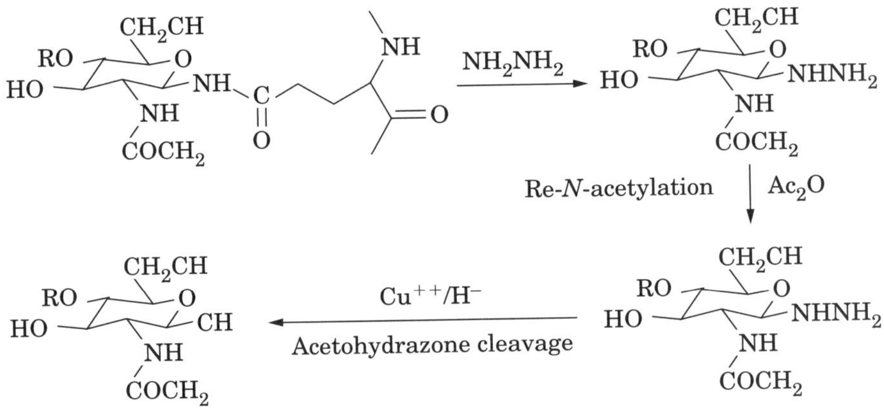

Figure 3. Hydrazinolysis reaction used to release glycans from glycoproteins.

Exoglycosidase Arrays

One of the most powerful methods for the elucidation of glycan structure uses an array of exoglycosidases (Table 2). These enzymes recognize predominantly the terminal nonreducing monosaccharide residues of glycans, glycopeptides, or whole glycoproteins and can be used to determine the anomeric configuration and glycosidic linkage of this terminal. Time can be saved by constructing a parallel array of these enzymes in combination (63), and the resultant mixtures can be analyzed by mass spectrometry (30,49) or by HPAE (64). Here, differences in mass or resolution time corresponding to the loss of specific sugar residues are determined.

Sialic Acid Analysis

Because of the importance of sialic acid in influencing half-life and protein stability, sialic acid analysis acquires a special significance, particularly for biopharmaceuticals destined for human therapy. Sialic acid is a collective term used to describe several derivatives of neuraminic acid. The neuraminic acid occurring in glycoproteins is usually substituted with an *N*-acetyl or *N*-glycolyl group in a species-specific fashion (2). Sialic acids can also be derivatized with *O*-acetyl, *O*-lactyl, or *O*-methyl groups, but these are rarer in natural glycoproteins. Each type of sialic acid has a characteristic mass/charge ratio that is recognizable by mass spectrometry; however, for detailed quan-

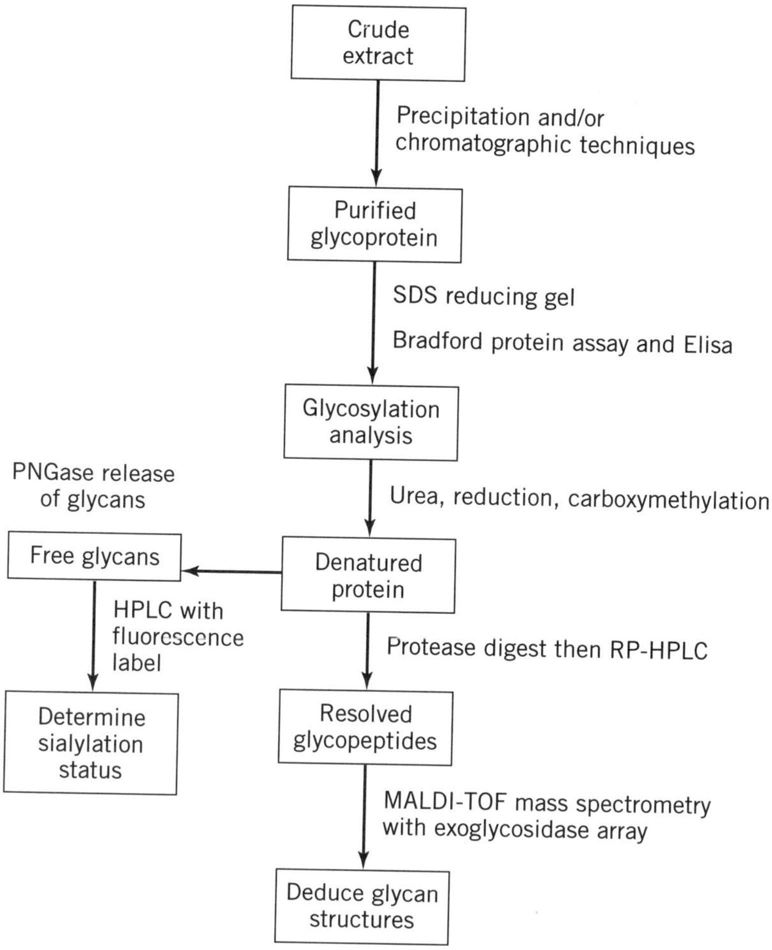

Figure 4. A strategy for glycan analysis at each glycosylation site.

tification it is often necessary to release the sialic acids from the glycoprotein by acid hydrolysis or by linkage-specific neuraminidases (65). Released sialic acids can then be analyzed by HPLC or HPAE (see earlier).

CONCLUSIONS

Glycobiology has advanced dramatically as people become aware of the biological implications of carbohydrate heterogeneity. Over the past five years, there has been a move away from simple monosaccharide composition analysis or gel estimates of glycan heterogeneity toward more sophisticated forms of analyses using techniques such as CE, mass spectroscopy, fluorescence HPLC, and HPAE. Regulatory authorities such as the U.S. Food and Drug Administration are demanding increasingly sophisticated carbohydrate analyses as part of the product or process validation (66). In addition to these analytical trends, there is increasing awareness of the relevance of glycosylation in producing a product with a good therapeutic profile and a high degree of batch consistency.

ACKNOWLEDGMENTS

The authors have been supported by the Biotechnology and Biological Research Council of the United Kingdom, and the European Union Biotechnology Framework Programme (DGXII).

BIBLIOGRAPHY

1. C. Abeijon and C.B. Hirschberg, *Trends Biochem. Sci.* **17**, 32–36 (1992).
2. N. Jenkins, R.B. Parekh, and D.C. James, *Nat. Biotechnol.* **14**, 975–981 (1996).
3. C.G. Gahmberg and M. Tolvanen, *Trends Biochem. Sci.* **21**, 308–311 (1996).
4. B.K. Hayes and G.W. Hart, *Curr. Opin. Struct. Biol.* **4**, 692–696 (1994).
5. M.C. Field and L.J. Wainwright, *Glycobiology* **5**, 463–472 (1995).
6. A. Varki, *Glycobiology* **3**, 97–130 (1993).
7. N. Jenkins and E.M. Curling, *Enzyme Microb. Technol.* **16**, 354–364 (1994).
8. R.J. Harris, V.T. Ling, and M.W. Spellman, *J. Biol. Chem.* **267**, 5102–5107 (1992).
9. S.B. Yan, Y.B. Chao, and H. Vanhalbeek, *Glycobiology* **3**, 597–608 (1993).
10. S. Allen and N.J. Bulleid, *Biochem. J.* **328**, 113–119 (1997).
11. R. Wilson, A.J. Allen, J. Oliver, J.L. Brookman, S. High, and N.J. Bulleid, *Biochem. J.* **307**, 679–687 (1995).
12. E.M. Curling, P.M. Hayter, A.J. Baines, A.T. Bull, K. Gull, P.G. Strange, and N. Jenkins, *Biochem. J.* **272**, 333–337 (1990).

13. T.A. Hsu, N. Takahashi, Y. Tsukamoto, K. Kato, I. Shimada, K. Masuda, E.M. Whiteley, J.Q. Fan, Y.C. Lee, and M.J. Betenbaugh, *J. Biol. Chem.* **272**, 9062–9070 (1997).
14. M.R. Lifely, C. Hale, S. Boyce, M.J. Keen, and J. Phillips, *Glycobiology* **5**, 813–822 (1995).
15. A.A. Bergwerff, J. Vanoostrum, J.P. Kamerling, and J.F.G. Vliegenthart, *Eur. J. Biochem.* **228**, 1009–1019 (1995).
16. P.L. Smith and J.U. Baenziger, *Proc. Natl. Acad. Sci. U.S.A.* **89**, 329–333 (1992).
17. E.F. Hounsell, M.J. Davies, and D.V. Renouf, *Glycoconjugate J.* **13**, 19–26 (1996).
18. A. Allen and J. Feehally, *Adv. Exp. Med. Biol.* **435**, 175–183 (1998).
19. J. Takeda and T. Kinoshita, *Trends Biochem. Sci.* **20**, 367–371 (1995).
20. L. Lehle, A. Eiden, K. Lehnert, A. Haselbeck, and E. Kopetzki, *FEBS Lett.* **370**, 41–45 (1995).
21. A.O. Herscovics and P. Orlean, *FASEB J.* **7**, 540–550 (1993).
22. G. Garcia-Casado, R. Sanchezmonge, M.J. Chrispeels, A. Armentia, G. Salcedo, and L. Gomez, *Glycobiology* **6**, 471–477 (1996).
23. F.J. Castellino and D.J.C. Davidson, *Am. Chem. Soc. Symp. Ser.* **202**, 20–20 (1991).
24. O.W. Ogonah, R.B. Freedman, N. Jenkins, K. Patel, and B.C. Rooney, *Biotechnology* **14**, 197–202 (1996).
25. T.R. Davis and H.A. Wood, *In Vitro Cell. Dev. Biol.* **31**, 659–663 (1995).
26. A. Noguchi, C.J. Mukuria, E. Suzuki, and M. Naiki, *J. Biochem.* **117**, 59–62 (1995).
27. E.S. Cole, E. Higgins, R. Bernasconi, L. Garone, and T. Edmunds, *J. Cell Biochem.* **265**, S18D (1994).
28. P.A. Prieto, P. Mukerji, B. Kelder, R. Erney, D. Gonzalez, J.S. Yun, D.F. Smith, K.W. Moremen, C. Nardelli, M. Pierce, Y.S. Li, X. Chen, T.E. Wagner, R.D. Cummings, and J.J. Kopchick, *J. Biol. Chem.* **270**, 29515–29519 (1995).
29. K. Yamashita, N. Koide, T. Endo, Y. Iwaki, and A. Kobata, *J. Biol. Chem.* **264**, 2415–2423 (1989).
30. D.C. James, R.B. Freedman, M. Hoare, O.W. Ogonah, B.C. Rooney, O.A. Larionov, V.N. Dobrovolsky, O.V. Lagutin, and N. Jenkins, *Biotechnology* **13**, 592–596 (1995).
31. M. Gawlitzek, U. Valley, M. Nimtz, R. Wagner, and H.S. Conradt, *J. Biotechnol.* **42**, 117–131 (1995).
32. H. Tachibana, K. Taniguchi, Y. Ushio, K. Teruya, K. Osada, and H. Murakami, *Cytotechnology* **16**, 151–157 (1994).
33. P.M. Hayter, E.M. Curling, A.J. Baines, N. Jenkins, I. Salmon, P.G. Strange, J.M. Tong, and A.T. Bull, *Biotechnol. Bioeng.* **39**, 327–335 (1992).
34. N. Jenkins, P.M.L. Castro, S. Menon, A.P. Ison, and A.T. Bull, *Cytotechnology* **15**, 209–215 (1994).
35. W.R.P. Rijcken, B. Overdijk, D.H. Vandeneijnden, and W. Ferwerda, *Biochem. J.* **305**, 865–870 (1995).
36. C.A. Gebert and P.P. Gray, *Cytotechnology* **17**, 13–19 (1995).
37. C.C. Yu-Ip, W.J. Miller, M. Silberklang, G.E. Mark, R.W. Ellis, L.H. Huang, J. Glushka, H. van Halbeek, J. Zhu, and J.A. Alhadeff, *Arch. Biochem. Biophys.* **308**, 387–399 (1994).
38. A.D. Hooker, M.H. Goldman, N.H. Markham, D.C. James, A.P. Ison, A.T. Bull, P.G. Strange, I. Salmon, A.J. Baines, and N. Jenkins, *Biotechnol. Bioeng.* **48**, 639–648 (1995).
39. A.A. Lin, R. Kimura, and W.M. Miller, *Biotechnol. Bioeng.* **42**, 339–350 (1993).
40. W. Chotigeat, Y. Watanapokasin, S. Mahler, and P.P. Gray, *Cytotechnology* **15**, 217–221 (1994).
41. M.C. Borys, D.J.H. Linzer, and E.T. Papoutsakis, *Bio/Technology* **11**, 720–724 (1993).
42. D.C. Andersen, C.F. Goochee, G. Cooper, and M. Weitzhandler, *Glycobiology* **4**, 459–467 (1994).
43. M.J. Gramer, C.F. Goochee, V. Chock, D.T. Brousseau, and M.B. Sliwkowski, *Bio/Technology* **13**, 692–698 (1995).
44. T.G. Warner, J. Chang, J. Ferrari, R. Harris, T. Mcnerney, G. Bennett, J. Burnier, and M.B. Sliwkowski, *Glycobiology* **3**, 455–463 (1993).
45. M.J. Gramer, D.V. Schaffer, M.B. Sliwkowski, and C.F. Goochee, *Glycobiology* **4**, 611–616 (1994).
46. M.J. Gramer and C.F. Goochee, *Biotechnol. Bioeng.* **43**, 423–428 (1994).
47. A. Varki, *FASEB J.* **11**, 248–255 (1997).
48. R.D. Cummings, *Methods Enzymol.* **230**, 66–86 (1994).
49. C.W. Sutton, J.A. O'Neill, and J.S. Cottrell, *Anal. Biochem.* **218**, 34–46 (1994).
50. D.C. James, M.H. Goldman, M. Hoare, N. Jenkins, R.W.A. Oliver, B.N. Green, and R.B. Freedman, *Protein Sci.* **5**, 331–340 (1996).
51. K. Yamamoto, *J. Biochem.* **116**, 229–235 (1994).
52. G.D. Roberts, W.P. Johnson, S. Burman, K.R. Anumula, and S.A. Carr, *Anal. Chem.* **67**, 3613–3625 (1995).
53. J.H.G. Mutsaers, J.P. Kamerling, R. Devos, Y. Guisez, W. Friers, and J.F.G. Vliegenthart, *Eur. J. Biochem.* **156**, 651–654 (1986).
54. M. Weitzhandler, C. Pohl, J. Rohrer, L. Narayanan, R. Slingsby, and N. Avdalovic, *Anal. Biochem.* **241**, 128–134 (1996).
55. P. Hermentin, R. Witzel, R. Doenges, R. Bauer, H. Haupt, T. Patel, R.B. Parekh, and D. Brazel, *Anal. Biochem.* **206**, 419–429 (1992).
56. P.L. Weber, T. Kornfelt, N.K. Klausen, and S.M. Lunte, *Anal. Biochem.* **225**, 135–142 (1997).
57. D.C. James, R.B. Freedman, M. Hoare, and N. Jenkins, *Anal. Biochem.* **222**, 315–322 (1994).
58. R.S. Rush, H.J. Boss, V. Katta, and M.F. Rohde, *Electrophoresis* **18**, 751–756 (1997).
59. X.C. Le, Y. Zhang, N.J. Dovichi, C.A. Compston, M.M. Palcic, R.J. Beever, and O. Hindsgaul, *J. Chromatogr. A* **781**, 515–522 (1997).
60. G.J. Rademaker, S.A. Pergantis, L. BlokTip, J.I. Langridge, A. Kleen, and J.E. Thomasoates, *Anal. Biochem.* **257**, 149–160 (1998).
61. P. Camilleri, D. Tolson, and H. Birrell, *Rapid Commun. Mass Spectrom.* **12**, 144–148 (1998).
62. D.I. Papac, A. Wong, and A.J.S. Jones, *Anal. Chem.* **68**, 3215–3223 (1996).
63. C.J. Edge, R.B. Parekh, T.W. Rademacher, M. Wormald, and R.A. Dwek, *Nature* **358**, 693–694 (1992).
64. F.H. Routier, M.J. Davies, K. Bergemann, and E.F. Hounsell, *Glycoconjugate J.* **14**, 201–207 (1997).
65. L. Monaco, A. Marc, A. EonDuval, G. Acerbis, G. Distefano, D. Lamotte, J.M. Engasser, M. Soria, and N. Jenkins, *Cytotechnology* **22**, 197–203 (1996).
66. D.T.Y. Liu, *Trends Biotechnol.* **10**, 114–120 (1992).

See also GLYCOSYLATION OF RECOMBINANT PROTEINS.

PROTEIN PURIFICATION, AQUEOUS LIQUID EXTRACTION

Maria-Regina Kula
Klaus Selber
Heinrich Heine University Düsseldorf
Jülich, Germany

KEY WORDS

Affinity extraction
Aqueous two-phase systems
Detergent-based systems
Large-scale
Protein extraction
Protein purification
Recycling
Reverse micelles
Thermodynamics
Thermoseparating polymers

OUTLINE

INTRODUCTION

For the large-scale production of proteins, complex separation procedures are necessary after protein synthesis by living cells is completed. In comparison to downstream processing in the chemical industry, the separation of proteins must be very gentle to conserve the native structure, which means that in general near-neutral pH values and moderate temperatures (≤30 °C) must be maintained. Important objectives in separation are high selectivity and yield because multistep procedures are usually required. Because of the polar character and high molecular mass of proteins, several unit operations common in the chemical industries cannot be applied, for example, distillation or extraction into organic solvents. Aqueous two-phase systems, however, can be used with advantage for protein isolation by liquid–liquid extraction. The technique is based on the incompatibility of two different hydrophilic polymers in the common solvent water or the reversible hydration of oxygen in polyethers.

In an aqueous liquid–liquid extraction process, cells and cell debris and other unwanted substances should remain in one phase while the desired protein is partitioned into the other phase, preferentially the top phase. The phases are separated and further processed until the desired level of product purity is reached. The extraction of proteins is performed preferentially at early stages of an isolation process using cell homogenates or whole fermentation broth because this allows the concomitant removal of solids, a product enrichment and concentration in a single step.

Partitioning of proteins by aqueous liquid–liquid extraction has been well described (1–6) and finds increasing application for large-scale processes in industry. For further reading the monographs by Albertsson (4) and especially by Walter et al. (5) are recommended. The latter gives detailed information about handling principles and applications. The Aqueous Two-Phase (a.k.a. Biphasic) Systems Community home page has a bibliography compiled by Harry Walter that contains more than 1,000 files of journal publications on aqueous two-phase systems published between 1956 and 1996: *http://radar.ch.ua.edu/~aq2phase/*

BIOCHEMICAL FUNDAMENTALS

Miscibility gaps exist in different aqueous systems; these are based on polymer–polymer interactions, polymer–salt interactions, or thermoseparation. A special case is affinity extraction, which can be viewed as a reactive extraction process. For further descriptions see "Alternative Two-Phase Systems".

Principles of Aqueous Two-Phase Extraction

Most of the hydrophilic, natural, and synthetic polymers are miscible with water and show phase separation in a mixture with another polymer or salt. The phase diagram of polyethylene glycol (PEG) and potassium phosphate in Figure 1 demonstrates the miscibility of polymer and salt phases. Above the curve (binodal) the system separates into two liquid phases. If PEG and salt are mixed in a percentage represented by a point M above the binodal, the mixture separates along the tie-line. The composition of the two resulting phases are represented by the cross-points of the tie-line and the binodal. The volume ratio of the PEG-rich phase to the salt-rich phase is given approximately by the ratio of the tie-line sections $\overline{MB}$ and $\overline{TM}$. The critical point is reached when the tie-line is represented by only one point on the binodal. At the critical point, the addition of an infinitesimally small amount of water results in two phases of equal volume and composition; therefore, any added component should have a partition coefficient K of 1, where K is defined as the ratio of concentration in the top ($c_{i,\mathrm{T}}$) and bottom phase ($c_{i,\mathrm{B}}$)

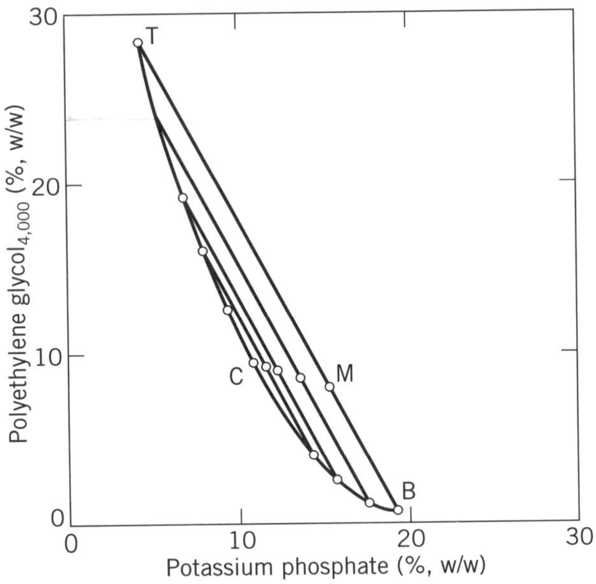

Figure 1. Phase diagram of PEG 4,000–potassium phosphate at 20 °C. *Source*: Ref. 4.

$$K = \frac{c_{i,T}}{c_{i,B}} \quad (1)$$

With increasing length of the tie-line the composition of the phases at equilibrium diverge and consequently K changes.

It must be noted that phase diagrams are altered by addition of other components in sufficient concentration such as cell homogenates and different salts (7,8). Nevertheless, the phase diagram is useful for the selection of phase systems. Many phase diagrams of polymer–polymer and polymer–salt systems can be found in Refs. 4 and 6. An extraction system is usually designed such that the wanted protein partitions predominantly to the top phase (PEG, in this case) and the undesired material including cells and cell debris partitions to the bottom phase (salt). The separation power of a system is characterized by the yield Y, the separation factor G, depending on the volume ratio R, and K. The yield in the top phase is defined as

$$Y_T = \frac{n_T}{n_T + n_B} \quad (2)$$

or consequently as

$$Y_T = \frac{1}{1 + \left[\frac{V_B}{V_T} \cdot \frac{1}{K}\right]} \quad (3)$$

where n is the molar quantity of the desired protein in the volume of the top (V_T) and bottom phase (V_B), respectively. G is defined as the product of the partition coefficient K and the volume ratio R. The definition of the volume ratio is given in equation 5.

$$G = K \cdot R \quad (4)$$

$$R = \frac{V_T}{V_B} \quad (5)$$

Y, K, and R affect each other. Usually Y_T increases if K is increased at a constant R or if R is increased at a constant K (Figure 2).

K does not depend on the concentration of the protein i, so long as the concentration c_i is much smaller than the concentration of the phase-forming polymers, and no complex formation takes place. According to Bronstedt (9), there exists an exponential relation between K and the surface area (molecular mass) of i:

$$\ln K_i = \frac{\alpha \cdot A_i}{k \cdot T} \quad (6)$$

α is a factor related to the chemical potential, A_i is the surface for interaction, k is the Boltzmann constant, and T is the absolute temperature. Thus, cells and cell debris should exhibit one-sided partition approaching zero or infinity even at low values of α because of their large value of A, while metabolites should have K values approximately 1 and proteins usually between 0.1 and 10. The relation between ln$K$ and A has been capitalized in a process to isolate hepatitis B surface antigen particles (molecular weight, ≃ 1 to 2 million Da) as a hepatitis B vaccine from recombinant yeast cultures (10). Equation 1 is valid for the equilibrium between two liquid phases only; however, sometimes a solid phase seems to be involved and rather large values of K may result as a consequence of preferential solubility of i under extraction conditions in the top phase of PEG–salt systems (11). In such a case, K will increase with c_i.

Factors Determining Partitioning

The forces that cause partitioning can be explained in a first approximation by splitting the partition coefficient into increments:

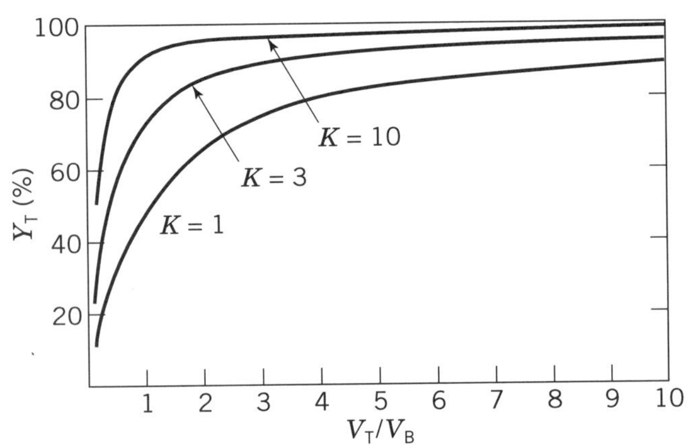

Figure 2. Interdependence of yield (Y), partition coefficient (K), and volume ratio (R). *Source*: Ref. 8.

$$\ln K = \ln K_{\text{ion}} + \ln K_{\text{hphob}} + \ln K_{\text{hphil}} + \ln K_{\text{conf}} + \ln K_{\text{lig}} \quad (7)$$

where K_{ij} denotes the partition coefficient from charges or hydrophobic or hydrophilic forces or its dependence on conformation or ligand interaction, respectively (12). Only $\ln K_{\text{lig}}$ is to some extent predictable. For this reason, affinity extraction is much studied, because the affinity interaction can be made to dominate ln *K*. Even in the absence of affinity interactions many parameters will influence the partitioning of proteins in aqueous two-phase systems, affecting the partition coefficient and the selectivity.

These factors are as follows:

- Selection of component(s) to form a two-phase system
- Concentration of polymers used
- Molecular weight of the polymers
- Type and concentration of ions
- pH
- Temperature
- Addition of other agents (e.g., chaotropic agents) (13,14)
- Concentration of cells and cell debris added

The strong influence on the partition of only small changes in the system composition is shown in Figure 3, where the partition coefficient for L-2-hydroxyisocaproate dehydrogenase is changed significantly by addition of potassium phosphate and the resulting change of the tie-line.

The molecular mass of a given polymer can strongly influence the partition of a protein. The partition coefficient can be increased by increasing the molecular mass of the

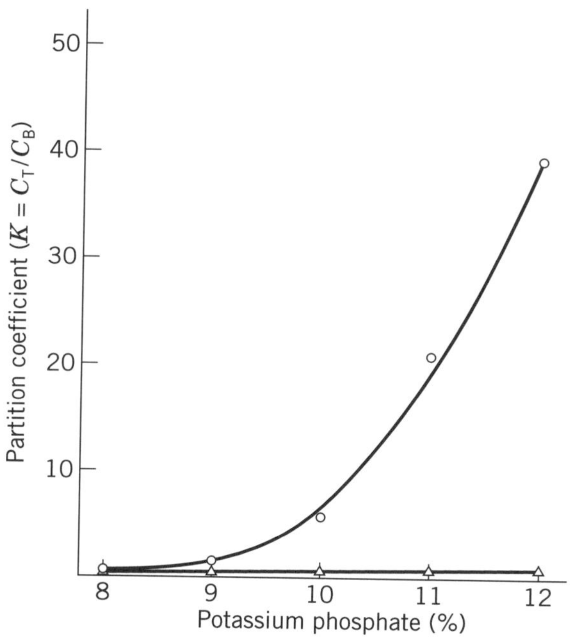

Figure 3. Partition of L-2-hydroxyisocaproate dehydrogenase (○) and D-lactate dehydrogenase (△) as a function of potassium phosphate concentration. *Source*: Ref. 15.

bottom phase forming polymer (for example, dextran 1) or by lowering the average molecular mass of the top-phase-forming polymer, most often PEG (1,4). The latter represents a fast and convenient way to increase protein yield in the top phase and is aided by the fact that the polymerization of ethyleneoxid can be easily adjusted to yield different molecular weight fractions of PEG. Several fractions are commercially available and are included in the pharmacopoeia of many countries. Changes in pH affect the ionization of side chains of the protein and therefore the charge and polar/hydrophobic balance of the surface; it will also alter the ionization of some weak and multivalent acids present in the extraction mixture leading to changes in *K*. pH values are easily manipulated in the range of protein stability. The addition of ions has a remarkable effect on *K* (1,4) by altering the length of the tie-line and the solvating power in PEG–salt systems as well as changes in the electrochemical potential across the interface in polymer–polymer systems.

The search for a separating system of a new protein is usually carried out by trial and error guided by the heuristic rules discussed previously. A series of phase systems differing in one parameter are produced in the scale of 5.0–10.0 g using a statistical design. A nice example of a statistical approach is shown by Hart et al. for the isolation of nonnative IGF-I (16). *Y*, *K*, and *R* are determined as well as protein concentration. From this information, a useful extraction system can be derived that can be further optimized with regard to specific goals, such as product yield and quality. Other criteria for optimization are biodegradability, ease of recycling, toxicity of phase-forming chemicals, physical properties of extraction systems, and last although not least the total costs (see "Economic Aspects").

For research purposes, the most investigated system is the PEG–dextran (1) combination. Both polymers are commercially available in different molecular weight fractions. For technical applications, fractionated dextran is too expensive; crude dextran may be used in low concentrations (17,18), but results in high-viscosity bottom phases. A less expensive possibility is PEG–salt systems, which are commonly used for large-scale applications (8). Other alternatives are reported (19,20). When substituting other polymers for dextran, for example, the limiting concentration for phase separation needs to be considered as well as product yield and quality. Combinations of the different systems are possible, like polymers and surfactants (21,22) or affinity partitioning combined with temperature-induced phase separation (23).

For large-scale operations, questions of waste disposal and the possibility of recycling auxiliary chemicals are of increasing interest for environmental as well as economic reasons. These topics are discussed separately under "Applications".

Thermodynamics

Phase separation in aqueous systems and partitioning of proteins are governed by thermodynamic forces. Because thermodynamics including mathematical models of aqueous two-phase systems cannot be explained in a few sentences, the reader is referred to the literature. For a broader introduction, the review by Cabezas (24) should be consulted.

Phase separation is based on the Gibbs free energy of mixing and will occur if ΔG^{mix} of two phases becomes energetically more favorable than ΔG^{mix} of a single phase:

$$\Delta G^{mix} = \Delta H^{mix} - T\Delta S^{mix} \quad (8)$$

where ΔH^{mix} and ΔS^{mix} are the enthalpy and entropy of mixing and T is the temperature. ΔG^{mix} is related to the excess enthalpy G^e by

$$\Delta G^{mix} = G^e + RT \sum_i x_i \ln x_i \quad (9)$$

where x_i is the mole fraction of the different phase components i.

There is a large diversity of polymer–polymer models available, such as Edmond and Ogston-based models using McMillan–Mayer or Hill virial expansions (25,26), Flory–Huggins–based lattice models using the classical polymer solution theory (27,28), the UNIQUAC lattice model (29), integral equation theory-based models (30), the VERS model based on group contributions of salt and polymer (31), and the excluded volume theory based on statistical geometric arguments (32). A comprehensive discussion can be found in a recent paper by Johansson et al. (33). The problem for most of these models is that they include theories derived for gases or crystalline solids. They have advantages and disadvantages, such as too many parameters, time-consuming experiments to obtain parameters, and derived parameters that are valid only for a certain range of conditions.

An interesting approach to solve binodales for PEG–dextran systems is shown by Baughman and Liu (34). They used about 100 different binodales and protein partitioning data from other investigators as raw data for an expert system based on a neural network, resulting in quantitative predictions for unknown phase diagrams. To predict protein partition away from the isoelectric point the results were not satisfactory, however. In polymer–salt systems most of the electrostatic interactions are screened by the high salt concentration of the phases, which makes modeling more complicated. Only a few models have been developed based on virial expansions (31), integral equation theory (30), and a potential of mean force model including fluctuation solution theory (35).

Until now a reliable prediction of protein partitioning in aqueous two-phase systems for a random protein has not been possible. The systems are too complex, depending on pH, salt, polymer, protein structure, temperature, specific interactions, and charge distribution on the protein surface. The aim to minimize experiments for certain systems and predict trends is advancing. The problem with the model experiments is that they rely upon commercially available cheap proteins, which have strong specific interactions, however, such as lysozyme or α-chymotrypsin in which dimerization or self-digestion occurs. These effects in aqueous two-phase systems have not been studied extensively. Work has been done to predict peptide partitioning (27,36–38). Predictions have improved over recent years for certain peptides and amino acids, but depending on the sequence, predictions do not work well yet in general.

The diverse models describe phase separation reasonably well but a state-of-the-art model is not available (24). The interaction between proteins and solutions needs more attention to understand protein partition and describe it in quantitative terms based on fundamental principles. This goal will be elusive for quite some time to come. Fortunately it is very easy to determine K values experimentally so long as an appropriate specific assay is available.

ALTERNATIVE TWO-PHASE SYSTEMS

In addition to the commonly used polymer–polymer and polymer–salt systems, there are some alternative, also well-investigated, aqueous two-phase systems.

Affinity Extraction

Ligands covalently bound to one of the phase-forming components are useful to direct the extraction of certain proteins because the ligand will predominantly partition with the polymer in one phase. There exists a linked equilibrium of partition and complex formation, as shown in Figure 4. The extraction efficiency depends on the number of binding sites of the protein and the dissociation constant of the complex. The latter may be different in the top and bottom phase (39,40).

Affinity ligands can be cofactors or inhibitors of the target protein. General ligands based on textile dyes are more often used than natural ligands. Besides textile dyes, metal chelators are attractive as general ligands. These are robust, cheap chemicals applicable in PEG–salt systems and offering useful selectivity. The syntheses of PEG derivatives has been described by Harris (41). Johansson et al. (42) investigated the partition of phosphofructokinase from yeast. The partition coefficient increased by a factor of 10^4 compared to physical extractions using the pseudo-affinity ligand cibacron-blue in an affinity process (Fig. 5).

The effectivity of affinity extraction was also demonstrated by Kroner et al. (43). With increase of the amount of PEG blue, the partitioning coefficient K increases. A yield of 98% could be obtained thereby with a single extraction step. Other parameters than the ligand concentration (43) are the polymer concentration (44), the kind of salt and its concentration (42), the pH value (45), the temperature (45), and the addition of free ligands (44,45). Useful models describing affinity extraction have been derived by Cordes et al. (40) and Suh and Arnold (46).

An example of screening for a suitable dye can be found in a patent of Guliano and Szlag (47) describing experiments with the triazin dyes Cibacron-Blue FGF, Procion Turquoise H-A, Procion Green HE-4BDA, and Procion Red HE-3B to separate alcohol dehydrogenase. Large-scale applications of affinity extraction are listed later. Reviews of polymer–ligand systems have been published by Johansson (48) and Johansson and Tjerneld (49). It can be expected that specific ligands originally selected from libraries of suitable synthetic chemicals or peptides for affinity

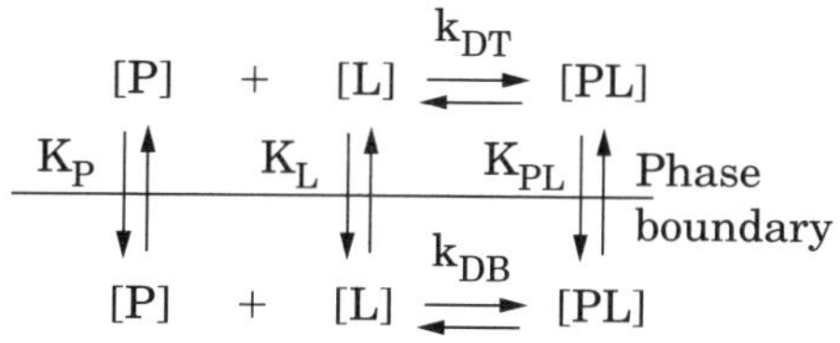

[P] Concentration of protein
[L] Concentration of ligand
[PL] Concentration of protein–ligand complex
K_P Partition coefficient of the protein
K_{PL} Partition coefficient of the ligand
K_L Partition coefficient of the complex
k_{DT} Equilibrium constant of the complex in the top phase
k_{DB} Equilibrium constant of the complex of the bottom phase

Figure 4. Equilibrium of partition and complex formation in affinity extraction processes. [P], concentration of protein; [L], concentration of ligand; [PL], concentration of protein–ligand complex; K_P, partition coefficient of the protein; K_{PL}, partition coefficient of the ligand; K_L, partition coefficient of the complex; k_{DT}, equilibrium constant of the complex in the top phase; and k_{DB}, equilibrium constant of the complex in the bottom phase.

chromatography will find applications also in affinity extraction.

Detergent-Based System

A recent development is the two-phase extraction in aqueous two-phase systems based on nonionic detergents carrying ethylenoxide groups as the hydrophilic moiety. The protein is partitioned using only one auxiliary chemical. The principle is shown in Figure 6. At low temperature, a homogeneous phase containing micelles of the detergent and fermentation broth exists. The detergent interacts preferably with hydrophobic proteins. As soon as the temperature is increased above a certain system-dependent temperature, turbidity develops as an indicator of phase separation. The so-called cloud point depends mainly on the average number of ethylene oxide groups; a narrow range is desirable. A C12EO5 detergent, for example, has a cloud point between 23 and 25 °C. The phase separation depends on the reversible, temperature-induced dehydration of the ether oxygens in the detergent and leads to a detergent-rich and a detergent-depleted phase in equilibrium. The micelles are believed to aggregate into laminar structures carrying hydrophobic proteins into the so-called coazervate phase, which may still contain 70–80% water (50).

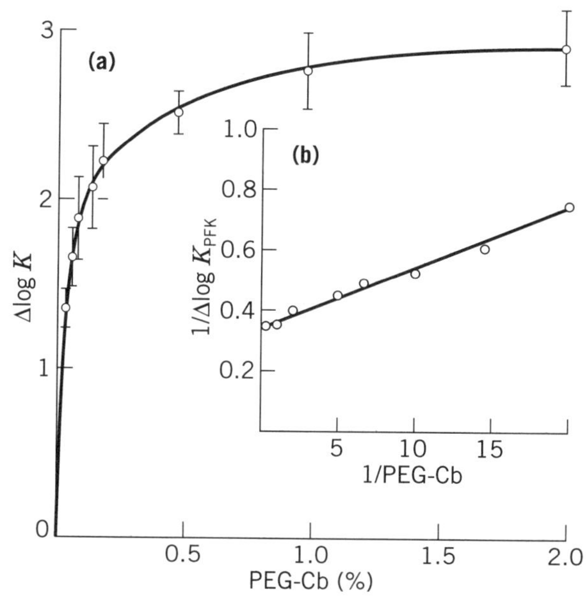

Figure 5. Changes of the partition coefficient of phosphofructokinase depending on the amount of added ligand. *Source*: Ref. 42.

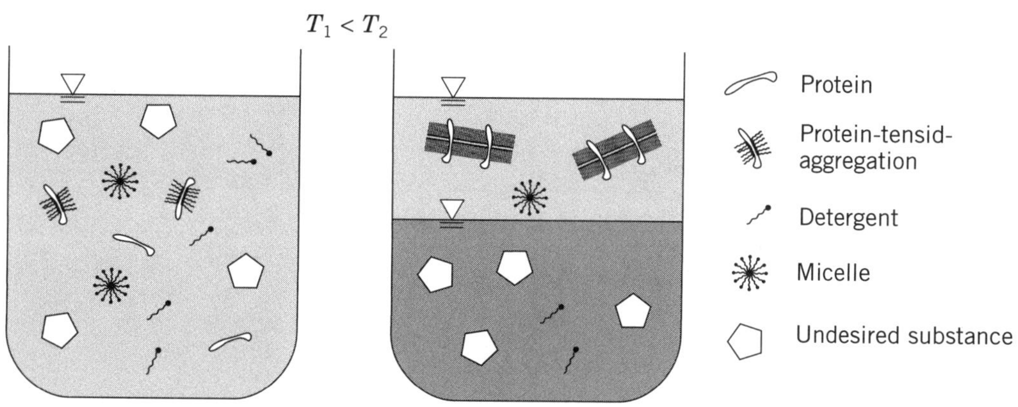

Figure 6. Schematic illustration of aqueous detergent solutions below and above the cloud point.

Bordier (51), Minuth et al. (52), and others investigated cloud-point extraction of proteins. Nonionic or zwitterionic detergents are most useful for the separation of membrane proteins (53,54). A complete process design for the purification of cholesterol oxidase is presented in Figure 7. A scale-up study indicated that the rather low density difference in detergent-based systems limits phase separation by centrifugation in the initial step (56).

It has been shown that the detergent for extraction of hydrophilic proteins and membrane-bound proteins should be chosen with respect to its hydrophile–lipophile balance (HLB) (57). Table 1 shows partitioning results using Tritons with different HLBs for the separation of an endopeptidase in the presence of sodium chloride. Yield, concentration factor, and partitioning coefficient increased significantly with an increasing HLB.

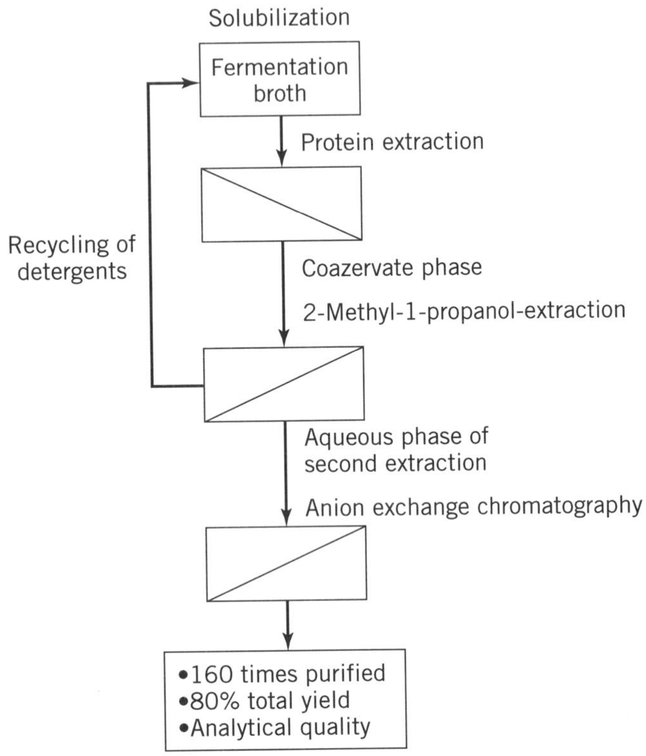

Figure 7. Optimal process design for the extraction of cholesterol oxidase from *Nocardia rhodochrous* using detergent-based systems in the initial step. (*Source*: Ref. 55.)

Table 1. Influence of the Surfactant Hydrophile–Lipophile Balance (HLB) on the Partition Coefficient, Concentration Factor, and Yield of the Purification Process for the Example of Endopeptidase

Surfactant	HLB	Partition coefficient	Concentration factor	Yield (%)
Triton X-100	13.5	0.81	0.7	15
Triton X-165	15.8	4.17	3.65	49
Triton X-305	17.3	8.93	3.1	69
Triton X-405	17.9	53	6.63	100

Source: Ref. 57.

Reversed Micelles

Because it is similar to aqueous liquid–liquid extraction, reversed micellar protein extraction is mentioned here briefly. Reversed micelles extract proteins into the water pool contained in the micelles present in an organic solvent depending on the isoelectric point of the protein and therefore pH and ionic strength in the process solution. The principle is shown in Figure 8. The organic solvent allows the organization of the reversed micelles and the separation of the bulk water from the water pool in the micelles.

Extraction into reversed micelles has been investigated for several industrially relevant proteins, but these are apparently not yet used in industry (58,59). The major remaining problem is the difficult reextraction of the bioactive protein into an aqueous phase and the reuse of reversed micelles.

Thermoseparating Polymers

Certain polyethylene–polypropylenoxide copolymers also show reversible dehydration of the oxygen groups in an interesting temperature range and have been investigated by Alred et al. (60), Johansson et al. (61), and recently by Persson et al. (62) for application in aqueous two-phase systems. Usually a polymer–polymer system is used in the first step selected to partition the protein in the copolymer-rich phase. After primary phase separation, the solution is heated above the cloud point, which results in phase separation under conditions in which the protein is almost exclusively found in the polymer-depleted phase.

APPLICATIONS

Technical Aspects

Aqueous liquid–liquid extraction uses the same principles as the well-known chemical process of liquid–liquid ex-

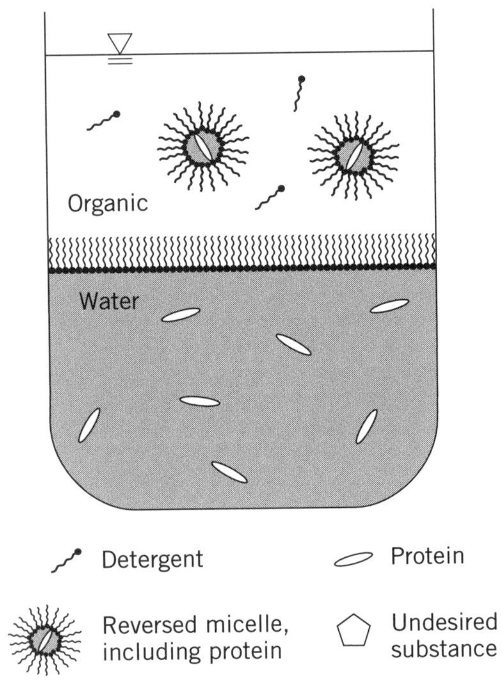

Figure 8. Operating principle of extraction into reversed micelles.

traction. Therefore similar, commercially available equipment may be used for mixing and phase separation. The physical properties of aqueous two-phase systems, however, are very different from those of solvent–water systems. The most notable difference is the dramatically lower interfacial tension (Table 2).

In general, for the purification of an intracellular protein the cells are disrupted, the phase-forming components added, and the pH adjusted; the suspension is then mixed to reach equilibrium. In large-scale processes this is done in a stirred vessel with baffles or using static mixers (63). An extremely low energy input is needed because of the low interfacial tension of the whole system. The approach to equilibrium is surprisingly fast ($\leq$30 s), as demonstrated by Fauqueux et al. (64), provided turbulent mixing is ensured. After equilibration is reached the phases should separate into a top and bottom phase. This separation can be done under unit gravity or can be speeded up by centrifugation. The latter is recommended if the viscosity is high or the density difference between the phases is low (8). Centrifugation gives higher yields as entrainment is avoided at higher g forces; optimal feed rates of the suspension up to 3 L/min corresponding to residence times of about 16 s have been reported using small disk stack separators (8).

After the initial separation step, the top phase is often mixed with salts to generate a secondary phase system. Conditions are selected such that the protein of interest is partitioned to the bottom phase and can be further processed after phase separation (Fig. 9). Other possible procedures for the isolation of proteins from polymer phases are centrifugation, adsorption, ultrafiltration, or electrophoreses (66). In large-scale processes, ultrafiltration or diafiltration is used to concentrate and condition the product from the secondary bottom phase. The top phase may be recycled (52,67). A secondary bottom phase may be applied under certain conditions to a chromatographic support, such as hydrophobic interaction matrices.

Extraction usually is performed in the batch mode, but often it may be more cost-effective to use a continuous process. In general there are three possibilities of continuous processing in production lines: the co-, cross-, and countercurrent flow. The three modes are illustrated in Figure 10, and a comparison of specific properties is shown in Table 3. In large-scale processes the counter-current method should be the most efficient because of a higher yield and lower consumption of chemicals (63). Countercurrent extraction has been studied for the multistep procedures discussed next. The continuous cross-current extraction has been investigated experimentally in pilot scale. Enzyme extraction has been successfully carried out using more than 500 kg yeast per day in a computer-controlled process (68). The procedure could also be applied to extract enzymes continuously from bacterial cell homogenates (69).

To achieve a higher purification, multiple partition steps would be required. Not every multistage apparatus used in the chemical industry for liquid–liquid extraction is suitable for operation with aqueous phases as the physical properties of the two phases are too similar and the interfacial tension too low. The performance of the Graesser contactor has been best investigated. For the separation of α-lactalbumin from whey, a space time yield of 29 g/L reactor volume and day was achieved (70). The speed of rotation in the "raining bucket contactor" had to be reduced from the usual 20 U/min to 2–5 U/min, working with PEG–salt systems (71). In addition, the rotational direction was reversed to reduce the intensity of mixing and to avoid flooding. Other apparatus investigated for multistage extractions are mixer-settler, Kühni columns (72), the Podbielniak extractor (1), and spray columns (73). All can be operated successfully with specific precautions against flooding. Changing operating conditions such as pH or the length of the tie-line, however, may be more effective for protein purification than multiple partition steps.

At Miles Inc., a conventional downstream process was compared to an operation using aqueous two-phase systems and *Bacillus* alpha-amylase as an example (74). The latter was carried out in a 50,000-L large-scale process (Fig. 11). In this example the extraction process in aqueous two-phase-systems required fewer steps, resulting in higher yield and lower costs.

Recycling of Process Chemicals

To lower high chemical and waste disposal costs, recycling or recirculation of the auxiliary chemicals is recommended. Recycling of phosphate by crystallization at 6 °C has been described from solid-free process streams (8). The successful recirculation of most of the PEG has been demonstrated by Papamichael et al. (75) and of detergents by Minuth et al. (55) (see Fig. 7). The recovery of salt from the primary bottom phase was accomplished by Greve and Kula (76) making use of miscibility gaps in the ternary-phase systems—water–salt–lower aliphatic alcohols (Fig. 9). The authors also presented an economic analysis (76). In principle, the material balance of the auxiliary chemicals introduced into the process stream to accomplish phase separation can be closed more than 90% by recycling, yielding cost-effective and environmentally acceptable procedures.

Large-Scale Applications

Examples of Large-Scale Applications. Direct information about industrial applications of aqueous two-phase extrac-

Table 2. Physical Properties of Aqueous Two-Phase Systems

System composition	$\Delta\rho$ (g/cm^3)	η_T (mPa s)	η_B (mPa s)	$\eta_{dispersion}$ (mPa s)	δ (mN/m)	Reference
9% PEG 4000, 2% dextran T-500	1,010	3	94	3.4	—	16
7% PEG 4000, 1.25% crude dextran	1,046	2.7	2,000	4	—	16
16% PEG 4000, 13% phosphate	1,085	12	2.1	—	1.25	58

Source: PEG, polyethylene glycol. Ref. 8.

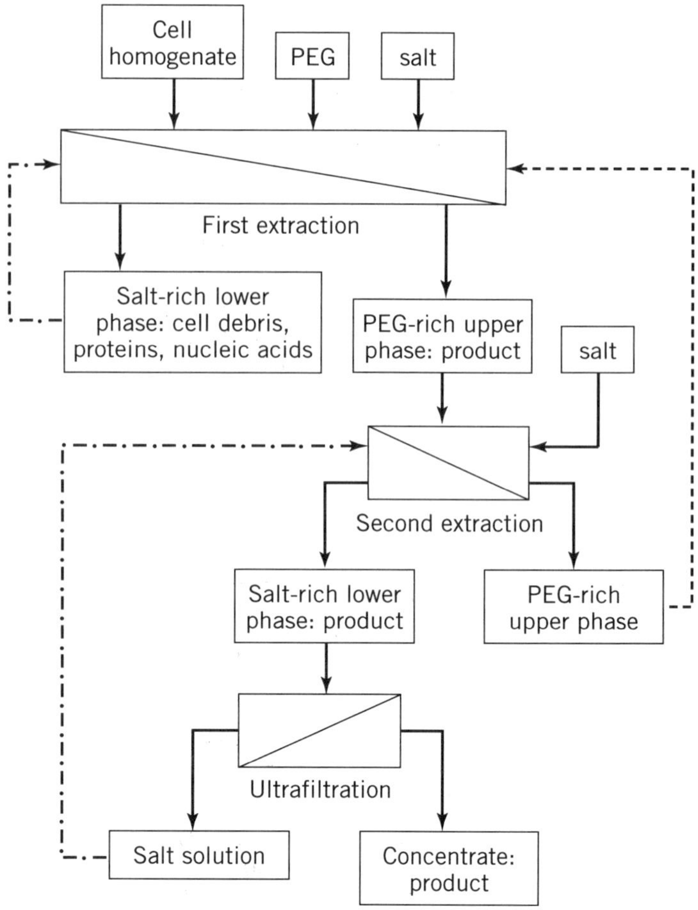

Figure 9. Flow sheet for product, chemicals, and waste flows in an extraction process using a PEG–salt aqueous two-phase system. ·—·—·—·, recycling of salt; ————, recycling of PEG. *Source*: Ref. 65.

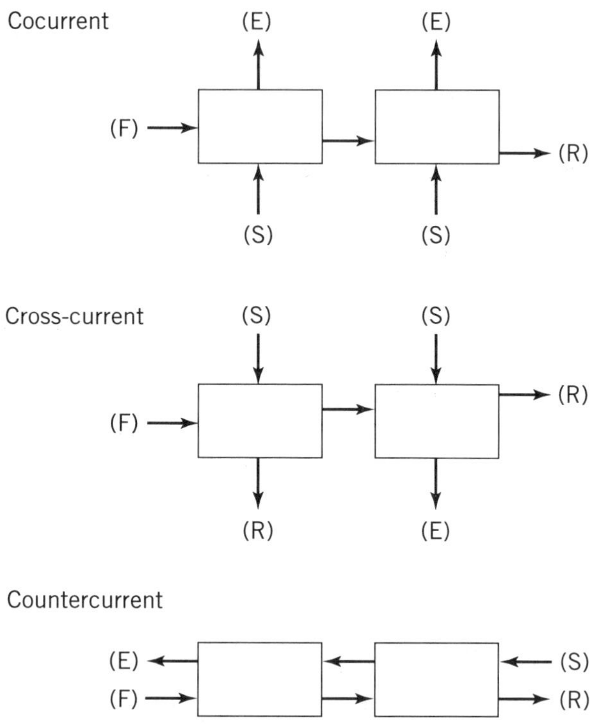

Figure 10. Principles of co-, cross-, and counter-current flow. F, feed; S, solvent; E, extract; R, raffinade.

tion is scarce. In Table 4, a list of patents and large-scale applications (beginning in 1982; minimum, 24-kg cells) described in the literature is compiled. The size of the operation reached 50,000 L or more (74). Usually centrifugation is preferred in the initial phase separation. (An example is illustrated in Fig. 11.)

Scale-Up. One big advantage of aqueous liquid–liquid extraction is the ease of scaling up. Hart et al. (89), Schütte et al. (93), and Kroner et al. (94) showed that scale-up is feasible by proportional linear increase of the amount of ingredients. The necessary centrifugation equipment is commercially available and can be selected using the Sigma concept. In Figure 12, scale-up data of Kroner et al. (94) are presented for the purification of formate dehydrogenase from *Candida boidinii* in PEG–potassium phosphate. A scale-up of nearly 40,000 was carried out for different separation steps without significant changes in yield.

Economic Aspects. The costs of the protein recovery and enrichment by aqueous liquid–liquid extraction is of course highly dependent on the actual system, scale, etc. As an example Kroner et al. (95) analyzed the cost structure for extraction and compared the economics with other downstream separation methods employed as alternatives for the purification of intracellular fumarase from *Brevibacterium ammoniagenes* in a 100-kg batch process. With a

Table 3. Advantages and Disadvantages of Different Flow Regimes in Continuous Processing

	Suitable for	Consumption requirements	Yield	Purification factor
Cocurrent	Low K	As cross-current	Medium	Medium
Cross-current	High K	As co-current	Lowest	Highest
Countercurrent	Low K	Lowest	Highest	Lowest

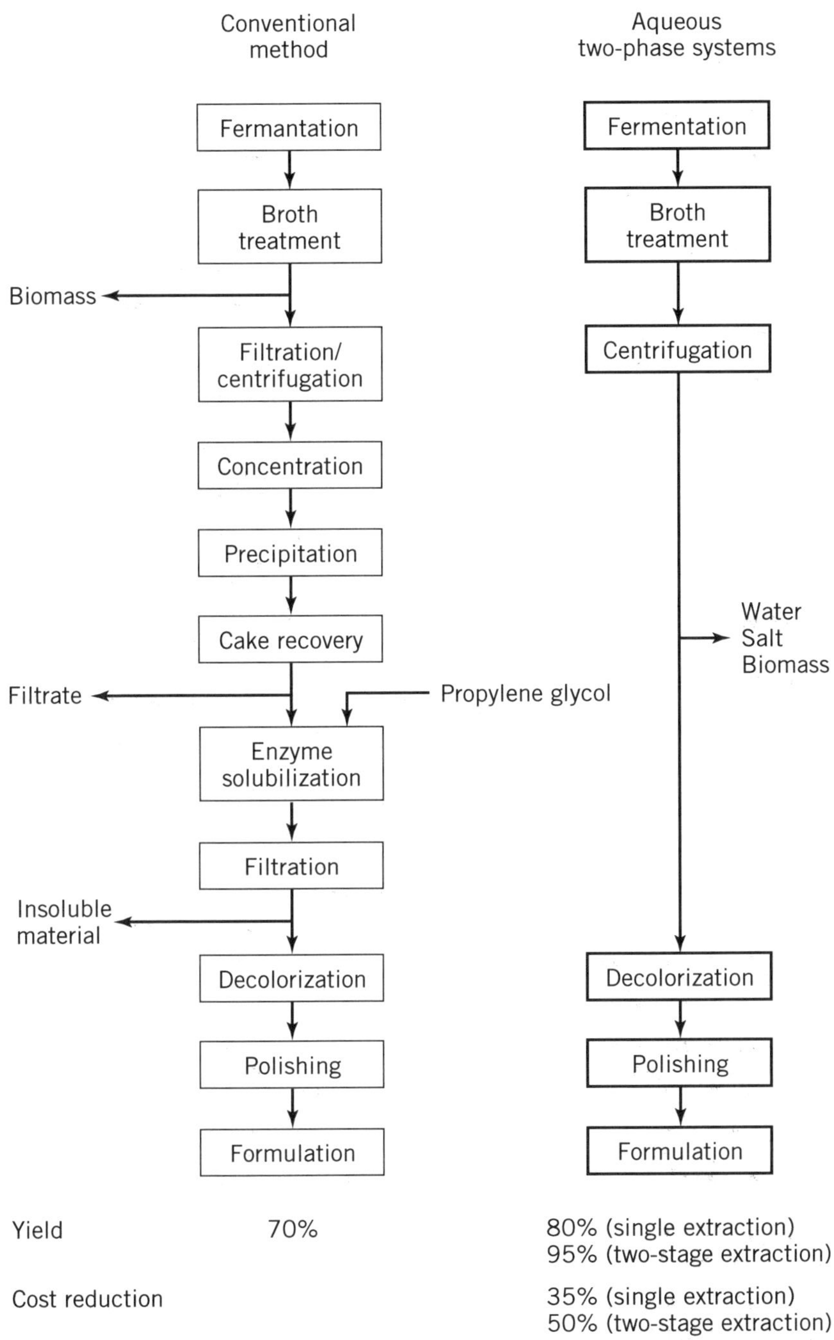

Figure 11. Comparison of the downstream process of alpha-amylase, using a conventional method (left column) and using aqueous two-phase systems (right column). *Source*: Ref. 74.

Table 4. Patents and Large-Scale Applications of Protein Extraction Using Aqueous Two-Phase Systems

Literature		Author	Date	Enzyme or protein	System used	Ligand	Origin
77	Patent	Kula	1982	Interferon	PEG–dextran		Raw material
15	Journal article	Hummel et al.	1983	D-Lactate dehydrogenase	PEG–phosphate		*Lactobacillus cellubiosus*
78	Patent	Kim et al.	1985	Protease, amylase	PEG–dextran		*Aspergillus orizae*
79	Journal article	Schütte et al.	1985	L-Leucin dehydrogenase	PEG–phosphate		*Bacillus cereus*
80	Patent	Gustafsson et al.	1986	ADH, hexokinase	PEG–K_2HPO_4, KH_2PO_4	Yes[c]	Yeast
80	Patent	Gustafsson et al.	1986	Transferrin	PEG–K_2HPO_4, KH_2PO_4	Yes	Blood plasma
81	Patent	Paul et al.	1986	Dextran-sucrase	PEG–dextran		*Leuconostoc mesenteroides*
82	Patent	van Wijnendaele et al.	1986	Hepatitis B antigen (HBsAg)	PEG–$(NHY_4)_2SO_4$		Yeast
82	Patent	van Wijnendaele et al.	1986	Alpha-1-antitrypsin	PEG–$(NH_4)_2SO_4$		Yeast
83	Patent	Dove et al.	1987	Therapeutically active proteins[a]	PEG–KH_2PO_4		Blood plasma
44	Journal article	Tjerneld et al.	1987	Lactate dehydrogenase	PEG–aquaphase PPT	Yes	Pig muscle
84	Patent	Ananthapadmanabhar	1988	Alcaline protease	PEG–Na_2SO_4	Yes	
85	Patent	Brewer et al.	1988	Protease	PEG–Na_2SO_4		Whole fermentation beer
14	Patent	Sieron et al.	1991	Recombinant proteins	PEG–polyvinylalcohol	[b]	
86	Patent	Enfors et al.	1992	Human immunoglobulin G	PEG–phosphate		Staphylococcus
47	Patent	Giuliano et al.	1992	Alcohol dehydrogenase (ADH)	PVP–maltodextrin	Yes	Baker's yeast
87	Patent	Heinsohn et al.	1992	Chymosin	PEG–Na_2SO_4		*Aspergillus niger* var. *awamori*
88	Patent	Kirchberger et al.	1992	Alkalische phosphatase	PEG–dextran		Calf intestine
71	Journal article	Coimbra et al.	1994	Beta-lactoglobulin	PEG–phosphate		Cheese whey
71	Journal article	Coimbra et al.	1995	Alpha-lactalbumin	PEG–phosphate		Cheese whey
18	Journal article	Cordes et al.	1994	Formate dehydrogenase	PEG–phosphate	Yes	*Candida boidinii*
89	Journal article	Hart et al.	1994	IGF[a]	PEG–Na_2SO_4		*E. coli*
90	Patent	Lorch et al.	1994	EG[c]	PEG–$(NH_4)_2SO_4$		Cellulase mixture
13	Patent	Builder et al.	1995	IGF-I or mammalian polypeptide	PEG–citrate	[b]	*E. coli*
91	Patent	Heinssohn et al.	1995	Chymosin	PEG–Na_2SO_4		Bovine stomach
92	Patent	Lee and Kahn	1995	Hemoglobin	PEG–phosphate buffer		Bovine blood
57	Patent	Braunstein et al.	1995	Different lipases and proteases	Detergents		Different organisms

Patents often cover wide ranges; therefore, in some cases only examples from the patent are given.

[a]Albumin, IgM, IgG, alpha-1-antitrypsin.

[b]Use of chaotropic substances.

[c]Adsorption at Sepharose particles.

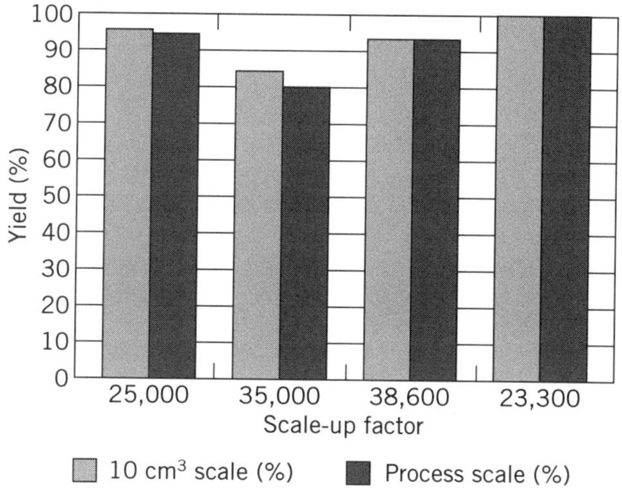

Figure 12. Scale-up of various extraction steps during the isolation and purification of formate dehydrogenase from *Candida boidinii*. *Source*: Ref. 94.

system of PEG and potassium phosphate, an overall yield of 70% was achieved by a two-step extraction. The cost structure of the extraction is shown in Table 5. Even if prices have changed meanwhile, the table still gives some idea about the relative cost distribution of material, labor, energy, and investment for the different process steps.

In Figure 13, a comparison of the cost factors and the total costs of the initial recovery operations is shown (95). Data in Table 5 and in Figure 13 are calculated from equation 10, which is derived from coefficients used for the cost calculation in the chemical industry and normalized for the yield:

$$T = \frac{1.13M + 2.6L + 1.13E + 0.13I}{Y} \tag{10}$$

where the letters represent costs for T, total production; M, material; L, labor; E, energy; I, investment (capital); and Y, yield.

Aqueous two-phase extraction is a low-energy, labor-requiring technique with comparatively low investment costs and high material costs. These high costs for auxiliary chemicals can be lowered by a good choice of volume

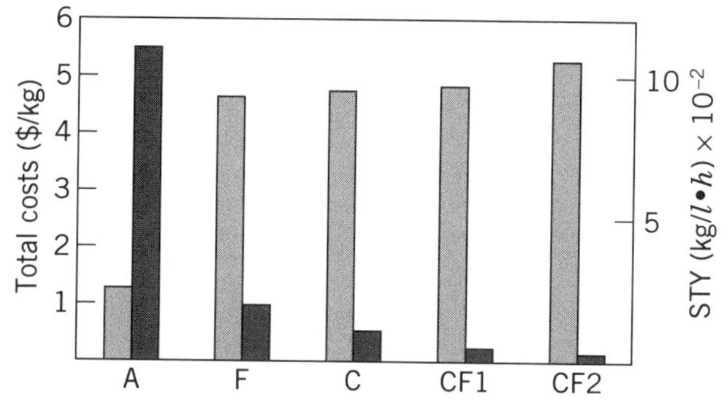

Figure 13. Cost factors and total costs comparing different initial isolation procedures. A, aqueous-phase system operation; F, filtration; C, centrifugation; CF, cross-flow filtration (1, retention coefficient $R = 0$; 2, $R = 0.7$). *Source*: Ref. 95.

ratio and by new phase systems and especially by the development of recycling processes (65). The process time can be decreased dramatically by centrifugal separation, resulting not only in high space time yields but also in improved product quality because of the absence of significant proteolytic degradation (96,97). Economic aspects of affinity extraction have been discussed by Johansson et al. (45) and Cordes et al. (18). In such cases, economics rely heavily on the initial costs and the reusability of the affinity ligand.

For a valid evaluation, it is important to realize that extraction constitutes an integrative technology; it achieves removal of solids and an increase in the specific activity of the product in a single step. In addition, in most cases the product can be separated from high molecular weight nucleic acids and its concentration can be increased, improving the performance of subsequent separation steps. An industrial total cost comparison between a downstream process using aqueous two-phase systems and a conventional one is shown in Figure 11. A cost reduction of 35% to 50% was achieved for the separation of *Bacillus* alpha-amylase.

CONCLUSION

Liquid–liquid extraction of proteins using aqueous two-phase systems is a comparatively new but nevertheless

Table 5. Cost Structure of an Extraction Process

Step	Feedstock ($/kg)	Material ($/kg)	Labor ($/kg)	Energy ($/kg)	Instrumentation ($/kg)	Total ($/kg)	Percentage of total
Fermentation	5.80					6.55	40.1
Disruption		0.03	0.92	0.03	2.76	2.82	17.2
1. Extraction		1.89	0.48	<0.01	0.94	3.52	21.5
2. Extraction		0.17	0.29	<0.01	0.94	1.08	6.6
Ultrafiltration		0.20	0.48	0.01	0.90	1.60	9.8
Waste water						0.78	4.8
Sum ($/kg)	5.80	2.29	2.17	<0.06	5.54	16.35	
Percentage of total	35.5	14.0	13.3	0.4	33.9		

Source: Ref. 95.

well-investigated technique for the isolation and separation of proteins. It is apparently not yet widely used in industry, but the number of industrial applications seems to have increased in recent years due to the effectiveness and ease of operation on a large scale. Intensive research is going on worldwide with respect to practical applications and theoretical understanding. Together with increases in production size for pharmaceutical proteins, this will make liquid–liquid extraction in aqueous two-phase systems a more common unit operation in the biotechnological industry.

BIBLIOGRAPHY

1. M.-R. Kula, K.H. Kroner, and H. Hustedt, *Adv. Biochem. Eng.* **24**, 73–118 (1982).
2. M.-R. Kula, *Bioseparation* **1**, 181–189 (1990).
3. H. Walter and G. Johansson, *Methods in Enzymology*, vol. 228, Academic Press, San Diego, Calif., 1994.
4. P.A. Albertsson, *Partition of Cell Particles and Macromolecules*, Wiley, New York, 1986.
5. H. Walter, D.E. Brooks, and D. Fisher, *Partitioning in Aqueous Two-Phase Systems*, Academic Press, Orlando, Fla., 1985.
6. B.Y. Zaslavsky, *Aqueous Two-Phase Partitioning*, Dekker, New York, 1994.
7. M.-R. Kula, in A. Humphrey, C.L. Cooney eds., *Comprehensive Biotechnology*, vol. 2, Pergamon Press, New York, 1985, pp. 451–471.
8. H. Hustedt, K.H. Kroner, and M.-R. Kula, in H. Walter, D.E. Brooks, and D. Fisher eds., *Partitioning in Aqueous Two-Phase Systems*, Academic Press, Orlando, Fla., 1985, pp. 529–587.
9. J.N. Bronsted, *Z. Phys. Chem. Abt. A* **157**, 257 (1931).
10. D.J. Kubek, in R.G. Harrison ed., *Protein Purification Process Engineering*, Dekker, New York, 1994, pp. 87–114.
11. U. Menge, M. Morr, U. Mayr, and M.-R. Kula, *J. Appl. Biochem.* **5**, 75–90 (1983).
12. P.A. Albertsson, *Endeavour* **1**, 69 (1977).
13. U.S. Pat. 5,407,810 (April 18, 1995), S. Builder, R. Hart, P. Lester, J. Ogez, and D. Reifsnyder (to Genentech, Inc.).
14. GDR Pat. DD 288 837 (April 11, 1994), R. Sieron, R. Wondraczek, K. Binder, H. Dittmar, K. Lambrecht, and H. Thessa (to Zentralinstitut für Mikrobiologie und experimentelle Therapie).
15. H. Schütte, W. Hummel, and M.-R. Kula, *Appl. Microbiol. Biotechnol.* **19**, 167–176 (1984).
16. R.A. Hart, J.R. Ogez, and S.E. Builder, *Bioseparation* **5**, 113–121 (1995).
17. K.H. Kroner, H. Hustedt, and M.-R. Kula, *Biotechnol. Bioeng.* **24**, 1015–1045 (1982).
18. A. Cordes and M.-R. Kula, in H. Walter and G. Johansson eds., *Methods of Enzymology*, vol. 228, Academic Press, San Diego, Calif., 1994, pp. 600–617.
19. D.C. Szlag, K.A. Giuliano, and S.M. Snyder, J.F.P. Hamel, J. Hunter, and S.K. Sikdar eds., *Downstream Processing and Bioseparation*, ACS Symposium Series, vol. 419, American Chemical Soc., Washington, D.C., 1990, pp. 71–86.
20. M.-R. Kula, in H. Brauer ed., *Biotechnology*, vol. 2, VCH-Verlagsgesellschaft, Weinheim, 1985, pp. 725–760.
21. U. Sivars, K. Bergfeld, L. Piculell, and F. Tjerneld, *J. Chromatogr. B* **680**, 43–53 (1996).
22. H.-O. Johansson, G. Karlström, and F. Tjerneld, *Macromolecules* **26**, 4478 (1993).
23. P.A. Alred, A. Koslowski, and J.M. Harris, *Bioseparation* **2**, 363–373 (1992).
24. H. Cabezas, Jr., *J. Chromatogr. B* **680**, 3–30 (1996).
25. D. Forciniti and C.K. Hall, *ACS Symp. Ser.* **419**, 53–70 (1990).
26. F. Döbert, A. Pfennig, and M. Stumpf, *Macromolecules* **28**, 7860–7868 (1995).
27. A.D. Diamond, K. Yu, and J.T. Hsu, in *Protein Purification*, ACS Symposium Series, American Chemical Soc., Washington, D.C., 1990, pp. 52–65.
28. A. Gustafsson, H. Wernerström, and F. Tjerneld, *Polymer* **27**, 1768–1770 (1986).
29. H. Hartounian and S.I. Sandler, *Biotechnol. Prog.* **7**, 279 (1991).
30. C.A. Haynes, F.J. Benitez, H.W. Blanch, and J.M. Prausnitz, *AIChE J.* **39**, 1539 (1993).
31. C. Großmann and G. Maurer, *Fluid Phase Equilib.* **106**, 17–25 (1995).
32. Y. Guan, T.H. Lilley, and T.E. Treffry, *Macromolecules* **26**, 3971 (1993).
33. H.-O. Johansson, G. Karlsström, F. Tjerneld, and C.A. Haynes, *J. Chromatogr. B* **711**, 3–17 (1998).
34. D.R. Baughman and Y.A. Liu, *Ind. Eng. Chem. Res.* **33**, 2668–2687 (1994).
35. M. Kabiri-Badr and H. Cabezas, Jr., *Fluid Phase Equilib.* **115**, 39–58 (1996).
36. J. Bringmann, B. Keil, and A. Pfennig, *Fluid Phase Equilib.* **101**, 211–225 (1994).
37. W.-Y. Chen, C.-G. Shu, J.Y. Chen, and J.-F. Lee, *J. Chem. Eng. Jpn.* **27**, 688–690 (1994).
38. M.A. Eiteman, C. Hassinen, and A. Veide, *Biotechnol. Prog.* **10**, 513–519 (1994).
39. M.-R. Kula, A. Walsdorf, and A. Cordes, *Ber. Bunsen-Ges. Phys. Chem.* **93**, 968–970 (1989).
40. A. Cordes, J. Flossdorf, and M.-R. Kula, *Biotechnol. Bioeng.* **30**, 514–520 (1987).
41. J.M. Harris, *J. Macromol. Sci.* **C-25**, 325 (1985).
42. G. Johansson, G. Kopperschläger, and P.A. Albertsson, *Eur. J. Biochem.* **131**, 589–594 (1983).
43. K.H. Kroner, A. Cordes, A. Schelper, M. Morr, A.F. Bückmann, and M.-R. Kula, in T.C.J. Gribnau, J. Visser, and R.J.F. Nivard eds., *Affinity Chromatography and Related Techniques*, Elsevier, Amsterdam, 1982, pp. 451–501.
44. F. Tjerneld, G. Johansson, and M. Joelsson, *Biotechnol. Bioeng.* **30**, 809–816 (1987).
45. G. Johansson and M. Andersson, *J. Chromatogr.* **303**, 39 (1984).
46. S.-S. Suh and F.H. Arnold, *Biotechnol. Bioeng.* **35**, 682–690 (1990).
47. U.S. Pat. 5,093,254 (March 3, 1992), K.A. Giuliano and D.C. Szlag (to the United States).
48. G. Johansson, in S.K. Sikdar and M. Bier eds., *Frontiers in Bioprocessing*, CRC Press, Boca Raton, Fla., 1989, pp. 271–284.
49. G. Johansson and H. Tjerneld, *J. Biotechnol.* **11**, 135–142 (1989).
50. G.C. Terstappen, R.A. Ramelmeier, and M.-R. Kula, *J. Biotechnol.* **28**, 263–275 (1993).
51. C. Bordier, *J. Biol. Chem.* **250**, 1604–1606 (1981).

52. T. Minuth, J. Thömmes, and M.-R. Kula, *J. Biotechnol.* **38**, 151–164 (1995).
53. R.P. Frankewich and W.L. Hinze, *Anal. Chem.* **66**, 944–954 (1994).
54. W.L. Hinze and E. Pramauro, *Crit. Rev. Anal. Chem.* **24**, 133–177 (1993).
55. T. Minuth, J. Thömmes, and M.-R. Kula, *Biotechnol. Appl. Biochem.* **23**, 107–116 (1996).
56. T. Minuth, H. Gieren, H. Pape, C. Raths, J. Thömmes, and M.-R. Kula, *Biotechnol. Bioeng.* **55**, 339–347 (1997).
57. Int. Pat. WO 96/23061 (January 27, 1995) E.L. Braunstein, N.T. Becker, G. Ganshaw, and T.P. Graycar (to Genencor International, Inc.).
58. B.D. Kelley, D.I.C. Wang, and T.A. Hatton, *Biotechnol. Bioeng.* **42**, 1199–1208 (1993).
59. M.J. Pires and J.M.S. Cabral, *J. Chem. Technol. Biotechnol.* **61**, 219–224 (1994).
60. P.A. Alred, A. Kozlowski, J.M. Harris, and F. Tjerneld, *J. Chromatogr. A* **659**, 289–298 (1994).
61. H.-O. Johansson, G. Karlsström, B. Mattiasson, and F. Tjerneld, *Bioseparation* **5**, 269–279 (1995).
62. J. Persson, L. Nyström, H. Ageland, and F. Tjerneld, *J. Chromatogr. B* **711**, 97–109 (1998).
63. N. Papamichael and H. Hustedt, *Methods Enzymol.* **228**, 573–584 (1994).
64. P.-F. Fauqueux, H. Hustedt, and M.-R. Kula, *J. Chem. Technol. Biotechnol.* **35B**, 51–59 (1985).
65. A. Greve and M.-R. Kula, *J. Chem. Technol. Biotechnol.* **50**, 27–42 (1991).
66. G. Johansson, in H. Walter and G. Johansson eds., *Methods of Enzymology*, vol. 228, Academic Press, San Diego, Calif., 1994, pp. 569–573.
67. H. Hustedt, K.-H. Kroner, U. Menge, and M.-R. Kula, *Trends Biotechnol.* **3**, 139–144 (1985).
68. H. Hustedt, B. Börner, K.H. Kroner, and N. Papamichael, *Biotechnol. Tech.* **1**, 49–54 (1987).
69. H. Hustedt and N. Papamichael, *Enzyme Eng.* **9**, 135–139 (1988).
70. J. Dos Reis Coimbra, J. Thömmes, A. Meirelles, and M.-R. Kula, *Bioseparation* **5**, 259–268, (1995).
71. J. Dos Reis Coimbra, J. Thömmes, and M.-R. Kula, *J. Chromatogr. A* **668**, 85–94 (1994).
72. H. Hustedt, K.-H. Kroner, U. Menge, and M.-R. Kula, *Enzyme Eng.* **5**, 45–47 (1988).
73. K.R. Jafarabad, S.B. Sawant, J.B. Joshi, and S.K. Sikdar, *Chem. Eng. Science* **47**, 57–68 (1992).
74. C.Y. Kim, J.W. Brewer, C.E. Brothers, T.F. Farver, and E.K. Lee, *3rd Chemical Cong. of North America*, Toronto, Canada, June 5–10, 1988.
75. N. Papamichael, B. Börner, and H. Hustedt, *J. Chem. Technol. Biotechnol.* **54**, 47–55 (1992).
76. A. Greve and M.-R. Kula, *Bioprocess Eng.* **6**, 173–177 (1990).
77. FRG Pat. DE 2943016 C2 (September 6, 1984), U. Menge, M. Morr, and M.-R. Kula (to the Gesellschaft für Biotechnologische Forschung mbH).
78. U.S. Pat. 4,508,825 (April 2, 1985), C.Y. Kim, T.F. Farver, and J.W. Brewer (to Miles Laboratories, Inc.).
79. H. Schütte, W. Hummel, H. Tsai, and M.-R. Kula, *Appl. Microbiol. Biotechnol.* **22**, 306–317 (1985).
80. U.S. Pat. 4,579,661 (April 1, 1986), S.J. Gustafsson, P.O. Hedman, T.G.I. Ling, and B.G. Mattiasson (to Pharmacia AB, Sweden).
81. U.S. Pat. 4,591563 (May 27, 1986), F. Paul, P. Monsan and D. Auriol (to Societe Nationale Elf Aquitaine, France).
82. E.U. Pat. 0 199 698 B1 (August 28, 1991), F. van Wijnendaele, D. Gilles, and G. Simonet (to Smithkline Biologicals S.A.).
83. U.S. Pat. 4,684,723 (August 4, 1987), G.B. Dove and G. Mitra (to Miles Laboratories, Inc.).
84. U.S. Pat. 4,743,550 (May 10, 1988), K.P. Ananthapadmanabhan and E.D. Goddard (to Union Carbide Corporation).
85. U.S. Pat. 4,728,613 (March 1, 1988), J.W. Brewer, C.E. Brothers, T.F. Farver, C.Y. Kim, and E. Lee (to Miles Laboratories, Inc.).
86. Int. Pat. WO 92/07868 (May 14, 1992), S.-O. Enfors, K. Köhler, Ch. Ljundquist, B. Nilsson, and A. Veide (to Pharmacia AB, Sweden).
87. U.S. Pat. 5,139,943 (August 18, 1992), H.G. Heinsohn, J.D. Lorch, and R.E. Arnold (to Genencor International, Inc.).
88. FRG Pat. DD 298 424 (February 20, 1992), J. Kirchberger, G. Kopperschläger, M. Rockstroh, and K. Eisenbrandt (to Universität Leipzig, Germany).
89. R.A. Hart, P.M. Lester, H. Riefsnyder, J.R. Ogez, and S.E. Builder, *BioTechnology* **12**, 1113–1117 (1994).
90. U.S. Pat. 5,328,841 (July 12, 1994), J.D. Lorch K.A. Clarkson, E. Larenas, B.S. Bower, and G.L. Weiss (to Genencor International, Inc.).
91. E.U. Pat. 0 477 284 B1 (August 16, 1995), H.G. Heinssohn and K.J. Hayenga (to Smithkline Biologicals S.A.).
92. U.S. Pat. 5,407,579 (April 18, 1995) C.-J. Lee and P. Khan (to National Science Council, Taipei).
93. H. Schütte, K.H. Kroner, W. Hummel, and M.-R. Kula, *Ann. N.Y. Acad. Sci.* **413**, 270–282 (1983).
94. K.H. Kroner, H. Schütte, W. Stach, and M.-R. Kula, *J. Chem. Technol. Biotechnol.* **32**, 130–137 (1982).
95. K.H. Kroner, H. Hustedt, and M.-R. Kula, *Process Biochem.* **19**, 170–179 (1984).
96. T.R. Pulliam, S. Winston, and W.E. Bentley, *Enzyme Microb. Technol.* **20**, 46–51 (1997).
97. H. Hellebust, A. Veide, and S. Enfors, *J. Biotechnol.* **7**, 185–198 (1988).

See also CELL DISRUPTION AND LYSIS; CRYSTALLIZATION, PROTEIN, KINETICS.

PROTEIN SECRETION, *SACCHAROMYCES CEREVISIAE*

K.D. WITTRUP
Kyowa Hakko Kogyo Co., Ltd.
Tokyo, Japan

KEY WORDS

Cell engineering
Chaperones
Expression vector
Foldases
Heterologous secretion
Protein expression

Saccharomyces cerevisiae
Yeast

OUTLINE

INTRODUCTION

The yeast *Saccharomyces cerevisiae* is an excellent host for production of nonglycosylated secreted proteins possessing disulfide bonds. As a microbial eukaryote, yeast possesses the protein folding and conformational proofreading apparatus of the endoplasmic reticulum and Golgi apparatus, yet it is amenable to high cell density fermentation in large, process-scale fermenters. Effective exploitation of the yeast cell's secretory capacity can produce yields in the gram per liter range of secreted, active protein.

YEASTS

The most widely studied yeast is *S. cerevisiae*, which is used as a model system for many studies of eukaryotic cell biology (1). As a result, a rich knowledge base is available concerning the physiology and genetics of *S. cerevisiae*, for example, the entire genome sequence has been obtained. Alternative yeasts have also been exploited for protein expression, such as *Pichia pastoris, Kluyveromyces lactis*, and *Hansenula polymorpha*. In particular, use of *P. pastoris* is widespread, and a kit is commercially available. In most cases, these nontraditional yeasts have been found to secrete many-fold higher levels of heterologous protein than *S. cerevisiae* (2,3). More recent studies have indicated, however, that the intrinsic secretory capacity of *S. cerevisiae* is inefficiently exploited by the most commonly used multicopy plasmid vectors; when chromosomal integrating vectors are used, the secretory capacity of *S. cerevisiae* is comparable to or greater than that of *P. pastoris* (4,5). Reviews of *P. pastoris* are available elsewhere (2); this article focuses on *S. cerevisiae*. Rather than exhaustively reviewing all facets of *S. cerevisiae* expression, the approaches expected to be of most practical value are highlighted. An excellent, more comprehensive review of yeast expression is available elsewhere (6).

PROTEINS SUITABLE FOR SECRETION IN YEAST

Yeast expression is used to best advantage for proteins whose native route of synthesis is the eukaryotic secretory pathway, a classification that includes most pharmaceutical proteins. Yeast is not well suited for producing glycosylated proteins destined for therapeutic administration, because of the introduction of immunogenic sugar linkages and the absence of sialic acid in yeast glycosylation. Nonmammalian glycosylation does not preclude yeast expression of glycoproteins for analytical, diagnostic, or other in vitro applications, however. A partial list of proteins industrially produced in *S. cerevisiae* for clinical trials is presented in Table 1 (7). Noteworthy yeast-produced commercial products are human insulin, granulocyte-macrophage colony-stimulating factor, and hepatitis B vaccine. A potentially important product class for yeast expression is single-chain antibody fragments and immunotoxins; many such therapeutics are currently in development (8).

VECTORS

Although the most commonly used *S. cerevisiae* vectors are based on the 2μ multicopy plasmid, amplifiable integrating vectors can offer strong advantages and are generally preferable for high-level secretion.

2μ Plasmids

Most laboratory strains of *S. cerevisiae* possess 50 to 100 copies of a circular plasmid called the ***2μ plasmid*** because of its dimensions observed by electron microscopy. The native 2μ plasmid does not confer a measurable growth phenotype, but represents parasitic selfish DNA and is apparently maintained simply as a result of effective amplification, replication, and partitioning mechanisms. Numerous expression vectors containing all or part of the 2μ plasmid have been constructed (9). 2μ-Based vectors exhibit broad clonal variation in copy number and are mitotically unstable. Controlled amplification of 2μ-based plasmids has been accomplished by overexpression of the FLP recombinase that converts θ replication intermediates to a rolling circle mode of amplification (9). 2μ vector copy numbers can also be amplified by use of faulty selection genes such as *leu2-d* or *ura3-d*, which are required at several hundred copies per cell to support growth (10,11).

Integrating Vectors

Chromosomal integration of heterologous expression cassettes takes advantage of the extremely high fidelity of the

Table 1. Proteins Expressed in *Saccharomyces cerevisiae* by Pharmaceutical and Biotechnology Companies for Clinical Trials

Immunomodulators
Interferon-α
Interferon-γ
Interleukin-1
Interleukin-2
Interleukin-3
Tumor necrosis factor
Growth factors
Granulocyte colony-stimulating factor
Macrophage colony-stimulating factor
Granulocyte-macrophage colony stimulating factor
Fibroblast growth factor
Transforming growth factor-α
Transforming growth factor-β
Epidermal growth factor
Platelet-derived growth factor
Fibronectin
Hormones
Insulin-like growth factor-1
Insulin-like growth factor-2
Human growth hormone
Human growth hormone releasing factor
Calcitonin
Human chorionic gonadotropin
Insulin
Proinsulin
Blood proteins
Human serum albumin
Hemoglobin
Factor VII
Factor XIII
Tissue-type plasminogen activator
Hirudin
α-1-Antitrypsin
Enzymes
Lysozyme
Superoxide dismutase
Gastric lipase protein
Vaccines
Hepatitis B
Malaria

Source: Data from Hodgson (7).

replication and partitioning apparatus used by the yeast cell to maintain its own genome. Vectors can be integrated either by homologous recombination or retrotransposition.

Homologous Recombination. The predominant majority of recombination events in *S. cerevisiae* are homologous, and vectors can be precisely targeted to particular chromosomal locations by restriction digestion in a segment of the vector identical to the chromosomal target sequence (12). Several integrating vectors have been developed that are targeted to multiple chromosomal targets. These vectors are typically integrated at one to three locations in iterated repeats, a configuration that can lead to instability if gene amplification reduces growth rate due to constitutive heterologous gene expression. The ribosomal DNA repeats have been used as integration targets (13), and Ty retrotransposon sequences have also been targeted. Several groups have developed δ vectors that are targeted to the 150 to 200 copies of the Ty long terminal repeat that are dispersed throughout the yeast chromosomes (14–19). The δ vectors have shown particular promise, increasing heterologous secretion 5 to 10 fold over levels achieved using 2μ plasmids (15,16). The recombination event that integrates the δ vector is reversed at a frequency of approximately 10^{-4}, looping out the vector sequences from the chromosome. If constitutive expression is used, clones with reduced copy number can quickly outgrow the amplified clones, resulting in substantial instability (18,20). If a strongly repressible promoter such as the GAL promoter is used instead, amplified expression cassettes are completely stable for more than 50 generations in the absence of antibiotic selection, because there is no growth selection against the amplified clones (16,20).

Retrotransposition. Expression cassettes can be inserted within inducible yeast retrotransposons and chromosomally integrated at dispersed sites via the processes of transcription, reverse transcription, and integration used by native yeast retrotransposons. Vectors of this type have been constructed from Ty1 and Ty3 retrotransposons (21,22). A key advantage of this approach is the dispersed nature of gene amplification, which reduces instability caused by recombination between iterated repeats. One disadvantage of this approach is that relatively low numbers of integration events occur at each round of amplification. A potential regulatory difficulty with this approach is the error-prone nature of the reverse transcription process, which could potentially introduce novel mutations in each integrated expression cassette.

SELECTION MARKERS

Nutritional Markers

Almost all yeast genetic studies not related to heterologous protein production use recessive chromosomal mutations with nutritional deficiency phenotypes. These mutations are complemented by plasmid-borne wild-type genes that allow transformants to propagate on medium lacking key amino acids or nucleotides. The genes and corresponding nutrients most commonly used are *URA3* (uracil), *TRP1* (tryptophan), *HIS3* (histidine), *LEU2* (leucine), and *ADE2* (adenine.) These genetic selection markers are often inserted into 2μ plasmids to select for plasmid maintenance. A refinement of this approach is to use a promoter-deleted version of the selection gene that is required at high amplification to support growth. The *leu2-d* and *ura3-d* alleles have been used in this way to select for very high 2μ copy numbers (10,11).

Dominant Markers

Dominant markers have two key advantages: the capability to transform polyploid industrial yeast strains and the capability to select for increasing gene dosage by increasing levels of the toxic selection compound. Several antibiotic resistance genes have been used as dominant markers for yeast transformation, but the most commonly used is the *G418* resistance gene from the *E. coli* transposon Tn903 (23). When used without a yeast promoter, expression of *G418*^{R} is weak enough to provide a selectable advantage by increasing gene dosage. *G418* resistance is approximately proportional to the number of integrated copies of the Tn903 *G418*^{R} gene (16). The *CUP1* copper metallothionein gene has also been used as a dominant selection marker, with the phenotype of increased copper resistance (24).

Autoselection

An essential gene can be used as the selectable marker in a strain with the chromosomal copy of that gene deleted. This approach allows the use of complex media without antibiotics, and has been used with *ura3fur1* (25) and Δ*tpi* (26) strains. Cells that lose the plasmid and its associated essential gene by segregational instability simply stop growing.

TRANSCRIPTIONAL PROMOTERS

Constitutive

Given that individual glycolytic enzymes can constitute several percent of the cell's total protein, the strong promoters from these genes are often used for constitutive heterologous expression. The most commonly used constitutive promoters are from the *PGK* and *GAP* genes (27,28). However, the use of a constitutive promoter for high level heterologous secretion is inadvisable because of inevitable negative growth phenotypes associated with high-level secretion, which in numerous studies have been shown to select for outgrowth of nonexpressing cells or cells with reduced expression (6,18,29,30).

Inducible

The GAL1 promoter is particularly well suited for heterologous expression because transcription is very tightly repressed in the presence of glucose, yet it is induced more than 1,000-fold by switching the carbon source to galactose. The wild-type levels of the Gal4p transcriptional activation protein can become limiting when multiple GAL1 promoters are induced, but this limitation can be overcome by single-copy inducible overexpression of Gal4p from a GAL promoter (31). Glucose-repressible promoters such as ADH2 and hybrid GAP/ADH2 promoters have also been used (32). However, difficulties may arise with such promoters during high cell density fermentations in balancing the central carbon metabolism against the requirement to repress heterologous expression during the growth phase. Glucose concentration is typically kept low during high cell density fermentation to prevent overflow metabolic production of ethanol resulting from the Crabtree effect, but this may lead to early induction of gene expression and growth defects. The CUP1 copper inducible promoter can be tuned across a 25-fold induction range by varying copper concentration in the growth medium (33). However, maximal induction occurs at copper concentrations that can have toxic effects on cell physiology, and basal expression in the absence of added copper could produce counterselection against gene amplification.

SIGNAL PEPTIDES

The most widely used leader peptide for heterologous secretion in *S. cerevisiae* is that of pre-pro α factor (34). This leader peptide is often found to be the most effective when compared with other signal peptides. The signal peptide is cleaved by signal peptidase after endoplasmic reticulum membrane translocation, and the pro region is cleaved C-terminal to a Lys-Arg site by the Kex2 protease in the Golgi apparatus. Another signal peptide that has been used frequently is the invertase signal peptide, which does not have an associated pro region.

POST-TRANSLATIONAL PROCESSING

Secreted proteins undergo a number of processing steps after translocation into the lumen of the endoplasmic reticulum: disulfide bond formation, folding, subunit assembly, glycosylation, and site-specific endoproteolytic processing. The endoplasmic reticulum and Golgi apparatus contain high concentrations of specialized chaperones, foldases, and enzymes dedicated to the task of conformational maturation of secreted proteins before release to the extracellular space (35). Many of these processing steps are performed inefficiently or not at all by bacteria, but yeast's secretory processes closely mirror those of mammalian cells (with the exception of glycosylation).

Protein Folding and Conformational Proofreading

Proteins do not exit the endoplasmic reticulum until they have folded correctly (36). The mechanistic details of this conformational proofreading process have not been completely elucidated, but they involve addition and trimming of glucose residues for glycoproteins and binding to the lectinlike chaperone calnexin. Two practical consequences of this proofreading step are of critical significance for heterologous protein production by secretion: (*1*) only correctly folded protein appears in the growth medium, obviating the need for refolding procedures; and (*2*) protein folding in the endoplasmic reticulum constitutes the key rate- and yield-limiting step in high-level heterologous protein secretion. Consequence *1* provides one of the major advantages of using yeast for protein production. Consequence *2* has motivated researchers to intensively study and manipulate the endoplasmic reticulum protein folding machinery in order to increase secretory capacity.

Most secreted eukaryotic proteins possess disulfide bonds. These covalent cross-links between cysteine side chains stabilize the folded conformation of a protein. As the number of disulfides in a protein increases, the combinatorial possibilities for incorrect cysteine pairings during folding grows geometrically. The endoplasmic-reticulum-resident foldase enzyme protein disulfide isomerase (PDI) catalyzes thiol–disulfide exchange to shuffle among possible disulfide pairings until the most stable folded conformation is attained. Overexpression of PDI in yeast can significantly improve secretion of some heavily disulfide bonded proteins (37,38); it is not a panacea, however. Secretion of some proteins is unaffected by PDI overexpression despite possession of disulfide bonds (37).

Another endoplasmic-reticulum-resident chaperone whose function in heterologous secretion has been closely examined is BiP, an hsp70 homologue. Although fulfilling an essential role in membrane translocation and subunit assembly, BiP levels are clearly not limiting for heterologous secretion in yeast or mammalian cells. Less clear is whether reduction of BiP levels is beneficial for secretion, because secretion of three heterologous proteins was decreased in yeast by reduction of BiP levels (39), whereas secretion of some proteins has been increased in mammalian cells by reducing BiP (40). The likely source of this discrepancy lies in the proteins studied; those proteins whose secretion is increased at low BiP levels are either mutant proteins with folding defects or subunits lacking stabilizing partners for a complex. It is possible that overexpression of other endoplasmic-reticulum-resident soluble or membrane bound chaperones and foldases could prove fruitful for heterologous secretion; such candidates include GRP94 chaperones, peptidyl prolyl isomerases, and redox foldases related to PDI such as EUG1 and ERP72.

Kex2p Endoproteolytic Processing

The Golgi-resident Kex2p endoprotease cleaves polypeptides C-terminal to Lys-Arg in unstructured regions of the protein. Kex2p removes the α-factor pro region from the α-factor peptide and heterologous fusion proteins. The sequence requirements for the amino acids adjacent to the Lys-Arg cleavage site have not been completely determined. In many cases, cleavage is efficient and the protein's native N-terminus is produced. For some proteins, however, Kex2 cleavage is inefficient, and a dipeptide spacer must be inserted at the junction to achieve efficient cleavage (41). This dipeptide may be tolerable as part of the protein product, or alternatively it may be removed by cathepsin-C processing (42). Undesirable Kex2 cleavage may also occur at internal Lys-Arg sites of a protein (43). This problem may be circumvented either by deletion of the *KEX2* gene or by alteration of the Lys-Arg sites by site-directed mutagenesis.

Glycosylation

S. cerevisiae covalently attaches polysaccharide structures at the same Asn-X-Ser/Thr sites recognized by mammalian cells. Although the core glycosylation structures initially added in the ER are identical for yeast and mammals, yeast subsequently adds extended mannose chains that are not capped with sialic acid and can be immunogenic. Although hypermannosylation can be reduced or eliminated by use of a *mnnl mnn9* strain (44), the absence of mammalian-type outer-chain glycosylation makes all yeasts undesirable hosts for expression of therapeutic glycoproteins.

SUPERSECRETION MUTATIONS

A classical genetic approach to strain improvement is to screen for spontaneous mutations that increase secretion of the protein of interest. Although such mutants may be difficult to analyze because of mutations at multiple loci or dominant alleles, the practical value of this strategy can be significant (45–48). One of the best characterized supersecretion mutants is the *pmr1* mutation, which involves loss of function of a calcium pump localized in the Golgi (45). How this causes substantial increases in heterologous secretion is not clear.

HIGH CELL DENSITY FERMENTATION

S. cerevisiae can be grown to cell densities of more than 100 g cell dry weight per liter in industrial-scale fed-batch fermentation (49,50). In such fermentations, glucose feed can be controlled by measurement of the respiratory quotient to prevent overfeeding and generation of ethanol. In the absence of instrumentation to measure the respiratory

Table 2. Genetic Components of *S. cerevisiae* Expression/Secretion Systems

Vector	Selection marker	Promoter	Leader peptide
Integrating	Dominant	Inducible	Pre-pro α
Recombination	***G418*^{R}**	**GAL1**	Invertase
Retrotransposition	*CUP1*	ADH2	Acid phosphatase
2μ plasmid	Recessive	CUP1	
	URA3	Constitutive	
	TRP1	GAP	
	HIS3	PGK	
	LEU2	TPI	
	ADE2		
	Autoselection		
	ura3fur1		
	Δtpi		

quotient, an exponentially increasing feed rate can also achieve high cell densities.

CONCLUSIONS AND RECOMMENDATIONS

Some of the most commonly used genetic components of heterologous secretion systems for *S. cerevisiae* are summarized in Table 2, and the recommended approach is highlighted in boldface. Amplifiable chromosomal integration vectors provide excellent mitotic stability and can improve secretion yields by an order of magnitude relative to 2μ plasmids. The *G418*^{R} dominant selectable marker enables tuning of gene amplification via increasing antibiotic resistance and allows use and optimization of complex media without auxotrophic nutritional constraints. Constitutive transcriptional promoters are not recommended because of counterselection against high-level expression and consequent strain instability. The GAL promoter provides a combination of near complete repression during a glucose fed growth phase, followed by strong induction upon feeding of galactose. The pre-pro α-factor leader peptide generally provides maximal targeting efficiency, and there is generally little benefit in examining several different leader peptides for comparison. The recommended combination of a δ integrating vector with *G418*^{R} selection, a GAL promoter, and pre-pro α-type leader peptide has been applied to secrete 400 mg/L of active bovine pancreatic trypsin inhibitor (BPTI) in shake flask cultures of *S. cerevisiae*, by comparison with 10 to 40 mg/L secreted with a constitutive promoter and 2μ plasmid (16). This specific productivity exceeds that reported for BPTI secretion in *P. pastoris* (4), demonstrating that efficient exploitation of *S. cerevisiae* can eliminate apparent differences among yeast species in their capacity for heterologous protein secretion.

BIBLIOGRAPHY

1. J.R. Broach, J.R. Pringle, and E.W. Jones eds., *The Molecular and Cellular Biology of the Yeast* Saccharomyces. Cold Spring Harbor Laboratory Press, Cold Spring Harbor, N.Y., 1991.
2. M. Romanos, *Curr. Opin. Biotechnol.* **6**, 527–533 (1995).
3. K.N. Faber, W. Harder, G. Ab, and M. Veenhuis, *Yeast* **11**, 1331–1344 (1995).
4. T. Vedvick, R.G. Buckholz, M. Engel, M. Urcan, J. Kinney, S. Provow, R.S. Siegel, and G.P. Thill, *J. Ind. Microb.* **7**, 197–202 (1991).
5. R.N. Parekh, M.R. Shaw, and K.D. Wittrup, *Biotechnol. Prog.* **12**, 16–21 (1996).
6. M.A. Romanos, C.A. Scorer, and J.J. Clare, *Yeast* **8**, 423–488 (1992).
7. J. Hodgson, *Bio / Technology* **11**, 887–893 (1993).
8. D.M. Goldenberg, *Am. J. Med.* **94**, 297–312 (1993).
9. A.B. Rose and J.R. Broach, *Methods Enzymol.* **185**, 234–279 (1990).
10. E. Erhart and C.P. Hollenberg, *J. Bacteriol.* **156**, 625–635 (1983).
11. G. Loison, A. Vidal, A. Findeli, C. Roitsch, J.M. Balloul, and Y. Lemoine, *Yeast* **5**, 497–507 (1989).
12. R. Rothstein, *Methods Enzymol.* **194**, 281–301 (1991).
13. T.S. Lopes, J. Klootwijk, A.E. Veenstra, P.C. van der Aar, H. van Heerikhuizen, H.A. Raue, and R.J. Planta, *Gene* **79**, 199–206 (1989).
14. J.R. Shuster, H. Lee, and D.L. Moyer, *Yeast* **6**, 579 (1990).
15. A. Sakai, F. Ozawa, T. Higashizaki, Y. Shimizu, and F. Hishinuma, *Bio / Technology* **9**, 1382–1385 (1991).
16. R.N. Parekh, M.R. Shaw, and K.D. Wittrup, *Biotechnol. Prog.* **12**, 16–21 (1996).
17. N. Shiomi, H. Fukuda, F. Murata, and A. Kimura, *Appl. Microbiol. Biotechnol.* **42**, 730–733 (1995).
18. X. Wang, Z. Wang, and N.A. DaSilva, *Biotechnol. Bioeng.* **49**, 45–51 (1996).
19. D. Mochizuki, K. Miyahara, D. Hirata, H. Matsuzaki, T. Hatano, S. Fukui, and T. Miyakawa, *J. Ferment. Bioeng.* **77**, 468–473 (1994).
20. T. Fujii, K. Kondo, F. Shimizu, H. Sone, J.-I. Tanaka, and T. Inoue, *Appl. Env. Microbiol.* **56**, 997–1003 (1990).
21. J.D. Boeke, H. Xu, and G.R. Fink, *Science* **239**, 280–282 (1988).
22. X. Wang and N.A. DaSilva, *Biotechnol. Bioeng.* **51**, 703–713 (1996).
23. T.D. Webster and R.C. Dickson, *Gene* **26**, 243–252 (1983).
24. R.C.A. Henderson, B.S. Cox, and R. Tubb, *Curr. Genet.* **9**, 133–138 (1985).
25. G. Loison, M. Nguyen-Juilleret, F. Alouani, and M. Marquet, *Bio / Technology* **4**, 433–437 (1986).
26. Eur. Pat. 017,114 (1986), G. Kawasaki.
27. S.M. Kingsman, D. Cousens, C.A. Stanway, A. Chambers, M. Wilson, and A.J. Kingsman, *Methods Enzymol.* **185**, 329–341 (1990).
28. S. Rosenberg, D. Coit, and P. Tekamp-Olson, *Methods Enzymol.* **185**, 341–351 (1990).
29. S. Elliott, J. Giffin, S. Suggs, E.P. Lau, and A.R. Banks, *Gene* **79**, 167–180 (1989).
30. A.S. Robinson and K.D. Wittrup, *Biotechnol. Prog.* **11**, 171–177 (1995).
31. L.M. Mylin, K.J. Hofmann, L.D. Schultz, and J.E. Hopper, *Methods Enzymol.* **185**, 297–308 (1990).
32. V.L. Price, W.E. Taylor, W. Clevenger, M. Worthington, and E.T. Young, *Methods Enzymol.* **185**, 308–318 (1990).
33. T. Etcheverry, *Methods Enzymol.* **185**, 319–329 (1990).
34. A.J. Brake, *Methods Enzymol.* **185**, 408–420 (1990).
35. A. Helenius, T. Marquardt, and I. Braakman, *Trends Cell Biol.* **2**, 227–231 (1992).
36. C. Hammond and A. Helenius, *Curr. Opin. Cell Biol.* **7**, 523–529 (1995).
37. A. Robinson, V. Hines, and K.D. Wittrup, *Bio / Technology* **12**, 381–384 (1994).
38. L. Schultz, H. Markus, K. Hofmann, D. Montgomery, C. Dunwiddie, P. Kniskern, R. Freedman, R. Ellis, and M. Tuite, *Ann. N.Y. Acad. Sci.* **721**, 148–157 (1994).
39. A.S. Robinson, J.A. Bockhaus, A.C. Voegler, and K.D. Wittrup, *J. Biol. Chem.* **271**, 10017–10022 (1996).
40. A.J. Dorner and R.J. Kaufman, *Biologicals*, **22**, 103–112 (1994).
41. R.N. Parekh, K.J. Forrester, and K.D. Wittrup, *Protein Exp. Purif.* **6**, 537–545 (1995).
42. C. Lauritzen, E. Tuchsen, P.E. Hansen, and O. Skovgaard, *Protein Exp. Purif.* **2**, 372–378 (1991).
43. O.S. Gabrielsen, R. Reppe, O. Saether, O.R. Blingsmo, K. Sletten, J.O. Gordeladze, A. Hogset, V.T. Gautvik, P. Alestrom, T.B. Oyen, and K.M. Gautvik, *Gene* **90**, 255–262 (1990).

44. C.E. Ballou, *Methods Enzymol.* **185**, 440–470 (1990).
45. R.A. Smith, M.J. Duncan, and D.T. Moir, *Science* **230**, 1219–1224 (1985).
46. J.R. Shuster, D.L. Moyer, H. Lee, A. Dennis, B. Smith, and J.P. Merryweather, *Gene* **83**, 47–55 (1989).
47. G.S. Gill, P.G. Zaworski, K.R. Marotti, and E.F. Renberg, *Bio / Technology* **8**, 956–958 (1990).
48. V. Chisholm, C.Y. Chen, N.J. Simpson, and R.A. Hitzeman, *Methods Enzymol.* **185**, 471–482 (1990).
49. G.M. O'Connor, F. Sanchez-Riera, and C.L. Cooney, *Biotechnol. Bioeng.* **39**, 293–304 (1992).
50. O. Mendoza-Vega, J. Sabatie, and S.W. Brown, *FEMS Microb. Rev.* **15**, 369–410 (1994).

See also ENZYMES, DIRECTED EVOLUTION; ENZYMES, EXTREMELY THERMOSTABLE; SECRETION FROM ANIMAL CELLS.

PROTEIN ULTRAFILTRATION

ROBERT VAN REIS
Genentech, Inc.
South San Francisco, California
ANDREW L. ZYDNEY
University of Delaware
Newark, Delaware

KEY WORDS

Buffer Exchange
Concentration
Cross Flow Filtration
Desalting
Diafiltration
Filtration
Membrane
Protein Purification
Tangential Flow Filtration
Ultrafiltration

OUTLINE

INTRODUCTION

Protein ultrafiltration (UF) is a pressure-driven membrane process used for the concentration or purification of protein solutions. UF membranes typically have mean pore size between 10 and 500 Å, which is intermediate between reverse osmosis (RO) and microfiltration (MF). Although UF has often been viewed as a purely size-based separation, with species larger than the membrane pores being fully retained while smaller species pass freely, separation in UF actually occurs because of differences in the rate of filtration of different components across the membrane in response to a given pressure driving force. Solute filtration rates, and thus the membrane selectivity, are determined by both thermodynamic (e.g., partitioning) and hydrodynamic (e.g., transport) interactions. UF is currently used throughout downstream processing for protein concentration, buffer exchange and desalting, protein purification, virus clearance, and clarification. Buffer exchange and desalting are typically accomplished using a diafiltration (DF) mode in which the low molecular weight components are washed away from the protein by simultaneously adding fresh buffer (or solvent) to the feed during UF. Protein purification is accomplished using high-performance tangential flow filtration (HPTFF), with the product collected in either the retentate or filtrate depending on the relative filtration rates. Virus filtration (VF) uses larger pore UF membranes that are permeable to protein but provide significant virus removal.

PRINCIPLES OF ULTRAFILTRATION AND HIGH-PERFORMANCE TANGENTIAL FLOW FILTRATION

Background and Theory

General Operating Principles. Although protein UF has been used in the laboratory since the turn of the century, large-scale industrial applications of UF only date back to about 1970. The key breakthrough was the development of the asymmetric membrane by Loeb and Sourirajan and the subsequent extension of this technique to the production of UF membranes by Michaels and colleagues at Amicon. These asymmetric membranes have a very thin skin (approximately 0.5 μm thick) that provides the membrane its selectivity and a more macroporous substructure that provides the required mechanical and structural integrity. The thin skin results in much higher filtration rates than obtainable with homogeneous membranes, significantly reducing the required membrane area or process time. Typical process fluxes in UF range from 25 to 250 L m^{-2} h^{-1}. Operating pressures are generally in the range of 0.2 to 4 bar, which are much lower than the 25 to 80 bar required in RO to overcome the large osmotic pressure of the retained salts. UF membranes are available with a variety of pore sizes and can be cast from a range of base polymers, each having unique physical and chemical characteristics.

Most large-scale UF devices use tangential flow filtration (TFF), also referred to as a cross-flow configuration, in which the feed flow is parallel to the membrane and thus perpendicular to the filtrate flow (1). This allows retained species to be swept along the membrane surface and out the device exit, significantly increasing the process flux compared to that obtained with dead-end operation. A concentration boundary layer consisting of a high concentration of retained solutes typically develops at the upstream surface of the membrane. This highly concentrated region reduces the effective pressure driving force and can cause extensive membrane fouling from protein adsorption, denaturation, precipitation, or aggregation. The performance of many TFF systems is determined almost entirely by the rate at which these retained solutes are transported away from the membrane and back into the bulk solution, a phenomenon referred to as concentration polarization. A variety of membrane modules have been developed to achieve the desired mass transport rates while maintaining high membrane packing densities, and these are discussed in some detail in "Modules and Devices". Most units are produced in modular form to allow easy replacement and scale-up. Membrane devices are available with surface areas as small as 50 cm^2 for bench-top operation to more than 200 m^2 for large-scale processing. Dead-end (or normal flow) filtration is used primarily for laboratory-scale separations and for systems in which the retained species are present at very low concentration. For example, dead-end MF cartridges have been used extensively for sterile filtration, in which the retained species are present at very low concentration. Similar modules can be used for virus removal.

Conventional UF has traditionally been used to separate solutes that differ by more than 10-fold in size, making it ideal for protein concentration and buffer exchange (2). UF processes are thus used throughout the downstream purification, often to condition the product before or immediately after a chromatographic column. The fairly limited selectivity of UF has generally been attributed to the wide pore size distributions in commercial membranes, the presence of significant bulk mass transfer limitations in existing modules, and membrane fouling phenomena. Recent work (3) has demonstrated that membrane systems can be used for protein–protein separations, even for proteins with very similar molecular weight, using a process known as HPTFF. High selectivity values are obtained in HPTFF by careful choice of both membrane and buffer conditions, with the latter chosen to maximize the difference in hydrodynamic volume for the two proteins. Concentration polarization is exploited to enhance, rather than limit, solute separation. Membrane fouling is controlled using appropriate fluid dynamic and mass transfer conditions. High yields and purification factors are obtained using a diafiltration mode. The net result is that HPTFF provides a competitive separation process that complements chromatographic techniques for protein purification.

Membrane Transport. Theoretical analyses of solute and solvent transport through UF membranes are generally developed using either the Kedem-Katchalsky or Stefan-Maxwell descriptions of irreversible thermodynamics (4). The filtrate flux J (equal to the volumetric filtration rate normalized by the membrane surface area) is proportional to the effective pressure driving force (5):

$$J = L_p[\Delta P - \Sigma(\sigma_i \Delta \Pi_i)] \tag{1}$$

where L_p is the overall permeability (equal to the reciprocal of the overall resistance), and ΔP is the transmembrane pressure. L_p is only equal to the clean membrane permeability (evaluated from water flux data obtained with the clean membrane) in the absence of irreversible fouling. The osmotic reflection coefficient (σ_i) provides a measure of the membrane selectivity: $\sigma_i = 1$ for a solute that is completely retained and $\sigma_i = 0$ for a solute that is completely permeable. The osmotic pressure difference across the membrane ($\Delta\Pi_i = \Pi_w - \Pi_f$) is a function of the solute concentration at the upstream surface (C_w) and in the filtrate solution (C_f). Although protein osmotic pressures are typically small for $C < 10$ g/L, $\Delta\Pi_i$ can be comparable to ΔP during UF because of the buildup of retained protein at the membrane surface (see "Concentration Polarization").

The solute flux N_s (in units of grams protein collected in the filtrate per membrane surface area per time) is conveniently expressed as (6)

$$N_s = JS_\infty C_w \tag{2}$$

where S_∞ is the intrinsic (or asymptotic) sieving coefficient, which is approximately equal to $1 - \sigma_i$ based on Onsager reciprocity. Equation 2 is valid only under conditions where the solute flux is dominated by convection; a more detailed analysis accounting for the coupling between convection and diffusion is given by Opong and Zydney (6). The asymptotic sieving coefficient is determined by thermodynamic partitioning of the solute into the membrane and the relative rate of solute transport through the pore (7):

$$S_\infty = \phi K_c = C_f/C_w \tag{3}$$

where K_c is the hydrodynamic hindrance factor, C_f is the filtrate concentration, and C_w is the retentate concentration at the membrane wall. The hydrodynamic hindrance factor accounts for the additional drag on the solute because of the presence of the pore wall. Theoretical analyses for a hard sphere in a cylindrical pore indicate that K_c only varies between 1.00 and 1.47 (8); thus S_∞ is determined primarily by the equilibrium partition coefficient (ϕ), which is a function of both steric (size) and long-range (e.g., electrostatic) interactions. The electrostatic energy of interaction between a protein and a pore has three distinct contributions (9). In UF systems, the dominant factor is typically the charge on the solute (σ_s). The ϕ decreases with increasing σ_s because of the increase in free energy associated with the distortion of the electrical double layer surrounding the protein caused by the presence of the pore wall. This can be interpreted as an increase in effective protein size (or hydrodynamic volume) associated with the diffuse ion cloud (10). This effect is the basis for the 100-fold decrease in S_∞ with decreasing salt concentration seen for bovine serum albumin (11). There are also contributions from the double layer surrounding the pore wall (which reduces the effective pore size) and from direct charge–charge interactions, with the latter causing an increase in ϕ when the solute and pore have opposite charge.

In the absence of long-range interactions, the sieving coefficient for a spherical solute in a cylindrical pore is (7)

$$S_\infty = (1 - \lambda)^2 \quad (4)$$

where λ is the ratio of the solute (r_s) to pore radius (r_p). Thus, even a membrane with a single pore size has a gradual decrease in S_∞ because of the reduction in accessible pore volume with increasing r_s. The retention characteristics of commercial UF membranes are more complex because of the broad distribution of irregularly shaped pores. A simple approximate analysis yields (12)

$$S_\infty = \exp(-r_s/s) \quad (5)$$

where s provides a measure of the mean pore size. In this case, S_∞ approaches zero asymptotically because of the presence of a small number of very large pores. The effects of different pore size distributions on S_∞ are discussed elsewhere (12).

Concentration Polarization. One of the critical factors determining the overall performance of any UF device is the rate of protein transport in the bulk solution adjacent to the membrane. The filtrate flow causes an accumulation of partially (or completely) retained solute at the upstream surface of the membrane, a phenomenon referred to as concentration polarization. The solute concentration thus varies from its maximum value at the membrane surface (C_w) to its bulk value (C_b) over the thickness of the concentration boundary layer (δ). Most analyses of concentration polarization have used the simple stagnant film model originally presented by Michaels (13):

$$J = k \ln([C_w - C_f]/[C_b - C_f]) \quad (6)$$

where the solute mass transfer coefficient (k) is equal to the ratio of the solute diffusivity (D) to the boundary layer thickness (δ). More rigorous analyses of solute transport indicate that equation 6 is valid at low degrees of concentration polarization, with the actual concentration dependence becoming more complex at high C_w/C_b (14). The wall concentration increases with increasing filtrate flux, causing an increase in $\Delta\Pi$ and a reduction in the effective pressure driving force (equation 1). This negative feedback balances any increase in applied pressure, with the net result that the flux approaches a nearly constant, pressure-independent value at high ΔP (Fig. 1). High wall concentrations can also reduce L_p through irreversible fouling or the formation of a protein gel on the membrane surface. In the latter case, the pressure-independent flux is determined by the maximum solubility or gel concentration for the protein. Although this gel polarization model has been used quite extensively, the calculated values of C_w are device-dependent and are often in poor agreement with independent estimates of protein solubility (8).

Theoretical analyses of solute mass transfer coefficients (k) in laminar flow are generally based on the Leveque solution (1):

$$k = 0.807(D^2\gamma/L)^{1/3} \quad (7)$$

which predicts a two-thirds power dependence on the protein diffusivity (D) and a one-third power dependence on

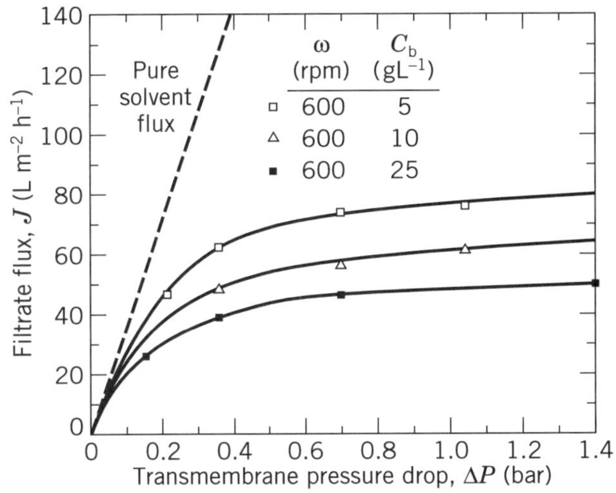

Figure 1. Filtrate flux as a function of transmembrane pressure for stirred-cell UF of BSA solutions of different bulk concentrations (C_b) using a constant stir speed (ω). Solid curves are model calculations based on equations 1 and 6. *Source*: Figure adapted from Opong and Zydney (6).

shear rate (γ). Empirical correlations for k have been developed from pressure drop data using the Chilton-Colburn analogy with a power law dependence on feed flow rate around 0.5 for spacer-filled channels and near 1.0 for turbulent flow (8). The most accurate method for evaluating the mass transfer coefficient, particularly for modules with complex fluid flow characteristics (see "Modules and Devices") is to fit equation 6 to actual filtrate flux data. The simplest approach is to evaluate k from the slope of a plot of the pressure-independent flux as a function of $\ln(C_b)$ (assuming a fully retentive membrane). This requires minimal experimental data, but the results can be in considerable error because of uncertainties in the limiting flux. In addition, the resulting value for k is only valid under conditions of very high polarization; the behavior at lower degrees of polarization can be considerably different because of the concentration dependence of the solution viscosity and protein diffusivity (8). The mass transfer coefficient can also be evaluated by fitting filtrate flux versus ΔP data to equations 1 and 6 simultaneously, which was the approach used to generate the solid curves in Figure 1. This requires an estimate of the process membrane permeability, which can be evaluated from buffer flux data obtained with the fouled membrane, and an expression for the protein osmotic pressure as a function of C_w. The osmotic pressure can be evaluated independently, or it can be described using a simple polynomial expansion, in which case k and the osmotic coefficients are fit simultaneously. van Reis et al. (15) developed a novel approach that eliminates the need to assume a specific functional relationship for $\Delta\Pi$ by simultaneously analyzing J versus ΔP data obtained at different C_b.

The accumulation of retained solute at the membrane also affects protein retention. Equation 6 can be rearranged to evaluate the observed protein sieving coefficient using $N_s = JC_f$, yielding (6)

$$S_o = C_f/C_b = S_\infty/((1 - S_\infty)\exp(-J/k) + S_\infty) \quad (8)$$

At low filtrate flux, $S_o = S_\infty$ because C_w is approximately equal to C_b. The observed sieving coefficient increases with increasing flux, approaching a value of 1 (i.e., $C_f = C_b$) at very high flux as long as the membrane is at least partially permeable to the solute and the required wall concentration ($C_w = C_b/S_\infty$) is physically attainable. Typical experimental results for bovine serum albumin are shown in Figure 2. The solid curve is the model calculation using equation 8. The slight increase in S_o at very low filtrate flux is because of the diffusive contribution to protein transport through the membrane (6).

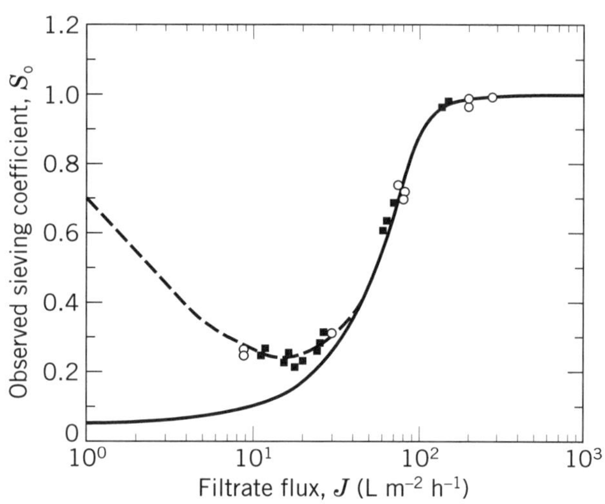

Figure 2. Observed sieving coefficient as a function of filtrate flux for the UF of BSA through a 100-kD poly(ether sulfone) membrane. Solid curve is model calculation based on equation 8. Dashed curve accounts for diffusive contribution. *Source*: Figure adapted from Opong and Zydney (6).

Membrane Properties

Membrane Chemistry and Morphology. Almost all UF membranes have an anisotropic or asymmetric structure, with the membrane selectivity determined by a thin skin layer and the mechanical integrity provided by a much thicker macroporous support. Most asymmetric UF membranes are prepared by immersion casting (8). A thin film of a polymer solution is spread on a suitable porous or woven support (e.g., polyethylene). The cast film is then immersed in a nonsolvent (often water). Diffusional exchange occurs between the solvent and nonsolvent, causing phase separation of the polymer solution and subsequent gelation of the polymer-rich phase. This gives the membrane a nodular morphology with the pore size determined by the size and spacing of the polymer-rich globules. The asymmetric structure develops because of mass transfer (kinetic) effects, such as growth and coalescence of polymer-rich globules before gelation (8). Large tear-shaped macrovoids often form beneath the membrane skin because of interfacial instabilities or the growth of polymer-lean regions in the film (16). Membranes with macrovoids have high permeability but lower maximum pressure limits. More importantly, defects can form if the macrovoids extend up to (or through) the skin. Hollow fiber membranes are also made by phase inversion with the polymer solution extruded through an orifice into a gelation bath while a second gelation medium is pumped into the fiber lumen. Skin layer(s) develop on the inner or outer surface, depending on the properties of the gelation media. Composite membranes are produced using multiple casting steps. For example, very uniform UF membranes that are free of macrovoids can be produced by casting a thin (5 to 10 μm) polymer film on a MF substrate (17). These membranes are particularly attractive for virus filtration where the presence of even trace defects leads to poor virus removal.

UF membranes can be cast from a wide range of polymers, including polysulfone, poly(ethersulfone), polyamides, polyimides, poly(vinyl chloride), polyacrylonitrile, poly(vinylidene fluoride), regenerated cellulose, and cellulose acetate. Cellulosic membranes bind the least amount of protein but can be damaged by harsh cleaning protocols. The synthetic polymers have much greater chemical and thermal stability but are more susceptible to protein fouling. Ceramic membranes, produced by slip-casting of Al_2O_3 or acid leaching of SiO_2 (18), have very high thermal and chemical stability but tend to be considerably more expensive than polymeric membranes. Many membranes are modified during or after casting by inclusion of polyvinylpyrrolidone in the initial polymer solution, sulfonation of cast polysulfone, radiation-grafting of methacrylate monomers to polyamides, or coating with hydroxypropyl cellulose (19). These modifications increase membrane hydrophilicity and alter membrane surface charge, but they can reduce long-term chemical and thermal stability.

Membrane Characterization. Information on UF membrane characterization most commonly provided to end users consists of water flux measurements, solute retention tests, selected protein binding data, chemical compatibility, and thermal limits. Water flux data provide a measure of the membrane permeability as determined by evaluating the filtrate flux (filtration rate normalized for surface area) as a function of the transmembrane pressure (ΔP). Normal flow filtration devices yield a linear relationship between flux (J) and transmembrane pressure as long as membrane compaction is avoided. The situation is more complex in TFF because the conversion of feed flow to filtrate flow creates a nonlinear pressure drop through the feed channel. A correct value for the permeability of a membrane in a TFF device can be obtained by measuring flux versus transmembrane pressure and extrapolating $J/\Delta P$ versus ΔP to $\Delta P = 0$. Membrane permeability is primarily determined by pore size distribution, porosity (pore density), membrane thickness, and solvent viscosity. Several issues should be considered when judging membrane permeability. The type and quality of water used for the tests can have a significant influence on the results. The pH and ionic strength of the water can influence the flux because of charge interactions, and contaminants can foul the membrane during the test. It is also important to carefully control and report the temperature of the test because permeability is linear with viscosity. As previously mentioned, membrane compaction and device configuration can also play a significant role in these tests. It should be noted that membrane water permeability data are of minimal value unless used in conjunction with other membrane characterization information. Water permeability data are not predictive of process fluxes because of significant differences in fouling behavior of various membrane chemistries. Polysulfone membranes typically have much higher water permeability values than regenerated cellulose membranes, whereas the latter membranes typically have higher process permeability values as a result of their inherent low fouling characteristics with most biotechnology feedstreams.

It is important to consider the intended use of membranes when selecting a retention test. The process objectives, and hence the criteria for membrane selection, are very different for protein concentration, buffer exchange, virus–protein separations, clarification, or protein purification. Retention testing of UF membranes involves measurements of solute transmission. Sieving coefficients (S) or retention coefficients (R) are determined by analyzing the solute in the feed and filtrate streams under specific process conditions.

$$S = C_f/C_F \tag{9}$$

$$R = 1 - S \tag{10}$$

Tests are carried out using either single solutes or multiple solutes in separate experiments. The most commonly used single solutes are polymers [such as poly(vinylpyrrolidone)], proteins and viruses. The purpose of the retention tests are to provide nominal molecular weight cutoff values for each membrane. Single-solute retention tests are the easiest to perform and provide either the least or most useful data, depending on the chosen solute and test conditions. Single-solute tests performed by membrane manufacturers typically involve the use of one or several experiments with solutes that are partially retained by the membrane. This information is of limited value for protein concentration and buffer exchange applications because the coefficients of partially retained solutes are not predictive of the retention values one would obtain for a highly retained solute. The retention of solutes with $R = 0.3$ and $R = 0.9$, for example, do not provide guidance on the molecular weight of a solute with a retention coefficient of $R = 0.999$, which is what one would strive for in a protein concentration–diafiltration operation. In all applications (UF, VF, and HPTFF), retention coefficients will also be highly dependent on fouling, buffer chemistry, and fluid dynamics. Furthermore, there is a lack of standardization of retention testing protocols between manufacturers. Different solutes, buffers, and fluid dynamic conditions are used. In addition, the protocols for assigning nominal molecular weight designations (for example, the molecular weight of a solute that has a retention coefficient of $R = 0.9$) are not even standardized within a given company. It is therefore imperative that the end users perform their own retention testing on the solutes of interest under conditions that are a linear scale-down of the intended process.

The mixed dextran test (20) provides an alternative methodology for membrane characterization. A mixture of several polydisperse dextrans are used with the feed and filtrate streams analyzed using size exclusion chromatography (SEC) with an on-line refractive index (RI) detector. Column calibration is performed using several narrow molecular weight distribution standards. The SEC chromatograms represent the solute concentrations (RI absorbance) as a function of dextran molecular weight (elution time). A continuous retention curve (retention versus molecular weight) can hence be generated by constructing a curve using equations 9 and 10 with the feed and filtrate concentrations represented by the respective chromatograms. Because the SEC chromatograms provide elution times for every molecular weight fraction, it is not necessary to use dextrans with tight molecular weight distributions. Mixed dextran tests performed under conditions of constant mass transfer coefficient and flux are a useful tool in comparing the relative retention characteristics of various UF membranes. The method is a good quality-control test for both membrane manufacturers and end users. The method can also be used to derive information on pore size distributions (21). Some predictive power can also be derived from this methodology by converting both dextran and protein molecular weights to equivalent hydrodynamic volumes (22).

A more sensitive method of characterizing UF membranes involves the principle of liquid–liquid pore intrusion. Two highly immiscible liquids, such as solutions of a sulfate salt and a poly(ethylene glycol) are contacted through mixing to reach equilibrium partitioning. The membrane to be tested is primed with one of the liquids so that all the pores are filled. After draining the feed channels, the second fluid is introduced into the system. The first fluid is then displaced out of the pores by the second fluid, and the flow rate is measured as a function of transmembrane pressure. The resulting data provide information on pore size distribution and can be correlated with the nominal molecular weight cutoff. It has also been demonstrated that this methodology can be used as an excellent quality-control test for membranes used in HPTFF applications because performance can be correlated with protein sieving data (23).

Protein binding studies are often performed to contrast the relative behavior of various membrane chemistries. The data generated in these studies can be useful as a general guide. It is important, however, to recognize that protein adsorption will be highly dependent on pH, ionic strength, and the specific isoelectric point and hydrophobicity of a particular protein. It is also useful to note that most membranes will not cause significant protein losses from adsorption because of the relatively low binding capacities (1 to 10 mg/m^2), the small surface areas required (0.01 m^2/g), and the high protein concentrations (0.05 to 1 g/L) obtained in modern biotechnology processes. This is even true for processes such as MF, VF, and HPTFF, in which the product may be exposed to the internal surface area, which is 10 to 500 times the frontal membrane area. Chemical and thermal limits for membranes are determined by characterizing the membranes (using the techniques described herein) both before and after controlled exposure. Membrane manufacturers often perform additional tests to determine pore size distribution, porosity, morphology, thickness, charge, process flux, and mechanical strength (compaction, tensile strength, and delamination). Pore size distribution and porosity can be determined by either mixed dextran, liquid–liquid intrusion, mercury intrusion, air–liquid intrusion (for large pore sizes), thermoporometry, scanning electron microscopy, or atomic force microscopy. Membrane morphology and thickness are determined by either scanning electron microscopy or transmission electron microscopy. Membrane charge can be determined by streaming potential measurements and is of particular use in HPTFF. Process fluxes provide some complementary information to the water permeability values provided by manufacturers but are best carried out by the end user on the specific application of interest. Determination of mechanical strength should include testing of membrane compaction, tensile strength, and ability to withstand reverse pressure. Soft membranes are susceptible to membrane compaction under process conditions. Membrane compaction can easily be determined by measuring water permeability as a function of transmembrane pressure throughout the desirable ΔP range. Tensile strength is most relevant to the membrane manufacturers and is analyzed using specialized equipment that measures the elongation of the membrane as a function of applied force. The reverse pressure limits of membranes are important to the end user, because unintended pressure spikes may occur during processing with resultant delamination of the membrane and subsequent product loss. Testing is done by subjecting the membrane to reverse pressure cycles at various pressures, followed by retention testing as outlined earlier.

Integrity testing of UF membranes can be performed by bubble-point measurement, air-diffusion flow measurement, or the liquid–liquid intrusion method just described. Bubble-point determinations involve measuring a gas flow rate through a membrane at increasing levels of pressure. The bubble point is determined by an inflection in the flow rate versus pressure curve indicative of liquid displacement out of the pores. Bubble-point measurements are primarily used for MF membranes that have bubble point values in the 1 to 6 bar range (for 0.65 to 0.1 μm pore sizes) using water as solvent. Bubble-point measurements of UF membranes are only practical when using much lower surface tension solvents, because the bubble point values using water as a solvent often exceed the pressure ratings of either the membrane or device. UF membrane integrity is therefore commonly performed using air (or nitrogen) diffusion methods. Air diffusion measurements are done at a single pressure and strive to detect abnormally high flow rates because of convective flow through membrane defects that are large enough to permit the bubble point to be exceeded.

Membrane Fouling and Regeneration. Membrane fouling is a critical factor in almost all UF processes. Fouling refers to the irreversible alteration in membrane properties caused by specific physical or chemical interactions between the membrane and the various components present in the feed. Fouling can occur because of protein adsorption, deposition, precipitation, denaturation, and aggregation, all of which affect membrane performance as well as product yield and quality. The rate and extent of fouling are determined by the morphological and chemical characteristics of the membrane, the fluid mechanics in the module, and the buffer chemistry (24). In addition, protein aggregation or denaturation can occur during repeated pump passes, with these species then deposited on the membrane during UF. Static protein adsorption (in the absence of filtration) generally attains monolayer values on the accessible membrane surface (25), with less adsorption on hydrophilic membranes. As discussed previously, adsorptive losses are small even when the protein has access to the internal pore surface. The convective filtrate flow during UF causes significant additional fouling that can be well beyond that associated with simple adsorption. Thus, considerable care must be taken in predicting fouling levels from static protein binding data. The initial fouling during MF is often caused by physical deposition of large protein aggregates on the membrane surface (26). Protein fouling in UF is strongly influenced by the magnitude of the local protein concentration at the membrane surface, with high values of C_w leading to increased adsorption, precipitation, aggregation, or denaturation. Protein fouling is generally maximum near the protein isoelectric point (27) and is enhanced under conditions where the protein and membrane have opposite charge. The extent of fouling is very sensitive to start-up conditions, with less fouling seen when the system is slowly ramped to the desired filtrate flux or transmembrane pressure (3).

Membrane cleaning or regeneration is performed whenever the flux or selectivity drop below some minimally acceptable level. In addition, membrane disinfection and sterilization must be performed between process runs even in the absence of significant fouling. Cleaning can be accomplished by physically removing the foulants from the membrane, for example, by fluid recirculation at zero transmembrane pressure, backflushing with negative ΔP, or mechanical scrubbing (generally limited to large diameter tubular modules). Chemical cleaning is done using appropriate detergents, acids, alkalis, enzymes, or chelating compounds. These cleaners must effectively displace, solubilize, or chemically modify the foulants while not exceeding the mechanical, thermal, or chemical limits of the membrane polymer or module. For example, cellulosic membranes are highly susceptible to damage by extremes of pH and temperature (28). Polysulfone and poly(ether sulfone) membranes are much more stable, with some manufacturers reporting acceptable performance even after limited exposure to temperatures as high as 125 °C (29). These membranes can also be operated over a wide range of pH (1 to 13) and can tolerate low levels of chlorine (up to 200 ppm). Alkali cleaners (e.g., NaOH) are particularly effective for removing biological foulants (30). Many alkali cleaners are used in combination with chelating agents (e.g., citrate or EDTA) to remove free Ca^{2+} and Mg^{2+} to prevent precipitation of saponified fats and oils (8). Detergents such as SDS are also effective in many systems. Enzymatic cleaners can be used for membranes that are unable to withstand elevated temperatures, strong chemicals, or pH extremes, but these solutions tend to be very expensive and care must be taken to ensure complete removal of residual proteases. Membrane cleaning remains very much an art, with the development of an optimal cleaning cycle for a given application determined largely by trial and error.

Sodium hypochlorite (NaOCl) is used extensively for chemical disinfection of membrane systems. Hypochlorite is most effective at low pH, but it is very corrosive under these conditions (31). Disinfection is typically performed at pH≈10 using fairly concentrated solutions (0.25 to 4% NaOCl). Other oxidizing agents used for membrane disinfection include hydrogen peroxide (H_2O_2) and peracetic acid ($CH_3COCOOH$) (32). Peracetic acid is attractive because it is fast-acting, and it decomposes into the nontoxic acetic acid. Sodium bisulfite ($NaHSO_3$), a strong reducing agent, can be used for disinfection of membrane systems that would be degraded by strong oxidizing agents such as hypochlorite (33). Steam sterilization can be performed on disassembled membrane systems in an autoclave, but the membrane unit must then be aseptically attached to the remainder of the presterilized system (34). Steam-in-place (SIP) is the preferred method of sterilization, with the entire unit exposed to flowing steam as part of the completely assembled filtration system. Minimum requirements for an effective steam sterilization are 15 min exposure to steam at 121 °C and 1 bar pressure. Effective validation of SIP procedures is essential (35). Membrane systems can also be sterilized using ethylene oxide (EtO) or chlorofluorocarbons (CFCs), but these gas sterilization procedures require fairly long incubation periods to ensure complete deactivation of all microorganisms and spores (35). Gamma irradiation can also be used, although some membrane materials (e.g., PTFE and cellulosics) are damaged on exposure to high radiation doses (35).

Modules and Devices

The membrane module is the physical unit that houses the UF membrane in an appropriately designed configuration. The module must provide physical separation of the retentate and filtrate streams, mechanical support for the membrane, high membrane packing densities, easy access for cleaning or replacement, and scalability. In addition, the feed channel must provide adequate bulk mass transfer rates without excessive feed flow rates. Many of these criteria are to a large extent contradictory, for example, devices with large membrane packing densities are generally more susceptible to particulate fouling and more difficult to clean. A number of commercial UF modules have been developed, differing primarily in the size and shape of the feed (and filtrate) flow channels. These are conveniently classified into five distinct types: flat-sheet cassettes, spiral-wound modules, hollow-fiber cartridges, tubular modules, and enhanced mass transfer modules. The latter category includes a wide range of system geometries and operating strategies. Additional information on the economics, fluid mechanics, and mass transfer characteristics of these modules are available in Zeman and Zydney (8).

Flat-Sheet Cassettes. Flat-sheet (also known as plate-and-frame) cassettes were among the earliest configurations developed for large-scale commercial applications. The module uses a support plate (which also defines the filtrate flow path), membrane, and channel spacer (which defines the retentate flow path) in a sandwich configuration (Fig. 3). The membranes can be sealed to the plates using gaskets, in which case the module (stack) is hydraulically clamped to form a tight fit. Alternatively, the membranes can be directly bonded or glued to the plates using heat sealing or an appropriate adhesive to form an integral membrane element. Several of these elements or plates are then stacked together and clamped to form a complete module. The feed flow is distributed among the different channels at one end of the device, and the retentate is collected at the opposite end. The filtrate is manifolded separately from the feedstream and can be combined within or outside the module depending on the particular device design. The feed channel is typically 0.03 to 0.1 cm in height. The channel can either be fully open or it can use an appropriate screen (typically a polypropylene mesh) to promote local mixing and improve mass transfer. The small channel height leads to relatively high membrane packing densities (generally 300 to 500 m^{-1}) and low hold-up volumes, which is particularly attractive for recovery of high-value products. Open channel cassettes are less susceptible to plugging and operate with smaller pressure drops (feed minus retentate pressure). However, back-flushing may not be possible because the membranes are effectively supported only on one side.

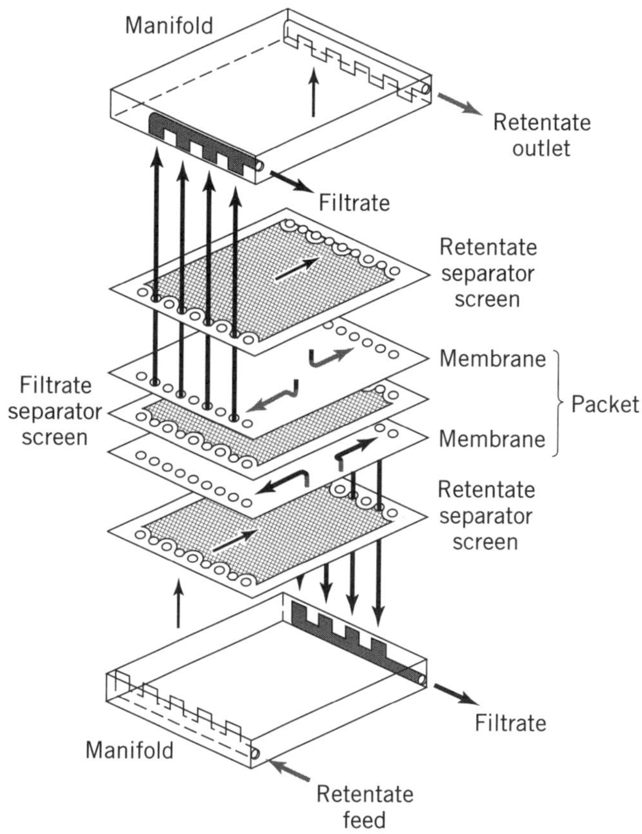

Figure 3. Schematic drawing of flat-sheet cassette. *Source*: Adapted from Ref. 8. Courtesy of Marcel Dekker, Inc.

Spiral Cartridges. Spiral-wound cartridges are constructed using flat-sheet membranes in the form of a pocket consisting of two membrane sheets separated by a highly porous permeable mesh that provides physical support for the membrane and defines the filtrate flow path (Fig. 4). The membranes are sealed along three edges to form a pocket using an appropriate epoxy or polyurethane adhesive. The open side of the pocket is attached (glued) to a central perforated tube that collects the filtrate flow. Several pockets are wound in a spiral configuration around a single collecting tube using a feed-side (polypropylene) mesh as a spacer between the pockets to establish the desired feed channel thickness. The use of multiple pockets reduces the total path length and pressure drop for the filtrate flow. The entire spiral is generally wrapped with fiberglass tape or shrink wrap for added mechanical strength and to prevent unwrapping of the spiral under pressure. The spiral-wound element is fitted into a tubular steel or plastic pressure vessel with an antitelescoping device placed at both ends to prevent distortion of the spiral during operation. This also allows several spiral elements to be inserted into a single tube. Pressurized feed solution flows through the plastic screens along the surface of the membranes with the retentate collected out the other end of the device. Filtrate flows into the closed membrane pockets and flows in a spiral path inward to the central tube. The region between the outer edge of the spiral element and the inside of the cylindrical housing is sealed to avoid fluid bypass. Spiral cartridges that direct the feed flow in the spiral direction and the filtrate flow in the axial direction have also been produced. Spiral-wound cartridges have relatively high membrane packing densities and very effective mass transfer characteristics because of the feed-side spacers. The cartridges also have fairly low energy costs because they provide effective mass transfer at relatively low flow rates. The primary disadvantage of the spiral wound modules is that they are highly susceptible to particulate fouling because of the narrow and irregular flow path through the spacers. In addition, the dead space between the outer edge of the element and the cylindrical housing is difficult to clean and highly susceptible to bacterial contamination. This is sometimes overcome by allowing a very limited feed bypass to continuously flush the annular space. There can also be problems with the integrity and chemical resistance of the seals used to form the pockets, particularly for repeated applications using strong cleaning solutions.

Hollow-Fiber Cartridges. Hollow-fiber cartridges use an array of narrow bore (self-supporting) fibers, typically with an anisotropic structure. Fiber diameters range from about 200 to 2,500 μm, with the fiber wall around 200 μm in thickness. The dense skin layer is typically at the lumen side of the fiber, although it can also be placed on the outside. A parallel array of 50 to 10,000 fibers is potted at the ends in an epoxy or polyurethane resin within a cylindrical cartridge of plastic or steel to form a tube that is machined to expose the open bores (lumens) of the fibers (Fig. 5). These modules are generally operated with the feed flow into the fiber lumens, with the filtrate moving in a radial direction outward through the fiber walls. The cartridge is typically equipped with a single feed inlet, a retentate outlet, and two filtrate ports (one near each end of the unit). The feed and filtrate flows can be cocurrent or countercurrent. Hollow fiber devices have high packing density because of the narrow diameter of the individual fibers and the ability to pack the fiber bundles tightly in the cylindrical cartridge (shell). The use of small diameter fibers

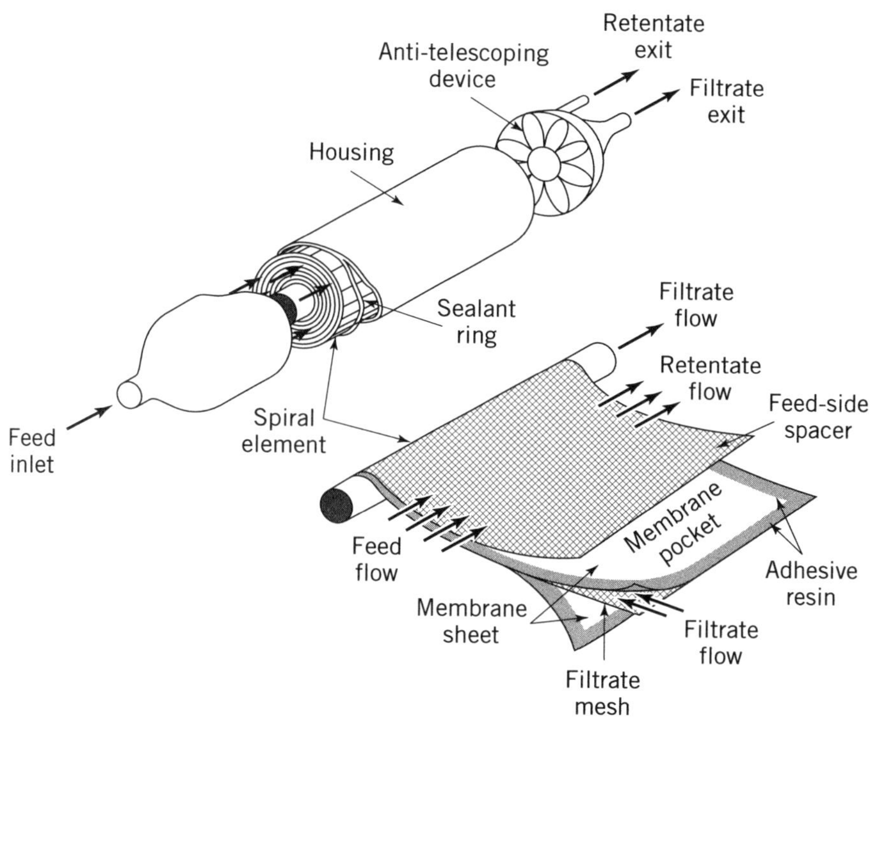

Figure 4. Schematic drawing of spiral-wound module showing membrane pocket. *Source*: Adapted from Ref. 8. Courtesy of Marcel Dekker, Inc.

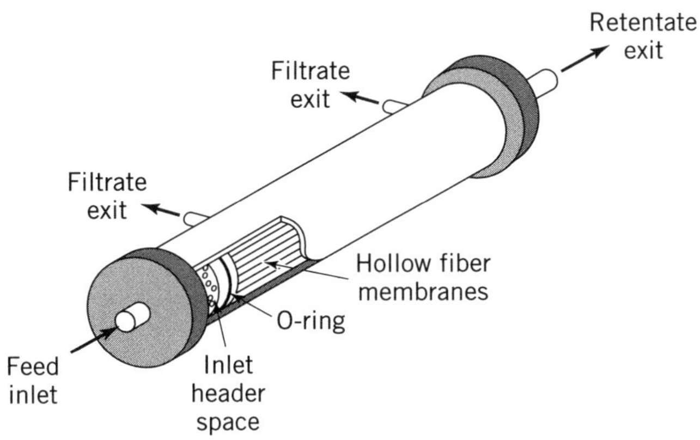

Figure 5. Schematic drawing of hollow-fiber module. *Source:* Adapted from Ref. 8. Courtesy of Marcel Dekker, Inc.

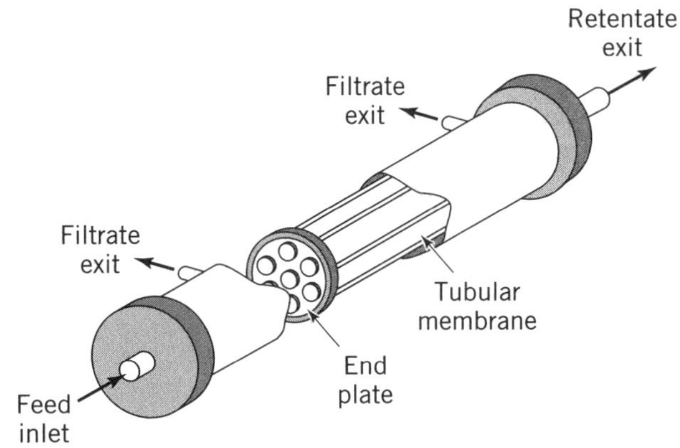

Figure 6. Schematic drawing of tubular module. *Source:* Adapted from Ref. 8. Courtesy of Marcel Dekker, Inc.

can provide effective mass transfer at relatively low feed flow rates. The absence of any spacers or multiple sealing procedures reduces labor costs involved in assembly. The hollow fiber membranes are self-supporting, which often allows them to be cleaned by backflushing of the device either by reversing the direction of the filtrate flow or by reversing the feed flow with the filtrate ports closed. One disadvantage of the hollow fiber modules is that they are susceptible to particulate plugging. Fiber breakage can also be a problem, so it is critical to follow the manufacturer's recommendations with respect to operating pressures. Fiber breakage can often be difficult to detect. Several manufacturers recommend using a modified bubble-point test to detect ruptured fibers.

Tubular Modules. Tubular modules are very similar in design to the hollow-fiber cartridges described in the previous section but use much larger diameter tubes (typically 0.3 to 2.5 cm). The tubular membranes are not self-supporting (with the exception of some inorganic membranes). The membranes are usually cast in place within a porous support tube made of fiberglass, ceramic, plastic, or stainless steel. The individual tubes can be placed inside a plastic or stainless steel sleeve to form a single tube cartridge, or they can be packed together in small bundles that are kept in place using an appropriate end plate (Fig. 6). The feed flows through the bore of the tubes, while the filtrate flows in a radial direction outward across the membrane and support tube where it is collected from the filtrate outlet ports. The two streams may be cocurrent or countercurrent. Inorganic membranes are typically made in a honeycomb monolith (18) in which the membranes are arranged in a parallel array. The unit is housed in an appropriate plastic or stainless steel cartridge. Tubular modules are typically operated in the turbulent flow regime to obtain effective mass transfer in the large diameter tubes. The primary advantages of the tubular modules are their resistance to particulate plugging and ease of chemical or mechanical cleaning. Physical cleaning of the membranes can be affected by passage of a rubber sponge, a scouring ball, or a narrow rod through the bore of the tubes. It is also possible to replace individual tubes in the multielement units in the field, resulting in a considerable reduction in membrane replacement costs and an increase in the overall unit lifetime. The primary disadvantages of the tubular modules are the low membrane packing densities ($<100\ m^{-1}$), high hold-up volumes, large feed flow rates, and high capital costs (including high floor space requirements).

Enhanced Mass Transfer Devices. Enhanced mass transfer modules exploit flow instabilities or turbulence to increase the transport of retained solutes. A variety of devices have been designed using pulsatile flow or membranes with pronounced surface corrugations (36). Alternatively, systems have been developed in which the feed flow generates secondary flow patterns or vortices. The most successful of these to date has been the rotating cylinder device originally proposed by Hallstrom and Lopez-Leiva (37). The feed flows into a thin (0.05 to 0.10 cm) annular region between two concentric cylinders, one (or both) of which is porous and has a membrane bound to the surface. The filtrate is collected in the central chamber (for a device with membrane on the inner cylinder) or in a separate annular filtrate region adjacent to the outer porous cylinder. The inner cylinder in these devices is rotated at high speed (typically >3,000 rpm) to induce Taylor vortices, which are caused by the centrifugal forces acting on the rotating fluid as it moves around the cylindrical flow path. The vortices effectively mix the fluid near the surface of the membrane, increasing the feed-side mass transfer coefficients. Mass transfer in the rotating system is determined almost entirely by the rotation rate of the inner cylinder, which effectively decouples the mass transfer characteristics from the feed flow rate. Consequently, rotating devices can be operated at very low feed flow rates. The use of low flow rates can be particularly attractive for many high viscosity applications. However, the rotating systems have very large capital costs, they are difficult to scale-up, and there is concern about long-time performance of a device with so many moving parts. In addition, membrane packing densities are low, and it is difficult to develop and maintain effective seals in such a rotating system.

Belfort et al. (38) and Winzeler (39) independently proposed using a vortex-flow system based on Dean, rather than Taylor, vortices. Dean vortices are induced by centrifugal forces exerted on a fluid as it flows inside a highly curved channel. This can be accomplished using spiral wound flat sheet membranes, with the channel curvature adjusted to maintain the Dean vortices throughout the system, or with hollow-fiber membranes wound around a central core at an appropriate pitch in a screwlike geometry (40). Both systems generate counter-rotating vortices that effectively depolarize the concentration boundary layer, but without the need for any moving parts or seals. Available data on the Dean vortex systems are very encouraging (38). These units have not yet been commercialized, so it is impossible to evaluate their economics and overall performance characteristics.

Another approach to achieving high mass transfer is to move the membrane, with the bulk of the fluid remaining largely stationary. For example, Vigo et al. (41) developed a module in which a cylindrical membrane is vibrated (longitudinally) inside a concentric cylindrical housing. Culkin and Armando (42) developed vibratory shear enhanced processing (VSEP) in which a stack of parallel circular membrane disks is spun at high speed inside a concentric cylindrical housing using a torsional spring. Shear rates at the membrane-feed interface can be as high as 150,000 s^{-1} for a 60-Hz oscillation. The high shear rates obtained in these systems can result in dramatic improvements in filtrate flux; however, important questions remain about the scalability and robustness of these modules as well as the overall economics.

High-frequency transmembrane pressure pulsing can also be used to enhance mass transfer. In contrast to classical backflushing, in which the reverse filtration occurs after several minutes or hours of normal filtration, transmembrane pressure pulsing is performed with a frequency around 0.1 to 1 Hz (43). The back pressure pulse is applied using nitrogen or air pressurization, with the frequency and duration of the pulse set using computer-controlled solenoid valves (43). Very short forward filtration times are used to prevent significant concentration polarization or fouling, thereby maintaining high average filtrate flux. Data for filtration of albumin–immunoglobulin mixtures showed more than an order of magnitude increase in flux (44), but a corresponding reduction in protein retention. Although transmembrane pressure pulsing systems appear to have excellent performance characteristics, they have not yet been used in large-scale commercial applications.

Process Configurations

Batch. UF and HPTFF processes consist of various combinations of concentration and diafiltration steps. A concentration step is either the main objective of the process or may be used as a means to optimize a subsequent diafiltration step (see "Design Equations" and "Process Optimization"). Concentration steps are carried out in either batch or fed-batch mode. In a batch mode, the entire volume of feed is contained within the recycle tank, as illustrated in Figure 7. Batch mode operations use a minimum of hardware, provide simple manual or automatic control, and provide the highest flux rates. Although the initial and final retentate concentrations are the same regardless of process configuration, the batch operation gives higher flux rates than a fed-batch operation because the bulk concentration follows a more dilute path from initial to final concentration. The disadvantages of a batch configuration can include less flexibility in using the same system for multiple processes, increased difficulty in designing a well-mixed system, and difficulties in obtaining high concentration factors. Mixing is important both to keep the bulk concentration at a minimum and for efficient buffer exchange during diafiltration. High concentration factors can in some cases be obtained in a batch mode using tanks with two or more sections of progressively decreasing diameters.

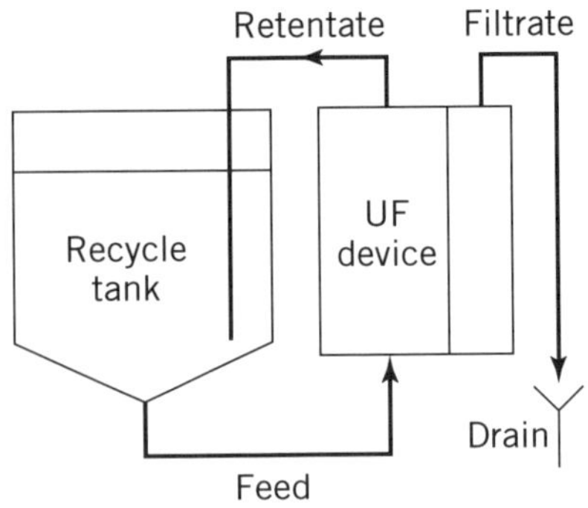

Figure 7. Batch-mode UF for a retained product.

Fed Batch. A fed-batch configuration uses an additional tank to feed into the recycle tank, as shown in Figure 8. Fed-batch configurations are commonly used in industry to obtain high concentration factors, provide flexibility for use in multiple processes, and provide well-mixed low hold-up retentate reservoirs for efficient diafiltration. In choosing a fed-batch configuration, several trade-offs must be carefully considered. For a given membrane area, process time will increase with the use of a fed-batch configuration because of the increase in bulk concentration compared to a batch operation. In addition, the number of times the retained solute passes through pumps and valves in the recycle line will increase because of the longer

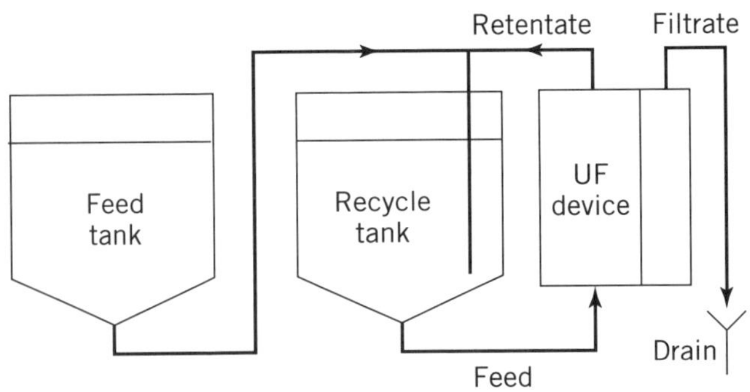

Figure 8. Fed-batch mode UF for a retained product.

processing time and the smaller retentate volume. An increased number of pump and valve passes can lead to shear-induced cell lysis and protein denaturation and aggregation. Protein denaturation in pumps and valves is believed to be the result of protein–gas interfaces (45) formed as a result of microcavitation.

Diafiltration. Diafiltration is an operation in which a solvent or solution (typically a buffer) is added to the recycle tank simultaneously with filtration (Fig. 9). The most common form is constant retentate volume diafiltration in which the buffer is added at the same rate as the filtrate rate. In UF processes, diafiltration is used to perform buffer exchange if the product is in the retentate, or to enhance yields if the product is in the filtrate. In HPTFF, diafiltration is used to achieve purification. In certain processes, it can be advantageous to perform simultaneous dilution or concentration with diafiltration (46). Optimization of diafiltration processes will be described in "Fluid Dynamics and Process Control" for UF and in "Process Optimization" for HPTFF.

Cascade Systems. In several plant designs, it may be advantageous to use cascaded process configurations. These configurations include both open-loop and closed-loop designs. In both configurations, the filtrate from the first stage is used as a feed to the second stage. In the open-loop design, the filtrate from the second stage is directed to drain while a replacement buffer is added to the first stage. In the closed-loop design (Fig. 10), the filtrate from the second stage is used as the diafiltration buffer for the first stage, thereby significantly reducing the amount of buffer required for the process. Cascade systems can be used to couple multiple process steps into a single operation. For example, a MF step used to harvest protein from a culture can be cascaded with a second-stage UF step to achieve simultaneous harvest and protein concentration. In HPTFF, two-stage cascade systems can be used in a similar manner when the product of interest is in the first-stage filtrate. Once a normal diafiltration has been performed on the first stage to achieve optimum buffer conditions (see "Buffer Chemistry"), a closed-loop configuration can be used to obtain a very large number of diavolumes by reusing a volume of buffer that is equal to the volume of retentate. It should be noted that the retentate volumes in the first- and second-stage tanks do not have to be the same. In cases where the buffer cost is very high, one might also consider using such a configuration even when the product is in the first-stage retentate if economically justified. Cascade configurations with more than two stages can also be used in processes where it is desirable to separate a feedstream into more than two fractions.

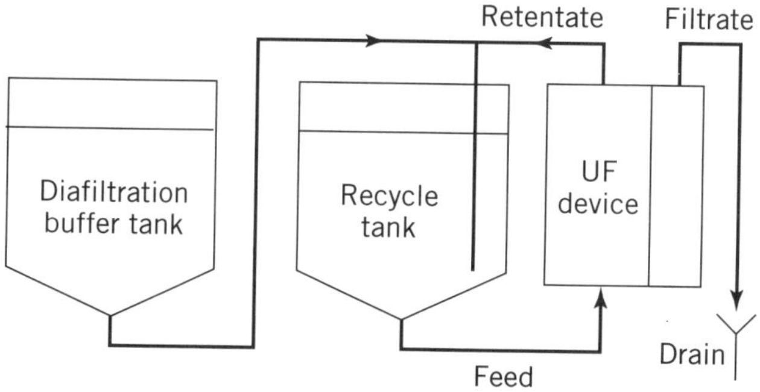

Figure 9. Diafiltration mode UF for a retained product.

PROCESS DESIGN

Ultrafiltration

Buffer Chemistry. Buffer exchange is often used to condition a product for subsequent loading onto a chromatographic column or for final formulation. The buffer chemistries used in ultrafiltration are hence often dictated by the preceding and subsequent steps in the process. Because of the very large amounts of protein processed per surface area of membrane, ultrafiltration can be used with a wide range of ionic strengths and pH values without significant adsorptive product loss. As discussed previously, adsorptive yield losses are usually insignificant ($<0.01\%$) for products in the retentate and can be limited to less than 5% for products in the filtrate. Although UF is routinely used to provide buffer exchange in preparation for subsequent processing, it can be advantageous to consider selecting buffer conditions for the sake of optimizing the UF step. Changes in pH and ionic strength can be made to affect protein solubility, stability, adsorption, and hydrodynamic volume. These parameters are directly linked to yield losses incurred at the membrane surface (see "Fluid Dynamics and Process Control"). In addition, buffer chemistry plays an essential role in membrane fouling, which can have a significant impact on process fluxes even when adsorptive product losses are minimized. The mass transfer coefficient can also be significantly enhanced by changes in buffer composition resulting in enhanced fluxes, reduced membrane area, and minimal volumetric hold-up losses. In general, higher fluxes are obtained at pH values either above or below the protein isoelectric point and at low to moderate ionic strength.

The theoretical removal of low molecular weight components during diafiltration is discussed in "Design Equations." It is important to note that one cannot assume that the retention coefficient for these components is zero despite the fact that the membrane may have a molecular weight cutoff an order of magnitude larger than the molecular weights of the components. Deviations in theoretical clearance of low molecular weight components can result from poor mixing in the recycle tank but can also result from buffer conditions. The use of detergents in either the feed material or in the diafiltration buffer can cause incomplete diafiltration if the critical micelle concentration is exceeded. Buffer components are entrained in the micelles, which may be retained by the membrane depending on micelle size and membrane molecular weight cutoff. It is generally advisable to avoid the use of detergents in UF for these reasons. Charged membranes may cause Donnan exclusion effects that result in increased retention of ionic solutes of all sizes. It has also been observed that low molecular weight ionic solutes can have

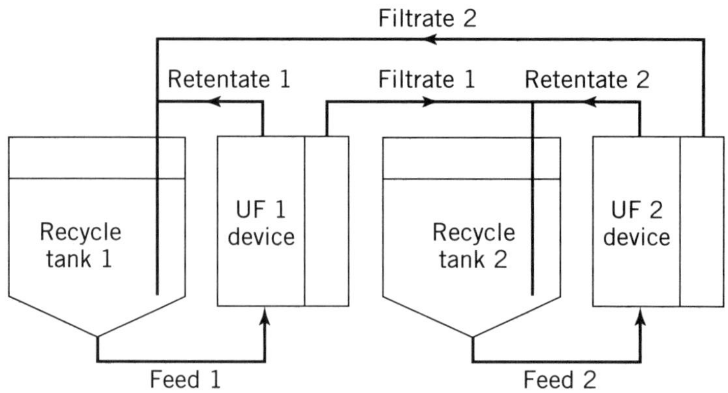

Figure 10. Two-stage closed-loop cascade configuration.

increased clearance, possibly as a result of Donnan exclusion by retained solutes.

Membrane Selection. Membrane selection should start with the choice of a high-quality vendor because robustness, reliability, and reproducibility of biotechnology manufacturing operations are of paramount importance. Consistent membrane and device characteristics can be as important to product quality, yield, and economics as are the inherent differences between various membranes and devices. The most important engineering parameters to consider during membrane selection are product retention, process permeability, mechanical strength, and chemical compatibility. Adsorptive properties play an important role in determining process permeability but are often less important with regard to product loss. Membrane chemistry should therefore be chosen in conjunction with the choice of buffer chemistry. Membrane charge should also be considered with regard to potential Donnan exclusion effects, as discussed previously. Water permeability values should not be used as a selection criteria since they are not predictive of process performance. Pore size distributions, porosity, morphology, and membrane thickness are only of indirect importance, because these parameters affect the ultimate retention, permeability, and mechanical characteristics of the membrane. As previously discussed ("Membrane Characterization"), retention properties should be determined on the actual solute under process conditions because marker studies can be completely misleading. Rules of thumb for selection of a membrane molecular weight cutoff relative to a protein molecular weight should also be avoided because protein hydrodynamic volumes vary drastically with pH and ionic strength (10) and because membrane manufacturers have not standardized their membrane characterization and nomenclature for UF membranes. The mechanical strength of a membrane can be important with regard to membrane compaction and reverse pressure tolerance. Membrane compaction can influence process permeability if retentate pressure limits are exceeded. The reverse pressure limits of a membrane are also important, because reverse pressure spikes are very difficult to avoid in industrial UF systems. The resulting membrane delamination can have a catastrophic impact on process yields. Chemical compatibility needs to be considered for the intended feedstream and the regeneration chemicals. Polysulfone and modified polysulfone membranes tend to have much broader chemical compatibility than regenerated cellulose membranes, but they also require much harsher chemical conditions for regeneration because of their inherent fouling characteristics. Long-term stability of membranes in bacteriostatic storage solution must also be carefully considered for economic reasons. The long-term stability of UF membranes should be evaluated by studying product retention as a function of membrane storage time in the chosen storage solution.

Fluid Dynamics and Process Control. Ultrafiltration processes can been performed using constant transmembrane pressure (ΔP), constant flux (J), or constant retentate wall concentration (C_w). The primary benefit of constant transmembrane pressure UF processes is the inherent simplicity of control, especially in manual systems. The feed rate is ramped up to the set point value, and a retentate valve is partially closed to achieve a specified retentate or transmembrane pressure:

$$\Delta P = (P_F + P_R)/2 - P_f \qquad (11)$$

where P_F is the feed pressure, P_R is the retentate pressure, and P_f is the filtrate pressure. Note that this equation is an approximation because the pressure drop along the feed channel is not truly linear because of conversion of feed to filtrate. If the feed stock viscosity changes during the process, one can adjust the retentate valve closure to maintain either constant retentate or transmembrane pressure. This is easily accomplished in both manual and automatic systems and often requires minimal adjustments after the initial set point has been achieved. The retentate pressure can also be controlled by using a gas overlay on the recycle tank. The feed rate and transmembrane pressure are empirically optimized by measuring the flux as a function of bulk solute concentration at various feed rates and transmembrane pressures. It is also important to measure solute sieving as a function of bulk concentration and flux. Even very small sieving coefficients may result in substantial product loss when a large number of diavolumes are required in DF processes. For example, a sieving coefficient of $S_o = 0.01$ will result in a 10% product loss during a 10-diavolume buffer exchange process (see equation 17).

In certain applications, it is not possible to maintain constant transmembrane pressure without severe reductions in filtrate rate. It has been demonstrated that one can obtain higher overall throughput in these applications (such as clarification) by operating at constant flux (47). This can be done by regulating the retentate pressure control valve or by using a pump on the filtrate line. The latter alternative provides more precise control and is easier to use in manual systems. Optimization is typically done through empirical studies varying feed rate and flux and determining process throughput, filtrate clarity, and product sieving.

In a third method of process control, the flux is varied with both retentate concentration and mass transfer coefficient to maintain a constant wall concentration of retained species (evaluated from equation 6). Control is performed using a control loop that measures flux and controls the transmembrane pressure with a retentate control valve to maintain a constant wall concentration (Fig. 11) throughout the concentration (15):

$$J = k \ln(C_w V / C_0 V_0) \tag{12}$$

where J is the flux, k is the mass transfer coefficient, C_w is the wall concentration of the retained solute, V is the retentate volume, V_0 is the initial feed volume, and C_0 is the initial retentate concentration. The benefits of constant C_w control are that product yield is maximized, product quality is ensured, membrane area is minimized, and process time is consistent and independent of membrane permeability. Process control for constant C_w during a concentration phase requires an automated system. Constant retentate volume DF using constant C_w, however, can easily be performed using either manual or automatic control because the C_w set point is maintained at a constant flux. Adjustments for changing mass transfer coefficients can be made during the DF process. The optimum C_w is determined by minimizing yield losses attributable to sieving, adsorption, solubility, and volumetric hold-up. (Fig. 12). The feed rate is optimized by evaluating a tradeoff between

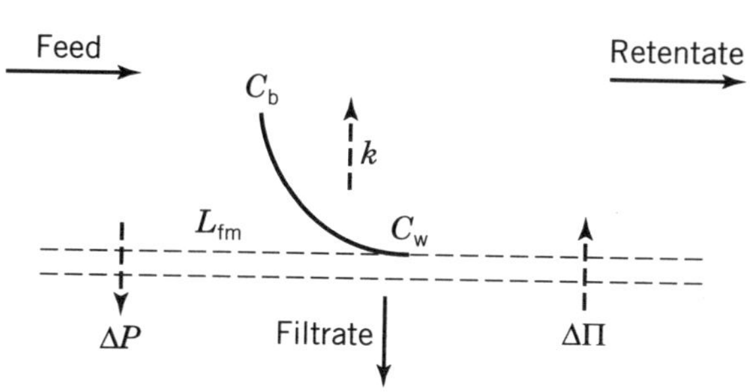

Figure 11. The concentration of a fully retained solute at the membrane wall (C_w) depends on the bulk concentration (C_b), the transmembrane pressure (ΔP), the fouled membrane resistance (L_{fm}), the osmotic pressure ($\Delta\Pi$), and the mass transfer coefficient (k). *Source*: Reprinted from R. van Reis et al. (15). With kind permission of Elsevier Science–NL, Amsterdam, The Netherlands.

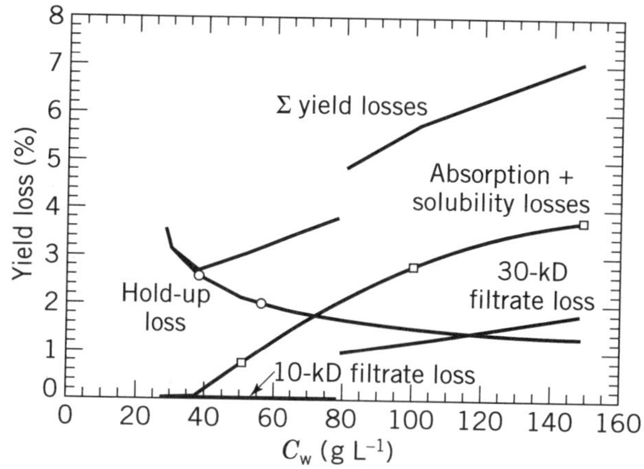

Figure 12. Yield losses in UF as a function of the wall concentration (C_w). Filtrate, adsorption, and solubility losses were determined with 0.1 to 0.4 m^2 linear scale-down devices. Hold-up losses were derived from data obtained with 5-, 21-, 63-, and 85-m^2 systems. *Source*: Reprinted from R. van Reis et al. (15). With kind permission of Elsevier Science–NL, Amsterdam, The Netherlands.

increased mass transfer coefficient (which increases flux and reduces membrane area for a given C_w) and an increased number of pump/valve passes (which may increase protein denaturation). The mass transfer coefficient can be calculated as:

$$k = aQ^b D^{2/3} \tag{13}$$

where a and b are coefficients determined for the specific UF module, Q is the feed flow rate normalized for membrane area, and D is the solute diffusivity. A typical constant C_w process is shown in Figure 13.

Design Equations The yield and purification obtained in UF can be evaluated from a mass balance, typically assuming a constant sieving coefficient. The fractional retentate product yield in a batch process is

$$Y_r = (V_0/V_F)^{-S_0} \tag{14}$$

where V_F is the final retentate volume, V_0 is the initial retentate volume, and S is the sieving coefficient. The analogous yield for a product in the filtrate can simply be calculated as

$$Y_f = 1 - Y_r \tag{15}$$

For a fed-batch process, the retentate yield is given by

$$Y_r = [1 - (1 - S_o)\exp(-S_o(V_0/V_F - 1))]/S_o(V_0/V_F) \tag{16}$$

and the corresponding filtrate yield can be calculated by combining equations 15 and 16. The retentate yield during constant retentate volume diafiltration is derived as

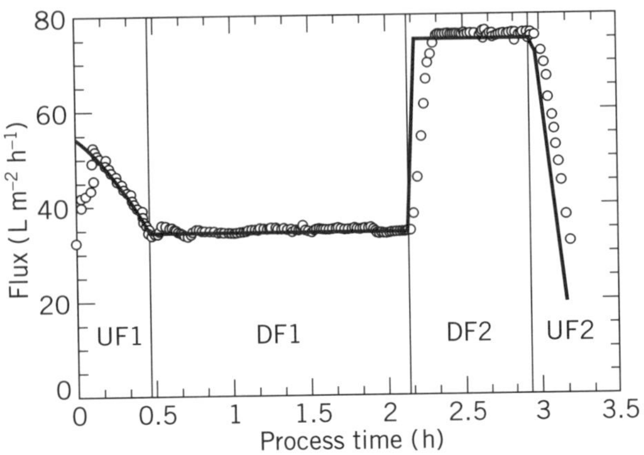

Figure 13. Predicted filtrate flux (solid line) and actual filtrate flux (open circles) for a constant C_w UF/DF process as a function of process time. Concentration of a recombinant antibody from 4 to 10 g L^{-1}, followed by a 10-diavolume buffer exchange and final concentration to 45 g L^{-1}. The increased fluxes during DF2 and UF2 process steps are the result of an increase in mass transfer coefficient and optimum C_w after the initial buffer exchange (DF1).

$$Y_r = \exp(-S_o N) \quad (17)$$

where the number of diavolumes is given by

$$N = V_D/V \quad (18)$$

where V_D is the DF buffer volume, and V is the constant retentate volume. It is often useful to combine and rearrange equations 14 and 17 to determine the retention coefficient required to obtain a desired retentate yield for a batch UF and constant volume DF process as

$$R = 1 + \ln Y/(\ln(V/V_0) + N) \quad (19)$$

For a filtrate product, the required retention coefficient is

$$R = 1 + \ln(1 - Y)/((\ln(V/V_0) + N) \quad (20)$$

DF processes are often used to remove undesirable low molecular weight components. The residual concentration can be calculated as

$$C = C_0 \exp(-S_o N) \quad (21)$$

Buffer exchange may also involve the addition of a new component:

$$C = C_i(1 - \exp(-N)) \quad (22)$$

where C_i is the bulk concentration of the component that is being added during diafiltration. Simultaneous addition and removal of a component may also occur in cases where the initial buffer and the DF buffer have a common component. The concentration of such a component can be calculated by adding equations 21 and 22. Process time and membrane area can be minimized by operating DF processes at an optimum bulk concentration (48):

$$C_b^* = C_w/e \quad (23)$$

By combining equations 23 and 6, with $C_f = 0$ for a fully retentive membrane, it has been shown (15) that the optimum diafiltration flux at C_b^* can be derived as

$$J^* = k \quad (24)$$

High-Performance Tangential-Flow Filtration

UF has been limited to separating solutes that are at least 10-fold different in size. In contrast, HPTFF enables the separation of solutes without limit to their relative size. HPTFF can even be used to perform reverse separations in which the higher molecular weight solute passes through the membrane while the lower molecular weight solute is highly retained. High-resolution separations are accomplished by exploiting concentration polarization, optimizing buffer chemistry, and using highly selective UF membranes. The performance of HPTFF is quantified by two dimensionless numbers, Ψ and $N\Delta S$:

$$\Psi = S_2/S_1 \quad (25)$$

$$N\Delta S = (JAt/V)(S_2 - S_1) \quad (26)$$

where Ψ is the selectivity, S_1 and S_2 are the respective sieving coefficients for the less and more highly retained species, N is number of diavolumes, A is membrane area, t is process time, and V is retentate volume. Membrane area and process time can be chosen to satisfy economic and manufacturing criteria independent of other variables. The retentate volume can be optimized separately. It is therefore useful to initially evaluate the selectivity (equation 25) and $J\Delta S$ values, which only depend on three simple experimental parameters:

$$J\Delta S = J(S_2 - S_1) \quad (27)$$

The next three sections will describe how buffer chemistry, membrane selection, and fluid dynamics are optimized. The final section will describe the mathematical basis and use of optimization equations and diagrams.

Buffer Chemistry. Selecting buffer pH and ionic strength are of paramount importance to achieving optimum separation performance in HPTFF. By operating close to the pI of the lower molecular weight solute and far away from the pI of the larger molecular weight solute, one can take advantage of differences in hydrodynamic volume and charge (3). The electric double layer that surrounds charged solutes increases the effective hydrodynamic volume as determined by SEC (10). Direct charge effects can also be exploited by using a membrane that has the same charge as the retained solute. In addition, the conformation of a molecule, and hence the hydrodynamic volume, may be changed under highly charged conditions. Use of charge effects are optimized by operating at low ionic strength (Fig. 14). The optimum ionic strength is generally achieved

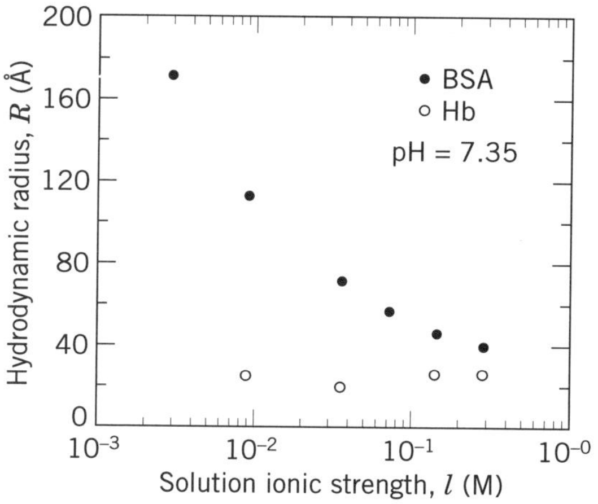

Figure 14. Effective hydrodynamic radius of BSA and hemoglobin (Hb) as a function of solution ionic strength. Calculated values of R were determined from retention time measurements in SEC using narrow molecular weight distribution dextran standards for calibration (10).

by using the lowest possible value while still maintaining an adequate pH buffering capacity. The buffering capacity of proteins at high concentrations near the membrane wall must be taken into account. Even in cases where there is no difference in pI between the solutes, it is advantageous to optimize both pH and ionic strength because the hydrodynamic volumes can be adjusted to increase selectivity and $J\Delta S$ for a given membrane molecular weight cutoff (3). This also reduces the need to have a large selection of different molecular weight cutoff membranes for HPTFF. Hydrodynamic volume and charge effects can also be altered by using buffering ions that have different protein binding characteristics (49). Rapid process development can, however, often be accomplished by testing just a few pH values around the pI of the lower molecular weight solute with a single buffering species at a concentration of 10 mM. As in all purification process development, it is important to ensure that the buffer composition does not cause undesirable biochemical alterations to the product and that product solubility is maintained at the concentrations incurred at the membrane wall (see "Fluid Dynamics and Process Control").

Membrane Selection. The general considerations for UF membrane selection outlined in "Membrane Selection" also apply to HPTFF. The selectivity and $J\Delta S$ that can be achieved with a membrane is determined by the pore-size distribution and membrane chemistry. Pore-size distribution affects both selectivity and $J\Delta S$ by altering the solute sieving coefficients and the filtrate flow distribution (8). Membrane chemistry has an indirect impact on the pore-size distribution because membrane formation conditions are linked to the casting solution chemistry. Membrane chemistry also plays a significant role in fouling, which in turn influences the selectivity and $J\Delta S$ values. Finally, the charge versus pH profile of the membrane can be selected to influence both charge repulsion and fouling. It should be noted that selective fouling of a membrane can actually be used to enhance separations because the pore-size distribution can be reduced and the surface charge can be altered. Using these guidelines, it is often possible to narrow down the choice to one or two membranes before proceeding with fluid dynamic optimization.

Fluid Dynamics and Process Control. The intrinsic sieving coefficient for a solute is defined by equation 3. The intrinsic sieving coefficient depends on the size and charge characteristics of both the solute and the membrane pores. Buffer optimization and membrane selection have hence focused on establishing conditions that provide the best differential intrinsic sieving coefficients for the solutes to be separated. The observed sieving coefficient of a solute, however, depends not only on the intrinsic sieving coefficient but also on the concentration of solute at the membrane wall. The wall concentration is determined by the mass transfer coefficient and the filtrate flux. The observed sieving coefficient is related to the intrinsic sieving coefficient (S_∞), mass transfer coefficient (k) and flux (J) by equation 8. It is therefore clear that the optimization of HPTFF is strongly influenced by concentration polarization effects. Concentration polarization has often been cited in the literature as one of the reasons why UF has not been successfully used for high-resolution separations. It has been demonstrated (3), however, that concentration polarization can be used to enhance, rather than limit, the resolving power of UF membranes. Key differences used in HPTFF methodology (50,51) include operating below the transition point (Fig. 15) and minimizing the transmembrane pressure gradient along the length of the filtration module. Although conventional UF has often sought to maximize flux by operating above the transition point, it has been shown that the optimum selectivity and $J\Delta S$ (a differential mass flux) occur below the transition point (3). It has also been demonstrated that membrane fouling is more consistent and can be controlled below the transition point (15). Optimization of HPTFF fluid dynamics can therefore be reduced to a set of simple experiments in which module geometry, feed flow rate, and filtrate flux are investigated. The results are evaluated by measuring selectivity (Ψ) and differential mass flux ($J\Delta S$) as a function of these variables.

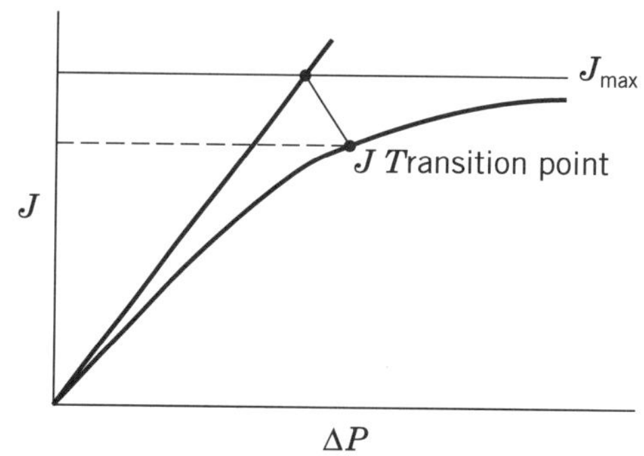

Figure 15. Typical flux versus transmembrane pressure curve showing the transition point flux at the inflection point (50). The optimum selectivity (Ψ) and differential mass flux ($J\Delta S$) occur below the transition point flux.

Process Optimization. The goal of process optimization is to achieve product quality specifications with the best combination of yield and purification factor using an economic amount of membrane area and providing an acceptable process time. Yield, purification factor, membrane area, and process time can be evaluated by using an optimization diagram (52) for either retentate product (Fig. 16) or filtrate product (Fig. 17). The optimization diagrams consist of a family of curves with the dimensionless parameters selectivity and $N\Delta S$. The family of curves in the optimization diagrams are mathematical expressions that link the yield and the purification factor to selectivity and $N\Delta S$ values. The equations are derived (52) as

$$P_R = Y_R^{1-\Psi} = \exp(N\Delta S) \tag{28}$$

$$\begin{aligned} P_F &= Y_F/(1 - (1 - Y_F)^{1/\Psi}) \\ &= Y_F/(1 + (Y_F - 1)\exp(N\Delta S)) \end{aligned} \tag{29}$$

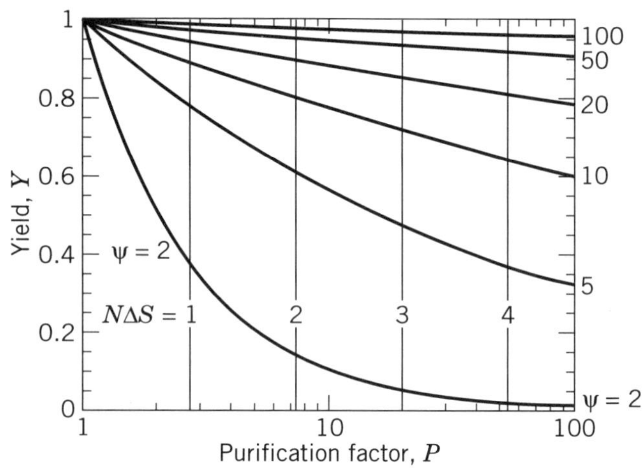

Figure 16. Optimization diagram for product in the retentate. The process follows a constant selectivity (Ψ) curve starting at a fractional yield $Y = 1$ and can be operated up to a maximum $N\Delta S$, which depends on the experimental $J\Delta S$ value and the ratio of membrane area (A), process time (t), and retentate volume (V). *Source*: Adapted from R. van Reis and S. Saksena (52). With kind permission of Elsevier Science–NL, Amsterdam, The Netherlands.

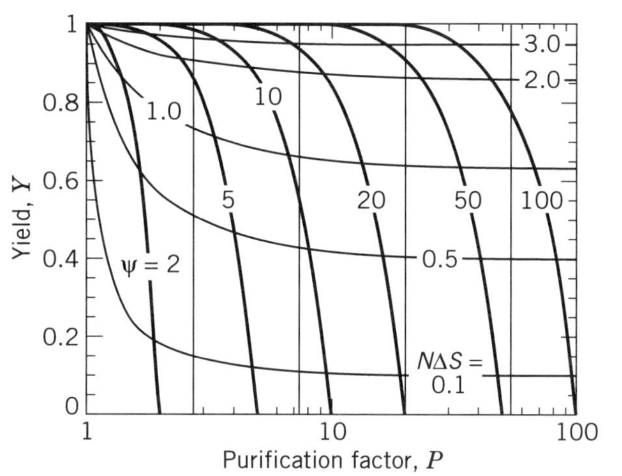

Figure 17. Optimization diagram for product in the filtrate. The process follows a constant selectivity (Y) curve starting at a fractional yield $Y = 0$ and can be operated up to a maximum $N\Delta S$, which depends on the experimental $J\Delta S$ value and the ratio of membrane area (A), process time (t), and retentate volume (V). *Source*: Adapted from R. van Reis and S. Saksena (52). With kind permission of Elsevier Science–NL, Amsterdam, The Netherlands.

where P is the purification factor, and Y is the fractional yield in the retentate (R) and filtrate (F). Both retentate and filtrate product processes operate along lines of constant selectivity. In the case of a retentate product, the process starts at a purification factor of 1 and a fractional yield of 1 and proceeds along a constant selectivity line up to the limit of $N\Delta S$ with an increasing purification factor and a decreasing yield. Conversely, the filtrate process starts at zero yield and a purification factor that numerically equals the selectivity (52) and proceeds along a constant selectivity curve with increasing yield and decreasing purification factor up to a limiting $N\Delta S$ value. Different experimental conditions (buffer chemistry, membrane chemistry and pore size, and fluid dynamics) can now be compared by evaluating the impact of selectivity and $J\Delta S$ on yield and purification factor. Selectivity and $J\Delta S$ often reach maxima under different experimental conditions. The best combination of selectivity and $J\Delta S$ depends on the overall process objectives and can be evaluated by comparing yield and purification factors for different combinations of these parameters using the optimization diagrams. As part of this analysis, $J\Delta S$ is converted to $N\Delta S$ by including membrane area, process time, and retentate volume according to equation 26. The membrane area and process times are chosen to provide acceptable economics, system hold-up volume, and process time. The retentate volume must be optimized. Decreasing the retentate volume (while maintaining a constant mass of solute) decreases the amount of buffer required for the process and can also decrease the amount of membrane area required for a given process time. This must be balanced against the increase in the number of pump and valve passes that occurs at increased concentrations, which can adversely impact product quality (see "Fluid Dynamics and Process Control").

Scale-Up

Effective commercial implementation of UF and HPTFF requires accurate and dependable scale-up of these membrane processes. It is also important to be able to scale-down existing manufacturing processes for further optimization, validation, and troubleshooting. Scale-up in UF and HPTFF obviously requires use of the same membrane material and pore size as well as the same module configuration and channel height at all scales of operation, with the membrane area scaled proportional to the filtrate volume. The most effective approach to scale-up is linear scaling, in which the pressure, fluid flow, and concentration profiles along the length of the filtration module are all kept constant when changing scale of operation (53). Because the local retentate velocity, solute concentration, and fluid pressure vary with position in the channel because of fluid removal and frictional pressure losses, linear scaling can only be achieved by keeping the channel length constant. Thus, the increase in membrane surface area required for large filtrate volume operation must be accomplished by increasing the number of parallel channels or fibers. It is also possible to increase the channel width in flat-sheet cassettes, although some care must be taken to ensure that this does not alter the fluid flow profiles in different size modules. Linear scaling of hollow fiber systems is relatively straightforward using modules with equal length but different numbers of fibers. Equal flow distribution between fibers and between fiber cartridges is easily accomplished with appropriate cartridge housings and piping manifolds. Linear scaling of flat-sheet designs has also been achieved using careful system design to ensure equivalent flow distribution, channel height compression, and entrance and exit effects at the small and large scale (53). Linear scale-up of hollow-fiber and flat-sheet devices over at least a 400-fold range of membrane area has been done successfully without intermediate pilot scale tests (53). Spiral-wound modules have a more limited range of linear scalability because pressure losses on the filtrate side will vary as the membrane area is changed, particularly in the design of very low area (<0.1 m^2) modules. Linear scaling can also be developed for spiral devices in which the feed flow is in the spiral direction (with the filtrate flow in the axial direction), and these modules should provide enhanced mass transfer through the formation of Dean vortices. Linear scaling of the other enhanced mass transfer modules discussed in "Modules and Devices" can often be difficult to achieve over a wide range of membrane surface area because of practical limitations on system geometry and design.

BIBLIOGRAPHY

1. W.F. Blatt, A. Dravid, A.S. Michaels, and L. Nelsen, in J.E. Flinn ed., *Membrane Science and Technology*, Plenum Press, New York, 1970, pp. 47–97.
2. R.T. Kurnik, A.W. Yu, G.S. Blank, A.R. Burton, D. Smith, A.M. Athalye, and R. van Reis, *Biotech. Bioeng.* **45**, 149–157 (1995).
3. R. van Reis, S. Gadam, L.N. Frautschy, S. Orlando, E.M. Goodrich, S. Saksena, R. Kuriyel, C.M. Simpson, S. Pearl, and A.L. Zydney, *Biotech. Bioeng.* **56**, 71–82 (1997).
4. E.N. Lightfoot, *Transport Phenomena in Living Systems*, Wiley, New York, 1974.
5. O. Kedem and A. Katchalsky, *Biochim. Biophys. Acta.* **27**, 229–246 (1958).
6. W.S. Opong and A.L. Zydney, *AIChE J.* **37**, 1497–1510 (1991).
7. W.M. Deen, *AIChE J.* **33**, 1409–1425 (1987).
8. L.J. Zeman and A.L. Zydney, *Microfiltration and Ultrafiltration: Principles and Applications*, Dekker, New York, 1996.
9. N.S. Pujar and A.L. Zydney, *J. Colloid Interface Sci.* **192**, 338–349 (1997).
10. N.S. Pujar and A.L. Zydney, *J. Chromatography A*, **796**, 229–238 (1998).
11. N.S. Pujar and A.L. Zydney, *Ind. Eng. Chem. Res.* **33**, 2473–2482 (1994).
12. S. Mochizuki and A.L. Zydney, *J. Membrane Sci.* **82**, 211–228 (1993).
13. A.S. Michaels, *Chem. Eng. Prog.* **64**, 31–44 (1968).
14. A.L. Zydney, *J. Membrane Sci.* **23**, 275–282 (1997).
15. R. van Reis, E.M. Goodrich, C.L. Yson, L.N. Frautschy, R. Whiteley, and A.L. Zydney, *J. Membrane Sci.* **130**, 123–140 (1997).
16. F.G. Paulsen, S.S. Shojaie, and W.B. Krantz, *J. Membrane Sci.* **91**, 265–282 (1994).
17. A.J. DiLeo and A.E. Allegrezza, *Nature* **351**, 420–421 (1991).
18. H.P. Hsieh, *AIChE Symp. Ser.* **84**, 1–18 (1988).
19. G. Belfort and A.L. Zydney, in M. Malmsten ed., *Biopolymers at Interfaces*, Dekker, New York, 1998, pp. 513–559.
20. G. Tkacik and S. Michaels, *Bio / Technol.* **9**, 941–946 (1991).
21. M. Meireles, P. Aimar, and V. Sanchez, *J. Membrane Sci.* **56**, 13–28 (1991).
22. M. Meireles, A. Bessieres, I. Rogissart, P. Aimar, and V. Sanchez, *J. Membrane Sci.* **103**, 105–115 (1995).
23. S. Gadam, M. Phillips, S. Orlando, R. Kuriyel, S. Pearl, and A.L. Zydney, *J. Membrane Sci.* **133**, 111–125 (1997).
24. G. Belfort, R.H. Davis, and A.L. Zydney, *J. Membrane Sci.* **96**, 1–58 (1994).
25. B.C. Robertson and A.L. Zydney, *J. Colloid Interface Sci.* **134**, 563–575 (1990).
26. S.T. Kelly, W.S. Opong, and A.L. Zydney, *J. Membrane Sci.* **80**, 175–187 (1993).
27. J.H. Hanemaajer, T. Robbertsen, T. van den Boomgard, and J.W. Gunnik, *J. Membrane Sci.* **40**, 199–212 (1989).
28. S.B. McCray and J. Glater, in S. Sourirajan and S. Matsuura eds., *Reverse Osmosis and Ultrafiltration*, ACS Symposium Series, Washington, D.C., 1985.
29. M. Cheryan, *Ultrafiltration Handbook*, Technomic, Lancaster, Penn., 1986.
30. G. Tragardh, *Desalination* **71**, 325–335 (1989).
31. C.A. Lawrence and S.S. Block, *Disinfection, Sterilization, and Preservation*, Lea and Febiger, Philadelphia, 1968.
32. V. Gekas, G. Tragardh, and B. Hallstrom, *Ultrafiltration Membrane Performance Fundamentals: 10 Years of Characterization and Fouling Studies*, The Swedish Foundation for Membrane Technology, Lund University, Sweden, 1993.
33. K.E. Smith and R.L. Bradley, in D. Lund, E. Plett, and C. Sandu eds., *Fouling and Cleaning in Food Processing*, Extension Duplicating, University of Wisconsin, Madison, Wis., 1985, pp. 410–423.
34. D. Berman, T. Meyers, and S. Chrai, *J. Parenteral Sci. Tech.* **40**, 119–121 (1986).
35. V. Goel, M.A. Accomazzo, A.J. DiLeo, P. Meier, A. Pitt, M. Pluskal, and R. Kaiser, in W.S.W. Ho and K.K. Sirkar eds., *Membrane Handbook*, Chapman and Hall, New York, 1992, pp. 506–570.
36. G. Belfort, *J. Membrane Sci.* **35**, 245–270 (1988).
37. B. Hallstrom and M. Lopez-Leiva, *Desalination* **24**, 273–279 (1978).
38. K.-Y. Chung, R. Bates, and G. Belfort, *J. Membrane Sci.* **81**, 139–150 (1993).
39. H.B. Winzeler and G. Belfort, *J. Membrane Sci.* **80**, 35–47 (1993).
40. C.H. Gooding, *Membrane Technology and Separations Planning Conference*, Newton, Mass., October 1997.
41. F. Vigo, C. Uliana, and E. Ravina, *Sep. Sci. Technol.* **25**, 63–82 (1990).

42. B. Culkin and A.D. Armando, *Filtr. Sep.* **29**, 376–378 (1992).
43. S.G. Redkar and R.H. Davis, *AIChE J.* **41**, 501–508 (1995).
44. V.G.J. Rodgers and R.E. Sparks, *J. Membrane Sci.* **68**, 149–168 (1992).
45. T.J. Narendranathan and P. Dunnill, *Biotech. Bioeng.* **24**, 2103–2107 (1982).
46. U.S. Pat. 5,597,486 (Jan. 28, 1997), H. Lutz (to Millipore Investment Holdings Limited).
47. J.J. Sheehan, B.K. Hamilton, and P.F. Levy, *ACS Symp. Ser.* **419**, 130–155 (1990).
48. P. Ng, J. Lundblad, and G. Mitra, *Sep. Sci.* **2**, 499–502 (1976).
49. M. Menon and A.L. Zydney, *American Chemical Society Annual Meeting*, San Francisco, 1997.
50. U.S. Pat. 5,256,294 (Oct. 26, 1993), R. van Reis (to Genentech Inc.).
51. U.S. Pat. 5,490,937 (Feb. 13, 1996), R. van Reis (to Genentech Inc.).
52. R. van Reis and S. Saksena, *J. Membrane Sci.* **129**, 19–29 (1997).
53. R. van Reis, E. Goodrich, C.L. Yson, L.N. Frautschy, S. Dzengeleski, and H. Lutz, *Biotech. Bioeng.* **55**, 737–746 (1997).

See also MEMBRANE CHROMATOGRAPHY; MEMBRANE SEPARATIONS.

PROTEOLYTIC CLEAVAGE, REACTION MECHANISMS

THOMAS P. GRAYCAR
Genencor International Inc.
Palo Alto, California

KEY WORDS

Aspartic proteinase
Cysteine protease
Endopeptidase
Exopeptidase
Metallopeptidase
Proteolytic cleavage
Reaction mechansim
Serine protease
Site-specific mutagenesis
X-ray crystallography

OUTLINE

INTRODUCTION

Proteolytic cleavage, the hydrolysis of peptide bonds in a protein, is a ubiquitous reaction in nature catalyzed by enzymes called proteases. Whether the purpose served is digestive scavenging of nutrients or regulation of a critical physiological process such as blood coagulation, proteases play an essential role in the growth and survival of all living organisms. Because of their abundance and importance, the reaction mechanisms of proteases have been studied more extensively than any other class of enzyme. Many of the first high-resolution X-ray crystal structures of proteins were obtained for proteases. Recently, our understanding of the mechanism of rate acceleration used by proteases and the structural basis for their specificity has improved dramatically through the use of a powerful recombinant DNA technique called site-specific mutagenesis. Developed in the early 1980s, site-specific mutagenesis is a method whereby one or more specific residues in the amino acid sequence of a protein can be removed or, more commonly, replaced with another amino acid via mutagenesis of its corresponding cloned gene (1). The contribution to catalysis of an amino acid functional group in the active site of an enzyme can then be determined directly (2). Because of the valuable insight into the reaction mechanism of proteases afforded by site-specific mutagenesis studies and high-resolution X-ray crystallography, there has been significant progress made in the design of new protease-inhibitor drugs as well as the development of proteases with improved properties for medicinal and industrial applications.

REACTION MECHANISMS

Mechanistic Classification of Proteases

Enzymes that hydrolyze peptide bonds can be grouped into two subclasses, exopeptidases and endopeptidases, depending on where the reaction takes place in the polypeptide substrate. Exopeptidases cleave peptide bonds at the amino or carboxyl ends of the polypeptide chain, whereas endopeptidases cleave internal peptide bonds. Based on their reaction mechanism, proteases may be further classified into four groups: serine, cysteine, aspartic, and metallo (3). The name of each class is derived from a distinct catalytic group involved in the reaction. The active sites of serine and cysteine proteases use serine hydroxyl and cysteine thiol side groups, respectively, as the attacking nucleophiles during catalysis. Their reaction pathway entails formation of a covalent intermediate between the nucleophile and the carbonyl carbon atom of the scissile peptide bond. By contrast, the mechanism of hydrolysis for aspartic and metallo proteases does not involve a covalent intermediate because the nucleophile for both of these enzymes is a water molecule. The different functional groups for each protease reaction mechanism affect their active pH range as well as the types of inhibitors suitable for inactivation. Indeed, measurements of the pH dependence for activity and reactivity with a set of class-specific inhibitors are typical laboratory methods for determining the mechanistic class of an unknown protease (4). Before describing each mechanism of catalysis in detail, it should be noted that the reaction pathways in all four protease classes share multiple common elements:

- Binding of the substrate polypeptide into a channel or pocket on the enzyme surface
- Nucleophilic attack on the carbonyl carbon of the scissile peptide bond by either oxygen or sulfur
- General base-assisted catalysis for removal of a proton from the nucleophile
- Stabilization of the tetrahedral transition state intermediate formed at the carbonyl carbon of the scissile peptide bond
- General acid-assisted catalysis for transfer of a proton to the amine leaving group

A description of substrate binding and the contribution of binding forces to catalysis and specificity will be discussed in a later section. The focus of this section will be the catalytic groups responsible for carrying out the chemical steps in the reaction.

Serine Proteases

The name *serine protease* refers to the nucleophilic serine residue located in the enzyme active site. In addition to the catalytic serine residue, this protease class is distinguished by having essential aspartate and histidine residues that together form a catalytic triad (Fig. 1). Serine proteases are prevalent in both prokaryotic and eukaryotic organisms and can be further classified into three families according to their structural homology to chymotrypsin, subtilisin, and wheat serine carboxypeptidase II (5,6). A fourth family of serine proteases has recently been identified based on amino acid sequence homology to prolyl oligopetidase (7). The catalytic serine and histidine residues in this enzyme have been identified by covalent modification with active-site-specific chemical reagents (8,9). Although it is expected that a catalytic triad is present in the prolyl oligopetidase family of serine proteases, identification of the catalytic aspartate awaits having a three-dimensional structure of the active site.

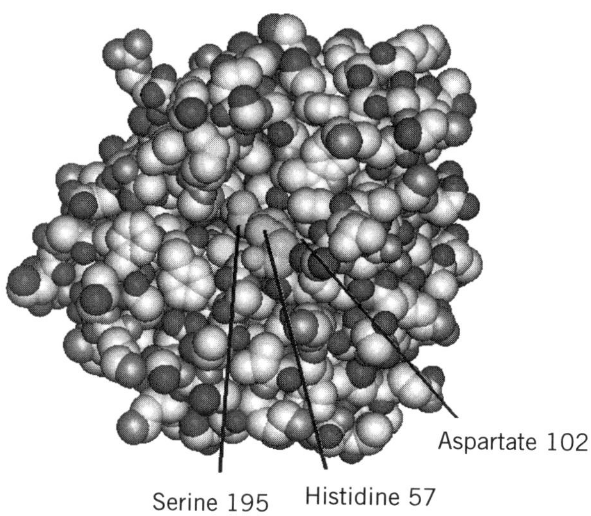

Figure 1. Three-dimensional structure of α-chymotrypsin. Residues forming the catalytic triad—serine 195, histidine 57, and aspartic acid 102—are located at the base of the substrate binding cleft. *Source*: Courtesy of Dr. Richard Bott, Genencor International Inc.

Chymotrypsin, a mammalian digestive enzyme, was the first protease to have its three-dimensional structure determined by X-ray crystallography (10). Other mammalian proteases structurally homologous to chymotrypsin but with different specificity are the digestive enzymes trypsin and elastase, and thrombin, which has specificity similar to trypsin and plays a critical role in blood clot formation. Subtilisin is a bacterial serine protease secreted extracellularly for the purpose of scavenging nutrients. Not all subtilisin-like serine proteases are bacterial in origin. Precursors of bioactive proteins such as hormones and neuropeptides are cleaved by serine proteases with greater structural homology to subtilisin than to chymotrypsin (11). The structure of wheat serine carboxypeptidase II has only recently been determined (12). Not only are the tertiary protein folds completely different for each structural family, the order of the catalytic triad residues in the amino acid sequence are also different. The arrangements are histidine-aspartate-serine, aspartate-histidine-serine, and serine-aspartate-histidine for chymotrypsin-like, subtilisin-like, and serine-carboxypeptidase-II-like serine proteases, respectively. Despite these differences, however, their reaction mechanism and geometry of catalytic residues are nearly identical and are believed to have arisen from different ancestral proteins through a process of convergent evolution.

Hydrolysis of amide and ester substrates by serine proteases follows an acyl transfer reaction pathway:

$$E + S \rightleftharpoons ES \xrightarrow{\nearrow P_1} EP_2 \xrightarrow{H_2O} E + P_2$$

where ES represents the noncovalent enzyme-substrate Michaelis complex, EP_2 is the acylenzyme intermediate, and P_1 and P_2 are the reaction products generated during acylation and deacylation, respectively. Evidence for the acyl transfer mechanism was obtained from numerous kinetic studies using activated ester substrates where deacylation is the rate-determining step in the reaction such that accumulation and partitioning of the intermediate could be monitored (13). Further support came from direct crystallographic observation of a stable elastase acylenzyme intermediate formed at low temperature (14). X-ray crystal structures also revealed key structural features related to the reaction mechanism of serine proteases: the catalytic triad comprising serine 195, histidine 57, and aspartate 102 (residue numbers are from the chymotrypsin amino acid sequence), and the oxyanion hole formed by the main chain NH groups of serine 195 and glycine 193 (15). A schematic of the reaction mechanism for serine proteases is illustrated in Figure 2. Formation of the covalent acylenzyme intermediate involves the nucleophilic attack of serine 195 on the substrate carbonyl carbon atom and release of the product P_1. This step is believed to proceed via general base-catalyzed transfer of a proton from the side chain hydroxyl group of serine 195 to the imidazole nitrogen of histidine 57, stabilization of the tetrahedral transition state by hydrogen bonds formed between the negatively charged carbonyl oxygen atom and the main chain NH groups of serine 195 and glycine 193, and general acid-catalyzed transfer of the proton of histidine 57 to the amine leaving group. Deacylation is essentially the same reaction except the attacking nucleophile is a water molecule bound at the site vacated by the P_1-leaving group. The pH dependence for catalysis by serine proteases follows the protonation state of the catalytic histidine residue. Chymotrypsin and subtilisin are active at neutral pH and above where the histidine side group is deprotonated. In contrast, wheat serine carboxypeptidase II is most active in the pH range of 4.5 to 5.5, suggesting an unusually low pK_a for its catalytic histidine residue. Although the role of the catalytic triad and oxyanion hole in the acyl transfer reaction mechanism of serine proteases was de-

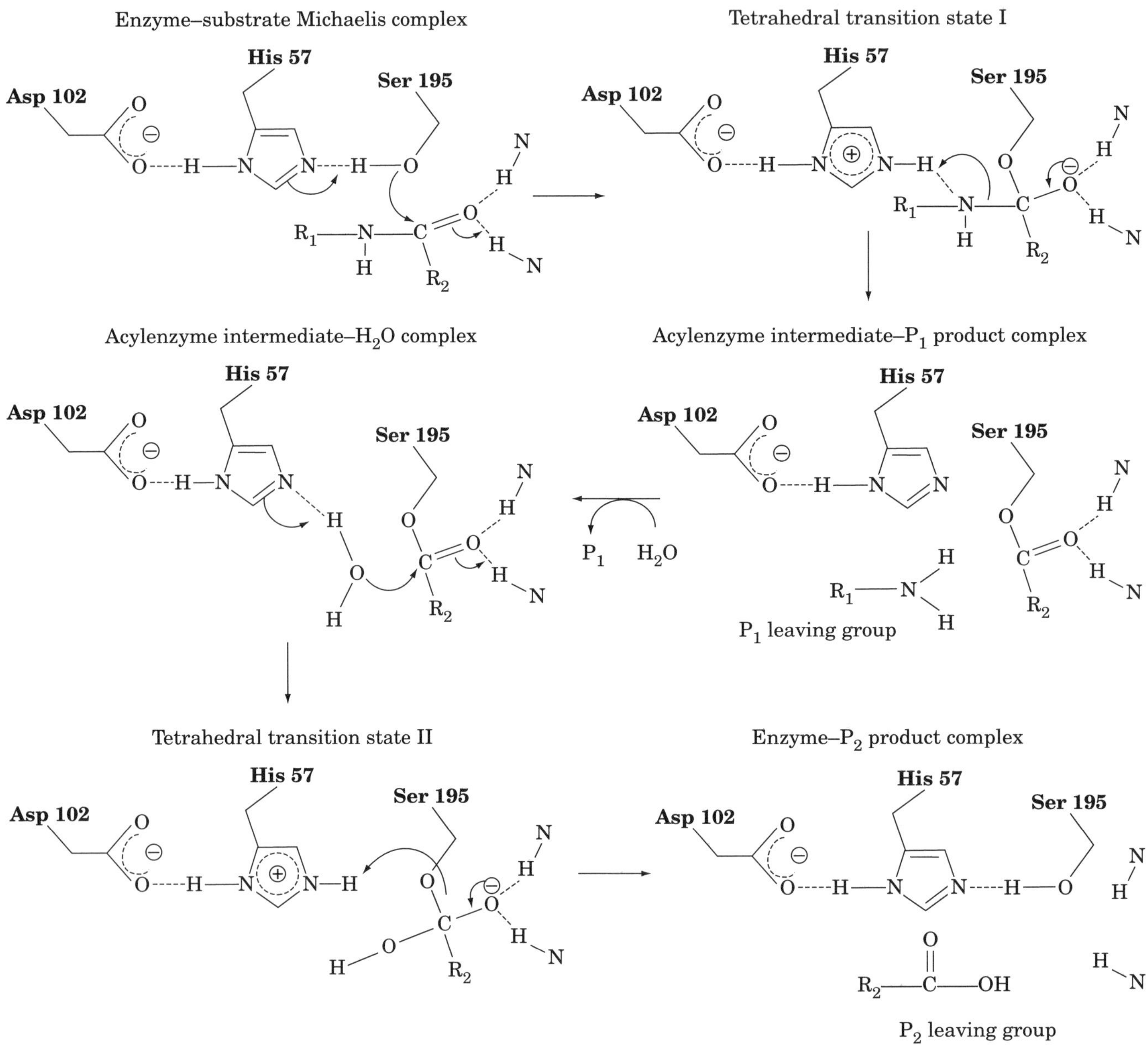

Figure 2. Schematic of the reaction mechanism in serine proteases. Nucleophilic attack of serine 195 on the peptide substrate bound in the substrate binding cleft forms an acylenzyme intermediate that is then hydrolyzed by a solvent water.

scribed more than 20 years ago, there continues to be considerable interest in understanding the fundamental source of their catalytic power.

Site-specific mutagenesis has been used extensively as a tool to better understand the catalytic mechanism of serine proteases. To assess the importance of transition-state stabilization, this technique was first applied to the oxyanion hole of subtilisin BPN' (16,17). Unlike chymotrypsin, where both hydrogen bonds stabilizing the substrate oxyanion are formed with main chain NH groups, the oxyanion hole in subtilisin BPN' consists of the main chain amide group of the catalytic serine 221 and one of the two protons on the side chain nitrogen atom of asparagine 155. Removal of the latter hydrogen bond via substitution of asparagine 155 with an amino acid lacking the necessary functional group or side chain geometry to form a good hydrogen bond to the oxyanion results in a reduction in the rate of catalysis by two to three orders of magnitude. In a similar fashion, the catalytic significance of aspartate 102 in trypsin was investigated by substitution with asparagine and replacement of each residue in the catalytic triad of subtilisin BPN' (serine 221, histidine 64, and aspartate 32) with alanine (18,19). The substitution of alanine for serine and histidine separately in subtilisin BPN' reduced the rate of catalysis by a factor of 10^6, whereas replacement of aspartate 32 with alanine led to a 10^4-fold loss in activity. Simultaneous substitution of all three amino acids in the catalytic triad of subtilisin with alanine also reduces catalysis by a factor of 10^6, illustrating the cooperative nature of these residues in catalysis. Assigning a functional role for aspartate 102 in the mechanism of serine proteases has been less straightforward than for either serine 195 or histidine 57. Initially, it was believed the high nucleophilicity of serine 195 was generated via a charge relay network in which histidine 57 accepts a proton from serine 195 and transfers another proton to aspartate 102 (20). This double-proton transfer reaction, however, has not been supported experimentally. It is now believed that aspartate 102, through a hydrogen bond to the side chain of histidine 57, plays the crucial roles of orienting histidine 57 in the correct tautomeric form to accept a proton from serine 195 and stabilizing the positive charge formed on histidine 57 in the transition state. Recently, it has been debated whether formation of a low-barrier hydrogen bond between the catalytic aspartate and histidine residues in the reaction transition state is an important feature in the catalytic mechanism of serine proteases (21–25). Although this issue remains to be resolved, it is evident that interactions between multiple residues work synergistically to stabilize the oxyanion transition state formed in the reaction mechanism of serine proteases (26).

Cysteine Proteases

The functional roles of cysteine proteases are quite broad, and members of this enzyme class are found in animals, plants, bacteria, and viruses. Like serine proteases, hydrolysis of amide and ester substrates by cysteine proteases follows an acyl transfer reaction pathway (equation 1), but the attacking nucleophile is the thiol group of an active-site cysteine residue instead of a serine hydroxyl group. The plant enzyme papain is the most extensively studied cysteine protease and the first to have its three-dimensional structure determined (27). It is a globular protein with the active site located in a cleft at the interface of two domains. The pH activity profile of papain is a bell-shaped curve centered about pH 6, with activity dependent on formation of a thiolate-imidazolium ion pair between the catalytic residues cysteine 25 and histidine 159. The loss in activity in the acidic pH range is attributed to protonation of cysteine 25 with an apparent pKa of 4, whereas the drop in activity in the basic pH range is attributed to deprotonation of histidine 159 with an apparent pKa of 8.5 (28). The side chain of asparagine 175 is hydrogen bonded to the imidazole of histidine 159, forming a cysteine-histidine-asparagine triad similar to the serine-histidine-aspartate catalytic triad found in serine proteases. The functional role attributed to asparagine 175 is that of orienting the imidazole ring of histidine 159. The importance of this residue for catalysis has been confirmed by screening of site-specific random amino acid replacements at position 175 in papain for catalytic activity (29). Glutamine at position 175 was found to have weak activity whereas no activity was observed for 11 other amino acids incorporated at position 175. Although a complete description of the acylation and deacylation steps in the reaction pathway of papain remains to be defined, it is clear there are significant differences in the mechanisms of cysteine and serine proteases.

Cysteine proteases have been referred to as activated enzymes because their thiolate-imidazolium ion pair present in the ground state obviates the need for general base-catalyzed removal of a proton to generate the nucleophilic thiolate anion in acylation and for general acid catalyzed proton transfer from histidine 159 during deacylation (30). A schematic of the reaction pathway for papain is shown in Figure 3. After formation of the noncovalent enzyme-substrate Michaelis complex, acylation proceeds via nucleophilic attack of the cysteine 25 thiolate anion on the carbonyl carbon of the substrate scissile bond. Breakdown of the tetrahedral transition state into a covalent acyl–enzyme intermediate is facilitated by rotation of the imidazole ring of histidine 159 and general acid-catalyzed transfer of its proton to the amide nitrogen of the leaving group. Hydrolysis of the acyl–enzyme intermediate during deacylation is base catalyzed with transfer of a proton from a water molecule to histidine 159 to generate the OH- nucleophile. Until recently, the existence of an oxyanion transition state in both acylation and deacylation has been the subject of considerable debate. A putative oxyanion hole in papain has been described comprising the main chain amide proton of cysteine 25 and the side chain NH_2 group of glutamine 19. Substitution of alanine and serine for glutamine at position 19 in papain via site-directed mutagenesis resulted in mutant enzymes having 60- and 600-fold decreases in catalytic efficiency, respectively, compared to wild-type enzyme (31). The effect of these mutations on the catalytic activity of papain are somewhat smaller than that observed for analogous mutations in the oxyanion hole of subtilisin. However, these results clearly demonstrate the importance of oxyanion stabilization in the reaction mechanism of cysteine proteases.

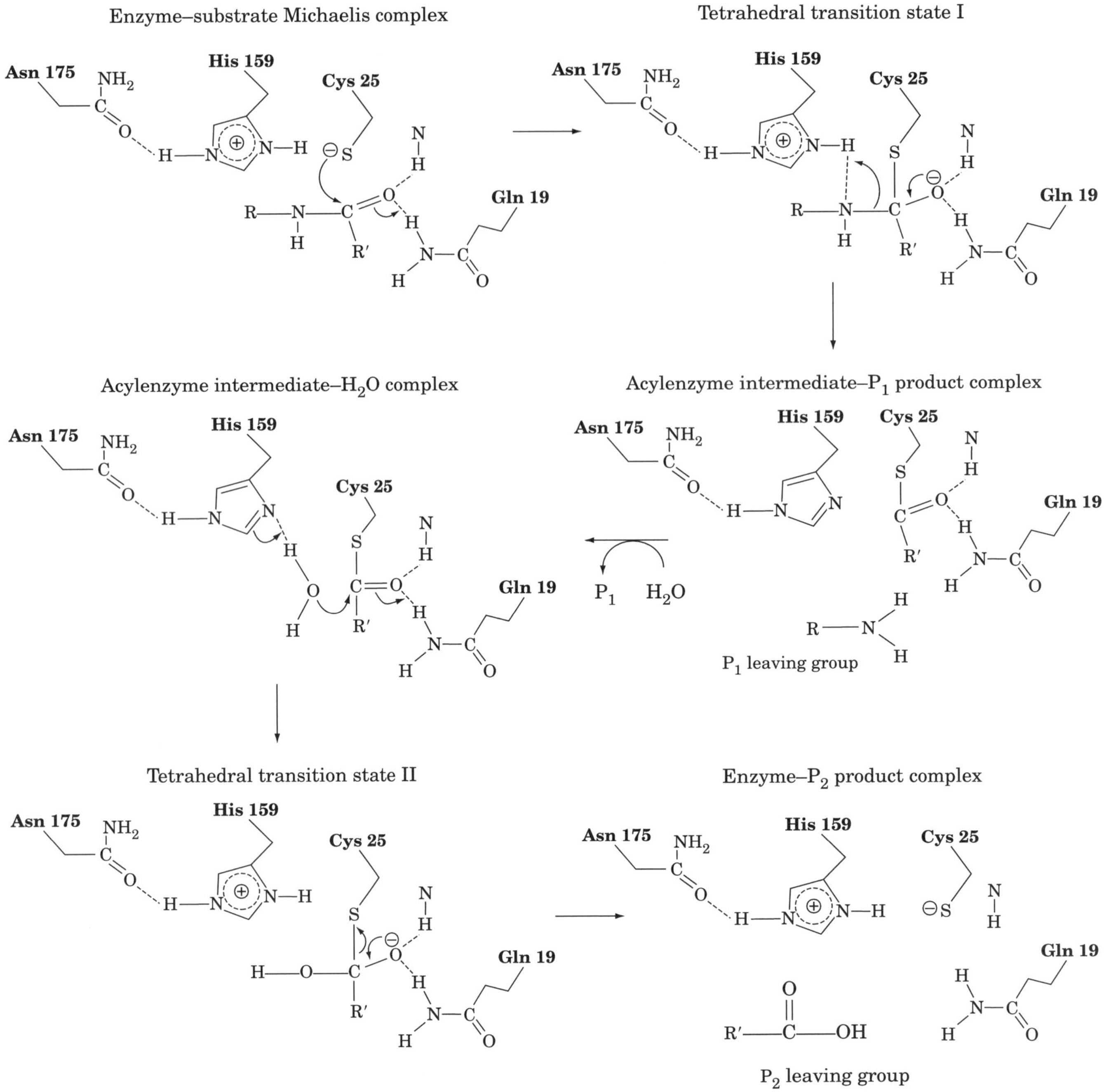

Figure 3. Schematic of the reaction mechanism in cysteine proteases. Nucleophilic attack of the thiolate anion of cysteine 25 on the peptide substrate bound in the substrate binding cleft forms an acylenzyme intermediate that is then hydrolyzed by a solvent water.

Aspartic Proteinases

The aspartic proteinases are named for the essential catalytic role of two aspartic acid residues. The three-dimensional structures for aspartic proteases of mammalian, fungal, and viral origin are quite comparable to each other. These enzymes are composed of two structurally similar domains with the active site formed within the cleft between each domain, much like the active site in cysteine proteases. Further, each domain contributes one of the two aspartate residues, aspartate 32 and aspartate 215, critical for catalysis (residue numbers are from the pepsin amino acid sequence) (32). The pH activity profile of aspartic proteinases is a bell-shaped curve, with activity dependent on protonation and ionization of the carboxylate group of aspartates 32 and 215, respectively. Most aspartic proteases are active in the range of pH 2 to 4, and in the past, this class of enzymes was referred to as acid proteases. Renin is one exception where activity is in the pH range of 5.5 to 7.5. Inhibition by pepstatin is an important feature common to all aspartic proteases. Pepstatin, a natural microbial hexapeptide containing a statine residue, is believed to be a mimic of the reaction-transition state. Crystal structures of aspartic proteases with various analogues of pepstatin have provided valuable insight in elucidating the reaction mechanism and for the purpose of designing inhibitor drug molecules (33).

Unlike serine and thiol proteases, proteolytic cleavage of peptide bonds by aspartic proteases does not proceed via a covalent acyl intermediate. A water molecule bound at the active site between the carboxylates of aspartates 32 and 215 is believed to be the attacking nucleophile. Attack of the water on the scissile carbonyl is facilitated by the general base and general acid functions of aspartates 215 and 32, respectively. As shown in Figure 4, the ionized aspartate 215 abstracts a proton from the water nucleophile, and aspartate 32 transfers its proton to the carbonyl oxygen. Stabilization of the tetrahedral intermediate is achieved through hydrogen bonds formed between the gem-diol and the negatively charged aspartate 32 and between the aspartate 215 and one of the hydroxyl oxygens. Breakdown of the intermediate and release of products is facilitated by transfer of the proton from aspartate 215 to the amine nitrogen of the leaving group and deprotonation of the statinelike hydroxyl by aspartate 32. Additional stabilization of the tetrahedral intermediate via hydrogen bonds from the side chain of tyrosine 75 has also been proposed. Substitution of tyrosine 75 with other amino acid residues in calf stomach chymosin and *Rhizomucor pusillus* pepsin by site-directed mutagenesis results in modified enzymes with dramatically altered catalytic activity, suggesting a critical role of this residue for catalysis, presumably via formation of hydrogen bonds to one of the hydroxyls formed in the transition state (34).

Metallopeptidases

The catalytic activity of metallopeptidases is dependent on the presence of one or two metal ions, most commonly zinc, bound at the active site. The reaction mechanism for the two-metal ion enzymes remains to be fully understood, although they clearly comprise a class of metallopeptidases distinct from that of the single-metal ion enzymes. Significant differences in amino acid sequence and tertiary structure are observed both for single- and two-metal ion metallopeptidases isolated from bacteria, fungi, and higher organisms. Crystal structures have recently been determined for three different aminopeptidases having two-metal ions at their active site (35–37). The identification of catalytic groups and elucidation of a reaction mechanism for single-metal ion metallopeptidases is based largely on crystallographic structural analyses of the bacterial endopeptidase thermolysin and the pancreatic exopeptidase carboxypeptidase A (38). Although the three-dimensional structures of these two enzymes are very different, they share a similar orientation of reactive groups, suggesting convergent evolution of a common reaction mechanism. Like cysteine and aspartic proteases, single-metal ion metallopeptidases are typically bilobal proteins with the active site located at a cleft formed at the junction of the two domains. Another feature in common with aspartic proteases is formation of a noncovalent intermediate during catalysis, because the attacking nucleophile in the single-metal ion metallopeptidases is a zinc-bound water molecule.

Figure 5 shows a schematic of the reaction mechanism for thermolysin. Binding of substrate at the active site displaces the water nucleophile toward glutamate 143, with the carbonyl oxygen of the scissile bond polarized by the positively charged zinc ion. Glutamate 143 functions as a proton shuttle during catalysis, similar to the role of histidine in the catalytic triad of serine proteases. General base-catalyzed transfer of a proton from the nucleophilic water to glutamate 143 facilitates attack of the hydroxyl anion on the substrate carbonyl carbon atom. This proton is then passed to the amide nitrogen of the leaving group. Not surprisingly, substitution via site-specific mutagenesis of glutamate 143 with alanine in the metallopeptidase from *Bacillus subtilis* results in complete loss of enzyme activity (39). The negatively charged oxyanion formed in the transition state is believed to be stabilized via hydrogen bonds to the side groups of histidine 231 and tyrosine 157. Replacement of histidine 231 with alanine or phenylalanine by site-specific mutagenesis of the thermolysin-like endopeptidase isolated from *B. stearothermophilus* leads to a large reduction in k_{cat} and little change in K_M

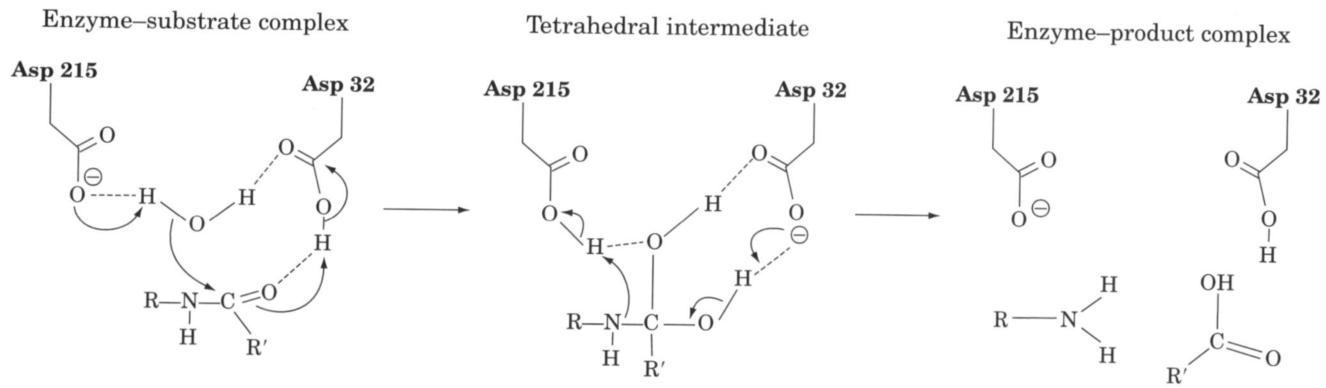

Figure 4. Schematic of the reaction mechanism in aspartic proteinases.

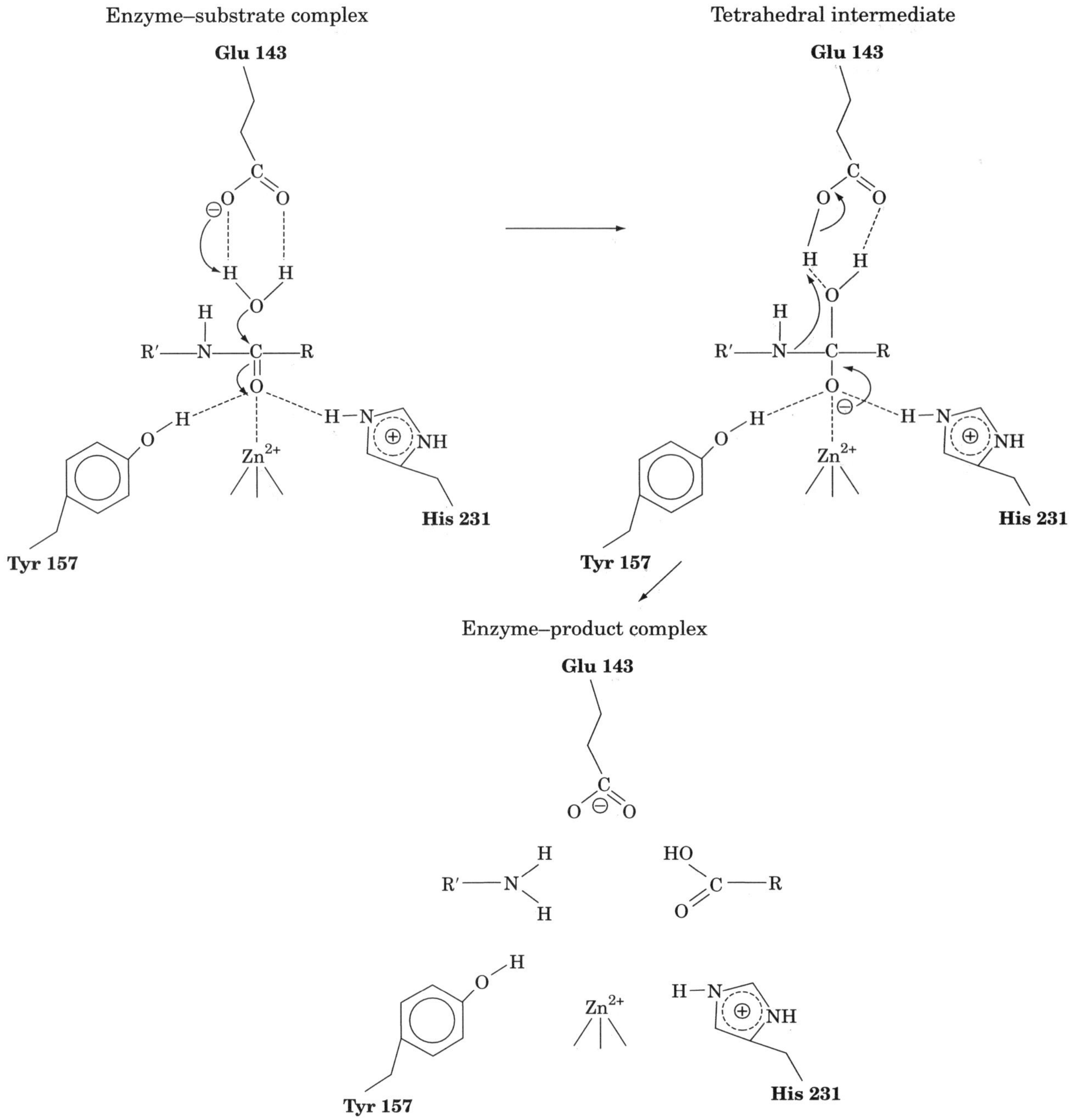

Figure 5. Schematic of the reaction mechanism of metallopeptidases.

(40). This result is consistent with assigning the role of transition state stabilization for histidine 231. A similar effect is obtained where arginine 127 in carboxypeptidase A, a mammalian exopeptidase, is substituted by site-specific mutagenesis with either alanine or methionine (41). The X-ray crystal structure of carboxypeptidase A reveals that arginine 127 is located at an active-site position equivalent to histidine 231 in thermolysin. Stabilization of the transition state in carboxypeptidase A is achieved via electrostatic interaction between the oxyanion and the positively charged guanidinium side group of arginine 127.

BINDING ENERGY IN CATALYSIS

A key feature used for rate acceleration by all four protease families is lowering of the reaction activation barrier through stabilization of the tetrahedral intermediate. Removal of the catalytic triad in the serine protease subtilisin via site-specific mutagenesis leads to an enzyme still capable of a 1,000-fold enhancement in the rate of peptide bond hydrolysis compared to the nonenzymatic reaction (16). Although much of the rate acceleration observed in the catalytically impaired subtilisin enzyme is likely attributable to the presence of a functional oxyanion hole, the contribution of additional binding energy to catalysis is also possible. The γ-hydroxyl group of threonine 220 in subtilisin has been shown to stabilize the oxyanion transition state independent of the oxyanion hole (42). In addition, the role of long-range electrostatic interactions in subtilisin catalysis has been explored via measurement of the contribution of charged residues on the enzyme surface to binding of a transition state analogue inhibitor (43). Interaction of the substrate peptide chain with residues along the protease binding cleft is another source of stabilization energy.

The substrate binding cleft of proteases is often quite large, with numerous regions of contact, called subsites, for binding amino acid residues of the polypeptide substrate (44). Figure 6 shows the structure of a peptide substrate bound into the binding cleft of the serine protease subtilisin. Hydrogen bonding, salt bridge formation, and hydrophobic interactions between substrate and subsite residues create significant levels of binding energy important in determining substrate specificity and in facilitating catalysis. The rate of catalysis for the aspartic proteinase pepsin is observed to increase progressively as the length of the substrate polypeptide chain increases (45). Dramatic alterations in the substrate specificity profiles and catalytic activity of numerous proteases have been achieved via

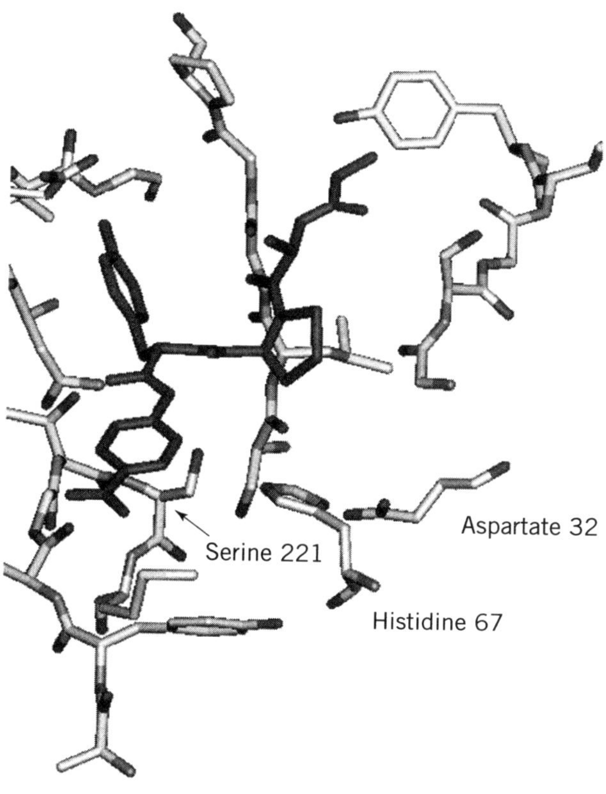

Figure 6. Close-up X-ray crystal structure view of the tetrapeptide succinyl-alanine-alanine-proline-phenylalanine-*p*-nitroanilide (solid dark lines) bound into the substrate binding cleft of subtilisin BPN′. The catalytic triad residues serine 221, histidine 64, and aspartic acid 32 are also shown. *Source*: Courtesy of Dr. Richard Bott, Genencor International Inc.

substitution of a limited number of subsite residues by site-directed mutagenesis (46). Protein structure outside of the binding cleft has also been found to affect the rate of peptide bond hydrolysis. For example, a recent site-specific mutagenesis experiment that altered the substrate specificity of trypsin to match that of chymotrypsin revealed that in addition to changes in the specificity pocket of the binding cleft, recruitment was required of two surface loop sequences that do not contact bound substrate (47). Although a description of the structural mechanism for how these loops influence catalysis remains to be determined, the importance of long-range enzyme–substrate interactions for transition-state stabilization is clearly apparent.

FUTURE DIRECTIONS

More than 50 years ago, Linus Pauling proposed that enzymes accelerate chemical reactions by providing a flexible template for binding the transition-state structure of a substrate molecule rather than its ground state (48). The structure-activity relationships elucidated from investigations on the reaction mechanisms of proteases provide an experimental framework that supports Pauling's theory on how enzymes catalyze chemical reactions. Although protease reaction mechanisms are still not completely understood, it is evident that essential catalytic groups can be readily identified from X-ray crystallographic and site-specific mutagenesis studies. The importance of having detailed information about the active-site geometry and chemical bonds stabilizing the reaction transition state has been established for the development of inhibitor drugs to treat protease-related diseases (49). Our knowledge of protease structure and function will also enable the facile discovery of new proteases from genetic sequence databases. The amount of genomic information available to researchers is growing at a fast rate. There are now 45 in-progress and 20 complete prokaryotic genome sequences, and 17 in-progress and 1 complete eukaryotic genome sequences available for analysis on the Internet (*www.mcs.anl.gov/home/gaasterl/genomes.html*). Computer searches for homologous sequence alignments and functional motifs such as metal–ion binding sites can rapidly identify protein sequences having similarity to a particular protease family. Monitoring of expression or genetic variability of the newly discovered gene via high-density DNA array technology can further our understanding of the physiological role of the new protease and its possible clinical significance (60).

The wealth of information known about protease reaction mechanisms has also led to exciting new applications of protein engineering in the creation of novel biocatalysts. For example, the activity of serine proteases for hydrolysis of amide and ester bonds can be modified by altering their active site structure via chemical modification or site-specific mutagenesis (51–55). Such a change in reaction specificity minimizes product degradation in protease-catalyzed peptide synthesis reactions. Incorporation of selenium at the active center of subtilisin results in an enzyme having significant peroxidase activity (56). More recently, a protease having novel specificity was created via incorporating the catalytic triad of serine proteases into the substrate binding site of a nonprotease enzyme (57). The success of these and other experiments with proteases demonstrate the feasibility of redesigning the structure of a protein to produce a catalytic activity or specificity unobtainable from a naturally occurring enzyme.

BIBLIOGRAPHY

1. P. Carter, *Methods Enzymol.* **154**, 382–403, (1987).
2. R.J. Leatherbarrow and A.R. Fersht, *Protein Eng.* **1**, 7–16 (1986).
3. A.J. Barrett, *Methods Enzymol.* **244**, 1–15, (1994).
4. B.M. Dunn, in R.J. Benyon and J.S. Bond eds., *Proteolytic Enzymes: A Practical Approach*, IRL Press at Oxford University Press, Oxford, 1989, pp. 57–81.
5. B.W. Mathews, in H. Neurath and R.L. Hill eds., *The Proteins*, vol. 3, Academic Press, New York, 1977, pp. 404–590.
6. D.-I. Liao and S.J. Remington, *J. Biol. Chem.* **265**, 6528–6531 (1990).
7. N.D. Rawlings, L. Polgar, and A.J. Barrett, *Biochem. J.* **279**, 907–908 (1991).
8. D. Rennex, B.A. Hemmings, J. Hoffsteenge, and S.R. Stone, *Biochemistry* **30**, 2195–2203 (1991).
9. S.R. Stone, D. Rennex, P. Wikstrom, E. Shaw, and J. Hoffsteenge, *Biochem. J.* **276**, 837–840 (1991).
10. B.W. Mathews, P.B. Sigler, R. Henderson, D.M. Blow, *Nature* **214**, 652–656 (1967).
11. W.J.M. Van de Ven, A.J.M. Roebroek, and H.L.P. van Duijnhoven, *Crit. Rev. Oncol.* **4**, 115–136 (1993).
12. D.-I. Liao, K. Breddam, R.M. Sweet, T. Bullock, and S.J. Remington, *Biochemistry* **31**, 9796–9812 (1992).
13. M.L. Bender and J.V. Killheffer, *CRC Crit. Rev. Biochem.* **1**, 149–199 (1973).
14. T. Alber, G.A. Petsko, and D. Tsemoglou, *Nature* **263**, 297–300 (1976).
15. J. Kraut, *Annu. Rev. Biochem.* **46**, 331–358 (1977).
16. J.A. Wells, B.C. Cunningham, T.P. Graycar, and D.A. Estell, *Philos. Trans. R. Soc. Lond., Ser. A* **317**, 415–423 (1986).
17. P. Bryan, M.W. Pantoliano, S.G. Quill, H.-Y. Hsiao, and T. Poulos, *Proc. Natl. Acad. Sci. USA* **83**, 3743–3745 (1986).
18. C.S. Craik, S. Roczniak, C. Largman, and W.J. Rutter, *Science* **237**, 909–913 (1987).
19. P. Carter and J.A. Wells, *Nature* **332**, 564–568 (1988).
20. D.M. Blow, J.J. Birktoft, and B.S. Hartley, *Nature* **221**, 337–340 (1969).
21. P.A. Frey, S.A. Whitt, and J.B. Tobin, *Science* **264**, 1927–1930 (1994).
22. A. Warshel, A. Papazyan, and P.A. Kollman, *Science* **269**, 102–104 (1995).
23. W.W. Cleland and M.M. Kreevoy, *Science* **269**, 104 (1995).
24. P.A. Frey, *Science* **269**, 104–106 (1995).
25. S.-O. Shan, S. Loh, and D. Herschlag, *Science* **272**, 97–101 (1996).
26. A. Warshel, G. Naray-Szabo, F. Sussman, and J.-K. Hwang, *Biochemistry* **28**, 3629–3637 (1989).
27. I.G. Kamphuls, K.H. Kalk, M.B.A. Swarte, and J. Drenth, *J. Mol. Biol.* **179**, 233–256 (1984).
28. A.C. Storer and R. Menard, *Methods Enzymol.* **244**, 486–500 (1994).
29. T. Vernet, J. Chatellier, D.C. Tessler, and D.Y. Thomas, *Protein Eng.* **6**, 213–219 (1993).
30. L. Polgar and B. Asboth, *J. Theor. Biol.* **121**, 323–326 (1986).
31. R. Menard, J. Carriere, P. Laflamme, C. Plouffe, H.E. Khouri, T. Vernet, D.C. Tessler, D.Y. Thomas, and A.C. Storer, *Biochemistry* **30**, 8924–8928 (1991).
32. D.R. Davies, *Annu. Rev. Biophys. Biophys. Chem.* **19**, 189–215 (1990).
33. B. Veerapandian, J.B. Cooper, A. Sall, T.L. Blundell, R.L. Rosati, B.W. Dominy, D.B. Damon, and D.J. Hoover, *Protein Sci.* **1**, 322–328 (1992).
34. Y.-N. Park, J.-I. Aikawa, M. Nishiyama, S. Horinouchi, and T. Beppu, *Protein Eng.* **9**, 869–875 (1996).
35. N. Strater and W.N. Lipscomb, *Biochemistry* **34**, 14792–14800 (1995).
36. S.L. Roderick and B.W. Mathews, *Biochemistry* **32**, 3907–3912 (1993).
37. B. Chevrier, C. Schalk, H. D'Orchymont, J.-M. Rondeau, D. Moras, and C. Tamus, *Structure* **2**, 283–291 (1994).
38. B.W. Mathews, *Acc. Chem. Res.* **21**, 333–340 (1988).
39. S. Toma, S. Campagnoli, E. De Gregorils, R. Gianna, I. Margarit, M. Zamal, and G. Grandi, *Protein Eng.* **2**, 359–364 (1989).
40. A. Beaumont, M.J. O'Donohue, N. Paredes, N. Rousselet, M. Assicot, C. Bohuon, M.-C. Fournie-Zaluski, and B.P. Roques, *J. Biol. Chem.* **270**, 16803–16808 (1995).
41. M.A. Phillips, R. Fletterick, and W.J. Rutter, *J. Biol. Chem.* **265**, 20692–20698 (1990).
42. S. Braxton and J.A. Wells, *J. Biol. Chem.* **266**, 11797–11800 (1991).
43. S.E. Jackson and A.R. Fersht, *Biochemistry* **32**, 13909–13916 (1993).
44. I. Schechter and A. Berger, *Biochem. Biophys. Res. Commun.* **27**, 157–162 (1967).
45. K. Medzihradzky, I.M. Voynick, H.M. Schweiger, and J.S. Fruton, *Biochemistry* **9**, 1154–1162 (1970).
46. J.J. Perona and C.S. Craik, *Protein Sci.* **4**, 337–360 (1995).
47. L. Hedstrom, L. Szllagyl, and W.J. Rutter, *Science* **255**, 1249–1253 (1992).
48. L. Pauling, *Chem. Eng. News* **24**, 1375–1377 (1946); see also *Nature* **161**, 707–709 (1948) and *Am. Sci.* **38**, 51–58 (1948).
49. E.W. Petrillo and M.A. Ondetti, *Med. Res. Rev.* **2**, 1–41 (1982).
50. L. Wodicka, H. Dong, M. Mittmann, M.-H. Ho, and D.J. Lockhart, *Nature Biotechnol.* **15**, 1359–1367 (1997).
51. J.B. West, W.J. Hennen, J.A. Bibbs, J.L. Lalonde, Z. Zhong, E.F. Meyer, and C.-H. Wong, *J. Am. Chem. Soc.* **112**, 5313–5320 (1990).
52. T. Nakatsuka, T. Sasaki, and E.T. Kaiser, *J. Am. Chem. Soc.* **109**, 3808–3810 (1987).
53. Z.-P. Wu and D. Hilvert, *J. Am. Chem. Soc.* **111**, 4513–4514 (1989).
54. P.R. Bonneau, T.P. Graycar, D.A. Estell, and J.B. Jones, *J. Am. Chem. Soc.* **113**, 1026–1030, (1991).
55. T. Graycar, R. Bott, R. Caldwell, J. Dauberman, P. Lad, S. Power, H. Sagar, R. Silva, G. Weiss, L. Woodhouse, and D. Estell, *Ann. N.Y. Acad. Sci.* **672**, 71–79 (1992).
56. Z.-P. Wu and D. Hilvert, *J. Am. Chem. Soc.* **112**, 5647–5648 (1989).
57. E. Quememeur, M. Moutiez, J.-B. Charbonnier, and A. Menez, *Nature* **391**, 301–304 (1998).

PSEUDOMONAS, PROCESS APPLICATIONS

ANDREW SCHMID
MARCEL WUBBOLTS
MARTIN HELD
BERNARD WITHOLT
Swiss Federal Institute of Technology
Zurich, Switzerland

KEY WORDS

Biocatalysis
Bioconversion
Bioprocesses
Biotransformation
Industrial processes
Pseudomonas

OUTLINE

INTRODUCTION

Pseudomonads are known to be widely encountered in nature, occupying niches ranging from contaminated soil to infected mammalian tissues. The extraordinary metabolic versatility of these species has been noticed since the early days of their study (1,2), probably because of the capability of many pseudomonads to share and acquire genetic information required to degrade a large number of compounds (3,4).

Because of the wide variety of degradation reactions and metabolic pathways existing in nature, it is not surprising that screening programs have led to the isolation of numerous bacterial strains, capable of using countless substrates. Interestingly, many substrates that can be degraded by *Pseudomonas* species include not only water-soluble chemicals but also highly hydrophobic and toxic compounds. The observation that (partially) purified enzymes and even intact cells can be efficiently used in mixed-solvent systems or in the presence of anhydrous solvents has led to widespread study and application of biocatalysts in organic synthesis during the past two decades.

The growing demand for optically pure drugs and agrochemicals, devoid of enantiomeric ballast, and the improvement of chemical catalysts have initiated development toward enantiospecific synthesis routes for chiral substances. However, the production of enantiopure products using chemical catalysts is not always feasible, and alternative chemical routes usually tend to be tedious, involving costly and dirty process steps. Therefore, biocatalysis exploiting the enzymatic activity of cells evolved in nature presents a valuable tool, with a large economical potential, for organic synthesis (Fig. 1).

BIOSYNTHESIS

A well-known biotechnological production process for a natural compound on an industrial scale with the help of *Pseudomonas* strains is the production of vitamin B-12 by *P. denitrificans* (7), which is especially efficient in media (such as beet molasses) that are rich in betaine (8,9). The production of vitamin B-12 by methylotrophic *Pseudomonas* sp. AM-1 on the cheap carbon source methanol (10) has not been able to replace the successful original industrial process (7).

Microbial polysaccharides are important biotechnological products that have found applications in the food, agricultural, and chemical industries (11–13). Exopolysaccharide production is common to *P. fluorescens*, *P. putida*, *P. aeruginosa*, *P. cepacia*, and other *Pseudomonas* (14–16). These compounds are believed to serve as a defense mechanism against the host immune system and as protection against bacteriophages or dehydration. Plant pathogens may use it as an anchoring material (17). Surfactants excreted by hydrocarbon degrading species improve the mobilization and bioavailability of mineral oils in soil (18,19).

The production of *Pseudomonas* exopolysaccharides such as alginate from *P. aeruginosa* and gellan gum from *P. elodea* have been commercialized, but these polysaccharides have not reached the level of application of xanthan gum, the principally used microbial polysaccharide from *Xanthomonas campestris* (17). The role of alginates in cystic fibrosis has triggered numerous scientific efforts to elucidate the molecular biology of their synthesis (20–22), which may well result in more efficient ways to produce this polysaccharide that is frequently applied to immobilize cells and enzymes in biotransformation procedures.

Other biopolymers that are of industrial importance are the bacterial polyesters: poly-3-hydroxyalkanoates (23,24). These biopolymers are biodegradable and provide an environmentally friendly alternative to common plastics. The commercialization of Biopol, a poly-3-hydroxybutyrate-poly-3-hydroxyvalerate copolymer from *Alcaligenes eutrophus* by Zeneca (ICI) has been realized (25), and the development of other biodegradable plastics is underway.

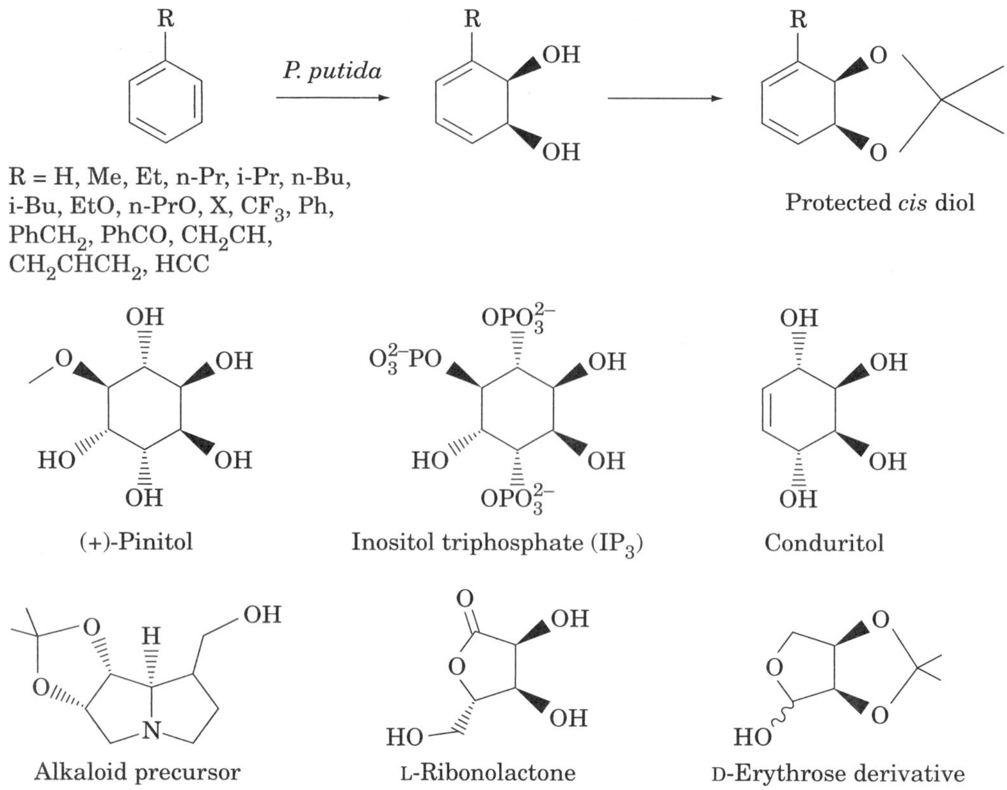

Figure 1. Synthesis of variously substituted *cis*-dihydrodiols by *P. putida* and several examples of complex molecules made using this synthon (5,6).

Poly-3-hydroxyalkanoates from *P. oleovorans* (26), an alkane-degrading microorganism (27,28), are of particular interest, because these bacterial polyesters can be produced with variable side groups, depending on the (co)substrate added to the culture (24,29,30). Biopolymers with variable-length side chains, unsaturated side chains, and aromatic pendant groups have been produced by this microorganism (24), but large-scale commercial application of these medium-chain-length poly-3-hydroxyalkanoates has yet to be realized.

BIOTRANSFORMATIONS

The use of enzymes in biocatalytic processes, either in (partially) purified form or in whole cell systems, is not evenly distributed over the different enzyme classes, defined by the Nomenclature Committee of the International Union of Biochemistry and Molecular Biology (formerly Enzyme Committee [EC]) (31).

Biocatalysts based on the hydrolases (EC class 3) are most used in synthetic reactions, because these enzymes are cofactor independent; are often secreted from the cells into the extracellular medium, facilitating their isolation, and in some cases are commercially available. Furthermore, many hydrolases have been shown to maintain their activity in media containing bulk amounts of hydrophobic, organic solvents. Oxidoreductases (EC class 1) are relatively complex enzymes that require cofactors and frequently consist of more than one protein component and are therefore mostly used as whole cell biocatalysts. Despite their complexity and difficulty of handling, these biocatalysts are important tools by virtue of their regio- and stereoselectivity. The enzymes that catalyze addition–elimination reactions, lyases (EC class 4), are underrepresented in biotransformations, although some biocatalysts of this type are used on a large scale in industrial applications, such as for the production of acrylamide (32). The development of biocatalysts that catalyze group transfer reactions (transferases, EC class 2) or isomerizations (isomerases, EC class 5) as well as ligase-based biotransformation reactions (EC class 6) is in a relatively early stage, with few realized industrial applications. Enzymes of these classes generally are highly substrate specific, which makes them less suited for wide-range biocatalytic applications.

Redox Reactions

Oxidation and reduction reactions catalyzed by oxidoreductases require cofactors, such as $NAD(P)^+$ or NAD(P)H, as electron acceptors or donors. Because the addition of cofactor to an enzyme reaction mixture in stoichiometric quantities is economically unfeasible, oxidative and reductive bioconversions are usually performed using whole cell cultures, where metabolically active cells efficiently recycle the cofactor. However, recent developments of efficient in situ cofactor regeneration systems (5,33) have stimulated the use of dehydrogenases in enzymatic biocatalyses.

Useful oxidoreductases encompass predominantly dehydrogenases, which catalyse the addition or elimination of hydrogen, and oxygenases, catalyzing the direct incorporation of molecular oxygen into organic molecules. Known hydrogenation reactions catalyzed by pseudomonads include the reduction of various aliphatic, cyclic, and aromatic ketones and aldehydes (34–36). *Pseudomonas*-catalyzed oxidations of hydroxy groups have been reported for substrates such as morphine, codeine, and (*S*)-malic acid (37,38).

Monooxygenases are a class of oxygenases that catalyze the addition of single oxygen atoms introducing hydroxy groups or epoxy moieties to aliphatic and olefinic compounds as well as aromatic rings. Additional reactivities include Baeyer-Villiger reactions, sulfoxidation reactions, and oxidative dehalogenations (5,39). The former two reaction types appear as promising alternatives to chemical synthesis pathways because of their (stereo-) specificity. Monooxygenase-catalyzed dehalogenation reactions performed by pseudomonads have not proved useful for synthetic biotransformation reactions. On the other hand, these reactions are of importance for biodegradation purposes.

Dioxygenases, a second class of oxygenases, drive the dihydroxylation of aromatic compounds to optically active *cis*-diols (Fig. 1). Often, these monooxygenase and dioxygenase-catalyzed hydroxylation reactions represent the initial steps in microbial degradation pathways of organic compounds. With the necessary genes in hand, recombinant strains have been developed that catalyze these initial oxidation reactions only, without further degradation, yielding specifically oxidized organic bioconversion products (40,41).

Generally, these enzyme systems possess a broad substrate range as well as a high regio- and stereospecificity (Fig. 2). In addition, different enzyme systems have been shown to ideally complement each other, such as the alkane hydroxylase system (EC 1.14.15.3) from *P. oleovorans* (43) and xylene oxygenase (EC 1.14.15.-) from *P. putida* mt-2 (44). For example, xylene oxidizes pendant methyl groups on aromatic rings (45), whereas alkane hydroxylase does not accept toluene derivatives as substrates but will introduce a terminal hydroxy group on ethyl side chains (46,47).

Because of their versatility and specificity, these enzyme systems represent valuable biocatalysts for the synthesis of chiral organic compounds, and they are being exploited in industrial applications for the production of fine chemicals (48,49) and pharmaceuticals (50).

Group Transfer Reactions

Transferases catalyze group transfer reactions. Although some acyltransferases, alkyltransferases, amino transferases, glycosyl transferases and phosphoryl transferases have found their way into biocatalytic processes, the application of transferases from pseudomonads on an industrial scale is rare.

Transferase-catalyzed reactions reported in the academic and patent literature include amino- (51,52), glycosyl- (53,54), and phospho-transfer reactions (55). The alkyl transfer reaction for the formation of L-serine from the achiral amino acid glycine and methanol was realized using serine hydroxymethyl transferase (EC 2.1.2.1) from *Pseudomonas* MS31, depicted in Figure 3 (56).

Hydrolyzations

Most industrial and laboratory enzyme processes are based on hydrolases. These enzymes do not use cofactors, usually consist of one catalytic component, and are readily available from commercial suppliers. Furthermore, the ability of hydrolases to catalyze synthesis reactions in nonaqueous environments has been thoroughly investigated and has provided paths to synthesize complex esters and amides.

Processes with pseudomonad hydrolases include reactions for the cleavage of ester, amide, and glycosidic bonds, the hydrolysis of epoxides to diols and nitriles to carboxylic acids, and the dehalogenation of chlorinated organic compounds (5,57). Their principle value lies in their contribution to the enantiopure synthesis of chemicals, often achieved by resolving racemic starting compounds or intermediates readily obtained from inexpensive precursors by chemical synthesis (Fig. 4). Also, many enzymatic reactions proceed under milder conditions and produce less-hazardous waste products compared to the chemical reaction alternative (5), which explains why especially nitrilases are chosen even when chemo-, regio- or enantioselectivity is not required for the synthesis of carboxylic acids from nitriles.

Hydrolytic reactions catalyzed by *Pseudomonas* enzymes are used as intermediate steps in various industrial processes for the synthesis of compounds such as natural and unnatural amino acids and peptides, organic acids, alcohols, and esters. Applications of these products range from food additives and agrochemicals to pharmaceuticals, including antibiotics, anticancer drugs, anti-inflammatory drugs, and β-blockers.

Lipases are the most applied hydrolases in biotransformations because these enzymes are highly enantioselective and are readily available from commercial suppliers. The pseudomonads are particularly rich in catalytically useful lipases. Lipases (EC 3.1.1.3) from *P. fluorescens*, *Pseudomonas* sp., *P. cepacia*, and *P. aeruginosa* have been and continue to be major players in the field of enantioselective ester hydrolysis and esterification reactions with technical or commercial applications (59–61). (Fig. 5)

Interestingly, proteases (EC 3.4.-.-) from pseudomonads have not received much attention, and they are hardly commercially available, although these hydrolytic enzymes obtained from other sources (e.g. subtilisin, chymotrypsin, papain) are very frequently used in many industrial and laboratory applications.

The discovery that many hydrolytic enzymes function well in organic solvents has lead to an even broader application spectrum of these biocatalysts, because it is possible to use substrates and to produce products unstable in water systems, and synthesis reactions have become feasible in low-water systems by eliminating the hydrolytic reaction dominant in aqueous media (62). For example lipases can be directed to catalyze esterifications or transesterifi-

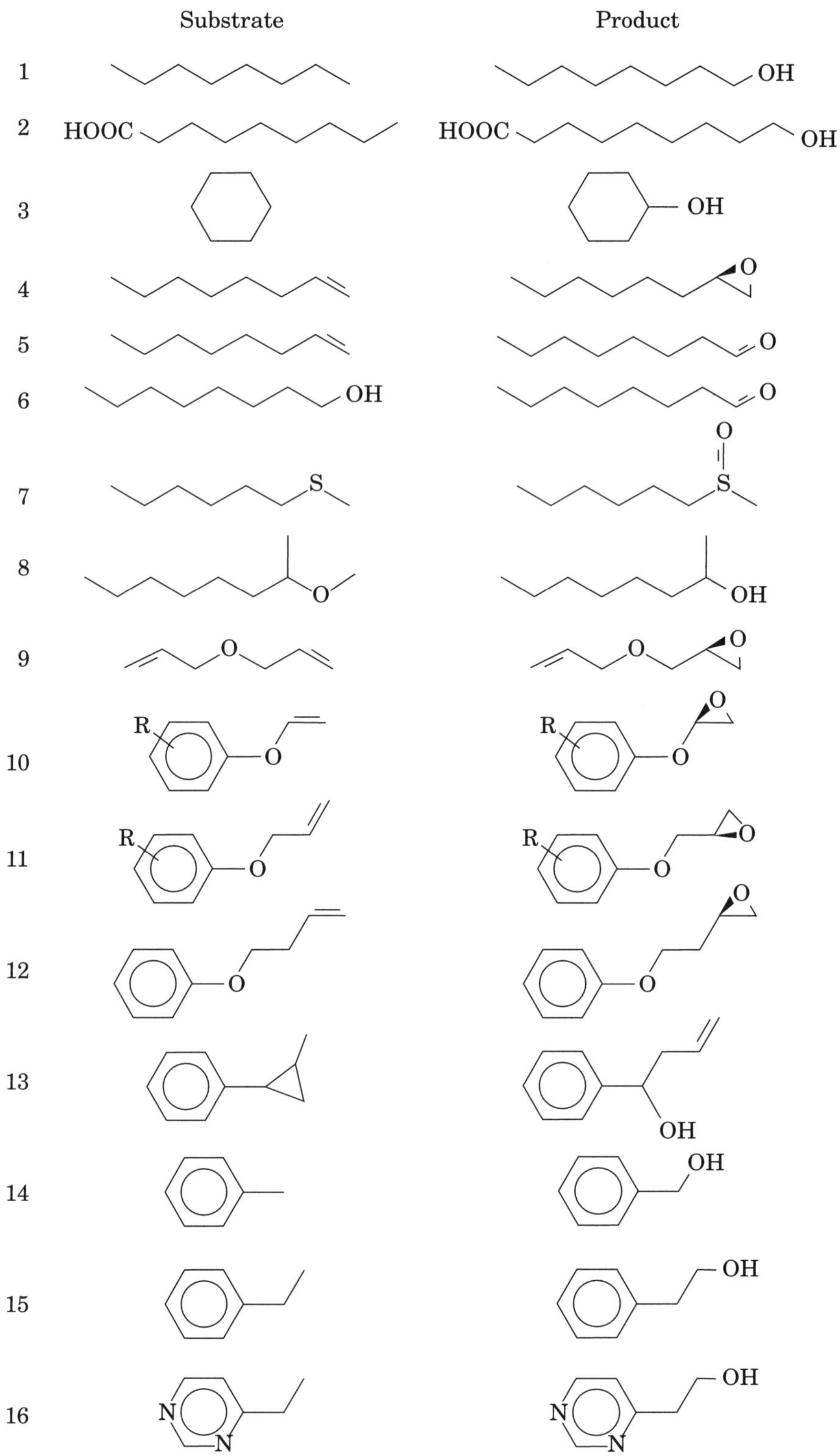

Figure 2. Substrate range of the alkane hydroxylase system (EC, 1.14.15.3) (42). Examples of the classes of substrates are shown. 1, C6–C14 alkanes to corresponding 1-alkanols; 2, C6–C12 fatty acids to ω-hydroxy fatty acids; 3, cyclohexane to cyclohexanol; 4, C3–C12 1-alkenes to 1,2-epoxyalkanes; 5, 1-octene to 1-octanal; 6, 1-octanol to 1-octanal; 7, methyl hexyl thioether to corresponding sulfoxide (and heptylthiol); 8, branched ethers to alcohol (demethylation); 9, diallyl ether to corresponding epoxide; 10, ring-substituted allylbenzene to 3-phenyl-1,2-epoxypropane; 11, ring-substituted allyl phenyl ethers to corresponding epoxides; 12, substituted allyl benzyl ethers to corresponding epoxides; 13, (*R,S*)-*trans*-2-phenyl-1-methylcyclopropane to 1-phenyl-3-butene-1-ol; 14, toluene to benzylalcohol; 15, ethylbenzene to 2-phenylethanol; 16, ethyl-substituted heterocyclic 5- and 6-membered aromatic rings to corresponding alcohols.

cation reactions such as acylation or alcoholysis. Increased conformational and thermal stability of water-soluble enzymes in hydrophobic solvents allows features such as molecular imprinting, by lyophilizing the enzyme from an aqueous solution at an ideal pH or in the presence of the preferred substrate, enzyme activity and specificity can be influenced (63).

Addition and Elimination Reactions

Lyases catalyze additions to double bonds or eliminations that generate double bonds and are specific for carbon–carbon, carbon–oxygen, carbon–nitrogen, carbon–sulfur, carbon–halide, or phosphorus–oxygen bonds. A strong enantioface discrimination during addition reactions is typ-

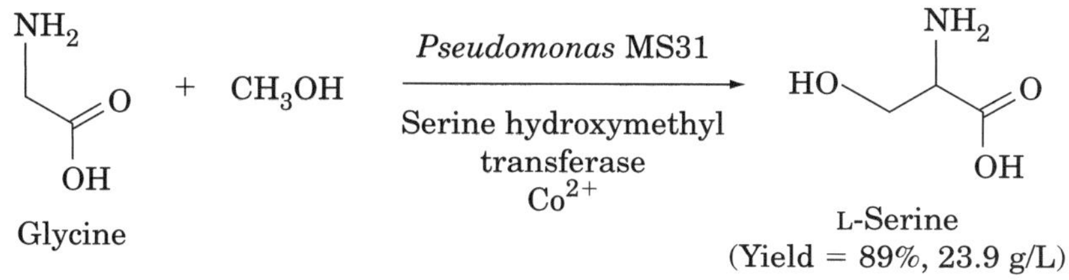

Figure 3. Biotransformation of glycine to L-serine using *Pseudomonas* (56).

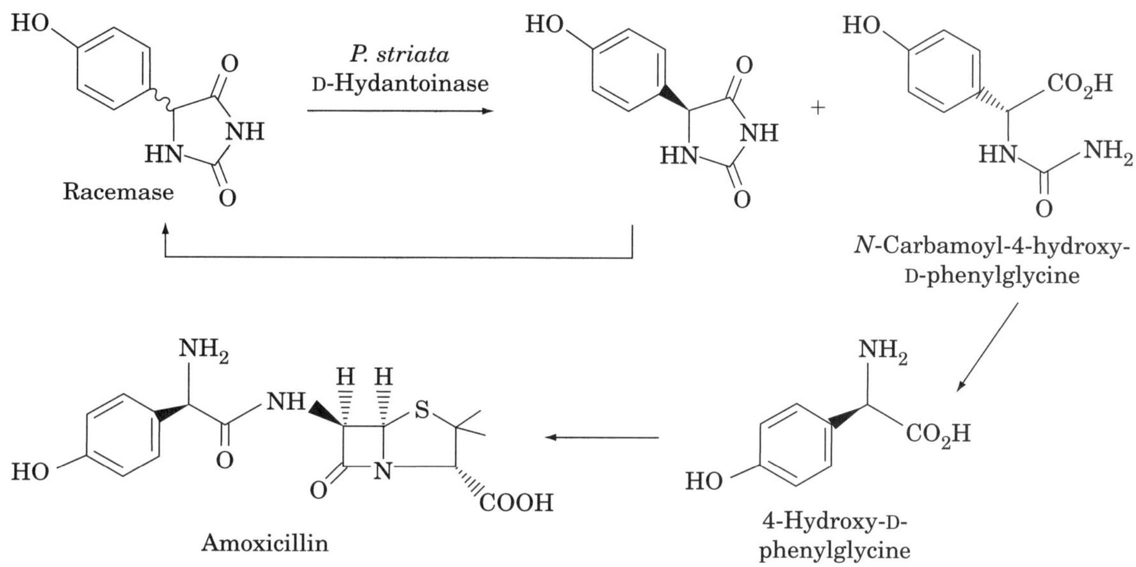

Figure 4. Amoxicillin production with a *P. striata* hydantoinase, a hydrolase that cleaves hydantoin rings to linear hydantoic acids (58).

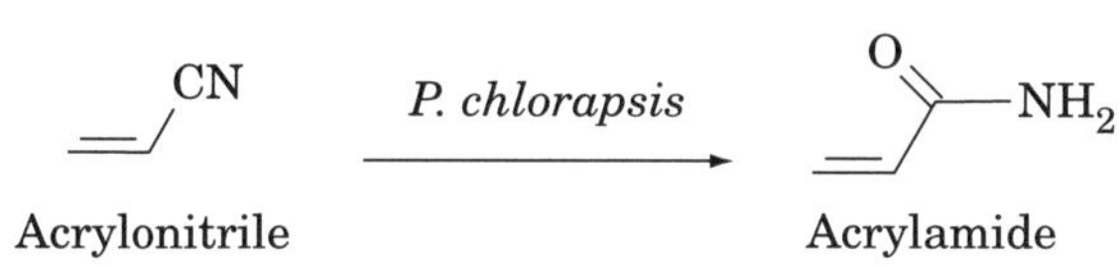

Figure 5. The production of acrylamide from acrylonitrile by Nitto using *P. chlorapsis* or *R. rhodochrous* J1 acrylonitrile hydrolyase (nitrile hydratase) (32).

ical of lyases, and lyase-catalyzed eliminations are as a rule also highly enantioselective, which makes them useful tools for chiral synthesis (64). Unfortunately, most lyases have a narrow substrate spectrum, which limits the applicability of these enzymes in biocatalytic processes (5).

Nevertheless, a number of industrial applications exploiting *Pseudomonas* lyases do exist (57). These include aminolyases for the synthesis of the artificial sweetener aspartame, L-aspartyl-L-phenylalanine methyl ester (51,65,66); carnitine hydrolyase (EC 4.2.1.89), containing whole cells of *Pseudomonas* sp. used by Lonza to produce the food additive L-carnitine (67,68), and haloyases, used for the synthesis of D-cysteine derivatives, important building blocks for the semisynthetic production of cephalosporin, (*S*)-carboxymethxy-D-cysteine. Originally, the hydrolyase acrylonitrile hydrolase from *P. chlororaphis* B23 (EC 4.2.1.84) was used as a catalyst for the production of acrylamide from acrylonitrile by the Nitto company on a scale of 30,000 tons per year (59), but since 1991 *R. rhodochrous* J1 has been used for production because of its superior reaction characteristics (69).

Isomerizations

Amino acid racemases (EC 5.1.1.10) catalyze the interconversion of D- and L-amino acids and play an important role in the industrial production of some amino acids, because racemization of the undesired or nonconverted enantiomer makes a 100% yield of the desired enantiomer attainable (59). In a number of processes, multienzyme reaction mixtures that include racemases have been used for the stereoselective synthesis of various amino acids (70,71) (Fig. 6).

The use of carbohydrate isomerases from pseudomonads is limited, but some processes have been described for the synthesis of fructose, mannose, and psicose (53,72,73).

Ligations

Ligases catalyze the joining of two molecules at the expense of the hydrolysis of a high-energy phosphate bond and are of particular use in molecular biology. In the bio-

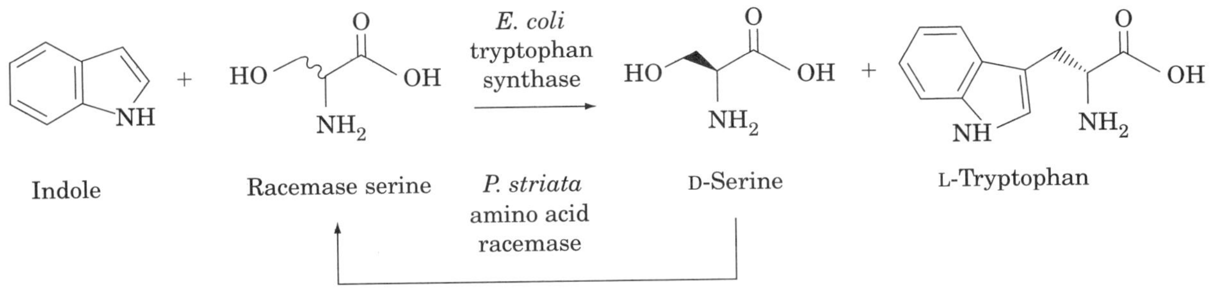

Figure 6. The Mitsui Toatsu process for production of L-tryptophan using serine racemase from *P. striata* (69).

transformation field, however, these enzymes are applied to a very limited extent (31).

Phenyl acetyl CoA ligase (EC 6.2.1.21) from *P. putida* M has been used in conjunction with *Penicillum chrysogenum* acyl-CoA:6-APA acyltransferase to produce a variety of penicillin analogues (74). Because of the wide substrate range of the enzymes involved (75), this system allows the in vitro synthesis of a wide variety of β-lactam antibiotics such as 3-furylmethylpenicillin and several ketoalkylpenicillins (76,77).

BIOPROCESS TECHNOLOGY

Toxicity of Apolar Organic Solvents

Many substrates and products involved in biosynthesis and biotransformation reactions inhibit the catalytic activity of cells or enzymes above critical concentrations. Which organic solvents cause inhibitory effects is most reliably predicted by the logarithm of the octanol-water partitioning coefficient (log*P*) (78). In general, complete biocatalyst inhibition is observed when apolar compounds with a log*P* below 2 are added to liquid cell suspension cultures at or above water-saturating concentrations. Above values of 4 to 5, inhibitory effects are markedly reduced. Gram-positive bacteria, which remain active in the presence of solvents with log*P* values above 4.5 to 6, show a higher sensitivity to solvents than Gram-negative bacteria, of which *Pseudomonas* species belong to the most resistant, where growth is observed in the presence of solvents with log*P* values as low as 3 (79–81).

Unless specific interactions with enzymes exist that limit growth, apolar chemicals generally harm cells as a result of their partitioning into the cell membrane bilayers. As a consequence, higher membrane fluidity and permeability results, leading to the loss of cell functionality and eventually cell lysis (82–84).

To withstand the exposure to organic solvents, cells react by increasing their membrane hydrophobicity and by decreasing membrane fluidity. The latter compensates the fluidizing effects of solvent molecules intercalated in the lipid bilayers, whereas changes in hydrophobicity have been observed to reduce the partitioning of solvents to the membranes (85,86). These effects are achieved by the incorporation of longer C_{18} fatty acids and proteins into the cell membrane, by modulation of the ratio between saturated and unsaturated fatty acids, by changes in the phospholipid composition, and by conversion of *cis*-unsaturated fatty acids to their *trans*-isomers (87–90). For a comprehensive review on the subject of cellular adaptation to organic solvents, see the work by Weber and de Bont (91).

Strain selection (92,93) and genetic engineering (94–96) are increasingly providing strains with improved resistance against elevated concentrations of organic chemicals. Alternative strategies to minimize cell damage and growth inhibition include cell immobilization (97–99). A range of carrier materials has been used with cells either adsorbed to the surface or entrapped within beads and gels. Examples of such materials are Celite, calcium alginate, κ-carrageenan, polyurethane, polyacrylamide, silicon, and chitosan (100,101). However, it must be considered that cell immobilization introduces an additional diffusion barrier, which can lead to mass transfer limitations (102).

Despite these advances, special procedures during bioprocess operations are nevertheless required to avoid toxic effects by harmful substrates and products when high productivities and high final product titers are to be attained in commercial applications. Strategies to attain this goal are described in the next section.

Controlled Substrate Feed

To avoid substrate inhibition, it is necessary to feed toxic starting compounds to a cell culture or an enzyme-containing reaction mixture in a controlled way, at a rate that avoids substrate accumulation to inhibitory levels. This is achieved either by supplying volatile compounds via the gas phase (103,104) or via controlled liquid feeds (105,106) in laboratory as well as production-scale reactors.

An alternative way—limited to laboratory studies—to eliminate substrate inhibition is to add toxic starting material dissolved in apolar carrier solvents, such as hexadecane and other high log*P* solvents. Because hydrophobic substrates mainly partition into the apolar solvent, they are less exposed but still well available to the biocatalyst residing in the aqueous phase of the two-liquid phase mixture (Fig. 7). Thus, higher quantities of toxic material can be added to the reaction mixture without encountering inhibitory effects. An example where such a two-liquid phase operation may in the future be scaled up for production purposes is the biosynthesis of functionalized poly-3-hydroxyalkanoates (PHA) from toxic precursors (24,107).

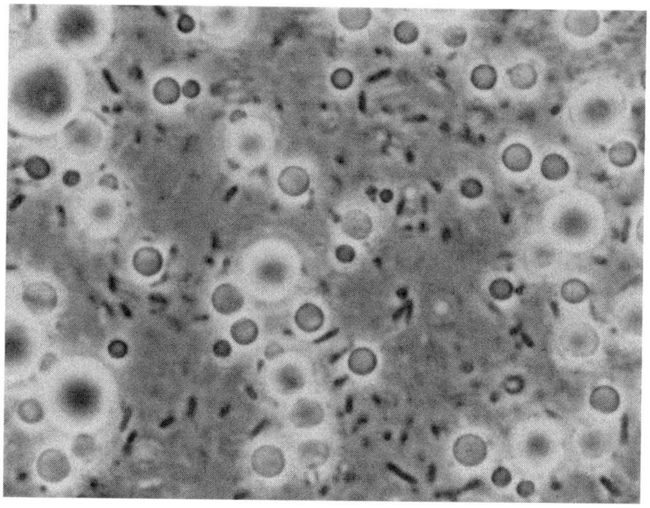

Figure 7. Micrograph of two-liquid phase medium. *P. oleovorans* cells (dark rods) are suspended throughout the continuous aqueous phase. The organic solvent *n*-octane is dispersed as fine droplets, with an average diameter in the order of 10 μm.

These precursors must be present in concentrations below inhibiting levels, but above limiting concentrations, because cells accumulate PHA as a carbon storage material only in abundance of the carbon source (108). Because the second liquid phase will buffer fluctuations of substrate uptake rate, as a result of continuously varying substrate uptake rates in batch and fed-batch processes, concentrations should be controlled more easily within the required narrow range.

In Situ Product Extraction

To obtain high titers of toxic or inhibitory products, it is necessary to continuously remove product from the reaction vessel. Depending on the nature of the product and the bioprocess involved, various methods are feasible.

Two-Liquid Phase Operation. By operating a two-liquid, aqueous-organic phase system, it is possible to continuously extract hydrophobic products formed by the biocatalyst residing in the aqueous phase (Fig. 7). As a result, cells and enzymes see less of the toxic products accumulating mainly in the organic phase during the course of the bioprocess, thus enabling higher final product titers to be attained. Compounds successfully produced in such in situ extraction processes include aliphatic and aromatic epoxides (41,109,110), alcohols (111,112) optically pure derivatives of citronellol (113) and organic acids (114). These biotransformations have been limited to laboratory scales; however, various industries are developing related processes.

Model studies have shown that the extractant can be separated from the aqueous phase by thermal and physical means (115) or, as recently developed, by continuously removing the organic solvent from the reaction mixture during the biotransformation step by in situ centrifugation (116). Product isolation and purification from the organic phase proceeds via conventional, chemical downstream processing techniques (117,118).

Solid Phase Extraction. Solid adsorbents have been successfully used to recover small molecules such as acetone, butanol, terpenes, aromatic amino acids, and steroids from fermentation broth. In these procedures, cells are first removed from the medium (119,120) or they are immobilized on solid carrier material (121). Solid adsorbent in situ recovery procedures have also been used in the presence of living cells, for example, to achieve enhanced production of antibiotics by streptomycetes (122), or mycobacteria (123). The addition of solid adsorption material such as XAD-4 to cultures was especially useful for adsorption of alkaloids from plant cell cultures (124–126). Solid adsorbents can be added into the bioreactor either directly or separated from the cells by a membrane. Alternatively, culture medium can be cycled from the bioreactor via an external loop through a packed or fluidized adsorption bed (127).

Membrane Separation. When using (partly) purified enzymes in biotransformation processes, it is often important to recycle the biocatalyst to optimize productivity and to minimize operation costs. This may be achieved by immobilizing the enzymes to a solid support and filtering the reaction mixture either continuously during the biotransformation reaction or batchwise at the end of the bioconversion step. Alternatively, soluble enzymes may be used in a homogeneous reaction mixture with ultrafiltration membranes used to separate the high molecular weight enzymes from the low molecular weight metabolites (128).

In oxidoreductase catalyzed-reactions, cofactors such as $NAD(P)^+$ and NAD(P)H are required as electron acceptors or donors. Because the addition of these metabolites in stoichiometric amounts is uneconomical, they must be regenerated. This can be achieved by supplementing the reaction mixture with a second enzyme, which catalyzes a redox reaction running in the opposite direction (129). For example, formate dehydrogenase (EC 1.2.1.2) added together with formate catalyzes the oxidation to carbon dioxide, thereby recycling the electron acceptor NAD^+ to NADH, which can again donate electrons to the desired enzymatic reaction (35). Because the native cofactor permeates through ultrafiltration membranes, NAD has been covalently bound to a water-soluble polymer, such as poly(ethylene glycol) (PEG).

OUTLOOK

Ongoing studies to elucidate anabolic and catabolic pathways and to characterize the enzymes that catalyze the individual reactions will, together with a multitude of screening programs, undoubtedly lead to further discoveries and applications of new and useful biotransformation reactions. In this process, genetic engineering will continue to play an important role for the characterization of enzymes as well as their production by cultures of cloned enzyme producing organisms. Genetically engineered whole cell biocatalysts can be beneficial when undesired side reactions of the host organism hamper a process and when space and time yields are insufficient. Modern techniques promise to allow an even more effective construction of biocatalytically active cell strains than is achieved today. By exerting a more sophisticated control over enzyme expression and regulation as well as by designing whole cell biocatalysts containing genetic information from various sources, whole cell biocatalysts tailor-made for specific processes will evolve.

New process technologies, useful for biotransformation reactions, that have been developed during recent years include in situ product removal (127), membrane reactor systems (128), and two-liquid phase bioprocesses (130). These techniques significantly widen the potential of biocatalysis, opening applications involving cofactor-dependent enzyme processes, bioconversions of toxic compounds, and syntheses of organic molecules in organic media. Thus, reactions catalyzed by biological systems, which formerly were restricted to the field of chemistry, have become feasible. Chemistry and biology are moving together to provide interchangeable tools for the efficient production of important chemicals useful in diverse applications in medicine, agriculture, and nutrition.

BIBLIOGRAPHY

1. L.E. den Dooren de Jong, Ph.D. Thesis, Leiden University, The Netherlands, 1926.
2. R.Y. Stanier, N.J. Palleroni, and M. Doudoroff, *J. Gen. Microbiol.* **43**, 159–271 (1966).
3. B. Holloway, M. Escuadra, A. Morgan, R. Saffery, and V. Krishnapillai, *FEMS Microbiol. Lett.* **100**, 101–105 (1992).
4. J.R. van der Meer, *FEMS Microbiol. Rev* **15**, 239–249 (1994).
5. K. Faber, *Biotransformations in Organic Chemistry*, Springer, Berlin, 1997.
6. T. Hudlicky, H.F. Olivo, and B. McKibben, *J. Am. Chem. Soc.* **116**, 5108–5115 (1994).
7. J. Florent, in H.-J. Rehm, and G. Reed, eds., Biotechnology, vol. 4, VCH Verlaggesellschaft, Weinheim, Germany, 1986, pp. 117–158.
8. Y.H. Fa, J.P. Kusel, and A.L. Demain, *Appl. Environ. Microbiol.* **47**, 1067–1069 (1984).
9. J.P. Kusel, Y.H. Fa, and A.L. Demain, *J. Gen. Microbiol.* **130**, 835–841 (1984).
10. Y. Tsuchiya, and N. Nishio, *J. Ferment. Technol.* **58**, 485–487 (1980).
11. J.D. Desai and I.M. Banat, *Microbiol. Mol. Biol. Rev.* **61**, 47–64 (1997).
12. V.J. Morris, *Agro Food Industry Hi-Tech* **3**, 3–8 (1992).
13. I.W. Sutherland, *Biotechnology of Microbial Exopolysaccharides*, Cambridge Univ. Press, Cambridge, U.K. 1991.
14. D.G. Allison and M.J. Goldsbrough, *J. Basic. Microbiol.* **34**, 3–10 (1994).
15. A.M. Marques, I. Estanol, J.M. Alsina, C. Fuste, and D. Simon-Pujol, *Appl. Environ. Microbiol.* **52**, 1221–1223 (1986).
16. I.W. Sutherland, E. Conti, A. Flaibani, and M. O'Regan, *Microbiology* **140**, 1125–1132 (1994).
17. A.N. Glazer, and H. Nikaido, *Microbial Biotechnology: Fundamentals of Applied Microbiology*, Freeman, New York, 1995.
18. D.S. Francy, J.M. Thomas, R.L. Raymond, and C.H. Ward, *J. Ind. Microbiol.* **8**, 237–245 (1991).
19. Y. Zhang, and R.M. Miller, *Appl. Environ. Microbiol.* **61**, 2247–2251 (1995).
20. V. Deretic, D.W. Martin, M.J. Schurr, M.H. Mudd, N.S. Hibler, et al., *Biotechnology* **11**, 1133–1136 (1993).
21. L.O. Martins and I. Sa-Correia, *Enzyme Microbial Technol.* **13**, 385–389 (1991).
22. M.J. Schurr, D.W. Martin, M.H. Mudd, and V. Deretic, *J. Bacteriol.* **176**, 3375–3382 (1994).
23. A.J. Anderson and E.A. Dawes, *Microbiol. Rev* **54**, 450–472 (1990).
24. A. Steinbüchel, and H.E. Valentin, *FEMS Microbiol. Lett.* **128**, 219–228 (1995).
25. *Eur. Plastic News* **12**, 38 (1985).
26. M.J. de Smet, G. Eggink, B. Witholt, J. Kingma and H. Wynberg, *J. Bacteriol.* **154**, 870–878 (1983).
27. R.D. Schwartz, *Appl. Microbiol.* **25**, 574–577 (1973).
28. R.D. Schwartz, and C.J. McCoy, *Appl. Microbiol.* **26**, 217–218 (1973).
29. R.G. Lageveen, G.W. Huisman, H. Preusting, P. Ketelaar, G. Eggink, and B. Witholt, *Appl. Environ. Microbiol.* **54**, 2924–2932 (1988).
30. R.W. Lenz, Y.B. Kim, and R.C. Fuller, *FEMS Microbiol. Rev.* **103**, 207–214 (1992).
31. WWW site *http://alpha.gmw.ac.uk/~ugca000/iupac/enzyme*.
32. Eur. Pat. 93782 (1983), H. Yamada, and Y. Tani.
33. C. Wandrey, and B. Bossow, *Biotechnol. Bioind.* **3**, 8–13 (1986).
34. C.W. Bradshaw, H. Fu, G.J. Shen, and C.H. Wong, *J. Org. Chem.* **57**, 1526–1532 (1992).
35. W.O. Pat. 9318138 (Sept. 16, 1993) M.R. Kula and J. Peters, (to Forschungszentrum Juelich GmbH).
36. S. Shimizu, and H. Yamada, *J. Syn. Org. Chem. Japan* **49**, 52–70 (1991).
37. M. Deamici, C. Demicheli, G. Molteni, D. Pitre, G. Carrea, S. Riva, S. Spezia, and L. Zetta, *J. Org. Chemistry* **56**, 67–72 (1991).
38. S.I. Suye, S. Yokoyama, and A. Obayashi, *J. Ferment. Bioeng.* **68**, 301–304 (1989).
39. D.B. Janssen, F. Pries, and J.R. Vanderploeg, *Annu. Rev. Microbiol* **48**, 163–191 (1994).
40. N. Mermod, J.L. Ramos, and K.N. Timmis, *J. Bacteriol.* **167**, 447–454 (1986).
41. M.G. Wubbolts, O. Favre-Bulle, and B. Witholt, *Biotechnol. Bioeng.* **52**, 301–308 (1996).
42. J.B. van Beilen, Ph.D. Thesis, Rijksuniversiteit Groningen, 1994.
43. M. Lee and A.C. Chandler, *J. Bacteriol.* **41**, 373–386 (1941).
44. D.A. Kunz and P.J. Chapman, *J. Bacteriol.* **146**, 179–191 (1981).
45. M.G. Wubbolts, P. Reuvekamp, and B. Witholt, *Enzyme Microb. Technol.* **16**, 608–615 (1994).
46. J.B. van Beilen, J. Kingma, and B. Witholt, *Enzyme Microb. Technol.* **16**, 904–911 (1994).
47. M.G. Wubbolts, R. Noordman, J.B. van Beilen, and B. Witholt, *Recl. Trav. Chim. Pays-Bas* **114**, 139–144 (1995).
48. A. Kiener, *Angew. Chem., Int. Ed. Engl.* **31**, 774–775 (1992).
49. A. Kiener *Chemtech* 31–35 (Sept. 1995).

50. J.B. Johnston, and V. Renganathan, *Enzyme Microb. Technol.* **9**, 706–708 (1987).
51. Eur. Pat. 289,846 (Nov. 9, 1988), R. Marquardt, J. Then, B. Braeu, P. Praeve, and G. Woehner, (to Hoecst).
52. H.I. Rhee, K. Murata, and A. Mimura, *Agric. Biol. Chem.* **51**, 1701–1702 (1987).
53. U.S. Pat. 5,049,494 (Sept. 17, 1991), P. Allenza, (to Allied-Signal Inc.)
54. S. Maruo, H. Yamamoto, M. Toda, N. Tachikake, M. Kojima, and Y. Ezure, *Biosci. Biotechnol. Biochem.* **57**, 499–501 (1993).
55. J. Alvarez-Jacobs, D. Court, and G. Guarneros, *Biotechnol. Lett.* **12**, 425–430 (1990).
56. M. Watanabe, Y. Morinaga, and H. Enei, *J. Ferment. Technol.* **65**, 617–620 (1987).
57. M.G. Wubbolts, submitted for publication.
58. S. Takahashi, *Hakko Kogaku Kaishi* **61**, 139–151 (1983).
59. K. Drauz and H. Waldmann, *Enzyme Catalysis in Organic Synthesis: A Comprehensive Handbook*. VCH Verlagsgesellschaft, Weinheim, Germany, 1995.
60. K.E. Jaeger and S. Wohlfarth, *Bioengineering* **9**, 39–46 (1993).
61. F. Theil, *Chem. Rev.* **95**, 2203–2227 (1995).
62. A.L. Gutman and M. Shapira, *Adv. Biochem. Eng. Biotechnol.* **52**, 87–128 (1995).
63. J.S. Dordick, *Curr. Opin. Biotechnol.* **2**, 401–407 (1991).
64. J. van der Werf, W.J.J. van den Tweel, J. Kamphuis, S. Hartmans, and J.A.M. de Bont, *Trends Biotechnol.* **12**, 95–103 (1994).
65. J.S. Takagi, R. Fukunaga, M. Tokushige, and H. Katsuki, *J. Biochem.* **96**, 545–552 (1984).
66. Eur. Pat. B 1 0102529 (May 10, 1989), H. Tuneo, T. Hisao, and I. Tatsuo, (to Toso Corp.).
67. G.H. Kulla, *Chimia* **86**, 295–323 (1991).
68. Eur. Pat. 410,430 (Jan. 30, 1991), F. Hocks (to the Lonza AG, Gampel/Wallis, Switzerland).
69. T. Nagasawa, H. Shimizu, and H. Yamada, *Appl. Microbiol. Biotechnol.* **40**, 189–195 (1993).
70. K.I. Ishiwata, M. Fukuhara, M. Shimada, N. Makiguchi, and K. Soda, *Biotechnol. Appl. Biochem.* **12**, 141–149 (1990).
71. U.S. Pat. 4,605,625 (Aug. 12, 1986), H. Yamada, S. Shimizu, and S. Shiozaki, (to Nippon Zeon).
72. H. Itoh, T. Sato, and K. Izumori, *J. Ferment. Bioeng.* **80**, 101–103 (1995).
73. Y. Takasaki, K. Hinoki, Y. Kataoka, S. Fukuyama, and N. Nishimura, *J. Ferment. Bioeng.* **76**, 237–239 (1993).
74. H. Martinez-Blanco, A. Reglero, J. Martin-Villacorta, and J.M. Luengo, *FEMS Microbiol. Rev.* **72**, 113–116 (1990).
75. M. Fernandez-Valverde, A. Reglero, M.-B. H., and J.M. Luengo, *Appl. Environ. Microbiol.* **59**, 1149–1154 (1993).
76. M.A. Ferrero, A. Reglero, H. Martinez-Blanco, M. Fernandez-Valverde, and J.M. Luengo, *Antimicrob. Agents Chemother.* **35**, 1931–1932 (1991).
77. H. Martinez-Blanco, A. Reglero, and J.M. Luengo, *J. Antibiotics* **44**, 1252–1258 (1991).
78. C. Laane, S. Boeren, K. Vos, and C. Vergeer, *Biotechnol. Bioeng.* **30**, 81–87 (1987).
79. A. Inoue and K. Horikoshi, *J. Ferment. Bioeng.* **71**, 194–196 (1991).
80. A.N. Rajagopal, *Enzyme Microb. Technol.* **19**, 606–613 (1996).
81. M. Vermue, J. Sikkema, A. Verheul, R. Bakker, and J. Tramper, *Biotechnol. Bioeng.* **42**, 747–758 (1993).
82. Q. Chen, A. Nijenhuis, H. Preusting, J. Dolfing, D.B. Janssen, and B. Witholt, *Enzyme Microb. Technol.* **17**, 647–652 (1995).
83. J. Sikkema, J.A.M. de Bont, and B. Poolman, *J. Biol. Chem.* **269**, 8022–8028 (1994).
84. J. Sikkema, J.A.M. de Bont, and B. Poolman, *Microbiol. Rev.* **59**, 201–222 (1995).
85. M.C. Antunes-Madeira, and V.M.C. Madeira, *Biochim. Biophys. Acta* **901**, 61–66 (1987).
86. M.A. Ascon-Cabrera and J.M. Lebeault, *J. Ferm. Bioeng.* **80**, 270–275 (1995).
87. H. Alexandre, B. Mathieu, and C. Charpentier, *Microbiology* **142**, 469–475 (1996).
88. Q. Chen, D.B. Janssen, and B. Witholt, *J. Bacteriol.* **177**, 6894–6901 (1995).
89. H.C. Pinkart, J.W. Wolfram, R. Rogers, and D.C. White, *Appl. Environ. Microbiol.* **62** 1129–1132 (1996).
90. F.J. Weber and J.A.M. de Bont, *Meded. Fac. Landbouwwet. Rijksuniv. Gent* **59**, 2295–2392 (1994).
91. F.J. Weber and J.A.M. de Bont, *Biochim. Biophys. Acta* **1286**, 225–245 (1996).
92. K. Moriya, S. Yanagitani, R. Usami, and K. Horikoshi, *J. Mar. Biotechnol.* **2**, 131–133 (1995).
93. J.L. Ramos, E. Duque, M.J. Huertas, and A. Haidour, *J. Bacteriol.* **177**, 3911–3916 (1995).
94. R. Aono, T. Negishi and H. Nakajima, *Appl. Environ. Microbiol.* **60**, 4624–4626 (1994).
95. H. Asako, H. Nakajima, K. Kobayashi, M. Kobayashi, and R. Aono, *Appl. Environ. Microbiol.* **63**, 1428–1433 (1997).
96. H. Nakajima, H. Kobayashi, T. Negishi, and R. Aono, *Biosci. Biotech. Biochem.* **59**, 1323–1325 (1995).
97. R. Bar and J.L. Gainer, *Biotechnol. Prog.* **3**, 109–114 (1987).
98. L.E.S. Brink, J. Tramper, K.C.A.M. Luyben, and K. Van't Riet, *Enzyme Microb. Technol.* **10**, 736–743 (1988).
99. M.D. Hocknull and M.D. Lilly, *Appl. Microbiol. Biotechnol.* **33**, 148–153 (1990).
100. A.C.P. Dias, J.M.S. Cabral, and H.M. Pinheiro, *Enzyme Microb. Technol.* **16**, 708–714 (1994).
101. J.Y. Houng, W.P. Chiang, K.C. Chen, and C. Tiu, *Enzyme Microb. Technol.* **16**, 485–491 (1994).
102. E.G. Ceen, J.P.R. Herrmann, and P. Dunnill, *Appl. Microbiol. Biotechnol.* **25**, 491–494 (1987).
103. U.S. Pat. 4,833,078 (May 23, 1989), J.H. Hsieh (to the Celgene Corp.).
104. U.S. Pat. 5,236,832 (Sept. 7, 1993), A. Kiener (to Lonza Ltd.).
105. D.G.H. Ballard, A.J. Blacker, J.M. Woodley, and S.C. Taylor in D.P. Mobley ed., *Plastics from Microbes: Microbial Synthesis of Polymers and Polymer Precursors*, Hanser Publishers, Munich, 1994, pp. 139–168.
106. K. Gbewonyo, B.C. Buckland, and M.D. Lilly, *Biotechnol. Bioeng.* **37**, 1101–1107 (1991).
107. H. Preusting, R. van Houten, A. Hoefs, E.K. van Langenberghe, O. Favre Bulle, and B. Witholt, *Biotechnol. Bioeng.* **41**, 550–556 (1993).
108. H.G. Schlegel, G. Gottschalk, and R. von Bartha, *Nature* **191**, 463–465 (1961).

109. M.J. de Smet, J. Kingma, H. Wynberg, and B. Witholt, *Enzyme Microb. Technol.* **5**, 352–360 (1983).
110. K. Kawakami, in H., Okada, A. Tanaka, and H.W. Blanch, eds., *Enzyme Engineering X*, vol. 613, New York Academy of Sciences, New York, 1990, pp. 707–711.
111. A. Bosetti, J.B. van Beilen, H. Preusting, R.G. Lageveen, and B. Witholt, *Enzyme Microb. Technol.* **14**, 702–708 (1992).
112. S.R. Roffler, T.W. Randolph, D.A. Miller, H.W. Blanch, and J.M. Prausnitz, in B. Mattiason and O. Holst, eds., *Extractive Bioconversions*, Dekker, New York, 1990, pp. 133–172.
113. S. Oda, Y. Inada, A. Kato, N. Matsudomi, and H. Ohta, *J. Ferment. Bioeng.* **80**, 559–564 (1995).
114. V.M. Yabannavar and D.I.C. Wang, *Biotechnol. Bioeng.* **37**, 1095–1100 (1991).
115. R.G. Mathys, Ph.D. Thesis, Eidgenössische Technische Hochschule Zürich, Switzerland, 1997.
116. A. Kollmer, Ph.D. Thesis, Eidgenössische Technische Hochschule Zürich, Switzerland, 1997.
117. R.G. Mathys, O.M. Kut, and B. Witholt, *J. Chem. Technol. Biotechnol.*, in press (1998).
118. R.G. Mathys, A. Schmid, O.M. Kut, and B. Witholt, *J. Chem. Technol. Biotechnol.*, in press (1998).
119. J.P. Corry, W.L. Reed, and W.R. Curtis, *Biotechnol. Bioeng.* **42**, 503–508 (1993).
120. L. Nielsen, M. Larsson, O. Holst, and B. Mattiasson, *Appl. Microbiol. Biotechnol.* **28**, 335–339 (1988).
121. M.H.L. Ribeiro, D.M.F. Prazeres, J.M.S. Cabral, and M.M.R. da Fonseca, *Bioprocess Eng.* **12**, 95–102 (1995).
122. H.P. Fielder, M. Nega, C. Pfefferle, I. Groth, C. Kempter, H. Stephan, and J.W. Metzger, *J. Antibiot.* **49**, 758–764 (1996).
123. V. Hecht, J. Vorlop, H. Kalbitz, K. Gerth, and J. Lehmann, *Biotechnol. Bioeng.* **29**, 222–227 (1987).
124. K.D. Green, N.H. Thomas, and J.A. Callow, *Biotechnol. Bioeng.* **39**, 195–202 (1992).
125. R.J. Robins, A.J. Parr, and M.J.C. Rhodes, *Biochem. Soc. Trans.* **16**, 67–71 (1988).
126. R.D. Williams, N. Chauret, C. Bedard, and J. Archambault, *Biotechnol. Bioeng.* **40**, 971–977 (1992).
127. A. Freeman, J.M. Woodley, and M.D. Lilly, *Bio/Technology* **11**, 1007–1012 (1993).
128. A.S. Bommarius, K. Drauz, U. Groeger, and S. Wandrey, in A.N. Collins, G.N. Sheldrake, and J. Crosby eds, *Chirality in Industry*, Wiley, Chichester, U.K., 1992, pp. 371–397.
129. P. Adlercreutz, *Biocat. Biotransf.* **14**, 1–30 (1996).
130. B. Witholt, M.J. de Smet, J. Kingma, J.B. van Beilen, M. Kok, R.G. Lageveen, and G. Eggink, *Tibtech* **8**, 46–52 (1990).

PULLULAN, MICROBIAL PRODUCTION METHODS

JULES THIBAULT
ANH LEDUY
Laval University
Sainte-Foy, Canada

KEY WORDS

Aureobasidium pullulans
Fermentation
Hyperbaric
Microbial polysaccharide
Mixing
Oxygen transfer
Pigmentation
Pullulan
Pullulanolysis
Reciprocating plate bioreactor
Rheology

OUTLINE

WHAT IS PULLULAN?

Pullulan is a water-soluble extracellular polysaccharide synthesized by the yeastlike fungus *Aureobasidium pullulans* (formerly named *Pullularia pullulans*) from a variety of mono- and disaccharides (1). It is generally accepted that pullulan is a neutral glucan that consists of either a linear chain of D-glucopyranosyl units that alternate regularly between one (1→6)-α-D and two (1→4)-α-D linkages, or a linear polymer of maltotriosyl units connected by (1→6)-α-D linkages (Fig. 1). However, pullulan containing 5–7% maltotetraosyl units has also been reported for several strains of *A. pullulans* (2,3).

Freeze-dried pullulan is white in color, tasteless, odorless, nonhygroscopic, nontoxic, biodegradable, and edible. It is insoluble in many solvents including methanol, ethanol, and acetone, but soluble in water to form a transparent, colorless, viscous, and adhesive solution. Shaped bodies as well as very thin pullulan films 0.01 mm thick can be formed. Thin films of pullulan are colorless, transparent, tasteless, odorless, tenacious, resistant to oil and grease, unaffected by small thermal variations, impermeable to oxygen, nontoxic, biodegradable, and edible (4,5).

Numerous applications of pullulan are claimed in a large number of patents. A comprehensive review of the applications of pullulan in food and drugs, as well as on the potential uses of pullulan in industrial and medical applications, is reported elsewhere (6).

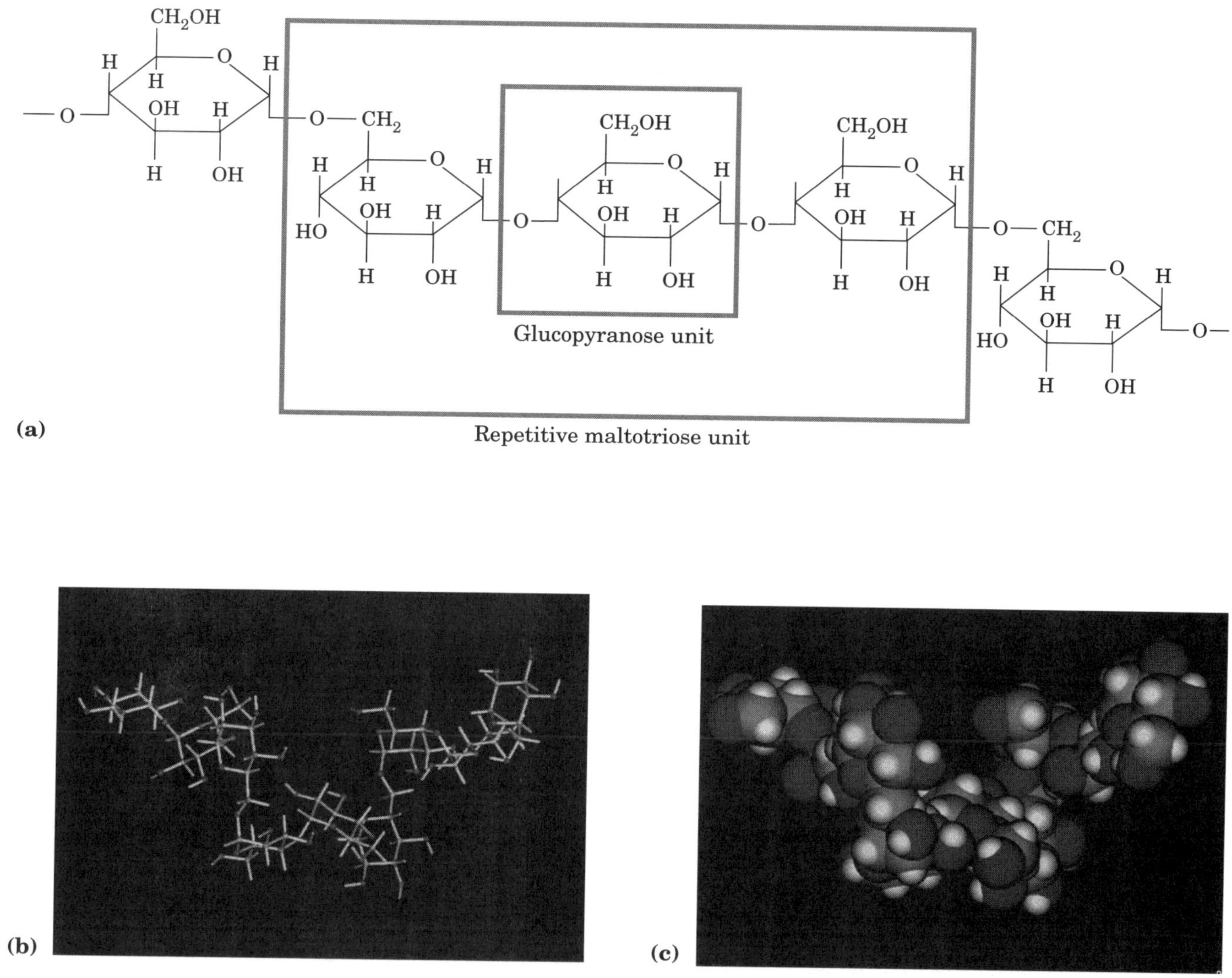

Figure 1. Generally accepted chemical structure of pullulan: (**a**) chemical formula; (**b**) ball-and-stick model for one hypothetical conformation of three maltotriose units; (**c**) space-filled model for one hypothetical conformation of three maltotriose units. *Source*: The ball-and-stick and space-filled models were kindly provided by Professor Josée Brisson from the Chemistry Department, Laval University.

PROCESS TECHNOLOGY FOR THE PRODUCTION OF PULLULAN

The general process flowchart in Figure 2 illustrates the five main steps in the production of pullulan by fermentation: culture medium preparation, inoculum preparation, fermentation, product separation and purification, and product alteration. This five-step general process flowchart for the production of pullulan by fermentation applies for small-scale laboratory experiments as well as large-scale commercial production of pullulan.

Step A. During the culture medium preparation step, appropriate quantities of raw materials to be used as culture medium ingredients are to be formulated, weighed, and dissolved in water. The ingredients are chosen and formulated in such a way to provide the exact amounts of carbon, nitrogen, and mineral salts required for cell growth and pullulan synthesis during the fermentation step. The pH of the solution is adjusted to the desired value; the medium is then sterilized, either in a fermenter or in a separate feed tank.

Step B. The inoculum preparation step consists of producing a large quantity of microorganism to seed or to inoculate the sterilized culture medium obtained at the end of step A in order to start the fermentation (step C). The appropriate strain of microorganism is selected, then propagated by successive fermentations from a small volume of a pure culture to obtain a large volume of microbial cell suspension required for inoculum purposes. In the industrial production of pullulan, the amount of inoculum required to seed the production fermenter may vary from hundreds to thousands of liters of microbial cell suspension. The inoculum culture-to-medium ratio used may also vary from 0.1% v/v to 10% v/v. Usually, larger inoculum

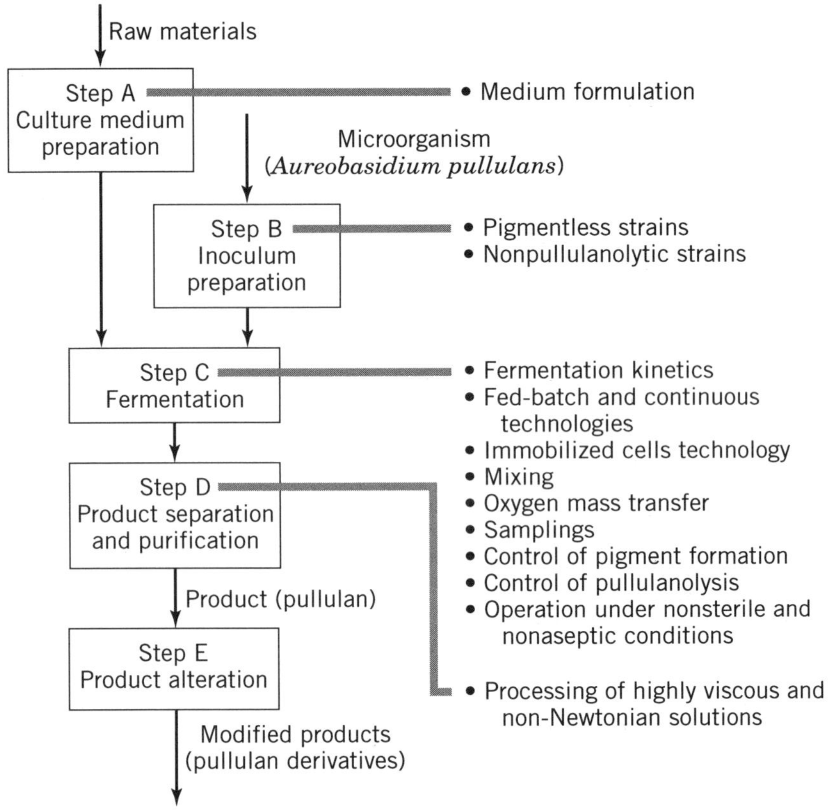

Figure 2. Five-step general process flowchart for the production of pullulan by fermentation. Research and development requirements for technology improvement are indicated for each step.

ratios are suitable for laboratory-scale fermentations, and smaller ratios are more practical for industrial-scale productions.

Step C. The fermentation step is the biosynthesis of pullulan from a carbon source by living and growing microbial cells in the fermenter under sterile and aseptic conditions. The fermentation technology for pullulan production may be batch mode, fed-batch mode, or continuous mode, using either free or immobilized cells. Batch fermentation using free cells of *A. pullulans* is still the most popular technology used in pullulan production in small as well as in large scales. The duration and the performance of the fermentation step are determined by the kinetic behavior of the combined system of microorganism–medium–fermentation conditions. It means that each specific strain of microorganism, in a specific culture medium (ingredients and composition), and under a specific set of fermentation conditions will respond with a specific fermentation kinetic behavior. Therefore, knowledge of fermentation kinetics is important for choosing the best fermentation technology and conditions for each specific strain of microorganism in a specific culture medium. During the fermentation step, operating conditions such as pH, temperature, oxygenation (by aeration), and mixing (by agitation) are either artificially controlled by maintaining each one at a certain value for a specific time period, or simply uncontrolled.

Step D. When the fermentation step is completed, the resulting fermented broth is a mixture of solids in suspension (which consists of microbial cells and cellular debries) together with dissolved solids in the aqueous solution (which comprise water-soluble pullulan synthesized during the fermentation, residual ingredients from the culture medium, and extracellular metabolites produced and excreted during the fermentation). In the product separation and purification step, solids in suspension are first separated from the culture medium by centrifugation or filtration. The obtained liquid phase containing soluble pullulan is then decolorized using activated carbon to remove the black pigmentation that is produced by *A. pullulans* during fermentation. The decolorized liquid phase is further purified to remove small molecular compounds (residual ingredients, extracellular metabolites), then dewatered and dried to obtain pullulan as a final product in solid form. The dewatering operation usually consists of the precipitation of pullulan from the aqueous solution using acetone, methanol, or ethanol as a solvent (Fig. 3).

Step E. The product alteration step is necessary only when modified pullulan or pullulan derivatives are desired as final products. Technologies are available to chemically modify pullulan in order to improve its existing properties and/or to impart newly desired properties to pullulan. Examples of such pullulan derivatives are presented in Table 1 (5,7–15).

COMMERCIAL PRODUCTION OF PULLULAN

At the present time, the only known commercial producer of pullulan by the fermentation process is the Hayashibara Co., Ltd. in Japan. Figure 4 illustrates the detailed flowchart for the commercial production of pullulan by this particular company (16). Note that it matches very well the

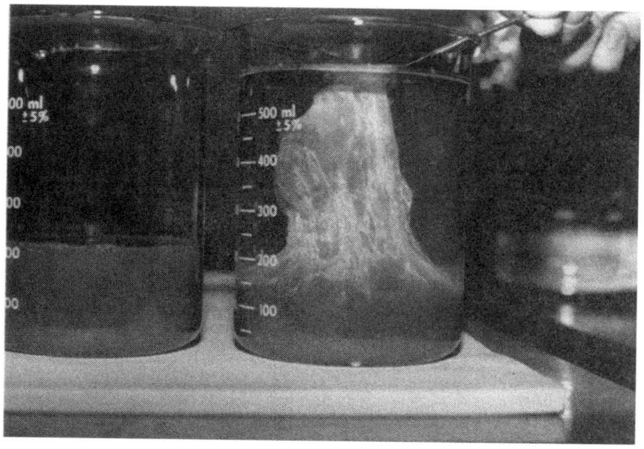

Figure 3. Solvent precipitation of pullulan: Pullulan, which is soluble in the fermentation broth (yellow solution), becomes insoluble in methanol (clear liquid). The fibrous texture appearance of the precipitated pullulan indicates that this polysaccharide has very high molecular weight.

five-step general process flowchart for the production of pullulan by fermentation shown in Figure 2.

The culture medium used consists of a carbon source (partial starch hydrolyzate with dextrose equivalent [DE] around 50% at a concentration from 10 to 20%), a nitrogen source (formulated from corn gluten, soybean protein, peptone, nitrate salts, and ammonium salts), and other mineral salts. The initial pH is within 5–6.8.

The inoculum is prepared from selected mutant strains of *A. pullulans* AHU 9553, *Dematium pullulans* IFO 4464, and *Pullularia fermentans* var. *fermentans* IFO 6401. The mutant strains used as production strains must have the least black pigment, require the shortest duration of fermentation, and give the highest yield of pullulan.

The fermentation is carried on in batch mode for approximately four days in a conventional fermenter, with special consideration to ensure adequate aeration and agitation of the highly viscous fermentation broth. The temperature is maintained at 30 °C and aeration at 0.5 VVM (volume of air per volume of culture medium per minute). The kinetic behavior of commercial pullulan fermentation corresponds very well with those reported in the scientific literature (17–20). The pH of the culture medium decreases sharply during the first day and remains constant at around pH 3.5 to the end of the fermentation. Cell growth and pullulan synthesis occur during this low and constant pH period. The concentrations of cell mass and pullulan continue to increase in the fermentation broth until the end of the fermentation. However, the molecular weight of pullulan in the fermentation broth increases to a maximum value of around 2,000,000–3,000,000 in the early stage of fermentation, then drastically decreases to very low molecular weight toward the end of the fermentation. This decrease is presumably caused by the action of a pullulan-decomposing enzyme, which is also synthesized by the microorganism during the fermentation period. Consequently, the viscosity of the fermentation broth, which results from both pullulan concentration and pullulan molecular weight, increases rapidly in the early stage of fermentation to over 300 cP, then reduces to less than 100 cP at the end of the fermentation. By controlling the fermentation conditions (phosphate content and initial pH), Hayashibara Co., Ltd. obtains a pullulan product that ranges from 50,000 to 500,000 MW.

The pullulan recovery and purification operations depend on the desired quality of pullulan product for specific applications (Fig. 4). Hayashibara Co., Ltd. defines three different pullulan grades: (*1*) Product I (crude pullulan culture and its concentrate) for use as an industrial adhesive, dispersant, and coagulant; (*2*) Product II (cell-free, decolorized pullulan, containing some minerals, mostly sodium chloride) for use as a raw material for various industries, including the food processing industries, as well as a base material for the production of pullulan derivatives; and (*3*) Product III (cell-free, decolorized, and desalted pullulan), which is highly purified pullulan in white powder for making films and shaped articles.

The annual production of pullulan by Hayashibara Co., Ltd. is 200 metric tons and the ex-factory sale price of pullulan PF-20 (Product III) is 2300 JP_l (approximately 17 US$) per kilogram (personal communication with Dr. K. Okada, Overseas Business Development, Hayashibara Co., Ltd., April 6, 1998).

RESEARCH AND DEVELOPMENT REQUIREMENTS FOR TECHNOLOGY IMPROVEMENT

Pullulan fermentation, like numerous other extracellular polysaccharide fermentations, is one of the most challeng-

Table 1. Pullulan Derivatives

Modified pullulan	Imported properties	References
Esterified pullulan	Various degrees of water insolubility	5
Etherified pullulan	Various degrees of water insolubility	5
Hydrogenated pullulan	High heat resistance	7
Pullulan sulfate and its salts	Anti–peptidic ulcer; low blood anticoagulation	8
Carboxylated pullulan	High solubility in cold water	9
Cross-linked pullulan	Hydrophilic gel; various degrees of water absorbability	10
Dialdehyde pullulan	Water resistance	11
Pullulan aminoalkyl ether	Cationic polymer	12
Ionic pullulan gel	Charged gel	13
Cyanoethylated pullulan	High heat resistance; insolubility in water; high dielectric constant	14
Cyanoethyldihydroxylakylated pullulan	High dielectric constant	15

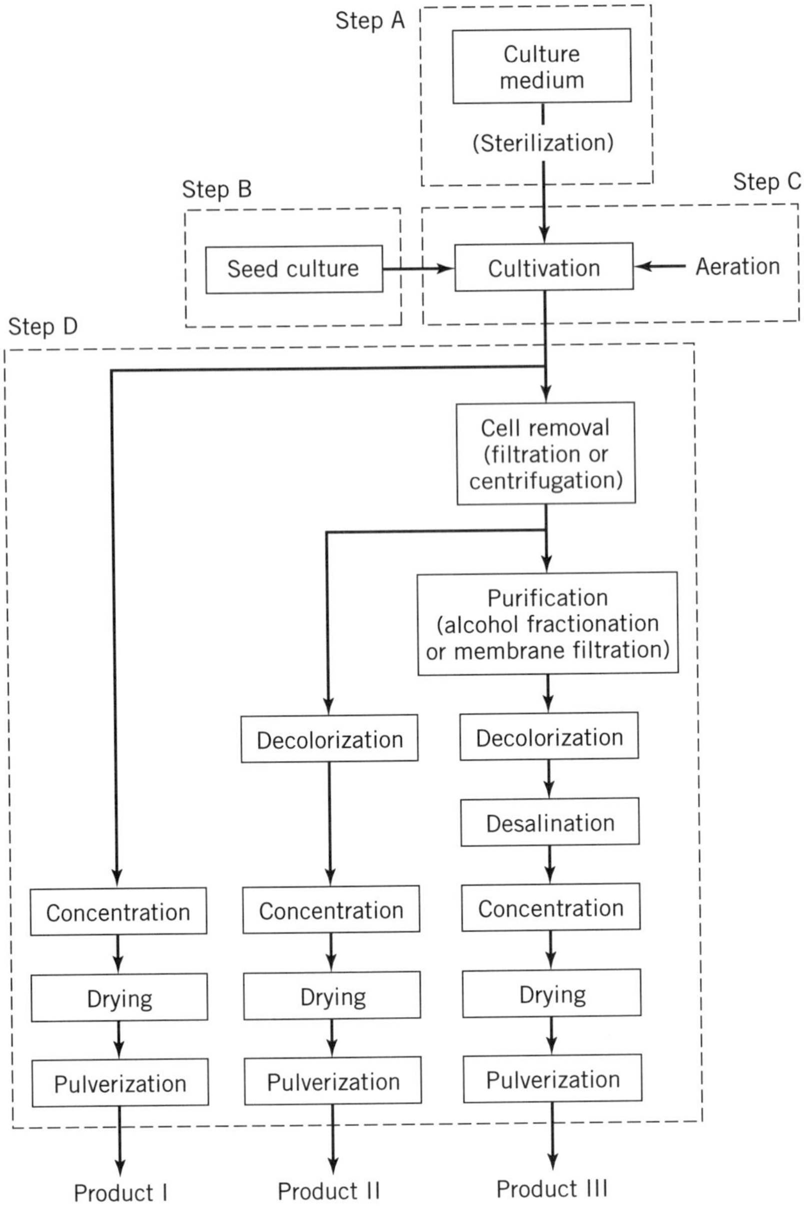

Figure 4. Process flowchart for the commercial production of pullulan by Hayashibara Co., Ltd. *Source*: Adapted from Ref. 16, reproduced with permission.

ing bioprocesses for engineers and scientists seeking ways to improve fermentation technology. Four groups of problems associated with pullulan fermentation still require better solutions in order to increase the performance of the fermentation process and lower product recovery costs.

Problems Due to Viscosity of the Fermentation Broth

The evolution of rheological properties of the fermentation broth during batch fermentation of *A. pullulans* has been studied in detail (17). In batch fermentation, the culture medium initially exhibits Newtonian behavior, and then the rheology of the broth progressively changes to a highly viscous and non-Newtonian behavior. The rheological properties of the fermentation broth may be described by two different parameters: the field apparent viscosity and the non-Newtonian flow behavior index, which can be expressed with the following rheological model:

$$\eta = K\dot{\gamma}^{n-1} \tag{1}$$

where η is the apparent viscosity, K is the consistency index, n is the flow behavior index, and $\dot{\gamma}$ is the shear rate. This equation shows that, in non-Newtonian fluids, the value of viscosity obtained depends on the rate of shear at which the viscosity is measured. The viscosity of a non-Newtonian fermentation broth is commonly standardized by its field value, which is the value measured at the unit rate of shear. Such values of the field apparent viscosity are independent of either rotational speed or geometrical configuration of the viscometer used.

It would be tempting to believe that the viscosity of the fermentation broth increases steadily throughout the fermentation to follow the increase in pullulan concentration. In fact, in the case of pullulan, the field apparent viscosity of the fermentation broth increases slowly during the initial phase of fermentation, then rises rapidly to its maximum value, and finally decreases toward its initial value. The maximum values of the apparent field viscosity, as well as the time when these maximum values occur, depend on fermentation conditions such as initial pH of the culture medium and age of inoculum (Fig. 5). The results of Figure 5 are very informative in the sense that the large differences observed for the maximum values of the field apparent viscosity can be directly attributed to the length of the pullulan chain or to the average molecular weight of pullulan, because the concentrations of pullulan in the fermentation broths are approximately the same around the fifth day of fermentation. The decrease in the field apparent viscosity during later stages of each fermentation, when the polysaccharide concentration has leveled off, is called pullulanolysis, which relates to the depolymerase secreted by *A. pullulans*. The evolution of the field apparent viscosity of the fermentation broth depends also on other factors such as the aeration rate (VVM) and the level of dissolved oxygen concentration (mg/L) during pullulan fermentation (21).

The flow behavior index characterizes very well the average length of the pullulan chain (22). It is therefore interesting to monitor this rheological parameter during the course of the fermentation to observe changes in pullulan average molecular weight. At the beginning of a typical pullulan fermentation, the culture medium is Newtonian (non-Newtonian index $n = 1$). The non-Newtonian index of the fermentation broth decreases slowly during the initial phase of fermentation, then drops rapidly to its minimum value, and finally rises toward its initial value. Therefore, throughout the fermentation the broth becomes more and more pseudoplastic (non-Newtonian index <1), then returns toward its initial Newtonian behavior. The minimum values of the non-Newtonian index as well as the time when these minimum values occur depend on fermentation conditions such as initial pH of the culture medium and the age of inoculum (Fig. 6). The rheological behavior of such non-Newtonian solutions has a profound impact on the flow characteristics within the fermenter. The apparent viscosity decreases when the fluid is in motion. This reduction in apparent viscosity with the velocity of the fluid is more pronounced as the flow behavior index is reduced.

The observed rheological properties of the fermentation broth in Figures 5 and 6 result from the combined contribution of the pullulan concentration and the average length of the pullulan chain. Both pullulan concentration and pullulan average chain length in the fermentation broth are continuously changing during batch fermentation. Therefore, from the results in Figures 5 and 6, it is not possible to say how and how much these two parameters contribute to the observed rheological properties. LeDuy et al. (17) have recovered pullulan from fermentation broths at 1, 3, 5, and 8 days of a batch fermentation

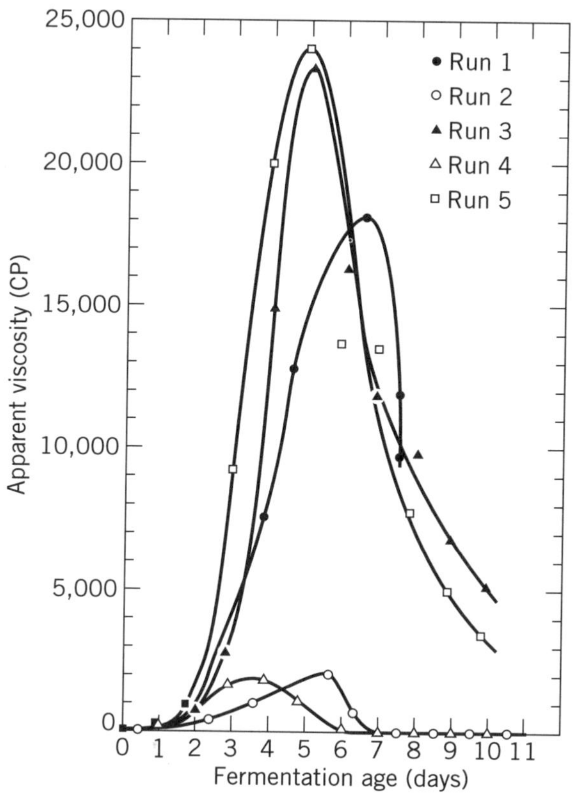

Figure 5. Evolution of the field apparent viscosity of the fermentation broths under various conditions of initial pH and inoculum age. *Source*: From Ref. 17, reproduced with permission.

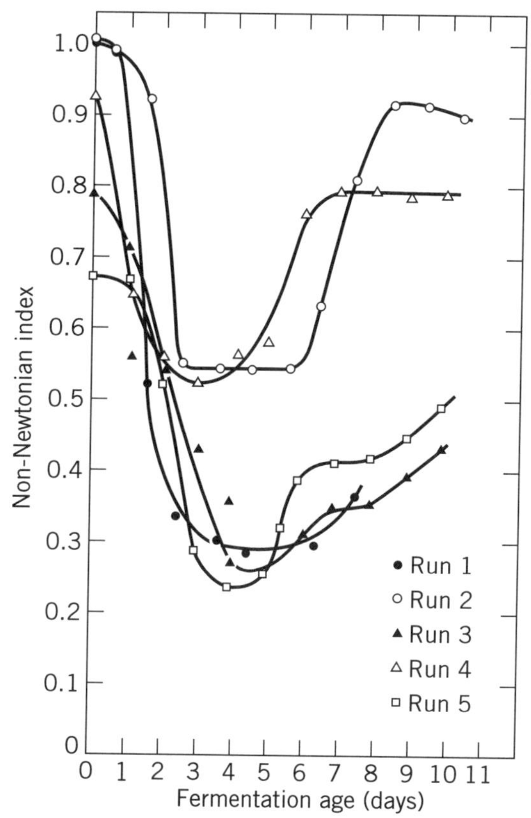

Figure 6. Evolution of the non-Newtonian index of the fermentation broths under various conditions of initial pH and inoculum age. *Source*: From Ref. 17, reproduced with permission.

(run 3 from Figs. 5 and 6), freeze-dried it, and then dissolved the recovered pullulan in water to form solutions having a pullulan concentration varying from 0.125 to 2.0 g/L. The field apparent viscosities and the non-Newtonian indices of these solutions were measured and are presented in Figure 7. At the same concentration of pullulan, the field apparent viscosity of pullulan solution recovered from the early stage of fermentation is higher than that recovered from the later stage, indicating that the pullulan recovered in the first days of fermentation has higher molecular weight than in subsequent days. The opposite phenomenon is observed for the non-Newtonian index. At the same concentration of pullulan, the non-Newtonian index of pullulan solution recovered from the early stage of fermentation is lower than that recovered from the later stage, indicating that the pseudoplasticity of the pullulan solution is directly related to the high molecular weight of pullulan. Each set of curves prepared with pullulan recovered after the same fermentation time in Figure 7 shows the very important contribution of pullulan concentration on the field apparent viscosity as well as on the non-Newtonian index of the pullulan solution. Higher pullulan concentration leads to higher field apparent viscosity and larger pseudoplasticity.

Thus, any fermentation process aiming to achieve high-yield production of high molecular weight pullulan still provides tremendous engineering challenges in mixing, oxygenation, and sampling during the fermentation, as well as the major challenge of removal of microbial cells from the highly viscous fermentation broth in the product separation and purification step. To clearly illustrate the difficulties involved in dealing with a highly viscous fermentation broth, Figure 8 shows a typical fermentation broth flowing out of a fully opened large-size sampling tube under the compression of the aeration gas inside the fermenter. During pullulan fermentation, the nonhomogeneous mixing problem is always observed, even in a laboratory-scale fermenter. The fermentation broth inside the fermenter tank segregates into two regions having different characteristics, with an active core around the impeller where the culture medium is well mixed and properly aerated. This core is separated from the innermost zone by a fictitious boundary, a stagnant fluid region characterized by a total lack of agitation and aeration (Fig. 9). Some authors even discuss the formation of three distinct mixing regions: a small central region (often referred to as a cavern) in the vicinity of the impellers where the turbulence is important, an annular section characterized by slow fluid motion, and finally, a stagnant outer region where the

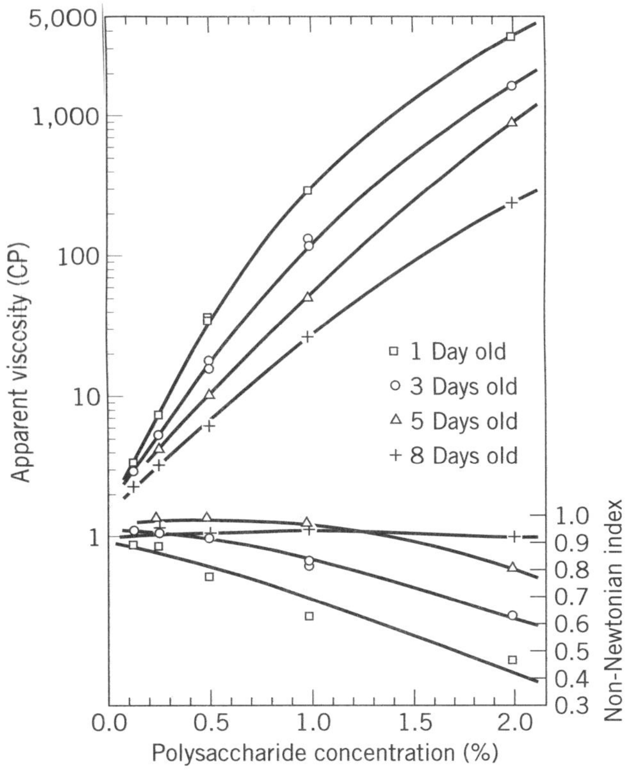

Figure 7. Variations of field apparent viscosity and non-Newtonian index as a function of the concentration of pullulan in a solution and the pullulan chain length. Pullulan molecules have longest chain lengths after one day and shortest chain lengths after eight days of batch fermentation. *Source*: From Ref. 17, reproduced with permission.

Figure 8. Highly viscous fermentation broth of *A. pullulans* in a sucrose medium: the broth slowly flowing out of a fully opened large-size sampling tube under the compressed air inside the fermenter.

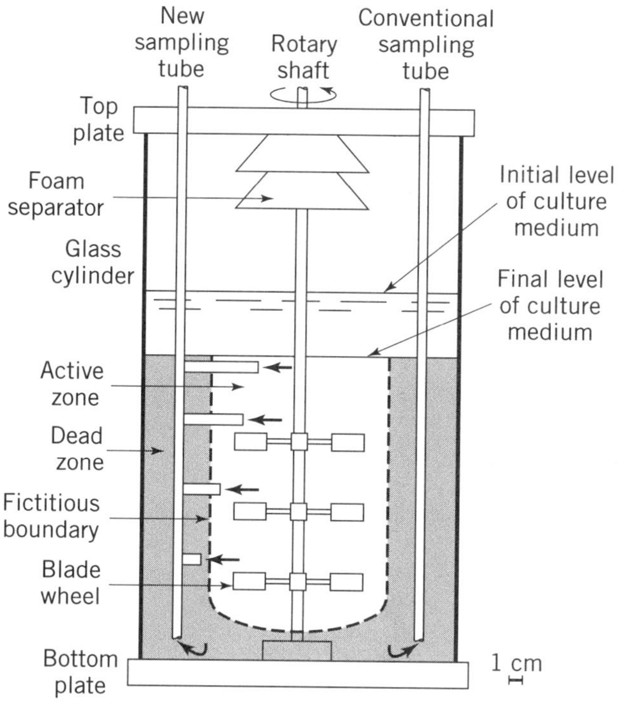

Figure 9. Active zone and stagnant zone of the highly viscous fermentation broth in the fermenter during the pullulan fermentation. Configurations of both conventional and newly improved sampling devices are shown. *Source*: From Ref. 23, reproduced with permission.

fluid is motionless and the apparent viscosity very high (24). The energy imparted to the fluid by the rotating impellers is dissipated in a restricted region of the total volume of the fermentation broth. The high viscosity combined with the non-Newtonian behavior creates major difficulties in sampling, oxygen mass transfer, mixing, and downstream processing. These are briefly analyzed in turn.

Process Monitoring and Sampling. With the absence of homogeneous conditions throughout the fermentation vessel, it is only possible to monitor locally the dissolved oxygen concentration and the pH. This also applies to any novel sensor for monitoring other fermentation process variables such as the *A. pullulans* population (25). These values will therefore depend greatly on the position of the various probes. The same phenomenon applies also to the fermentation broth sampling port. The geometry and the location of the conventional sampling tube, as provided in fermenters by manufacturers (Fig. 9), shows that the samples are taken very frequently from the stagnant zone, where the data are not representative for the entire culture medium content. A newly improved sampling device was designed and tested in a 7-L fermenter (23) and enables one to obtain a more representative sample because it draws a sample from different locations within the culture vessel (Fig. 9). Nevertheless, because of the evolving and shear-dependent viscosity, it is not guaranteed that a uniform sampling in all branches of the sampling tube will be achieved. Thus, the sampling of representative samples and process monitoring of highly viscous fermentation broth in larger-scale fermenters having nonhomogeneous mixing remains a challenge that one should not overlook.

Oxygen Transfer. Pullulan fermentation is an aerobic fermentation. It was demonstrated that oxygen is essential for the biosynthesis of pullulan by *A. pullulans*. In a growth medium, pullulan yield and synthesis rates are proportional to the oxygen availability. However, under a controlled oxygen environment in a nongrowth (nitrogen-free) medium, pullulan yield and synthesis rates are inversely proportional to the oxygen concentration in the culture medium (26). During the fermentation, air is continuously provided to the culture medium to supply oxygen to the microbial cells. However, oxygen molecules in the air bubbles have to move into the fermentation broth and then diffuse through this liquid phase to reach the microbial cells in order to be used for the growth and biosynthesis of pullulan. With the highly viscous fermentation broth in pullulan fermentation, the oxygen mass transfer process is very difficult and remains a major engineering challenge for the future. Indeed, as the fermentation broth becomes increasingly viscous, the gas bubbles in the peripheral stagnant zone of the fermenter remain trapped in the motionless fluid and are progressively depleted from their oxygen content. Figure 10 shows a photograph taken during a typical pullulan production where the fermentation broth and gas bubbles in the outer region remained stagnant for long periods.

The evaluation and comparison of the performance of oxygen mass transfer within a mixing vessel is usually quantified by the volumetric mass transfer coefficient K_La. The overall oxygen transfer rate from air bubbles to the fermentation broth (dC_L/dt) is the product of the volumetric oxygen transfer coefficient (K_La) and the mean concen-

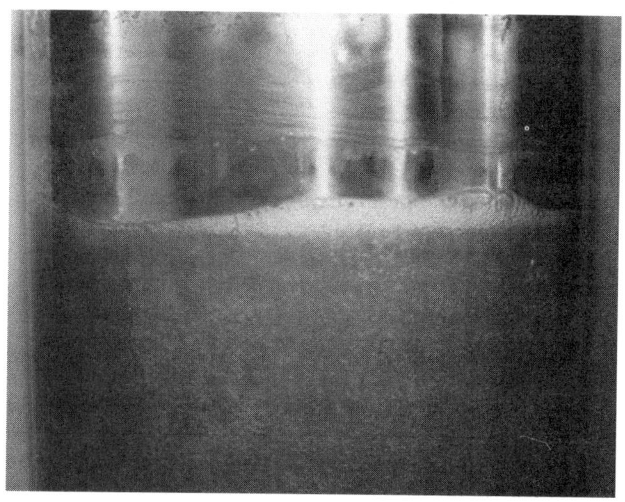

Figure 10. Mixing and mass transfer challenges in pullulan fermentation. The picture is taken when the agitator equipped with Ruston turbines is rotating at 500 rpm. The layer of fermentation broth adjacent to the fermenter wall remains stagnant with trapped air bubbles.

tration driving force of oxygen in the fermentation broth ($C^* - C_L$):

$$\frac{dC_L}{dt} = K_La(C^* - C_L) \qquad (2)$$

where C_L is the dissolved oxygen concentration in the fermentation broth and C^* is the saturated dissolved oxygen concentration in the fermentation broth that is the solubility of oxygen under the fermentation conditions. According to equation 2, one possible way to improve the oxygen mass transfer rate is to improve the mean concentration driving force of oxygen in the fermentation broth ($C^* - C_L$) by increasing the solubility of oxygen in the liquid phase. The solubility of oxygen in the fermentation broth, C^*, during the fermentation can be enhanced either by using oxygen enriched gas, up to pure oxygen, in aeration or by pressurizing the fermenter headspace (commonly referred to as hyperbaric fermentation). For practical and economic reasons, the use of pure oxygen is not feasible in large-scale fermenters. On the other hand, enriched gas would be used only when air would not suffice to maintain the desired dissolved oxygen concentration within the fermenter, because high oxygen concentration in the fermentation broth may enhance the biosynthesis of the black melanin pigment, which is undesirable (26).

The effects of pressure on the growth of *A. pullulans* and the synthesis of pullulan have been studied in a specially designed fermenter and fermentation system (Figs. 11 and 12) (27). It is reported that for all volumetric airflow rates in the range of 0.2 to 2.0 VVM, the production of biomass increases with pressure up to a critical value, which ranges from 0.50 MPa to 0.75 MPa, at which point a drastic decrease in biomass production and a change in cellular morphology occur. Above the critical pressure, *A. pullulans* forms aggregates of irregular shape that are fairly uniform in size under each set of fermentation conditions. In general, larger aggregates of approximately 8 mm are associated with a higher pressure level (1.48 MPa) and aeration rate (2 VVM). The details of *A. pullulans* aggregates are shown in the scanning photomicrographs in Figure 13. The same pattern is also observed for pullulan synthesis: it increases with pressure up to approximately the same critical pressure, then decreases drastically beyond this value. Much more investigation is required to better understand the behavior of hyperbaric pullulan fermentation before reaching a practical and economic solution for this technology.

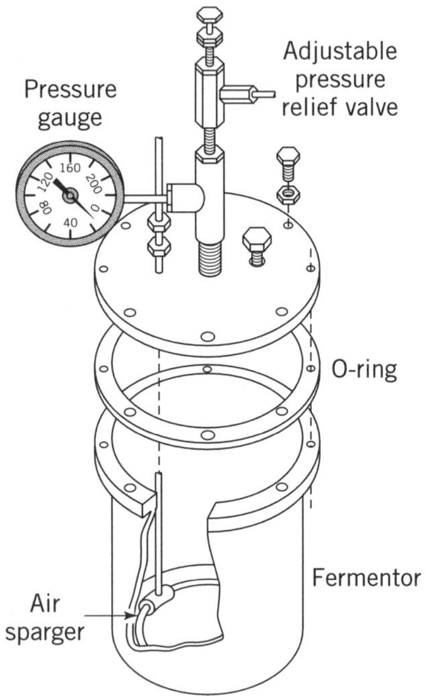

Figure 11. The specially designed fermenter for studying pullulan fermentation under hyperbaric conditions. *Source*: From Ref. 27, reproduced with permission.

A second way to improve the oxygen mass transfer rate is to increase the volumetric oxygen transfer coefficient (K_La) by using better mixing devices. It has been clearly shown that K_La is not uniform throughout most mixing vessels. It is usually large in the vicinity of the mixing element and lower further away. It is evident from Figures 9 and 10 that any mixing device that can reduce or even eliminate the dead zones in the fermenter will have a chance to provide better volumetric oxygen transfer to the pullulan fermentation broth and in addition create more homogeneous operating conditions throughout the fermentation vessel. An adequate uniform mixing would also contribute to the maintenance of gas bubbles with high oxygen content and therefore keep the oxygen gradient at a high level. Several unconventional mixing devices have been designed and studied. These are the reciprocating plate mixer, the helical ribbon mixer, and the helical ribbon screw mixer.

Although the physical properties of the fluid and operating conditions, used by many authors, are hardly comparable over a wide range, their results have a common trend of increasing oxygen mass transfer coefficients with increasing agitation and gasflow rate. The efficiency of mixing devices, evaluated in terms of K_La, is usually correlated in terms of power input per unit volume (P_G/V_L) and superficial gas velocity (U_G), using the following expression:

$$K_La = C\left(\frac{P_G}{V_L}\right)^{\alpha}(U_G)^{\beta} \qquad (3)$$

The correlation coefficients C, α, and β depend on the nature of the fluid, the geometrical configuration of the mixing element, and the hydrodynamic regimes of the fluid (28,29).

The biochemical engineering group at Laval University has been interested for the last decade in developing a relatively new type of bioreactor, the reciprocating plate bioreactor (RPB). The RPB (Fig. 14a) consists of a stack of perforated plates that undergoes reciprocating vertical motions inside the fermenter. The RPB, with its particular geometry, possesses the desired attributes to produce efficient mixing, with the result that microorganisms, nutrients, and air bubbles are properly distributed (31). In the RPB, the perforated plates occupy the total cross section

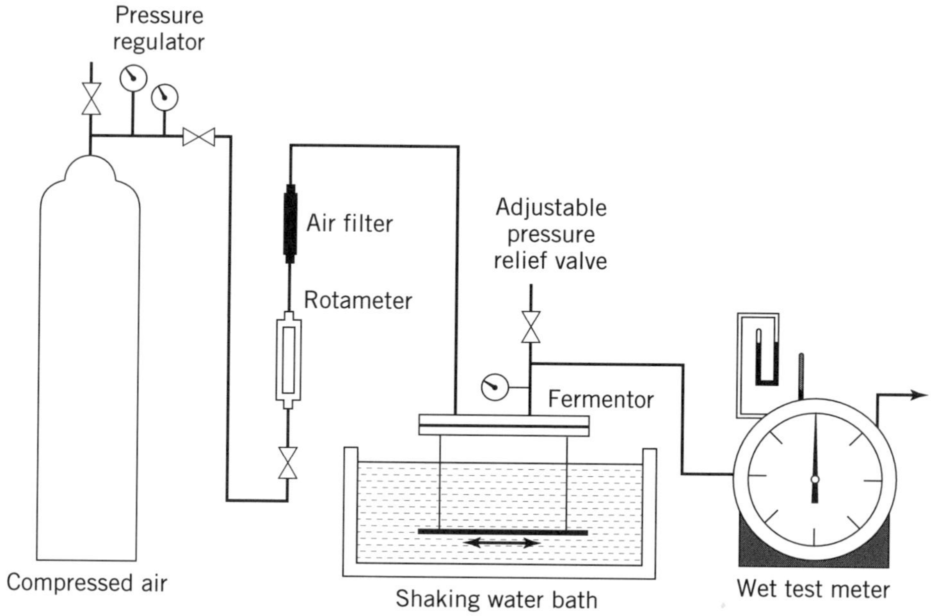

Figure 12. The fermentation setup for studying pullulan fermentation under hyperbaric conditions. *Source*: From Ref. 27, reproduced with permission.

of the fermentation vessel so that the upward and downward movements of the plate stack lead to uniform mixing. Three different RPBs with aspect ratios ranging from 0.08 to 0.5 have been characterized with fluid models and pullulan fermentation (21,28,30). Audet et al. (29) have performed experiments to evaluate the influence of polysaccharide concentration and molecular weight on oxygen mass transfer in the RPB. They used model fluids consisting of different concentrations of dextran (another water-soluble microbial polysaccharide) of different molecular weights to simulate the pullulan fermentation broth at different stages of fermentation. They correlated their results using equation 3 and reported that the correlation coefficients C, α, and β in this equation also depend on the polysaccharide concentration, the polysaccharide molecular weight, and the hydrodynamic regime of the solution.

The most recently constructed RPB was designed to be identical, except for the mixing mechanism, to a stirred-tank (radial-type mixing device) bioreactor that was built simultaneously. The schematic diagram of the experimental setup of the RPB is shown in Figure 15. Each bioreactor has a total volume of 22.5 L and a working volume of 18 L. The two bioreactors, using a series of mixing elements, were initially characterized in water for their ability to produce high K_La values. A total of five plate stacks were used in the RPB. Each plate stack consists of 6 (plate stacks PA, PB, and PC) or 12 (plate stacks PD and PE) perforated stainless steel plates, 221 mm in diameter and 1.25 mm thick, mounted on a central shaft and uniformly distributed with a distance of, respectively, 50 and 25 mm between adjacent plates. The diameters of perforations were 6.35 mm for plate stacks PA and PD, 12.7 mm for plate stacks PB and PE, and 19.05 mm for plate stack PE. A total of five mixing elements were used with the radial-type bioreactor: assemblies containing one, two, or three identical Rushton turbines (RT1, RT2, and RT3, Fig. 14b), and a helical ribbon impeller with or without surface baffles (HR and HRB, Fig. 14c). A more complete description is presented by Gagnon et al. (32).

These bioreactors were first characterized in terms of oxygen mass transfer (32). Typical plots of the oxygen mass transfer coefficient in water as a function of the power input per unit volume are presented in Figure 16 for all mixing devices at one superficial gas velocity. All mixing devices show identical patterns. For lower power inputs per unit volume, the oxygen mass transfer, K_La, is constant and varies only with the gas superficial velocity. At a certain power input per unit volume, K_La starts to increase rapidly with the power input per unit volume. Perez and Sandall (33) and Nishikawa et al. (34) previously observed these two distinct regimes of operation with Rushton turbines.

The two bioreactors were then used in an investigation undertaken with the objective of comparing experimentally the performance of four different mixing devices for the production of pullulan (30). The four mixing devices include two different reciprocating plate stacks (PA and PB), the assembly of three Rushton turbines (RTB), and the helical ribbon impeller (HR). These four fermentations were performed with an equal level of power input per unit volume (1,000 W/m^3) and gasflow rate (0.5 VVM or 9 L/min) during the entire fermentation. The evolution of the concentrations of biomass, pullulan, and dissolved oxygen during each fermentation are presented in Figures 17–19. Under identical power inputs per unit volume, better cell growth and pullulan synthesis were obtained with the reciprocating plate mixer that had larger perforations. The helical ribbon mixer gave the poorest cell growth and pullulan synthesis. However, pullulan of the highest molecular weight, as indicated by its highest fermentation broth viscosity, was obtained with the helical ribbon mixer.

The evolution of dissolved oxygen (Fig. 19) clearly shows that the RPBs are more efficient at maintaining a high and homogeneous level of dissolved oxygen within the fermen-

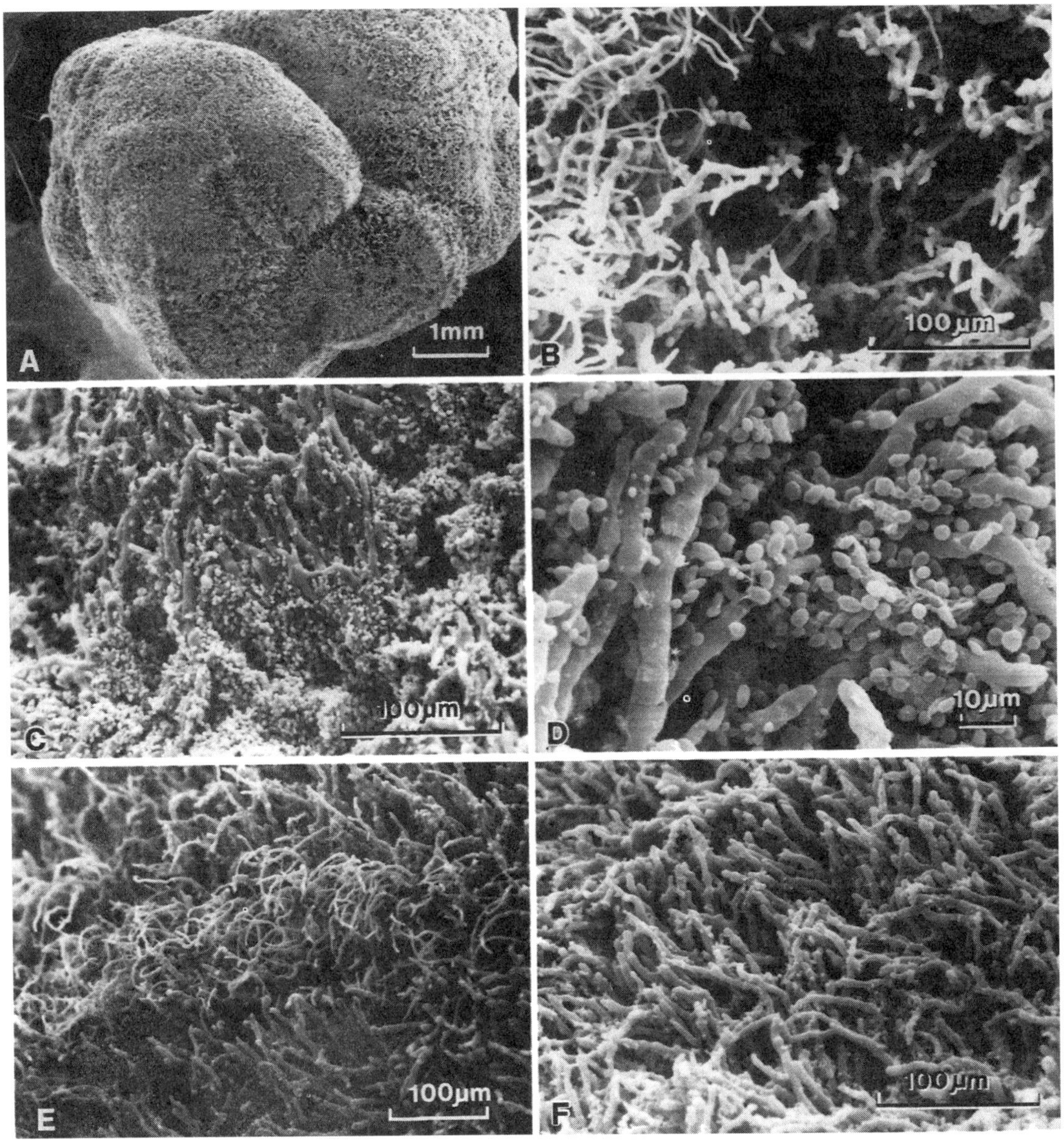

Figure 13. Agglomerates of *A. pullulans* in pullulan fermentation under 1.13 MPa and 0.2 VVM: (**a**) Global view of agglomerate; (**b**) cavity region; (**c**) and (**d**) regions where yeast cells dominate; (**e**) and (**f**) regions where mycelial cells dominate. *Source*: From Ref. 27, reproduced with permission.

tation broth. For the other two mixing devices, it can be stated that the local dissolved oxygen varied greatly and was always very low in the vicinity of the dissolved oxygen probe.

Guerinik (35) has modified the helical ribbon impeller by adding a central screw installed on the rotating shaft to provide the pumping action on the fermentation broth in order to enhance its circulation inside the fermenter. He also found that pullulan fermentations using the helical ribbon screw mixer resulted in a higher concentration of pullulan of higher molecular weight as compared with that obtained using standard Ruston turbines.

In a recent study, Tanguy et al. (36) proposed a dual coaxial impeller composed of a Rushton turbine and a helical ribbon impeller, both rotating clockwise but at different speeds. They have shown that the hybrid mixer system outperforms the standard helical ribbon impeller when the fluid rheology changes with time. These hybrid systems could be used advantageously in pullulan fermentation.

Product Separation and Recovery. The first operation in the product separation and purification step (Step D in Figures 2 and 4) is the removal of *A. pullulans* cells from the highly viscous fermentation broth. This is done by either centrifugation or filtration. The fermentation broth has to be diluted many fold to lower its viscosity prior to the centrifugation or filtration operation. The resulting large volume of diluted processing broth leads to higher operating

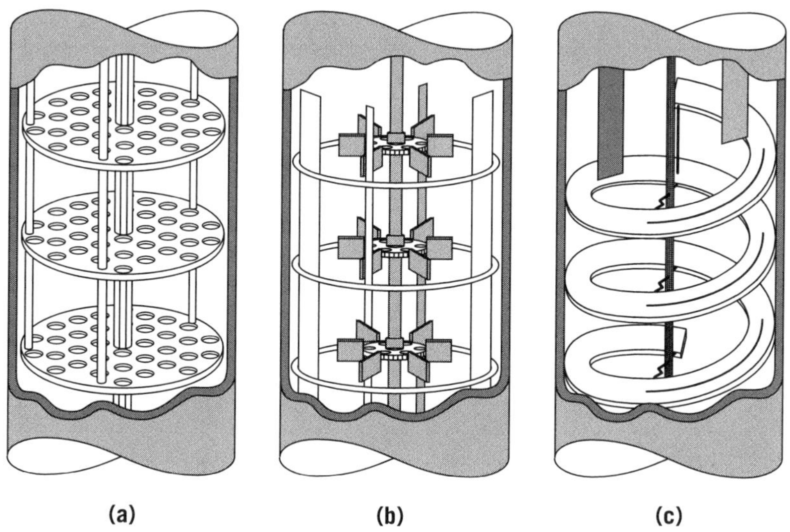

Figure 14. Mixers used in pullulan fermentations: (**a**) the reciprocating plate mixer; (**b**) the Rushton turbines; (**c**) the helical ribbon mixer. *Source*: From Ref. 30, reproduced with permission.

costs in the subsequent steps of pullulan purification, especially in the solvent precipitation step. In order to eliminate the presence of cells in the highly viscous fermentation broth, attempts to immobilize *A. pullulans* cells on solid supports and to use immobilized cells in pullulan fermentation has been reported. The utilization of immobilized cells for this purpose is not feasible because considerable numbers of cells leak into the fermentation broth (37). In addition, because pullulan synthesis is partially growth associated, it is inevitable that some cells will be present in the fermentation broth.

Problems Due to the Sterile and Aseptic Conditions Requirement

At the present time, pullulan production by fermentation is carried under sterile and aseptic conditions, which require very high capital and operating costs as in the case of pharmaceutical production. If pullulan is to be produced for use as bulk chemicals and/or bulk commodities, its production costs have to be greatly lowered. The technological challenge is to develop a nonsterile and nonaseptic fermentation process for the production of pullulan. A bistaged pH process has been developed for this purpose (18). In batch fermentation, the pH of the culture medium decreases rapidly from its initial value of 5.0 to the final value of 2.5 within the first 12–36 h and then self-stabilizes at this value until the end of the fermentation. It is observed that most of the polysaccharide from the batch fermentation process was synthesized during the self-stabilized acidic phases. Experiments on the effect of the initial pH on the fermentation reveal that at very low initial pH values, such as pH 2, the polysaccharide production is insignificant. However, the biomass concentration obtained under these conditions is very high. This interesting phenomenon served as the basic principle for the development of the bistaged pH fermentation process. In this process, the first stage of fermentation is conducted at a very acidic pH for the best production of biomass. When the biomass concentration reaches its maximum value, the second stage of fermentation is initiated by adjusting the medium pH to a higher value for promoting the synthesis of pullulan. The bistaged pH process enhances the pullulan concentration in the medium, influences the rheological properties of the fermentation broth, and has a potential of operation under nonsterile and nonaseptic conditions. During the first stage of fermentation, the microbial contamination could be eliminated due to the very low pH of the medium. During the second stage of fermentation, the contamination could be minimized to an insignificant level due to the high concentration of *A. pullulans* already present in the culture medium. This bistaged pH process could be applied to batch or continuous fermentation processes.

Problems Due to Melanine Pigmentation

Aureobasidium pullulans is also known as a black yeast because of its characteristic black pigmentation associated with mature cells (Fig. 20). The pigment, melanin, is synthesized from the pentaketide pathway in which 1,8-dihydroxynaphthalene is the direct precursor (38). Both intracellular and extracellular melanin are synthesized during the later stage of fermentation where the cell morphology changes from swollen cells to true chlamydospores (39). Thus, melanin production occurs simultaneously with pullulan synthesis, resulting in a dark green to black fermentation broth at the end of the fermentation. This is an undesirable feature for industrial production of pullulan because decolorization (melanin removal) is required during the product separation and purification operations (Step D of Figures 2 and 4). The synthesis of melanin by *A. pullulans* depends on the culture conditions, such as carbon source and concentration, nitrogen source and concentration, oxygen concentration, and trace minerals (26,40,41) (Fig. 20). Therefore, such conditions can serve as control variables to minimize melanin synthesis during the pullulan fermentation process. However, the most promising solution to the problem of melanin production during fermentation may be the use of nonpigmented strains of *A. pullulans*, which may be obtained by mutation (42).

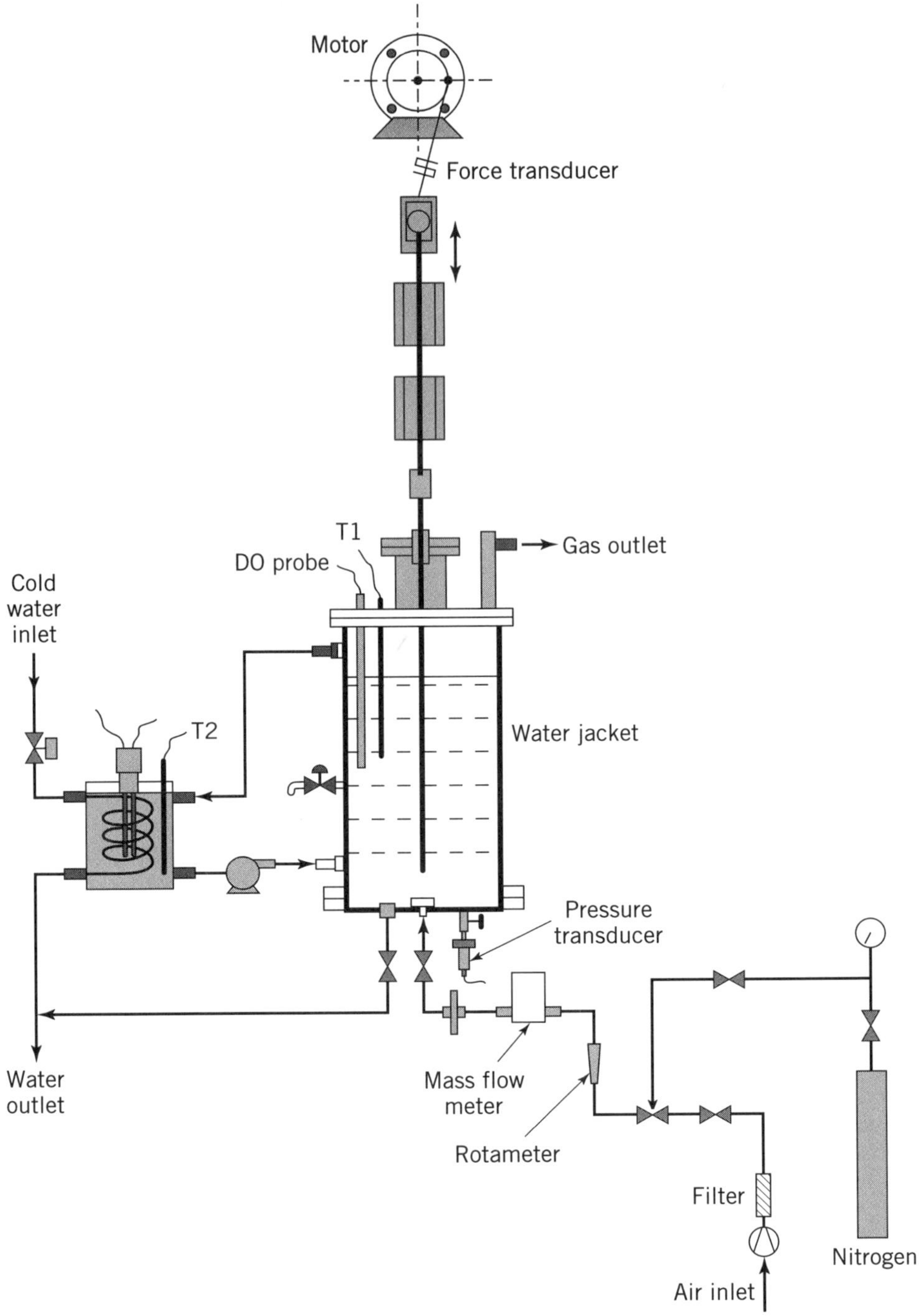

Figure 15. Schematic diagram of the experimental system showing the reciprocating plate bioreactor. *Source*: From Ref. 32, reproduced with permission.

Problems Due to Pullulanolysis during Fermentation

Pullulanolysis is the breakdown of high molecular weight pullulan in the fermentation broth into lower molecular weight pullulan by the depolymerase activity of *A. pullulans*. The pullulanolysis always occurs during the progress of fermentation and is responsible for the decrease of viscosity of the fermentation broth during the later stage of fermentation. The presence of an extracellular endoamylase is suggested as the cause of the observed phenomenon (43). It is not clearly understood why *A. pullulans* produces the pullulan-destructive endoamylase during the synthesis of pullulan. The resulting consequence of the pullulanolysis is that only low molecular weight pullulan is obtained at the end of the fermentation. The pullulanolytic breakdown of the pullulan product is worsen when a high-shearing mixing device such as the Rushton turbine is employed in the fermenter (30,35). Thus, if the high molecular weight pullulan is desired as a final product, the fermentation has to be stopped after approximately 24 h, when the molecular weight of pullulan is at its highest value (44,45). This is not, however, an economically feasible solution for industrial practice because the fermentation

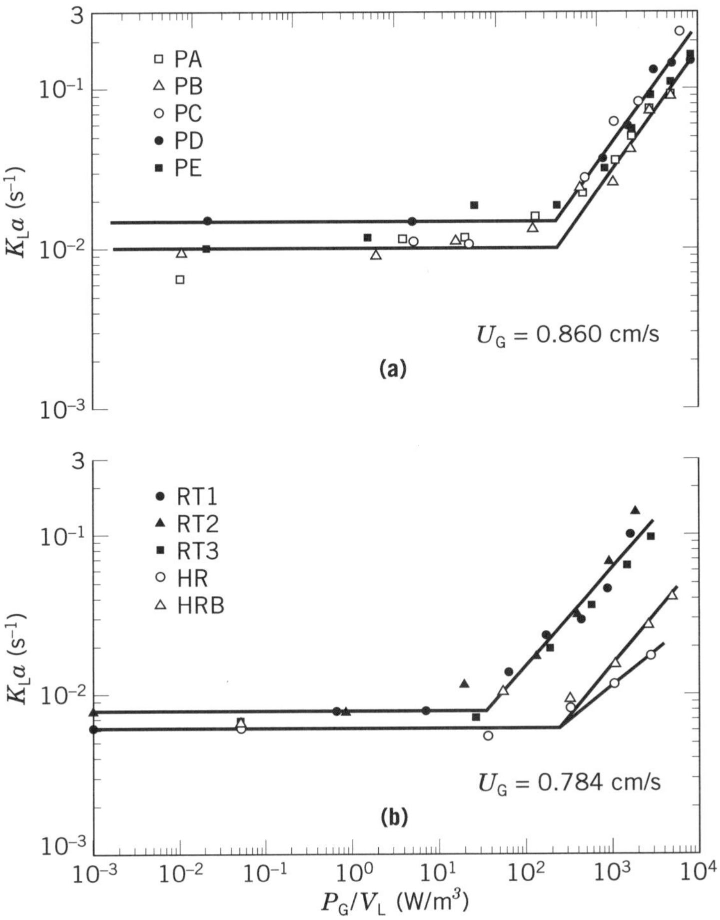

Figure 16. K_La as a function of the power input per unit volume for four mixing devices. *Source*: From Ref. 32, reproduced with permission.

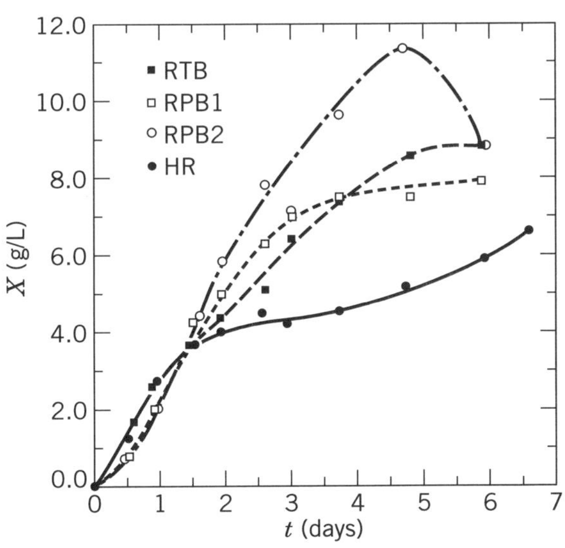

Figure 17. Evolution of biomass dry weight concentration as a function of fermentation time for four mising devices. *Source*: From Ref. 30, reproduced with permission.

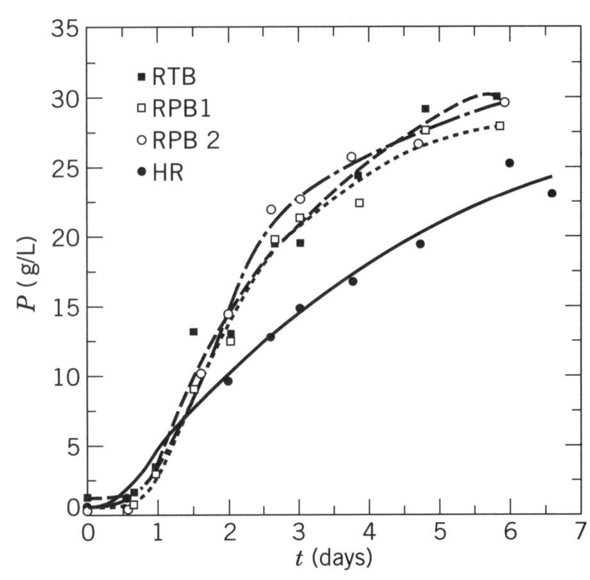

Figure 18. Evolution of pullulan concentration as a function of fermentation time for four mising devices. *Source*: From Ref. 30, reproduced with permission.

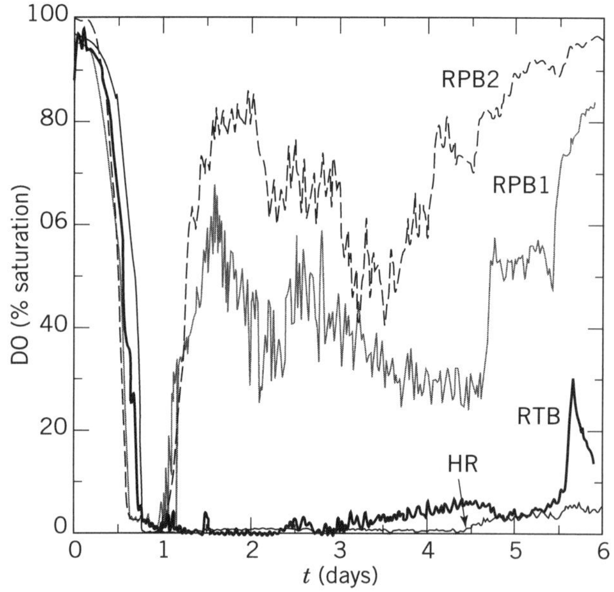

Figure 19. Evolution of dissolved oxygen concentration as a function of fermentation time for four mising devices. *Source*: From Ref. 30, reproduced with permission.

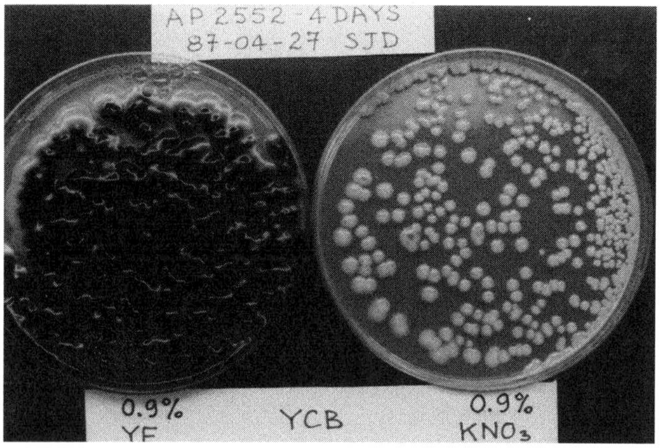

Figure 20. Melanin pigment of *A. pullulans* can be controlled by the nitrogen source. Potassium nitrate as nitrogen source instead of yeast extract in an yeast-carbon-base medium greatly reduces the melanin synthesis (*A. pullulans* 2552 after 4 days of incubation).

efficiency (expressed as grams of pullulan produced per 100 g of supplied substrate) is very low after this short period of fermentation. A more effective but challenging alternative solution is to acquire more knowledge on the pullulanolysis phenomenon by research, then to develop the technology for the suppression or the inhibition of endoamylase synthesis by *A. pullulans* during the pullulan fermentation.

BIBLIOGRAPHY

1. H. Bender, J. Lehmann, and K. Wallenfels, *Biochim. Biophys. Acta* **36**, 309–316 (1959).
2. R. Taguchi, Y. Kikuchi, Y. Sakano, and T. Kobayashi, *Agric. Biol. Chem.* **37**, 1583–1588 (1973).
3. B.J. Catley and W.J. Whelan, *Arch. Biochem. Biophys.* **143**, 138–142 (1971).
4. Ger. Pat. 1,096,850 (January 12, 1961), K. Wallenfels and H. Bender (to Farbwerke Hoechst AG).
5. S. Yuen, *Process Biochem.* **9**(9), 7–9, 22 (1974).
6. A. LeDuy, L. Choplin, J.E. Zajic, and J.H.T. Luong. in H.F. Mark, N.M. Bikales, C.G. Overberger, and G. Menges eds., *Encyclopedia of Polymer Science and Engineering*, 2nd ed., vol. 13, Wiley, New York, 1988, pp. 650–660.
7. U.S. Pat. 3,931,146 (January 6, 1976), K. Kato and M. Shiosaka (to Hayashibara Biochemical Laboratories, Inc.).
8. French Pat. 2,287,911 (May 14, 1976), Anonymous (to Sumitomo Chemical Co., Ltd. and Hayashibara Biochemical Laboratories, Inc.).
9. U.S. Pat. 4,090,016 (May 16, 1978), K. Tsuji, M. Fujimoto, F. Masuko, and T. Nagase (to Sumitomo Chemical Co., Ltd. and Hayashibara Biochemical Laboratories, Inc.).
10. U.S. Pat. 4,152,170 (May 1, 1979), T. Nagase, K. Tsuji, M. Fujimoto, and F. Masuko (to Sumitomo Chemical Co., Ltd. and Hayashibara Biochemical Laboratories, Inc.).
11. U.S. Pat. 4,186,024 (January 29, 1980), M. Fujimoto, K. Fukami, K. Tsuji, and T. Nagase (to Sumitomo Chemical Co., Ltd. and Hayashibara Biochemical Laboratories, Inc.).
12. Can. Pat. 1,078,830 (June 3, 1980), F. Fujita, K. Fukami, and M. Fujimoto (to Sumitomo Chemical Co., Ltd. and Hayashibara Biochemical Laboratories, Inc.).
13. Can. Pat. 1,094,550 (January 27, 1981), F. Fujita, K. Fukami, M. Fujimoto, and T. Nagase (to Sumitomo Chemical Co., Ltd.).
14. French Pat. 2,461,718 (February 6, 1981), H. Muto and H. Suzuki (to Shin-Etsu Chemical Co., Ltd.).
15. European Pat. 76,698 (April 3, 1983), I. Murase, F. Fujita, T. Ohnishi, and T. Tamura (to Sumitomo Chemical Co., Ltd.).
16. K. Sugimoto, *Ferment. Ind.* **36**(2), 98–108 (1978).
17. A. LeDuy, A.A. Marsan, and B. Coupal, *Biotechnol. Bioeng.* **16**, 61–76, (1974).
18. C. Lacroix, A. LeDuy, G. Noël, and L. Choplin, *Biotechnol. Bioeng.* **27**, 202–207 (1985).
19. A. Mulchandani, J.H.T. Luong, and A. LeDuy, *Biotechnol. Bioeng.* **32**, 639–646 (1988).
20. J.M. Boa and A. LeDuy, *Biotechnol. Bioeng.* **30**, 463–470 (1987).
21. J. Audet, M. Lounes, and J. Thibault, *Bioprocess Eng.* **15**, 209–214 (1996).
22. S. Onogi, T. Masuda, and K. Kitagawa, *Macromolecules* **3**, 109–116 (1970).
23. A. LeDuy and J.E. Zajic, *Biotechnol. Bioeng.* **15**, 579–581 (1973).
24. G.B. Tatterson, *Fluid Mixing and Gas Dispersion in Agitated Tanks*, McGraw-Hill, New York, 1991, pp. 417–528.
25. H. Guterman and Y. Shabtai, *Biotechnol. Bioeng.* **51**, 501–510 (1996).
26. D. Rho, A. Mulchandani, J.H.T. Luong, and A. LeDuy, *Appl. Microbiol. Biotechnol.* **28**, 361–366 (1988).
27. R. Dufresne, J. Thibault, A. LeDuy, and R. Lencki, *Appl. Microbiol. Biotechnol.* **32**, 526–532 (1990).
28. M. Lounes, J. Audet, J. Thibault, and A. LeDuy, *Bioprocess Eng.* **13**, 1–11 (1995).
29. J. Audet, J. Thibault, and A. LeDuy, *Biotechnol. Bioeng.* **52**, 507–517 (1996).

30. J. Audet, H. Gagnon, M. Lounes, and J. Thibault, *Bioprocess Eng.* **19**, 45–52 (1998).
31. H. Brauer, *Bioprocess Eng.* **6**, 939–944 (1991).
32. H. Gagnon, M. Lounes, and J. Thibault, *Can. J. Chem. Eng.*, **76**, 379–389 (1998).
33. J.F. Perez and O.C. Sandall, *AIChEJ.* **20**, 770–775 (1974).
34. M. Nishikawa, M. Nakamura, H. Yagi, and K. Hashimoto, *J. Chem. Eng. Jpn.* **14**, 219–226 (1981).
35. K. Guerinik, Ph.D. Thesis, Université Laval, Sainte-Foy, Canada, 1989.
36. P.A. Tanguy, F. Thibault, E. Brito de la Fuente, T. Espinosa-Solares, and A. Tecante, *Chem. Eng. Sci.* **52**, 1733–1741 (1997).
37. A. Mulchandani, J.H.T. Luong, and A. LeDuy, *Biotechnol. Bioeng.* **33**, 306–312 (1989).
38. D.J. Siehr, *J. Coatings Technol.* **53**, 23–25 (1981).
39. G.M. Gadd, *FEMS Microbiol. Lett.* **9**, 237–240 (1980).
40. Y. Shabtai and I. Mukmenev, *Appl. Microbiol. Biotechnol.* **43**, 595–603 (1995).
41. T.P. West and B. Strohfus, *World J. Microbiol. Biotechnol.* **13**, 233–235 (1997).
42. T.J. Pollock, L. Thorne, and R.W. Armentrout, *Appl. Environ. Microbiol.* **58**, 877–883 (1992).
43. B.J. Catley, *FEBS Lett.* **3**, 190–193 (1970).
44. B. McNeil and B. Kristiansen, *Biotechnol. Lett.* **9**, 101–104 (1987).
45. K.Y. Lee and Y.J. Yoo, *Biotechnol. Lett.* **15**, 1021–1024 (1993).

PULP AND PAPER PROCESSING. See ENZYMES, PULP AND PAPER PROCESSING; HEMICELLULASES; HEMICELLULOSE CONVERSION.

PUMPS, INDUSTRIAL

BOB STOVER
ED DOMANICO
Tri-Clover
Valencia, California

KEY WORDS

Cavitation
Centrifugal
Head
Horsepower
Mechanical seals
Motors
Net positive suction head (NPSH)
Positive displacement
Specific gravity
Viscosity

OUTLINE

INTRODUCTION

Liquid transfer is a vital component of most biotechnology processes. There is possibly more information available on pumps than any other single piece of equipment (pumps being a very common piece of rotating equipment, second only to the electric motor) existing in a typical plant, so this section must be a rough introduction to hydraulic theory. This article provides a basic understanding of pump mechanics, nomenclature, and principles and sufficient information to begin the selection process of a new pump.

THEORY

Before a specific selection can be initiated, the operating conditions the pump will be expected to meet must be understood. Breaking down the system into clearly defined and agreed-upon components will not only simplify calculations, and help guarantee nothing is overlooked, but also make it easier to evaluate the consequences of changes to any part of the system. There are universally understood and applied ways of breaking down a system, recognized by all pump manufacturers and users.

Flow

Capacity is expressed in flow/unit of time, commonly gallons per minute (gpm) or per hour (gph), or as cubic meters per hour (m^3/h).

To convert from volume to gallons:

$$1\ ft^3 = 7.48\ \text{gallons}$$
$$1\ m^3 = 264.2\ \text{gallons}$$

It is important to be realistic with estimates, because accurate pressure loss calculations depend on an accurate flow rate. Overestimating will result in equipment that will need to be valved back or slowed down to produce the demands actually required or in oversized/overpriced pumps.

Head

Head is the accepted hydraulic term for pressure. It is most commonly expressed in pounds/inch² (psi), or feet of head. The concept of a linear unit being used to express pressure can be most easily visualized as taking a resistance to flow and mentally converting it to a vertical column of water, such that the weight of this imaginary column is equivalent to the resistance of the considered section. Any pump that can push fluid against a column of water of the height thus calculated will be able to deliver flow against the system used to make the calculation.

To convert from psi to feet:

$$1\ \text{psi} = 2.31\ \text{ft of water}$$
$$1\ \text{ft of water} = .433\ \text{psi}$$

Most pipe friction loss charts express their losses as feet of head per 100 feet run of pipe, and all losses given as psi can be converted to feet by the above conversion factors (see Specific Gravity for adjustments given liquids heavier or lighter than water).

This linear unit has become the most common choice for hydraulic pressure (in the sense of resistance to flow) because:

- It is easily visualized.
- A linear unit fits into most hydraulic formulas (velocity, horsepower, specific speed, etc.).
- It compensates for various specific gravities before inclusion.
- It provides a common unit for all components of the total head that can then be added together to arrive at total resistance.

Types of Head. The total head referred to above is made up of four components. Any resistance to flow will fall into one of the following categories.

Static Head (ΔZ). Static head refers to the elevation differential between the source level of the liquid and the elevation at the destination. In other words, the elevation differential the pump must overcome. Caveats to this are:

- If a line enters at or near the bottom of a destination tank, the liquid level *in* the tank is considered the appropriate one, not where the pipe makes its entry (the pump must overcome the weight of the column *inside* the tank as part of its pressure capability).
- The vertical ups and downs are irrelevant (they will be accounted for later under Friction Head); the elevations at the beginning and end of the line are the ones used for ΔZ.
- The pump's position in relationship vertically to either the source or destination point is also irrelevant.

Static head is easily expressed in feet.

Pressure Head (ΔP). Pressure head is the differential in pressure at the ends of the system, the inlet and outlet, expressed in feet. An example would be a vacuum tank at the source; it provides resistance to flow (the vacuum holds back the fluid) and must be overcome by the pump. Resistance to flow from vacuum can be converted to feet of head by the following:

$$1\ \text{inch mercury (Hg)} = 1.133\ \text{ft of water}$$

Pressure at the discharge end (a pressurized tank or injection into a pressurized line) also needs to be added into the total and is included in ΔP.

Velocity Head (H_V). Velocity head is that energy necessary to overcome inertia. It is calculated by

$$H_V = V^2/2g$$

where V = velocity through the piping lines (in feet/second), and g = 32.2 ft/s², the acceleration caused by gravity at sea level.

Velocity head is reasonably estimated by taking the average velocity in a system and applying the formula. However, given a maximum average velocity of less than 8 ft/s in most systems, H_V is usually a relatively small number (less than 1 ft in the average 8 ft/s example above) in a typical hydraulic calculation and almost always ignored.

Friction Head (H_F). Friction loss is usually the largest individual component of total head for most calculations. There are formulas available for its calculation given various types of pipe, fittings, valves, and filters in texts, but fortunately there are charts in existence giving the calculated heads. Pipe losses themselves are usually expressed in feet of head/unit of length (usually 100 ft). Fitting losses are expressed either in equivalent feet of straight pipe to be added to the run of actual straight pipe for what is called an *equivalent feet of pipe* to be used in the calculations. Or, charts may give the actual loss per each fitting. Losses through valves, filters, and such are often expressed in psi or feet of loss per valve. Each component in a line must be considered, and a grand total for the system is arrived at as a total of the individual pieces.

The friction loss numbers will always vary considering the flow through the line; higher flows mean more drag and therefore more resistance. Filters will vary with flow and also condition of the filter element. It is common practice to use worst-case conditions.

Total Head (H_M). The four individual elements listed previously are added together for the *total mechanical* *head* (H_M, also known as mechanical head, total dynamic head [TDH], or simply total head). They are the components for Bernoulli's formula, modified for the pump industry, given as:

$$H_M = \Delta Z + \Delta P + H_V + H_F$$

Adjustments to Head. There are qualities a liquid can possess that will affect the H_M.

Specific Gravity. Specific gravity (SG) is the direct numeric ratio of the weight of a particular liquid to the weight of water. Given that at certain standard conditions water weighs approximately 8.34 lb/gallon, anything heavier than that would have a specific gravity greater than 1.0 (for example, a liquid weighing 16.68 lb/gallon would have an SG of 2.0), anything lighter less than 1.0. Specific gravity will affect a conversion from psi to feet as follows:

$$\text{psi} = (\text{feet of water} \times \text{SG})/2.31$$

or

$$\text{feet of water} = (\text{psi} \times 2.31)/\text{SG}$$

Specific gravity will also directly affect horsepower because of the extra torque needed to overcome the additional weight of the liquid (as will be shown under Horsepower).

The second quality a liquid possesses that directly affects total head calculations is its *viscosity*. Most pump selection charts assume waterlike viscosity (1 centipoise [cps] or 31 Saybolt second universal [SSU]). The viscosity of liquids thicker than water affects pumps in several ways:

- It affects friction loss calculations; losses are higher with more viscous liquids. There are separate charts used for liquids more viscous than water.
- It affects horsepower. Manufacturers often provide specific charts to give the added torque needed to overcome viscosity.
- It affects the very capability of a pump to even *deal* with a product. Centrifugal pumps especially have limited abilities to handle viscous products.

Viscosity is usually expressed in the referenced units (cps or SSU), but there are many others. Viscosity can vary under dynamic conditions. Specifically:

- If a liquid becomes thinner under shear, it is *thixotropic* (example: paint).
- If it becomes thicker under shear, it is *dilatant* (example: taffy).
- If it becomes thinner only after exceeding a certain minimum threshold of shear, it is *plastic* (example: ketchup).
- If a liquid is unaffected by shear, it is *Newtonian* (example: water).

As mentioned above, viscosity affects centrifugal pumps in their very ability to pump a liquid; with viscous fluids their pressure capability falls off, the flow capability falls off even more, and their efficiency may fall off to a point where their use is totally unacceptable. Positive displacement pumps are affected in the speed they can turn and still move product without starving at the inlet. Manufacturers provide charts giving maximum speed for varying viscosities, for adjustments needed to horsepower to drive the unit with a thicker product, and perhaps for minimum port sizes necessary to allow product to get in quickly enough.

If the viscosity of a product is unknown, many pump manufacturers provide services wherein a liquid's viscosity will be tested, often at no charge, in exchange for the opportunity to quote their pumps.

Horsepower

Given the results of the total head calculations above, it is possible to make accurate calculations of the horsepower necessary to drive a particular pump.

The basic horsepower formula is

$$\text{HP} = (\text{gpm} \times H_M \times \text{SG}) \div (3{,}960 \times \text{eff.})$$

where gpm = flow (in gallons/min); H_M = total mechanical head (in feet); SG = specific gravity; eff. = efficiency (expressed as a decimal; e.g., 50% would be .50). *Note*: 3,960 is a constant, derived from the horsepower units and the weight of water.

A variation of this formula, commonly used for positive displacement pumps wherein specific gravity has already been calculated into the total head before being expressed as psi, is

$$\text{HP} = (\text{gpm} \times \text{psi}) \div (1{,}715 \times \text{eff.})$$

The 3,960 constant has been changed to 1,715 to provide for psi as a pressure unit by dividing by 2.31.

Often, viscosity will affect the horsepower as calculated above; in those cases manufacturers will provide correction factors for differing viscosity. The Hydraulic Institute has generic charts for centrifugal pumps that seem to work fairly well for almost all manufacturers, although some especially high-efficiency pumps claim to outperform the charts and therefore provide their own correction factors.

Net Positive Suction Head

Net positive suction head (NPSH) is the hydraulic term referring to the pressure under which liquid enters the pump, not only in terms of what the existing system is able to provide but also in reference to what any particular pump needs in order to function properly. It is an absolute unit, always expressed relative to absolute zero (as opposed to gauge units, whose values are always in reference to normal atmospheric pressure, called zero gauge). If a pump does not receive the liquid under sufficient pressure, it will cavitate.

Cavitation. Liquids will stay in liquid form only if sufficient atmospheric or system pressure is available to prevent it from going off to vapor. Even cold water, for example, will flash off into vapor in a strong enough vacuum, outer space as an obvious example. All liquids have a predictable vapor pressure necessary to keep them in their liquid state, varying up or down respectively as the liquid is heated or cooled.

This is of particular interest in hydraulics because of the often high vacuum present at the inlet of a pump. If the pressure provided by the system is less than the liquid's vapor pressure, the pump will undergo a physical process called cavitation. This action varies in specifics from centrifugal to positive displacement type pumps, but the general process is always approximately the same:

- The liquid finds itself in an area of low pressure, the pump's inlet port.
- If the pressure is *too* low (less than the liquid's vapor pressure at the operating temperature), it undergoes a state change from liquid to vapor. Because the vapor is not free to escape, it stays entrained in the liquid and assumes a bubble (spherical) shape.
- The vapor bubbles are carried further into the pump with the flow of surrounding liquid.
- Eventually the low-pressure bubbles and surrounding liquid find themselves in a higher pressure area as they near the pump discharge.
- The higher pressure causes the vapor bubbles to implode (collapse). This implosion is called cavitation.

The implosion of the low pressure vapor bubbles makes a very identifiable sound; in centrifugal pumps it produces a rattle, sounding like gravel is being carried through the housing; in positive displacement pumps the sound is more percussive, often *very* loud. In both cases the implosion of the vapor bubbles is damaging to the pump itself and often to the product. Centrifugal impellers and positive displacement pump rotors will actually have material removed by the very powerful implosions, shafts can be deflected to the point of fatigue, higher horsepower is required, and performance will be negatively affected.

Net Positive Suction Head Required. Manufacturers can test their pumps to see exactly how much positive suction pressure is necessary at various flows to keep their pumps from cavitating. This is called the net positive suction head required ($NPSH_R$). Typically, the greater the flow, the more pressure necessary because of the higher velocities going through the housing. A possible exception: at extremely low flows, the slippage caused by recirculation within the housing in some centrifugal pumps may create low pressure areas, sometimes low enough to fall under a liquid's vapor pressure and allow cavitation. This is called either suction or discharge recirculation cavitation, depending on where it physically occurs in the pump housing. Manufacturers determine the $NPSH_R$ from empirical data. They set up a test skid wherein they can gradually restrict flow into a pump, constantly monitoring the performance with pressure gauges and flow meters. Eventually they find a point at which the restrictions affect performance an agreed-upon minimum amount (3% below normal curve is the current standard). The pump is considered to be cavitating, and this point is registered on a chart. The same process is performed for a number of flows, and an $NPSH_R$ curve is generated. $NPSH_R$ curves are assumed to be drawn using 60 °F water; adjustments can be made later for other liquids.

Net Positive Suction Head Available. Net positive suction head available ($NPSH_A$) is the pressure any particular system makes available at a pump's inlet under operating conditions. It is a total of several factors. After calculation (or determination by an actual physical test), it will be compared to the pump's $NPSH_R$. Also by previous agreed-upon standard, $NPSH_A$ must exceed $NPSH_R$ by a minimum of 2 ft for a pump to be considered acceptable. This is a controversial issue; some people believe that this differential is insufficient, that the $NPSH_R$ chart represents the point at which the pump is *already* cavitating, a 2-ft differential is often too low; and that a sliding scale of minimal percentages by which $NPSH_A$ should exceed $NPSH_R$ would serve the user better. However, at this time the 2-ft differential is still accepted by most hydraulic engineers.

$NPSH_A$ is calculated by the formula:

$$NPSH_A = H_{ATM} \pm \Delta Z - H_F - H_{VAP}$$

where H_{ATM} = atmospheric pressure in absolute units, expressed in feet (atmospheric pressure at sea level = 14.7 psi = 33.9 ft); ΔZ = elevation differential from liquid level to pump's inlet (in feet). If the liquid level is above the pump inlet, the number will be positive (+), if below it will be considered a negative (−); H_F = friction loss in suction line only (in feet); and H_{VAP} = vapor pressure of liquid at operating temperature (in feet).

H_{ATM} and H_{VAP} are available from various reference books; H_F and ΔZ are calculated as in the section on Friction Head.

$NPSH_A$ can be determined on an operating pump by installing either a compound or vacuum gauge at the pump's inlet and by taking this reading and subtracting the liquid's vapor pressure (which will not be indicated on the gauge: H_{ATM}, H_F, and ΔZ will all be indicated by the gauge reading).

As stated earlier, the results of the calculation (or gauge reading $-H_{VAP}$) should be compared to the manufacturer's $NPSH_R$, a minimum 2-ft differential necessary to guarantee a pump will not cavitate. The $NPSH_A$ calculation is valuable for several reasons:

- It can show us beforehand whether a pump will cavitate after installation.
- It indicates the area most easily altered to eliminate existing cavitation and how much it needs to be altered. For example, if H_F is high, it may be lowered with larger pipe or fewer elbows. If ΔZ is low (or even negative), we can pick up positive head by raising the source tank.

If a system is still cavitating, and the $NPSH_A$ calculation does not indicate an easy correction, it may be that a pump with a lower $NPSH_R$ is needed. There are whole classes of pumps considered low NPSH that typically have requirements below 1 ft or 2 ft.

CENTRIFUGAL PUMPS

Pumps generally are divided into two main classifications, centrifugal and positive displacement (PD). There are many subclassifications. Several will be listed here later according to their suitability for use in bioprocessing, including a few that seem to cross the boundaries of centrifugal and PD. The vast majority of pumps in use are centrifugal because of the many advantages of centrifugal design (to be listed later), but bioprocessing has unusual demands, and the percentages of positive displacement pumps in use is significantly higher than in other industries.

Principals of Operation of Centrifugal Pumps

Centrifugal pumps (Fig. 1) consist of

- A rotating impeller
- An impeller housing (volute) to channel the flow from the suction port to the discharge port
- A power source, usually an electric motor
- A connection from the motor to the impeller; either the motor shaft itself, or sometimes a separate shaft extension
- A mechanical seal assembly to seal liquid inside the liquid-end assembly at the point where the shaft goes through the backplate.

There are many designs of impeller, having to do with whether the vanes are enclosed by shrouds or not (open or enclosed; open is cleaner, but less efficient; enclosed has less slip and is therefore more efficient, but is much harder to clean and to inspect); number and angle of the vanes themselves (influencing the shape of the pump curve); and diameter and thickness of the vanes themselves (deeper vanes give more flow, larger diameters mean more pressure capability). Manufacturers can create impeller designs for very specific sets of conditions. A centrifugal pump's simple design means that

- Centrifugal pumps are usually less expensive (often *considerably* less expensive) than PD pumps.
- Centrifugal pumps can be close-coupled, meaning they can be mounted directly on the face of specially designed close-coupled motors, a rare feature on PD pumps, helping keep them inexpensive (and also compact).
- Their flow is relatively pulsation free.
- Because of the inherent slip caused by the generally open nature of the impeller/volute package, there is a maximum pressure (shut-off head) a centrifugal pump can develop. This also means they can be easily controlled by regulating valves and often do not need pressure relief valves or controls in the system because HP loads actually decrease when flow is valved back.
- They are easy to maintain and repair.
- They are easily adapted to sanitary standards.

There are also disadvantages possessed by centrifugal pumps that are often especially significant in bioprocessing.

- Their inherent slip means they do not handle viscous materials well. The rule of thumb is anything over 2,000 SSU (about 500 cps) is usually considered too viscous for a centrifugal pump.
- Because pressure is generated by vane tip speeds, centrifugal pumps are not by nature easily adapted to high-pressure applications. This can be overcome to a degree by running them especially fast (designs exist that run at over 10,000 rpm) or by incorporating multistage designs wherein a chain of impellers is mounted on a common shaft, each impeller boosting the pressure of the impeller preceding it.
- Centrifugal pumps must be run at relatively fast speeds to develop sufficient impeller vane tip velocity to generate reasonable head (usually a minimum of 1,200 rpm). PD pumps, on the other hand, are often run efficiently at well under 100 rpm. This minimum speed means that centrifugal pumps are not usually good in low shear applications.
- The minimum speed also means that centrifugal pumps are not efficient with low-flow applications; turning slowly to limit flow means all pressure capability is gone. Even very small centrifugal pumps are often rated at 10 gpm or more; they will run at lesser flows by regulating them back in any of various fashions (regulating valves, inverters, bypass lines), but they are not efficient at very low flows.
- The presence of a mechanical seal means that abrasive products are a potential problem, not only because of the limited life of the components, but also for the danger of materials wearing off the faces and

Figure 1. Centrifugal pump. *Source*: Courtesy of Tri-Clover Inc., Kenosha, Wis.

entering the product stream. There are designs that remove seals from the product flow (double mechanical seals) or eliminate seals altogether (magnetically driven pumps, or pumps with gas injection seals), but they are often expensive and sometimes not available in sanitary designs.

Types of Centrifugal Pumps

End Suction. By far the most popular general classification of centrifugal pump is the end-suction design. Liquid enters through the suction port in the center of the housing, is picked up by the leading edge of the impeller, has energy imparted by the impeller's rotation, and exits tangentially at the discharge port on the volute's periphery. End-suction centrifugal pumps are available as close coupled (mounted directly onto a motor's face) or pedestal mounted (the pump is a separate entity with its own shaft/ bearing housing, connected to the motor by a flexible coupling, the assembled package bolted to a common base). Close-coupled design is generally preferred because of lower cost, its more compact design, and its inherent alignment between components, but pedestal-mounted pumps often have more space for special seal options and often will adapt more easily to any special motor requirements.

Magnetically Coupled. One way to arrive at a seal-less design and remove the difficulties of seal compatibility (see Mechanical Seals) is to adapt the standard end-suction design. Rather than the motor shaft extending directly through the mechanical seal and into the liquid end of the pump, it drives a large, hollow rotating magnet. The liquid end of the pump is in a sealed housing and fits up against the motor with a magnetized impeller extended into the hollowed-out area of the drive magnet. As the motor shaft rotates the hollow drive magnet, the impeller is pulled along. Because of the lack of a direct connection between the motor and impeller, magnetically driven pumps can handle only limited viscosity or specific gravity before the impeller decouples (uncouples). They can be very inexpensive in smaller designs, but the price escalates quickly as they get larger. Often, soft-start motors are necessary to prevent decoupling. At present, there are no sanitary-design magnetically driven pumps, but they will undoubtedly be available soon.

Multistage. Centrifugal design is normally best suited for medium-to-high flow, low-head applications such as liquid transfer. When the many advantages of centrifugal pumps makes them the design of preference, but the total head necessary exceeds that available from standard models, multistage pumps may provide an alternate to PD pumps. Impellers are staged, one after another in a line, and are driven by a long common shaft. Internal channels see that the first impeller's discharge is fed to the inlet side of the second, which discharge feeds the third, and so forth. Pressure is cumulative, but the flow is limited to that which can be carried by any individual impeller. The only restriction to the number of stages, and therefore the total pressure possible, is the housing-pressure limitations. This design carries all the advantages of standard centrifugal pumps, except they are not as easy to repair (impellers must be removed in series, and clearances can be difficult to adjust), and they are not necessarily inexpensive. There are sanitary designs available.

POSITIVE DISPLACEMENT PUMPS

Design and Principles of Operation

Positive displacement pumps make-up the second major classification of pumps used in the bioprocess industry today. In contrast to centrifugal pumps, discussed earlier, PD pumps function on a very different principle: that compressing a fluid volume will increase the ability of the fluid to do work. This change in volume takes place within the pump cavity as the pump drives or pushes that volume to the outlet, thereby increasing the pressure on that volume of fluid to be equal to the pressure at the pump outlet or discharge. An example of this principle is a simple syringe. As the syringe plunger is pushed in, the fluid within the syringe will begin to be exposed to a higher pressure as the volume holding the fluid decreases. The higher pressure then drives the fluid out the end of the syringe. In both the syringe and a pump, the pressure that the pump will generate is controlled by the resistance to flow that the fluid is exposed to as it leaves the pump discharge. Unlike dynamic or centrifugal pumps, this creates a situation whereby the PD pump can become overpressurized. The PD pump will continue to drive fluid forward regardless of the level of resistance. This overpressurization will only be relieved by either a pressure relief device on the discharge side of the pump or through the slip that is experienced as the pump operates.

Slip, in a positive displacement pump, is defined as an efficiency calculated by taking the amount of fluid that escapes the encapsulated volume as the pressure of the fluid volume increases and dividing by the amount of fluid delivered. The escaping fluid is driven back to the inlet or low-pressure side of the pump through clearances within the pump. By knowing the amount of slip that the pump is experiencing, one can calculate the pump efficiency. Typically, PD pump efficiencies are relatively high, 80 to 90%, and will increase as the fluid viscosity, discussed earlier, increases.

Advantages. Advantages that can be found in using PD pumps in the bioprocess industry include:

- The ability to deliver a constant flow rate; the discharge rate for a PD pump is relatively constant, against a varying discharge pressure.
- They function well at a low-flow rate; the PD pump can deliver a low-flow rate at a high discharge pressure and still maintain a high efficiency as compared to centrifugal pumps.
- They function well at high viscosity; the PD pump can easily be adjusted to handle high viscosity liquids and, in fact, will run at a higher efficiency as the higher viscosity acts to reduce the amount of slip experienced by the pump.

- They have the ability to self-prime; most PD pumps will act to create a vacuum at the pump inlet. This vacuum will then pull liquid into the pump cavity.
- PD pumps can easily be made to sanitary design specifications.

Disadvantages. Disadvantages that can be found in using PD pumps in the bioprocess industry include:

- Overpressurization can occur; a flow restriction on a PD pump will build pressure until that pressure is relieved. An internal or external pressure relief valve may be required for safe operation of the pump.
- Pulsing flow will occur; the design of PD pumps is such that usually they will cause pressure pulsation as they maintain flow rate. This pressure pulsation can act to damage discharge piping caused by excessive vibration. A pulsation dampener may be required.
- High flow rates are hard to maintain; high flow rates may be hard to achieve even with large pumps because the maximum pump speed may be limited by the viscosity and suction port size.
- Power limitation; because the discharge pressure is directly proportional to the horsepower required, very high discharge pressures may require a very high horsepower, which may exceed the physical strength of the pump.

Types of Positive Displacement Pumps

Gear Pumps. Gear pumps (Fig. 2) comprise, perhaps, the most common PD pump in industry. They have wide acceptance because of low cost and simple construction. In general, gear pumps are made in two very distinct designs. The first is the external gear pump. This design uses two meshed gears that rotate in opposite directions. The first gear is driven by the pump motor, referred to as the driven gear, and the second gear is driven by the first and is called the pump gear. Fluid enters and exits the gear housing perpendicular to the gear mesh, with the incoming fluid being trapped within the gear teeth as the opening mesh creates a vacuum pulling liquid in. This trapped liquid rotates on the outside of the gears and is expelled at the pump discharge port as the gear teeth again start to mesh. The fluid is forced out of the gear housing at the higher pressure of the discharge side of the pump.

The other type of gear pump used in industry is the internal gear pump. The internal gear pump works under the same principle as the external gear pump; however, the gears are configured such that a smaller, off-center, internal gear is placed inside a planetary gear. The internal gear meshes with the inner teeth of the planetary gear to create the pumping action. In this design, the larger planetary gear is the motor driven gear, with the inner gear acting as an idler. The gap created by mounting the internal gear off center is filled with a crescent-shaped scraper. This scraper acts to seal the inlet port from the discharge port and reduces the amount of slip in the pump. This pump is slightly more expensive, but is quieter and often longer lived.

Air-Driven Double-Diaphragm Pumps. Double-diaphragm pumps (Fig. 3) are a unique type of PD pump. The design of a double-diaphragm pump is made up through the porting of four check valves and two dia-

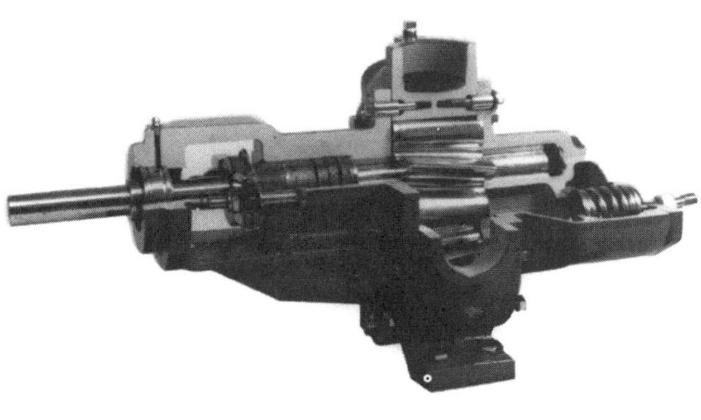

Figure 2. Gear pump. *Source*: Photo used by permission of Roper Pump Company.

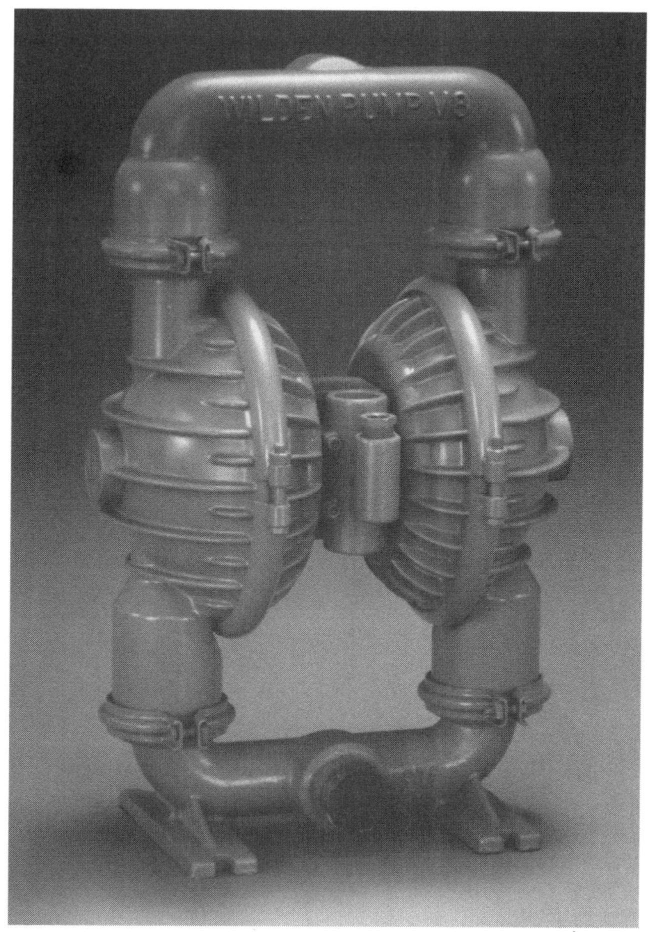

Figure 3. Double-diaphragm pump. *Source*: Courtesy of Wilden Pump Company.

phragms mounted in parallel. The two diaphragms are connected to an air-driven piston. The operation of the pump has liquid entering a common inlet port and the liquid separating and alternately flowing through one of two check valves, the check valve that is exposed to the opening diaphragm. At the same time, on the opposite side of the closed check valve, the closing diaphragm is expelling fluid at a discharge pressure (limited by the air supply pressure that powers the pump). The fluid being expelled pushes open the check valve exiting the closing diaphragm section and pushes closed the check valve on the opening diaphragm. This action cycles back and forth as the air piston that drives the two diaphragms moves in and out. Because both diaphragms are attached to the same shaft, when one diaphragm is opening, the other is closing.

Progressing Cavity Pumps. Progressing cavity pumps (Fig. 4) have a special design that allows for very gentle pumping of highly viscous liquids. Most progressing cavity pumps are made up from a single helical (metal) rotor rotating inside a (elastomeric) double helical stator. Cavities formed by this combination of rotor and stator progress down the length of the rotating shaft as the shaft turns. The function of the rotor is very similar to that of a screw pump slowly pushing the pumped liquid forward toward the discharge of the pump. Pressure is developed because close clearances between the rotor and stator minimize the amount of slip that is found in the pump. Further, because the pump is well suited to pumping high-viscosity liquids, this action is even better utilized.

Flexible Impeller Pumps. Flexible impeller pumps (Fig. 5) make up a class of very inexpensive PD pumps that can be easily adapted to many applications. Flexible impeller pumps work on the principle of a multivaned rubber impeller rotating in a housing that is cammed such that the vanes are pushed close as the impeller rotates. This closed impeller expels fluid out the discharge of the pump and at the same time pulling liquid into the suction port. This action creates a vacuum at the suction port. Vacuum values as high as 22 in Hg are easily achievable. The major drawback of this style of PD pump is the inability of the

Figure 4. Progressing cavity pumps. *Source*: Photo used by permission of Roper Pump Company.

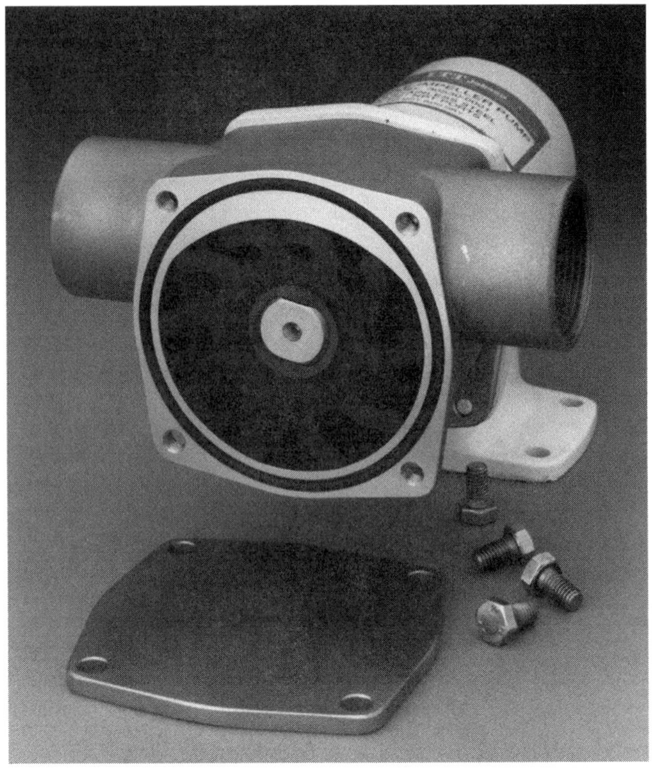

Figure 5. Flexible impeller pump. *Source*: Courtesy of ITT Jabsco.

pump to run dry. Dry running of this type of pump will quickly erode the rotor.

Circumferential Piston Pumps. Circumferential piston pumps (Fig. 6a) are used in the bioprocess industry to transfer a wide range of both raw material and finished products. The circumferential piston pump works with two opposing hemispherical rotors spaced 180° apart. These two rotors turn in opposite directions much like that of the external gear pump; however, the action of the circular piston moving past the suction port creates a vacuum that draws fluid into the pump cavity. This fluid, which is trapped between the sides of the rotating piston and the pump housing, travels around the pump casing and is discharged at the pump outlet as the piston of the opposing rotor pushes the trapped liquid into the outlet of the pump. This action alternates from pump rotor to driven rotor as trapped fluid is pushed into the pump outlet. This pump design requires that the rotors of the pump not come in contact with one another. To accomplish this, the rotors are driven with a gear housing that contains supporting shafts, bearings, and timing gears. It is therefore very clean and ideally adds no metal to the product.

Lobe Pumps. Very similar in design to circumferential piston pumps, lobe pumps (Fig. 6b) also use a gear housing containing supporting shafts, bearings, and timing gears. The rotors of the lobe pump rotate in opposite directions, creating a low pressure or vacuum at the pump inlet as the lobes separate. The fluid drawn into the pump cavity

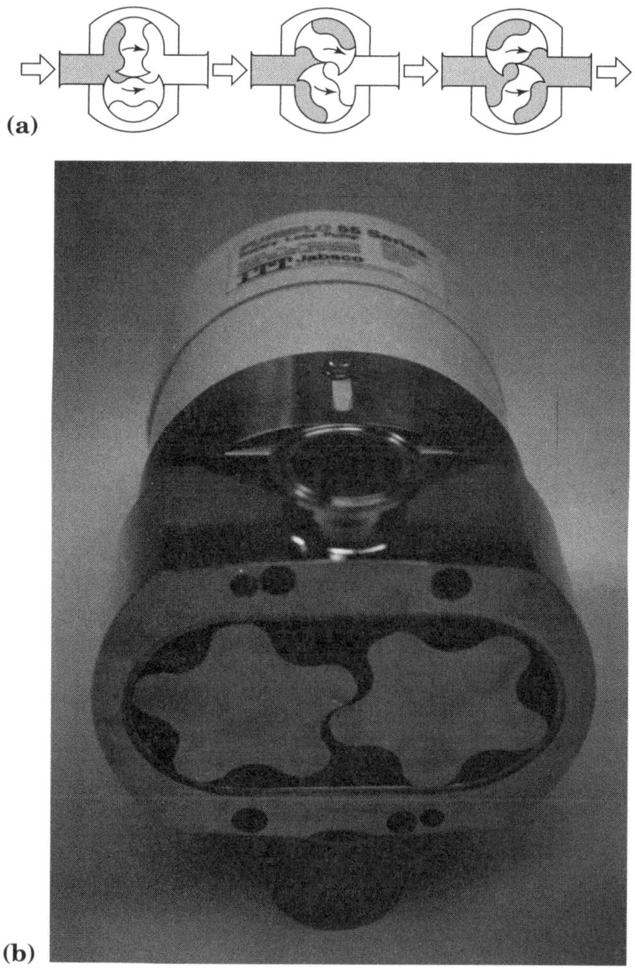

(a)

(b)

Figure 6. (**a**) Circumferential piston pump; (**b**) Lobe pump. *Source*: (**a**) courtesy of Waukesha Fluid Handling; (**b**) courtesy of ITT Sherotec.

is trapped in the valley of the lobe and rotated to the outlet of the pump where the trapped fluid is forced into the pump outlet as the opposing rotor lobe rotates into the lobe valley containing the trapped liquid. This action alternates from rotor to rotor similar to that found on the circumferential piston pump. The lobe pump is somewhat less shear intensive as compared to the circumferential piston pump and is more easily cleaned when cleaning a pump in place.

Peristaltic Pumps. Peristaltic pumps (Fig. 7) are very common throughout the bioprocess industry. They can be found in all size ranges, handling flows on the laboratory bench of 3 to 4 mL/min to flow rates on the production floor of 30 to 40 gpm. The basic design of the pump functions through the progressive squeezing of tubing by rollers mounted to a rotating plate. The rollers trap fluid within the tubing and rotate it to the outlet. They are self-priming and have the major advantage that the fluid only contacts the pump tubing so sealing the pump is not required. The action also minimizes the amount of shear that the pump fluid is exposed to.

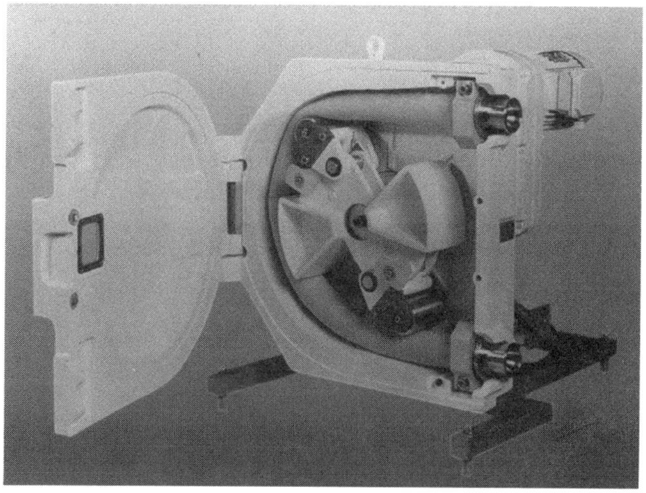

Figure 7. Peristaltic pump. *Source*: Courtesy of Watson-Marlow Inc.

Nutating Disk-Type Pumps. Nutating disk-type pumps (Fig. 8) are new to the bioprocess pump market and provide for advantages typically found in several different types of older pump designs. The principle that drives this class of PD pumps is the action of a wobbling disk within the pump cavity. This disk can be used to trap a volume of fluid between it and the stationary pump housing. The trapped fluid is pushed toward the pump outlet through this wobbling action and the use of a scraper gate that rides on the width of the rotating disk.

DRIVERS

The most popular choice of drivers for pumps, by far, is the AC motor. They have become the standard because they

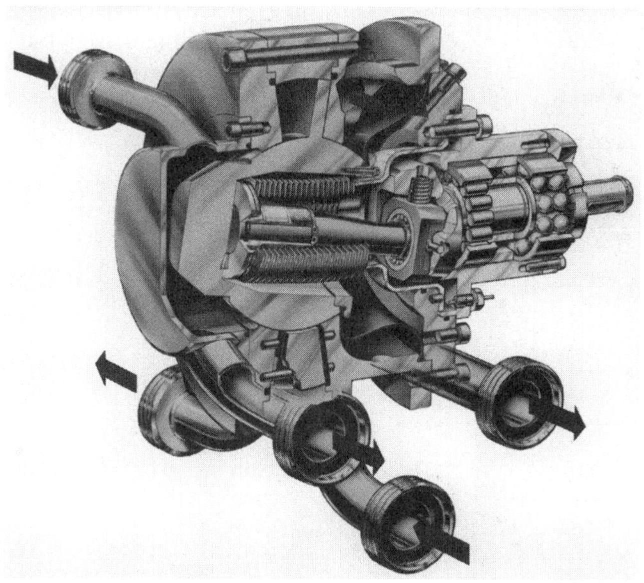

Figure 8. Nutating disk-type pump. *Source*: Courtesy of KSI.

are inexpensive and long lived, have relatively few moving parts, and are available from a large variety of sources, in a large variety of styles and enclosures. They are relatively quiet (typically between 70 and 85 decibels), efficient (often over 90%), and institutions exist (National Electrical Manufacturer Association [NEMA]) that guarantee a certain degree of interchangeability from manufacturer to manufacturer. Typically, only if an AC motor is unacceptable are other choices considered. Considerations include style, electrical specifications, and enclosure.

Style

Close coupled, foot-mounted motors are common for centrifugal pumps that will accept this. Standard horizontal foot-mounted motors are needed when pedestal-mounted centrifugal pumps have been chosen. PD pumps usually need motors with gear boxes integrally mounted to achieve the low speed necessary. Because most pump suppliers sell their pumps as motor and pump packages, the user is not usually concerned with physically matching the components, but should understand the principles.

Electrical Specifications

Unlike the general motor style, it is the user's responsibility to know the electrical systems the drive will be expected to operate under. Concerns the supplier must address will be:

- *Phase*: Single phase is available for small horsepower drivers (below 5 hp), and common for fractional horsepower, but three phase is available for fractional and large integral motors.
- *Voltage*: in the United States, standards are 115 or 230 V for single phase and 230 or 460 V for three phase, but others voltages exist. Because motors are good for $\pm 10\%$, there is some flexibility.
- *Frequency*: 60 Hz is the standard in the United States, but 50 Hz exists all over the world. Motors tolerate only $\pm 5\%$ frequency variation.

Enclosure

The desired enclosure is an important consideration. Standard types available are open drip-proof for environments where the motor will be protected from atmospheric dangers; totally enclosed, fan cooled (TEFC) for areas where the motor will be exposed to light moisture; explosion proof where a UL rating is necessary and the motor's ability to keep sparks from escaping is crucial for safety considerations; and common in bioprocessing, the wash-down enclosure where the area the unit is to be mounted in will be periodically hosed down.

NEMA is an organization that has written a standardized list of motor dimensions such that a user may substitute one brand for another in confidence that the new motor will physically interchange. This is done by a series of frame designations, such that a brand A motor in, for example, frame 145TC will exactly fit in place of a brand B motor of the same frame.

An accessory often considered today is the frequency inverter, which varies frequency and voltage in such a fashion as to allow an AC motor to act as a variable speed device. This has many advantages in addition to allowing a user to dial in more precisely the desired flow, such as the ability to control speed remotely, and to be able to start up a pump slowly to minimize start-up damage to the pump and the motor.

Other options to an AC motor are gasoline or diesel engines for areas where electrical power is unavailable, and more and more commonly, compressed air drives. Air motors have the advantages of being inherently explosion proof and being able to stall without incurring damage, but have the disadvantages of needing very dry compressed air and of being relatively inefficient. DC power is used in some very small pump packages, but is uncommon.

SPECIAL CONSIDERATIONS FOR BIOPROCESSING PUMPS

Sanitary Standards

The main standard for the sanitary design and construction of pumps to be used in the bioprocess industry is the 3A Standards (Code of Federal Register Title 21, part 177). These standards govern certain design modifications that are key elements in maintaining sanitary conditions within a pump. These guidelines cover surface finish, materials of construction, fabrication, and other areas. The sanitary design of a pump must also comply with all other applicable standards, including FDA, cGMP, and the ANSI/ASME standards for food, drug, and beverage equipment.

Materials of Construction and Surface Finishes

The most commonly found material of construction for pumps in the bioprocess industry is austenitic stainless steel alloys. Austenitic stainless steel alloys used in the construction of sanitary stainless steel pumps include AISI types 304, 304L, 316, and 316L. Analysis of the makeup of these alloys reveals that a major element in all stainless steel alloys is iron. Iron alone is quickly attacked by most fluids, but in the presence of chromium in correct combinations of Ni, C, Mn, P, Mo, Si, and S, the ability of stainless steel to withstand corrosion is excellent. This, plus its low carbon content and its ability to be easily welded, machined, and polished, makes 316L the most popular material of construction for pumps in bioprocessing. Product-contact surface finish is critical in most bioprocess systems because of the need to preserve sterile or sanitary conditions. All product contact surfaces must be free of cracks, crevices, and occlusions that can harbor and promote bacteria growth. Surface finish in most pumps is obtained through the use of mechanical polishing and electropolishing. This surface treatment will result in a surface finish that would measure at approximately 15 Ra surface roughness.

Mechanical Seals

Theory. Mechanical seals (Fig. 9) are a standard product used in the manufacture of most pumps to seal between

Figure 9. Mechanical seals. *Source*: Courtesy of the Durametallic Company.

the rotating pump shaft and the pump cavity. The type of mechanical seal used to seal the pump cavity is a key component of most pumps used in the bioprocess industry. The mechanical seal that is placed into pumps must serve two functions:

- Provide a barrier to the atmosphere
- Contain the fluid in the pump cavity without affecting fluid quality

Bioprocess pump seals are designed to be easily cleansed and not harbor bacteria. However, for them to be used in pumps for sterile fluids, particular modifications are required. These modifications include changes in the seal design, placement of the seal within the pump housing and selection of seal materials.

A discussion on mechanical seals is best started with a brief background on how mechanical seals work. A single mechanical seal works with two very flat surfaces, one a rotating face element and the other a stationary face element. These faces ride on one another with an ultrathin layer of fluid between the seal faces. When the fluid enters the zone between the two faces, the fluid temperature rises owing to the frictional resistance of one face on the other. If the seal is functioning correctly, the fluid will vaporize as it exits this zone and enters the atmosphere. This principle applies to most single mechanical seals in bioprocess pumps. The key to this functioning correctly is to have proper force pushing the rotating element on to the stationary seat. The force can be obtained either mechanically or hydraulically. The mechanical force is provided by a compressed spring. The hydraulic force, found in a hydraulically balanced seal, uses the pressure inside the pump head to push the rotating element to the stationary seat. Both of these methods are used in bioprocess pumps seals to produce a force that will generate enough friction between the faces to vaporize the fluid just before it leaves the seal faces without causing the faces to wear too quickly or generate too much heat.

If the mechanical seal is to function as described and provide a sterile barrier between the atmosphere and the fluid within the pump cavity, certain modifications are required to the seal design. First, a secondary mechanical seal is needed behind the primary inboard seal. This combination is referred to as a double mechanical seal. Second, both of these seals are placed into a stuffing box or gland that is attached to the pump where the rotating shaft enters the pump head. This is done so that a flushing fluid can be run between the two seals. The fluid barrier will guard against any contaminants from the atmosphere penetrating into the fluid within the pump cavity. The flushing fluid is pushed through the stuffing box under pressure so that a positive pressure on the outboard seal stops contaminants from entering the stuffing box. A secondary function of the seal flush liquid is to provide a certain amount of cooling to the seal faces.

The requirement that the seal contain the fluid within the pump cavity without affecting the fluid quality necessitates changes in the way the inboard seal functions. Fluid in the pump cavity is kept pure by ensuring that a positive pressure is maintained on the pump side of the inboard seal. This pressure functions in the same manner as that on the outboard seal; however, it should be greater than the stuffing box pressure to ensure that any leakage through the seal passes from inside the pump into the stuffing box. This will prevent the passage of contaminants from the stuffing box into the pump cavity. The seal designed for this configuration creates some problems.

First, if the inboard seal fails, product can be lost though the flush outlet of the stuffing box. Second, if the fluid in the pump contains active organisms that must be contained, these organisms could pass into the seal flush. The common flushing procedure for double mechanical seals in this situation is to maintain the stuffing box at a higher pressure than the pump side inboard seal pressure. In the case of seal failure, the flush fluid would move into the pump head, and no product would be lost. This arrangement can present problems if particulate, worn from the seal faces, is forced into the pump cavity by the high stuffing box pressure, thereby contaminating the fluid in the pump. Applications should be reviewed to determine which flushing configuration is most applicable.

The mechanical seal chosen for an application should be constructed of the highest quality FDA-approved materials of construction to provide a reliable, long-lasting seal. The face and rotating element materials should be selected for fluid compatibility, hardness, and flatness. Commonly used materials include tungsten carbide, silicone carbide, ceramics, carbon graphite, and stainless steel. The supporting items of the seal, including the springs, cups, O-rings, drive collar and set screws, should also be selected for fluid compatibility and performance characteristics. These pieces will function as a system to provide a reliable seal.

Static Seals

Static seals within pumps used in the bioprocess industry fall into two categories. These categories are elastomer O-rings and flat gaskets. O-rings are more commonly used in modern pump designs because they can be sized to properly seal flat or round pump components within the pump design. Further, O-rings and their corresponding O-ring grooves can be easily configured to minimize any fluid entrapment and provide a cleanable crevice-free seal. O-rings are commonly used to seal pump shafts to mechanical seals and pump coverplates to pump housings. The O-ring material is selected based on the required durometer (hardness) and material compatibility with the sealed fluid. Commonly found materials include Viton®, EPDM, Buna-N, Silicone, and Kalrez®.

Flat gaskets are less common in modern pump designs; however, they are used when maintaining clearances within the pump design is critical. As with O-rings, they can also be constructed of an elastomeric compound, but it is also common to see them made from such materials as paper or PTFE (Teflon®).

Aseptic Modifications

Sterile fluids require systems that are capable of being steamed-in-place (SIP). To minimize downtime and preserve the integrity of the sterile atmosphere within the pumping system, pumps can be modified so they can be fully steamed-in-place. Most pumps can be steamed-in-place if they are designed in accordance with 3A standards. These standards, as mentioned earlier, will specify proper surface finish, impeller, and seal design. Further, the pump must utilize a seal that will allow for full steam penetration under pressure. Because proper steam flow is required to maintain saturated steam within the pump being sterilized, steam flow is enhanced in pumps by providing a means of condensate removal. As steam sterilizing the pump cools, it will form condensate. For proper sterilization to occur, this condensate must be removed from the pump quickly, preventing cold spots from forming. Hence, there is often the need for the pump to be self-draining by adding a tangential bottom drain.

Water-for-Injection Modifications (Centrifugal Pumps)

The design of centrifugal pumps make them very well suited to the application of pumping sterile water for injection (WFI). Because WFI must be kept moving at high temperature and high velocity within a piping system, a standard sanitary centrifugal pump can be modified to handle this application. The first modification is to the surface finish and materials of construction. The pump should be constructed of AISI Type 316L stainless steel that has been finished so that the surface roughness measures approximately 15 Ra. Second, all elastomeric components of the pump should be selected to be compatible with the WFI. This requires matching compatibility for temperature. Typical materials include steam resistant Viton® or EPDM. The third measure is the correct configuration of the mechanical seal to be used. This should be selected to reduce the chances of contamination. The last typical modification includes the addition of a casing drain to promote the removal of condensate during the steam sterilization of the WFI pump.

TROUBLESHOOTING

There is a general theme in successful troubleshooting. Specifically:

1. Make sure the system is fully understood (to eliminate oversights).
2. Narrow down or isolate potential sources of the problem.
3. Determine all possible causes within these areas.
4. In order of probability, go through them one by one.

The difficult step is often number 3, but all four hold potential hazards and solutions.

1. System analysis can be tedious. Fear of asking what appear to be obvious questions can tempt one to give this step short shrift. However, if problems began after a change in the system, system analysis will usually hold the answer. Walking through a typical flow pattern and questioning each step is often helpful. Keep in mind that flow at any division seeks equilibrium; flow will divide such that resistance to flow on one side given its gallonage will exactly equal the resistance to flow on the other side with the remainder of flow. Double checking that the pump in question will, in theory, meet the system conditions may be also be appropriate.

2. In many cases, narrowing down whether a problem is in the system or the pump can be affected by switching pumps in an installation containing one that is working well. If the problem follows the pump, it is in the pump; if it stays with the line, it is in the system. In other cases, noises can be isolated by separating various components of a system to determine if they continue or cease. The motor sometimes can be separated from the pump, for example, or suspect components temporarily removed. Valving can sometimes isolate pumps or piping sections to check the results.

3. There are usually more possible explanations for a problem than a typical manufacturers' lists provides. Following is a sampling of potential causes for a few specific complaints. Creativity is essential in this step, with nothing *ever* taken for granted. Often, a marginal explanation tentatively or casually offered up will be immediately recognized by an engineer and the problem solved.

4. There will always be possibilities (listed in the following section) that are more likely to occur than the others, depending on the symptoms. Of these, the most easily checked should be the first inspected, and then the more difficult. At times there are two simultaneous flaws in a system, making checking of individual items fruitless. In these cases, logic will be of more value than experiment.

Suggestions for a few common problems follow.

Low Flow

Centrifugal Pumps Only

1. Reverse rotation (*Note*: a centrifugal impeller flings the liquid as opposed to scoops it, if there is ever a question as to rotational arrows. A centrifugal pump will pump if turning in reverse, but with low flow and pressure and higher horsepower because of lowered efficiency).

2. More head than originally calculated.
3. Clogged impeller.
4. Wrong impeller diameter for conditions.
5. Cavitation (see earlier section).
6. Impeller clearances worn too large.
7. Motor speed slower than nameplate indicates.
8. Impeller spins loosely on shaft.
9. Broken coupling.
10. Viscosity has increased.
11. Vortexing in source tank.
12. Air in tank introduced from splashing or churning close enough to tank outlet to allow air to enter with stream.
13. Wrong pump for conditions.
14. If pump is above source tank, air leak in suction pipe.

Positive Displacement Pumps Only

1. Excess wear on components.
2. Stuck internal check valves, if any.
3. Torn diaphragms, if present.
4. Stuck internal (or external) pressure relief valve.
5. Speed too slow.
6. Cavitation (see preceding section).
7. Wrong pump for conditions.
8. If belt driven, a slipping belt.
9. Broken coupling.
10. Speed not slow enough for viscosity, leading to cavitation.
11. If pump is above source, air leak in inlet pipe.

Cavitation

Note: Cavitation is indicated by crackling noise for centrifugal pumps, by banging for most PD pumps, or by finding that TDH has fallen at least 3% below the standard curve for the particular centrifugal pump.

1. Blocked inlet.
2. Too much friction loss in suction line.
3. Elbow too close to pump suction port (*Note*: The rule-of-thumb is that the elbow should be located no closer than 6 × the suction port diameter).
4. For centrifugal pumps only, running at extreme left end of the curve causing low pressure areas to develop.
5. Suction line with high spot (as opposed to continually rising), therefore, air trap to disturb or restrict flow.
6. Liquid too hot and therefore vapor pressure too high.
7. Elevation differential too great (too much lift).
8. If drawing from closed tank, vacuum level too high, or vacuum level that builds to an unacceptable point as liquid is pumped out.
9. Excess prerotation at inlet of centrifugal, causing flow to enter in wrong pattern for design (*Note*: Rotation against impeller rotation is a positive factor and will move the curve upward, but the more common prerotation in the same direction as the impeller is harmful and may drop the curve).

Motor Overheats

Centrifugal Pumps Only.

1. Too much flow; therefore, further out on curve than calculated.
2. Wrong rotation.
3. Oversized impeller.
4. Unbalanced impeller.

Positive Displacement Pumps Only.

1. Too much backpressure.
2. Binding or rubbing components.
3. In case of progressive cavity, flexible impeller, or other pump with rubbing components, insufficient motor starting torque to get up to full speed

Any Pumps

1. Voltage above ±10% allowed by manufacturers.
2. Frequency beyond ±5% allowed by manufacturers.
3. Bad motor bearing(s).
4. Three-phase motor not getting power to all three lines (single phasing).
5. Cavitation.
6. Misaligned motor/pump shaft, or misaligned pulleys.
7. Undersized cable to motor.
8. Loose connection at motor outlet box or control box.
9. Insulation has broken down through time and use.
10. Cable too long to handle amperage.
11. Ambient temperature too high.
12. Overgreased bearings, causing binding when bearings heat up.
13. On TEFC design, broken or slipping fan.
14. Speed too high for pump.
15. Not enough air movement to keep motor cool because of confined or poorly vented area.
16. If single phase motor, starting windings do not disengage.
17. Viscosity higher than calculated.
18. Pipe torque, causing misalignment and binding.
19. Motor stops and starts too often (consult manufacturer for maximum stops and starts for applicable horsepower, it is probably lower than you think!).
20. Dirty surfaces not allowing efficient cooling.
21. On drip-proof design, internal diffuser or fans not channeling air flow properly.

Noise

Note: Various designs have various degrees of noise considered standard and acceptable. The manufacturer will be able to supply the numbers if given operating conditions.

1. Motor or pump bearings worn.
2. Cavitation.
3. If a PD pump, components hitting or rubbing.
4. If rigidly piped, amplification of what might be considered normal noise through lines into abnormal amounts.
5. If mounted on channel steel base, amplification of noise from pocket of air beneath motor/pump package.
6. Coupling hitting coupling guard.
7. Motor fan hitting fan cover guard.
8. Check valves chattering as they stick.
9. Water hammer (valves must close slowly enough for momentum in line to be absorbed). A rough rule of thumb is $t = 2L/a$, where t = time in seconds for valve to close (minimum); L = length of pipe via longest route, in feet; a = speed of pressure wave in feet/second (an average for water is 4,000 ft/s). An example would be a 200-ft length of pipe handling water, which would demand a valve closing in no less than $2 \times 200/4{,}000 = 0.1$ s to avoid water hammer.

Abnormal Seal Wear

Note: A mechanical seal in a continuous-duty, clean application can be expected to last several years, as a minimum, to as many as 10 years or longer if under absolutely ideal conditions.

Seat Fracture

1. Thermal shock from dry running followed by sudden introduction of cooler liquid, or any sudden drastic change in the temperature or the liquid. Usually appears as a hairline fracture.
2. Seal head installed incorrectly, causing the head to strike the seat.
3. Physical shock from a blow such as being dropped.
4. Water hammer.
5. Pipe misalignment, twisting the bracket holding the seat.

Carbon Washers Wear Prematurely

1. Abrasives (clearly visible by scored faces).
2. Chemical incompatibility between the chemical and either the washer itself, or the binder holding the carbon together.
3. Improper installation, usually with the seal head cocked.

Elastomer Problems

1. Chemical incompatibility. Elastomer may become brittle, softened, swollen, or even totally eaten away.
2. Dry running. Elastomer may look burned or cracked or become brittle and hard.

ADDITIONAL READING

Hydraulic Institute Standards, Hydraulic Institute.

I. Karassik, W. Krutzsch, W. Fraser, and J. Messina eds. *The Pump Handbook*, 2nd ed., McGraw Hill, New York, 1985.

R. Stover, in B. Lydersen, N. D'Elia, and K. Nelson eds. *Bioprocess Engineering: Systems, Equipment and Facilities*, Wiley, New York, 1994, pp. 253–315.

V. Streeter, *Handbook of Fluid Dynamics*, 1961.

C.R. Westaway and A.W. Loomis, *Cameron Hydraulic Data*, 16th ed., Ingersol Rand, 1984.